Biological Effects and Physics of Solar and Galactic Cosmic Radiation

Part B

NATO ASI Series

Advanced Science Institutes Series

A series presenting the results of activities sponsored by the NATO Science Committee, which aims at the dissemination of advanced scientific and technological knowledge, with a view to strengthening links between scientific communities.

The series is published by an international board of publishers in conjunction with the NATO Scientific Affairs Division

A	Life Sciences	Plenum Publishing Corporation
B	Physics	New York and London
C	Mathematical and Physical Sciences	Kluwer Academic Publishers
D	Behavioral and Social Sciences	Dordrecht, Boston, and London
E	Applied Sciences	
F	Computer and Systems Sciences	Springer-Verlag
G	Ecological Sciences	Berlin, Heidelberg, New York, London,
H	Cell Biology	Paris, Tokyo, Hong Kong, and Barcelona
I	Global Environmental Change	

Recent Volumes in this Series

Volume 243A —Biological Effects and Physics of Solar and Galactic Cosmic Radiation, Part A
edited by Charles E. Swenberg, Gerda Horneck, and E. G. Stassinopoulos

Volume 243B —Biological Effects and Physics of Solar and Galactic Cosmic Radiation, Part B
edited by Charles E. Swenberg, Gerda Horneck, and E. G. Stassinopoulos

Volume 244 —Forest Development in Cold Climates
edited by John Alden, J. Louise Mastrantonio, and Søren Ødum

Volume 245 —Biology of *Salmonella*
edited by Felipe Cabello

Volume 246 —New Developments in Lipid–Protein Interactions and Receptor Function
edited by K. W. A. Wirtz, L. Packer, J. Å. Gustafsson, A. E. Evangelopoulos, and J. P. Changeux

Volume 247 —Bone Circulation and Vascularization in Normal and Pathological Conditions
edited by A. Schoutens, J. Arlet, J. W. M. Gardeniers, and S. P. F. Hughes

Series A: Life Sciences

Biological Effects and Physics of Solar and Galactic Cosmic Radiation

Part B

Edited by

Charles E. Swenberg

Armed Forces Radiobiology Research Institute
Bethesda, Maryland
and Complexity Incorporated
Potomac, Maryland

Gerda Horneck

DLR Institute of Aerospace Medicine
Cologne, Germany

and

E. G. Stassinopoulos

NASA-Goddard Space Flight Center
Greenbelt, Maryland

Springer Science+Business Media, LLC

Proceedings of a NATO Advanced Study Institute on
Biological Effects and Physics of Solar and Galactic Cosmic Radiation,
held October 13–23, 1991,
in Algarve, Portugal

NATO-PCO-DATA BASE

The electronic index to the NATO ASI Series provides full bibliographical references (with keywords and/or abstracts) to more than 30,000 contributions from international scientists published in all sections of the NATO ASI Series. Access to the NATO-PCO-DATA BASE is possible in two ways:

—via online FILE 128 (NATO-PCO-DATA BASE) hosted by ESRIN, Via Galileo Galilei, I-00044 Frascati, Italy

Library of Congress Cataloging-in-Publication Data

NATO Advanced Study Institute on Biological Effects and Physics of
  Solar and Galactic Cosmic Radiation (1991 : Algarve, Portugal)
    Biological effects and physics of solar and galactic cosmic
  radiation. Part B / edited by Charles E. Swenberg, Gerda Horneck,
  and E.G. Stassinopoulos.
        p.   cm. -- (NATO advanced science institutes series. Series
  A, Life sciences ; v. 243B)
    "Proceedings of a NATO Advanced Study Institute on Biological
  Effects and Physics of Solar and Galactic Cosmic Radiation, held
  October 13-23, 1991 in Algarve, Portugal"--T.p. verso.
    Includes bibliographical references and index.

    1. Cosmic rays--Physiological effect--Congresses. 2. Solar
  radiation--Physiological effect--Congresses. 3. Radiation
  dosimetry--Congresses. 4. Outer space--Exploration--Health aspects-
  -Congresses.   I. Swenberg, Charles E. II. Horneck, G. (Gerda)
  III. Stassinopoulos, E. G. IV. Title. V. Series.
  RC1151.R33N38 1991a
  616.9'897--dc20   93-8449
                                                                  CIP

Additional material to this book can be downloaded from http://extra.springer.com.

ISBN 978-1-4613-6265-4   ISBN 978-1-4615-2916-3 (eBook)
DOI 10.1007/978-1-4615-2916-3

©1993 Springer Science+Business Media New York
Originally published by Plenum Press, New York in 1993
Softcover reprint of the hardcover 1st edition 1993

PREFACE

Space missions subject human beings or any other target of a spacecraft to a radiation environment of an intensity and composition not available on earth. Whereas for missions in low earth orbit (LEO), such as those using the Space Shuttle or Space Station scenario, radiation exposure guidelines have been developed and have been adopted by spacefaring agencies, for exploratory class missions that will take the space travellers outside the protective confines of the geomagnetic field sufficient guidelines for radiation protection are still outstanding. For a piloted Mars mission, the whole concept of radiation protection needs to be reconsidered. Since there is an increasing interest of many nations and space agencies in establishing a lunar base and /or exploring Mars by manned missions, it is both, timely and important to develop appropriate risk estimates and radiation protection guidelines which will have an influence on the design and structure of space vehicles and habitation areas of the extraterrestrial settlements.

This book is the result of a multidisciplinary effort to assess the state of art in our knowledge on the radiation situation during deep space missions and on the impact of this complex radiation environment on the space traveller. It comprises the lectures by the faculty members as well as short contributions by the students given at the NATO Advanced Study Institute "Biological Effects and Physics of Solar and Galactic Cosmic Radiation" held in Armacao de Pera, Portugal, 12-23 October, 1991. The following scientists served on the Organizing Committee:

C. E. Swenberg, Armed Forces Radiobiology Research Institute, Bethesda, Maryland, USA

G. Horneck, Deutsche Forschungsanstalt für Luft- und Raumfahrt, Köln, Germany

E.G. Stassinopoulos, NASA Goddard Space Flight Center, Greenbelt, Maryland, USA

P.D. McCormack, US Naval Medical Center, Washington D.C., USA

The participants, coming from various countries including Russia, Ukraine, Czechoslovakia, Bulgaria are listed at the end of this book. The event is in many respects a sequel to the NATO Advanced Study Institute "Terrestrial Space Radiation and its Biological Effects", Corfu, Greece, October 1987, which was mainly concerned with radiation problems for manned missions in Low Earth Orbit (LEO).

During this meeting, it was emphasised that in order to safeguard future human enterprises in space, especially those of very long duration or beyond our geomagnetic shield, an intense research program has to be initiated. The program objectives should include:

(1)   with respect to the radiation environment in deep space missions: development of a better physical model of the galactic cosmic radiation modulation as a function of solar cycle observables; development of models of HZE particles propagation in the interplanetary medium; a better understanding on the periodicity (magnitude, duration) of a solar cycle, in order to make a prediction several years in advance; a standardized approach on an

international scale for predicting/forecasting solar flares that give rise to proton events and an efficient warning system; microdosimetric approaches in future development of space dosimetry; determination of the dose contributing products as a function of shielding; testing of new space technologies with respect to their vulnerability to space radiation;

(2)   with respect to biological responses to the radiation in space: selection of appropriate biological test systems for radiobiological space experiments in order to quantify and qualify various long-term biological radiation effects; analysis of the biological responses to single particle traversals; development of biological dosimeters; a better understanding of the radiobiological chain of events as a function of radiation characteristics from the initial interactions altering the essential chemical processes, such as DNA strand breaks to the biological response, e.g. cellular lethality, mutagenesis, transformation; investigation of transeffects, i. e. radiation lesions in the DNA that may lead to changes at remote sites; development of radiobiological models appropriate for determining biological effects of HZE particles; international cooperation in the analysis of radiation effects in higher organisms and humans including data obtained in space;

(3)   with respect to risk estimates for deep space missions: development of new and more appropriate concepts to quantitate the radiation risk in space missions; development of shielding concepts; development of countermeasures including radioprotectants, nutritional supplements; determination of the radiation tolerances of different individuals.

This list is by no means complete. It reflects the need for a long-term program where ground based studies will be augmented by flight experiments, especially in high inclination orbits or on precursor missions to Moon and Mars. It was considered to be extremely important to reach a standardisation on an international level with respect to data collection, protocol comparison and formulation of guidelines for future exploratory class missions.

The committee is most grateful to the North Atlantic Treaty Organization for the outstanding support provided for this meeting and for the production of this monograph. It also acknowledges substantial financial support provided by German Aerospace Research Establishment DLR, The US Armed Forces Radiobiology Research Institute, Bethesda MD, the US Department of Energy, and the Committee on Interagency Radiation Research and Policy Coordination in cooperation with Oak Ridge Associated Universities. We thank Lisa Steimel for her valuable and efficient assistance in assuring that the meeting was truly successful. The editors acknowledge the advise and guidance of Mr. Gregory Safford of Plenum Press and the assistance of Dr. Mei-Lie Swenberg in typing and reorganizing the manuscripts to the final version.

Charles E. Swenberg

Gerda Horneck

E. G. Stassinopoulos

CONTENTS

RADIATION ENVIRONMENT, DOSIMETRY, SHIELDING EFFECTS

RADIATION EXPOSURE IN MANNED SPACE FLIGHT, RISK ESTIMATES, PROTECTION

RADIATION ENVIRONMENT DURING THE LONG SPACE MISSION (MARS) DUE TO GALACTIC COSMIC RAYS

N.F. Pissarenko
Space Research Institute
Moscow, 117810, USSR

ABSTRACT

Galactic cosmic radiation {GCR} mostly determines dose equivalents inside the spacecraft during long-term manned missions in space. In this paper some new results are collected concerning different characteristics of GCR's. Together with earlier obtained data they show that during most part of the solar cycle such spaceflights are not possible. Attention is drawn to very great errors in the estimates of dose equivalent and shielding thickness.

GALACTIC COSMIC RADIATION

The galactic cosmic radiation (GCR) is one of the main factors which determine the radiation situation during long-term (longer than one year) manned space missions. Energetic charged particles of galactic and extragalactic origin observed in the interplanetary medium are called galactic cosmic rays. Observations show that inside a spacecraft this radiation can be assumed to be isotropic, in practice, though due to the presence of the magnetic field of solar origin in the heliosphere there is some weak anisotropy of GCR towards the Sun which does not exceed several percent of their total flux.

Galactic cosmic rays are characterized by a wide energy spectrum from several tens of MeV to 10^{20} eV, and even greater. The GCR integral flux with $E > 30$ MeV observed in the interplanetary medium near the Earth (inside our heliosphere) depends on the place in the solar activity cycle being during minimum-activity years equal to $N = 4.5$ part•cm^{-2} s^{-1} and $N = 2$ part•cm^{-2}•s^{-1} during maximum-activity years. This galactic radiation intensity modulation in the near-earth space is caused by the 11-year solar activity cycle (see Fig. 1; Mavromishalaki et al, 1989). The solar activity cycle is characterized by the variation of the solar spot number (Wolf numbers) and by the appearance of different forms of the solar activity mainly associated with the amplification of local magnetic fields in the photosphere and atmosphere of the Sun (active regions, flares, transients and so on).

The GCR consists of 83% protons, 13% α-particles, and about 1% nuclei with $Z>2$; the electron component is about 3% of the total flux. It should be noted that electrons with energy $E < 20$ MeV are mostly of Jupiter origin. Table 1 (Smart and Shea, 1985) illustrates in detail the composition of galactic cosmic rays. GCR nuclei with $Z>2$ are classified as several charge groups: L- (light nuclei, $3<Z<5$), M- (medium nuclei, $6<Z<8$), LM- (semi-heavy nuclei, $9<Z<14$),

Biological Effects and Physics of Solar and Galactic Cosmic Radiation,
Part B, Edited by C.E. Swenberg *et al.,* Plenum Press, New York, 1993

1

H- (heavy nuclei, 15<Z<19), and vH- (very heavy nuclei, 20<Z<28). Often the charge group from manganese (Mn, Z = 25) to nickel (Ni, Z = 27) is called the iron group.

According to L.I. Dorman (1977) the GCR nucleon spectrum observed in the quiet time (i.e., in absence of flares and Forbush decreases) can be subdivided into five intervals:

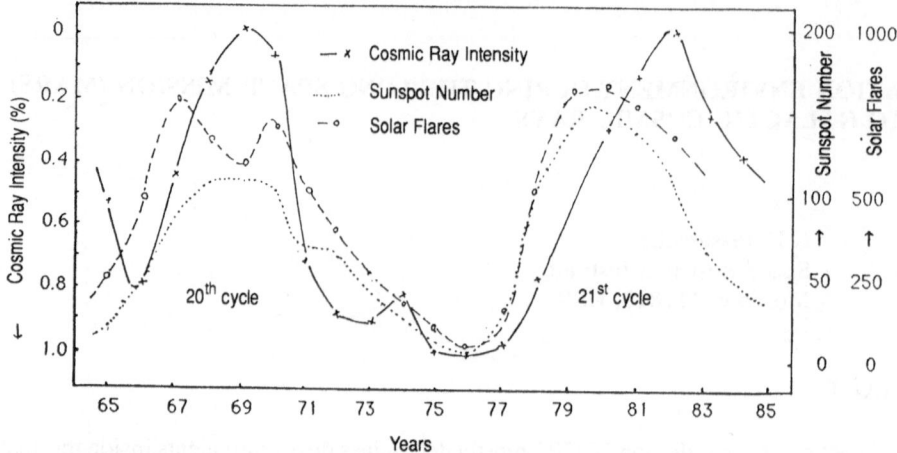

Fig. 1 Neutron monitor data, sunspot number and solar flares during the 20 and 21 solar cycles.

Table 1 Composition of Galactic Cosmic Rays

Ions	Z	Group	Abundance for E > 450 MeV/Nuc.
He	2	α	44700 ± 500
Li	3		192 ± 4
Be	4	L	94 ± 2.5
B	5		329 ± 5
C	6		1130 ± 12
N	7	M	278 ± 5
O	8		1000
F	9		24 ± 1.5
Ne	10		158 ± 3
Na	11	LH	29 ± 1.5
Mg	12		203± 3
Al	13		36 ± 1.5
Si	14		141 ± 3
P	15		7.5 ± 0.6
S	16		34 ± 1.5
Cl	17	H	9.0 ± 0.6
Ar	18		14.2 ± 0.9
K	19		10.1 ± 0.7
Ca	20		26 ± 1.3
Sc	21		6.3 ± 0.6
Ti	22	vH	14.4 ± 0.9
V	23		9.5 ± 0.7
Cr	24		15.1 ± 0.9
Mn	25		11.6 ± 1.0
Fe	26	Fe	103 ± 2.5
Ni	27		5.6 ± 0.6

I	$3 \cdot 10^5 eV \leq E_1 \leq 3 \cdot 10^{20} eV,$
II	$3 \cdot 10^{11} eV/nucleon \leq E_2 \leq 3 \cdot 10^{15} eV/nucleon,$
III	$30\ MeV/nucleon \leq E_3 \leq 3 \cdot 10^{11} eV/nucleon,$
IV	$1\ MeV/nucleon \leq E_4 \leq 30\ MeV/nucleon,$
V	$0.01\ MeV/nucleon \leq E_5 \leq 1 MeV/nucleon.$

Within the first interval the upper boundary of a spectrum is at the energy $E_B > 10^{19} eV$ for protons. Protons of metagalactic origin are observed in this interval, and their spectral cutoff is associated with scattering of cosmic rays on the relic electromagnetic radiation with T=2.7K and on photons of other origin. The boundary between the intervals I and II is characterized by the change in the GCR differential spectrum slope with $\gamma_1 = 3.2$-3.3 to $\gamma_2 = 2.7$ in the power spectral presentation. It can be also noted that the boundaries of the interval I are given in the total energy per particle rather than in energy per nucleon due to the scarce information about the chemical composition of particles in this interval. The interval II contain particles of both galactic and metagalactic origin. The boundary between the intervals II and III corresponds to the upper limit of the GCR modulation region inside the Solar system ($E_m \approx 200$ GeV/nucleon). The interval III contains particles of galactic origin as the analysis of their chemical and isotopic composition demonstrates. This interval is modulated most highly by magnetic fields of solar wind. The boundary between the intervals III and IV corresponds to the minimum GCR energy (energy per nucleon) observed in the near-earth environment. The origination of particles within this interval is not still clear. The significant portion of them is the so-called "anomalous" component. And finally, the boundary between the intervals IV and V is rather conventional (with the solar activity variation it shifts from several tenths of MeV/nucleon to several MeV/nucleon). Particles of this interval differ from those in other intervals by their chemical composition, energy spectrum, and character of time variations. The lower boundary of this interval, in principle, coincides with the upper energy boundary of solar wind.

Energy spectra of GCR particles (electrons, protons, and nuclei) for undisturbed periods can be estimated using the following equation (Nimmik and Suslov, 1990):

$$N_i(E,t) = F_i(R,t)\ \frac{A_i}{Z_i}\ \frac{10^{-3}}{\beta} \tag{1}$$

where

N_i in $s^{-1}\ m^{-1}\ st^{-1}\ MeV^{-1}$ for particles with $A_i = 1$;
F_i in $s^{-1}\ m^{-1}\ st^{-1}\ (MeV/nucleon)^{-1}$ for particles with $A_i > 2$
A_i is the mass number;
Z_i is the particle charge in units of electron charge;
β is the ratio of the particle velocity to the light speed in vacuum:

$$\beta = \frac{R}{[R^2 + (A_i\ M_{0i}/Z_i)^2]^{1/2}} \tag{2}$$

where

M_{0i} is the particle rest mass;
$M_{0i} = 0.0511 \cdot 10^{-3} GeV$ for electrons;
$M_{0i} = 0.938$ GeV for protons;
$M_{0i} = 0.939$ GeV/nucleon for nuclei;
R is the particle rigidity, GV.

With the given values of the kinetic energy E of GCR electrons, protons, and nuclei the rigidity and β can be determined using the equations:

$$R = \frac{A^i}{Z_i} [E(E+2M_{0i})]^{\frac{1}{2}} \quad ; \quad \beta = \frac{[E(E+2M_{0i})]^{\frac{1}{2}}}{E + M_{0i}} \tag{3}$$

and

$$F_i(R,t) = \frac{D_i \, \beta^{\alpha i}}{R^{\gamma i}} \left(\frac{R}{R + R_0(t)} \right)^{\Delta i(t)} \tag{4}$$

where $\Delta_i(t)$ is the dimensionless parameter evaluated using the equation:

$$\Delta_i(t) = 5.5 \left[1 - b_i \exp\left(- \frac{\beta \, R^2}{d_i} \right) \right] +$$

$$1.13 \, \frac{Z_i \, \beta \, R}{Z_i \, R_0(t)} [\sin(2\pi/22(t-t_0))]^{\frac{1}{2}} \cdot \exp\left(- \frac{\beta \, R}{R_0(t)} \right). \tag{5}$$

Here $R_0(t) = 0.375 + 3.10^{-4} \, W^{1.445}$ where R_0 is the modulation potential of the heliosphere for the time t; $W_{t\text{-}to}$ is the smoothed monthly-average Wolf number dated from the time t- t_0 and calculated from the equation:

$$\tilde{W}_{t\text{-}\delta t} = \frac{{}_{k=1}\Sigma^n \, k \, W_k \; + \; {}_{k=n+1}\Sigma^{2n-1} \, (2n - k) \, W_k}{n^2} \tag{6}$$

The parameters D_i, α_i, γ_i, b_i, and t_{0i} depend on the particle kind (see Table 2); d_i =0.106 GV (for e^-), d_i = 0.012 GV (for p^+), d_i = A_i /Z_i $\bullet$ 0.012 GV (for nuclei).

Table 2   The parameters for the determination of energy spectra GCR.

GCR	CHARGE Z_i	MASS A_i	D_i	α_i	γ_i	b_i	t_{0i}, yr	RIGITY RANGE GV
e^-	-1	1	1.7E2	-	γ_e	0.9	1980.5	E(-2)-E2
p^+	1	1	2.E4	3.0	2.75	1.2	1982.5	1.4E(-1)-E2
He	2	4	3.5E3	3.0	2.75	1.2	1982.5	2.8E(-1)-2E2
C	6	12	9.6E1	3.1	2.75	1.2	1982.5	2.8E(-1)-2E2
Fe	26	55.8	9.2E0	3.1	2.6	1.2	1982.5	2.8E(-1)-2E2

For GCR nuclei with the charge Z > 2 the values of D_i, α_i, b_i, and t_{0i} refer to the mixture of isotopes of appropriate elements. For electrons $\gamma_e = 3.0-1.4 \bullet \exp(-R/R_i)$ where $R_i = 1GV$, Figures 2, 3, 4, and 5 show examples of spectra (Nimmik and Suslov, 1990). In this model the

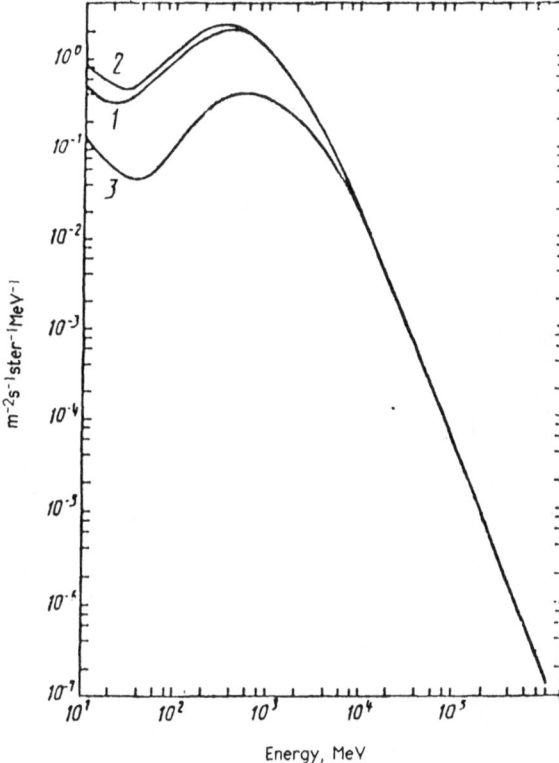

Fig. 2 GCR proton energy spectra for minimum even (negative) -(1), odd (positive)-(2) and for maximum 11-year solar cycles.

time of delay of flux GCR variations relative to solar activity variations δt(months) does not depend on particle energy and can be estimated using the equation:

$$\delta t = 10 + 5[\sin(2\pi/22)(t-t_0)]^{\frac{1}{2}} \tag{7}$$

where t is the time instant for which the calculation is made and which is expressed in years $t_0 = 1978.5$. For example, March 15, 1991 = 1991.2. To calculate the modulation potential of the heliosphere $R_0(t)$ one can take as initial data the monthly average Wolf numbers W_k which correspond to the time instants t or to the instants K-months back from them k = 1,2, ..., n, ..., 2n-1 where n is the time of delay t rounded to the integer. During calculations of energy spectra for the past time instant t one employs the real monthly average values of Wolf numbers, however, for the future time instants t one does the monthly average Wolf numbers which are determined using linear interpolation of quarterly average Wolf numbers. For all nuclei the detailed information about parameters and the FLXGCR calculation program was published by Nimmik and Suslov (1990).

It is seen from spectra shown in Figures 2 to 5 that with the energy lower than several GeV/nucleon differential spectra of GCR nucleons become flatter and with energies of about several hundreds of MeV/nucleon the maximum can be singled out on spectra. As it has been mentioned above the GCR intensity variation correlates with the solar activity variation. The physical reason of this modulation is based on the filling of the Solar system up by the flows of completely ionized plasma propagating from the Sun (solar wind). The solar wind includes and carries the so-called "frozen-in" magnetic field. The galactic cosmic radiation outside the heliosphere is diffusing inside the Solar system scattering on inhomogeneities ("disturbances")

5

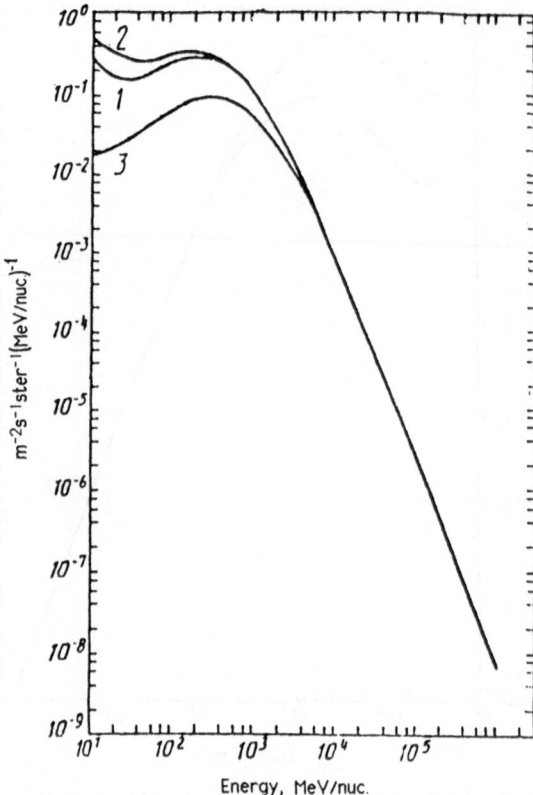

Fig. 3 GCR α-particles energy spectra for minimum even (1), odd - (2) and for maximum - (3) 11- year solar cycles.

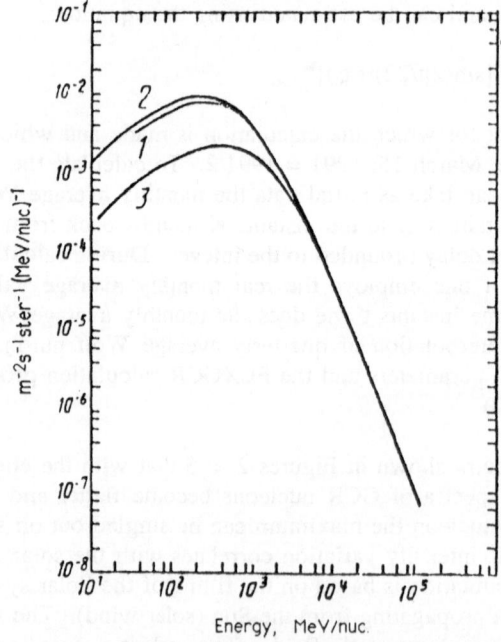

Fig. 4 GCR carbon nuclei energy spectra for minimum even (1), odd - (2) and for maximum - (3) 11-year solar cycles.

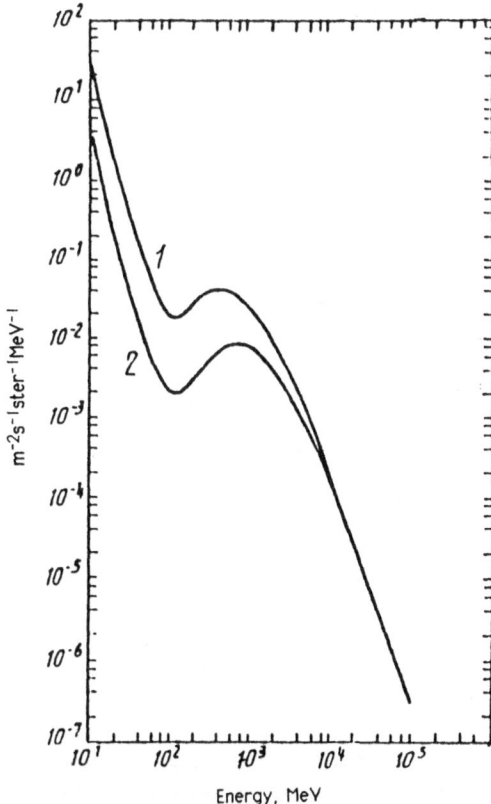

Fig. 5 GCR electron energy spectra for minimum - (1) and for maximum - (2) 11 -year solar cycles.

of the solar wind. During this movement GCR particles are decelerated in the process of adiabatic cooling due to the expansion of scattering structures of the solar wind. The GCR spectrum outside the heliosphere contains evidently the much greater number of particles with energies E < 10 GeV. Several authors characterizing the GCR spectrum outside the modulation region simply draw a straight line to the small energy region thus retaining the spectrum slope observed within 10 to 100 GeV. It should be noted that the validity of such a procedure is rather doubtful.

The total energy density for cosmic rays in Galaxy is about 1eV cm^{-3} that corresponds to the GCR total energy equal to about 10^{56} erg for the Galaxy volume V_G = 10^{68}cm^3. Nuclei of the GCR radiation during their propagation in space will interact with nuclei of the environment (ρ=10^{-2} cm^{-3}). This interaction is characterized by the following cross-sections and paths (Ginzburg and Syrovatskii, 1966, see Table 3).

This process of the interaction between GCR nuclei with the environmental matter determines the mean lifetime for a particle τ = 10^{17} to 10^{18} s. The estimates show that supernovae outbursts can be the source of GCR generation. Both nuclei (accelerated nucleus and target nucleus) can disintegrate during the interaction of space radiation nuclei with the interstellar gas. Table 4 lists the coefficients of fragmentation, i.e., the average number of K-group nuclei generated during absorption of one i-group nucleus (the lower index means the heavier nucleus) (Ginzburg and Syrovatskii, 1964).

Table 3 Cross-sections and free paths for different groups of nuclei.

NUCLEI	p	α	L	M	H	vH
cross-section, 10^{-26} cm^{-2}	2.8	11.2	20.7	29.7	50.5	74.3
free path, g cm^{-2}	7.2	18	9.8	8.0	6.0	2.7

Table 4 Fragmentation coefficients.

$\underset{i}{\overset{k}{}}$	H	M	L	α
H	.31±.07	.36±.07	.12±.04	1.35±.18
M		.11±.02	.28±.04	1.22±.11
L			.15±.05	1.09±.18
α				.41±.03

In a first approximation during fragmentation for high-energy particles the energy-per-nucleon is conserved, hence, energy spectra for primary particles and products of fragmentation are identical. In the process of fragmentation unstable secondary particles are also formed: K-mesons, π°-mesons, $\pi^\pm$-mesons, neutrons whose decay ultimately leads to generation of gamma-quanta, electrons, and neutrinos ($\pi^\circ \Rightarrow 2\gamma$; $\pi^\circ \Rightarrow \mu^\pm \Rightarrow e^\pm$). Thus the generation of secondary particles during the GCR interaction in the interstellar matter will result in the following fluxes:

$$I(E)dE = KS \cdot 10^{-4} E^{-2.64} dE \text{ cm}^{-2} \text{ st}^{-1} \text{ s}^{-1}.$$

Here K = 3.27 for gamma-quanta; K = 3.05 for neutrino; K = 1.11 for electrons; E is given in GeV; S is the matter thickness in g•cm^{-2} on the line of sight.

Besides the nuclear interaction nuclei of the primary, cosmic radiation interact with the relic radiation (T = 2.7K) and the Sun's photon radiation. The former interaction leads to the GCR spectrum limitation from the high energy wing (E>10^{19}eV); the second-type interaction results in photodisintegration of nuclei. Therefore, the primary cosmic radiation spectrum is the result of several processes of acceleration and deceleration of particles in the interstellar medium. In this case the significant role, besides the processes mentioned above, may be betatron acceleration or deceleration in the increasing or decreasing magnetic fields, acceleration in electric fields appearing due to dissipation of the magnetic field, as well as mechanisms of particle acceleration near shock-wave fronts and stochastical mechanisms related to the motion of magnetic inhomogeneities, wave processes and other events. In general, our Galaxy (halo and galactic disk) forms a specific trap whose characteristic sizes amount to tens of thousands of parsec. This trap keeps rather well particles of moderate or high energy, however, it is transparent for superhigh-energy particles. If the characteristic size of a trap is L = 10^4 parsec = 3.10^{22}cm and $v_{particle}$ = 3.10^{10} cm s^{-1} the transport path of GCR particles is L = 3 to 30 parsec or 10^{19} to 10^{20}cm for scattering and the mean time for particles being inside a trap is 10^7 to 10^8 years that is comparable with the GCR particle lifetime determined by nuclear interaction.

In recent years the doses caused by the GSR were estimated many times. The accuracy of these estimates is based on the knowledge of several basic factors:

1. the absolute differential fluxes of all components of the primary radiation;

2. transport processes occurring in spacecraft walls or in a special protection screen which transform the primary radiation (ionization losses, elastic and inelastic collisions, nuclear fragmentation, and reactions associated with electromagnetic radiation dissociation). The data on effective cross-sections of reactions of nuclear absorption and nuclear fragmentation are especially scarce;

3. the relative biological effectiveness for all components and the dependence of the integral relative biological effectiveness on a protection-screen thickness that determines the behavior of an equivalent dose in rems (cSv).

The accuracy with which components of the primary space radiation are measured does not exceed 10%, however, together with the uncertainty of nuclear fragmentation models this lead to the factor 2 for the inaccuracy in determining the dose (Letaw et al, 1987; Wilson et al, 1991). Due to this we should be extremely careful when estimating the dose rate dependence on the shielding thickness. The shielding thickness dependence on the dose rate level being rather strong, even a slight inaccuracy in dosage estimation leads to a very considerable increase in the shielding thickness. For instance, a 15 percent underestimation of the dose, which may result when meson production is neglected or nuclei fragmentation is not fully taken into account causes a 100 percent increase in the shielding mass (Letaw et al, 1987). Recently, to get compatible estimates of the shielding efficiency and dose estimates, dose calculations for blood-forming organs (BFO) are used, which corresponds to the dose taken at 5 cm depth in the tissue equivalent material whereas tables and plots present higher values of shielding thickness. Fig. 6 illustrates compatible estimates obtained in recent years. The estimates made by J.K. Letaw et al (1987) and J.W. Wilson et al (1991) rely upon the well known CREME model (Adams et al, 1981) to estimate GCR fluxes whereas the estimates made by J.H. Adams et al (1981) and G.D. Badhvar (1991) use new models for GCRs. Among these new models the one developed at the Moscow State University (Nimmik et al, 1991) should be mentioned as well as the model of G.D. Badhvar (1991). This model ensures a comprehensive description of GCR spectra from hydrogen to nickel. It employs a smooth and weighted average of monthly Wolf numbers to evaluate the so-called solar modulation potential. The model yields a much better agreement with the experiment than the well-known CREME model. Soon we should expect dose estimates as well as the predictions in the radiation situation based on that model. Elements of the model were described at the beginning of this paper within the discussion of energy spectra of GCR.

The model relied upon the solution of a spherically symmetric equation of diffusion for estimating the differential spectra of GCR particles. It successfully explains the previous periods of solar activity; however not so evident are its capabilities to predict the radiation condition. More recent results have shown that, compared with tolerant doses of 50 rem/year (Badhvar et al, 1991) genuine doses - even if there is a radiation- resistant shelter with a protective 20 to 30 g/cm^2 screen aboard the spacecraft - cannot reduce the dose to tolerance values. It thus prevents long-term (2 to 3 years) flights of manned spacecraft during the periods of minimum solar activity. How long could such "closed" periods last during a solar activity cycle? If the duration of a such period is equal the time when the GCR intensity varies by not more than 30 percent of its maximum value it should be 5 to 6 years. Though periods have been observed (see Fig. 7) when the intensity of GCR (reflected in the data of the neutron monitor) according above stated criteria lasted almost 8 years. The duration of a given specific minimum of solar activity can be predicted from the characteristics of the previous solar activity maximum. If that maximum was not sufficiently 'active', as for instance, in the case of a maximum of cycle 20,

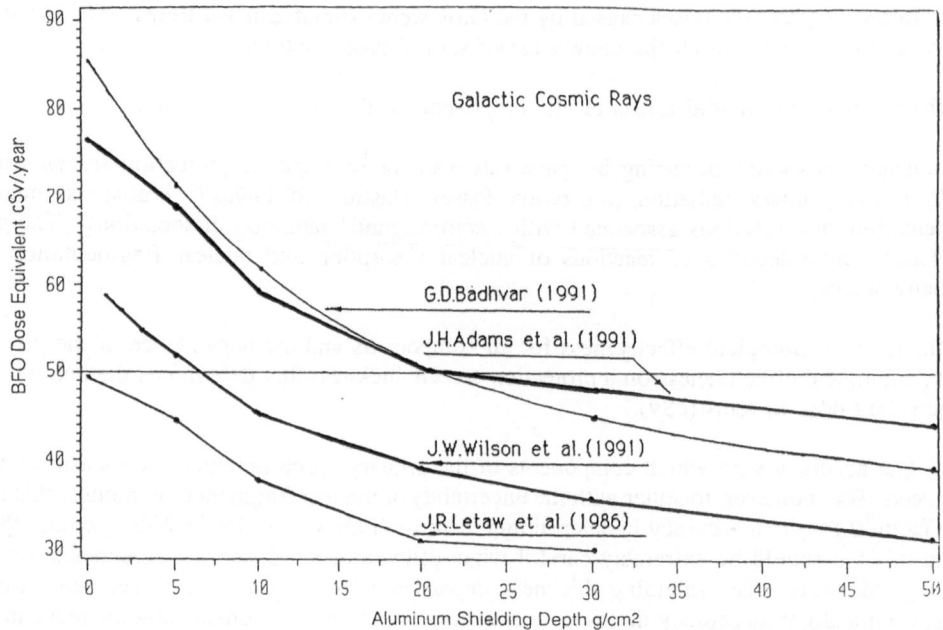

Fig. 6 BFO dose equivalent behind Al shield for GCR.

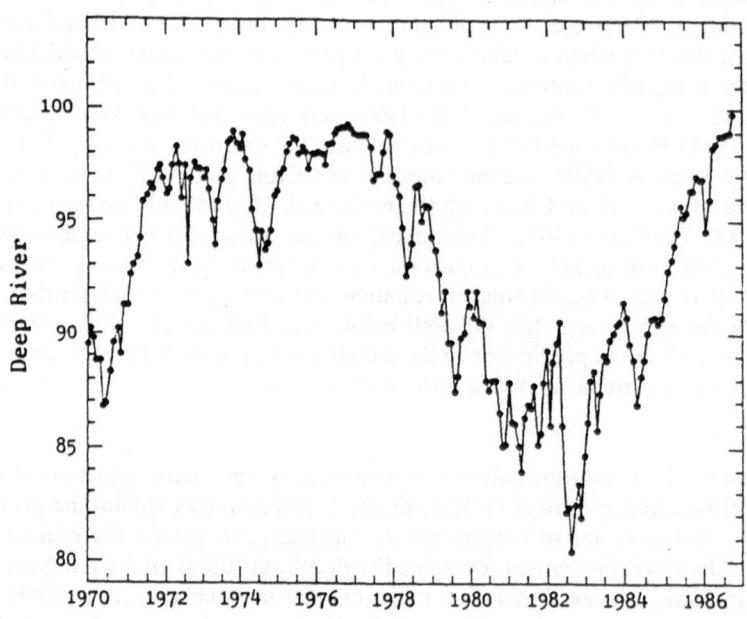

Fig. 7 Monthly average values of the Deep River neutron monitor counting rate normalized to 100% in May 1965 (solar minimum).

then the next minimum should surely be "deeper" than others. In reality, however, this situation is more complicated: in particular it is necessary to estimate in detail the 'hysteresis' of GCR respective to solar activity as well as to account for major flares that occur during the fall and rise phases of solar activity (see Fig 8, NCRP 98, 1989). In the future this problem should be more thoroughly considered. The radiation levels and doses induced by GCRs during the maxima of solar activity reduce by a factor of 2 to 2.7 depending on particle energy (2.3 times on the average) and reach values of 30 cSv per year inside modules with about 1 to 5 g•cm^{-2} Al protective screen and 20 cSv per year for a shelter 20 to 30 g•cm^{-2} Al screen. For the flight of a spacecraft with a shelter aboard such a value might be acceptable from the viewpoint of radiation hazard (if flares do not add more radiation). Though the above cited ideas of different authors concerning the accuracy of dose estimates also make eventually hazardous the value given for the solar activity maximum. The two-times change of the accuracy of the dose estimate in the shelter already puts to doubt the radiation safety of the crew even in the absence of radiation-hazardous flares over the larger part of the solar cycle, not to mention the fact that - in the context of the mission plan - the spacecraft crew cannot stay in the shelter during the entire flight. Thus from the radiation safety point of view, the general problem (concerning flight to Mars) shifts to questions of selecting the flight program (determination of the flight period within a solar activity cycle, duration of trajectory portions, and time of stay on the surface of the planet; development of a system for predicting major flares, procedures of protection against flare-induced emissions, etc.). It should be mentioned that whatever the flight option might be, the GCRs would generate a certain 'background' upon which doses caused by other factors would be superimposed. Table 5 summarizes doses induced by GCRs passed through several different typical shieldings (service module $d_1 = 1$ g•cm^{-2}Al; habitable module for crew $d_2 = 5$ g•cm^{-2}Al; shelter $d_3 = 20$ to 30 g•cm^{-2} Al) always for blood-forming organs (screening 5 g•cm^{-2} -tissue) in the maximum and minimum periods of solar activity. For reference the same Table 5 offers equivalent doses from the radiation of three strongest solar flares behind the same shieldings as in GCR-case.

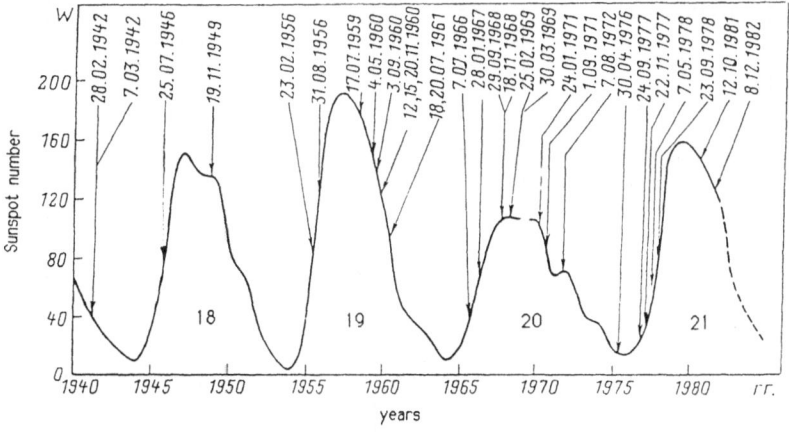

Fig. 8 Major solar particle events with energy E >500 MeV of the last four solar cycles.

Table 5 makes clear that the situation with long-term flights of manned spacecraft (even with the radiation shelter 20 to 30 g•cm^{-2}Al thick) is such that the permanent radiation level due to galactic space radiation prevents any flights lasting 12 months or longer. Hardly could this situation considerably improve in the nearest years in the course of more detailed studies of the effect of HZE-nuclei and very-high-energy, nuclei on biological objects. Most probably the permissible radiation doses (50 cSv per year in the USA and about 60 cSv per year in the Soviet Union) will very soon be reduced as we are learning more about the aftereffects of the Chernobyl accident, as well as about the influence of radiation on the immune system of man's organism. The prediction of the period of the first manned flight to Mars would most probably include a prohibition to carry a flight during the minimum solar activity period and during some years adjacent to the minimum. Thus, the manned flight to Mars will have to be conducted during the solar activity maximum or in the intermediate period between the minimum and the maximum.

The effect of radiations due to large flares affects the decision. It is clear, on the one hand, that in this case the radiation situation will depend on the protection of the spacecraft and shelter. The extremely thick protection (up to 50 g•cm^{-2}) could decrease the dose inside the habitable module or shelter down to acceptable values of the order of several cSv even for large flares of the type "February 1956", "August 1972" or "October 1989". The point is whether it is possible to design such a protection from the viewpoint of the structure itself and the implementation of the flight program. To evaluate possible hazard caused by flares the so-called worst case solar flare (most unfavorable for us) was simulated by Wilson et al (1991). This flare was designed by two ways: A) two integral spectra of the largest flares (February 23, 1956 and August 4, 1972) were superimposed to obtain one common spectrum and B) the spectral shape of the other "hard" flare was attributed to the flare with most intense fluxes of moderate-energy particles (10 MeV). The method of building these spectra is shown on Figure 9 (Wilson et al, 1991). The A-type spectrum yields the reasonable estimates of doses behind different protection screens. We built the differential spectrum of protons due to 28 largest solar flares ion the maximum of intensity in each energy range using the method of the A-type spectrum design.

Table 5 Doses induced by GCRs passing through different shielding thicknesses.

Solar Minimum			
	Service Modul 1 g·cm^{-2} Al	Habitable Space 5 g·cm^{-2} Al	Shelter 20 - 30 g·cm^{-2}
J.Letaw et al. 1988	48	37	31 - 28
S.Curtis et al. 1989	58	42	32 -
IMBP 1989	60	50	38 - 34
J.Adams et al. 1991	75	59	50 - 48
G.Badhvar 1991	82	72	51 - 45
Solar Maximum			
IMBP	32	28	23 - 20
Solar Flares			
23 February 1956	60	48	30 - 24
4 August 1972	400	140	10 - 5

Figure 10 demonstrates this spectrum. To compare with spectra on Figure 8 it is necessary to make time integration, however, this procedure needs specific assumptions and estimates of the time profile of the particle intensity in each energy interval. This work is under process now. As to the B-type spectrum the dose rates determined by such a spectrum turn out to be extremely great. So, for example, the dose is 20 cSv•hour^{-1} behind the protection screen from Al 40 g•cm^{-2} thick and 15 cSv•hour^{-1} behind the protection screen 70 g•cm^{-2} thick. For a burst duration of several hours the total doses become close to or exceed the critical value 50 cSv. It is practically impossible to protect from such flares. However, the assumptions taken into account in the process of the B-type flare building cannot be substantiated reliably. Hence, the probability that it adequately exists is low.

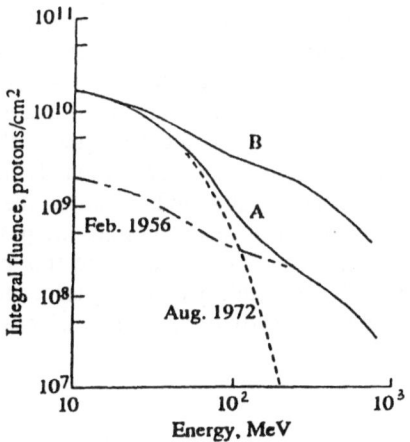

Fig. 9 Fluence spectra for hypothetical worst - case solar particle events compared with February 23, 1956 and August 4, 1972 cases. Case A is superposition both events. Case B is August 1972 event with February 1956 energy spectrum.

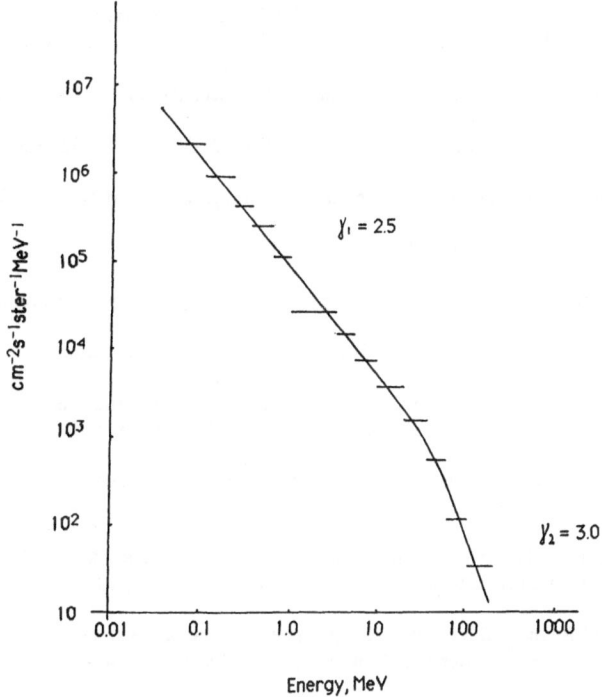

Fig. 10 Maximum flux spectra for hypothetical giant flare. Spectrum is superposition maximum fluxes in subsequent energy intervals observed during 28 different flares.

From the viewpoint of flares the question at issue is that substantiated assumptions about the limit of maximum energy release in a flare do not exist. The flares mentioned above were characterized by the total energy of the order of several units per 10^{32} erg. Observations made up to now show no evidence that the flares had higher energy release, i.e., we may say that during 50 years of more or less regular observations of flares no flare with energy release higher than 10^{33}erg occurred. Unfortunately we cannot still state that in future there will be no flare with such or even higher energy release. Therefore, the fact that we cannot explain why such flares were not observed during 50 years of observations causes anxiety for us. It is most probable that our Sun has such a structure that its size, mass, and emittance determine the level of turbulence which is responsible for those maximum magnetic disturbances in the convective zone; thus the photosphere and atmosphere of the Sun ultimately determine the maximum flare energy release. This is the problem necessitating fundamental studies in the nearest future. Thus, in years of maximum solar activity long-term space missions are possible only using a spacecraft with a very thick protection screen. Thus the emphasis of this problem is transferred to the design and manufacturing aspects. The possibility of a long-term space mission in years between the maximum and minimum of the solar activity needs special analysis.

REFERENCES

Adams, J.H., Report OG 5.2.7. on the Workshop "Galactic Cosmic Rays: Constraints on Space Exploration". Dublin, Aug.6 1991.

Adams, J.H., Silberberg, R. and Tsao, C.H., 1981, "Cosmic Ray Effects on Microelectronics", Part 1, The Near-Earth Particle Environment, NRL Memorandum, Rep. 4506, Naval Research Laboratory, Washington DC.

Badhvar, G.D., 1991, "Model of Galactic Cosmic Radiation for Space Exploration Missions", Report on the Scientific Meeting "Radiation Safety of Manned Mars Mission", Oct. 1-3, 1991, Dubna, Moscow Region, USSR.

Dorman, L.I., 1977, Proceedings of the 15nd Cosmic Ray Conference, Plovdiv, vol.1, p.405.

"Galactic Cosmic Rays", the Model of the Particle Fluxes", GOST 25645.150-90, Ed. Standard", Moscow, (1991).

NCRP 98, National Council on Radiation Protection and Measurements, Report No 98, (1989).

Ginzburg, V.L. and Syrovatskii, S.I., 1966, UFN, vol. 88, p.485.

Ginzburg, V.L. and Syrovatskii, S.I., 1964, "The Origin of Cosmic Rays", Pergamon Press.

Letaw, J.R., Silberberg, R. and Tsao, C.H., 1987, Radiation Hazard on Space, Nature, vol. 330, p. 709.

Mavromishalaki, H., Marmatsouri, L. and Vasilaki, A., 1989, "Some Characteristics Feature of Solar Activity During Solar Cycle 21". Proceedings of the 20th Cosmic Ray Conference, Moscow, vol.3 p.357.

Miroshnichenko, L.I. and Petrov, V.M., 1985, "Dynamics of the Radiation Environment in Space", Ed. Energoatomizdat, Moscow.

Nimmik, R.A., Panasjuk, M.I., Pervaya, T.I. and Suslov, A.A., "Model GCR Fluxes", Report on the Workkshop "Galactic Cosmic Rays: Constraints on Space Exploration", Dublin, Aug. 16, 1991.

von Rosenwinge, T.T. and Reames, D.V., 1989, "Reappearence of the Anomalous Oxygen Component at 1A.U.", Proceedings of the 20 Cosmic Ray Conference, Moscow, Aug. 1989, vol.3, p.434.

Smart, D.F. and Shea, M.A., 1985, Galactic Cosmic Radiation and Solar Energetic Particles in: Handbook of Geophysics and the Space Environment", A.S.Jursa ed. AFGL, USA.

Nimmik, R.A. and Suslov, A.A., 1990, Semi-empirical Model for the large Scale Modulation of the Galactic Cosmic Rays Energy Spectra, Proceedings of the 21st Cosmic Rays Conference, Adelaida, vol. p.33.

Wilson, J.W., Townsend, L.W., Schimmerling, R., Khandelwal, G.S., Khan, F., Nealy, J.E., Cucinotta, A., Simonsen, L.C., Shinn, J.L. and Norbury, J.W., 1991, "Transport Methods and Interactions for Space Radiation". NASA Ref. Pub. 1257; USA.

INFLUENCE OF THE GEOMAGNETIC FIELD AND OF THE SOLAR ACTIVITY CYCLE ON THE COSMIC RAY ENERGY SPECTRUM

R. Beaujean

Institute for Nuclear Physics
University Kiel, Germany

INTRODUCTION

Energetic charged particles impinging on low altitude earth-orbiting satellites originate from four different sources: a) galactic cosmic rays from outside the heliosphere, b) the singly ionized anomalous component, c) solar energetic particles from solar flares and d) trapped particles from the earth's radiation belts. The topic of this lecture is the modulation of the galactic cosmic ray energy spectrum while the particles propagate through the heliosphere and the earth's magnetic field until they reach the spacecraft. Additional variation by the shielding material of the spacecraft and of the radiation sensitive volume is not discussed here. It is shown that the flux of energetic particles strongly depends on the solar activity and on the orbit parameters.

GALACTIC COSMIC RAYS

The galactic cosmic ray particle radiation inside the heliosphere is composed of about 98% nuclei and 2% electrons and positrons. The nuclear component consists of about 87% hydrogen, about 12% helium and about 1% for all heavier nuclei. These nuclei are stripped of all their orbital electrons and have travelled about 10 million years in our galaxy before they arrived at earth. All elements of the periodic system are present in this matter from outside our solar system. However, the relative abundances of the elements vary orders of magnitude (Fig.1). There is no conclusive proof on the source and the mechanism accelerating the nuclei to cosmic ray energies after injection. At earth the highest intensity of the galactic cosmic ray nuclei (as the differential flux of particles per m² s sr MeV/nuc) is in the energy range 0.1 - 10 GeV/nuc. Above 10 GeV/nucleon the energy spectrum is represented by an inverse power law with an exponent of 2.5-2.7 depending on the nuclear charge. The spectra of nuclei extend to energies in excess of 10^{20} eV with a steepening of the slope at about 10^{16} eV. It appears that the ultra-high-energy nuclei probably are of extra-galactic origin. Due to the irregular deflection in interstellar magnetic fields in our galaxy we observe a highly isotropic arrival at the solar system. This means that we can not deduce information about the location of the source from their arrival direction. For a more detailed review on the properties of the galactic cosmic rays see Simpson (1983) and Wefel (1988).

MOTION OF CHARGED PARTICLES IN MAGNETIC FIELDS

When a particle with charge Ze moves with velocity v in a uniform magnetic field B it

Biological Effects and Physics of Solar and Galactic Cosmic Radiation,
Part B, Edited by C.E. Swenberg *et al.*, Plenum Press, New York, 1993

15

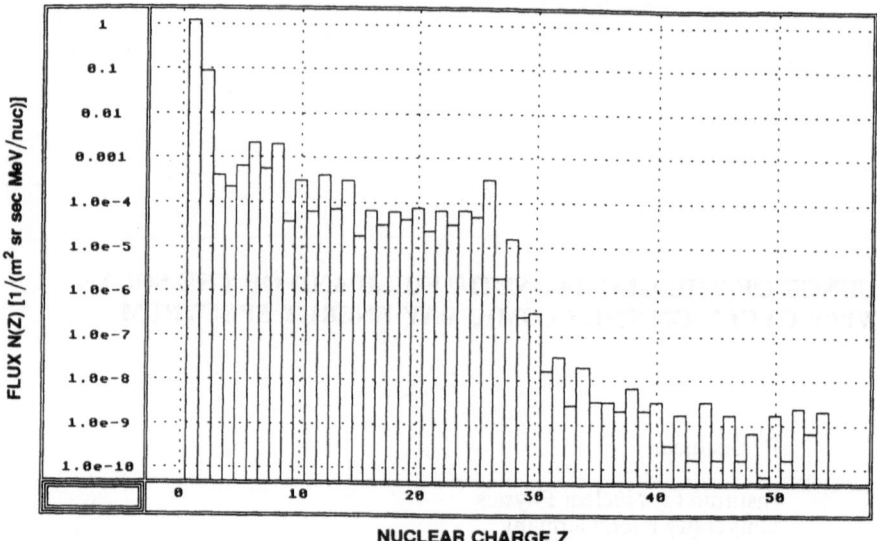

FLUX N(Z) [1/(m² sr sec MeV/nuc)]

NUCLEAR CHARGE Z

Figure 1. Flux of the Galactic Cosmic Ray Nuclei (Elemental Abundances) at 1 GeV/nuc in 1984 as Calculated from Adams et al. (1981) and Simpson (1983).

is deflected by the Lorentz force according to the relation

$$m \, (\, dv \, / \, dt \,) = (\, Ze \, / \, c \,)(\, \underline{v} \times \underline{B} \,) \tag{1}$$

The direction of this force is perpendicular to the plane defined by the vectors $\underline{v}$ and $\underline{B}$. The angle between the two vectors $\underline{v}$ and $\underline{B}$ is called the pitch angle α. Due to this force the moving direction is changed whereas the kinetic energy of the particle remains constant. In general, $\underline{v}$ has components perpendicular and parallel to $\underline{B}$, and the path of the particle is a helix with radius $r=f(v,B)$, which is called the gyroradius or cyclotron radius (Fig. 2a). If this radius is small compared with the dimension of the magnetic field curvature, the particle essentially follows a line of force which defines the direction of motion of its guiding center - the point about which the particle executes its cyclotronic motion. Fig. 2b shows the drift effect for particles moving in a nonuniform field. Due to the different gyroradii the particle drifts perpendicular to the field lines.

If B is perpendicular to v, the force is mutually perpendicular to both, and in an uniform field the particle describes a circle with radius r and the acceleration is given by

$$dv \, / \, dt \, = \, v^2/ \, r \, = \, (\, Ze \, / \, mc \,) \, (v^*B_{\perp}) \tag{2}$$

With the relativistic momentum p this can be written as

$$r^*B_{\perp} = (pc) \, / \, (Ze) \, = \, R \tag{3}$$

defining the magnetic rigidity R. Particles with the same rigidity R will have the same trajectory in a given magnetic field. However, the time needed for this trajectory depends on the velocity v of the particle. The following relation may be used to convert a given rigidity R (in Volt) into kinetic energy per nucleon E_k (in eV/nuc)

$$E_k \, = \, (\, (\, mc^2 \,)^2 + (\, R^* \, (Ze) \, / \, A \,)^2 \,)^{1/2} \, - \, mc^2 \tag{4}$$

where mc^2 is the rest mass energy per nucleon, A is the number of nucleons and Ze is the charge (Smart and Shea, 1984). This relation is shown in Fig. 3 for protons (A/Z=1) and particles with A/Z=2.

In an arbitrary, nonuniform magnetic field the charged particle follows a complex

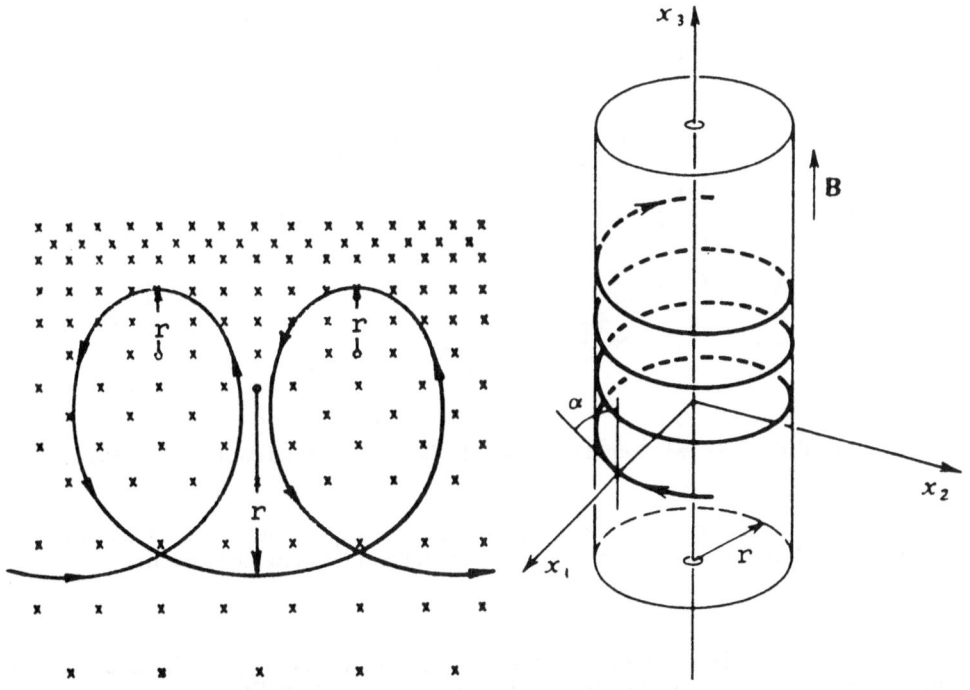

Figure 2. Motion of Charged Particles in Magnetic Fields.

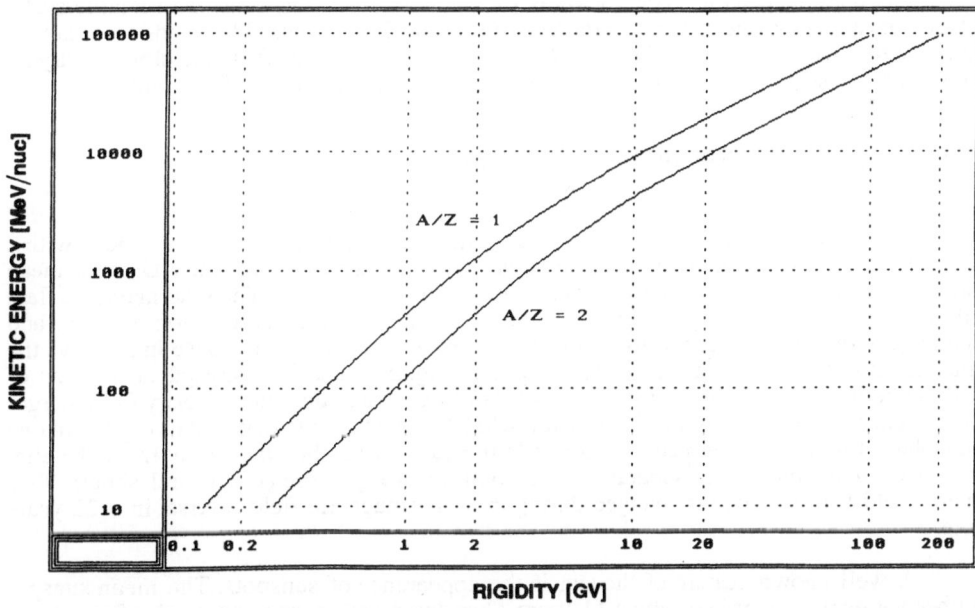

Figure 3. Kinetic Energy per Nucleon Versus Rigidity for Protons (A/Z=1) and A/Z=2 Particles.

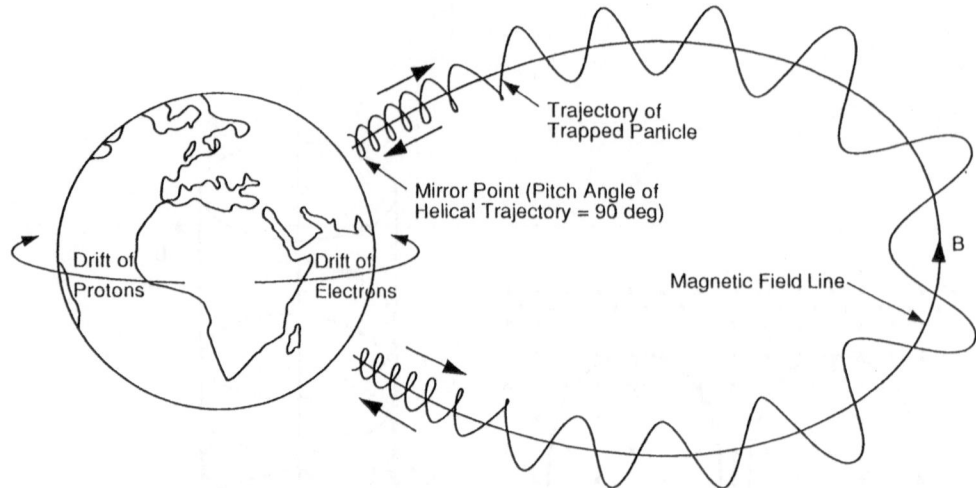

Figure 4. Path of Trapped Charged Particles in the Geomagnetic Field (Watts et al., 1989).

three dimensional trajectory. If the variation of the B field within the dimension of the gyroradius is small the conservation of the magnetic moment leads to the relation

$$\sin^2(\alpha) / B = \text{const.} \tag{5}$$

This relation describes the focussing effect of a divergent magnetic field. Particles moving into sites of decreasing fields will be focussed in the direction of the field line (the pitch angle decreases). If the particle moves into an increasing field the pitch angle will increase up to 90 degrees where the particle is reflected (mirror point). An example for this motion is shown in Fig. 4.

The interplanetary magnetic field (as well as the interstellar) consists of a large scale, slowly varying component with overlaid shorter scale fluctuations. This leads to a particle motion with two components: helical motion with systematic drift in the slowly varying fields, and stochastic diffusion like processes in the fluctuating portion of the field.

MODULATION IN THE SOLAR SYSTEM

The solar system, a small part of our galaxy, may be described as a sphere with a radius of 50 -100 AU with the sun located in the center (1 astronomical unit, AU, is the mean distance between Sun and Earth and equals to about 150 million km). The solar magnetic field within the heliosphere is coupled to the solar wind, a plasma emerging from the sun and streaming outward at a mean velocity of 400 km/s. As the sun rotates once in 27 days the solar magnetic field lines form archimedes spirals (Fig. 5a). The field strength at the orbit of Earth is about 5×10^{-5} Oe and it decreases to $1-2 \times 10^{-6}$ Oe when the interplanetary field merges in the interstellar magnetic field. The general field line orientation is depicted in Fig.5b: in one hemisphere the field line direction is towards the sun and in the other away from the sun. These two hemispheres are separated by a complex wavy plane (the neutral sheet). This general field line direction is changed during the solar magnetic field reversal in a 22 years cycle.

A well known feature of the sun is the appearance of sunspots. The mean sunspot number varies with a cycle of about 11 years. Correlated with sunspots are solar flares from which solar material is ejected into the interplanetary space. Associated with the ejection of high speed plasma, shock waves and magnetic disturbances move outward to the boundary. When the galactic cosmic rays penetrate the heliosphere they interact with the general

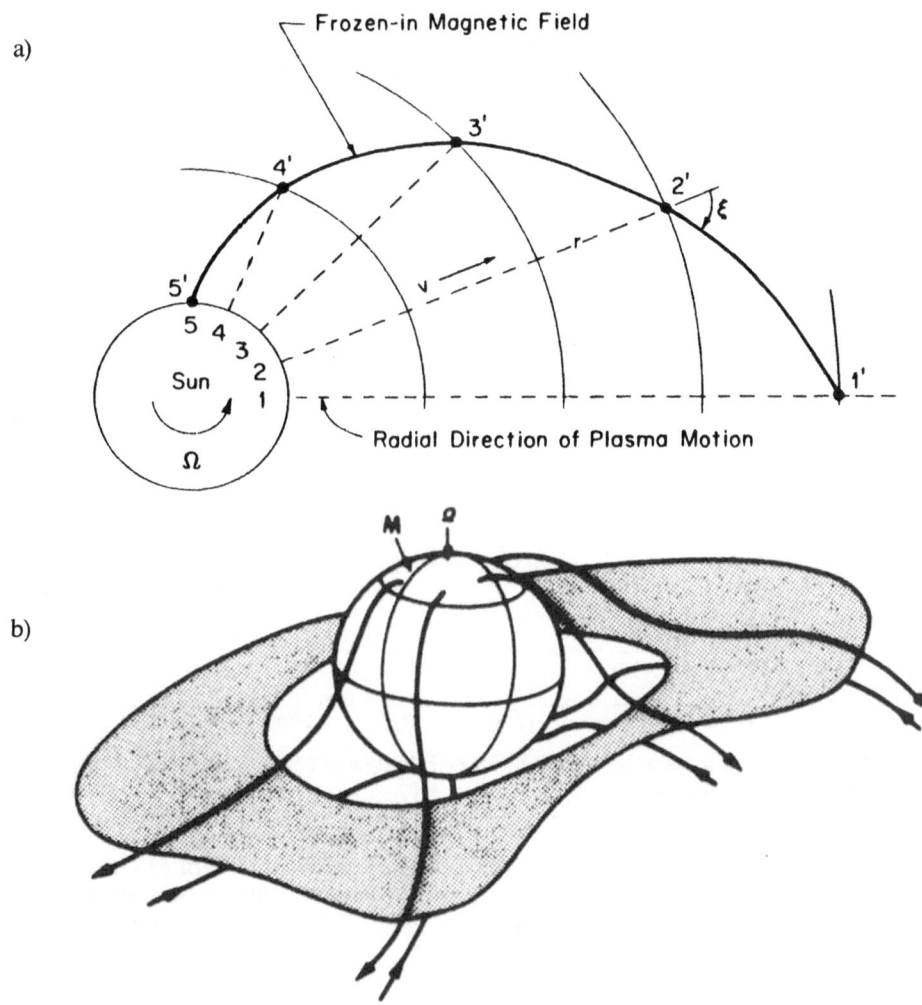

a)

Frozen-in Magnetic Field

Sun

Ω

ξ

r

v

Radial Direction of Plasma Motion

b)

M   Ω

Figure 5. Sketch of the Solar Magnetic Field.

magnetic field and the shorter scale disturbances. The propagation is described by the Fokker-Planck transport equation taking into account diffusion, convection, adiabatic deceleration and drift. However, a grand unified model is not yet established.

Let us now consider how these effects influence the cosmic ray particle spectra at the orbit of Earth. Outside the geomagnetic field the intensity of the different nuclei depending on their energy can be measured at different times within the 11 years solar activity cycle. Another integral method is to study the production of secondary neutrons produced by the cosmic rays in collisions with the atoms of the atmosphere. This is done by a network of ground based neutron monitors.

Fig. 6 shows the time dependence of the mean sunspot numbers and neutron intensity measured by the Climax neutron monitor (located in Colorado, USA). Obviously the neutron intensity is anticorrelated with the solar activity. Furthermore the shape of the intensity profile is correlated with the solar magnetic polarity which can be understood in terms of drift effects.

Fig. 7 shows recent results for the differential energy spectra of protons and helium

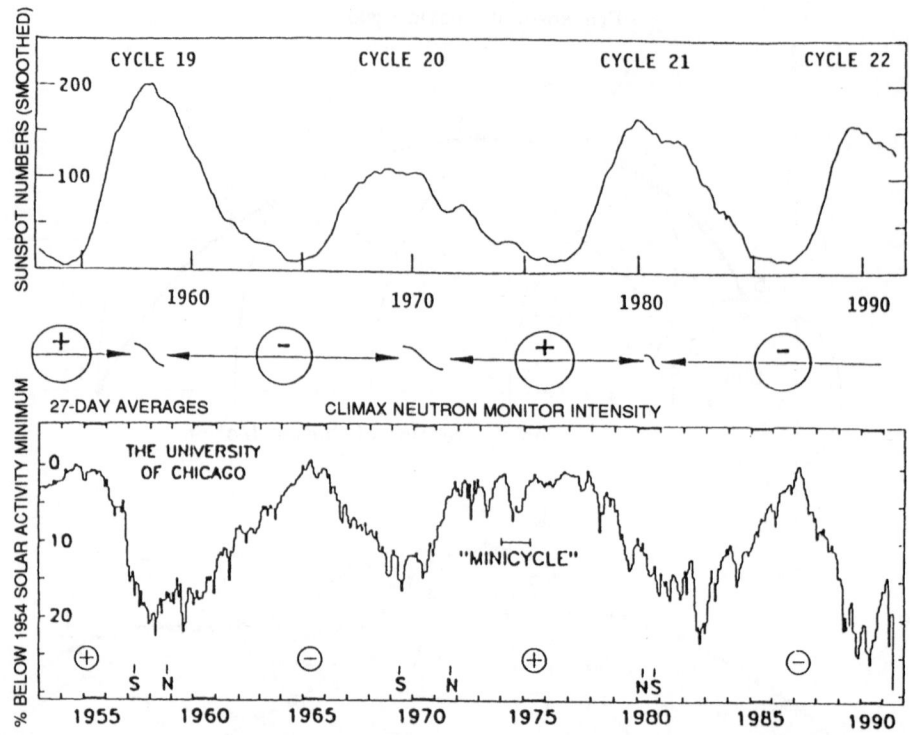

Figure 6. Smoothed Sunspot Numbers, Solar Magnetic Polarity and 27-Day Averages of Climax Neutron Monitor Intensity (Flückiger, 1991).

presented by Seo et al. (1991). Using this data the local interstellar spectra LIS were deduced from the symmetric solution of the transport equation. Measurements on deep space missions like Pioneer, Voyager and Ulysses contribute to the understanding of the three dimensional modulation of the charged particles in the heliosphere.

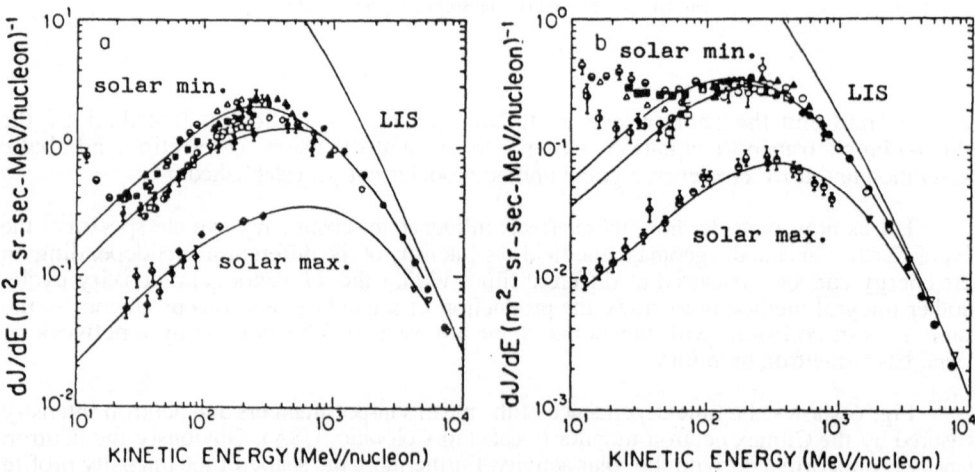

Figure 7. Measured Proton (a) and Helium (b) Spectra during the 1987, 1977 and 1965 Solar Minima and the 1969 Solar Maximum. The Curves Represent Modulation Theory Fits to the 1977, 1965 and 1969 Measurements (Top to Bottom) (Seo et al.,1991).

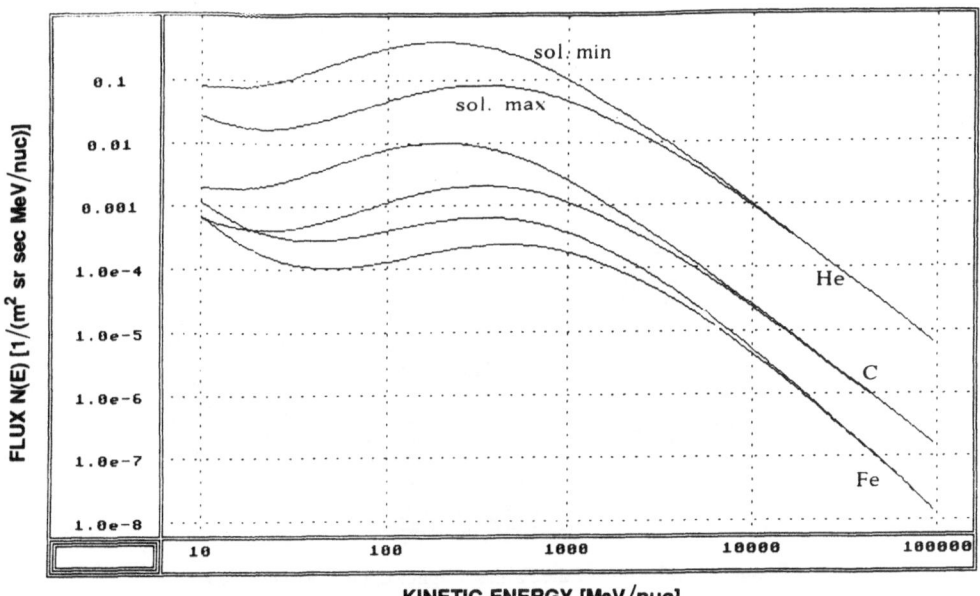

Figure 8. Modulated Energy Spectra as Calculated after Adams et al., 1981.

Attempts have been made to develop a numerical code for the time dependence of the cosmic ray flux (Adams et al.,1981, Badhwar and O'Neill,1991, Nimmik et al.,1991). The spectra of Fig. 8 were calculated using the model of Adams. They show that the solar modulation can be neglected above 10 GeV/nuc and the maximum variation at energies around 100 MeV/nuc is about a factor of ten.

In conclusion the features of the solar modulation are summarized. Charged particles entering the heliosphere from the interstellar medium interact with the outbound streaming solar wind and the interplanetary magnetic field. The large scale structure, the tilt angle of the mean solar dipole, the polarity if this field and the form of the wavy neutral sheet vary during the solar activity cycle. Short scale magnetic disturbances and interplanetary shock waves cause "minicycles" on the general modulation. The modulation effect on the LIS decreases with increasing radial distance from the sun, the integral radial gradient is about 3% / AU. Due to drift effects electrons and positive charged particles show different modulation. At the orbit of Earth a high (low) solar activity is correlated with a low (high) intensity of galactic cosmic rays respectively.

MODEL OF THE EARTH MAGNETIC FIELD

After passing the interplanetary space the cosmic ray nuclei enter the earth magnetic field on their way to a low altitude orbiting satellite. Again they are deflected depending on their magnetic rigidity $R=(pc)/(Ze)$ (equation 3).

The complex structure of the earth magnetic field (Figs. 9 & 10) can be divided into two parts,which are both time dependent and together form the magnetosphere which is embedded in the interplanetary magnetic field: a) The inner (or main) field is produced by sources inside the Earth and can be described by an inclined, eccentric dipole field with multipole additions. b) The outer field is the result of the interaction between the earth magnetic field and the solar wind. This field is associated with a system of currents which strongly depends on the solar wind conditions.

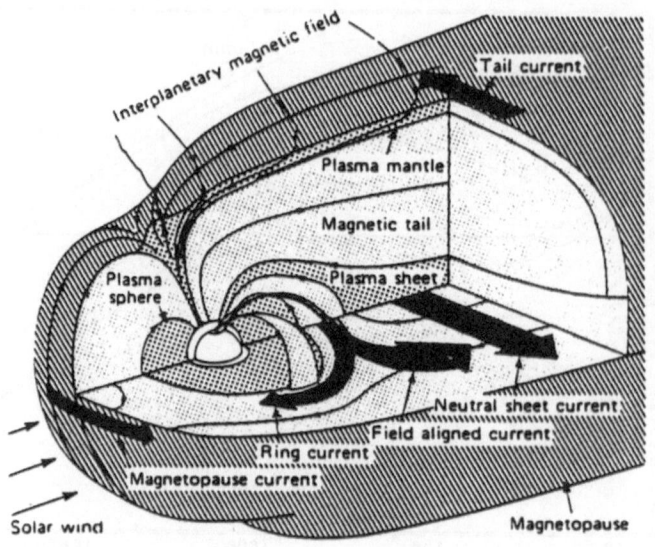

Figure 9. Structure of the Earth's Magnetosphere (Kobel, 1989).

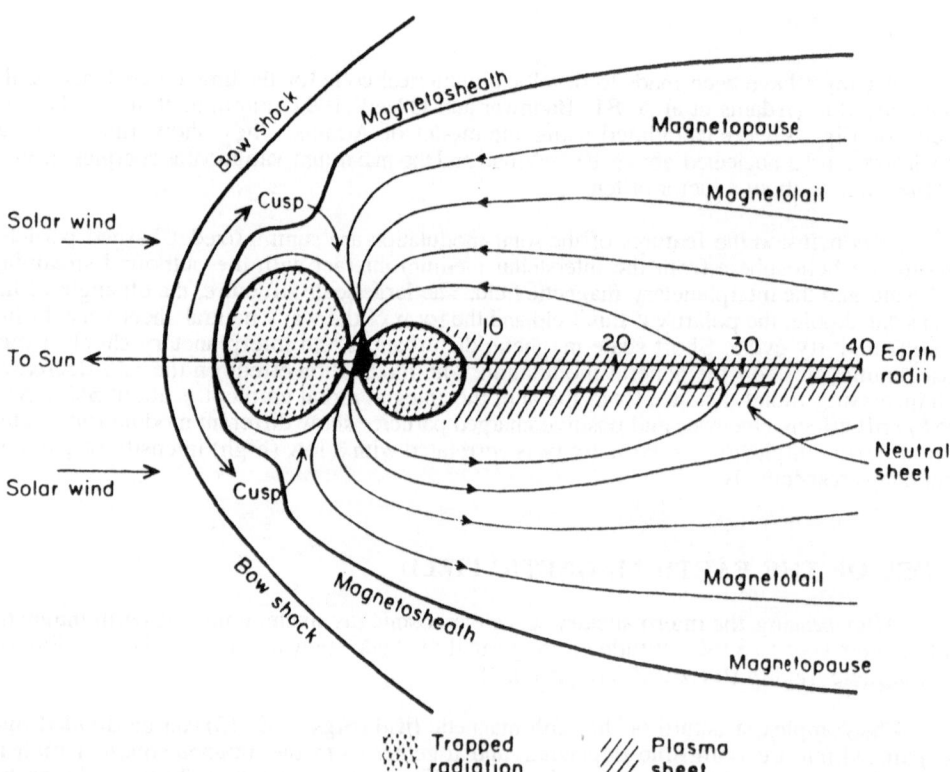

Figure 10. The Different Regions of the Magnetosphere in the Noon-Midnight Meridian Plane (Merrill and McElhinny, 1983).

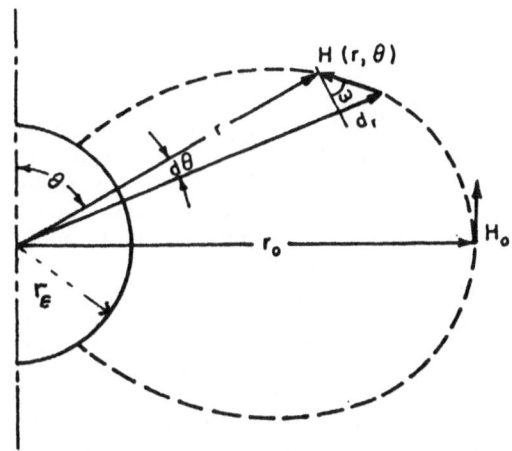

Figure 11. The Geometry of a Dipole Magnetic Field Line

Let us first consider a pure dipole field (Fig. 11). The geometry of the field line is

$$r = r_o * \sin^2(\theta) \tag{6}$$

where θ is the magnetic co-latitude. The inclination of the field line is

$$\tan(\omega) = 2*\cot(\theta). \tag{7}$$

Using the distance r_o where the field line crosses the magnetic equator, the McIlwain-parameter L is defined by

$$L = r_o / r_e = r / (r_e * \sin^2(\theta)) \tag{8}$$

in units of the earth radius r_e=6378 km. When this definition of L is transferred to the real field it defines the invariant latitude Λ of the field line on the ground by

$$\cos^2(\Lambda) = 1 / L \tag{9}$$

which is the most appropriate latitude for geomagnetic effects.

A dipole with a magnetic moment $M=8.05*10^{25}$ Gauss*cm^3 and tilt angle of 11.5 degrees with respect to the geographic axis is a good first approximation for the inner field. The field strength at the surface of the Earth is 0.3 Oe at the magnetic equator and twice as much at the poles.

Measurements of the real field have shown that a better approximation is achieved when the center of this dipole is shifted about 500 km out of the center of the earth towards about 160 degree east. Final adjustments are done by multipole additions in a standard field model for the main field. This international geomagnetic reference field IGRF is defined by a set of coefficients for spherical harmonics taking into account the secular variation of about 0.1% per year. Using the published coefficients the main field can be calculated for a given epoch (year).

One important feature of the tilted, eccentric dipole field is the South Atlantic Anomaly SAA. Due to the asymmetry of the field with respect to the earth surface the mirror points of the inner radiation belt are located at lower altitudes in the SAA which leads to a higher trapped particle flux in this area compared to other positions in a low altitude orbit(see Fig. 4).

GEOMAGNETIC SHIELDING

The geomagnetic field acts as a huge magnetic spectrometer. Fig. 12 shows the calculated trajectory of a cosmic ray particle with A/Z=2 at 4 GV. The arrival direction at Earth is quite different from the original direction of approach far away from earth (the asymptotic direction). The field forms a complicated lens which imposes restrictions on the directions of viewing. Another consequence is a filtering effect whereby particles which do not meet a certain minimum requirement (depending on the location inside the field and on the arrival direction) are not admitted. This requirement is called the cutoff rigidity.

Störmer's theory dealt with the problem of determining the regions of space around a pure dipole that are accessible to particles of any specified magnetic rigidity. The calculation showed that for a given location particles with a given rigidity R can reach a specific location only outside a solid angle region called the forbidden cone. The remainder solid angle contains the allowed cone (where all arrival directions are allowed) and the penumbral region where the access is allowed for some directions. However, some regions of the allowed cone and the penumbra are excluded due to short range earth intersections of the approaching trajectory (shadow cone, see Cooke et al, 1985).As an example of Störmer's analysis let us consider positive charged particles with a magnetic rigidity of R=10 GV. These particles cannot reach the geomagnetic equator but begin to arrive within a narrow cone about the western horizon at some critical latitude. The cone broadens out until, at another critical latitude, it fills the entire hemisphere.

The Störmer treatment is restricted to a pure dipole field. For the real field the allowed arrival directions can be calculated by the trajectory tracing method TTM. In this method an inverse charged particle is injected from the specific location into the field and the inverse trajectory is calculated by solving the equation of motion (1) in the field model. The trajectory of Fig. 12 was calculated this way for the 1975 IGRF.

Let us now restrict to vertical arrival, i.e. looking to the zenith. All particles with rigidities above the critical cutoff rigidity are allowed. The flux of these particles is the same as it would have been without the presence of the magnetic field. As we move towards the

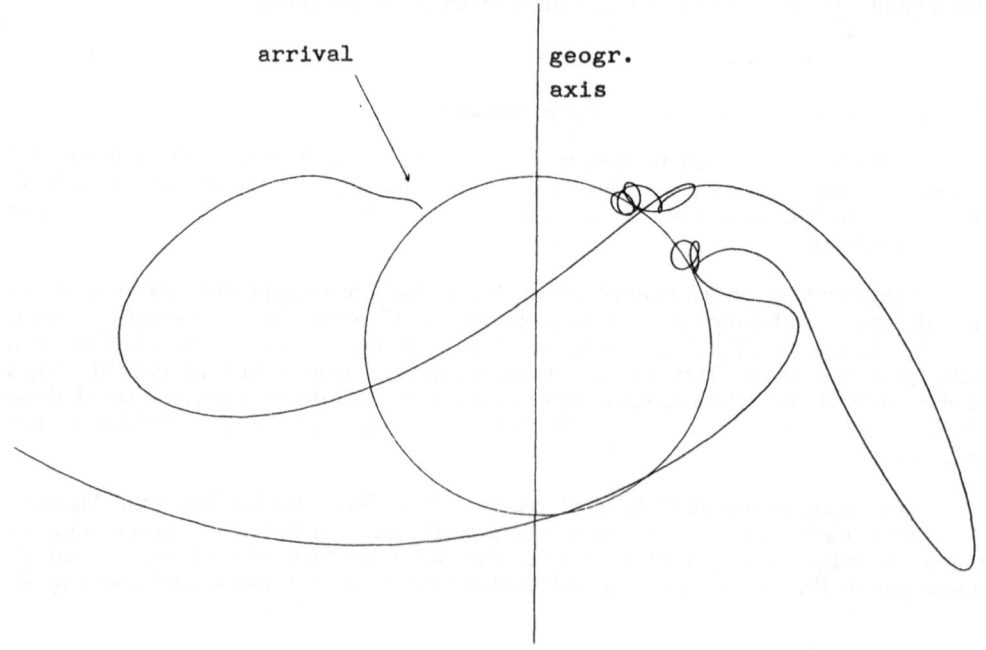

arrival
geogr.
axis

Figure 12. Calculated Trajectory of a Cosmic Ray Particle in the Earth's Magnetic Field.

pole the threshold rigidity decreases. At a given geomagnetic latitude, no particle with less than the Störmer cutoff rigidity can enter in the vertical direction, but all particles having rigidities exceeding the main cutoff have access. Some particles with rigidities between these two limits may enter (depending on their individual trajectories). The penumbra for a specific location is shown in Fig. 13, calculated by the TTM (Kobel 1990). The Störmer treatment for the dipole field leads to an analytic expression for the vertical Störmer cutoff rigidity (in GV) as a function of the geomagnetic latitude λ and the distance r from the dipole center (in earth radii) (or of the L-value according to equation 9)

$$Rs = 14.9 * \cos^4 (\lambda) / (r / r_e)^2 = 14.9 / L^2 \tag{10}$$

Fig. 14 shows a world map of effective vertical cutoff rigidities calculated from the IGRF model of the main field using the trajectory tracing method (Shea et al., 1987). This map can be used for a rough estimate of the geomagnetic shielding at a given location.

SATELLITE ORBITS

A set of five orbital parameters defines the motion of a satellite around the Earth. In general the trajectory is an ellipse with given eccentricity and semi major axis. The angle between the orbital and the equatorial plane is called the inclination. Furthermore the longitudinal position of the satellite above the equator at a certain time and the angle between this position and the perigee have to be known in order to calculate the geocentric position of the satellite at any specific time. For longer missions at low altitude, the atmospheric drag has to be taken into account.

The revolution time of low altitude orbits is close to 90 minutes. During 90 minutes the Earth revolves for 22.5 degrees and thus successive satellite ground tracks are shifted by this value. After completing 16 revolutions the satellite is above the same geographic location. Fig. 15 shows some ground tracks for Spacelab 1 (almost circular orbit at 249 km altitude). The inclination of 57 degrees determines the highest latitude reached in this orbit. The variable magnetic shielding conditions for the spacecraft can be obtained by combining these ground tracks with cutoff rigidities of Fig. 14.

Another description of the varying magnetic conditions on orbit is depicted in Fig. 16. For any given orbital position the corresponding L-value is defined by the local field line (field line tracing to the magnetic equator). Successive peaks of Fig. 16 correspond to maximum northern and southern latitudes respectively, equator crossings at about L=1 occur every 45 minutes. The variation of the peak heights is caused by the tilt angle of the dipole axis with respect to the geographic axis: For a specific orbit the maximum latitude position is close to the magnetic pole, while eight orbits later this distance is at maximum.

MISSION AVERAGED CUTOFF PROBABILITIES

As a consequence of Liouville's theorem (well known from mechanics), the particle

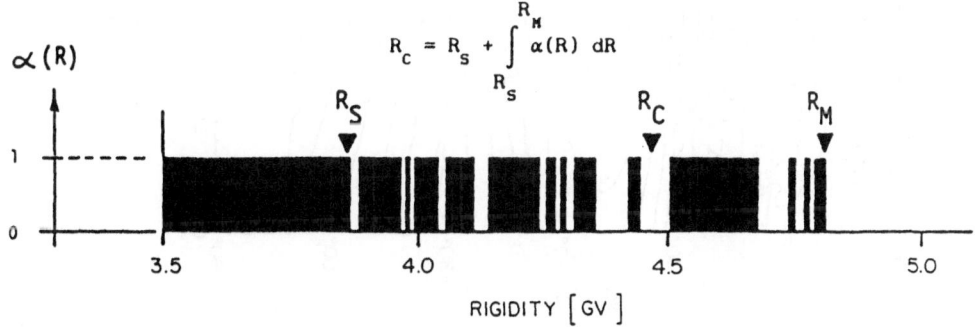

Figure 13. Penumbra at Jungfraujoch (Switzerland) for Vertical Arrival Showing Störmer Cutoff Rs, Main Cutoff Rm and Effective Cutoff Rc (Kobel, 1989).

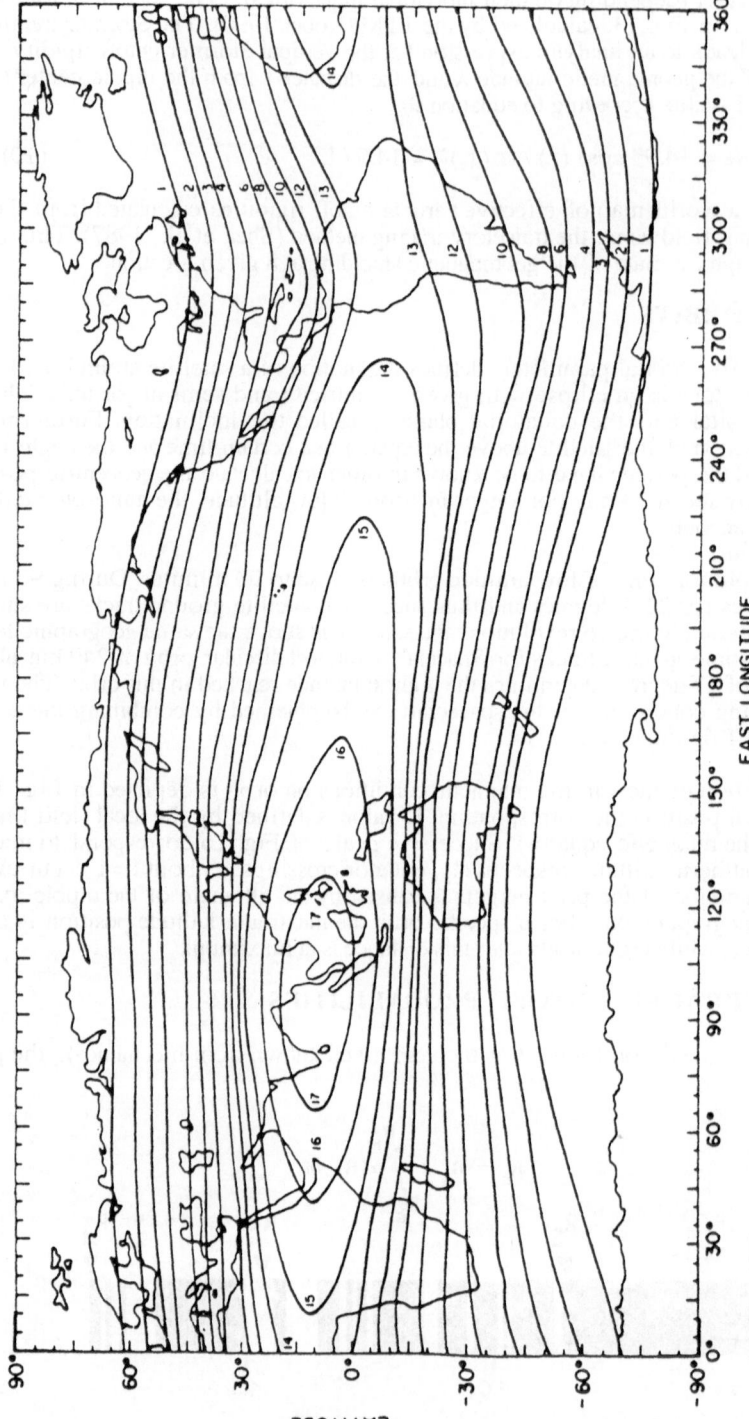

Figure 14. Contours of Vertical Cutoff Rigidities (in Units of GV) as Calculated for Epoch 1955.0 (Shea et al., 1987).

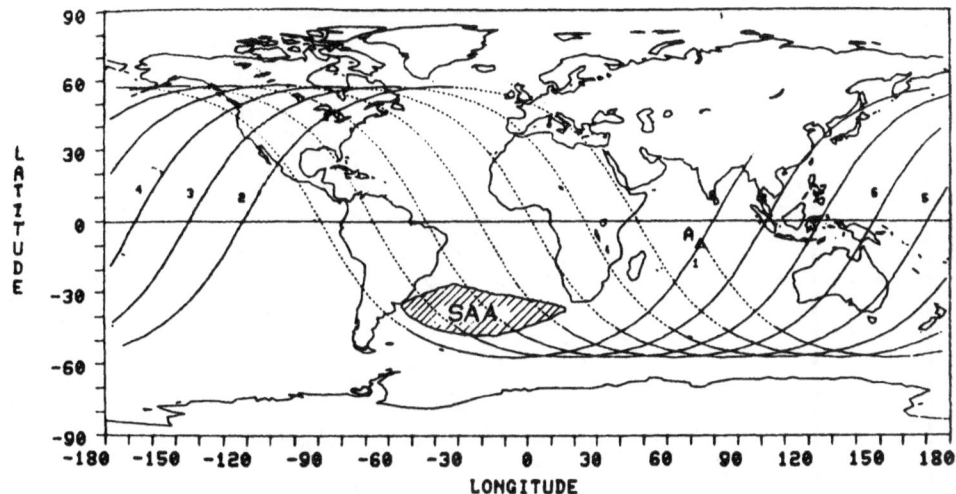

Figure 15. Ground Tracks for Spacelab 1 (A = insertion into 248x249 km orbit) (NASA SL-1 FDRD,MSFC,1978).

flux inside the magnetic field is the same at all allowed places and zero elsewhere. Due to the continuously changing cutoff conditions on orbit, the time interval spent at allowed places depends on the particle rigidity. We need to know the time history of cutoff versus mission elapsed time to compute the effect of the geomagnetic modulation on the mission averaged spectrum.

In a first approach to this problem, we consider only zenith viewing, i.e. vertical cutoff values. Following Fig. 13 the effective vertical cutoff Rc (averaging the penumbra) is used to define the transmission function T(R) at a specific location which is a step function with two values: T(R)=0 for all rigidities R below Rc and T(R)=1 for all rigidities above Rc.

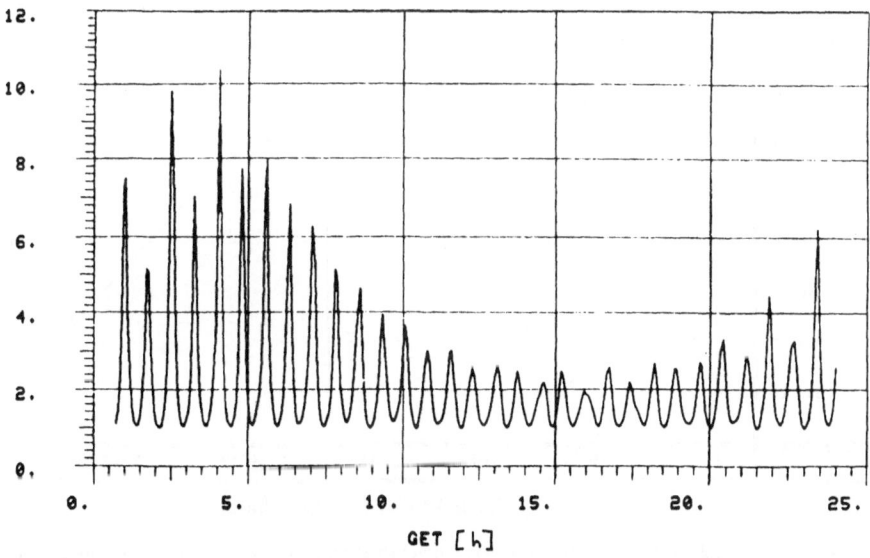

Figure 16. McIlwain Parameter Versus Time for First Day of SL-1 at 57 Degree Inclination (NASA SL-1 FDRD,MSFC,1978).

27

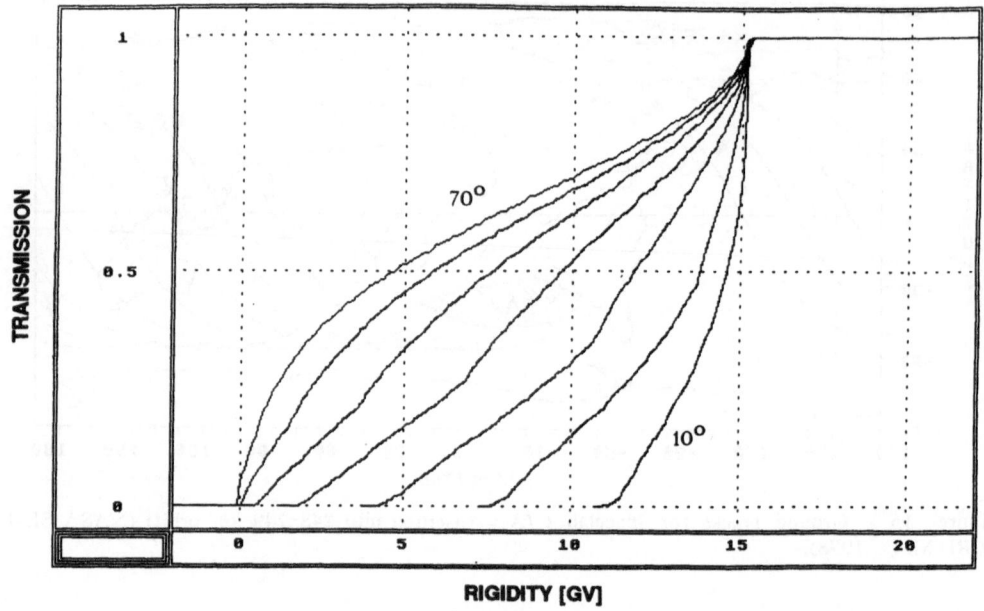

RIGIDITY [GV]

Figure 17.   Transmission Functions for Circular Orbits at 223 km Altitude and Different Inclinations. Effective Cutoff Values are Calculated from Equation (11), the Inclination Increment is 10 Degree.

Knowing the effective cutoff values along the orbit trajectory, the mission averaged access probability for particles with a certain rigidity R can be calculated as the ratio of the mission time, spent at positions with effective cutoff values less than R, to the total mission

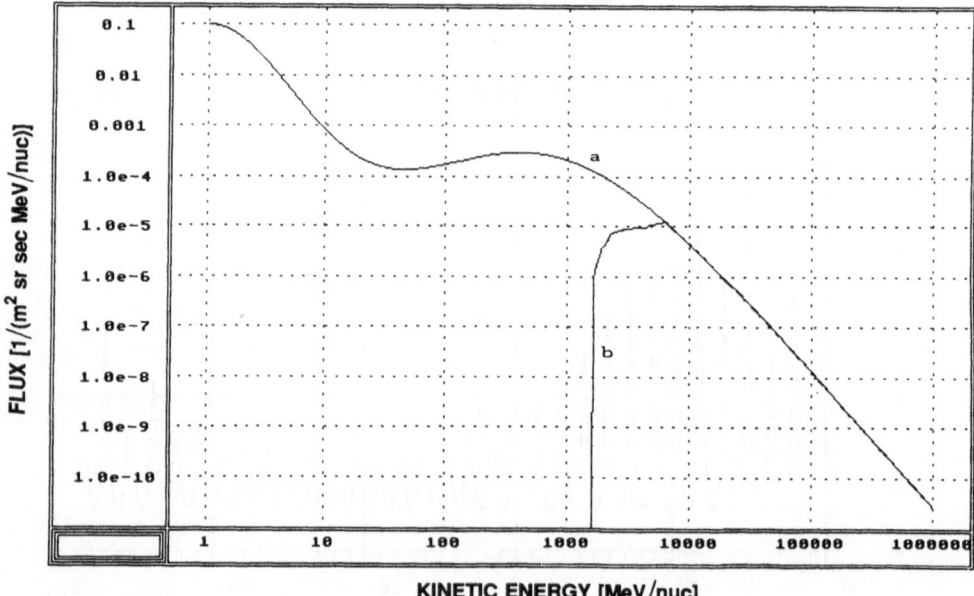

KINETIC ENERGY [MeV/nuc]

Figure 18.   Differential Energy Spectrum for Iron Nuclei in 1984.3, a) Outside the Geomagnetic Field, b) 24 hours Average for LDEF Orbit (28 Degree Inclination, Circular Orbit at 463 km Altitude).

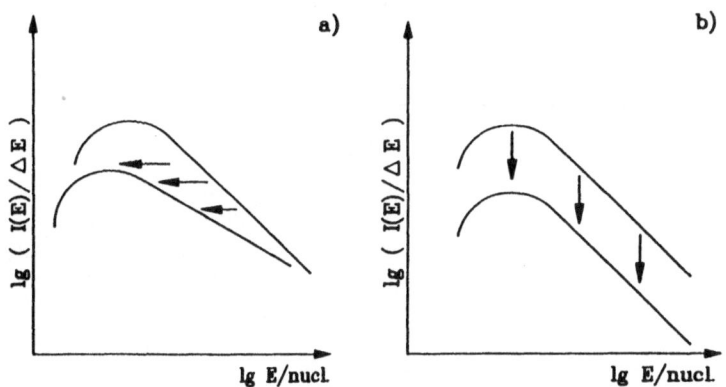

Figure 19. Effects of Energy Loss (a) and Fragmentation (b) in Shielding Material on the Energy Spectrum (Heinrich 1988).

time. The result is a transmission function growing from T(R)=0 at the minimum cutoff value to T(R)=1 at the maximum value for the specific orbit. The access probabilities are different for successive revolutions (see Fig. 16), but in most cases it is sufficient to average over 16 revolutions. The required vertical cutoff values may be taken from data at 20 km altitude, published by Shea and Smart (1983a). Data are only available for points in a world grid because the calculation, based on the trajectory-tracing method, is very time consuming. Cutoff rigidities for intermediate locations must be obtained by interpolation between these values, whereas values for different altitudes may be obtained using the relation $R(h1)*L^2(h1)=R(h2)*L^2(h2)=const$ (equation 10).

Another approach is based on equation (10) which contains the relation between the vertical cutoff rigidity and the McIlwain parameter L for a dipole field. Shea et al.(1987) have

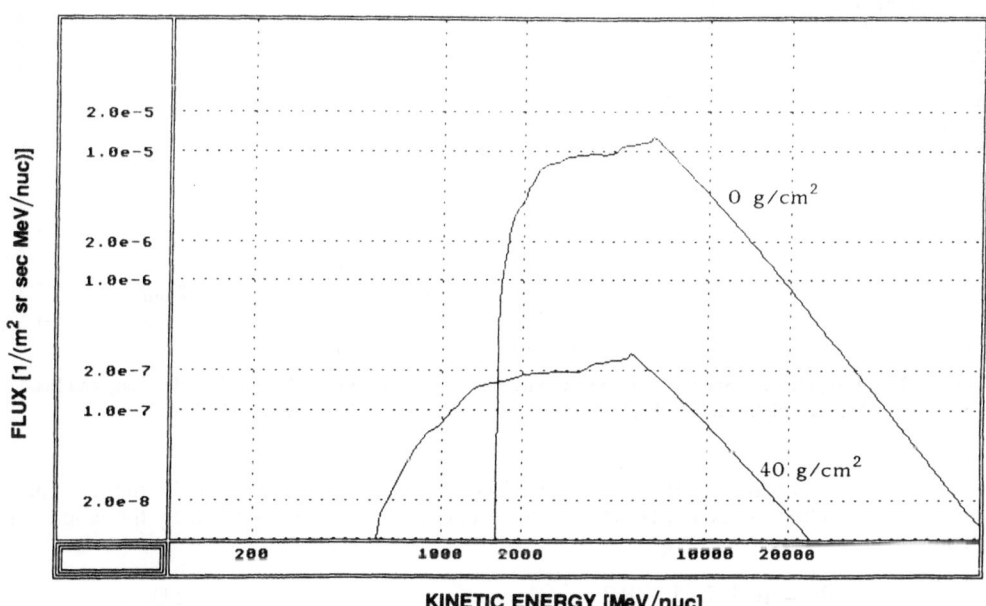

Figure 20. Influence of 40 g/cm² Shielding Material on theDifferential Energy Spectrum for Iron Nuclei (24 h Average on the LDEF Orbit, See Fig. 18).

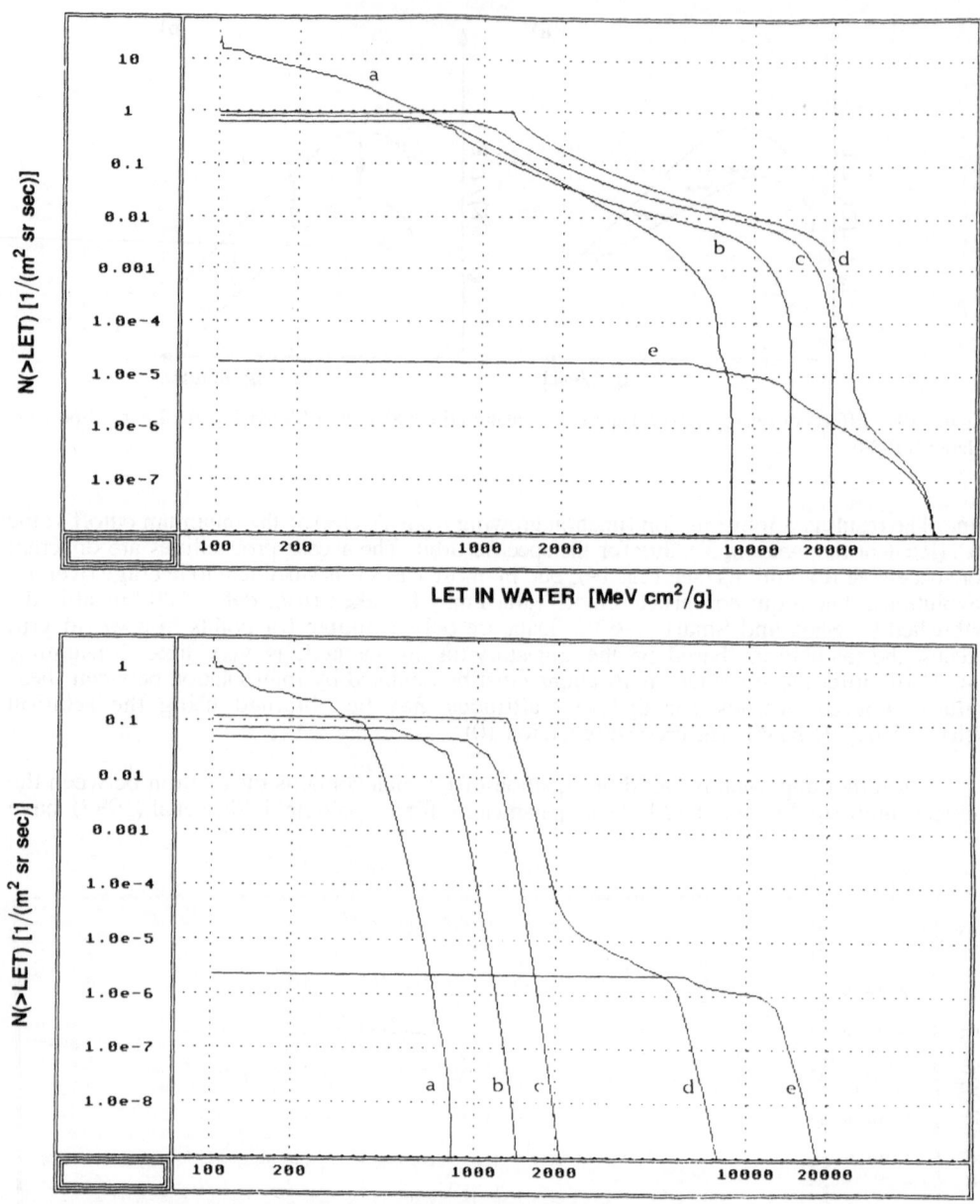

Figure 21. Integral LET Spectra of Different Nuclear Charge Groups in Water Outside (Top) and Inside (Bottom) the Magnetosphere: a) Z=1-15, b) Z=16-20, c) Z=21-25, d) Z=26-54, e) Z=55-92.

analyzed this relation for cutoff rigidities and L-values calculated for the IGRF model. They came to the conclusion that the effective vertical cutoff Rc can be described by the empirical relation

$$Rc = 16.222 / L^{2.0441} \qquad (11)$$

for the epoch 1980. The equation is valid only for regions outside +-5 degrees of the cosmic ray or the minimum L equators.

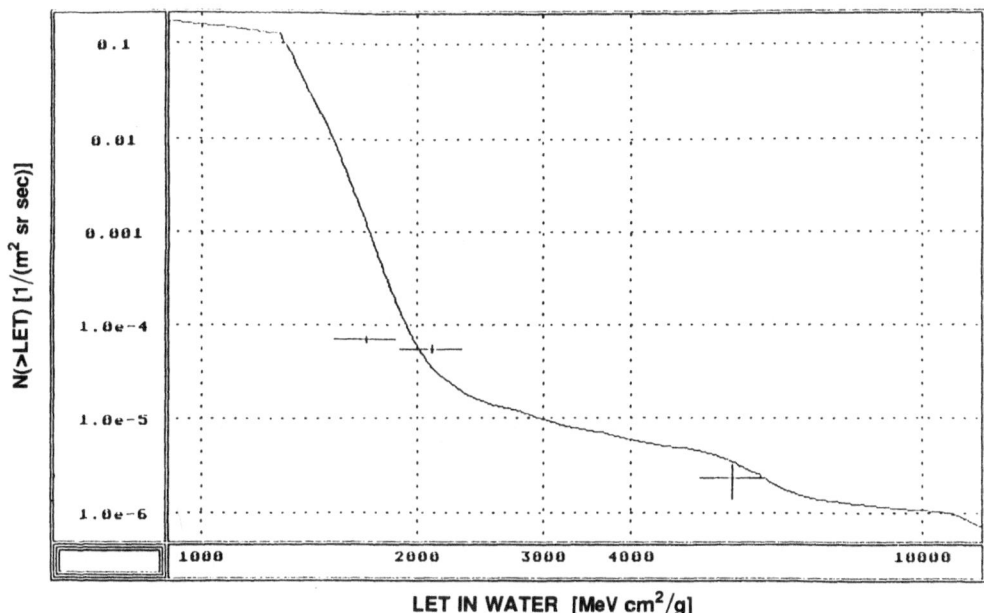

Figure 22. Integral LET Spectrum for the LDEF Mission. The Curve was Calculated for 0.1 g/cm2 (no shielding material).

Equation (11) provides an easy way for the calculation of cutoff rigidities on small computers. The computation of the field, using the IGRF coefficients, and the field line tracing for the L value is faster than the trajectory-tracing technique for the cutoff values. The transmission functions in Fig. 17 were calculated using equation (11) with L values of the 1985 IGRF model. The strong dependence on the orbit inclination is apparent: rigidities below a certain threshold value have no access to low inclination orbits.

The mission averaged modulation effect on the energy spectrum of iron nuclei is shown in Fig. 18 for the LDEF orbit. It is the quantitative description of the filtering effect discussed above. The modulated spectra are obtained by multiplying the free space spectra with the transmission function after converting rigidity to kinetic energy (equation 4).Additional modulation is caused by the shielding material as depicted in Figs. 19 & 20. For a more detailed discussion of energy loss and fragmentation effects in shielding material see Heinrich (1988).

The modulated energy spectra for the individual cosmic ray nuclei can be used to calculate the linear energy transfer (LET) spectrum of HZE particles in a detector or in tissue. The partial contributions in groups of elements to the integral LET spectrum outside the magnetic field is shown in Fig. 21a, whereas Fig. 21b shows the modulated spectra for the LDEF orbit. Due to the magnetic cutoff the high LET (=low energy) particles are strongly decreased. The calculation is confirmed by preliminary measurements in plastic nuclear track detectors (Fig.22).

IMPROVEMENT OF CUTOFF RIGIDITY CALCULATIONS

The discussed approach to the problem of geomagnetic shielding does not take into account the variation of the cutoff value with a) the viewing direction and b) the solar wind conditions. The calculations were restricted to vertical viewing and contributions of the inner field.

As already mentioned above the cutoff value at a given geographic location depends on the zenith and azimuthal angle (i.e. the final arrival direction of the particle). The Störmer

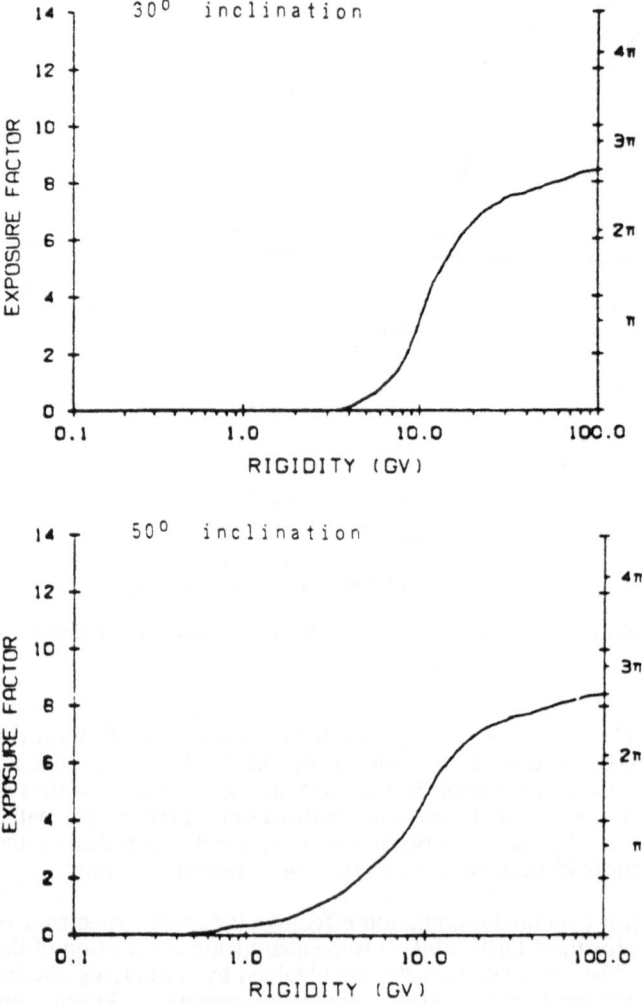

Figure 23. Geomagnetic Transmission Functions for a 400 km Altitude Satellite (Smart and Shea, 1983)

treatment for a dipole field leads to the cutoff rigidity relation

$$R = 4*R_c* (1+ (1-\cos (aa) \sin (za) \cos^3(\lambda))^{1/2})^{-2} \tag{12}$$

with $R_c = M*\cos^4(\lambda) / (2r)^2$ being the vertical cutoff at distance r from the dipole center, where λ is the magnetic latitude, za the zenith angle, and aa the azimuthal angle measured from magnetic east.

Smart and Shea (1983) have calculated geomagnetic transmission functions including zenith and azimuth dependence by the trajectory tracing method for the 1975 IGRF model. The basic procedure was to construct a cosmic ray sphere of access at each grid location. The result is an exposure factor that is a function of rigidity for the specified point with units of solid angle (steradians). The obtained mission averaged exposure factors (Fig. 23) include the shadow cone of the solid earth and arrival directions below the optical horizon.

The IGRF model describes the inner field (emerging from interior sources) taking into account the secular variation. It does not include the complex structure of the magnetosphere with its dependence on the solar wind conditions. However, asymptotic directions and cutoff

32

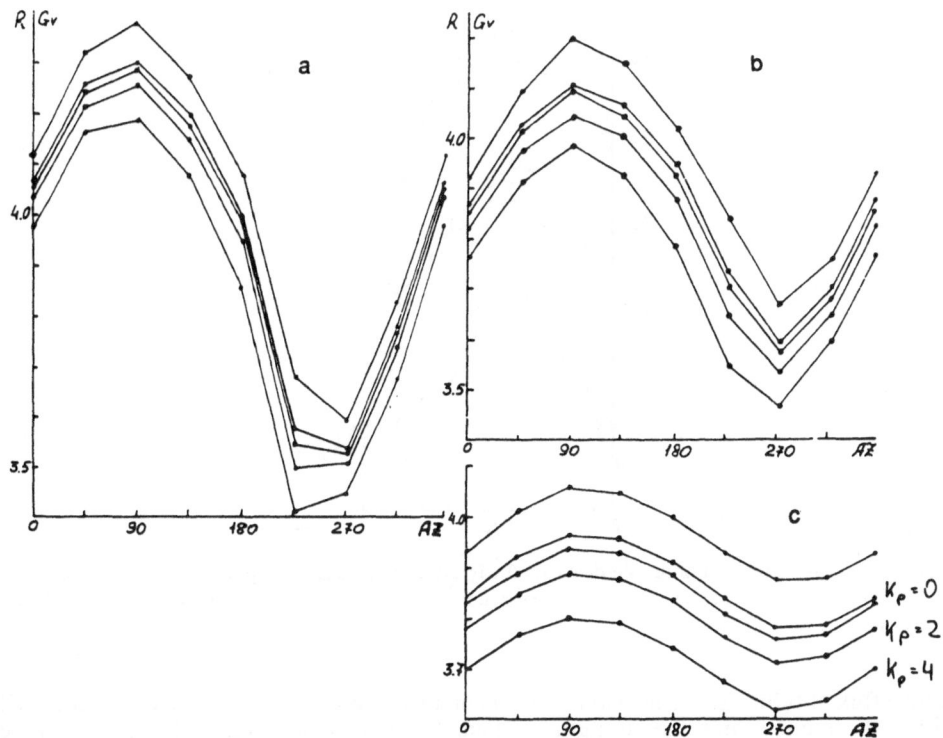

Figure 24. Cosmic Ray Cutoff Rigidities as Function of Azimuth Angle for 40 (a), 24 (b) and 8 (c) Degree Zenith Angle. The Topmost Curve in Each Panel Corresponds to Cutoff Rigidities in the Main field, the Lower Curves were Computed Using the Tsyganenko-Usmanov Magnetospheric Field for kp=0,1,2,4 (Top to Bottom). (Danilova and Tyasto, 1987).

rigidities of cosmic rays are influenced not only by the main field but also by the external sources. The disturbance of the outer field during solar events is described by the planetary magnetic indices kp. The result of the disturbance are decreasing cutoff values at increasing kp values (quiet condition at kp=0, highly disturbed condition at kp=9).

An improved quantitative representation of the field in the magnetosphere was developed by Tsyganenko (1989). Based on a similar model Danilova and Tyasto (1987) have calculated the non vertical cutoff rigidities as a function of geomagnetic activity (kp indices). Their results for the station Irkutsk demonstrate the significant role of the asymmetric magnetosphere (Fig. 24). The influence of the external sources reduces the cutoff for all considered arrival directions compared to the main field cutoff values. Additional calculations showed that the asymmetry of the real magnetosphere causes the daily variation of cutoff rigidities at the middle latitudes (Danilova and Tyasto, 1990). The amplitude of this variations is increasing with the rise of the geomagnetic activity (kp indices) and has the maximum seasonal modulation in summer and the minimum in winter (Fig. 25).

CONCLUSION

It has been shown that the number of particles, impinging on a low earth orbit, depends on the solar modulation in the interplanetary space and on the terrestrial modulation in the geomagnetic field.

The basic modulation can be understood in terms of solar wind interactions and magnetic field deflections. However, at the present time it is not possible to calculate the exact

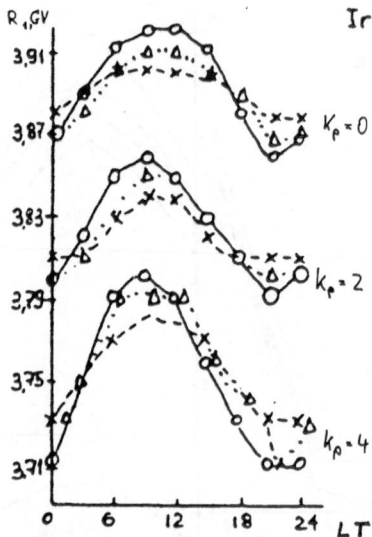

Figure 25. Daily Variation of the Vertical Cutoff Rigidities in the Magnetospheric Field for Irkutsk Station (52.47 N, 104.03 E) at Summer solstice (Circle), Equinox (tTriangle) and Winter Solstice (Cross). (Danilova and Tyasto, 1990).

particle flux at future space missions. This is mainly due to the unpredictable solar activity: neither the number nor the strength of solar flares can be predicted with our present knowledge of the solar physics. Furthermore we need to improve the cutoff calculations by using the best available time dependent magnetospheric field model.

The additional effect of shielding by matter was not discussed here. Only limited data exist for nuclear fragmentation cross sections. Future experimental and theoretical studies are needed to improve the accuracy of the propagation calculation in matter.

One motivation for the work on the modulation model is to access the potential hazard posed by galactic cosmic rays to manned space missions, i.e. to keep the radiation dose equivalent within prescribed limits. Future studies may lead to the recognition of a pattern in the solar activity that would give more confidence in predicting future cosmic ray flux levels from which doses inside manned spacecraft can be calculated.

ACKNOWLEDGEMENT

Part of this work was supported by the Bundesministerium für Forschung und Technologie (BMFT) under grants 01 QV 297 and 50 QV 8566.

REFERENCES

Abarbanell, T.; 1991: Energieverlust schwerer Ionen der Kosmischen Strahlung in der LDEF-Weltraummission,Diploma thesis, University Kiel.

Adams, J.H. Jr., Silberberg, R.; and Tsao, C.H; 1981: Cosmic ray effects on microelectronics, part I: The near-earth particle environment, NRL Washington, memorandum report no. 4506.

Badhwar,G.D.; and O'Neill,P.M.; 1991: An improved model of galactic cosmic radiation for space exploration missions, paper OG 5.2.13 at the 22nd ICRC Dublin, to appear in the Conf. Proc.

Cooke,D.J.; Humble,J.E.; Shea, M.A.; Smart,D.F.; Lund, N.;Rasmussen, I.L.; Byrnak, B.; Goret,P.; and Petrou,N.; 1985: Re-evaluation of cosmic ray cutoff terminology, Proc.19th ICRC La Jolla,   Vol.5, 328-331.

Danilova,O.A. and Tyasto, M.I.;1987: Nonvertical cosmic ray cutoff rigidities in the asymptotic magnetosphere, Proc.20th ICRC Moscow, Vol. 4, 208-211.

Danilova,O.A. and Tyasto, M.I.; 1990: Variations of cosmic ray rigidities at the mid-latitude stations due to the asymmetric Magnetosphere, Proc.21st ICRC Adelaide, Vol.7, 6-9

Flückiger, E.O.; 1991: Solar and terrestrial modulation, rapporteur paper at the 22nd ICRC Dublin, , to appear in the Conf. Proc.

Heinrich,W., 1988, Variation of galactic cosmic radiation by solar modulation, geomagnetic shielding and shielding by material, Terrestrial Space Radiation and its Biological Effects, Plenum Publishing Corp.

Kobel,E., 1989: Bestimmung der Grenzsteifigkeiten und der asymptotischen Richtungen der kosmischen Strahlung, Ph.D. thesis, University Bern

Merril,R.T. and McElhinny, M.W.; 1983: The Earth's magnetic field, Academic Press, London

Nimmik, R.A.; Panasyuk, M.I.; Pervaya,T.I.; and Suslov, A.A.;1991: Spatial and temporal relationship of galactic cosmic rays and anomalous component energy spectra in the near heliosphere, Workshop on "Constraints on space exploration", 22nd ICRC Dublin, to appear in the Conf. Proc..

Seo,E.S.,Ormes, J.F., Streitmatter, R.E.;Stochaj, S.J.; Jones, W.V.;Stephens,S.A. and Bowen,T.; 1991: Cosmic ray proton and helium spectra during the 1987 solar minimum, paper OG 5.2.5 at the 22nd ICRC Dublin, to appear in the Conf. Proc..

Shea,M.A.; and Smart, D.F.; 1983a: A world grid of calculated cosmic ray vertical cutoff rigidities for 1980.0, Proc. 18th ICRC, Bangalore,Vol.3, 415-418.

Shea,M.A.; Smart, D.F.; and Gentile, L.C.; 1987: Estimating cosmic ray vertical cutoff rigidities as a function of the McIlwain L-parameter for different epochs of the geomagnetic field, Phys. Earth Plan. Int.,48, 200-205.

Simpson,J.A.; 1983: Elemental and isotopic composition of the galactic cosmic rays, Ann.Rev.Nuc.Part.Sci, 33, 323-381.

Smart,D.F.; and Shea, M.A.; 1983b: Geomagnetic transmission functions for a 400 km altitude satellite, Proc. 18th ICRC., Bangalore, Vol. 3, 419-422.

Smart,D.F.; and Shea,M.A.; 1984: Cosmic ray exposure factors for shuttle altitudes derived from calculated cut-off rigidities, Adv. Space Res., Vol.10,161-164.

Tsyganenko, N.A.; 1989: A magnetospheric magnetic field model with a warped tail current sheet,Planet.Space Sci.,Vol.1,5-20.

Watts, J.W.Jr.; Parnell, T.A.; and Heckman, H.H.; 1989: Approximate angular distribution and spectra for geomagnetically trapped protons in low-earth orbit, Proc. AIP Conf., Sanibel Island, Vol.186, 75-85.

Wefel,J.P.; 1988: An overview of cosmic ray research, in: Genesis and Propagation of Cosmic Rays, 1-40 D. Reidel Publ. Comp..

HISTORY OF ENERGETIC SOLAR PROTONS FOR THE PAST THREE SOLAR CYCLES INCLUDING CYCLE 22 UPDATE

M.A. Shea and D.F. Smart

Space Physics Division
Geophysics Directorate/PL
Hanscom AFB, Bedford, MA 01731-5000, USA

ABSTRACT

Solar proton events have been recorded at the earth since 1942 although the detection techniques varied considerably over the past 50 years. From 1942 to 1957 the identification of solar proton events was limited to very high energy events. This situation improved over the next decade after which lower energy solar proton events became routinely identified by satellite measurements. Even though the detection threshold differed between the 19th and more recent cycles, more than 200 solar proton events with a flux of over 10 protons/cm^2-sec-ster above 10 MeV have been recorded at the earth in the last three solar cycles. From the composite record of major solar proton events that have occurred during each of the last five solar cycles, it appears that the 20th and 21st solar cycles are deficient in extremely high energy long duration solar particle events. A summary of solar proton events for the past 50 years is given together with information on the flux of solar protons to various spacecraft orbits.

INTRODUCTION

Major solar flares are often associated with the acceleration of energetic particles at the sun and their injection into the interplanetary medium where the particles can be detected by a variety of techniques. These solar proton events have been recorded at the earth for 50 years although the routine monitoring of these events by spacecraft did not commence until 1965. Between 1942 and the beginning of the space era in 1957 only extremely high energy solar particle events could be detected. These particles had enough energy to penetrate the earth's magnetic shield and into the atmosphere where they were occasionally recorded by balloon borne sensors. In rare cases very high energy particles contained sufficient energy to generate a nuclear cascade in the atmosphere that resulted in an increase in the radiation intensity as monitored by cosmic ray ground-level detectors. Proton events could also be inferred using selected ionospheric techniques. Svestka and Simon (1975) give a summary of measurement techniques used to detect proton events from 1955-1969.

Over the years these solar particle events have been referred to by a number of descriptive names such as solar cosmic ray events, solar proton events, ground-level events, and polar cap absorption events. These terms are still in use. Since the advent of the space era, data obtained from particle sensors on near-earth satellites and on space probes

Biological Effects and Physics of Solar and Galactic Cosmic Radiation,
Part B, Edited by C.E. Swenberg *et al.,* Plenum Press, New York, 1993

37

throughout the heliosphere coupled with improved balloon and ground-based instrumentation have greatly increased our understanding of solar particles and their propagation in the solar system.

Since solar proton events can adversely affect the terrestrial environment it is essential we learn as much as possible about these events in order to accurately predict their occurrence and severity. This objective assumes greater importance in planning for the safety of astronauts during long term missions for space exploration. We already know that the solar proton flux to locations in space for current and projected manned space missions is completely dependent on the characteristics of the spacecraft orbit. For example, the flux of solar protons to earth-orbiting spacecraft will be limited to those particles that can penetrate through the earth's magnetic shielding to the spacecraft position. However, this situation changes dramatically for lunar and interplanetary missions where the spacecraft will be subjected to the full solar proton flux in the interplanetary medium.

Investigation of solar proton events over the past 50 years has been primarily devoted to statistical studies or research on the characteristics of individual events. Although we do not yet understand how the sun accelerates ions to relativistic energies, nor how to predict the fluence from an individual flare, we have assembled enough data to be able to place some preliminary, but nevertheless realistic, limits on the extent and severity of these events. This paper presents a summary of our knowledge of solar proton events as gained over the past half century. Hopefully this information may provide broad guidelines to space exploration planners.

SOLAR FLARES AND ASSOCIATED EMISSIONS

Solar Particle Source

Although the solar flare process is the most commonly assumed source of solar protons, recent research indicates that the coronal mass ejection may be the phenomenon that is associated with the release of solar protons into the interplanetary medium (Kahler et al., 1984). Since most major solar flares are associated with solar mass ejections, it is still customary to refer to solar proton events as emanating from solar flares, and we will continue to use this nomenclature throughout the remainder of this paper.

Solar Electromagnetic Emissions

During a solar flare, electromagnetic radiation such as X-ray and radio emission, is generated by the hot plasma and travels at the speed of light through interplanetary space. This type of radiation takes ~ 8.3 minutes to reach the earth and is usually the first indication that a major solar flare has occurred. The onset of an increase in solar X-ray emission detected by sensors on earth-orbiting satellites is approximately simultaneous with the visual observation of a solar flare usually made in the H-alpha wavelength. Although the detection of X-rays may actually precede the optical observation by a minute or two, this is primarily the result of instrumentation sensitivity.

Solar Energetic Particle Emissions

The transit of solar energetic particles to a point in space is dependent upon the energy of the particle and the location of the flare on the sun with respect to the detection point. Under idealized circumstances, from well connected solar flares, relativistic solar protons can reach the orbit of the earth (i.e. one Astronomical Unit) within 10-15 minutes of the onset of the flare; 10 MeV protons take approximately 80 minutes to reach the same distance. X-rays, having no charge or mass, travel rectilinearly; solar electrons and ions, being charged particles, spiral along the interplanetary magnetic field lines between the sun and the detection point in space.

Solar Plasma Emission

When a major solar flare occurs there is also an associated release of enhanced solar

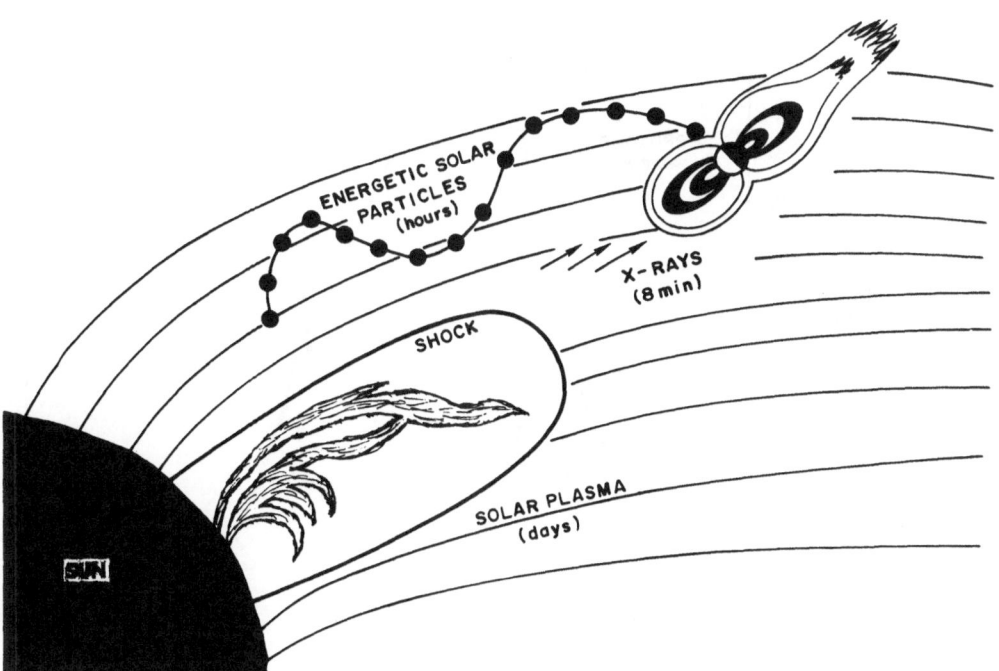

Figure 1. Pictorial representation of solar emissions from a solar flare. The time scales for the various emissions to reach the earth are also shown.

plasma into the interplanetary medium. Occasionally this dense plasma will propagate far into the heliosphere travelling at speeds of 1-3 days per Astronomical Unit. When these plasmas interact with the earth's magnetic field the resulting energy transfer manifests itself by the occurrence of aurora and geomagnetic disturbances, the magnitude of which are dependent upon the interplanetary plasma and magnetic field characteristics at the time of the arrival of the plasma at the earth. These "travelling interplanetary plasma discontinuities" can severely disrupt the ambient particle environment; occasionally, for major solar proton events, the ambient flux can be re-accelerated by interaction with the shock front (Lee, 1988). In severe cases the solar particle flux can be greatly intensified over a period of a few hours (Smart et al., 1990). Figure 1 is a pictorial representation of emissions from a solar flare; Figure 2 illustrates the relative time scales of solar particle emissions as detected at the earth.

Non-flare Related Energetic Particle Increases

In the past two decades it has been recognized that other phenomena such as "disparitions brusques" (i.e. disappearing solar filaments) may be associated with increases in the energetic particle flux as measured by spacecraft (Kahler, et al., 1986). These non-flare related increases are relatively small and would not pose a significant hazard for manned space missions.

SOLAR PARTICLE COMPOSITION AND NOMENCLATURE

Composition of Solar Particle Events

Solar particle events are composed primarily of solar protons and solar electrons. However, the relative abundance of the heavy ions produced by solar flares is of considerable importance both to equipment in space and to manned space missions. The energy deposition in matter is proportional to the square of the atomic charge number, and consequently heavy ions such as iron will deposit 676 (i.e. 26^2) times as much energy as protons of the same kinetic energy per nucleon.

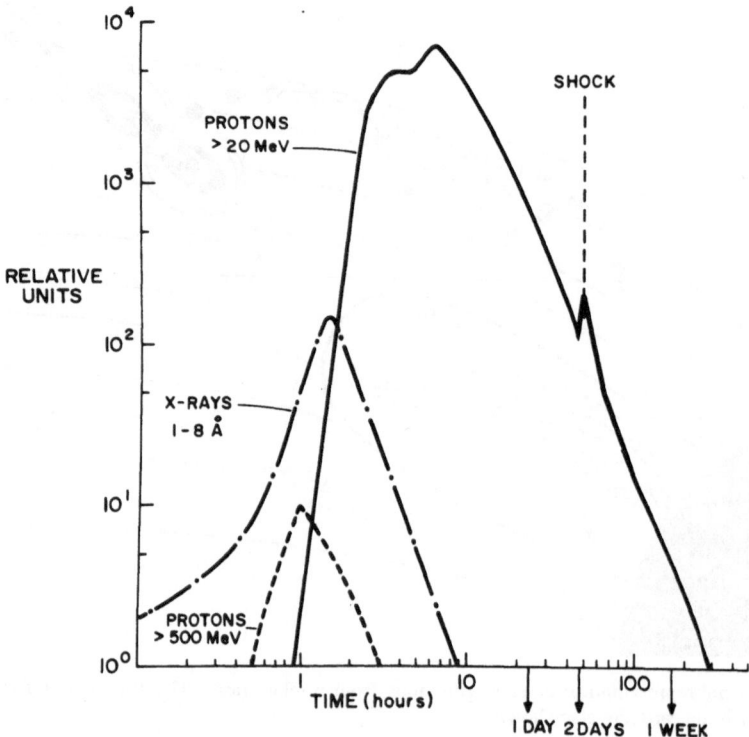

Figure 2. Relative time scales of solar particle emissions (at 1 AU). The increase in particle flux at the time of the arrival of the interplanetary shock is from the additional acceleration of the ambient particle flux caused by particle interaction with the shock front.

There have been a number of papers reporting research on the elemental abundance of solar particle events in the last decade; see Lin (1987), and Mason (1987) for recent reviews. A general summary of the results is that "small" events may have the greatest variability in elemental composition and "large" events tend toward "normal" composition. The hydrogen to helium ratios are the most variable even for "large" events.

There is not yet universal agreement among researchers of what is "normal" composition. Stone (1989) and co-workers consider "normal" composition to be the elemental abundance ratios expected if the composition of the solar corona is selected by the first ionization potential of each element. Breneman and Stone (1985) reached this conclusion from the analyses of large events in solar cycle 21. Garrard and Stone (1991) consider the measurements of elements ranging from carbon (Z = 6) to nickel (Z = 28) obtained during the October 1989 sequence of events as convincing proof of this hypothesis. Other researchers, also using measurements obtained during the October 1989 event sequence (Reedy, et al., 1991), compared their measured abundance ratios with the elemental abundance ratios in the solar photosphere.

From these contrasting examples, it is apparent that the definitions of "normal" composition depend on the criteria being used for evaluation. Our opinion is that the elemental abundance ratios for elements with atomic charge >3 seem to be in general agreement with the ratios expected from normal coronal material organized by the first ionization potential. Table 1 presents solar particle element abundance ratios normalized to hydrogen originally developed by Smart (1988). Also, as discussed elsewhere in this volume

Table 1. Normalized Elemental Abundances of Solar Energetic Particle Events.

		Adams et al. (1981) Mason et al. (1980) ~ 1 MeV	Gloeckler (1979) 1-20 MeV	Cook et al. (1984) 10 MeV	McGuire et al. (1986) 6.7-15 MeV
1	H	1.0	1.0	1.0	1.0
2	He	2.2×10^{-2}	1.5×10^{-2}		1.5×10^{-2}
3	Li		1.0×10^{-7}	4.8×10^{-8}	2.8×10^{-6}
4	Be		1.5×10^{-7}	6.0×10^{-9}	1.4×10^{-7}
5	B		1.5×10^{-7}	1.2×10^{-8}	1.4×10^{-7}
6	C	1.6×10^{-4}	1.2×10^{-4}	9.6×10^{-5}	1.3×10^{-4}
7	N	3.8×10^{-5}	2.8×10^{-5}	2.7×10^{-5}	3.7×10^{-5}
8	O	3.2×10^{-4}	2.2×10^{-4}	2.2×10^{-4}	2.8×10^{-4}
9	F		4.3×10^{-7}	1.0×10^{-8}	1.4×10^{-7}
10	Ne	5.1×10^{-5}	3.5×10^{-5}	3.1×10^{-5}	3.6×10^{-5}
11	Na	1.6×10^{-6}	3.5×10^{-6}	2.6×10^{-6}	2.4×10^{-6}
12	Mg	4.8×10^{-5}	3.9×10^{-5}	4.3×10^{-5}	5.2×10^{-5}
13	Al	3.5×10^{-6}	3.5×10^{-6}	3.1×10^{-6}	3.3×10^{-6}
14	Si	3.8×10^{-5}	2.8×10^{-5}	3.5×10^{-5}	4.2×10^{-5}
15	P	2.3×10^{-7}	4.3×10^{-7}	1.7×10^{-7}	4.0×10^{-7}
16	S	1.8×10^{-5}	5.7×10^{-6}	7.8×10^{-6}	6.5×10^{-6}
17	Cl	1.7×10^{-7}		7.1×10^{-8}	
18	Ar	3.9×10^{-6}	8.7×10^{-7}	7.3×10^{-7}	4.6×10^{-6}
19	K	1.3×10^{-7}		1.0×10^{-7}	
20	Ca	2.3×10^{-6}	2.6×10^{-6}	3.1×10^{-6}	3.2×10^{-6}
21	Sc			7.8×10^{-9}	
22	Ti	1.0×10^{-7}		1.2×10^{-7}	
23	V			1.2×10^{-8}	
24	Cr	5.7×10^{-7}		5.0×10^{-7}	
25	Mn	4.2×10^{-7}		1.8×10^{-7}	
26	Fe	4.1×10^{-5}	3.3×10^{-5}	3.4×10^{-5}	
27	Co	1.0×10^{-7}		4.8×10^{-7}	
28	Ni	2.2×10^{-6}		1.2×10^{-6}	
29				1.4×10^{-8}	
30				3.8×10^{-8}	

there are significant unresolved questions as to the relative biological effectiveness of heavy ions compared to protons.

Flux and Fluence

In the following sections the terms "flux" and "fluence" will be used. Particle physicists usually refer to the peak flux observed in a specific channel of a solar particle detector. This can be either an integral flux above a specified energy level, in units of particles/cm2-sec-ster or a differential measurement which specifies the flux at a specific energy in units of particles/cm2-sec-ster-MeV. Individual events are usually compared using identical channels. The peak flux specifies the maximum particle flux from which can be derived a corresponding interaction rate. Peak flux is usually used to describe individual solar particle events.

Fluence is the total number of particles above a selected energy that is experienced throughout an individual event or episode of particle intensity. Fluence may be given in either directional units of particles/cm2-ster or omni-directional units of particles/cm2. The fluence is generally of concern for the total radiation exposure.

SOLAR PROTON PROPAGATION

Solar Proton Propagation to the Earth

To understand solar proton propagation it is necessary to briefly describe the basic magnetic topology of the interplanetary medium. The flow of the supersonic ionized plasma coming from the solar corona is popularly called the solar wind. (See Hundhausen, 1972, and Feynman, 1985, for a more detailed definition of the solar wind.) This plasma flows radially from the sun into the interplanetary medium and is primarily composed of low energy electrons (~ 100 eV) and protons (~ 1000 eV). Contained within this plasma is the interplanetary magnetic field. In interplanetary space the energy density of this plasma is much larger than the energy density of the magnetic field, and so for practical purposes the interplanetary magnetic field may be considered to be "frozen" in the plasma. This means that the topology of the interplanetary medium is dominated by the plasma flow. Although the plasma flows radially from the sun, the interplanetary magnetic field lines would appear to be curved as viewed from the earth - an effect resulting from the continuous plasma flow and the rotation of the sun. This is often called the "garden hose" effect since it is analogous to the drops of water that individually stream radially from a garden hose, but "appear" (to a spectator getting wet) to come along a curved path as the person holding the hose turns.

In much the same way, the interplanetary magnetic field lines from the sun to the earth "appear" to be curved in this Archimedean spiral nature. Solar protons, being charged particles, are guided by the topology of the interplanetary magnetic field lines as they propagate in the interplanetary medium to a detection point in space. Under nominal conditions, from the viewpoint of an observer at the earth, the apparent particle flow is from a direction west of the sun-earth line along an Archimedean spiral between the sun and the earth. The nominal "footpoint" of this spiral path is located at approximately 60° West of the sun-earth line.

The amount of solar rotation the sun has undergone between the time the solar wind has reached supersonic velocities and has left the inner solar corona (assumed to be 18 solar radii) and the time the solar wind is detected is calculated as follows:
  (1) From spacecraft measurements obtain the speed of the solar wind.
  (2) Divide the radial distance between the sun and the detection point by the speed of the solar wind. This gives the transit time.
  (3) Convert the transit time into days.
  (4) Knowing that the sun rotates approximately 13.3 degrees per day, determine how many degrees the sun has rotated during the elapsed time.

For the earth, the nominal "footpoint" of approximately 60° West of the sun-earth line is obtained from using a constant solar wind speed of 385 km/sec. A fast solar wind speed (e.g. 650 km/sec) will have a "connection longitude" closer to the earth-sun line to about 35° West; a slow solar wind speed (e.g. 300 km/sec) will move the connection longitude beyond 60° West to 77° West. However, in reality, any major perturbation to the interplanetary magnetic field such as an increase in the solar wind velocity associated with a coronal hole or the passage of a major interplanetary shock, can considerably alter the idealized Archimedean spiral structure between the sun and the earth in addition to rapidly shifting the location of the "footpoint" many degrees in solar longitude.

The maximum possible solar proton flux is presumed to be at the solar flare site. If a solar flare occurs at the "footpoint" of the interplanetary magnetic field line connecting the sun with the earth, the solar particle flux will propagate rapidly along the interplanetary magnetic field lines to the earth, and the earth will receive the maximum flux. If the solar flare occurs at any other location on the sun, the proton flux must first propagate through the solar corona to the "footpoint" of the interplanetary magnetic field line connecting the sun with the earth after which the particles move outward into the interplanetary medium to the earth. The solar particle flux that reaches the footpoint of the Archimedean spiral is attenuated as a function of the angular distance from the flare site. (See Smart and Shea, 1985, 1992a, for a more detailed description of the solar particle propagation between the sun and the earth.)

Thus, solar protons appear to propagate into the interplanetary medium through two independent phases. The first phase is diffusion from the flare site through the solar corona to the "foot" of the idealized Archimedean spiral path formed by the interplanetary magnetic field line between the sun and the detection point. The second phase is the propagation in the interplanetary medium from the sun to the detection point along the interplanetary magnetic field lines. Figure 3 illustrates the idealized interplanetary magnetic field lines between the sun and the earth and the "most favorable propagation path" for solar particles to reach the earth independent of the location of the flare on the sun.

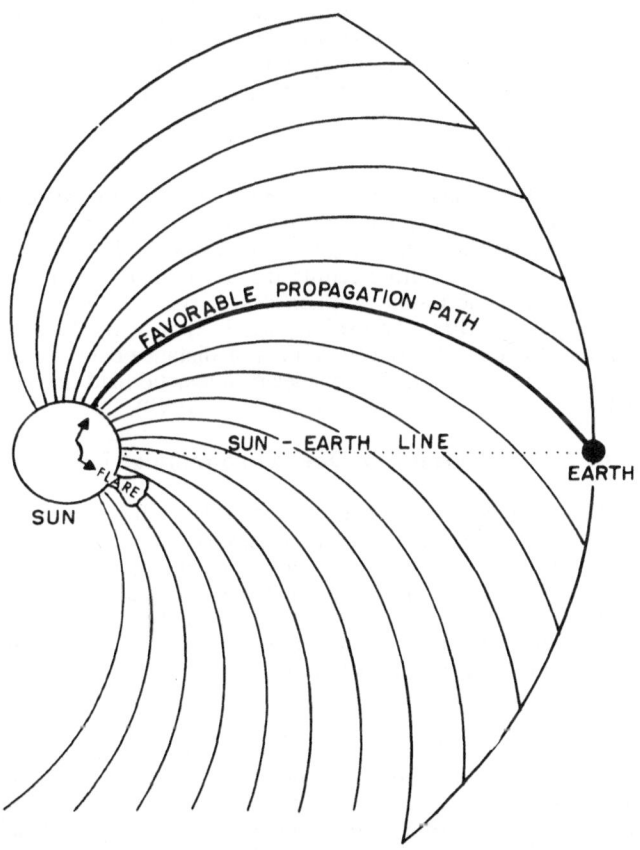

Figure 3. Conceptual figure illustrating solar proton propagation from the sun to the earth. The coronal propagation distance from the site of the flare to the location of the favorable propagation path for solar particles to intercept the earth, is illustrated by the arc between the arrows shown on the sun. Interplanetary propagation proceeds along the Archimedean spiral path from the sun to the earth.

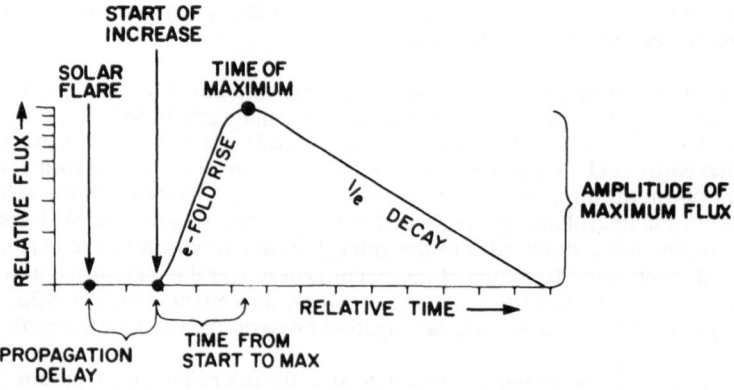

Figure 4. Characteristic profile of the solar particle intensity as a function of time at one Astronomical Unit.

Solar Proton Event Characteristics

Solar proton events, as measured at the orbit of the earth, have a characteristic intensity-time profile as illustrated in Figure 4. Although the general shape of the intensity-time profile will differ from event to event and also with respect to the location of the flare on the sun and the detection point in the heliosphere, particle events can be characterized by the following: a propagation delay between the onset of the solar flare in electromagnetic emission and the onset of the particle increase, a relatively rapid rise in intensity to a maximum value, and a slow decay to the background level. Although actual event profiles can be complicated by multiple particle injections or interplanetary perturbations, this simplified picture is appropriate for any one isolated event.

For a solar flare on the western portion of the sun, the solar particle flux usually rises and decays fairly rapidly; for an identical solar particle event from a flare on the eastern hemisphere of the sun as viewed from the earth, the attenuation as the flux propagates through the solar corona together with the time required for the particles to propagate to the "favorable" interplanetary field line results in much slower rise and decay profiles in addition to a smaller maximum particle flux. The solar particle flux at the flare site is attenuated approximately one order of magnitude for each radian of distance between the location of the flare and location of the "footpoint" of the interplanetary magnetic field line connecting the sun with the eventual detection point in space (Smart and Shea, 1979).

Solar Proton Event Characteristics at One Astronomical Unit

Figures 5 and 6 illustrate typical intensity-time profiles that would be recorded at one Astronomical Unit from identical solar flare energetic proton sources at different locations on the sun. The intensity-time profile shown on the lower right side of Figure 5 is typical for a flare that occurred at the "footpoint" of the interplanetary magnetic field line connecting the sun with the earth. Notice the rapid rise to maximum intensity. The particle flux would be maximum along the favorable propagation path (shown by the larger dots) whereas particles that diffuse through the solar corona to other field lines would have a smaller flux (shown by smaller dots).

Figure 6 illustrates the particle flux from a flare to the east of the earth-sun line. The left side of this figure shows that the maximum flux (large dots) would be along the interplanetary magnetic field line from the flare location to the hypothetical spacecraft located at one Astronomical Unit. While particles from this flare are propagating along the field line to the spacecraft, they are also propagating, albeit with a reduced intensity, through the solar corona to other field lines. Those particles which reach the interplanetary magnetic field line connecting the sun with the earth propagate along the field line to the earth. The lower section of this figure shows the intensity-time profiles which would have been recorded by both

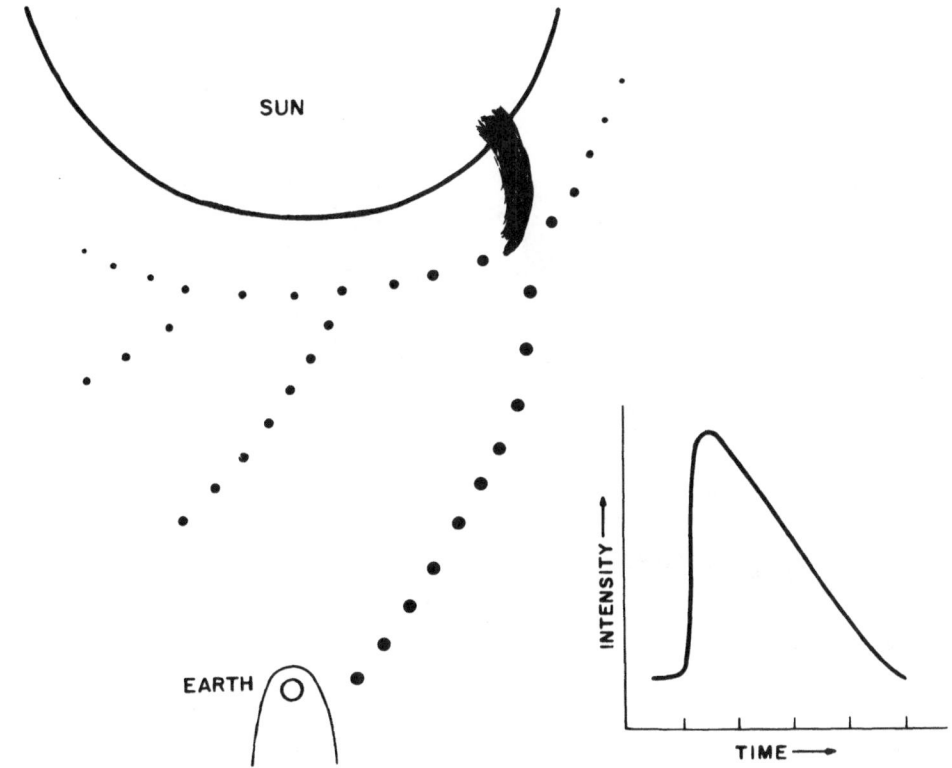

Figure 5. Solar particle propagation along the interplanetary magnetic field lines and in the inner heliosphere from a hypothetical flare west of the sun-earth line. Coronal propagation and subsequent particle transport along other field lines are indicated by dots; the larger the dots, the larger the flux. The insert at the lower right shows a typical intensity-time profile at an earth-orbiting spacecraft in the interplanetary medium.

spacecraft during this event. The spacecraft at the earth would have recorded a later onset time, slower rise time, smaller maximum flux, and longer decay time than the spacecraft located along the field line connected to the flare site.

In addition to solar protons from flares that are visually observed from the earth, approximately 20% of the solar proton events detected at the earth have been associated with flares from beyond the western limb of the sun and hence invisible from the earth. These events pose a special problem for an interplanetary mission since earth-based observations can only see one half of the sun.

Major solar flares can populate the entire inner heliosphere with particles although the flux at the various detection points will be determined by the angular distance from the flare location to the footpoint of the interplanetary magnetic field line connecting the sun with the detection location. An example of a solar proton event detected at vastly different locations in space is shown in Figure 7. In this figure the relative solar proton intensity (on a log scale) as measured on 8 and 9 August 1970 by the Pioneers 8 and 9 space probes and the earth-orbiting IMP 5 is represented by the bars placed at the location of the indicated spacecraft. Pioneer 9 observed the largest increase on 8 August, and Pioneer 8 had a smaller increase with maximum intensity on 9 August. A flare located approximately 40° behind the east limb was assumed to be the source of this particle event, an assumption that was strengthened when a very active solar region rotated around the eastern hemisphere of the sun a few days later (Dodson-Prince et al., 1977). The small increase observed on IMP 5 is consistent with this flare location. The spacecraft which detected this event were all close to a radial distance of

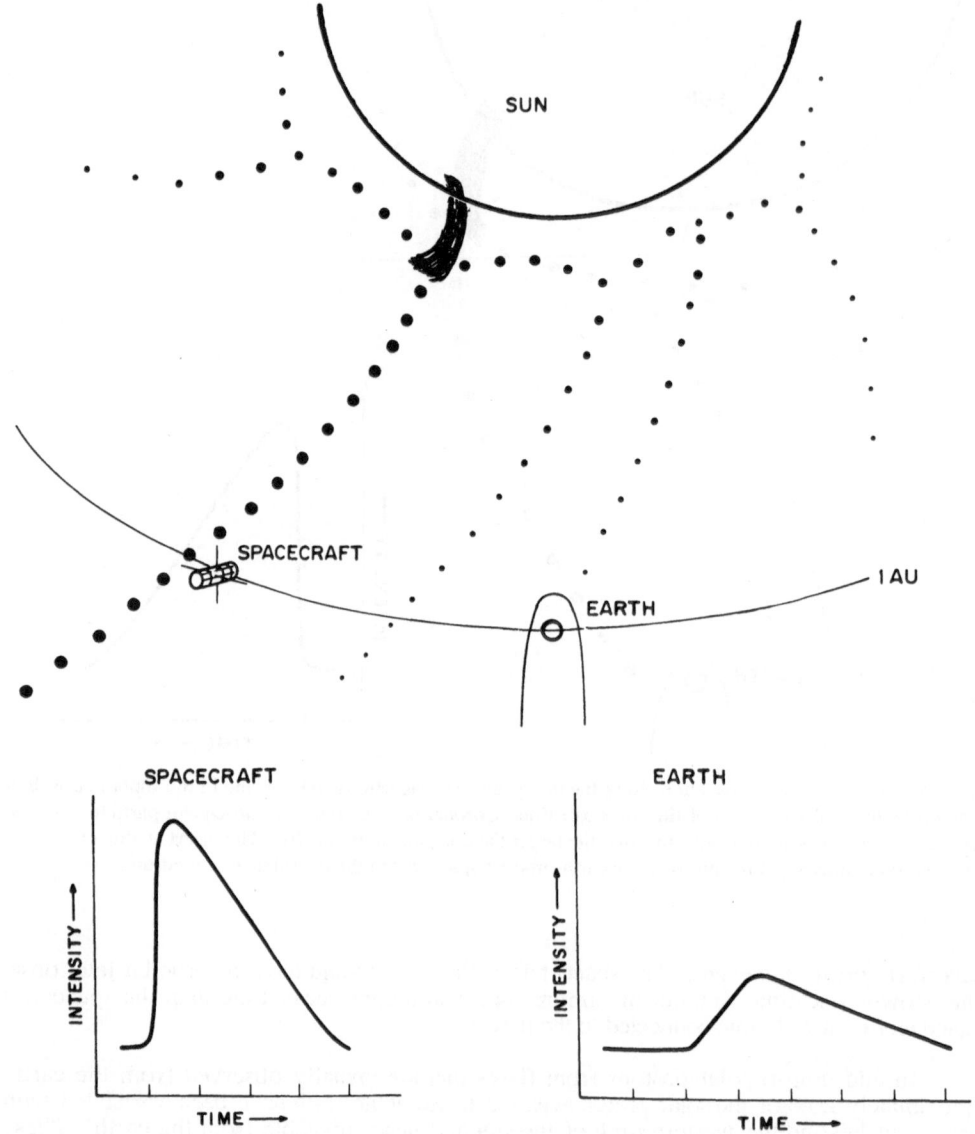

Figure 6. Solar particle propagation along the interplanetary magnetic field lines and in the inner heliosphere from a hypothetical flare east of the sun-earth line. Coronal propagation and subsequent particle transport along other field lines are indicated by dots; the larger the dots, the larger the flux. The insert at the lower left shows the intensity-time profile for the solar proton flux for the entire event as measured at the spacecraft directly connected to the flare site via the interplanetary magnetic field line; the insert at the lower right shows the intensity-time profile for the solar proton flux for the entire event as measured by identical instrumentation on an earth-orbiting spacecraft in the interplanetary medium.

one Astronomical Unit albeit at different heliolongitudes. Had an interplanetary mission been located on the "far" side of the sun (as viewed from the earth) and been "connected" to the flaring region via the interplanetary magnetic field lines, the spacecraft would have been subjected to a major proton event which would not have been anticipated by any earth-based measurements. For this reason, on-board solar optical, X-ray, and particle monitors would be almost essential equipment for a manned interplanetary mission.

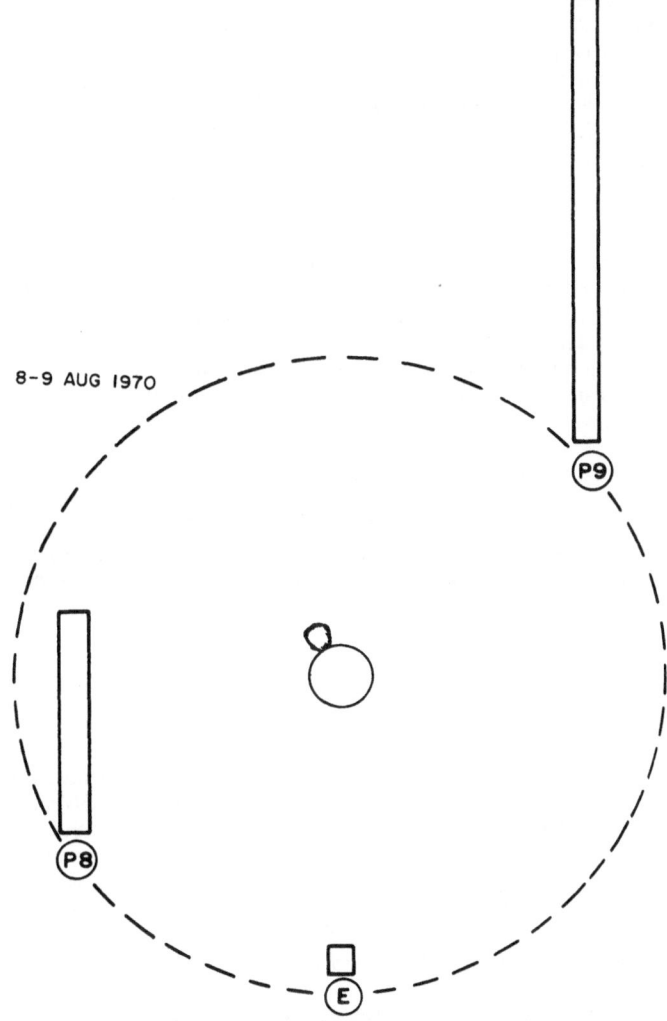

Figure 7. Relative solar proton amplitude vs. spacecraft position.

Solar Proton Propagation to Other Locations in the Heliosphere

The principals of solar proton propagation to the earth, as described in the previous sections, can be applied to any point in the inner heliosphere. The solar particles first propagate through the solar corona to the "footpoint" of the interplanetary magnetic field lines connecting the sun to the detection point in space. The particles then propagate along the magnetic field lines to this detection point. If Earth and Mars were radially aligned, the "nominal" location of the footpoint of the Archimedean spiral path connecting the sun to Mars would be at ~ 90° West heliolongitude. In comparison, the "nominal" location of the footpoint of the Archimedean spiral path connecting the sun to Earth is located at ~ 60° West heliolongitude. As the planets rotate about the sun, the relative location of the nominal footpoint of the Archimedean spiral path to Earth and to Mars continually changes.

If a spacecraft were traveling between Earth and Mars, the "footpoint" of the idealized Archimedean spiral path between the spacecraft and the sun would also be continually changing. As a spacecraft moves radially away from the sun, the "connection longitude" on the sun of the idealized Archimedean spiral between the spacecraft and the sun moves further and further to the west of the spacecraft-sun radial direction. Beyond the orbit of Mars, the idealized "connection longitude" would be behind the western limb of the sun as viewed from the spacecraft. The principals of determining the longitude on the sun at which the idealized interplanetary magnetic field line would be connected, as described in the section "propagation to the earth", can be used for other locations within the orbit of Mars.

Solar Proton Prediction

A full description of our ability to predict the occurrence and magnitude of solar proton events is beyond the scope of this paper and will be discussed elsewhere in this volume (Smart and Shea, 1992b). At the present time it is not possible to predict when a solar proton-producing flare is likely to occur. Even flare prediction is in its infancy. Certain magnetic configurations within the solar active region, indicative of magnetic stress, have proved to be useful parameters for identifying a region that might produce a significant flare; however, there is no unique indicator that the flare will produce copious solar particles.

Solar proton prediction is based on observables such as the location of the flare on the sun, the electromagnetic emission, and the status of the interplanetary medium. The techniques used by the forecast centers are all based on observations made at the earth. Although the general principals would remain the same, additional techniques would have to be developed to predict an intensity-time profile for an interplanetary mission.

SOLAR CYCLE STUDY

The Solar Cycle

The sun is a controlling factor for many aspects of our terrestrial environment. The solar activity cycle averages about 11 years (Schwabe, 1843; Wolf, 1856) although there can be large deviations from this average. For the past century the "measure" of the solar cycle has been the number of sunspots as observed by astronomers around the world. Because the sunspot numbers as developed by Wolf in 1848, and modified to the present method in 1882, vary considerably from one month to the next, a "smoothed" 13-month running mean of monthly means was derived to typify the solar cycle. (See Waldmeier, 1961, for a description of how the smoothed sunspot number is calculated.)

The latitude of the location of sunspots as a function of time, the so-called "butterfly diagram", is plotted in the top part of Figure 8 for the past 10 solar cycles. The first spots for each new cycle are seen at high solar latitudes. Successive spot groups erupt at slightly lower latitudes in each hemisphere and with increasing (and then decreasing) frequency to form a pattern suggestive of pressed butterfly wings. Toward solar minimum, sunspots for the next cycle appear at higher latitudes while the spots for the waning cycle are at low latitudes. The smoothed sunspot number for this time period is illustrated in the bottom part of this figure.

Although there is no one-to-one correspondence between solar flares and sunspots, major solar flares usually occur in large sunspot groups. Each sunspot group is identified by a region number. During solar minimum there are relatively few flares and relatively few solar proton events. During solar maximum there are many more solar flares and also many more solar proton events. Not all solar flares produce major solar proton events; the search for an observable that will identify which active region or which solar flare will produce copious protons is an ongoing subject of intense research. At the present time the most promising parameter appears to be associated with the magnetic field configuration within and around the active region.

Solar Proton Detection

The detection of solar proton events covers a span of 50 years. Our ability to detect

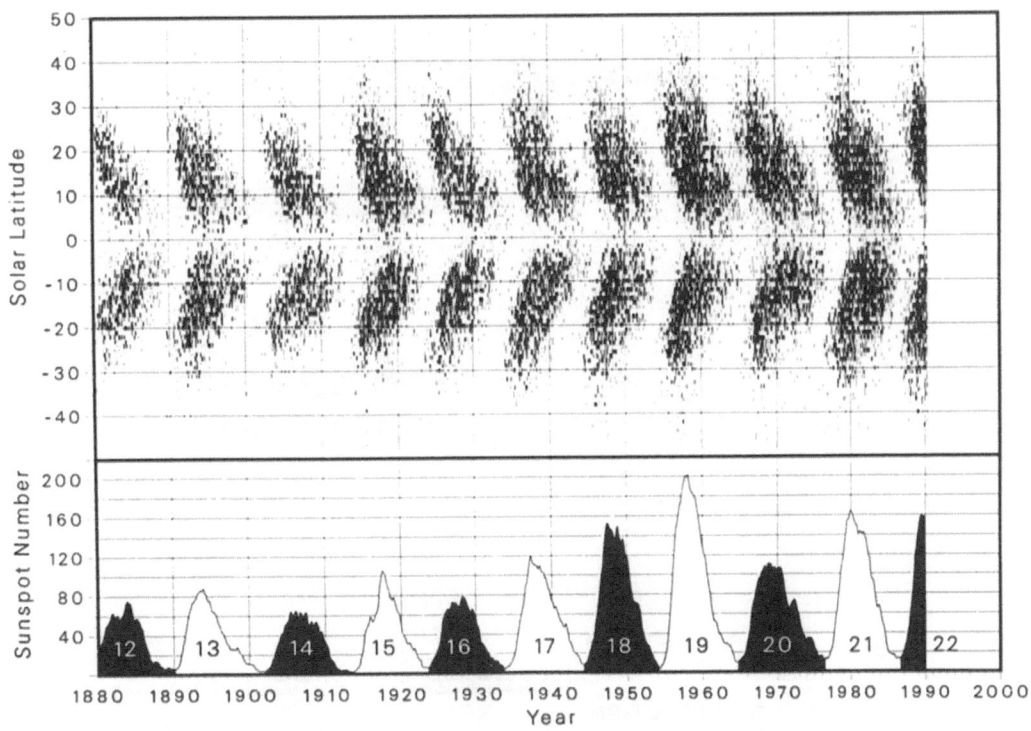

Figure 8. Solar Butterfly Diagram. (Top) The solar latitude of sunspots plotted as a function of time for the last 10 solar cycles. (Bottom) The "smoothed" sunspot number from 1880 until 1990.

these events started with the operation of cosmic ray muon detectors (i.e. ionization chambers) in 1932. The first instances where the sun was unambiguously identified as the source of particles detected at the earth were on 28 February and 7 March 1942. However, it was not until the 25 July 1946 ground-level event detected by the Godhavn and Cheltenham ionization chambers that Forbush (1946) published both these measurements and his earlier measurements of the 1942 events and suggested that these sudden increases were associated with the emission of energetic particles by a solar flare (Elliott, 1985).

The initial observations of "solar cosmic rays" relied on ground-based measurements of secondary particles generated at the "top" of the earth's atmosphere. The events in the 17th and 18th solar cycles were detected by ground-based ionization chambers or muon detectors. These detectors respond to secondary particles generated in the atmosphere from incident solar protons having energies greater than ~ 4 GeV. Most of the solar particle flux and fluence data available from the 19th solar cycle were derived from riometer measurements in the earth's polar regions as these detectors proved to be very sensitive to particles above a few MeV depositing energy in the ionosphere directly above the instrument. High energy solar particle events were also detected during the 19th solar cycle by neutron monitors which measure cosmic radiation from an atmospheric detection threshold of about 450 MeV, if located at high latitudes, to a geomagnetic cutoff threshold of approximately 15 GeV, if located in the equatorial latitudes. Cosmic ray neutron monitors have been in continuous operation since 1953.

Solar proton events have also been detected by sensors flown on balloons since the International Geophysical Year (1957-1958), and on spacecraft since the early 60's. In addition, the ionosphere responds to an influx of solar protons and is, in itself, a means of detecting the presence of solar particles. Figure 9 gives a conceptual history of the availability and energy threshold of each technique used to detect solar proton events since 1933. At the present time our most reliable method of identifying solar proton events is by spacecraft measurements.

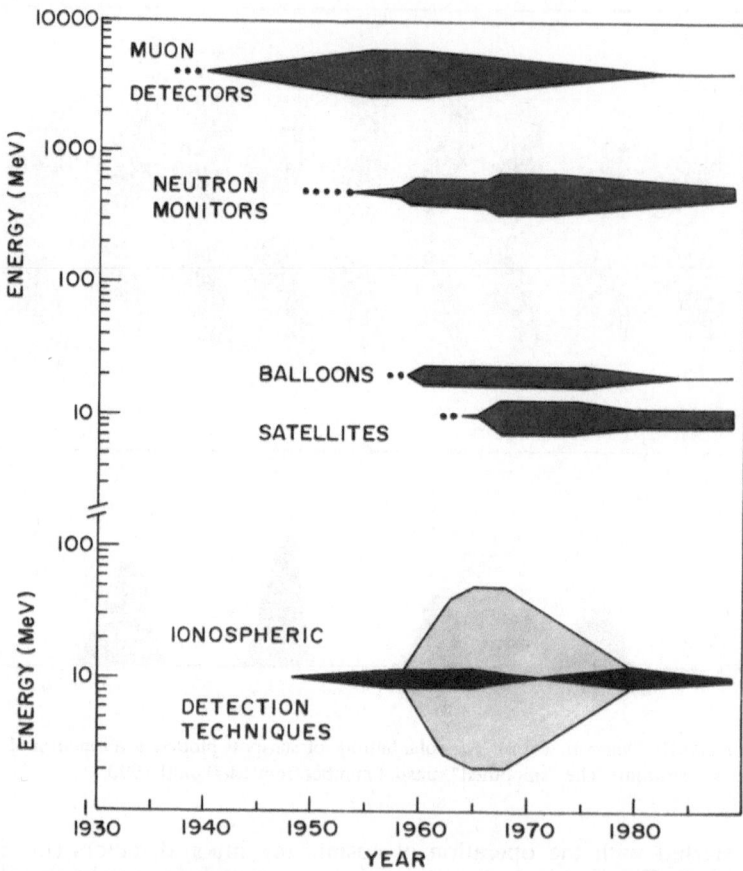

Figure 9 Conceptual history of the detection thresholds of solar proton events. The thickness of the lines indicates the relative number of each type of detector in use. The differences in shading in the ionospheric section indicates changes in detection technique.

Solar Proton Events During Solar Cycles 19-21 (1954-1986)

Using the various data sets that extend over the past three solar cycles (1954-1986), Shea and Smart (1990a) identified 218 "significant" solar proton events that had been detected at the earth. Because of the different measurement techniques employed over this period, it was not possible to assemble a completely homogeneous list of solar proton events. However, the Shea and Smart (1990a) list is as homogeneous as possible using the criterion that each proton event must have a 10 MeV particle flux above 10 particles/cm^2-sec-ster. This criterion is the presently accepted value for identification of a proton event that has the potential of producing perturbations to the geophysical environment.

In deriving this list, each separate solar particle injection was counted as an individual event even if the particle flux was enhanced from a previous increase. However, an additional particle increase related to a geomagnetic storm sudden commencement/interplanetary shock acceleration was not identified as a separate event. A detailed description of the methods for selection and a listing of these events including peak flux and fluence is given by Shea and Smart (1990a).

Using the monthly mean sunspot number as the major ordering parameter (McKinnon, 1987) with the smoothed sunspot number to define the length of each cycle, the 218 events were separated into individual solar cycles with the start of each cycle being the

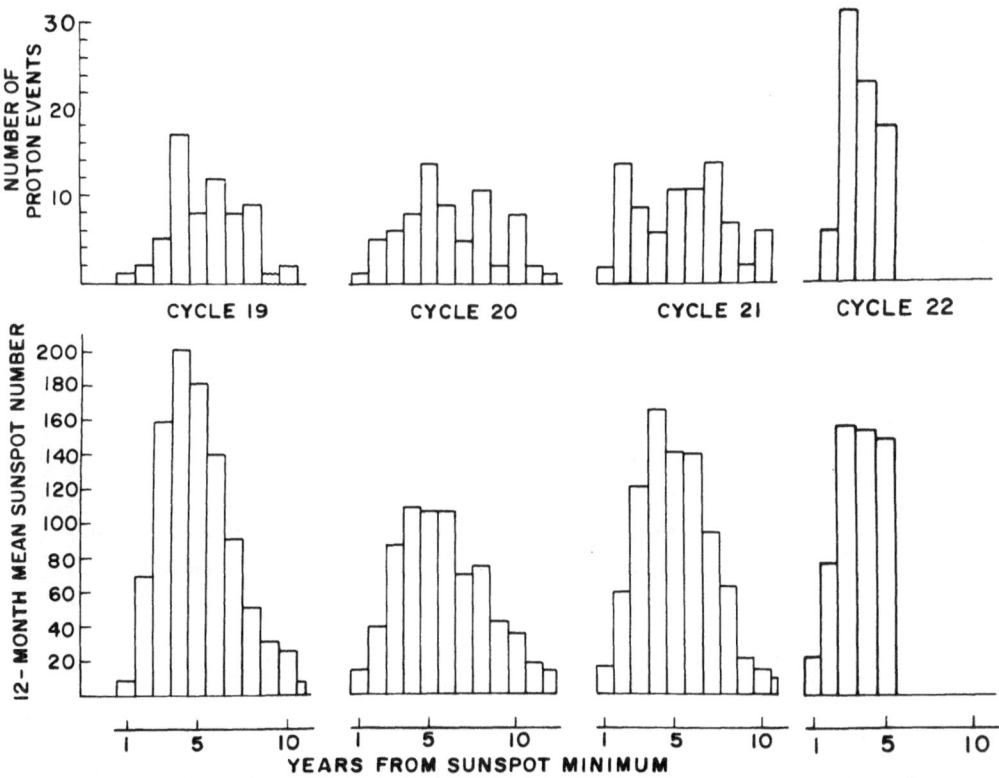

Figure 10. The number of significant discrete solar proton events for each 12-month period after solar minimum (top) and the 12-month mean sunspot number for the corresponding period (bottom) for the completed solar cycles 19-21 and the first five years of solar cycle 22. The data for the present solar cycle are preliminary.

first month after the minimum defined by the smoothed sunspot number. The top portion of Figure 10 shows the number of solar proton events that occurred each 12-month period after sunspot minimum since the start of cycle 19. The 12-month mean sunspot numbers for the same time intervals are shown in the bottom of the figure. Figure 11 is another representation of the same data where the number of discrete solar proton events are shown as individual dots and the 12-month mean sunspot number for the corresponding period is given by the histograms. From these data Shea and Smart (1990a) concluded that although the number of proton events increased during the years of maximum sunspot number, there was no predictable pattern from one solar cycle to the next.

Figure 12 shows the distribution of proton events when the data for the 19th, 20th, and 21st solar cycles were combined into one set. These data, also organized in 12-month periods beginning with the month after sunspot minimum as defined by the statistically smoothed sunspot number, indicate that proton events occur primarily from the 2nd through the 8th year of the solar cycle. For unexplained reasons, there is also an increase in the 10th year of each of the three solar cycles which combine to give an increase in events for year 10.

Solar Proton Events During the First Five Years of Solar Cycle 22 (1986-1991)

Solar cycle 22 started in October 1986 and has defied all predictions of a benign uninteresting cycle. The solar activity, as measured by the sunspot number is the third largest value recorded exceeded only by solar cycles 19 and 21. It now appears that July 1989 will be the maximum of this solar cycle with a smoothed sunspot number of 159. For comparison

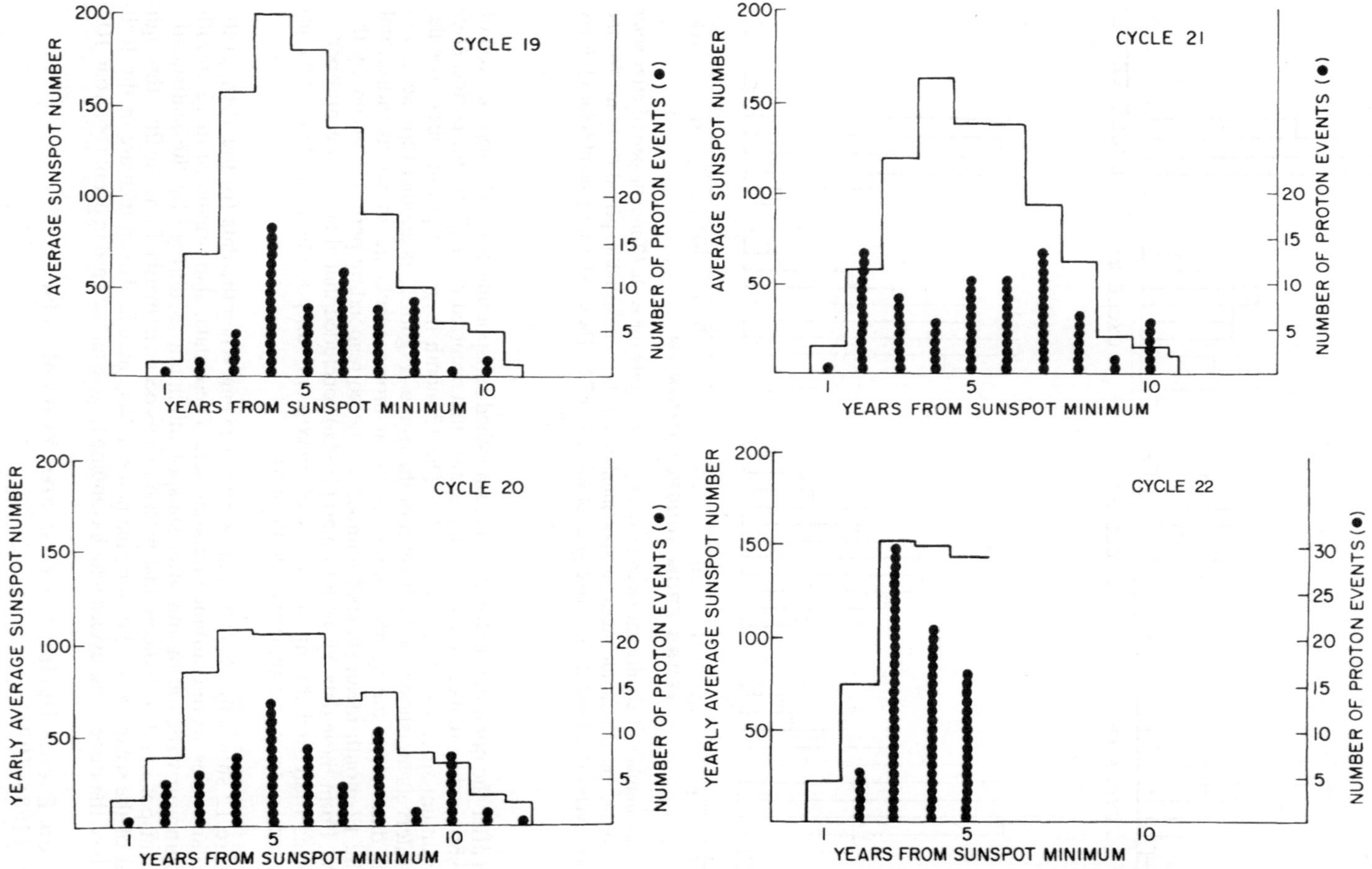

Figure 11. The number of significant discrete solar proton events for each 12-month period after solar minimum (circles) and the 12-month mean sunspot number for the corresponding period (histograms) for solar cycles 19-21 and the first five years of solar cycle 22. The data for the present solar cycle are preliminary.

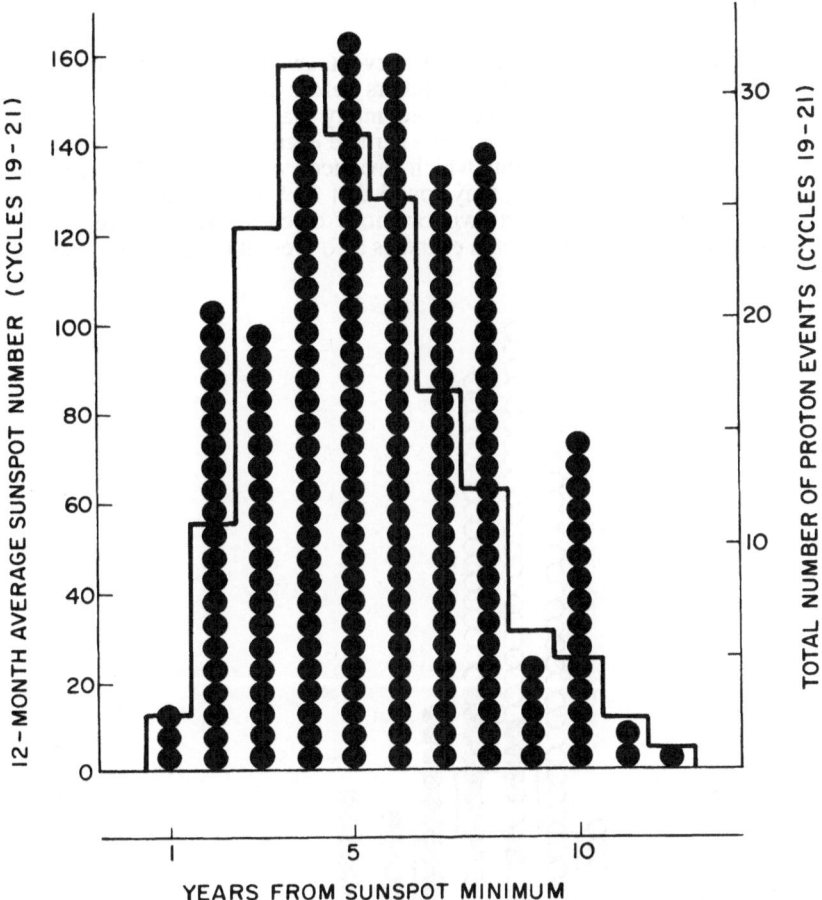

Figure 12. Summation of significant discrete solar proton events for cycles 19-21 (solid circles) and the corresponding 12-month average sunspot number (histogram). The data are organized in 12-month periods beginning with the month after sunspot minimum as defined by the statistically smoothed sunspot number.

purposes the maximum smoothed sunspot number and corresponding month for each of the past three solar cycles were: Cycle 19: 201 (March 1958); Cycle 20: 111 (November 1968); Cycle 2l: 164 (December 1979).

There has been a plethora of proton events this solar cycle. Using the same criteria as used for solar cycles 19-21, the preliminary total number of events for the first five years of this cycle already exceeds the total number of events for the 20th solar cycle and is only a few events less than the total number recorded during the 21st cycle. The far right portion of Figure 10 and the lower right side of Figure 11 shows the number of proton events for the first five years of this cycle; a record number of events was recorded during the third year of this cycle.

The open circles in Figure 13 show the proton events for the first five years of cycle 22 superimposed on the summation of the events for the previous three solar cycles. No events are shown for the 6th year which will not be complete until September 1992. It will be of interest to see if there is an increase in the number of proton events during the 10th year (October 1995 - September 1996).

Sequences of Solar Proton Events

In compiling the list of significant proton events over the past three solar cycles, Shea and Smart (1990a) found, as did many researchers before them, occasions where one active region passing across the solar disk produced many major solar flares several of which released particles associated with the aggregate particle event observed at the earth. From an identification of the individual regions which produced each of the 218 significant solar proton events for solar cycles 19-21, they found that 22% of the proton events were from solar regions that produced at least two or more discrete proton events. This same phenomenon of sequences of solar proton events is also present in the 22nd solar cycle.

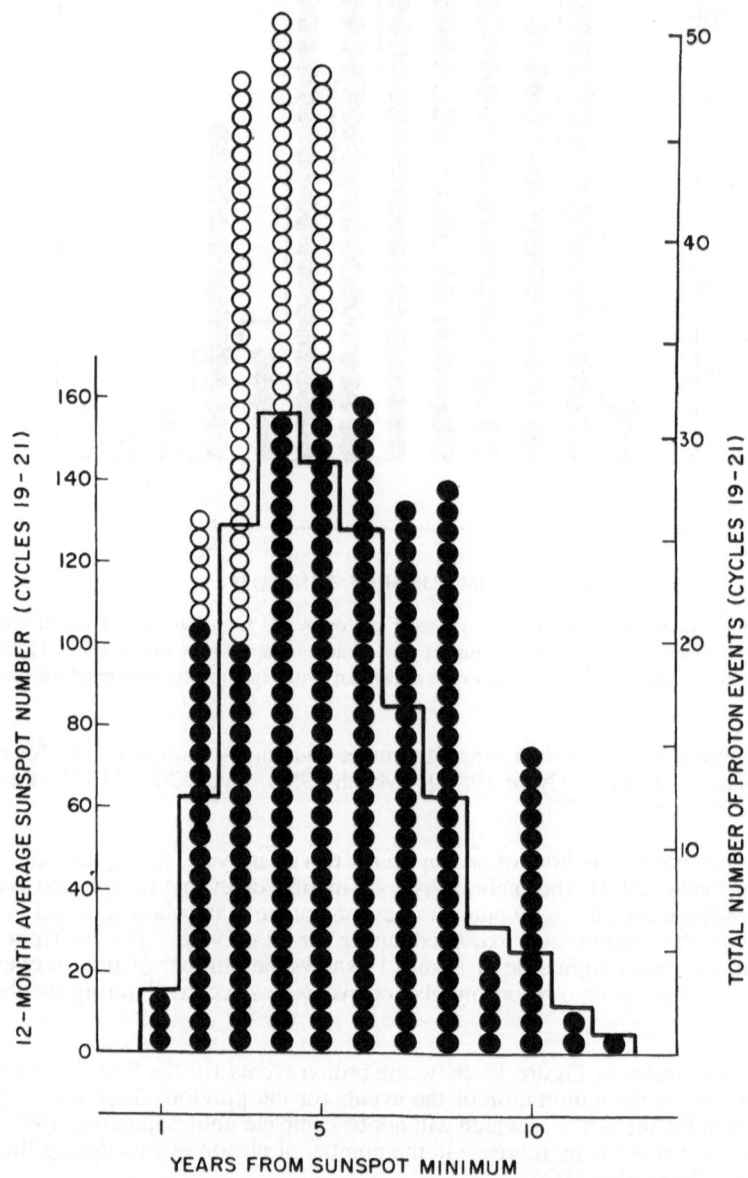

Figure 13. Summation of significant discrete solar proton events for cycles 19-21 (solid circles) and the first five years of cycle 22 (open circles). The corresponding 12-month average sunspot number is shown by the histogram. The data are organized in 12-month periods beginning with the month after sunspot minimum as defined by the statistically smoothed sunspot number. The data for the present solar cycle are preliminary.

RELATIVISTIC SOLAR PROTON EVENTS

Frequency of Relativistic Solar Proton Events

The most homogeneous extended set of solar proton events derived from a standard observational technique is the list of relativistic solar proton events or "ground-level events" as detected by neutron monitors. The sensitivity of this instrument has been essentially unchanged since its inception in 1953 although some small events may not have been identified during the 19th solar cycle because of a sparsity of detectors. Protons having energies greater than approximately 450 MeV can be detected by neutron monitors in the earth's polar regions (above a geomagnetic latitude of ~ 60°). For detectors at mid and equatorial latitudes, the threshold energy is determined by the earth's magnetic field. Protons must have energies of approximately 15 GeV or greater to be detected in the vertical direction in the equatorial regions.

Although folklore was established in the 19th solar cycle that relativistic solar proton events only occur on the increasing or decreasing side of the solar cycle, the improved instrumentation coverage of the more recent cycles has shown that these events can occur at any time including solar minimum. Most of the flares accelerating relativistic protons detected at the earth are on the western hemisphere of the sun, although flares on the eastern hemisphere as well as assumed flares behind the western limb of the sun have occasionally been associated with relativistic proton events.

Figure 14 shows the temporal distribution of these relativistic solar proton events since 1954. Figure 15 shows the location of the flare associated with each event; ten of the 48 events have been associated with solar activity (i.e. an assumed flare) behind the west limb of the sun. Since 1955 more than 17% of the significant solar proton events recorded at the

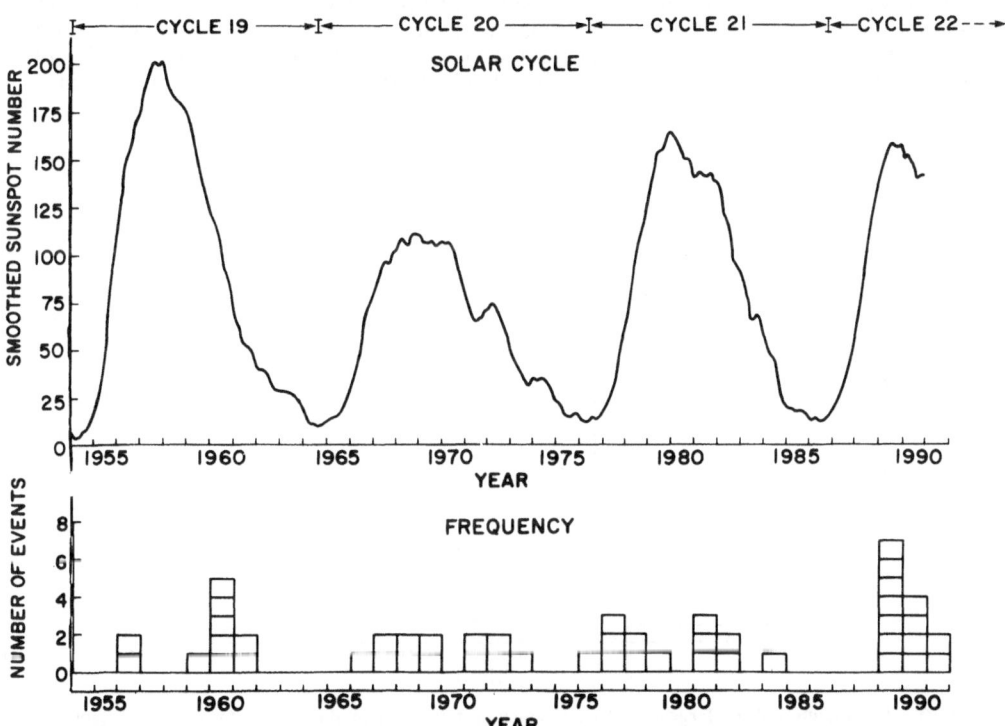

Figure 14. The relativistic solar proton events observed since 1955. The top part of the figure shows the smoothed sunspot number; the bottom part of the figure shows the number of high energy solar proton events each year.

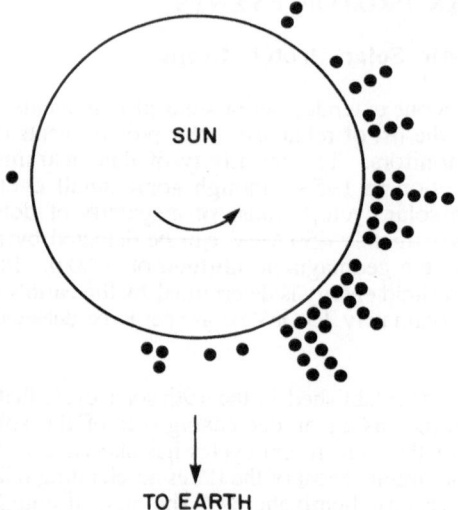

SUN

TO EARTH

Figure 15. The solar longitude (as viewed from the earth) of each flare considered to be the source region of the high energy solar proton acceleration for events between 1956 and 1991.

earth contained relativistic solar protons (i.e. protons with energies > 450 MeV).

Duration of Relativistic Solar Proton Events

Most relativistic solar proton events are relatively short lived with the high energy particle flux passing the earth within a few hours. The higher the energy of particles present, the shorter the exposure time to particles having a specific energy. There is currently no way to predict the duration of relativistic solar proton events since the duration depends primarily on the rate at which the particles diffuse in the interplanetary medium, and this diffusion is

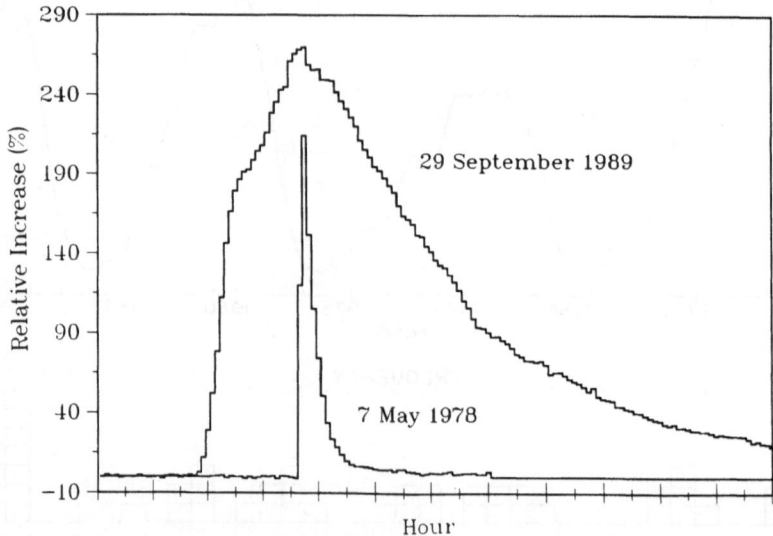

Figure 16. The duration of the largest relativistic solar proton event of the 22nd solar cycle (29 September 1989) compared with the largest event of the 20th and 21st solar cycles (7 May 1978). The individual tic marks on the x-axis represent 30 minutes. The Kerguelen Island neutron monitor is located at 49.35° S, 70.27° E and measures the cosmic radiation above approximately 508 MeV.

dependent upon many factors. Figure 16 illustrates the duration of two comparable events as recorded by the neutron monitor located at sea level on Kerguelen Island in the Indian Ocean. This monitor is located at a vertical geomagnetic cutoff rigidity of 1.1 GV (i.e. equivalent proton energy of 508 MeV) which is essentially the atmospheric cutoff. The event on 7 May 1978 had the highest increase for all relativistic solar proton events in either the 20th or 21st solar cycles with a recorded increase of 214% for protons above 508 MeV. The event had a duration of just over an hour. The event on 29 September 1989 with its comparable increase lasted more than 10 hours. For unknown reasons, the events of this 22nd solar cycle have been of considerably longer duration than similar events of the previous two solar cycles.

Anisotropy of Relativistic Solar Proton Events

Most relativistic proton events are quite anisotropic as viewed at a distance of one Astronomical Unit. It is now generally accepted that these particles usually have long mean free path lengths (>0.3 Astronomical Units) with a variability of a factor of three within the range of normal expectations. It is possible that our observation location of one Astronomical Unit may be close enough to the sun that we may be sampling the solar injection profile.

The anisotropy of particles travelling along the interplanetary magnetic field line is a function of the number of scattering centers between the sun and the earth. When the interplanetary magnetic conditions are quiet such that the magnetic turbulence is relatively minimal, the interplanetary magnetic lines will approximate the "idealized Archimedean spiral shape". Under these conditions, a solar particle flux travelling along the interplanetary magnetic field lines can be extremely well collimated giving rise to a very anisotropic distribution at the earth. In some cases, the anisotropy, defined as the ratio of the flux in the forward quadrant to the average flux, can be as large as 10 to 1 especially during the onset of a well connected event. If there are many scattering centers in the interplanetary medium between the sun and the earth the particles will undergo more scattering enroute to the earth thereby decreasing the amount of anisotropy.

The procedure for analyzing relativistic solar proton events using neutron monitor data is a complex function of solar particle spectrum, pitch angle distribution, asymptotic cone of acceptance for each station, and geomagnetic cutoff rigidity (Shea and Smart, 1982; McCracken, 1962a). In brief, the earth and its geomagnetic field act as a type of magnetic analyzer on "spacecraft Earth". These solar protons are very difficult to measure on satellites because of their high energy; thus neutron monitor measurements are extremely valuable for the analysis of these events.

When high energy solar protons enter the earth's magnetosphere their paths are deflected by the geomagnetic field with the lower energy particles being "bent" more than the higher energy particles. This phenomenon has lead to the concept of "asymptotic cones of acceptance" for cosmic ray detectors on the earth (Brunberg, 1958, McCracken, 1962b). Thus each cosmic ray detector has a unique viewing direction in space which is a function of energy. This viewing direction rotates as the earth rotates. If a station is "viewing" into the direction of the interplanetary magnetic field connecting the sun with the earth when a relativistic solar proton event occurs, the station will detect a maximum particle increase. If a station is "viewing" in a completely different direction it will record a smaller increase depending upon the anisotropy (i.e. pitch angle distribution about the interplanetary magnetic field) of the event. A conceptual illustration of asymptotic cones and an anisotropic particle flux travelling along the interplanetary magnetic field line from the sun to the earth is shown in Figure 17. In this example the station with asymptotic cone "A" viewing into the interplanetary magnetic field direction would record a higher increase than the station with asymptotic cone "B" viewing in a completely different direction. If the event occurred a few hours later, when the station with asymptotic cone "B" had rotated such that this cone was viewing in the direction of the interplanetary magnetic field leading away from the sun, then station "B" would have recorded the larger increase.

Generally the anisotropy is the most extreme during the onset of an event. The longer the duration of an event, the more likely scattering processes will influence particle transport, and the more likely the degree of anisotropy will be reduced. Figure 18 is an example of an extreme anisotropy that was present during the onset of the relativistic solar proton event on

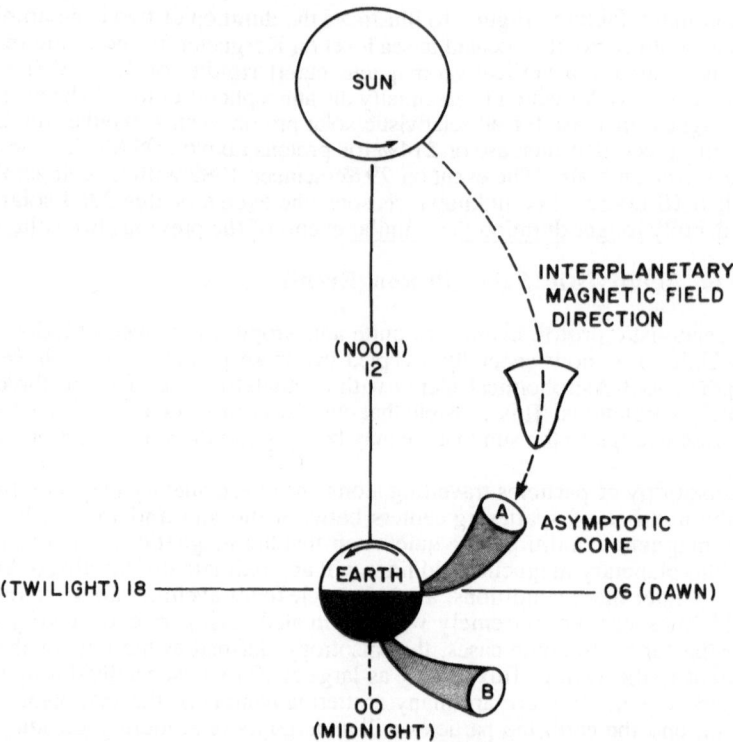

Figure 17. Conceptual illustration of asymptotic cones for two cosmic ray stations and an anisotropic particle flux travelling along the interplanetary magnetic field line from the sun to the earth.

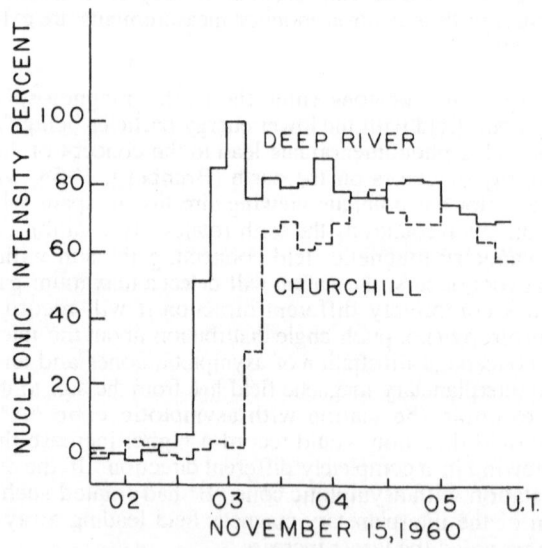

Figure 18. Relative increase in cosmic ray intensity as recorded by two Canadian neutron monitors during the 15 November 1960 relativistic solar proton event. This is an example of an extreme anisotropy. The Deep River monitor was located at 46.10° N, 282.50° E; the Churchill monitor at 58.75° N, 265.92° E. Both stations would have recorded the solar proton intensity above the atmospheric cutoff of ~ 500 MeV.

15 November 1960. In this case the associated flare at 35º W heliolongitude was relatively well connected to interplanetary magnetic field lines that might intersect the earth. At the time of the event, the asymptotic cone of acceptance of the Deep River, Canada neutron monitor, which is elongated over a wide range of longitudes, extended into the probable interplanetary magnetic field direction. This monitor recorded an increased particle flux for 30 minutes before the neutron monitor at Churchill, Canada with a comparable cutoff rigidity but having a very narrow cone of acceptance viewing in a completely different direction of space. Approximately 90 minutes after the onset of the particle increase, both detectors were recording the same increase in intensity indicting that the particle flux had become relatively isotropic.

Figure 19 is another example of an anisotropic particle distribution as recorded by neutron monitors located in the polar regions. The associated flare for the 29 September 1989 relativistic solar proton event was assumed to be at ~105° West heliolongitude - 15° behind the western limb of the sun. Although a major event, the anisotropy was not as pronounced as for the event on 15 November 1960; however, the stations at Thule, Greenland and McMurdo, Antarctica, recorded vastly different intensity-time profiles for the first five hours of the event. In this case the different profiles were the result of a moderate anisotropy along the interplanetary magnetic field direction coupled with changes in the orientation of the interplanetary magnetic field line (relative to the earth-sun line) throughout the event (Smart et al., 1991). For reasons such as these it is necessary to utilize the world network of neutron monitor stations to analyze relativistic solar proton events.

Although we have used neutron monitor data to illustrate examples of an anisotropic particle flux, spacecraft measurements have also shown that the lower energy particles follow a similar pattern.

Magnitude and Spectra of Relativistic Solar Proton Events

The largest relativistic solar proton event recorded at the earth was on 23 February 1956 with a 4000% increase in the counting rate of the neutron monitor at Leeds, England. The event on 29 September 1989 was the largest relativistic solar proton event since 23

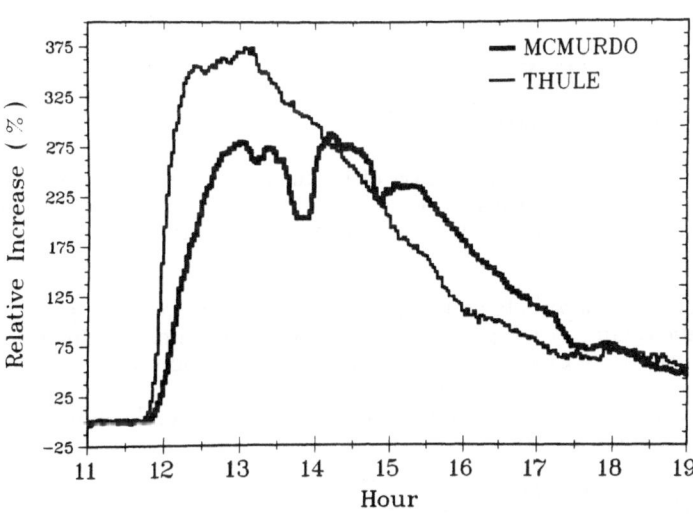

Figure 19. Relative increase in cosmic ray intensity as recorded by two polar neutron monitors during the 29 September 1989 relativistic solar proton event. This is an example of a moderate anisotropy where the interplanetary magnetic field line was also undergoing perturbations. The Thule, Greenland monitor is located at 76.50º N, 291.30º E; the McMurdo, Antarctica monitor is located at 77.85º S, 166.72º E. Both stations are at sea level and record the solar proton intensity above the atmospheric cutoff of ~ 500 MeV.

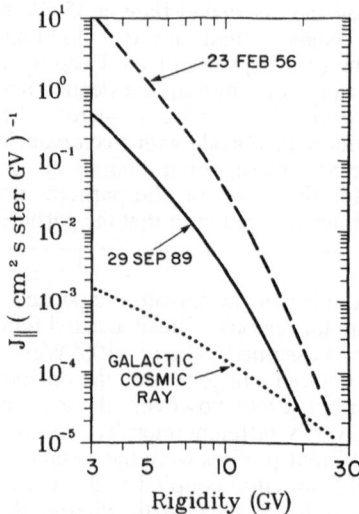

Figure 20. Differential solar proton spectra for the relativistic solar proton events of 23 February 1956 and 29 September 1989. The galactic cosmic ray spectrum is also shown.

February 1956, and was the third largest since 1942 (Smart and Shea, 1991). Figure 20 illustrates the solar proton spectrum derived at the high energy event maximum for the 29 September 1989 event (Smart et al., 1991) compared with the spectrum derived at the high energy event maximum for the 23 February 1956 event (Smart and Shea, 1990). The spectrum for galactic cosmic radiation is also shown for comparison. The area between each event curve and the galactic cosmic ray spectrum represents the excess solar proton flux above the normal cosmic ray background. The x-axis is given in units of rigidity (momentum per unit charge) which is a common unit used by cosmic ray scientists. A proton having a rigidity of 1 GV has an energy of 433 MeV; a proton having a rigidity of 10 GV has an energy of 9.11 GeV. Figure 21 illustrates the energy to rigidity conversion for protons, electrons and alpha particles.

GEOMAGNETIC CUTOFF RIGIDITIES

Basic Cutoff Rigidity Equation

A complete and detailed discussion of geomagnetic cutoff rigidities is beyond the scope of this paper. Nevertheless, a description of the geomagnetic shielding effect is necessary to understand the particle flux that can penetrate to various spacecraft orbits within the magnetosphere.

The detection of a particle at any specific point in the magnetosphere is dependent upon the magnetic rigidity of the particle, the geographic coordinates, altitude, and angle of incidence at the detection location, and the geomagnetic field conditions.

The basic equation for calculating the lowest rigidity particle allowed from a specific direction in a dipole field is

$$R_s = \frac{M \cos^4 \lambda}{r^2[1 - (1 - \sin\varepsilon \, \sin\phi \, \cos^3 \lambda)^{1/2}]^2} \tag{1}$$

where M is the effective dipole moment which has a normalized value of 59.6 when "r" is expressed in units of earth radii, λ is an appropriate magnetic latitude, "r" is the distance from

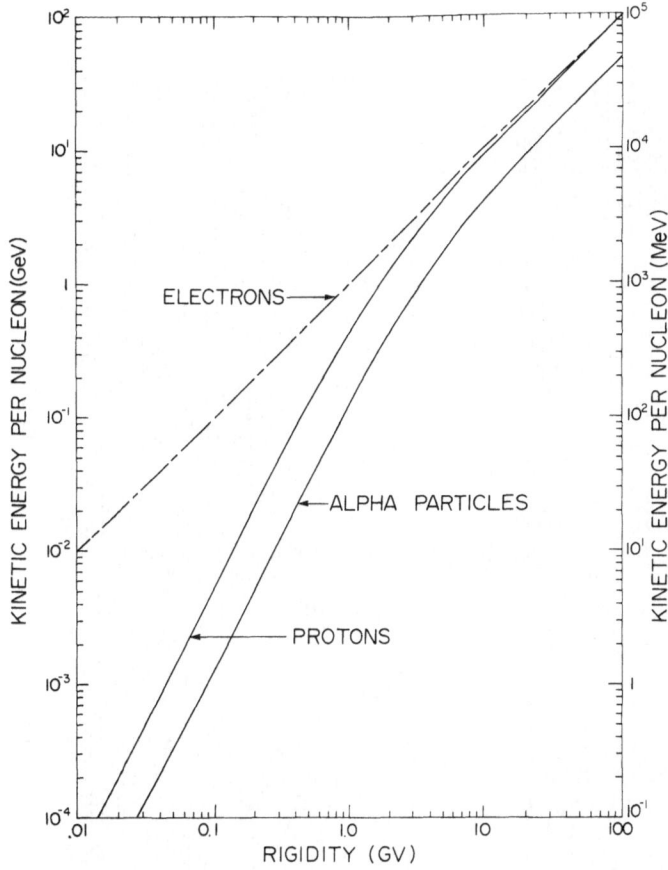

Figure 21. Conversion from magnetic rigidity to kinetic energy per nucleon for electrons, protons and alpha particles. Since the ratio of neutrons to protons for all elements ≥ 2 is similar, the alpha particle conversion curve can be used for all heavier elements.

the dipole in earth radii, ϕ is the azimuthal angle measured clockwise from the geomagnetic east direction (for positive particles), and ε is the angle from the local magnetic zenith direction. Since the earth's magnetic field only approximates a dipole, the use of the above equation gives a crude approximation of the cutoff rigidity. For more accurate values, detailed particle trajectory calculations using a mathematical model of the geomagnetic field are necessary.

The terminology surrounding the cosmic ray cutoff rigidity problem has evolved considerably in the last 50 years; this entire process and the current terminology is documented by Cooke et al. (1991).

Vertical Cutoff Rigidity Values

The calculation of the cutoff rigidity for any one direction in space involves a complex set of calculations of the trajectory of charged particles in the geomagnetic field (Shea et al., 1965). These calculations take a prodigious amount of high speed computational time depending upon the degree of precision of the geomagnetic field model utilized. Because of these complexities, most calculations are done for the vertical direction. For the purposes of this paper it is sufficient to know that the effective cutoff rigidity, which allows for the transparency of the cosmic ray penumbra, is the most commonly used value of cutoff rigidity (Shea et al., 1965). In today's nomenclature the term "vertical cutoff rigidity" usually denotes the "effective cutoff rigidity in the vertical direction".

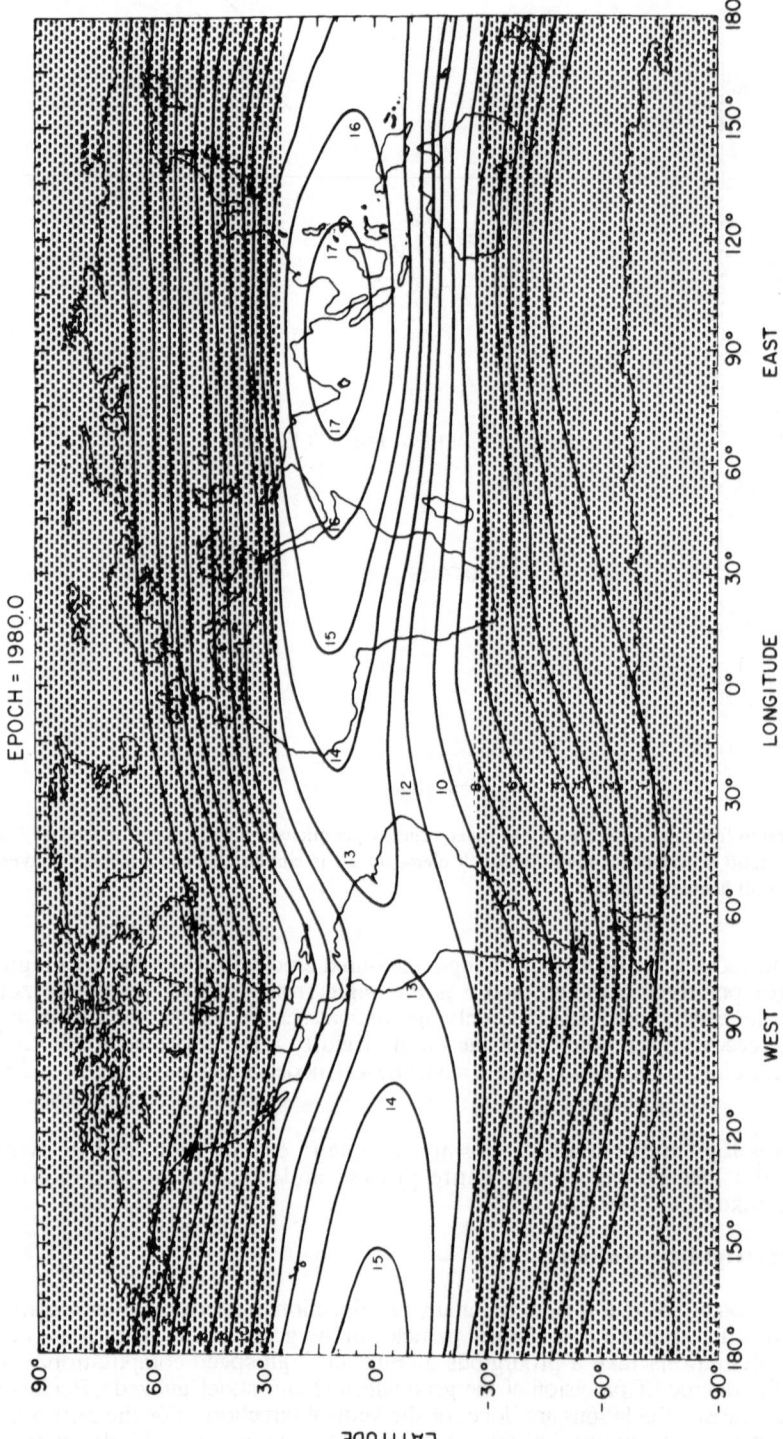

Figure 22. World map of effective vertical cutoff iso-rigidity contours. The unshaded areas, between 30° north and south latitudes represent the latitude ranges covered by low inclination shuttle flights. The cutoff rigidity contours are given for the following intervals: 1, 2, 3, 4, 6, 8, 10, 12, 13, 14, 15, 16, and 17 GV.

Figure 22 shows the iso-rigidity contours of the effective vertical cutoff rigidity as determined from a large number of calculations made around the world (Smart and Shea, 1985; Shea and Smart, 1990b). The high precision cutoff rigidity values were calculated using a 10th order geomagnetic field model appropriate for Epoch 1980.0. The values given represent the rigidity, in GV, needed for a particle to reach the top of the atmosphere at a specific location in the vertical direction. In other words, all particles having higher rigidity can reach that location from the specified direction. Because of the off-set of the earth's dipole and higher order moments in the earth's magnetic field, the contour lines are not symmetric. The unshaded area between 30° north and south latitudes represents the latitude excursions of most shuttle flights and the planned space station. From this figure it can be seen that the minimum rigidity required to reach the shuttle in the vertical direction (under quiescent conditions) is 4 GV for an extremely limited portion of the orbit over Florida. Incoming protons would require higher rigidities to reach lower latitudes in this orbit. Over India, for example, particles would need a rigidity of > 17 GV to reach the spacecraft from the vertical direction.

The threat of solar protons to this orbit is minimal. Only relativistic solar protons can penetrate to low latitude spacecraft having a 30° inclination, and, as mentioned previously, these events are relatively rare and are usually short lived. However, each solar cycle there are one or two solar proton events having high energy particles that penetrate to the top of the atmosphere at equatorial latitudes.

Figure 23 is similar to Figure 22 except that the unshaded area between 60° north and south latitudes represent the approximate latitude excursions of selected shuttle flights and the initial Soviet space station. If the spacecraft is in a 60° inclination orbit, solar protons must have a rigidity above 1 GV to impact the spacecraft, except for a limited portion of the orbit over North America and south of Australia where lower rigidities can penetrate.

The iso-rigidity contour maps shown in Figures 22 and 23 were prepared from calculations for an altitude of 20 km which is roughly equivalent to the top of the atmosphere. As the altitude increases the vertical cutoff rigidity decreases in proportion to the inverse square of the radial distance from the effective magnetic center of the earth. For more accurate values, particularly for higher altitudes, the use of the McIlwain (1961) L-parameter is recommended as an interpolation aid (Smart and Shea, 1973). The L-parameter was originally used to bring order to trapped radiation data.

It has been shown (Smart and Shea, 1967; Shea et al., 1987) that the effective vertical cutoff rigidity is well correlated with the L-parameter since the calculation of the L-parameter accounts for the higher order moments present in the earth's magnetic field. The use of the L-parameter, calculated for various Epochs of the geomagnetic field, also accounts for the changes observed in the geomagnetic field for the past 30 years. For practical applications, a satellite within the domain of L = 4 may be considered to be "shielded" from the solar cosmic ray flux, but a spacecraft beyond L = 4 is considered to be "exposed" to the solar cosmic ray flux. The L = 4 position is equivalent to a corrected geomagnetic latitude of 60° and also a vertical cutoff rigidity of 1 GV.

For both high-inclination and polar orbits most spacecraft spend only a small fraction of orbital time at high latitudes. For the remainder of the orbit, the spacecraft traverses lower latitudes where the geomagnetic field effectively shields all but the most energetic protons. Therefore, even for a major low energy solar proton event with the full interplanetary flux present in the polar cap, a spacecraft in a high inclination orbit would be impacted by solar protons for only a portion of the orbit.

Cutoff Rigidities in the Non-Vertical Direction

As mentioned previously, the calculation of the cutoff rigidity value is usually made for the vertical direction. The use of vertical cutoff rigidity values, extrapolated to low altitude spacecraft, is useful for estimating the dose expected from cosmic rays during quiescent times for a particular orbit within the magnetosphere.

EPOCH = 1980.0

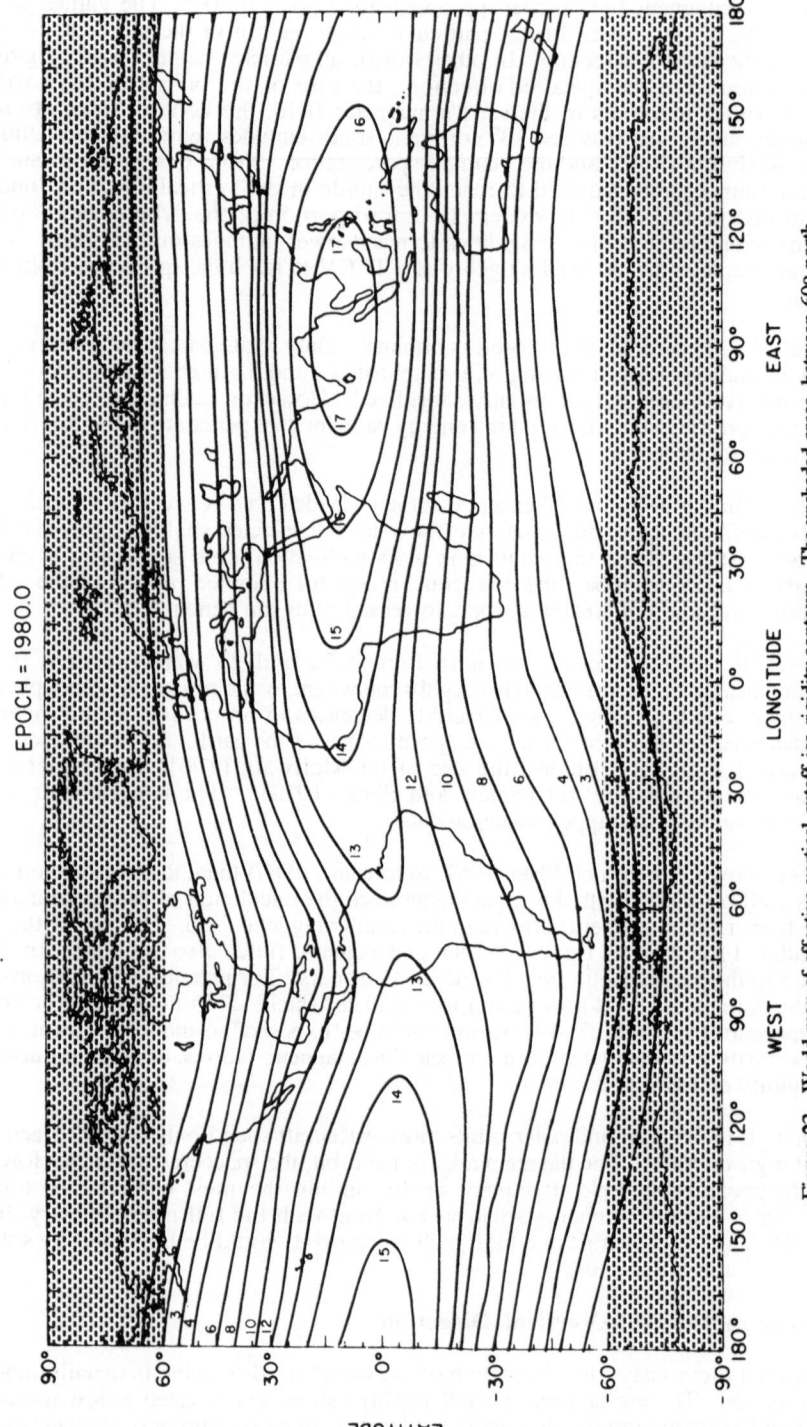

Figure 23. World map of effective vertical cutoff iso-rigidity contours. The unshaded areas, between 60° north and south latitudes represent the range in latitudes covered by selected high inclination shuttle flights. The cutoff rigidity contours are given for the following intervals: 1, 2, 3, 4, 6, 8, 10, 12, 13, 14, 15, 16, and 17 GV.

However, as seen from equation (1) given above, the cutoff rigidity is a function of both zenith and azimuthal directions. It is lower for protons arriving from the west direction and higher for particles arriving from the east direction. The cutoff in the north-south magnetic meridian is the same as the vertical cutoff. The east-west effect, which is minimal for high latitude locations, assumes greater importance at mid and equatorial latitudes. The greater the zenith angle, the larger the east-west difference. Therefore, to determine the lowest energy particle that could penetrate to a specific location on a spacecraft orbit the cutoff should be calculated for a direction 90° from the zenith in the magnetic west direction (i.e. azimuthal direction 270° from magnetic north measured clockwise).

The angular cutoff problem, because of its intrinsic complexity relies on approximations for solutions. Smart and Shea (1983) calculated a 10° in latitude by 30° in longitude world grid of angular cutoffs where the trajectory-tracing method was used to calculate cutoffs in 97 directions for each grid location. These cutoffs were combined to give a "transmission function" which describes the cosmic ray access at each position. The more recent computer-based models for calculating cosmic-ray access (Adams et al., 1983, and Wilson et al., 1990) demonstrate an equivalence to the Smart and Shea (1983) transmission functions at 400 km. These more recent computer based approximations can be applied to other altitudes or satellite orbits.

The Effect of Geomagnetic Disturbances on Cutoff Rigidity

The cutoff rigidity values shown in Figures 22 and 23 were calculated for quiescent magnetic field conditions. When the geomagnetic field is highly disturbed, such as during a major geomagnetic storm, the geomagnetic cutoff can be appreciably lowered in a predictable manner (Flückiger et al., 1986). Figure 24 is a north polar view of the quiescent iso-rigidity contours. Under extremely disturbed magnetic conditions, for example when aurora are visible at mid and low latitudes, the entire polar cap enlarges and the cutoff contours move equatorward. During these times, which usually persists for only a day or two, lower rigidity particles can penetrate to much lower latitudes than during quiescent conditions. During episodes of solar activity when there is a series of solar proton events combined with geomagnetic storms, the particle flux for spacecraft within the magnetosphere would be larger than if the particle events had occurred without a coincidental geomagnetic storm.

EXTREMELY LARGE EVENTS

Occasionally solar and interplanetary conditions will combine to produce an unusually large solar proton event. These events have been called "Unusually Large" or "Anomalously Large" events; however, as shown by Feynman, et al. (1990), they are at the high end of a log-normal distribution and should not be called anomalous.

These events typically occur when a series of major proton-producing flares combine with one or more interplanetary shock waves in such a manner that the ambient particle flux is accelerated to energies considerably higher than the initial energy. Most of these large fluence events are also associated with great geomagnetic storms since a series of large flares, particularly near the central meridian of the sun (as viewed from the earth) often produce the interplanetary shock waves that are associated with both geomagnetic storms at the earth and interplanetary particle acceleration. Examples of such events are those of July 1959, July 1961, and August 1972. The events of August 1972, with a particle fluence of 1.1×10^{10} protons/cm^2, has often been used as a fiducial mark for a "worse case" solar proton fluence above 10 MeV, primarily because it was the first extremely large event for which we had comprehensive spacecraft measurements. However, there have been other events since August 1972 that had larger fluences at specific energies such as the major episode of activity in October 1989.

These extremely large events provide the greatest hazard to manned space missions. Although the radiation from this type of activity sequence would warrant cancellation of Extra Vehicular Activity (EVA) during the major portion of the particle event for a low altitude spacecraft in the magnetosphere, the mission could probably continue to its planned termination without debilitating effects to the astronauts. The spacecraft shielding plus the additional shielding provided by the geomagnetic field would dramatically attenuate the charged particle flux so that it will be considerably less than the interplanetary solar particle flux. Even for a major solar proton event such as occurred in September 1989, the geomagnetic shielding provides very effective protection to a 60° inclination orbit. Interplanetary radiation dose calculations by Lett et al. (1990) based on solar proton measurements for the 29 September 1989 event (Sauer et al., 1990) indicate that astronauts shielded by only 2 gm of material would have experienced a dose of 0.24 Gy to the blood forming organs. The actual dose measurements during this event under about 10 gm absorber on the MIR Space Station (orbital inclination of 51.6°) indicate that the cosmonauts experienced an actual dose of 0.525 rad (0.00525 Gy) as measured by a tissue-equivalent ionization chamber (Teltsov and Tverskaya, 1991).

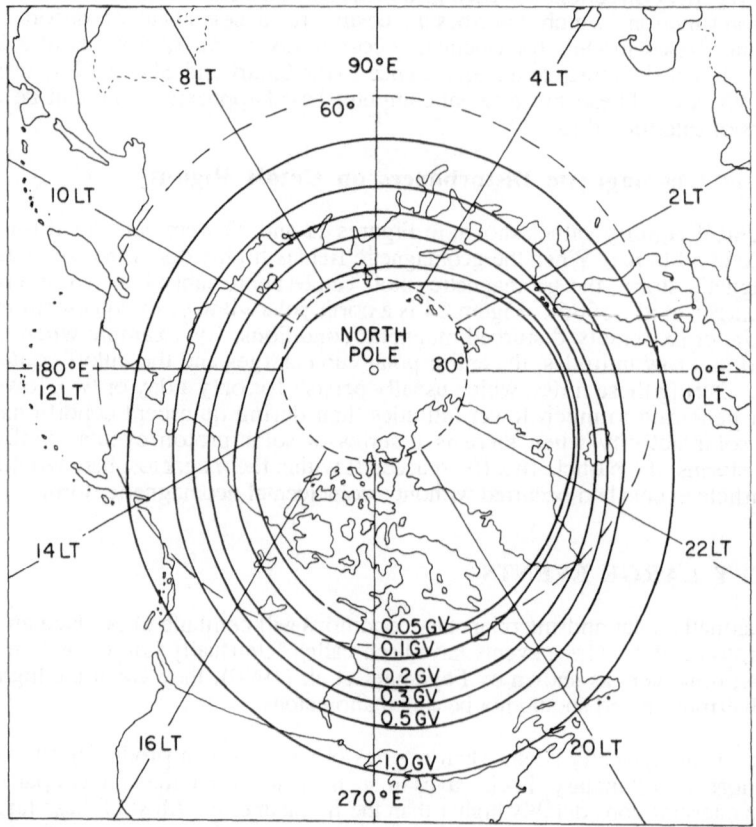

Figure 24. Effective vertical cutoff iso-rigidity contours for quiescent conditions on a north polar projection. During geomagnetic disturbances the contours move equatorward.

The effect of a decrease in cutoff rigidity coupled with a major solar proton event can be more hazardous than a single isolated solar proton event. This is demonstrated by comparing the measurements on the MIR Space Station for the 29 September 1989 event and the major sequence of activity between 19 and 24 October 1989. The radiation increased by 0.375 rad (0.00375 Gy) during 16 hours of the 29 September event; it increased by 1.5 rad (0.015 Gy) during three hours of the major geomagnetic disturbance that occurred when the proton event of 19 October was still in progress. Although part of this difference must be attributed to the fact that the MIR was under maximum geomagnetic shielding during the maximum of the 29 September event, the effect of a coincidental proton event and major geomagnetic disturbance can not be underestimated.

Space missions to the moon will be subjected to the full solar particle flux present in interplanetary space in the vicinity of the earth. Although the moon spends a portion of its orbit in the magnetospheric tail, the magnetic field in this region is insufficient to shield the lunar surface from solar protons.

For a mission to Mars, the spacecraft will also be vulnerable to all energies from solar proton events. The possible exposure to a very large solar proton event during a Mars mission should be viewed as an operational constraint which can be mitigated by proper contingency planning.

COMPARISON OF FLUENCE: 1955-1991

Table 2 summarizes the number of events and the particle fluence above 10 and 30 MeV for solar cycles 19-21. Also listed are the > 10 and > 30 MeV fluence for the major sequences of activity that have occurred within the first five years of solar cycle 22. The fluence for solar cycle 22 has already exceeded that for either cycle 20 or 21; it is approximately half the value estimated for cycle 19. Of special interest is the fact that the sequence of activity in October 1989 generated more fluence above 10 and 30 MeV than was measured at the earth during the entire 21st solar cycle! Since each cycle has produced major solar activity during the declining portion of the cycle, it is possible that the fluence of cycle 19 will be equalled if not exceeded.

SUMMARY

This review of energetic solar proton events for the past 50 years can be summarized as follows:

(a) Energetic solar proton events are characterized by a rapid rise in intensity followed by a slower decay.

(b) The magnitude of a solar proton event at any position in space is a function of location of the flare on the sun with respect to the detection position. The flux is attenuated a factor of 10 per radian as measured by the angular distance between the heliolongitude of the solar activity (i.e. assumed to be a solar flare) and the heliolongitude of the "footpoint" of the interplanetary magnetic field line connecting the sun with the detection point.

(c) Major solar flares can populate the entire inner heliosphere with solar particles.

(d) Energetic solar proton events usually occur from the 2nd through the 8th year of the solar cycle; however, they can occur at any time including solar minimum. Beyond the general relationship of increased solar proton events with increased solar activity, no repeatable pattern could be found between the occurrence of solar proton events and the solar cycle.

(e) Occasionally, one active solar region passing across the solar disk will produce many major solar flares, associated solar particle events and interplanetary magnetic shocks. When the ambient solar particle flux is re-accelerated by interaction with the shock front, an unexpectedly large particle flux may be recorded. These extremely large events provide the greatest hazard to manned space missions since the proton fluence can exceed 10^{10} particles/cm^2 above 10 MeV.

(f) The earth's magnetic shield provides considerable protection against solar protons at low and mid latitudes. Even a 57° inclination low altitude polar orbiting spacecraft is offered considerable protection since the spacecraft orbit continually crosses the equatorial regions between high latitude passes.

(g) If a major solar proton event coincides with a major geomagnetic storm (with a resulting sudden decrease in cutoff rigidity) earth-orbiting spacecraft will experience an increased particle flux exposure for a few hours.

Table 2. Summary of Solar Proton Events for Solar Cycles 19, 20, and 21.

Cycle	Start*	End	No. of Months in Cycle	No. of Discrete Proton Events	No. of Discrete Proton Producing Regions	Solar Cycle Integrated Solar Proton Fluence (cm^{-2}) > 10 MeV	> 30 MeV
19	May 1954	Oct 1964	126	65	47	7.2×10^{10}	1.8×10^{10}
20	Nov 1964	Jun 1976	140	72	56	2.2×10^{10}	6.9×10^{9}
21	Jul 1976	Sep 1986	123	81	57	1.8×10^{10}	2.8×10^{9}

****

22	Mar 7-25, 1989					0.12×10^{10}	0.03×10^{9}
	Aug 12-18, 1989					0.76×10^{10}	1.4×10^{9}
	Sept 29 - Oct 2, 1989					0.38×10^{10}	1.4×10^{9}
	Oct 19-30, 1989					1.9×10^{10}	4.2×10^{9}
	Dec 30, 1989 - Jan 2, 1990					0.21×10^{10}	0.13×10^{9}
	May 21-31, 1990					0.04×10^{10}	0.14×10^{9}
	Mar 22-26, 1991					0.96×10^{10}	1.8×10^{9}
	June 4-21, 1991					0.32×10^{10}	0.79×10^{9}
					Totals:	4.7×10^{10}	9.9×10^{9}

****

* The start of each solar cycle was selected as the month after the minimum in the smoothed sunspot number (McKinnon, 1987).

(h) Space vehicles on the moon or in the interplanetary medium would be subjected to the full fluence of each solar proton event as present in the interplanetary medium.

(i) For an interplanetary mission, we recommend that optical, X-ray, and particle sensors be onboard the spacecraft.

ACKNOWLEDGMENTS

The authors thank D. C. Wilkinson for providing Figure 8, L.C. Gentile for Figure 16, and J. Collins for artwork.

REFERENCES

Adams, J.H., Jr., Silberberg, R., and Taso, C.H., 1981, "Cosmic Ray Effects on Microelectronics, Part I: The Near Earth Particle Environment", NRL Memorandum Report 4506, Naval Research Laboratory, Washington, D.C.

Adams, J.H., Jr., Letaw, J.R., and Smart, D.F., 1983, "Cosmic Ray Effects on Microelectronics, Part II: The Geomagnetic Cutoff Effects", NRL Memorandum Report 5099, Naval Research Laboratory, Washington, D.C.

Breneman, H.H., and Stone, E.C., 1985, Solar coronal and photospheric abundances from solar energetic particle measurements, Astrophys. J. Lett., 299: L57.

Brunberg, E.-A., 1958, The optics of cosmic ray telescopes, Arkiv för Fysik, 14: 195.

Cook, W.R., Stone, E.C., and Vogt, R.E., 1984, Elemental composition of solar energetic particles, Astrophys. J., 297: 827.

Cooke, D.J., Humble, J.E., Shea, M.A., Smart, D.F., Lund, N., Rasmussen, I.L., Byrnak, B., Goret, P., and Petrou, N., 1991, On cosmic-ray cut-off terminology, Il Nuovo Cimento, 14C: 213.

Dodson-Prince, H.W., Hedeman, E.R., and Mohler, O.D., 1977, "Survey and Comparison of Solar Activity and Energetic Particle Emission in 1970", Air Force Geophysics Laboratory Technical Report, AFGL-TR-77-0222.

Elliot, H., 1985, Cosmic ray intensity variations and the Manchester school of cosmic ray physics, in: "Early History of Cosmic Ray Studies," Y. Sekido and H. Elliot, eds., Astrophysics and Space Science Library, Vol. 118, D. Reidel Publ. Co., Dordrecht, Holland, 375.

Feynman, J., 1985, The solar wind, Chapter 3 in "Handbook of Geophysics and the Space Environment", A.S. Jursa, ed., Air Force Geophysics Laboratory, Bedford, MA.

Feynman, J., Armstrong, T.P., Dao-Gibner, L., and Silverman, S., 1990, New interplanetary proton fluence model, J. Spacecraft and Rockets, 27: 403.

Flückiger, E.O., Smart, D.F., and Shea, M.A., 1986, A procedure for estimating the changes in cosmic ray cutoff rigidities and asymptotic directions at low and middle latitudes during periods of enhanced geomagnetic activity, J. Geophys. Res., 91: 7925.

Forbush, S.E., 1946, Three unusual cosmic-ray intensity increases due to charged particles from the sun, Phys. Rev., 70: 771.

Garrard, T.L., and Stone, E.C., 1991, Heavy ions in the October 1989 solar flares observed on the Galileo spacecraft, 22nd Intl. Cosmic Ray Conf., Contributed Papers, The Dublin Institute for Advanced Studies, Dublin, Ireland, 3: 331.

Gloeckler, G., 1979, Composition of energetic particle population in interplanetary space, Rev. of Geophys., 17: 569.

Hundhausen, A.J., 1972, "Coronal Expansion and Solar Wind", Springer-Verlag, New York.

Kahler, S.W., Sheeley, N.R., Jr., Howard, R.A., Koomen, M.J., Michels, D.J., McGuire, R.E., von Rosenvinge, T.T., and Reames, D.F., 1984, Associations between coronal mass ejections and solar energetic proton events, J. Geophys. Res., 89: 9683.

Kahler, S.W., Cliver, E.W., Cane, H.V., McGuire, R.E., Stone, R.G., and Sheeley, N.R., Jr., 1986, Solar filament eruptions and energetic particle events, Astrophys. J., 302: 504.

Lee, M.A., 1988, Particles accelerated by shocks in the heliosphere, in: "Interplanetary Particle Environment," Joan Feynman and Stephen Gabriel, eds., JPL Publication 88-28, Jet Propulsion Laboratory, California Institute of Technology, Pasadena: 111.

Lett, J.T., Atwell, W., and Golightly, M.J., 1990, Radiation hazards to humans in deep space: A summary with special reference to large solar particle events, in: "Solar-Terrestrial Predictions: Proceedings of a Workshop at Leura, Australia, October 16-20, 1989," 1, R.J. Thompson, D.G. Cole, P.J. Wilkinson, M.A. Shea, D.F.Smart, and G.R. Heckman, Eds., NOAA, Boulder, Colorado, 140.

Lin, R.P., 1987, Solar particle acceleration and propagation, Rev. of Geophys., 25: 676.

Mason, G.M., 1987, The composition of galactic cosmic rays and solar energetic particles, Rev. of Geophys., 25: 685.

Mason, G.M., Fisk, L.A., Hovestadt, D., and Gloeckler, G., 1980, A survey of ~ 1 MeV nucleon-1 solar flare particle abundances, 1 < Z < 26, during the 1973-1977 solar minimum period, Astrophys. J., 239: 1070.

McGuire, R.E., Von Rosenvinge, T.T., and McDonald, F.B., 1986, The composition of solar energetic particles, Astrophys. J., 301: 938.

McKinnon, J.A., 1987, "Sunspot Numbers 1610-1986 Based on the Sunspot Activity in the Years 1610-1960," UAG-95, NOAA, Nat. Geophys. Data Center, Boulder, Colorado.

McCracken, K.G., 1962a, The cosmic-ray flare effect. II. The flare effects of May 4, November 12, and November 15, 1960, J. Geophys. Res., 67: 435.

McCracken, K.G., 1962b, The cosmic-ray flare effect. I. Some new methods of analysis, J. Geophys. Res., 67: 423.

McIlwain, C.E., 1961, Coordinates for mapping the distribution of magnetically trapped particles, J. Geophys. Res., 66: 3681.

Reedy, R.C., Cayton, T.E., Belian, R.D., Gary, S.P., Gisler, G.R., Reeves, G.D., Fritz, T.A., and Christensen, R.A., 1991, Solar particle events during the rising phase of solar cycle 22, in "Workshop on Ionizing Radiation Environment Models and Methods", 1, unnumbered report, NASA Marshall Space Flight Center, Huntsville, Alabama, 227.

Sauer, H.H., Zwickl, R.D., and Ness, M.J., 1990, "Summary Data for the Solar Energetic Particle Events of August through December 1989," Space Environment Laboratory, unnumbered report, NOAA, Boulder, Colorado.

Schwabe, H., 1843, Verzeichniss der Gruppen und fleckenfreien Tage für 1826 bis 1843, Astron. Nachr., 21 (No. 495): 234.

Shea, M.A., and Smart, D.F., 1982, Possible evidence for a rigidity-dependent release of relativistic protons from the solar corona, Space Sci. Rev., 32: 251.

Shea, M.A., and Smart, D.F., 1990a, A summary of major solar proton events, Solar Phys., 127: 297.

Shea, M.A., and Smart, D.F., 1990b, The influence of the changing geomagnetic field on cosmic ray measurements, J. Geomag. Geoelect., 42: 1107.

Shea, M.A., Smart, D.F., and McCracken, K.G., 1965, A study of vertical cutoff rigidities using sixth degree simulations of the geomagnetic field, J. Geophys. Res., 70: 4117.

Shea, M.A., Smart, D.F., and Gentile, L.C., 1987, Estimating cosmic ray vertical cutoff rigidities as a function of the McIlwain L-parameter for different epochs of the geomagnetic field, Phys. Earth Planet. Inter., 48: 200.

Smart, D.F., 1988, Predicting the arrival times of solar particles, in: "Interplanetary Particle Environment," Joan Feynman and Stephen Gabriel, eds., JPL Publication 88-28, Jet Propulsion Laboratory, California Institute of Technology, Pasadena: 101.

Smart, D.F., and Shea, M.A., 1967, A study of the effectiveness of the McIlwain coordinates in estimating cosmic-ray vertical cutoff rigidities, J. Geophys. Res., 72: 3447.

Smart, D.F., and Shea, M.A., 1973, An empirical method of estimating cutoff rigidities at satellite altitudes, 13th Intl. Cosmic Ray Conf., Conference Papers, 2: 1070.

Smart, D.F., and Shea, M.A., 1979, PPS76 - A computerized "event mode" solar proton forecasting technique, in "Solar-Terrestrial Predictions Proceedings," Vol. 1, Richard F. Donnelly, ed., Environmental Research Laboratories, National Oceanic and Atmospheric Administration, U.S. Department of Commerce, Boulder, Colorado, 406.

Smart, D.F., and Shea, M.A., 1983, Geomagnetic transmission functions for a 400 km satellite, 18th Intl. Cosmic Ray Conference, Conference Papers, 3: 419.

Smart, D.F., and Shea, M.A., 1985, Galactic cosmic radiation and solar energetic particles, Chapter 6 in "Handbook of Geophysics and the Space Environment", A.S. Jursa, ed., Air Force Geophysics Laboratory, Bedford, MA.

Smart, D.F., and Shea, M.A., 1990, Probable pitch angle distribution and spectra of the 23 February 1956 solar cosmic ray event, 21st Intl. Cosmic Ray Conf., Conference Papers, 5, The University of Adelaide, Adelaide, Australia, 257.

Smart, D.F., and Shea, M.A., 1991, A comparison of the magnitude of the 29 September 1989 high energy event with solar cycle 17, 18 and 19 events, 22nd Intl. Cosmic Ray Conf., Contributed Papers, The Dublin Institute for Advanced Studies, Dublin, Ireland, 3: 101.

Smart, D.F., and Shea, M.A., 1992a, Modeling the time-intensity profile of solar flare generated particle fluxes in the inner heliosphere, Adv. Space Res., 12: (2)303.

Smart, D.F., and Shea, M.A., 1992b, Predicting and modeling solar flare generated proton fluxes in the inner heliosphere, this volume.

Smart, D.F., Shea, M.A., and Webber, W.R., 1990, Study of the August 1972 solar proton events: A flux intensity paradox, 21st Intl. Cosmic Ray Conf., Conference Papers, 5, The University of Adelaide, Adelaide, Australia, 324.

Smart, D.F., Shea, M.A., Wilson, M.D., and Gentile, L.C., 1991, Solar cosmic rays on 29 September 1989; An analysis using the world-wide network of cosmic ray stations, 22nd Intl. Cosmic Ray Conf., Contributed Papers, The Dublin Institute for Advanced Studies, Dublin, Ireland, 3: 97.

Stone, E.C., 1989, Solar abundances as derived from solar energetic particles, in "Symposium on Cosmic Abundances of Matter", C.J. Waddington ed., AIP Conference Proceedings 183, American Physical Society, New York, 72.

Svestka, Z., and Simon, P., 1975, "Catalog of Solar Particle Events 1955-1969", D. Reidel Publishing Co., Dordrecht, Holland.

Teltsov, M.V., and Tverskaya, L.V., 1991, "Radiation dose measurements onboard station "MIR" during solar proton flares in September-October 1989," Preprint, Presented at 1st SOLTIP Symposium, Liblice, Czechoslovakia.

Waldmeier, M., 1961, "The Sunspot-Activity in the Years 1610-1960", Schulthess and Company, Zurich.

Wilson, J.W., Khandelwal, G.S., Shinn, J.L., Nealy, J.E., Townsend, L.W., and Cucinotta, F.A., 1990, "Simplified Models for Solar Cosmic Ray Exposure in Manned Earth Orbital Flights", NASA TN-4182.

Wolf, J.R., 1856, "Neue Untersuchunger Uber die Periode der Sonnenflecken", Astronomische Mitteilungen der Eidgenossischen Sternwarte Zurich, No. 1.

ORIGINS AND EFFECTS OF SOLAR FLARES

D. M. Rust

The Johns Hopkins University Applied Physics Laboratory
Laurel, Maryland 20723 USA

From veils of haze now comes the gleam,
Here to a tender scarf it tapers,
Here gushes forth a vivid stream;
Then threads of light in a network surging
Their silver veins through valleys run,
Till, gathered by the hills converging,
The sundered filaments are one.

--Goethe: *Faust*

Walpurgis Night came to Montreal and Quebec City, Canada, just before 3 A.M. on March 13 in 1989, when the Quebec power grid crashed, after a series of extraordinary solar flares. The crash was quickly followed by the collapse of other generating capacity as the networked power grid protected itself from the excessive demands caused by the massive failure. The unnatural night lasted nine hours and many Quebecquois were without electricity for much longer than that.

Other solar-terrestrial storms the same year destroyed two transformers at the Oyster Creek Nuclear Station in New Jersey and a transformer at the Salem Nuclear Generating Station in New Hampshire. An electrical generator caught fire at a nuclear plant in Maine during another storm. The Nuclear Regulatory Commission investigated the incidents and is trying to match previous failures at nuclear plants against other solar-terrestrial events.

The same solar flares that brought down the Quebec power system caused severe problems with seven commercial geostationary communications satellites. Some 177 manual operator interventions were required during the storms to keep the satellites' antennas pointed at their earthly receptors. That was more than normally required in a year of routine operations.

Ultraviolet and X-radiation from the flares and the geomagnetic storms they induced heated the upper atmosphere and disrupted satellite orbits. On a normal day, satellite trackers temporarily lose about 1000 orbiting targets. Because of the increased atmospheric drag on March 13, the number of "lost" targets rose to 2000 and increased further each day until 6000 "lost" targets were reported on March 18. One of them was NASA's SMM satellite which dropped three miles in its orbit during the disturbed period.

HF communications were severely disrupted by the March 1989 storms. In Australia, polar and mid-latitude circuits were "useless" and equatorial circuits were "very weak and noisy ".

Biological Effects and Physics of Solar and Galactic Cosmic Radiation,
Part B, Edited by C.E. Swenberg *et al.*, Plenum Press, New York, 1993

73

PROTON STORMS

The hail of protons from the 1989 storms was the heaviest since the legendary outbursts of 1972. In three of the storms even an astronaut shielded by a quarter-inch of aluminum plate would have absorbed a dose of radiation higher than the National Council of Radiation Protection and Measurement (NCRP) recommends for a lifetime.

Flares, which accelerate the protons, are much more frequent at the maximum of the 11-year sunspot cycle than at minimum. The next maximum will be in the year 2000. No telescope will show the protons as they spiral into space. They will come down on space travellers and Earth-orbiting satellites with unpredictable savagery. Helium, electrons and heavy nuclei will accompany the protons in approximately their usual proportions (Gloeckler, 1979), but the shear numbers of protons will make them the most dangerous.

Paradoxically, 2000 be a safer year to venture into space than, say, 2005, the next sunspot minimum. In 2000, magnetic fields throughout the solar system will be tangled from the effects of many solar shocks (Burlaga 1967). They will efficiently deflect the cosmic rays from the inner solar system, just as Earth's magnetic fields deflect solar protons to shield astronauts and satellites in low Earth orbit. Cosmic rays between Earth and Mars are two to ten times more numerous when the Sun is quiet, making that a dangerous time for deep-space

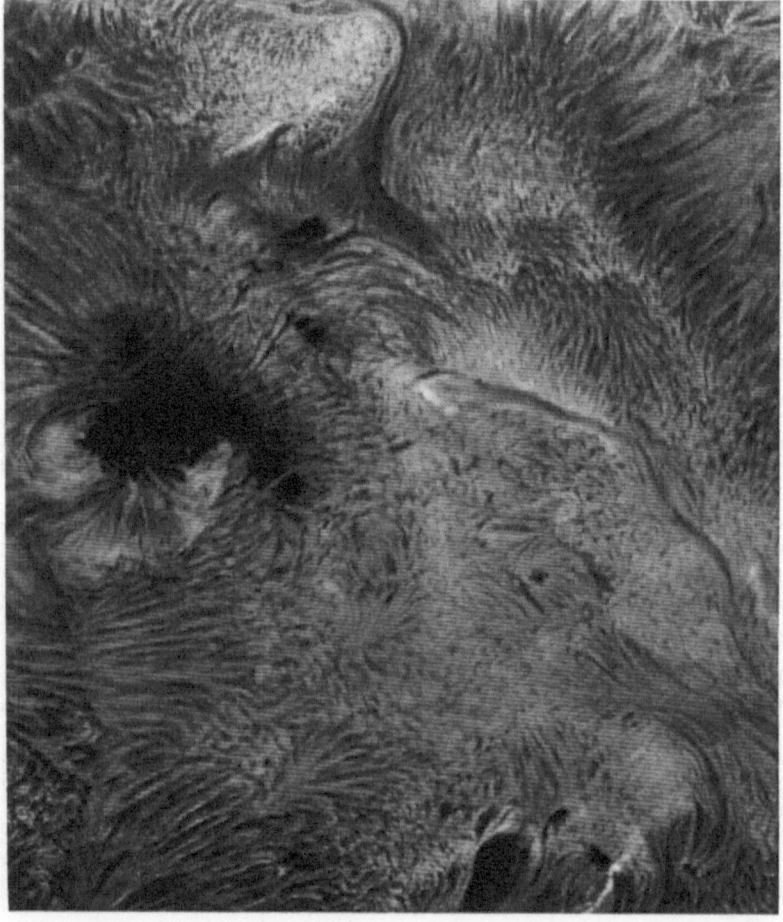

Figure 1.   An Active Region on 17 May 1990 Recorded at the Big Bear Solar Observatory with a Filter Tuned to the Hydrogen Balmer Line at 6563 Ångstroms.

travellers (Toptygin 1985). Given adequate warning, space travelers *can* to find temporary shelter from proton storms behind inch and a half thick aluminum plate or in a one-inch water jacket, but there is no practical protection from the far more penetrating cosmic rays (Curtis 1973).

Thus, the need to make accurate forecasts of solar flares puts a practical hue on a modern astronomical mystery: How do stars like the Sun, and galaxies, and supernovas accelerate atoms (at nearly the speed of light)?

THROUGH THE TELESCOPE

Only one star, the Sun, is so close that we can imagine a day when the apparatus of its cosmic accelerators is revealed. Most solar physicists think the accelerators are magnetic, just as the man-made ones are. There is indirect evidence for 100-kilometer magnetic structures on the Sun's surface (Stenflo 1973). Unfortunately, they are just beyond the resolving power of today's best telescopes. Better resolution might be achieved in the next decade, but meanwhile, solar physicists are using a global network of the best available telescopes and some modern inventions such as adaptive optics, or "rubber mirrors," to see what they can of the magnetic elements and flares. Theorists are taking ideas from the fusion laboratories to try to learn how magnetic instabilities on the Sun's surface accelerate atoms.

An "active region" may have many large sunspots and dark filaments (upper part of Figure 1) or dense clouds suspended by magnetic fields in the solar atmosphere. There are also thin, dark fibrils. Like iron filings near a magnet, they trace the lines of magnetic force. The fibrils trace the force fields because they are made of ionized gases that can flow smoothly along the lines of force, but cannot move across them.

Most of the emission visible in flares comes from the chromosphere, which is the pink

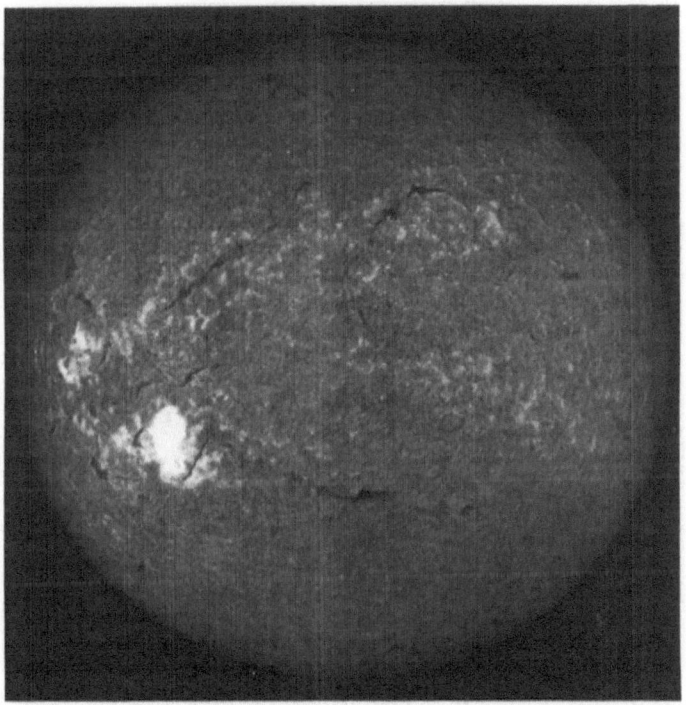

Figure 2.   The Sun at the Peak of the Great Flare of 7 August 1972.

layer just above the familiar pale yellow photosphere. The pink color comes from the red and powder blue emissions of hydrogen heated to 10000 degrees. The Air Weather Service's Solar Observing Optical Network (SOON) takes chromospheric pictures around the clock and few flares are overlooked (Figure 2). The pictures help to document the brightness and area of flares and are useful in forecasting proton storms since bigger, brighter flares normally produce more protons.

Flares often start with a few intensely bright knots near sunspots. The bright knots expand to form two or more ribbons that engulf most of the active region around the spots. After about an hour, the ribbons fade, and the active region returns to its appearance beforehand. But sometimes there is a notable difference. One of the dark filaments may be missing, and an examination of "flare patrol" films will show that the filament darkened half an hour before the flare and erupted into the Sun's outer atmosphere (Figure 3), the corona, at 200 - 1000 kilometers per second. Films of the corona will show a corresponding expansion of the bright arches that overlay the active region and the filament (Figure 4). It is a remarkable fact that the filament eruption and the coronal expansion, called a mass ejection,start before any brightening in the active region.

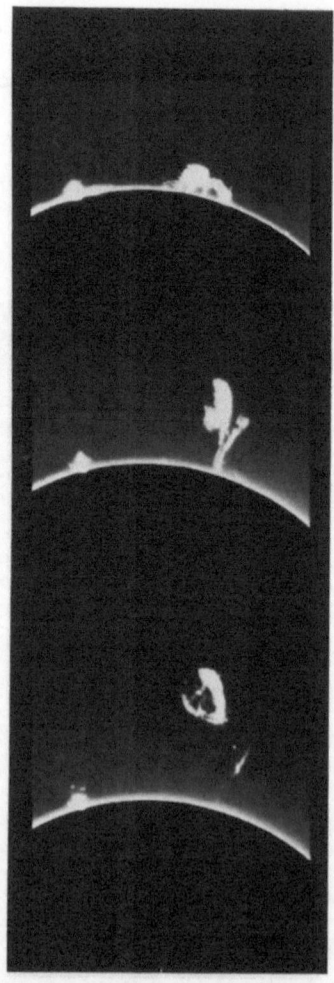

Figure 3.   (Top to Bottom) Eruption of an Active Region Filament, as Photographed with an Artificial Moon Eclipsing the Photosphere.

Figure 4.   Ejection of Mass from the Corona. The Loops Shown here were Expanding Away from the Sun at ~ 500 Kilometers per Second.

As flare pictures achieve higher and higher resolution, they show smaller and smaller features. Many solar physicists believe that the fundamental elements of flares are only 100 kilometers wide. But this opinion is not universally shared. Those who favor the view that the important instabilities are small point out that flare X-ray bursts and radiowave bursts are composed of microbursts, each lasting only a few milliseconds. It is also true that theorists have so far had little success in explaining how any magnetic structures larger than about 100 kilometers in diameter could become unstable and release so much energy in such short bursts (Spicer et al. 1986).

Does it follow that a flare is a fusillade of microflares? Not necessarily, because there is no good explanation for the propagation of flaring, or destabilization, from one microstructure to another. And the microbursts may be a red herring, because the energy in each microburst is a very small fraction of all the energy in a flare. Yet, tens of thousands of unresolved microbursts could blur together to produce the smooth and relatively slow rise and fall of emission seen in most flares.

Most solar physicists believe that flares are large because the unstable magnetic loops, photographed in X-rays are large. The loops emit copious X-rays because the temperature of the plasma confined there is 10 to 40 million degrees (Dennis 1988).

The first trace of a flare usually appears at two intersecting magnetic loops, and one of the loops is usually expanding outward at 10 kilometers per second or faster. Then an entire arcade of loops may appear as the brightening spreads to more of the active region (Figure 5). The flare stops when most of the loops in the arcade have pulled back, or 'reconnected', after a mass ejection phase. In the end most of the energy will have been spent on the eruptive filament and the mass ejection, not on the protons or on the showy emissions. Thus, despite

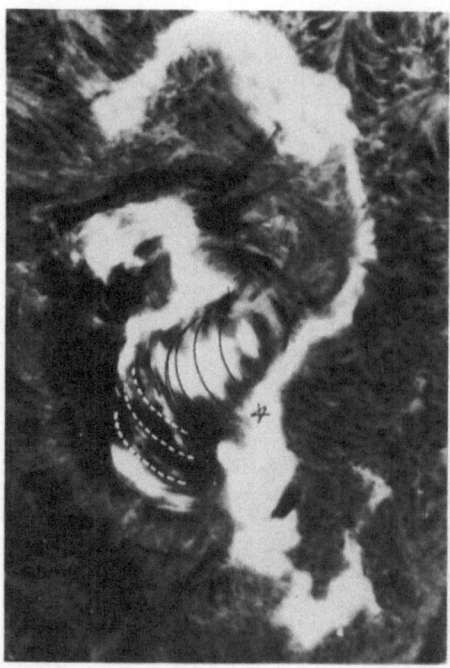

Figure 5. The Great Flare of 7 August 1972 (close up). The Bright Ribbons Show where the Chromosphere was Heated. The Loops (Dark Threads and Dashed and Solid Lines) Show where the Magnetic Fields Arch Through the Corona between the Ribbons.

the interesting microbursts, the brilliant ribbons, the diaphanous X-ray loops and the invisible burst of protons, the real muscle of a flare is usually in a slow outward expansion of the corona. One such massive ejection, on March 10, 1989, started Walpurgis Night in Quebec.

PROTON ACCELERATORS

In the chromosphere, a flare is glistening red rivulets flowing for a few minutes around a cluster of sunspots. In an X-ray imager, it is a blinding loopy cloud. But a flare is also an atomic accelerator. The emissions signal the onset of proton acceleration The protons can arrive at Earth in twenty minutes or in several hours (Zirin and Tang 1990). A plot (Figure 6) of the X-ray emission from a series of big flares shows the Sun becoming a hundred to a thousand times brighter before each corresponding proton storm at Earth. Often the flux of protons, each with an energy of more than ten million electron volts or MeV, stays at a dangerously high level for hours.

By monitoring flares with the SOON telescopes and satellite-borne detectors, forecasters at the NOAA Space Environment Forecast Center try to give warnings of impending proton storms. They base their forecasts on statistical records from thousands of past flares. But chance is not destiny so the false alarm rate and the nasty surprise rate are both too high (Simon et al. 1986). A large flare can produce either no proton storm or one aimed at another part of the solar system. Or a flare may ambush the forecasters and aim a storm directly at Earth as in the memorable events of August, 1972 (Figure 2). The storm tracks are invisible.

The exact location of a flare is important, since flares on the western hemisphere of the Sun, as viewed from Earth, are more likely to lie at the base of the spiralling lines of magnetic force that run directly from the Sun to Earth (Figure 7). Protons injected from the west will usually reach Earth more rapidly than protons from the east. For example, the 21 - 22 May

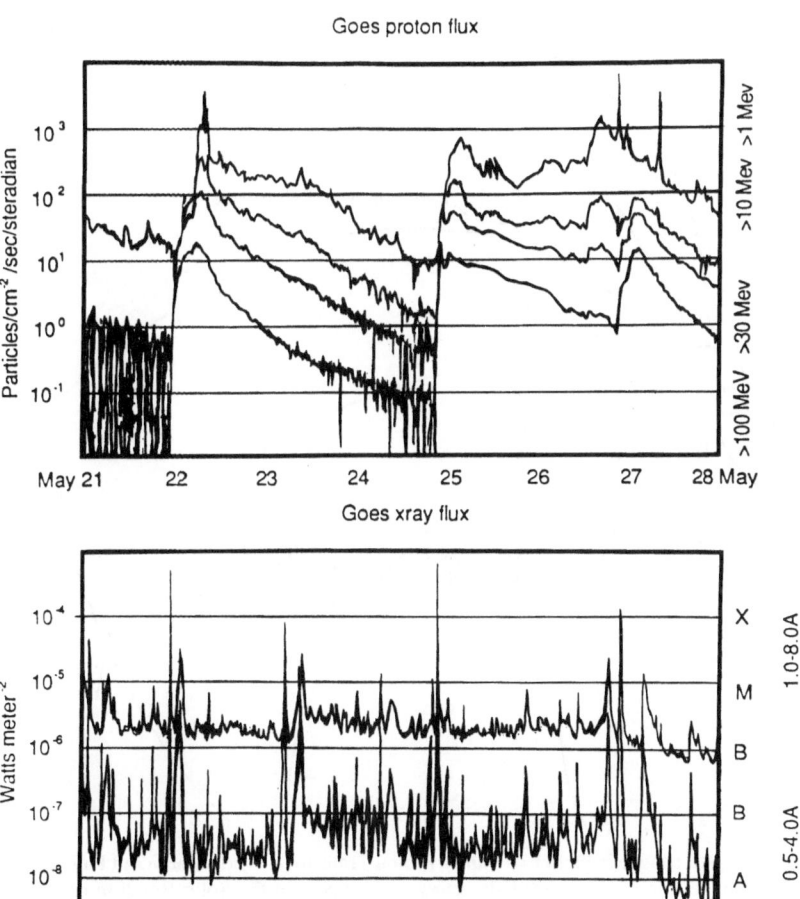

Figure 6. Five-Minute Averaged X-Ray and Proton Fluxes for 21 - 28 May 1990, as Measured at the NOAA Goes Satellite. The Proton Flares on 22, 25 and 26 May were all Produced by the Same Sunspot Group.

1990 flare in Figure 1 occurred at 36° West. The delay was 96 minutes. The delay for the 24 - 25 May flare at 78° West was only 34 minutes. Protons from the earlier flare took longer to reach Earth because they had to diffuse across the interplanetary magnetic fields. The effects of the diffusion process is hard to forecast because it depends on the twists and turns in the fields made by shocks from previous flares. It also depends on how the magnetic fields were kinked and stretched in the photosphere over the previous several months (Cane and Reames 1990).

NASA's Solar Maximum Mission answered many questions about the proton acceleration site (Chupp 1987). Since proton storms generally outlast flares by many hours it had been unclear whether they originate in similarly long-lived, but invisible acceleration regions, or whether their long duration simply reflected the delays caused by their slow diffusion through interplanetary space.

The SMM had a sensitive gamma-ray detector on board. The only way some of the gamma rays in flares can be produced is by the collisions of protons with the photosphere, that is, the accelerator is also an atom smasher. The gamma-ray data showed that the spectrum of protons striking the photosphere is the same as for the protons arriving at Earth, provided one makes proper allowance for the effects of diffusion through the interplanetary

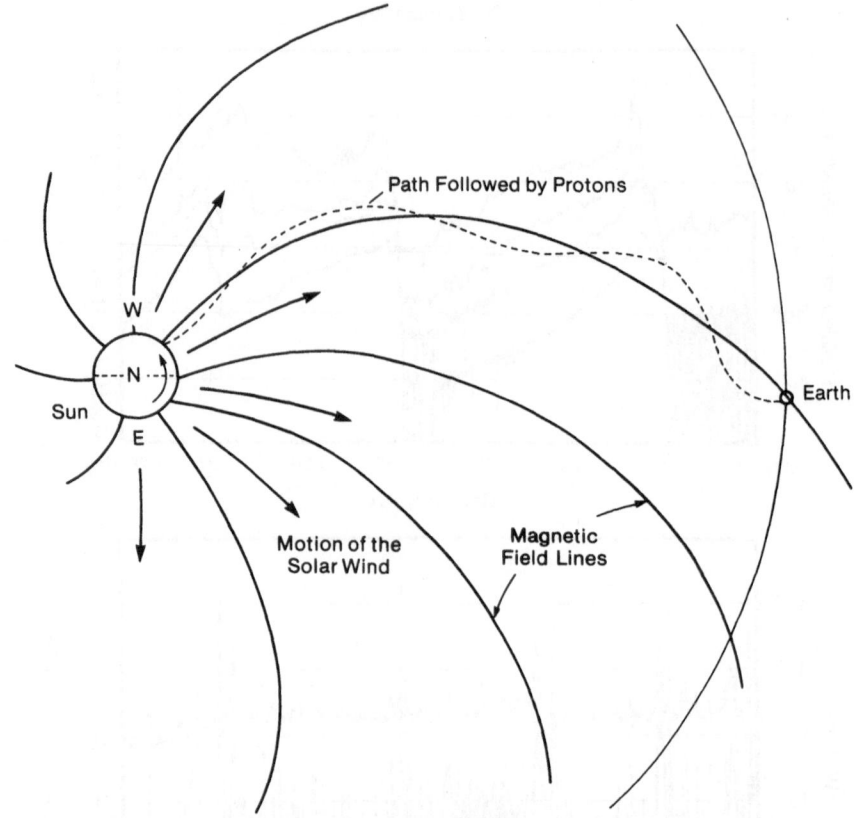

Figure 7. Solar Wind, Magnetic Field Lines and a Proton Trajectory. The Solar Wind Pulls Magnetic Fields Radially away from the Sun at about 400 Kilometers per Second, but the Solar Rotation of 2 Kilometers per Second Twists the Fields into a Gradual Spiral. Protons Released at the Base of Magnetic Fields 20° to 80° West of the Central Meridian can Easily Follow the Fieldlines to Earth.

magnetic fields (Vlahos et al. 1986, Forman et al. 1986). This indicated that a single population of protons is released at one instant. SMM measurements also showed that the protons are accelerated at the same time as the electrons that heat the chromosphere. It is an exciting result because high-resolution pictures may eventually show enough detail of the collision points, which must be near the proton accelerator, to betray the acceleration process. Telescopes that can resolve the gamma-ray sites are planned for NASA's High-Energy Solar Physics mission which plans to launch it in 2002.

MAGNETIC FIELDS

Only the magnetic fields in active regions have enough concentrated energy to accelerate particles and to produce all the other effects of flares. In fact, all solar activity depends on magnetic action. Any of the instabilities seen in the fusion laboratories could happen, but no one knows which ones actually do. It is not even clear whether the most important instabilities are large, say 100,000 kilometers in scale, or small, because flares include mysterious phenomena on every scale.

Before virtually every flare, some magnetic fields in the host active region begin to strengthen and move about. The fields are said to undergo shear, emergence and even cancellation. The physical meaning of these terms is not universally agreed upon. But most agree that the fields are frequently sheared by the action of surface flows. Magnetic fields can act very much like rubber bands. If they are held at two points, where they intersect the photosphere, any twist or movement of the points propagates from the photosphere into the unconstrained part of the field, in the corona. Because the photospheric gases are constantly churning, the magnetic fields will eventually be severely stretched, a condition that solar observers call shear, because of the resemblance of maps of the stretched regions to geological shear layers.

Although observations have yet to show an unambiguous reduction in shear, and hence a loss of built-up energy, during a flare, theorists believe that sheared fields could supply the energy for even very large flares (Low 1989). A popular scenario is that the sheared fields develop an unstable kink and erupt outward, like the fields in unstable fusion machines (Schmidt, 1966).

The magnetic fields in established sunspot groups are quite stable. The greatest changes occur when fresh magnetic flux emerges from beneath the photosphere. Enough flux can emerge to make a large sunspot in seven days (Bray and Loughhead 1965). Many solar physicists think that new flux emerging into an established active region can destabilize a dark filament (Figure 8) and the coronal arcade above it. A venerable model (Heyvaerts et al. 1977) of such flare triggering is based on this view and on observations that the magnetic fields in and around filaments obviously become unstable and open into the corona. Martin et al. (1985) showed that some *small* flares, at least, begin when magnetic flux disappears from the photosphere. The interpretation of the observations is not clear, so they call the process flux cancellation. Their examples of positive (upward directed) and negative (downward directed) fields merging and apparently producing transient brightenings are the clearest illustrations yet that magnetic fields are annihilated in flares.

The coronal fields arching over a sunspot group are generally part of a structure called a coronal streamer. Streamers extend from one-half to ten solar radii into space. The loops at the base of the streamer often begin to expand outward about an hour before a flare (Harrison 1986). This expansion is usually accompanied by twisting and destabilization of the low coronal fields (Rust, Nakagawa and Neuport 1975; Kahler et al. 1988). An X-ray telescope on the Apollo Skylab mission and later observations with other satellites confirmed that coronal destabilization precedes the first flashes of light, even in relatively small flares (Kahler et al. 1990).

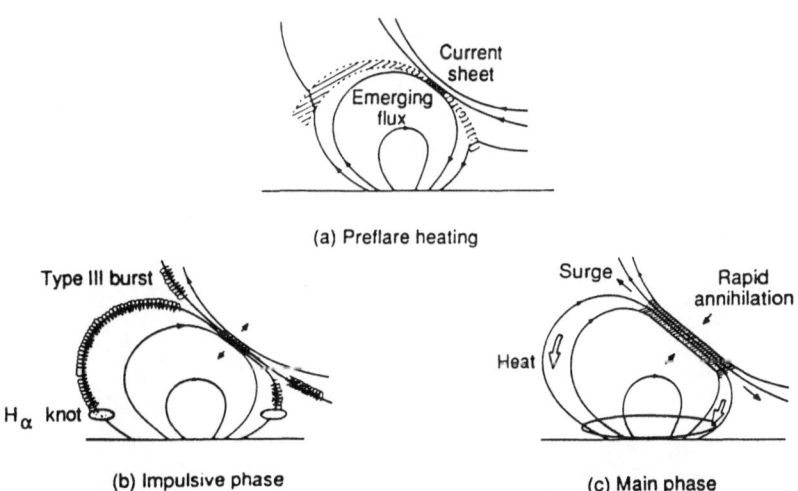

Figure 8. Emerging Flux Model of Solar Flares. An Electric Current Sheet Forms at the Intersection of the New Flux and the Overlying Magnetic Fields. Particles Accelerated there Escape upward to Produce Radiowave Bursts and downward to Produce the Optical Flare. Energy for Electromagnetic Radiation and Fast Particles Comes from Annihilation of the Magnetic Fields.

The eruption produces open magnetic fields which the protons can easily follow into interplanetary space. The acceleration mechanism is still not understood but it may take place in shocks generated where the opened fields begin to reconnect and collapse back toward the surface. Or it may occur at the intersection of two flux systems as shown in Figure 8. Intense heating, which can drive the coronal plasma temperature to 100 million degrees, may be the signature of fields reconnecting. Thermal conduction fronts (Rust, Simnett and Smith 1985) or beams of electrons (Petrosian 1990) carry energy to the chromosphere where the optical and ultraviolet emissions appear

The developments that lead to coronal destabilization may occur on a small scale (flux cancellation), an intermediate scale (flux emergence) or a large scale (shear). No one knows whether any of these sketchy concepts describe a process that is both necessary and sufficient to destabilize filaments and coronal arcades. Turbulence in the Earth's atmosphere blurs all current flare pictures, and because the turbulence varies during the day, a subtle buildup of shear and the beginnings of flux emergence usually escape detection.

The whole subject of whether flare instabilities develop from shearing motions is highly controversial. Klimchuk and Sturrock (1989) claim that the catastrophic eruptive behavior of a sheared field theorized by Low (1987) was an artifact of his method of solving his equations. McClymont and Fisher (1989) found that actual transfer of magnetic energy to the corona by mass motions in the photosphere is probably a very inefficient process. They believe that only emerging magnetic fields can provide enough energy for the largest flares. According to them, erupting magnetic flux can give up energy as it expands and an average-size magnetic feature can easily contain enough free energy to power any flare.

There is some fresh evidence for this view in observations made in New Mexico with a solar vector magnetograph (Rust, O'Byrne and Harris 1988). A vector magnetograph maps all components of the fields in an active region every few minutes. Maps were being made on April 2, 1991 when a two-ribbon flare started at 2251 Universal Time (UT). The flare reached maximum brightness at 2321 UT (Figure 9). On the basis of its three-hour life and

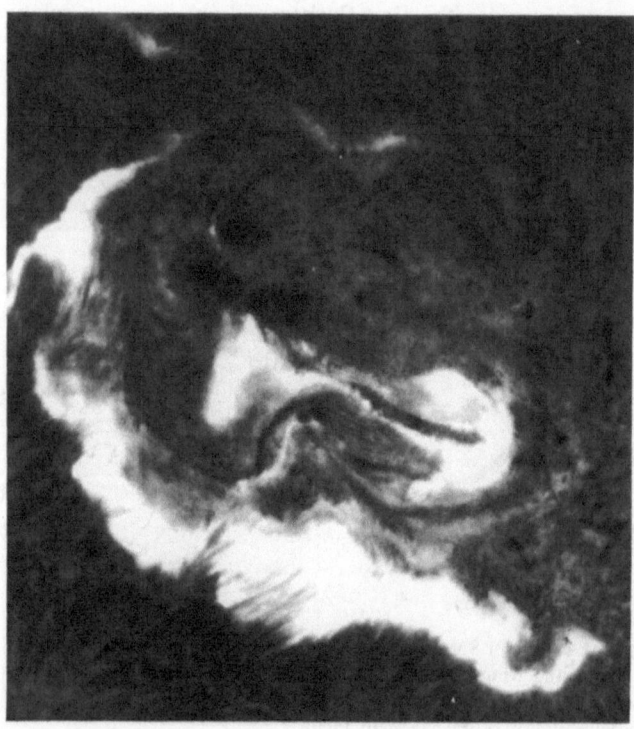

Figure 9. The Flare of 2 April 1991 at 2321 UT.

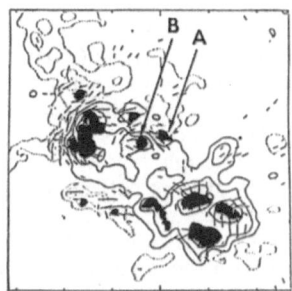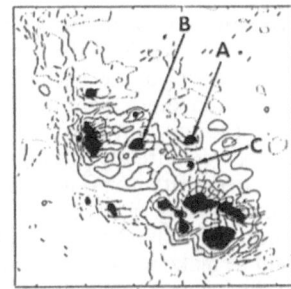

Figure 10. Maps of the Magnetic Fields in the Active Region where the Flare Shown in Figure 9 Occurred, at 1455 UT (left) and 2330 UT (right). Solar North is at the Top, East is on the Left. Arrows A and B Show the Emerging Flux Region. The Contours Denote Longitudinal Magnetic Strength at 100, 400 and 800 Gauss Levels. Positive, upward-Directed, Fields Have Solid Contours, Negative Fields Have Dashed Contours. Slashes Represent Transverse Fields up to ~ 1600 G. The Black Hand-Drawn Patches Represent the Major Sunspots.

the formation of prominent X-ray loops, we think it was a classic eruptive flare (Svestka 1976). The flare produced a three-day proton event at Earth. On April 4 the flux of 10 MeV protons peaked at 52 proton-flux units.

We examined the maps for sheared (i.e., stretched) fields and moving or emerging magnetic elements. We compared maps made 8 hours before and 45 minutes after flare onset (Figure 10). The principal features of the active region were the southwest cluster of spots with positive polarity and the east cluster, also with positive polarity. The two clusters were very different, according to our vector magnetograms. Fields in the southwest cluster resembled unstretched, untwisted fields, i.e., they point radially outward from the sunspots (Figure 10). The field direction near the east spots was nearly parallel to the boundary between the positive spots and the surrounding negative fields. They looked as though the spots had rotated and twisted the fields up. We expected the flare to begin in this region.

The flare began, or was triggered, in smaller sunspots that developed that morning. Arrows in Figure 10 show the two spots, A and B, of the affected part of the active region. The spots separated at ~ 0.2 kilometers per second in the eight hours before the flare. Spot A, with negative fields, moved southwestward while its size and magnetic strength grew rapidly. Gesztelyi et al. (1989), Wang (1991), Rust, Nakagawa, and Neupert (1975), Hoyng et al. (1981) and others had already flagged rapid growth and drift in sunspots as precursors of major flares.

The vector magnetograms uncovered effects of the growth and sideways drift of the new sunspots that had only been suspected before. Most magnetographs installed earlier than the one in New Mexico measure only the longitudinal or "line-of-sight" component of the magnetic fields. The new instrument shows the transverse fields as well, that is, the part of the field that lies flat in the surface, perpendicular to the line of sight. Maps made eight hours before the flare showed transverse fields that resembled fields without electric currents, which means they had no energy available to fuel a flare. But just 36 minutes before flare onset, we noticed that the transverse field had changed course by 90° in some parts of the active region near spot A (Figure 11). Thus, the maps were showing a build up of electric currents near spot A - currents that could be carrying enough energy for a flare. Maps made after the flare had started show less "shear" in the A spot region, but we have not yet been able to estimate the size of the pre-flare energy buildup or the change after the flare started.

Just before the flare, the magnetograms revealed a curious new transverse field development on one side of spot A (third panel from the top in Figure 11). There, a strong transverse field appeared, as though a previously upright or line-of-sight field had bent over and was lying horizontally in the photosphere. It was not seen in other magnetograms, so its reality may never be firmly established. But could be evidence for an upthrust of new flux just before the flare.

The chromospheric ribbons in the April 2 flare stretched across the active region and nearly covered some sheared fields in the eastern cluster of spots (Rust and Cauzzi, 1991). To that extent our observations agree with the earlier vector magnetograph data obtained by Hagyard et al. (1984). And there were no flare ribbons near the western spots, where the vector fields most resembled current-free fields. Yet, observations on April 1 showed that the sheared fields in the east were already present then, although they were weaker. So, we think sheared fields may be necessary for flares, but they are not sufficient. Some perturbation must destabilize them, probably from below the visible surface. In the case of the April 2 flare, the destabilization may have been related to the emerging and bending magnetic fields near spot A. We want to make more observations with the magnetograph in New Mexico to test these ideas. We think it is possible that fields from below the surface could emerge in ~ 45 minutes and disturb the previous equilibrium and that a magnetograph operating under ideal conditions could solidly detect them.

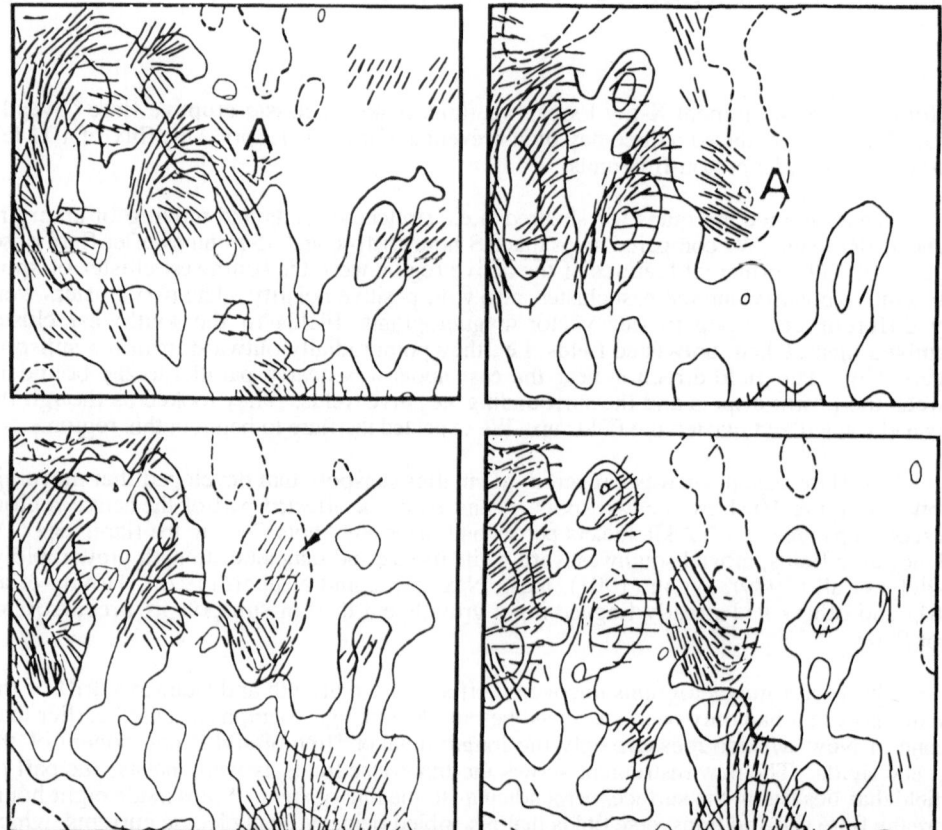

Figure 11. A Sequence of Vector Magnetograms from 2 April 1991. Observation Times were (Top to Bottom) 1452, 1815, 2210 and 2335 UT. Note the Sudden Appearance of A Strong, New Transverse Field at 2210 UT, which is 45 Minutes Before the Onset of the Flare Shown in Figure 9.

The April 2 data shed some light on earlier observations of sunspot motion and morphological changes noticed in strong flares that were obtained before the development of vector magnetographs. They show that emerging magnetic fields can create a sheared field in less than 8 hours. Movies of the sunspots, fibrils and filaments seem to show that magnetic fields bubble up, stretch and twist with particular vehemence near flares, but vital hourly magnetograms to measure it were not available.

A BIGGER TELESCOPE

Even though the new vector magnetograph in New Mexico and others now being built will help flare research, solar physicists have not been able to greatly improve the pictures obtained with ground-based telescopes beyond what they had thirty years ago. NASA was planning an orbiting solar laboratory that would have resolved the 100-kilometer magnetic elements, but it died in the 1991 budget cuts. While the new vector magnetographs will undoubtedly produce useful results, past experience suggests that the maps may be too blurry to make positive identifications of unstable currents.

There is an acute need for much sharper images than any ground-based telescope can make. A telescope in space or in the stratosphere could get much better pictures and magnetic maps and make the dramatic progress in flare research that future missions to the moon and Mars will need. So we are building a vector magnetograph that can be lifted by a balloon 100,000 feet into the stratosphere. Its 80-centimeter mirror will make it the largest solar telescope ever flown. Our goal, of course, is to measure the fundamental magnetic elements in active regions. The balloon-borne magnetograph will provide ten times better resolution and magnetic sensitivity than the best ground-based instruments.

The first flight is scheduled for Antarctica in 1994. Antarctica has been called Earth's window on outer space because telescopes there can do many jobs that can otherwise only be done in space. And the cost is about one percent of a space mission's cost. Our planned circumpolar flight in the Antarctic summer will produce ten to twenty days of uninterrupted measurements of the magnetic fields, sunspots, filaments and fibrils, and flares. At the end of one trip around the pole, a parachute will carry the magnetograph to an ice shelf where an airplane can retrieve it. It will be returned by ship to the United States for refurbishment and flight the next year.

SUMMARY

During the 1989-1991 peak of solar activity, geomagnetic storms from interplanetary shocks caused a massive failure in the Canadian power grid, minor failures in other power equipment, and many communications disruptions and satellite malfunctions. The proton storms would have been lethal for unshielded space travellers.

Had the power company managers been given a credible, timely forecast of the solar storm, they could have protected their generating equipment and the grid. They do not keep protective circuits in place full-time because that reduces efficiency and increases the cost of power distribution. Nor will astronauts on the moon or in deep space confine themselves full-time to thick-walled, radiation-resistant closets. To enable manned deep space exploration we have to find a way to determine what happens in solar flares. Only this will improve the forecasts. Expensive and restrictive protective measures would then have to be applied only when a major flare is clearly imminent.

There is no generally accepted flare theory or description of the pre-flare state or of the instabilities. The Solar Maximum Mission cleared up many questions about electromagnetic flare emissions and the structure of the flaring atmosphere, but the dynamic of the magnetic fields is still a mystery.

ACKNOWLEDGEMENTS

I am grateful to Craig Gullixson and John O'Byrne of the magnetograph project and to Gianna Cauzzi for help in analyzing the magnetograms. This work was sponsored by the Air Force Office of Scientific Research, grant AFOSR-90-0102.

REFERENCES

Bray, R. J. and Loughhead, R. E. 1965. Sunspots, Wiley, New York, p. 215.

Burlaga, L. F. 1967. Anisotropic Distribution of Solar Cosmic Rays, J. Geophys Res. 72: 4449.

Cane, H. V. and Reames, D. V. 1990. The Relationship Between Energetic Particles and Flare Properties for Impulsive Solar Flares, Astrophys. J. Suppl. 73: 253.

Chupp, E. L. 1987. Gamma Ray Emission in Solar Flares, Physica Scripta T18: 5.

Curtis, S. B. 1973. Frequency of Heavy Ions in Space and their Biologically Important Characteristics, Life Sci. Space Res. 11: 209.

Dennis, B. 1988. Solar Flare Hard X-Ray Observations, Solar Physics 118: 49.

Forman, M. A., Ramaty, R. and Zweibel, E. G. 1986. The Acceleration and Propagation of Solar Flare Energetic Particles, in: Physics of the Sun II, (P. A. Sturrock, T. E. Holzer, D. M. Mihalas and R. K. Ulrich, Eds.), D. Reidel, Dordrecht, p. 249.

Gesztelyi, L, Karlicky, M., Farnik, F., Gerlei, O., and Valnicek, B. 1986. White-Light Flare of 26 July 1981, in The Lower Atmosphere of Solar Flares (D. F. Neidig, ed.), Natl. Solar Obs., Sunspot, NM, p. 163.

Gloeckler, G. 1979. Composition of Energetic Particle Populations in Interplanetary Space, Rev. Geophys. Space Phys. 17: 569.

Hagyard, M. J., Smith, J. B., Teuber, D., and West, E. A. 1984. A Quantitative Study Relating Observed Shear in Photospheric Magnetic Fields to Repeated Flaring, Solar Phys. 91: 115.

Hagyard, M. J. and Rabin, D. M. 1986. Measurement and Interpretation of Magnetic Shear in Solar Active Regions, Adv. Space Res. 6: 7.

Harrison, R. A. 1986. Coronal Mass Ejections and Flares, Astron. Astrophys. 162: 283.

Heyvaerts, J., Priest, E. R., and Rust, D. M. 1977. An Emerging Flux Model for Solar Flares, Astrophys. J. 216: 123.

Hoyng, P., Frost, K. J., Woodgate, B. E., and the SMM HXIS Team 1981. Origin and Location of the Hard X-Ray Emission in a Two-Ribbon Flare, Astrophys. J. Lett. 246: L155.

Kahler, S. W., Moore, R. L., Kane, S. R., and Zirin, H. 1988. Filament Eruptions and the Impulsive Phase of Solar Flares, Astrophys. J. 328: 824.

Kahler, S. W., Sheeley, N. R., Jr., and Liggett, M. 1990. Coronal Mass Ejections and Associated X-Ray Flare Durations, Astrophys. J. 344: 1026.

Klimchuk, J. A. and Sturrock, P. A. 1989. Force-Free Magnetic Fields: Is There a Loss of Equilibrium?, Astrophys. J. 345: 1034.

Low, B. C. 1987. Electric Current Sheet Formation in a Magnetic Field Induced by Continuous Magnetic Footpoint Displacements, Astrophys. J. 323: 358.

Low, B. C. 1989. Magnetic Free-Energy in the Solar Atmosphere, in Solar System Plasma Physics, (J. H. Waite, Jr., J. L. Burch, and R. L. Moore Eds.,) Amer. Geophys. Union Monograph 54, Washington, p. 21.

Martin, S. F., Livi, S. H. B., and Wang, J. 1985. The Cancellation of Magnetic Flux. II In a Decaying Active Region, Australian J. Phys. 38: 929.

McClymont, A. N. and Fisher, G. H. 1989. On the Mechanical Energy Available to Drive Solar Flares, in Solar System Plasma Physics (J. H. Waite, Jr., J. L. Burch, and R. L. Moore, Eds.), Amer. Geophys. Union Monograph 54, Washington, p. 219.

Petrosian, V. 1990. Correlation of Hard X-Ray and Type III Bursts in Solar Flares, in Basic Plasma Processes on the Sun, (E. R. Priest and V. Krishnan, eds.), D. Reidel, Dordrecht, p. 391.

Rust, D. M. and Cauzzi, G. 1991. Variation of the Vector Magnetic Field in an Eruptive Flare, in Eruptive Solar Flares, Proc. IAU Colloq. 133 (B. V. Jackson, M. E. Machado and Z. Svestka, eds.), Springer Verlag, p. 46.

Rust, D. M., Nakagawa, Y., and Neupert, W. 1975. EUV Emission, Filament Activation and Magnetic Fields in a Slow-Rise Flare, Solar Phys. 41: 397.

Rust, D. M., O'Byrne, J. W. and Harris. T. J. 1988. An Optical Instrument for Measuring Solar Magnetism, Johns Hopkins APL Tech. Dig. 9 349.

Rust, D. M., Simnett, G. M., and Smith, D. F. 1985. Thermal Wave Fronts in Solar Flares, Astrophys. J. 288: 401.

Schmidt, G. 1966. Physics of High Temperature Plasmas, Academic Press, New York, p. 134.

Simon, P., Heckman, G. and Shea, M. 1986. Solar Terrestrial Predictions, U.S. Dept. of Commerce Pub.

Spicer, D. S., Mariska, J. T. and Boris, J. P. 1986. Magnetic Energy Storage and Conversion in the Solar Atmosphere, in Physics of the Sun II, (P. A. Sturrock, T. E. Holzer, D. M. Mihalas and R. K. Ulrich Eds.), D. Reidel, Dordrecht, p. 181.

Stenflo, J. O. 1973. Magnetic Field Structure of the Photospheric Network, Solar Physics 32: 41.

Svestka, Z. 1976. Solar Flares, Reidel, Dordrecht, p. 6.

Toptygin, I. N. 1985. Cosmic Rays in Interplanetary Magnetic Fields, D. Reidel, Dordrecht.

Vlahos, L., Machado, M. E., Ramaty, R. and Murphy, R. J. 1986. Particle Acceleration, in Energetic Phenomena on the Sun, (M. Kundu and B. Woodgate, eds.), NASA Conf. Pub. 2439: 2-1.

Wang, H. 1991. Evolution of Vector Magnetic Fields and the August 27 1990 X-3 Flare, Solar. Phys. 140: 85.

Zirin, H., and Tang, F. 1990. Optical Properties of Impulsive Flares, Astrophys. J. Suppl. 73: 111.

PREDICTION OF SOLAR PARTICLE EVENTS FOR EXPLORATION CLASS MISSIONS

G. Heckman

National Oceanic and Atmospheric Administration
Boulder, Co 80303

ABSTRACT

Manned space missions beyond the Earth's magnetosphere require forecasts of solar activity to insure that crews are safe enough to perform their duties and live normal lives after they complete their missions. Solar flares and associated activity produce temporary increases in the number of ionized particles in interplanetary space near Earth, Mars, and the Moon. These increases, called Solar Particle Events (SPE), typically last a few hours and are a source of radiation intense enough to degrade people's ability to perform physical activity, to cause lingering after effects such as cancer and cataracts, and in extreme cases, to endanger lives. Crews on missions beyond the protective shield of Earth's atmosphere and magnetic field can go into storm shelters or take protective chemicals if they have timely forecasts of SPEs. A forecast and observing system for SPEs and associated solar activity is already in place. In this paper, comparisons of forecasts and observations for the past several years are analyzed to identify strengths and shortcomings of the present program. This data indicates that forecasts made tens of minutes to a few hours in advance are rather reliable in forecasting whether SPEs will occur but are less reliable in forecasting the intensity of SPEs. Longer term forecasts, made one to three days in advance, would be useful for planning exploration trips away from protective shelters. Though such forecasts are moderately reliable for solar flares, they are less reliable in forecasting whether a SPE will follow.

This evaluation provides guidance for planning a forecasting system for future missions. The principal components can be copied in a Martian system. Improvements needed include better observations and research to incorporate the effects of shocks in producing and moderating SPEs and observations and research to improve the prediction of solar activity that will produce SPEs before the source activity occurs.

INTRODUCTION

Accurately forecasting solar particle events (SPEs) for astronauts on a mission to Mars or on a long stay on the Moon may mean the difference between success and failure of the mission. The lives of the astronauts and their ability to do their work can be jeopardized if they are exposed to a rapidly increasing, large radiation dose from a solar particle event. A SPE is a transient increase in the flux of high-energy ionized particles in the heliosphere; this increase is produced by solar flares and other types of solar activity. Development of a capability to forecast solar activity and the ensuing SPEs at Mars is part of the preparation necessary for future space missions when astronauts will spend months or years outside the protective shield of Earth's atmosphere and magnetic field. Long-duration missions to the

Biological Effects and Physics of Solar and Galactic Cosmic Radiation,
Part B, Edited by C.E. Swenberg *et al.*, Plenum Press, New York, 1993

moon will require a similar warning system. If astronauts have been provided with shelter or other protection from the radiation hazard posed by SPE, forecasts will allow them to seek life-saving shelter before the radiation increases to dangerous levels.

This report describes a current program for forecasting SPEs and discusses problems and areas where improvements are needed, especially if this program becomes the basis for a program to support future exploration missions. The report begins with a brief description of aspects of solar-terrestrial activity relevant to forecasting SPEs. It describes the type of forecasts useful to long-duration missions and summarize the techniques now used for making forecasts. It shows some limitations of the current techniques and suggest areas of research and development necessary to achieve the accuracy necessary for future, long-duration missions.

Detailed descriptions of solar-terrestrial activity and SPEs are beyond the scope of this paper. Foukal (1990) provided a comprehensive description of the physics of solar activity and its consequences. Shea and Smart (1992) have reviewed the characteristics and variety of SPEs. Techniques for making solar flare and SPE forecasts have been discussed by Rust (1982, 1992) and Smart and Shea (1979, 1992) respectively. Clark (1992) has described alternatives for radiation shelters that astronauts on a long-duration mission could use when dangerous SPEs are forecast.

SOLAR PARTICLE EVENTS

How serious is the radiation hazard from the Sun for astronauts in deep space? Bursts of energetic ionized particles in SPEs can produce short, intense radiation doses exceeding 0.1 Sv/hour to the blood-forming organs (BFO) of astronauts in a space ship with typical shielding. If vehicles for exploration missions have the same amount of shielding as current vehicles (20 kg m-2), three large SPEs in August, September, and October of 1989 would have produced a radiation dose of 1.3 Sv to the BFO of the astronauts unless they had taken refuge in a shelter that offered further shielding. For comparison, the limit of allowable radiation dose for astronauts on current space missions is 0.5 Sv annual dose to the BFO (National Council on Radiation Protection and Measurements, 1989).

SPEs are measured by their (1) intensity, (2) duration, and (3) spectra of energetic particles. (1) Intensity is the flux of particles per unit area. In this paper one particle flux unit (p.f.u.) equals one particle cm-2 s-1 sr-1, which in turn equals 10-4 particles m-2 s-1 sr-1. (2) Durations of SPEs are a few hours to several days. (3) Energies of the particles of interest fall in the range from 10 Million electron Volts (MeV) to a few hundred MeV per nucleon.

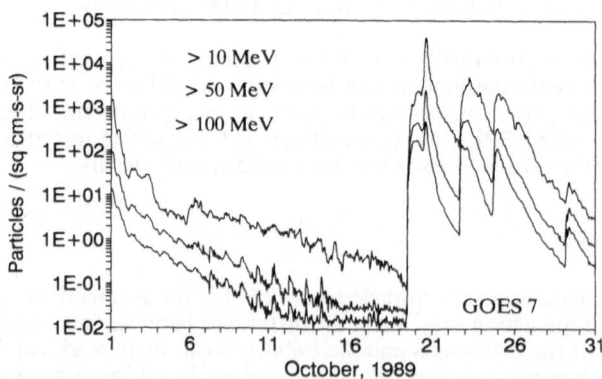

Figure 1. Energetic Particle Fluxes for an Active Period in 1989. Fluxes are Declining from an Earlier Event that Began in September. On October 19, a Series of Three Very Large SPEs Began. The Fluxes (Hourly Average Proton Flux) are for Particles with Energies > 10 Mev, > 50 Mev, and > 100 MeV, Plotted from Top to Bottom. The Data are from the GOES Satellite and are Processed in the NOAA Space Environment Laboratory (after Zwickl and Kunches, 1989).

The primary particles in SPEs are protons. For space radiation shielding studies, other ions of interest in SPEs include nuclei of helium, iron, and carbon (Shea and Smart, 1992). Spectra of SPEs can be approximated with a power law in energy. Typical solar particle fluxes are plotted for an active period of 1989 in Figure 1 (Zwickl and Kunches, 1989).

Observers have long associated SPEs of interest to space missions with solar flares. More recent work shows that SPEs most likely occur in solar flares or other forms of solar activity where a solar mass ejection (an ejection outward of a self-contained solar mass and magnetic field) occurs accompanied by a shock wave that runs ahead of the mass ejection (Rust, 1982), (Cane, Reames, and von Rosenvinge, 1988). After initial acceleration in the original solar activity, the particles in SPEs move through the solar corona out into the solar wind.

The solar wind, a plasma traveling at velocities of several hundred km s^{-1}, constantly carries magnetic field from the Sun out past Earth and into the distant heliosphere. The Sun rotates constantly from left (east) to right (west) as viewed by a person standing in the northern hemisphere on Earth. If visible markers were placed in the solar wind as it escaped the Sun, a section of plasma and magnetic field approaching Earth at any given moment would appear to have originated toward the west limb of the Sun (Figure 2). The solar wind velocity waxes and wanes with solar activity, and the source point for earth can vary from near the center of the Sun to near or beyond the west limb.

When charged particles reach the solar wind, they will have been scattered sufficiently that they reach an extended area covering several tens of degrees. In the solar wind, the particles tend to follow magnetic field lines with additional scattering from irregularities in the magnetic field. Shocks may continue to accelerate the charged particles as both move outward through the solar atmosphere into interplanetary space.

Superposed on this solar wind-magnetic field structure is the effect of solar mass ejections. Such ejections may travel several times faster than the solar wind around them. They push their way through the prior solar wind, rearranging its magnetic field, meanwhile carrying their own magnetic field with them. The resulting structure of the magnetic field that guides the energetic charged particles in SPEs may be a simple archimedean spiral, or it may be a complex mixture of radial, spiral field and looped, mass-ejection field.

The temporal profile of particles in a SPE after they have transversed this intervening structure to reach Earth is typically a rise to maximum over several hours and a decay back to background covering several days. The profiles are a combination of the effects of solar

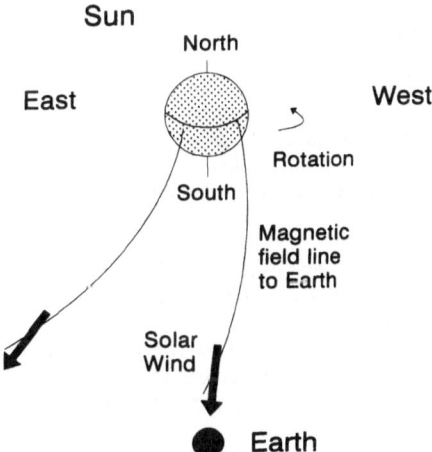

Figure 2. The Sun-Earth Coordinate System Showing a Magnetic Field Line Connected from the Sun to Earth and Another Extending East of the Earth. The Magnetic Field is Carried Radially Outward from the Sun by the Solar Wind.

rotation, guiding and scattering by the interplanetary field, and acceleration by shocks.

USES OF SPE FORECASTS FOR MARS MISSIONS

Scenarios for Mars missions include the provision of a storm shelter where astronauts could go during a SPE. The walls of the shelter would contain shielding such as water or food stores consisting largely of water. The shielding of the water would reduce the high radiation dose of a large SPE to tolerable levels (Clark, 1992). Escape to the shelter could be combined with intake of chemical protection prior to radiation exposure (Maisin, 1989). These actions would require a few minutes to a few hours advance notice of the SPE, depending on the distance of the astronauts from the shelter and the difficulty getting to it. Such a warning time is consistent with that available with the current forecast technique of observing a solar flare in electromagnetic wavelengths on the Sun and then making a forecast of an ensuing SPE. If the astronauts are a day or more removed from their shelter, as is possible during an exploration sortie on the surface of Mars, then a longer lead-time forecast is required. Forecasts a day or more in advance require a capability to forecast a solar flare before it has occurred, and at the same time, determine whether the flare will be one of the one percent of all flares that produce SPEs at Earth. A potential use of the one-day or longer forecasts would be to reliably forecast no chance of a SPE occurrence. Such all-clear forecasts would allow the scheduling of safe exploration sorties away from a fixed shelter.

A FORECASTING PROGRAM IN CURRENT USE

A space radiation forecasting and monitoring system is currently in use to support U. S. space missions (Heckman, et al., 1992). Primary elements of the system are the NOAA Space Environment Services Center (SESC) in Boulder, Colorado, ground-based solar observatories around the world (especially the Solar Electronic Observing Network (SEON)), and the NOAA Geosynchronous Environmental Observing Satellites (GOES).

The GOES satellites carry charged particle sensors to measure the in-situ flux of protons, electrons, and alpha particles over the approximate range of one MeV to one BeV for protons and alpha particles, and of a few MeV for electrons (Grubb, 1975). The proton data are processed by spectral fitting to obtain estimates of the particle fluxes for arbitrary proton energies, as shown in Figure 1 (Zwickl and Kunches, 1989).

The GOES satellites also provide constant monitoring of solar X-ray radiation, which gives rapid, quantitative information on the state of solar activity, especially the occurrence of solar flares (Grubb, 1975). The GOES X-ray data are used to determine the start and maximum time of solar flares and the energy emitted by a solar flare, estimated by summing the x-ray emission over the duration of the event.

Optical telescopes in the SEON network provide the location, size, and complexity of sunspot groups on the Sun; maps of the magnetic field structures in active regions; and the size, intensity, and location of solar flares as they occur. Radio telescopes provide the time, intensity, and duration of bursts of solar radio noise associated with solar flares and other forms of solar activity, especially shocks near the Sun. Thompson and Secan (1979) described the SEON network.

The SESC in Boulder receives GOES and SEON data and space environment data from additional satellites and ground-based observatories. With these data, space scientists monitor the level of various types of solar activity, make forecasts of solar activity and SPEs, issue alerts of events when they occur, and provide information about events in progress. The SESC provides these data, alerts, and forecasts in real time to the NASA Space Radiation Analysis Group (SRAG) during manned U.S. space missions.

SHORT-TERM FORECASTS

One forecast in the present system is for time and intensity of a SPE following the

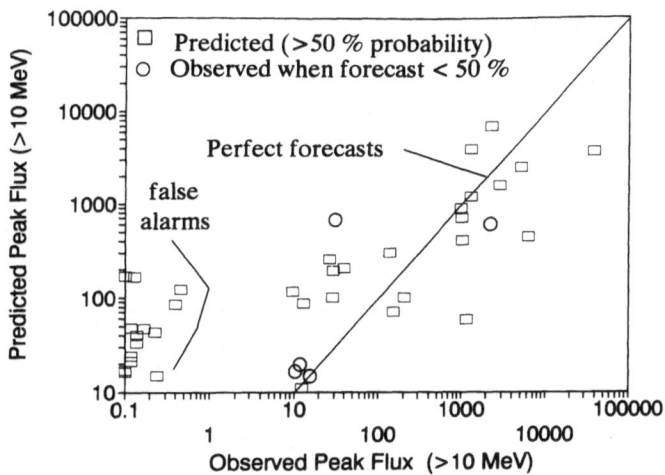

Figure 3. Comparison of the Forecasts and Observed Peak Particle Fluxes for All the Proton Events that Occurred in 1989. Not Shown are the 63 Cases When the Forecast Probability was Less than 50 Percent and No Event Occurred.

observation of a solar flare. Less than one percent of all flares produce a SPE at Earth large enough to be a risk for space flight crews. Therefore, the first task when a flare is observed is to make a <u>yes</u> or <u>no</u> forecast whether a SPE is expected from a flare even while it is still occurring. X-rays, visible light, and radio emission from the flare reach the Earth in slightly more than eight minutes. Energetic particles travel slower than the speed of light and have their route to the Earth modulated by the magnetic field in the solar atmosphere and solar wind. The energetic charged particles from a flare, if they reach Earth at all, require some tens of minutes to hours to arrive. The delay time until particles first arrive and increase to maximum intensity can be used to take protective actions.

The forecasts use a simple hybrid/empirical model called PROTONS; it assumes the flux of energetic charged particles that will reach Earth from a flare can be estimated to first order from the intensity and duration of electromagnetic flare emissions, especially the x-rays. The effect of preferred particle propagation along magnetic fields is included by down-scaling the size of the predicted event at Earth if the flare is removed from the magnetic field lines that connect most directly between the Sun and Earth. The predicted particle flux at Earth is scaled up if there is evidence from previous flares or radio emissions that very energetic events have already produced a mass ejection or shock wave that may be moving outward through the solar atmosphere as part of the flare and associated processes. Specific data are used to run PROTONS:

-- Time-integrated 0.1-0.8 nm x-ray energy (Joules m-2)
-- Heliographic latitude and longitude of the flare
-- Peak 0.1-0.8 nm x-ray flux (Watts m-2)
-- Type II and Type IV radio burst information
-- X-ray energy from the same solar location in the preceding 24-48 hours

PROTONS produces a set of forecasts about the occurrence of a <u>proton event</u> at Earth[1]. It produces a probability that a proton event will occur, an expected maximum intensity (peak particle flux) of the particle flux for the event for several energy ranges, the delay time until the event begins, and the time for the >10 MeV particle flux to reach maximum. Details of the PROTONS model are available in Balch and Kunches (1986),

---

[1] A <u>proton event</u> is defined when the flux of particles (primarily protons) from an SPE reaches or exceeds 10 p.f.u. for at least 15 minutes for particles with energies equal or greater than 10 MeV as measured by the GOES Energetic Particle Sensor.

Table 1. Comparison of Yes/No SPE Forecasts to Observations for 1989.

		Forecast	
		No event	Yes event
Observation	No event	63 (hits)	16 (False alarms)
	Yes event	4 (+1) (missed events)	17 (hits)

Kunches et al. (1991), and Heckman, et al.(1992). In actual practice, the decision to run the PROTONS model is left to the forecaster on duty in the SESC. The forecaster, using experience and other judgmental factors, may modify the forecast made by the PROTONS model. Changes to the PROTONS forecast made by the forecaster are not included in this evaluation.

Results of the Short-Term Forecasts in 1989

The results from using PROTONS in real time in 1989 are summarized in Table 1 and Figure 3. In that year the forecaster decided that rapid flare reports warranted running PROTONS exactly 100 times. The results of all 100 runs are included in the analysis. In addition, one proton event occurred when no significant flare had occurred and the forecaster had not run the model.

The PROTONS program produces a probability between zero and 100 percent that a proton event will occur at earth. The probability forecast for each event in 1989 was converted into a yes/no forecast by assuming a yes forecast if the probability of occurrence was greater than 50 percent and a no forecast if the probability was less than 50 percent. The reliability of the resulting forecasts is shown in Figure 3. Missed events are those when a forecast was for no event but an event did occur (four times). False alarms are those events when the forecast was yes but no event subsequently occurred (16 times).

PROTONS also produces a forecast of the peak flux of protons with energy greater than 10 MeV. Forecasts of peak flux intensity as compared to observed intensities are included in Figure 3 if they were forecast to occur with a probability of 50 percent or greater and if an event occurred when the predicted probability of occurrence was less than 50 percent. If the forecast probability was less than 50 percent and no event occurred (the no event/no event category in Table 1), no point for that flare is shown in Figure 3.

What is Needed to Improve the Forecasts?

Our operational experience with PROTONS in 1989 suggests that omission of the effects of interplanetary shocks is a major deficiency with the present version of the model. In some cases the effects were confined to particles with energies near 10 MeV. In the large event on October 10 (Figure 4), the probable shock effect increased the peak particle flux >10 MeV by a factor of ten. We are unable to assess the source of the forecast errors in Figure 3 beyond the order-of-magnitude effects of the shocks. We know that solar mass ejections are important in understanding the acceleration and escape of particles from the flare (Rust, 1982). We have no ready explanation for the separate grouping of the false alarm forecasts in Figure 3.

Assumption of uniform propagation in the corona, as is done in PROTONS, appears

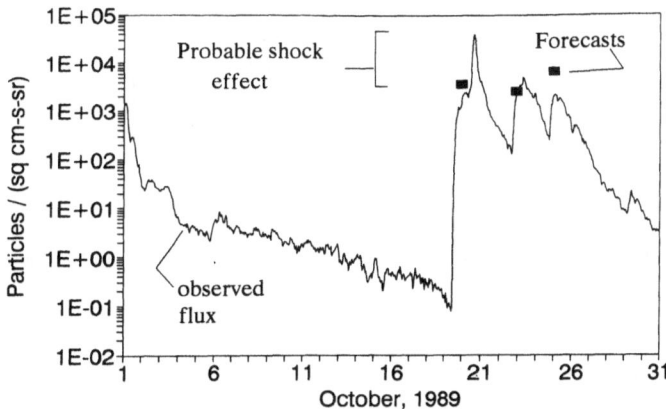

Figure 4. Comparison of the Forecast and Observed Peak Particle Fluxes for the Three Very Large Events in October 1989. Data from Solar Geophysical Data (1989) Indicate an Interplanetary Shock Passed the Earth at the Time of the Abrupt Spike-like Increase of Fluxes on October 20. Proton Flux (>10 MeV) Hourly Average.

simplistic, given the complex structures that are apparent in coronal images recorded during the Skylab mission. Two major areas of observations would improve the use of the model immediately and would provide data for further research and development to improve the model. X-ray image patrol observations of the corona (Wagner, et al., 1987) would improve coverage of flare observations including detection of limb events. Direct observations of the solar wind would improve present techniques of inferring solar wind velocities. These new observations will require research to develop a more complete understanding of solar mass ejections, coronal structures, and the propagation and effects of shocks. Research efforts need to address the unsolved problems of the acceleration of charged particles at the Sun and their propagation in the solar corona.

FORECASTS MADE ONE TO THREE DAYS IN ADVANCE

The present forecast and monitoring service provides daily forecasts of solar flares and proton events. Flares are classified according to the peak emission of their x-rays measured in the 0.1-0.8 nm band on the GOES satellites. Flares whose peak x-ray emission exceeds 10^{-5} W m^{-2} are defined as Class M, and those whose peak exceeds 10^{-4} W m^{-2} are defined as Class X. The definition of proton events for these forecasts is the same as given previously, that the flux of particles with energy >10 MeV reaches or exceeds 10 p.f.u. for at least 15 minutes. Forecasts are made daily for 24 hour periods of the probability of occurrence of Class M flares, Class X flares, and proton events. There are separate forecasts for each of the three days following the day of the forecast. Less than one flare in ten reaches Class M level, and not all Class M or Class X flares produce proton events. In 1989, there were 620 Class M flares and 59 Class X flares.

How the One-, Two-, and Three-day forecasts are Made

Rust (1992) has discussed available methods for flare forecasting and those that might be used in the future. Flare forecasts are made daily in the SESC using a combination of objective guidelines and subjective decisions by a duty forecaster. Some basic criteria, as described by Rust, are the complexity of solar active regions, including the structure of sunspots and magnetic fields, and the history of flare production within each active region visible on the sun. Following is a list of the criteria used by forecasters, ranked in approximate priority (Heckman, 1979):

- Presence of a high level of sheared magnetic field in the sunspot group/center of activity on the sun
- Presence of strong gradients in the magnetic fields

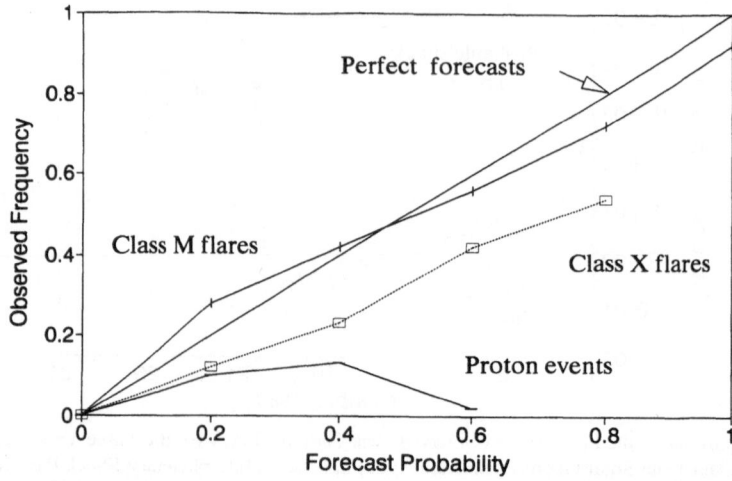

Figure 5. Reliability of the Forecasts of Class M and Class X Flares and of Proton Events for All the One-Day-in-Advance Forecasts Made in 1989. If the Forecasts were Perfect, All Points Would Lie Along the Diagonal Line. Class M Forecasts Fell near the Line. Class X Flare and Proton Event Forecasts Fell Below, Indicating a Trend toward Over-Forecasting of these Events.

- Complex magnetic field structure
- Previous moderate or major flare activity
- Relative motions of sunspots that will increase the magnetic field shear
- Many minor flares with elevated x-ray emissions
- Formal sunspot classification (complexity, size, compactness) (McIntosh, 1990)
- Emergence of new sunspots or magnetic field into stable or decaying centers of activity
- Rapid sunspot growth
- Colliding sunspot regions
- Structural similarity to past flare-producing regions
- Existence of bright emission of solar plage in the active center
- Compact plage structures

Once the forecaster produces a flare forecast for each day, it is then necessary to forecast whether any flares that do occur will be likely to produce proton events. Additional considerations at this step are whether the region has a history of producing solar particle events, the relative duration of the flares that have occurred, and the location of the region on the sun.

Evaluation of the One-, Two-, and Three-Day Forecasts

For this report, we include and compare the results of the forecasts for flares and for proton events. Reliability plots provide a measure of reliability of the forecasts. Reliability in this instance is akin to consistency. If the forecasts are highly reliable, the observed frequency of flare occurrence will equal the forecast probability. If a collection of forecasts are for 90 percent probability of flare occurrence and they are reliable, flares will occur in 90 percent of the observed cases. If a collection of forecasts are for 10 percent probability of occurrence and are reliable, then flares should be observed in 10 percent of the cases. The reliability of the one-day forecasts are plotted in Figure 5 for those forecasts with a lead time of one day. These forecasts are the probability that an event of the type specified will occur during the 24-hour period on the following day. Forecasts for the second day in the future are shown in Figure 6 and for the third day in the future in Figure 7. The differences in the forecasts are apparent. Forecasters are more proficient at forecasting both Class M and Class X flares and less proficient at forecasting proton events one or two days in advance. The reliability decreases in going from one day to two days lead time and the trend continues for three-day forecasts. The forecasters can forecast flares in advance with some reliability, but they are unable to forecast which flares will produce proton events.

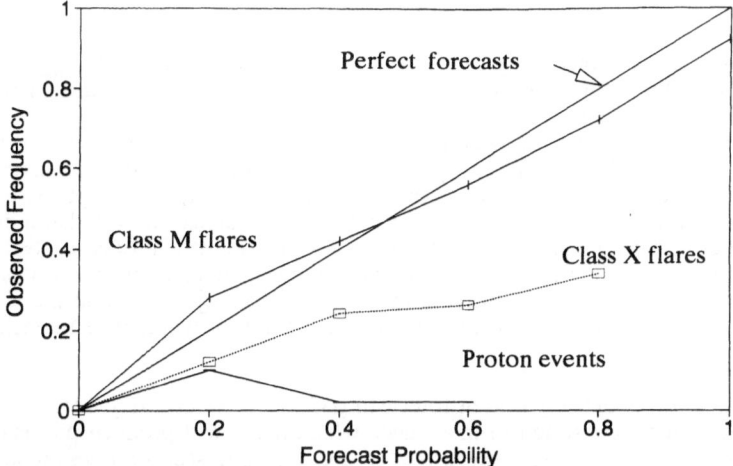

Figure 6. Reliability of the Forecasts of Class M and Class X Flares and of Proton Events for All the Two-Day-in-Advance Forecasts Made in 1989. If the Forecasts were Perfect, All Points would Again Lie along the Diagonal Line. Class M Forecasts were Less Reliable than for the One-Day Forecasts. Class X Flare and Proton Event Forecasts Fell Further Below the Perfect-Forecast Line than in the One-Day Forecasts, Indicating an Increasing Trend Toward Over-Forecasting these Events.

What is Needed to Improve the Forecasts?

The present daily forecasts are based on little physical understanding of the flare and particle acceleration process. They rely on forecaster judgement to evaluate the morphology of solar active regions and to identify factors indicative of a flare. The parameters and observations currently emphasized in Class M forecasting are dominated by small-scale structures. This approach appears to be more useful for anticipating the release of the X-ray energy. Proton event forecasts probably cannot be improved without (1) gaining significantly improved understanding of the physical process through research and (2) providing observations that show the larger scale processes.

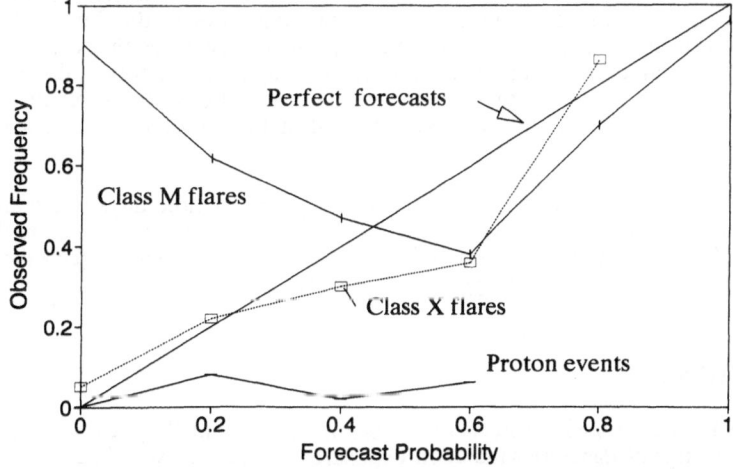

Figure 7. Reliability of the Forecasts of Class M and Class X Flares and of Proton Events for All the Three-Day-in-Advance Forecasts Made in 1989. All Forecasts Departed Further from the Perfect-Forecast Diagonal Line Compared to the One- and Two-Day Forecasts.

DISCUSSION

We have examined forecasting capability for the subset of SPEs called proton events for space crews spending long periods in space. As the lead times of forecasts increase from a few tens of minutes to days, the accuracy of the forecasts decreases. Forecasts of a proton event made after a solar flare has occurred will allow some tens of minutes to several hours for astronauts to seek shelter. The capacity to make such forecasts is comparatively good. A major aspect of the present technique is the ability to identify the small proportion of solar flares that will produce proton events. Besides the yes/no forecast of whether a proton event will occur, there are techniques to forecast the size of the event if it occurs. Such forecasts can be in error by factors of ten over the observed range of maximum intensity of ten to 10^5 p.f.u. Monitoring, modeling, and forecasting of solar mass ejections and the shock waves associated with them would appear to result in the most improvement of intensity for these forecasts.

Forecasters do moderately well at forecasting Class M flares a day or more in advance. However, techniques to distinguish in advance which flares will produce proton events near Earth are missing, and the proton forecasts at comparable time scales are less reliable than the flare forecasts. Improvement in these proton event forecasts requires improved understanding of the physics of the processes by which energy stored in the sun's magnetic fields is released and converted into the energy that leads to the acceleration of charged particles. While research continues to improve physical understanding of these processes, development of improved observational models of the evolution of solar active regions (Bornmann, 1989) could improve forecasts of flares and proton events up to a few days in advance.

SPE FORECASTS FOR MARS MISSIONS

The present Earth-based forecast system will not be optimal to protect space crews on the far side of the solar system, where they will be during portions of a Martian mission. Earth-based telescopes cannot see all the solar flares and other activity that can produce SPEs on the other side of sun. The capability to make these observations from the vicinity of a Martian spacecraft will be necessary. A further complication is the time necessary for radio signals to travel from Mars, when it is on the opposite side of the Sun, to Earth and back to Mars. A significant portion of the warning time for SPEs (up to 30 minutes) would be lost while radio signals were in transit. One solution would be to provide the crew of a Martian mission with observations from a set of telescopes similar to those now used to make the SPE forecasts but with additions and improvements to eliminate the present observational gaps. The telescopes could be placed in a separate orbit around the Sun, carried on the spacecraft itself, or both. Solar wind observations from several points around the Sun would provide a significant improvement in understanding the large-scale structures that can affect particle propagation. Martian crews would then also need the capability to collect the observations, recognize and correct errors, synthesize the various observations into a real time picture of solar activity and its effects in the solar system, and make the forecasts necessary for their own protection.

Reliable forecasts that no SPEs will occur, made one to three days in advance, would allow astronauts on the surface of Mars leave their base and its shelter to explore the Martian surface. The formulation of the forecasts for such an approach would require the joining of the one- and two-day forecasts of no event derived from the forecasts of event occurrences.

CONCLUSIONS

Our experience with SPE forecasting systems shows that they are capable of giving short-term warnings of dangerous radiation exposures to space crews. Eighty percent of the time they were correct in forecasting the occurrence or non-occurrence of the subset of SPEs called proton events. Forecasting models similar to current models but using better observations and improved physical understanding of the relationship between solar mass ejections, shocks, and SPEs will allow completion of missions with less risk of exposure and improved real-time planning of mission operations. They would reduce the loss of time inherent in passive monitoring and response systems that would respond to all radiation increases.

Forecasts made one or more days in advance need significantly improved physical understanding of the process of energy storage and release in solar flares. The SPE/proton forecasts show that major advances are needed in observing and understanding the structures in the solar magnetic field that contribute to particle acceleration even before energetic acceleration events begin.

A forecasting system for exploration missions to Mars requires an ability to observe, analyze, and forecast solar and interplanetary activity from the Martian side of the solar system.

ACKNOWLEDGMENTS

The author thanks Herbert Sauer, Ron Zwickl, and the other staff of the Space Environment Laboratory for their work in making the GOES energetic particle data as reliable and well calibrated as it is. Allan Murphy and Barbara Brown designed and developed the forecast evaluation system used in the Space Environment Services Center. Judy Stephenson did the work to necessary to implement it on the Space Environment Laboratory computers. Joseph Hirman and his forecast staff have preserved all the relevant information used in their forecasts and have provided the information for open examination.

REFERENCES

Balch, C. and J., Kunches, 1986. SESC methods for flare forecasts, in: "Sol. Terr. Predict. Proc.", Simon, P., Shea, M.A. and Heckman, G.R., ed., NOAA, Boulder, CO.

Bornmann, P., D.Kalmbach, D. Kulhanek, and A. Casale, A study of the evolution of solar active regions for improving solar flare forecasts, in: "Sol. Terr. Predict. Proc.", Thompson, R.J., D. G.Cole, P. J. Wilkenson, M. A. Shea, D. F. Smart, and G. R. Heckman, ed., NOAA, Boulder, CO.

Cane, H.V., D. V Reames,and T. T.von Rosenvinge,1988. The role of interplanetary shocks in the longitude distribution of solar energetic particles, J. Geophys. Res., 93, #A9: 9555-9567.

Clark, B., 1992. Radiation protection for human interplanetary spaceflight and planetary surface operations, this Volume

Foukal, P. V., 1990. "Solar Astrophysics", John Wyley and Sons Inc., New York.

Grubb, R.N., 1975. The SMS/GOES Space Environment Monitor Subsystems, NOAA Tech. Memo. ERL SEL-42, NOAA, Boulder, CO.

Heckman, G. R., Kunches, J.M., and Allen, J.H., 1992. Prediction and evaluation of solar particle events based on precursor information, Adv. Space Res., 12, 2-3: (2)313-(2)320.

Heckman, G.R., 1979, Products and services of the Space Environment Services Center, in: "Sol. Terr. Predict. Proc.", R. Donnelly, ed., NOAA, Boulder, CO.

Hirman, J. W., 1988. Solar proton event forecasts, in: "Terrestrial Space Radiation and Its Biological Effects", McCormack, P. D., Swenberg, C. E., and Bücker, H., ed., Plenum Press, New York and London.

Kunches, J. M., G. R.Heckman, E. Hildner,and S.T. Suess, 1991. Solar radiation forecasting and research to support the Space Exploration Initiative, NOAA Space Environment Laboratory Special Report, NOAA, Boulder, CO.

Lett, J.T., W. Atwell, and M. J. Golightly, 1991. Radiation hazards to humans in deep space: a summary with special reference to large solar particle events. in: "Sol. Terr. Predict. Proc.", Thompson, R.J., Cole, D.G.,

Wilkenson, P.J., Shea, M.A., Smart, D.F., and Heckman, G.R., ed., NOAA, Boulder, Co.

National Council on Radiation Protection and Measurements, 1989. "Guidance on Radiation Received in Space", NCRP Report No. 98, National Council on Radiation Protection and Measurements, Bethesda, MD.Maisin, J.R., 1989, Chemical protection against ionizing radiation, Adv. Space Res., 9, 10 : (10)205-(10)212.

McIntosh, P.S., 1990. The classification of sunspot groups, Sol. Phys., 125: 251-267.

Rust, D.M., 1982. Solar flares, proton showers, and the space shuttle, Science, 216: 4549.

Rust, D.M., 1992. Solar flares and their effects, this Volume.

Shea, M.A. and D.F. Smart, 1992, History of energetic solar particles for the past three solar cycles including Cycle 22 update, this Volume.

Smart, D.F., and M.A.Shea, 1979, PPS76-A computerized "Event Mode" solar proton forecasting technique, in: "Sol. Terr. Predict. Proc.", R. Donnelly, ed., NOAA, Boulder, CO.

Smart, D.F. and M.A. Shea, 1992, Predicting and modeling the time-intensity profile of solar flare generated particle fluxes in the inner heliosphere, this Volume.

Solar Geophysical Data, 1989, Part 1 (ISSN 0038-0911), U.S. Department of Commerce, Boulder, CO.

Thompson, R.L. and J.A. Secan, 1979, Geophysical forecasting at AFGWC, in: "Sol. Terr. Predict. Proc.", R. Donnelly, ed., NOAA, Boulder, CO.

Wagner, W.J., R.N. Grubb, Heckman, G.R., and Mulligan, P.J., 1987. The Solar X-ray Imagers (SXI) on NOAA's GOES, Bull. Am. Astron. Soc., 19: 923.

Zwickl, R. and J.M. Kunches, 1989. Energetic particle events observed by NOAA/GOES during Solar Cycle 22, Trans. Am. Geophys. Un., 70, 43: 1258.

PREDICTING AND MODELING SOLAR FLARE GENERATED PROTON FLUXES IN THE INNER HELIOSPHERE

D.F. Smart and M.A. Shea

Space Physics Division, Geophysics Directorate
Phillips Laboratory (OL-AA)
Hanscom AFB
Bedford, Massachusetts, 01731 USA

ABSTRACT

Solar energetic particles are assumed to be accelerated above the solar active regions from the available coronal material. The composition of "large" solar particle events is consistent with an ion selection process based on the first ionization potential of the elements in the solar corona. The transport of solar protons in interplanetary space is controlled by the topology and characteristics of the interplanetary magnetic field. The topology of the magnetic field lines in interplanetary space is controlled by the flow speed of the ionized plasma and the rotation rate of the sun, resulting in the so called "Archimedean spiral" configuration. The particle flux longitudinal gradients observed in the inner heliosphere are variable, and local interplanetary conditions and structures greatly influence the time-intensity profiles observed. The most extensive solar particle measurements are those observed by earth-orbiting spacecraft, and forecast and prediction procedures are best for the position of the earth. These earth-based models can be extended to other heliolongitudes or to more distant locations in the inner heliosphere.

OVERVIEW OF CONCEPTS INVOLVED

The Solar Flare Source

Solar energetic particles are accelerated from the available coronal material during the solar flare process. (See Svestka, 1976, for a detailed discussion of the solar flare phenomena.) After the initial solar acceleration there may be further acceleration of the energetic particle population by interactions with interplanetary shocks, but these subjects are beyond the scope of this paper. The solar flare emissions that we observe at Earth are indicators, probably secondary manifestations, that particle acceleration is occurring. The particle acceleration site is probably high in the solar corona where we have no method of direct observation. The various emissions we associate with solar flares occur as the result of some of the accelerated particles travelling down from the probable high coronal acceleration site into the chromosphere and photosphere.

The English term "solar flare" is the name given to the sudden energy release of about 10^{32} ergs of energy in a relatively small volume of the solar atmosphere. The Russian term, which translates into English as chromospheric brightening is more descriptive of the observations, occasionally seen in white light, but usually observed in the hydrogen-alpha

Biological Effects and Physics of Solar and Galactic Cosmic Radiation,
Part B, Edited by C.E. Swenberg *et al.*, Plenum Press, New York, 1993

101

line. The radio emissions result from energetic electrons at positions where the magnetic field is intense enough to confine the gyration radius of these electrons to a diameter consistent with the emission wavelengths. The solar X-rays are emitted from the hot plasma generated in the solar flare process, and gamma ray emissions occur at heights where there is sufficient density for the flare energized particles to interact with the ambient solar atmosphere.

The solar particles are apparently accelerated high in the solar corona. The composition of solar particles observed at the various interplanetary spacecraft is consistent with the particles having passed through less than 30 mg cm^{-2} of matter from the acceleration site to their detection location since the elemental and isotopic composition of the solar particles observed in space do not appear to have undergone fractionization due to interaction with significant mass.

Solar Particle Transport

Once the solar particles have been accelerated by the solar flare process, they must still escape the confines of the solar magnetic fields and be transported through the solar corona into space. Coronal propagation is a concept developed to explain the apparent transport of solar particles around the sun. It is assumed that there are intensity gradients in the solar flare particle distribution in the solar corona. Current ideas suggest that the magnetic structures in the corona extend into space and that the intensity in the solar corona can be mapped into interplanetary space (Reinhard et al., 1986). The coronal propagation characteristics are probably dependent on the coronal magnetic fields, which are poorly observed but are thought to be quite variable. The old ideas developed during the 60's that most of the particle diffusion occurred in space proved to be inconsistent with the satellite observations of later years from which apparent mean free path lengths derived for solar particles in interplanetary space ranged between approximately 0.1 to more than 1.0 AU, depending on the event. These large variations in the mean free path lengths are thought to be due to variations in the

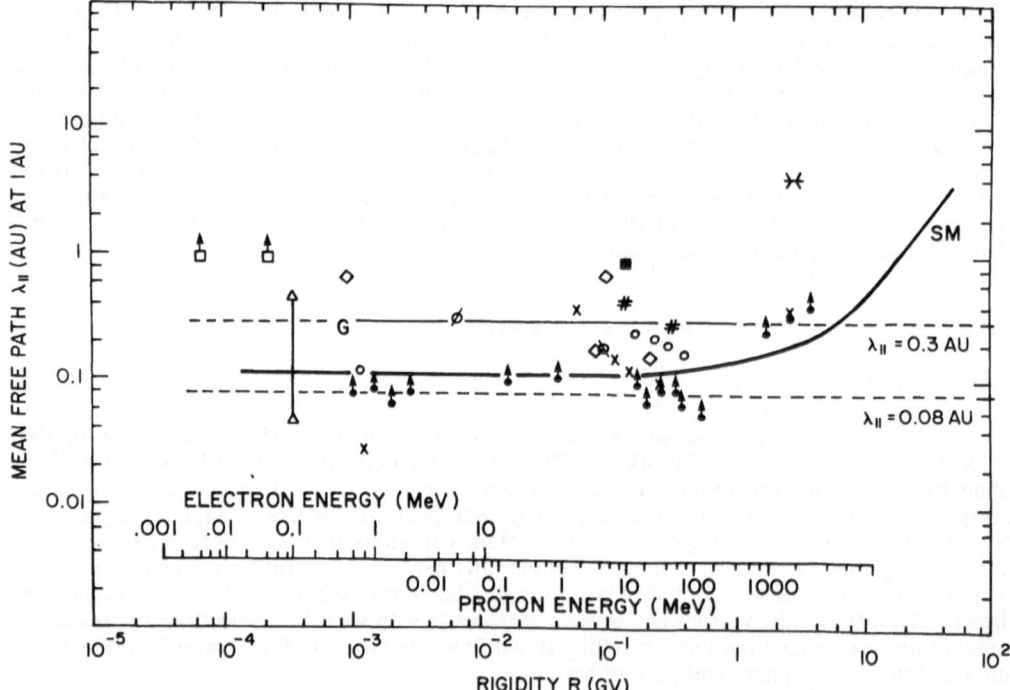

Figure 1. The variability in the mean free path (the distance between scattering events) for a number of interplanetary solar particle events. The solid curve labeled "SM" is from Morfill et al. (1976). A energy-to-rigidity conversion curve is given at the bottom of the figure for electrons and protons. After Valdes-Galica (1991).

turbulence of the interplanetary magnetic field. The variability in the mean free path (the distance between scattering events) for solar particles in the interplanetary medium is illustrated in Figure 1.

Interplanetary magnetic field topology. Charged particle transport in space is controlled by the interplanetary magnetic field topology. In organizing solar energetic particle data it is very useful to use the gross topology expected for the interplanetary magnetic field (Roelof 1973, 1975, 1976, Roelof and Krimigis 1973, Reinhard et al., 1986) as illustrated in Figure 2. The interplanetary medium in the vicinity of the earth is dominated by the plasma flowing out into space from the solar corona called the "solar wind". This plasma is highly ionized and therefore highly conducting; for most practical purposes the interplanetary magnetic field may be considered to be "frozen" in the solar wind. As a result of the outward flowing solar wind from a rotating source, in the regions of space not too far from the ecliptic plane, the idealized interplanetary magnetic field lines carried away from the sun into space have a characteristic curved shape, the mathematical form of an Archimedean spiral. Application of this mathematical form makes it possible to estimate the probable solar connection longitude of the interplanetary field passing near the earth or a spacecraft in the inner heliosphere. This simplified method of estimating the probable solar source location was formalized by Nolte and Roelof (1973). The basic procedure is to divide the distance from the sun to the observation location by the observed solar wind speed to obtain the travel time and then compute the amount of solar rotation that has occurred in that transit time. This method is still used because of its simplicity, and because there is no assurance that more complex formulations generate results that can be used with more confidence.

The portion of space where the plasma outflow from the sun dominates the characteristic of space is the heliosphere. The dimensions of the heliosphere are still a matter of scientific speculation, but the heliosphere is estimated to extend beyond 100 Astronomical Units (AU). (An Astronomical Unit is the distance from the sun to the earth, 149.6 million km.) The inner heliosphere is generally considered the domain from the sun to the earth and

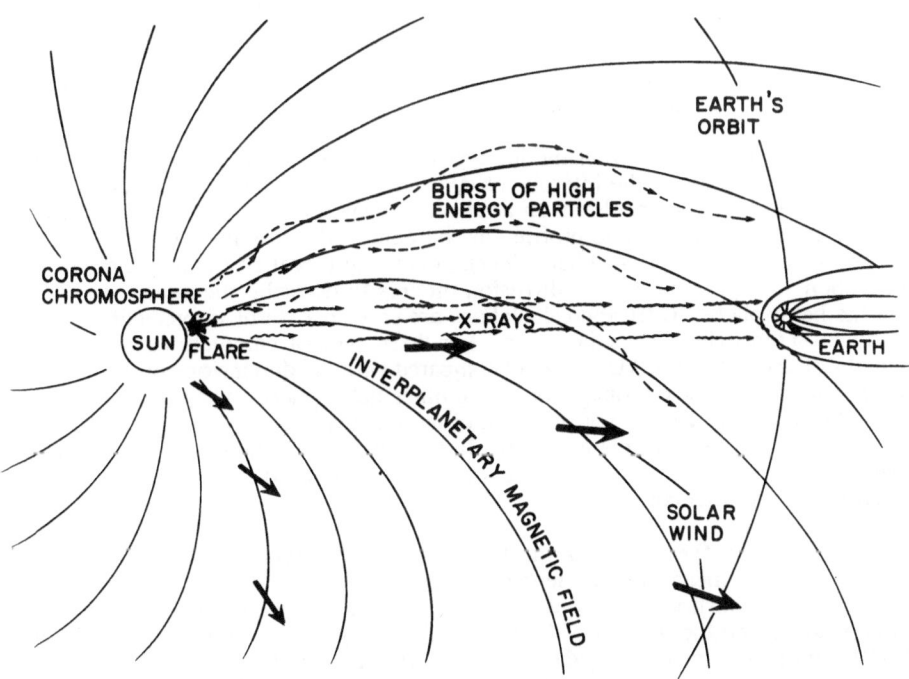

Figure 2. Illustration of the sun and the gross topology of the idealized structure of the interplanetary medium.

may include the orbit of Mars at ~1.5 AU. The outer heliosphere may begin about the distance of Jupiter and continue to the termination shock at the heliospheric boundary, the distance where the solar wind flow is hypothesized to change from super-sonic flow to sub-sonic flow. In the outer heliosphere, diffusion dominates the processes controlling energetic charged particle transport. In the inner heliosphere, at distances from "near sun" to about 1 AU, during "quiet" times, the transport processes controlling energetic charged particle motion is not diffusion dominated. However, the characteristics of the inner heliosphere are best described as "highly variable" and dominated by solar activity.

Particle transport by diffusion - elementary diffusion theory. From a theoretical standpoint the problem of particle diffusion in the interplanetary medium should be solvable with known equations. There are a large number of theoretical papers (not enumerated here) on this subject. The fundamental equations for transport of solar energetic particles were derived by Parker (1965). The spherical transport equation given below includes the effects of diffusive convection and adiabatic energy loss in the expanding solar wind:

$$\frac{\partial U}{\partial t} + \frac{1}{r^2}\frac{\partial U}{\partial r} \left\{ r^2 UV - r^2 K_r \frac{\partial U}{\partial r} \right\} - \frac{2V}{3r}\frac{\partial U}{\partial t}(\alpha TU) = 0. \tag{1}$$

In this equation, U(r, T, t) specifies the differential number density, V the solar wind velocity, K_r the radial diffusion gradient, and T the particle kinetic energy. The parameter α is defined by

$$\alpha = (T + 2m_0 c^2)/(T + m_0 c^2). \tag{2}$$

When solving this equation, it is generally assumed that the diffusion constant, K_r, behaves as:

$$K_r = K_0 r^b, \tag{3}$$

where r is the radial distance from the sun, K_0 is the diffusion coefficient at 1 AU, and b is an exponent describing the interplanetary turbulence spectrum. This type of equation can be simplified and solved for the time of maximum particle flux resulting in:

$$t_{(max)} = \frac{r^{(2-b)}}{3 K_0 (2 - b)}. \tag{4}$$

The fundamental problem using this type of equation in a prediction sense is the total dependance of the result on the interplanetary scattering parameters. It is precisely these parameters that are not known. From an academic viewpoint, after the occurrence of a solar particle event, the intensity-time profile can be analyzed to deduce what the scattering parameters were (Zwickl and Webber, 1977). Once these scattering parameters have been deduced, then it is possible to use diffusion models to describe the intensity-time profile through the heliosphere. Diffusion calculations are very successful in predicting the cosmic ray modulation throughout the solar cycle. For these type of calculations, even mean free path lengths of the order of 1 AU are small compared with the dimensions of the heliosphere. However, for solar particle propagation in the inner heliosphere, when the mean free path length is relatively long, there is not sufficient distance between the sun and the earth to permit diffusion to dominate the particle transport. Indeed, there are a number of suggestions that the solar particle event profile observed at 1 AU may be dominated by the injection and release of particles from the solar corona.

The "scatter free" approximation for solar particle transport. Since the apparent solar proton mean free path length derived from spacecraft observations is relatively long (a significant portion of an Astronomical Unit in many cases), Roelof (1975) and co-workers developed the concept of "scatter free" propagation where the solar particles are essentially confined to travel along the interplanetary magnetic field lines. This concept has been very useful in "mapping" particles back to the sun and deriving probable source locations. In the inner heliosphere, the "scatter free" approximation is often as successful as the much more complex diffusion calculations. The utility and demonstrated simplicity of the "scatter free" concept makes it simple to apply to the solar particle prediction problem in the inner heliosphere. At large distances in the heliosphere, say beyond the orbit of Jupiter (5 AU), it has been demonstrated that diffusion dominates the transport of particles in space.

PREDICTION OF SOLAR PARTICLE EVENTS

The prediction of solar particle events depends on observable parameters on which to base a calculation. The questions to be answered by a prediction procedure are:

(1) When will the event start?
(2) When will the particle flux reach maximum intensity?
(3) How large will the event be?
(4) How long will the event last (at a specific energy or above a specific flux level)?

Indicators of a Solar Proton Event

There are general relations between the electromagnetic emissions from a solar flare and the number of protons observed at the earth. The X-ray, radio and optical emissions during the solar flare event are the indicators (perhaps secondary manifestations) that proton acceleration is occurring. There are correlations between the observed proton flux in the ~10 MeV energy range and the electromagnetic emission from the solar flare. The best correlation coefficients for these "observables" are approximately 0.7 in both the microwave frequencies or the soft X-ray wave lengths. More relationships have been developed for radio emissions than for X-ray emissions. This is probably a result of the length of time these measurements have been available to researchers rather than any reflection on the physical processes involved. Algorithms that relate the emission power at various radio frequencies to the expected proton flux at 1 AU have been developed by several researchers including Cliver (1976), Cliver et al. (1978), Akun'yan et al. (1977), Akinyan et al. (1979), Smart and Shea (1979), and Chertok and Fomichev, (1984). Algorithms that relate the emission power at X-ray wave lengths to the expected proton flux at 1 AU have been developed by Kuck et al. (1971), Smart and Shea (1979), and Kuck and Hudson (1990). In a comparison of prediction methods Lantos (1990) found that the radio and X-ray peak flux proton prediction methods were equivalent.

Parameters Necessary for Solar Proton Event Model

In a real-time prediction mode, it is necessary to construct a model based on the information available at the time of the solar flare. We require some indication of the occurrence of a significant solar flare that is likely to accelerate and release particles into the interplanetary medium. We know of no unique parameter that indicates this situation. The best indicators, the big flare syndrome (Kahler, 1982), or the U-shaped peak power spectra (Castelli et al., 1967; Castelli and Guidice, 1976; Cliver et al.,1985), or the long duration soft X-ray event, all correlate at about the same level. Kuck and Hudson (1989) concluded that the correlation between the 1-8Å integrated soft X-ray emission and the peak >10 MeV proton flux is about 0.7. Mel'nikov et al. (1990) reported a similar correlation between the integrated microwave radio emission and peak >10 MeV proton flux. Once a likely solar flare is identified, it is essential to determine the position of the solar flare on the sun and calculate the energy in the solar flare electromagnetic emission. The peak proton flux expected at locations "well connected" to the solar fare site via the interplanetary magnetic field can be computed from selected algorithms that convert one of the electromagnetic emission parameters to expected peak proton flux.

It is possible to calculate (using the methods detailed below) the probable onset time and the expected time of maximum flux. This is done by separating the propagation of solar protons from the flare site to the observation location into two distinct and independent phases. Both of these phases, coronal propagation and interplanetary propagation, are illustrated in Figure 3. The first phase, coronal propagation, transports the particles

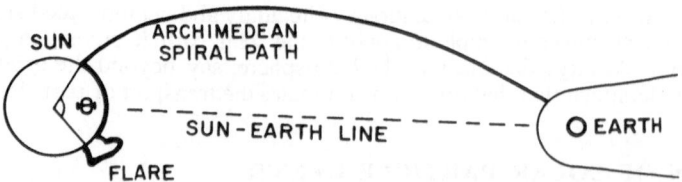

Figure 3. Illustration of the concept of two phases of solar particle propagation from the flare site on the sun to the earth. The coronal propagation distance on the sun is illustrated by the heavy arc. Interplanetary propagation proceeds along the interplanetary magnetic field lines which, for a constant speed solar wind, forms an Archimedean spiral path.

accelerated by the solar flare from the flare location to the "foot" of the interplanetary magnetic field line (Archimedean spiral path) between the sun and the earth. The coronal propagation phase is only weakly energy dependant, at least in the MeV energy range, and for our modeling purposes is assumed to be energy independent. The maximum possible prompt proton flux is presumed to be at the solar flare site, and it is further assumed that there is a gradient in the solar corona extending from the flare site. This gradient attenuates the maximum particle intensity one order of magnitude per radian of heliocentric angular distance from the flare site. The second phase, interplanetary propagation, transports the particles through the interplanetary medium from the sun to the earth along the interplanetary magnetic field. The interplanetary propagation phase is velocity dependant. For the particle onset phase, we use the "scatter free" approximation. Many of the concepts described here were first used by Smart and Shea (1979) and are summarized by Smart and Shea (1985).

Coronal propagation - theoretical expectations. The concepts we have used for the propagation of solar protons in the solar corona are similar to those originally advanced by Reinhard and Wibberenz (1974). We utilize the fundamental elements of solar particle diffusion theory as developed by early researchers (Reid, 1964; Axford, 1965; Krimigis; 1965; Burlaga, 1967; Wibberenz, 1974) and assume that almost all of the major diffusive effects occur in the solar corona. We make very few assumptions as to the manner of coronal transport except that stochastic processes dominate the particle transport between their source at the flare site and their release point along an interplanetary magnetic field line. We consider that the time required for coronal propagation is a function of heliocentric angular distance θ. From diffusion theory we would expect it to be proportional to θ^2. [See Wibberenz (1974) for a discussion of diffusion theory relating to coronal propagation.] For large coronal propagation distances, the propagation delay time would be dominated by the coronal diffusion time rather than interplanetary propagation time.

We expect that there is a solar particle gradient existing in the solar corona such that the proton intensity decreases as a function of distance from the flare site. There has been considerable observational evidence of a coronal gradient (McCracken and Rao, 1970; McCracken et al., 1971; Roelof, 1973; Roelof et al., 1975; Gold et al., 1975; McGuire et al., 1983). The observations suggest that the gradient varies from case to case. We use an average gradient of a factor of 10 per radian of heliocentric propagation distance. This means that the expected flux observed at some position in space removed from the Archimedean spiral path of the interplanetary magnetic field line from the presumed particle source (i.e. the flare location) is expected to be reduced by a factor of 10 per radian of heliocentric angular distance.

Coronal propagation - comparison with observations. The onset times of particle events at the earth have been catalogued since the early satellite observations, (Barouch et al., 1971; Lanzerotti, 1973). When these earth-acquired data sets are organized in a heliographic coordinate system they show that the minimum time from the flare onset to particle detection occurs in a broad range of heliolongitudes around 60° west of solar central meridian, and that the longest times between the associated flare and the onset of particles observed at the earth are for flares on the eastern side of the visible solar disk. More recent data sets tend to confirm the general trends noted by the earlier investigators.

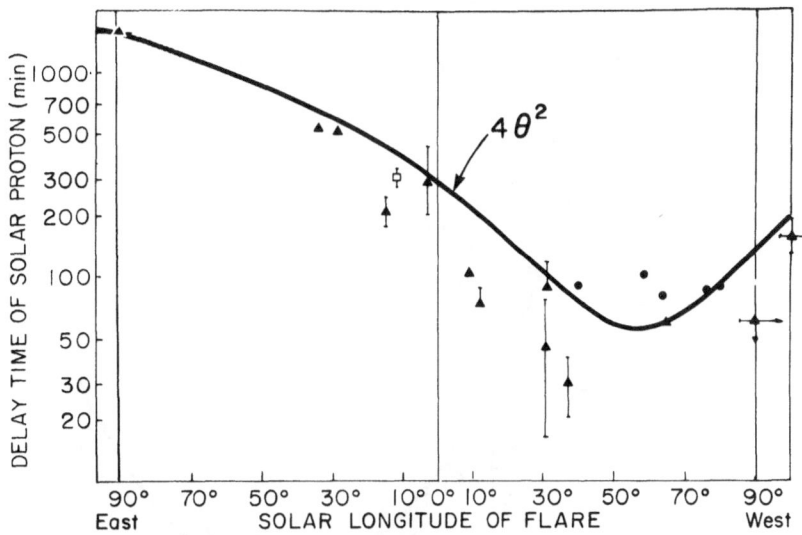

Figure 4. Distribution of the onset time of 30 MeV protons observed at the earth as a function of solar longitude. The data points are the measurements of Barouch et al. (1971).

The distribution of onset times at the earth for 30 MeV protons is shown in Figure 4. The data points shown on the figure indicate typical variations that may be expected. The minimum in the figure corresponds to a flare at the "foot point" of the Archimedean spiral path between the sun and the earth (57° west of central meridian). To our prejudiced eye, a reasonable fit to the onset data at the earth for this specific energy has the functional form of $4\theta^2$.

The distribution observed at the earth for time of maximum as a function of heliolongitude is illustrated in Figure 5. The data points taken from Van Hollebeke et al., (1975) show the typical range of variations that can be expected. The minimum in the curve corresponds to a flare at the "foot point" of 57° for the Archimedean spiral path between the earth and the sun computed from a nominal solar wind of 404 km/sec. It is our opinion that a reasonable fit to the onset data at the earth for this specific energy has the functional form of $8\theta^2$. Other data sets can be plotted in this manner and illustrate the same general characteristics.

Propagation in the interplanetary medium. After the particles propagate through the solar corona and are released into the interplanetary medium, we assume that they propagate along the interplanetary magnetic field lines. In the inner heliosphere the minimum interplanetary propagation time will be for particles that essentially travel along the interplanetary magnetic field lines with very little scattering, so for "scatter free" (Roelof, 1975) onsets the propagation time from the sun will be the distance traveled (i.e. the length along the Archimedean spiral path divided by the particle velocity). After the initial onset it is reasonable to expect that some scattering does occur and that some aspects of diffusion theory are applicable. The time for the propagation of any specified ion along this path is the Archimedean spiral path distance divided by the velocity which is determined by the kinetic energy of the ion.

We make the simplest possible assumptions regarding transport in the interplanetary medium as follows:

     a. Diffusion perpendicular to the interplanetary magnetic field is assumed to be negligible.

     b. The particles travel essentially along the interplanetary magnetic field lines with a

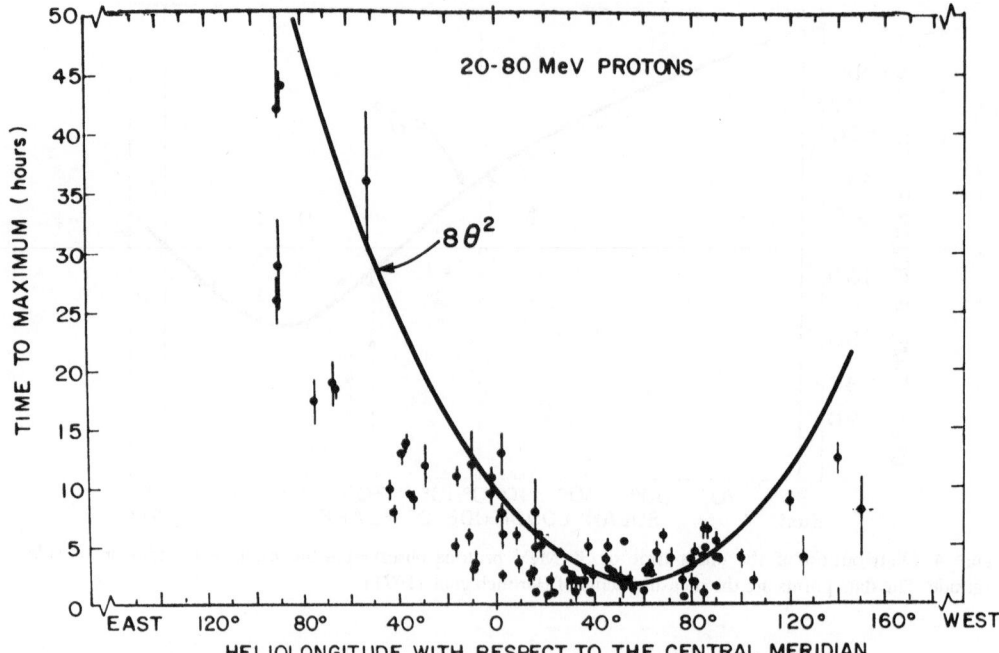

Figure 5. The time to the maximum 20 - 80 MeV proton flux observed at the earth as a function of the heliolongitude of the associated solar flare. The data points are from Van Hollebeke et al. (1975) and the heavy solid line is the 8 θ^2 curve for a nominal solar wind speed.

velocity which is a function of the particle energy. We will use the symbol β to denote the speed of a solar proton as a ratio of the speed of light. β is given by

$$\beta = \{ 1 / (1 - [(E/mc^2) + 1]^2) \}^{0.5} \qquad (5)$$

where E is the kinetic energy of the particle in MeV, and mc^2 is the mass-energy equivalence of a proton which is 938.323 MeV.

    c. The distance traveled from the sun to the observation location in space is the distance along the Archimedean spiral path. The length of the Archimedean spiral path can be obtained by integration of the polar form of the Archimedean spiral equation.

From these assumptions, detailed below, we can calculate the expected propagation delay time and the time to peak particle flux.

Event decay. The decaying portion of the event can be modeled after the principles of collimated convection originally developed by Roelof (1973). After making a number of simplifying assumptions (some of which are that the particle flux can be represented by a simple power law, the anisotropy of the particle flux is small, the magnitude of the interplanetary magnetic field falls off as r^{-2}, and the particle flux gradient is field aligned and small), a 1/e decay constant can be derived which is a function of the distance along the Archimedean spiral path, the solar wind velocity, and the magnitude of the differential energy spectral exponent as follows:

$$T_d = 3D / [4V_{sw} (\gamma + 1)] \qquad (6)$$

where T_d is the 1/e decay constant, D is the distance along the Archimedean spiral path, V_{sw} is the solar wind velocity, and γ is the solar proton energy differential spectral exponent.

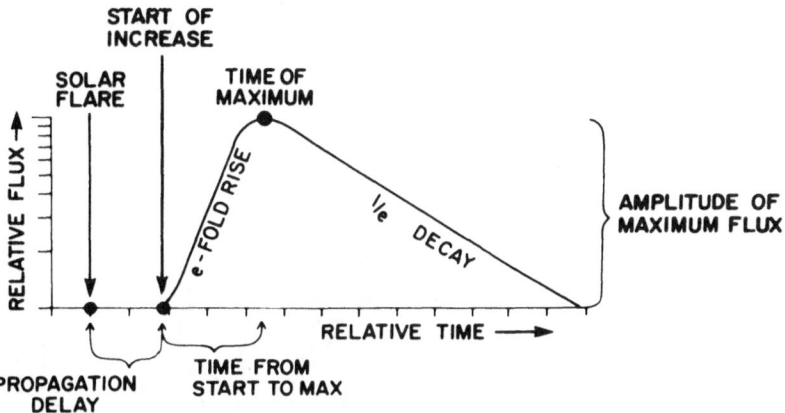

Figure 6. Illustration of the general characteristics of solar proton events. The large dots indicate the critical information that must be known to construct this type of event profile.

Solar Proton Event Model Construction

We now have the necessary information to construct a prediction for the proton intensity as a function of time. We can define the critical points needed to construct a predicted intensity-time profile for the specified energy. These are: the propagation delay between the solar flare and the proton onset at the earth, the time of the predicted flux maximum, the amplitude of the predicted flux maximum, and the predicted decay rate of the maximum flux as illustrated by large dots in Figure 6. An exponential curve is fitted between the proton onset and the maximum. Then the maximum flux is predicted to decay at the 1/e decay rate obtained from equation 5.

Solar proton onset time. The solar proton onset at the earth will be the solar flare onset time plus the propagation delay time. In our model, the propagation delay time is the sum of the coronal propagation time, and the interplanetary propagation time along the length of the Archimedean spiral from the sun to the earth for the fastest (highest energy, scatter-free) protons being detected. This propagation delay time, $T_{(pd)}$, after the solar flare onset time can be represented by:

$$T_{(pd)} = 4 \, \theta^2 + 0.1333 \, D / \beta. \qquad (7)$$

In this and the following equations, time is specified in hours, the coronal propagation heliocentric distance θ is in radians and the distance along the Archimedean spiral from the sun to the earth, D, is in Astronomical Units.

Solar proton maximum time. The time of the solar proton flux maximum at the earth will again be the sum of the coronal propagation time and the interplanetary propagation time along the length of the Archimedean spiral from the sun to the earth. For this computation, several additional factors influence the interplanetary propagation calculation. Almost all theories involving differential transport show that the time of maximum is proportional to the square of the distance traveled (See Wibberenz, 1974). Also, the "steep" solar proton energy spectral slope results in most of the flux in a specific energy interval to be at the lowest energy being detected. Hence, for the time of maximum we compute the particle velocity for the lowest energy protons being detected. Since we also assume some scattering is applicable, we do not use the "scatter free approximation, but assume that the "average pitch angle is 60° to the interplanetary magnetic field direction, and hence the average "forward" speed is one-half the proton velocity around the interplanetary magnetic field line.

The time of maximum intensity $T_{(m)}$, can be calculated by:

$$T_{(m)} = 8 \, \theta^2 + 2 \, (\, 0.1333 \, D^2 / \beta \,). \qquad (8)$$

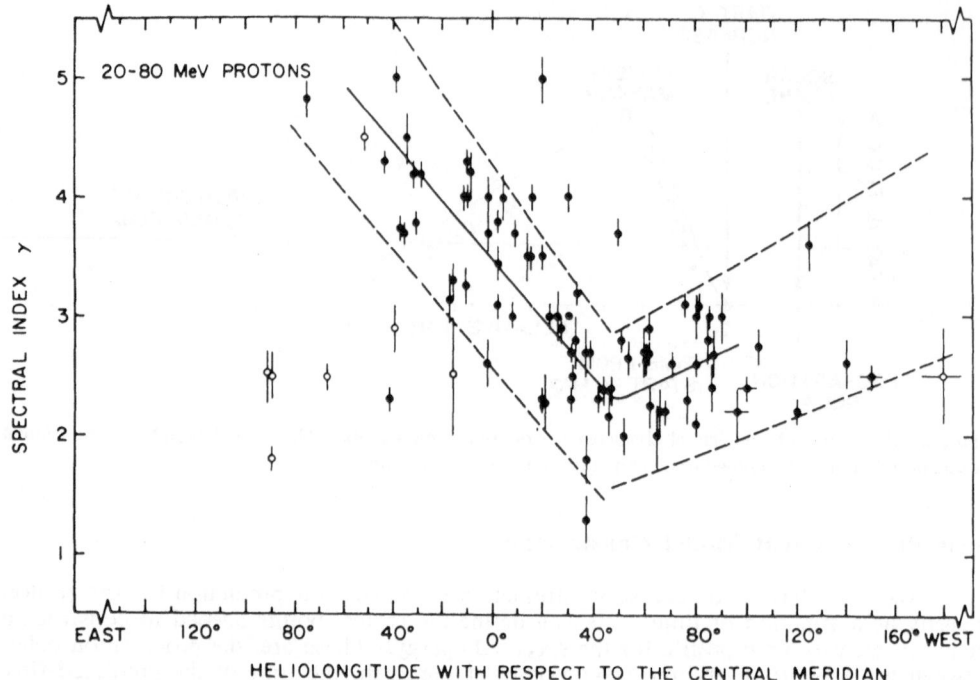

Figure 7. The variation of the solar proton spectral exponent in the 20 - 80 MeV energy range as a function of heliolongitude. The data points are from Van Hollebeke et al. (1975).

Solar proton maximum flux amplitude. The maximum proton flux expected is obtained by converting the selected electromagnetic emission parameters to predicted proton flux along the Archimedean spiral path leading from the solar flare position. By assuming an average coronal gradient of a factor of 10 per radian from the presumed particle source (i.e. the flare location) to the release point of solar protons "favorable" to the observation location (i.e. the "foot" of the Archimedean spiral of the interplanetary magnetic field line between the observation position and the sun) it is possible to estimate particle intensity.

Solar proton energy spectrum exponent prediction. There are few predictors of the slope of the solar proton spectrum. Some of the early work attempting this prediction from the maximum and minimum flux values in the radio peak power spectrum (Bakshi and Barron 1974, 1975, 1979) has not proven extremely reliable. The statistical tendencies of the exponent of the proton energy spectral slope noted by Van Hollebeke et al. (1975) shown in Figure 7 have demonstrated usefulness. These data can be ordered by the same parameter used to order the other solar proton flux parameters, the heliocentric angular distance from the solar flare location to the probable "root" of the Archimedean spiral path (interplanetary magnetic field lines) from the detection location to the sun. A prediction of the magnitude of the exponent of the solar proton differential energy spectral slope, γ is given by the equation:

$$\gamma = 2.7 \, [\, 1 + \theta \, / \, 2 \,] \qquad\qquad (9)$$

EXTENDING PROTON PREDICTED FLUXES TO HEAVIER ELEMENTS

The same principles involved for organizing and estimating the proton (ions with Z=1) arrival and time-intensity profiles are also applicable to heavy ions. It is reasonable to assume

Table 1. Normalized Elemental Abundances of Solar Particle Events

		Adams et al. (1981) Mason et al. (1980) ~ 1 MeV	Gloeckler (1979) 1-20 MeV	Cook et al. (1984) 10 MeV	McGuire et al. (1986) 6.7-15 MeV
1	H	1.0	1.0	1.0	1.0
2	He	2.2×10^{-2}	1.5×10^{-2}		1.5×10^{-2}
3	Li		1.0×10^{-7}	4.8×10^{-8}	2.8×10^{-6}
4	Be		1.5×10^{-7}	6.0×10^{-9}	1.4×10^{-7}
5	B		1.5×10^{-7}	1.2×10^{-8}	1.4×10^{-7}
6	C	1.6×10^{-4}	1.2×10^{-4}	9.6×10^{-5}	1.3×10^{-4}
7	N	3.8×10^{-5}	2.8×10^{-5}	2.7×10^{-5}	3.7×10^{-5}
8	O	3.2×10^{-4}	2.2×10^{-4}	2.2×10^{-4}	2.8×10^{-4}
9	F		4.3×10^{-7}	1.0×10^{-8}	1.4×10^{-7}
10	Ne	5.1×10^{-5}	3.5×10^{-5}	3.1×10^{-5}	3.6×10^{-5}
11	Na	1.6×10^{-6}	3.5×10^{-6}	2.6×10^{-6}	2.4×10^{-6}
12	Mg	4.8×10^{-5}	3.9×10^{-5}	4.3×10^{-5}	5.2×10^{-5}
13	Al	3.5×10^{-6}	3.5×10^{-6}	3.1×10^{-6}	3.3×10^{-6}
14	Si	3.8×10^{-5}	2.8×10^{-5}	3.5×10^{-5}	4.2×10^{-5}
15	P	2.3×10^{-7}	4.3×10^{-7}	1.7×10^{-7}	4.0×10^{-7}
16	S	1.8×10^{-5}	5.7×10^{-6}	7.8×10^{-6}	6.5×10^{-6}
17	Cl	1.7×10^{-7}		7.1×10^{-8}	
18	Ar	3.9×10^{-6}	8.7×10^{-7}	7.3×10^{-7}	4.6×10^{-6}
19	K	1.3×10^{-7}		1.0×10^{-7}	
20	Ca	2.3×10^{-6}	2.6×10^{-6}	3.1×10^{-6}	3.2×10^{-6}
21	Sc			7.8×10^{-9}	
22	Ti	1.0×10^{-7}		1.2×10^{-7}	
23	V			1.2×10^{-8}	
24	Cr	5.7×10^{-7}		5.0×10^{-7}	
25	Mn	4.2×10^{-7}		1.8×10^{-7}	
26	Fe	4.1×10^{-5}	3.3×10^{-5}	3.4×10^{-5}	
27	Co	1.0×10^{-7}		4.8×10^{-7}	
28	Ni	2.2×10^{-6}		1.2×10^{-6}	
29				1.4×10^{-8}	
30				3.8×10^{-8}	

that the same principles of coronal and interplanetary propagation apply to all ions independent of the mass or atomic charge. There is a major problem in anticipating the flux or fluence in finding a simple common factor for the elemental abundance ratios. There have been a number of papers reporting the variation of the elemental abundances in solar particle events; see Lin (1987) and Mason (1987) for recent reviews. A general summary is that "small" events may have the greatest variability in elemental composition. The elemental abundance ratios seem to have a slight variation according to the energy of the measurement. This may be a reflection of the "size" of the particle event since small particle events would not have many heavy ions at high energies. The hydrogen to helium ratios are the most variable even for "large" events; the heavier elemental abundance ratios seem to be in general agreement with the ratios expected from normal coronal material organized by the first ionization potential. Since the prediction parameters available convert from electromagnetic emission parameters to proton flux, it is necessary to extrapolate these proton predictions, by some expected ratio, to heavier elements. Initial estimates of these ratios have been done by Smart (1988) and are summarized in Table 1.

EXTRAPOLATION OF EARTH-BASED PREDICTION METHODS TO OTHER LOCATIONS IN SPACE

We assume that the maximum possible prompt solar proton flux would be at the position that is "well connected" to the solar flare source region. Using the intrinsic assumptions that the coronal particle intensity gradients control the particle flux observed around the sun it is possible to estimate the particle flux at any heliographic longitude. The arguments used for extrapolation of proton fluxes to other heliocentric distances rely on the assumption that the diffusion across magnetic field lines is negligible, and that the volume of the magnetic flux tube as the distance from the sun increases expands in the manner expected from classical geometry. In this case a power law function of the form R^{-3} can be used to extrapolate to other distances. Hamilton (1988) analyzed the probable effects of diffusion and suggested that $R^{-3.3}$ would be an appropriate factor.

Extrapolation at 1 AU Radial Distance to Other Heliocentric Angles

To extrapolate a prediction to other locations at 1 AU, it is necessary to use the Archimedean spiral and the coronal gradient concept. First, compute the longitude on the sun from which the interplanetary magnetic field line passing through the spacecraft position would originate. Then determine the heliocentric angular distance between the location of the solar flare and the solar longitude of the "root" of the idealized spiral field line passing through the spacecraft. Next multiply the coronal gradient per radian by the heliocentric angular distance between the two positions in order to estimate the flux diminution. Finally, multiply the peak proton flux expected at the "favorable" propagation path by this flux reduction factor.

Radial Dependence of Protons

The extrapolation arguments to other heliocentric distances rely on the volume of the magnetic flux tube behaving in a "classical" manner as the distance from the sun increases. If "classical" behavior is assumed, then a power law function can be used to extrapolate to other distances. Any distortion to the magnetic flux tubes are an unknown that we have no way of accurately estimating. Because of this there is no consensus view on the proper method for extrapolating solar particle fluxes and fluences from 1 AU to other distances in the heliosphere. The existing meager measurements are from comparisons of earth-orbiting satellite measured proton fluxes compared with space-probe measurements of the same event in the energy range of 10 to 70 MeV from 1 to 5 AU. These investigations of the radial dependence of the solar energetic particle flux have been done by Hamilton (1977, 1981, 1988) and Beeck et al. (1987).

Radial flux extrapolation from 1 AU. For distances greater than 1 AU extrapolate the expected proton flux at 1 AU using a functional form of $R^{-3.3}$ where R is the radial distance from the sun. This is the average solar proton radial gradient derived by Hamilton (1988) from a combination of Voyager and earth-satellite data. The limited measurements available suggest we should expect variations ranging from R^{-3} to R^{-4}. For distances less than 1 AU extrapolate the expected proton flux at 1 AU using a functional form of R^{-3}. Again, the limited measurements available suggest that variations ranging from R^{-3} to R^{-2} should be expected.

Radial fluence extrapolation from 1 AU. To extrapolate proton fluence from 1 AU to other distances in the heliosphere, use a functional form of $R^{-2.5}$ and expect variations ranging from R^{-3} to R^{-2}.

EXTRAPOLATION OF EARTH-BASED PREDICTION METHODS TO A MARS MISSION

In a mission to Mars, the radial distance will vary according to the spacecraft trajectory chosen. The shielding provided by the spacecraft must be consistent with constructing a minimum mass vehicle to minimize the exposure to galactic cosmic ray secondaries generated in the vehicle, and yet provide adequate protection against a very large solar proton event. As discussed elsewhere in this volume, approximately 15 g cm^{-2} of shielding appear to be sufficient to reduce the dose of a "major" solar proton event to a "tolerable level". (A much more detailed discussion of this issue is given by Wilson et al., 1991.)

The solar proton particle flux is expected to vary as a power law with radial distance from the sun. As discussed in the previous section, a power law exponent of -3 would be expected from magnetic flux tube geometry. Since the radial distance to Mars is ~1.5 AU, then the flux at the orbit of Mars would be expected to be about 1/3 of the flux at 1.0 AU along the same spiral path. This variation should be contrasted with the average helio-longitudinal gradient of one order of magnitude per radian of heliocentric angular distance. A consideration of these expected variations suggest that the proton prediction problem for Mars is not dramatically different from the earth. Sensors on-board the spacecraft viewing in the optical, radio and soft X-ray wavelengths should be able to provide useful prediction information.

The probability of a "surprise", a solar proton event being detected when there is no visible preceding solar activity, is significantly larger for the Mars radial distance. At the earth, about 20% of the recorded solar proton events are not associated with visually observed solar flares. It is presumed that the origin of "major" proton events not associated with visual solar flares have their source from solar activity from behind the western limb of the sun as viewed from the earth. See Figure 15 of Shea and Smart (1992) for assumed source location on the sun for solar cosmic ray events from 1956. This same type of distribution is present for major non-relativisitic solar proton events. Similarly, for the position of Mars, we would expect that 1/2 of the detected solar proton events would have their source on the portion of the sun that is not observable from Mars. Consider the probable "favorably connected" heliolongitude for Mars. At 1.5 AU distance the sun-Mars transit time for a 400 km sec^{-1} solar wind would be about 6.5 days. During that time the sun would have rotated ~86°. This is essentially at the western limb of the solar disk visible from Mars. Assuming that the solar proton flare distribution is symmetrical in heliolongitude, then approximately 1/2 of the source solar proton flares cannot be observed from the Mars orbital distance.

This situation strongly argues for on-board particle and radiation sensors on a Mars mission. If we consider the intensity-time profile of a solar particle event, then the critical factor is the time from event onset to "hazardous" radiation levels. Depending on the propagation conditions even for "well connected" events, this is likely to be of the order of an hour. The most "dangerous" particle radiation will be the ions that penetrate the shielding and stop in blood forming organs (thus depositing most of their energy in these organs). These will be the protons between 70 and 150 MeV assuming that there is ~5 g cm^{-2} of shielding provided by the body structure. The typical intensity-time profile observed in this energy range at 1 AU and expected at 1.5 AU provides for about an hour from particle onset until the maximum proton flux will be observed. We suggest that prudent mission planning would allow for movement of personnel to a more heavily shielded area or the re-distribution of mass on this time scale.

SUMMARY

We have discussed the procedures we use to model the intensity-time profile of solar protons expected in space after the occurrence of a significant solar flare on the sun. The particle flux detected at any point in space is a function of the location of the flare on the sun with respect to the detection position. The procedures and techniques used for predicting solar proton fluxes at the earth can be extrapolated to help predict solar particle fluxes at other locations in space including the orbit of Mars.

REFERENCES

Adams, J.H., Jr., Silberberg, R., and Taso, C.H., 1981, "Cosmic Ray Effects on Microelectronics, Part I: The Near Earth Particle Environment", NRL Memorandum Report 4506, Naval Research Laboratory, Washington, D.C.

Akun'yan, S.T., Fomichev, V.V., and Chertok, I.M., 1977, Determination of the parameters of solar protons in the neighborhood of the earth from radio bursts, 1 Intensity functions, Geomag. and Aeron., 17: 5.

Akinyan, S.T., Fomichev, V.V., and Chertok, I.M., 1979, Quantitative forecasts of solar protons based on solar flare radio data, 3:D14, in: "Solar-Terrestrial Prediction Proceedings", R.F. Donnelly, ed., U. S. Department of Commerce, NOAA/ERL, Boulder, Colorado.

Axford, W.I., 1965, Anisotropic diffusion of solar cosmic rays, Planet. Space Sci., 13: 1301.

Bakshi, P., and Barron, W., 1974, Spectral Correlations Between Solar Fare Radio Bursts and Associated Proton Fluxes, I, AFCRL-TR-74-0508, Air Force Cambridge Research Laboratories, Hanscom Air Force Base, Massachusetts.

Bakshi, P., and Barron, W.R., 1975, Spectral Correlation Between Solar Flare Radio Bursts and Associated Proton Fluxes, II, AFCRL-TR-75-0579, Air Force Cambridge Research Laboratories, Hanscom Air Force Base, Massachusetts.

Bakshi, P., and Barron, W.R., 1979, Prediction of solar flare proton spectral slope from radio burst data, J. Geophys. Res., 84: 131.

Beeck, J., Mason, G.M., Hamilton, D.C., Wibberenz, G., Kunow, H., Hovestadt, D., and Klecker, B., 1987, A multispacecraft study of the injection and transport of solar energetic particles, Astrophys. J., 322: 1052.

Barouch, E., Gros, M., and Masse, P., 1971, The solar longitude dependence of proton event delay, Sol. Phys., 19: 483.

Burlaga, L.F., 1967, Anisotropic diffusion of solar cosmic rays, J. Geophys. Res., 72: 4449.

Castelli, J.P., Aarons, J., and Michael, G.A., 1967, Flux density measurements of radio bursts of proton-producing flares and nonproton flares, J. Geophys. Res., 72: 5491.

Castelli, J.P., and Guidice, D.A., 1976, Impact of current solar radio patrol observations, Vistas in Astronomy, 19: 355.

Chertok, I.M., and Fomichev, V.V., 1984, Development of the quantitative proton flare diagnostics technique by radio burst data, p. 270 in:"Solar-Terrestrial Prediction Proceedings: Proceedings of a Workshop at Meudon, France, June 18-22, 1984", P.A. Simon, G. Heckman and M.A. Shea, ed., U. S. Department of Commerce, NOAA/ERL, Boulder, Colorado.

Cliver, E.W., 1976, Parent Flare Emission at 2. 8 GHz As A Predictor of the Peak Absorption of Polar-Cap Events, NELC-TR-2015, Naval Electronics Laboratory, San Diego, California.

Cliver, E.W., Secan, J.A., Beard E.D., and Manley, J.A., 1978, Prediction of solar proton events at the Air Force Global Weather Central's Space Environment Forecasting Facility, p. 393, In: "Proceedings of the NRL Symposium on the Effect of the Ionosphere on Space and Terrestrial Systems", Naval Research Laboratory, Washington D. C.

Cliver, E.W., McNamara, L.F., and Gentile, L.C., 1985, Peak flux density spectra of large solar radio bursts and proton emission from flares", J. Geophys. Res., 90: 6251.

Cook, W.R., Stone, E.C., and Vogt, R.E., 1984, Elemental composition of solar energetic particles, Astrophys. J., 297: 827.

Gloeckler, G., 1979, Composition of energetic particle population in interplanetary space, Rev. of Geophys., 17: 569.

Gold, R.E., Roelof, E.C., Nolte, J.T., and Krieger, A.S., 1975, Relation of large-scale coronal x-ray structure and cosmic rays: 5,. Solar wind and coronal influence on a Forbrush decrease lasting one solar rotation, Proc. 14th International Cosmic Ray Conference, 3: 1095.

Hamilton, D.C., 1977, Radial transport of energetic solar flare particles from 1 to 6 AU, J. Geophys. Res., 82: 2159.

Hamilton, D.C., 1981, Dynamics of solar cosmic ray bursts at large heliocentric distances ($\geq$ 1 AU), Adv. Space. Res., 1: 25.

Hamilton, D.C., 1988, The radial dependence of the solar energetic particle flux, in: "Proceedings of the JPL Workshop in the Interplanetary Charged Particle Environment", 1: 86, J. Feynman and S. Gabriel ed., NASA JPL Publication 88-28, Jet Propulsion Laboratory, Pasadena, California.

Kahler, S.W., 1982, The role of the big flare syndrome in correlations of solar energetic protons and associated microwave parameters, J. Geophys. Res., 87: 3439.

Krimigis, S.M., 1965, Interplanetary diffusion model for the time behavior of intensity in a solar cosmic ray event, J. Geophys. Res., 70: 2943.

Kuck, G.A., Davis, S.R., and Krause, G.J., 1971, Prediction of Polar Cap Absorption Events, AFWL-TR-71-1, Air Force Weapons Laboratory, Kirtland AFB, New Mexico.

Kuck, G.A., and Hudson, S., 1990, Prediction of solar proton fluxes from X-ray Signatures, 1: 422, in: "Solar-Terrestrial Prediction Proceedings: Proceedingsofa Workshop at Leura, Australia, October 16-20, 1989", R.J. Thompson, D.G. Cole, P.J. Wilkinson, M.A. Shea, D.F.Smart, and G.R. Heckman, Eds., U. S. Department of Commerce, NOAA/ERL, Boulder, Colorado.

Lantos, P., 1990, Evaluation of proton events prediction, 1:487, in: "Solar-Terrestrial Prediction Proceedings: Proceedings of a Workshop at Leura, Australia, October 16-20, 1989", R.J. Thompson, D.G. Cole, P.J. Wilkinson, M.A. Shea, D.F.Smart, and G.R. Heckman, Eds., U. S. Department of Commerce, NOAA/ERL, Boulder, Colorado.

Lanzerotti, L.J., 1973, Coronal propagation of low-energy solar protons, J.Geophys. Res., 78: 3942.

Lin, R.P., 1987, Solar particle acceleration and propagation, Reviews of Geophysics, 25: 676.

Mason, G.M., 1987, The composition of galactic cosmic rays and solar energetic particles, Reviews of Geophysics, 25: 685.

Mason, G.M., Fisk, L.A., Hovestadt, D., and Gloeckler, G., 1980, A survey of ~1 MeV nucleon-1 solar flare particle abundances, $1 < Z < 26$, during the 1973-1977 solar minimum period, Astrophys. J., 239: 1070.

McCracken, K.G., and Rao, U.R., 1970, Solar cosmic ray phenomena, Space Science Reviews, 11: 155.

McCracken, K.G., Rao, U.R., Bukata, R.P., and Keath, E.P., 1971, The decay phase of solar flare events, Sol. Phys., 18: 100.

McGuire, R.E., van Hollebeke, M.A.I., and Lau, N., 1983, A multi-spacecraft study of the coronal and interplanetary transport of solar cosmic rays, Proc. 18th International Cosmic Ray Conference, 10: 353.

Morfill, G., Volk, H., and Lee, M.A., 1976, On the effect of directional medium-scale interplanetary variations on the diffusion of galactic cosmic rays and their solar cycle variations, J. Geophys. Res., 81: 5841.

McGuire, R.E., Von Rosenvinge, T.T., and McDonald, F.B., 1986, The composition of solar energetic particles, Astrophys. J., 301: 938.

Nolte, J.T., and Roelof, E.C., 1973, Large scale structure of the interplanetary medium, 1: High coronal structure and the source of the solar wind, Solar Physics, 33: 241.

Mel'nikov, V.F., Podstrigach, T.S., Dajbog, E.I., Logachev, Yu.I., and Stolpovskij, V.G., 1990, 1: 533, in: "Solar-Terrestrial Prediction Proceedings: Proceedings of a Workshop at Leura, Australia, October 16-20, 1989", R.J. Thompson, D.G. Cole, P.J. Wilkinson, M.A. Shea, D.F.Smart, and G.R. Heckman, Eds., U. S. Department of Commerce, NOAA/ERL, Boulder, Colorado.

Parker, E.N., 1965, The passage of energetic charged particles through interplanetary space, Planet. Space Sci., 13: 9

Reid, G.C., 1964, A diffusive model for the initial phase of a solar proton event, J. Geophys. Res., 69: 2659.

Reinhard, R., Roelof, E.C., and Gold, R.E., 1986, Separation and analysis of temporal and spatial variations in the 10 April 1969 solar flare particle event, in: "The Sun and the Heliosphere in Three Dimensions", p. 297, R.G. Marsden, ed., Proceedings of the XIX ESLAB Symposium, Astrophysics and Space Science Library, D. Reidel Publishing Co., Dordecht.

Reinhard, R., and Wibberenz, G., 1974, Propagation of flare protons in the solar atmosphere, Sol. Phys., 36:473.

Roelof, E.C., 1973, New aspects of interplanetary propagation revealed by 0. 3 MeV solar proton events in 1967, p. 411, in: Solar-Terrestrial Relations,University of Calgary, Canada.

Roelof, E.C., 1975, Scatter-free collimated convection and cosmic-ray transport at 1 AU, Proc. 14th International Cosmic Ray Conference, 5:1716.

Roelof, E.C., 1976, Solar Particle Emissions, in: "Physics of Solar Planetary Environments", p. 214, D.J. Williams, ed., American Geophysical Union, Washington, D. C., USA.

Roelof, E.C., Gold, R.E., Krimigis, S.M., Krieger, A.S., Nolte, T.J., McIntosh, P.S., Lazarus, A.J., and Sullivan, J.D., 1975, Relation of large-scale coronal x-ray structure and cosmic rays: 2, Coronal control of interplanetary injection of 300 keV solar protons, Proc. 14th International Cosmic Ray Conference, 5: 1704.

Roelof, E.C., and Krimgis, S.M., 1973, Analysis and synthesis of coronal and interplanetary energetic particle, plasma and magnetic field observations over three solar rotations, J. Geophys. Res., 78: 5375.

Shea, M.A., and Smart, D.F., 1992, History of energetic solar protons for the past three solar cycles including Cycle 22 update, in "Biological Effects and Physics of Solar and Galactic Cosmic Radiation", C.E. Swenberg, G. Horneck and E. G. Stassinpoulos, ed., this volume.

Smart, D.F., 1988, Predicting the arrival times of solar particles, in:"Proceedings of the JPL Workshop in the Interplanetary Charged Particle Environment", J. Feynman and S. Gabriel, ed., NASA JPL Publication 88-28, 1: 101, Jet Propulsion Laboratory, Pasadena, California.

Smart, D.F., and Shea, M.A., 1979, PPS76 - a computerized "event mode" solar proton forecasting technique, 1:406, in: "Solar-Terrestrial Prediction Proceedings", R.F. Donnelly, ed., U. S. Department of Commerce, NOAA/ERL, Boulder Colorado.

Smart, D.F., and Shea, M.A., 1985, Galactic cosmic radiation and solar energetic particles, Chapter 6 in: "Handbook of Geophysics and the Space Environment", A.S. Jursa ed., Air Force Geophysics Laboratory, Bedford, MA.

Svestka, Z., 1976, "Solar Flares", Volume 8, Geophysics and Astrophysics Monographs, D. Reidel Publishing Co., Dordrecht, Holland.

Valdes-Galica, J.F., 1991, Transport of energetic particles in the interplanetary medium, in "Solar and Galactic Cosmic Rays, Proceedings of the 12th European Cosmic Ray Symposium", P.R. Blake and W.F. Nash, ed., Nuclear Physics B (Proc. Suppl.) 22B:46, North Holland, Amsterdam.

Van Hollebeke, M.A.I., Ma Sung, L.S., and McDonald, F.B., 1975, The variation of solar proton energy spectra and size distribution with heliolongitude, Sol. Phys., 41: 189.

Wibberenz, G., 1974, Interplanetary magnetic fields and the propagation of cosmic rays, J. Geophys., 40: 667.

Wilson, J.W., Townsend, L.W., Schimmerling, W., Khandelwal, G.S., Ferdous, K., Nealy, J.E., Cucinotta, F.A., Simonsen, L.C., Shinn, J.L., andNorbury, J.W., 1991, Application to space exploration, Chapter 12 in "Transport Methods and Interactions for Space Radiation", NASA Reference Publication 1257, NASA, Washington, D. C.

Zwickl, R.D., and Webber, W.R., 1977, Solar particle propagation from 1 to 5 AU, Sol. Phys., 54: 457.

SOLAR ENERGY AND ITS INTERACTION WITH EARTH'S ATMOSPHERE

Yurdanur Tulunay

Department of Aeronautical Engineering
Middle East Technical University
ODTÜ, Ankara, Turkey

ABSTRACT

The Sun is responsible for many of the phenomena on Earth, including the maintenance of life. In addition, magnetic storms, capable of disrupting radio communication, and auroral displays are associated with solar events. Man-made electrical, satellite, and communication systems are affected strongly by the near-Earth space environments. The purpose of this paper is to review briefly the interaction of solar activity with the near-Earth environment. These processes can be studied by examining two sets of interactions. That is, the interaction of the solar electromagnetic output with the Earth's neutral atmosphere, and the solar corpuscular output with the geomagnetic field. In order to understand the types of interactions one needs to know more details of the interacting components. Therefore, the near-Earth environments which comprise neutral atmospheric, ionospheric and magnetospheric regions will be discussed in relation to the direct and indirect influences of solar activity.

INTRODUCTION

Since the beginning of this century there has evolved a picture of two separate types of solar energy: the electromagnetic energy which is almost constant and photoionize the Earth's upper atmosphere over the whole dayside of the Earth; whereas the other is the corpuscular energy which is emitted from the sun and interacts with the Earth via the geomagnetic field by day and night but only around the high latitudes produces the aurorae. The corpuscular radiation contains energetic particles, solar wind plasma etc... In order to understand the interactions of the solar activity with the near-Earth environment one needs to know more about the interacting components. Figure 1 shows the spectral distribution of solar energy. It covers from the frequencies and wavelengths between 1020 Hz, or 10^{-3} A° and hundreds of MHz or tens of meters. 70% of the solar energy lies between near ultraviolet, visible, and near infrared, corresponding wavelengths of 0.32 μm to 1 μm. 2% of the solar energy appears at ultraviolet and X-ray wavelenths shorter than 0.32 μm. The remainder of the solar energy appears at infrared and radio wavelenths longer than 1.0 μm. The solar energy corresponding to wavelengths shorter than 0.32 μm is strongly absorbed in the upper atmosphere where it drives several important photochemical reactions, including ozone production. Much of the energy corresponding to wavelengths longer than 0.32 μm is strongly absorbed by atmospheric water vapor and carbon dioxide and thus contributes to the energy budget of the lower atmosphere (see Altrock et al. 1985). The height to which solar radiation with different wavelengths penetrate before being attenuated by a factor (exp)-1 is illustrated in Figure 2. It is evident that at altitudes greater than about 100 km most of the energetic radiation is absorbed , resulting in photoionization, to form atomic oxygen. The

Biological Effects and Physics of Solar and Galactic Cosmic Radiation,
Part B, Edited by C.E. Swenberg *et al.,* Plenum Press, New York, 1993

119

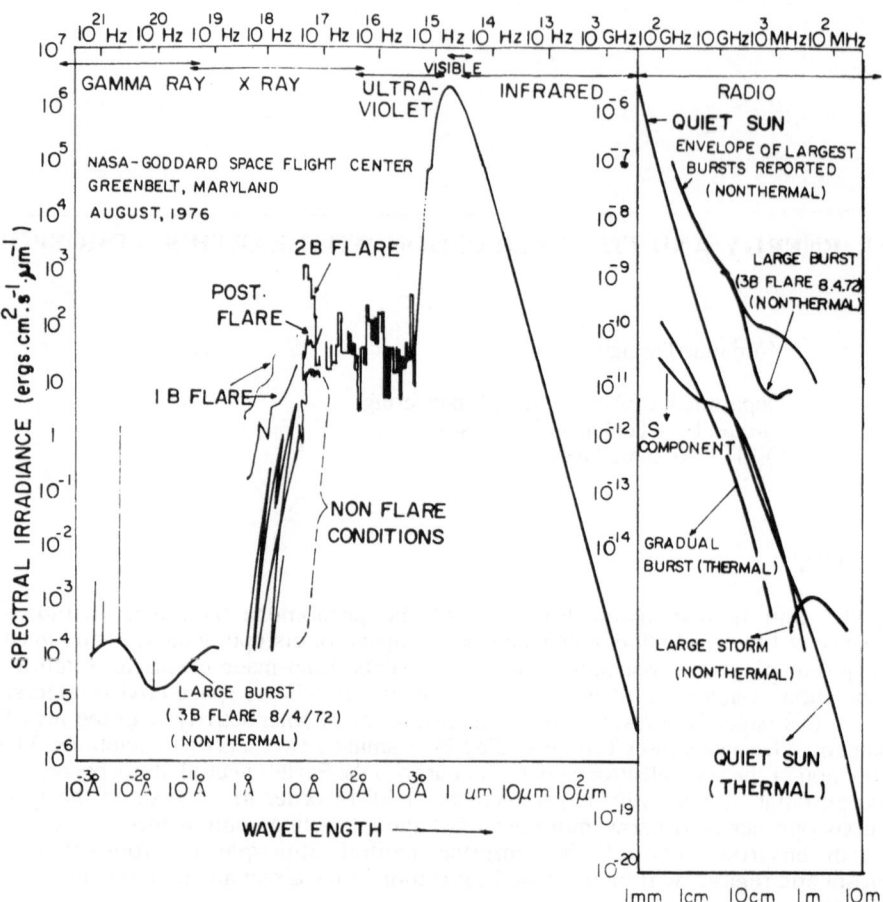

Figure 1. Spectral Distribution of Solar Electromagnetic Radiation (or Energy) (from Altrock, R.C., et al., 1985).

neutral temperature increases up to 1500K or 2000K depending on the solar cycle. Between about 50km and 100km there is little radiation absorbed. At altitudes less than 50km very little energetic solar radiation can be absorbed and the ozone formation is the most important event at those altitudes. The layer warms up at around 50 km altitude. The next figure (Figure 3) represents a more complete picture of the different sources which contribute to the ionization of the upper atmosphere (Richmond, 1987). At higher altitudes (>200km) where molecular species become less and less in number concentration photochemical reactions become less probable. Instead, transport phenomena become more compatible. For example, diffusion of plasma, electromagnetic drifts, neutral winds etc. can be named as the major transport processes. At altitudes where photoionization competes with transport phonomena there forms a peak in ionization between 200-450km altitude.

Figure 4 is the height profile of electron density at the ionosphere. The ionosphere is the region of the Earth's atmosphere where the propagation of radio waves can be affected by the presence of free electrons. The ionosphere can be viewed as a variable shell of plasma surrounding the Earth(Richmond, 1987). To facilitate a comparison, the neutral gas profile is also illustrated by a dashed line in Figure 4. Below the main peak there often appear density maxima, sometimes distinct enough to form secondary peaks. Historically, the earliest radar

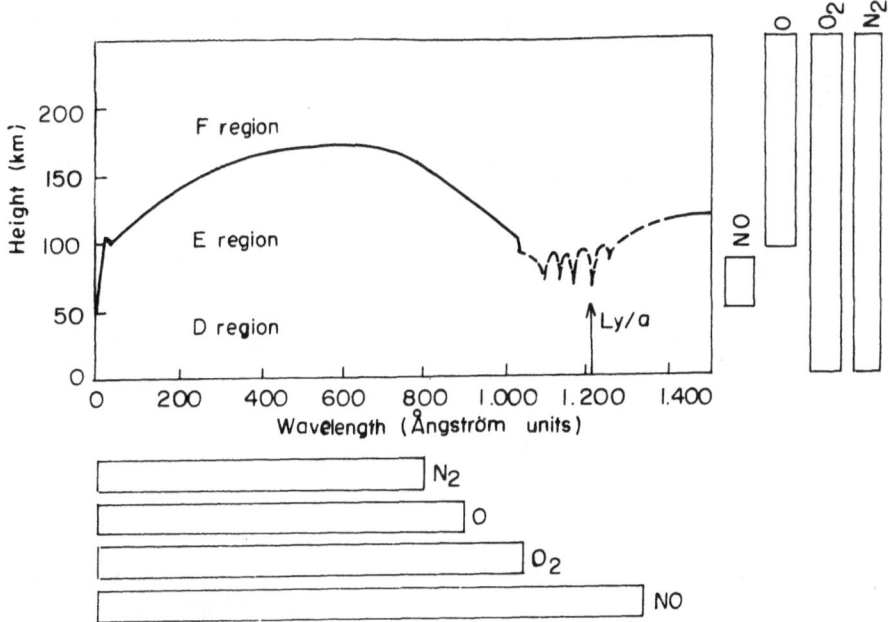

Figure 2.   Altitude at which the Intensity of Solar Radiation Drops to 1/exp of its Value Outside the Earth's Atmosphere, for Vertical Incidence (from Ratcliffe, 1970).

measurements detected reflections from around 100 km. The ionosphere at this height was named the "E Layer" (E for electric). For alphabetical continuity, the ionosphere around the main peak higher up was named the "F Layer". Since a secondary "bump" in the profile sometimes appears in the lower part of the F layer, this layer is sometimes divided into an "F_1 layer" (the lower "bump"), and an "F_2 layer" (the main peak). Equivalently, we speak of F, F_1 and F_2 "regions" rather than "layers". Below the E region is a highly variable region of lesser electron density called the D region, around 60 to 90km in altitude. The electron density in all of these regions varies with time of day, altitude, season, strength of solar ionizing radiation, and level of magnetospheric activity (Richmond, 1987). Typical electron densities (or plasma densities) in the ionosphere range from 10^3 to 10^6 cm^{-3} (as compared with neutral gas densities about 10^7 to 10^{16} cm^{-3}). The ionosphere extends into the protonosphere where the major constituent is the singly ionized hydrogen atoms, H+. Figure 5 shows how the electron density, Ne, total ion density Ni, H+ and 0+ vary in invariant magnetic latitude, at two particular altitudes of two satellites, 0G0-4 and Ariel 3. In this figure the transition between 0+ (reflecting Ne) and H+ is also well illustrated (Taylor and Tulunay, 1973).

Radio waves are refracted or reflected by the ionospheric plasma. Of particular importance is the peak density of the ionospheric plasma. The largest frequency of a radio wave which the F- layer reflects back to a vertical incidence sounder is called "critical frequency". The critical frequency fc is related to the F-region peak, electron density, Nmax, by

$$fc = 9.0 \sqrt{Nmax} \text{ (electr./m}^3).$$

Radio waves of sufficiently high frequency, greater than fc, can pass through the ionosphere. A vertically transmitted radio wave with a frequency below fc will normally be reflected by the ionosphere. Because of the reflective properties of the ionosphere, radio waves from very low frequency (VLF) to high frequency (HF) (broadly 3kHz to 30MHz) can be propagated to great distances. Figure 6 shows vertical profiles of electron density and how radio wave propagation is influenced by the ionosphere. As illustrated in Figure 6, at sufficiently low radio frequencies, the ionosphere can be used for long-distance communication. In the short-wave band, signals are reflected from the F region or from the

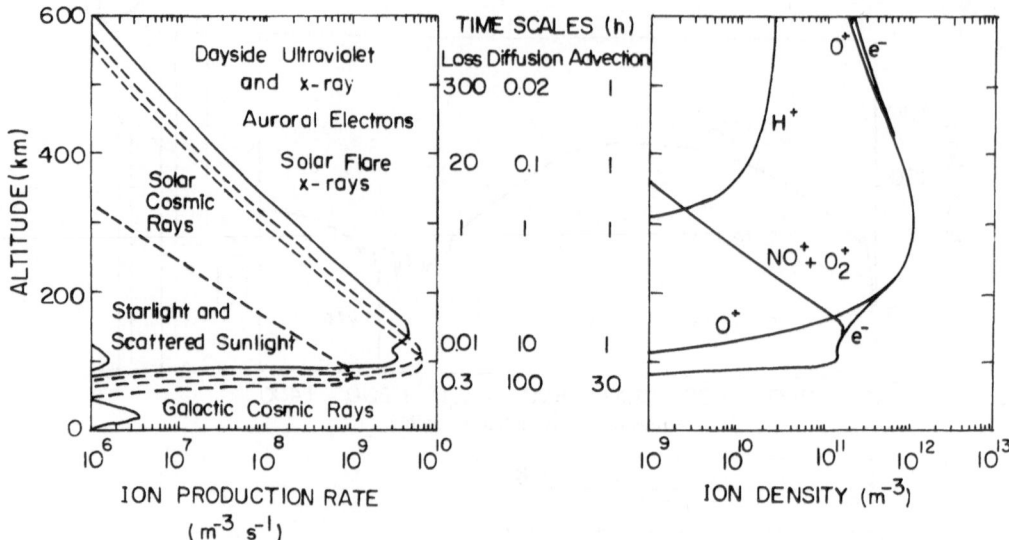

Figure 3. (Left) Different Sources Contribute to the Ionization of the Upper Atmosphere. Solid Lines Show Regular Sources. During the Day, Direct Solar Extreme Utraviolet (EUV) Light is the Main Ionization Source. At Night, EUV Scattered from the Geocorona as Well as that from Stars Helps Maintain the E-regionionosphere. Other Sources are Highly Variable in Tme, and are Shown with Dashed Lines: Solar Flare X rays and EUV (Day Side only) and Auroral Electrons and Solar Cosmic Rays (High Latitudes only). (Middle). The Table of Representative Time Scales for Different Physical Processes Effecting the Daytime Ionosphere Shows the Relative Importance of these Processes at Various Altitude. The Smaller A Time Scale, the More Rapidly that Process Acts. The Shortest Time Scale at A Given Altitude therefore Points to the Dominant Process. For Example, in the E Region Ions are Lost Rapidly so they are Influenced Little by Diffusion and Wind Transport. In the Upper Ionosphere Ions Diffuse Rapidly and tend to Approach a State of Hydrostatic Equilibrium, with Density Falling of Exponentially with Altitude. (Right). The Height Profiles of Ion and Electron Densities are A Result of the Various Processes Affecting the Ionosphere: Production, Charge Exchange, Loss, Diffusion, and Transport. O^+ Ions are Dominant in the Fregion, while NO^+ and O_2^+ are Dominant in the E Region. At Very High Altitudes H^+ ions (i.e., Protons) Eventually Dominate (Richmond, 1987).

dayside E region, while at lower frequencies they are reflected from the D or lower E regions. Waves are partly absorbed by the ionosphere as the vibrated electrons collide with air molecules. Because air is most dense at the lower ionosphere, wave absorption in the D region can be particularly important especially when the electron density increases sharply during solar flares or solar cosmic ray events. In addition to sporadic absorption, ionospheric radio propagation is also degraded by fading and interference caused by temporal and spatial ionospheric variations. Higher-frequency satellite communications overcomes most of these problems since they are little affected by ionospheric refraction and absorption, but they nonetheless are subject to signal distortion and modulation due to small-scale ionospheric irregularities (Richmond, 1987).

The sun is also the source of the solar wind and energetic particles. The solar wind is a "wind" not a static phenomena. In simple terms, it is the extension of the solar atmosphere into the interplanletary medium. It is electrically highly conducting and low density (5-10 particles /cm³), low energy (10-200eV) plasma. The solar wind mainly consists of protons, H+, and electrons with speeds between 250 km/s to 2000 km/s. Near the orbit of Earth its average density is about 8 ion-electron pairs per cubic centimeter (Neugebauer, 1991). The dynamic pressure of the solar wind is greater than its magnetic pressure. Therefore, one may envisage that, as the solar wind plasma is being ejected from the solar atmosphere it also imprisons the solar magnetic field, and thus the solar wind introduces the interplanetary magnetic field in the interplanetary space. Depending on where it originates on the sun the solar wind can be a "fast"-"quasi stationary" wind; or it can be a "slow" - "transient" wind.

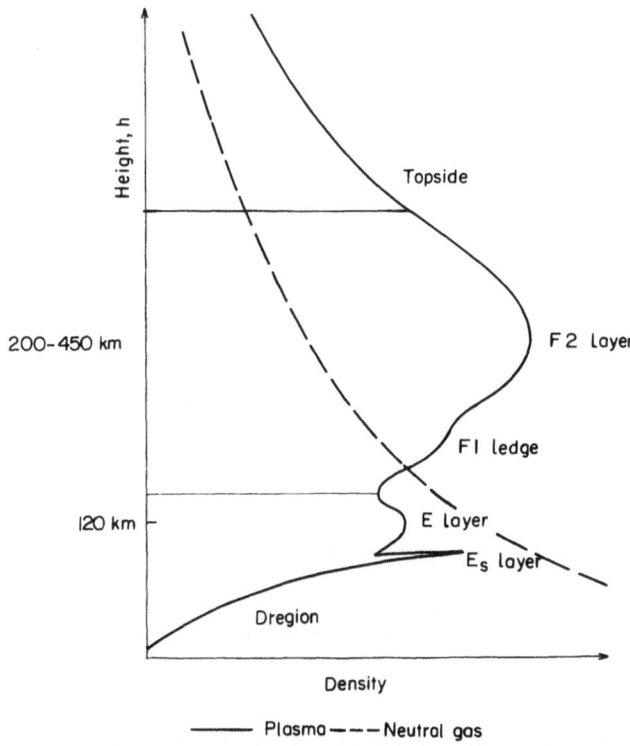

Figure 4. A Typical Ionospheric Electron Density Height Distribution (Vertical Profie). Different Altitude Regions of the Ionosphere are Labeled D.E.F1 and F2. (Diagram: Rutherford Appleton Laboratory)

The fast solar wind originates from the solar coronal holes. From coronal holes single polarity, weak solar magnetic fields open out into the interplanetary space. The boundary between out going field and incoming field is called sector boundary (Neugebauer, 1991). On the other hand, the slow solar wind originates from the more active regions of the Sun. The slow wind is highly variable (Dryer, 1987). The solar rotational velocity couples with the outward velocity of the solar wind producing the Archimedean spiral pattern in the interplanetary magnetic field (Figure 7; Gorney, 1990). It is now common knowledge that the key link between the solar atmosphere and the Earth system is the solar wind (Schwenn, 1990). Therefore the solar wind travels in space at supersonic speeds. When the supersonic solar wind encounters an obstacle, a shock wave is produced. Due to the interaction between the interplanetary and geomagnetic fields the geomagnetic field is pushed into the dayside, and it is swept backwards from the nightside to form a long tail which extends hundreds of Earth radii; well beyond lunar orbit (Gorney, 1990). The interaction between the solar wind and the geomagnetic field is maximal when the southward component of the interplanetary magnetic field attracts the northward component of the geomagnetic field (Ely, 1984; Gorney 1990; Heikkila, 1990). Such merging forces the solar wind particles into the Earth system (Russell,1987; Schwenn, 1990; Rodger, 1990). Thus, as a result of this interaction the solar wind particles can directly enter, dump their kinetic energy or trigger the relase of different energies which are stored at other places within the magnetosphere. As a consequence geomagnetic activity may increase. The high latitude ionosphere is strongly linked to the magnetosphere by the geomagnetic field, and the variations that take place in that region are mainly driven by geomagnetic activity whose origins lie within the magnetosphere. The typical signatures of geomagnetic activity are the geomagnetic storms, visible aurorae, ionospheric disturbances, and others (Gorney, 1990).

The shocked solar wind flows around the magnetosphere in the region called the magnetosheath. A small fraction of the solar wind plasma enters the magnetosphere through

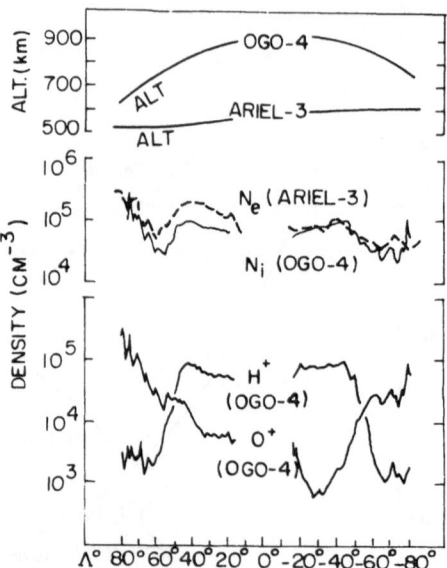

Figure 5. Comparison of Ion Composition Results Obtained from 0G0-4 with Electron Density Measurements from Ariel-3, Using Orbits Nearly Coincident in Local Time and Season. (Taylor and Tulunay,1973). Sept. 10,1967   OGO-4- 1512-1602UT ARIEL-3 1357-1441UT.

the polar cusp. Some of this entering plasma forms a boundary layer called the plasma mantle, and some of it drifts to a region of the neutral point where it is accelerated and forms the plasma sheet (Russell, 1987). It has been estimated that only 1% of the incident solar wind

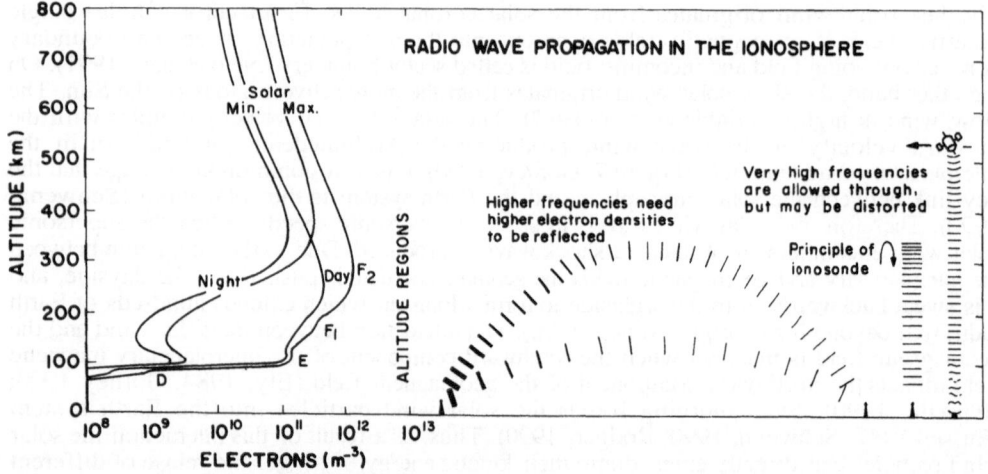

Figure 6. (Left). Typical Midlatitude Ionospheric Electron Density Profiles for Sunspot Maximum and Minimum, Day and Night. Different Altitude Regions of the Ionosphere are Labeled D,E,F1 and F2. (Right). Radio Wave Propagation is Influenced by the Ionosphere. At Frequences Below about 30 MHz (Radio Wavelengths Longer than 10m) Waves can be Reflected back to Earth. The Higher the Frequency, the Greater the Electron Density Required to Reflect the Signal. Higher Frequencies Tend to be Reflected from the F Region, while Lower Frequencies are Reflected from the E and D Regions. Part of the Wave Energy is Absorbed in Passing through the Ionosphere. Transmissions to and from Satellites Use Much Higher Frequencies, which are Deflected Only Slightly by the Ionosphere. These Transmission can Nevertheless be Distorted by Ionospheric Irregularities.(from Richmond,1987,1987; Diagram:Rutherford Appleton Laboratory).

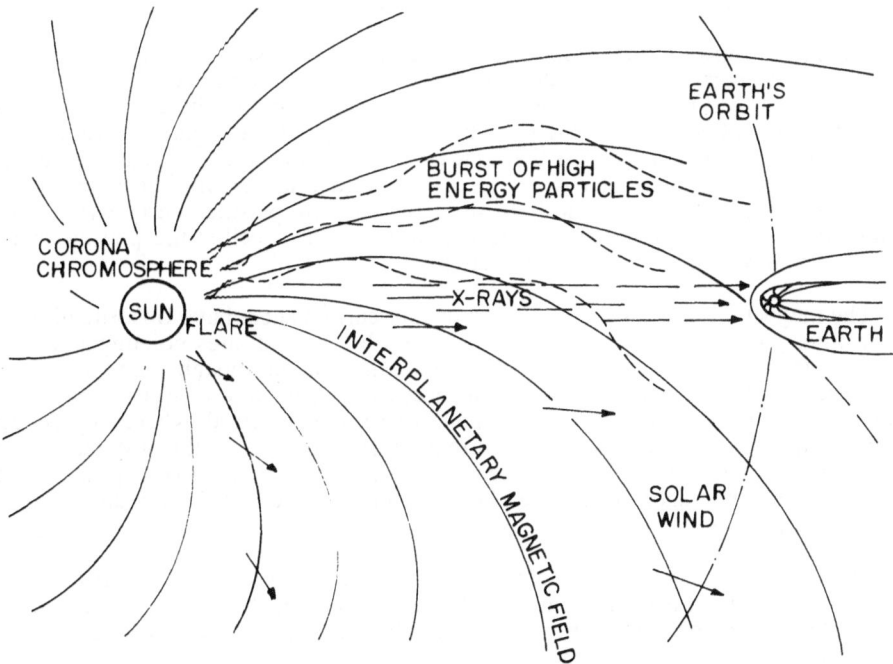

Figure 7. Polar View of Interplanetary Space in the Ecliptic Plane, Showing the Propagation of Solar X rays, Solar Energetic Particles, and the Solar Wind from the Sun to the Vicinity of Earth. The Sun's Rotational Velocity Coupled with the Outward Velocity of the Solar Wind produces the Archimedean Spiral Pattern of the Interplanetary Magnetic Field (From Gorney, 1990).

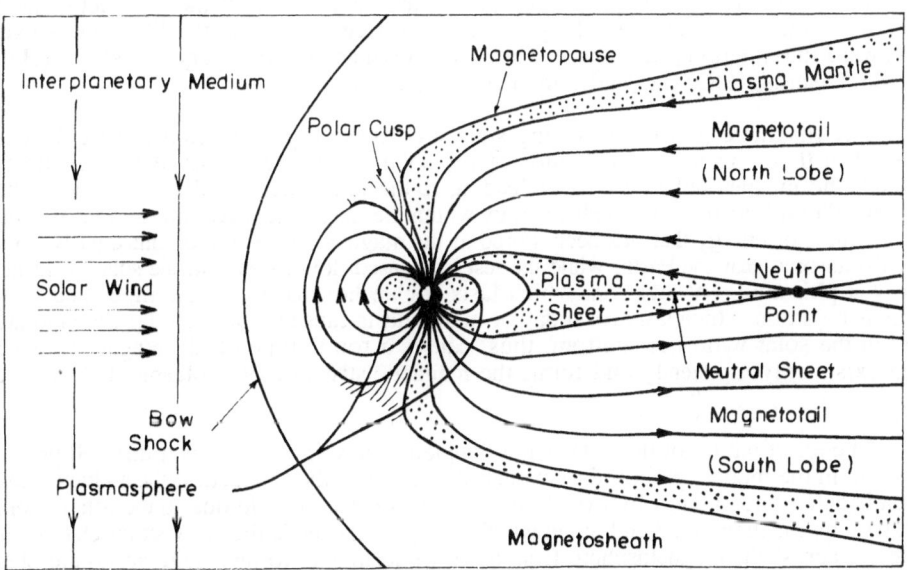

Figure 8. A Simplified Cross-section of the Noon-midnight Meridian of the Terrestrial Magnetosphere showing Schematically the Interaction of a Southward Interplanetary Magnetic Field with the Geomagnetic Field (from Russell, 1987).

energy flux is ultimately dissipated within the magnetosphere (Gorney, 1990).

Figure 8 shows a noon-midnight meridian cross section of the magnetosphere and labels its different morphological feautures (Russell, 1987). The polar cusps are two regions at the interface between the day and night, one north, and one south, where the magnetic field lines form a funnel shaped geometry. The solar wind plasma gains entrance to the magnetosphere at the polar cusps. As the solar wind flows downstream to the plasma mantle it is subjected to an ExB drift force which forces it toward a neutral sheet at the mid-plane of the tail where there is no magnetic field since the flowing plasmas carry with them the opposing, frozen in magnetic fields. The neutral sheet, also called the plasma sheet, is occupied by a current made up ions and electrons.

The magnetic fields on the two sides of this current sheet have opposite directions. The field lines pointing sunward arrive from the north lobe and lines pointing antisunward arrive from the south lobe. A situation of this kind can be unstable and under some circumstances the energy stored in the current sheet can be passed to charged particles. They can travel along lines of force towards the polar regions of the Earth where they appear as energetic particles. If the energy of the current sheet is passed in the way suggested to energetic particles the shape of the magnetic field altered. The field lines of opposite direction begin to touch and reconnect across the plasma sheet at about 100 Earth radii downstream from the Earth (Hones, Jr., 1986).

As a result of reconnections the magnetotail has three types of magnetic field line. The first consists of the field lines in the lobes. One end of each of these lines is attached to the Earth; the other end extends downstream into the solar wind. Such lines are called open field lines.

The second type is found in the plasma sheet on the earthward side of the magnetic neutral sheet. Here each field line comprises the earthward ends of two lobe field lines that have reconnected. Thus each field line in the plasma sheet on the Earthward side of the magnetic neutral line is a loop, both ends of which are attached to the Earth. Such lines are called closed field lines.

The third type is found in the plasma sheet downstream from the magnetic neutral sheet. Here each field line comprises the downstream ends of two of the lobe field lines that have reconnected; thus they are loops completely free of the Earth. Their ends extend antisunward into interplanetary space. They are known as interplanetary field lines (Hones, Jr., 1986; Nishida, 1982; Haerendel and Paschmann, 1982).

The magnetic pattern consisting simply of these types of magnetic field lines is disturbed at times when the interaction of the solar wind and the Earth's magnetic field overloads the magnetotail with energy, leading to the phenomenan called a magnetospheric substorm. The substorm is the mechanism by which the magnetosphere intermittently releases large amounts of energy that has been stored in the magnetotail. Some of this energy goes to create the aurorae near the Earth, while the rest is released downstream to the solar wind in the form of a plasmoid. (A substrom which lasts for an hour or so, is distinguished from a geomagnetic storm which lasts for a day or more and is caused when a solar flare initiates a shock in the solar wind.) A subtrom, thus, causes a fourth type of tail magnetic field line which exists only transiently and forms the magnetic structure of a plasmoid (Hones, Jr., 1986).

The exact nature of the solar wind-magnetosphere interaction is a difficult problem. However, in the models proposed to explain the nature of the interaction the ambient plasma motion is determined mainly by two electric fields. An electric field due to the rotation of the Earth with its magnetic field and, upon which is superimposed is the convection electric field E directed across the magnetosphere from dawn to dusk. The former field causes the plasma close to the Earth to co-rotate while the dawn-dusk field causes the plasma at greater distances from the Earth to convect towards the magnetopause in the direction of the sun with a convective velocity. The boundary between the region where convection dominates and the region where co-rotation is of major importance, is the plasmapause. The plasma contained within the plasmapause is in diffusive equilibrium with the ionosphere. Outside the boundary no such equilibrium can be established since the plasma is continously being convected away. A consequence of this is that there is a sharp discontinuuity in plasma density as one crosses the plasmapause. Equatorial densities of 100-1000 cm^{-3} can occur within the plasmasphere, compared with densities of 0.1-10 cm^{-3} in the outer magnetosphere. Since the rotational flow of the field lines (or plasma) have no radial velocity, while the convective flow has a radial component inward on the night side, and outward on the dayside the flow lines in the combined picture have a minimum geocentric distance at dawn and a maximum geocentric distance at dusk. The ionospheric projection of the plasmapause is the ionospheric electron density trough (see Figure 9, Tulunay and Grebowsky 1987).

The convection cycle is completed in about 12-20 hours. Periods of southward interplanetary magnetic field and high solar wind velocities are well correlated with enhancements of magnetospheric convection velocities. The enhanced activity periods result in enhanced fluxes of plasma within the magnetosphere and in enhanced precipitation of plasma into the high latitude ionosphere and atmosphere (Gorney, 1990).

During geomagnetically quite conditions the plasmasphere can extend beyond 5-6 Earth radii. During disturbed conditions due to the enhanced magnetospheric convection the plasmasphere is eroded considerably. The erosion of the outer plasmasphere during magnetically active periods can take place within an hour of the onset of enhanced convection,while it can take days for the refilling of the plasmasphere. Thus, the distribution

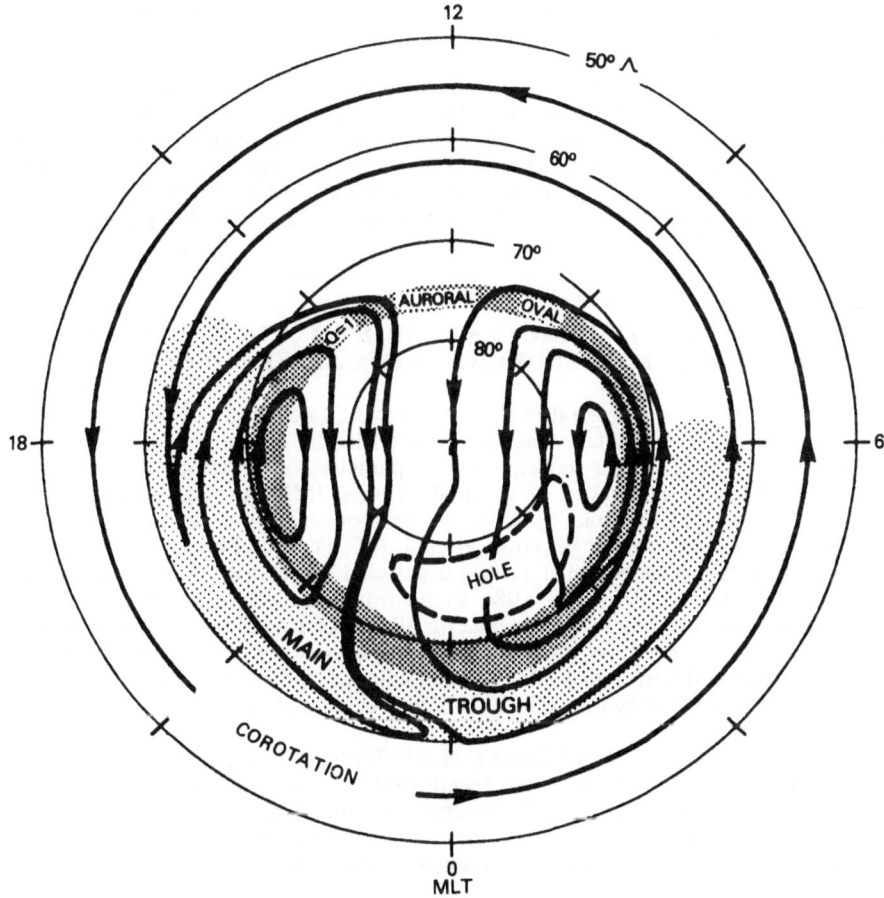

Figure 9. Average Locations of the High Latitude Ionization "Hole" and the Main Trough in Relation to a Representative Plasma Drift Configuration (From Brinton et al. 1978; see Heppner, 1977) and the Quiet-Time Auoral Oval of Feldstein and Starkov (1967).

of plasma tends to be quite variable, and the shape of the plasmapause or trough depends both on the magnitude and time history of geomagnetic activity. Because of infinite conductivity along field lines ExB drifts of flux tubes (or plasma) in the plasmasphere produce changes in the F region of the ionosphere.The convection patterns imposed upon the ionosphere by the coupling of the solar wind to the magnetosphere during periods of northward interplanetary magnetic field are very poorly described (Rodger, 1990).

Correlations between solar activity and disturbances in the near-Earth magnetosphere, ionosphere and atmosphere are well documented. Unfortunately, because of the complex and sometimes indirect interactions between the Sun and near-Earth space environment very few long-term quantitative predictions can be made regarding the effects of an extreme solar maximum on the near-Earth environment or on the complex systems operating in that environment (Gorney, 1990; Akasofu, 1981; Crooker and Siscoe, 1986)

The system composed of the Earth's magnetosphere, ionosphere and solar wind responds to a variety of processes on time scales ranging from about 10^{-6} s^{-1}, the electron gyro-frequency, to 3×10^8 s^{-1} a solar cycle. Feedback between micro-structure of the ionosphere (<10 km) and the macro structure of geospace implies that realistic modelling should have very high spatial and temporal resolution on a global scale (Rodger, 1990). Therefore, it is difficult to obtain an exact model which can cope with such a wide range of variation. Currently the model studies are mainly at macro scale state of physical parameter under investigation in terms of temporal and spatial structure. However, progress is being made for short-term forecasting of the occurrence of geomagnetic storms based on real-time observations of solar activity, and a number of important qualitative predictions can be made with high confidence (Gorney, 1990; see also Feynman and Gu, 1986)

The following section will be composed of some of the findings of a survey based on hourly averages of observations of the interplanetary medium, made by satellites at about 1 A.U. in the years 1963-1986, and the possible effects of the interplanetary magnetic field has on the mid-latitude ionosphere. This section will provide an example of the possible influence of solar activity on the near-Earth space environment directly and also implies indirectly some of the expected possible effects of solar activity on systems which operate within that environment.

THE INTERPLANETARY MEDIUM AT ABOUT 1 A.U. AND ITS POSSIBLE INFLUENCE ON THE IONOSPHERIC CRITICAL FREQUENCIES

Tulunay (1991) and Hapgood et al. (1991) investigated some of the questions about the variability of both interplanetary magnetic field (IMF) Bz component and the solar wind dynamic pressure using a 24-year sequence of interplanetary observations between 1963 and 1986. Hapgood et al, (1991) showed the distribution of the IMF and solar wind values for the whole data set (Figure 10). The solar cycle variations in out-of-ecliptic component of the IMF and the solar wind are represented in Hapgood et al, (1991) second and third figures (Figure 11 and Figure 12 here). Hapgood et al. (1991) concluded that distributions of solar-wind and IMF parameters between 1963 and 1986 to be much the same as the corresponding distributions that had been presented from data for all part of cycle 20 (references in Hapgood et al., 1991). The most notable of the differences they reported between solar cycles 20 and 21 is that the clear anti-correlation of annual means of solar-wind density and sunspot numbers, as reported previously for cycle 20 is hardly evident at all in cycle 21.

Hapgood et al. (1991) quoted Bame et al. (1976) in ascribing the increase in the mean and spread of the distribution of solar-wind speeds in the declining and minimum phase of the solar cycle to increased occurrence of high-speed streams. The solar cycle variation of the averaged magnitude of the IMF-Bz and the the corresponding standard deviations are similar in both solar cycles 20 and 21, and both parameters reflect the larger maximum of the solar cycle 21.

Hapgood et al. (1991) investigated the major reversals of the IMF Bz component since these reversals are important events which should cause significant changes in the Earth's

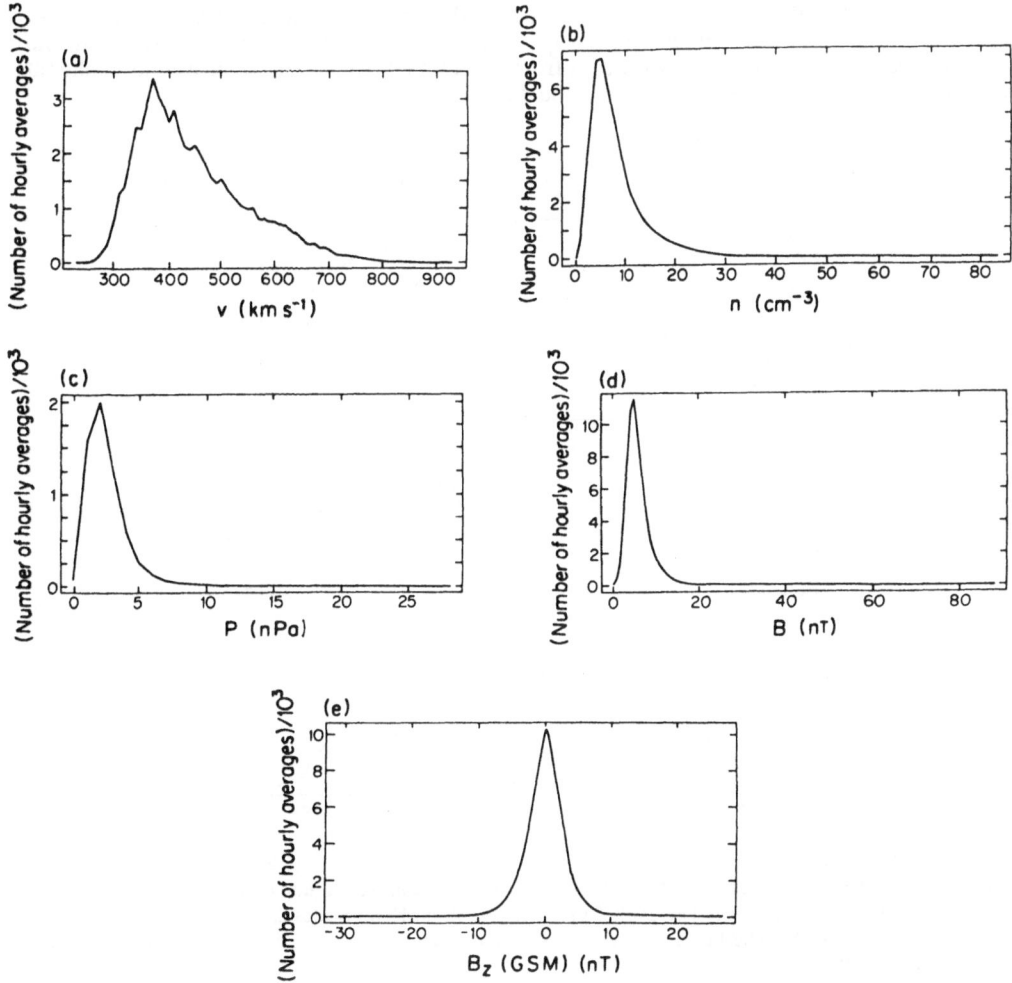

Figure 10.   Distributions of IMF and Solar-Wind Values for the Complete Dataset (a) The Solar-Wind Speed v; (b) the Solar-Wind Density. n: (c) the Solar-Wind Dynamic Pressure P; (d) the IMF Field Strength B; and (e) the Northward Component of the IMF (GSM coordinate) Bz (From Hapgood et al., 1991).

magnetosphere. They identified "events" as times when there was a reversal of the polarity of Bz between adjacent hourly mean values. As a secondary condition they required that the magnitude of Bz was greater than 1nT for both hourly values (Tulunay et al. 1991). To estimate the duration of stable IMF conditions Hapgood et al. (1991) determined the interval following each event for which Bz maintained the same polarity (see Figure 4 and 5 in Hapgood et al. 1991). The distribution of the duration of periods of stable polarity of the IMF-Bz component showed that the Earth's magnetosphere could achieve steady state for only a small fraction of the time and there was some evidence for a solar-cycle variation in this fraction. These authors also found that the polarity changes in the IMF- Bz fall into two classes : one with an associated change in solar wind dynamic pressure, the other without such a change. However, in only 20% of cases does the dynamic pressure change exceed 50%.

Figure 13 and 14 are taken from a paper by Tulunay et al. (1991) in which they searched the possible effects of the orientation of the IMF on the Slough (51.48°N, 0.57°W) critical frequencies, f0F2, in the period 1967 to 1986. Figure 13 shows the diurnal variations of the critical frequencies, after the "quiet-time" diurnal variation was subtracted (δf0F2), for a northward IMF Bz polarity change, (δBz>0) for both Toward (Bx>0), and Away (Bx<0) sectors. Both the Toward and Away cases show the diurnal variation present in the average case (dashed line). The deviations from quiet time values seem to be a little smaller at all times for the Toward (Bx>0) subset of the data.

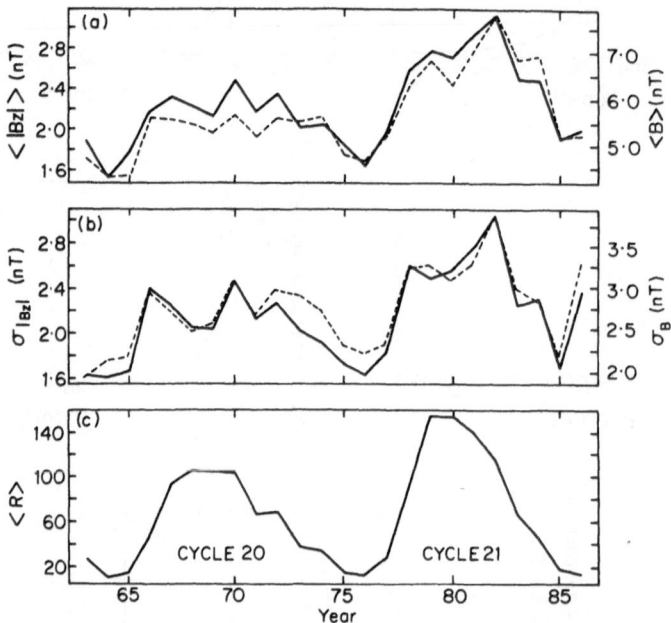

Figure 11. Solar-cycle Variations in Out-of-Ecliptic Component of the IMF (GSM coordinates). The Solid Lines and Left-Hand Scales are for Annual Values of (a) the Mean Magnitude of Bz and (b) the Standard Deviation of the Magnitude of Bz. The Dashed Lines and the Right-Hand Scales are the Equivalent Variations for the IMF Strength, B. The Lowest Panel (c) gives the mean International Sunspot Number, R (From Hapgood et al., 1991).

Figure 14 is the same as Figure 13, but for the southward IMF Bz polarity dataset. For the night time data there is no significant differences between the Toward and Away sectors and the all-data average variation (shown by the dashed line). However, for the dayside there is a considerable and significant difference between the two cases. The difference between the Toward (Bx>0) and Away (Bx<0) sectors on the noonside is associated with the southward IMF Bz polarity. Tulunay et al. (1991) concluded that larger deviations from quiet-time foF2 values may exist on the dayside during Toward (Bx>0) than Away (Bx<0) sectors, when the IMF is southward. This effect is not seen when the IMF is northward. The magnitude of this effect is of order 0.8 MHz, which is a sizable part of the day-to-day variability, as shown by the distribution of the critical frequencies in their second figure. The results do suggest the possibility that dayside auroral heating in the northern hemisphere is greater for IMF By<0(Bx>0) than for IMF By>0 (Bx<0) as invoked by Lockwood and Fuller-Rowell (1987 a;b).

CONCLUSIONS

The interaction of solar energy with the near-Earth environment has been attracting scientific interest both in its long-term behavior of solar activity and in the physics of such

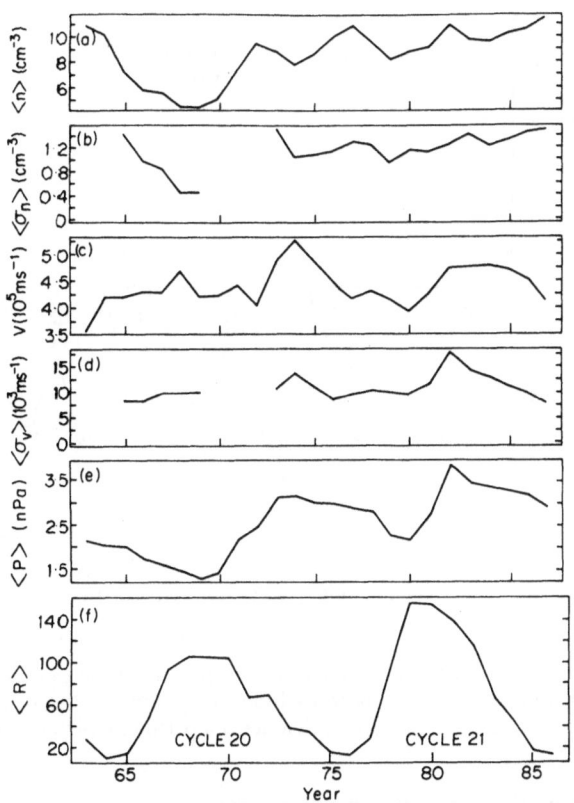

Figure 12. Solar-cycle Variations in the Solar Wind. Annual means of: (a) the Solar-wind Density, n; (b) the Hourly Standard Deviation of n values, σ_n ; (c) the Solar-wind Speed, v; (d) the Hourly Standard Deviation of v Values, σ_v ; (e) the Dynamic Pressure, P; and (f) the International Sunspot Number, R (From Hapgood et al., 1991).

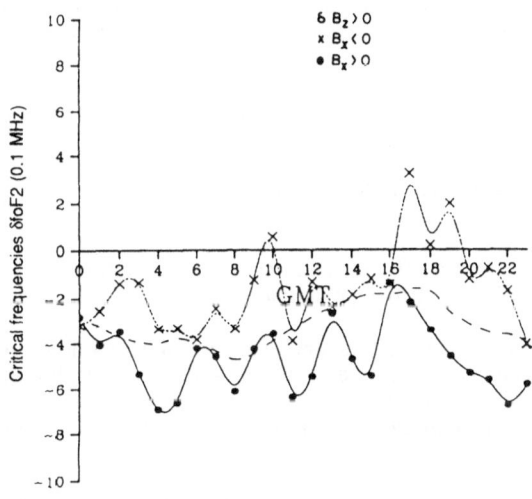

Figure 13.   Diurnal Variation of $\delta t0F2$ for $Bz>0$ Dataset for $Bx<0$ and $Bx>0$. The Dashed Line Gives a Curve for Comparison (From Tulunay et al., 1991).

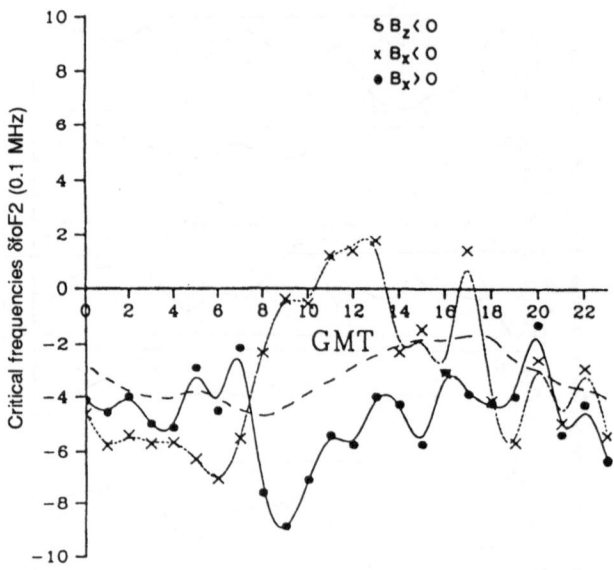

Figure 14.   Diurnal Variation of δt0F2, but for Bz<0 (From Tulunay et al., 1991).

interaction. It has been also attracting considerable interest from the operational points of view concerned with man-made systems. These include, for example, the importance of the STP monitoring program (courtesy of the Space Science Department of the S.E.R.C. Rutherford Appleton Laboratory) :

(i) The reliability of worldwide communication, broadcast, navigation and over-the-horizon radar systems depends on the variability of the ionosphere.

(ii) Ionospheric currents cause destructive voltage surges in power lines.

(iii) Ionospheric currents cause corrosion in pipelines.

(iv) Upper atmospheric heating causes additional drag on satellites and hence orbital decay and premature reentry. For example, the lifetime of a satellite at an initial altitude of 500 km during quiet solar conditions is about thirty years, although the lifetime of the vehicle is limited to just over two years during active solar conditions (Gorney, 1990, see also Walterscheid, 1989).

(v) Energetic particles cause temporary faults or permanent damage in electronic systems on satellites.

(vi) Energetic particles represent a radiation hazard to astronauts and high-altitude aeroplane pilots.

        In addition the adverse effects of the solar and geophysical events may further be illustrated with the help of the following examples :

(vii) At altitudes between 500 and 800 km the highly reactive atomic oxygen concentration can vary over the solar cycle as much as a factor of thousand. High concentration of atomic oxygen can react chemically with various surfaces of a satellite or sensor, leading to mass loss from external structures and degraded sensor performance (Gorney, 1990, see also Visentine, 1988).

(viii) The plasma sheet can cause electrical charging of the surfaces of satellites at about 5-6 Earth radii. The probability of occurrence and the severity of spacecraft charging events are directly correlated with periods of enhanced geomagnetic activity (see Gorney, 1990).

This general review of this very broad and multidisciplinary subject on the physics of solar radiation and its effect on the Earth's atmosphere constitutes only an incomplete introduction to this active research area. Hopefully it will create some interest among the scientists who are working in the subject of the biological effects and physics of solar and galactic cosmic radiation.

ACKNOWLEDGEMENT

The author would like to record her thank to Ms. G. Elden for the typing of this text.

REFERENCES

Akasofu, S. -I., 1981, Energy coupling between the solar wind and the magnetosphere, Space Science Reviews, 20:121.

Altrock, R.C., Neidig, D.F., DeMastus, H.I., Radick, R.R., Evans, J.W., Simon, G.W. and Keil, S.I.,1985, The Sun, in: "Handbook of Geophysics and the Space Environment", A.S. Jursa ed., National Technical Information Servic, Spring field, VA, U.S.A.

Bame, S.J., Asbridge, J.R., Feldman, W.C. and Gosling J.T., 1976, Solar cycle evolution of high speed solar wind streams, Astrophys. J., 207 : 977.

Brinton, H.C., Grebowsky J.M. and Brace, L.H., 1978, J. Geophys, Res., 83: 4767.

Crooker, N.U. and Siscoe, G.L., 1986, the effect of the solar wind on the terrestrial environment, in: "Physics of the Sun", vol.3, P.A. Sturrock, ed., D.Reidel, Hingham, Mass.

Dryer, M., 1987, Solar Wind and Heliosphere, in: "The Solar Wind and the Earth", S,-I. Akasofu and Y.Kamide, eds., Terra Scientific Publishing Company, Tokyo.

Ely, J.T.A., 1984, "Cosmic Rays, Solar Activity, Magnetic Coupling, and Lightning Incidence, 1984", NASA Contractor Report 3812, NASA.

Feldstein, Y.T., and Starkov, G.V., 1967, Dynamics of auroral belt and polar geomagnetic distrubances, Planet Space. Sci., 15 : 209.

Feynman, J., Gu, Y.Y., 1986, Prediction of geomagnetic activity on time scales of one to ten years, Rev. Geophys., 24: 650.

Gorney, D.J., 1990, Solar-cycle effects on the near-Earth space environment, Rev. Geophys., 28.: 315.

Haerendel, G. and Paschmann, G., 1982, Interaction of the solar wind with the dayside magnetosphere, in: "Magnetospheric Plasma Physics", A. Nishida, ed., D.Reidel Publishing Co., Dordrecht. Boston. London.

Hapgood, M.A., Lockwood, M., Bowe G.A., Willis, D.M. and Tulunay Y.K., 1991, Variability of the interplanetary medium at 1 a.u. over 24 years 1963 - 1986, Planet Space Sci., 39 : 411.

Heikkila, W.J., 1990, Magnetic reconnection, merging, and viscous interaction in the magnetosphere, Space Science Reviews, 53:1.

Heppner, J.P, 1977, Empirical models of high latitude electric fields, J.Geophys. Res., 82 : 1115.

Hones, Jr., E.W., 1986, the Earth's magnetotail, Scientific American(march) : 32.

Lockwood, M. and Fuller-Rowell T.J., 1987 a, Geophys. Res. Lett. 14 : 371

Lockwood, M. and Fuller-Rowell T.J., 1987 b, Geophys. Res. Lett. 14 : 581

Neugebauer, M., 1991, The Quasi-stationary and transient states of the solar wind, Science, 252: 405.

Nishida, A., 1982, Origin of magnetospheric plasma, in: "Magnetospheric Plasma Physics", A. Nishida, ed., D.Reidel Publishing Co., Dordrecht. Boston. London.

Ratcllife, J.A., 1970 "Sun, Earth and Radio" world University Library, Weidenfeld and Nicolson, London.

Richmond, A.D., 1987, The ionosphere, in: "The Solar Wind and the Earth,"S-I. Akasofu and Y. Kamide, eds., Terra Scientific Publishing Company, Tokyo.

Rodger, A.S., 1990, Recent progress In understanding the magnetospheric environment,: in "Solar-Terrestrial Predictions : Proceedings of a Workshop at Leura, Australia, October 16-20, 1989", National, Oceanic Boulder, Colorado, U.S.A. and Atmospheric Administration Environmental Research Lab. Boulder, Colorado, U.S.A.

Russell, C.T., 1987, The Magnetosphere, in: "The Solar Wind and the Earth", S,-I. Akasofu and Y.Kamide, eds., Terra Scientific Publishing Company, Tokyo.

Schwenn, R., 1990, What does the solar wind tell the prediction community?, in : "Solar-Terrestrial Predictions : Proceedings of a Workshop at Leura, Australia, October 16-20, 1989", National Oceanic and Atmospheric Administration Environmental Research Lab. Boulder, Colorado U.S.A.

Taylor, H.A. and Tulunay, Y.K., 1973, Near-simultaneous measurements of the plasma trough and plasmatail from OGO-4 and Ariel-3, Proc. Chapman memorial Symposium June 18-22, Boulder, USA.

Tulunay, Y.T., Willis, D.M., Hapgood, M.A. and Lockwood, M., 1991, Influence of the interplanetary medium on mid-latitude ionospheric variability,"Final Report", NATO Collaborative Research Project, NATO, 0753/87.

Tulunay, Y.K. and Grebowsky, J.M., 1987, Hemispheric differences in the morphology of the high latitude ionosphere measured at ~ 550 km, Planet. Space Sci., 35:821.

Visentine, J.T. (Ed.), 1988, "Atomic Oxygen effects measurements for shuttle missions STS-8 and 41-G", NASA Tech. Memo., TM-100459,I-III.

Walterscheid, R.L., 1989, Solar cycle effects on the upper atmosphere : Implication for satellite drag, J.Spacecr. Rockets, 26 : 439.

SPACE RADIATION DOSIMETRY

F. A. Hanser and B. K. Dichter

Panametrics, Inc.
221 Crescent Street
Waltham, MA 02154, USA

INTRODUCTION

Dosimetry is the measurement of the energy deposited in matter by various forms of radiation. In space the radiation is primarily energetic electrons, protons and heavier ions from planetary radiation belts, solar flares, and interstellar cosmic rays. Experimentally, dose is frequently obtained by summing the individual energy deposits in a solid state detector. If the detector is calibrated and the sensitive mass is known, the energy sum can be converted directly to accumulated radiation dose in Gy (J/kg). Such detectors can also be used to provide an approximate separation of dose into the components due to electrons, protons, and heavier ions, which is useful if it is desired to convert the measured dose into a biological effective dose (Sv) for manned spaceflight purposes. The output can also be used to provide an essentially instantaneous dose rate for use as warning devices. This is the primary type of space radiation dosimeter to be discussed here. The MOS-type dosimeter is another solid state sensor which can be of small size and low power. These devices integrate the total dose once through, can not separate particle types, and are not suitable for instantaneous dose rate measurement at low levels. There are several additional methods of measuring space radiation dose using scintillators, etc., but these are not discussed in detail. In this paper emphasis is given to descriptions of active solid state detector instruments which have successfully worked in space. Some results of in-orbit dose measurements are presented.

GENERAL DISCUSSION OF RADIATION DOSIMETRY

A radiation dose is energy deposited in matter by ionization produced by incident radiation. For neutral particles (x-rays, neutrons) the first step is an interaction that produces an energetic charged particle, since energy deposition is generally by ionization energy loss by charged particles. X-rays produce energetic electrons from photo-absorption, Compton scattering, and pair production (above 1.022 MeV), and the electrons (+positrons from pair production) produce the ionization. Initially dose was measured in Roentgens (R) which is the dose required to produce a charge of 1 statcoulomb/cm^3 in air at STP, and is strictly applicable only to X-rays (see Evans (1955), Chapter 25; Greening (1985)). Using a value of 34 eV/(electron-ion pair) (see e.g., Attix (1986), p 31) for air this converts to 88 erg/(gram air). The rad is defined as an ionization energy deposit of 100 erg/g of the material being irradiated, and has been replaced by the MKS unit of the Gray (Gy) = 1 J/kg = 100 rad. For charged particle energy loss in matter, energy losses are usually calculated in terms of the electron volt (eV), which is equal to 1.602 x 10^{-19} J.

Protons, neutrons, and heavy ions have additional dose effects resulting from nuclear collisions. These massive particles can displace atoms and thus introduce crystal defects or molecular damage. Nuclear cross sections are generally much lower than the ionization cross sections, so the interaction rate per incident particle is small. However, the damage from nuclear displacement can be more severe, particularly in very pure crystals, such as some types of solid state detectors and electronic chips. Some nuclear collisions result in nuclear interactions which can produce a large energy deposit in a small volume. A shattered nucleus can emit several highly ionizing fragments which produce a large amount of local ionization, with attendant destructive effects.

Radiation exposure for humans has an additional quality factor (Q) which varies with the particle ionization rate. One set of Q factors was given by the ICRP (1977), with the ionization rate in water being the defining base. These Q factors are listed in Table 1 along with the corresponding proton energies and ionization rates in silicon (Si). The ionization

Table 1. Radiation Dose Q Factors

Ionization energy loss rate in water			proton energy	Energy loss rate in Si	Ratio
keV/μm	MeV-cm^2/g	Q	(MeV)	(Mev-cm^2/g)	water/Si
≤3.5	35.	1	13.8	27.2	1.29
7.	70.	2	5.7	52.6	1.33
23.	230.	5	1.17	158.	1.46
53.	530.	10	0.35	303.	1.75
≥175.	1750.	20	-	-	-

rates (also called stopping powers, and frequently expressed as MeV-cm^2/g) for Si are taken from the tabulations of Janni (1982). The maximum stopping power for protons in water is 996 MeV-cm^2/g at an energy of 0.065 MeV. Protons of 13.8 MeV with a Q = 2 (ICRP (1977) have a range in water of 2.23 mm which is easily shielded. Thus most proton irradiation is from high energy particles which have a Q = 1 except for the (usually small) fraction of locally stopping protons. A 10 keV electron has a stopping power in water of 22.6 MeV-cm^2/g and a range of 2.5 microns (Berger and Seltzer (1982)), so electrons (and x-rays) generally have a Q factor of 1. As is noted in other articles in these Proceedings, there is some controversy about the use of Q factors, and the 1977 ICRP values have already been updated.

The use of Si solid state detectors to measure dose requires some correction to obtain the dose in human tissue (effectively water for most cases). The water/Si stopping power ratios in Table 1 show that for protons above about 1 MeV this ratio is nearly constant at 1.3. The measured dose in Gy(Si) is thus multiplied by 1.3 to get an effective Gy (water - or tissue), and this dose is in turn multiplied by the Q factor to get the dose in Sievert (Sv). As will be shown in the following sections of this paper, the measurement of dose in Gy(Si) is quite direct, and can be readily calibrated once for all types of radiation. The conversion to Sv, using Q factors, is where the largest uncertainty lies. As is discussed in several other papers in these Proceedings, the Q values have some uncertainty, and there is some question of their applicability, since radiation susceptibility varies for different organs and organisms.

The stopping power of charged particles varies with energy, having a low energy peak and a high energy minimum. The general form for the ionization energy loss for a particle with mass m and charge q (in electron charges e) in a target material with atomic number A, nuclear charge Z and density ρ is

$$\frac{1}{\rho}\frac{dE}{dx} = \frac{2\pi N_0 q^2 e^4 Z}{mc^2 \beta^2 A}\left[\ln\frac{2m c^2 \beta^2 W}{I_{adj}(1-\beta^2)} - 2\beta^2 - (correction\ terms) \right]$$

where

$$W = \frac{2m\,c^2\,\beta^2}{1 - \beta^2}\left[1 + \frac{2m}{M\,(1 - \beta^2)^{1/2}} + \frac{m^2}{M^2}\right]$$

N_o is Avogadro's number, ß is v/c (the particle velocity divided by the speed of light), I_{adj} is the target approximate effective ionization energy and M is the target atomic mass (to a good approximation, given by A x mass of proton). For very low energies (below about 10 keV/nucleon for protons) the stopping power is reduced because of partial charge neutralization from electron capture, and has the form

$$\frac{1}{\rho}\,\frac{dE}{dx} \propto E^{1/2}$$

Electrons must have the bremsstrahlung energy loss added to the ionization energy loss. Bremsstrahlung is the energy loss suffered by a decelerating or accelerating electric charge. It is much larger for low mass particles like electrons than it is for the more massive protons. The above equations must also be modified because of the quantum mechanical identity of the incident electron and the target atom electrons. More detailed discussion of the proton stopping power is given by Janni (1982), and of that for electrons and positrons by Berger and Seltzer (1982).

The stopping power of Si for protons, taken from Janni (1982), is shown in Fig. 1. This plot illustrates the general form of the stopping power-incident energy relationship, with a peak at low energy and a high energy minimum at a relativistic energy (kinetic energy at the minimum is about twice the rest mass energy).

Fig. 2 shows the stopping power of Si for electrons, including the bremsstrahlung effects, taken from Berger and Seltzer (1982). The bremsstrahlung part of the stopping power becomes dominant in Si at energies above 50 MeV. The bremsstrahlung stopping power is strongly dependent on the Z of the target material, and it becomes dominant at about 10 MeV for lead and 90 MeV for water (and tissue). It should also be noted that electrons scatter very readily, so their path in the material is not a straight line.

Heavy ion stopping powers are illustrated by the ^{16}O and ^{56}Fe curves for Si in Fig. 3, from the tabulations of Hubert et al.(1990). The absolute values are much larger than for protons or electrons, and show why heavy ions can cause a large amount of damage to both electronic devices and to living tissue.

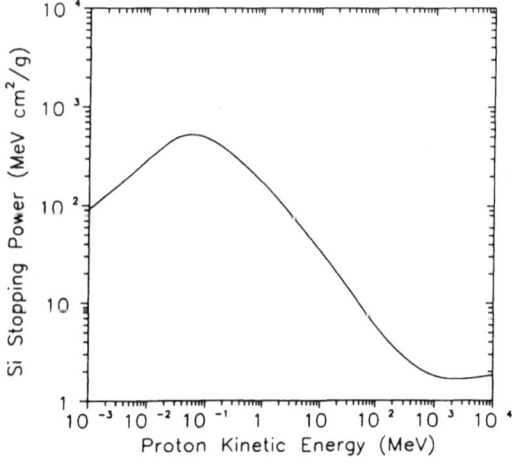

Figure 1.   Stopping Power Curve of Silicon for Protons.

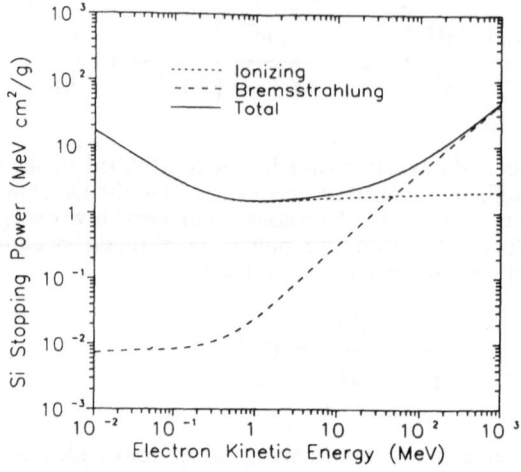

Figure 2. Stopping Power Curves of Silicon for Electrons, Including Bremsstrahlung effects.

The effects of radiation on electronic devices occur through a number of processes:

1. Ionization and nuclear collisions can produce atomic dislocations and resulting crystal damage. This leads to degradation of electrical and physical properties, and ultimately to electrical or physical breakdown.

2. Ionization can produce trapped charges in insulating materials. This can lead to changes in the electrical properties of devices, such as threshold shifts, and result in malfunction.

3. High ionization densities can produce severe effects. The large charge concentration can result in device state changes (single event upsets, or SEU's), or more serious latchups where a device must be powered off and back on to allow proper operation. Some devices can be destroyed by a current surge along an ionization path between parts of the device which are at different voltages and have insufficient current limitation.

The effects of radiation on living tissue are similar. The ionization provides energy for

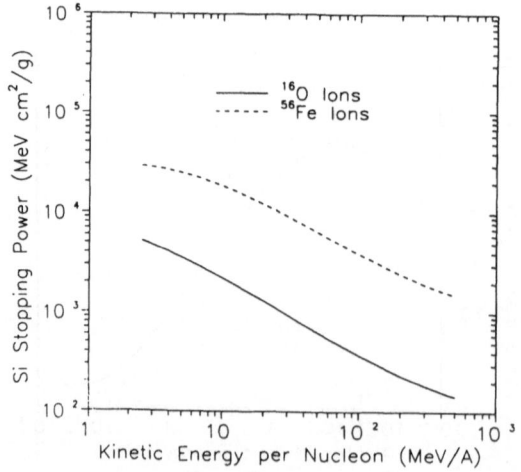

Figure 3. Stopping Power Curves of Silicon for ^{16}O and ^{56}Fe.

chemical changes which can ultimately produce sickness and death. Some changes can be genetic and these can lead to a long term risk increase for cancer. These effects of radiation are discussed in more detail in other papers in these Proceedings.

Radiation dose measurements can be made with a wide variety of instruments. Any device which produces a signal or undergoes some form of property change from radiation may be useable. The important factors are that the device is stable, the measurements are repeatable, and an absolute calibration is possible. Several types of device are: 1. Solid state detectors. These are most commonly silicon or germanium, with the device producing an output proportional to the deposited energy. 2. Electronic device property changes. The MOS dosimeter is the most common of this type. 3. Ion chambers. These can be used for high dose rate applications. 4. Scintillator/PMT combinations. These can use a wide variety of available scintillators, including organic scintillators designed to simulate tissue in their radiation response. 5. Geiger tubes. These do not produce an output proportional to the deposited energy, and so must be separately calibrated for different radiation types. 6. Thermoluminescent detectors. These store radiation energy stably and then release it as light when heated. Measurement of the light with a PMT provides the exposed dose. 7. Film. This can provide a dose measure by developing the film after exposure. The above can provide dose measurements of varying degrees of accuracy for various radiation environments. Some are not practical for spacecraft use because of their bulk and complex processing requirements. Here we will consider primarily the Si detectors of (1), with a brief discussion of (2).

SPACE RADIATION CHARACTERISTICS

Radiation in space consists primarily of three components: 1. The trapped radiation belts of planetary magnetospheres, with that of the earth being of most immediate interest. 2. Solar cosmic rays (mostly proton) which come from solar flares. These can be very intense and of high particle energy, so they are an important radiation hazard when planetary magnetospheric shielding is not available. 3. Galactic Cosmic Rays (GCR), which are high energy particles, mostly protons but with an important heavy ion component. The fluxes are weak, but they are always present, and the heavy ion component may be important for long term human exposure. These radiation components are discussed in great detail in several other papers in these Proceedings, so only a brief summary is given here.

The trapped radiation of the earth's magnetosphere is confined by the dipole magnetic field. The radiation is concentrated at the magnetic equator, with the polar regions being open to external radiation (solar protons and GCR). The particle distributions are usually described as a function of the "L" value, the approximate equatorial radius, in earth radii, of the magnetic shell of the trapped particles. A value of $L = 1$ corresponds to the earth's surface at the equator. The trapped radiation forms an "inner belt" for L about 1.1 to 2, which consists primarily of high energy protons, up to several hundred MeV. The fluxes are intense, with about 10^5 p/(cm^2-s) above 50 MeV at $L = 1.5$. Shielding is difficult, so any spacecraft operation near $L = 1.5$ must allow for a substantial radiation dose rate. Most manned space missions have thus far avoided this region, except for a brief passage during the moon flights.

An "outer belt" for L about 3 to 6 consists primarily of electrons up to several MeV. This belt is quite variable in intensity, particularly at the high L values, and can have intensities in excess of 10^7 e/(cm^2-s) above 0.5 MeV. Shielding from the electrons themselves is comparatively easy, but the bremsstrahlung from the stopped electrons produces an x-ray dose rate that is more difficult to shield against. Solar activity producing magnetic storms at the earth is responsible for the variability of the trapped radiation intensities.

Solar flares can on occasion produce intense fluxes of protons with energies in excess of 1 GeV. These events can produce significant proton fluxes in the outer radiation belt and over the polar caps. Intense events can produce radiation dose rates behind a shield of 1 g/cm^2 aluminum in excess of 0.01 Gy/s. While such events are rare, they can be fatal for manned operation outside the earth's magnetosphere (or over the polar caps) if there is insufficient shielding.

GCR fluxes are comparatively low in intensity, about 1 p/(cm^2-s) above 500 MeV, but they have high energies (to many GeV/nucleon) and include heavy ions. The spectra peak at 0.1 - 1 GeV/nucleon, with the low energy intensity being modulated by solar activity. During solar particle events the solar proton fluxes can be dominant below 1 GeV. The heavy ion components are about 10^{-3} (C - O) to 10^{-4} (Ne - Si; Fe group) of the proton intensity, but their much larger stopping powers can produce significantly more damage. The radiation dose rate from GCR is not significant in terms of Gy/s, but the damage to individual electronic devices, or to cells and neurons, may be significant.

Other planetary magnetospheres can also be important radiation hazards. Jupiter has intense fluxes of trapped electrons in the many MeV range and this must be considered in the design of electronics for spacecraft headed for that environment. Most of the large outer planets have been found to have significant fluxes of trapped radiation, and this must be considered in designing spacecraft for their exploration.

SPACE RADIATION DOSIMETER INSTRUMENTS

There are two radiation dose monitor designs that have been commonly used in space applications: the MOS type dosimeter and the solid state detector dosimeter. The MOS dosimeter was born of the realization that the very radiation susceptibility of a device can be used to measure the radiation dose that the device had received (Adams and Holmes-Siedle (1978)). The solid state detector (SSD) dosimeter directly measures the radiation dose delivered to the detector by each particle that traverses it.

The principle of the dose measurement is the same in both cases. A detector, sensitive to ionizing radiation, is placed behind a known amount of shielding. Detector output is monitored to determine the radiation dose that it receives. This information is then used to infer the radiation dose to a human crew member or electronic devices behind a similar amount of shielding. Typically, several different shielding thicknesses are used.

MOS Dosimeter

In a MOS dosimeter, the parameter measured to determine the radiation dose is the threshold voltage (or the applied voltage at which the transistor "turns on") of a radiation-soft PMOS transistor. The threshold voltage increases monotonically with increasing radiation dose. There are two possible modes of operation of such a device, one suitable for measuring the dose up to 500 Gy, the other up 25,000 Gy (August (1984)).

The first mode is to irradiate the transistor while holding the gate voltage at a high positive value, such that the electric field strength in the gate oxide is $\geq$ 1 MV/cm. Under this condition, the dominant radiation damage mechanism is the trapping of the holes created by the ionization track in hole trapping sites that result from the production process. The change in threshold voltage is then, to a good approximation, linearly related to the dose. A typical change in threshold voltage, measured by August (1984), is 15 V for a 500 Gy dose. This mode of operation is not suitable for doses much larger than 500 Gy because the increase in the threshold voltage would make the applied voltage large enough to cause breakdown in the transistor during the measurement.

The second, high dose, mode studied by August (1984) is to expose the transistor to radiation while holding all transistor elements (source, drain and gate) at ground and only bias the transistor during the threshold voltage measurement. August (1984) found that a 15 V increase in the threshold voltage corresponds to a dose of approximately 25,000 Gy.

The advantages of a MOS dosimeter are its low weight and power and very small data processing requirements. A space qualified MOS dosimeter, with four PMOS transistors located behind various thicknesses of shielding, weighs less than half a pound and has an average power consumption of 120 mW (August et al. (1983)). The MOS dosimeter is a passive detector, in that the change in threshold voltage occurs automatically due to the radiation bombardment. As a result, the only data processing activity is the measurement of the "turn on" voltage. The disadvantages of a MOS dosimeter are that it is not well suited to be used to give dose rate alerts and its threshold voltage-dose relationship cannot be calibrated directly.

The changes in the threshold voltage are gradual and cumulative and can only be detected following a read out, which is typically done once per day (due to the nature of the voltage increase, more frequent readouts may not yield meaningful data). A sudden influx of energetic particles which delivers a full daily dose over a period of minutes may not be noted by the MOS dosimeter until many hours later. Even if the dosimeter is read out more frequently than once per day, the increase in the dose due to a brief period of high flux may be lost in the uncertainty of the measurement. Finally, since the radiation damage to the transistor is permanent, a flight PMOS transistor cannot be calibrated with a radiation source prior to flight. Instead, an identical device from the same wafer and production batch must be calibrated and that calibration curve must be assumed to represent the behavior of the flight transistor.

Solid State Detector Dosimeter

This type of dosimeter utilizes solid state detectors, for example silicon p-i-n photodiodes, to detect each particle that traverses the detector and to measure the energy that it deposits in the detector. The passage of a charged particle through the detector, or an x-ray which knocks out an electron from an atom in the detector, generates electron-hole pairs in the detector material. The number of pairs is proportional to the energy that the incident particle deposited in the detector. For Si SSD's the rate is one pair per 3.6 eV of deposited energy. An electric field applied to the detector sweeps out the electrons and holes and the resulting current flows into a charge sensitive pre-amplifier. The total charge collected after a passage of an ionizing particle is then proportional to the deposited energy. If the volume of the detector is known precisely, the total dose in Gy (Si) can be calculated precisely from the measured deposited energy.

The SSD dosimeter has a number of advantages.

1. Since the detectors are made of the same material, Si, as most of the spacecraft electronic devices, the dose measured by the dosimeter can be directly applied to evaluate the dose received by those devices.
2. Each incident particle produces a signal, which is processed by the detector electronics. Therefore, the instantaneous particle flux and dose rate are available from the instrument in real time. This feature is very useful if dose rate information is needed in real time, for example for manned space missions.
3. Radiation is not inherently destructive to the dosimeter detectors (except after very large fluences, $>10^{13}/cm^2$ for electrons or $>10^{10}/cm^2$ for protons) and, therefore, the flight instrument response to electrons and protons of various energies can be calibrated with accelerator beams or radiation sources.
4. Since the energy deposition rates of electrons, protons and heavy ions are very different, the total dose may be divided into two or more components; one due to protons, one primarily due to electrons, and the heavy ion component, split as desired. This feature is of use in determining the biologically effective dose, since electrons, protons, and heavy ions have very different Q factors.

Examples of this type of dosimeter design are the nearly identical instruments flown aboard the DMSP/F7 and CRRES spacecraft (Morel et al., 1989). A cutaway view of the CRRES instrument is shown in Figure 4. A p-i-n photodiode detector is located under each of the four domes, each of different thickness. This geometry, shown in Fig. 5, allows the instrument to have a 2π sr field of view. The material in the domes (Al) serves as an energy degrader for the incident particles. It stops particles with energies below the threshold energies needed to penetrate the domes and prevents them from reaching the photodiodes. The four domes have energy thresholds of 20, 35, 51 and 75 MeV for protons and 1.0, 2.5, 5.0 and 10.0 MeV for electrons. The characteristics of the DMSP/F7 and CRRES dosimeter domes (channels) are listed in Tables 2 and 3. All SSD's are 400 microns thick, and are shielded from the rear by a minimum of 1.27 cm of tungsten.

The advantage of using p-i-n photodiodes, rather than surface barrier, silicon SSD's,

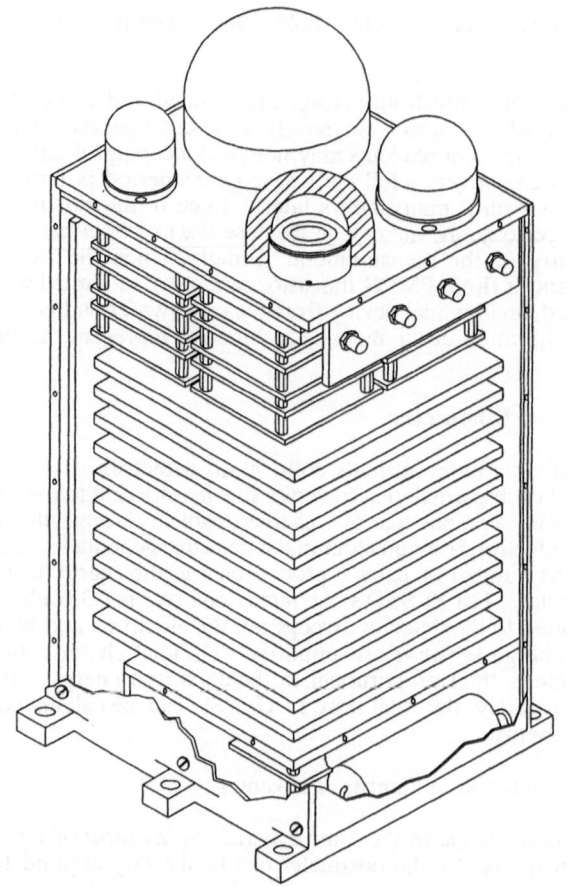

Figure 4. Isometric, Cross-Sectional View of the CRRES Dosimeter.

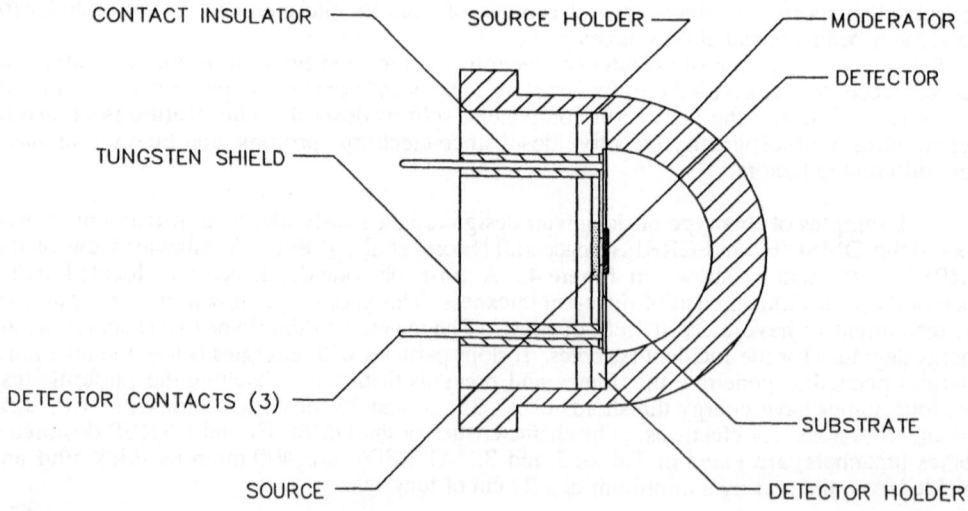

Figure 5. Cross-Section View of a Typical Dosimeter Dome, Showing the Detector Mounting Geometry.

Table 2. SSD Dosimeter Design Used on the DMSP/F7 and CRRES Satellites

Dome No.	Al dome thickness (g/cm^2)	SSD area (cm^2) DMSP/F7	SSD area (cm^2) CRRES	Energy deposition ranges (MeV) LOLET	Energy deposition ranges (MeV) HILET	Energy deposition ranges (MeV) VHLET[a]
1	0.57	0.051	0.008	0.05-1	1-10	>40
2	1.59	1.000	0.051	0.05-1	1-10	>40
3	3.14	1.000	0.051	0.05-1	1-10	>75/40
4	6.08	1.000	1.000	0.05-1	1-10	>40/75

[a]Values listed for domes 3 and 4 are for DMSP/F7 and CRRES, respectively.

Table 3. Primary Particles Measured by the DMSP/F7 and CRRES SSD Dosimeters

Dome No.	LOLET Electrons	LOLET Protons	HILET Protons	VHLET Protons[a]
1	>1.0	>120	20-120	>45
2	>2.5	>125	35-125	>55
3	>5.0	>130	51-130	>95/70
4	>10.0	>140	75-140	>90/110

[a]This refers to the minimum proton energy which could produce a count after traversing the aluminum shield. The values listed for domes 3 and 4 are for DMSP/F7 and CRRES, respectively.

is that in the "guard ring" photodiode geometry the sensitive volume of the detector can be determined much more accurately than the sensitive volume of the standard silicon surface barrier detector. Diagrams showing the typical construction of the p-i-n photodiode and the silicon surface barrier detectors are shown in Figures 6 and 7, respectively. Use of the guard ring, surrounding the signal collecting electrode, on the p-i-n photodiode detector results in a very well defined sensitive detector volume. Only the electron-hole pairs produced inside the sharply defined volume (dashed lines in Figure 6) can induce a signal on the collecting electrode. Those produced outside of the volume produce a signal on the guard ring only.

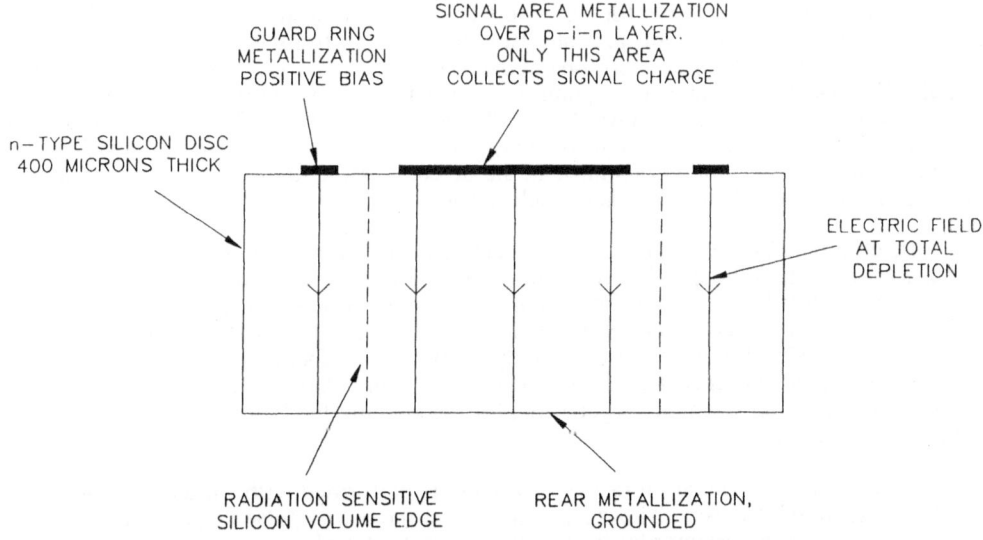

Figure 6. Silicon P-i-n Photodiode Construction Showing Precise Definition of Radiation Sensitive Silicon Volume.

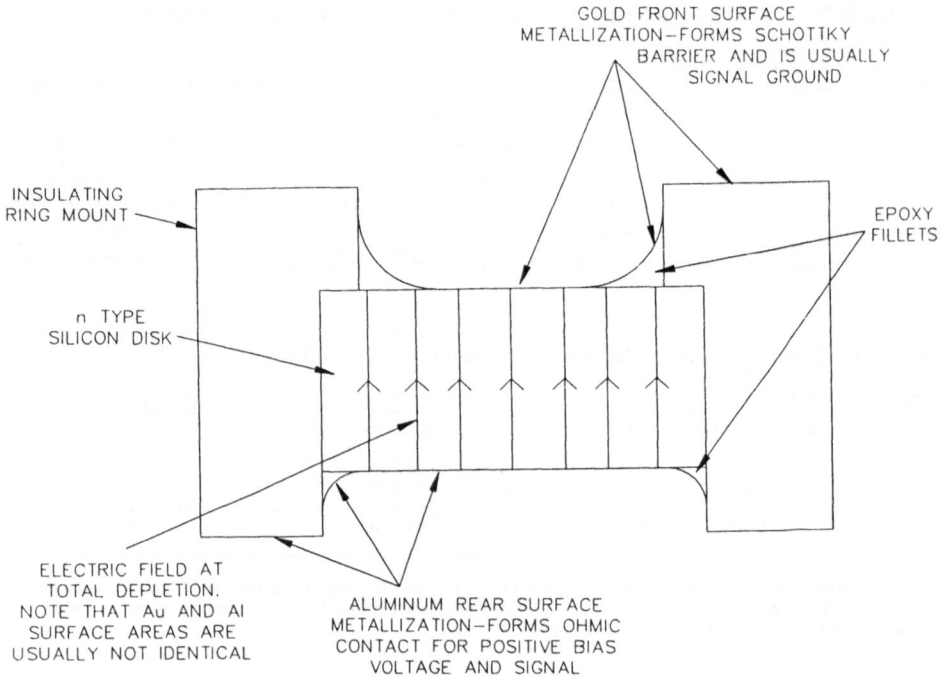

GOLD FRONT SURFACE
METALLIZATION–FORMS SCHOTTKY
BARRIER AND IS USUALLY
SIGNAL GROUND

INSULATING
RING MOUNT

EPOXY
FILLETS

n TYPE
SILICON DISK

ELECTRIC FIELD AT
TOTAL DEPLETION.
NOTE THAT Au AND Al
SURFACE AREAS ARE
USUALLY NOT IDENTICAL

ALUMINUM REAR SURFACE
METALLIZATION–FORMS OHMIC
CONTACT FOR POSITIVE BIAS
VOLTAGE AND SIGNAL

Figure 7. Silicon Surface Barrier Detector Construction, Showing the Imprecise Sensitive Volume Definitions. Note that the Gold and Aluminum Metallizations do not have the Same Area.

The sensitive volume of the silicon surface barrier detector is not very precisely defined because of the different sizes of the Al and Au metallization areas and the size variability in the epoxy fillets (Fig. 7). The precise knowledge of the sensitive volume of a detector is necessary in order to convert the energy deposited by incident radiation into a radiation dose.

The electronic circuitry of the instrument is set up to count separately the incident flux and radiation dose due to particles that deposit between 50 keV and 1 MeV of energy (LOLET), between 1 and 10 MeV (HILET) and above 40 MeV (VHLET). As can be seen from the energy deposition curve for the lowest energy threshold dome, shown in Figure 8, the LOLET flux and dose counts are integral counts for electrons with energies above the electron threshold (1 MeV) and protons above 120 MeV (at a 60° angle to the detector normal). The HILET flux and dose counts are counts for protons with energies between the proton threshold (20 MeV) and 120 MeV. The very rare VHLET flux counts are due to events that deposit very large amounts of energy in the detector; these can be caused by the passage of high energy heavy ions through the detector or the high energy proton-induced nuclear reactions in the detector volume (nuclear stars).

In effect the LOLET counts, and associated dose, are due to incident particles with dose Q factors of 1, while the HILET counts and dose are due to particles (mostly protons) with dose Q factors larger than 1 (see Table 1). In this particular instrument, the incident particles are divided into two energy deposition (and hence 2 Q factor) ranges, HILET and LOLET (a third VHLET channel measures only flux), but in principle this type of instrument can subdivide the total energy deposition range into as many channels as are desirable. This feature makes the SSD dosimeter a good tool for determining the biological effectiveness of the incident radiation.

Radiation dose calibration of the instrument is performed with the aid of radioactive sources and is straightforward. The correspondence of the output pulse height voltages to the energy deposited in the detector by the incident particles (the electronic gain), is determined with a precision pulser, calibrated with 5.48 MeV alpha particles from 241Am and the 477

144

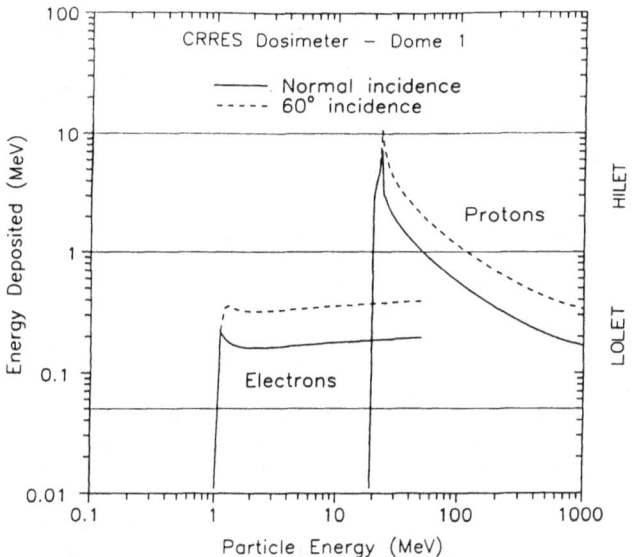

Figure 8.   Energy Loss Curves for the DMSP/F7 and CRRES Dosimeters Dome 1 Detector.

keV Compton edge of the 662 keV ^{137}Cs gamma rays. The thickness of the detector is measured accurately by X-ray transmission methods (Hanser and Sellers (1974)), while the sensitive area is determined from the guard ring geometry. This allows a determination of the sensitive volume of the detector. The electronic gain calibration and the knowledge of the detector volume allow the determination of the dose calibration of each detector.

The instrument can also be used to measure the flux of the incident protons and electrons. While measuring dose, any energy deposition in the photodiodes is counted and added to the total dose (regardless of energy and identity of the incident particle: proton, electron or photon). A flux measurement requires that the incident proton or electron be correctly identified and its initial energy (before the degrader domes) be determined. The knowledge of incident proton and electron flux is useful because radiation transport calculations can then be employed to determine the effect of the particular radiation environment on biological and electronic systems behind shielding of very different thickness than that of the four instrument domes.

The response of the instrument to protons (HILET counts) can be calculated easily, because energetic protons lose energy while traversing the Al domes and the photodiodes without a great deal of angular scattering. Calibration work done at a proton accelerator with monoenergetic proton beams with energies between 20 and 144 MeV verified the accuracy of the calculations.

The electron response of the instrument is complicated by the fact that electrons undergo severe angular scattering while traversing the degrader domes and sub threshold energy electrons can give rise to bremsstrahlung photons. When absorbed by the photodiode these photons produce pulses that exceed the LOLET energy deposition threshold and so are counted as much higher energy electrons. The response of the instrument to energetic electrons has been measured and the results reported in Gussenhoven et al. (1986).

SPACE RADIATION DOSE MEASUREMENTS

DMSP / F7 Dosimeter Measurements

The first SSD dosimeter of the design just discussed was launched on the DMSP/F7 satellite in November, 1983. The DMSP/F7 satellite was in a polar orbit (98.8° inclination) at

840 km altitude, with a 101 minute sun-synchronous orbit (Gussenhoven et al. (1987)). The satellite operated to October, 1987. Incorrect thermal design by the spacecraft manufacturer resulted in the dosimeter operating at 50° C to 60° C, well above the nominal maximum of 40° C. Initial operation was proper, but after about one year in orbit the dome 3 and 4 LOLET channels showed some noise degradation at the end of the sunlit half of the orbit, when the temperature was highest. Beyond this minor degradation, the dosimeter operated properly for the entire satellite lifetime.

A significant component of the dose measured came from the several daily passes through the South Atlantic Anomaly (SAA). The SAA is a region just east of the South American continent near 30° south geographic (15° south geomagnetic) latitude where the inner belt of trapped particles approaches closest to the earth. It is caused primarily by the offset of the earth's dipole magnetic field from the earth's center. The dose measured comes almost entirely from high energy protons. There are secondary belts of dose near 60° geomagnetic latitude, north and south, from the low altitude edges of the outer belt electrons. A plot of the dome 1 LOLET dose contours averaged over 1 year is shown in Fig. 9. Corrected geomagnetic coordinates are used, since they yield a much greater regularity in the trapped particle data by reducing the distortions produced in geographic plots by the offset, tilted magnetic dipole of the earth. This is illustrated in Fig. 10, which shows in exaggerated form the effect of the offset dipole on the radiation belts in a half-plane through the SAA. The SAA is the closest approach of the inner (primarily proton) belt to the earth's surface (and upper atmosphere). The outer (mostly electron) belt has almost continuous particle precipitation and so is usually detectable at both the north and south extremes.

The outer zone electrons show significant time variations as is shown in Fig. 11, where the dome 2 LOLET channel (electrons >2.5 MeV) count rate average for the north and south outer zone bands near 55° - 60° latitude are plotted for 1984. The time axis is numbered in increments of the 27-day solar rotation period, and some correlation with solar rotation can be seen. The variations for this period cover two orders of magnitude in electron flux.

A plot of the dome 4 HILET (protons >75 MeV) flux count rate for 9 November 1984

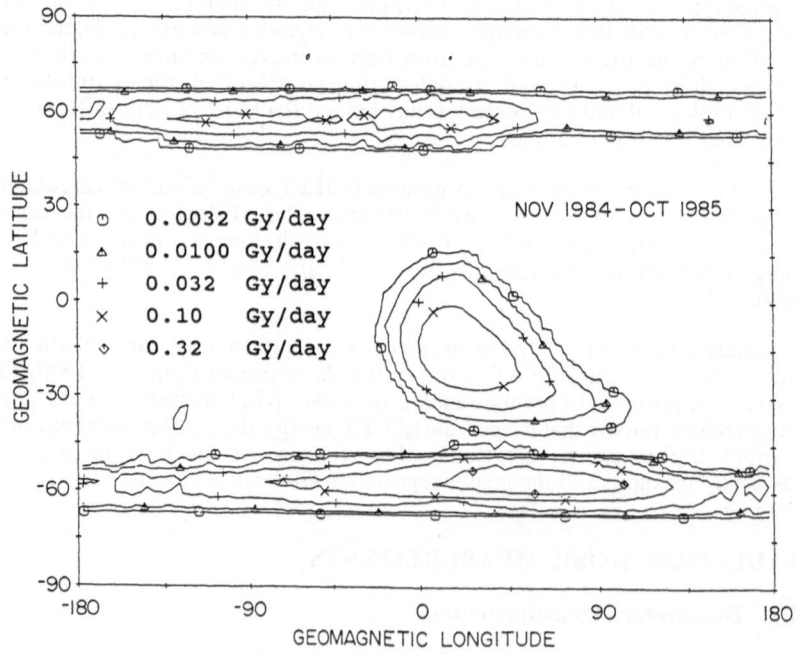

Figure 9.   One Year Average Dose Contours for the DMSP/F7 Dosimeter Dome 1 LOLET (0.05 - 1 MeV Energy Deposits).

146

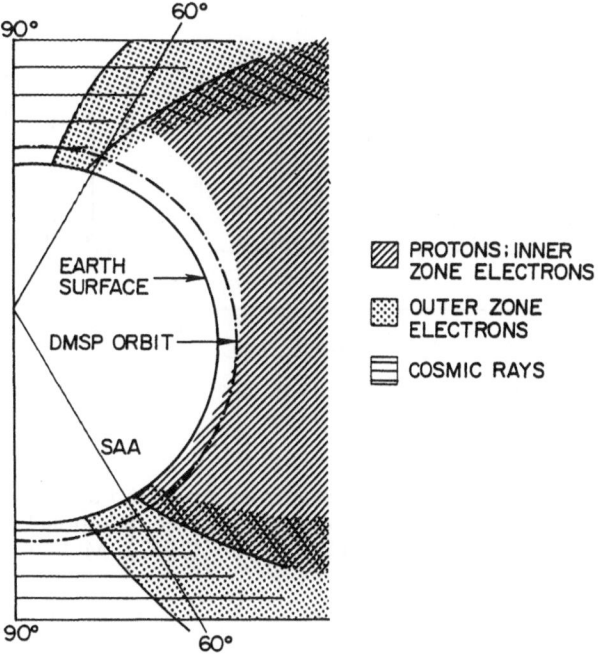

Figure 10. Diagram Illustrating how the Offset of the Earth's Magnetic Dipole Field Produces the SAA. Dimensions are Exaggerated for Clarity.

is shown in Fig. 12. This shows only protons from the SAA, measured on either the north or south going part of the orbit depending on the position of the earth in its daily rotation.

During solar proton events the dosimeter also responds to protons over the polar cap. The event of 26 April 1984 produced a moderately soft (rapidly falling off with increasing energy) proton spectrum, with a strong variation of dose with shielding thickness. This is shown by the dose data in Fig. 13, which is for a north polar cap pass during the peak of the event. The dose comes primarily from the HILET channels, and decreases by two orders of magnitude for a shielding increase from 0.57 g/cm^2 aluminum to 6.08 g/cm^2 aluminum.

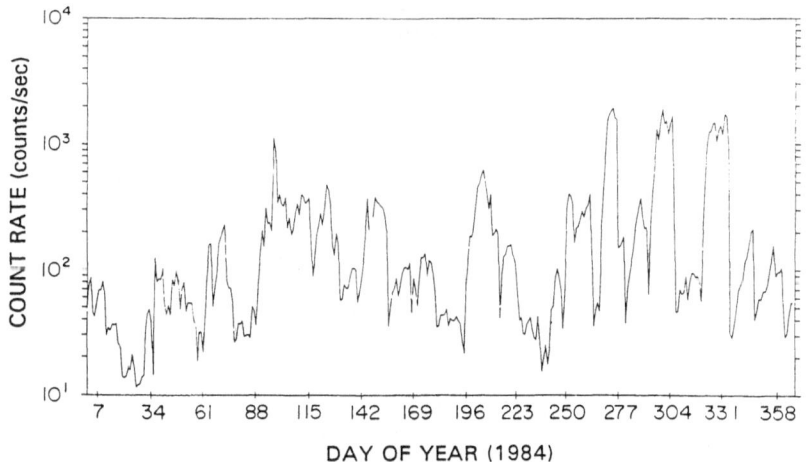

Figure 11. DMSP / F7 Dome 2 LOLET Flux (electrons > 2.5 MeV) Daily Count Rate Average for Outer Zone Electrons During 1984.

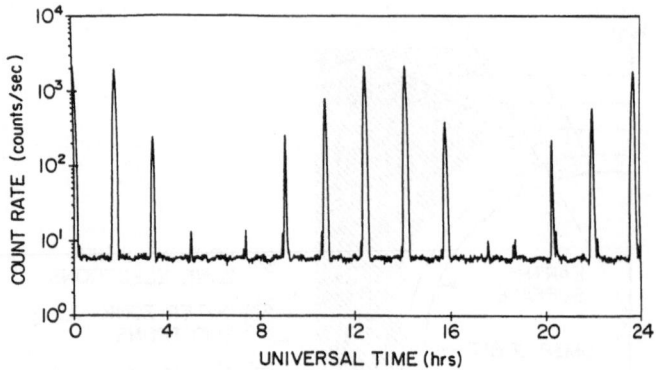

Figure 12. DMSP/F7 Dome 4 HILET Flux (Protons > 75 MeV) for 9 November 1984 Showing the SAA Protons.

The lowest shielding dose is about 3 times that from a pass through the peak of the SAA, while the highest shielding dose is a factor of 10 lower. For additional information on the DMSP/F7 dosimeter results see Gussenhoven et al. (1987), Mullen et al. (1987) and Gussenhoven et al. (1988).

CRRES Dosimeter Results

The CRRES (Combined Release and Radiation Effects) satellite has an SSD type dosimeter, and several MOS type dosimeters in the Microelectronics Test Package (MEP) (see Gussenhoven et al. (1985) and Gussenhoven et al. (1991)). The satellite was launched on 25 July 1990 into an 18° inclination orbit, with a 350 km perigee and 33,500 km apogee. CRRES operated properly until 12 October 1991, when a satellite communications problem occurred. The dosimeter has operated properly since its turn-on on 26 July 1990.

The orbit average dose rates measured by the CRRES dosimeters have shown a significant change with time, arising primarily from electron increases in the outer belt caused

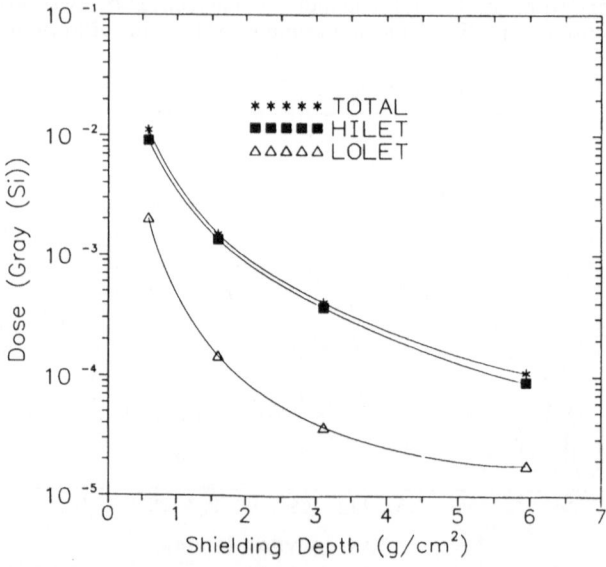

Figure 13. DMSP/F7 Depth/dose Measurement for a North Polar Cap Pass (53817 to 54969 Seconds UT) for the Peak of the 26 April 1984 Solar Proton Event.

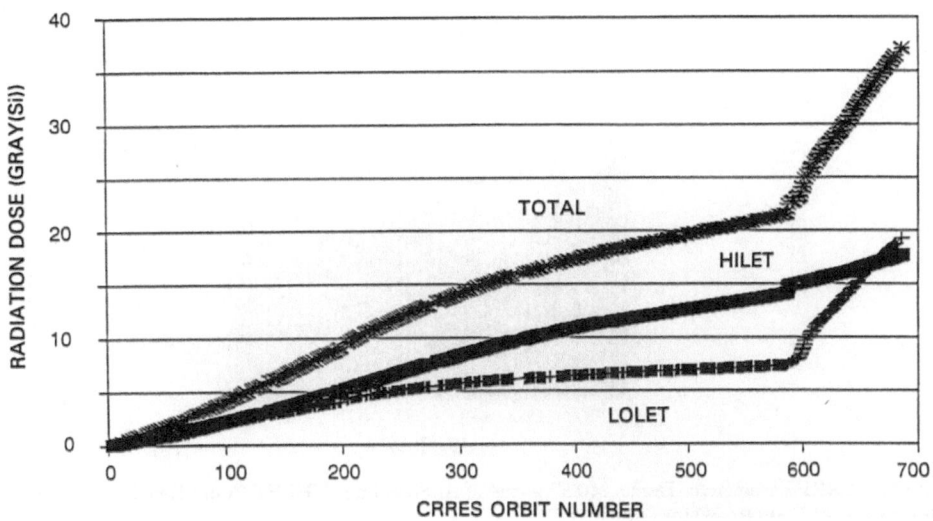

Figure 14. CRRES Dosimeter Measurements for Dome 2 (1.59 g/cm2 Al Shielding) for the First 700 Orbits. (0 = 25 July 1990; 700 = 9 May 1991).

by solar activity. The dome 2 accumulated doses from turn-on (orbit 2; 27 July 1990) to near orbit 700 (9 May 1991) are shown in Fig. 14. A similar plot for three MOS dosimeters in the MEP are shown in Fig. 15, which includes data to past orbit 900. Note that the dose rate increased significantly after the solar proton events of late March and early April 1991. Fig. 15 shows that another increase occurred near orbit 800, and this is associated with a number of strong solar proton events in June 1991. The GOES-7 solar x-ray data (Coffey (1991)) show several very intense x-ray flares in the first part of June 1991.

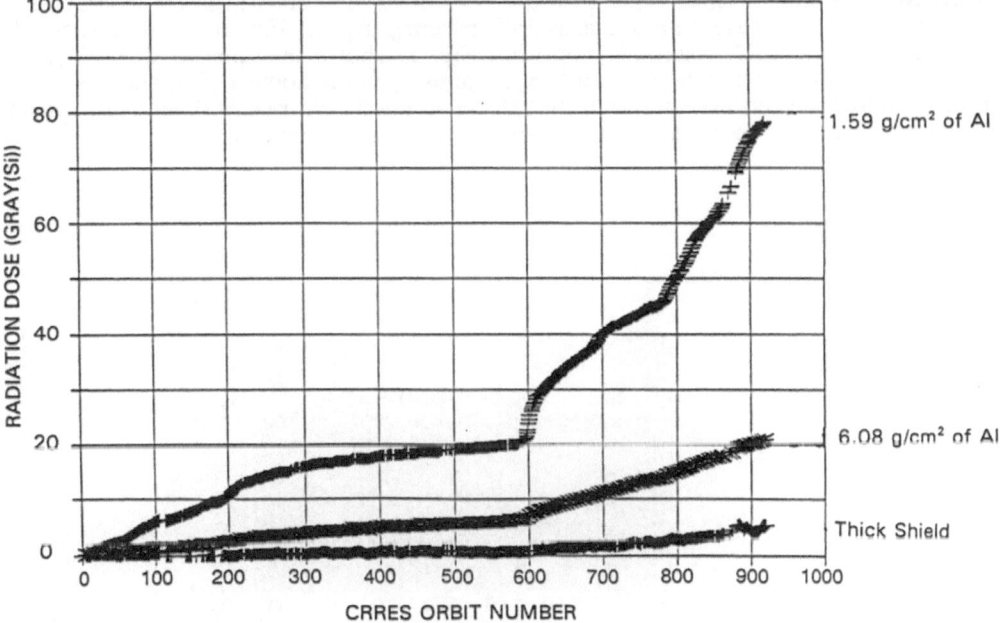

Figure 15. CRRES MEP MOS Dosimeter Measurements for Various Levels of Shielding for the First 900 Orbits (0 = 25 July 1990; 1000 = 9   September 1991).

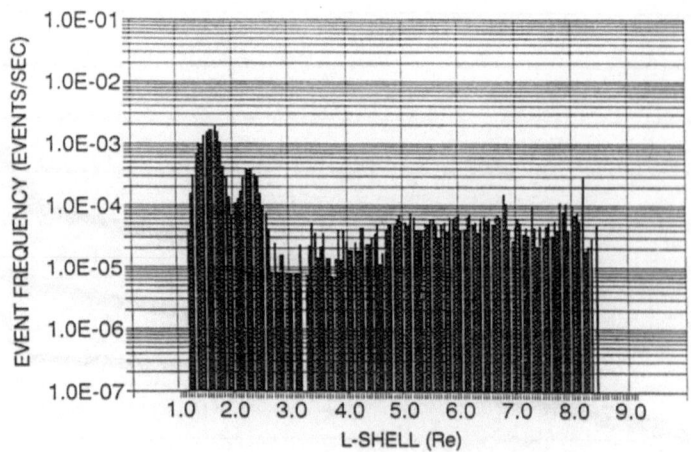

Figure 16. CRRES Dosimeter Dome 1 (0.57 g/cm² of Al Shielding) VHLET Count Rate Distribution in L-shell (in Units of Earth Radii) for Orbits 3 - 838.

The data in Fig. 14 show that the dose properties changed near orbit 600, with the LOLET dose rate from the outer belt electrons becoming dominant. The MEP MOS dosimeter results in Fig. 15 show that the dose accumulated in orbits 600-900 is about three times that accumulated in orbit 0-600, a factor of six increase in dose rate. The large variability seen by the DMSP/F7 dosimeter in the outer belt electron flux (Fig. 11) is consistent with the CRRES data, and shows how solar activity effects can change satellite dose exposures in the trapped radiation belts by orders of magnitude.

A final set of data from the CRRES dosimeter is shown in Figs. 16, 17 and 18, where the VHLET (energy deposition in the detectors of >40 MeV) count rates are plotted against L-shell. These extreme energy deposition events are most likely the cause of single event upsets (SEU's) in electronic devices, and may cause severe damage in living tissue. The data are averaged for orbits 3 - 838 (Figs. 16 and 17) or orbits 1 - 500 (Fig. 18) to provide a sufficiently large data base to make a statistically meaningful plot. Fig. 16 shows the dome 1 (0.57 g/cm² Al) VHLET count rate plot, while Fig. 17 shows the same for dome 2 (1.59 g/cm² Al). A trapped particle belt near L = 2.3 appears in the dome 1 plot, and this is the result of trapped belt changes in March 1991 (these occurred near orbit 600; see Figs. 14 and 15), and appears to be the result of the formation of a second proton belt.

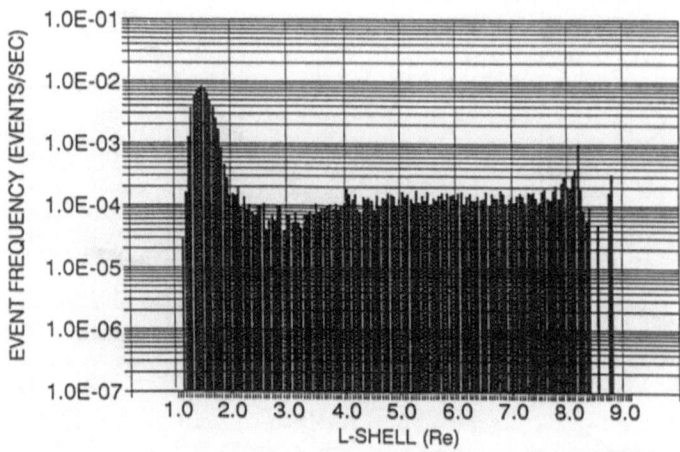

Figure 17. CRRES Dosimeter Dome 2 (1.59 g/cm² of Al Shielding) VHLET Count Rate Distribution in L-shell for Orbits 3 - 838.

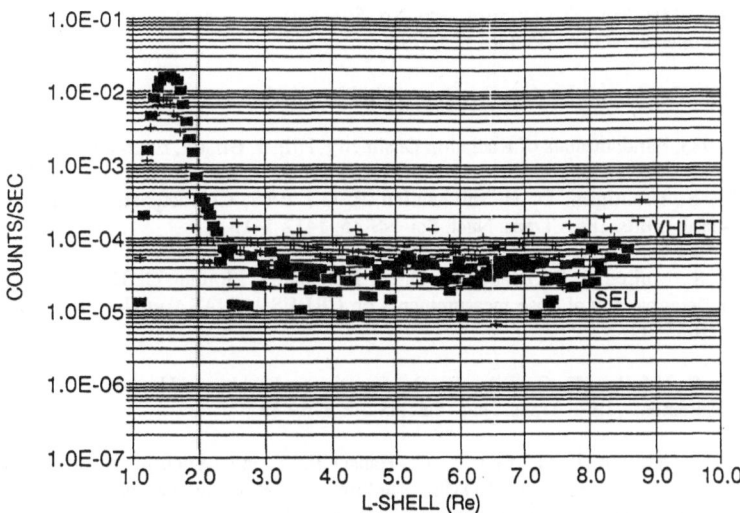

Figure 18. CRRES Dosimeter Domes 2 and 3 VHLET and MEP SEU Count Rate Distribution in L-shell for Orbits 1 - 500.

Fig. 18 shows the L-shell distributions for the dosimeter domes 2 + 3 VHLET count rates, and the MEP SEU count rates from about 500,000 tested bits. The two distributions track quite well. Above L = 2.5 they are equal to within a factor of 2, and show a very slight increase toward large L, showing the response to high energy cosmic rays. The relation changes by about a factor of four for the inner belt protons, and is probably caused by the proton spectrum change (GCR spectrum is harder in the few hundred MeV range) as well as the higher Z component of the GCR. The VHLET measurements of the dosimeter are thus a good predictor of SEU type phenomena in electronic devices.

ACKNOWLEDGMENT

We wish to thank E. G. Mullen, M. S. Gussenhoven, and D. A. Hardy of the Phillips Laboratory, Hanscom AFB, for their help in various aspects of the DMSP and CRRES dosimeter programs; they also provided some of the data summaries necessary for the figures. F. B. Sellers of Panametrics, Inc. was involved in much of the original design work on the DMSP and CRRES dosimeters, and provided some useful comments on this paper.

REFERENCES

Adams, L. and Holmes-Siedle, A., 1978, The development of an MOS dosimetry unit for use in space, IEEE Trans. Nucl. Sci. NS-25:1607-1612.

Attix, F. H., 1986, Introduction to Radiological Physics and Radiation Dosimetry, Wiley, New York.

August, L. S., 1984, Design criteria for a high-dose MOS dosimeter for use in space, IEEE Trans. Nucl. Sci. NS-31:801-803.

August, L. S., Circle, R. and Ritter, J. C., 1983, An MOS dosimeter for use in space, IEEE Trans.Nucl. Sci. NS-30:508-511.

Berger, M. J. and Seltzer, S. M., 1982, Stopping Powers and Ranges of Electrons and Positrons,NBSIR 82-2550, U. S. DOT, NBS, Washington, D. C.

Coffey, H. E. (Ed.), 1991, Solar-Geophysical Data, No. 563, Part 1, National Geophysical Data Center, U. S. Department of Commerce, NOAA, Boulder, CO, USA.

Evans, R. D., 1955, The Atomic Nucleus, McGraw-Hill, New York.

Greening, J. R., 1985, Fundamentals of Radiation Dosimetry, Hilger, Bristol, England.

Gussenhoven, M. S., Brautigam, D. H. and Mullen, E. G., 1988, Characterizing solar flare high energy particles in near-earth orbits, IEEE Trans. Nucl. Sci. NS-35: 1412-1419.

Gussenhoven, M. S., Mullen, E. G., Brautigam, D. H., Holeman, E., Jordan, C., Hanser, F. and Dichter, B., 1991, Preliminary comparison of dose measurements on CRRES to NASA model predictions,IEEE Trans. Nucl. Sci.NS-38: 1655-1662.

Gussenhoven, M. S., Mullen, E. G., Filz, R. C., Brautigam, D. H. and Hanser, F. A., 1987, New low altitude dose measurements, IEEE Trans. Nucl. Sci. NS-34: 676-683.

Gussenhoven, M. S., Mullen, E. G., Filz, R. C., Hanser F. A. and Lynch K. A., 1986, Space radiation dosimeter SSJ* for the block 5D/flight 7 DMSP satellite: calibration and data presentation, AFGL-TR-86-0065, Phillips Laboratory, Geophysics Directorate, Hanscom AFB, MA.

Gussenhoven, M. S., Mullen, E. G., Sagalyn, R. C., editors, 1985, CRRES/SPACERAD experiment descriptions, AFGL-TR-85-0017, Environmental Research Papers, No. 906, Air Force Geophysics Laboratory, Hanscom AFB, MA.

Hanser, F. A. and Sellers, B., 1974 Measurement of totally depleted silicon solid state detector thickness by x-ray attenuation, Rev. Sci. Instrum. 45: 226-231.

Hubert, F., Bimbot, R. and Gauvin, H., 1990, Range and Stopping Power Tables for 2.5-500 MeV/Nucleon Heavy Ions in Solids, Atomic Data and Nuclear Data Tables, 46: 1-213.

ICRP, 1977, Publ. 26, Recommendations of the Intl. Comm. on Radiological Protection, Ann. ICRP 1, No. 3.

Janni, J. F., 1982, Proton Range-Energy Tables, 1 keV - 10 GeV, Atomic Data and Nuclear Data Tables, 27:Part 1. Compounds 147-339; Part 2. Elements 341-529.

Morel, P. R., Hanser, F., Sellers, B., Hunerwadel, J. H., Cohen, R., Kane, B. D. and Dichter, B. K., 1989, Fabricate, calibrate and test a dosimeter for integration into the CRRES satellite, GL-TR-89-0152, Phillips Laboratory, Geophysics Directorate, Hanscom AFB, MA.

Mullen, E. G., Gussenhoven, M. S., Lynch, K. A. and Brautigam, D. H., 1987, DMSP dosimetry data: a space measurement and mapping of upset causing phenomena, IEEE Trans. Nucl. Sci. NS-34:1251-1255.

TIME RESOLVING DETECTOR SYSTEMS FOR RADIOBIOLOGICAL INVESTIGATIONS OF EFFECTS OF SINGLE HEAVY IONS

J.U. Schott

DLR FF-ME, Linder Höhe
5000 Köln 90, FRG

ABSTRACT

The combination of the properties of Charge Coupled Devices (CCDs) as image sensor and as detector for energetic charged particles makes then unique for time resolved radiation experiments in which single particles are to be correlated either to their radiation effects in individual objects, or to their sources or origins respectively. The experimental set up for the calibration of a frame transfer CCD type VALVO NXA 1011 with heavy ions from accelerators is described. For low energetic heavy ions, the single particle effects observed in the pixels of the sensor show a fairly linear response with the Linear Energy Transfer (LET) of the particle. As an example for the application of these detector systems in radiobiology, radiation effects of single ionizing particles in the meristem of moving biologigal objects have been investigated.

INTRODUCTION

The knowledge of the radiobiological effectiveness of single heavy ions represents a substantiated base for models and estimates of radiation risk to men and material in ionizing particle fields, such as in orbital and galactic spaceflight environment. In individual resting biological specimen of different state of organization, significant radiation effects clearly have been correlated to the passage of single heavy ions of the orbital and lunar radiation field. A quantification of the radiation parameters to different endpoints of damage has been performed in the Biostack experiments. Increased radiation loads being expected in future space missions require dose equivalent estimates, based on more reliable data of radiation effects in active metabolizing biological specimens of a significantly higher radio-sensitivity, or in their distinguished organs or tissues, respectively. The investigation in those developing and/or moving systems requires either an online correlation of any particle transversal to the distinguished biological target, or a time resolved record of the particle events, and of the actual position of the object. In both cases, the ionization parameters, the coordinates and the spatial direction of the any particle and the actual position of the target are to be detected.

DETECTION OF CHARGED PARTICLES

In general, energetic charged particles are detected by observing their interaction processes or products when they loose kinetic energy on their passage through absorbing matter. For energetic heavy ions, the main contributions of primary interactions to an observable quantity are inelastic scattering with the electrons and elastic scattering with the nuclei of the absorber atoms. Inelastic collisions with the electrons result in ionization processes and the generation of energetic electrons (δ-electrons) with the consequence of secondary ionizations within the range of the δ-electrons. The energy loss of charged particles through electronic processes being proportional to Z^2/b^2 has been quantified in the Bethe Bloch formula and its derivates. The local dissipation of the energy deposited depends strongly on the macroscopic data of the absorbing material, first of all its electronic properties. Following a theory of Kobetich and Katz (1968), the lateral distribution of the ionizing energy has been calculated for several materials of concern (i.e. Stapor, McDonald, 1988).

Compared to the inelastic scattering with electrons, elastic interactions with nuclei are rare events, as long as the penetrating particle is not close to its stopping point. However, in solid matter they result in displacements of atoms and lattice defects which, in many cases, can be detected as a permanent damage of the matrix. Both scattering processes and endpoints are not independent to each other, but cross linked to some extend.

Other mechanisms of interaction, like Bremsstrahlung, are of less concern for heavy particles (proportional to $1/M^2$), as their contribution to the total energy loss less than 0.1% even at high energies and atomic charge numbers, as in the case of Cerenkov radiation.

Materials and mechanisms for particle detection

The transformation of the primary effects of heavy charged particles into a quantity suitable for quantification is determined by a long chain of secondary processes in the absorbing material and of processes aiming at an amplification of the radiation damage up to the necessary level for detection. The endpoints of the secondary processes are mainly determined by the lifetime and mobility of the charge carriers.

In materials with a high electronic conductivity, the separated electrons and holes undergo fast recombination before any appreciable dissipation of energy to the lattice can occur. For single lattice disturbancies in solids no amplification processes are known which make them accessible for an easy read out. Their only practical use for particle detection is to act as material for a faraday cup from which the charge of an ion beam is directed to an charge sensitive amplifier in order to determine the product of the number of particles and its mean charge state. In materials with low or medium conductivity, a variety of secondary processes lead to either the manifestation of detectable atomic displacement, to the excitation of the atoms of the absorber or to the production of free electrons and holes with lifetimes that permit their separation in an electric field.

In materials with low conductivity the radiation effects result in either excitations or, in solids, in the establishment of stable atomic displacements along the trajectory of the particle predominantly. Using adequate physical or chemical processes, these disturbancies get amplified up to optical visibility or the feasibility of electrical scanning. Thus, the fundamental properties of these materials for particle detection are to accumulate trajectories and, by quantification of the observable, to yield information on the energy deposition of the particle per unit path length, the linear energy transfer (LET) and its lateral distribution. These data can be used for the determination of the atomic charge number and the energy of the particle.

The best known groups of these Solid State Nuclear Track Detectors (SSNTD) make use of either the visualization of phase precipitations in crystals (grains of nuclear emulsions or silver chloride single crystals with chemical or physical redox processing of silver) or on the visualisation of molecular damage in glasses or dielectrics with etching techniques (etching detectors: i.e. cellulose nitrate, CR39, polycarbonates). SSNTDs have found a broad field of applications due to some excellent features: Their typical spatial resolution, including read out, is in the order of a few tenth of one micron, they are completely passive i.e. they accumulate transversals of particles without any support. Thus they are predestined for the detection of rare events and for the analysis of complex nuclear interactions. The discovery of high energetic particles in cosmic rays in 1948 (Freier et al, 1948), the investigation of nuclear disintegrations and, in nearer past, the clear-cut correlation of biological effects in individual organisms with the passage of a single charged particle from the earth orbital radiation field in the Biostack experiments (Bücker et al, 1973) are a few examples, only.

However, the later example demonstrates the intrinsic limitations of SSNTDs as well. With experiments, in which particle transversals get correlated to any other systems, it is essential that the relative position of transversal and object must not change between the exposure to particles and their evaluation. Since SSNTDs do not provide online signals and the visualization of particle track in SSNTDs and their evaluation is still very time consuming, in spite of computer assisted read out and evaluation techniques, the method of the Biostack experiments applying SSNTDs is restricted to the investigation of biological objects with a long shelf life in dormant states. Higher radiobiological sensitivity of metabolizing and active moving biological objects and higher relevance of the relative biological effectiveness (RBE) from these systems for radiation risk estimates for man in spaceflights calls for an extension of this method towards investigations with moving objects. This requires detecting systems with spatial resolution and online read out for localization. Online systems either make use of the detection of excitations (scintillation counters), or on the separation of charge carriers in an electric field with subsequent transformation into a voltage signal by a charge sensitive amplifier. Representatives for gaseous systems of the later type are ionization chambers with its different working conditions as proportional counter or Geiger Müller counter. The development of solid state semiconducting structures based on germanium and silicon (surface barrier detector, PIN diode) started, after McKay observed an electric signal from a pn junction in germanium being exposed to alpha particles (McKay, 1951). With an intrinsic time resolution of some 10^{-9} seconds, all these detectors are worthwhile for applications in radiobiology.

Their spatial resolution, however, is given by the overall size of the volume from which electric charge is collected (sensitive volume). In order to overcome the lack of spatial resolution these detectors have been developed further, making use of two different techniques: to subdivide the sensitive volume into smaller volumes by replacing the electrodes of the external electric field by sets of parallel wires, each of which hooked onto a single charge sensitive reading system (multi wire proportional counter (Melissinos, 1966), or micro strip detector, respectively (Hyams et al, 1983)), or, to calculate the position of charge generation from the charge carrier mobilities and the time delay between the detection of electrons and holes (gaseous and solid state drift chambers (Gatti and Rehak, 1984)). In spite of some skilful measures for the reduction of read out electronics, for instance by using capacitive or resistive coupling of the single electrodes (Gerber et al, 1977), these spatially resolving detectors, as well as the scintillation counters, require extensive and fast online read out electronics. With solid systems, spatial resolutions of a few microns have been achieved, gaseous systems are limited by the diameter and the mechanical stability of the wires to tens of millimeters.

In the last two decades high integrated silicon devices have been developed for several purposes of electronic techniques. Charge sensitive structures especially in microprocessors, memory devices and optical sensors are candidates for radiation induced errors. In memory devices, the critical charge to setting a bit flip is in the order of some 10^{-13} C, thus, the tranversal of a single particle with an LET $> 5 \times 10^3$ MeV cm^2/g in silicon through the sensitive volume of a typical storage element can change its electronic status and thus cause Single Event Upsets (SEUs) (Pickel and Blandford, 1978). The occurrence of SEUs - in electronics raise severe problems with proper function and reliability in particle radiation fields (Binder et al, 1975) - can be used for particle detection with a spatial resolution determined by the size of the sensitive volume.

For a quantification of the parameters of the particles, charge sensitive structures with an analog response to charge deposition are of higher benefit than digital circuits. The sensitivity of Charge Coupled Devices (CCDs) as optical imaging sensor is limited by a few thermally free electrons and the noise of the read out electronic, only, and thus higher by a factor of about 10^2. They have been proposed for particle detection in 1970 by Boyle and Smith (1970). A detailed description of CCDs and their use for tracking of particles in high energy physics has been given by Damerell et al, 1981 and Bailey et al, 1983. Their application in radiobiological investigations, following the concept of the Biostack experiments, with single particle correlation on moving biological objects has been demonstrated in accelerator experiments with seedlings of *Arabidopsis thaliana* (Schott, 1988).

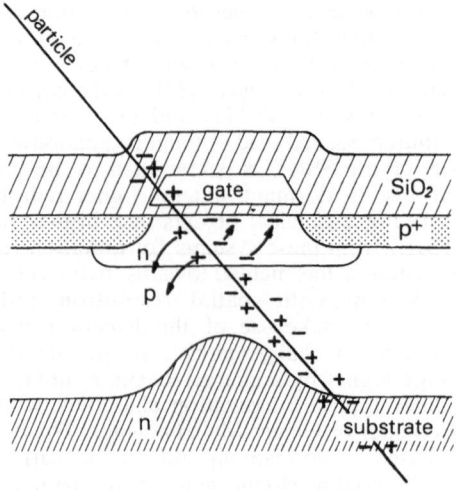

Figure 1.   Section through a Pixel of a CCD.

PARTICLE DETECTION WITH CHARGE COUPLED DEVICES

Charge coupled devices consist of a matrix of MOS diodes (pixels). Compared to other semiconducting detectors with spatial resolution, the electrons being produced by the passage of a charges particle are stored in the sensing elements. Subsequent read out with TV techniques yields information on the position and stored charge and thereby on the penetration of single particles.

For imaging purpose, three technical concepts for the storage and read out design have been discussed, an Interline-, XY- and Frame Transfer-type (Collet et al, 1986). For particle detection and localization, the frame transfer concept guarantees a full area charge sensitivity with a clear-cut distinction between charge production and its local storage in the corresponding pixel element. Fig. 1 shows the structure of a pixel. A transversing relativistic proton generates about 100 electron hole pairs per micron track length. Within the depletion volume, the electrons are swept into the interface to the insulating silicon dioxide of the positive gate, whereas the holes migrate to the substrate. Most of the electrons generated in the non depleted volume of the p doped layer undergo either recombination with holes or will be lost in the substrate. Only a few of them might contribute to the stored quantity at the interface. Due to very short diffusion lengths of the charge carriers, the high doped p+ substrate does not contribute to charge collection. Its function is to isolate neighbouring depletion volumes and thus to define the size of one pixel along a TV line. The border towards the other direction is defined by a pattern of electrodes set on external potentials. For read out purpose, these potentials are switched sequentially. The stored electrons are swept line by line from one pixel through the next into a register and transformed into a sequential TV-signal.

The bended substrate below each pixel in Fig. 1 supports the function of the p+ substrate and the potential pattern. In case of an overloading by high ionizing rates it acts as a sink for charge carriers and restricts non controlled charge spreading over the entire neighbouring pixels, a technological concept, which avoids blooming effects in optical applications. High charge deposition might cause bias voltage shifts as well and contribute, due to the formation of electron trapping sites by atomic displacement of silicon atoms, to an increase of dark current. Both, bias shift and dark current increase, are important contributions onto the decrease of reliability and lifetime of these sensors. They are more dominantly caused by sparsely ionizing radiations at doses higher than 10^2 Gy. For the detection of single particles in low intensive radiation fields these effects are of minor concern. High charge depositions by single particles, however, may cause permanent destruction of the insulating layers and result in a burn up of single pixels.

Since a quantitative correlation of the signals from CCDs to the parameters of transversing particles are not available for modern frame transfer sensors with high spatial resolution, calibration work and its results are reported in more detail.

We decided to investigate the radiosensitivity of the VALVO frame transfer sensor type NXA 1011 using standard TV techniques and commercial image analyzing equipment. The 1/2" format sensor consists of 600 x 576 pixel elements corresponding to a pixel size of 10 x 15 mm^2. It is operated without cover glass in a TV-camera type AQUA HR 600 hooked on an image analyzing system BM 901 with a storage capacity of 8 frames with 8 bit resolution in real time. Fig. 2 shows the experimental set up for calibration work at accelerators.

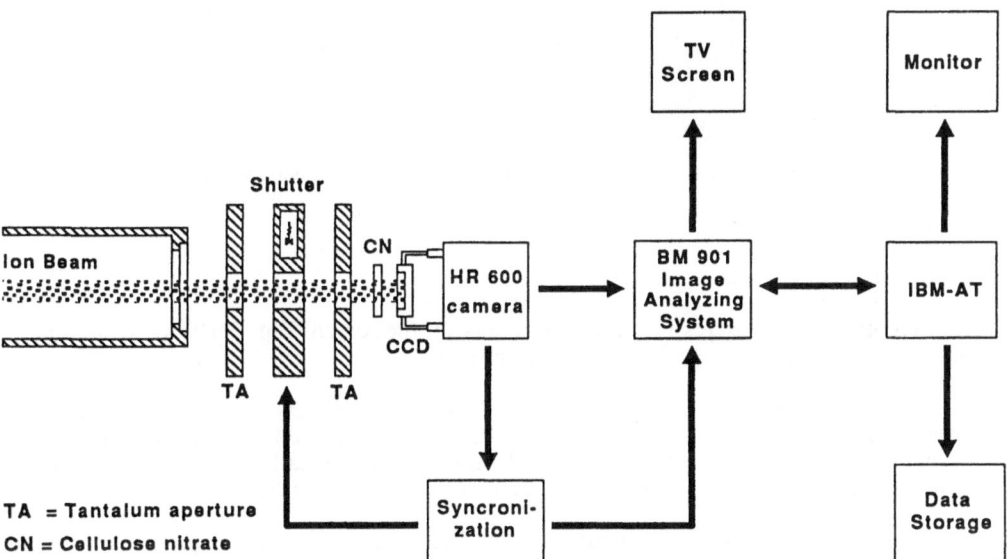

Figure 2.   Experimental Set up for CCD Calibration Work and, with Some Modifications, for Radiobiological Investigations of Single Particle Effects in Metabolizing Objects.

In order to avoid potential permanent damage of the pixel elements, the particle exposures have been limited to a single integration phase of one TV-frame only. The vertical synchronization signals of the camera have been used for a synchronized single shot operation of a 3 mm thick tantalum particle shutter. Aside the storage of this frame (particle image), subsequent frames are taken at closed shutter position in an exponential time scale and are stored for the off-line analysis of permanent damage of single pixel elements. The pixelwise correction of these frames with those taken shortly prior to particle exposure, permit the elimination of detection and calibration errors caused by potential permanent set up from previous detecting cycles as well as from slightly different sensitivity of single pixels and read out electronics. For exposures with particles of sufficiently high energy and LET, a 100 μm thick cellulose nitrate detector (CN) was placed in front of the CCD surface to clearly discriminate heavy ion effects from those of any other sources, like electronic noise or background radiation.

Exposures have been performed with protons, alpha particles and low energetic heavier ions at the accelerator of the University of Frankfurt, with heavy ions of medium energy at GSI, and at GANIL with even higher energies at typical fluxes of 10^4 - 10^5 particle/cm^2 s.

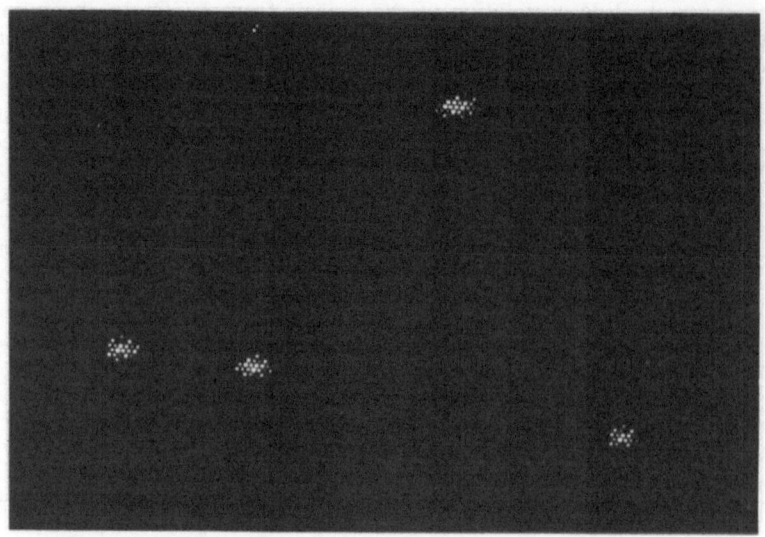

Figure 3.   Radiation Effects of Protons, Visualized on a Colour TV Screen.

Evaluation and results

The particle exposures result in bright dots on the screen. Fig. 3 shows sectors of a colour TV screen with the visualized effects of protons.

For the quantitative evaluation of particle radiation effects in the CCD pixels, an extended software package for image analysis has been developed (Lichtenberg, 1989). It recognizes particle radiation effects, reduces the image data to small data packages of single particle effects and permits its statistic evaluation with the following options:

Pixelwise correction of the particle image with an image taken prior to the particle image, in order to exclude potentially pre damaged pixels;

Calculation of the mean value of those pixels obviously not belonging to a particle effect (background), using an iterating mathematical algorithm;

Definition of pixels with a significant higher amplitude than the background;

Putting together neighbouring significant pixels as single particle events;

Setting up a reduced data file, including all particle events of one image.

Table 1 shows an event from an uranium ion of 15 MeV/u in an 8 bit digitized half image. The low background is due to a correction of the pixel matrix with an image taken before particle exposure. Subsequent software can evaluate statistically the events of up to 8 reduced data files with respect of selectable criteria for the exclusion of other events than proper single events, such as minimum extension in x or y, minimum sum of all pixels or minimum sum of all pixel values.

Fig. 4 shows the spectra of the measured sum of the pixel elements belonging to single particle effects (signal/event) of protons with an energy of 0.5 MeV, 2.0 MeV and 5.0 MeV and of Argon ions with an energy of 0.5 MeV/u. In order to exclude contributions of some noisy single pixels, events with less than two pixels have been excluded. Fig. 5 shows the electronic signal summarized over all pixels belonging to one particle event against the LET of the particle or the number of free electrons being produced in the sensitive volume of the sensor, respectively, for low energetic particles (< 5 MeV/u) at normal incidence.

Table 1. Single Particle Event in a NXA 1011 Charge Coupled Device, caused by an Uranium ion of 15 MeV/u (Numbers give the 8 Bit Digitized Read out of the Pixels of the Corrected Densor Matrix).

Line	Column								
	610	611	612	613	614	615	616	617	618
324	0	0	0	0	0	0	1	0	0
322	0	0	1	0	0	0	0	0	0
320	0	0	0	0	0	0	0	0	0
318	0	0	30	156	134	48	36	0	0
316	0	2	48	196	161	48	37	0	0
314	2	0	21	128	115	39	17	0	0
312	0	0	0	4	4	0	0	0	0
310	0	0	0	0	0	1	0	0	1
308	0	0	1	0	0	0	0	0	0

Discussion

Assuming, that the effect of charged particles is based on ionization and charge separation only (damage on the semiconducting matrix, the insulation layers and dynamic effects being neglected), the pixel elements should show a linear response over a broad range of LET. Its lower limitation is given by the reset noise of thermal electrons and corresponds to a particle LET of about 6 MeV cm^2/g at room temperature. The upper limit is given by the storage capacity for electrons at an LET of some 10^4 MeV cm^2/g.

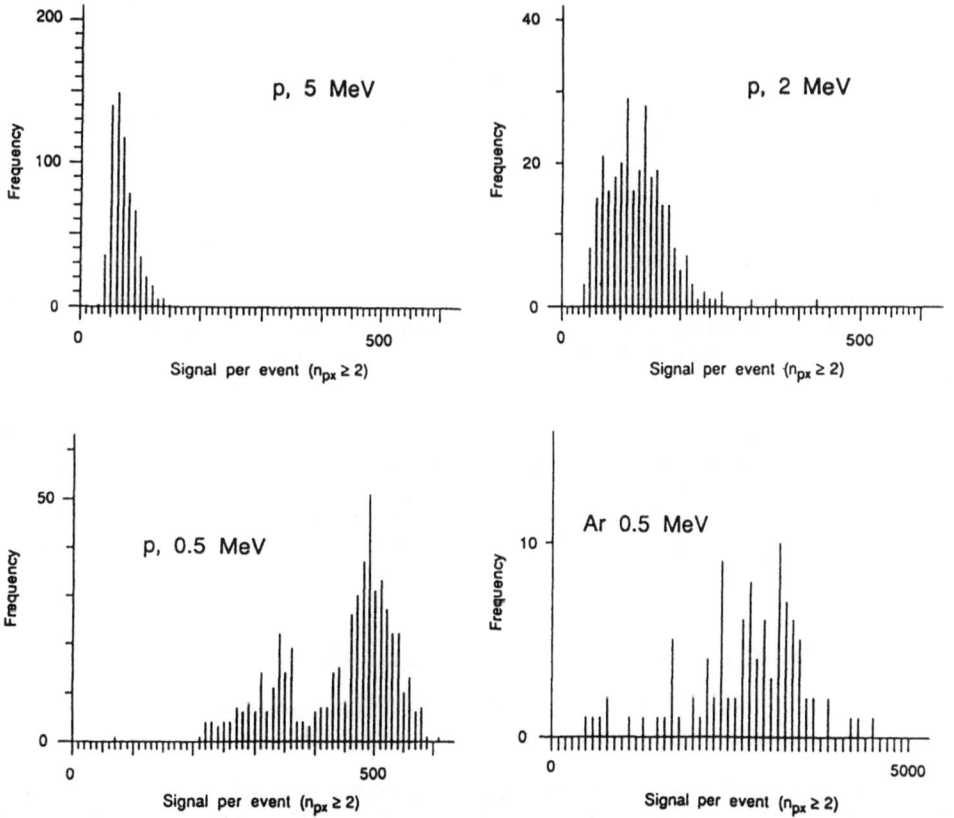

Figure 4. Spectrum of the Measured Signal per Particle Event for Protons & Argon Ions of Different Energy .

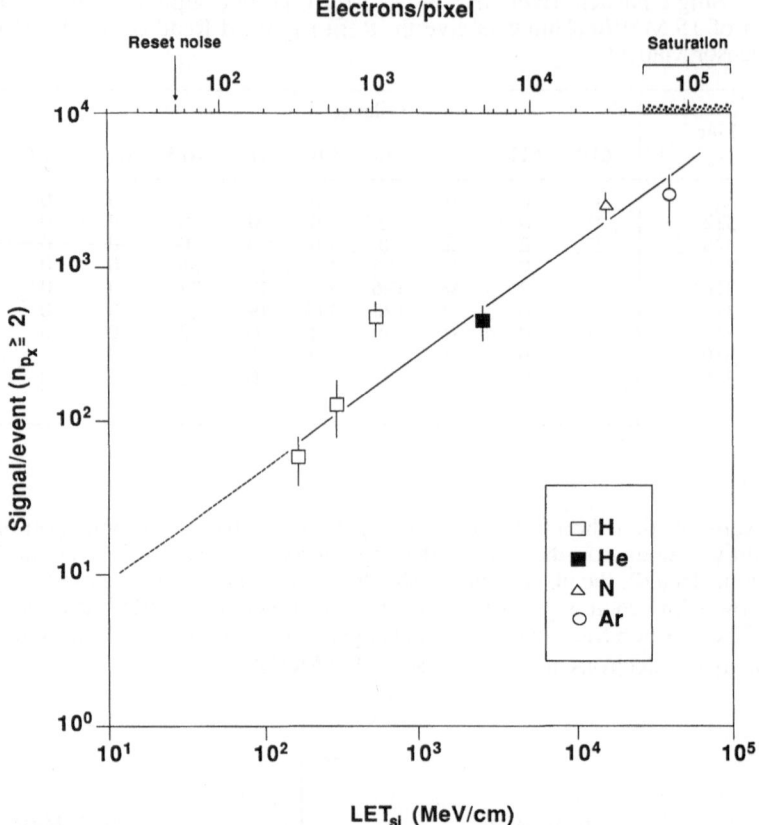

Figure 5.   Response of CCD Type NXA 1011 Plotted against LET. The Upper Abscissa Gives the Transformation of LET into the Number of Electrons per Pixel Element Neglecting other Mechanisms than the Production of Free Electrons.

For low energetic heavy ions, the linearity of the response with LET seems to be reasonable. However, first results of exposures with high energetic Xe-ions of 40 Mev/u and 400 Mev/u at GANIL and the SIS at GSI show a reduced response and give rise to doubt, that the radiation effects are to be described by the LET of the particle, only.

For single particles with LET < 2×10^4 MeV cm^2/g no permanent damage has been observed in subsequent frames (Dt = 20 ms) at room temperature. In precursor experiments, with a less sophisticated data acquisition system, permanent damage has been observed (Schott, 1988). Vanadium and Krypton ions with LET values of 3×10^4 and 4×10^4 MeV cm^2/g as well as uranium of 10^5 MeV cm^2/g set permanent damage with a lifetime of several days at room temperature.

RADIOBIOLOGICAL INVESTIGATIONS WITH CCDs

The combination of optical imaging and ionizing particle detecting make CCDs an easy tool for single particle correlation experiments with time resolution.

The experimental concept is based on a simultaneous or nearly simultaneous online detection and correlation of particle transversals and the spatial position of the biological object. For this reason, one or a few biological objects get prepared on the light sensitive surface of the CCD. In order to keep adequate environmental conditions for the objects of investigation i.e. humidity, the sensors are integrated in a hermetically sealed container with a 10 mm thick Nickel window for particle penetration. The exposure to particles is performed with the set up being similar to that shown in Fig.2. For a better quality of the particle beam (lower number of secondary particles), no cellulose nitrate foil is used. Additionally a light emitting diode close to the beam axis is used for imaging the shadow of the specimen on top of the CCD surface.

Experiment operation

The operation of the experiment is fully software controlled. After a synchronized half image particle exposure of the biologically prepared CCD in total darkness, one single image, or an averaged image of up to 256 frames, respectively, at a low intensive illumination of the chip is digitized and stored in the image analyzing system. The particle image displays single bright dots composed of one or more responding pixels, the subsequent biological image shows the shadow of the objects. For a fast evaluation both images get binarized, the biological image is transverted to an outline image, additionally. A print of the overlay of the binary particle image and the biological outlines in different colours provides a fast dosimetric mapping of any individual. This minimizes the delay with the replacement of the samples into an optimum growing environment and guaranties minimum impacts from stress factors, other than ionizing radiation.

In cooperation with the biogenetic group of A.R.Kranz at the University of Frankfurt, seedlings of *Arabidopsis thaliana* have been exposed to Ar, Xe and Pb-ions at energies between 29 MeV/amu and 45 MeV/amu (Schott et al, 1992). Depending on their size, 7 to 10 seedlings have been prepared on the 4x6 mm² large light sensitive area of the CCD per single exposure.

The seeds were kept in petri dishes on an aseptic synthetic growth medium at 2°C for 72 hours of vernalisation (Bork et al, 1986). During their pre-exposure germination phase of 48 to 60 hours, they have been kept in a growth cabinet at a controlled temperature of 25 ± 2 °C, and a periodical light/dark cycle of 16/8 hours. The typical time for preparing the samples onto the CCDs, the exposure to particles and subsequent disassembly is in the order of 20 minutes. Thereafter, the seedlings were transplanted onto new aseptic growth medium for further investigations of biological damage under the same environmental conditions for 28 days. Under these conditions of handling the biological objects, cell transformation (tumor formation) is the most reliable biological endpoint for the correlation to heavy ion effects.

Results

The experiments performed have been pilot experiments with small sample sizes.In spite of low level confidence, the preliminary results are presented, in order to demonstrate the experimental method and to give an example for a typical biological endpoint of investigation in heavy ion radiation experiments. Table 2 shows the tumor induction frequency in seedlings, irradiated with Pb ions of 29 MeV/amu. Even at very low numbers of particle hits, spread over the entire volume of the seedling, tumor induction gets significant. Most of the tumours originate in the vicinity of the apex meristem. Fig. 6a shows a picture of a typical apex tumor taken 14 days after particle exposure. The non-differentiated cell proliferation mainly covers the apex and the bases of the cotyledons.

Table 2. Tumor Induction Frequency of Pb-Ions of 29 MeV/n in Seedlings of *Arabidopsis thaliana*

particle hits per seedling	tumor frequency (%)	sample size
0	0	7
1-10	7.7	13
11-20	4.5	22
21-30	18.2	11

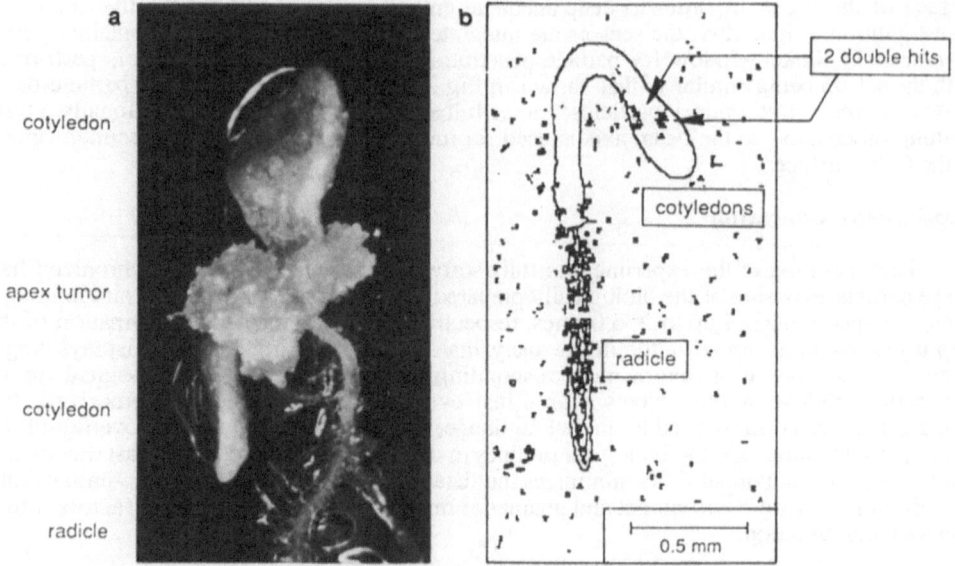

Figure 6.   Seedlings of *Arabidopsis thaliana*, Hit by Pb-ions of 29 MeV/amu. a) Apex Tumor 14 Days after Exposure (Courtesy of A.R.Kranz, University Frankfurt). b) Hit Map (overlay Image) of the Exposure.

The corresponding binary overlay image of the outline of this seedling and the inverted image of particle transversals shows the compact lines of the cotyledon and the hypocotyl and the scattered plot of the radicle (Fig.6b). The pixel clusters from particle transversals are spread over the entire area of exposure, remarkable are the two double hits. Their localization close to the base of the cotyledons suggests a direct correlation to tumor induction.

CONCLUSIONS

CCDs represent an easy to handle compact detector system for energetic particles with spatial resolution and near real time read out characteristics. It combines the detection of ionization processes with the high sensitivity of silicon devices, with the localized storage of the information and the reduction of read out channels on one chip. Together with its properties as image sensor it realizes a simple and cheap experimental approach for the investigation of single particle correlated effects. Their resolution in time and space meets the requirements for many applications, their electronic response on the particle LET is fairly linear for low energetic particles. However, for experiments in unknown radiation fields in which a particle identification is requested, further calibration work is necessary with particles over the entire range of atomic numbers and LET. For radiobiological experiments with moving objects in space, hybrid designs of electronic and solid state nuclear track detectors improve the identification of particles.

ACKNOWLEDGEMENTS

The support with charge coupled devices by VALVO-Bauelemente GmbH(Hamburg) and with exposure facilities at GSI (Darmstadt), GANIL (Caen), and IKF (Frankfurt) is gratefully acknowledged.

REFERENCES

Bailey, R., Damerell, C.J.S., English, R.L., Gillman, A.R., Lintern, A.L., Watts, S.J. and Wickens, F.J., 1983. First measurements of efficiency and precision of CCD detectors for high energy physics, Nucl. Instr. Meth., 213: 201-215.

Binder, D., Smith, E.C.. and Holman, A.B, 1975. Satellite anomalies from galactic cosmic rays, IEEE Trans. Nucl. Sci., NS-22, No. 6, 2675-2680.

Boyle, W.S. and Smith, G.E., 1970. Charge coupled semicoductor devices, Bell System Technical Journal 49: 587-593.

Bork U., Gartenbach K., Koch C. and Kranz A.R., 1986. Biological effects of heavy ions in *Arabidopsis thaliana*, Adv. Space Res. 6-XXI: 149-152.

Bücker, H., Horneck, G., Reinholz, E., Scheuermann, W., Rüther, W., Graul, E.H., Planel, H., Soleilhavoup, J.P., Cuer, P., Kaiser, R., Massue, J.P., Pfohl, R., Schmitt, R., Enge, E., Bartolomä, K.P., Beaujean, R., Fukui, K., Allkofer, O.C., Heinrich, W., Francois, H., Portal, G., Kühn, H., Wollenhaupt, H. and Bowman, G.H., 1973. NASA Apollo 16 Preliminary Science Report, NASA SP-315, 27-1 - 27-20.

Collet, M., Gabler, L., Kürzinger, W. and Euler, G., 1986. Vergleich von Halbleiterbildaufnehmern: Interline-, XY- und Frame Transfer-Konzept, VALVO, Technische Information No. 860310, 1-8.

Damerell, C.J.S, Farley, F.J.M., Gillman, A.R. and Wickens, F.J., 1981. Charge coupled devices for particle detection with high spatial resolution, Nucl. Instr. Meth.: 185: 33-42.

Freier, P., Lofgren, E.J., Ney, E.P. and Oppenheimer, F., 1948. The heavy component of primary cosmic rays, Phys. Rev. 74(No. 12): 1818-1827.

Gatti, E. and Rehak, P., 1984. Semiconductor drift chamber - An application of a novel charge transport scheme, Nucl. Instr. Meth., 225: 608-614.

Gerber, M.S., Miller, D.W., Schlosser, P.A., Steidley, J.W. and Deutchman, A.H., 1977. Position sensitive gamma ray detectors using resistive charge division readout, IEEE Trans. Nucl. Sci., NS-24, No. 1, 182-187.

Hyams, B., Kötz, U., Belau, E., Klanner, R., Lutz, G., Neugebauer, E., Wylie, A. and Kemmer, J., 1983. A silicon counter telescope to study short lived particles in high energy hadronic interactions, Nucl. Instr. Meth.: 205: 99-105.

Kobetich, E.J., and Katz, R., 1968. Energy deposition by electron beams and delta rays, Phys. Rev. 170: 391-396.

Lichtenberg, G., 1989. Erfassung und Auswertung schwerioneninduzierter Ladungsbilder in CCD-Elementen, Diploma Thesis, Fachhochschule Giessen/DLR.

McKay, K.G., 1951. Electron-hole production in germanium by alpha-particles, Phys. Rev. 84(No. 4): 829-832.

Melissinos, 1966. Experiments in Modern Physics, Academic Press, New York.

Pickel, J.C. and Blandford, J.T. Jr., 1978. Cosmic ray induced errors in MOS memory cells, IEEE Trans. Nucl. Sci., NS-25(No. 6): 1166-1171.

Schott, J.U., 1988. Charge Coupled Devices (CCDs) - A detector system for particles with time resolution and local assignment with particle trajectories, Nucl. Tracks Radiat. Meas. 15, Nos. 1-4, 81-89.

Schott, J.U., Kranz, A.R., Gartenbach, K., Zimmermann, M., 1992. Investigation of single particle effects in active metabolizing seedlings of *Arabidopsis thaliana*, Nuclear Physics at GANIL, Scientific Report 1989-1991.

Stapor, W.J. and McDonald, P.T., 1988. Practical approach to ion track energy distribution, J. Appl. Phys. 64 (9): 4430-4434.

MICRODOSIMETRY IN SPACE USING MICROELECTRONIC CIRCUITS

P.J. McNulty, D.R. Roth, W.J. Beauvais, W.G. Abdel-Kader [1] and
E. G. Stassinopoulos [2]

1 Department of Physics and Astronomy
  Clemson University
  Clemson, SC 29634-1911
2 NASA Goddard Space Flight Center
  Greenbelt, MD 20771

ABSTRACT

Single event phenomena in biological organisms are compared with similar phenomena in microelectronic circuits and the similarities and differences are discussed. Both types of phenomena appear to obey the same model, at least to first order. The general applicability of this first-order model appears to reflect the need to operate despite the noise inherent in the storage and processing of information within microscopic volume elements. Arrays of p-n junctions, each having dimensions of a few microns, are proposed as solid-state microdosimeters.

INTRODUCTION

Modern microelectronic circuits respond to the heavily ionizing radiations of space in ways which resemble the effects the same radiations have on biological organisms (McNulty, 1983, 1988 and to be published). These similarities first became pronounced when circuit designers began to design spacecraft systems around microelectronic chips whose transistors had dimensions in the size range typical of biological cell nuclei. At that point, circuits exhibited single-event phenomena initiated either by the traversal of sensitive microstructures by individual cosmic-ray ions or by one or more particles from a nearby nuclear reaction (McNulty, 1983). Moreover, there are similarities in the response curves when the cross sections for single-hit phenomena are plotted versus the LET of the incident heavy ion. In this paper, we explore the possibility that these similarities in response may form the basis for using microelectronic circuits as microdosimeters for astronauts flying on future space missions. We will also demonstrate that accelerator exposures of microstructures can generate the data necessary to test those models which predict the energy deposition within sensitive microvolumes.

A circuit is proposed for measuring the energy-deposition events in an array of as many as two million junctions where each volume element has dimensions typical of biological cell nuclei. Furthermore, commercial lithographic techniques can be used to customize the dimensions and shape to represent different target structures.

Biological Effects and Physics of Solar and Galactic Cosmic Radiation,
Part B, Edited by C.E. Swenberg *et al.*, Plenum Press, New York, 1993

Analogies with Radiobiological Effects

Single-event phenomena are those which are initiated by a single energetic particle. The normal macroscopic units of exposure, dose and dose equivalent, cannot easily be applied to single-event phenomena because they do not predict the probability that a given microstructure will be traversed by an ionizing particle or the probability that its being traversed will produce an effect, with the trivial exception of predicting the exposure to identical monoenergetic particles whose effects have already been measured. In other words, the measured response from exposure to one type of heavily ionizing radiation cannot be used to predict the response to another type of radiation. Single-event phenomena involving damage or a permanent change in at least one microstructure on a device are called hard errors. Chromosome aberrations are biological single-event hard errors. Errors are said to be soft when information or logic flow is altered without any direct damage to any part of the circuit. The device works as well after a soft error as before, only it is now processing erroneous information. Since more data is available for soft errors than for damaging events, they will be emphasized in what follows.

Two types of soft errors are known to have biological analogies. A single event upset (SEU) is said to occur when the information stored in a memory cell is altered without direct damage to the microstructures involved. Its biological equivalent is the somatic mutation which involves changes in the genetic code stored in the DNA. Both types of soft error can have catastrophic effects on the system. Mutations in cells which remain healthy can be harmless or can lead to serious organic illnesses like cancer. Similarly, the loss of information stored in a memory can be important if it is part of a critical instruction. The second type of soft error known to have a biological equivalent is the single-event transient (SET). The SET is a transient signal generated by a radiation event. The SET is treated as a valid logic signal by the system and may lead to some unforeseen consequence. Both SEUs and SETs have been known to trigger catastrophic results such as unscheduled rocket firings, system shutdowns, and loss of contact with the ground station (McNulty, 1991). The visual phenomena experienced by Apollo astronauts are examples of biological SETs (Pinsky et al, 1974).

The propagation of errors resulting from an isolated SEU or SET in a logic network is

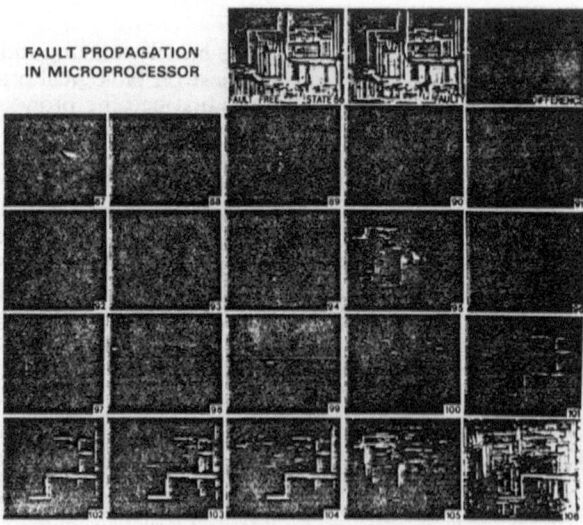

Figure 1.   Error Propagation in the Intel 806186 Picroprocessor. Light and Dark Areas Represent Different Voltage States on the First Two SEM Photographs (Pinsky et al, 1974).

illustrated for the Intel 80186 microprocessor in a series of electron microscope photographs (Fig.1) taken by Tim May (May, 1984). The voltage levels and, therefore, the logic states of the gates of the microprocessor are represented as light and dark areas on the photographs of Fig. 1. The fault free micrograph of the device taken after machine cycle 86 of a programmed sequence is shown in the image on the left in the top row. The next image shows the micrograph taken after step 86 and a single error has been introduced. No difference between such complicated photographs is noticeable to the eye. The third image on the top row is the difference between the first two, i.e., lighted points in the third image represent those pixels which appear lighted in one of the preceding images but not the other. The light spot in the third image on the top row, then, represents the location of the original error. Unfortunately, a single point is barely distinguishable from the photographic reproducing noise in Fig. 1. The microprocessor then operated normally and micrographs were taken after each machine cycle. Each image numbered above 86 in Fig. 1 represents the difference between the fault image and the fault-free image after the same number of machine cycles. The error points spread over the images, contract, then spread again in complicated patterns. It takes a large number of machine cycles, in this case twenty, before any of the errors reach a bond pad . Only after an error reaches the edge of the die is the problem observable to the world outside the chip. The outside world sees a pattern of erroneous information on the pins of the device that bear little resemblance to the simple change that was first introduced. The complications to the system caused by an error of this type are not the direct result of the error but rather the result of the system operating on the false information subsequently generated.

Astronauts experienced SETs in the form of visual experiences both while in deep space during the Apollo program (Pinsky et al, 1974) and while in low earth orbit on Skylab and Shuttle missions (Pinsky et al, 1975). One feature of the light flash phenomena which puzzled workers during the Apollo program can now be seen as similar to what has just been described for logic circuits. Descriptions of the visual experiences given by astronauts on Apollo and those given by scientists exposed under controlled conditions at accelerators typically included features, both physical and temporal, which clearly differed from the physical events at the retina. Figure 2 shows the two types of particle events known to induce visual experiences in space. Traversals of the retina by cosmic rays in deep space generated streaks and large bright flashes (Budinger, Lyman and Tobias, 1972; McNulty et al, 1972) while nuclear reactions induced by protons in the portion of the radiation belt known as the South Atlantic Anomaly produced point flashes (Pinsky, 1975; Rothwell, Filz and McNulty,

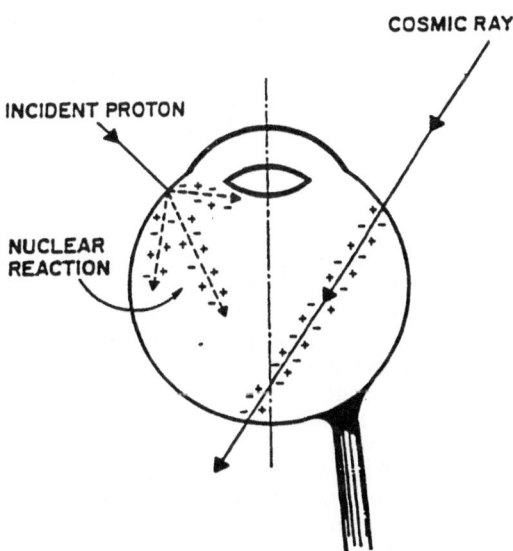

Figure 2.   Cosmic Ray Incident on the Retina of the Human Eye and a Nuclear Spallation Reaction. These are the Two Physical Events Known to Initiate Visual Phenomena in Astronauts Flying in Space.

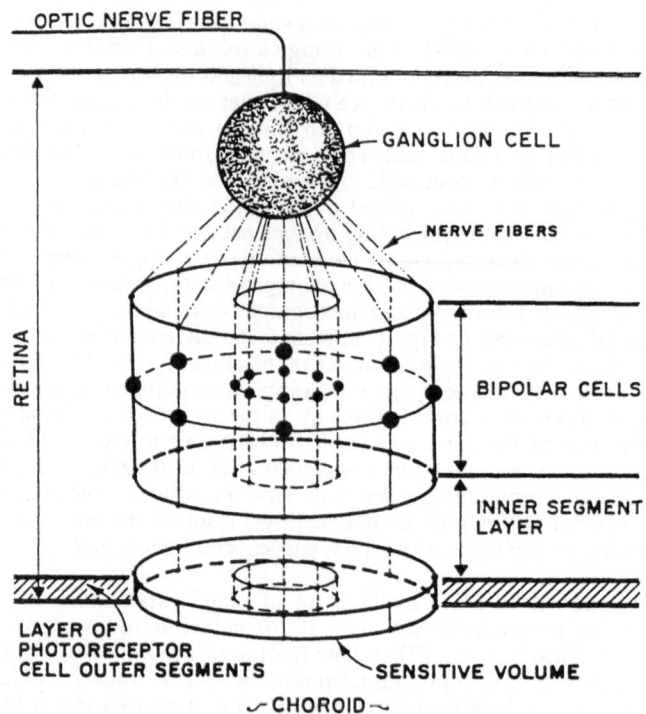

OPTIC NERVE FIBER

GANGLION CELL

NERVE FIBERS

RETINA

BIPOLAR CELLS

INNER SEGMENT
LAYER

LAYER OF
PHOTORECEPTOR
CELL OUTER SEGMENTS

SENSITIVE VOLUME

CHOROID

Figure 3. Schematic of the Functional Visual Signal Processing Unit on the Peripheral Retina. The Axons of the Ganglion Cells Form the Optic Nerve Fibers which Lead to the Central Nervous System.

1976). The nuclear events generated short-range recoiling nuclei which deposited a relatively large amount of energy within a small region of the retina. The fact that such localized generations of ionizations result in the experience of point flashes of light was predictable. However, the streaks observed in deep space and at accelerators did not appear to correlate so nicely with the physics of the inducing events. The streaks often had curvature and were broken in two or three places. Large bright flashes sometimes were followed by slower streaks elsewhere. This temporal sequence was consistently reported despite the fact that the particle events were obviously instantaneous compared to the time constants of any neural processing. Moreover, their trajectories were straight and intersected the retina at one, or at most, two places. These apparent discrepancies clearly result from neural processing by the retina and the central nervous system.

The First-Order Model

The prediction of soft-error rates in space began with the light-flash phenomena. The organization of the retina and its response to optical light at near threshold intensities were well known and this provided a basis for developing a model of its response to ionizing radiation. The functional unit of the peripheral retinal, shown schematically in Fig. 3, corresponds roughly to the region monitored by a single ganglion cell. The lateral dimensions of this region, known as a summation area, can be determined by probing with small spots of light. The thickness of this sensitive volume is just that of the layer of the rod cell outer segments. These outer segments contain the rhodopsin molecules which absorb the photons and initiate the electrical response. Photons absorbed within this sensitive volume contribute to the detection of light; those absorbed outside this volume do not contribute and, if they are absorbed within adjacent sensitive volumes, can inhibit detection.

The first-order model assumes that a visual experience occurs if a threshold number rhodopsin isomerizations occur within the sensitive volume and not otherwise. The isomerizations must be generated within the time constants of the unit, called the summation

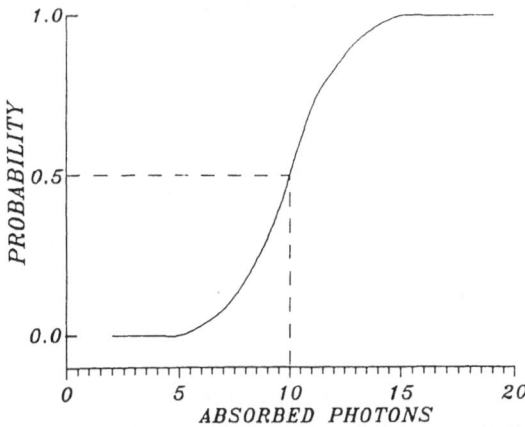

Figure 4. Response Curve for the Probability that a Human Observer Will Detect a Small Spot of Light on the Retina Versus the Average Number of Photons within the Sensitive Volume at that Intensity.

time, which is about 0.1 sec for a dark-adapted retina. This model is known to work well for optical vision. A response curve typical of the threshold or ``S'' shaped curves measured for vision is shown in Fig. 4.

The response of the visual unit to any type of ionizing radiation localized to within a single sensitive volume could be predicted knowing that the the visual response to ionizing radiation also follows a threshold response curve (Lipetz, 1953) and assuming that the relationship between large area and small area exposures is the same for ionizing radiation as it is for optical light. The results were in good quantitative agreement with accelerator data and the more limited data available from space (Rothwell, Filz and McNulty, 1976). This was the first time that the response of a biological system to quite different types of ionizing radiation could be explained quantitatively using a single set of parameters, the dimensions of the sensitive volume and the threshold number of ionizations (more correctly, the threshold amount of energy deposited) required for a visual response. This appears to be the essential test for all soft-error modeling, whether in biological systems or in circuits - that a single set of parameters can be used to predict the response to radiations which are quite different. In the visual system, the geometry of the sensitive volume was known from studies with visible light, and the value of threshold energy which had to be deposited in the sensitive volume could be estimated from the X-ray data. Another important characteristic of all soft errors was illustrated by the light flashes: it is a systems effect and the consequences of the soft error can only be measured in a complete organism.

SINGLE EVENT UPSETS

Since microelectronics fly in the same regions of space as the astronauts, the physical mechanisms leading to their soft errors are the same: cosmic-ray traversals and nuclear spallation reactions. The SEU-sensitive structures in microcircuits are small reverse-biased n-p junctions. Besides being the building blocks of microelectronic circuit elements, these junctions are also used as the sensors of silicon particle detector systems. One such system is shown schematically in Fig. 5. The sensitive volume in a detector is the region in the silicon immediately surrounding the depletion region formed at the reverse-biased junction. In a fully depleted detector, it is the entire volume of the silicon. Pulses are generated at the junction and shaped by the nearby circuit elements, and those which exceed some threshold value set by the discriminator are counted while lesser pulses are not. This operation is very similar to the first-order model for the retina's detection of photons and, as a result, it is not surprising that two independent groups attempting to model SEUs, one using the retina as a model (Wyatt et al, 1979; McNulty et al, 1980) and the other silicon detectors (Pickel and Blandford, 1980; 1981), arrived at essentially the same set of assumptions, now known as the first-order model.

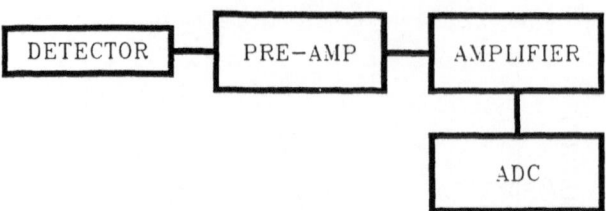

Figure 5. Schematic of a Silicon Particle Detector System.

First-Order Model for SEUs

The first assumption is that there is a sensitive volume associated with each SEU-sensitive junction on a microelectronic circuit whose dimensions are chosen such that the charge generated within this volume equals the charge collected across the junction. The dimensions of the sensitive volumes on the retina and the one within the fully depleted detector are known because they correspond to actual structures. Unfortunately, the sensitive volumes in microelectronic devices often differ in at least one dimension from identifiable microstructures, usually the thickness measured normal to the crystal surface. Moreover, it has only recently been shown that this dimension is independent of LET (McNulty, Beauvais and Roth, 1991) for circuit elements, a necessary condition for the concept of a sensitive volume to be useful quantitatively. Of course, the lateral dimensions of the sensitive volume should be only slightly larger than those of the junction but the value of the thickness could, until only recently, be estimated for most devices.

The second assumption is that there is a threshold number of ionizations which must be generated within the sensitive volume to induce an upset. This value is known as the critical charge. Since the number of ionizations is proportional to the energy deposited, the value of the critical charge is often listed in energy units where the conversion is the W value for silicon, 3.6 eV.

Luckily, both the area and the thickness of the sensitive volume can now be estimated from pulse-height measurements made on signals generated at the junction. Figure 6 shows the circuit used for these measurements. It is a modified version of the pulse-height system used in nuclear spectroscopy. The time constants of the charge-sensitive preamplifier should approximate the time constants of the circuit in order to collect only that charge which would

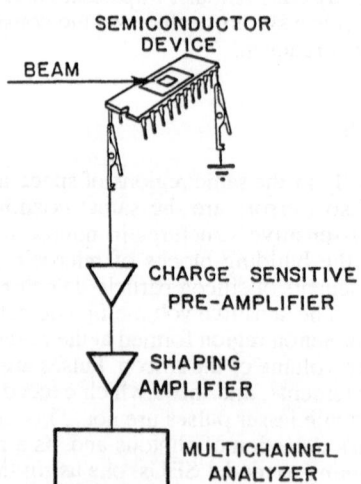

Figure 6. Schematic of Circuit Used for Pulse-height Measurements between the Power Pins of Static Memories.

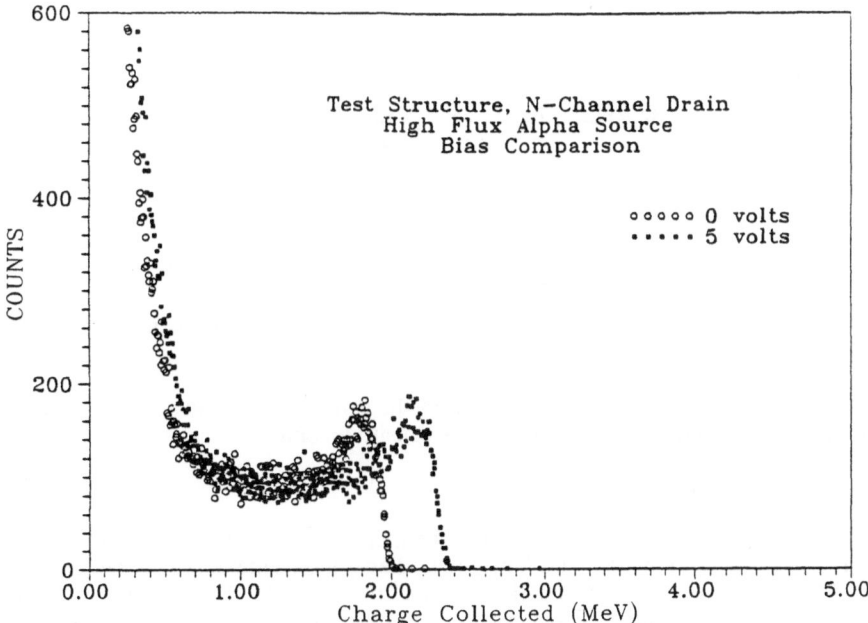

Figure 7. Pulse-height Spectra Measured across a Single n-p Junction of the Type Used in Microelectonic Circuits. Measurements are Plotted for Two Values of the Applied Bias, 0 V and 5 V.

contribute to upsetting the device. Figure 7 shows the pulse-height spectrum measured from a single n-p junction exposed to 4.8 MeV alphas. The junction has dimensions similar to the SEU-sensitive junctions in memory cells. Measurements are plotted for two values of the applied bias: 0 V and 5 V. A spectrum is obtained at zero bias because of the built-in potential across the metallurgical junction resulting from the differences in doping levels on either side. Increasing the bias to a working voltage of 5 V produces only a 20% shift in the spectrum toward higher energies with little or no change in the shape. This is important because the dimensions of the sensitive volume can be obtained from the pulse-height spectrum and, for reasons given below, they are usually estimated at zero bias.

Since particles initiate pulses by traversing the sensitive volume, the area can be estimated from the ratio of the number of events under the peak and the fluence. The thickness can be estimated from the position of the peak in the spectrum. If the energy of the incident particle is E and the peak position corresponds to a deposition of energy ΔE in the sensitive volume, then the thickness t of the sensitive volume can be estimated using range-energy tables and the formula:

$$t = R(E) - R(E - \Delta E)$$

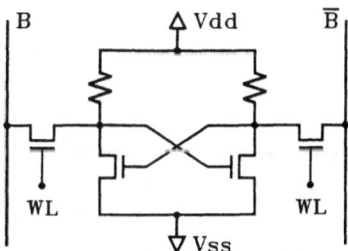

Figure 8. Memory Element of a Resistor-loaded NMOS Static Random Acess Memory (SRAM) Device.

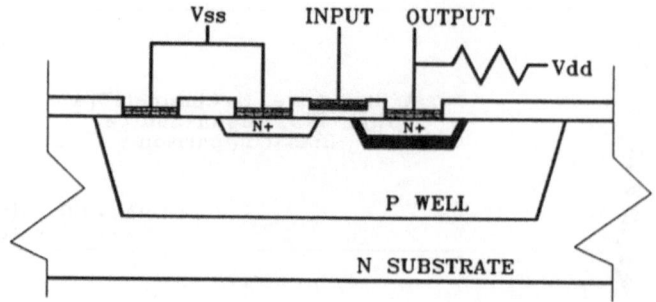

Figure 9.   Cross Section of an Inverter in a Resistor-loaded NMOS SRAM.

where R(E) is the range of the particle entering the sensitive volume and R(E - ΔE) is the range of the incident particle after traversing the sensitive volume.

SEUs in Static RAMs

The resistor-loaded NMOS SRAM is the memory circuit best suited to illustrate the analysis of SEUs in working devices. The circuit diagram of a memory element of an NMOS static RAM is shown in Fig. 8. The cell consists of two cross-latched inverters held between high voltage V_{DD} and ground. The sensitive junction for this device is the reverse-biased drain-substrate junction. There are two drains shared by the four transistor cell. When the cell is powered up in either state, only one of these junctions is reverse-biased, but at zero bias both junctions in the cell are reverse-biased as a result of the built-in potential. The structure of a single inverter is shown in cross section in Fig. 9. When the chip is irradiated, the pulses at the drain-substrate junction can be monitored through the power (V_{DD}) and the ground lines of the device. The experimental configuration for measuring these pulses is shown in Fig. 6. By selecting only those pulses of the appropriate polarity, the spectrum obtained should be dominated by a peak due to particles traversing the drain-substrate junctions. Pulses due to hits on other junctions have a different polarity and are ignored. Figure 10 shows the pulse-height spectrum measured between the V_{DD} and ground pins of an IDT 6116V, a 16K

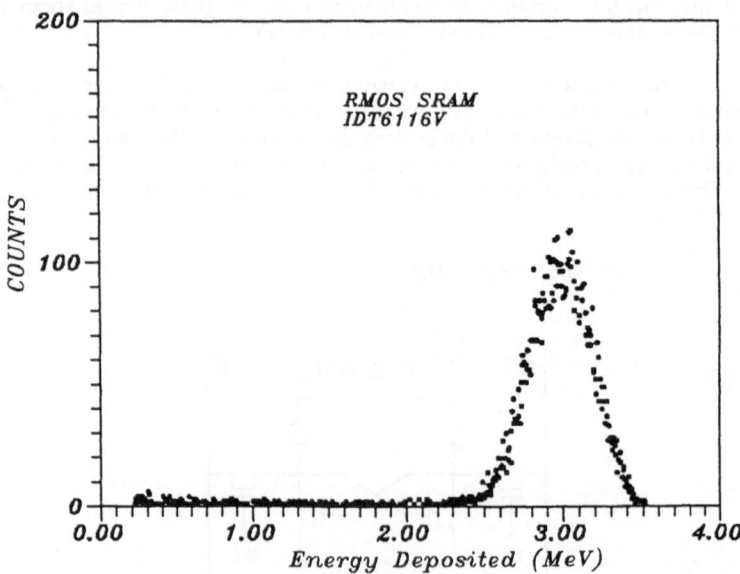

Figure 10.   Pulse-height spectrum measured between the power and ground pins of an IDT 6116V resistor-loaded NMOS SRAM.

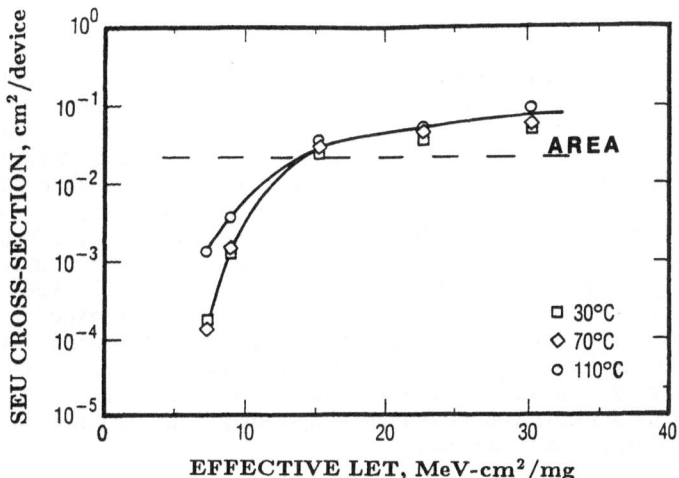

Figure 11.   SEU Cross Section Versus LET for Heavy Ions Incident on the IDT 6116V at Different Temperatures. The Horizontal Curve is the Cross Sectional Area of the Sensitive Volume Determined from Fig. 10.

memory device with over 32,000 reverse-biased drain-substrate junctions in its unbiased state. As with test structures, the area of the junction can be found from the ratio of the number of events under the peak to the product of the fluence and the number of junctions. This area is also the cross sectional area of the target which must be hit to upset the memory cell. The thickness of the equivalent sensitive volume for this junction is obtained from the position of the peak in the spectrum as described earlier.

SEU Cross Section Measurements

For heavy ions the probability that an individual energetic particle will induce an SEU is expressed in terms of the per-bit SEU cross section which is typically determined from experimental measurements of the number of upsets induced by a parallel beam of accelerator particles. (See McNulty, 1991 for a recent review of the literature.) The per-bit cross section is defined to be the ratio of the number of SEUs generated to the product of the fluence and the number of memory cells in the device. Since all circuits are designed to work error free in the natural environment at sea level, they do not upset to particles incident with very low LET. At sufficiently high values of LET, any commercially available microcircuit will upset every time an SEU-sensitive junction is traversed. Then, the response curve of an SRAM exposed to particles of intermediate LET should be similar to the response curve of the retina to increasing number of photons, i.e., at low numbers of ionizations (low LET) the SEU cross section is zero, rising quickly at threshold to approximately the junction area and then continuing to rise slowly with LET as the cells begin to upset even with near misses (McNulty, 1991).

The response curve for the IDT 6116V was measured by Koga et al (1988) at different temperatures, and the results are plotted in Fig. 11. The horizontal line is the estimate of the junction area obtained from the pulse-height measurements described above. This value is in excellent agreement with the value of the SEU cross section at the point of transition between fast and slow rise in cross section. If this data, or any of the more complete data sets we are familiar with, were replotted as the probability that a heavy ion traversing one of the junctions caused an upset versus the number of ionizations generated by the ion as it traversed the sensitive volume (obtained by dividing the product of the LET, the density and the thickness of the sensitive volume by the average energy deposited per ion pair for silicon), the resulting curve would have the shape of the response curve of Fig. 4 except that the units of the abscissa would be number of ionizations instead of photon absorption.

SEU Measurements for Protons

A Monte-Carlo code called CUPID (McNulty, Farrell and Tucker, 1981) simulates the energy deposition in silicon microvolumes as a result of nearby proton-induced nuclear spallation reactions. It was developed following the approach of our earlier code for predicting proton-induced events in the retina (Rothwell, Filz and McNulty, 1976). The results of calculations for a sensitive volume having the dimensions obtained earlier for the IDT 6116V and exposed to 148 MeV protons are compared in Fig. 12 to experimental pulse-height measurements between power and ground on that device (McNulty et al, 1991). The plots are of integral cross section versus energy, i.e., the cross section for depositing at least some energy E in the microvolume versus E. The agreement between theory and the experimental curve is quite good. The horizontal line in Fig. 12 is the experimental SEU cross section. The locations on the abscissa which mark where the cross-section line intersects the curves are theoretical and experimental estimates of the value of the threshold energy deposition required to upset the device. The values are very close. The ratio of either value to the thickness of the sensitive volume is in good agreement with the threshold LET measured by Koga et al (1988) for heavy-ion data.

Universality of the First-Order Model

Similar studies have been carried out for CMOS static memories (McNulty, Beauvais and Roth, 1991) and NMOS dynamic memories (McNulty, Abdel-Kader and Lynch, 1991). The spectra for these technologies are more complicated, but the peak due to ion hits on the SEU-sensitive junctions can usually be identified and the dimensions of the SEU sensitive volumes determined. The proton SEU cross sections for more complicated devices must be compared to simulations of the energy-deposition in the sensitive volume rather than experimental measurements because, with spallation reactions, events on the SEU-sensitive junctions cannot be differentiated from other events. It has been shown for a variety of devices having different microstructures that the heavy-ion SEU cross sections and the proton SEU cross sections can be fit by a single set of parameters: the dimensions of the sensitive volume and the value of threshold for upset (critical charge). Events which generate fewer than a critical number of ionizations, or deposit less than a threshold amount of energy, do not upset the device. Events which exceed threshold always result in upsets. The shape of the response curve of cross section versus LET has been explained as being due primarily to the spread of energy-deposition events observed at a single LET (McNulty, Abdel-Kader and Lynch, 1991) just as the shape of the visual response curve has been attributed to quantal fluctuations at low intensities.

Bond and Varma have developed a variation on the first-order model for somatic mutations in biological cells (Bond and Varma, 1983). Since the dimensions of the sensitive

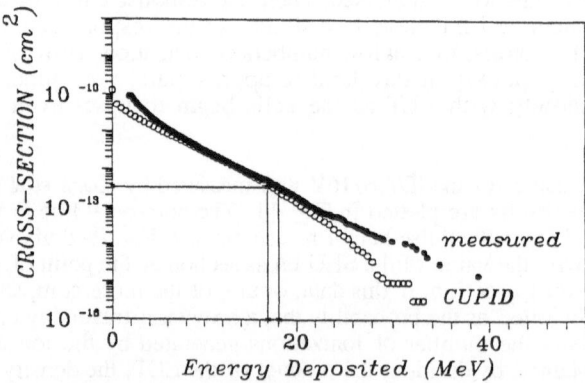

Figure 12. Cross Section as Calculated by CUPID for Depositing at Least Energy E within the Sensitive Volume of the IDT 6116V Plotted Versus E. The Experimental Curve Represents the Values Obtained from the Pulse-height Spectrum Measured between the Power and Ground Pins while the Unpowered Device is Irradiated by 148 MeV Protons.

volume are poorly defined for mutations, available data defining a response curve (probability of a mutation versus LET) cannot be used with confidence to specify the value of the threshold energy which must be deposited. It has been shown (Sondhaus, Bond and Feinendegen,1990) that a wide range of mutations and chromosome aberrations have response curves which are similar in shape to the curve of Fig. 12, again with ionizations within the sensitive volume replacing photon absorption on the abscissa. An interesting exercise, then, would be to try estimating the size of the sensitive volume following the procedure outlined earlier for SEUs. The per-cell cross sections for mutations are needed to do this. They are obtained for heavy ions at various values of LET from the ratio of the number of mutations to the product of the fluence of incident particles and the number of cells at risk. The values of the per-cell cross sections would have to be plotted versus LET as the SEU data plotted in Fig. 11. The value of the cross section that marks the transition between the region of fast rise in cross section with LET and the region of slower increases with LET is the cross sectional area of the sensitive volume, just as the corresponding value of the SEU cross section in Fig. 11 was the junction area. Assuming that the sensitive volume is a sphere, the radius of the sensitive volume and the the pathlength distribution through the sensitive volume can also be calculated.

It is not clear that the model of Bond and Varma can be adequately tested without at least approximate dimensions for the sensitive volume being known. If the sensitive volumes have submicron feature sizes, the energy-deposition spectra may be dominated by events where the particle missed the target (Dicello, private communication). This is also true for large volumes when there is a low value of threshold. In this regime, the plateau in the cross section may reflect track structure effects more than the dimensions of the sensitive volume. Our experiences with the retina and microcircuits suggest that, if the first-order model is valid for mutations, the proper dimensions of the sensitive volume would be those which when combined with the corresponding value of the threshold would fit data for two different radiation types. Nuclear reactions and traversals by heavy ions might be the logical place to start, but they are not the only radiations which should give consistent results.

We know why the SEU response of microelectronic circuits follows the first-order model: circuits are deliberately designed not to upset in the natural environment at ground level. This environment includes thermal fluctuations as well as background radiations. So circuits are designed not to upset at low values of LET, but the information stored in random access memories is meant to be stored and recalled easily at low power. This makes the circuits sensitive to ion hits at higher LET. It seems clear that evolutionary processes also designed a threshold into the human visual system. Individual photoreceptors respond electrically to individual photons but, if a visual experience resulted from individual photon absorptions, the visual system would be swamped by the infrared radiation of the thermal background. Both types of systems have adjusted to a signal to noise problem. If there is a threshold in radiation induced mutations, perhaps it is also an evolutionary response to a signal to noise problem. Mutations are necessary for evolutionary progress, but to be too sensitive would make organisms vulnerable.

Circuits are designed with different threshold values depending on how harsh the intended environment is expected to be. Increasing the level of protection typically involves trade offs in performance: usually speed. Often, protection is designed in at the system level rather than by hardening individual components. Redundancy of stored information and voting logic are frequently used. The genetic code also incorporates redundancy. Comparisons in this area may be fruitful.

APPLICATIONS TO MICRODOSIMETRY

Parallel arrays of p-n junctions have been proposed as a microdosimetry tool for characterizing complex radiation environments for microelectronic circuits (McNulty et al, 1990). The pulse-height spectra measured from an array containing over two million junctions is shown in Fig. 13. The pulse-height spectrum measured at a microjunction having a sensitive volume with dimensions similar to the circuit of interest would certainly be a better predictor of the risk of SEUs in a given environment than any other form of dosimetry including detailed measurements of the charge and energy of the particles making up the

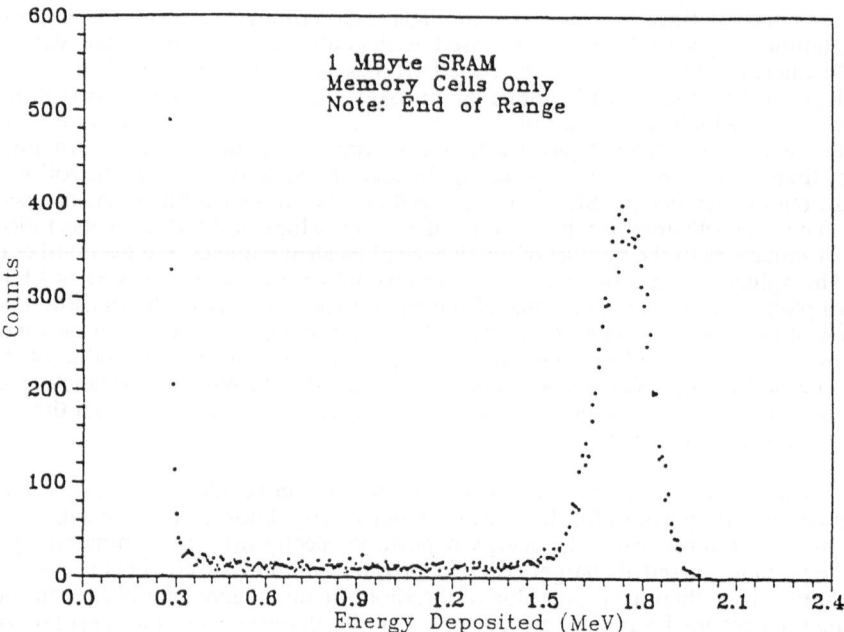

Figure 13. Pulse-height Spectrum Obtained from an Array of over Two Million Junctions in Bulk CMOS.

environment. Clearly, such arrays would also be useful in predicting light flash rates if proper corrections for the differences in composition were taken into account in selecting the dimensions to use for the sensitive volume. The possibility of using microelectronics to characterize a radiation environment for risk of mutations is less obvious. We do not know the shape or size of the tissue equivalent volumes that must be approximated. Although not essential for the space radiations mentioned above, a good general purpose microdosimeter should have all the individual sensitive volumes isolated from one another. Technology is now available to do just that. Figure 14 shows the pulse-height spectrum obtained with a parallel array of 2,316 junctions from a Texas Instruments test chip where the sensitive volume associated with each junction is completely surrounded by dielectric except where the electrical contacts are made.

Even in the absence of clearly specified dimensions for the sensitive volumes for biological mutations, microelectronic junction arrays still have important potential advantages in characterizing a radiation environment in terms of radiobiological risk. Sensitive volumes can be scaled to the size of biological cell nuclei, a probable upper limit to any sensitive volume. The pulse-height spectra would separate the high-energy-deposition events from the low-energy ones. These spectra could then be used to calculate dose equivalents based on quality factors or equivalent quantities determined by the NCRP or ICRP. Battery operated personnel dosimeters small enough to fit into shirt pockets can be made with commercially available devices and dosimeters as small as a film badge can be constructed using modern lithographic techniques.

CURRENT TESTING OF MODELS IN SPACE

The accuracy of predictions of single event phenomena strongly depends on the accuracy of the environmental models and the transport calculations for handling the effects of shielding. Solar flares and magnetic disturbances result in dramatic temporary changes in the radiation incident on spacecraft and consequently the error rates. There is a slow variation in the environments from solar minimum to solar maximum which should result in corresponding increases in SEU rates over the poles and decreases in the portions of the low earth orbits where spacecraft penetrate the radiation belts. This is an eleven year cycle and

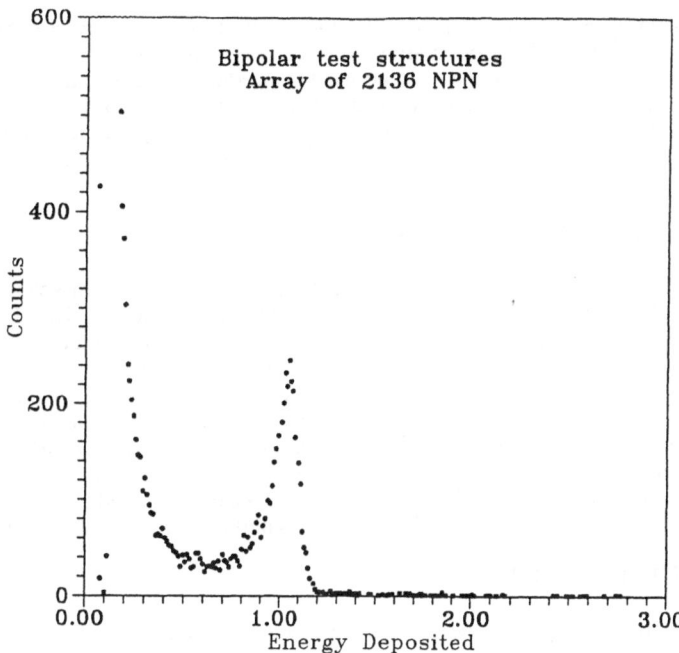

Figure 14.   Pulse-height Spectrum Obtained with an Array of 2316 Microjunctions where Sensitive Volume is Surrounded by a Layer of Dielectric Isolation. Energy in MeV.

longer trends cannot be ruled out. Current NASA models of the radiation environments are static models which only incorporate the dynamic variations described above by long-term averaging. The model for the cosmic-ray environment is probably valid to within a factor of two for averages of the environment in excess of a year. It is necessary, of course, to provide the proper mix of the solar maximum and the solar minimum in calculations. Also, AP8, the NASA code of the trapped proton environment, may have an error of a factor of two in the high energy proton flux for low-earth orbits. The J* dosimeter recently characterized the low-earth orbit of the DMSP satellite in terms of the energetic nuclear reactions which would lead to SEUs in microelectronic devices (Gussenhoven et al, 1987). The data was found to agree with predictions of the AP8 model of the proton environment combined with CUPID's predictions of the number of spallation events which deposit at least 40 MeV and 75 MeV in the detectors (Beauvais et al, to be published).

Relatively good agreement between predictions based on heavy-ion accelerator data and cosmic-ray spaceflight data have been reported by Binder (1988), Smith and Simpson (1987) and Blake and Mandel (1986). These predictions were for a small number of devices in different satellite programs and included geosynchronous orbits and low-earth polar orbits.

The CRRES satellite is an entire satellite dedicated to measuring the radiation environments and correlating them with the effects on circuit components. The orbit is a highly elliptical orbit typical of the transfer orbits used to shift satellites from low-earth orbit to a geostationary orbit. Analysis is just beginning, but data sufficient for careful test of environmental and transport codes is being made available.

PREDICTING PROTON RESPONSE FROM HEAVY ION DATA

The procedure for using heavy-ion data to predict the proton cross section versus incident energy response function is discussed in Bisgrove et al, 1986. As described earlier, for this the work, the dimensions of the sensitive volume must be determined from charge collection measurements. The critical charge can then be estimated from the threshold LET measured in the heavy ion data. The agreement obtained using CUPID was excellent for the device reported on by Bisgrove et al , (1986) and for a number of devices tested later. Bion and Bourrieau (1989) have developed simulation codes of the same general type as CUPID. They also find that the heavy-ion data can be used to predict the proton data. It is important to note that, in all of these calculations, the detailed shape of the response curve must be taken into account for accurate calculations. Errors of an order of magnitude result from ignoring the shape of the response curve in predicting SEU rates in space Brucker and Stassinopoulos, 1991).

SUMMARY AND CONCLUSIONS

The single event upset phenomena induced in circuits flown in space have response curves in plots of SEU cross section versus LET which exhibit a threshold behavior similar to the response curves reported for the light flash phenomena experienced by astronauts and mutations in biological cells. It is possible to independently determine the dimensions of the sensitive volume for a visual unit on the retina and for a memory cell in a circuit, but not for mutations. Knowledge of these dimensions and the threshold energy that must be deposited in the sensitive volume is sufficient to predict the response to a variety of radiation types. The first-order model assumes that if more than a threshold amount of energy is deposited within the sensitive volume, an event is induced, but not otherwise. Knowledge of the dimensions of the sensitive volumes would be essential for testing whether mutations obey similar models.

ACKNOWLEDGEMENTS

Conversations with V.P. Bond, J. Booz, and L.E. Feinendegen are gratefully acknowledged.

REFERENCES

Beauvais, W.J. , P.J. McNulty, S. El-Teleaty, W. G. Abdel-Kader, M.S. Gussenhoven, E.G. Mullen, and G.E. Farrell, Comparison of Spallation-Reaction Simulations with DMSP Satellite Data , IEEE Trans. Nucl. Sci., to be published.

Binder, D., 1988, Analytic SEU Rate Calculation Compared to Space Data, IEEE Trans. Nucl. Sci., NS-35: 1570-1572.

Bion, T. and J. Bourrieau, 1989, A Model for Proton-Induced SEU, IEEE Trans. Nucl. Sci., NS-36: 2281-2286.

Bisgrove, J.M., J.E. Lynch, P.J. McNulty, W. Abdel-Kader, V. Kletnieks, and W.A. Kolasinski, 1986, Comparison of Soft Errors Induced by Heavy ions and Protons, IEEE Trans. Nucl. Sci. , NS-33: 1571-1576.

Blake, J.B. and R. Mandel, 1986On-Orbit Observations of Single Event Upsets in Harris HM-6508 1K SRAMs, IEEE Trans. Nucl. Sci., NS-34: 1616-1619.

Bond, V.P., and M.N. Varma, 1983, Low-Level Radiation Response Explained in Terms of Fluence and Cell Critical Volume Dose, in Eight Symposium on Microdosimetry, (Julich West Germany; Commission of the European Communities) pp 423-439.

Brucker, G. J. and E.G. Stassinopoulos, 1991, Prediction of Error Rates in Dose-Imprinted Memories on Board CRESS by Two Different Methods, IEEE Trans. Nucl, Sci., NS-38: 913-922.

Budinger, T.F., J.T. Lyman, and C.A. Tobias, 1972, Visual Perception of Accelerated Nitrogen Nuclei Interacting with the Human Retina, Nature 239: 209-211.

Gussenhoven, M.S., E.G. Mullen, R.C. Filz, D.H. Brautigam, and F.A. Hanser, 1987, New Low-Altitude Dose Measurements, IEEE Trans. Nucl. Sci., NS-34: 676-683.

Koga, R., W.A. Kolasinski, J.V. Osborne, J.H. Elder, and R. Chitty, 1988, SEU Test Techniques for 256K Static RAMS and Comparison of Upsets Induced by Heavy Ions and Protons, IEEE Trans. Nucl. Sci., NS-35: 1638-1643.

Lipetz, L.E., Ph.D. Thesis, 1953, University of California Radiation Laboratory Report No. 2056.

May, Tim; 1984, Dynamic Fault Imaging of VLSI Logic Devices, presented at the International Reliability Symposium, Los Vegas, Nevada April 2-5, 1984.

McNulty, P.J. ; 1983, Charged Particles Cause Microelectronic Malfunction in Space , Physics Today (Guest Comment) 36: 9.

McNulty, P.J.; 1988, New Directions in Space Dosimetry; In Terrestrial Space Radiation and Its Biological Effects P.D. McCormack, C.E. Swenberg, and H. Bucker, Eds. (Plenum Publishing Co.) pp 819 - 840.

McNulty, P.J. ; Single Event Upsets in Microelectronics, Radiation Research, to be published.

McNulty, P. C.; Predicting Single Event Phenomena in the Natural Space Environments ; in Microelectronics for the Natural Radiation Environments of Space P.J. McNulty, ed., (IEEE Short Course for the 1991 International Nuclear and Space Radiation Effects Conference, Reno, Nevada) pp 3-1 to 3-93.

McNulty, P.J. , W.G. Abdel-Kader, and J.E. Lynch, 1991, Modeling Charge Collection and Single Event Upsets in Microelectronics," Nuclear Instruments and Methods in Physics Research, B61, No. 1: 52-60.

McNulty, P.J., W.J. Beauvais, and D.R. Roth,1991 Determination of SEU Parameters of CMOS SRAMS by Charge Collection Measurements on SRAMS, IEEE Trans. Nucl. Sci., NS-38: 1463-1470, Dec..

McNulty, P.J., G.E. Farrell, and W.P. Tucker, 1981, Proton-Induced Nuclear Reactions in Silicon, IEEE Trans. Nucl. Sci., NS-28: 4007-4012.

McNulty, P.J. , V.P. Pease, L.S. Pinsky, V.P.Bond, W. Schimmerling and K.G. Vosburgh; 1972, Visual Sensations Induced by Relativistic Nitrogen Nuclei, Science 178: 160-162.

McNulty, P.J. , D.R. Roth, W.J. Beauvais, W.G. Abdel-Kader, and D.C. Dinge, 1991, Comparison of the Charge Collecting Properties of Junctions and the SEU Response of Microelectronic Circuits,Int. J. Radiat. Instrum., Nuclear Tracks and Radiation Measurement, 19, Nos. 1-4: 929-938.

McNulty, P.J., D.R. Roth, E.G. Stassinopoulos, and W.J. Stapor, 1990, Characterizing Complex Radiation Environments Using MORE (Monitor of Radiation Effects), in Proceedings of The Symposium on Detector Research and Development for The Superconducting Super Collider, (Fort Worth, Texas, October 15-18, 1990) pp 690-692.

Pickel, J.C. and J.T. Blandford, Jr., 1980, Cosmic Ray Induced Errors In MOS Memory Cells, IEEE. Trans. Nucl. Sci., NS-27: 1006-1015; 1981, CMOS RAM Cosmic Ray Induced Error Rate Analysis,IEEE Trans. Nucl. Sci., NS-28: 3962-3967.

Pinsky, L.S., W.Z. Osborne, J.V. Bailey, R.E. Benson, and L.F. Thompson; 1974, Light Flashes Observed by Astronauts On Apollo 11 Through Apollo 17, Science 183: 957 - 959.

Pinsky, L.S. ,W.Z. Osborne, R.A. Hoffman, and J.V. Bailey, 1975 Science 188: 928.

Rothwell, P. , R. Filz, and P.J. McNulty; 1976, Light Flashes Observed on Skylab IV - The Role of Nuclear Stars, Science 193: 1002.

Smith, E.C. and T.R. Simpson, 1987, Prediction of Cosmic-Radiation-Induced Single-Event Upsets in Digital Logic Devices On Geostationary Orbit,TRW Final Report, November 21, 1987.

Sondhaus, C.A., V.P. Bond and L.E. Feinendegen, 1990, Cell Oriented Alternatives to Dose, Quality Factor, and Dose Equivalent for Low-Level Radiation, Health Physics 59: 35-48.

Wyatt, R.C. , P.J. McNulty, P. Toumbas, P.L. Rothwell, and R.C. Filz, 1979, Soft Errors Induced by Energetic Protons, IEEE Trans. Nuc. Sci., NS-26: 4905-4910; McNulty, P.J. , G.E. Farrell, R.C. Wyatt, P.L. Rothwell, R.C. Filz, and J.N. Bradford, 1980, Upset Phenomena Induced by Energetic Protons and Electrons, IEEE Trans. Nucl. Sci., NS-27: 1516-1522.

HEAVY-ION FRAGMENTATION STUDIES IN THICK WATER ABSORBERS

M. R. Shavers[1],*, J. Miller*, W. Schimmerling*, J. W. Wilson**, and L. W. Townsend**

*Research Medicine and Radiation Biophysics Division, Lawrence Berkeley Laboratory, 29/100 #1 Cyclotron Road, Berkeley, CA 94720
**High Energy Science Branch, M. S. #493 NASA Langley Research Center, Hampton, VA 23665

INTRODUCTION

Proposed long-term space missions could expose crewmembers to significant fluxes of galactic cosmic radiation (GCR) particles and secondary particles created from nuclear collisions. An assessment of radiobiological risks is dependent upon an accurate description of the charged-particle radiation field inside the human body.

The unshielded GCR environment in near-Earth free space is fairly well known, and is described in detail in this publication and elsewhere (Adams et al. 1981). A recent analysis of available experimental measurements and the development of a theoretical model which describes the cyclic influence of solar-surface activity on the interplanetary magnetic field have improved the accuracy with which we can predict GCR charged-particle flux in free space near Earth to within 7% to 30% error (Badhwar and O'Neill 1991).

Adequate protection from GCR must be attained with massive shields and strategies for using the shields to reduce exposure to tolerable levels. Use of an artificial magnetic field (active shield) for this purpose is not feasible, even with superconductor technology (Townsend and Wilson 1984, Townsend et al. 1990a). Mission scheduling, to take advantage of reduced GCR flux at maximum solar activity, probably is not possible for long-term space activities such as a return to the Moon or manned exploration of Mars. Once on the surface of the moon or Mars, GCR flux levels are reduced by essentially 2π shielding. Exposure on Mars is further reduced by a weak planetary magnetic field and thin atmosphere (~ 27 g/cm^2), but, at an orbit nearly twice the distance of the Earth from the sun, the interplanetary magnetic field is less effective at providing protection from GCR.[2] Additional protection will be required.

The shielding effectiveness of various materials has been demonstrated by dose calculations of space workers shielded by thick layers of aluminum, regolith (lunar soil) and hydrogenous material for various geometries (Wilson et al. 1987a; Townsend et al. 1988,

---

[1]Current address: Department of Nuclear Engineering, Texas A&M University, College Station, TX 77843-3133.
[2]G. Badhwar, personal communication. Solar Systems Exploration Division, SN-3, NASA-Johnson Space Center, Houston, TX 77058.

Biological Effects and Physics of Solar and Galactic Cosmic Radiation,
Part B, Edited by C.E. Swenberg *et al.*, Plenum Press, New York, 1993

1990b; Nealy et al. 1990; Simonsen et al. 1990). Although optimization of shields will minimize overall launch mass and fuel needs, the present large uncertainty in required shielding imposes expensive and unnecessary launch mass penalties.

As shield thickness increases and the incident ions are slowed, the production of secondary particles contributes an increasingly significant fraction of the total dose until eventually secondary particles become more important than the primary particles. The nuclear mean free path of the GCR ions (which usually have nuclear charge between 1 (protons) and 26 (iron), both inclusive) are comparable with thicknesses typical of spacecraft structures and the human body. Collisions in these media will create *projectile* and *target fragments* with charge less than that of the primary particle, and each interaction event can have a multiplicity of of more than one emerging interaction product. *Projectile fragments* usually continue on with very nearly the velocity of the primary ion (the so-called straightahead approximation). Having sufficient energy, the fragments may collide with atomic nuclei in thick shields and create a second generation of fragments, and so on. *Target fragments* are emitted from a struck nucleus, usually with much lower energy than projectile fragments and nearly isotropically in the rest frame of the absorbing medium. The resulting spectrum of particles and their energy loss rates will be very different from that in the unshielded environment, will determine the radiological impact on exposed living tissues---whether in space or in ground-based radiobiology experiments---and will play an important role in radiation effects on microelectronics.

Although the physics of the nuclear fragmentation phenomenon is not well understood (Townsend et al. 1987), a semiempirical model has been developed from existing data (Silberberg et al. 1973). More recently, an energy-independent fragmentation model was developed from the classical, geometric abrasion-ablation model of Bowman et al. (1973), and modified to include semiempirical higher order corrections for the abraded "prefragment" excitation energies (Wilson et al. 1987b). Though probably more accurate than earlier models, the latter methodology has not yet been incorporated into the "accelerator beam" transport model described below.

Very few data are available for the production of nuclear fragments from relativistic heavy-ion collisions, and, because of the complexity of such measurements and the number of relevant projectile-target-and-energy combinations, it is not likely that an adequate data set will be available in the foreseeable future.

NEON EXPERIMENT

In a program designed to measure radiation fields produced by high-energy heavy-ions incident on thick absorbers, detailed measurements were made of 670A MeV accelerated neon beams interacting in water using a time-of-flight telescope to measure velocity and a set of silicon detectors to measure energy loss of each particle. Fluence spectra of identified projectile fragment nuclei between Be (Z=4) and Ne (Z=10), both inclusive, were measured along the central axis of the beam-line downstream from absorbers of thicknesses between 0 and 38.5 g/cm^2 (Schimmerling et al. 1989). The data set was compared with theoretical fluence spectra predicted by the heavy-ion transport code HZESEC, a model based on "first principles" which calculates only the projectile fragments created directly from nuclear collisions of the the primary beam. Multiple Coulombic-scattering effects on the particles emerging from the water column were separately considered for each element, as were detector resolution and particle energy cutoff effects. The comparison of the calculation with measurements and approximations inherent in the analysis have been described in detail (Shavers et al. 1990). The study revealed aspects of the analysis which are sensitive to error, and indicated that higher-order reaction products are probably significant at depth.

HEAVY ION TRANSPORT MODEL LBLBEAM

The heavy-ion transport model LBLBEAM, developed at NASA-Langley Research Center, is used to calculate integral fluence and differential LET spectra of the primary neon beam, of the secondary projectile fragments created from neon interactions, and of tertiary particles created by first-generation secondaries. The calculation is based on an analytical solution of a one-dimensional Boltzmann transport equation using the straight-ahead approximation. Assuming energy-independent fragmentation parameters and neglecting the target fragments, the equation is

$$[\frac{\partial}{\partial x} - \frac{\partial}{\partial E} S_j(E) + \sigma_j]\phi_j (X, E) = \sum_{k>j} m_{jk}\sigma_k\phi_j (X, E) .$$

The flux of fragment ions $\phi_j(x,e)$ of type j with atomic mass A_j at x move along the x-axis with kinetic energy A_jE in units of MeV. The production cross section for type j ions by collisions of type k ions with a target nuclei is given by $m_{jk}\sigma_k$. σ_j are the macroscopic nuclear absorption cross sections (in cm^{-1}); fragmentation cross sections. The incident ion beam flux is described by the Gaussian distribution

$$\phi_j (X=0, E) = \frac{1}{\sqrt{2\pi}\,\Delta} \exp[\frac{-(E-E_0)^2}{2\,\Delta^2}]$$

where the central energy $E_0 \approx 636.28A$ MeV is the nominal beam energy per nucleon and $\Delta = 0.0024E_0$ is the standard deviation of the extracted neon beam. The analytical solutions for first through third successive generations of fragments are straightforward, though lengthy, and are published in these proceedings and elsewhere (Wilson et al. 1989, 1990). The method of Barkas and Berger, as described by Schimmerling et al. (1986), is used to calculate range-energy relations and LET spectra from both calculated and measured energy spectra. In this work, multiple scattering effect on particles emerging from the water column is separately considered for each isotope.

RESULTS AND DISCUSSION

Theoretical differential LET spectra of first- and second-generation fluorine (Z=9) and boron (Z=5) fragments predicted by LBLBEAM, and with detector effects folded in, are compared with the data in Figs. 1 and 2. The water thickness (in g/cm^2) is shown in the upper right-hand corner of each panel. The data are represented by filled circles, and statistical error

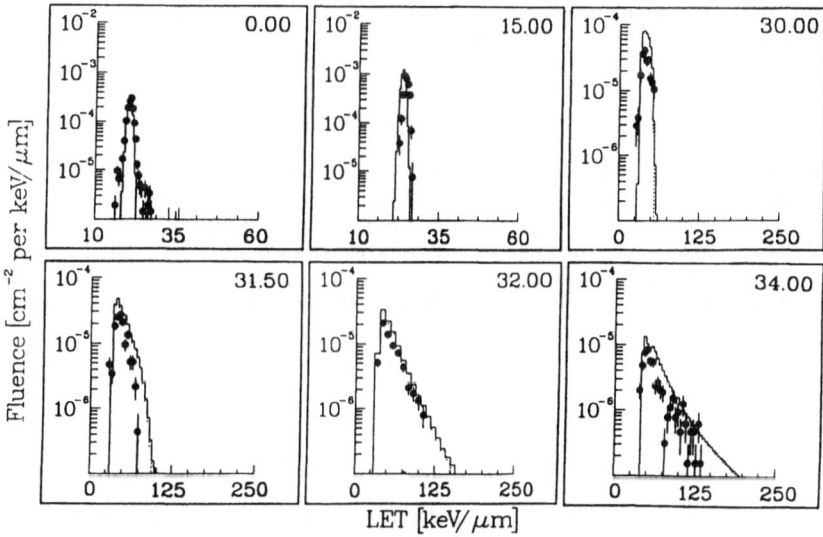

Figure 1. LBLBEAM vs EXPERIMENT Ne + H$_2$O → F + X Yield per Unit Fluence of the Incident Beam. Differential fluence spectra for fluorine. The filled circles represent the data; the dotted-line represents first-generation fragments; the summation of first- and second-generation fragments is shown with a solid-line. The thickness of water in g/cm^2 is indicated in the upper right corner of each panel.

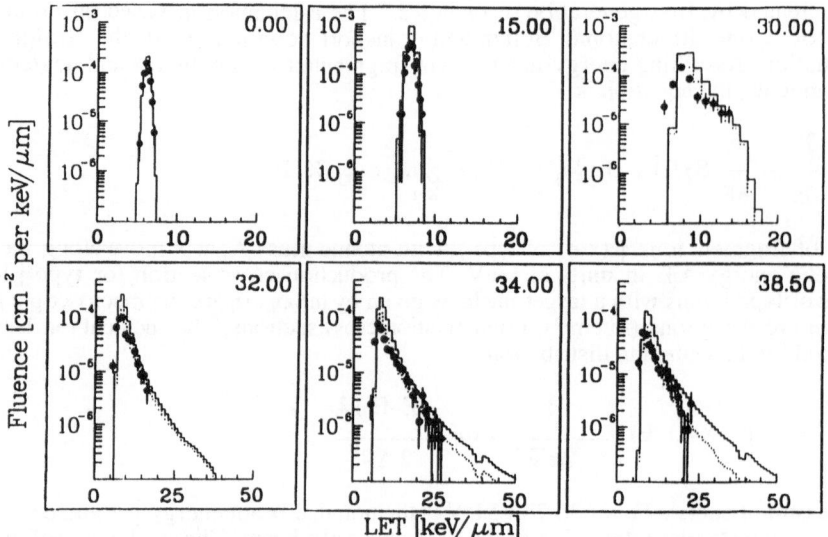

Figure 2. LBLBEAM vs EXPERIMENT Ne + $H_2O \rightarrow$ B + X Yield per Unit Fluence of the Incident Beam. Differential fluence spectra for boron, as in Fig. 1.

bars are shown when larger than the symbol. The first-generation fragments predicted by LBLBEAM are represented by the dotted-line histogram. The sum of first- and second-generation fragments is shown as a solid line. Spectral shapes of the data are accurately reproduced by the model, although the theory systematically over-estimates the integral fluence. Agreement is reasonably good, especially for elements dominated by fragments generated directly from neon interactions, such as fluorine. Uncertainties in fragmentation cross sections, multiple Coulombic scattering, and energy threshold effects account for most of the observed discrepancies, which can be a factor of two or more for boron.

The comparison between experiment and theory demonstrates increasing importance of second-generation fragments for the thickest targets, especially for the lightest fragments (all heavier fragments can undergo collisions and produce these) and at the high-LET "tail" of each spectrum. Whereas few neon projectile fragments can collide and produce second-generation fluorine fragments, this is not the case for boron. Nearly 50% of all boron fragments downstream from 34 g/cm^2 of water are tertiary fragments. This order of magnitude was expected from discrepancies observed in the earlier work.

The results demonstrate the significance of considering second-order fragmentation reaction products at depth, and the importance of an accurate knowledge of inclusive nuclear fragmentation parameters. The experimental methods and the procedure for comparing accelerated heavy-ion beams with theory are shown to be quite useful for validating transport models. A detailed comparison of theory with detected neon and neon projectile fragments with charge greater than or equal to four (beryllium) will be described in a future paper.

REFERENCES

Adams, J. H.; R. Silberberg, C.H. Tsao, 1981, Cosmic ray effects on microelectronics, Part I: The near-Earth particle environment. Naval Research Laboratory, NRL Memo. Rep. 4506.

Badhwar, G. D.; P.M. O'Neill, An improved model of galactic cosmic radiation for space exploration missions. In: 22nd International Cosmic Ray Conference (paper OG 5-2), Dublin, Ireland, August, 1991 (to be published).

Bowman, J. B.; W. J. Swiatecki, C.F. Tsang, 1973, Abrasion and ablation of heavy ions. Lawrence Berkeley Laboratory Rep. LBL-2908.

Nealy, J. E.; J.W. Wilson, L.W. Townsend, 1990, Preliminary analyses of space radiation protection for lunar base surface systems. In: 20th Annual Intersociety Conference on Environmental Systems, Williamsburg, VA. SAE Technical Paper Series #891487.

Schimmerling, W.; M. Rapkin, M. Wong, J. Howard, 1986, The propagation of relativistic heavy ions in multielement beam lines. Med. Phys. 13: 217-228.

Schimmerling, W.; J. Miller, M. Wong, M. Rapkin, J. Howard, H.G. Spieler, B.V. Jarret, 1989, The fragmentation of 670A MeV Neon-20 as a function of depth in water. I. Experiment. Radiat. Res. 120: 36-71.

Shavers, M. R.; S.B. Curtis, J. Miller, W. Schimmerling, 1990, The fragmentation of 670A MeV Neon-20 as a function of depth in water. II. One-generation transport theory. Radiat. Res. 124:117-130.

Silberberg, R.; C.H. Tsao, 1973, Partial cross sections in high energy nuclear reactions and astrophysical applications. Astrophys. J. Suppl. 25: 315-368.

Simonsen, L. C.; L. W. Townsend, J.W. Wilson, J. E. Nealy, J. E., 1990, Space radiation shielding for Martian habitats. In: 20th Annual Intersociety Conference on Environmental Systems, Williamsburg, VA. SAE Technical Paper Series #901346.

Townsend, L. W.; J. W. Wilson, 1984, Galactic heavy-ion shielding using electrostatic fields. Hampton VA: NASA TM 86265.

Townsend, L. W.; M. Wong, M. Schimmerling, J.W. Wilson, 1987, Development of a nuclear data base for relativistic ion beams. In: Busick and Swanson, eds. Health physics of radiation generating machines, 20th Midyear Topical Symposium of the Health Physics Society. CONF-8602106, National Technical Information Service, Springfield, VA.

Townsend, L. W.; J.W. Wilson, J.E. Nealy, 1988, Preliminary estimates of galactic cosmic ray shielding requirements for manned interplanetary missions. Hampton, VA: NASA TM 101516.

Townsend, L. W. ; J.W. Wilson, J.E. Nealy, 1990b, Space radiation shielding strategies and requirements for deep space missions. In: 20th Annual Intersociety Conference on Environmental Systems, Williamsburg, VA: SAE Technical Paper Series #891433.

Townsend, L. W.; J.W. Wilson, J.L. Shinn, J.E. Nealy, L. C. Simonsen, 1990a Radiation protection effectiveness of a proposed magnetic shielding concept for manned Mars missions. In: 20th Annual Intersociety Conference on Environmental Systems, Williamsburg, VA. SAE Technical Paper Series #901343.

Wilson, J. W.; L.T. Townsend, F.F. Badavi, 1987b, A semiempirical nuclear fragmentation model. Nucl. Instrum. & Methods Phys. Res. B18, no. 3:225-231.

Wilson, J. W.; L.W. Townsend, W. Atwell, 1987a, Preliminary estimates of galactic cosmic ray exposures for manned interplanetary missions. Hampton, VA: NASA TM 100519.

Wilson, J. W.; S.L. Lamkin,H. Farhat,B.D. Ganapol, L.W. Townsend, 1989, A hierarchy of transport approximations for high energy heavy (HZE) ions. Hampton, VA: NASA TM 4118.

Wilson, J. W.; L.W. Townsend, S.L. Lamkin, B.D. Ganapol, 1990, A closed-form solution to HZE transport. Radiat. Res. 22:223-228. 1990.

TRANSPORT METHODS AND INTERACTIONS

FOR SPACE RADIATIONS

John W. Wilson and Lawrence W. Townsend
Langley Research Center
Hampton, Virginia

Walter Schimmerling
Lawrence Berkeley Laboratory
Berkeley, California

Govind S. Khandelwal and Ferdous Khan
Old Dominion University
Norfolk, Virginia

John E. Nealy, Francis A. Cucinotta,
Lisa C. Simonsen, and Judy L. Shinn
Langley Research Center
Hampton, Virginia

John W. Norbury
Rider College
Lawrenceville, New Jersey

ABSTRACT

This report presents a brief history leading to the involvement of the Langley Research Center, of the National Aeronautics and Space Administration (NASA) in space-radiation physics and protection. Indeed, a relatively complete summary of technical capability as of the summer of 1990 is given. The Boltzmann equations for coupled ionic and neutronic fields are presented and inversion techniques for the Boltzmann operator are discussed. Errors generated by the straight ahead approximation are derived and are shown to be negligible for most problems of space-radiation protection. A decoupling of projectile propagation from the target fields greatly simplifies the Boltzmann equations and allows an analytic solution of the target fragment transport. Analytic and numerical methods of solving the projectile transport equations are discussed. The study shows that explicit numerical techniques can develop unstable roots that require some care in applying discrete numerical methods. A second class of numerical methods is derived by first inverting the Boltzmann operator to form a Volterra equation from which an unconditionally stable numerical marching procedure is derived. Error propagation in the marching procedure is studied. Local relative errors must be on the order of h^2 for adequate control of propagated errors, where h is the step size.

The nuclear physics underlying the coefficients in the Boltzmann equation is discussed. A coupled-channel optical model is found as a consequence of the loose binding of nuclear matter and closure of the nuclear states in high-energy reactions. An abrasion optical model is derived that agrees well with experiment if the two-body interaction matrix is properly symmetrized. The optical model is found to be a good approximation to the elastic channel in the coherent approximation. Noncoherent effects are explicitly evaluated in a bordered matrix approximation. A complete elastic channel data base is presented. Inelastic and nonelastic processes are treated in the bordered matrix approximation with encouraging comparisons with experiment. The theory of electromagnetic dissociation is reviewed, and a model for single-nucleon and two-nucleon knockouts is presented and compared with experimental data. A semiempirical nuclear-fragmentation model is presented for the generation of a nuclear reaction data base and compared with experimental data. A relatively complete nuclear reaction data base is presented.

Transport solutions with the developed data base are used with laboratory experiments to validate both the transport code and the data base. Numerical benchmarks and comparison with Monte Carlo calculations are also used for code validation.

The analytic methods and data base are used to study coupling of the local radiation fields to electronic devices, dosimeters, and biological systems. Energy deposition fluctuations in thin silicon detectors caused by target fragmentation in the silicon device are shown and compared with experiment. Energy fluctuation in microscopic volumes is studied and the relation to tissue equivalent microdosimeters is described. The bone-tissue interface is examined for possible damage enhancement effects in the transition region near the interface. Target fragmentation corrections to damage coefficients of biological experimental data are discussed. Comparisons are made with results obtained by others using different nuclear data bases.

Approximate solutions to the Boltzmann equation in arbitrary convex geometry are found in preparation for application to space radiations. A buildup factor formalism is derived for space use, and example calculations for the human geometry in a space vehicle or on the surface of the Moon and Mars are given. The heavy ion transport code is used to study the shielding requirements for lunar or martian missions. Future needs of the NASA radiation physics program are discussed.

1. INTRODUCTION

1.1. Pre-NASA History

The panel meeting (Armstrong, 1949) on "Aero Medical Problems of Space Travel," sponsored by the School of Aviation Medicine, Wright Field, Ohio, was held in 1949, the year following the first published account of the existence of heavy ions in the galactic cosmic rays (GCR) observed at high altitude in the Earth's atmosphere (Freier et al., 1948). It was C. F. Gell, a member of the panel, who suggested that space radiation may be life threatening despite stratospheric radiation studies indicating the contrary. He gave two reasons for this possibility: (1) The cosmic radiation that is unable to penetrate to the stratosphere may be important and (2) the geomagnetic field deflects many of the particles away from the Earth; therefore, they are not observed in current stratospheric flight experiments. He proposed the need to further investigate space-radiation protection and the subject plunged into immediate controversy. In the following year, H. J. Schaefer (1950), of the Naval School of Aviation Medicine, provided a review of atmospheric radiations. He reported that cosmic rays are greatly diminished at the Earth's surface (0.1 mR/day (1 R corresponds to the exposure unit of formation of 1 esu/cm^3 of dry air at standard conditions)), increase to a maximum of 15 mR/day at 70 000 ft, and decrease beyond the transition maximum formed by the well-known transition effect (fig. 1.1). The transition effect results from interactions of the most penetrating radiations producing secondary particles in sufficient numbers to increase the dose. A belief that the ionization rates would decline to the free space values, where only the primary particles were present, had been generally accepted. However, as Schaefer notes, the discovery of "heavy nuclei rays" with their low penetrating power leads one to expect the decline beyond 70 000 ft to reach a minimum followed by a rise in ionization at higher altitudes. He credits C. F. Gell for first suggesting this possibility. Schaefer further suggests that the unusually high specific ionization of these energetic heavy nuclei indicates they may pose a significant health hazard and stresses the importance of further study.

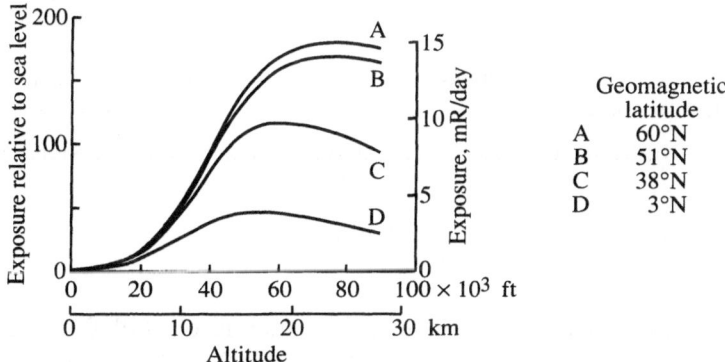

Figure 1.1. Altitude dependence of ionization in tissue from cosmic radiation (from Schaefer, 1950).

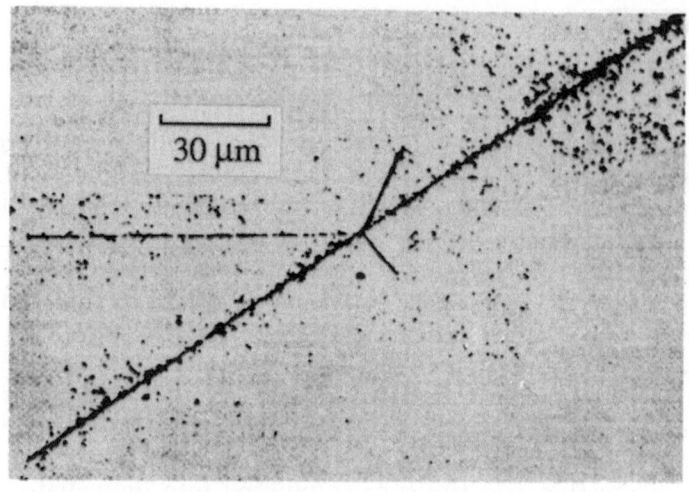

(a) $Z \approx 20$ projectile, low-energy transfer event.

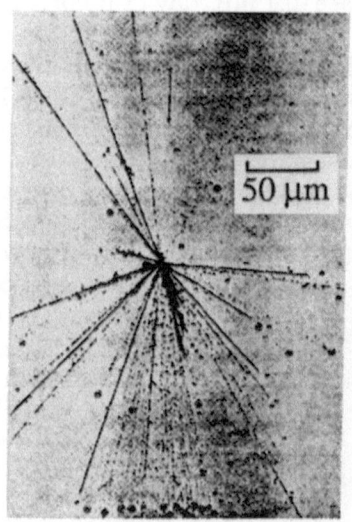

(b) Neutral primary, 27 relativistic charged prongs.

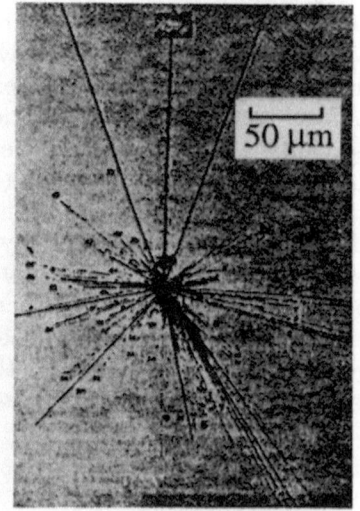

(c) $Z \approx 17$ projectile, high-energy transfer event.

Figure 1.2. Nuclear-star events observed in nuclear emulsion (from Krebs, 1950).

After the publication of the findings of the panel, Krebs (1950) at the Field Research Laboratory of the Army Medical Service, Fort Knox, Kentucky, described his work on biological experiments with cosmic air showers by emphasizing the importance of nuclear-star contributions. Krebs suggests in particular that the "explosive (or 'nuclear') stars," assumed to be created by "heavy nuclei coming from outside of the atmosphere," are a novel physical process (fig. 1.2) with potential for biological effects that "cannot be overemphasized." Clearly, Krebs' emphasis on the nuclear-star contribution is to be distinguished from Schaefer's concern over the direct ionization of cosmic heavy nuclei. Although nuclear-star effects in tissue remains an important biological issue (see chapter 11), it was the insight of Schaefer on the nature of high energy and charge (HZE) ions 40 years ago that is having a lasting impact on radiation physics and biology.

The Symposium on Space Medicine at the 23rd Annual Meeting of the Aero Medical Association, held in March of 1952, was a watershed for space-radiation biology and protection. Schaefer (1952) argued eloquently that delayed effects are the likely consequence of cosmic heavy nuclei exposure and that we cannot extrapolate from well-established dose response curves for common radiations. The nature of atmospheric ionization exposure was discussed as was the problem of extrapolation to free space (fig. 1.3) with the limiting effects of geomagnetic cutoff, solar modulation, and the uncertainty in the radiobiology. Schaefer then looked at the issue of track structure (fig. 1.4) and described a model of injury near the end of the heavy nuclei tracks (microbeams), which was a small linear lesion somewhat

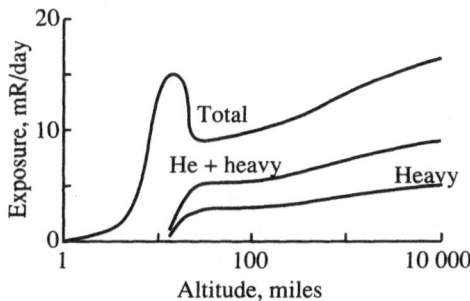

Figure 1.3. Ionization dosage from cosmic radiation for distances from 1 to 10 000 miles from Earth at higher latitudes (from Schaefer, 1952).

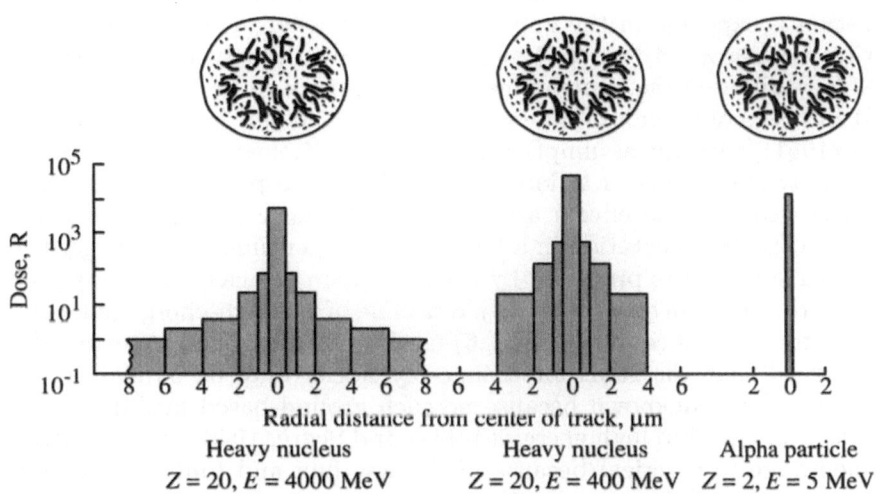

Figure 1.4. Radial spread of ionization about heavy nucleus and alpha tracks to be compared with size of human cell at top (from Schaefer, 1952).

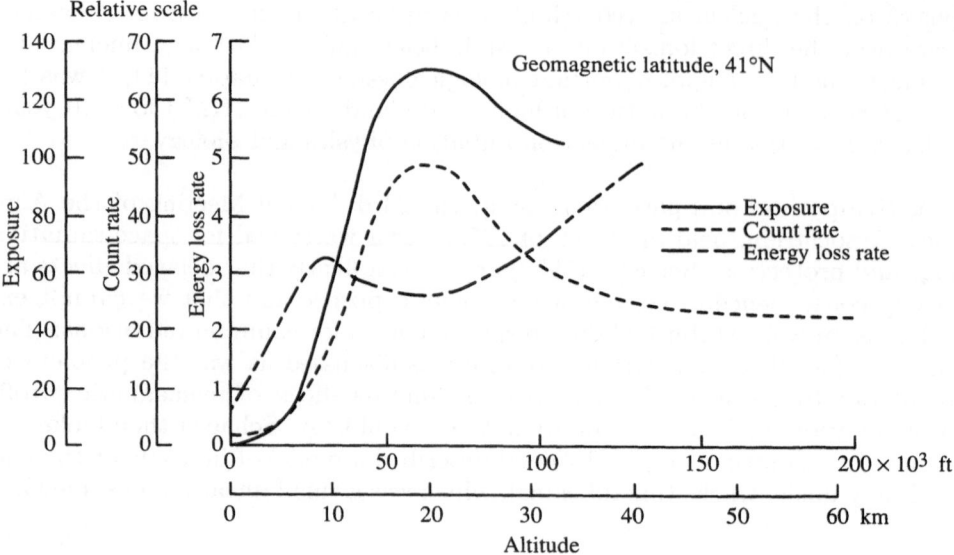

Figure 1.5. Ionization, counting rate, and average rate of energy loss (proportional to specific ionization) (from Tobias, 1952).

similar to Todd's microlesion but without the carcinogenic interpretation Todd (1983) suggests. It was Schaefer (1952) who first suggested that linear energy transfer (LET) may not be a good predictor of biological response because the track width also controls the biochemistry.

Also presented at the Symposium on Space Medicine was a particularly lucid paper by C. A. Tobias (1952), of the Donner Laboratory, on the radiation hazards in high-altitude aviation. Like Schaefer, Tobias argued that a rapid change in GCR composition is expected in the upper atmosphere where particles of high specific ionization are absorbed (fig. 1.5) and are converted partly to particles of lesser charge. He estimated the neutron-biological exposure to be 10 mrem/day (1 rem (= 0.01 Sv) is an older unit of dose equivalent) at 45 000 ft from the measurements of Yuan (1951), with the assumption of a Relative Biological Effectiveness (RBE) of 10, and stressed the need to look for low-energy, cosmic nuclei near the North Pole where geomagnetic effects are minimal. Because primary iron nuclei will undergo nuclear fragmentation in a few grams per centimeter^2, he suggested that less ionizing secondaries produced by fragmentation of heavy ions may pose a reduced hazard. One can observe the rapid decline of the high-energy pulse events in gas-filled proportional counters (fig. 1.6) used by McClure and Pomerantz (1950). Tobias argued that the RBE for cosmic-ray nuclei may be as high as 100 but that the values are unknown because no such ground-based facilities are able to accelerate iron nuclei to high energy (Tobias and Segrè, 1946) and biological flight experiments are impractical because of the low flux and limited exposure time. He estimated the exposure to be 26 rem per year at the top of the atmosphere or about 50 rem per year in free space. He then surveyed the biological data available and proposed a radiobiological program that would be the mold for the

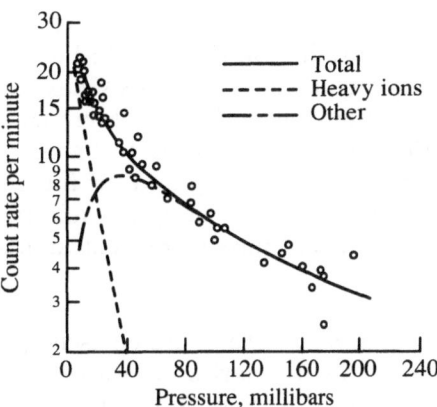

Figure 1.6. Bursts produced by protons, neutrons, α-particles, and heavy ions (from McClure and Pomerantz, 1950).

next 40 years of heavy ion, space-radiation biology. He predicted the possible direct observations of light flashes from heavy ion exposure in dark-adapted eyes, which were observed by Apollo astronauts nearly 20 years later.

New emphasis was given to space radiation after the occurrence of an enormous solar flare on February 23, 1956, which was summarized by Schaefer (1957) and further detailed in 1958 (Schaefer, 1958). After the successful launch of an unmanned satellite by the USSR, NASA was formed out of the older civil aeronautical agency (National Advisory Committee for Aeronautics (NACA)) and elements of the military space effort in 1958. In this same year, Van Allen discovered the trapped radiation belts. In July 1959, Schaefer (1959) began to explore the possibilities of space travel despite the presence of the Van Allen belts.

1.2. History of Langley Program

In June 1960, a conference on radiation problems in manned space flight, organized by the Office of Life Science Programs, NASA (Jacobs, 1960), was convened to address the problem of potential acute and chronic radiation damage. A background paper on space radiation was presented by J. A. Winckler of the University of Minnesota in which the 1956 solar flare and subsequent events through 1959 were discussed. We now know that an even bigger solar flare event occurred on November 12–13, 1960, 5 months after the June 1960 conference. Nearly all factors important to solar flare events were identified at the NASA conference: important locations on the Sun for active regions to affect the Earth, propagation and storage effects, geomagnetic effects including magnetic disturbances, the significance of type-IV radio noise as a signature of particle events, importance of riometer and ground-level neutron monitors, and an estimate of the required shielding thickness. Winckler suggested that GCR exposures were probably unimportant for short-duration missions. Winckler's review noted that the inner Van Allen zone was reasonably stable with dose rates of 30 R/hr compared with the dynamic outer zone with peak rates near 10 R/hr. (Shielding of the ion chambers was not specified.) An attempt was made to establish a rationale to define acceptable risk. J. E. Pickering, of the Air Force School of Aviation Medicine, Wright Field, Ohio, suggested radiation risk should be in line with other mission risks; this then became a dominant theme in NASA's exposure-limits assessments. The main conclusions to be drawn were that the Mercury program, which flew at 100 n.mi. at low inclination, was not expected to have a radiation problem, but a vigorous radiation program would be required for future NASA missions. This is the historical context of the beginning of space-radiation protection at NASA Langley Research Center which continues at present and is the main focus of the rest of this report.

The Langley effort began in 1958 with Trutz Foelsche (1959) evaluating specific ionization caused by cosmic-ray primaries in water or tissue. Aware of the high altitudes projected by the U.S. Supersonic Transport (SST) Program, the potential impact on commercial operations was brought to the forefront (Foelsche, 1961). A major concern in Foelsche's estimates was uncertainty in neutron and target-recoil contributions (Foelsche, 1962a and 1962b) which would be a dominant issue at Langley for the next decade. His estimates of space-radiation doses were a prime contribution to the Conference on Environmental Problems of Space Flight Structures, convened under the Advisory Committee on Missile and Space Vehicle Structures (Vosteen, 1962).

The first Symposium on the Protection Against Radiation Hazards in Space, held in Gatlinburg, Tennessee (first Gatlinburg Conference, Anon., 1962), was a coming together of the diverse elements working on various aspects of the space-radiation problem. At the conference, plans for the Space Radiation Effects Laboratory (SREL) at Langley were unveiled with its central 600-MeV synchrocyclotron and various other low-energy machines. (The site is the current location of the Continuous Electron Beam Acceleration Facility (CEBAF).) The main experimental thrust of the radiation-protection group was secondary particle production from collisions of energetic protons and α-particles (Orr, 1972; Beck and Powell, 1976). Also presented at the conference by Kinney, Coveyou, and Zerby (1962) were the beginnings of the High-Energy Transport Code (HETC). The most surprising feature of the conference in retrospect was the lack of papers on energetic heavy ions, except for the biological experiments of H. J. Curtis, who used deuteron microbeams to simulate the high-energy, heavy ion microlesions suggested by Schaefer (1952) as a potential biological hazard a decade before.

The second Gatlinburg conference, held in Gatlinburg, Tennessee, 2 years later (Reetz, 1965), showed considerable maturation. Foelsche identified major uncertainties in neutron-exposure rates that justified an atmospheric-measurements program starting in 1965, which ran out of funds just 7 months before the now-famous solar event of August 1972 (Korff et al., 1979). Unlike the earlier conference, there were three papers in reference to high-energy heavy ions. One was written by P. Todd (1965) presenting a host of cell survival data for heavy ion beams measured at the Donner Laboratory of the Lawrence Radiation Laboratory at Berkeley (LRLB) and the other was written by S. B. Curtis (Curtis, Dye, and Sheldon, 1965), who later joined the staff at the Donner Laboratory. Several fundamental papers appeared using HETC, including one by R. G. Alsmiller, Jr., et al. (1965), showing the validity of the straight ahead approximation for high-energy nucleon transport as applied to space radiations. This paper had an important impact on transport theory development at Langley. Probably the most surprising paper in retrospect was the four-part paper written by J. Billingham, D. E. Robbins, J. L. Modisette, and P. W. Higgins (1965). This paper described the dose limits for design purposes in the Apollo mission as 200 rem (blood-forming organ, ocular lens), 700 rem (skin), and 980 rem (hands and feet), which were adequate to return the astronauts to Earth for proper medical attention (Reetz, 1965). A clear balance was established between radiation risk and other mission risks for this exploratory high-risk mission.

In December 1964, the FAA requested that NASA resolve the issue of radiation exposure for the commercial supersonic transport as had been so elegantly discussed by Foelsche at the second Gatlinburg conference. A detailed measurements program began the following year by combining efforts at Langley with the work of the Korff group at New York University (NYU). The flight experiment package included tissue-equivalent ion chambers, a fast neutron spectrometer (1–10 MeV), and nuclear emulsion. Over the next several years, there were hundreds of high-altitude balloon and airplane flights, a world latitude survey on a Boeing 707 airplane, and high-altitude studies, especially for solar flare events, in U2 and RB-57F flights. (See fig. 1.7.) The main limitation of the Langley experimental effort was the lack of neutron spectrum measurements outside the fast region. The year after the Langley measurements program began, the International Commission for Radiological Protection (ICRP) Task Group (1966) for SST exposure published their conclusion that the biological exposure from atmospheric neutrons was nearly negligible.

The Langley effort was to extend the measured neutron spectrum to both lower and especially to higher energies by using the Monte Carlo work at the Oak Ridge National Laboratory (ORNL), namely, the HETC. The importance of

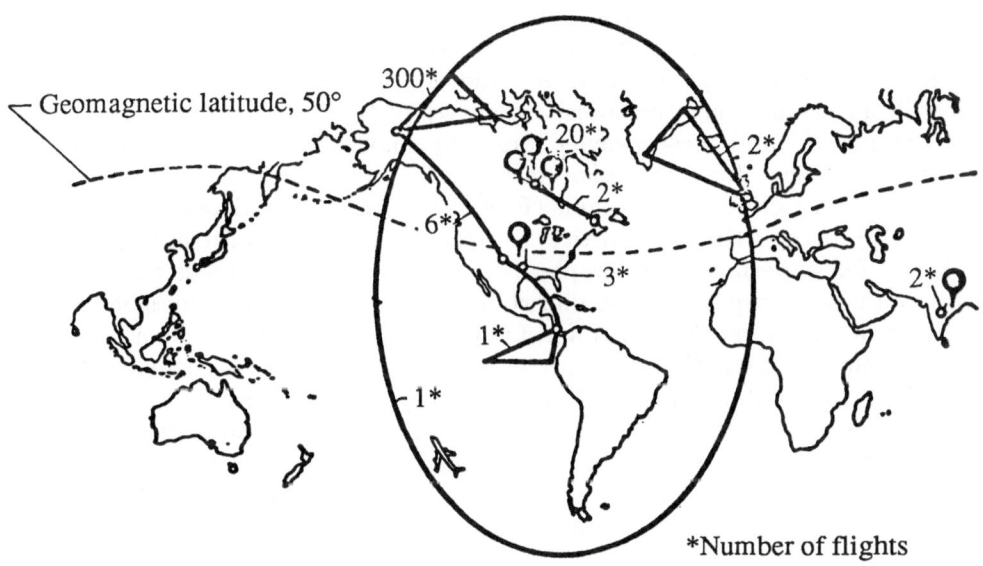

Figure 1.7. High-altitude radiation measurements from 1965 to 1971.

195

the development of the Bertini (1967) nuclear reaction code and the associated shielding code (HETC) cannot be overemphasized. An outstanding feature of the code is the inclusion of the intranuclear cascade code (Bertini, 1967) as part of the internuclear cascade calculation (HETC) reducing the reliance on external nuclear data bases. At the same time, this feature made the complete code computationally inefficient in the midst of demands and requests for results for various disciplines. The Langley atmospheric program found itself standing in line with accelerator, biomedical, dosimetry, and space programs waiting to be serviced by the HETC code. The decision was made at Langley to develop an in-house capability.

Code development was undertaken by physicists within the computational division at Langley with great vigor; for after all, when the supply of nuclear data is exhausted, a real opportunity to develop nuclear theory exists. The code (PROPER-C) chosen for Langley development was written by Leimdorfer and Crawford (1968) for applications at energies below pion-production threshold. This code was extended to high energies (PROPER-3C) by incorporating the recently published Bertini (1967) data (Wilson, 1972b) and making a high-energy extrapolation (Lambiotte, Wilson, and Filippas, 1971). There were critical meetings concerning the SST in early 1969, and results from the Langley code were the only available results to fill the gap (Foelsche et al., 1969; Foelsche and Wilson, 1969; Wilson, Lambiotte, and Foelsche, 1969). The Langley code was extremely fast because the intranuclear nucleon cascade was represented by a numerical data set and yet required over $80,000 of computer time (1968 dollars) to make the extension of the fast neutron spectrum to high energies. The results predicted the transition curve (fig. 1.8) measured for fast neutrons (Foelsche et al., 1969; Foelsche and Wilson, 1969), the importance of high-energy neutrons (Foelsche and Wilson, 1969; Wilson, 1969) in contributing to biological dose (fig. 1.9), and an interesting structure in the atmospheric neutron spectrum (Wilson, Lambiotte, and Foelsche, 1969). These results were confirmed by later calculations at NYU (Korff et al., 1979) and ORNL (Foelsche et al., 1974). A summary of the atmospheric radiation program is given by Foelsche et al. (1974) and Korff et al. (1979). From these studies, the background radiation levels were still uncertain, since the transition curves of the other heavier primary ions were not known, and these heavier ions may make important contributions to the dose equivalent for some solar flare events. Therefore, preliminary studies for heavy ion reactions were begun (Foelsche et al., 1974; Skoski, Merker, and Shen, 1973) at the Princeton Particle Accelerator (PPA). As further justification for heavy ion experiments, a simple model of visual impairment by heavy ion exposure revealed required shield uncertainties for a 3-year Mars mission of 4.5 to 29 g/cm^2 of aluminum (fig. 1.10) and further emphasized the value of a vigorous heavy ion physics and radiobiology program (memorandum to the Langley director concerning continuation and modification of the Princeton Particle Accelerator by NASA in 1970). Experiments began at a meager level and were later moved to the Lawrence Berkeley Laboratory (Schimmerling, Kast, and Ortendahl, 1979), where they continue to this day at a very modest funding level (Schimmerling, Curtis, and Vosburgh, 1977; Schimmerling et al., 1987 and 1989).

Two accomplishments resulted from the PROPER-3C code (Lambiotte, Wilson, and Filippas, 1971): The available nuclear data were exhausted, laying the

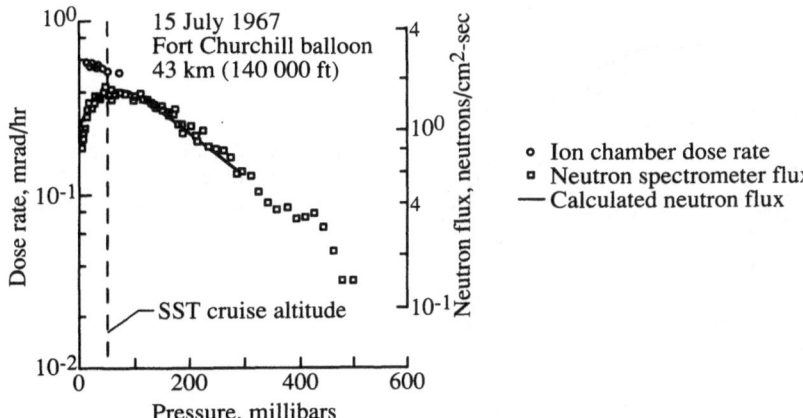

Figure 1.8. Galactic cosmic rays 2 years after galactic cosmic-ray maximum (July 1967, Fort Churchill, geomagnetic latitude $\approx 69°$). The neutron flux and ion chamber dose rate have both decreased about 25 to 30 percent at SST altitudes (solar modulation). (From Foelsche et al., 1974.)

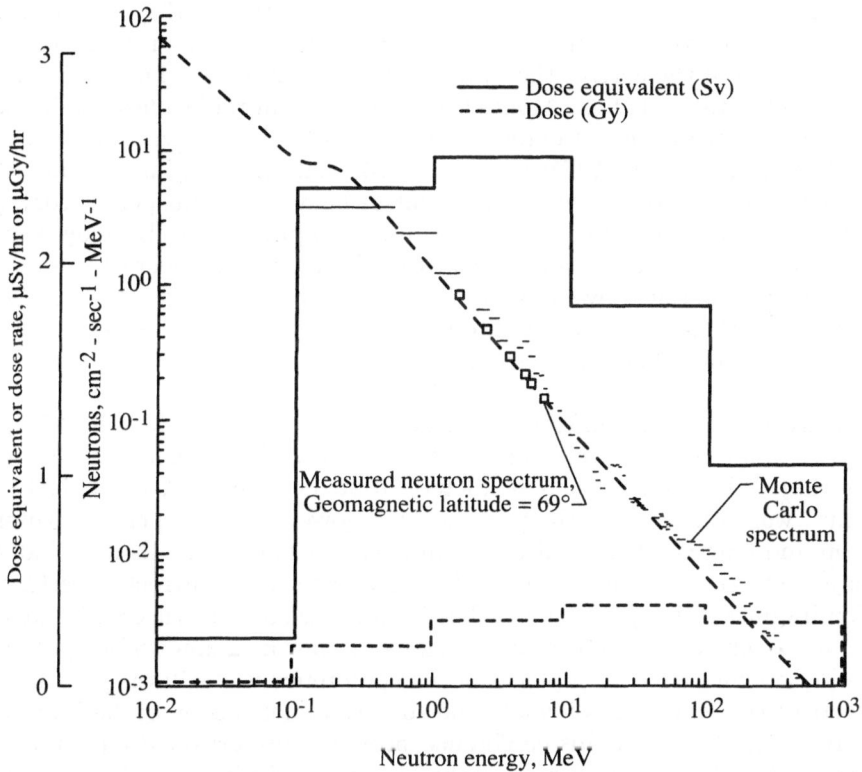

Figure 1.9. The high-latitude (geomagnetic latitude $\approx 69°$) neutron spectrum measured at SST altitudes (≈ 50 g/cm^2) on August 3, 1965, by Korff et al. (1979), with its extension to lower and higher energies compared with the shape of the Monte Carlo spectrum. (From Foelsche et al., 1974.)

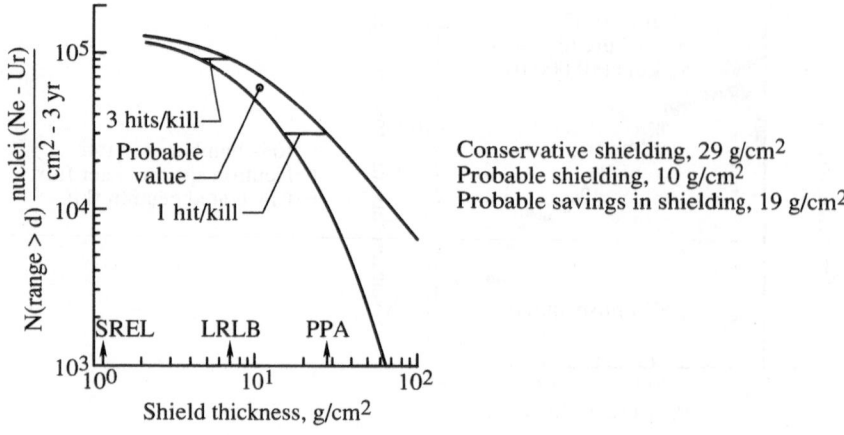

Figure 1.10. Shield analysis of Wilson.

groundwork for a theoretical nuclear program, and an appreciation of the tedious details of the Monte Carlo method, not to mention the intense computer requirements. Consequently, the first fruits of a nuclear theory program produced new skills in multiple scattering theory (Wilson, 1972a, 1973, and 1974b), a fundamental theory of heavy ion reactions (Wilson, 1974a), and the first Langley-developed data base for heavy ion cross sections (Wilson and Costner, 1975). These theories provided the framework for nuclear model development for the next 15 years and continue to provide the core of the Langley nuclear program. The greater appreciation of the limitations of the Monte Carlo methods in radiation shielding led to the development of a series of deterministic codes beginning with nucleon transport (Wilson and Lamkin, 1975; Lamkin, 1974; Wilson and Khandelwal, 1976b) and moving onward toward the development of heavy ion transport theory (Wilson, 1977a, 1977b, and 1978). The deterministic approach at Langley was seen as the necessary means of obtaining codes useful for an engineering design environment. A more detailed overview of the Langley program was given by Wilson (1978) at the workshop on the satellite power system, held at Lawrence Berkeley Laboratory (Schimmerling and Curtis, 1978).

The third Gatlinburg conference was held in Las Vegas in March 1971 (Warman, 1972). Reported at this conference were the light flashes in the Apollo missions that Tobias (1952) had predicted 20 years earlier. A great deal of the symposium was concerned with space nuclear power. Two important papers by Wilkinson and Curtis (1972) and Curtis and Wilkinson (1972) showed that there were major uncertainties in shield requirements caused by current uncertainty in heavy ion fragmentation parameters. The importance of galactic cosmic-ray exposure was a concern for long-duration missions in view of unknown but potentially large biological effects. The emphasis in the conference was still the proton shielding aspects of the Apollo mission and the successful conclusion of the Man on the Moon Program. Although this conference provided important documentation of the previous decade of work, it also marked the rapid decline in radiation research funding within NASA. In spite of the total lack of funding from 1973 to 1980, Langley maintained its files on radiation interaction and managed to perform radiation related tasks on a time-available basis. Fundamental work on dosimetry (Khandelwal, Costner, and Wilson, 1974; Khandelwal and Wilson, 1974; Wilson,

1975b), new methods in radiation transport (Wilson and Khandelwal, 1976a and 1976b; Wilson, 1975a; Wilson and Denn, 1977b and 1977c), and analysis of space-radiation-protection issues (Wilson and Denn, 1976 and 1977a; Wilson, 1981) were completed.

It was natural in these intermediate years to work in closely related disciplines. The first such area was nuclear-induced plasmas and nuclear pumped lasers (Wilson and De Young, 1978a and 1978b; Harries and Wilson, 1979; De Young and Wilson, 1979; Wilson, De Young, and Harries, 1979; Wilson and Shapiro, 1980; Wilson, 1980). The nuclear flash-lamp-pumped laser work (Wilson, 1980) was a natural lead into direct solar-pumped laser systems (Wilson and Lee, 1980; Harries and Wilson, 1981; Wilson, Raju, and Shiu, 1983; Wilson et al., 1984). With new skills in nuclear-induced plasma chemistry, nuclear interactions in materials became a natural work area more closely akin to space radiations. A small amount of funds was available allowing work on structural materials (Wilson and Kamaratos, 1981; Wilson and Xu, 1982; Wilson et al., 1982; Kamaratos et al., 1982; Xu, Khandelwal, and Wilson, 1984a and 1984b; Rustgi et al., 1988) and electronic materials (Wilson, Stith, and Stock, 1983; Wilson and Stock, 1984). The space-radiation-protection research was restored under a proposal to the Life Sciences Division entitled "Space Radiation Protection Methods," submitted July 31, 1979, by John Wilson. The proposal contained a local theoretical effort at Langley and experiments at the Lawrence Berkeley Laboratory as an augmentation of experiments funded by the National Cancer Institute conducted by Walter Schimmerling.

The present report gives an account of the methods and underlying data bases currently in use at the Langley Research Center. It is the goal of the Langley program to go beyond progress in fundamental methods to provide analysis tools that can be easily used by the nonexpert in engineering and experimental design applications. Such tools are not only to be convenient to use but are also to have been validated by laboratory experiments so that their domain of applicability is clearly delineated. Although such a goal was barely conceivable 20 years ago when we embarked on this course, this report demonstrates great progress toward this goal. We look forward to its successful completion in the coming decades.

1.3. Overview of Space-Radiation Interactions

An overview of the space environment and its interaction with materials was given by Wilson (1978). A number of details could be added but very little change in the basic protection requirements would result, and considerable uncertainty in radiation-protection practice remains even today. Here, we present a pedestrian view of space-radiation interaction and refer to the earlier review (Wilson, 1978) for somewhat expanded detail. The present document contains the interaction description in greater detail and our aim in this section is to give an overview to the processes described herein.

The energetic particles in space consist mainly of atomic constituents covering a very broad energy spectrum and flux values as shown in figure 1.11 (Wilson, 1978). The particles themselves are small ($\approx 10^{-13}$ cm) but are electrically charged

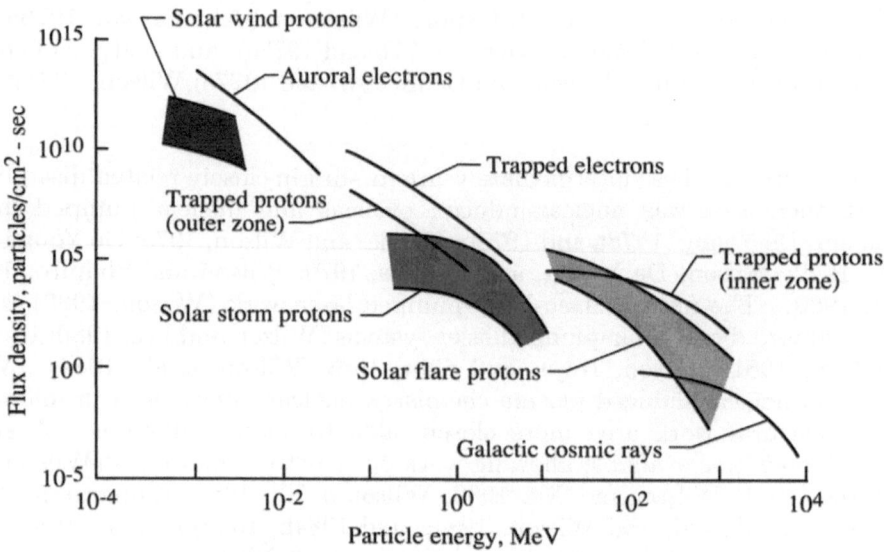

Figure 1.11. Space-radiation environment.

resulting in a long-range force component. A casual look at condensed matter reveals mostly the structure of the electron clouds which contain only 0.05 percent of the mass but occupy virtually all the space within the material. Embedded within these electron clouds are the atomic nuclei whose dimensions are 10^{-5} times smaller than the complete atom but contain 99.95 percent of the mass of the atom. Clearly, an energetic particle passing through such a material will mainly interact with the electrons in the cloud and seldom strike a nucleus.

We now discuss some of the physical parameters related to shielding calculations using elementary concepts. The dominant term in a shielding calculation is energy loss through ionization, that is, a collision between the incoming charged particle (whether it is a proton, electron, or heavy ion) and the orbital electrons of the shielding material (fig. 1.12). They interact through coulomb scattering, and the energy transferred from an ion of energy E and charge Z_P to a target particle of charge Z_T is labeled Q. The cross section σ has an inverse Q^2 dependence, and therefore the energy transfer is usually quite small. In the figure, μ is reduced mass for the projectile target system of masses M_P and M_T.

When the target is an electron bound in an atomic orbital, there are two options of either producing excitation when specific energy transfers ($\varepsilon_i - \varepsilon_j$, where ε_i and ε_j denote atomic energy levels) are made or ionization where the energy transferred must be greater than the ionization potential (fig. 1.13). The cross section is related to this energy transfer and goes like the inverse of Q^2. Another process that is extremely important, especially for incident electrons, is coulomb interaction with the atomic nucleus which results in multiple-scattering effects. These multiple-scattering effects are important for electron shielding or for laboratory ion experiments.

The cross sections for secondary electrons produced from impacts of ions with atoms as described in figures 1.12 and 1.13 are shown in figure 1.14. This

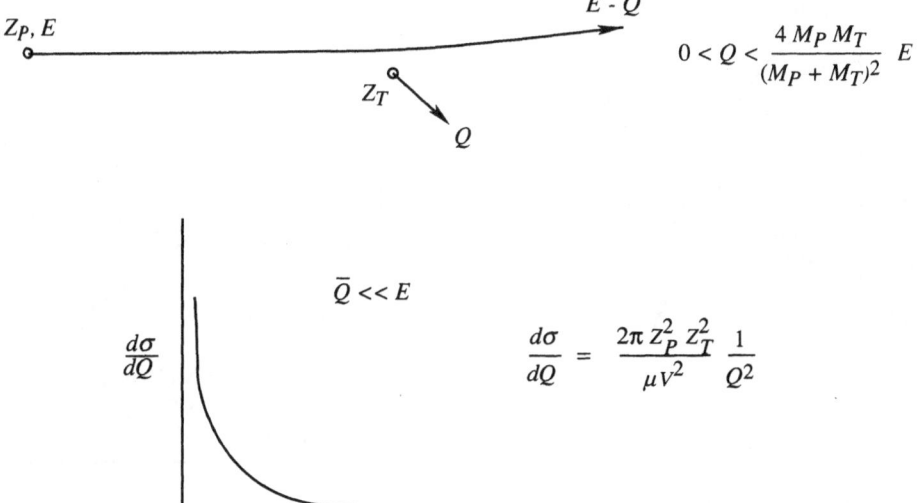

$$0 < Q < \frac{4\,M_P\,M_T}{(M_P + M_T)^2}\,E$$

$$\bar{Q} \ll E$$

$$\frac{d\sigma}{dQ} = \frac{2\pi\,z_P^2\,z_T^2}{\mu v^2}\,\frac{1}{Q^2}$$

Figure 1.12. Coulomb scattering.

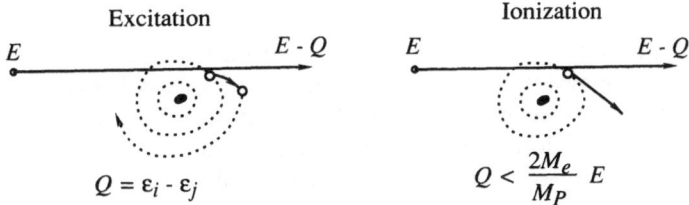

Excitation

$$Q = \varepsilon_i - \varepsilon_j$$

Ionization

$$Q < \frac{2M_e}{M_P}\,E$$

Coulomb interactions with atomic electrons

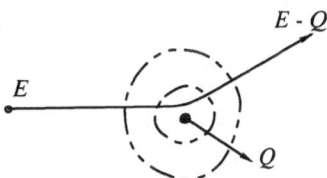

Coulomb interaction with atomic nucleus

Figure 1.13. Schematic of coulomb interactions with atomic electrons and atomic nucleus.

figure shows curve fits to the experimental data (Manson et al., 1975) at 1 and 5-MeV proton impacts, and the inverse Q^2 dependence above about 20 eV for the secondary electron energy is again evident. The corrections below 20 eV are due to binding effects which can only be treated quantum mechanically. The electron is actually bound in an atom, and these binding effects become important when the energy transfer is on the order of the binding energy. These types of data are important in giving the lateral spread of the energy from the track as the particle passes through a material.

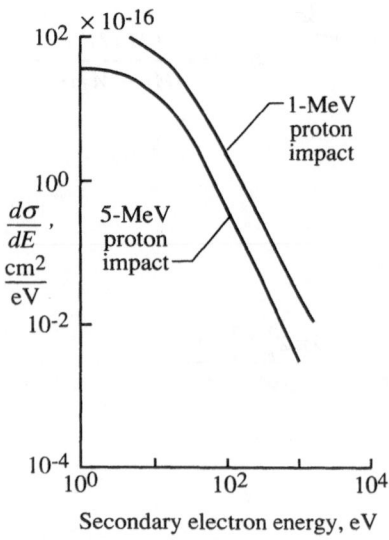

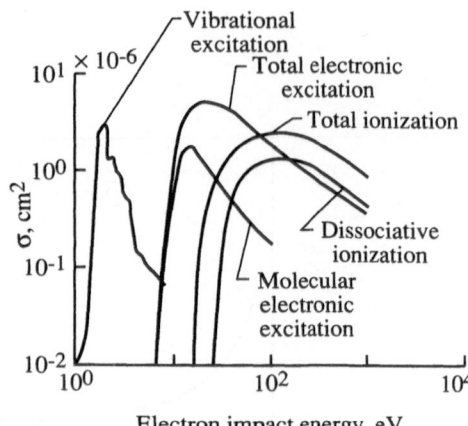

Figure 1.14. Secondary electron production spectra from proton impact with helium.

Figure 1.15. Electron impact cross section with N_2 molecules.

There are a number of other degrees of freedom that one contends with when looking at molecular systems. Shown in figure 1.15 is a collection of data for N_2 molecules, which we chose as a typical molecule mainly because we could find the most data for it. Vibrational excitation is important for electron energies below about 10 eV. Once the electronic excitation or ionization threshold is exceeded, everything becomes heavily dominated by those two processes alone. In about one half the cases, ionization results in dissociation; and according to the data we have been able to collect, most molecules undergoing electronic excitation result in dissociation. There are, however, considerable differences in the dissociation cross section for the two processes as seen in figure 1.15. Those differences are probably due to the small number of molecular states observed in the experiments. The dissociative excitation cross section will probably change as future experiments are performed, and total dissociative cross section will probably show the same energy dependence as the ionization cross section at high energy. The data are taken from Schulz (1976), Cartwright et al. (1977), Köllmann (1975), and Wight, Van der Wiel, and Brion (1976). The problem of molecular binding effects is difficult to treat using quantum theory but local plasma models have shown some success in treating both the molecular binding problem (Wilson and Kamaratos, 1981; Kamaratos, 1982; Xu, Khandelwal, and Wilson, 1984a and 1984b) and condensed phase effects (Wilson et al., 1984; Xu, Khandelwal, and Wilson, 1985).

Although most collisions in the material are with orbital electrons, the rare nuclear collisions are of importance because of the large energy transferred in the collision and the generation of new energetic particles. This process of transferring kinetic energy into new secondary radiations occurs through several different processes, such as direct knockout of nuclear constituents, resonant excitation followed by particle emission, pair production, and possible coherent effects within the nucleus. Through these processes, a single-particle incident on the shield

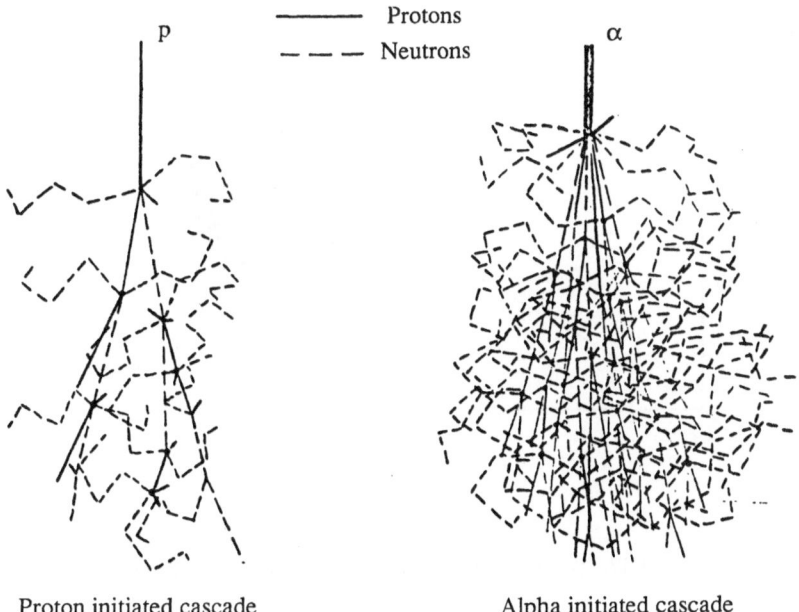

Figure 1.16. Cascade development in matter.

may attenuate through energy transfer to electrons of the media or generate a multitude of secondaries causing an increase in exposure (transition effect). The process that dominates depends on energy, particle type, and material composition. This development of cascading particles is depicted in figure 1.16 as a relative comparison between high-energy proton and α-particle cascades in the Earth's atmosphere. Note the similarities displayed in figure 1.16 for individual reaction events and the nuclear-star events shown in figure 1.2 for nuclear emulsion.

The relevant transport equations are derived on the basis of conservation principles. Consider a region of space filled by matter described by appropriate atomic and nuclear cross sections. In figure 1.17, we show a small portion of the region enclosed by a sphere of radius δ. The number of particles of type j leaving a surface element $\delta^2 \, d\vec{\Omega}$ is given as $\phi_j(\vec{x} + \delta\vec{\Omega}, \vec{\Omega}, E)\delta^2 \, d\vec{\Omega}$, where $\phi_j(\vec{x}, \vec{\Omega}, E)$ is the particle flux density, $\vec{x}$ is a vector to the center of the sphere, $\vec{\Omega}$ is normal to the surface element, and E is the particle energy. The projection of the surface element through the sphere center to the opposite side of the sphere defines a flux tube through which pass a number of particles of type j given as $\phi_j(\vec{x} - \delta\vec{\Omega}, \vec{\Omega}, E)\delta^2 \, d\vec{\Omega}$, which would equal the number leaving the opposite face if the tube defined by the projection were a vacuum. The two numbers of particles, in fact, differ by the gains and the losses created by atomic and nuclear collisions as follows:

$$
\begin{aligned}
\phi_j&(\vec{x} + \delta\vec{\Omega}, \vec{\Omega}, E)\delta^2 \, d\vec{\Omega} \\
&= \phi_j(\vec{x} - \delta\vec{\Omega}, \vec{\Omega}, E)\delta^2 \, d\vec{\Omega} \\
&\quad + \delta^2 \, d\vec{\Omega} \int_{-\delta}^{\delta} dl \sum_k \int \sigma_{jk}(\vec{\Omega}, \vec{\Omega}', E, E')\phi_k(\vec{x} + l\vec{\Omega}, \vec{\Omega}', E') \, d\vec{\Omega}' dE' \\
&\quad - \delta^2 \, d\vec{\Omega} \int_{-\delta}^{\delta} dl \, \sigma_j(E) \, \phi_j(\vec{x} + l\vec{\Omega}, \vec{\Omega}, E) \qquad (1.1)
\end{aligned}
$$

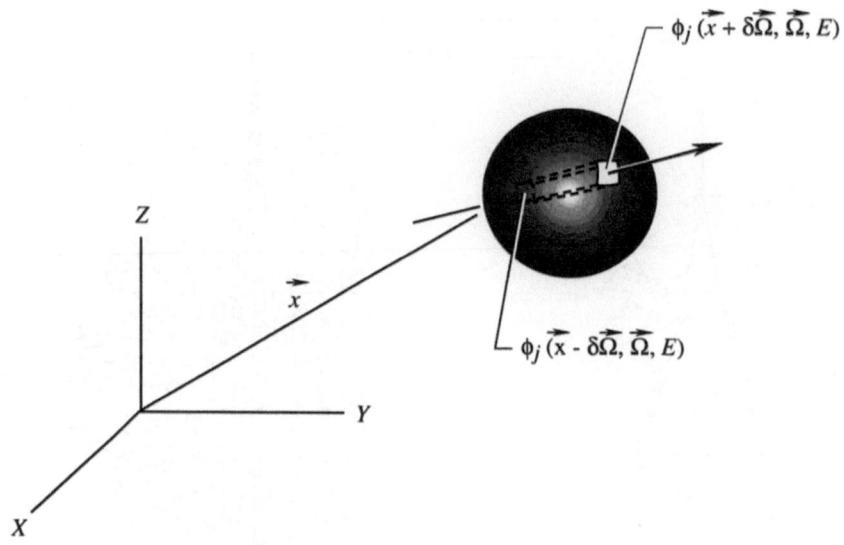

Figure 1.17. Transport of particles through spherical region.

where $\sigma_j(E)$ and $\sigma_{jk}(\vec{\Omega}, \vec{\Omega}', E, E')$ are the media macroscopic cross sections. The cross section $\sigma_{jk}(\vec{\Omega}, \vec{\Omega}', E, E')$ represents all those processes by which type k particles moving in direction $\vec{\Omega}'$ with energy E' produce a type j particle in direction $\vec{\Omega}$ with energy E. Note, there may be several reactions which may accomplish this result and the appropriate cross sections of equation (1.1) are the inclusive ones. Note that the second term on the right-hand side of equation (1.1) is the source of secondary particles integrated over the total volume $2\delta^3\,d\vec{\Omega}$ and the third term is the loss through nuclear reaction integrated over the same volume. We expand the terms of each side and retain terms to order δ^3 explicitly as

$$\delta^2\,d\vec{\Omega}\left[\phi_j(\vec{x}, \vec{\Omega}, E) + \delta\vec{\Omega}\cdot\nabla\phi_j(\vec{x}, \vec{\Omega}, E)\right]$$

$$= \delta^2 d\vec{\Omega}\left[\phi_j(\vec{x}, \vec{\Omega}, E) - \delta\vec{\Omega}\cdot\nabla\phi_j(\vec{x}, \vec{\Omega}, E)\right.$$

$$+ 2\delta\sum_k\int\sigma_{jk}(\vec{\Omega}, \vec{\Omega}', E, E')\phi_k(\vec{x}, \vec{\Omega}', E')\,d\vec{\Omega}'\,dE'$$

$$\left. - 2\delta\sigma_j(E)\,\phi_j(\vec{x}, \vec{\Omega}, E)\right] + O(\delta^4) \tag{1.2}$$

which may be divided by the cylindrical volume $2\delta(\delta^2\,d\vec{\Omega})$ and written as

$$\vec{\Omega}\cdot\nabla\phi_j(\vec{x}, \vec{\Omega}, E) = \sum_k\int\sigma_{jk}(\vec{\Omega}, \vec{\Omega}', E, E')\,\phi_k(\vec{x}, \vec{\Omega}', E')\,d\vec{\Omega}'dE'$$

$$- \sigma_j(E)\,\phi_j(\vec{x}, \vec{\Omega}, E) + O(\delta) \tag{1.3}$$

for which the last term $O(\delta)$ approaches zero in the limit as $\delta \to 0$. Equation (1.3) is recognized as a time independent form of the Boltzmann equation for a tenuous gas. Atomic collisions (i.e., collisions with atomic electrons) preserve the identity of the particle and two terms of the right-hand side of equation (1.3) contribute. The differential cross sections have the approximate form for atomic processes

$$\sigma_{jk}^{at}(\vec{\Omega},\vec{\Omega}',E,E') = \sum_n \sigma_{jn}^{at}(E)\,\delta(\vec{\Omega}\cdot\vec{\Omega}'-1)\,\delta_{jk}\,\delta(E+\varepsilon_n-E') \qquad (1.4)$$

where n labels the electronic excitation levels and ε_n are the corresponding excitation energies which are small (1–100 eV in most cases) compared with the particle energy E. The atomic terms may then be written as

$$\sum_k \int \sigma_{jk}^{at}(\vec{\Omega},\vec{\Omega}',E,E')\,\phi_k(\vec{x},\vec{\Omega}',E')\,d\vec{\Omega}'\,dE' - \sigma_j^{at}(E)\,\phi_j(\vec{x},\vec{\Omega},E)$$

$$= \sum_n \sigma_{jn}^{at}(E)\,\phi_j(\vec{x},\vec{\Omega},E+\varepsilon_n) - \sigma_j^{at}(E)\,\phi_j(\vec{x},\vec{\Omega},E)$$

$$\approx \sum_n \sigma_{jn}^{at}(E)\left[\phi_j(\vec{x},\vec{\Omega},E) + \sum_n \varepsilon_n\,\frac{\partial}{\partial E}\,\sigma_{jn}^{at}(E)\,\phi_j(\vec{x},\vec{\Omega},E)\right]$$

$$- \sigma_j^{at}(E)\,\phi_j(\vec{x},\vec{\Omega},E)$$

$$= \frac{\partial}{\partial E}\left[S_j(E)\phi_j(\vec{x},\vec{\Omega},E)\right] \qquad (1.5)$$

since the stopping power is

$$S_j(E) = \sum_n \sigma_{jn}(E)\varepsilon_n \qquad (1.6)$$

and the total atomic cross section is

$$\sigma_j^{at}(E) = \sum_n \sigma_{jn}^{at}(E) \qquad (1.7)$$

Equations (1.5) to (1.7) allow us to rewrite equation (1.3) in the usual continuous slowing down approximation as

$$\vec{\Omega}\cdot\nabla\phi_j(\vec{x},\vec{\Omega},E) - \frac{\partial}{\partial E}\left[S_j(E)\,\phi_j(\vec{x},\vec{\Omega},E)\right] + \sigma_j(E)\,\phi_j(\vec{x},\vec{\Omega},E)$$

$$= \int \sum_k \sigma_{jk}(\vec{\Omega},\vec{\Omega}',E,E')\,\psi_k(\vec{x},\vec{\Omega}',E')\,d\vec{\Omega}'\,dE' \qquad (1.8)$$

where the cross sections of equation (1.8) now contain only the nuclear contributions.

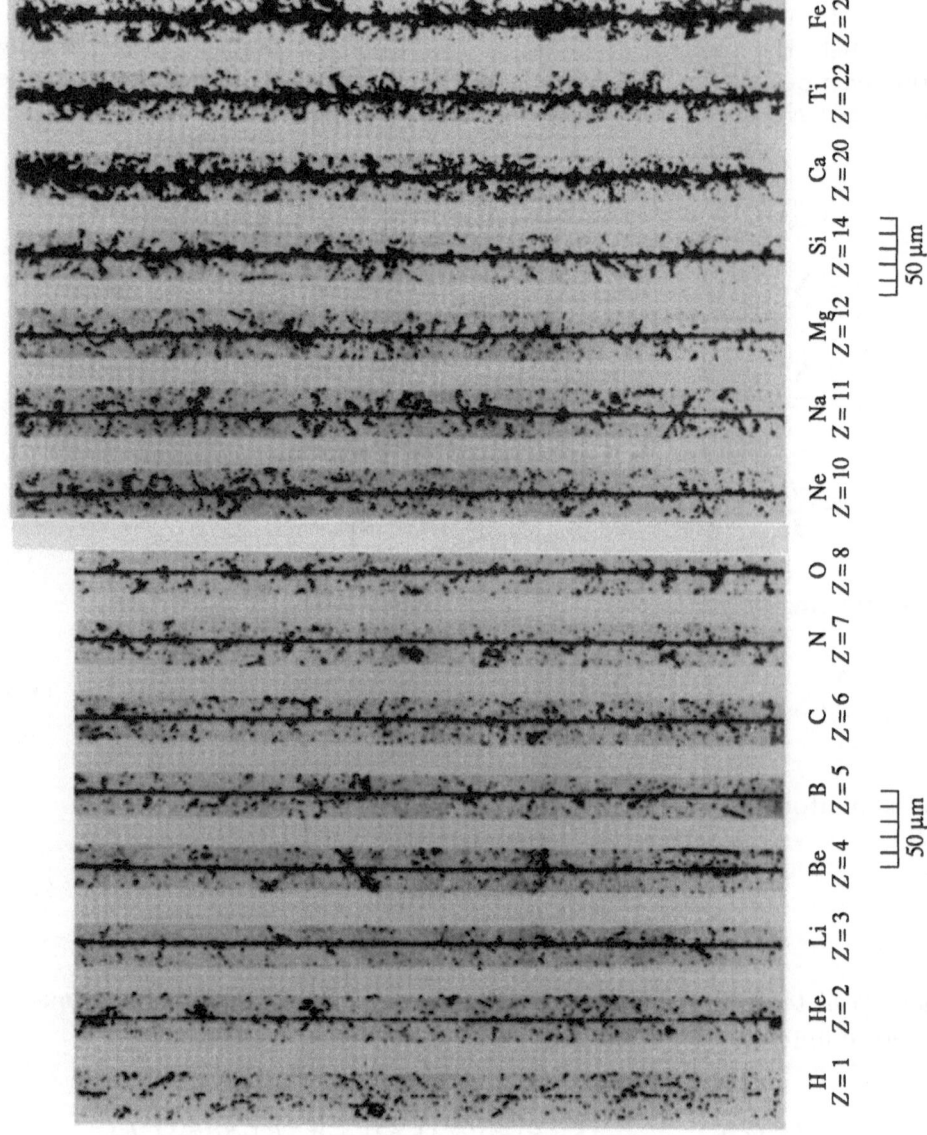

Figure 1.18. Cosmic-ray ion tracks in nuclear emulsion. (Taken from McDonald, 1965.)

The rest of this report concerns finding values for the atomic and nuclear cross sections, evaluating solutions to equation (1.8) for various boundary conditions, and making application to various radiation-protection issues.

The response of materials to ionizing radiation is related to the amount of local energy deposited and the manner in which that energy is deposited. The energy given up to nuclear emulsion (McDonald, 1965) is shown for several ions in figure 1.18. The figure registers developable crystals caused by the passage of the particle directly by ionization or indirectly by the ionization of secondary electrons (δ-rays). These δ-rays appear as hairs emanating from the particle track. Note that the scale of the δ-ray track is on the order of biological cell dimensions (2–10 μm). Many of the modern large integrated circuits are even of the 0.5-μm scale. For this scale, track structure effects become important as interruptive events as a particle passes through active elements of such circuits.

From the radiation-protection perspective, the issues of shielding are somewhat clearly drawn. Given the complex external environment, the shield properties alter the internal environment within the spacecraft structure as shown in figure 1.19. The internal environment interacts with onboard personnel or equipment. If sufficient knowledge is known about specific devices and biological responses, then the shield properties can be altered to minimize adverse effects. Since the shield is intimately connected to the overall engineering systems and often impacts launch cost, the minimization of radiation risk is not independent of other risk factors and mission costs. Even mission objectives are at times impacted by radiation-protection requirements (e.g., Viking solar cell design to ensure sufficient solar

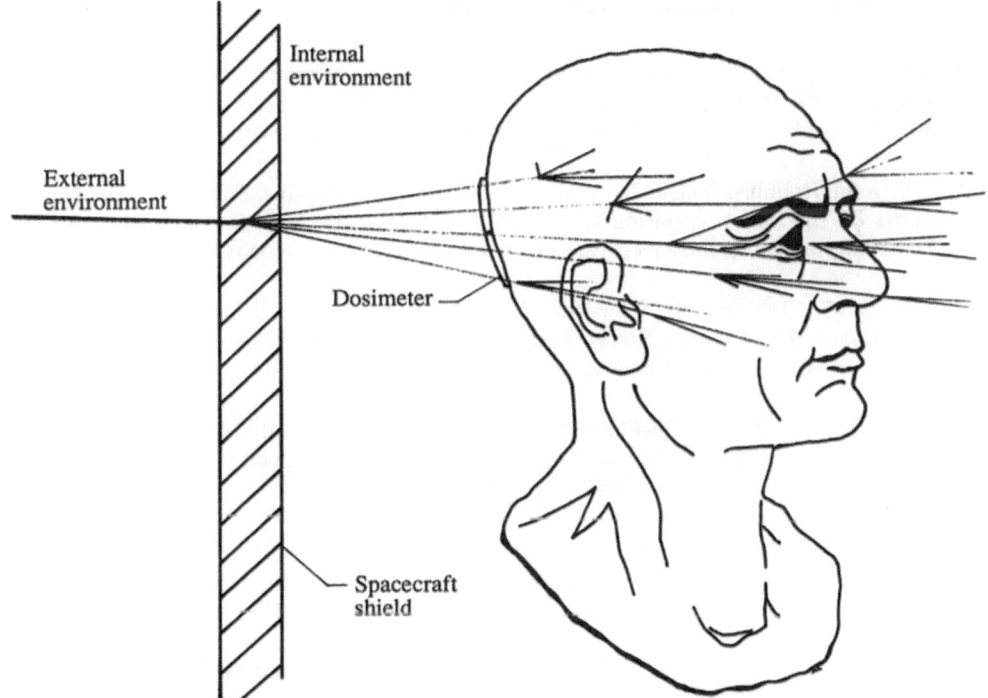

Figure 1.19. Schematic of space-radiation-protection problem.

power in the event of decreased performance caused by a large solar flare during the mission affected the weight allowed the experiments package). Clearly, the uncertainty in shield specification is an important factor when such critical issues are being addressed. There is uncertainty in subsystem response which can be easily (more or less) obtained for electronic or structural devices. The uncertainty in response of biological systems is complicated by the long delay times (up to 30 years) before system response occurs and the unusually small signal-to-noise ratio in biological response. Clearly, a difficult task remains before risk assignments can be made for long-duration deep space missions.

1.4. References

Alsmiller, R. G., Jr.; Irving, D. C.; Kinney, W. E.; and Moran, H. S., 1965: The Validity of the Straightahead Approximation in Space Vehicle Shielding Studies. *Second Symposium on Protection Against Radiations in Space,* Arthur Reetz, Jr., ed., NASA SP-71, pp. 177–181.

Anon., 1962: *Proceedings of the Symposium on the Protection Against Radiation Hazards in Space,* Books 1 and 2. TID-7652, U.S. Atomic Energy Commission.

Armstrong, Harry; Haber, Heinz; and Strughold, Hubertus, 1949: Aero Medical Problems of Space Travel—Panel Meeting, School of Aviation Medicine. *J. Aviation Med.,* vol. 20, no. 6, pp. 383–417.

Beck, Sherwin M.; and Powell, Clemans A., 1976: *Proton and Deuteron Double Differential Cross Sections at Angles From 10° to 60° From* Be, C, Al, Fe, Cu, Ge, W, *and* Pb *Under 558*-MeV-*Proton Irradiation.* NASA TN D-8119.

Bertini, Hugo W., 1967: *Preliminary Data From Intranuclear-Cascade Calculations of 0.75-, 1-, and 2-GeV Protons on Oxygen, Aluminum, and Lead, and 1-GeV Neutrons on the Same Elements.* ORNL-TM-1996, U.S. Atomic Energy Commission.

Billingham, John; Robbins, Donald E.; Modisette, Jerry L.; and Higgins, Peter W., 1965: Status Report on the Space Radiation Effects on the Apollo Mission. *Second Symposium on Protection Against Radiations in Space,* Arthur Reetz, Jr., ed., NASA SP-71, pp. 139–156.

Cartwright, D. C.; Trajmar, S.; Chutjian, A.; and Williams, W., 1977: Electron Impact Excitation of the Electronic States of N_2. II. Integral Cross Sections at Incident Energies From 10 to 50 eV. *Phys. Review A,* vol. 16, no. 3, pp. 1041–1051.

Curtis, S. B.; Dye, D. L.; and Sheldon, W. R., 1965: Fractional Cell Lethality Approach to Space Radiation Hazards. *Second Symposium on Protection Against Radiations in Space,* Arthur Reetz, Jr., ed., NASA SP-71, pp. 219–223.

Curtis, S. B.; and Wilkinson, M. C., 1972: The Heavy Particle Hazard—What Physical Data Are Needed? *Proceedings of the National Symposium on Natural and Manmade Radiation in Space,* E. A. Warman, ed., NASA TM X-2440, pp. 1007–1015.

DeYoung, R. J.; and Wilson, J. W., 1979: Population Inversion Mechanisms Producing Nuclear Lasing in He-3-Ar, Xe, Kr, Cl, and UF6. *Proceedings of the 1st International Symposium on Nuclear Induced Plasmas and Nuclear Pumped Lasers,* Les Editions de Physique Z. I. de Courtaboeuf (Orsay, Essonne, France), pp. 25–32.

Foelsche, T., 1959: Estimate of the Specific Ionization Caused by Heavy Cosmic Ray Primaries in Tissue or Water. *J. Astronaut. Sci.,* vol. VI, no. 4, pp. 57–62.

Foelsche, T., 1961: Radiation Exposure in Supersonic Transports. *Symposium on Supersonic Air Transport,* Conf. 14/WP-SYMP/49, International Air Transportation Assoc.

Foelsche, Trutz, 1962a: Protection Against Solar Flare Protons. *Advances in the Astronautical Sciences,* Volume 8, Plenum Press, Inc., pp. 357–374.

Foelsche, Trutz, 1962b: *Radiation Exposure in Supersonic Transports.* NASA TN D-1383.

Foelsche, Trutz; and Wilson, John W., 1969: Results of NASA SST-Radiation Studies Including Experimental Results on Solar Flare Events. Minutes of the Standing Committee on Radiobiology Aspects of the SST, Federal Aviation Adm.

Foelsche, T.; Mendell, Rosalind; Adams, Richard R.; and Wilson, John W., 1969: Measured and Calculated Radiation Levels Produced by Galactic and Solar Cosmic Rays in SST Altitudes and Precaution Measures To Minimize Implications at Commercial SST-Operations. NASA paper presented at French-Anglo United States Supersonic Transport Meeting (Paris, France).

Foelsche, Trutz; Mendell, Rosalind B.; Wilson, John W.; and Adams, Richard R., 1974: *Measured and Calculated Neutron Spectra and Dose Equivalent Rates at High Altitudes; Relevance to SST Operations and Space Research.* NASA TN D-7715.

Freier, Phyllis; Lofgren, E. J.; Ney, E. P.; and Oppenheimer, F., 1948: The Heavy Component of Primary Cosmic Rays. *Phys. Review,* vol. 74, second ser., no. 12, pp. 1818–1827.

Harries, W. L.; and Wilson, J. W., 1979: Simplified Model of a Volumetric Direct Nuclear Pumped He-3-Ar Laser. *Proceedings of the 1st International Symposium on Nuclear Induced Plasmas and Nuclear Pumped Lasers,* Les Editions de Physique Z. I. de Courtaboeuf (Orsay, Essonne, France), pp. 33–42.

Harries, W. L.; and Wilson, J. W., 1981: Solar-Pumped Electronic-to-Vibrational Energy Transfer Lasers. *Space Sol. Power Review,* vol. 2, no. 4, pp. 367–381.

ICRP Task Group, 1966: Radiobiological Aspects of the Supersonic Transport. *Health Phys.,* vol. 12, no. 2, pp. 209–226.

Jacobs, George J., ed. (appendix A by J. R. Winckler), 1960: *Proceedings of Conference on Radiation Problems in Manned Space Flight.* NASA TN D-588.

Kamaratos, E.; Chang, C. K.; Wilson, J. W.; and Xu, Y. J., 1982: Valence Bond Effects on Mean Excitation Energies for Stopping Power in Metals. *Phys. Lett.,* vol. 92A, no. 7, pp. 363–365.

Khandelwal, G. S.; Costner, C. M.; and Wilson, J. W., 1974: Geometric Correction for Spherical Ion Chambers. *Trans. American Nuclear Soc.,* vol. 19, p. 481.

Khandelwal, G. S.; and Wilson, John W., 1974: *Proton Tissue Dose for the Blood Forming Organ in Human Geometry: Isotropic Radiation.* NASA TM X-3089.

Kinney, W. E.; Coveyou, R. R.; and Zerby, C. D., 1962: A Series of Monte Carlo Codes To Transport Nucleons Through Matter. *Proceedings of the Symposium on the Protection Against Radiation Hazards in Space,* Book 2, TID-7652, U.S. Atomic Energy Commission, pp. 608–618.

Köllmann, K., 1975: Dissociative Ionization of H_2, N_2 and CO by Electron Impact—Measurement of Kinetic Energy, Angular Distributions, and Appearance Potentials. *Int. J. Mass Spectrom. & Ion Phys.,* vol. 17, pp. 261–285.

Korff, Serge A.; Mendell, Rosalind B.; Merker, Milton; Light, Edward S.; Verschell, Howard J.; and Sandie, William S., 1979: *Atmospheric Neutrons.* NASA CR-3126.

Krebs, A. T., 1950: Possibility of Biological Effects of Cosmic Rays in High Altitudes, Stratosphere and Space. *J. Aviation Med.,* vol. 21, no. 6, pp. 481–494.

Lambiotte, Jules J., Jr.; Wilson, John W.; and Filippas, Tassos A., 1971: Proper 3C: *A Nucleon-Pion Transport Code.* NASA TM X-2158.

Lamkin, Stanley Lee, 1974: A Theory for High-Energy Nucleon Transport in One Dimension. M.S. Thesis, Old Dominion Univ.

Leimdorfer, Martin; and Crawford, George W., eds., 1968: *Penetration and Interaction of Protons With Matter—Part I. Theoretical Studies Using Monte Carlo Techniques.* NASA CR-108228.

Manson, Steven T.; Toburen, L. H.; Madison, D. H.; and Stolterfoht, N., 1975: Energy and Angular Distribution of Electrons Ejected From Helium by Fast Protons and Electrons: Theory and Experiment. *Phys. Review A,* vol. 12, third ser., no. 1, pp. 60–79.

McClure, G. W.; and Pomerantz, M. A., 1950: Ionization Chamber Bursts at Very High Altitudes. *Phys. Review,* vol. 79, second ser., no. 5, pp. 911–912.

McDonald, F. B., 1965: Review of Galactic and Solar Cosmic Rays. *Second Symposium on Protection Against Radiations in Space,* Arthur Reetz, Jr., ed., NASA SP-71, pp. 19–29.

Orr, Harry D., III, 1972: Low Energy Protons From C, AL, NI, CU, and AU Under 600 MeV Proton Bombardment. Ph.D. Thesis, Univ. of South Carolina.

Reetz, Arthur, Jr., ed., 1965: *Second Symposium on Protection Against Radiations in Space.* NASA SP-71.

Rustgi, M. L.; Pandey, L. N.; Wilson, J. W.; Long, S. A. T.; and Zhu, G., 1988: Distribution of Energy in Polymers Due to Incident Electrons and Protons. *Radiat. Eff.,* vol. 105, pp. 303–311.

Schaefer, Hermann J., 1950: Evaluation of Present-Day Knowledge of Cosmic Radiation at Extreme Altitude in Terms of the Hazard to Health. *J. Aviation Med.,* vol. 21, no. 5, pp. 375–394, 418.

Schaefer, Hermann J., 1952: Exposure Hazards From Cosmic Radiation Beyond the Stratosphere and in Free Space. *J. Aviation Med.,* vol. 23, no. 4, pp. 334–344.

Schaefer, Hermann J., 1957: Cosmic Ray Dosage During the Giant Solar Flare of February 26, 1956. *J. Aviation Med.,* vol. 28, no. 4, pp. 387–396.

Schaefer, Hermann J., 1958: New Knowledge of the Extra-Atmospheric Radiation Field. *J. Aviation Med.,* vol. 29, no. 7, pp. 492–500.

Schaefer, Hermann J., 1959: Radiation Dosage in Flight Through the Van Allen Belt. *Aerosp. Med.,* vol. 30, no. 9, pp. 631–639.

Schimmerling, Walter; Curtis, Stanley B.; and Vosburgh, Kirby G., 1977: Velocity Spectrometry of 3.5-GeV Nitrogen Ions. *Radiat. Res.,* vol. 72, no. 1, pp. 1–17.

Schimmerling, Walter; and Curtis, Stanley B., eds., 1978: *Workshop on the Radiation Environment of the Satellite Power System.* LBL-8581 (Contract W-7405-ENG-48), Lawrence Berkeley Lab., Univ. of California.

Schimmerling, Walter; Kast, John W.; and Ortendahl, Douglas, 1979: Measurement of the Inclusive Neutron Production by Relativistic Neon Ions on Uranium. *Phys. Review Lett.,* vol. 43, no. 27, pp. 1985–1987.

Schimmerling, Walter; Alpen, Edward L.; Powers-Risius, Patricia; Wong, Mervyn; DeGuzman, Randy J.; and Rapkin, Marwin, 1987: The Relative Biological Effectiveness of 670 MeV/A Neon as a Function of Depth in Water for a Tissue Model. *Radiat. Res.,* vol. 112, pp. 436–448.

Schimmerling, Walter; Miller, Jack; Wong, Mervyn; Rapkin, Marwin; Howard, Jerry; Spieler, Helmut G.; and Jarret, Blair V., 1989: The Fragmentation of 670A MeV Neon-20 as a Function of Depth in Water. *Radiat. Res.,* vol. 120, pp. 36–71.

Schulz, George J., 1976: A Review of Vibrational Excitation of Molecules by Electron Impact at Low Energies. *Principles of Laser Plasmas,* George Bekefi, ed., John Wiley & Sons, Inc., pp. 33–88.

Skoski, L.; Merker, M.; and Shen, B. S. P., 1973: Absolute Cross Section for Producing ^{11}C From Carbon by 270-MeV/Nucleon ^{14}N Ions. *Phys. Review Lett.,* vol. 30, no. 2, pp. 51–54.

Tobias, Cornelius; and Segrè, Emilio, 1946: High Energy Carbon Nuclei. *Phys. Review,* vol. 70, second ser., nos. 1 and 2, p. 89.

Tobias, Cornelius A., 1952: Radiation Hazards in High Altitude Aviation. *J. Aviation Med.,* vol. 23, no. 4, pp. 345–372.

Todd, Paul, 1965: Biological Effects of Heavy Ions. *Second Symposium on Protection Against Radiations in Space,* Arthur Reetz, Jr., ed., NASA SP-71, pp. 105–114.

Todd, Paul, 1983: Unique Biological Aspects of Radiation Hazards—An Overview. *Adv. Space Res.,* vol. 3, no. 8, pp. 187–194.

Vosteen, Louis F., 1962: *Environmental Problems of Space Flight Structures—I. Ionizing Radiation in Space and Its Influence on Spacecraft Design.* NASA TN D-1474.

Warman, E. A., ed., 1972: *Proceedings of the National Symposium on Natural and Manmade Radiation in Space.* NASA TM X-2440.

Wight, G. R.; Van der Wiel, M. J.; and Brion, C. E., 1976: Dipole Excitation, Ionization and Fragmentation of N_2 and CO in the 10–60 eV Region. *J. Phys. B: At. Mol. Phys.*, vol. 9, no. 4, pp. 675–689.

Wilkinson, M. C.; and Curtis, S. B., 1972: Galactic Cosmic Ray Heavy Primary Secondary Doses. *Proceedings of the National Symposium on Natural and Manmade Radiation in Space,* E. A. Warman, ed., NASA TM X-2440, pp. 104–107.

Wilson, John W., 1969: Description of Transport Calculations. Minutes of the Standing Committee on Radiobiology Aspects of the SST, Federal Aviation Adm.

Wilson, John W.; Lambiotte, Jules J.; and Foelsche, T., 1969: Structure in the Fast Spectra of Atmospheric Neutrons. *J. Geophys. Res.,* vol. 74, no. 26, pp. 6494–6496.

Wilson, John W., 1972a: *Comparison of Exact and Approximate Evaluations of the Single-Scattering Integral in Nucleon-Deuteron Elastic Scattering.* NASA TN D-6884.

Wilson, John W., 1972b: *Isosinglet Approximation for Nonelastic Reactions.* NASA TN D-6942.

Wilson, John W., 1973: Intermediate Energy Nucleon-Deuteron Elastic Scattering. *Nuclear Phys.,* vol. B66, pp. 221–244.

Wilson, John W., 1974a: Multiple Scattering of Heavy Ions, Glauber Theory, and Optical Model. *Phys. Lett.,* vol. B52, no. 2, pp. 149–152.

Wilson, John W., 1974b: Proton-Deuteron Double Scattering. *Phys. Review C,* vol. 10, no. 1, pp. 369–376.

Wilson, John W., 1975a: Composite Particle Reaction Theory. Ph.D. Diss., College of William and Mary in Virginia.

Wilson, John W., 1975b: Weight Optimization Methods in Space Radiation Shield Design. *J. Spacecr. & Rockets,* vol. 12, no. 12, pp. 770–773.

Wilson, John W.; and Costner, Christopher M., 1975: *Nucleon and Heavy-Ion Total and Absorption Cross Section for Selected Nuclei.* NASA TN D-8107.

Wilson, John W.; and Lamkin, Stanley L., 1975: Perturbation Theory for Charged-Particle Transport in One Dimension. *Nuclear Sci. & Eng.,* vol. 57, no. 4, pp. 292–299.

Wilson, John W.; and Denn, Fred M., 1976: *Preliminary Analysis of the Implications of Natural Radiations on Geostationary Operations.* NASA TN D-8290.

Wilson, John W.; and Khandelwal, G. S., 1976a: *Computer Subroutines for the Estimation of Nuclear Reaction Effects in Proton-Tissue-Dose Calculations.* NASA TM X-3388.

Wilson, John W.; and Khandelwal, Govind S., 1976b: Proton-Tissue Dose Buildup Factors. *Health Phys.,* vol. 31, no. 2, pp. 115–118.

Wilson, John W., 1977a: *Analysis of the Theory of High-Energy Ion Transport.* NASA TN D-8381.

Wilson, J. W., 1977b: Depth-Dose Relations for Heavy Ion Beams. *Virginia J. Sci.*, vol. 28, no. 3, pp. 136–138.

Wilson, John W.; and Denn, Fred M., 1977a: *Implications of Outer-Zone Radiations on Operations in the Geostationary Region Utilizing the AE4 Environmental Model.* NASA TN D-8416.

Wilson, John W.; and Denn, Fred M., 1977b: *Improved Analysis of Electron Penetration and Numerical Procedures for Space Radiation Shielding.* NASA TN D-8526.

Wilson, John W.; and Denn, Fred M., 1977c: Methods of Shield Analysis for Protection Against Electrons in Space. *Nuclear Technol.*, vol. 35, no. 1, pp. 178–183.

Wilson, John W., 1978: Environmental Geophysics and SPS Shielding. *Workshop on the Radiation Environment of the Satellite Power System,* Walter Schimmerling and Stanley B. Curtis, eds., LBL-8581, UC-41 (Contract W-7405-ENG-48), Univ. of California, pp. 33–116.

Wilson, J. W.; and DeYoung, R. J., 1978a: Power Density in Direct Nuclear-Pumped ^{3}He Lasers. *J. Appl. Phys.*, vol. 49, no. 3, pt. I, pp. 980–988.

Wilson, John W.; and DeYoung, R. J., 1978b: Power Deposition in Volumetric ^{235}UF$_6$-He Fission-Pumped Nuclear Lasers. *J. Appl. Phys.*, vol. 49, no. 3, pt. I, pp. 989–993.

Wilson, J. W.; DeYoung, R. J.; and Harries, W. L., 1979: Nuclear-Pumped ^{3}He-Ar Laser Modeling. *J. Appl. Phys.*, vol. 50, no. 3, pt. I, pp. 1226–1235.

Wilson, J. W., 1980: Nuclear-Induced Xe-Br* Photolytic Laser Model. *Appl. Phys. Lett.*, vol. 37, no. 8, pp. 695–697.

Wilson, J. W.; and Lee, J. H., 1980: Modeling of a Solar-Pumped Iodine Laser. *Virginia J. Sci.*, vol. 31, pp. 34–38.

Wilson, J. W.; and Shapiro, A., 1980: Nuclear-Induced Excimer Fluorescence. *J. Appl. Phys.*, vol. 51, no. 5, pp. 2387–2393.

Wilson, J. W., 1981: Solar Radiation Monitoring for High Altitude Aircraft. *Health Phys.*, vol. 41, no. 4, pp. 607–617.

Wilson, J. W.; and Kamaratos, E., 1981: Mean Excitation Energy for Molecules of Hydrogen and Carbon. *Phys. Lett.*, vol. 85A, no. 1, pp. 27–29.

Wilson, J. W.; Chang, C. K.; Xu, Y. J.; and Kamaratos, E., 1982: Ionic Bond Effects on the Mean Excitation Energy for Stopping Power. *J. Appl. Phys.*, vol. 53, no. 2, pp. 828–830.

Wilson, J. W.; and Xu, Y. J., 1982: Metallic Bond Effects on Mean Excitation Energies for Stopping Powers. *Phys. Lett.*, vol. 90A, no. 5, pp. 253–255.

Wilson, John W.; Stith, John J.; and Stock, L., 1983: *A Simple Model of Space Radiation Damage in GaAs Solar Cells.* NASA TP-2242.

Wilson, John W.; Raju, S.; and Shiu, Y. J., 1983: *Solar-Simulator-Pumped Atomic Iodine Laser Kinetics.* NASA TP-2182.

Wilson, John W.; and Stock, L. V., 1984: Equivalent Electron Fluence for Space Qualification of Shallow Junction Heteroface GaAs Solar Cells. *IEEE Trans. Electron Devices,* vol. ED-31, no. 5, pp. 622–625.

Wilson, John W.; Lee, Yeunggil; Weaver, Willard R.; Humes, Donald H.; and Lee, Ja H., 1984: *Threshold Kinetics of a Solar-Simulator-Pumped Iodine Laser.* NASA TP-2241.

Xu, Y. J.; Khandelwal, G. S.; and Wilson, J. W., 1984a: Intermediate Energy Proton Stopping Power for Hydrogen Molecules and Monoatomic Helium Gas. *Phys. Lett.,* vol. 100A, no. 3, pp. 137–140.

Xu, Y. J.; Khandelwal, G. S.; and Wilson, J. W., 1984b: Low-Energy Proton Stopping Power of N_2, O_2, and Water Vapor, and Deviations From Bragg's Rule. *Phys. Review A,* vol. 29, third ser., no. 6, pp. 3419–3422.

Xu, Y. J.; Khandelwal, G. S.; and Wilson, J. W., 1985: Proton Stopping Cross Sections of Liquid Water. *Phys. Review A,* vol. 32, third ser., no. 1, pp. 629–636.

Yuan, Luke C. L., 1951: Distribution of Slow Neutrons in Free Atmosphere up to 100,000 Feet. *Phys. Review,* vol. 81, second ser., no. 2, pp. 175–184.

2.COULOMB INTERACTIONS IN ATOMS AND MOLECULES

2.1. Introduction

In deriving the Boltzmann equation in chapter 1, we included atomic/molecular and nuclear collision processes. The total cross section $\sigma_j(E)$ with the medium for each particle type of energy E may be expanded as

$$\sigma_j(E) = \sigma_j^{at}(E) + \sigma_j^{el}(E) + \sigma_j^{r}(E) \tag{2.1}$$

where the first term refers to collision with atomic electrons, the second term is for elastic nuclear scattering, and the third term describes nuclear reactions. The microscopic cross sections are ordered as follows:

$$\sigma_j^{at}(E) \sim 10^{-16} \text{ cm}^2 \tag{2.2}$$

$$\sigma_j^{el}(E) \sim 10^{-19} \text{ cm}^2 \tag{2.3}$$

$$\sigma_j^{r}(E) \sim 10^{-24} \text{ cm}^2 \tag{2.4}$$

to allow flexibility in expanding solutions to the Boltzmann equation as a sequence of perturbative approximations, for example, the continuous slowing down approximation is one such approach. It is clear that many atomic collisions ($\sim 10^6$) occur in a cm of ordinary matter, whereas $\sim 10^3$ nuclear elastic collisions occur per cm. In distinction, nuclear reactions are separated by many cm. We shall further elaborate this point of view and indicate important atomic and molecular quantities required for transport theory development. In particular, we will examine a more general formulation than that presented in equation (1.5).

The Boltzmann equation, ignoring terms associated with equations (2.3) and (2.4), can be written with the aid of equation (1.4) as

$$\vec{\Omega} \cdot \nabla \phi_j(\vec{x}, \vec{\Omega}, E) = \sum_n \sigma_n^{at}(E + \epsilon_n) \, \phi_j\left(\vec{x}, \vec{\Omega}, E + \epsilon_n\right) - \sigma_j^{at}(E) \, \phi_j(\vec{x}, \vec{\Omega}, E) \tag{2.5}$$

where ϵ_n is the atomic/molecular excitation energy. Equation (2.5) is equivalent to one-dimensional transport along the ray directed by $\vec{\Omega}$. For simplicity of notation we use a one-dimensional equation as

$$\frac{\partial}{\partial z} \phi_j(z, E) = \sum_n \sigma_n^{at}(E + \epsilon_n) \, \phi_j(z, E + \epsilon_n) - \sigma_j^{at}(E) \, \phi_j(z, E) \tag{2.6}$$

where we drop the superscript at and subscript j in the rest of this section. The boundary condition is taken as

$$\phi(0, E) = \delta(E - E_0) \tag{2.7}$$

The solution can be written with perturbation theory as

$$\phi^{(0)}(z, E) = \exp\left(-\sigma z\right)\delta(E - E_0) \tag{2.8}$$

$$\phi^{(1)}(z, E) = \sigma z \exp(-\sigma z) \sum_n g_n \, \delta(E + \epsilon_n - E_0) \qquad (2.9)$$

$$\phi^{(2)}(z, E) = \frac{(\sigma z)^2}{2!} \exp(-\sigma z) \sum_n g_m g_n \, \delta(E + \epsilon_n + \epsilon_m - E_0) \qquad (2.10)$$

and similarly for higher order terms, where $\delta()$ is the Dirac delta function, and $E_0 \gg \epsilon_n$ has been assumed so that σ and $g_n = \sigma_n/\sigma$ are evaluated at E_0. The average energy after penetration to a distance z is given by

$$\langle E \rangle = E_0 - \bar{\epsilon} \sigma z \qquad (2.11)$$

where the average excitation energy is

$$\bar{\epsilon} = \sum_n g_n \epsilon_n \qquad (2.12)$$

and the sum over n contains both discrete and continuous terms. The standard deviation about the average energy is similarly found to be

$$s^2 \equiv \left\langle (E - \langle E \rangle)^2 \right\rangle = \overline{\epsilon^2} \sigma z \qquad (2.13)$$

where

$$\overline{\epsilon^2} = \sum_n g_n \epsilon_n^2 \qquad (2.14)$$

Similar results can be derived for the higher moments of the energy distribution, which depend on atomic quantities through the g_n terms. Considering the nonlinear dependence of the transported spectrum on the atomic cross sections σ_n, it is somewhat surprising that the transported spectral parameters depend linearly on g_n. Equations (2.11) and (2.13) apply when $\bar{\epsilon} \sigma z \ll E_0$ so that the energy variations in the cross sections can be ignored. The expressions are easily generalized to deep penetrations as

$$\overline{E}(z) = E_0 - \int_0^z S[\overline{E}(y)] \, dy \qquad (2.15)$$

and

$$\overline{E^2}(z) = \overline{E^2}(0) + \int_0^z \frac{S_1[\overline{E}(y)]}{S[\overline{E}(y)]} \, dy \qquad (2.16)$$

where the stopping power is given by

$$S(E) = \sum_n \sigma_n(E) \, \epsilon_n \qquad (2.17)$$

and the straggling is related to

$$S_1(E) = \sum_n \sigma_n(E) \, \epsilon_n^2 \qquad (2.18)$$

The degrading particle energy $\overline{E}(y)$ is found through the usual range-energy relations

$$R(E) = \int_0^E \frac{dE'}{S(E')} \tag{2.19}$$

It is clear that $R(E)$ is the average stopping path length for the ions. The corresponding spectrum is taken as

$$\phi(z, E) = \frac{1}{\sqrt{2\pi}s} \exp\left[\frac{-(E - \overline{E})^2}{2s^2}\right] \tag{2.20}$$

where the standard deviation s is given by equation (2.13). The usual continuous slowing down approximation is found in the limit as $s \to 0$. The evaluation requires knowledge of the appropriate atomic cross sections σ_n.

The atomic collisions occur quite frequently in ordinary matter (10^6–10^7 per cm). Less frequent are the elastic nuclear collisions, the largest contribution of which is coulomb scattering. The elastic cross section for scattering from the nucleus is represented as $\sigma_s(\theta)$ with expansion in terms of Legendre polynomials $P_n(x)$ as

$$\sigma_s(\theta) = \sum a_n \, P_n(\cos\theta) \tag{2.21}$$

where the coefficients are given as

$$a_n = \frac{2n+1}{2} \int_{-1}^1 \sigma_s(\theta) \, P_n(\cos\theta) \, d\cos\theta \tag{2.22}$$

and the corresponding equation for transport through a slab

$$\cos\theta \, \frac{\partial}{\partial z} \, \phi(z,\theta) + \sigma\phi(z,\theta) = \int \sigma_s(\gamma) \, \phi(z,\theta') \, d\cos\theta' d\varphi'$$

$$= 2\pi \sum a_n \, P_n(\cos\theta) \int P_n(\cos\theta') \, \phi(z,\theta') \, d\cos\theta' \tag{2.23}$$

where

$$\cos\gamma = \cos\theta \cos\theta' + \sin\theta \sin\theta' \cos(\varphi - \varphi') \tag{2.24}$$

and we have used the addition theorem

$$P_n(\cos\gamma) = P_n(\cos\theta) \, P_n(\cos\theta') + 2\sum_l P_n^l(\cos\theta) \, P_n^l(\cos\theta') \cos l(\varphi - \varphi') \tag{2.25}$$

The differential operator can be inverted in equation (2.23) to obtain

$$\phi(z,\theta) = \exp\left(\frac{-\sigma z}{\cos\theta}\right) \phi(0,\theta)$$

$$+ \sum_n 2\pi \int_0^z \exp\left[\frac{-\sigma(z-\tau)}{\cos\theta}\right] a_n \, P_n(\cos\theta) \int P_n(\cos\theta') \, \phi(\tau,\theta') \, d\cos\theta' \, d\tau \tag{2.26}$$

The scattering is strongly peaked in the forward direction and the integral kernel varies slowly for forward propagation along off-axis rays (Breitenberger, 1959). The attenuation kernel $\exp[-\sigma(z - \tau)/\cos\theta]$ is shown in figure 2.1. We simplify the propagation equation as

$$\phi(z, \theta) = \exp(-\sigma z)\ \phi(0,\theta)$$

$$+ 2\pi \sum_n \int_0^z \exp[-\sigma(z - \tau)] a_n\ P_n(\cos\theta)\ P_n(\cos\theta')\ \phi(\tau,\theta')\ d\cos\theta' d\tau \qquad (2.27)$$

The approximate multiple-scattering equation may be solved by expanding the flux as

$$\phi(z, \theta) = \sum A_n(z)\ P_n(\cos\theta) \qquad (2.28)$$

where

$$A_n(z) = \frac{2n+1}{2} \int \phi(z,\theta)\ P_n(\cos\theta)\ d\cos\theta \qquad (2.29)$$

The coefficients then satisfy

$$A_n(z) = \exp(-\sigma z) A_n(0) + \frac{4\pi}{2n+1} a_n \int_0^z \exp[-\sigma(z - \tau)] A_n(\tau)\ d\tau \qquad (2.30)$$

Let the boundary condition be

$$\phi(0, \theta) = \frac{1}{2\pi}\ \delta(\cos\theta - 1) \qquad (2.31)$$

where $\delta(\ )$ is the Dirac delta function. Then $A_n(0) = (2n + 1)/4\pi$. The iterated solution of equation (2.30) may be written as

$$A_n(z) = \frac{2n+1}{4\pi} \exp\left[-\left(\sigma - \frac{4\pi}{2n+1} a_n\right) z\right] \qquad (2.32)$$

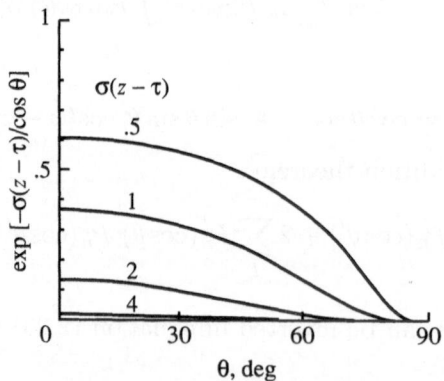

Figure 2.1. Transport kernel as a function of angle of propagation.

In the absence of absorptive processes, the forward isotropic term ($n = 0$) shows no spatial dependence. In distinction, the higher order ($n > 0$) terms display spatial attenuation at greater depths of penetration. The angular distribution is now characterized in terms of the mean cosine of the zenith angle as

$$\langle \cos \theta \rangle = \int \cos \theta \phi(z, \theta) \, d\Omega$$

$$= \frac{2}{3} A_1(z)$$

$$= \exp \left[-\left(\sigma - \frac{4}{3} \pi a_1 \right) z \right] \tag{2.33}$$

and is related to the average angular deflection as

$$\langle \theta^2 \rangle \approx 2[1 - \langle \cos \theta \rangle]$$

$$= 2 \left\{ 1 - \exp \left[-\left(\sigma - \frac{4}{3} \pi a_1 \right) z \right] \right\} \tag{2.34}$$

Initially the root-mean-square angle is zero as expected for the unidirectional boundary condition (eq. (2.31)) and increases to a value approaching 90° at very large depths. In cases of interest to us, we find that the asymptotic value is never reached since energy loss due to atomic collision or nuclear reaction processes limits the beam propagation before this occurs (e.g., see Janni, 1982a and 1982b).

2.2. Extremely Rarefied Gas Interactions

In passing through matter, an ion loses a large fraction of its energy to atomic/molecular excitation of the material. Although a satisfactory theory of high-energy interaction exists in the form of Bethe's theory (Bethe, 1950) using the Born approximation or more exact calculations using transitions from specific atomic shells (Merzbacher and Lewis, 1958; Khandelwal, 1968), an equally satisfactory theory for low-energy collisions is not available. In the rest of this chapter. we give a brief overview of the theory of stopping power and the formalism used in our transport calculations. Future directions of research to allow more accurate evaluation of these transport parameters are discussed.

In an extremely rarefied gas, we may assume that the passing ion interacts singly with the media molecules. This is an extreme simplification but is an idealization which still leaves many challenges to theoretical treatment. Even so, our aim is to treat the noble gases in fair detail, but even the interaction for the extremely rarefied noble gases cannot as yet be fully calculated with great confidence.

The gas atom can, for practical purposes, be taken as in the ground state before interacting with the passing particle. At the lowest energies, the gas molecule or atom interacts through adiabatic processes for which the Born-Oppenheimer approximation is appropriate. The electronic portion of the total Hamiltonian appears as part of the potential through which the massive nuclei move. The exchange of electrons between the moving particle and target molecule or atom can change the charge state of the projectile as it passes through the media. The exchange of electrons leads to potential curve crossing which is usually treated in the Landau-Zener model (Landau and Lifshitz, 1958; Zener, 1932) or by a molecular orbital approximation (Suzuki, Nakamura, and Ishiguro, 1984; Xu, Khandelwal, and Wilson, 1989). In the media, the projectile state is not well-defined and changes randomly in charge state and excitation level. The charge state is usually described by some equilibrium distribution with some mean charge that depends on both the kinetic energy of the projectile and the character of the media. These charge changing cross sections are quite large, and equilibrium values are achieved over relatively short distances (less than 1 mg/cm^2 of material).

2.2.1. Stopping at low energies. At the lowest energies, the projectile is hardly able to penetrate the atomic orbitals of the media, and the media atoms or molecules recoil in tack. The stopping cross section has been calculated by Firsov with the Thomas-Fermi model (Martynenko, 1970) to be

$$S_n(E) = \left(\frac{2\alpha NC}{E} \right) \ln \left(1 + \sqrt{\lambda} \frac{E}{C} \right) \tag{2.35}$$

where

$$\alpha = \frac{\pi^2}{8} \frac{Z_P Z_T C^2 0.8853 a_o}{\left(Z_P^{1/2} + Z_T^{1/2} \right)^{2/3}} \left(\frac{M_P}{M_T} \right)^{1/2} \tag{2.36}$$

$$C = \frac{8e^2 Z_P Z_T \left(Z_P^{1/2} + Z_T^{1/2} \right)^{2/3} (M_P/M_T)^{1/2}}{\pi 0.8853 a_o} \tag{2.37}$$

$$\lambda = \frac{4 M_P M_T}{(M_P + M_T)^2} \tag{2.38}$$

where Z_P and Z_T are projectile and target atomic numbers and M_P and M_T are their atomic weights, a_o is the Bohr radius, e is the electron charge, and N is the number of target atoms per unit volume.

At somewhat higher kinetic energy, the outer electron cloud is penetrated and the nuclear electric repulsion becomes more effective giving rise to Rutherford scattering described by

$$S_n(E) \approx \frac{C'}{E} \ln E \tag{2.39}$$

where C' is a constant and the raising of electrons to higher orbitals is possible.

2.2.2. Bethe stopping theory. At sufficiently high energies, the Born approximation is applicable, which may be used for both molecules and atoms. In practice, the molecular electronic wave functions are not known, and such calculations are limited to interaction with atoms (noble gases in practice). The differential cross section for the nonrelativistic case in lowest order (Born) approximation is given by the formula

$$\frac{d\sigma_n}{dQ} = \frac{2\pi Z_P^2 e^4}{mv^2} Z_T \frac{|F_n(\vec{q})|^2}{Q^2} \tag{2.40}$$

and the mean energy loss (see eq. (2.17)

$$-\frac{dE}{dx} = \frac{2\pi Z_P^2 e^4 N Z_T}{mv^2} \sum_n (E_n - E_o) \int_{\frac{(E_n-E_o)^2}{2mv^2}}^{2mv^2} |F_n(\vec{q})|^2 \frac{dQ}{Q^2} \qquad (2.41)$$

In equations (2.40) and (2.41), $\vec{q}$ is the momentum transferred to the electron, and $Q = q^2/2m$, the energy absorbed by a free electron of mass m at rest (Xu, Khandelwal, and Wilson, 1986), Z_T is the target atomic number, and N is the number of target atoms per unit volume. The quantity $(E_n - E_o)$ is the excitation energy, and the form factor $F_n(\vec{q})$ is defined as

$$F_n(\vec{q}) = \sum_{j=1}^{Z_T} \left\langle \Psi_n \left| \exp\left(\frac{i\vec{q} \cdot \vec{r}_j}{\hbar}\right) \right| \Psi_o \right\rangle \qquad (2.42)$$

in which $\vec{r}_j$ denotes the position of the jth atomic electron relative to the nucleus, and Ψ_n and Ψ_o are the final- and the initial-state wave functions of the target.

The Bethe method (Fano, 1963; Bethe, 1933; Livingston and Bethe, 1937; Bethe and Ashkin, 1953) depends on rewriting equation (2.41) by dividing the integration over Q into two parts: low Q and high Q in which the intermediate value is Q_o. Thus consider the following term of equation (2.41):

$$\sum_n (E_n - E_o) \int_{Q_{\min}}^{Q_{\max}} |F_n(\vec{q})|^2 \frac{dQ}{Q^2}$$

$$= \sum_n (E_n - E_o) \int_{\frac{(E_n-E_o)^2}{2mv^2}}^{Q_o} |F_n(\vec{q})|^2 \frac{dQ}{Q^2}$$

$$+ \sum_n (E_n - E_o) \int_{Q_o}^{2mv^2} |F_n(\vec{q})|^2 \frac{dQ}{Q^2} \qquad (2.43)$$

Consider the first term in equation (2.43). Expanding $|F_n(q)|^2$ for low q in equation (2.42) and retaining only the first nonvanishing term give:

$$|F_n(\vec{q})|^2 \approx \frac{q^2}{\hbar^2} \left| \left\langle \Psi_n \left| \sum_j x_j \right| \Psi_o \right\rangle \right|^2 \qquad (2.44)$$

Notice that the other higher order terms are neglected in this approximation. Thus, the first term in equation (2.43) becomes

$$\sum_n (E_n - E_o) \int_{\frac{(E_n-E_o)^2}{2mv^2}}^{Q_o} |F_n(\vec{q})|^2 \frac{dQ}{Q^2} = \ln\left(\frac{Q_o 2mv^2}{I_T^2}\right) \qquad (2.45)$$

where

$$\ln I_T = \sum_n f_n \ln(E_n - E_o) \qquad (2.46)$$

and

$$f_n = \frac{2m}{\hbar^2}(E_n - E_o)\left|\sum_j \langle n|x_j|0\rangle\right|^2 \qquad (2.47)$$

In equations (2.46) and (2.47), I_T is known as mean excitation energy of the medium and f_n is the optical oscillator strength.

The second term in equation (2.43) can be written as

$$\sum_n (E_n - E_o)\int_{Q_o}^{2mv^2} |F_n(\vec{q})|^2 \frac{dQ}{Q^2} \approx \ln Q|_{Q_o}^{2mv^2} \qquad (2.48)$$

where the Bethe sum rule

$$\sum_n (E_n - E_o)|F_n(\vec{q})|^2 = Q \qquad (2.49)$$

has been used. Equation (2.41) with equations (2.43), (2.45), and (2.48) becomes

$$-\frac{dE}{dx} = \frac{4\pi Z_P^2 e^4 N Z_T}{mv^2} \ln \frac{2mv^2}{I_T} \qquad (2.50)$$

which is the celebrated Bethe stopping power equation.

The derivation of equation (2.50) from equation (2.41) depends on the sum rule (eq. (2.49)), the upper limit $2mv^2$ in equation (2.41), and the intermediate value Q_o. The main thrust of these assumptions is to treat all the electrons as essentially free electrons. This assumption fails for innershell electrons which are tightly bound to the atom. To incorporate the correct treatment of these innershell electrons, one introduces a "shell correction" term C in equation (2.50). Basically, the treatment of the correction involves the exact evaluation of the form factor $|F_n|^2$ of equation (2.42). The equation for energy loss per unit path length then reads as

$$-\frac{dE}{dx} = \frac{4\pi Z_P^2 e^4 N Z_T}{mv^2}\left[\ln \frac{2mv^2}{I_T} - \frac{C}{Z_T}\right] \qquad (2.51)$$

The evaluation of mean excitation energy I_T from equations (2.46) and (2.47) has been studied intently for the last several decades. Extensive calculations for many atoms using the Hartree-Slater potential model have been performed recently by many authors (Dehmer, Inokuti, and Saxon, 1975; Inokuti, Baer, and Dehmer, 1978; Inokuti and Turner, 1978; Inokuti et al., 1981). These are later compared with the values obtained with the local plasma model.

Similarly, shell corrections have been studied by various authors (Bethe, 1933; Livingston and Bethe, 1937; Bethe, Brown, and Walske, 1950; Brown, 1950; Walske, 1952 and 1956; Khandelwal, 1968 and 1982; Janni, 1966; Merzbacher and Lewis, 1958; Bichsel, 1966[1]) for the last 60 years. Basically, one employs

---

[1] Research on the L-shell correction in stopping power done by Hans Bichsel at the Nuclear Physics Laboratory, University of Southern California, and supported by the National Cancer Institute and the U.S. Atomic Energy Commission.

the screened hydrogenic approximation (only one parameter for screened nuclear charge Z_i for both the initial- and final-state electron is used) and calculations are made for a particular shell. The total shell correction C, in principle, can be obtained by summing the contributions shell by shell.

As noted in section 2.1, the fluctuation in energy loss is also related to the atomic cross sections as (see eq. (2.18))

$$S_1(E) = \frac{2\pi Z_P^2 e^4 N Z_T}{mv^2} \sum_n (E_n - E_o)^2 \int_{Q_{\min}}^{Q_{\max}} |F_n(q)|^2 \frac{dQ}{Q^2} \qquad (2.52)$$

where the limits on Q are those discussed in connection with equation (2.41). By arguments similar to those leading to equation (2.50), one finds

$$S_1(E) \approx \frac{4\pi Z_P^2 e^4 N Z_T}{mv^2} \ln \frac{2mv^2}{\Delta} \qquad (2.53)$$

where Δ is given as

$$\ln \Delta = \sum_n f_n(E_n - E_o) \ln(E_n - E_o) \qquad (2.54)$$

These quantities are important in calculating the energy spectra of slowing ions within a medium. (See eq. (2.20).)

As is evident from equation (2.51), the determination of the energy loss per unit path length depends upon the accurate knowledge of the mean excitation energy I and the shell corrections C. In practice one invokes some sort of parameter fitting involving the experimental data on stopping power and the quantities I and C. Quite often (Bichsel, 1963; Janni, 1966) the theoretical values are used in conjunction with the experimental values for parameter fitting. It would thus be desirable to obtain stopping power without the need to have access to the parameters I and C.

We have initiated such an attempt which is described as follows. The main thrust of the approach is to calculate exactly the one-electron form factor within a screened hydrogenic model. As is known for an atom with more than one electron, the form factor given by equation (2.41) within the one-electron model can be approximated as

$$F_{n'n}(q) \approx \left\langle \psi_{n'}(\vec{r}) \middle| \exp(i\vec{q} \cdot \vec{r}) \middle| \psi_n(\vec{r}) \right\rangle \qquad (2.55)$$

where ψ_n and $\psi_{n'}$ are wave functions with a single electron (henceforth, we use natural units in which $\hbar$ and c are unity). Historically, equation (2.55) has been justified on the basis of Hartree-Fock approximation. The knowledge of the form factor of equation (2.55) thus depends on knowing the radial integrals for the process of excitation as well as ionization when a projectile passes through matter. We have recently calculated the radial integral for the optically allowed transitions in He atoms and helium-like ions under the screened hydrogenic model. The model describes the atom by single-particle hydrogenic wave functions and treats the initial state and the final state by two different effective charge parameters Z_i and Z_f, respectively. The generalized radial integral (corresponding to the expansion of eq. (2.55) in a power series in q) is presented in the following section and the dipole term is discussed.

2.2.3. Optical oscillator strengths within screened hydrogenic model.
The generalized radial integral $R_\beta(n'l' - kl)$ of concern is the following:

$$R_\beta(n'l' - kl) = \int_0^\infty [R(n', l'; r)] r^{\beta+1} R(k, l; r) r^2 \, dr \tag{2.56}$$

where $R(n', l'; r)$ and $R(k, l; r)$ are the bound and the free-state radial wave functions, respectively. These wave functions in terms of Z_i and Z_f are

$$R(n', l'; r) = \frac{(-1)^{n'+l'+1} 2^n Z_i^{n+1/2}}{[(n'+1)!]^{1/2}[(n'-l'-1)!]^{1/2} n'^{n'+1}}$$

$$\times \left[\sum_{j=0}^{n'-l'-1} \frac{1}{j!} \left[-\frac{n'}{2Z_i} \right]^j \prod_{i=0}^j (n'+l'-i)(n'-l'-1-i) \right.$$

$$\left. \times \exp\left(\frac{-Z_i r}{n'} \right) r^{n'-1-j} \right] \tag{2.57}$$

$$R(k, l; r) = \frac{(-1)^{l+1} \left(2\sqrt{Z_f}\right)}{\left[1 - \exp\left(-2\frac{\pi Z_f}{k}\right)\right]^{\frac{1}{2}}} \prod_{s=0}^l \left(s^2 + \frac{Z_f^2}{k^2}\right)^{1/2} \left(\frac{1}{2k}\right)^{l+1}$$

$$\times \frac{r^{-(l+1)}}{2\pi} \oint \exp\left(-2ikr\xi\right) \left(\xi + \frac{1}{2}\right)^{-i\frac{Z_f}{k}-l-1}$$

$$\times \left(\xi - \frac{1}{2}\right)^{i\frac{Z_f}{k}-l-1} d\xi \tag{2.58}$$

When equations (2.57) and (2.58) are substituted into equation (2.56), one obtains (Khandelwal et al., 1989)

$$R_\beta(n'l'-kl) = F_1 \frac{\sqrt{k}\exp\left\{-2(Z_f/k)\,\tan^{-1}[k(n'/Z_i)]\right\}}{\left(1 - \exp\left(-2\pi Z_f/k\right)\right)^{1/2} \left[k^2(n')^2 \Big/ Z_i^2 + 1\right]^{(\beta+n'+2)}} \left[\prod_{s=1}^l \left(\frac{k^2}{Z_f^2}s^2 + 1\right)^{1/2}\right] G_1 \tag{2.59}$$

where the quantities F_1 and G_1 are defined by

$$F_1 \equiv \frac{Z_f^{l+1/2} 2^{n'+l+1} (n')^{\beta+l+2} (-1)^{l+l'-\beta}}{Z_i^{\beta+l+5/2} i^{\beta+n'-l+1} [(n'+l')!]^{1/2} [(n'-l'-1)!]^{1/2}} \tag{2.60}$$

$$G_1 \equiv \sum_{j=0}^{n'-l'-1} \left\{ (i/2)^j (k/Z_f)^{j-n'-1-\beta+1} \left[1 + \frac{k^2 (n')^2}{Z_i^2} \right]^j \frac{(p-1)!}{j!} \right.$$

$$\times \prod_{i=0}^{j-1} (n'+l'-i)(n'-l'-1-i) \right\} \sum_{m=1}^{p} \left\{ \left(i + \frac{kn'}{Z_i} \right)^{p-m} \left(i - \frac{kn'}{Z_i} \right)^{m-1} \right.$$

$$\times \prod_{\gamma=1}^{m-1} \left(-i - l\frac{k}{Z_f} - \gamma\frac{k}{Z_f} \right) [(p-m)! \, (m-1)!]^{-1} \prod_{\kappa=1}^{p-m} \left(i - l\frac{k}{Z_f} - \kappa\frac{k}{Z_f} \right) \right\} \tag{2.61}$$

with

$$p = \beta + n' - l - j + 2 \tag{2.62}$$

The square of the radial matrix element is given by

$$R_\beta^2(n'l' - kl) \, dk = |F_1|^2 \frac{k \exp[(-4Z_f/k) \, \tan^{-1}(kn'/Z_i)]}{[1 - \exp(-2\pi Z_f/k)][(k^2 n'^2/Z_i^2) + 1]^{2(\beta+n'+2)}}$$

$$\times \prod_{s=1}^{l} \left(\frac{k^2}{Z_f^2} s^2 + 1 \right) |G_1|^2 \, dk \tag{2.63}$$

The radial integral for bound-bound transitions can be obtained by substituting the bound wave functions into equation (2.56). However, it is easy to accomplish the same task if one recognizes the fact that a continuous spectrum of positive eigenvalues adjoins the discrete levels of negative energy (Bethe and Salpeter, 1957). This implies the calculations of the residue of the bound-free matrix element at

$$k = i\frac{Z_f}{n} \tag{2.64}$$

Furthermore, $[1 - \exp(-2\pi Z_f/k)] \to 1$. Such a prescription has been tested by various authors (Khandelwal et al., 1989). Thus from equation (2.63), one obtains

$$R_\beta^2(n'l' - nl) \, dn = \frac{Z_f^2 |F_1|^2}{n^3 \left\{ 1 - [(n')^2/n^2] \left(Z_f^2/Z_i^2 \right) \right\}^{2(\beta+n'+2)}}$$

$$\times \left(\frac{nZ_i - n'Z_f}{nZ_i + n'Z_f} \right)^{2n} \left[\prod_{s=1}^{l} \left(1 - \frac{s^2}{n^2} \right) \right] |G_2|^2 \, dn \tag{2.65}$$

225

where the quantity G_2 stands for the following:

$$
\begin{aligned}
G_2 \equiv & \sum_{j=0}^{n'-l'-1} \left\{ \left(\frac{1}{2}\right)^j (i)^{j+p-1} \left(\frac{1}{n}\right)^{j-n'-1-\beta+l} \left[1 - \frac{Z_f^2}{Z_i^2}\frac{(n')^2}{n^2}\right]^j \frac{(p-1)!}{j!} \right. \\
& \times \left. \prod_{i=0}^{j-1} (n'+l'-i)(n'-l'-1-i) \right\} \\
& \times \sum_{m=1}^{p} \left\{ \frac{[1+(Z_f/Z_i)(n'/n)]^{p-m}\,[1-(Z_f/Z_i)(n'/n)]^{m-1}}{(p-m)!\,(m-1)!} \right. \\
& \times \left. \prod_{\gamma=1}^{m-1}\left(-1-\frac{l}{n}-\frac{\gamma}{n}\right) \prod_{\kappa=1}^{p-m}\left(1-\frac{l}{n}-\frac{\kappa}{n}\right) \right\}
\end{aligned}
\tag{2.66}
$$

with $p = \beta + n' - l - j + 2$.

The discrete dipole oscillator strength f_n and the differential oscillator strength $df/d\epsilon$ for ejected energy ϵ are important in various physical applications (Khandelwal, Khan, and Wilson, 1989; Khan, Khandelwal, and Wilson, 1988a, 1988b, and 1990). These can be obtained from equations (2.66) and (2.63) for $1s - np$ or k transitions for $\beta = 0$ as:

$$
f_n = E_n \frac{2^9}{3} n^7 (n^2 - 1) Z_i^3 Z_f^5 (2Z_i - Z_f)^2 \frac{(nZ_i - Z_f)^{2n-6}}{(nZ_i + Z_f)^{2n+6}}
\tag{2.67}
$$

and

$$
\frac{df}{d\epsilon} = \left(\epsilon + 2Z_i^2 - Z^2\right) \left(\frac{Z_f}{Z-1}\right)^2 \frac{1}{3k} R^2 (1s - k)
\tag{2.68}
$$

where

$$
k^2 = \frac{Z_f^2 \epsilon}{(Z-1)^2}
\tag{2.69}
$$

$$
E_n = 2Z_i^2 - Z^2 - \frac{Z_f^2}{n^2}
\tag{2.70}
$$

and

$$
R^2(1s-k)\, dk = \frac{2^8 k Z_i^3 Z_f \left(Z_f^2 + k^2\right)(2Z_i - Z_f)^2 \exp\left[-4(Z_f/k)\tan^{-1}(k/Z_i)\right]}{\left(Z_i^2 + k^2\right)^6 \left[1 - \exp\left(2\pi Z_f/k\right)\right]}\, dk
\tag{2.71}
$$

Recently, we have applied equations (2.67) and (2.68) to helium atoms and to helium-like ions (Khandelwal, Khan, and Wilson, 1989; Khan, Khandelwal, and Wilson, 1988a, 1988b, and 1990). We find that the screened hydrogenic model reasonably reproduces the existing dipole oscillator-strength values with little effort, and nonrelativistic numerical values for bound-bound and for bound-continuum transitions are available for many target He-like ions. The model has also been successful in reproducing the known dipole polarizability values and in predicting the other unknown values. Moments of dipole oscillator-strength distribution (Khan, Khandelwal, and Wilson, 1990) for the helium sequence have also recently been calculated under the screened hydrogenic model. This approach has resulted in values which are in reasonable agreement with the various moment values of other authors (including the mean excitation energy parameter I_T).

In order to obtain the stopping power, one has to include all momentum transfers in the form factor. Khandelwal and coworkers at Old Dominion University, Norfolk, Virginia, under sponsorship of the NASA Langley Research Center, have recently calculated the related radial integral (generalized oscillator strength) for the $1s$ to nl transitions. Thus, it would be an easy matter to obtain stopping power of a helium atom for a projectile such as a proton or a heavy ion. This work is currently underway. This is an ambitious undertaking but is more satisfying in that the calculations are done directly for each atom from first principles, thereby avoiding the inherent approximations such as the underlying Bethe energy loss formula (involving I_T and C).

2.3. Stopping in Molecular Gases

In an extremely rarefied atomic gas, charge particle interactions can occur singly with individual gas constituents leading to great simplification in theoretical treatment. Two physical effects occur as the gas density increases: (1) The projectile no longer reaches asymptotic states in subsequent reactions and (2) the interaction is modified by the presence of the surrounding medium. In addition, for low-energy collisions, the charge state of the projectile is likewise altered by these same physical effects and new states of the partially charged projectile states become important since radiative and Auger transition times become on the order of or greater than the mean free time between collisions. Although Bethe's theory for ordinary matter has questionable applicability it has been shown to be useful in estimating stopping powers provided empirical mean excitation energies are used. This is further discussed in section 2.6.

2.3.1. Historical perspective. Early in the classical treatment of charged particle slowing down it was recognized that the free-electron, long-range coulomb interaction leads to divergencies in the energy-loss rate. These divergencies indicate that there is a need for a long-range saturation effect. The saturation in gases was discussed by Bohr (1915) in terms of Ehrenfest's principle. Bohr proposed that the saturation in gases is caused by the bonding of the electrons. To effect energy transfer, the interaction time $\tau = b/v$ (where b is the impact parameter and v is the ion velocity) must be short compared with the oscillating period of the bonded electron. Hence, the adiabatic long-range collisions provide the necessary saturation, and an upper limit is established for the effective impact parameters. Most of our modern understanding stems from Bethe's detailed quantum theory (1930) based on the Born approximation. Stopping power for gaseous media with this approximation is given by

$$S = \frac{4\pi N Z_P^2 Z_T e^4}{m v^2} \left\{ \ln\left[\frac{2mv^2}{(1-\beta^2)I_T} \right] - \beta^2 - \frac{C}{Z_T} \right\} \tag{2.72}$$

where Z_P is the projectile charge, N is the number of targets per unit volume, Z_T is the number of electrons per target, m is the electron mass, v is the projectile velocity, $\beta = v/c$, c is the velocity of light, C is the velocity-dependent shell-correction term (Walske and Bethe, 1951), and I_T is the mean excitation energy given by solving

$$Z_T \ln I_T = \sum_n f_n \ln E_n \tag{2.73}$$

where f_n is the electric dipole oscillator strength of the target and E_n is the corresponding excitation energy. The sum in equation (2.73) includes discrete and continuum levels. Empirically, it was observed that molecular stopping power is reasonably approximated by the sum of the corresponding empirically derived "atomic" stopping powers (Bragg and Kleeman, 1905). Equations (2.72) and (2.73) imply

$$Z_T \ln I_T = \sum_j n_j Z_j \ln I_j \tag{2.74}$$

where Z_T and I_T pertain to the molecule, Z_j and I_j are the corresponding atomic values, and n_j represents the stoichiometric coefficients. This additivity rule, given by equation (2.74), is called Bragg's rule.

Sources of deviations from Bragg's additivity rule for molecules and the condensed phase are discussed by Platzman (1952a and 1952b). Aside from shifts in excitation energies and adjustments in line strengths as a result of molecular bonding, new terms in the stopping power are caused by the coupling of vibrational and rotational modes. Additionally, in the condensed phase, some discrete transitions are moved into the continuum, and collective modes among valence electrons in adjacent atoms produce new terms to be dealt with in the absorption spectrum. Platzman proposed that the experimentally observed additivity rule may not show that molecular stopping power is the sum of atomic processes but rather it demonstrates that molecular bond shifts for covalent-bonded molecules are relatively independent of the molecular combination. On the basis of such arguments, Platzman suggested that ionic-bonded substances should be studied as a rigid test of the additivity rule because of the radical difference in bonding type. He further estimated that ionic-bond shifts could change the stopping power by as much as 50 percent.

Among the early indicators of the violation of the Bragg rule was the calculation of 15 eV for the mean excitation energy of atomic hydrogen (using eq. (2.72) with the exactly known oscillator strengths and excitation levels) compared with a rather firmly established experimental value for molecular hydrogen of about 18 eV. Since accurate values of atomic mean excitation energies have been calculated for numerous elements by Inokuti and coworkers (Dehmer, Inokuti, and Saxon, 1975; Inokuti, et al., 1981) for the purpose of evaluating chemical bonding effects in molecules, empirical values have been substantially perturbed by effects of the chemical bonds. Although the mean excitation energy for gas molecules could be evaluated in principle from equation (2.72), the lack of knowledge of the excitation levels and corresponding oscillator strengths is the main hindrance.

It was suggested by Dalgarno (1960) that the oscillator strength distributions could be determined empirically from the photoabsorption spectra (aside from experimental uncertainty). Much of these data are obtained by energy-loss experiments by electron impact scattering at forward angles. Values of mean excitation energy for a number of simple molecules have in this way been estimated and demonstrate the shift in atomic values caused by chemical bonding (Zeiss and Meath, 1975; Zeiss et al., 1977).

Theoretical calculation of mean excitation energies is hindered by the difficulty of solving for the complete excitation spectrum of complex quantum systems. Dalgarno (1963) was able to simplify the calculation by introducing a generalized function, which is related to the excitation spectrum as follows:

$$F_D(\omega) = \sum_n \frac{f_n}{E_n + \omega} \tag{2.75}$$

However, this function can be evaluated without explicitly forming the indicated sum. Thus, Dalgarno was able to reduce equation (2.75) to

$$F_D(\omega) = \frac{2}{3} \left(\vec{X}, \sum_{i=1}^{Z_T} \nabla_i \psi_o \right) \tag{2.76}$$

with

$$(H - E_o + \omega)\vec{X} + \sum_{i=1}^{Z_T} \vec{r}_i \psi_o = 0 \tag{2.77}$$

where ψ_o is the ground-state wave function, E_o is the corresponding energy, ω is an energy eigenvalue, and $\vec{X}$ is the corresponding eigenvector. Chan and Dalgarno (1965) calculated I as 42 eV for helium and Kamikawai, Watanabe, and Amemiya (1969) calculated 18.2 eV for molecular hydrogen by the same method. These values are in excellent agreement with experiments.

Simultaneous with the development of the microscopic theory of stopping power was the macroscopic electrodynamic description of energy loss as required for the description of the long-range part of the interaction in the condensed phase. This is because the interaction is simultaneous among many constituents. The slowing down is through the force exerted on the passing particle by the electric field induced in the medium by the passage (Landau and Lifshitz, 1960). It is customary to assume that the electric displacement vector is linearly related to the time-varying electric field as

$$\vec{D}(t) = \vec{E}(t) + \int_0^t g(\tau) \, \vec{E}(t - \tau) \, d\tau \tag{2.78}$$

for which the dielectric constant is

$$\epsilon(\omega) = 1 + \int_0^\infty g(\tau) \exp(i\omega\tau) \, d\tau \tag{2.79}$$

The short-range collisions are still treated by Bethe theory with the result for total stopping power (see Ahlen (1980) for details) of

$$S = \frac{4\pi N Z_P^2 Z_T e^4}{mv^2} \left\{ \ln \left[\frac{2mv^2}{(1-\beta^2)I_T} \right] - \beta^2 - \frac{\delta}{2} \right\} \tag{2.80}$$

where δ is a density-effect correction applicable at high energies ($\beta^2 > 1/\epsilon(0)$). Also,

$$Z_T \ln I_T = \frac{m}{2\pi^2 N e^2} \int_0^\infty \omega \operatorname{Im}\left[\frac{-1}{\epsilon(\omega)} \right] \ln(\hbar\omega) \, d\omega \tag{2.81}$$

where $\operatorname{Im}(Z)$ denotes the imaginary part of Z and $\hbar$ is Planck's constant. A result of dispersion theory is

$$\frac{m}{2\pi^2 N e^2} \omega \operatorname{Im}[\epsilon(\omega)] = f(\omega) \tag{2.82}$$

where $f(\omega)$ is the dipole oscillator strength per unit cell of the medium, and

$$Z_T \ln I_T = \int_0^\infty \frac{f(\omega)}{|\epsilon(\omega)|^2} \ln(\hbar\omega) \, d\omega \tag{2.83}$$

which reduces to the usual Bethe expression (eq. (2.73)) in a sparse gas for which $\epsilon(\omega) \approx 1$.

If the long-range saturation effect is in terms of adiabatic limits for a gas and in terms of the medium polarization response for condensed dielectrics, the saturation effect for a free-electron gas is related to the tendency of a neutral plasma to screen a local charge imbalance at large distances (Kramers, 1947). The dielectric function of a free-electron gas is derived by Lindhard (1954) and applied to the stopping power problem for a classical electron gas and for the interaction-free Sommerfeld electron gas model. For a free-electron gas at rest, Lindhard arrives at the equation

$$S = \frac{4\pi Z_P^2 e^4 \rho}{mv^2} \ln\left(\frac{2mv^2}{\hbar\omega_p} \right) \tag{2.84}$$

where ρ is the electron density and ω_p is the classical plasma frequency given by

$$\omega_p^2 = \frac{4\pi e^2}{m} \rho \tag{2.85}$$

Strictly speaking, equation (2.84) applies only when the electron gas is at rest, but it also applies in the limit of high projectile velocity compared with the average motion of the electrons.

A discovery which paralleled the Lindhard investigations was made by Bohm and Pines (Bohm and Pines, 1951 and 1953; Pines and Bohm, 1952), in which collective long-range interactions in a quantum electron gas were separated from individual electron motion through a canonical transformation, after which the normal coordinates of collective oscillation appear. This separation of the Hamiltonian into collective and individual electron motions is accomplished because of the effective screening of the coulomb fields of individual electrons for distances greater than the screening distance λ_c. For collective motion to give a major contribution to the Hamiltonian, the individual electron wavelength must be greater than λ_c. Bohm and Pines (1953) found the average collective plasma frequency to be

$$\langle \omega \rangle = \left[1 + \frac{3}{2} \frac{\chi^2}{\lambda_s} \left(1 + \frac{3}{10} \chi^2 \right) \right] \omega_p \qquad (2.86)$$

where λ_s is the average electron separation and χ is the ratio of the average electron wavelength to the screening distance. Pines (1953) suggests that the screening parameter χ should be chosen to minimize the electron long-range correlation energy (that is, the electronic coulomb energy), which, for plane-wave states appropriate to their degenerate electron gas model, is given by

$$E_{lr,\text{corr}} = \frac{0.866\chi^3}{\lambda_s^{1.5}} - \frac{0.458\chi^2}{\lambda_s} + \frac{0.019\chi^4}{\lambda_s} \qquad (2.87)$$

Pines (1953) derived the stopping power in this degenerate electron gas and showed that the usual classical plasma frequency ω_p is replaced by $\langle \omega \rangle$, which includes corrections for individual electron motion.

A rather bold suggestion was made by Lindhard and Scharff (1960) that equation (2.84) could be applied on the atomic scale if the appropriate average over the atomic electron density was made. They further suggested that the effects of individual bonding of the electrons in their atomic orbitals could be incorporated through the added factor $\gamma \approx \sqrt{2}$ as

$$S = \frac{4\pi e^4 Z_P^2 N}{mv^2} \int d^3r \; \rho(\vec{r}) \; \ln \left(\frac{2mv^2}{\gamma \hbar \omega_p} \right) \qquad (2.88)$$

From equation (2.88), the mean excitation energy is given by

$$Z_T \; \ln \; I_T = \int d^3r \; \rho(\vec{r}) \; \ln(\gamma \hbar \omega_p) \qquad (2.89)$$

Lindhard and Scharff estimated the mean excitation energy for atomic Hg as 768 eV compared with ≈ 800 eV from experiment. For He, they got 37 eV compared with 35 eV from quoted experiments (more modern experiments yield 42 eV). They further approximated molecular hydrogen by taking the effective charge to be $Z = 1.2$ and obtained 16 eV.

Following this initial success of treating atoms as localized electron plasmas, Lindhard and Winther (1964) extended equation (2.88) by using the more general velocity-dependent dielectric function derived by Lindhard (1954), and demonstrated the ability of the Lindhard theory to predict tight bonding corrections of similar character to those of Walske (1952) in connection with the Bethe theory.

Chu and Powers (1972) made extensive use of the work of Lindhard and Scharff (1960) to demonstrate Z_2 oscillations in the mean excitation energy. This work gave rise to corresponding Z_2 oscillations in stopping power from which periodic variations are associated with the atomic shell structure (Chu and Powers, 1972). The more detailed calculations of Rousseau, Chu, and Powers (1971) utilized the velocity-dependent Lindhard-Winther theory and Bonderup's (1967) simplified form of the Lindhard theory and showed good agreement with 2-MeV α-particle stopping power data (Chu and Powers, 1969). Throughout these efforts, the parameter γ is taken as the square root of 2, as suggested by Lindhard and Scharff (1960).

Chu, Moruzzi, and Ziegler (1975), using the theory of Lindhard and Winther in which individual electron corrections to the local collective excitation were treated empirically by taking γ as an adjustable parameter, evaluated the aggregation effects for condensed noble gases and metals. The condensed-gas calculations determined electron densities according to atomic Hartree-Fock densities, including overlap from the nearest neighbors in the condensed phase. Metallic wave functions were taken from the muffin-tin model calculations of Moruzzi, Janak, and Williams (1978). In most cases, the empirically determined γ was in the range from 1.2 to 1.3. (See Ziegler, 1980.)

As noted by Dehmer, Inokuti, and Saxon (1975), equation (2.89) may be rewritten as

$$Z_T \ln I_T = \int d\omega \left[\int d^3r \, \delta(\omega - \gamma\omega_p) \, \rho(r) \right] \ln(\hbar\omega) \qquad (2.90)$$

from which can be obtained

$$f(\omega) = \int d^3r \, \delta(\omega - \gamma\omega_p) \, \rho(r) \qquad (2.91)$$

where $\delta(x)$ is the Dirac delta function. It is seen from equation (2.91) that, in the local plasma approximation, the volume of plasma with cutoff frequency $\gamma\omega_p = \omega$ approximates the total oscillator strength of the system at frequency ω. No exact equivalence is implied between the oscillator frequency distribution given by equation (2.91) and the oscillator frequency distribution of a quantum system. (This is true because equation (2.91) exhibits a continuous spectrum, although quantum systems generally exhibit a series of poles associated with the discrete quantum levels as well as a continuum at higher frequencies.) Some insight may be gained by comparing dispersion relations for atomic systems with those for a related plasma. The dispersion relation for a classical plasma is given by the dielectric constant $\epsilon(\omega)$ as

$$\epsilon(\omega) = 1 - \frac{\omega_p^2}{\omega^2} \qquad (2.92)$$

232

where ω_p is the usual plasma frequency and equation (2.92) results from the plasma conductivity (Hubbard, 1955). Indeed, the same pole term in $\epsilon(\omega)^{-1}$ appears in metals as the result of the conduction electrons that give metals their characteristic optical properties (Hubbard, 1955; Fröhlich and Pelzer, 1955). The more general dispersion relation, derived from equations (2.78) and (2.81), is

$$\epsilon(\omega) = 1 - \frac{4\pi Z_T e^2}{m} P \int_0^\infty \frac{f(x)}{x^2 - \omega^2} \, dx \qquad (2.93)$$

where P denotes the principal value at the singularity. In atomic systems, the oscillator strengths are broadly separated in frequencies according to shells; the outer shells appear at the lowest frequencies, and the innermost shell appears at the highest frequencies. The lack of oscillator strength at frequencies between shells results in large gaps in the spectrum. Let ω be a frequency in the broad gap between two successive shells—the first centered at ω_1 and the second at ω_2. Then the dispersion relation (eq. (2.93)) becomes

$$\epsilon(\omega) \approx 1 - \frac{\omega_{p,1}^2}{\omega^2} \qquad (2.94)$$

where

$$\omega_{p,1}^2 \approx \frac{4\pi Z_T e^2}{m} \int_0^{\omega_1} f(x) \, dx \qquad (2.95)$$

so that $\omega_{p,1}$ is the plasma frequency associated with the electrons of the outermost shell. Although equations (2.94) and (2.95) provide motivation (Wilson et al., 1984b) for using the local plasma approximation (eq. (2.91)), there is plenty of room for a more complete understanding as to why the model works as well as it does in practical calculations (Wilson and Kamaratos, 1981; Wilson et al., 1984b).

In previous investigations, we considered the use of the local plasma model to evaluate molecular bonding effects on the mean excitation energy of molecules of covalent-bonded hydrogen and carbon (Wilson and Kamaratos, 1981) as well as ionic crystals and gases (Wilson et al., 1982), in which quite sensible corrections to the usual Bragg's rule were obtained. The chemical-bond shifts were unambiguously defined in terms of atomic integrals and molecular parameters. In the usual implementation of the local plasma model (eq. (2.89)), γ corrects for a shift in the local plasma frequency caused by individual electron effects. Lindhard and Scharff (1960) suggest $\gamma = \sqrt{2}$; however, $\gamma \approx 1.2$ yields atomic mean excitation energies from the local plasma model in better agreement with the accurate atomic values calculated by Dehmer, Inokuti, and Saxon (1975). The fact that the larger value $\left(\gamma = \sqrt{2}\right)$ gives better agreement with empirical data suggests that this larger value corrects (in addition to individual electron shifts) for the chemical shifts as well. Such chemical shifts were estimated separately for covalent and ionic bonds by Wilson and Kamaratos (1981) and Wilson et al. (1982).

Encouraged by the smallness (<30 percent) of the empirical individual electron corrections to the collective plasma frequency (Ziegler, 1980; Wilson and Kamaratos, 1981; Wilson et al., 1982), a calculation (Wilson and Xu, 1982) in which individual electron shifts were estimated according to the theory for plane-wave states in an extended plasma, as calculated by Pines (1953), yields results that are in good agreement with Dehmer, Inokuti, and Saxon (1975). Consequently, the local plasma model is placed on a parameter-free basis (Wilson et al., 1984a and 1984b) in which chemical shifts are determined from atomic/molecular parameters alone, and effects of individual electron motion are evaluated in terms of the Pines correction, the combined effects of which are on the order of the plasma frequency shift of $\gamma \approx \sqrt{2}$ suggested by Lindhard and Scharff.

The Pines correction makes a remarkable improvement in the prediction of the local plasma model, and further adjustments in the theory to account for the plasma frequency shifts resulting from the atomic shell structure should bring the model into predictive capability. To further elucidate the relationship between the local plasma model and the more exact quantum treatment of bonded systems, related quantities of both theories in the case of one- and two-electron systems are examined in section 2.3.2. Atomic mean excitation energies and straggling parameters, based on the local plasma model, are compared with accurate calculations of Inokuti et al. (Dehmer, Inokuti, and Saxon, 1975; Inokuti, Baer, and Dehmer, 1978; Inokuti et al., 1981) in section 2.3.3. The use of the Gordon-Kim electron gas model of molecular bonding (Gordon and Kim, 1972) to determine the effects of covalent chemical-bond shifts of mean excitation energy for elements of the first two rows is presented in section 2.3.4. Calculations of mean excitation energies of ionic-bonded substances are discussed in section 2.3.5, and the mean excitation energies of metals are discussed in section 2.3.6.

2.3.2. Excitation spectra of one- and two-electron systems. The hydrogen atomic excitation spectrum in the dipole approximation is well-known as

$$
f_{\rm H}(\omega) = \left\{
\begin{array}{ll}
\displaystyle\sum_{n=2} \frac{2^8}{3}\left(1 - \frac{1}{n^2}\right) n^7 \frac{(n-1)^{2n-5}}{(n+1)^{2n+5}} \, \delta(\omega - \omega_n) & (\hbar\omega < R) \\[2ex]
\displaystyle\frac{1}{2}\sqrt{\frac{\hbar}{R\omega}} \frac{2^8}{3}\hbar\omega \frac{k}{(1+k^2)^5} \frac{\exp[(-4/k)\tan^{-1} k]}{1 - \exp(-2\pi/k)} & (\hbar\omega > R)
\end{array}
\right\}
\tag{2.96}
$$

where n is the principal quantum number, R is the Rydberg constant, ω_n is given by

$$
\hbar\omega_n = R\left(1 - \frac{1}{n^2}\right)
\tag{2.97}
$$

and

$$
Rk^2 = \hbar\omega - R
\tag{2.98}
$$

The corresponding spectrum for the local plasma model (eq. (2.91)) is given as

$$
f_p(\omega) = \left\{
\begin{array}{ll}
4\omega\omega_o^{-2} \, \ln^2(\omega/\omega_o) & (\omega < \omega_o) \\[2ex]
0 & (\omega > \omega_o)
\end{array}
\right\}
\tag{2.99}
$$

where $\omega_o = 55.12$ eV. The cumulative oscillator strength

$$
F(\omega) = \int_0^\omega f(\omega') \, d\omega'
\tag{2.100}
$$

234

is shown in figure 2.2 for each of the two models. Similarly, the excitation spectrum of the helium atom has been evaluated for screened wave functions and is shown in figure 2.2. The fractional excitations of the two models never differ by more than ≈ 15 percent above the excitation threshold. As noted by Dehmer, Inokuti, and Saxon (1975), the main error in the local plasma model is the contribution to absorption below excitation threshold all the way down to zero. This error is also evident in the energy moments of the plasma model. The moments of the energy spectrum for the hydrogen atom are shown in figure 2.3, where

$$\langle (E/R)^m \rangle = \int_0^\infty \left(\frac{\hbar\omega}{R} \right)^m f(\omega) \, d\omega \tag{2.101}$$

and m is a continuous parameter. The low-frequency contributions associated with the local plasma model cause a divergence in equation (2.101) at $m \approx -2$ which is not present in the quantum system. Atomic polarizability and the low-frequency refractive index are affected the most. Other atomic properties, such as the total inelastic cross section, the mean excitation energy, the straggling parameter, and the mean electronic kinetic energy, are reasonably presented by the plasma model. Also shown in figure 2.3 are results, including the Pines (1953) correction to the plasma frequency, which indicate substantial improvement in the prediction of atomic properties, although low-energy atomic properties are still beyond the scope of the model.

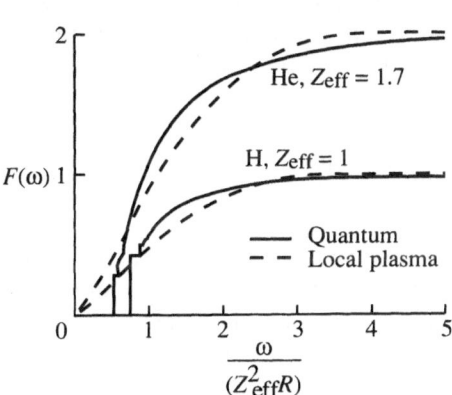

Figure 2.2. Cumulative oscillator strength distribution for atomic hydrogen and helium.

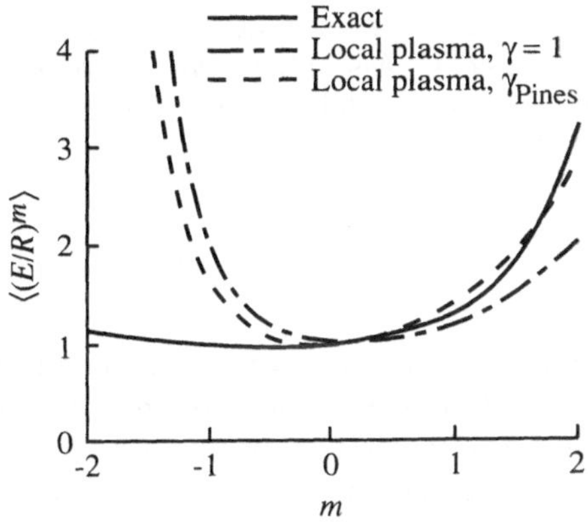

Figure 2.3. Moments of oscillator strengths of hydrogen atom.

The plasma model is expected to be more accurate as more electrons are added to the system. This occurs in two ways, as seen in figure 2.2. First, a greater contribution comes from the continuum, which is most like the plasma. Second, the excitation thresholds shift to relatively lower energies and fill in the low-frequency region, for which the plasma model normally tends to err. A considerable improvement in the energy moments of helium for the local plasma approximation is clearly shown in figure 2.4.

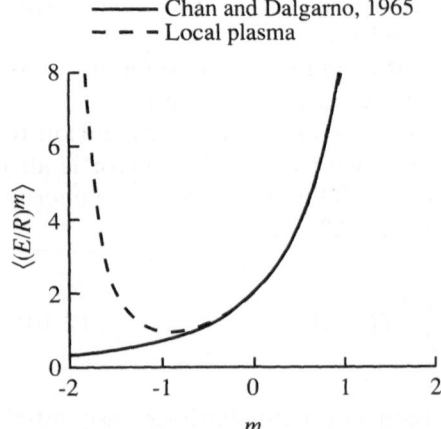

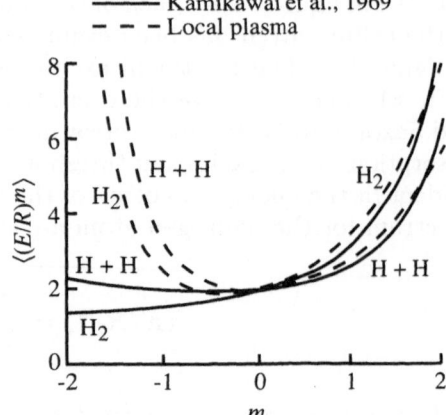

Figure 2.4. Moments of oscillator strengths of helium atom according to quantum oscillator strengths using screened wave functions. $Z_{\text{eff}} = 1.7$.

Figure 2.5. Moments of oscillator strengths of hydrogen molecule for several models.

The moments of the excitation spectrum of H_2 have been evaluated empirically by using experimental oscillator strengths (Dalgarno and Williams, 1965) and theoretically (Kamikawai, Watanabe, and Amemiya, 1969) using the Dalgarno (1963) sum rules (eqs. (2.75) and (2.76)). These are compared in figure 2.5 with an "atomic" approximation to H_2 taken as a generalization of Bragg's rule (Zeiss et al., 1977). Also shown in figure 2.5 are values for H_2 calculated by using the local plasma model with the Pines correction and with the Gordon-Kim model of the molecular wave functions given as

$$\rho_{H_2}(\vec{r}) = \rho_H(\vec{r}) + \rho_H(\vec{r} - \vec{R}) \qquad (2.102)$$

where $\rho_H(\vec{r})$ is the atomic hydrogen electron density and $\vec{R}$ is the displacement vector of length $1.4a_o$ between the two centers. It is clear from figure 2.5 that, even with the simple Gordon-Kim approximation, the plasma model is a considerable improvement over the Bragg rule, except for the lowest-energy molecular properties (i.e., $m < -0.5$). Figure 2.5 also shows that the Gordon-Kim approximation introduces minor errors compared with the inherent limitations of the local plasma model.

The mean excitation energy for stopping power may likewise be evaluated. Atomic hydrogen and molecular hydrogen are presented in table 2.1 along with a recent compilation of experimental data (Seltzer and Berger, 1982). Quite reasonable estimates of atomic and molecular properties of importance to ionizing radiation are obtained by this local plasma model if the Pines correction is included. Optical and other low-frequency properties, however, are poorly represented. The plasma model should become more accurate for more complex many-electron systems, especially those in which the optical properties are more in line with those predicted by the plasma model.

Table 2.1. Hydrogen Mean Excitation Energy for Stopping Power

| Chemical species | Hydrogen mean excitation energy, eV, for stopping power for— | | | |
	Quantum model	Oscillator strength distribution	Local plasma model (a)	Experiment
H	[b]14.99		14.69	
H_2	[c]18.2	[d]18.4	18.9	[e]18.5 ± 0.5

[a]With Pines correction.
[b]With oscillator strengths of equation (2.96).
[c]Kamikawai, Watanabe, and Amemiya, 1969.
[d]Dalgarno and Williams, 1965.
[e]Seltzer and Berger, 1982.

With the present results, it is now clear what approach should be taken to improve the plasma model applications. Clearly, a correction factor similar to that of Pines should be introduced to suppress absorption below excitation threshold and, correspondingly, to enhance frequencies just above threshold. A number of possibilities are open to implement such a correction, which would appear as a first-order quantum correction for the discrete spectrum. Preliminary work by Walecka (1976) on the study of collective atomic oscillations may be a starting point for further development.

2.3.3. Stopping and straggling parameters of atoms. In this section, parameters are considered for atoms associated with the stopping of charged particles and fluctuations in their energy transfer. The energy moment is

$$S(m) = \langle (E/R)^m \rangle \tag{2.103}$$

and the related quantity is

$$L(m) = \frac{dS(m)}{dm} \tag{2.104}$$

In terms of these quantities, the mean excitation energy is

$$\ln I = \frac{L(0)}{S(0)} \tag{2.105}$$

and the straggling parameter related to fluctuations in energy loss is

$$\ln \Delta = \frac{L(1)}{S(1)} \tag{2.106}$$

The mean excitation energy (eq. (2.105)) has been evaluated in the context of the local plasma model and is presented in figure 2.6 along with the values computed by Inokuti and coworkers for atoms through krypton and the compilation of experimental data by Seltzer and Berger (1982). Hartree-Fock wave functions (Clementi and Roetti, 1974) have been used for elements helium through neon and sodium through krypton are represented by screened wave functions (Clementi and Raimondi, 1963).

The values for the straggling parameter were similarly evaluated and are presented in figure 2.7 with the values obtained by Inokuti and coworkers. Also shown are values for noble gases compiled by Inokuti et al. (1981) and values obtained by Zeiss et al. (1977). The present values tend to be about 25 percent low at $Z \approx 36$, with improvements at lower values of Z, which may be caused by the lack of shell structure corrections in the plasma frequencies of the K and L shells.

It is clear from these atomic calculations that the plasma model with the Pines correction generally provides good results for mean excitation energy and reasonable estimates for the straggling parameter. Although the Hartree-Fock wave functions are required for low atomic numbers, reasonable results are obtained using screened wave functions for atoms heavier than argon. The low-energy atomic properties mainly require improvements beyond the Pines correction. These properties emphasize the need for a first-order quantum correction to the atomic structure.

2.3.4. Covalent-bond effects.

2.3.4. Covalent-bond effects. Early experimental work with ionization energy loss was conducted in covalent-bonded gases (also noble gases) from which Bragg's rule was derived. Although more recent experimental work, beginning with Thompson (1952), has shown systematic variation from Bragg's rule, such rules still seem appropriate for fixed molecular structures (Lodhi and Powers, 1974; Neuwirth, Pietsch, and Kreutz, 1978). As a result of the theoretical efforts of Inokuti and coworkers, it is clear that chemical-bond shifts in the mean excitation energy have occurred, and as suggested by Platzman (1952a), all covalent shifts are of similar magnitude.

In any molecular dynamic calculation, there is a trade-off between model accuracy and computation efficiency. As pertains to the radiolysis of large molecular structures, the most useful model is the lowest order possible. It is clear that the use of self-consistent field methods to determine molecular wave functions would seriously limit the ability to study systems of practical interest. Considering the relative success of the Gordon-Kim electron gas model of molecular bonding (Gordon and Kim, 1972; Tossell, 1979; Waldman and Gordon, 1979), a simple method for the calculation of chemical-bond effects on the mean excitation energies is suggested. As suggested by Gordon and Kim, the molecular electron density as a superposition of the unperturbed atomic states is given by

$$\rho(\vec{r}) = \rho_1(\vec{r}) + \rho_2\left(\vec{r} - \vec{R}_{12}\right) \tag{2.107}$$

for diatomic molecules. There is an obvious generalization of equation (2.107) for the polyatomic case. Whereas Gordon and Kim used equation (2.107) to calculate the molecular potential (see Tossell, 1979; Waldman and Gordon, 1979, for ionic and covalent applications) from which R_{12} is theoretically obtained, here R_{12} is taken from observed experimental bond distances. Substituting equation (2.107)

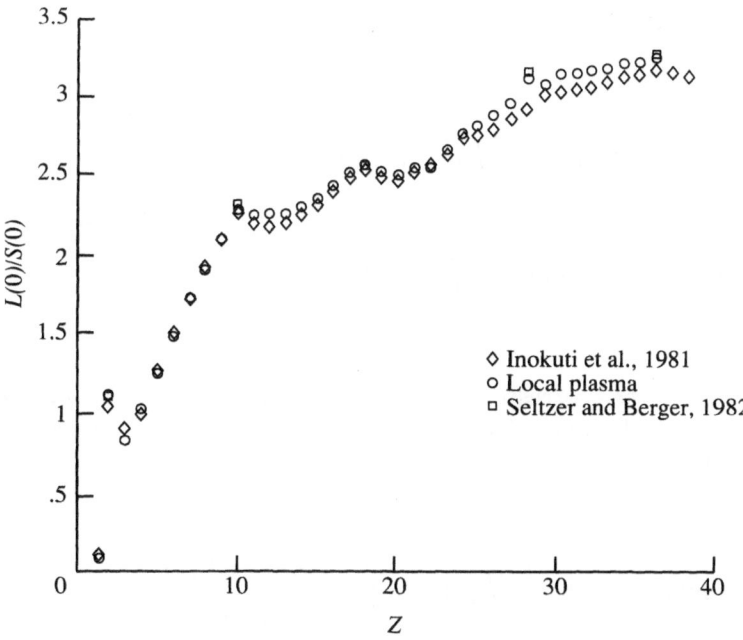

Figure 2.6. Atomic mean excitation energies from quantum calculations of Inokuti et al., 1981, and local plasma model. Empirical values are from Seltzer and Berger, 1982.

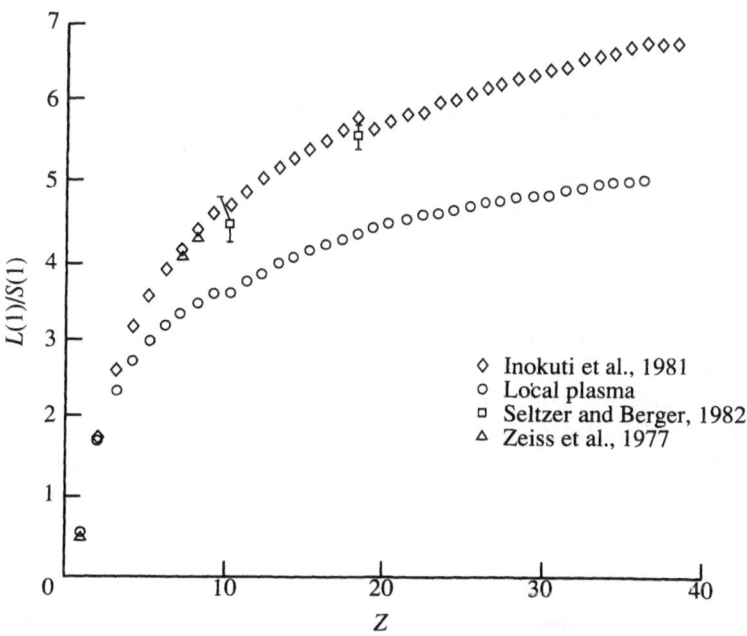

Figure 2.7. Atomic straggling parameters from quantum calculations of Inokuti et al., 1981, and local plasma model along with various experimental results.

into equations (2.85) and (2.89) and reducing results in

$$Z \ln(I) = Z_1 \ln(I_1) + \int \rho_1(\vec{r}) \ln \left[1 + \frac{\rho_2 \left(\vec{r} - \vec{R}_{12} \right)}{\rho_1(\vec{r})} \right]^{1/2} d^3r$$

$$+ Z_2 \ln(I_2) + \int \rho_2(\vec{r}) \ln \left[1 + \frac{\rho_1 \left(\vec{r} - \vec{R}_{21} \right)}{\rho_2(\vec{r})} \right]^{1/2} d^3r \qquad (2.108)$$

where I_1 and I_2 are the corresponding atomic values, which are accurately known (Inokuti, Baer, and Dehmer, 1978; Inokuti et al., 1981). The chemical-bonding correction is generally

$$\ln(1 + \delta_{ij}) = \frac{1}{Z_i} \int \rho_i(\vec{r}) \ln \left[1 + \frac{\rho_i \left(\vec{r} - \vec{R}_{ij} \right)}{\rho_i(\vec{r})} \right]^{1/2} d^3r \qquad (2.109)$$

Equation (2.108) for diatomic molecules is generalized for polyatomic systems as

$$Z \ln(I) = \sum_i Z_i \ln \left[\left(1 + \sum_j \delta_{ij} \right) I_i \right] \qquad (2.110)$$

where the sum over j includes every bond in which Z is attached in the molecule. Correction factors have been calculated (Wilson and Kamaratos, 1981) for hydrogen and carbon molecules with the bond parameters in table 2.2. Carbon sp^3 hybrid orbital wave functions were used in these calculations, although s^2p^2 values were only slightly different. The tetrahedral orbitals were spherically symmetrical in their electron densities. Therefore, spherical symmetry was assumed throughout subsequent calculations.

Recommended values of mean excitation energies (Seltzer and Berger, 1982) are presented in table 2.3 along with theoretical values calculated by using atomic mean excitation energies from Dehmer, Inokuti, and Saxon (1975) with the bond corrections in table 2.2. Bragg's rule is also used with the atomic values of Dehmer, Inokuti, and Saxon for comparison. Although the theoretical values are within 4 percent of the experimental and empirical values, Bragg's rule values are from 17 to 21 percent low, indicating a substantial adjustment as the result of chemical bonding.

Mean excitation energies have been calculated for covalent gases of the first two rows using the local plasma model and the Pines correction. Results of this calculation and the empirical values of Seltzer and Berger (1982) are given in table 2.4. Corresponding values for covalent solids are shown in table 2.5.

Table 2.2. Hydrogen and Carbon Molecular Parameters

Molecular parameter	H-H	H-C	C-H	C-C	C=C	C≡C	H-C benzene	C-H benzene	CC benzene	CC graphite
R_{AB}, bohrs	1.40	2.08	2.08	2.94	2.52	2.28	2.02	2.02	2.64	2.68
δ	0.261	0.432	0.044	0.062	0.087	0.105	0.453	0.045	0.079	0.076

Table 2.3. Molecular Mean Excitation Energy

Chemical species	Molecular mean excitation energy, eV, for—		
	Present theory	Seltzer and Berger, 1982	Bragg's rule
CH_4	44.7	42.8	35.1
$(CH_2)_x$	55.0	53.4	43.5
C_6H_6	60.6	61.4 ± 1.9	50.6
H_2	18.9	18.5 ± 0.5	15.0
Graphite	76.1	78.5 ± 1.5	62.0

Table 2.4. Molecular Mean Excitation Energies for Covalent Gases

Chemical species	R_{AB}, bohrs	I, eV	
		Local plasma model	Seltzer and Berger, 1982
H_2	1.40	18.9	[a]19.2 ± 0.4
N_2	2.08	85.0	[a]82 ± 1.6
O_2	2.34	99.6	[a]95 ± 1.9
F_2	2.67	114.2	115 ± 10
Cl_2	3.76	170.8	171 ± 14

[a] These values are strongly influenced by Zeiss et al. (1977).

Table 2.5. Mean Excitation Energies for Covalent-Bonded Crystals

Chemical species	R_{AB}, bohrs	I, eV	
		Local plasma model	Seltzer and Berger, 1982
B (tetragonal)	3.06	67.3	76 ± 7.6
C (diamond)	2.94	75.3	
C (graphite)	2.68	76.1	78 ± 2.3
Si (diamond)	4.42	151.0	173 ± 4
P (black)	4.16	155.7	[a]181 ± 14
S (rhombic)	3.85	162.7	[a]190 ± 15

[a]Unspecified allotropic form.

Moments for the N_2 molecule using the plasma model are presented in figure 2.8 with values calculated from the oscillator strengths compiled by Dalgarno, Degges, and Williams (1967). As can be seen, good agreement between the present simple plasma model calculations and the oscillator strength distribution of Dalgarno, Degges, and Williams is obtained except for the lowest frequency phenomena.

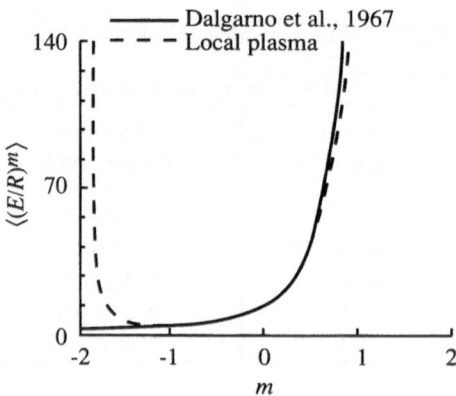

Figure 2.8. Moments of N_2 oscillator strengths from empirical values of Dalgarno, Degges, and Williams (1967) and local plasma model using Gordon-Kim molecular model densities.

2.3.5. Ionic-bond effects.

Although covalent-bond shifts were found to be relatively small corrections to atomic values, such a separation as in equation (2.108) in terms of neutral atomic values is not possible for ionic bonds. Using the Gordon-Kim model electron density of the partial ionic (diatomic) system,

$$\rho_p(\vec{r}) = \rho_{A(\pm p)}(\vec{r}) + \rho_{B(-p)}\left(\vec{r} - \vec{R}_{AB}\right) \tag{2.111}$$

where $A^{(+p)}$ and $B^{(-p)}$ refer to partially ionic states of the two constituents, $\vec{R}_{AB}$ is their nuclear separation, and p is the partial ionic fraction. The electron density of a partial ionic atom in equation (2.110) is

$$\rho_{A(\pm p)}(\vec{r}) = (1-p)\rho_A(\vec{r}) + p\rho_{A\pm}(\vec{r}) \tag{2.112}$$

where $\rho_A(\vec{r})$ is the electron density of the neutral atom and $\rho_{A\pm}(\vec{r})$ is the electron density of the atomic ion. With the aid of equations (2.111) and (2.112), shifts in the mean excitation energy caused by ionic and covalent effects can be evaluated. As shown by Wilson et al. (1982),

$$Z \ln(I) = Z_{A(+p)} \ln \left[I_{A(+p)} \right] + \int \rho_{A(+p)}(\vec{r}) \ln \left[1 + \frac{\rho_{B(-p)} \left(\vec{r} - \vec{R}_{AB} \right)}{\rho_{A(+p)}(\vec{r})} \right]^{1/2} d^3r$$

$$+ Z_{B(-p)} \ln \left[I_{B(-p)} \right] + \int \rho_{B(-p)}(\vec{r}) \ln \left[1 + \frac{\rho_{A(+p)} \left(\vec{r} - \vec{R}_{AB} \right)}{\rho_{B(-p)}(\vec{r})} \right]^{1/2} d^3r \qquad (2.113)$$

with

$$Z_{A(+p)} \ln \left[I_{A(+p)} \right] = \int \rho_{A(+p)}(\vec{r}) \ln \left[\gamma \hbar \omega_{A(+p)}(\vec{r}) \right] d^3r \qquad (2.114)$$

where γ is the Pines correction given by equation (2.85) or estimated empirically as given by Neuwirth, Pietsch, and Kreutz (1978). Mean excitation energies for various stages of ionization calculated with the Pines correction and the atomic wave functions of Clementi and Roetti (1974) are shown in figure 2.9. In addition to the ionic-bond shifts, there are shifts caused by covalent-like character, as given by

$$\ln \left[1 + \delta_{A(+p),B(-p)} \right] = \frac{1}{Z_{A(+p)}} \int \rho_{A(+p)}(\vec{r}) \ln \left[1 + \frac{\rho_{B(-p)}(\vec{r} - \vec{R})}{\rho_{A(+p)}(\vec{r})} \right]^{1/2} d^3r \quad (2.115)$$

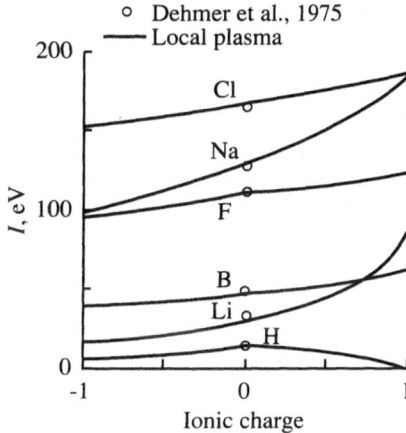

Figure 2.9. Mean excitation energies for partially ionic atoms.

Mean excitation energies for partial ionic-bonded substances are shown in table 2.6 with the corresponding bond parameters used in the model. Also shown are values for a pure covalent bond and Bragg's values using the neutral atomic mean excitation energies of Dehmer, Inokuti, and Saxon (1975), as well as Bragg's values of the corresponding partial ionic states. The ionic-bond fractions are taken from Pauling (1967) as experimental data for HF and LiH. Bond lengths are for ionic crystals except for the HF gas. Atomic mean excitation energies are shown for partial ionic states in figure 2.9 and differ from values of Wilson et al. (1982) because of the Pines correction.

Table 2.6. Ionic-Bond Parameters

Chemical species	R_{AB}, bohrs	p	I_c, eV	I_{IB}, eV	I_B, eV	I, eV
HF	1.72	0.50	97.6	91.7	91.0	96.4
LiH	3.85	0.25	27.8	25.2	25.9	26.7
LiF	3.85	0.90	83.4	92.6	81.6	93.6

It is clear from table 2.6 that the main contribution to corrections to the Bragg rule is the adjustment from atomic neutral to atomic ion mean excitation energies as proposed by Platzman (1952a). Indeed, when there is little difference between the usual Bragg value and the partially ionic Bragg value, the covalent value is in near agreement with the predicted value of I for HF and LiH in the table. For LiF, the relatively large adjustment from the usual Bragg value (81.6) to the partially ionic Bragg value (92.6) leaves a large difference between the covalent value (83.4) and the predicted value of I (93.6). The adjustment of the ionic-bond shift caused by the covalent-like character for LiF is 1 eV compared with adjustments of the neutral states caused by the pure covalent bond of 1.8 eV. This comparison shows the greater role of the coulomb attraction in forming the bond of the ionic molecules relative to the two-electron interaction in forming the covalent bond.

Calculated mean excitation energies for ionic crystals using the Pines correction are shown in table 2.7, along with recommended values of Seltzer and Berger (1982). The crystal parameter and fractional ionic charge have been taken from Pauling (1967). The LiF value is the only one with an experimental basis (Wilson et al., 1982).

Table 2.7. Mean Excitation Energies of Ionic Crystals

Chemical species	R_{AB}, bohrs	p	I, eV Local plasma model	I, eV Seltzer and Berger, 1982
LiF	3.80	0.90	92.8	94 ± 8
LiCl	4.86	0.73	139.1	144 ± 12
NaF	4.37	[a]0.91	131.5	147 ± 12
NaCl	5.31	0.75	159.1	181 ± 14

[a]Pauling partial ionic character function.

2.3.6. Metallic-bond effects. Our first approach to metals is similar to that taken by Chu, Moruzzi, and Ziegler (1975), in which they employed the muffin-tin wave functions (Moruzzi, Janak, and Williams, 1978) and stopping power theory according to Lindhard and Winther (1964). Individual electron corrections to the local plasma frequency are treated empirically through an adjustable parameter γ. (See table I of Ziegler (1980) and related discussion.) Unlike this previous work, the present work includes estimates of shifts in the plasma frequency according to the Pines correction in equation (2.86) and is in that sense completely deterministic.

The metallic wave functions for lithium metal approximated by the Wigner-Seitz model (Wigner and Seitz, 1934) are considered first. In deriving these wave functions, the lithium ion core potential was taken from the screened wave functions of Clementi and Raimondi (1963), and the calculated crystal-valence wave functions (aside from normalization) were found to be a slight perturbation (mainly due to boundary conditions) of the free hydrogenic ($2s$) orbital inside the Wigner-Seitz sphere (Wigner and Seitz, 1934). The final crystal wave functions used were constructed from unperturbed Hartree-Fock orbitals (Clementi and Roetti, 1974) in the core region with a small perturbation outside the core. This perturbation matched the boundary conditions on the surface of the Wigner-Seitz sphere. This was followed by normalization of the valence-shell wave functions (to make the valence electron density add up to give the correct number of valence electrons). These wave functions are quite similar to the muffin-tin model and yield mean excitation energies in substantial agreement with Ziegler (1980) when γ is taken as his empirical value. The mean excitation energies for metals of the second and third rows using Wigner-Seitz wave functions (treating all valence electrons as spatially equivalent) and the Pines correction are presented in table 2.8 along with empirical values from Seltzer and Berger (1982).

Table 2.8. Metallic Parameters for Selected Metal of First Two Rows

Chemical species	I_{at}, eV Dehmer, Inokuti, and Saxon, 1975	r_s, bohrs	I, eV	
			Wigner-Seitz model	Seltzer and Berger, 1982
Lithium	34.0	3.260	45	41.5 ± 3.7
Beryllium	38.6	2.375	60	63.7 ± 3.2
Sodium	123.6	3.99	140	162 ± 8
Magnesium	121.2	3.34	144	164 ± 8
Aluminum	124.3	2.991	149	166 ± 3

The present results clearly demonstrate that the effects of the metallic bond in lithium and beryllium are large and are mainly the result of collective oscillations in the free-electron gas formed by the valence electrons. Although similar good agreement should be expected for sodium and magnesium, it is emphasized here only that these empirical values are interpolations without an experimental basis, and smaller empirical values more in line with the present results should not be eliminated. The small value predicted for aluminum (149 eV) is in doubt, as the empirical value (166 ± 3 eV) is based on one of the most experimentally studied quantities since aluminum served as a standard in stopping power experiments for many years. The fault could well lie in the use of the Wigner-Seitz model for group III metals. It is well-known that the success of the Wigner-Seitz theory rests mainly on application to alkali metals. Although some hope for application to group II metals exists, treating the three valence electrons of group III as spatially equivalent is clearly in error. Correction to metals from an alternate model, proposed by Pauling (1967) for metallic orbitals and implemented here in simplified fashion, is considered next.

In X-ray diffraction experiments, even beryllium metal shows a considerable degree of covalent quality, as suspected from bulk material properties (Brown, 1972). In this view, a model is considered in which the valence-bond effects can be included explicitly. In the spirit of the Pauling valence-bond theory and the Gordon-Kim model of valence bonding, the electron density about the ion cores is assumed to be a superposition of partial ionic core states among nearest core neighbors. Additional contributions from the next nearest neighbors are assumed to add to the electron continuum states in a manner analogous to the Pauling unsynchronized resonances in lithium crystals (Pauling, 1967). The electron density of the partially ionic core of charge p is

$$\rho_{A(+p)}(\vec{r}) = \left(\frac{v-p}{v}\right)\rho_A(\vec{r}) + \frac{p}{v}\rho_{A(+v)}(\vec{r}) \tag{2.116}$$

where v is the number of valence electrons, $\rho_A(\vec{r})$ is the electron density of the atomic neutral state, and $\rho_{A(+v)}(\vec{r})$ is the electron density of the valence-stripped ion cores. We have used the observation by Slater that radial wave functions of the L shell are nearly the same for both values of l as a result of exchange interaction between the $(2s)$ and $(1s)$ orbitals. The same is true for the M shell. In the present calculation, each metal ion core has been placed into a Wigner-Seitz cell, and the electron density from nearest neighbors has been approximated by reflecting the exterior core density function across the cell boundary. The continuum electron density is then taken as

$$\rho_e = [p + (v-p)\delta]\frac{3}{4\pi r_s^3} \tag{2.117}$$

where δ is the next nearest neighbor contribution to the continuum. The value of δ is determined by requiring a full complement of v valence electrons per cell. The resultant electron density $\rho(r)$ was used to calculate the local plasma frequency and mean excitation energy per cell. The Pines correction was used for individual particle shifts. The radii r_s for the Wigner-Seitz cell are given in table 2.8. The ion-core wave functions were calculated from the Hartree-Fock wave functions of Clementi and Roetti. A slight dependence on the ion-core charge appears (Kamaratos et al., 1982) in which there is some increase in mean excitation to $I = 155$ eV for aluminum. However, there are some unresolved questions concerning periodicity at the cell boundaries, which leave the value of this model somewhat in doubt.

The mean excitation energy for aluminum requires the reconsideration of the data on which it is based and the corresponding analysis. In an analysis by Andersen and Ziegler (1977), 162 eV was assumed as the mean excitation energy for aluminum. A reduction in I to 150 eV results in a 3-percent increase in stopping power at 1 MeV, which leaves it within the stated uncertainty limits of the Andersen and Ziegler parametric curves. These curves correspond to the uncertainty in the experimental data used in the analysis. (See fig. 2.10.) Indeed, a number of authors have reported mean excitation energies for aluminum in line with the present results (Bakker and Segrè, 1951; Simmons, 1952; Mather and Segrè, 1951; Sachs and Richardson, 1953; Bogaardt and Koudijs, 1952; Vasilevskii and Prokoshkin, 1960), although more recent analyses are higher. A recent study of aluminum optical properties indicates that a value of I several electron volts lower than 166 is not inconsistent with the empirical dielectric function (Shiles et al., 1980). The shift of several electron volts is associated with polarization of the Al^{3+} core by the valence electrons in their metallic orbitals. Such core polarization effects are not calculated in the present model. Furthermore, quantum corrections to K- and L-shell discrete spectra may cause further small adjustments. In any case, the apparent discrepancy is due to the electronic wave functions used in the present calculation, to the inadequate treatment of corrections to the Bethe formula, from which I is extracted from the experimental data (see, for example, Andersen et al. (1977) and Khandelwal (1982)), to quantum corrections, or to a combination of these.

To further clarify the relationship between the mean excitation energy for aluminum and experimental data, a band is shown in figure 2.10 which brackets the experimental data of Andersen et al. (1977), Kahn (1953), Neilsen (1961), Leminen, Fontell, and Bister (1968), Nakata (1971), and Sorensen and Andersen (1973) for proton energies between 0.5 and 10 MeV. These energies are compared with the reduced stopping power calculated from the Andersen and Ziegler (1977) empirical shell corrections. The older data of Kahn (1953), which would have lowered the band considerably, were excluded from the figure. The mean excitation energies exhibited in the figure are 167 eV used as input to Shiles et al. (1980), 162 eV determined by Andersen and Ziegler (1977), 155 eV estimated using one form of valence-bond theory, and 149 eV calculated according to the present (simplified) Wigner-Seitz model. Although it is not clear that the curve for 167 eV is superior to the curve for 149 eV, a modest shift in the empirical shell corrections can bring any of the four curves into an equally good fit to the data. It is further emphasized that shell corrections are not exactly known, and, in empirical analysis, shell corrections are not usually differentiated from other corrections to the Bethe formula (eq. (2.72)).

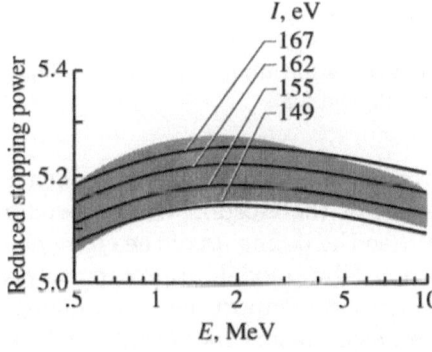

Figure 2.10. Reduced stopping power for aluminum for several mean excitation energies and range of experimental data. Shaded area is band of experimental data.

2.3.7. Discussion of results. The present results are combined in figure 2.11 with the evaluated data of Seltzer and Berger (1982). Care is taken when possible to model the same physical chemical state. (See specific tables for details.) Results for free atoms (Hartree-Fock wave functions for $Z \leq 10$ and screened wave functions elsewhere) and the accurate atomic values of Dehmer, Inokuti, and Saxon (1975) are presented in figure 2.11. It is clear that the trends in the first- and second-period elements are well approximated by the present application of the local plasma model, especially when the Pines correction is applied. The present results are generally in fair agreement with the compilation and recommendations of Seltzer and Berger (1982), although small discrepancies in the third period remain to be resolved.

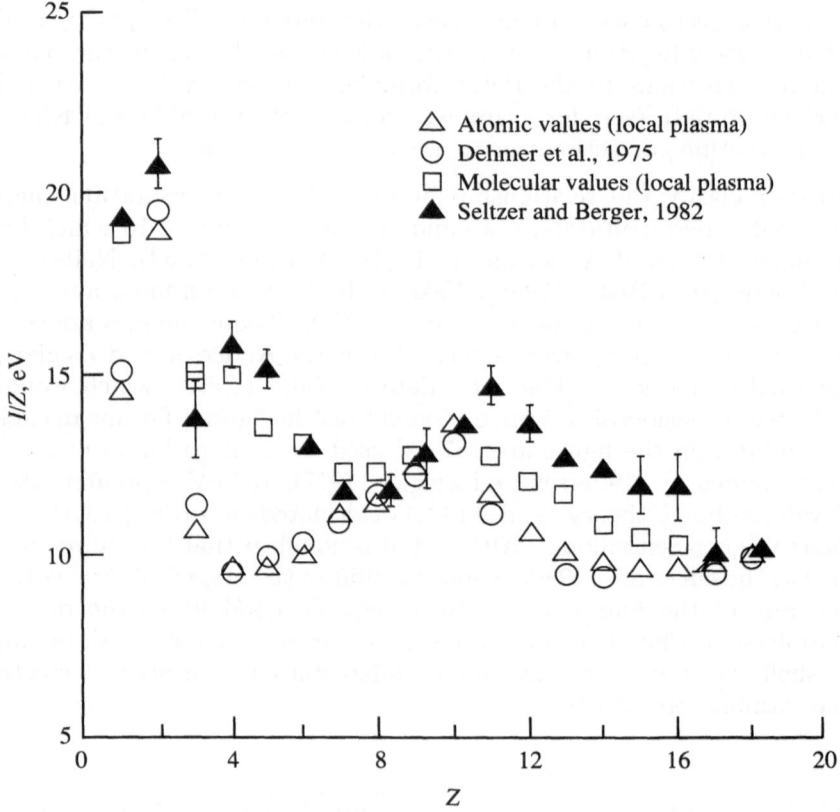

Figure 2.11. Mean excitation energies for atoms, molecules, solids, and metals. Specific data taken from tables 2.3, 2.4, 2.5, and 2.8.

Perhaps the greatest criticism of the present application of the local plasma model calculations is the use of the Gordon-Kim approximation to the covalent-bonded wave functions. When the moments of the energy spectrum are considered, it is clear that the Gordon-Kim model approximately adjusts the excitation spectrum in the region of greatest importance to ionizing radiation and appears no more in error than the basic plasma model in which it is used. (See fig. 2.5.) Of course, accurate use of the local plasma model implies the necessary use of the Pines correction, as demonstrated for the hydrogen atom in figure 2.3 and used throughout the present calculations. Although the Pines correction produces marked improvements in the predictive capability of the model, further quantum corrections for the discrete spectrum would produce additional corrections and would hopefully remove most of the remaining error in the plasma model. Further improvement in electronic wave functions would be helpful in identifying the remaining corrections required for the plasma model.

2.4. Molecular Stopping Cross Sections

In section 2.3, departures from Bragg's rule have been noticed in the theoretical calculations of the mean excitation energies of various molecular systems. Analysis of the experimental data on energy loss of low-energy α-particles in gases also indicates deviations from Bragg's rule (Bourland and Powers, 1971; Lodhi and Powers, 1974). In this section, the stopping power theory of Lindhard and Winther (1964) and the local plasma theory of Lindhard and Scharff (1960) are used to perform calculations in the low-energy region. Modifications are introduced through a simplifying model which incorporates the effects of the shell corrections and of the screening of the projectile (Xu, Khandelwal, and Wilson, 1984a and 1984b). The model is justified on the basis of fulfilling the more ambitious aim of obtaining the molecular stopping power. The Gordon-Kim electron density model of molecular wave functions (Gordon and Kim, 1972) is utilized in the calculations. As shown, such a model allows a successful method of calculating chemical-bond effects. Calculations done on N_2, O_2, and water vapor are found to be in fair agreement with experiments (Xu, Khandelwal, and Wilson, 1984b). Furthermore, departures from Bragg's rule are noticed for all these systems.

The celebrated stopping power formula for an energetic charged particle of charge Z_P and velocity v traversing matter of charge number Z_T is given by

$$-\frac{dE}{dx} = \frac{4\pi Z_P^2 e^4}{mv^2} N Z_T L \tag{2.118}$$

where m is the mass of an electron and N the number of atoms per unit volume of the medium.

The stopping number L of equation (2.118) has been a topic of considerable study. For instance, Lindhard and Winther (1964) have investigated the function L for a free-electron gas in the regions of low- and high-energy incident charged particles. For the high-energy case, these authors give the expression for L to order $1/v^2$ as

$$L = Z_T \ln Y - \frac{\langle T \rangle}{\frac{1}{2}mv^2} \tag{2.119}$$

where $Y = 2mv^2/\hbar\omega_p$, the classical plasma frequency $\omega_p = (4\pi\rho e^2/m)^{1/2}$, ρ is the electron density, and $\langle T \rangle$, the average kinetic electron energy, is given by

$$\langle T \rangle = \left(\frac{3}{10}\right) mv_F^2$$

where v_F is the Fermi velocity.

For the low-energy case, they give

$$L = \left(\frac{\chi^2}{3}\right)^{3/4} Y^{3/2} C_1(\chi) \tag{2.120}$$

where

$$C_1(\chi) = \frac{1}{2[1 - (\chi^2/3)]^2} \left(\ln \frac{1 + \frac{2}{3}\chi^2}{\chi^2} - \frac{1 - \frac{1}{3}\chi^2}{1 + \frac{2}{3}\chi^2}\right) \tag{2.121}$$

with

$$\chi^2 = \frac{e^2}{\pi \hbar v_F} \tag{2.122}$$

$$\hbar = \frac{h}{2\pi} \tag{2.123}$$

Equation (2.119) for the L function warrants some discussion. First, one notes that the L function of equations (2.119) and (2.120) is derived by Lindhard and Winther for a free-electron system. Transition to an atomic system of the first term of equation (2.119), as studied widely, is accomplished under the so-called local plasma model in which density $\rho(\vec{r})$ is evaluated by using quantum mechanical wave functions. The local plasma model is equivalent to replacing the molecular dipole oscillator strengths by the corresponding classical plasma absorption spectrum. The adequacy of such a replacement was recently shown by Johnson and Inokuti (1983) to be most accurate for evaluating atomic quantities associated with stopping power in spite of differences between the plasma spectrum and the actual oscillator strength distribution. A quantum mechanical analog of the second term of equation (2.119) would be of interest. In this context, a result first derived by Brown would prove to be useful. Brown (1950) studied the K-shell asymptotic stopping power of a hydrogenic system (with two K electrons) for a fast projectile, taking the maximum momentum transfer equal to $2mv$ as if the electron was free. (See Xu, Khandelwal, and Wilson, 1986.) The asymptotic stopping power equation obtained by Brown can be expressed in a form similar to equation (2.119). The first terms of both these equations, since they involve the mean excitation energy, can be assumed essentially equivalent within the local-plasma approximation. For the second term in equation (2.119) for a hydrogenic system, he obtained $1/\eta_s$, where $\eta_s = \frac{1}{2}mv^2/Z_s^2 R$, Z_s is the effective nuclear charge for the s shell ($s = K, L, \ldots$), and R is the Rydberg constant. Walske (1952), on the other hand, taking the upper limit for momentum transfer as infinity, overestimated the nuclear momentum recoiling and obtained $2/\eta_s$ instead. In reality, however, because of the recoiling of the nucleus, the result should be expected to fall somewhere between $1/\eta_s$ and $2/\eta_s$. This fact is incorporated into equation (2.126) as a parameter which we later estimate. At the present, however, for the sake of simplicity, combining Brown's result for the K shell with Walske's result for the L shell, but retaining the consistency with the free-electron model, we write the analogous second term (known as shell correction C) for a hydrogenic system with Z_T electrons as

$$C = C_{K,\text{total}} + C_L = \frac{1}{\eta_K} + \frac{1}{\eta_L}\left(\frac{Z_T - 2}{8}\right) \tag{2.124}$$

which can be rewritten for a real atom as

$$C = \frac{1}{2} \frac{\langle T \rangle}{\frac{1}{2}mv^2} \; \phi(Z_T) \tag{2.125}$$

where

$$\phi(Z_T) = Z_T f(Z_T)g \tag{2.126}$$

and

$$\langle T \rangle = \frac{1}{Z_T} \left[Z_K^2 R + (Z_T - 2)\frac{1}{4}Z_L^2 R \right] \tag{2.127}$$

In equation (2.126), a coefficient $f(Z_T)$ has been introduced to distinguish a real atom from a hydrogenic atom. The coefficient $f(Z_T)$ is known to be less than unity for L shells for targets with low atomic number. The coefficient g is introduced to incorporate the effect due to the recoiling of the nucleus.

At this stage, it is appropriate to discuss various features associated with the low-energy projectiles and the targets with low atomic number. First, in the low-energy region, the projectile's full charge Z_P will not be operational in the stopping process due to electron capture that is influenced mainly by the outer shell electrons of the medium. Second, Walske has pointed out that the coefficient $f(Z_T)$ is unreliable for the low atomic number $Z_T \leq 30$ due to use of the hydrogenic wave functions.

It is evident from the above observations that some sort of crude estimate of the quantity C is in order. This is justified since the usual incorporation of these effects involves fitting with experimental data. The inclusion of the effect of the projectile's effective charge should decrease the stopping number of all elements. The decrease should be the most for Li and the least for Ne. In order to incorporate this effect and the other problem of the need for an accurate value of the coefficient $f(Z_T)$, it is reasonable as a first approximation to assume a semiempirical constant value of the quantity $\phi(Z_T)$ equal to one half the total number of electrons in noble-gas atoms. Such a division should overestimate shell corrections for lithium and beryllium in decreasing fashion and underestimate that for helium, neon, carbon, nitrogen, oxygen, and fluorine also in a decreasing manner. Such a change in shell corrections is indeed needed to compensate for the effect of the effective charge of the projectile on the stopping power. In this paper since we are interested in the atoms with atomic number below 10, this assumption implies that

$$\phi(Z_T) = \left\{ \begin{array}{ll} 1 & (Z_T \leq 2) \\ 5 & (3 \leq Z_T \leq 10) \end{array} \right\} \tag{2.128}$$

Implicit in the above partition of ϕ (eq. (2.128)) is the fact that the quantity C no longer represents the so-called shell corrections only but presumably also some other effects including those of the projectile's effective charge and the neglect of the higher order terms in equation (2.119). One can now write equation (2.125) as

$$\frac{C}{Z_T} = \left\{ \begin{array}{ll} \dfrac{1}{2} \dfrac{\langle T \rangle}{\frac{1}{2}mv^2} \dfrac{1}{Z_T} & (Z_T \leq 2) \\[3ex] \dfrac{1}{2} \dfrac{\langle T \rangle}{\frac{1}{2}mv^2} \dfrac{5}{Z_T} & (3 \leq Z_T \leq 10) \end{array} \right\} \tag{2.129}$$

where $\langle T \rangle$ by virial theorem is just the average kinetic energy of the electron and should be averaged over all the Z_T electrons in the atom.

In order to make a transition to an atomic system, we assume the above results and accordingly replace equation (2.119) with

$$L = \left\{ \begin{array}{ll} \ln Y - \dfrac{3^{1.5}}{10\chi Z_T} \dfrac{1}{Y} & (Z_T \leq 2) \\[3ex] \ln Y - \dfrac{3^{1.5}}{2\chi Z_T} \dfrac{1}{Y} & (3 \leq Z_T \leq 10) \end{array} \right\} \tag{2.130}$$

The low- and high-energy L functions should now be combined to determine the appropriate dependence of the stopping power on energy. To do this, we used equations (2.120) and (2.130) for our desired results after replacing ω_p by $\gamma \omega_p$, where nonconstant values of γ were obtained from Wilson and Xu (1982). Bonderup (1967) had combined equations (2.119) and (2.120) and assumed a constant value of γ equal to $\sqrt{2}$. Unlike Bonderup, we tried to preserve the continuity between the low-energy stopping number function given by equation (2.120) and the high-energy function given by equation (2.130). In this way, stopping number values for a system can be obtained given the velocity of the projectile and the density $\rho(\vec{r})$.

For a diatomic molecule, the Gordon-Kim model gives the density as

$$\rho_{\text{molecule}} = \rho_a(\vec{r}) + \rho_b\left(\vec{r} - \vec{R}_{ab}\right) \tag{2.131}$$

where $\rho_a(\vec{r})$ is the atomic ground-state density and $\vec{R}_{ab}$ is the distance between the two atoms, which is known to be 1.094 Å for N_2 and 1.207 Å for the O_2 molecule. Equation (2.131) was generalized for water vapor including its partial ionic-bond nature and neglecting the overlap between the two H atoms. The distance between the O and H nuclei was taken as 0.958 Å. The molecular stopping power for protons was obtained by averaging the stopping number over $\vec{r}$ for N_2, O_2, and water vapor molecules. Hartree-Fock wave functions were employed in these calculations. Tables 2.9 and 2.10 list the results of the present work, together with curve-fitted results of Andersen and Ziegler (1977), and two sets of experimental data for the O_2 and N_2 molecules, respectively, (Reynolds et al., 1953; Langley, 1975). In table 2.11, the results for water vapor from the present work and experimental data for energies ranging from 40 to 500 keV are presented. Good agreement, within 10 percent, is found between the two sets of data.

Table 2.9. Proton Stopping Cross Sections for Oxygen Molecule

| E, keV | Proton stopping cross section, eV-cm^2/10^{15} atoms | | | |
| | Theoretical (a) | Curve fitted (b) | Experimental | |
			Reynolds et al., 1953	Langley, 1975
40	15.89	14.6	15.2 ± 2.6	
80	17.48	17.0	17.25 ± 2.6	
100	17.43	17.0	17.17 ± 2.6	
300	11.84	11.9	11.99 ± 1.7	
500	8.92	8.8	8.84 ± 1.7	
1037	5.64			5.25
2591	2.97			2.85

[a]Present paper.
[b]Andersen and Ziegler, 1977.

Table 2.10. Proton Stopping Cross Sections for Nitrogen Molecule

| E, keV | Proton stopping cross section, eV-cm^2/10^{15} atoms | | | |
| | Theoretical (a) | Curve fitted (b) | Experimental | |
			Reynolds et al., 1953	Langley, 1975
40	17.20	16.0	17.1 ± 2.6	
80	18.41	17.9	18.5 ± 2.6	
100	17.79	17.7	17.9 ± 2.6	
300	10.85	11.2	11.2 ± 1.7	
500	8.10	8.1	8.08 ± 1.7	
1037	5.20			4.78
2591	2.71			2.56

[a]Present paper.
[b]Andersen and Ziegler, 1977.

In order to discuss the departures from Bragg's rule, it would be relevant to cite a systematic study carried out in a series of experiments at Baylor University (Bourland and Powers, 1971; Powers et al., 1972; Lodhi and Powers, 1974). The study revealed that for low-energy projectiles there may exist a deviation from Bragg's rule depending on the physical state, but most importantly, on the chemical structure of the compounds. The confusing status of the dependence on the chemical structure can best be described by citing these studies in chronological order. First, in 1971 the Baylor group (Bourland and Powers, 1971) theorized that the compounds with single and double bonds should obey Bragg's rule. The compounds containing triple-bond structures were found to deviate from Bragg's rule by as much as 12.8 percent (α-particles of energy between 0.3 and 2.0 MeV often were the projectiles). In particular, these authors indicated that the molecular hydrogen (single-bonded molecule) should obey Bragg's rule. Later in 1972, the Baylor group (Powers et al., 1972) critically looked again at their previous conclusions. They indicated that perhaps the hydrogen atomic stopping cross section may be considerably different than one half the molecular stopping cross section and thus should cause considerable deviations. However, the Baylor group in 1974 (Lodhi and Powers, 1974) recognized the difficulty of obtaining atomic cross sections experimentally and based their analysis on the existence of some modified, but unique, atomic stopping cross sections.

Table 2.11. Proton Stopping Cross Sections for Water Vapor

Source	Proton stopping cross section, eV-cm^2/10^{15} molecules, at E, keV, of—				
	40	80	100	300	500
Present paper	28.81	27.8	26.8	17.1	12.6
Reynolds et al., 1953	25.0 ± 2.6	27.6 ± 2.6	27.3 ± 2.6	17.9 ± 1.7	13.0 ± 1.7

It is therefore imperative that in order to discuss the deviations from Bragg's rule, one must have access to the atomic and molecular stopping cross sections. We calculated both the atomic and the molecular stopping cross sections as a function of projectile energy of the O_2, N_2, and H_2 molecules. These results, together with the deviations from Bragg's rule, are exhibited in tables 2.12 through 2.14. One sees that the deviations from Bragg's rule become small as incident energy increases—in agreement with observations made by many workers including those at Baylor University. It is to be noted that N_2 is a triple-bonded molecule, O_2 is an approximately double-bonded molecule (from the bond energy point of view), and H_2 is a single-bonded molecule. The maximum deviations from Bragg's rule for energy of 100 keV and above are 6.1, 2.6, and 10 percent, respectively, for these molecules. Thus, the deviation depends on the chemical structure. When the Gordon-Kim model is used, the overlap of electron density determines the deviation or molecular bond effects. For instance, for the hydrogen molecule, the distance between nucleons is very small, 0.74 Å. It is expected that the overlap of electron density is large, thus explaining the considerable deviation from Bragg's rule. The stronger the bond energy, the shorter the distance will be. It is interesting to note that single-, double-, and triple-bonded carbon molecules have internuclear distances equal to 2.94, 2.52, and 2.24 bohrs, respectively. We may thus expect that the triple-bonded carbon will have more deviation from Bragg's rule than the single-bonded carbon.

Table 2.12. Deviations From Bragg's Rule for Oxygen Molecule

Stopping cross section	E, keV, of—						
	40	100	200	300	500	1037	100 000
ϵ (atomic)[a], eV-cm^2/10^{15} atoms . . .	17.44	17.48	14.65	12.15	9.1	5.72	0.1492
ϵ (molecule), eV-cm^2/10^{15} atoms . . .	15.89	17.43	14.36	11.84	8.92	5.64	0.1476
Deviation, percent	8.9	0.3	2	2.6	2	1.4	1.1

[a]Obtained from equations (2.116) and (2.126).

Table 2.13. Deviations From Bragg's Rule for Nitrogen Molecule

Stopping cross section	E, keV, of—						
	40	100	200	300	500	1037	100 000
ϵ (atomic)[a], eV-cm^2/10^{15} atoms . . .	19.33	18.57	14.32	11.46	8.53	5.30	0.1340
ϵ (molecule), eV-cm^2/10^{15} atoms . . .	17.20	17.79	13.75	10.85	8.10	5.20	0.1319
Deviation, percent	11	4.2	4.00	6.1	5.0	1.9	1.3

[a]Obtained from equations (2.116) and (2.126).

Table 2.14. Deviations From Bragg's Rule for Hydrogen Molecule

Stopping cross section	E, keV, of—						
	100	200	300	500	800	1037	2591
ϵ (atomic)[a], eV-cm^2/10^{15} molecules . .	12.7	8.13	6.1	4.17	2.89	2.36	1.11
ϵ (molecule), eV-cm^2/10^{15} molecules .	11.43	7.53	5.71	3.93	2.75	2.24	1.07
Deviation, percent	10	7.4	6.4	5.8	4.8	5.1	3.6

[a]Obtained from equations (2.116) and (2.126).

2.5. Stopping Cross Sections of Liquid Water. The stopping cross section of water is of interest in many fields but especially in radiation protection. Since the living tissues are basically composed of liquid water, a simple theoretical model for the calculation of the stopping power of water is of great practical interest. The importance of such a direct calculation has increased since various authors (Bourland and Powers, 1971; Powers et al., 1972; Lodhi and Powers, 1974) found that deviations from Bragg's rule may exist in the low-energy regime; meanwhile, the physical-state effect has also been observed in many experiments.

In previous work (Xu, Khandelwal, and Wilson, 1984a), we have established a modified local plasma model, based on the works of Lindhard and Winther (1964),

Brown (1950), and Walske (1952 and 1956). The model, applied to molecules using the Gordon-Kim molecular wave function (Gordon and Kim, 1972), has shown reasonable predictive capability for molecular bond effects as well as for atomic targets and covers a rather wide energy region.

In this section the modified local plasma model is applied to liquid water by employing a simple model of water molecules (Xü, Khandelwal, and Wilson, 1985). The calculated stopping cross section of liquid water is found to be about 5.5 to 14 percent lower than the calculated gas-state results for energies from 80 to 500 keV and is about 8.5 to 13.4 percent lower than gas-state results in the same energy range measured by Reynolds et al. (1953). The calculated liquid-water stopping power is within 2.5 percent of experimental values for ice in the energy range of 60 to 500 keV. It is proposed that for liquid water, this physical effect is due to interactions with neighbor molecules which confine each molecule to an effective close-packed sphere, thus causing the electrons to be more bonded and confined. Hence, the momentum transfer between projectile and electrons is reduced; this reduction decreases stopping power.

As is well-known, the structure of liquid water is complicated and far from completely known. As Pauling (1960) pointed out, "the structure for water that has received serious consideration for many years is the one proposed by Bernal and Fowler." Bernal and Fowler (1933) suggested that liquid water retains in part a hydrogen-bonded structure, similar to that of ice. They pointed out that as more and more hydrogen bonds are broken with increase in temperature, the oxygen atoms rearrange themselves into an approximately more and more close-packed structure. The rigidity of the hydrogen-bonded crystal structure is lost, allowing the motion between liquid molecules to be more flexible than that in the ice. A simple model of the close-packed structure of liquids was considered by Lennard-Jones and Devonshire (1937 and 1938) who derived the potential of the molecule in the liquid state from the interaction of all neighbors (fig. 2.12). Figure 2.12 shows clearly the effective volume to which the molecule is confined.

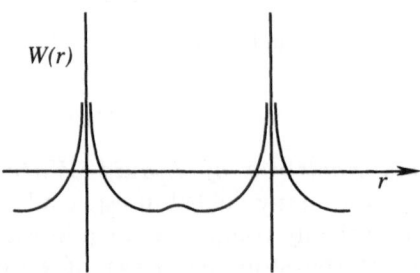

Figure 2.12. Intermolecular potential of molecule in liquid phase (Lennard-Jones and Devonshire, 1937 and 1938).

The hydrogen bond of the ice structure is rather weak compared with the molecular O-H bond. When ice melts, these hydrogen bonds of ice are distorted, and finally many are broken. Therefore, as a first-order approximation in the stopping power calculation, we can ignore the electronic overlap between molecules. As was pointed out by Xu, Khandelwal, and Wilson (1984a), the overlap of the electronic density in some sense expresses the bond energy or molecular bonding effect. The simplest model of this picture is to confine each molecule inside its close-packed sphere. Hence, the electronic density of this water molecule vanishes at distances exceeding the sphere radius, as shown in figure 2.13.

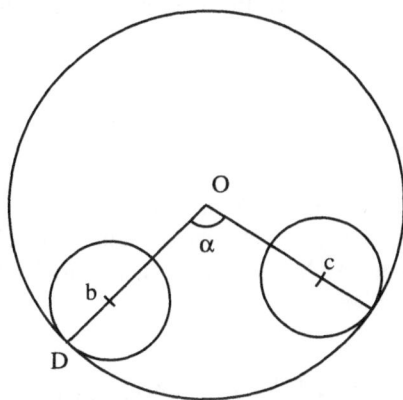

Figure 2.13. Liquid-water molecule in its close-packed spherical configuration.

In figure 2.13, O is the center of an oxygen atom, b and c are centers of two hydrogen atoms, α is the angle between two O-H bonds, and Ob is the distance between the center of the oxygen and the center of the hydrogen atom (Ob = 1.01 Å) obtained by neutron diffraction of deuterium oxide ice (Peterson and Levy, 1957). The radius OD is obtained from the effective volume per molecule V_{eff} as $R = (3V_{\text{eff}}/4\pi)^{1/3} = 2.992$ bohrs. Inside the sphere, the Gordon-Kim model is employed; this model assumes that the molecular electronic density to first-order approximation is simply the algebraic sum of corresponding atomic electron densities. The molecular density for H_2O can be expressed as follows:

$$\rho_{\text{mol}} = \rho_O(\vec{r}) + \rho_H(\vec{r} - \vec{R}_{Ob}) + \rho_H(\vec{r} - \vec{R}_{Oc}) \tag{2.132}$$

where $\vec{R}_{Ob}$ and $\vec{R}_{Oc}$ are displacement vectors of the nuclei at b and c, respectively (fig. 2.13). Since the radius of the hydrogen atom is much smaller than that of oxygen, the internuclear distance between the hydrogen atoms is large compared with their radii and the H-H interaction may be neglected within the H_2O molecule. Thus,

$$\rho_{\text{mol}} = \rho_O(\vec{r}) + 2\rho_H(\vec{r} - \vec{R}) \tag{2.133}$$

The partial ionic-bond effects are considered through

$$\rho_{\text{mol}} = \rho_O^+(\vec{r}) + 2\rho_H^-(\vec{r} - \vec{R}) \tag{2.134}$$

with $\rho^\pm = (1-p)\rho(\vec{r}) + p\rho^\pm(\vec{r})$, where p is the partial ionic fraction with $\rho(\vec{r})$ the neutral-atom electron density and $\rho^\pm(\vec{r})$ the ionic electron density. In the water molecule, according to Pauling (1960), p equals 0.33. The wave functions are obtained from Clementi and Roetti (1974). These wave functions are renormalized within the close-packed sphere according to

$$\int_{V_{\text{eff}}} \rho_{\text{mol}} \, d^3\vec{r} = 10 \tag{2.135}$$

The above electronic density is to be used with the formula for stopping power given by

$$-\frac{dE}{dx} = \frac{4\pi Z_P^2 e^4}{mv^2} N Z_T L \tag{2.136}$$

where m is the mass of an electron, N is the number of atoms per unit volume of the medium, Z_P is the charge of the projectile, and Z_T is the charge number of target; L is given elsewhere (Xu, Khandelwal, and Wilson, 1984a and 1984b). For the low projectile energies,

$$L = \left(\frac{\chi^2}{3}\right)^{3/4} Y^{3/2} C_1(\chi) \tag{2.137}$$

where

$$C_1(\chi) = \frac{1}{2(1 - \frac{\chi^2}{3})^2} \left(\ln \frac{1 + \frac{2\chi^2}{3}}{\chi^2} - \frac{1 - \frac{\chi^2}{3}}{1 + \frac{2\chi^2}{3}} \right) \tag{2.138}$$

where $\chi^2 = e^2/\pi\hbar v_F$, $\hbar = h/2\pi$, v_F is the Fermi velocity, $Y = 2mv^2/\gamma\hbar\omega_p$, $\omega_p^2 = 4\pi e^2 \rho/m$ is the classical plasma frequency, and ρ is the electronic density. For high projectile energies

$$L = \begin{cases} \ln Y - \dfrac{3\sqrt{3}}{10\chi}\dfrac{1}{Y} & (Z_T \leq 2) \\[3mm] \ln Y - \dfrac{3\sqrt{3}}{2\chi Z_T}\dfrac{1}{Y} & (3 \leq Z_T \leq 10) \end{cases} \tag{2.139}$$

The low- and high-energy L functions should be combined by joining them continuously (Xu, Khandelwal, and Wilson, 1984a and 1984b).

Table 2.15 shows the proton stopping cross section values of water vapor and liquid water together with the experimental results of water vapor measured by Reynolds et al. (1953). As can be seen, the agreement between the experiments and the theory is very good. There is a marked reduction in stopping power in the liquid phase by a few percent even at the highest energies shown. As noted by Thwaites (1981), there appears to be little difference between the stopping power of water in the liquid phase and of ice. (See, in particular, fig. 2 of Thwaites (1981).) The experimental data of Wenzel and Whaling (1952) for ice are shown in comparison with the calculated values for liquid water. The results given in table 2.15 are shown graphically in figure 2.14.

Table 2.15. Proton Stopping Cross Sections for Water

E, keV	Proton stopping cross section, eV-cm²/10¹⁵ molecules			
	Theoretical[a]		Experimental	
	Water vapor	Liquid water	Vapor[b]	D_2O ice[c]
40	28.7	21.6	25.0 ± 2.6	22.6
60	28.6	23.5	26.9 ± 2.6	24.0
80	27.8	23.9	27.6 ± 2.6	24.0
100	26.8	23.7	27.3 ± 2.6	23.7
200	21.1	19.5	22.0 ± 1.7	20.1
300	17.1	16.0	17.9 ± 1.7	16.0
400	14.4	13.6	15.0 ± 1.7	13.3
500	12.6	11.9	13.0 ± 1.7	11.6
600	11.2	10.7		
700	10.1	9.7		
800	9.3	8.9		
900	8.6	8.2		
1 000	8.0	7.6		
10 000	1.37	1.34		

[a]Present theory with partial ionic fraction of $p = 0.33$.
[b]Reynolds et al., 1953.
[c]Wenzel and Whaling, 1952.

We see from table 2.15 that the calculated stopping cross section of liquid water is about 5.6 to 14 percent lower than calculated gas-state results from 80 to 500 keV and is about 8.5 to 13.4 percent lower than measured gas-state results. In the same energy regime, Matteson, Powers, and Chau (1977) reported that in their experimental results for 0.3- to 2.0-MeV α-particles, the stopping cross section of H_2O vapor was found to be 4 to 12 percent higher than that of *ice*. This difference is less than that found previously for protons (10 to 14 percent) in the same velocity interval. Thwaites (1981) reported that for α-particles down to 1.8 MeV, the stopping cross section of H_2O vapor was found to be ~4 percent higher than that of *liquid water*. In the same velocity interval, our calculated results show that the proton stopping cross section of water vapor is about 5.6 percent higher than that of liquid water. The work of De Carvalho and Yagoda (1952) and Ellis, Rossi, and Failla (1955) found that the stopping cross section of 5- to 8-MeV α-particles in H_2O was the same in the vapor and condensed states. Wenzel and Whaling (1952) and Reynolds et al. (1953) found a greater stopping cross section for protons in the vapor states of H_2O than in the solid state, and Palmer (1966) observed the stopping cross section for α-particles to be less in the liquid than in the vapor state of H_2O. This physical-state effect is observed by most experimental physicists. The results of De Carvalho and Yagoda (1952) and Ellis, Rossi, and Failla (1955) may be explained since, in the high-energy regime, this effect becomes less important. This tendency is also exhibited in our calculated results. Palmer (1966) tried to explain this effect as a low-energy polarization screen effect. Matteson, Powers, and Chau (1977) explained it as an aggregation effect.

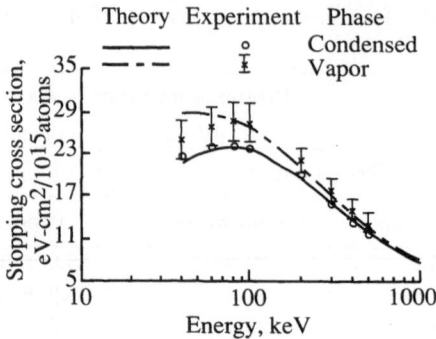

Figure 2.14. Proton stopping cross section of water molecules in vapor and condensed phases.

We now consider the effect according to the local plasma model. We will first explain the molecular effect or deviation from Bragg's rule. It is found that in the low-energy regime, the molecular stopping power may be lower than the atomic stopping power calculated by employing Bragg's rule. This is due to the fact that when an electron in a molecular gas is more bonded than in an atomic gas caused by the chemical bond, the momentum transfer between projectile and the electron becomes harder, causing decreased energy loss or stopping power of the projectile. In the formula

$$L = \ln Y - \frac{3\sqrt{3}}{10\chi Z_T}\frac{1}{Y} = \ln\left(\frac{2mv^2}{\gamma\hbar\omega_\rho}\right) - \frac{1}{Z_T}\frac{\langle T \rangle}{mv^2} \quad (Z_T \leq 2) \qquad (2.140)$$

where $\omega_\rho \approx \sqrt{\rho}$ and the average kinetic energy of electron $\langle T \rangle = \frac{3}{10}mv_F^2\rho^{2/3}$ are functions of electronic density. The chemical bond causes an increased electron density in the interatomic space and, hence, increased mean excitation energy and average kinetic energy. Physically, when an electron is more bonded, it is harder to excite, resulting in increased mean excitation energy. When an electron is more bonded, the magnitude of the total energy is increased (in bonded states), since $\langle T \rangle = |\langle E \rangle|$, the average kinetic energy is increased. Both effects tend to decrease the stopping power. In the low-energy regime, this effect becomes more important, mainly due to the contribution of the $\langle T \rangle$ term. In the high-energy regime, this term vanishes. The mean excitation term also becomes less important due to the influence of the factor $1/v^2$. Meanwhile, in liquid water, many of these hydrogen bonds between molecules are broken; also, compared with the partial ionic bond, the hydrogen bond is weaker. With this picture of molecular bond effects, how are we to understand the considerable difference of stopping cross section due to the physical state?

Matteson, Powers, and Chau (1977) correctly pointed out that the physical-state effects are due to the effects of aggregation upon the molecules. As mentioned previously, in liquid water these hydrogen bonds are not the main reason for this physical effect. Rather, it is a collective effect of the whole liquid-water molecule, since liquid molecules are more mobile than those in ice. Because of the interactions with all its neighbors each molecule is confined in an effective volume. This can be seen from figure 2.12, where we show $W(r)$ as the average potential of each molecule due to the interactions of all neighbors, according to the well-known Lennard-Jones and Devonshire theory (1937 and 1938). Thus, as a part of the molecule, electrons are more difficult to remove from this volume or more difficult to be excited. Simply said, the electrons are more bound because of the neighborhood interactions. This confinement of the electron causes less momentum transfer between the projectile and the electrons and decreases the stopping power. Our local plasma model describes this picture because the electronic density is more concentrated so that both the mean excitation energy and average kinetic energy terms are increased and cause a decrease in the stopping power. Moreover, we notice that the structure of ice is more open than that of liquid water. It is reasonable to expect a slight difference of stopping cross section between ice and liquid water.

2.6. Semiempirical Methods

In passing through an ordinary material, an ion loses the larger fraction of its energy to electronic excitation of the material. Although a satisfactory theory of high-energy ion-electron interaction is available in the form of Bethe's theory utilizing the Born approximation, an equally satisfactory theory for low energies is not available. Bethe's high-energy approximation to the energy loss per unit path (that is, stopping power) is given as

$$S_e = \frac{4\pi N Z_P^2 Z_T e^4}{mv^2} \left\{ \ln\left[\frac{2mv^2}{(1-\beta^2)I_T}\right] - \beta^2 - \frac{C}{Z_T} \right\} \qquad (2.141)$$

where Z_P is the projectile charge, N is the number of target molecules per unit volume, Z_T is the number of electrons per target molecule, m is the electron mass, v is the projectile velocity, $\beta = v/c$, c is the velocity of light, C is the velocity-dependent shell correction term (Walske and Bethe, 1951), and I_T is the mean excitation energy given by

$$Z_T \ln(I_T) = \sum_n f_n \ln(E_n) \qquad (2.142)$$

where f_n represents the electric dipole oscillator strengths of the target and E_n represents the corresponding excitation energies. Note that the sum in equation (2.142) includes discrete and continuum levels. Empirically it has been observed that molecular stopping power is reasonably approximated by the sum of the corresponding empirically derived atomic stopping powers for which equations (2.141) and (2.142) imply that

$$Z_T \ln(I_T) = \sum_j n_j Z_j \ln(I_j) \qquad (2.143)$$

where Z_T and I_T pertain to the molecule, Z_j and I_j are the corresponding atomic values, and n_j is the stoichiometric coefficient. This additive rule (eq. (2.143)), usually called Bragg's rule (Bragg and Kleeman, 1905), is the basis for providing stopping cross sections for arbitrary material compositions.

Sources of deviations from Bragg's additive rule for molecules and for the condensed phase are discussed. Aside from shifts in excitation energies and adjustments in line strengths as a result of molecular bonding, new terms in the stopping power appear due to coupling between vibrational and rotational modes. Additionally, in the condensed phase, some discrete transitions are moved into the continuum, and collective modes among valence electrons in adjacent atoms produce new terms in the absorption spectrum that needs to be dealt with. Platzman (1952a and 1952b) proposed that the experimentally observed additive rule may not show that molecular stopping power is the sum of atomic processes but rather demonstrates that molecular bond shifts for covalent-bonded molecules are relatively independent of the molecular combination as was theoretically demonstrated in section 2.3.4. On the basis of such arguments, Platzman suggested that ionic-bonded substances should be studied as a rigid test of the additive rule because of the radical difference in bonding type. He further estimated that ionic-bond shifts could change the stopping power by as much as 50 percent. Recent results on molecular bond shifts and condensed phase effects on mean excitation energies were discussed in section 2.5 (Wilson and Kamaratos, 1981; Wilson and Xu, 1982; Wilson et al., 1982).

The electron stopping power for protons is adequately described by equation (2.141) for energies above 500 keV for which the shell or "tight binding" correction C makes an important contribution below 10 MeV (Andersen and Ziegler, 1977; Janni, 1982a and 1982b). For proton energies below 500 keV, charge exchange (electron transfer) reactions alter the proton charge over much of its path; therefore, equation (2.141) is to be understood in terms of an average over the proton charge states. Normally an average over the charge states is introduced into equation (2.141) so that the effective charge is the root-mean-square ion charge and not the average ion charge. At any ion energy, charge equilibrium is established very quickly in all materials. Utilizing the effective charge in equation (2.141) appears to make only modest improvement below 500 keV, presumably an indication of the failure of this theory based on an empirical basis (Andersen et al., 1977; Janni, 1966, 1982a, and 1982b). The resultant stopping power for protons in water is shown along with the evaluated data of Bichsel (1963) in figure 2.15.

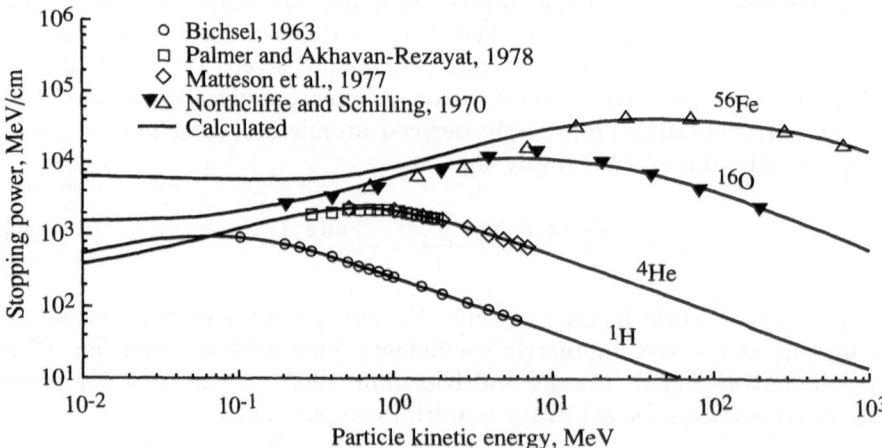

Figure 2.15. Calculated and experimental stopping powers in water for typical cosmic-ray ions as function of particle kinetic energy.

The electronic stopping power for α-particles requires terms in equation (2.141) of higher order in the projectile charge Z_P resulting from corrections to the Born approximation. The alpha and proton stopping powers cannot be related through their effective charges. Parametric fits to experimental data are given by Ziegler (1977) for all elements in both the gaseous and condensed phases.

The electronic stopping powers for heavier ions are related to the alpha stopping power through their corresponding effective charges. The effective charge suggested by Barkas (1963) is used:

$$Z^* = Z_P \left[1 - \exp \left(\frac{-125\beta}{Z_P^{2/3}} \right) \right] \tag{2.144}$$

where Z_P is the atomic number of the ion.

At sufficiently low energies, the energy lost by an ion in a nuclear collision becomes important. The nuclear stopping theory used in this report is a modification of the theory of Lindhard, Scharff, and Schiott (1963). The reduced energy is given as

$$\epsilon = \frac{32.53 A_P A_T E}{Z_P Z_T (A_P + A_T) \left(Z_P^{2/3} + Z_T^{2/3} \right)^{1/2}} \tag{2.145}$$

where E is in units of keV/nucleon and A_P and A_T are the atomic masses of the projectile and target. The nuclear stopping power in reduced units (Ziegler, 1977) is

$$S_n = \left\{ \begin{array}{ll} 1.59\epsilon^{1/2} & (\epsilon < 0.01) \\[2mm] \dfrac{1.7\epsilon^{1/2} \ln[\epsilon \, + \, \exp(1)]}{1 \, + \, 6.8\epsilon \, + \, 3.4\epsilon^{3/2}} & (0.01 < \epsilon < 10) \\[2mm] \dfrac{\ln(0.47\epsilon)}{2\epsilon} & (10 < \epsilon) \end{array} \right\} \tag{2.146}$$

and the conversion factor to units of eV-cm^2/10^{15} atoms is

$$f = \frac{8.426 Z_P A_T A_P}{(A_P + A_T) \left(Z_P^{2/3} + Z_T^{2/3} \right)^{1/2}} \tag{2.147}$$

The total stopping power S_j is obtained by summing the electronic and nuclear contributions. Other processes of energy transfer such as Bremsstrahlung and pair production are unimportant.

For energies above a few MeV per nucleon, Bethe's equation is adequate provided that appropriate corrections to Bragg's rule (Wilson et al., 1984a and 1984b), shell corrections (Janni, 1982a and 1982b), and an effective charge are included. Electronic stopping power for protons is calculated from the parametric formulas of Andersen and Ziegler (1977). The calculated stopping power for protons above a few MeV in water is shown in figure 2.15 along with data given by Bichsel (1963).

Because alpha stopping power is not derivable from the proton stopping power formula using the effective charge at low energy, the parametric fits to empirical alpha stopping powers given by Ziegler (1977) are used. Applying his results for condensed phase, water poorly represented the data of Matteson, Powers, and Chau (1977) and Palmer and Akhavan-Rezayat (1978). Considering that the physical-state and molecular binding effects are most important for hydrogen (Wilson and Kamaratos, 1981), the water stopping power was approximated by using the condensed phase parameters for hydrogen and the gas phase parameters for oxygen (which are known experimentally). These results are presented along with experimental data for condensed phase water in figure 2.15. It appears that Ziegler overestimated the condensed phase effects for oxygen since the gas phase oxygen data give satisfactory results as seen in figure 2.15.

Electronic stopping powers for ions with a charge greater than 2 are related to the alpha stopping power through the effective charge given by equation (2.144). For water, the condensed phase formula of Ziegler for α-particles probably gives the best stopping powers for heavier ions. Calculated results for ^{16}O and ^{56}Fe ions in water, shown in figure 2.15 along with the Northcliffe and Schilling (1970) results for ^{56}Fe ions, are especially important, since their data seem to agree with the range measured in General Electric Lexan plastic by J. H. Chan (Fleischer, Price, and Walker, 1975).

2.7. References

Ahlen, Steven P., 1980: Theoretical and Experimental Aspects of the Energy Loss of Relativistic Heavily Ionizing Particles. *Reviews Modern Phys.*, vol. 52, no. 1, pp. 121–173.

Andersen, H. H.; Bak, J. F.; Knudsen, H.; and Nielsen, B. R., 1977: Stopping Power of Al, Cu, Ag, and Au for MeV Hydrogen, Helium, and Lithium Ions Z_1^3 and Z_1^4 Proportional Deviations From the Bethe Formula. *Phys. Review A*, vol. 16, no. 5, pp. 1929–1940.

Andersen, H. H.; and Ziegler, J. F., 1977: *Hydrogen—Stopping Powers and Ranges in All Elements.* Volume 3 of *The Stopping and Ranges of Ions in Matter*, J. F. Ziegler, organizer, Pergamon Press, Inc.

Bakker, C. J.; and Segrè, E., 1951: Stopping Power and Energy Loss for Ion Pair Production for 340-Mev Protons. *Phys. Review*, vol. 81, second ser., no. 4, pp. 489–492.

Barkas, Walter H., 1963: *Nuclear Research Emulsion—I. Techniques and Theory.* Academic Press, Inc.

Bernal, J. D.; and Fowler, R. H., 1933: A Theory of Water and Ionic Solution, With Particular Reference to Hydrogen and Hydroxyl Ions. *J. Chem. Phys.*, vol. 1, no. 8, pp. 515–548.

Bethe, H., 1930: Zur Theorie des Durchgangs schneller Korpuskularstrahlen durch Materie. *Ann. Phys.*, 5 Folge, Bd. 5, pp. 325–400.

Bethe, H., 1933: Quantenmechanik der Ein- und Zwei-Elektronenprobleme. *Handbook of Physics*, Bd. XXIV, Kap. 3, Julius Springer (Berlin), pp. 273–560.

Bethe, H. A., 1950: The Range-Energy Relation for Slow Alpha-Particles and Protons in Air. *Reviews Modern Phys.*, vol. 22, no. 2, pp. 213–219.

Bethe, H. A.; Brown, L. M.; and Walske, M. C., 1950: Stopping Power of K-Electrons. *Phys. Review*, vol. 79, second ser., no. 2, p. 413.

Bethe, Hans A.; and Ashkin, Julius, 1953: Passage of Radiations Through Matter. *Experimental Nuclear Physics*, Volume I, E. Segrè, ed., John Wiley & Sons, Inc., pp. 166–357.

Bethe, Hans A.; and Salpeter, Edwin E., 1957: *Quantum Mechanics of One- and Two-Electron Atoms.* Academic Press.

Bichsel, Hans, 1963: Passage of Charged Particles Through Matter. *American Institute of Physics Handbook*, Second ed., Dwight E. Gray, ed., McGraw-Hill Book Co., Inc., pp. 8-20–8-47.

Bogaardt, M.; and Koudijs, B., 1952: On the Average Excitation Potentials and the Range-Energy Relations in Light Elements. *Physica*, vol. 18, pp. 249–264.

Bohm, David; and Pines, David, 1951: A Collective Description of Electron Interactions. I. Magnetic Interactions. *Phys. Review*, vol. 82, second ser., no. 5, pp. 625–634.

Bohm, David; and Pines, David, 1953: A Collective Description of Electron Interactions. III. Coulomb Interactions in a Degenerate Electron Gas. *Phys. Review*, vol. 92, second ser., no. 3, pp. 609–625.

Bohr, N., 1915: On the Decrease of Velocity of Swiftly Moving Electrified Particles in Passing Through Matter. *Philos. Mag. & J. Sci.*, ser. 6, vol. 30, no. 175, pp. 581–612.

Bonderup, E., 1967: Stopping of Swift Protons Evaluated From Statistical Atomic Model. *Mat.-Fys. Medd. —K. Dan. Vidensk. Selsk.*, vol. 35, no. 17.

Bourland, P. D.; and Powers, D., 1971: Bragg-Rule Applicability to Stopping Cross Sections of Gases for α Particles of Energy 0.3–2.0 MeV. *Phys. Review B*, vol. 3, third ser., no. 11, pp. 3635–3641.

Bragg, W. H.; and Kleeman, R., 1905: On the α Particles of Radium, and Their Loss of Range in Passing Through Various Atoms and Molecules. *Philos. Mag. & J. Sci.*, ser. 6, vol. 10, no. 57, pp. 318–340.

Breitenberger, E., 1959: Theory of Multiple Scattering. *Proc. Royal Soc. (London)*, ser. A, vol. 250, no. 1263, pp. 514–523.

Brown, L. M., 1950: Asymptotic Expression for the Stopping Power of K-Electrons. *Phys. Review*, vol. 79, second ser., no. 2, pp. 297 303.

Brown, P. J., 1972: A Study of Charge Density in Beryllium. *Philos. Mag.*, vol. 26, no. 6, pp. 1377 1394.

Chan, Y. M.; and Dalgarno, A., 1965: The Stopping Powers of Atoms and Molecules. *Proc. Phys. Soc. (London)*, vol. 85, pp. 457–459.

Chu, W. K.; and Powers, D., 1969: Alpha-Particle Stopping Cross Section in Solids From 400 KeV to 2 MeV. *Phys. Review*, vol. 187, second ser., no. 2, pp. 478–490.

Chu, W. K.; and Powers, D., 1972: Calculations of Mean Excitaton Energy for All Elements. *Phys. Lett.*, vol. 40A, no. 1, pp. 23–24.

Chu, W. K.; Moruzzi, V. L.; and Ziegler, J. F., 1975: Calculations of the Energy Loss of ^{4}He Ions in Solid Elements. *J. Appl. Phys.*, vol. 46, no. 7, pp. 2817–2820.

Clementi, E.; and Raimondi, D. L., 1963: Atomic Screening Constants From SCF Functions. *J. Chem. Phys.*, vol. 38, no. 11, pp. 2686–2689.

Clementi, Enrico; and Roetti, Carla, 1974: Roothaan-Hartree-Fock Atomic Wavefunctions. *At. Data & Nuclear Data Tables*, vol. 14, nos. 3–4, pp. 177–478.

Dalgarno, A., 1960: The Stopping Powers of Atoms. *Proc. Phys. Soc. (London)*, vol. 76, pt. 3, no. 489, pp. 422–424.

Dalgarno, A., 1963: Sum Rules and Atomic Structure. *Reviews Modern Phys.*, vol. 35, no. 3, pp. 522–524.

Dalgarno, A.; and Williams, D. A., 1965: Properties of the Hydrogen Molecule. *Proc. Phys. Soc. (London)*, vol. 85, pt. 4, no. 546, pp. 685–689.

Dalgarno, A.; Degges, T.; and Williams, D. A., 1967: Dipole Properties of Molecular Nitrogen. *Proc. Phys. Soc. (London)*, vol. 92, pt. 2, no. 576, pp. 291–295.

De Carvalho, Hervasio G.; and Yagoda, Herman, 1952: The Range of Alpha-Particles in Water. *Phys. Review*, vol. 88, no. 2, pp. 273–278.

Dehmer, J. L.; Inokuti, Mitio; and Saxon, R. P., 1975: Systematics of Moments of Dipole Oscillator-Strength Distributions for Atoms of the First and Second Row. *Phys. Review A*, vol. 12, third ser., no. 1, pp. 102–121.

Ellis, R. Hobart, Jr.; Rossi, H. H.; and Failla, G., 1955: Stopping Power of Water Films. *Phys. Review*, vol. 97, no. 4, pp. 1043–1047.

Fano, U., 1963: Penetration of Protons, Alpha Particles, and Mesons. *Annual Review of Nuclear Science*, Volume 13, Emilio Segrè, ed., Annual Reviews, Inc., pp. 1–66.

Fleischer, Robert L.; Price, P. Buford; and Walker, Robert M., 1975: *Nuclear Tracks in Solids— Principles and Applications.* Univ. of California Press.

Fröhlich, H.; and Pelzer, H., 1955: Plasma Oscillations and Energy Loss of Charged Particles in Solids. *Proc. Phys. Soc. (London)*, vol. 68, pt. 6, no. 426A, pp. 525–529.

Gordon, Roy G.; and Kim, Yung Sik, 1972: Theory for the Forces Between Closed-Shell Atoms and Molecules. *J. Chem. Phys.*, vol. 56, no. 6, pp. 3122–3133.

Hubbard, J., 1955: The Dielectric Theory of Electronic Interactions in Solids. *Proc. Phys. Soc. (London)*, vol. 68, pt. 11, no. 431A, pp. 976–986.

Inokuti, Mitio; Baer, T.; and Dehmer, J. L., 1978: Addendum: Systematics of Moments of Dipole Oscillator-Strength Distributions for Atoms in the First and Second Row. *Phys. Review A*, vol. 17, no. 3, pp. 1229–1231.

Inokuti, Mitio; and Turner, James E., 1978: Mean Excitation Energies for Stopping Power as Derived From Oscillator-Strength Distributions. *Sixth Symposium on Microdosimetry*, Volume I, J. Booz and H. G. Ebert, eds., Harwood Academic Publ., Ltd. (London), pp. 675–687.

Inokuti, M.; Dehmer, J. L.; Baer, T.; and Hanson, J. D., 1981: Oscillator-Strength Moments, Stopping Powers, and Total Inelastic-Scattering Cross Sections of All Atoms Through Strontium. *Phys. Review A*, vol. 23, no. 1, pp. 95–109.

Janni, Joseph F.; 1966: *Calculations of Energy Loss, Range, Pathlength, Straggling, Multiple Scattering, and the Probability of Inelastic Nuclear Collisions for 0.1- to 1000-Mev Protons.* AFWL-TR-65-150, U.S. Air Force. (Available from DTIC as AD 643 837.)

Janni, Joseph F., 1982a: Proton Range-Energy Tables 1 keV–10 GeV—Energy Loss, Range, Path Length, Time-of-Flight, Straggling, Multiple Scattering, and Nuclear Interaction Probability. Part 1. For 63 Compounds. *At. Data & Nuclear Data Tables,* vol. 27, nos. 2/3, pp. 147–339.

Janni, Joseph F., 1982b: Proton Range-Energy Tables, 1 keV–10 GeV—Energy Loss, Range, Path Length, Time-of-Flight, Straggling, Multiple Scattering, and Nuclear Interaction Probability. Part 2. For Elements $1 \leq Z \leq 92$. *At. Data & Nuclear Data Tables,* vol. 27, nos. 4/5, pp. 341–529.

Johnson, R. E.; and Inokuti, Mitio, 1983: The Local-Plasma Approximation to the Oscillator-Strength Spectrum: How Good Is it and Why? *Comments At. Mol. Phys.*, vol. 14, nos. 1 & 2, pp. 19–31.

Kahn, David, 1953: The Energy Loss of Protons in Metallic Foils and Mica. *Phys. Review,* vol. 90, second ser., no. 4, pp. 503–509.

Kamaratos, E.; Chang, C. K.; Wilson, J. W.; and Xu, Y. J., 1982: Valence Bond Effects on Mean Excitation Energies for Stopping Power in Metals. *Phys. Lett.*, vol. 92A, no. 7, pp. 363–365.

Kamikawai, Ryotaro; Watanabe, Tsutomu; and Amemiya, Ayao, 1969: Mean Excitation Energy for Stopping Power of Molecular Hydrogen. *Phys. Review,* vol. 184, second ser., no. 2, pp. 303–311.

Khan, F.; Khandelwal, G. S.; and Wilson, J. W., 1988a: Differential Oscillator Strengths and Dipole Polarizabilities for Transitions of the Helium Sequence. *Phys. Review A*, vol. 38, third ser., no. 12, pp. 6159–6164.

Khan, F.; Khandelwal, G. S.; and Wilson, J. W., 1988b: $1s^2$ ^{1}S-1s np ^{1}P Transitions of the Helium Isoelectronic Sequence Members up to $Z = 30$. *Astrophys. J.*, vol. 329, no. 1, pt. 1, pp. 493–497.

Khan, F.; Khandelwal, G. S.; and Wilson, J. W., 1990: Moments of Dipole Oscillator-Strength Distribution for the Helium Sequence. *J. Phys. B.: At., Mol. & Opt. Phys.*, vol. 23, no. 16, pp. 2717–2726.

Khandelwal, Govind S., 1968: Shell Corrections for K- and L-Electrons. *Nuclear Phys.*, vol. A116, no. 1, pp. 97–111.

Khandelwal, Govind S., 1982: Stopping Power of K and L Electrons. *Phys. Review A*, vol. 26, third ser., no. 5, pp. 2983–2986.

Khandelwal, G. S.; Khan, F.; and Wilson, J. W., 1989: An Asymptotic Expression for the Dipole Oscillator Strength for Transition of the He Sequence. *Astrophys. J.*, vol. 336, no. 1, pt. 1, pp. 504–506.

Khandelwal, G. S.; Pritchard, W. M.; Grubb, G.; and Khan, F., 1989: Screened Hydrogenic Radial Integrals. *Phys. Review A*, vol. 39, third ser., no. 8, pp. 3960–3963.

Kramers, H. A., 1947: The Stopping Power of a Metal for α-Particles. *Physica*, vol. 13, pp. 401–412.

Landau, L. D.; and Lifshitz, E. M. (J. B. Sykes and J. S. Bell, transl.), 1958: *Quantum Mechanics—Non-Relativistic Theory*. Addison-Wesley Publ. Co., Inc.

Landau, L. D.; and Lifshitz, E. M., 1960: *Electrodynamics of Continuous Media*. Addison-Wesley Publ. Co., Inc.

Langley, Robert A., 1975: Stopping Cross Sections for Helium and Hydrogen in H_2, N_2, O_2, and H_2S (0.3–2.5 MeV). *Phys. Review B,* vol. 12, no. 9, pp. 3575–3583.

Leminen, E.; Fontell, A.; and Bister, M., 1968: Stopping Power of Al, Zn, and In for 0.6–2.4 MeV Protons. *Ann. Acad. Sci. Fenn.*, ser. A, no. 281, pp. 4–12.

Lennard-Jones, J. E.; and Devonshire, A. F., 1937: Critical Phenomena in Gases—I. *Proc. Royal Soc. (London),* ser. A, vol. CLXIII, pp. 53–70.

Lennard-Jones, J. E.; and Devonshire, A. F., 1938: Critical Phenomena in Gases—II. Vapour Pressures and Boiling Points. *Proc. Royal Soc. (London),* ser. A, vol. CLXV, pp. 1–11.

Lindhard, J., 1954: On the Properties of a Gas of Charged Particles. *Mat.-Fys. Medd.—K. Dan. Vidensk. Selsk.,* vol. 28, no. 8, pp. 1–58.

Lindhard, Jens; and Scharff, Morten, 1960: Recent Developments in the Theory of Stopping Power. 1. Principles of the Statistical Method. *Penetration of Charged Particles in Matter,* Edwin Albrecht Uehling, ed., Nuclear Sci. Ser. Rep. No. 29 (Publ. 752), National Academy of Sciences, National Research Council, pp. 49–55.

Lindhard, J.; Scharff, M.; and Schiott, H. E., 1963: Range Concepts and Heavy Ion Ranges (Notes on Atomic Collisions, II). *Mat.-Fys. Medd.—K. Dan. Vidensk. Selsk.,* vol. 33, no. 14, pp. 1–42.

Lindhard, Jens; and Winther, Aage, 1964: Stopping Power of Electron Gas and Equipartition Rule. *Mat.-Fys. Medd.—K. Dan. Vidensk. Selsk.,* vol. 34, no. 4, pp. 3–22.

Livingston, M. Stanley; and Bethe, H. A., 1937: Nuclear Physics—C. Nuclear Dynamics, Experimental. *Reviews Modern Phys.*, vol. 9, no. 3, pp. 245–390.

Lodhi, A. S.; and Powers, D., 1974: Energy Loss of α Particles in Gaseous C-H and C-H-F Compounds. *Phys. Review A*, vol. 10, no. 6, pp. 2131–2140.

Martynenko, Yu. V., 1970: Radiation Damage of Crystals Irradiated by Atomic Particles. *Sov. Phys.—Solid State*, vol. 11, no. 7, pp. 1582–1585.

Mather, R.; and Segrè, E., 1951: Range Energy Relation for 340-Mev Protons. *Phys. Review*, vol. 84, second ser., no. 2, pp. 191–193.

Matteson, S.; Powers, D.; and Chau, E. K. L., 1977: Physical-State Effect in the Stopping Cross Section of H_2O Ice and Vapor for 0.3 to 2.0 MeV α Particles. *Phys. Review A*, vol. 15, ser. A, no. 3, pp. 856–864.

Merzbacher, E.; and Lewis, H. W., 1958: X-ray Production by Heavy Charged Particles. *Encyclopedia of Physics*, Volume XXXIV, S. Flugge, ed., Springer-Verlag, pp. 166–192.

Moruzzi, V. L.; Janak, J. F.; and Williams, A. R., 1978: *Calculated Electronic Properties of Metals.* Pergamon Press, Inc.

Nakata, Hiroshi, 1971: Analysis of Energy-Loss Data for 0.2–0.5 MeV/amu p, α, and N in Se. *Phys. Review B*, vol. 3, no. 9, pp. 2847–2851.

Neilsen, L. P., 1961: Energy Loss and Straggling of Protons and Deuterons. *Mat.-Fys. Medd.—K. Dan. Vidensk. Selsk.*, vol. 33, no. 6.

Neuwirth, Wolfgang; Pietsch, Werner; and Kreutz, Ronald, 1978: Chemical Influences on the Stopping Power. *Nuclear Instrum. & Methods*, vol. 149, nos. 1–3, pp. 105–113.

Northcliffe, L. C.; and Schilling, R. F., 1970: Range and Stopping-Power Tables for Heavy Ions. *Nuclear Data*, Sect. A, vol. 7, no. 3–4, pp. 233–463.

Palmer, Rita B. J., 1966: The Stopping Power of Hydrogen and Hydrocarbon Vapours for Alpha Particles Over the Energy Range 1 to 8 MeV. *Proc. Phys. Soc.*, vol. 87, pp. 681–688.

Palmer, Rita B. J.; and Akhavan-Rezayat, Ahmad, 1978: Range-Energy Relations and Stopping Power of Water, Water Vapour and Tissue Equivalent Liquid for α Particles Over the Energy Range 0.5 to 8 MeV. *Sixth Symposium on Microdosimetry*, Volume II, J. Booz and H. G. Ebert, eds., Harwood Academic Publ., Ltd. (London), pp. 739–748.

Pauling, Linus, 1960: *The Nature of the Chemical Bond*, Third ed. Cornell Univ. Press.

Pauling, Linus, 1967: *The Chemical Bond.* Cornell Univ. Press.

Peterson, S. W.; and Levy, Henri A., 1957: A Single-Crystal Neutron Diffraction Study of Heavy Ice. *Acta Crystallogr.*, vol. 10, pp. 70–76.

Pines, David; and Bohm, David, 1952: A Collective Description of Electron Interactions: II. Collective vs Individual Particle Aspects of the Interactions. *Phys. Review*, vol. 85, second ser., no. 2, pp. 338–353.

Pines, David, 1953: A Collective Description of Electron Interactions: IV. Electron Interaction in Metals. *Phys. Review*, vol. 92, second ser., no. 3, pp. 626–636.

Platzman, Robert L., 1952a: Influences on Details of Electronic Binding on Penetration Phenomena, and the Penetration of Energetic Charged Particles Through Liquid Water. *Symposium on Radiobiology—The Basic Aspects of Radiation Effects on Living Systems,* James J. Nickson, ed., John Wiley & Sons, Inc., pp. 139–176.

Platzman, Robert L., 1952b: On the Primary Processes in Radiation Chemistry and Biology. *Symposium on Radiobiology—The Basic Aspects of Radiation Effects on Living Systems,* James J. Nickson, ed., John Wiley & Sons, Inc., pp. 97–116.

Powers, D.; Chu, W. K.; Robinson, R. J.; and Lodhi, A. S., 1972: Measurement of Molecular Stopping Cross Sections of Halogen-Carbon Compounds and Calculation of Atomic Stopping Cross Sections of Halogens. *Phys. Review A,* vol. 6, no. 4, pp. 1425–1435.

Reynolds, H. K.; Dunbar, D. N. F.; Wenzel, W. A.; and Whaling, W., 1953: The Stopping Cross Section of Gases for Protons, 30–600 kev. *Phys. Review,* vol. 92, no. 3, pp. 742–748.

Rousseau, C. C.; Chu, W. K.; and Powers, D., 1971: Calculations of Stopping Cross Sections for 0.8- to 2.0-MeV Alpha Particles. *Phys. Review A,* vol. 4, third ser., no. 3, pp. 1066–1070.

Sachs, Donald C.; and Richardson, J. Reginald, 1953: Mean Excitation Potentials. *Phys. Review,* vol. 89, second ser., no. 6, pp. 1163–1164.

Seltzer, S. M.; and Berger, M. J., 1982: Evaluation of the Collision Stopping Power of Elements and Compounds for Electrons and Positrons. *Int. J. Appl. Radiat. & Isot.,* vol. 33, no. 11, pp. 1189–1218.

Shiles, E.; Sasaki, Taizo; Inokuti, Mitio; and Smith, D. Y., 1980: Self-Consistency and Sum-Rule Tests in the Kramers-Kronig Analysis of Optical Data: Applications to Aluminum. *Phys. Review B,* vol. 22, no. 4, pp. 1612–1628.

Simmons, D. H., 1952: The Range-Energy Relation for Protons in Aluminum. *Proc. Phys. Soc. (London),* vol. 65, pt. A, pp. 454–456.

Sorensen, H.; and Andersen, H. H., 1973: Stopping Power of Al, Cu, Ag, Au, Pb, and U for 5–18-MeV Protons and Deuterons. *Phys. Review B,* vol. 8, no. 5, pp. 1854–1863.

Suzuki, Reiko; Nakamura, Hiroki; and Ishiguro, Eiichi, 1984: Semiclassical Scattering Theory Based on the Dynamical-State Representation: Application to the $Li^+ + NA$ and $LI + Na^+$ Collisions. *Phys. Review A,* vol. 29, third ser., no. 6, pp. 3060–3070.

Thompson, Theos Jardin, 1952: Effect of Chemical Structure on Stopping Powers for High-Energy Protons. UCRL-1910, Univ. of California.

Thwaites, D. I., 1981: Stopping Cross-Sections of Liquid Water and Water Vapour for Alpha Particles Within the Energy Region 0.3 to 5.5 MeV. *Phys. Med. & Biol.,* vol. 26, no. 1, pp. 71–80.

Tossell, J. A., 1979: Calculation of Bond Distances and Cohesive Energies for Gaseous Halides Using the Modified Electron Gas Ionic Model. *Chem. Phys. Lett.,* vol. 67, no. 2–3, pp. 359–364.

Vasilevskii, I. M.; and Prokoshkin, Yu. D., 1960: *Ionizatsionnye Potentsialy Atomov (Ionization Potentials of Atoms).* Dubna:Ob'edinen. Inst. of Yadernykh Issledovanii.

Waldman, Marvin; and Gordon, Roy G., 1979: Scaled Electron Gas Approximation for Intermolecular Forces. *J. Chem. Phys.*, vol. 71, no. 3, pp. 1325–1339.

Walecka, J. D., 1976: Collective Excitations in Atoms. *Phys. Lett.*, vol. 58A, no. 2, pp. 83–86.

Walske, M. C.; and Bethe, H. A., 1951: Asymptotic Formula for Stopping Power of K-Electrons. *Phys. Review*, vol. 83, pp. 457–458.

Walske, M. C., 1952: The Stopping Power of K-Electrons. *Phys. Review*, vol. 88, second ser., no. 6, pp. 1283–1289.

Walske, M. C., 1956: Stopping Power of L-Electrons. *Phys. Review*, vol. 101, second ser., no. 3, pp. 940–944.

Wenzel, W. A.; and Whaling, Ward, 1952: The Stopping Cross Section of D_2O Ice. *Phys. Review,* vol. 87, no. 3, pp. 499–503.

Wigner, E. P.; and Seitz, F., 1934: On the Constitution of Metallic Sodium. II. *Phys. Review*, vol. 46, second ser., no. 6, pp. 509–524.

Wilson, J. W.; and Kamaratos, E., 1981: Mean Excitation Energy for Molecules of Hydrogen and Carbon. *Phys. Lett.*, vol. 85A, no. 1, pp. 27–29.

Wilson, J. W.; Chang, C. K.; Xu, Y. J.; and Kamaratos, E., 1982: Ionic Bond Effects on the Mean Excitation Energy for Stopping Power. *J. Appl. Phys.*, vol. 53, no. 2, pp. 828–830.

Wilson, J. W.; and Xu, Y. J., 1982: Metallic Bond Effects on Mean Excitation Energies for Stopping Powers. *Phys. Lett.*, vol. 90A, no. 5, pp. 253–255.

Wilson, J. W.; Xu, Y. J.; Chang, C. K.; and Kamaratos, E., 1984a: Atomic Mean Excitation Energies for Stopping Powers From Local Plasma Oscillator Strengths. *J. Appl. Phys.*, vol. 56, no. 3, pp. 860–861.

Wilson, J. W.; Xu, Y. J.; Kamaratos, E.; and Chang, C. K., 1984b: Mean Excitation Energies for Stopping Powers in Various Materials Composed of Elements Hydrogen Through Argon. *Canadian J. Phys.*, vol. 62, no. 7, pp. 646–660.

Xu, Y. J.; Khandelwal, G. S.; and Wilson, J. W., 1984a: Intermediate Energy Proton Stopping Power for Hydrogen Molecules and Monoatomic Helium Gas. *Phys. Lett.*, vol. 100A, no. 3, pp. 137–140.

Xu, Y. J.; Khandelwal, G. S.; and Wilson, J. W., 1984b: Low-Energy Proton Stopping Power of N_2, O_2, and Water Vapor, and Deviations From Bragg's Rule. *Phys. Review A,* vol. 29, third ser., no. 6, pp. 3419–3422.

Xu, Y. J.; Khandelwal, G. S.; and Wilson, J. W., 1985: Proton Stopping Cross Sections of Liquid Water. *Phys. Review A,* vol. 32, third ser., no. 1, pp. 629–632.

Xu, Y. J.; Khandelwal, G. S.; and Wilson, J. W., 1986: The Maximum Momentum Transfer in Proton-Hydrogen Collisions. *Phys. Lett. A,* vol. 115, no. 1, 2, pp. 37–38.

Xu, Y. J.; Khandelwal, G. S.; and Wilson, J. W., 1989: *Charge Exchange Transition Probability for Collisions Between Unlike Ions and Atoms Within the Adiabatic Approximation.* NASA TM-101559.

Zeiss, G. D.; and Meath, W. J., 1975: The H_2O-H_2O Dispersion Energy Constant and the Dispersion of the Specific Refractivity of Dilute Water Vapour. *Mol. Phys.*, vol. 30, no. 1, pp. 161–169.

Zeiss, G. D.; Meath, William J.; MacDonald, J. C. F.; and Dawson, D. J., 1977: Accurate Evaluation of Stopping and Straggling Mean Excitation Energies for N, O, H_2, N_2, O_2, NO, NH_3, H_2O, and N_2O Using Dipole Oscillator Strength Distributions—A Test of the Validity of Bragg's Rule. *Radiat. Res.*, vol. 70, no. 2, pp. 284–303.

Zener, Clarence, 1932: Non-Adiabatic Crossing of Energy Levels. *Proc. Royal Soc. (London)*, ser. A, vol. CXXXVII, pp. 696–702.

Ziegler, J. F., 1977: *Helium—Stopping Powers and Ranges in All Elemental Matter.* Volume 4 of *The Stopping and Ranges of Ions in Matter,* J. F. Ziegler, organizer, Pergamon Press, Inc.

Ziegler, J. F., 1980: The Stopping of Energetic Ions in Solids. *Nuclear Instrum. & Methods*, vol. 168, nos. 1–3, pp. 17–24.

3. HIGH-ENERGY INTERACTIONS

3.1. Introduction

We will not attempt to give any comprehensive review of nuclear physics but will touch only on the highlights of work directly related to the development of nuclear models at the Langley Research Center. We will attempt to place the Langley work into a cohesive framework and relate it to the early work on reaction model development.

Following the discovery of cosmic rays through air ionization phenomena, the development of the cloud chamber and nuclear emulsion began to reveal the details of cosmic-ray interactions. These observations showed the production of fast particles in the forward direction and isotropic slow particles.

Serber (1947) suggested that the fast particles produced in high-energy proton and neutron reactions are direct knockout products of scattering with individual nuclear constituents followed by the emission of slow particles in the evaporation decay of the residual excited nucleus. This two-step Serber model was implemented through the first step by Goldberger (1948) at the University of Chicago using semiclassical methods and Monte Carlo techniques whereby the importance of the Pauli exclusion principle is demonstrated. G. F. Chew (1951) who performed the Monte Carlo calculation as a student for Goldberger introduced a corresponding quantum mechanical model in the so-called impulse approximation. Watson (1953) derived a complete quantum description of the first Serber step as a multiple-scattering series in which the impulse approximation of Chew is the first term in the series. A great simplification in the quantum theory came with the introduction of the eikonal by Glauber (1955) for which a successful, yet simple, scattering theory was derived including the multiple-scattering series (Franco and Glauber, 1966). Remler (1968) later derived a formal relation between the Glauber multiple-scattering theory and the Watson multiple-scattering theory.

The second step of the Serber model assumed that the level density within a nucleus is large and that the high excitation energy is distributed in thermal equilibrium among the many states. The excitation energy is then given up to the most energetic nuclear particles that can escape the nuclear potential region. Such a model was based on low-energy reaction studies in which reactions proceed through a compound nuclear state. The formation of the compound nuclear state at low energies was taken as the absorption of the passing plane wave by the nucleus which was assumed to act like a "cloudy crystal ball." This crystal ball is referred to as the optical model (Fernbach, Serber, and Taylor, 1949) in which the nuclear interior is treated as a medium with a complex index of refraction. The relationship between multiple scattering and the optical model was given by Watson (1953).

There was rapid progress in the development of intranuclear cascade models with compound nuclear de-excitation after the introduction of large-scale scientific computers (Metropolis, et al., 1958; Bertini, 1969). These intranuclear cascade codes had a large impact on the Apollo mission and continue to be used even to this day (Santoro et al., 1986). The primary data base for shielding against high-energy protons and neutrons remains to be that derived from intranuclear cascade codes. An effort to extend the intranuclear cascade codes to include complex projectiles has met with some success but is computationally inefficient (Gabriel, Bishop, and Lillie, 1984).

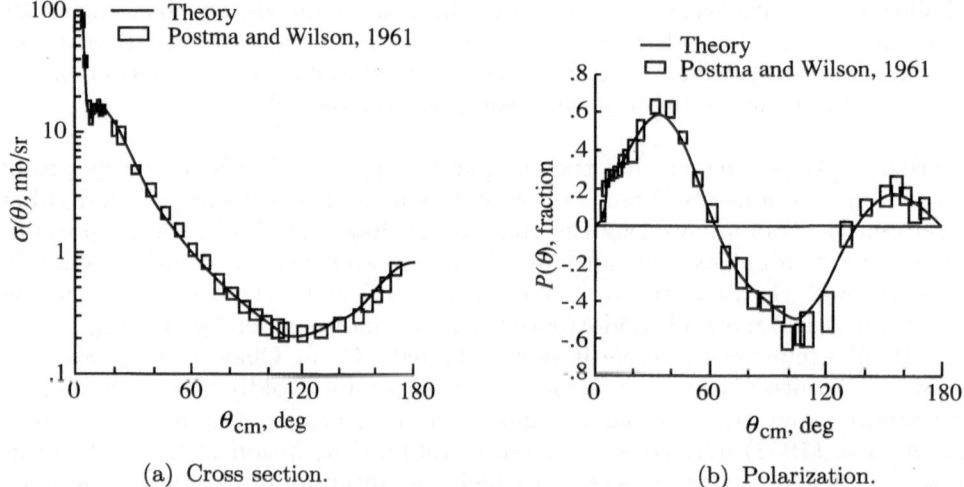

(a) Cross section.　　　　　　　　　　　(b) Polarization.

Figure 3.1. Comparison of S-wave plus P-wave fits using numerical integration for single scattering. D-state restricted to 6.93 ± 1 percent in search procedures and 5.93 at minimum $E = 146$ MeV.

The Langley program began with computations using the multiple-scattering formalism (Wilson, 1973a and 1973b) derived from the work of Mandelstam (1955). In accordance with Mandelstam, the transition amplitude is related to the residue at the pole of the propagator of the particles appearing in the asymptotic states. The propagator is fully symmetrized and can be formed into a multiple-scattering series by neglecting three-body terms and by projecting only the positive energy states. As Gross (1965) noted, the nuclear vertex can be related to nonrelativistic nuclear wave functions. The two-body scattering amplitudes were reconstructed from phase shift analysis and extrapolated to off-shell values by evaluating the one-pion exchange contributions directly and extrapolating the remaining contributions by assuming a two-pion exchange pole (Wilson, 1972). The S-state and D-state wave functions were taken from Humberston and Wallace (1970). The final calculation performed by Wilson (1973a, 1973b, and 1974a) was very successful (fig. 3.1) at describing the measured angular distributions of cross section and polarization measured for 146 MeV protons by Postma and Wilson (1961). With this success at applying the multiple-scattering theory to the three-body nuclear problem, we were encouraged to see these new skills help solve problems closer to NASA's interest.

3.2. Multiple-Scattering Theory

3.2.1. Glauber theory. The content of the Glauber (1955) theory is contained in the profile operator

$$\Gamma(\vec{b}) = 1 - \exp[-i\chi(\vec{b})] \tag{3.1}$$

where $\vec{b}$ is the impact parameter vector and the phase shift operator is

$$\chi(\vec{b}) = \frac{1}{\hbar v} \int\limits_{-\infty}^{\infty} V(\vec{b} + \vec{z}) \, dz \tag{3.2}$$

where v is the projectile velocity, V is the interaction potential, and z is the space variable in the direction of motion. The interaction potential for scattering an elementary projectile from a composite system (assuming only two-body potentials between projectile and constituents) is taken as

$$V(\vec{r}) = \sum_\alpha V_\alpha(\vec{r} - \vec{r}_\alpha) \tag{3.3}$$

which leads to the usual Glauber result. The usual extension to the scattering for composite systems is to take (Czyż and Maximon, 1969)

$$V(\vec{r}) = \sum_{j\alpha} V_{j\alpha}(\vec{r}_j - \vec{r}_\alpha) \tag{3.4}$$

where α constituents are located in the target of atomic weight A_T, and j constituents are located in the projectile of atomic weight A_P. The multiple-scattering form of the profile function is then (assuming that the potentials are commutative)

$$\Gamma(\vec{b}) = 1 - \prod_j \prod_\alpha [1 - \gamma_{j\alpha}(\vec{b})] \tag{3.5}$$

where $\gamma_{j\alpha}$ is the profile of the (j, α) colliding pair given as

$$\gamma_{j\alpha}(\vec{b}) = 1 - \exp[-i\frac{1}{\hbar v} \int\limits_{-\infty}^{\infty} V_{j\alpha}(\vec{r}_j - \vec{r}_\alpha) \, dz] \tag{3.6}$$

Equation (3.5) is expanded as

$$\Gamma(\vec{b}) = \sum_{j\alpha} \gamma_{j\alpha}(\vec{b}) - \sum_{j>k}^{(j,\alpha)\neq(k,\beta)} \gamma_{j\alpha}(\vec{b}) \, \gamma_{k\beta}(\vec{b}) + \ldots \tag{3.7}$$

The graphical representation of the single-scattering term of equation (3.7) is shown in figure 3.2. The double-scattering term contains two distinct types of graphs illustrated in figure 3.3. Note that the series in equation (3.7) ends after $(A_P \cdot A_T)$ terms.

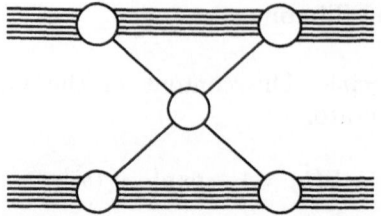

Figure 3.2. Single-scattering graph.

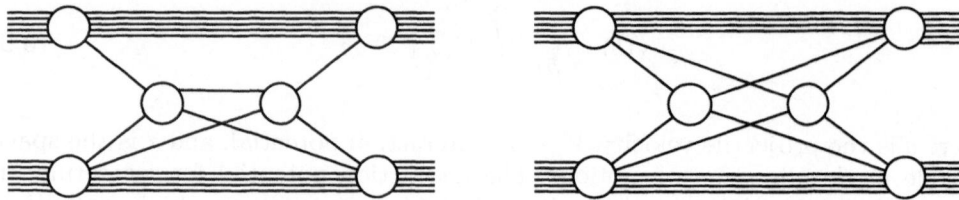

Figure 3.3. Rescattering graphs.

3.2.2. Multiple-scattering series.

The free projectile and target Hamiltonians, H_P and H_T, respectively, are taken together with the interaction potential V (assumed to be sums of two-body potentials between constituents) to form the full Hamiltonian

$$H = H_P + H_T + \sum_{\alpha j} V_{\alpha j} \tag{3.8}$$

The wave function in a remote region of space after the scattering satisfies

$$H\Psi = E\Psi \tag{3.9}$$

and consists of the superposition of the incident plane wave and the asymptotically scattered wave

$$\Psi = \varphi + \Psi_{sc} \tag{3.10}$$

where

$$(H_P + H_T)\varphi = E\varphi \tag{3.11}$$

and

$$\Psi_{sc} = GT\varphi \tag{3.12}$$

with Green's function given by

$$\lim_{\eta \to 0_+} (E - H_P - H_T + i\eta)G = 1 \tag{3.13}$$

and the transition operator by

$$T = V + VGT \tag{3.14}$$

In future equations we will assume that η is set to zero in the sense of the limit in equation (3.13). The usual wave operator Ω that transforms plane wave entering states to final scattered states

$$\Psi = \Omega\varphi \tag{3.15}$$

is given as

$$\Omega = 1 + GV\Omega \tag{3.16}$$

so that $\mathcal{T}$ is formally given as

$$\mathcal{T} = V\Omega \tag{3.17}$$

Our purpose is to find a series for $\mathcal{T}$ that is in terms of simpler functions. The development closely follows the original work of Watson (1953). The present derivation for heavy ions was made by Wilson (1974b).

To proceed with this purpose, the transition operator is defined for scattering the α constituent of the target with the j constituent of the projectile as

$$t_{\alpha j} = V_{\alpha j} + V_{\alpha j} G t_{\alpha j} \tag{3.18}$$

The wave operator that transforms the entering free state up to the collision of the α and j constituents is given by

$$\omega_{\alpha j} = 1 + \sum_{(\beta,k) \neq (\alpha,j)} G t_{\beta k} \omega_{\beta k} \tag{3.19}$$

Equation (3.19) is interpreted in the following way. The propagation to the time just before the α and j constituents scatter is the sum of an operator that brings the initial free state plus the scattered part from the scattering of all other β and k constituents. Clearly, the full wave operator consists of the wave operator that transforms the system to the α and j collision, plus the additional contribution caused by the scattering of the α and j constituents; that is,

$$\Omega = \omega_{\alpha j} + G t_{\alpha j} \omega_{\alpha j} \tag{3.20}$$

which, written symmetrically using equation (3.19), is

$$\Omega = 1 + \sum_{\alpha j} G t_{\alpha j} \omega_{\alpha j} \tag{3.21}$$

The series given by equations (3.18) through (3.21) constitutes an exact representation of the scattering process defined by equations (3.8) through (3.17). Consider the product

$$V_{\alpha j}\Omega = V_{\alpha j} \omega_{\alpha j} + V_{\alpha j} G t_{\alpha j} \omega_{\alpha j}$$
$$= (V_{\alpha j} + V_{\alpha j} G t_{\alpha j}) \omega_{\alpha j} = t_{\alpha j} \omega_{\alpha j} \tag{3.22}$$

Summing the α and j constituents gives

$$\mathcal{T} = \sum_{\alpha j} V_{\alpha j}\Omega = \sum_{\alpha j} t_{\alpha j} \omega_{\alpha j} \tag{3.23}$$

which shows equations (3.18), (3.19), and (3.21) as a solution to (3.16). By iteration of equations (3.23) and (3.19), the multiple-scattering series

$$\mathcal{T} = \sum_{\alpha j} t_{\alpha j} + \sum_{(\beta,k) \neq (\alpha,j)} t_{\alpha j} G t_{\beta k} + \cdots \tag{3.24}$$

is obtained, which constitutes a formal solution to the exact scattering problem (Wilson, 1974b). If the usual replacement (Watson, 1953; Wilson, 1974b) is made, that is,

$$G \to G_0 \equiv \frac{1}{E - \sum\limits_{j} T_j - \sum\limits_{\alpha} T_\alpha}$$

where G_0 is the free n-body Green's function given in terms of total energy and constituent kinetic energy operators, then $t_{\alpha j}$ becomes essentially two-body operators and equation (3.24) becomes a series of sequential two-body operators. The graphical representations of the terms of the series of equation (3.24) are the same as those shown in figures 3.2 and 3.3. The series (eq. (3.24)) reduces to the usual Watson series when the projectile consists of a single particle. When equation (3.24) is evaluated using the eikonal approximation, the Glauber theory is obtained, implying cancellation of an infinity of terms of equation (3.24) in the eikonal context. This type of cancellation was first noted by Remler (1968) and Harrington (1969).

3.2.3. Optical potential.

A potential operator V_{opt} must be found whose corresponding Born series for the **T**-matrix is equivalent to the multiple-scattering expansion (eq. (3.24)). Such an operator is closely related to the so-called optical potential (Ûlehla, Gomolčák, and Pluhař, 1964; Foldy and Walecka, 1969), which will be referred to as V_{opt}. The transition operator

$$\mathcal{T}_{\text{opt}} = V_{\text{opt}} + V_{\text{opt}} G \mathcal{T}_{\text{opt}} \tag{3.25}$$

will be defined by

$$V_{\text{opt}} = \sum_{\alpha j} t_{\alpha j} \tag{3.26}$$

from which

$$\mathcal{T} = \mathcal{T}_{\text{opt}} + \sum_{\alpha j} t_{\alpha j} G t_{\alpha j} + \ldots \tag{3.27}$$

The optical model is obtained by retaining the first term in equation (3.27), and the order of approximation is

$$\mathcal{T} - \mathcal{T}_{\text{opt}} \approx \frac{V_{\text{opt}} G V_{\text{opt}}}{A_T A_P} \tag{3.28}$$

because $t_{\alpha j} \approx V_{\text{opt}}/(A_T A_P)$ where A_T and A_P are the atomic weights of the target and projectile, respectively. The amplitude in equation (3.25) is a rather good approximation to the exact amplitude for light as well as heavy nuclei.

In summary, a multiple-scattering series for heavy ion scattering has been derived that appears as a natural extension to the Watson formalism. The structure of this series indicates that it reduces to the Glauber result within the eikonal context. A potential operator is found which shows that an optical model for heavy ion scattering is a good approximation for even rather light nuclei.

3.3. Heavy Ion Dynamical Equations

In the previous section, an optical potential equation was derived for use in the scattering of heavy ions. In this section, the coupled-channel equations for composite particle scattering are examined. Our method will be similar to that of Foldy and Walecka (1969) and has been presented elsewhere (Wilson, 1975). Particular attention will be given to the relation between the coherent elastic-scattered wave, the Born approximation, Chew's (1951) form of impulse approximation, the distorted-wave Born approximation (DWBA), and various approximation procedures to the coupled equations. Finally, the coupled equations will be solved by using the eikonal approximation. A simplified expression for the scattering amplitude is derived from that approximation, which includes the elastic- and all the inelastic-scattered amplitudes for small scattering angles. A discussion about the customary use of the optical theorem to estimate total cross sections from the coherent elastic-scattered wave will shed some light on the reasons that this estimate of total cross sections is successful.

3.3.1. Coupled-channel equations. The starting point for the present discussion is the coupled-channel (Schrödinger) equation relating the entrance channel to all excited states of the target and projectile. This equation was derived by Wilson (1974b and 1975) by assuming the kinetic energy to be large compared with the excitation energy of the target and projectile and closure for the accessible internal eigenstates. These coupled equations are given as

$$\left(\nabla_x^2 + \vec{k}^2\right)\psi_{m\mu}(\vec{x}) = \frac{2mA_PA_T}{N}\sum_{m'\mu'}V_{m\mu,m'\mu'}(\vec{x})\,\psi_{m'\mu'}(\vec{x}) \tag{3.29}$$

where subscripts m and μ label the eigenstates of the projectile and target; A_P and A_T are projectile and target mass number, respectively; m is constituent mass; $\vec{k}$ is projectile momentum relative to the center of mass; and $\vec{x}$ is the projectile position vector relative to the target, with

$$V_{m\mu,m'\mu'}(\vec{x}) = \left\langle g_{P,m}\left(\vec{\xi}_P\right)g_{T,\mu}\left(\vec{\xi}_T\right)\left|V_{\text{opt}}\left(\vec{\xi}_P,\vec{\xi}_T,\vec{x}\right)\right|g_{P,m'}\left(\vec{\xi}_P\right)g_{T,\mu'}\left(\vec{\xi}_T\right)\right\rangle \tag{3.30}$$

The quantities $g_{P,m}\left(\vec{\xi}_P\right)$ and $g_{T,\mu}\left(\vec{\xi}_T\right)$ are the projectile and target internal wave functions, respectively; $\vec{\xi}_P$ and $\vec{\xi}_T$ are collections of internal coordinates of the projectile and target constituents, respectively; and $V_{\text{opt}}\left(\vec{\xi}_P,\vec{\xi}_T,\vec{x}\right)$ is the effective potential operator derived in the previous section and is given by

$$V_{\text{opt}}\left(\vec{\xi}_P,\vec{\xi}_T,\vec{x}\right) = \sum_{\alpha j}t_{\alpha j}\left(\vec{x}_\alpha,\vec{x}_j\right) \tag{3.31}$$

Here, $t_{\alpha j}\left(\vec{x}_\alpha,\vec{x}_j\right)$ is the two-body transition operator for the j constituent of the projectile at position $\vec{x}_j$ and the α constituent of the target at $\vec{x}_\alpha$. The total constituent number N is defined as

$$N = A_P + A_T \tag{3.32}$$

The notation is simplified by introducing the wave vector

$$\psi(\vec{x}) = \begin{bmatrix} \psi_{00}(\vec{x}) \\ \psi_{01}(\vec{x}) \\ \psi_{10}(\vec{x}) \\ \psi_{11}(\vec{x}) \\ \vdots \end{bmatrix} \tag{3.33}$$

and the potential matrix

$$\boldsymbol{U}(\vec{x}) = \frac{2mA_T A_P}{N} \begin{bmatrix} V_{00,00}(\vec{x}) & V_{00,01}(\vec{x}) & V_{00,10}(\vec{x}) & \cdots \\ V_{01,00}(\vec{x}) & V_{01,01}(\vec{x}) & V_{01,10}(\vec{x}) & \cdots \\ V_{10,00}(\vec{x}) & V_{10,01}(\vec{x}) & V_{10,10}(\vec{x}) & \cdots \\ V_{11,00}(\vec{x}) & V_{11,01}(\vec{x}) & V_{11,10}(\vec{x}) & \cdots \\ \vdots & \vdots & \vdots & \vdots \end{bmatrix} \tag{3.34}$$

The coupled equations are then written in matrix form as

$$\left(\nabla_x^2 + \vec{k}^{\,2} \right) \boldsymbol{\psi}(\vec{x}) = \boldsymbol{U}(\vec{x}) \, \boldsymbol{\psi}(\vec{x}) \tag{3.35}$$

for which the approximate solution is considered.

The object of the solution of equation (3.35) is the calculation of the scattering amplitude given by

$$\boldsymbol{f}(\vec{q}) = -\sqrt{\frac{\pi}{2}} \int \exp\left(-i\vec{k}_f \cdot \vec{x} \right) \, \boldsymbol{U}(\vec{x}) \, \boldsymbol{\psi}(\vec{x}) \, d^3\vec{x} \tag{3.36}$$

where $\vec{k}_f$ is the final projectile momentum and $\vec{q}$ is the momentum transfer vector

$$\vec{q} = \vec{k}_f - \vec{k} \tag{3.37}$$

Because equation (3.35) cannot be solved in general, the rest of this chapter is devoted to the study of approximation procedures for the evaluation of equation (3.36). To gain insight, the simplest approximations are examined first and provide a basis for more accurate and complex procedures.

3.3.2. Born approximation. The Born approximation is obtained by approximating $\psi(\vec{x})$ by the incident plane wave. The coupled amplitude is then written as

$$\boldsymbol{f}^B(\vec{q}) = -\frac{1}{4\pi} \int \exp\left(-i\vec{q} \cdot \vec{x} \right) \, \boldsymbol{U}(\vec{x}) \, d^3\vec{x} \tag{3.38}$$

which is a matrix of approximate scattering amplitudes relating all possible entrance channels to all possible final channels. For example, diagonal elements relate to all possible elastic scatterings of the system where the elastic channel is defined by the entrance channel. Using the definition of the potential given in equations (3.30) and (3.31) results in

$$V_{m\mu,m'\mu'}(\vec{x}) = \sum_{\alpha j} \left\langle g_{P,m}\left(\vec{\xi}_P\right) \, g_{T,\mu}\left(\vec{\xi}_T\right) \left| t_{\alpha j}\left(\vec{x}_\alpha, \vec{x}_j\right) \right| g_{P,m'}\left(\vec{\xi}_P\right) \, g_{T,\mu'}\left(\vec{\xi}_T\right) \right\rangle$$

$$= \sum_{\alpha j} \int \rho_{T,\mu\mu'}\left(\vec{r}_\alpha\right) \, t_{\alpha j}\left(\vec{x}_\alpha, \vec{x}_j\right) \, \rho_{P,mm'}\left(\vec{r}_j\right) \, d^3\vec{r}_\alpha \, d^3\vec{r}_j \tag{3.39}$$

where

$$\rho_{P,mm'}\left(\vec{r}_j\right) = \int g^*_{P,m}\left(\vec{\xi}_P\right) \, \delta^3\left(\vec{r}_j - \vec{\xi}_{P,j}\right) \, g_{P,m'}\left(\vec{\xi}_P\right) \, d^3\vec{\xi}_P \tag{3.40}$$

and

$$\rho_{T,\mu\mu'}\left(\vec{r}_\alpha\right) = \int g^*_{T,\mu}\left(\vec{\xi}_T\right) \, \delta^3\left(\vec{r}_\alpha - \vec{\xi}_{T,\alpha}\right) \, g_{T,\mu'}\left(\vec{\xi}_T\right) \, d^3\vec{\xi}_T \tag{3.41}$$

where an asterisk denotes complex conjugation. The Fourier transform of equation (3.39) yields

$$\int V_{m\mu,m'\mu'}(\vec{x}) \, \exp(-i\vec{q} \cdot \vec{x}) \, d^3\vec{x}$$

$$= \sum_{\alpha j} \int \exp(-i\vec{q} \cdot \vec{x}) \left[\int \rho_{T,\mu\mu'}\left(\vec{r}_\alpha\right) \, \rho_{P,mm'}\left(\vec{r}_j\right) \, t_{\alpha j}\left(\vec{x}_\alpha, \vec{x}_j\right) \, d^3\vec{r}_\alpha \, d^3\vec{r}_j \right] d^3\vec{x}$$

$$= \sum_{\alpha j} t_{\alpha j}(k,\vec{q}) \, F_{T,\mu\mu'}(\vec{q}) \, F_{P,mm'}(-\vec{q}) \tag{3.42}$$

where the transition amplitudes $t_{\alpha j}\left(\vec{x}_\alpha, \vec{x}_j\right)$ used depend only on the relative position vector of the α and j constituents relative to one another. The form factors $F_{P,mm'}(-\vec{q})$ and $F_{T,\mu\mu'}(\vec{q})$ are the Fourier transforms of their corresponding single-particle transition densities given in equations (3.40) and (3.41), respectively. Using equations (3.34) and (3.42) in equation (3.38) results in the following form for the Born approximation:

$$f^B_{m'\mu',m\mu}(\vec{q}) = -\frac{1}{4\pi}\left(\frac{2mA_P^2 A_T^2}{N}\right) F_{T,\mu'\mu}(\vec{q}) \, F_{P,m'm}(-\vec{q}) \, \tilde{t}(k,\vec{q}) \tag{3.43}$$

where

$$\tilde{t}(k,\vec{q}) = \frac{1}{A_P A_T}\sum_{\alpha j} t_{\alpha j}(k,\vec{q}) \tag{3.44}$$

is the transition amplitude averaged over nuclear constituents.

Consider now the projectile form factor given by the Fourier transform of the single-particle densities as

$$F_{P,m'm}(\vec{q}) = \int \exp\left(i\vec{q} \cdot \vec{r}_\alpha\right) \rho_{P,m'm}\left(\vec{r}_\alpha\right) d^3\vec{r}_\alpha$$

$$= \int g^*_{P,m'}\left(\vec{\xi}_P\right) \exp\left(i\vec{q} \cdot \vec{\xi}_{P,\alpha}\right) g_{P,m}\left(\vec{\xi}_P\right) d^3\vec{\xi}_P \qquad (3.45)$$

Expanding the exponential factor as a power series results in

$$F_{P,m'm}(\vec{q}) = \delta_{m'm} + i\vec{a}_{P,1} \cdot \vec{q} - \frac{1}{2}\vec{q} \cdot \overset{\leftrightarrow}{a}_{P,2} \cdot \vec{q} + \dots \qquad (3.46)$$

where the first term in equation (3.46) corresponds to the normalization condition of the eigenstates; the second term contains the dipole transition moment given by

$$\vec{a}_{P,1} = \left\langle g_{P,m'}\left(\vec{\xi}_P\right) \left| \vec{\xi}_{P,\alpha} \right| g_{P,m}\left(\vec{\xi}_P\right) \right\rangle \qquad (3.47)$$

and the third term contains the dyadic quadrupole transition moment

$$\overset{\leftrightarrow}{a}_{P,2} = \left\langle g_{P,m'}\left(\vec{\xi}_P\right) \left| \vec{\xi}_{P,\alpha}\vec{\xi}_{P,\alpha} \right| g_{P,m}\left(\vec{\xi}_P\right) \right\rangle \qquad (3.48)$$

The higher order multipole transitions are indicated by dots in equation (3.46). The lowest order nonzero term in equation (3.46) depends on the properties of the internal wave functions involved. In general, the ℓth transition moment with magnitude given by

$$a_{P,\ell} = \left| \left\langle g_{P,m'}\left(\vec{\xi}_P\right) \left| \left(\vec{\xi}_{P,\alpha}\right)^\ell \right| g_{P,m}\left(\vec{\xi}_P\right) \right\rangle \right| \qquad (3.49)$$

is zero unless

$$\left| J_{m'} - J_m \right| \leq \ell \leq \left| J_{m'} + J_m \right| \qquad (3.50)$$

as a result of the Wigner-Eckart theorem where J_m and $J_{m'}$ are the projectile internal angular-momentum quantum numbers in the entering and final states, respectively. Because of the orthogonality, equation (3.49) reduces to

$$a_{P,0} = \delta_{m'm} \qquad (3.51)$$

for $\ell = 0$. It follows from relations (3.50) and (3.51) and for small momentum transfer that

$$F_{P,m'm}(\vec{q}) \approx \delta_{m'm} + \frac{a_{P,\ell_P} |\vec{q}|^{\ell_P}}{\ell_P!} \qquad (3.52)$$

where

$$\ell_P = \max\left\{\left| J_{m'} - J_m \right|, 1\right\} \qquad (3.53)$$

is the angular momentum associated with the lowest order transition moment. Similarly, for the target one obtains

$$F_{T,\mu'\mu}(\vec{q}) \approx \delta_{\mu'\mu} + \frac{a_{T,\ell_T}|\vec{q}|^{\ell_T}}{\ell_T!} \qquad (3.54)$$

where

$$\ell_T = \max\left\{\left|J_{\mu'} - J_\mu\right|, 1\right\} \qquad (3.55)$$

It follows from relations (3.43), (3.52), and (3.54) that the Born amplitude has proportionality given by

$$f^B_{m'\mu',m\mu}(\vec{q}) \propto \left(\delta_{m'm} + \frac{a_{P,\ell_P}|\vec{q}|^{\ell_P}}{\ell_P!}\right)\left(\delta_{\mu'\mu} + \frac{a_{T,\ell_T}|\vec{q}|^{\ell_T}}{\ell_T!}\right)\tilde{t}(k,\vec{q}) \qquad (3.56)$$

where a_{P,ℓ_P} and a_{T,ℓ_T} are the lowest order nonvanishing transition moments of the projectile and target corresponding to equations (3.53) and (3.55).

On the basis of the Born approximation, a very strong threshold effect on the various excitation processes is observed. This effect causes an ordering in the contribution of specific excitation channels in going from small to large momentum transfer. At zero momentum transfer, only the elastic channel is open. As the momentum transfer increases, the single dipole transitions for either the target or the projectile, but not both, are displayed first. Note that this condition severely restricts the accessible angular momentum states in the excitation process. At slightly higher momentum transfer, coincident dipole transitions in projectile and target and single quadrupole transitions are in competition with and may eventually dominate the single dipole transitions at sufficiently high momentum transfer. Similarly, at higher momentum transfer, transitions to higher angular-momentum states are possible.

3.3.3. Perturbation expansion and distorted-wave Born approximation.

According to the previous discussion, for a restricted range of momentum transfer, the off-diagonal elements of the "Born" matrix of scattering amplitudes are small compared with the elastic-scattering amplitudes for the various channels found along the diagonal. By noting that these amplitudes are proportional to the potential, a decomposition of the potential into large and small components (Wilson, 1975) may be made as

$$U(\vec{x}) = U_d(\vec{x}) + U_o(\vec{x}) \qquad (3.57)$$

where $U_d(\vec{x})$ denotes the diagonal parts of $U(\vec{x})$ and $U_o(\vec{x})$ denotes the corresponding off-diagonal parts. Clearly,

$$U_d(\vec{x}) >> U_o(\vec{x}) \qquad (3.58)$$

which is in accordance with the preceding discussion. Treating the off-diagonal contribution as a perturbation and considering the iterated solution will lead to substantial simplification (Wilson, 1975).

Rewriting equation (3.35) as

$$\left[\nabla_x^2 + \vec{k}^2 - U_d(\vec{x})\right] \psi(\vec{x}) = U_o(\vec{x}) \, \psi(\vec{x}) \tag{3.59}$$

and taking as a first approximation

$$\left[\nabla_x^2 + \vec{k}^2 - U_d(\vec{x})\right] \psi_0(\vec{x}) = 0 \tag{3.60}$$

leads to a solvable problem. The only nonzero component of $\psi_0(\vec{x})$ is the elastic coherent scattered wave. If the initial prepared nuclei are in their ground states, then the solution for the coherent elastic wave is obtained from

$$\left(\nabla_x^2 + \vec{k}^2\right) \psi_c(\vec{x}) = U_{00,00}(\vec{x}) \, \psi_c(\vec{x}) \tag{3.61}$$

and the first approximation to the coupled-channel problem is

$$\psi_0(\vec{x}) = \begin{bmatrix} \psi_c(\vec{x}) \\ 0 \\ 0 \\ 0 \\ \vdots \end{bmatrix} \tag{3.62}$$

Estimating the perturbation by using equation (3.62) yields the lowest order correction as

$$\left[\nabla_x^2 + \vec{k}^2 - U_d(\vec{x})\right] \psi^{(1)}(\vec{x}) = U_o(\vec{x}) \, \psi_0(\vec{x}) \tag{3.63}$$

The right-hand side is a term describing the source of excitation caused by the interaction of the coherent amplitude and is of the form

$$U_o(\vec{x}) \, \psi_0(\vec{x}) = \begin{bmatrix} 0 \\ U_{01,00} \\ U_{10,00} \\ U_{11,00} \\ \vdots \end{bmatrix} \psi_c(\vec{x}) \tag{3.64}$$

Because the first component of the source of excitation is zero, the equation for the first component of equation (3.35) is

$$\left[\nabla_x^2 + \vec{k}^2 - U_{00,00}(\vec{x})\right] \psi_{00}^{(1)}(\vec{x}) = 0 \tag{3.65}$$

and reveals that the iteration of the elastic channel yields again the coherent elastic amplitude

$$\psi_{00}^{(1)}(\vec{x}) = \psi_c(\vec{x}) \tag{3.66}$$

The remaining components of equation (3.63) are

$$\left[\nabla_x^2 + \vec{k}^2 - U_{m\mu,m\mu}(\vec{x})\right] \psi_{m\mu}^{(1)}(\vec{x}) = U_{m\mu,00}(\vec{x}) \, \psi_c(\vec{x}) \tag{3.67}$$

This process of successive iteration is equivalent to the series approximation

$$\psi(\vec{x}) = \psi_0(\vec{x}) + \psi_1(\vec{x}) + \psi_2(\vec{x}) + \dots \tag{3.68}$$

where

$$\left[\nabla_x^2 + \vec{k}^2 - U_d(\vec{x}) \right] \psi_0(\vec{x}) = 0 \tag{3.69}$$

and

$$\left[\nabla_x^2 + \vec{k}^2 - U_d(\vec{x}) \right] \psi_i(\vec{x}) = U_o(\vec{x}) \, \psi_{i-1}(\vec{x}) \tag{3.70}$$

The iterated solution and series solution are related as

$$\left. \begin{array}{c} \psi_i(\vec{x}) = \psi^{(i)}(\vec{x}) - \psi^{(i-1)}(\vec{x}) \\[2mm] \psi^{(-1)}(\vec{x}) \equiv 0 \end{array} \right\} \tag{3.71}$$

and the ith iterate $\psi^{(i)}(\vec{x})$ is the ith partial sum of the series.

Further insight can be gained by considering the formal solution to the coupled equations (3.69) and (3.70). Introducing the diagonal coherent propagator

$$G_c = \left[\nabla_x^2 + \vec{k}^2 - U_d(\vec{x}) \right]^{-1} \tag{3.72}$$

and the coherent wave operator

$$\Omega_c = 1 + \left(\nabla_x^2 + \vec{k}^2 \right)^{-1} U_d(\vec{x}) \tag{3.73}$$

produces the solution to equation (3.70) as

$$\psi_i(\vec{x}) = G_c \, U_o(\vec{x}) \, \psi_{i-1}(\vec{x}) \tag{3.74}$$

with

$$\psi_0 = \Omega_c \psi_p \tag{3.75}$$

where ψ_p is the entering plane-wave state. The series (eq. (3.68)) may now be written as

$$\psi = \Omega_c \psi_p + G_c U_o \Omega_c \psi_p + G_c U_o G_c U_o \Omega_c \psi_p + \dots \equiv \Omega \psi_p \tag{3.76}$$

The first term is the coherent elastic-scattered wave as noted in equation (3.75) and represents attenuation and propagation of the incident plane wave in matter. Since Ω_c is diagonal, this propagation is in undisturbed matter. The second term of equation (3.76) relates to the excitation caused by the presence of the coherent elastic wave followed by coherent propagation in disturbed matter. Note that the second term has no contribution in the elastic channel. The third term of equation (3.76) relates to further excitation caused by the presence of the scattered waves formed exclusively by coherent excitation and the first correction to the elastic channel caused by incoherent processes. Hence, the coherent elastic wave is correct up to second-order terms in off-diagonal elements of the potential matrix. These off-diagonal elements show considerable damping or suppression at small momentum transfer as shown in connection with equation (3.56). This may well be the reason that the coherent elastic amplitude has been so successful in nuclear applications (Wilson, 1975; Wilson and Costner, 1975; Best, 1972).

The structure of the second term in the series (eq. (3.76)) is either the usual distorted-wave Born approximation (Austern, 1963) or the single inelastic scattering approximation (Goldberger and Watson, 1964). The entire series could be aptly referred to as the distorted-wave Born series. However, recalling that the terms of the series correspond to a successively larger number of changes in states of excitation (that is, the first term contains no excitation, the second term transforms the coherent elastic wave to the excited states, the third term transforms the excited states of the second term to new excitation levels, and so on), a more appropriate name for the series would be the "multiple-excitation series."

3.4. Coupled-Channel Amplitudes

The coupled equations (3.35) are now solved within a small-angle approximation. This solution in effect sums the multiple-excitation series to all orders and, as a final result, gives expressions for the scattering amplitudes connecting all possible entrance channels to all possible final channels. By making the forward-scattering assumption, the boundary condition is given by

$$\lim_{z \to -\infty} \psi(\vec{x}) = \left(\frac{1}{2\pi}\right)^{3/2} \exp(i\vec{k} \cdot \vec{x}) \, \hat{\boldsymbol{\delta}} \tag{3.77}$$

where $-\hat{z}$ is the direction to the beam source and $\hat{\boldsymbol{\delta}}$ is a constant vector with a unit entry at the entrance channel element but zero elsewhere. Equation (3.77) simply states that no particles are scattered backward. Physically, this assumption is justified because the backward-scattered component for most high-energy scattering is many orders of magnitude less than the forward-scattered component. The form of the solution to equation (3.35) is taken as

$$\psi(\vec{x}) = \left(\frac{1}{2\pi}\right)^{3/2} \exp\left[i\phi(\vec{x})\right] \, \exp(i\vec{k} \cdot \vec{x}) \, \hat{\boldsymbol{\delta}} \tag{3.78}$$

where the boundary condition (3.77) implies that

$$\lim_{z \to -\infty} \phi(\vec{x}) = 0 \tag{3.79}$$

as a boundary condition on $\phi(\vec{x})$. By using equation (3.78), one may write an equation for $\phi(\vec{x})$ as

$$i\nabla_x^2 \, \phi(\vec{x}) - [\nabla_x \, \phi(\vec{x})]^2 - 2\vec{k} \cdot \nabla_x \phi(\vec{x}) - U(\vec{x}) = 0 \tag{3.80}$$

If $U(\vec{x})$ is small compared with the kinetic energy

$$U(\vec{x}) << k^2 \tag{3.81}$$

and if the change in $U(\vec{x})$ is small over one oscillation of the incident wave, that is,

$$\nabla_x U(\vec{x}) << \vec{k} \, U(\vec{x}) \tag{3.82}$$

where inequalities refer to magnitudes of elements on each side of equations (3.81) and (3.82), then equation (3.80) may be approximated by

$$2k \frac{\partial}{\partial z} \, \phi(\vec{x}) = -U(\vec{x}) \tag{3.83}$$

which has the solution

$$\phi(\vec{x}) = -\frac{1}{2k} \int_a^z U(\vec{x}') \, dz' \tag{3.84}$$

where the value of a is fixed by the boundary condition (eq. (3.79)) to be $-\infty$. The scattered wave (eq. (3.78)) may now be written as

$$\psi(\vec{x}) = \left(\frac{1}{2\pi}\right)^{3/2} \exp\left[-\frac{i}{2k} \int_{-\infty}^z U(\vec{x}') \, dz'\right] \exp(i\vec{k} \cdot \vec{x}) \, \hat{\delta} \tag{3.85}$$

Note that the wave operator is approximated by

$$\Omega \approx \exp\left[-\frac{i}{2k} \int_{-\infty}^z U(\vec{x}') \, dz'\right] \tag{3.86}$$

The eikonal result for the scattering amplitudes is given by

$$f(\vec{q}) \, \hat{\delta} = -\sqrt{\frac{\pi}{2}} \int \exp\left(-i\vec{k}_f \cdot \vec{x}\right) U(\vec{x}) \, \psi(\vec{x}) \, d^3\vec{x}$$

$$= -\frac{1}{4\pi} \int \exp(-i\vec{q} \cdot \vec{x}) U(\vec{x}) \exp\left[-\frac{i}{2k} \int_{-\infty}^z U(\vec{x}') \, dz'\right] \hat{\delta} \, d^3\vec{x} \tag{3.87}$$

where $\vec{k}_f$ is the final projectile momentum and $\vec{q}$ the momentum transfer given by

$$\vec{q} = \vec{k}_f - \vec{k} \tag{3.88}$$

The eikonal approximation to the coupled-channel amplitude (eq. (3.87)) can be further simplified by making an additional small-angle approximation as follows. By using a cylindrical coordinate system with cylinder axis along the beam direction and writing

$$\vec{x} = \vec{b} + \vec{z} \tag{3.89}$$

where $\vec{b}$ is the impact parameter vector, the product of $\vec{q}$ and $\vec{x}$ may be written as

$$\vec{q} \cdot \vec{x} = \vec{q} \cdot \vec{b} + O(\theta^2) \tag{3.90}$$

where θ is the scattering angle which is assumed to be small. This small-angle approximation allows equation (3.87) to be written as

$$f(\vec{q}) = -\frac{1}{4\pi} \int \exp(-i\vec{q} \cdot \vec{b}) \, U(\vec{b} + \vec{z}) \, \exp\left[-\frac{i}{2k} \int_{-\infty}^z U(\vec{b} + \vec{z}') \, dz'\right] d^2\vec{b} \, dz \tag{3.91}$$

where the integral over $\vec{z}$ can be performed exactly. Performing the integral over $\vec{z}$ in equation (3.91) yields the final simplified expression for the scattering amplitude as

$$f(\vec{q}) = -\frac{ik}{2\pi} \int \exp(-i\vec{q} \cdot \vec{b}) \left\{ \exp\left[i\, \chi(\vec{b}) \right] - 1 \right\} \, d^2\vec{b} \qquad (3.92)$$

where

$$\chi(\vec{b}) = -\frac{1}{2k} \int_{-\infty}^{\infty} U(\vec{b} + \vec{z}) \, d\vec{z} \qquad (3.93)$$

Equation (3.92) gives the matrix of scattering amplitudes of all possible entrance channels to all possible final channels of the system.

The relation between the eikonal result for the full scattering amplitude (eq. (3.92)) and the various approximate results discussed earlier in this section is now derived. First, consider the expansion in powers of χ of the integrand of equation (3.92):

$$f(\vec{q}) = \frac{ik}{2\pi} \int \exp(-i\vec{q} \cdot \vec{b}) \left(i\chi - \frac{1}{2!}\chi^2 - \frac{1}{3!}i\chi^3 + \dots \right) d^2\vec{b} \qquad (3.94)$$

The first term is the Born approximation at small angles. Higher order terms are multiple-scattering corrections to the Born result. Recall that the Born approximation for the optical potential is equivalent to Chew's impulse approximation. A more interesting result is obtained by separating the χ matrix into its diagonal and off-diagonal parts as

$$\chi(\vec{b}) = \chi_d(\vec{b}) + \chi_o(\vec{b}) \qquad (3.95)$$

which correspond to the diagonal and off-diagonal parts of the matrix potential $U(\vec{x})$. An expansion in powers of the off-diagonal part of χ in equation (3.92) yields

$$f(\vec{q}) = -\frac{ik}{2\pi} \int \exp(-i\vec{q} \cdot \vec{b}) \left\{ \exp\left[i\, \chi_d(\vec{b}) \right] - 1 \right\} \, d^2\vec{b}$$

$$- \frac{ik}{2\pi} \int \exp(-i\vec{q} \cdot \vec{b}) \exp\left[i\, \chi_d(\vec{b}) \right] \left(i\chi_o - \frac{1}{2!}\chi_o^2 - \frac{1}{3!}i\,\chi_o^3 + \dots \right) d^2\vec{b} \qquad (3.96)$$

The first integral is the elastic coherent amplitude, the first term of the second integral is the distorted-wave Born approximation, and the remaining terms are multiple-excitation corrections.

3.5. The Elastic Channel

Section 3.4 showed that within a small-angle approximation, the coupled-channel equations could be solved. The principal difficulty in calculating the full coupled-channel amplitude lies in the almost complete lack of knowledge of the internal wave functions for the colliding nuclei for all orders of excitation. On the very general principles for near forward scattering, transitions to the excited states are kinematically suppressed. This was the main motivation for expanding the solution in terms of off-diagonal matrix elements of the potential. Near forward scattering, the scattering amplitude is dominated by the diagonal elements. If elastic scattering is strongly forward, then a reasonable approximation to the elastic amplitude is obtained by neglecting the off-diagonal contribution (coherent approximation), and, in addition, the eikonal small-angle approximation should be accurate. In this vein, the elastic-channel amplitude is approximated by retaining only the first term in equation (3.96). Detailed comparisons with experimental data are made to justify this approximation.

Wilson (1975) showed that the elastic-channel potential (actually the coherent potential) can be reduced to

$$U_c(\vec{x}) = \frac{2mA_T^2 A_P^2}{N} \int d^3\vec{z}\; \rho_T(\vec{z}) \int d^3\vec{y}\; \rho_P(\vec{x} + \vec{y} + \vec{z})\; \tilde{t}(k,\vec{y}) \qquad (3.97)$$

where $\rho_T(\vec{z})$ and $\rho_P(\vec{z})$ are the target and projectile ground-state single-particle densities, respectively, and $\tilde{t}(k,\vec{y})$ is the energy- and space-dependent two-body transition amplitudes averaged over the projectile and target constituent types as

$$\tilde{t} = \frac{1}{A_P A_T} \left[N_P N_T t_{nn} + Z_P Z_T t_{pp} + (N_P Z_T + Z_P N_T) t_{np} \right] \qquad (3.98)$$

with N_P and N_T being the projectile and target neutron numbers, respectively, and Z_P and Z_T being the corresponding proton numbers. The normalization of the $\tilde{t}$ amplitude is given by

$$\tilde{t}(k,\vec{y}) = -\frac{1}{(2\pi)^2 \mu} \int \exp(i\vec{q} \cdot \vec{y})\; f(e,\vec{q})\; d^3\vec{q} \qquad (3.99)$$

with the usual expression for the spin-independent two-nucleon transition amplitudes as

$$f(e,\vec{q}) = \frac{\sigma(e)\sqrt{me}}{4\pi} \left[\alpha(e) + i\right]\; \exp\left[-\frac{1}{2}\; B(e)\; \vec{q}^{\,2} \right] \qquad (3.100)$$

where e is the kinetic energy in the two-body center-of-mass frame, $\mu = m/2$ is the two-body reduced mass, $\sigma(e)$ is the energy-dependent total two-body cross section, $\alpha(e)$ is the ratio of real to imaginary parts, and $B(e)$ is the slope parameter. The elastic-channel phase function may now be approximated by

$$\chi(\vec{b}) = -\frac{1}{2k} \int_{-\infty}^{\infty} U_c(\vec{b} + \vec{z})\; d\vec{z} \qquad (3.101)$$

from which the elastic-channel (coherent) amplitude may be calculated by

$$f_c(\vec{q}) = -ik \int_0^{\infty} b\; db\; J_0\left(2kb\; \sin\frac{\theta}{2} \right) \left\{ \exp\left[i\; \chi(\vec{b}) \right] - 1 \right\} \qquad (3.102)$$

where the property that the phase function is cylindrically symmetric about the $\hat{z}$-direction has been used and $J_0(\ )$ is the zeroth-order Bessel function. Applying now the optical theorem

$$\sigma_{\text{tot}} = \frac{4\pi}{k}\; \text{Im}\left[f_c(\vec{0}) \right] \qquad (3.103)$$

yields

$$\sigma_{\text{tot}} \approx 4\pi \int_0^{\infty} b\; db\; \left\{ 1 - \exp\left[-\chi_i(\vec{b}) \right] \cos\left[\chi_r(\vec{b}) \right] \right\} \qquad (3.104)$$

where χ_r and χ_i are the real and imaginary parts, respectively, of χ. Since the scattering is strongly forward, the total elastic cross section may be calculated by using the eikonal expression by

$$\sigma_s = \int |f(\vec{q})|^2 \, d\hat{q}$$

$$\approx 4\pi \int_0^\infty b \, db \left\{ 1 - \exp\left[-\chi_i(\vec{b})\right] \cos\left[\chi_r(\vec{b})\right] \right\}$$

$$- 2\pi \int_0^\infty b \, db \left\{ 1 - \exp\left[-2 \, \chi_i(\vec{b})\right] \right\} \tag{3.105}$$

from which it follows that

$$\sigma_{\text{abs}} = \sigma_{\text{tot}} - \sigma_s \approx 2\pi \int_0^\infty b \, db \left\{ 1 - \exp\left[-2 \, \chi_i(\vec{b})\right] \right\} \tag{3.106}$$

The use of the coherent wave as an approximation to the elastic channel has, at least in part, been justified by comparison with experiment (Wilson, 1975; Wilson and Costner, 1975). The formalism gave good agreement with the experiments of Schimmerling et al. (1971 and 1973), as shown in figures 3.4 and 3.5, and predicted oscillations in cross sections for nuclei corresponding to the shell structure of nuclei (Wilson, 1975) as shown in figures 3.4 to 3.6.

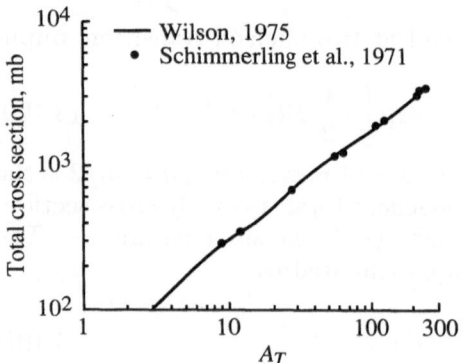

Figure 3.4. Total nucleon-nucleus cross section as a function of a nuclear mass number at 1.064 GeV.

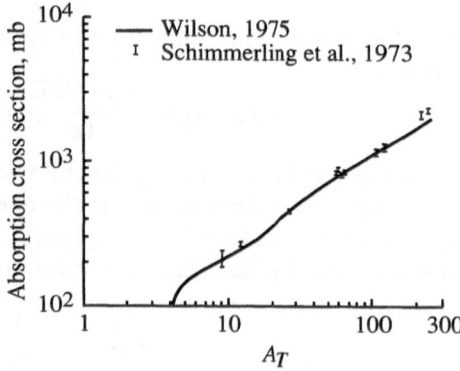

Figure 3.5. Nucleon-nucleus absorption cross section as a function of nuclear mass number at about 1 GeV.

3.6. Abrasion Theory

Abrasion theories developed in recent years have relied on Glauber theory as the basic formalism for the evaluation of probabilistic collision factors. Consequently, the inherent restrictions of Glauber theory are also limitations in these models. With the more powerful theoretical methods now available (Wilson, 1975), the development of a new abrasion theory is appropriate based on these more general results from current abrasion theories. The present development follows closely the work of Townsend (1983 and 1984).

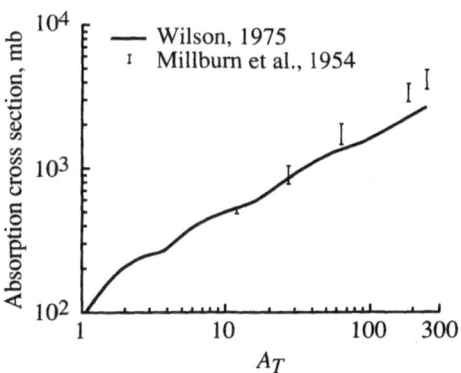

Figure 3.6. Triton-nucleus absorption cross section as a function of target mass at 100 MeV/nucleon.

In the abrasion-ablation collision model, projectile fragmentation is a three-step process. In the first step (abrasion), m nucleons are knocked out of the projectile nucleus of mass number A_P, leaving an excited prefragment nucleus of mass number

$$A_F = A_P - m \qquad (3.107)$$

In the next step, the prefragment is ablated by gamma emission, particle emission (usually nucleons or α-particles), or a combination of the two. The third and final phase involves interactions between the particles in the final state. These final-state interactions, although not unique to this collision formalism, are nevertheless significant experimentally and must be included in any complete theory.

3.6.1. Abrasion cross section. From Bleszynski and Sander (1979), the cross section for abrading m projectile nucleons is given by

$$\sigma_m = \binom{A_P}{m} 2\pi \int \left[1 - P(\vec{b}) \right]^m P(\vec{b})^{A_F} \, b \, db \qquad (3.108)$$

where $\binom{A_P}{m}$ is the binomial coefficient that reflects the number of possible combinations of m nucleons taken from an ensemble of A_P identical nucleons. The total absorption cross section

$$\sigma_{\text{abs}} = 2\pi \int \left[1 - P(\vec{b})^{A_P} \right] b \, db \qquad (3.109)$$

is obtained by summing over all values of m according to

$$\sigma_{\text{abs}} = \sum_{m=1}^{A_P} \sigma_m \qquad (3.110)$$

In equations (3.108) and (3.109), $P(\vec{b})$ is the probability as a function of impact parameter for not removing a single projectile nucleon in the abrasion process. Hence, $1 - P(\vec{b})$ is the probability for the removal of a nucleon.

The probability in Glauber theory is given by (Bleszynski and Sander, 1979)

$$P(\vec{b}) = 2\pi \int D_P(\vec{s}) \, \exp\left[-A_T \sigma_{NN} \, D_T(\vec{s} + \vec{b})\right] s \, ds \qquad (3.111)$$

where A_T is the mass number of the target and $D(\vec{s})$ denotes the single-particle densities summed along the beam direction (thickness functions)

$$D(\vec{s}) = \int_{-\infty}^{\infty} \rho(\vec{s} + \vec{z}) \, dz \qquad (3.112)$$

The abrasion theory is now extended to a more general collision theory that does not exhibit the convergence problems inherent with Glauber theory (Wilson, 1975; Wilson and Costner, 1975; Wilson and Townsend, 1981). An added feature of the extended abrasion theory, which gives symmetry to the final result, is that the projectile and target nuclei are treated on an equal basis.

3.6.2. Generalized abrasion theory. From the optical model derived in a previous section, the absorption cross section is expressed using the eikonal approximation

$$\sigma_{\rm abs} = 2\pi \int_0^{\infty} \left\{1 - \exp\left[-2 \, \text{Im} \, \chi(\vec{b})\right]\right\} b \, db \qquad (3.113)$$

where the eikonal phase function $\chi(\vec{b})$, with the optical-model potential approximation from Wilson and Costner (1975) incorporated, is written as

$$\chi(\vec{b}) = \frac{1}{2} A_P A_T \, \sigma(e) \, [\alpha(e) + i] \, I(\vec{b}) \qquad (3.114)$$

where

$$I(\vec{b}) = [2\pi \, B(e)]^{-3/2} \int d\vec{z} \int d^3\vec{\xi}_T \, \rho_T(\vec{\xi}_T)$$

$$\times \int d^3\vec{y} \, \rho_P(\vec{b} + \vec{z} + \vec{y} + \vec{\xi}_T) \, \exp\left[\frac{-\vec{y}^{\,2}}{2 \, B(e)}\right] \qquad (3.115)$$

In equations (3.114) and (3.115), $\sigma(e)$ is the energy-dependent nucleon-nucleon cross section, $\alpha(e)$ is the energy-dependent ratio of the real part to the imaginary part of the scattering amplitudes, $B(e)$ is the energy-dependent slope parameter, and ρ_P and ρ_T are the projectile and target single-particle nuclear densities, respectively. Townsend (1983) uses equations (3.109) and (3.113) to imply that

$$P(\vec{b})^{A_P} = \exp\left[-2 \, \text{Im} \, \chi(\vec{b})\right] \qquad (3.116)$$

Substitution of equation (3.114) into equation (3.116) yields

$$P(\vec{b}) = \exp\left[-A_T \, \sigma(e) \, I(\vec{b})\right] \qquad (3.117)$$

Finally, the cross section for abrading any m nucleons (eq. (3.108)) is written as

$$\sigma_m = \binom{A_P}{m} 2\pi \int \left\{ 1 - \exp\left[-A_T \, \sigma(e) \, I(\vec{b}) \right] \right\}^m \exp\left[-A_T A_F \, \sigma(e) \, I(\vec{b}) \right] b \, db$$

(3.118)

In evaluating equation (3.118), values for $\sigma(e)$ and $B(e)$ were taken from the compilations of Hellwege (1973) and Benary, Price, and Alexander (1970). The nuclear single-particle densities in equation (3.115) were extracted from the charge density data in De Jager, De Vries, and De Vries (1974) using the detailed procedure of Wilson and Costner (1975).

3.6.3. Isotope production cross section.

Up to this point, all nucleons have been treated as identical objects. To differentiate between protons and neutrons, equation (3.111) is replaced by (Hüfner, Schäfer, and Schürmann, 1975)

$$\sigma_{nz} = \binom{N_P}{n} \binom{Z_P}{z} 2\pi \int \left[1 - P(\vec{b}) \right]^{n+z} P(\vec{b})^{A_P - n - z} b \, db$$

(3.119)

where $P(\vec{b})$ is again given by equation (3.117). In equation (3.119), σ_{nz} is the cross section for abrading n out of N_P neutrons and z out of Z_P protons from the projectile nucleus. Implicit in this expression is the assumption that the neutron and proton distributions in the projectile nucleus are completely uncorrelated. This oversimplification of the actual complex nature of nucleon correlations in nuclei provides an analytically simple and convenient starting point for computing cross sections for specific fragment species.

3.6.4. Results.

Figure 3.7 displays results obtained from equation (3.118) for ^{16}O projectile nuclei colliding with various stationary target nuclei. The incident kinetic energy is 2.1 GeV/nucleon. The shapes of the curves are largely determined by the $2\pi b$ factor and the effect of the spatial variations of ρ_T and ρ_P on $P(\vec{b})$ in the integrand of equation (3.118). The comparatively large cross sections for abrading 1 or 2 nucleons are indicative of the dominance at large impact parameters of the $2\pi b$ factor. Were it not for the large degree of nuclear-matter transparency in this very low density region, these cross sections would be even larger in magnitude. Physically, these theoretical results are not unexpected. In peripheral interactions, the nucleons near the surface are the least tightly bound and are more easily removed than those in the nuclear interior. Because of the short finite range of the nuclear force, abrasion is possible even if the projectile and target densities do not physically overlap.

As the number of abraded nucleons increases, overlap between the projectile and target must occur. This increases the overlapping densities that do not, however, offset the initial decrease in the impact parameter. As a result, the cross sections initially decrease with increasing values of m. Between $m = 5$ and $m = 11$, the cross-section curves flatten as the increasing nuclear densities tend to balance the decrease in the $2\pi b$ factor. For $m \geq 11$, the curves display a marked dependence on the size of the target nucleus. The rapid decrease in σ_m for the ^9Be target indicates that abrasion of all, or nearly all, projectile nucleons by the smaller target is likely to occur only for very small impact parameters. If the target is pure hydrogen (curve not shown), the cross section for abrading all projectile nucleons in one collision, from equation (3.118), is less than 5 nanobarns (nb)—approximately a million times smaller than for the Be target. As target size increases, the abrasion cross sections increase as m increases. This results from the larger geometric area for which the projectile and target volumes completely overlap.

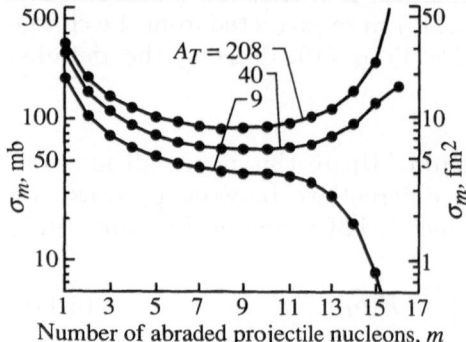

Figure 3.7. Oxygen-target abrasion cross sections σ_m as a function of the number of abraded projectile nucleons m. The lines are merely to guide the eye. Incident kinetic energy is 2.1 GeV/nucleon.

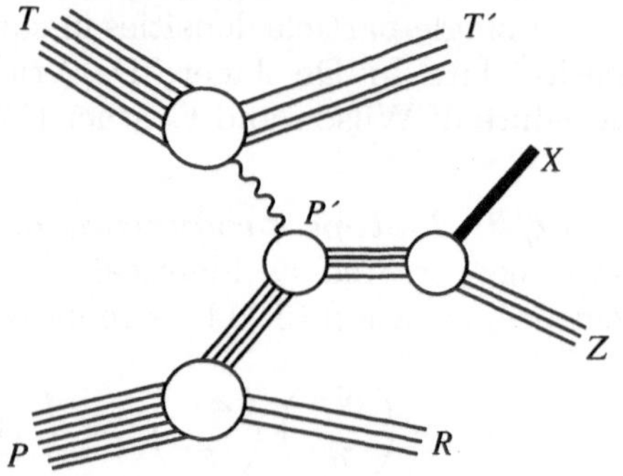

Figure 3.8. Equivalent Feynman diagram (lowest order; no time reversal) of projectile prefragmentation used in this work.

3.7. Abrasion-Ablation Model

In previous work (Townsend, 1984; Townsend et al., 1984b; Townsend, 1983), abrasion-ablation cross sections have been determined by calculating abrasion cross sections that are then multiplied by an ablation probability obtained from compound nucleus decay probabilities. This study demonstrates (Norbury, Townsend, and Deutchman, 1985) that the method of determining abrasion-ablation cross sections arises solely from particular approximations to the general formalism developed herein, and it is therefore only a special case of this more general formalism.

In terms of the transition rate, the total cross section is written as

$$\sigma = \frac{\nu}{v} w \tag{3.120}$$

where ν is the normalization volume and v is the incident velocity of the projectile. The transition rate is given as

$$w = \frac{2\pi}{\hbar} |T_{ki}^{AA}|^2 \, \rho(\epsilon_k) \tag{3.121}$$

where Planck's constant is denoted by $\hbar$ and the transition amplitude is given by

$$T_{ki}^{AA} = \sum_n T_{kn}^{abl} \, G_{ni} \, T_{ni}^{abr} \tag{3.122}$$

where T_{kn}^{abl} is the ablation amplitude, G_{ni} is the propagator, and T_{ni}^{abr} is the abrasion amplitude. The total abrasion-ablation cross section for the phase space associated with figure 3.8 is

$$\sigma(Z) = \frac{2\pi\nu}{\hbar v} \frac{\nu^3}{(2\pi\hbar)^9} \frac{d}{d\epsilon_{ZXRT'}}$$

$$\times \int \int \int |T_{ki}^{AA}|^2 \, d^3p_X \, d^3p_{T'} \, d^3p_Z \tag{3.123}$$

Using a phase space recurrence relation

$$\rho_4(\epsilon_{ZXRT}) = \int \int \int \rho_2(\epsilon_{ZXRT})$$

$$\times \rho_2(\epsilon_{ZXR}) \, \rho_2(\epsilon_{ZX}) \, d\epsilon_{ZX} \, d\epsilon_{ZXR} \tag{3.124}$$

demonstrates that d^3p_X can be replaced by $d^3p_{P'}$ in equation (3.123) where

$$d^3p_{P'} \equiv d^3p_{ZX} \tag{3.125}$$

This, together with equation (3.122), allows the cross section in equation (3.123) to be written as

$$\sigma(Z) = \frac{2\pi\nu}{\hbar v} \frac{\nu^3}{(2\pi\hbar)^9} \frac{d}{d\epsilon_{ZXRT'}} \int \int \int \left| \sum_n T_{kn}^{abl} \, G_{ni} \, T_{ni}^{abr} \right|^2$$

$$\times d^3p_{P'} \, d^3p_{T'} \, d^3p_Z \tag{3.126}$$

A major approximation is now introduced as

$$\left| \sum_n T_{kn}^{abl} \, G_{ni} \, T_{ni}^{abr} \right|^2 \approx \sum_n \left| T_{kn}^{abl} \right|^2 \left| G_{ni} \right|^2 \left| T_{ni}^{abr} \right|^2 \tag{3.127}$$

which will henceforth be referred to as the "classical probability approximation" because it involves the classical addition of probabilities (right-hand side) rather than the quantum mechanical addition of amplitudes (left-hand side). In essence, it involves ignoring the interference terms of the left-hand side of equation (3.127). We believe that the famous Bohr assumption for compound nucleus decay (Blatt and Weisskopf, 1959), which justifies the separation of a two-step cross section (such as compound nucleus formation and decay or abrasion-ablation) into a product of formation and decay (partial width) cross sections, is based upon this classical probability approximation. The Bohr assumption is so widely used because of the reasonableness of the classical argument. Equation (3.127) is sometimes justified quantum mechanically, especially when dealing with angular-momentum matrix elements (Brink and Satchler, 1968) where theorems on Clebsch-Gordan coefficients are available (Norbury, 1983). This is especially true, for example, for a single (one-level) resonant state involving several different angular-momentum projections M where the summation over n simply becomes a summation over M for the single resonance of a particular energy (Brink and Satchler, 1968). This was also the case for the pion production work of Townsend and Deutchman (1981); Deutchman and Townsend (1980 and 1982); Deutchman et al. (1983); Townsend et al. (1984a); and Norbury, Deutchman, and Townsend (1984) where there was only the single intermediate isobar Δ resonance at a fixed energy but with various spin and isotopic spin projections. Norbury (1983) has shown that equation (3.127) results from the spin-isospin Clebsch-Gordan algebra. Another example is the photonuclear excitation of a compound nucleus where the formation of a resonant state of a single energy, but with different spin projections (Norbury et al., 1978), justifies the use of the Bohr assumption when calculating (γ, n) cross sections via compound nucleus formation and decay. In general, however, the preceding simplifications that justify the classical probability assumption do not hold for the abrasion-ablation process. For example, a particular final projectile fragment could result from the ablation of numerous different prefragments, each with a quite different excitation energy.

The partial width, which is simply a transition rate multiplied by Planck's constant, is

$$\Gamma_n = 2\pi \frac{\nu}{(2\pi\hbar)^3} \frac{d}{d\epsilon_{P'}} \int \left| T_{kn}^{abl} \right|^2 d^3 p_Z \tag{3.128}$$

Substituting equations (3.127) and (3.128) into equation (3.126) yields

$$\sigma(Z) = \sum_n \frac{\nu}{\hbar v} \frac{\nu^2}{(2\pi\hbar)^6} \frac{d}{d\epsilon_{ZXRT'}} \int \int \int \Gamma_n \left| G_{ni} \right|^2 \left| T_{ni}^{abr} \right|^2$$
$$\times d^3 p_{P'} \, d^3 p_{T'} \, d\epsilon_{P'} \tag{3.129}$$

which can be rewritten as

$$\sigma(Z) = \sum_n \frac{\nu}{\hbar v} \int \int \Gamma_n \left| G_{ni} \right|^2 \left| T_{ni}^{abr} \right|^2$$
$$\times d \, \mathrm{Nips} \, (\epsilon_{P'RT'}; p_{P'}, p_{R'}, p_{T'}) \, d\epsilon_{P'} \tag{3.130}$$

where d Nips is the noninvariant phase space factor. The abrasion cross section is

$$\sigma_n(A) = \frac{2\pi\nu}{\hbar v} \int \left| T_{ni}^{abr} \right|^2$$
$$\times d \, \mathrm{Nips} \, (\epsilon_{P'RT'}; p_{P'}, p_R, p_{T'}) \tag{3.131}$$

where P' is approximated by the on-shell value. Equation (3.131) yields

$$\sigma(Z) = \frac{1}{2\pi} \sum_n \int \Gamma_n \left| G_{ni} \right|^2 \sigma_n(A) \, d\epsilon_{P'} \tag{3.132}$$

Inserting Green's function, the abrasion-ablation cross section is

$$\sigma(Z) = \frac{1}{2\pi} \sum_n \int \frac{\Gamma_n}{(\epsilon_n - \epsilon_i)^2 + (\Gamma/2)^2} \, \sigma_n(A) \, d\epsilon_{P'} \tag{3.133}$$

where the total Γ and partial widths are related by

$$\Gamma = \sum_n \Gamma_n \tag{3.134}$$

To evaluate the integral in equation (3.133), the zero-width approximation (Pilkuhn, 1967)

$$\lim_{\Gamma \to 0} \frac{\Gamma/2\pi}{(\epsilon_n - \epsilon_i)^2 + (\Gamma/2)^2} = \delta(\epsilon_n - \epsilon_i) \tag{3.135}$$

is introduced. If we write the energies explicitly as

$$\epsilon_n = \epsilon_{P'} + \epsilon_{T'} + \epsilon_R \tag{3.136}$$

with an initial-state energy given by

$$\epsilon_i = \epsilon_P + \epsilon_T \tag{3.137}$$

and the final-state energy as

$$\epsilon_k = \epsilon_X + \epsilon_Z + \epsilon_{T'} + \epsilon_R \tag{3.138}$$

then conservation of energy

$$\epsilon_k = \epsilon_i \tag{3.139}$$

yields

$$\epsilon_n - \epsilon_i = \epsilon_{P'} - (\epsilon_X + \epsilon_Z) \tag{3.140}$$

Inserting equation (3.140) into equation (3.133) indicates a variable, intermediate, virtual resonance energy $\epsilon_{P'}$ centered about $\epsilon_X + \epsilon_Z$, which is integrated over. The nature of the delta function in equation (3.135), however, destroys this quantum mechanical feature of virtual energy in the integral. The zero-width approximation, then, can be considered as another classical approximation. Inserting equations (3.135) and (3.140) into equation (3.133) yields

$$\sigma(Z) = \sum_n \frac{\Gamma_n}{\Gamma} \sigma_n(A) \tag{3.141}$$

If the branching ratio is defined as

$$g_n \equiv \frac{\Gamma_n}{\Gamma} \tag{3.142}$$

and is recognized as the usual ablation probability factor (Townsend et al., 1984b; Townsend, 1983)

$$\sigma(Z) = \sum_n g_n \, \sigma_n(A) \tag{3.143}$$

which is the standard abrasion-ablation cross-section result (Bleszynski and Sander, 1979; Hüfner, Schäfer, and Schürmann, 1975; Townsend et al., 1984b; Townsend, 1983; Bowman, Swiatecki, and Tsang, 1973).

This result (eq. (3.143)) can also be obtained from equation (3.133) by an alternative method. Since $\sigma_n(A)$ is obtained by integrating over all impact parameters, it is independent of $\epsilon_{P'}$. Taking it outside the integral enables equation (3.133) to be written as

$$\sigma(Z) = \frac{1}{2\pi} \sum_n \sigma_n(A) \int \frac{\Gamma_n}{(\epsilon_n - \epsilon_i)^2 + (\Gamma/2)^2} \, d\epsilon_{P'} \qquad (3.144)$$

Inserting

$$\frac{\Gamma}{\Gamma} = 1 \qquad (3.145)$$

inside the integral in equation (3.144) and substituting equation (3.142) yields

$$\sigma(Z) = \frac{1}{2\pi} \sum_n \sigma_n(A) \int g_n \frac{\Gamma}{(\epsilon_n - \epsilon_i)^2 + (\Gamma/2)^2} \, d\epsilon_{P'} \qquad (3.146)$$

If g_n is independent of $\epsilon_{P'}$ (which merely requires Γ_n and Γ to possess the same energy dependence), then it can be taken outside the integral to yield

$$\sigma(Z) = \frac{1}{2\pi} \sum_n g_n \, \sigma_n(A) \int \frac{\Gamma}{(\epsilon_n - \epsilon_i)^2 + (\Gamma/2)^2} \, d\epsilon_{P'} \qquad (3.147)$$

In principle, if the dependence of Γ on $\epsilon_{P'}$ is known, then the integral can be calculated numerically if not analytically. If the zero-width approximation is inserted from equation (3.135), equation (3.143) is again obtained.

Equation (3.143) is one of the central results of the present work. It represents a first-principles derivation of the usual abrasion-ablation cross section and results directly from the following: (1) the time-ordering approximation, (2) the classical probability approximation, and (3) the zero-width approximation. Clearly, then, the most obvious improvements to the abrasion-ablation theory would be to remove these assumptions. (The time-ordering approximation is the least important.)

3.8. Electromagnetic Interactions

So far we have discussed specifically nuclear interaction processes that dominate whenever the impact parameter is less than or equal to the sum of the nuclear radii. At larger impact parameters, the electromagnetic processes dominate because of the long-range interaction of the coulomb field. The elastic coulomb scattering contributes to the beam divergence with negligible energy loss (Rossi and Greisen, 1941; Highland, 1975). The inelastic coulomb scattering contributes to fragmentation of the projectile and target nuclei (Norbury and Townsend, 1990).

The ejection of the particles X from a nucleus by coulomb dissociation is given by

$$\sigma_{EM}(X) = \sum_{\pi\ell} \int_{E_o(x)}^{\infty} \sigma_\gamma^{\pi\ell}(E, X) \, N^{\pi\ell}(E) \, dE \qquad (3.148)$$

where $\pi\ell$ indicates the active electromagnetic moment of the transition ($\pi = E$ or $M, \ell = 1, 2, \ldots$), $N^{\pi\ell}(E)$ is the virtual photon density distribution generated by the passing heavy ion, and $\sigma_\gamma^{\pi\ell}(E, X)$ is the usual photonuclear cross section. The electric dipole (E1) contribution is related to the giant dipole resonance absorption cross section and the Weizsäcker-Williams virtual photon density function (Norbury and Townsend, 1986 and 1990; Norbury, Townsend, and Badavi, 1988; Norbury et al., 1988; Norbury, 1989a and 1989b; Cucinotta, Norbury, and Townsend, 1988). The electric quadrupole (E2) contributions are considered by Norbury (1990) and Norbury and Townsend (1990).

3.9. References

Austern, N., 1963: Direct Reaction Theories. *Fast Neutron Physics. Part II: Experiments and Theory*, J. B. Marion and J. L. Fowler, eds., Interscience Publ., pp. 1113–1216.

Benary, Odette; Price, Leroy R.; and Alexander, Gideon, 1970: *NN and ND Interactions (Above 0.5 GeV/c)—A Compilation.* UCRL-20000 NN, Lawrence Radiation Lab., Univ. of California.

Bertini, Hugo W., 1969: Intranuclear-Cascade Calculation of the Secondary Nucleon Spectra From Nucleon-Nucleus Interactions in the Energy Range 340 to 2900 MeV and Comparisons With Experiment. *Phys. Review*, vol. 188, second ser., no. 4, pp. 1711–1730.

Best, Melvyn E., 1972: Particle-Nucleus Scattering at Intermediate Energies. *Canadian J. Phys.*, vol. 50, no. 14, pp. 1609–1613.

Blatt, John M.; and Weisskopf, Victor F., 1959: *Theoretical Nuclear Physics.* Springer-Verlag.

Bleszynski, M.; and Sander, C., 1979: Geometrical Aspects of High-Energy Peripheral Nucleus-Nucleus Collisions. *Nuclear Phys. A,* vol. 326, nos. 2–3, pp. 525–535.

Bowman, J. D.; Swiatecki, W. J.; and Tsang, C. F., 1973: *Abrasion and Ablation of Heavy Ions.* LBL-2908, Lawrence Berkeley Lab., Univ. of California.

Brink, D. M.; and Satchler, G. R., 1968: *Angular Momentum, Second ed.* Clarendon Press.

Chew, Geoffrey F., 1951: High Energy Elastic Proton-Deuteron Scattering. *Phys. Review*, vol. 84, second ser., no. 5, pp. 1057–1058.

Cucinotta, Francis A.; Norbury, John W.; and Townsend, Lawrence W., 1988: *Multiple Nucleon Knockout by Coulomb Dissociation in Relativistic Heavy-Ion Collisions.* NASA TM-4070.

Czyż, W.; and Maximon, L. C., 1969: High Energy, Small Angle Elastic Scattering of Strongly Interacting Composite Particles. *Ann. Phys. (N.Y.)*, vol. 52, no. 1, pp. 59–121.

De Jager, C. W.; De Vries, H.; and De Vries, C., 1974: Nuclear Charge- and Magnetization-Density-Distribution Parameters From Elastic Electron Scattering. *At. Data & Nuclear Data Tables*, vol. 14, no. 5/6, pp. 479–508.

Deutchman, P. A.; and Townsend, L. W., 1980: Coherent Isobar Production in Peripheral Relativistic Heavy-Ion Collisions. *Phys. Review Lett.*, vol. 45, no. 20, pp. 1622–1625.

Deutchman, P. A.; and Townsend, L. W., 1982: Isobars and Isobaric Analog States. *Phys. Review*, vol. 25, ser. C, no. 2, pp. 1105–1107.

Deutchman, P. A.; Madigan, R. L.; Norbury, J. W.; and Townsend, L. W., 1983: Pion Production Through Coherent Isobar Formation in Heavy-Ion Collisions. *Phys. Lett.*, ser. B, vol. 132, no. 1, 2, 3, pp. 44–46.

Fernbach, S.; Serber, R.; and Taylor, T. B., 1949: The Scattering of High Energy Neutrons by Nuclei. *Phys. Review*, vol. 75, second ser., no. 9, pp. 1352–1355.

Foldy, L. L.; and Walecka, J. D., 1969: On the Theory of the Optical Potential. *Ann. Phys.*, vol. 54, no. 3, pp. 447–504.

Franco, V.; and Glauber, R. J., 1966: High-Energy Deuteron Cross Sections. *Phys. Review*, vol. 142, second ser., no. 4, pp. 1195–1214.

Gabriel, T. A.; Bishop, B. L.; and Lillie, R. A., 1984: *Shielding Considerations for Multi-GeV/Nucleon Heavy Ion Accelerators: The Introduction of a New Heavy Ion Transport Code,* HIT. ORNL/TM-8952 (Contract No. W-7405-eng-26), Oak Ridge National Lab.

Glauber, R. J., 1955: Cross Sections in Deuterium at High Energies. *Phys. Review*, vol. 100, second ser., no. 1, pp. 242–248.

Goldberger, M. L., 1948: The Interaction of High Energy Neutrons and Heavy Nuclei. *Phys. Review*, vol. 74, no. 10, pp. 1269–1277.

Goldberger, Marvin L.; and Watson, Kenneth M., 1964: *Collision Theory.* John Wiley & Sons, Inc.

Gross, Franz, 1965: Relativistic Treatment of Loosely Bound Systems in Scattering Theory. *Phys. Review*, vol. 140, second ser., no. 2B, pp. B410–B421.

Harrington, David R., 1969: Multiple Scattering, the Glauber Approximation, and the Off-Shell Eikonal Approximation. *Phys. Review*, vol. 184, second ser., no. 5, pp. 1745–1749.

Hellwege, K.-H., ed., 1973: *Landolt-Börnstein Numerical Data and Functional Relationships in Science and Technology—Group I: Nuclear and Particle Physics,* Volume 7, *Elastic and Charge Exchange Scattering of Elementary Particles.* Springer-Verlag.

Highland, Virgil L., 1975: Some Practical Remarks on Multiple Scattering. *Nuclear Instrum. & Methods*, vol. 129, no. 2, pp. 497–499.

Hüfner, J.; Schäfer, K.; and Schürmann, B., 1975: Abrasion-Ablation in Reactions Between Relativistic Heavy Ions. *Phys. Review*, vol. 12, ser. C, no. 6, pp. 1888–1898.

Humberston, J. W.; and Wallace, J. B. G., 1970: Deuteron Wave Functions for the Hamada-Johnston Potential. *Nuclear Phys.*, vol. A141, no. 2, pp. 362–368.

Mandelstam, S., 1955: Dynamical Variables in the Bethe-Salpeter Formalism. *Proc. Royal Soc. (London)*, ser. A, vol. 233, no. 1193, pp. 248–266.

Metropolis, N.; Bivins, R.; Storm, M.; Turkevich, Anthony; Miller, J. M.; and Friedlander, G.. 1958: Monte Carlo Calculations on Intranuclear Cascades. I. Low-Energy Studies. *Phys. Review*, vol. 110, second ser., no. 1, pp. 185–203.

Millburn, G. P.; Birnbaum, W.; Crandall, W. E.; and Schecter, L., 1954: Nuclear Radii From Inelastic Cross-Section Measurements. *Phys. Review*, vol. 95, second ser., no. 1, pp. 1268–1278.

Norbury, J. W.; Thompson, M. N.; Shoda, K.; and Tsubota, H., 1978: Photoneutron Cross Sections of ^{54}Fe. *Australian J. Phys.*, vol. 31, no. 6, pp. 471–475.

Norbury, John William, 1983: *Pion Production in Relativistic Heavy Ion Collisions.* Ph.D. Diss., Univ. of Idaho Graduate School.

Norbury, J. W.; Deutchman, P. A.; and Townsend, L. W., 1984: A Particle-Hole Formalism for Pion Production From Isobar Formation and Decay in Peripheral Heavy Ion Collisions. *Bull. American Phys. Soc.*, vol. 29, no. 4, p. 688.

Norbury, John W.; Townsend, Lawrence W.; and Deutchman, Philip A., 1985: *A T-Matrix Theory of Galactic Heavy-Ion Fragmentation.* NASA TP-2363.

Norbury, John W.; and Townsend, Lawrence W., 1986: *Electromagnetic Dissociation Effects in Galactic Heavy-Ion Fragmentation.* NASA TP-2527.

Norbury, John W.; Cucinotta, F. A.; Townsend, L. W.; and Badavi, F. F., 1988: Parameterized Cross Section for Coulomb Dissociation in Heavy-Ion Collisions. *Nuclear Instrum. & Methods Phys. Res.*, vol. B31, no. 4, pp. 535–537.

Norbury, John W.; Townsend, Lawrence W.; and Badavi, Forooz F., 1988: *Computer Program for Parameterization of Nucleus-Nucleus Electromagnetic Dissociation Cross Sections.* NASA TM-4038.

Norbury, John W., 1989a: Comment on "Electromagnetic Dissociation of ^{59}Co and ^{197}Au Targets by Relativistic ^{139}La Projectiles." *Phys. Review C*, vol. 39, third ser., no. 6, pp. 2472–2473.

Norbury, John W., 1989b: Nucleon Emission Via Electromagnetic Excitation in Relativistic Nucleus-Nucleus Collisions: Reanalysis of the Weizsäcker-Williams Method. *Phys. Review C*, vol. 40, third ser., no. 6, pp. 2621–2628.

Norbury, John W., 1990: Electric Quadrupole Excitations in the Interactions of ^{89}Y With Relativistic Nuclei. *Phys. Review C*, vol. 41, third ser., no. 1, pp. 372–373.

Norbury, John W.; and Townsend, Lawrence W., 1990: *Calculation of Two-Neutron Multiplicity in Photonuclear Reactions.* NASA TP-2968.

Pilkuhn, Hartmut, 1967: *The Interactions of Hadrons.* John Wiley & Sons, Inc.

Postma, Herman; and Wilson, Richard, 1961: Elastic Scattering of 146-MeV Polarized Protons by Deuterons. *Phys. Review*, vol. 121, second ser., no. 4, pp. 1229–1244.

Remler, Edward A., 1968: High-Energy Scattering by Nuclei. *Phys. Review*, vol. 176, second ser., no. 5, pp. 2108–2112.

Rossi, Bruno; and Greisen, Kenneth, 1941: Cosmic-Ray Theory. *Reviews Modern Phys.*, vol. 13, no. 4, pp. 240–309.

Santoro, R. T.; Alsmiller, R. G.; Barnes, J. M.; and Corbin, J. M., 1986: *Shielding of Manned Space Stations Against Van Allen Belt Protons: A Preliminary Scoping Study.* ORNL/TM-10040 (Contract No. IDOD3070-0000), Oak Ridge National Lab.

Schimmerling, W.; Devlin, T. J.; Johnson, W.; Vosburgh, K. G.; and Mischke, R. E., 1971: Neutron-Nucleus Total Cross Sections From 900 to 2600 MeV/c. *Phys. Lett.*, vol. 37B, no. 2, pp. 177–180.

Schimmerling, Walter; Devlin, Thomas J.; Johnson, Warren W.; Vosburgh, Kirby G.; and Mischke, Richard E., 1973: Neutron-Nucleus Total and Inelastic Cross Sections: 900 to 2600 MeV/c. *Phys. Review C,* vol. 7, third ser., no.1, pp. 248–262.

Serber, R., 1947: Nuclear Reactions at High Energies. *Phys. Review*, vol. 72, no. 11, pp. 1114–1115.

Townsend, L. W.; and Deutchman, P. A., 1981: Isobar Giant Resonance Formation in Self-Conjugate Nuclei. *Nuclear Phys.,* vol. A355, pp. 505–532.

Townsend, L. W., 1983: Abrasion Cross Sections for ^{20}Ne Projectiles at 2.1 GeV/Nucleon. *Canadian J. Phys.*, vol. 61, no. 1, pp. 93–98.

Townsend, Lawrence W., 1984: *Ablation Effects in Oxygen-Lead Fragmentation at 2.1 GeV/Nucleon.* NASA TM-85704.

Townsend, L. W.; Deutchman, P. A.; Madigan, R. L.; and Norbury, J. W., 1984a: Pion Production Via Isobar Giant Resonance Formation and Decay. *Nuclear Phys. A,* vol. 415, pp. 520–529.

Townsend, Lawrence W.; Wilson, John W.; Norbury, John W.; and Bidasaria, Hari B., 1984b: *An Abrasion-Ablation Model Description of Galactic Heavy-Ion Fragmentation.* NASA TP-2305.

Ûlehla, Ivan; Gomolčák, Ladislav; and Pluhař, Zdeněk (G. Alter, transl.), 1964: *Optical Model of the Atomic Nucleus.* Academic Press Inc.

Watson, Kenneth M., 1953: Multiple Scattering and the Many-Body Problem—Applications to Photomeson Production in Complex Nuclei. *Phys. Review*, vol. 89, second ser., no. 3, pp. 575–587.

Wilson, John W., 1972: *Comparison of Exact and Approximate Evaluations of the Single-Scattering Integral in Nucleon-Deuteron Elastic Scattering.* NASA TN D-6884.

Wilson, J. W., 1973a: Intermediate Energy Nucleon-Deuteron Scattering Theory. *Phys. Lett.,* vol. 44B, no. 2, pp. 125–126.

Wilson, John W., 1973b: Intermediate Energy Nucleon-Deuteron Elastic Scattering. *Nuclear Phys.,* vol. B66, pp. 221–244.

Wilson, John W., 1974a: Proton-Deuteron Double Scattering. *Phys. Review C,* vol. 10, no. 1, pp. 369–376.

Wilson, J. W., 1974b: Multiple Scattering of Heavy Ions, Glauber Theory, and Optical Model. *Phys. Lett.*, vol. B52, no. 2, pp. 149–152.

Wilson, John W., 1975: Composite Particle Reaction Theory. Ph.D. Diss., College of William and Mary in Virginia.

Wilson, John W.; and Costner, Christopher M., 1975: *Nucleon and Heavy-Ion Total and Absorption Cross Section for Selected Nuclei.* NASA TN D-8107.

Wilson, J. W.; and Townsend, L. W., 1981: An Optical Model for Composite Nuclear Scattering. *Canadian J. Phys.*, vol. 59, no. 11, pp. 1569–1576.

4. ELASTIC CHANNEL DATA BASE

4.1. Introduction

The development of the multiple-scattering theory and the corresponding optical model described in chapter 3 had a tremendous unifying effect on the Langley approach to nuclear scattering. The elastic channel amplitude could be reasonably represented by the free two-body scattering amplitudes and the ground state nuclear matter densities. From the elastic channel amplitude, one obtains the values of elastic differential cross sections, total elastic cross sections, and (by the optical theorem) the total cross sections. Armed with these new methods, a search for adequate nuclear matter density functions was undertaken (Wilson, 1975). Matter densities were derived from charge density distributions, and the Woods-Saxon distributions gave the best overall agreement with the neutron experiments of Schimmerling et al. (1971, 1973) and Palevsky et al. (1967) as seen in figures 4.1 and 4.2. Energy dependence was introduced through the usual analytic form for the two-body amplitudes, which is

$$f_{\text{NN}}(\vec{q}) = \frac{1}{4\pi} \, \sigma(e) k_{\text{NN}} [\alpha(e) + i] \exp\left[-\frac{1}{2} B(e) \vec{q}^2 \right] \qquad (4.1)$$

where $\vec{q}$ is the momentum transfer, $\sigma(e)$ is the total cross section at kinetic energy e, k_{NN} is the wave number, $\alpha(e)$ is the ratio of real to imaginary part, and $B(e)$ is the slope parameter. In the first data base derived by Wilson and Costner (1975), the nuclear matter densities below $A_T = 17$ were taken as Gaussian and densities above $A_T = 16$ as Woods-Saxon. Results for copper targets are shown in figures 4.3 and 4.4. Calculated absorption cross sections for various projectiles and targets are shown in figures 4.5 through 4.8 with experimental results (Lindstrom et al., 1975; Cheshire et al., 1974; Jakobsson and Kullberg, 1976; Antonchik et al., 1981). (See Wilson and Townsend (1981) for details.) The matter densities for light nuclei ($3 \leq Z \leq 8$) were subsequently replaced by Townsend (1982) with harmonic well functions, and Pauli correlations were added to modify the free two-body

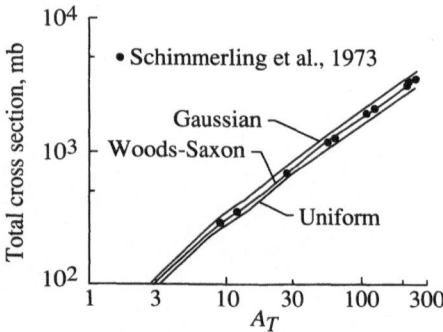

Figure 4.1. Total nucleon-nucleus cross section at ≈ 1 GeV as function of nuclear mass number for three model single-particle densities.

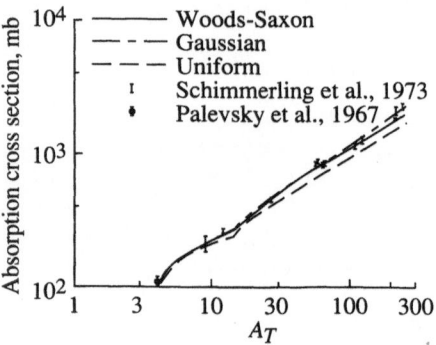

Figure 4.2. Nucleon-nucleus absorption cross section at ≈ 1 GeV function of nuclear mass number for three model single-particle densities.

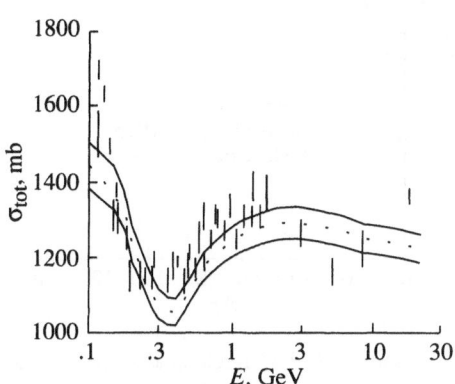

Figure 4.3. Nucleon-copper total cross section as function of laboratory energy. Curves give range of uncertainty.

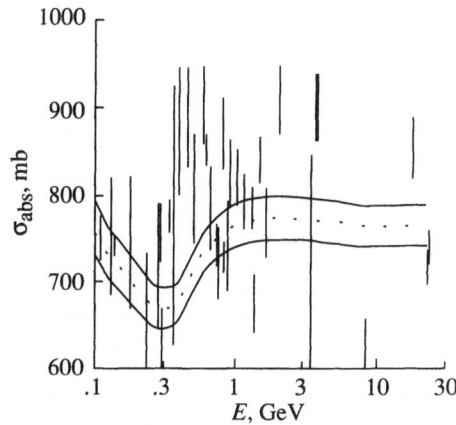

Figure 4.4. Nucleon-copper absorption cross section as function of laboratory energy. Curves give range of uncertainty.

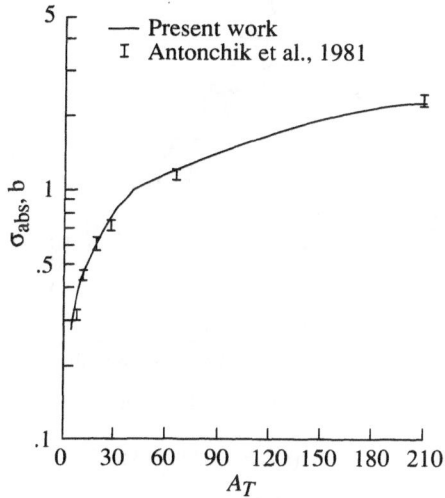

Figure 4.5. ^{4}He nucleus absorption cross sections at 3.6 GeV/nucleon.

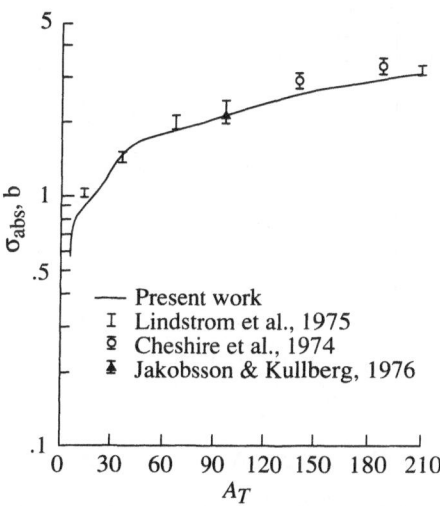

Figure 4.6. ^{16}O-nucleus absorption cross sections at 2.1 GeV/nucleon.

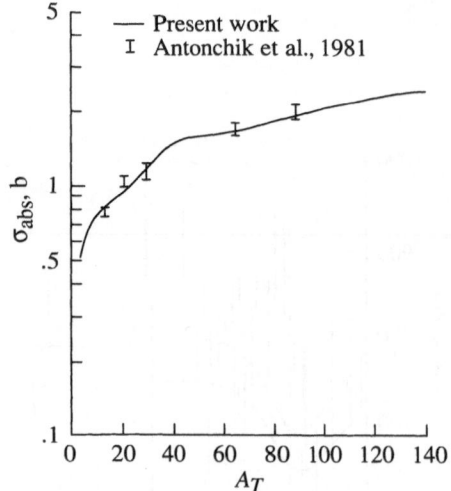

Figure 4.7. ^{12}C-nucleus absorption cross sections at 3.6 GeV/nucleon.

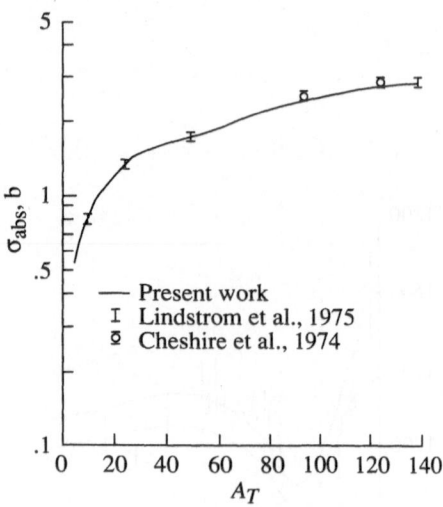

Figure 4.8. ^{12}C-nucleus absorption cross sections at 2.1 GeV/nucleon.

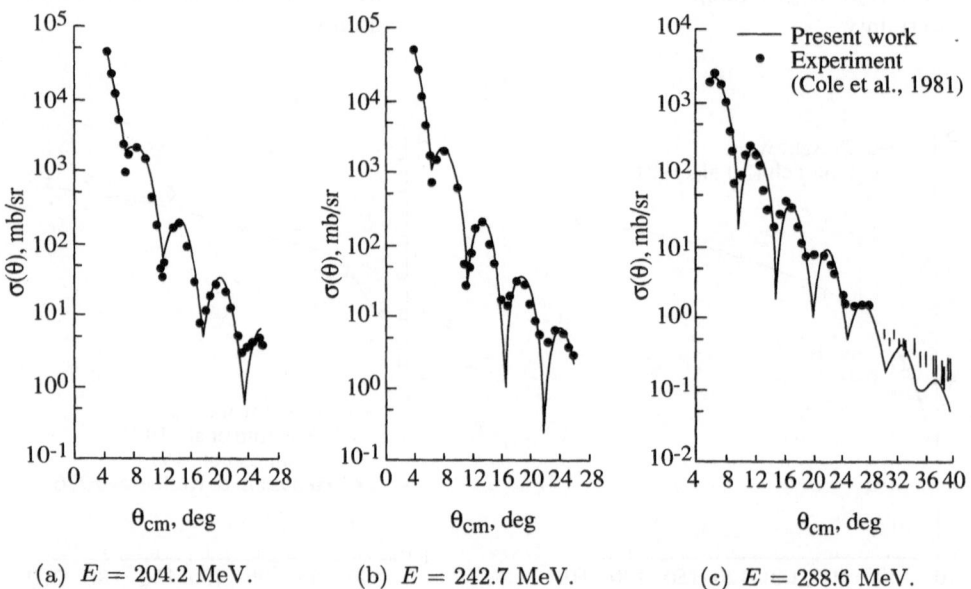

(a) $E = 204.2$ MeV.   (b) $E = 242.7$ MeV.   (c) $E = 288.6$ MeV.

Figure 4.9. ^{12}C-^{12}C elastic differential cross sections as function of center-of-mass scattering angle.

amplitudes. The low-energy elastic scattering required a partial wave analysis after which Bidasaria, Townsend, and Wilson (1983) found good agreement with scattering experiments (Cole et al., 1981) as shown in figure 4.9. The final data base uses the charge form factor data compiled by De Jager, De Vries, and De Vries (1974).

Although model developments for meson (Hong et al., 1989) and antinucleon (Buck et al., 1986 and 1987) data bases are underway, we will only discuss the nucleonic and heavy ion data as they now exist for space-radiation shielding.

4.2. Optical Model Cross Sections

The nucleus-nucleus potential (Wilson, 1975; Wilson and Townsend, 1981) including Pauli correlation effects (Townsend, 1982) is

$$W(\vec{x}) = A_P A_T \int d^3 \vec{\xi}_T \rho_T (\vec{\xi}_T) \int d^3 \vec{y} \ \rho_P(\vec{x} + \vec{y} + \vec{\xi}_T)$$

$$\times \tilde{t}(e, \vec{y})[1 - \tilde{C}(\vec{y})] \tag{4.2}$$

This potential was derived from an optical model potential approximation to the exact composite-particle multiple-scattering series.

The collision absorption (incoherent) cross sections are given by

$$\sigma_{\text{abs}} = 2\pi \int_0^\infty (1 - \exp\{-2 \ \text{Im}[\chi(\vec{b})]\}) b \ db \tag{4.3}$$

where the complex phase function, in terms of the reduced potential U, is

$$\chi(\vec{b}) = -\frac{1}{2k} \int_{-\infty}^\infty U(\vec{b}, z) \ dz \tag{4.4}$$

and the reduced (coherent) potential is

$$U(\vec{x}) = 2m A_P A_T (A_P + A_T)^{-1} W(\vec{x}) \tag{4.5}$$

where m is the nucleon mass, A_P is the nuclear mass number of the projectile, and A_T is the nuclear mass number of the target.

In equation (4.2), $\tilde{t}$ is the constituent-averaged, energy-dependent, two-body transition amplitude

$$\tilde{t}(e, \vec{y}) = -\left(\frac{e}{m}\right)^{1/2} \sigma(e)[\alpha(e) + i][2\pi B(e)]^{-3/2} \exp\left[\frac{-\vec{y}^2}{2B(e)}\right] \tag{4.6}$$

and the correlation function is taken to be

$$\tilde{C}(\vec{y}) = 0.25 \ \exp\left(\frac{-k_F^2 \vec{y}^2}{10}\right) \tag{4.7}$$

For the analyses of this work, the Fermi momentum is assumed to be that of infinite nuclear matter, $k_F = 1.36$ fm^{-1}.

4.2.1. Nuclear density distributions. The correct nuclear density distributions ρ_j $(j = P, T)$ to use in equation (4.2) are the nuclear ground state, single-particle number densities for the collision pair. Since these are not experimentally known, the number densities are obtained from their experimental charge density distributions by assuming that

$$\rho_c(\vec{r}) = \int \rho_p(\vec{r}\,')\rho_A(\vec{r} + \vec{r}\,')\, d^3\vec{r}\,' \tag{4.8}$$

where ρ_c is the nuclear charge distribution, ρ_p is the proton charge distribution, and ρ_A is the desired nuclear single-particle density. All density distributions in equation (4.8) are normalized to unity. The proton charge distribution is taken to be the usual Gaussian form and is

$$\rho_p(\vec{r}) = \left(\frac{3}{2\pi r_p^2}\right)^{3/2} \exp\left(\frac{-3r^2}{2r_p^2}\right) \tag{4.9}$$

where $r_p = 0.87$ fm is the proton root-mean-square charge radius (Borkowski et al., 1975).

When the projectile is a nucleon, equation (4.8) yields a delta function for ρ_A:

$$\rho_A(\vec{r} + \vec{r}\,') = \delta(\vec{r} + \vec{r}\,') \tag{4.10}$$

because ρ_c and ρ_p are identical.

For nuclei lighter than neon ($A < 20$), the nuclear charge distribution is the harmonic well (HW) form given by De Jager, De Vries, and De Vries (1974) as

$$\rho_c(\vec{r}) = \rho_o \left[1 + \gamma \left(\frac{r}{a}\right)^2\right] \exp\left(\frac{-r^2}{a^2}\right) \tag{4.11}$$

where ρ_o is the normalization constant, r is the radial coordinate, and a and γ are charge parameters. Values for a and γ used herein are given in table 4.1. Substituting equations (4.9) and (4.11) into equation (4.8) yields (Townsend, 1982)

$$\rho_A(\vec{r}) = \frac{\rho_o a^3}{8s^3} \left(1 + \frac{3\gamma}{2} - \frac{3\gamma a^2}{8s^2} + \frac{\gamma a^2 r^2}{16s^4}\right) \exp\left(\frac{-r^2}{4s^2}\right) \tag{4.12}$$

where

$$s^2 = \frac{a^2}{4} - \frac{r_p^2}{6} \tag{4.13}$$

For neon and heavier nuclei ($A > 20$), the nuclear charge distribution is taken to be the Woods-Saxon (WS) form

$$\rho_c(\vec{r}) = \frac{\rho_o}{1 + \exp[(r - R)/c]} \tag{4.14}$$

where R is the radius at half-density, and the surface diffuseness c is related to the nuclear skin thickness t through

$$c = \frac{t}{4.4} \tag{4.15}$$

Values for R and t used herein are given in table 4.1. Most values in table 4.1 are taken from De Jager, De Vries, and De Vries (1974). Inserting equations (4.9) and (4.14) into equation (4.8) yields, after some simplification (Wilson and Costner, 1975), a number density ρ_A that is of the WS form (see eq. (4.14)) with the same R, but different overall normalization factor ρ_o and surface thickness. The latter is given by

$$t_A = \frac{8.8r_p}{3^{1/2}} \left[\ln \left(\frac{3\beta - 1}{3 - \beta} \right) \right]^{-1} \tag{4.16}$$

where

$$\beta = \exp \left(\frac{4.4r_p}{t_c 3^{1/2}} \right) \tag{4.17}$$

with t_c noting the charge skin thickness obtained by using equation (4.15) and the charge distribution surface diffuseness values listed by De Jager, De Vries, and De Vries (1974).

Table 4.1. Nuclear Charge Distribution Parameters
From Electron Scattering Data

Nucleus	Distribution (a)	γ or t, fm (b)	a, fm or R, fm (b)
^{2}H	HW	0	1.71
^{4}He	HW	0	1.33
^{7}Li	HW	0.327	1.77
^{9}Be	HW	0.611	1.791
^{11}B	HW	0.811	1.69
^{12}C	HW	1.247	1.649
^{14}N	HW	1.291	1.729
^{16}O	HW	1.544	1.833
^{20}Ne	WS	2.517	2.74
^{27}Al	WS	2.504	3.05
^{40}Ar	WS	2.693	3.47
^{56}Fe	WS	2.611	3.971
^{64}Cu	WS	2.504	4.20
^{80}Br	WS	2.306	4.604
^{138}Ba	WS	2.621	5.517
^{108}Ag	WS	2.354	5.139
^{208}Pb	WS	2.416	6.624

[a]The harmonic well (HW) distribution (eq. (4.11)) is used for $A < 20$ and the Woods-Saxon (WS) distribution (eq. (4.14)) for $A \geq 20$.

[b]γ and a are for HW distributions and t and R are for WS distributions.

4.2.2. Nucleon-nucleon scattering parameters. The nucleon-nucleon cross sections $\sigma(e)$ used in the energy-dependent, two-body transition amplitude (eq. (4.6)) are obtained by performing a spline interpolation of values taken from various compilations (Benary, Price, and Alexander, 1970; Schopper, 1973 and 1980; Binstock, 1974). The results are displayed in figures 4.10 and 4.11 as a function of incident kinetic energy. No curve for neutron-neutron cross sections is displayed because only limited quantities of experimental data exist for these collisions. For computation purposes, we assumed that the proton-proton values for each energy listed adequately represented the neutron-neutron cross sections. Details of the constituent averaging for $\sigma(e)$ are given by Wilson and Costner (1975).

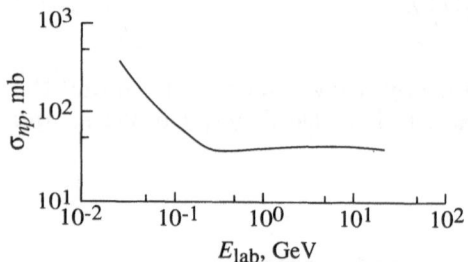 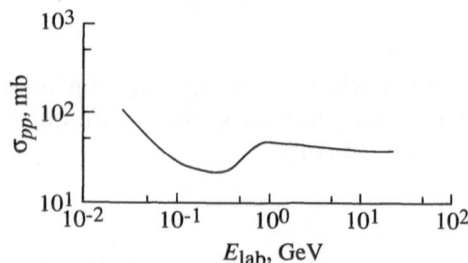

Figure 4.10. Neutron-proton total cross section as function of incident kinetic energy.

Figure 4.11. Proton-proton total cross section as function of incident kinetic energy.

Since scattering at these energies is mainly diffractive, the nucleon-nucleon slope parameters $B(e)$ are those appropriate to purely diffractive scattering. From Ringia et al. (1972) these are given by

$$B(e) = 10 + 0.5 \ \ln \left(\frac{s'}{s_o} \right) \tag{4.18}$$

where s' is the square of the nucleon-nucleon center-of-mass energy and $s_o = 1 \ (\text{GeV}/c)^{-2}$. Typical values from equation (4.18), displayed in figure 4.12, differ markedly from the nondiffractive compilation values of $B \approx 5 \ (\text{GeV}/c)^{-2}$ used previously by Townsend, Wilson, and Bidasaria (1983a and 1983b). The improved agreement between theory and experiment obtained with equation (4.18) is clearly demonstrated by Bidasaria and Townsend (1983). Values of the parameter $\alpha(e)$ are not required for these analyses, because only the imaginary part of equation (4.6) is used in equations (4.3) and (4.4).

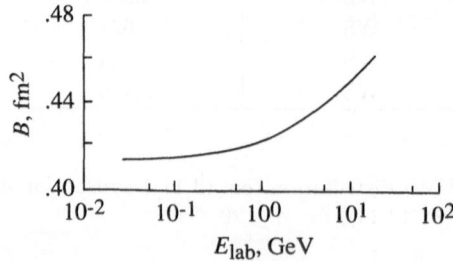

Figure 4.12. Nucleon-nucleon scattering slope parameter as function of incident kinetic energy.

4.2.3. Results. With the formalism described in sections 4.2.1 and 4.2.2, absorption cross sections for nucleons, deuterons, and selected heavy ions colliding with various target nuclei have been calculated.

Theoretical predictions for nucleon-nucleus scattering and representative experimental results of Schimmerling et al. (1973); Renberg et al. (1972); and Barashenkov, Gudima, and Toneev (1969) are presented in figures 4.13 through 4.18. Also displayed are the predictions using the empirical parameterization of Letaw, Silberberg, and Tsao (1983). The agreement between theory, empirical predictions, and experimental data is good.

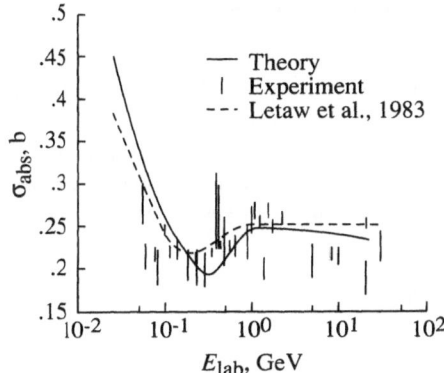

Figure 4.13. Nucleon-carbon absorption cross sections as function of incident nucleon kinetic energy.

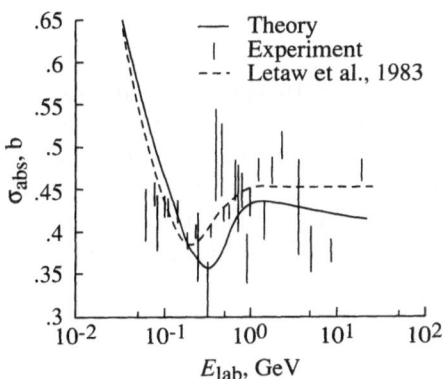

Figure 4.14. Nucleon-aluminum absorption cross sections as function of incident nucleon kinetic energy.

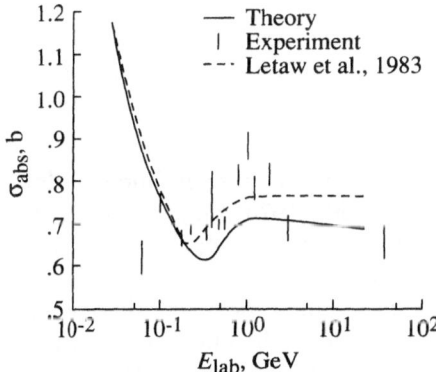

Figure 4.15. Neutron-iron absorption cross sections as function of incident nucleon kinetic energy.

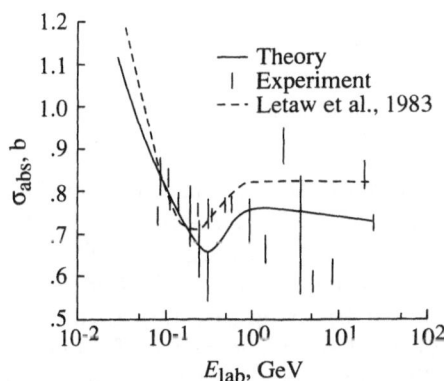

Figure 4.16. Nucleon-copper absorption cross sections as function of incident nucleon kinetic energy.

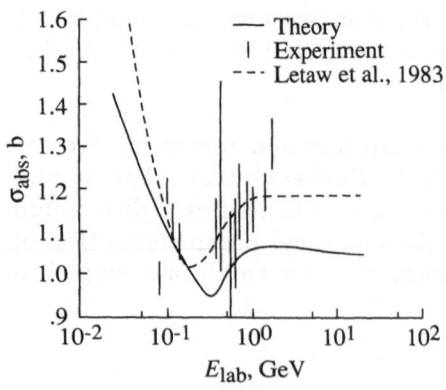

Figure 4.17. Nucleon-silver absorption cross sections as function of incident nucleon kinetic energy.

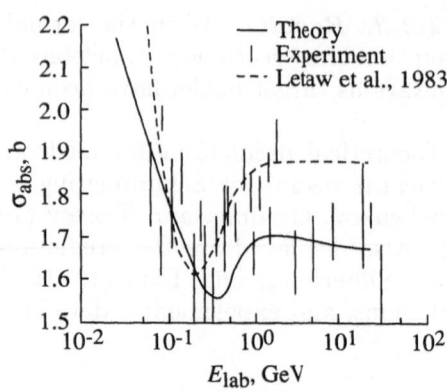

Figure 4.18. Nucleon-lead absorption cross sections as function of incident nucleon kinetic energy.

Figures 4.19 and 4.20 compare the theoretical predictions for deuteron-helium and deuteron-carbon scattering with experimental results from Jaros et al. (1978). For the helium target, theory and experiment agree to within 1 percent of the quoted cross sections, and the theory is well within the uncertainty in the experiment. For the carbon target, the disagreement between theory and experiment is less than 3 percent.

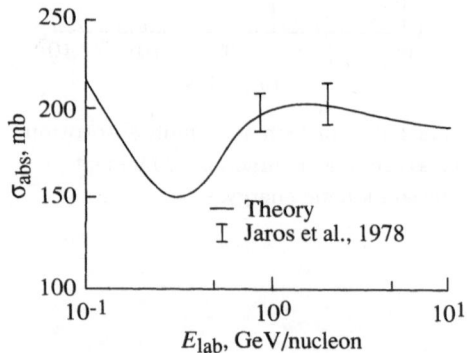

Figure 4.19. Absorption cross sections for deuteron-helium scattering as function of incident kinetic energy.

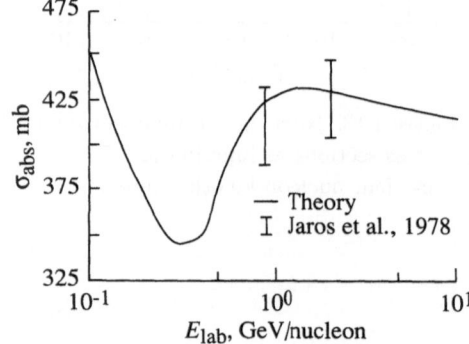

Figure 4.20. Absorption cross sections for deuteron-carbon scattering as function of incident kinetic energy.

Heavy ion absorption cross sections are presented along with experimental data (Jaros et al., 1978; Heckman et al., 1978; Cheshire et al., 1974; Skrzypczak, 1980; Jakobsson and Kullberg, 1976; Cole et al., 1981; Kox et al., 1984; Buenerd et al., 1984; Kullberg et al., 1977; Perrin et al., 1982; Antonchik et al., 1981; Westfall et al., 1979) in figures 4.21 through 4.30. The agreement between theory and experiment is excellent, even for energies lower than 100 MeV/nucleon, where the validity of the eikonal formalism is questionable (Vary and Dover, 1974). Further details are given by Townsend and Wilson (1985).

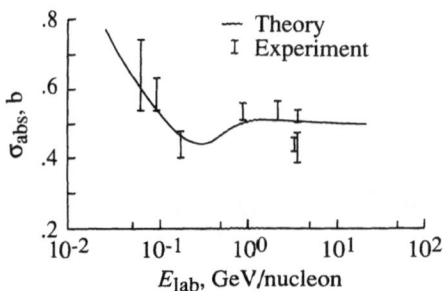

Figure 4.21. Absorption cross sections for helium-carbon scattering as function of incident kinetic energy.

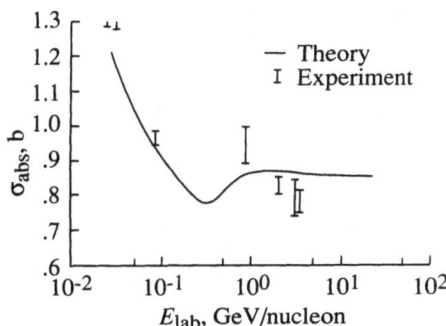

Figure 4.22. Absorption cross sections for carbon-carbon scattering as function of incident kinetic energy.

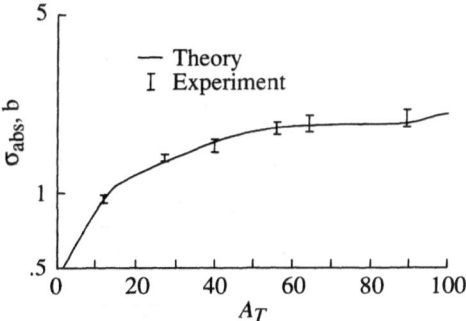

Figure 4.23. Absorption cross sections for carbon projectiles at 83 MeV/nucleon as function of target mass number.

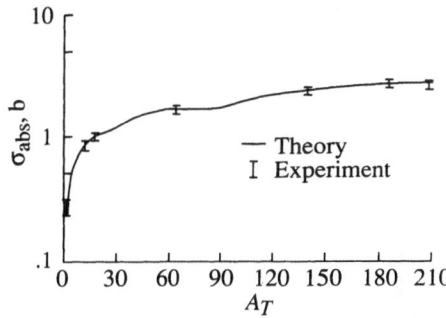

Figure 4.24. Absorption cross sections for carbon projectiles at 2.1 GeV/nucleon as function of target mass number.

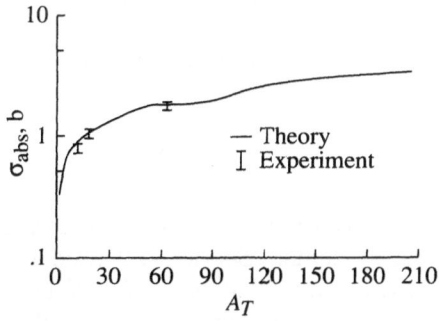

Figure 4.25. Absorption cross sections for carbon projectiles at 3.6 GeV/nucleon as function of target mass number.

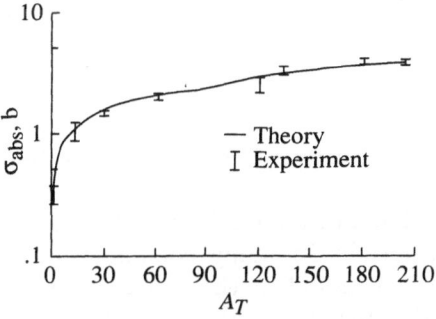

Figure 4.26. Absorption cross sections for oxygen projectiles at 2.1 GeV/nucleon as function of target mass number.

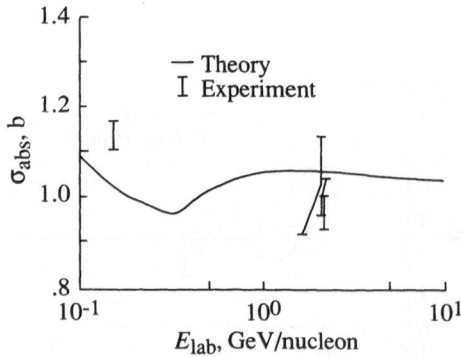

Figure 4.27. Absorption cross sections
for oxygen-emulsion scattering
as function of energy.

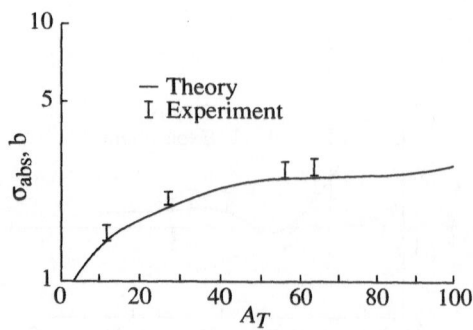

Figure 4.28. Absorption cross sections
for neon projectiles at 30 MeV/nucleon
as function of target mass number.

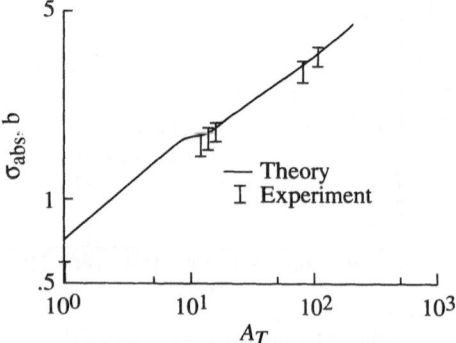

Figure 4.29. Absorption cross sections
for iron projectiles at 1.88 GeV/nucleon
as function of target mass number
with experimental data obtained
with emulsion target.

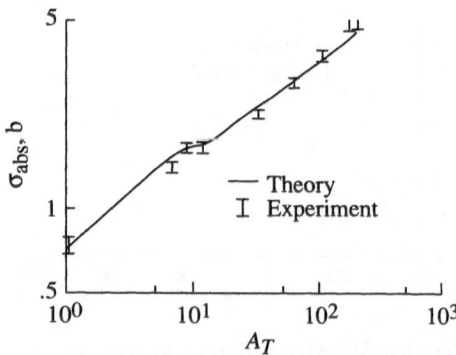

Figure 4.30. Absorption cross sections
for iron projectiles at 1.88 GeV/nucleon
as function of target mass number
with experimental data obtained
for removal of one or more nucleons.

4.3. Coupled-Channel Formalism

The optical model is extremely successful in describing the elastic scattering amplitude for many combinations of interacting systems. Section 4.2 used the optical model in the coherent amplitude approximation (Wilson, 1975; Wilson and Costner, 1975). This section represents the work of Cucinotta et al., 1989, and evaluates noncoherent contributions to the elastic scattering amplitude.

The coupled-channel (CC) Schrödinger equation for heavy ion scattering can be solved in the eikonal approximation (Wilson, 1975; Feshbach and Hüfner, 1970; Dadić, Martinis, and Pisk, 1971) resulting in the following matrix of scattering amplitudes

$$f(\vec{q}) = \frac{-ik}{2\pi} \int \exp(-i\vec{q} \cdot \vec{b})\{\exp[i\chi(\vec{b})] - 1\} \, d^2b \qquad (4.19)$$

where the boldface quantities represent matrices, k is the projectile momentum relative to the center of mass, $\vec{b}$ is the projectile impact-parameter vector, $\vec{q}$ is the momentum transfer, and $\boldsymbol{\chi}(\vec{b})$ is the eikonal phase matrix. For a projectile transition from quantum states n to n' and target transition from ν to ν', we write

$$\chi_{n\nu,n'\nu'}(\vec{b},z) = \sum_{\alpha j}^{A_P,A_T} \frac{-\mu}{2k} \int_{-\infty}^{z} dz' \left\langle n\nu|t_{\alpha j}|n'\nu'\right\rangle \tag{4.20}$$

where $t_{\alpha j}$ is the free-particle, two-body amplitude in the overall center-of-mass frame and μ is the projectile-target reduced mass. The matrix elements of $\boldsymbol{\chi}$ are given by equation (4.20) with $z \to \infty$. Equation (4.19) holds only if the commutator (Feshbach and Hüfner, 1970)

$$\left[\boldsymbol{\chi}(\vec{b},z), \frac{d\boldsymbol{\chi}(\vec{b},z)}{dz}\right] = 0 \tag{4.21}$$

Assuming this commutation relation will hold effectively eliminates all reflection terms and reduces the optical potential solution to Watson's form of the nucleus-nucleus multiple-scattering series (within small-angle and high-energy approximations) to the Glauber series (Wilson and Townsend, 1981). This can be seen by considering an element of $\boldsymbol{f}$ and expanding the exponential in equation (4.19):

$$f_{n\nu,n'\nu'}(\vec{q}) = \frac{-ik}{2\pi} \int \exp(-i\vec{q}\cdot\vec{b})\left(i\left\langle n\nu|\hat{\chi}|n'\nu'\right\rangle\right.$$

$$\left. -\frac{1}{2}\sum_{m,\mu}\left\langle n\nu|\hat{\chi}|m\mu\right\rangle\left\langle m\mu|\hat{\chi}|n'\nu'\right\rangle + \ldots\right) d^2b \tag{4.22}$$

which is equivalent to

$$f_{n\nu,n'\nu'}(\vec{q}) = \frac{ik}{2\pi} \int \exp(-i\vec{q}\cdot\vec{b})\left\langle n\nu|[1-\exp(i\hat{\chi})]|n'\nu'\right\rangle d^2b \tag{4.23}$$

Upon introduction of the two-body profile function, we arrive at the Glauber form for the nucleus-nucleus scattering amplitude (Franco and Nutt, 1978)

$$f_{n\nu,n'\nu'}(\vec{q}) = \frac{ik}{2\pi} \int \exp(-i\vec{q}\cdot\vec{b})\left\langle n\nu\left|\left(1 - \prod_{\alpha j}[1-\Gamma_{\alpha j}(\vec{b}-\vec{s}_\alpha-\vec{s}_j)]\right)\right|n'\nu'\right\rangle d^2b \tag{4.24}$$

We note that we have not considered the question of noncommutating interactions. Also, the eikonal CC approach is based on an ansatz for the optical-model CC equation wave function; therefore, we have not rigorously considered the connection between Watson's form of the nucleus-nucleus multiple-scattering series and the Glauber approximation. Such considerations can be found in Wallace (1975). Having shown the equivalence of the CC approach to the Glauber approximation, we next consider the second-order solution to equation (4.19) for the elastic channel, which we will compare with the second-order optical phase shift approximation to the Glauber amplitude.

The second-order approximation to the elastic amplitude is obtained by including all transitions between the ground and excited states and assuming that transitions between excited states are negligible. Furthermore, the densities of all excited states are approximated by an average excited-state density. The phase matrix is then of the bordered form

$$\boldsymbol{\chi}(\vec{b}) = \begin{pmatrix} \chi_{\text{el}} & \chi_{00,01} & \chi_{00,10} & \chi_{00,11} & \cdots \\ \chi_{01,00} & \chi_{\text{exc}} & 0 & 0 & \cdots \\ \chi_{10,00} & 0 & \chi_{\text{exc}} & 0 & \cdots \\ \chi_{11,00} & 0 & 0 & \chi_{\text{exc}} & \cdots \\ \vdots & \vdots & \vdots & \ddots & \vdots \end{pmatrix} \tag{4.25}$$

where $\chi_{\text{el}} = \chi_{00,00}$. The characteristic equation of this bordered matrix is

$$(\chi_{\text{exc}} - \lambda)^{N-2}[(\chi_{\text{el}} - \lambda)(\chi_{\text{exc}} - \lambda) - \Upsilon^2] = 0, \tag{4.26}$$

where N is the rank of χ, λ is the eigenvalue, and Υ^2 is defined by

$$\Upsilon^2(\vec{b}) = \sum_{n \text{ or } \nu \neq 0} \chi_{00,n\nu}\chi_{n\nu,00} \tag{4.27}$$

The eigenvalues are then given by

$$\lambda_{1,2} = \frac{1}{2}(\chi_{\text{el}} + \chi_{\text{exc}}) \pm \left\{ \left[\frac{1}{2}(\chi_{\text{el}} - \chi_{\text{exc}})\right]^2 + \Upsilon^2 \right\}^{1/2} \tag{4.28}$$

with all others taking the value χ_{exc}. The form of the eigenvalues allows us to treat the scattering system as an effective two-channel problem with

$$\chi = \begin{pmatrix} \chi_{\text{el}} & \Upsilon \\ \Upsilon & \chi_{\text{exc}} \end{pmatrix} \tag{4.29}$$

Then using Sylvester's theorem (Merzbacher, 1970), we find

$$f_{CC}^{(2)}(\vec{q}) = \frac{-ik}{2\pi} \int \exp(-i\vec{q} \cdot \vec{b}) \left\{ \exp\left[\frac{1}{2}i(\chi_{\text{el}} + \chi_{\text{exc}})\right] \left[\cos\left(\chi_{\text{dif}}^2 + \Upsilon^2\right)^{1/2} \right. \right.$$

$$\left. \left. + i\chi_{\text{dif}} \frac{\sin\left(\chi_{\text{dif}}^2 + \Upsilon^2\right)^{1/2}}{\left(\chi_{\text{dif}}^2 + \Upsilon^2\right)^{1/2}} \right] - 1 \right\} d^2b \tag{4.30}$$

where $\chi_{\text{dif}} = \frac{1}{2}(\chi_{\text{el}} - \chi_{\text{exc}})$. An examination of equation (4.30) reveals, as expected, that χ_{exc} appears only in third-order and higher order terms in $f_{\text{NN}}(\vec{q})$.

As discussed by Feshbach and Hüfner (1970) a reasonable approximation to χ_{exc} is to assume the ground-state density for the excited states. If χ_{exc} is set equal to χ_{el}, we find

$$f_{CC}^{(2)}(\vec{q}) \approx \frac{-ik}{2\pi} \int \exp(-i\vec{q} \cdot \vec{b})[\exp(i\chi_{\text{el}}) \cos \Upsilon - 1] \, d^2b \qquad (4.31)$$

The coherent approximation (Wilson, 1975; Wilson and Townsend, 1981) is recovered in the limit of small Υ.

Using closure to perform the summations in equation (4.27) and transforming from the overall center-of-mass (CM) frame to the nucleon-nucleon (NN) CM frame using nonrelativistic kinematics, Υ^2 is

$$\Upsilon^2(\vec{b}) = A_P A_T \left(\frac{1}{2\pi k_{\text{NN}}}\right)^2 \int d^2q \, d^2q' \, \exp(-i\vec{q} \cdot \vec{b}) \exp(-i\vec{q}' \cdot \vec{b})$$

$$\times f_{\text{NN}}(\vec{q}) f_{\text{NN}}(\vec{q}') \left[-A_P A_T F^{(1)}(\vec{q}) F^{(1)}(\vec{q}') G^{(1)}(-\vec{q}) G^{(1)}(-\vec{q}') \right.$$

$$+ (A_P - 1)(A_T - 1) F^{(2)}(\vec{q}, \vec{q}') G^{(2)}(-\vec{q}, -\vec{q}')$$

$$+ (A_T - 1) F^{(1)}(\vec{q} + \vec{q}') G^{(2)}(-\vec{q}, -\vec{q}')$$

$$+ (A_P - 1) F^{(2)}(\vec{q}, \vec{q}') G^{(1)}(-\vec{q} - \vec{q}')$$

$$\left. + F^{(1)}(\vec{q} + \vec{q}') G^{(1)}(-\vec{q} - \vec{q}') \right] \qquad (4.32)$$

where $F^{(1)}$ and $F^{(2)}$ $\left(G^{(1)} \text{ and } G^{(2)}\right)$ are one- and two-body ground-state form factors, respectively, for the projectile (target). The last term on the right-hand side of equation (4.32) is a self-correlation term that appears through the use of closure. The physical meanings of the other terms in equation (4.32) have been discussed by Franco and Nutt (1978).

The optical phase shift expansion given by Franco and Nutt (1978) to the Glauber approximation is written

$$f_{\text{Glauber}}(\vec{q}) = \frac{-ik}{2\pi} \int \exp(-i\vec{q} \cdot \vec{b}) \left[\exp(i\chi_{\text{opt}}) - 1\right] \, d^2b \qquad (4.33)$$

with

$$\chi_{\text{opt}} = \chi_1 + \chi_2 \dots. \qquad (4.34)$$

In comparison, we note that $\chi_1 = \chi_{\text{el}}$, and dropping the last term in Υ^2 yields $i\chi_2 = -\frac{1}{2}\Upsilon^2$. Approximating the density of all excited states by the ground-state density yields almost the same results for the coupled-channel and Glauber optical models:

$$f_{CC}^{(2)}(\vec{q}) = \frac{-ik}{2\pi} \int \exp(-i\vec{q} \cdot \vec{b}) \left[\exp\left(i\chi_{\text{el}}\right)\left(1 - \frac{1}{2}\Upsilon^2 + \frac{1}{24}\Upsilon^4 - \dots\right) - 1\right] d^2b$$

$$(4.35)$$

and

$$f_{\text{Glauber}}^{(2)}(\vec{q}) = \frac{-ik}{2\pi} \int \exp(-i\vec{q} \cdot \vec{b}) \left[\exp(iX_{\text{el}}) \left(1 - \frac{1}{2}\Upsilon^2 + \frac{1}{8}\Upsilon^4 - \cdots \right) - 1 \right] d^2 b \tag{4.36}$$

if $\frac{1}{2}\Upsilon^2 \ll 1$. Note that $\frac{1}{2}\Upsilon^2 \ll 1$ is found to be true for light collision pairs, whereas for large mass number nuclei, this condition should at least hold at large impact parameters, where most of the scattering occurs but may give rise to differences and should be further studied.

We now consider the evaluation of the elastic amplitude for α-α scattering. At high energies, only the central piece of the NN amplitude will be important in spin-0–spin-0 scattering. Therefore, we use the following parameterization:

$$f_{\text{NN}}(\vec{q}) = \frac{1}{4\pi} k_{\text{NN}}\sigma(\alpha + i)\exp\left(-\frac{1}{2}Bq^2\right) \tag{4.37}$$

The isospin-averaged values for the parameters σ, B, and α at the energies considered in this paper are listed in table 4.2. For the calculation of X_{el}, we use the following parameterization for the ^{4}He charge form factor (McCarthy, Sick, and Whitney, 1977):

$$F_{\text{ch}}(\vec{q}) = [1 - (aq)^{12}]\exp(-bq^2) \tag{4.38}$$

with $a = 0.316$ fm and $b = 0.681$ fm^2. The charge form factor F_{ch} is related to the matter form factor F by $F = F_{\text{ch}}/F_P$ with $F_P = \exp\left(-\frac{1}{6}r_P^2 q^2\right)$, where $r_P = 0.86$ fm. We also include coulomb effects in the usual way, assuming just the first term in equation (4.38).

Table 4.2. Nucleon-Nucleon Parameters

E, MeV	σ, mb	B, fm^{-2}	α
635	3.93	0.132	-0.39
1050	4.4	0.25	-0.28

The ^{4}He correlations caused by CM recoil are important. It is well-known that the CM motion can only be treated exactly for shell-model, harmonic-oscillator wave functions. Therefore, we use the harmonic-oscillator CM correction factor in our calculations such that the intrinsic one- and two-body form factors that appear in equation (4.32) are written in terms of model form factors F_M such as

$$F^{(1)}(\vec{q}) = \frac{F_M^{(1)}(\vec{q})}{F_{\text{CM}}(\vec{q})} \tag{4.39}$$

and

$$F^{(2)}(\vec{q}, \vec{q}\,') = \frac{F_M^{(2)}(\vec{q}, \vec{q}\,')}{F_{\mathrm{CM}}(\vec{q} + \vec{q}\,')} \tag{4.40}$$

with $F_{\mathrm{CM}}(\vec{q}) = \exp\left[-\left(R^2/4A\right)q^2\right]$. For ^{4}He, we use $R^2 = 1.94$ fm^2. The model two-particle density described below will be integrated to obtain these form factors.

In the Jastrow (1955) method (see Frullani and Mougey, 1984), if three-particle and higher particle correlations are ignored, we write

$$\rho_M^{(2)}(\vec{r}, \vec{r}\,') = N_n \rho_s^{(1)}(\vec{r})\rho_s^{(1)}(\vec{r}\,')|g(\vec{r}, \vec{r}\,')|^2 \tag{4.41}$$

where $\rho_s^{(1)}$ is the uncorrelated single-particle density, $\rho_s \propto \exp\left(-r^2/R^2\right)$, and N_n is the normalization constant. The correlation factor can be written in terms of the nucleon-nucleon relative momentum distribution as (Frullani and Mougey, 1984)

$$g(\vec{r}, \vec{r}\,') = 1 - \int \exp[i\vec{p}_r \cdot (\vec{r} - \vec{r}\,')]N(\vec{p}_r)\, d\vec{p}_r \tag{4.42}$$

where $\vec{p}_r$ is the NN relative momentum vector. We parameterize $N(\vec{p}_r)$ according to Akaishi (1984) as

$$N(\vec{p}_r) = C\left[\exp\left(\frac{-p_r^2}{a_1}\right) + S\exp\left(\frac{-p_r^2}{a_2}\right)\right] \tag{4.43}$$

with $a_1 = 5.4$ fm^{-2}, $a_2 = 4a_1$, $S = 0.015$, and C as the normalization. The first term on the right-hand side of equation (4.43) can be attributed to a Hartree-Fock-type correlation with the value of a_1, leading to a correlation length of about 0.8 fm, upon comparison with the usual Gaussian parameterization of the correlation factor. The higher momentum component in equation (4.43) should reflect the true dynamical correlations (Akaishi, 1984).

An average excited-state phase can be a complicated quantity to calculate. ^{4}He has many resonance states lying below an excitation of 40 MeV that should contribute. Some calculations are available (Liu, Zamick, and Jaqaman, 1985), but states higher in the continuum should be more dominant. Because this phase element only appears at third order in f_{NN}, a simple model will suffice to show that significant deviations from the ground-state phase are of negligible importance in the double scattering region. Since the form factor for this state must approach unity as $\vec{q} \to 0$, we choose a Gaussian and consider deviations from the ground state through

$$R_{\mathrm{exc}} = R(1 + \delta R) \tag{4.44}$$

In figures 4.31 and 4.32, we show our predictions compared with 635 MeV/nucleon (Berger et al., 1980) and 1050 MeV/nucleon (Satta et al., 1984) α-α scattering data. As can be seen, the Glauber and CC calculations, with excited-state phase approximated by the ground-state phase, are virtually indistinguishable at all momentum transfers. We find no appreciable differences between the results for

$\chi_{exc} = \chi_{el}$ and $\delta R = 0.5$ until the second minima at both energies. The effect of this phase on the double scattering region appears to have been overestimated by Feshbach and Hüfner (1970). No conclusions can be made at larger angles because three-body correlations should become important there. As noted by previous authors (Franco and Nutt, 1978) and as can be seen in figures 4.31 and 4.32, the differences between first- and second-order calculations become significant for increasing angles in α-α scattering. The second-order effects should be large enough to distinguish between models for the two-body density, for example, the Jastrow method used here and the more phenomenological Gaussian parameterization of the correlation factor that is used by Franco and Nutt (1978).

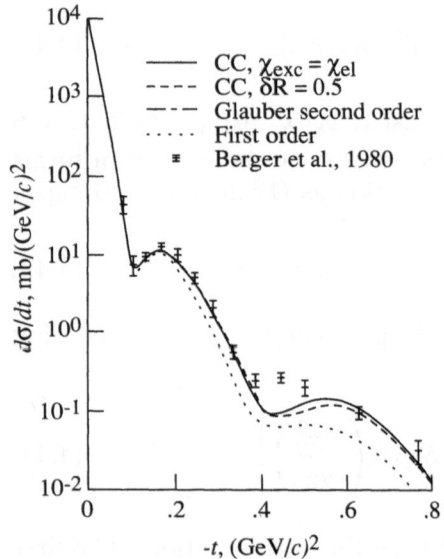

Figure 4.31. The α-α elastic differential cross section at 5.05 GeV/c.

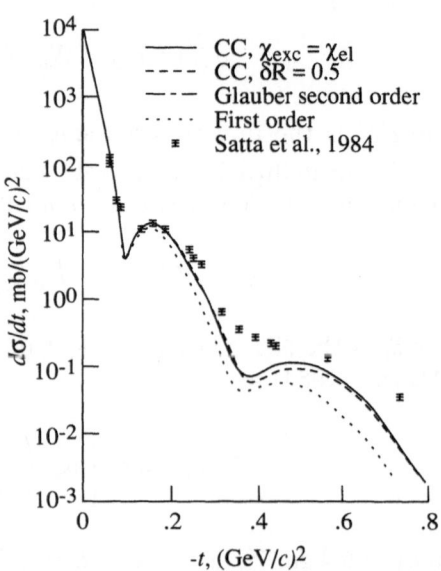

Figure 4.32. The α-α elastic differential cross section at 7 GeV/c.

In conclusion, the coupled-channels, semiclassical approximation of an optical-model solution to Watson's form of the nucleus-nucleus multiple-scattering series has been shown to be equivalent to the Glauber approximation. A second-order solution to the elastic channel obtained by neglecting all transitions between nuclear-excited states was found to be almost identical to the second-order optical phase shift expansion of the Glauber series. An average excited-state phase was seen to be of minor importance in studying the role of correlations in α-α scattering. The α-α scattering data studied is expected to be sensitive to how the two-body density is modeled and should warrant further study.

4.4. Parametric Cross Sections

In sections 4.2 and 4.3, we discussed basic theoretical issues required for an accurate understanding of the interaction process. This understanding is necessary to fill in gaps in experimental data and to further develop reactive theories. The rest of this chapter is a review of parametric representation of results for transport code input.

4.4.1. Total nuclear cross sections. After many decades of experimental activity at various accelerators with ever increasing energies, the cross sections for two-nucleon interactions are reasonably well-defined. Although recent advances in the theory of the two-nucleon interaction in terms of phenomenological meson exchange models (Gross, 1974) show considerable success, a simple parameterization of the experimental data is sufficient for our purposes. For $E \geq 25$ MeV, the proton-proton (pp) total cross section (mb) is found to be reasonably approximated by

$$\sigma_{pp}(E) = \left(1 + \frac{5}{E}\right)\left\{40 + 109\cos\left(0.199\sqrt{E}\right)\exp\left[-0.451\left(E - 25\right)^{0.258}\right]\right\} \quad (4.45)$$

and for lower energies, by

$$\sigma_{pp}(E) = \exp\left\{6.51\left[\exp\left(-\frac{E}{134}\right)^{0.7}\right]\right\} \quad (4.46)$$

These forms are compared with experiments above 50 MeV (Lock and Measday, 1970) shown in figure 4.33. For $E \geq 0.1$ MeV, the neutron-proton (np) cross section is taken as

$$\sigma_{np}(E) = 38 + 12\ 500\exp\left[-1.187(E - 0.1)^{0.35}\right] \quad (4.47)$$

and at lower energies, by

$$\sigma_{np}(E) = 26\ 000\exp\left[-\frac{E}{0.282}\right]^{0.3} \quad (4.48)$$

These forms are compared with experiments above 25 MeV (Lock and Measday, 1970) in figure 4.34.

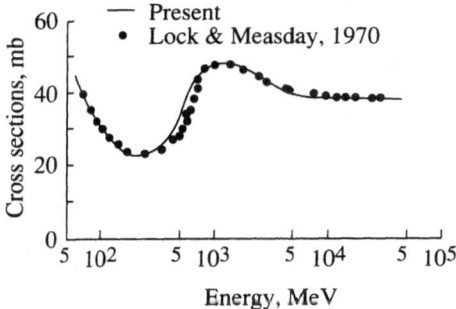

Figure 4.33. Total proton-proton cross sections.

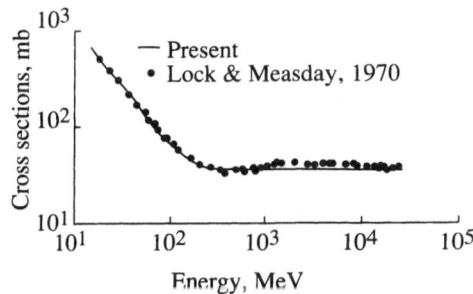

Figure 4.34. Total neutron-proton cross sections.

The low-energy, neutron-nucleus total cross sections exhibit a complicated fine resonance structure over a broad, slowly varying background. This background is marked by very broad Ramsauer resonances that persist even to neutron energies of 100 MeV. Although a simple fundamental theory for the Ramsauer resonances is not available, a semiempirical formalism is given by Angeli and Csikai (1970 and 1971). Their formalism starts with the usual partial wave expansion as

$$\sigma_{\text{tot}} = 2\pi\lambda^2 \sum_{\ell} (2\ell + 1)[1 - \text{Re}(\eta_\ell)] \tag{4.49}$$

with

$$\eta_\ell = \exp(i\delta_\ell) \tag{4.50}$$

where δ_ℓ is the complex phase shift for the ℓth partial wave and $\text{Re}(Z)$ denotes the real part of Z. In the opaque nucleus model, the fact that $n_\ell \approx 1$ for all values of $\ell > R/\lambda$, where R is the nuclear radius, leads Angeli and Csikai to assume that

$$\sigma_{\text{tot}} \approx 2\pi(R + \lambda)^2[1 - \text{Re}(\eta)] \tag{4.51}$$

where $\eta = 0$ gives the usual opaque nucleus result such that

$$\text{Re}(\eta) = \exp\left[-\text{Im}(\delta)\right] \cos\left[\text{Re}(\delta)\right]$$
$$\equiv p\cos\left(qA_T^{1/3} - r\right) \tag{4.52}$$

is a reasonable starting point to parameterize the total cross sections, where $\text{Im}(\delta)$ denotes the imaginary part of δ. Their complete parameterization is

$$\sigma_{\text{tot}} = 2\pi\left(r_0 A_T^{1/3} + \lambda\right)^2 \left[a - p\cos\left(qA_T^{1/3} - r\right)\right] \tag{4.53}$$

where $r_0 = 1.4$ fm, and the neutron wavelength is

$$\lambda = \frac{4.55}{\sqrt{E}} \frac{A_t + 1}{A_t} \tag{4.54}$$

The parameters of Angeli and Csikai (1970 and 1971) are adequately approximated by

$$a = \frac{1}{1 + \left[2/(3.8E + 0.1E\sqrt{E} + 0.1E^3\sqrt{E})\right]} \tag{4.55}$$

$$p = 0.15 - 0.0066\sqrt{E} \tag{4.56}$$

$$q = 2.72 - 0.203\sqrt{E} \tag{4.57}$$

$$r = \min\left\{-5.3 + 1.66\sqrt{E}; 1.3\right\} \tag{4.58}$$

Strictly speaking, equations (4.53) to (4.58) apply only to $A_T \geq 40$ and $0.5 \leq E \leq 40$ MeV. A simple extension to all values of A_T and $0.1 \leq E \leq 100$ MeV gives qualitatively similar results to the experimental data and provides a starting point to representing the total cross section. The cross sections given by equations (4.53) through (4.58) are shown in figure 4.35. These should be compared with the experimental data (Hughes and Schwartz, 1958) shown in figure 4.36. Note that the data in figure 4.36 have only the broad resonances shown. The very narrow resonances have been averaged. We now seek some pure empirical modification to the Angeli-Csikai cross sections to better approximate the total cross sections.

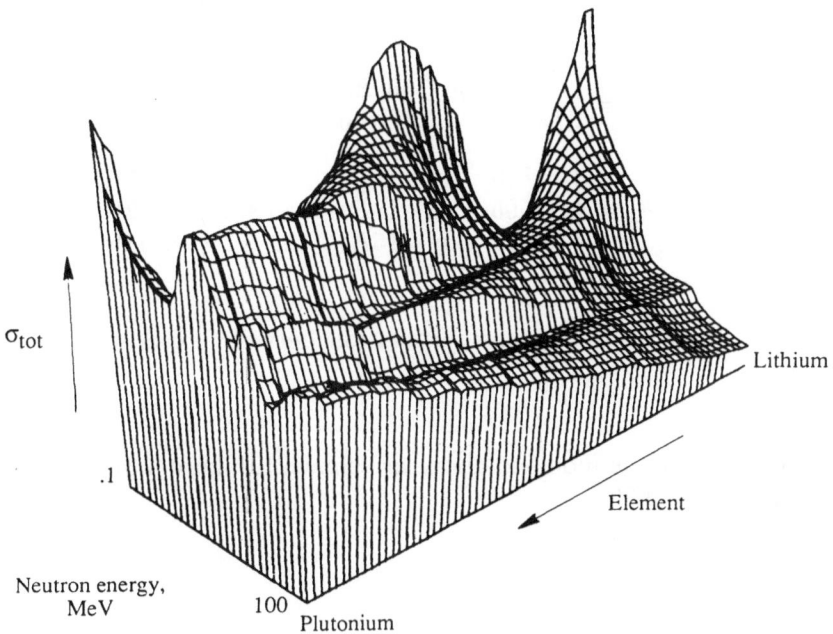

Figure 4.35. Total neutron-nucleus cross section according to Ramsauer resonance formalism.

Our modifications to the Angeli-Csikai formalism are as follows:

1. If $A_T > 75$, then a is taken as 0.18 for values of equation (4.55) less than 0.18

2. The value of p is taken to be greater than $0.4a$ unless $A_T > 76$ for which p can be as small as $0.3a$

3. A modifying factor of $1 + D \exp(-\alpha E)$ is used with

$$D = \begin{cases} 0.5 & (145 < A_T < 235) \\ 1.0 & \text{(Otherwise)} \end{cases}$$

and

$$\alpha = \begin{cases} 1.0 & (205 < A_T < 235) \\ 2.0 & \text{(Otherwise)} \end{cases}$$

4. An additional modifying factor is applied as

$$F_1\left\{1 - 0.5\exp\left[\frac{-(A_T - 63.54)^2}{20}\right]\right.$$

$$\left. - 0.45\exp\left[\frac{-(A_T - 58.71)^2}{4}\right]\exp(-2E) + F_2\right\}$$

where

$$F_1 = \begin{cases} 0.7 & (A_T \leq 63; E \leq 0.8) \\ 1.0 & \text{(Otherwise)} \end{cases}$$

$$F_2 = \begin{cases} 0 & (E > 0.5) \\ -4.95\ \exp(-18E) & (40 \leq A_T < 42) \\ -1.79\ \exp(-15E) & (32 \leq A_T < 34) \end{cases}$$

5. If $A_T < 30$, then numerical interpolation between experimental values is used

The final cross sections as modified above are shown in figure 4.37 and should be compared with figure 4.36

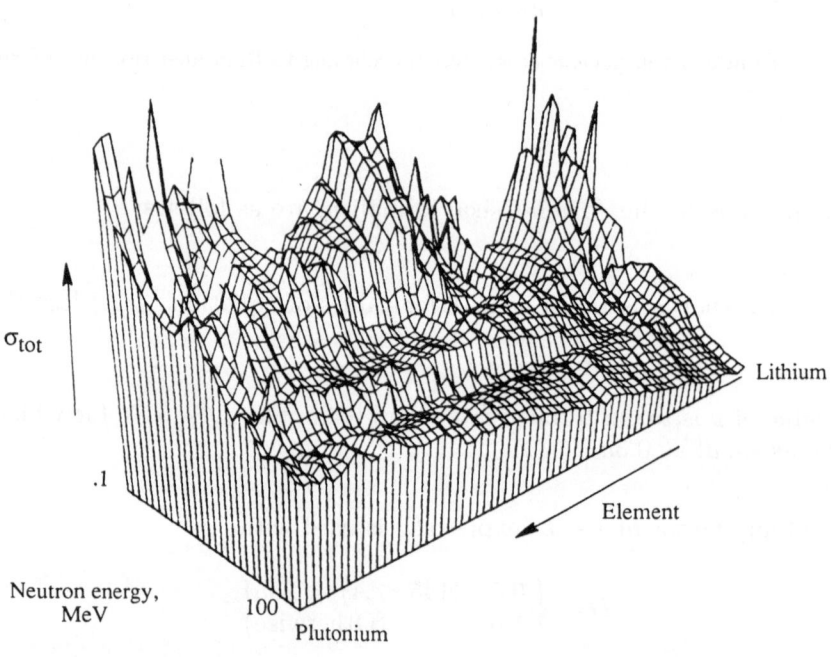

Figure 4.36. Total neutron cross section according to Hughes and Schwartz (1958).

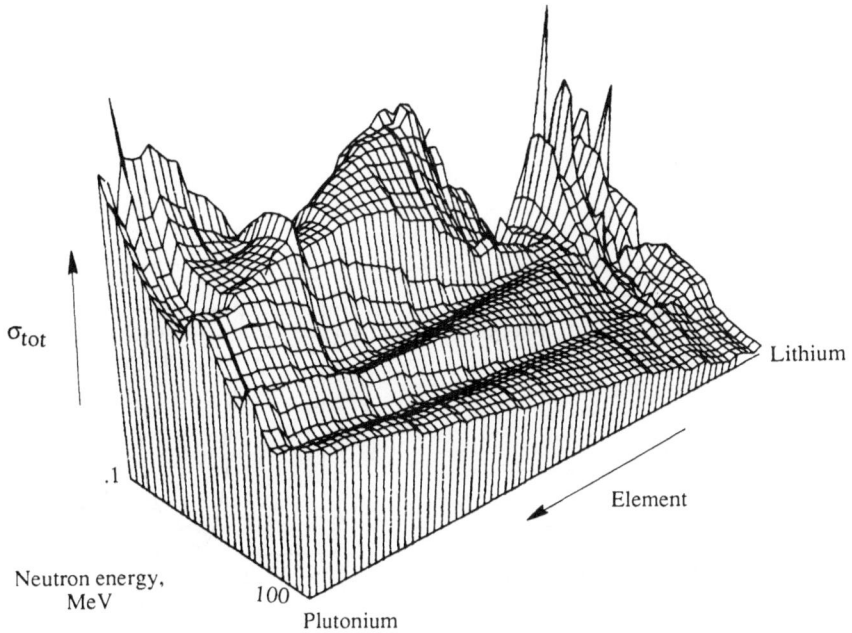

Figure 4.37. Total neutron-nucleus cross section according to present formalism.

The total cross sections above 100 MeV have been taken from Townsend, Wilson, and Bidasaria (1983b). The high-energy cross sections of Townsend, Wilson, and Bidasaria (1983b) have been approximated by

$$\sigma_{\text{tot}}(A_T, E) = 52.5 A_T^{0.758} \left[1 + \left[0.8 + 2.4 \, \exp\left(\frac{-A_T}{30}\right) \right] \, \exp\left(\frac{-E}{135}\right) \, \sin \Theta_E \right]$$
(4.59)

where the phase angle is given by

$$\Theta_E = \left\{ \begin{array}{ll} 14.41 & (E \leq 40 \text{ MeV}) \\ 1.29 \ln^2(E) - \pi & (E > 40 \text{ MeV}) \end{array} \right\}$$
(4.60)

The expressions (4.59) and (4.60) are shown along with the theory of Townsend, Wilson, and Bidasaria (1983b) and a compilation of experiments in figures 4.38 through 4.41. Equations (4.53) through (4.58) are connected smoothly at 70 MeV to the results of equations (4.59) and (4.60) at 130 MeV with an assumed exponential dependence on energy. The total cross section is used to calculate the scattering cross section as

$$\sigma_s(E) = \sigma_{\text{tot}}(E) - \sigma_{\text{abs}}(E)$$
(4.61)

The total (tot) neutron-nucleus cross section is shown with experimental data (Hughes and Schwartz, 1958) in figures 4.42 through 4.45.

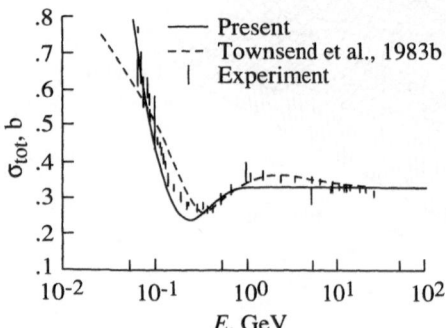

Figure 4.38. Total nucleon-carbon cross sections.

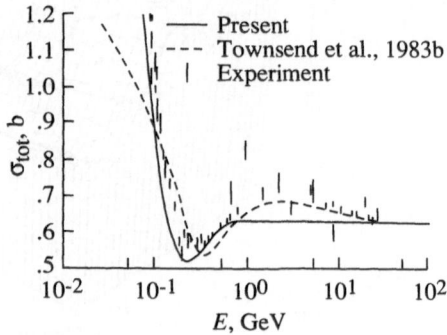

Figure 4.39. Total nucleon-aluminum cross sections.

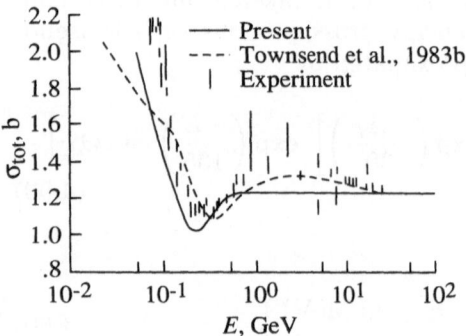

Figure 4.40. Total nucleon-copper cross sections.

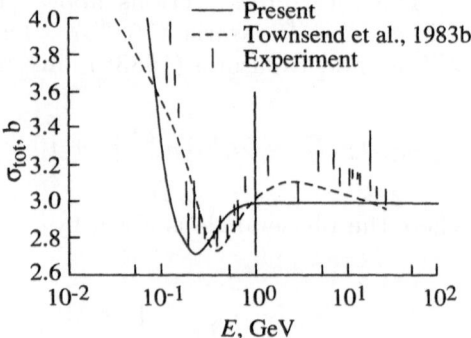

Figure 4.41. Total nucleon-lead cross sections.

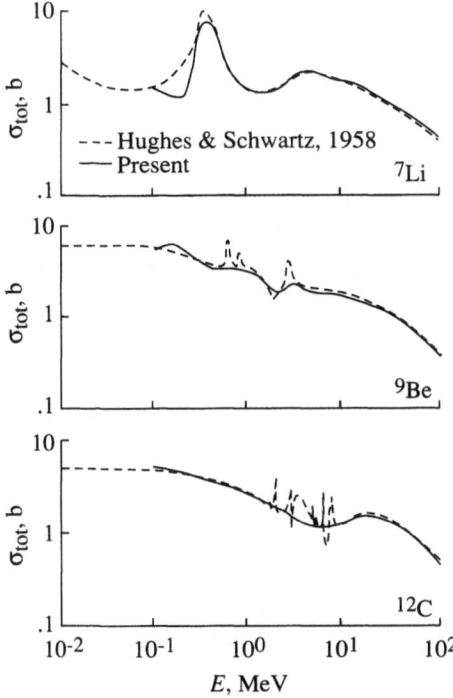

Figure 4.42. Total neutron-nucleus cross sections of ^{7}Li, ^{9}Be, and ^{12}C.

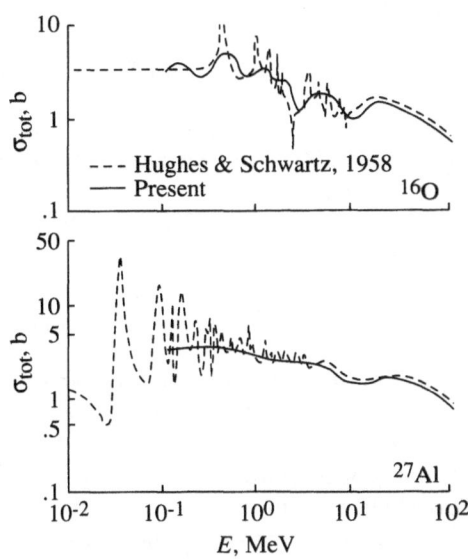

Figure 4.43. Total neutron-nucleus cross sections of ^{16}O and ^{27}Al.

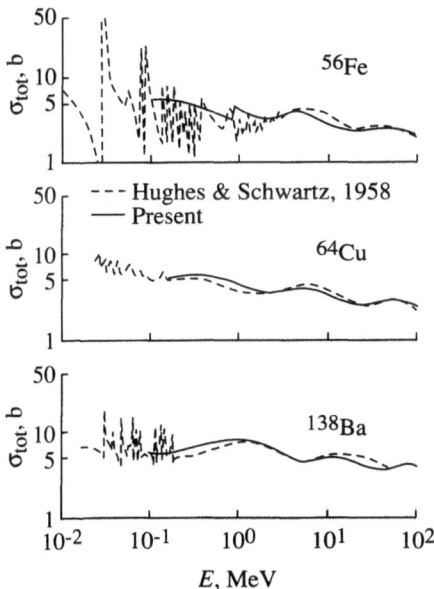

Figure 4.44. Total neutron-nucleus cross sections of ^{56}Fe, ^{64}Cu, and ^{138}Ba.

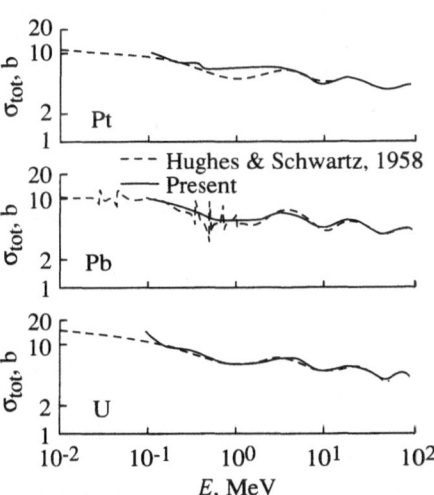

Figure 4.45. Total neutron-nucleus cross sections of Pt, Pb, and U.

4.4.2. Nuclear-absorption cross sections. Qualitatively, the nuclear-absorption cross sections show an energy dependence similar to that observed for the total nuclear cross sections. An analytic formula for protons was derived by Letaw, Silberberg, and Tsao (1983) by first fitting the cross sections of Bobchenko et al. (1979) with the formula

$$\sigma_{\text{abs}} = 45 A_T^{0.7}\{1 + 0.016 \sin[5.3 - 2.63 \ln(A_T)]\} \tag{4.62}$$

where A_T is the mass number of the target nucleus. Equation (4.62) reproduces the Bobchenko data to within ± 2 percent. A somewhat better fit to the Bobchenko data is given by

$$\sigma_{\text{abs}} = 45 A_T^{0.7}(1 - 0.018 \sin \Theta_A) \tag{4.63}$$

where the angle Θ_A is

$$\Theta_A = 2.94 \ln(A_T) + 0.63 \sin[3.92 \ln(A_T) - 2.329] - 0.176 \tag{4.64}$$

Equation (4.63) fits the Bobchenko data to within the 1.2-percent difference, which is on the order of the quoted experimental uncertainty. Although the Bobchenko data represent a consistent set of measurements for many different targets and probably well define the A-dependence of the high-energy cross sections, they may nonetheless be in error in absolute value as suggested by many other independent experiments (Townsend and Wilson, 1985).

Letaw, Silberberg, and Tsao (1983) assume the energy dependence for all nuclei to be the same and to be approximated by

$$f(E) = 1 - 0.62 \exp\left(\frac{-E}{200}\right) \sin\left(10.9 E^{-0.28}\right) \tag{4.65}$$

where the nucleon kinetic energy is in units of MeV. We observe oscillations according to the quantum mechanical calculations of Townsend, Wilson, and Bidasaria (1983b) with phase angle

$$\Theta_E = \begin{cases} 1.44 & (E < 25 \text{ MeV}) \\ 1.33 \ln(E) - 2.84 & (\text{Otherwise}) \end{cases} \tag{4.66}$$

but with an A-dependent amplitude given by

$$f(E) = 1 - \left[0.3 E^{-0.22} + 0.76 \exp\left(\frac{-E}{135}\right)\right]\left[0.4 + 0.9 \exp\left(\frac{-A_t}{30}\right)\right] \sin \Theta_E \tag{4.67}$$

The absorption cross section as given by equations (4.64), (4.66), and (4.67), the fit of Letaw, Silberberg, and Tsao, and various experimental results are given in figures 4.46 through 4.50. As one can see from the figures, a figure of merit is difficult to assign to the fit because great scatter in the data obscures the result. Generally, above 20 MeV the results are on the order of ± 10 percent accurate as estimated from the scatter in the experiments.

Below 20 MeV, the neutron cross sections are represented by numerical data sets at discrete energies of 1, 3, 5, 10, 14, and 20 MeV as taken from Hughes and Schwartz (1958), Stehn et al. (1964), and Brodsky (1978). Interpolated values

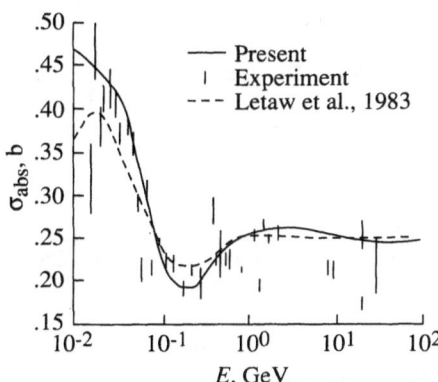

Figure 4.46. Neutron-carbon absorption cross sections.

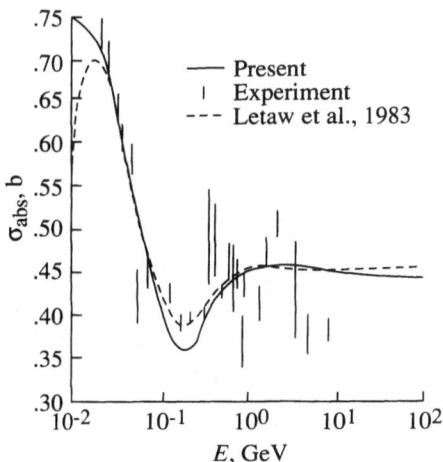

Figure 4.47. Neutron-aluminum absorption cross sections.

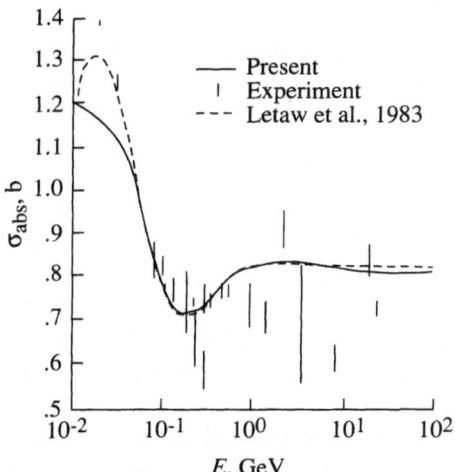

Figure 4.48. Neutron-copper absorption cross sections.

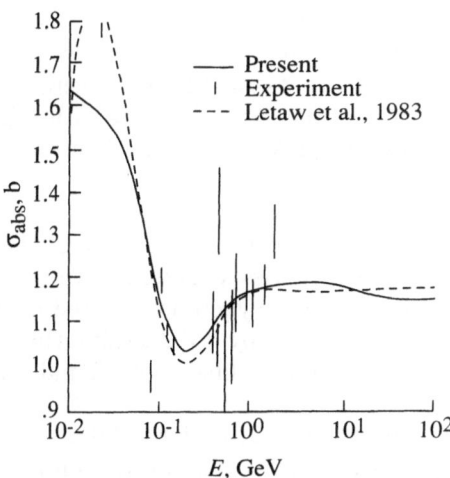

Figure 4.49. Neutron-silver absorption cross sections.

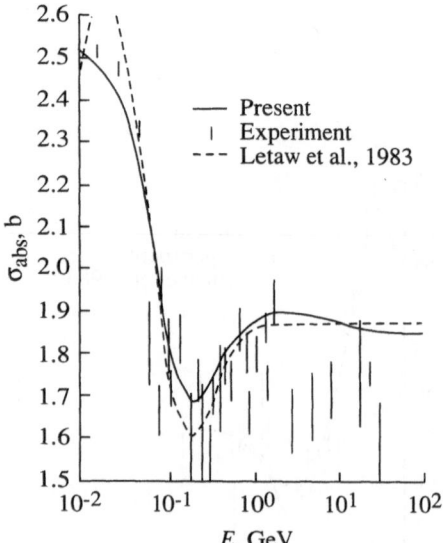

Figure 4.50. Neutron-lead absorption
cross sections.

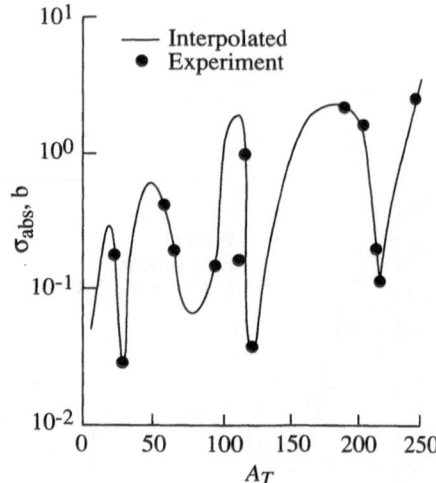

Figure 4.51. Interpolated neutron-nucleus
absorption cross sections at 1 MeV.

between data points at the available target masses are shown in figures 4.51 through 4.56. Intermediate energy values are found according to

$$\sigma(A_T, E) = \sigma(A_T, E_i) \ \exp[-a(E - E_i)] \tag{4.68}$$

where E_i and a are taken according to the appropriate subinterval. The cross sections are assumed to be zero at energies below 0.5 MeV. The absorption cross sections for elements from lithium to plutonium for energies between 1 and 100 MeV are displayed in figure 4.57.

The cross sections presented in this section are probably sufficiently accurate for most applications. Because of their special importance in evaluating radiation quantities in tissue systems, the low-energy neutron cross sections for carbon, nitrogen, and oxygen are treated on a special basis. These neutron cross sections are represented by a data table that was compared with the ENDF/B-V data file compiled by Brookhaven National Laboratory (1982) in figures 4.58 through 4.60.

In section 4.2, we formulated a fully energy-dependent optical model potential approximation to the exact composite particle, multiple-scattering series. The formulation includes the effects of the finite nuclear force, treats Pauli correlations in an approximate way, and has no arbitrarily adjusted parameters. It is applicable to any projectile nucleus of mass number A_P colliding with any target nucleus of mass number A_T at any energy above 25 MeV/nucleon. When used within the context of eikonal scattering theory, which has been shown to be valid (Townsend, Bidasaria, and Wilson, 1983) even at energies as low as 25 MeV/nucleon, the absorption cross sections can be calculated from

$$\sigma_{\mathrm{abs}} = \int d^2\vec{b}(1 - \exp\{-2 \ \mathrm{Im} \ [\chi(\vec{b})]\}) \tag{4.69}$$

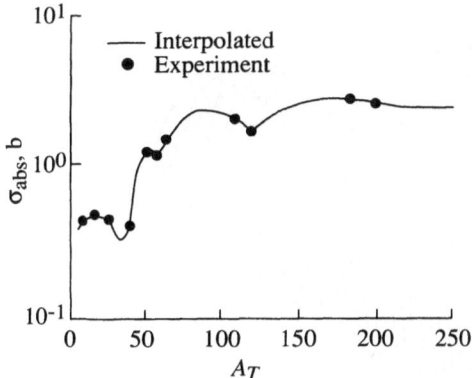

Figure 4.52. Interpolated neutron-nucleus absorption cross sections at 3 MeV.

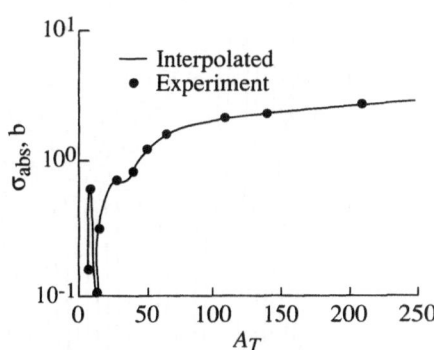

Figure 4.53. Interpolated neutron-nucleus absorption cross sections at 5 MeV.

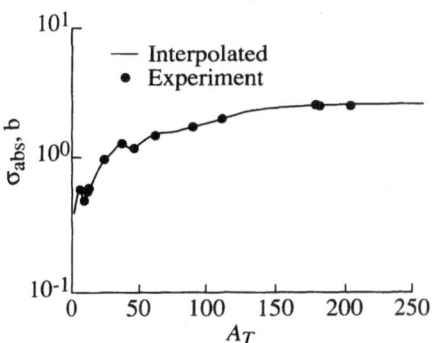

Figure 4.54. Interpolated neutron-nucleus absorption cross sections at 10 MeV.

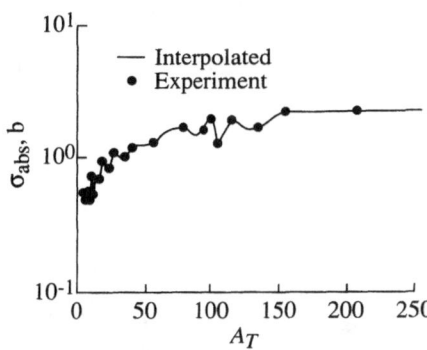

Figure 4.55. Interpolated neutron-nucleus absorption cross sections at 14 MeV.

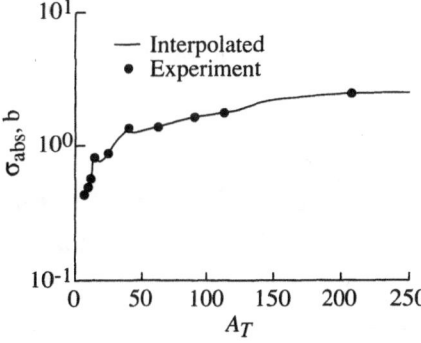

Figure 4.56. Interpolated neutron-nucleus absorption cross sections at 20 MeV.

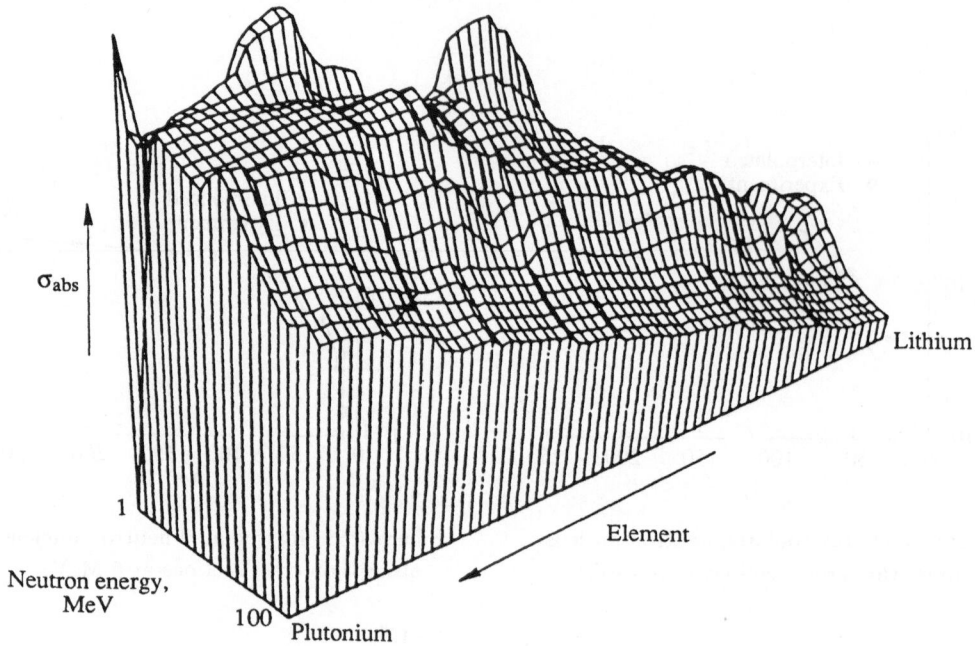

Figure 4.57. Neutron-nucleus absorption cross section.

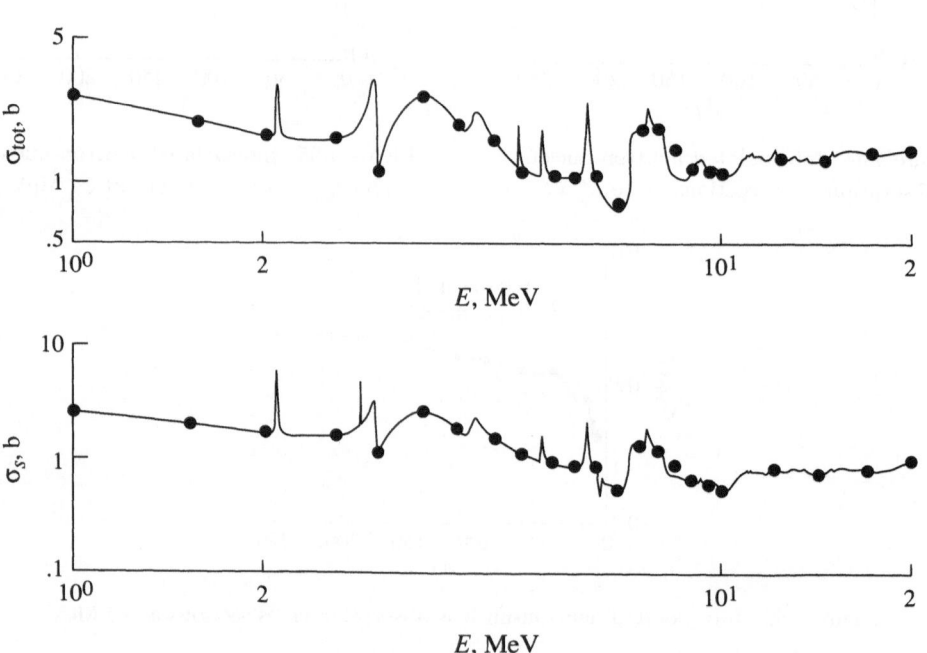

Figure 4.58. Nuclear cross sections for neutron projectiles onto carbon targets.

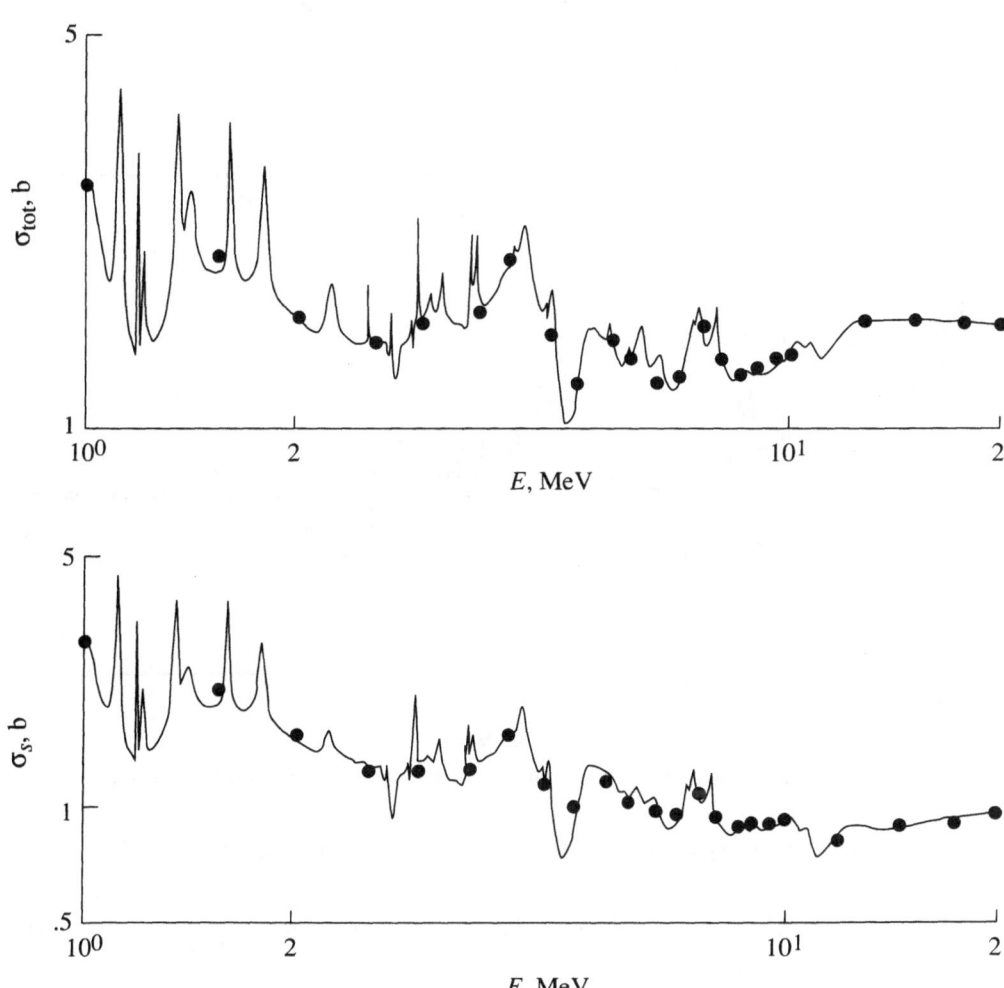

Figure 4.59. Nuclear cross sections for neutron projectiles onto nitrogen targets.

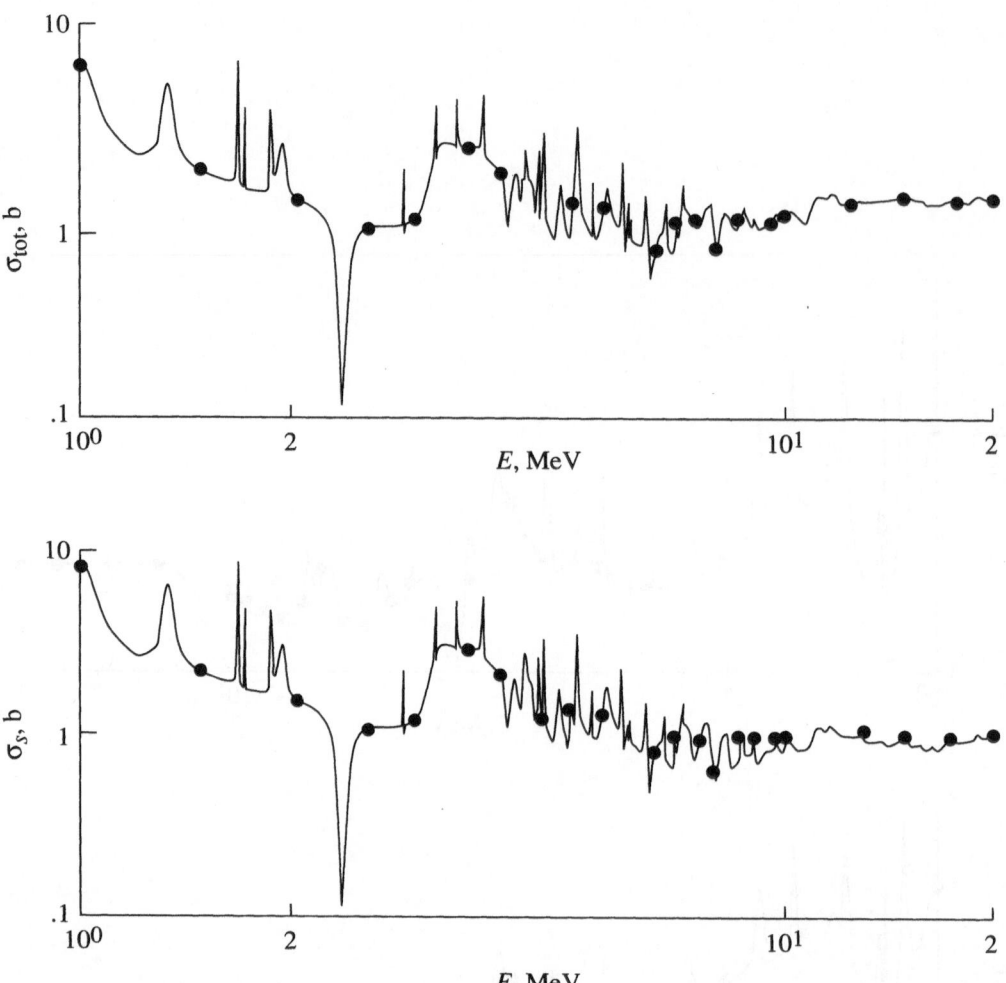

Figure 4.60. Nuclear cross sections for neutron projectiles onto oxygen targets.

where the complex phase function as a function of impact parameter $\vec{b}$ is

$$\chi(\vec{b}) = -mA_PA_Tk^{-1}(A_P + A_T)^{-1} \int V(\vec{b}, z) \, dz \qquad (4.70)$$

and the optical potential is

$$V(\vec{b}, z) = A_PA_T \int d^3\vec{r}d^3\vec{y} \; \rho_P(\vec{b} + \vec{z} + \vec{r} + \vec{y})\rho_T(\vec{r}) \; \tilde{t}(e, \vec{y}) \qquad (4.71)$$

In equations (4.69) and (4.70), m is the nucleon mass, k is the momentum wave number, and ρ_i $(i = P, T)$ is the respective number density distribution for the projectile and target nuclei. The constituent-averaged, two-nucleon transition amplitude $\tilde{t}$ is used to describe high-energy nucleon-nucleon scattering. Details can be found in Wilson and Townsend (1981) and Townsend and Wilson (1985). Typical results for carbon projectiles are displayed in figure 4.61 along with recent experimental data of Kox et al. (1984). Because these calculations are too complex to be repeatedly performed within a transport calculation, extensive tables, which can be easily stored on disk or magnetic tape for access as needed, have been published (Townsend and Wilson, 1985). Typical agreement between theory and experiment is within 10 percent for energies as low as 25 MeV/nucleon and within 3 percent for energies above 80 MeV/nucleon.

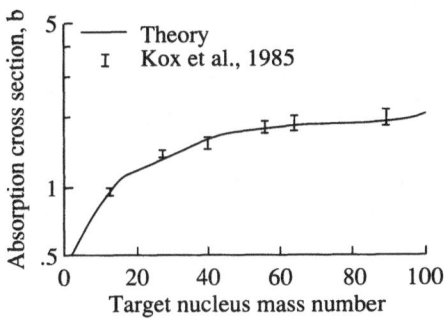

Figure 4.61. Absorption cross sections for carbon beams at 83 MeV/nucleon.

As an alternative to these tables, an energy-dependent parameterization of these tables has been formulated (Townsend and Wilson, 1986)

$$\sigma_{\text{abs}} = \pi r_o^2 \beta(E) \left(A_P^{1/3} + A_T^{1/3} - \delta \right)^2 \tag{4.72}$$

where

$$\delta = 0.200 + A_P^{-1} + A_T^{-1} - 0.292 \ \exp\left(\frac{-E}{792}\right) \ \cos\left(0.229 E^{0.453}\right) \tag{4.73}$$

with

$$\beta(E) = 1 + 5E^{-1} \tag{4.74}$$

$r_o = 1.26$ fm, and E expressed in units of MeV/nucleon. Note that for large values of E, $\beta(E) \to 1$ and δ becomes energy independent, so that a typical form from Bradt and Peters (1950) is reproduced. Nominal differences between the cross sections obtained with equations (4.69) and (4.72) are less than 5 percent for $A > 4$ and $E > 50$ MeV/nucleon. For $E < 50$ MeV/nucleon, the differences are less than 10 percent. For helium-helium collisions, differences of approximately 20 percent exist at all energies. Representative predictions for carbon-carbon scattering as a function of energy are displayed in figure 4.62 along with experimental results (Kox et al., 1984; Jaros et al., 1978; Aksinenko et al., 1980; Heckman et al., 1978; Kox et al., 1985) and estimates obtained from a recently proposed energy-independent parameterization (Silberberg et al., 1984). The agreement with experimental data is quite good for the energy-dependent predictions, whereas the energy-independent parameterization clearly breaks down at low energies.

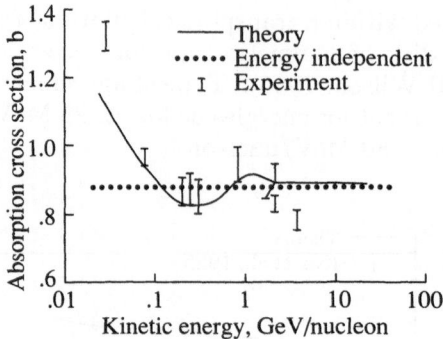

Figure 4.62. Absorption cross sections estimated by energy-dependent and energy-independent parameterizations and experiment for carbon-carbon scattering.

4.5. Parametric Differential Cross Sections

After the angular distribution in elastic scattering is sufficiently known, then the energy transferred to the target nucleus may be found as well as the new energy spectrum of the projectile. The differential energy and angle distributions are discussed in this section and simple parametric forms are given.

4.5.1. Nucleon-nucleon spectrum. The forward scattered nucleon differential cross section (Schopper, 1973) is well represented by

$$f_f(E, E') = B \, \frac{\exp\left[-B(E' - E)\right]}{1 - \exp(-BE')} \tag{4.75}$$

where

$$B = \frac{2mc^2 b}{10^6} \tag{4.76}$$

In equation (4.76), mc^2 is the nucleon rest energy (938 MeV), and b is the usual slope parameter given by (in units of GeV^{-2})

$$b = \left\{ \begin{array}{ll} 3 + 14 \, \exp\left(\frac{-E'}{200}\right) & \text{(For } pp\text{)} \\ 3.5 + 30 \, \exp\left(\frac{-E'}{200}\right) & \text{(For } pn\text{)} \end{array} \right\} \tag{4.77}$$

where E' (MeV) is the initial nucleon energy in the rest frame of the target. The backward scattering spectrum is similar in form

$$f_b(E, E') = \frac{B \, \exp(-BE)}{1 - \exp(-BE')} \tag{4.78}$$

where we assume the backward scatter slope parameter is the same as the forward value. This is strictly true for pp scattering, but the slope parameter for pn charge exchange scattering (Bertini, Guthrie, and Culkowski, 1972) would be more correct. The forward-to-backward ratio for np scattering is well represented by

$$F_B(E') = 0.12 - 0.015E' + \frac{0.41}{1 + \exp[4(E' - 1.2)]} \tag{4.79}$$

where E' in equation (4.79) has units of GeV. The full differential spectrum is then

$$f(E, E') = \frac{B \, \exp[-B(E' - E)] + F_B(E')B \, \exp(-BE)}{[1 - \exp(-BE')][1 + F_B(E')]} \tag{4.80}$$

where $F_B(E') = 1$ for pp scattering. The differential cross sections are normalized such that

$$\frac{d\sigma}{dE} = \sigma(E')f(E, E') \tag{4.81}$$

where $\sigma(E')$ is the appropriate nucleon-nucleon total cross section. Obviously, we have neglected the inelastic processes that must yet be included so that $\sigma(E')$ in equation (4.81) is currently set equal to total cross section to ensure conservation of energy, mass, and charge. The distribution of the center-of-mass angle θ_{cm} is related to the energy change in the laboratory frame of reference (relativistic kinematics are not yet included) by

$$\frac{d\sigma}{d\Omega} = \frac{E'}{4\pi} \frac{d\sigma}{dE} \tag{4.82}$$

where Ω denotes the solid angle element in the center-of-mass frame of reference. The center-of-mass angular distributions are compared with the compilation of experimental data (Hess, 1958) in figures 4.63 and 4.64.

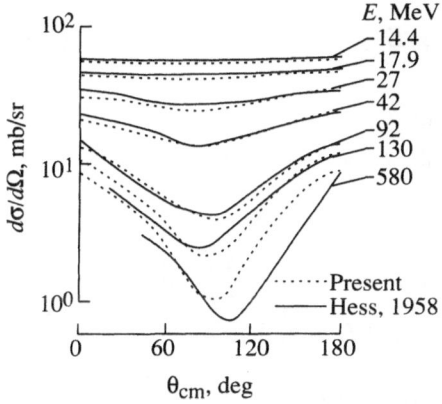

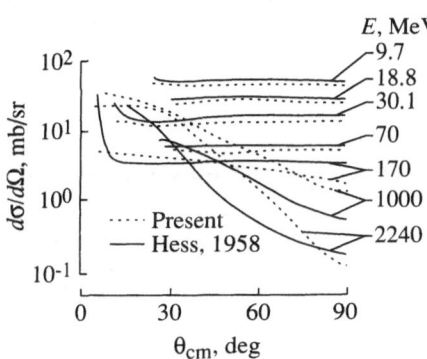

Figure 4.63. Neutron-proton differential elastic scattering cross section of present model and experiment.

Figure 4.64. Proton-proton differential elastic scattering cross sections.

4.5.2. Nucleon-nucleus spectrum.
The nucleon-nucleus differential cross section in Chew's form of the impulse approximation (note that this is just the Born term of the optical model) is given by

$$\frac{d\sigma_I}{dq^2} = c \, \exp(-2bq^2) \left| F_A(q^2) \right|^2$$

$$\approx c \, \exp(-2bq^2) \exp\left(\frac{-2a^2q^2}{3} \right) \tag{4.83}$$

where b is the slope parameter of equation (4.77) averaged among nuclear constituents, q is the magnitude of momentum transfer, and a is the nuclear root-mean-square (rms) radius. The nuclear rms radius (Wilson, 1975) in terms of the rms charge radius (in fermi) is given as

$$a = \left(a_c^2 - 0.64\right)^{1/2} \tag{4.84}$$

where the rms charge radius (in fermi) is

$$a_c = \begin{cases} 0.84 & (A_T = 1) \\ 2.17 & (A_T = 2) \\ 1.78 & (A_T = 3) \\ 1.63 & (A_T = 4) \\ 2.4 & (6 \le A_T \le 14) \\ 0.82 A_T^{1/3} + 0.58 & (A_T \ge 16) \end{cases} \tag{4.85}$$

the nuclear form factor is the Fourier transform of the nuclear-matter distribution. Note that the above equation assumes that the nuclear-matter distribution is a Gaussian function. Such an approximation is reasonable for the light-mass nuclei but is less valid for $A_T \gg 20$.

The energy transferred to the nucleus E_T is restricted by kinematics to

$$0 \le E_T \le (1 - \alpha)E' \tag{4.86}$$

where

$$\alpha = \frac{(A_T - 1)^2}{(A_T + 1)^2} \tag{4.87}$$

The energy-transfer spectrum is given as

$$f_I(E_T, E') = \frac{4A_T mc^2 \left(B + \frac{a^2}{3}\right) \exp\left[-4A_T mc^2 \left(B + \frac{a^2}{3}\right) E_t\right]}{1 - \exp\left[-4A_t mc^2 (1 - \alpha)\left(B + \frac{a^2}{3}\right) E'\right]} \tag{4.88}$$

Similarly, the scattered nucleon energy E is restricted to

$$\alpha E' \le E \le E' \tag{4.89}$$

The nucleon spectrum is given by

$$f(E, E') = \frac{4A_T mc^2 \left(B + \frac{a^2}{3}\right) \exp\left[-4A_T mc^2 \left(B + \frac{a^2}{3}\right)(E' - E)\right]}{1 - \exp\left[-4A_T mc^2 (1 - \alpha)\left(B + \frac{a^2}{3}\right) E'\right]} \tag{4.90}$$

One should note that both equations (4.88) and (4.90) reduce to the usual isotropic scattering results at low incident energy. The differential spectrum is normalized as

$$\frac{d\sigma}{dE} = \sigma_s(E') \, f(E, E') \tag{4.91}$$

where $\sigma_s(E')$ is the total scattering cross section obtained from equation (4.61).

The angular distribution of scattered nucleons is rather well-defined by equation (4.83) near the forward direction (Wilson et al., 1989). To approximate the cross section at large angles, we evaluate the $\ell = 0$ phase shift (Merzbacher, 1970) and introduce an energy-dependent parameter as follows. The S-wave phase shift δ_0 is related to the optical potential as

$$\tan \delta_0 \approx -k \int_0^\infty \left[j_0(kr') \right]^2 U(r')(r')^2 \, dr' \tag{4.92}$$

where

$$U(\vec{x}) = \frac{2m A_T^2 A_P^2}{(A_T + A_P)} \int d^3z \, d^3y \, \rho_T(\vec{z})\rho_P(\vec{x} + \vec{z} + \vec{y}) \, \tilde{t}(k, y) \tag{4.93}$$

as given in chapter 3. Because we assume that $\rho_T(\vec{z})$, $\rho_P(\vec{z})$, and $\tilde{t}(k, \vec{y})$ are Gaussian in coordinate space, the integrals are easily evaluated. The S-wave cross section is given as

$$\sigma_0 = \sqrt{\frac{mE'}{2}} \sin \delta_0 \tag{4.94}$$

with the corresponding differential contribution

$$\frac{d\sigma_0}{dE} = \frac{(A_P + A_T)^2}{A_P A_T E'} \frac{\sigma_0}{4\pi} \tag{4.95}$$

where $A_P = 1$ for neutron scattering.

The S-wave contribution is combined with the impulse approximation, with the interference terms neglected, as follows

$$\frac{d\sigma_s}{dE} = \left[\frac{d\sigma_I}{dE} + b_s(E') \frac{d\sigma_o}{dE} \right] N_s \tag{4.96}$$

where the renormalization factor N_s is chosen to preserve the relation

$$\sigma_s = \int_0^{(1-\alpha)E'} \frac{d\sigma_s}{dE} \, dE \tag{4.97}$$

for which N_s is found to be

$$N_s = \frac{\sigma_s(E')}{\sigma_s(E') + b_s(E')\sigma_0(E')} \tag{4.98}$$

The parameter $b_s(E')$ is taken as a function of energy

$$b_s(E') = \frac{1}{2} + \frac{3}{32} E' \tag{4.99}$$

The results are compared with the work of others (Fernbach, 1958; Goldberg, May, and Stehn, 1962) in figures 4.65 through 4.68. The scaled S-wave contribution used to represent the large angle scattering of neutrons shows improvement for most nuclei and gives satisfactory KERMA values as shown in chapter 10.

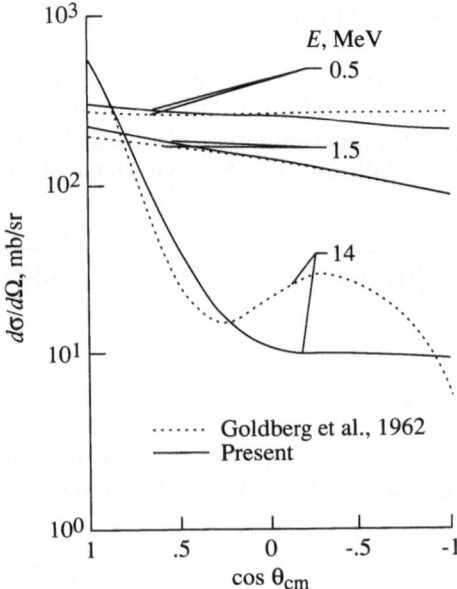

Figure 4.65. Neutron-carbon differential elastic scattering cross sections.

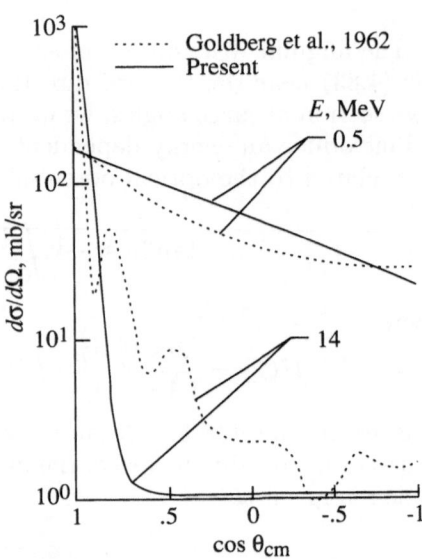

Figure 4.66. Neutron-uranium differential elastic scattering cross sections.

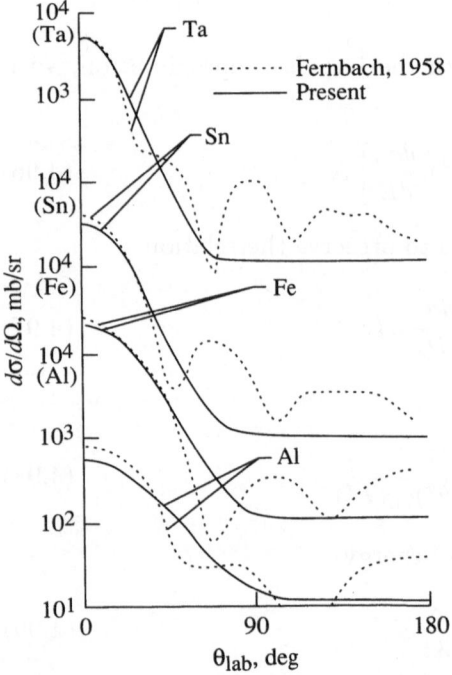

Figure 4.67. Neutron scattering cross section for several elements for 7-MeV neutrons.

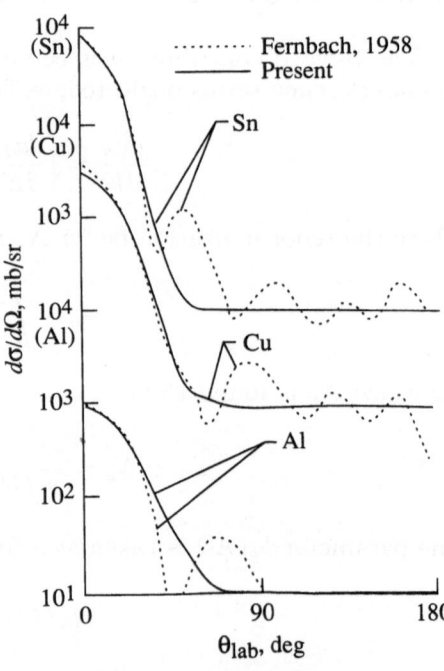

Figure 4.68. Neutron scattering cross section for several elements for 14-MeV neutrons.

4.6. Summary

A reasonably accurate data base is available to describe the elastic channel of nucleon and nucleus interactions. Future activity should concentrate on generating a meson and antinucleon data base.

4.7. References

Akaishi, Yoshinori, 1984: Random Number Method in Few Body Calculation. *Nuclear Phys.*, vol. A416, pp. 409c–419c.

Aksinenko, V. D.; Anikina, M. Kh.; Buttsev, V. S.; Chkaidze, L. V.; Glagoleva, N. S.; Golokhvastov, A. I.; Grachov, R. G.; Gudima, K. K.; Dementjev, E. A.; Kadikova, S. V.; Kaminski, N. I.; Khorozov, S. A.; Kuznetsova, E. S.; Lukstins, J.; Matyushin, A. T.; Matyushin, V. T.; Mescherjakov, M. G.; Musulmanbekov, Zh. Zh.; Nurgozhin, N. N.; Okonov, E. O.; Ostanevich, T. G.; Peisert, A.; Skrzypczak, E.; Szwed, R.; Shevchenko, E. A.; Sidorin, S. S.; Khusainov, E. K.; Toneev, V. D.; Vardenga, G. L.; Volodin, V. D.; Zakrzewski, J. A.; and Zuravleva, M. S., 1980: Streamer Chamber Study of the Cross Sections and Multiplicities in Nucleus-Nucleus Interactions at the Incident Momentum of 4.5 GeV/c per Nucleon. *Nuclear Phys.*, vol. A348, nos. 4,5, pp. 518–534.

Angeli, I.; and Csikai, J., 1970: Total Neutron Cross Sections and the Nuclear Ramsauer Effect. *Nuclear Phys.*, vol. A158, no. 2, pp. 389–392.

Angeli, I.; and Csikai, J., 1971: Total Neutron Cross Sections and the Nuclear Ramsauer Effect. (II). $E_n = 0.5$–42 MeV. *Nuclear Phys.*, vol. A170, no. 3, pp. 577–583.

Antonchik, V. A.; Bakaev, V. A.; Bogdanov, S. D.; Vikhrov, A. I.; Dudkin, V. E.; Nefedov, N. A.; Ostroumov, V. I.; and Potapov, Yu. V., 1981: Interaction of ^{56}Fe at 1.8 GeV Nucleon With the Nuclei C, N, and O and With Ag and Br. *Sov. J. Nuclear Phys.*, vol. 33, no. 4, pp. 558–560.

Barashenkov, V. S.; Gudima, K. K.; and Toneev, V. D., 1969: Cross Sections for Fast Particles and Atomic Nuclei. *Prog. Phys.*, vol. 17, no. 10, pp. 683–725.

Benary. Odette; Price, Leroy R.; and Alexander, Gideon, 1970: *NN and ND Interactions (Above 0.5 GeV/c)—A Compilation.* UCRL-20000 NN, Lawrence Radiation Lab., Univ. of California.

Berger, J.; Duflo, J.; Goldzahl, L.; Oostens, J.; Plouin, F.; Fabbri, F. L.; Picozza, P.; Satta, L.; Bizard, G.; Lefebvres, F.; Steckmeyer, J. C.; and Legrand, D., 1980: $\alpha\alpha$ Elastic Scattering at 4.32 GeV/c and 5.07 GeV/c. *Nuclear Phys.*, vol. A338, no. 2, pp. 421–428.

Bertini, Hugo W.; Guthrie, Miriam P.; and Culkowski, Arline H., 1972: *Nonelastic Interactions of Nucleons and π-Mesons With Complex Nuclei at Energies Below 3 GeV.* ORNL-TM-3148, U.S. Atomic Energy Commission.

Bidasaria, H. B.; and Townsend, L. W., 1983: Microscopic Optical Potential Analyses of Carbon-Carbon Elastic Scattering. *Canadian J. Phys.*, vol. 61, no. 12, pp. 1660–1662.

Bidasaria, H. B.; Townsend, L. W.; and Wilson, J. W., 1983: Theory of Carbon-Carbon Scattering From 200 to 290 MeV. *J. Phys. G: Nuclear Phys.*, vol. 9, no. 1, pp. L17–L20.

Binstock, Judith, 1974: Parameterization of σ_{tot}, $\sigma(\Theta), P(\Theta)$ for 25–100 MeV np Elastic Scattering. *Phys. Review*, vol. 10, ser. C, no. 1, pp. 19–23.

Bobchenko, B. M.; Buklei, A. E.; Viasov, A. V.; Vorob'ev, I. I.; Vorob'ev, L. S.; Goryainov, N. A.; Grishuk, Yu. G.; Gushchin, O. B.; Druzhinin, B. L.; Zhurkin, V. V.; Zavrazhnov, G. N.; Kosov, M. V.; Leksin, G. A.; Stolin, V. L.; Surin, V. P.; Fedorov, V. B.; Fominykh, B. A.; Shvartsman, B. B.; Shevchenko, S. V.; and Shuvalov, S. M., 1979: Measurement of Total Inelastic Cross Sections for Interaction of Protons With Nuclei in the Momentum Range From 5 to 9 GeV/c and for Interaction of π^- Mesons With Nuclei in the Momentum Range From 1.75 to 6.5 GeV/c. *Sov. J. Nuclear Phys.*, vol. 30, no. 6, pp. 805–813.

Borkowski, F.; Simon, G. G.; Walther, V. H.; and Wendling, R. D., 1975: On the Determination of the Proton RMS-Radius From Electron Scattering Data. *Z. Phys. A.*, vol. 275, no. 1, pp. 29–31.

Bradt, H. L.; and Peters, B., 1950: The Heavy Nuclei of the Primary Cosmic Radiation. *Phys. Review*, vol. 77, second ser., no. 1, pp. 54–70.

Brodsky, Allen, ed., 1978: *CRC Handbook of Radiation Measurement and Protection—General Scientific and Engineering Information, Volume I: Physical Science and Engineering Data.* CRC Press, Inc.

Brookhaven National Lab., 1982: *Guidebook for the ENDF/B-V Nuclear Data Files.* EPRI NP-2510 (Res. Proj. 975-1. BNL-NCS-31451, ENDF-328).

Buck, W. W.; Norbury, J. W.; Townsend, L. W.; and Wilson, J. W., 1986: Theoretical Antideuteron-Nucleus Absorptive Cross Sections. *Phys. Review*, vol. 33, third ser., no. 1, pp. 234–238.

Buck, Warren W.; Wilson, John W.; Townsend, Lawrence W.; and Norbury, John W., 1987: *Possible Complementary Cosmic-Ray Systems: Nuclei and Antinuclei.* NASA TP-2741.

Buenerd, M.; Lounis, A.; Chauvin, J.; Lebrun, D.; Martin, P.; Duhamel, G.; Gondrand, J. C.; and De Saintignon, P., 1984: Elastic and Inelastic Scattering of Carbon Ions at Intermediate Energies. *Nuclear Phys.*, vol. A424, no. 2, pp. 313–334.

Cheshire, D. L.; Huggett, R. W.; Johnson, D. P.; Jones, W. V.; Rountree, S. P.; Verma, S. D.; Schmidt, W. K. H.; Kurz, R. J.; Bowen, T.; and Krider, E. P., 1974: Fragmentation Cross Sections of 2.1-GeV/Nucleon ^{12}C and ^{16}O Ions. *Phys. Review*, vol. 10, ser. D, no. 1, pp. 25–31.

Cole, A. J.; Rae, W. D. M.; Brandan, M. E.; Dacal, A.; Harvey, B. G.; Legrain, R.; Murphy, M. J.; and Stokstad, R. G., 1981: ^{12}C + ^{12}C Reaction Cross Section Between 70 and 290 MeV Obtained From Elastic Scattering. *Phys. Review Lett.*, vol. 47, no. 24, pp. 1705–1708.

Cucinotta, F. A.; Khandelwal, G. S.; Townsend, L. W.; and Wilson, J. W., 1989: Correlations in α-α Scattering and Semi-Classical Optical Models. *Phys. Lett.*, vol. B223, no. 2, pp. 127–132.

Dadić, I.; Martinis, M.; and Pisk, K., 1971: Inelastic Processes and Backward Scattering in a Model of Multiple Scattering. *Ann. Phys.*, vol. 64, no. 2, pp. 647–671.

De Jager, C. W.; De Vries, H.; and De Vries, C., 1974: Nuclear Charge- and Magnetization-Density-Distribution Parameters From Elastic Electron Scattering. *At. Data & Nuclear Data Tables*, vol. 14, no. 5/6, pp. 479–508.

Fernbach, S., 1958: Nuclear Radii as Determined by Scattering of Neutrons. *Reviews Modern Phys.*, vol. 30, no. 2, pt. 1, pp. 414–418.

Feshbach, H.; and Hüfner, J., 1970: On Scattering by Nuclei at High Energies. *Ann. Phys.*, vol. 56, no. 1, pp. 268–294.

Franco, Victor; and Nutt, W. T., 1978: Short Range Correlations in High Energy Heavy Ion Collisions. *Phys. Review C*, vol. 17, third ser., no. 4, pp. 1347–1358.

Frullani, Salvatore; and Mougey, Jean, 1984: Single-Particle Properties of Nuclei Through $(e, e'p)$ Reactions. *Advances in Nuclear Physics*, Volume 14, J. W. Negele and Erich Vogt, eds., Plenum Press, pp. 1–289.

Goldberg, Murrey D.; May, Victoria M.; and Stehn, John R., 1962: *Angular Distributions in Neutron-Induced Reactions. Volume I, Z = 1 to 22*. BNL 400, Second ed., Vol. I, Sigma Center, Brookhaven National Lab.

Gross, Franz, 1974: New Theory of Nuclear Forces. Relativistic Origin of the Repulsive Core. *Phys. Review D,* vol. 10, no. 1, pp. 223–242.

Heckman, H. H.; Greiner, D. E.; Lindstrom, P. J.; and Shwe, H., 1978: Fragmentation of ^{4}He, ^{12}C, ^{14}N, and ^{16}O Nuclei in Nuclear Emulsion at 2.1 GeV/Nucleon. *Phys. Review*, vol. 17, ser. C, no. 5, pp. 1735–1747.

Hess, Wilmot N., 1958: Summary of High-Energy Nucleon-Nucleon Cross-Section Data. *Reviews Modern Phys.*, vol. 30, no. 2, pt. I, pp. 368–401.

Hong, Byungsik; Maung, Khin Maung; Wilson, John W.; and Buck, Warren W., 1989: *Kaon-Nucleus Scattering.* NASA TP-2920.

Hughes, Donald, J.; and Schwartz, Robert B., 1958: *Neutron Cross Sections.* BNL 325, Second ed., Brookhaven National Lab.

Jakobsson, B.; and Kullberg, R., 1976: Interactions of 2 GeV/Nucleon ^{16}O With Light and Heavy Emulsion Nuclei. *Phys. Scr.*, vol. 13, no. 6, pp. 327–338.

Jaros, J.; Wagner A.; Anderson, L.; Chamberlain, O.; Fuzesy, R. Z.; Gallup, J.; Gorn, U; Schroeder, L.; Shannon, S.; Shapiro, G.; and Steiner, H., 1978: Nucleus-Nucleus Total Cross Sections for Light Nuclei at 1.55 and 2.89 GeV/c per Nucleon. *Phys. Review*, vol. 18, ser. C, no. 5, pp. 2273–2292.

Jastrow, Robert, 1955: Many-Body Problem With Strong Forces. *Phys. Review*, vol. 98, second ser., no. 5, pp. 1479–1484.

Kox, S.; Gamp, A.; Cherkaoui, R.; Cole, A. J.; Longequeue, N.; Menet, J.; Perrin, C.; and Viano, J. B., 1984: Direct Measurements of Heavy-Ion Total Reaction Cross Sections at 30 and 83 MeV/Nucleon. *Nuclear Phys.*, vol. A420, no. 1, pp. 162–172.

Kox, S.; Gamp, A.; Perrin, C.; Arvieux, J.; Bertholet, R.; Barundet, J. F.; Buenerd, M.; El Masri, Y.; Longequeue, N.; and Merchez, F., 1985: Transparency Effects in Heavy-Ion Collisions Over the Energy Range 100–300 MeV/Nucleon. *Phys. Lett.*, vol. 159B, no. 1, pp. 15–18.

Kullberg, R.; Kristiansson, K.; Lindkvist, B.; and Otterlund, I., 1977: Production Cross Sections of Multiply Charged Fragments in Heavy Ion Interactions at 150–200 MeV/Nucleon. *Nuclear Phys.*, vol. A280, no. 2, pp. 491–497.

Letaw, John R.; Silberberg, R.; and Tsao, C. H., 1983: Proton-Nucleus Total Inelastic Cross Sections: An Empirical Formula for $E > 10$ MeV. *Astrophys. J.*, Suppl. ser., vol. 51, no. 3, pp. 271–276.

Lindstrom, P. J.; Greiner, D. E.; Heckman, H. H.; Cork, B.; and Bieser, F. S., 1975: *Isotope Production Cross Sections From the Fragmentation of ^{16}O and ^{12}C at Relativistic Energies.* LBL-3650 (NGR-05-003-513), Lawrence Berkeley Lab., Univ. of California.

Liu, H.; Zamick, L.; and Jaqaman, H., 1985: Excited States of ^{4}He. *Phys. Review C*, vol. 31, third ser., no. 6, pp. 2251–2255.

Lock, W. O.; and Measday, D. F., 1970: *Intermediate Energy Nuclear Physics.* Methuen & Co. Ltd. (London).

McCarthy, J. S.; Sick, I.; and Whitney, R. R., 1977: Electromagnetic Structure of the Helium Isotopes. *Phys. Review C*, vol. 15, third ser., no. 4, pp. 1396–1414.

Merzbacher, Eugen, 1970: *Quantum Mechanics*, Second ed. John Wiley & Sons, Inc.

Palevsky, H.; Friedes, J. L.; Sutter, R. J.; Bennett, G. W.; Igo, G. J.; Simpson, W. D.; Phillips, G. C.; Corley, D. M.; Wall, N. S.; Stearns, R. L.; and Gottschalk, B., 1967: Elastic Scattering of 1-BeV Protons From Hydrogen, Helium, Carbon, and Oxygen Nuclei. *Phys. Review Lett.*, vol. 18, no. 26, pp. 1200–1204.

Perrin, C.; Kox, S.; Longequeue, N.; Viano, J. B.; Buenerd, M.; Cherkaoui, R.; Cole, A. J.; Gamp, A.; Menet, J.; Ost, R.; Bertholet, R.; Guet, C.; and Pinston, J., 1982: Direct Measurement of the $^{12}C + {}^{12}C$ Reaction Cross Section Between 10 and 83 MeV/Nucleon. *Phys. Review Lett.*, vol. 49, no. 26, pp. 1905–1909.

Renberg, P. U.; Measday, D. F.; Pepin, M.; Schwaller, P.; Favier, B.; and Richard-Serre, C., 1972: Reaction Cross Sections for Protons in the Energy Range 220–570 MeV. *Nuclear Phys.*, vol. A183, no. 1, pp. 81–104.

Ringia, F. E.; Dobrowolski, T.; Gustafson, H. R.; Jones, L. W.; Longo, M. J.; Parker, E. F.; and Cork, Bruce, 1972: Differential Cross Sections for Small-Angle Neutron-Proton and Neutron-Nucleus Elastic Scattering at 4.8 GeV/c. *Phys. Review Lett.*, vol. 28, no. 3, pp. 185–188.

Satta, L.; Duflo, J.; Plouin, F.; Picozza, P.; Goldzahl, L.; Banaigs, J.; Frascaria, R.; Fabbri, F. L.; Codino, A.; Berger, J.; Boivin, M.; and Berthet, P., 1984: Elastic Scattering of α Particles on Light Nuclei at $P_\alpha = 7$ GeV/c. *Phys. Lett.*, vol. 139B, no. 4, pp. 263–266.

Schimmerling, W.; Devlin, T. J.; Johnson, W.; Vosburgh, K. G.; and Mischke, R. E., 1971: Neutron-Nucleus Total and Inelastic Cross Sections From 900 to 2600 MeV/c. *Phys. Lett.*, vol. 37B, no. 2, pp. 177–180.

Schimmerling, Walter; Devlin, Thomas J.; Johnson, Warren W.; Vosburgh, Kirby G.; and Mischke, Richard E., 1973: Neutron-Nucleus Total and Inelastic Cross Sections: 900 to 2600 MeV/c. *Phys. Review*, vol. 7, no. 1, pp. 248–262.

Schopper, H., ed., 1973: *Elastic and Charge Exchange Scattering of Elementary Particles. Landolt-Börnstein Numerical Data and Functional Relationships in Science and Technology*, Group I, Volume 7, Springer-Verlag.

Schopper, H., ed., 1980: *Elastic and Charge Exchange Scattering of Elementary Particles. Landolt-Börnstein Numerical Data and Functional Relationships in Science and Technology*, Group I, Volume 9, Springer-Verlag.

Silberberg, R.; Tsao, C. H.; Adams, J. H., Jr.; and Letaw, J. R., 1984: Radiation Doses and LET Distributions of Cosmic Rays. *Radiat. Res.*, vol. 98, pp. 209–226.

Skrzypczak, E., 1980: Cross-Sections for Inelastic ^{4}He and ^{12}C-Nucleus Collisions at 4.5 GeV/c/N Incident Momentum. *Proceedings of the International Conference on Nuclear Physics, Volume 1, Abstracts*, LBL-11118 (Contract No. W-7405-ENG-48), Lawrence Berkeley Lab., Univ. of California, p. 575.

Stehn, John R.; Goldberg, Murrey D.; Magurno, Benjamin A.; and Wiener-Chasman, Renate, 1964: *Neutron Cross Sections. Volume 1, Z = 1 to 20.* BNL 325, Second ed., Suppl. No. 2 (Physics—TID-4500, 32nd ed.), Sigma Center, Brookhaven National Lab. Associated Univ., Inc.

Townsend, Lawrence W., 1982: *Harmonic Well Matter Densities and Pauli Correlation Effects in Heavy Ion Collisions.* NASA TP-2003.

Townsend, L. W.; Bidasaria, H. B.; and Wilson, J. W., 1983: Eikonal Phase Shift Analyses of Carbon-Carbon Scattering. *Canadian J. Phys.*, vol. 61, no. 6, pp. 867–871.

Townsend, Lawrence W.; Wilson, John W.; and Bidasaria, Hari B.; 1983a: *Heavy-Ion Total and Absorption Cross Sections Above 25* MeV/*Nucleon.* NASA TP-2138.

Townsend, Lawrence W.; Wilson, John W.; and Bidasaria, Hari B., 1983b: *Nucleon and Deuteron Scattering Cross Sections From 25* MeV/*Nucleon to 22.5* GeV/*Nucleon.* NASA TM-84636.

Townsend, Lawrence W.; and Wilson, John W., 1985: *Tables of Nuclear Cross Sections for Galactic Cosmic Rays—Absorption Cross Sections.* NASA RP-1134.

Townsend, L. W.; and Wilson, J. W., 1986: Energy-Dependent Parameterization of Heavy-Ion Absorption Cross Sections. *Radiat. Res.*, vol. 106, pp. 283–287.

Vary, J. P.; and Dover, C. B., 1974: Microscopic Models for Heavy Ion Scattering at Low, Intermediate and High Energies. *2nd High Energy Heavy Ion Summer Study—July 15–26, 1974 at the Lawrence Berkeley Laboratory*, LBL-3675, Lawrence Berkeley Lab., Univ. of California, pp. 197–259.

Wallace, Stephen J., 1975: High-Energy Expansion for Nuclear Multiple Scattering. *Phys. Review C*, vol. 12, third ser., no. 1, pp. 179–193.

Westfall, G. D.; Wilson, Lance W.; Lindstrom, P. J.; Crawford, H. J.; Greiner, D. E.; and Heckman, H. H., 1979: Fragmentation of Relativistic ^{56}Fe. *Phys. Review*, vol. 19, ser. C, no. 4, pp. 1309–1323.

Wilson, John W., 1975: Composite Particle Reaction Theory. Ph.D. Diss., College of William and Mary in Virginia.

Wilson, John W.; and Costner, Christopher M., 1975: *Nucleon and Heavy-Ion Total and Absorption Cross Section for Selected Nuclei.* NASA TN D-8107.

Wilson, J. W.; and Townsend, L. W., 1981: An Optical Model for Composite Nuclear Scattering. *Canadian J. Phys.*, vol. 59, no. 11, pp. 1569–1576.

Wilson, John W.; Townsend, Lawrence W.; Nealy, John E.; Chun, Sang Y.; Hong, B. S.; Buck, Warren W.; Lamkin, S. L.; Ganapol, Barry D.; Khan, Ferdous; and Cucinotta, Francis A., 1989: BRYNTRN: *A Baryon Transport Model.* NASA-TP-2887.

5. REACTION CHANNEL DATA BASE

5.1. Introduction

After substantial improvements had been made to the description of the elastic channel (Wilson and Townsend, 1981), Townsend (1981) began the development of an abrasion reaction model for the absorptive processes observed in the elastic amplitude, Cucinotta (1988) began a theory for α-particle breakup and Khan et al. (1988) investigated heavy ion abrasion dynamics by using the optical model. It was in this development that the need for inclusion of Pauli correlation and more accurate density functions for light nuclei became apparent (Townsend, 1982). Development of abrasion theory was greatly encouraged by the work of Stevenson, Martinis, and Price (1981) (whose experiments measured directly the abrasion event). The α breakup model is required to further extend the nucleon transport code to light fragments. The first semiempirical code was also developed by Wilson, Townsend, and Badavi (1987a and 1987b) and Badavi et al. (1987) to provide the data base for heavy ion reactions.

5.2. Nuclear Abrasion Model

We now discuss the work of Townsend (1981) in deriving a nuclear abrasion model for the optical potential formalism.

5.2.1. Optical potential. The optical model potential operator (Wilson, 1974a) is

$$V_{\text{opt}} = \sum_{\alpha j} t_{\alpha j} \qquad (5.1)$$

where $t_{\alpha j}$ is the transition operator for scattering between the α constituent of the target and the j constituent of the projectile. The optical potential was derived as

$$W(\vec{x}) = A_P A_T \int d^3\vec{\xi}_T \, \rho_T(\vec{\xi}_T) \int d^3\vec{y} \, \rho_P(\vec{x} + \vec{y} + \vec{\xi}_T) \, t(e, \vec{y}) \qquad (5.2)$$

when couplings to various excited internal states were neglected. The development of equation (5.2) was made independent of the eikonal approximation (Wilson, 1975) and then subsequently used within that context. Note that Franco and Varma (1978) use this same expression to represent their single-scattering term. Differences between Wilson's and Glauber's approximations occur in the higher order terms. For example, unlike the Glauber theory, the Wilson (1974b) propagator includes target recoil and terms to order k^2. In equation (5.2), t is the two-body transition amplitude averaged over the constituent types, and ρ_T and ρ_P are the target and projectile single particle matter densities. Equation (5.2) does not include the correlation effects of the Pauli exclusion principle because only simple unsymmetrized product wave functions were used by Wilson and Townsend

(1981). The accuracy of the results of Wilson and Townsend (1981) supports the idea that exchange correlation effects were unimportant when determining total and absorption cross sections. This section confirms this idea. For abrasion predictions, however, correlation effects of Pauli are found to be important when there is a large overlap between the colliding nuclear volumes (i.e., for small residual mass fragments of the projectile nucleus).

Because equation (5.1) was derived independent of any assumptions regarding nuclear wave functions, it is the starting point. When rewriting equation (5.1) in second quantization notations, we have

$$V_{\mathrm{opt}} = \sum_{\beta k} \sum_{\alpha j} (\beta k|t|\alpha j) a_k^\dagger a_\beta^\dagger a_\alpha a_j \tag{5.3}$$

where the $a_i^\dagger$ and a_i are the usual creation and annihilation operators for the single-particle state i. After the usual operator manipulations, the optical potential reduces to

$$W(x) = \sum_\alpha^{A_T} \sum_j^{A_P} [(\alpha j|t|\alpha j) - (\alpha j|t|j\alpha)] \tag{5.4}$$

When assuming a correlation function C, which depends only upon the relative separation of the α- and j-constituents, Townsend (1982) derives

$$W(\vec{x}) = A_P A_T \int d^3\vec{\xi}_T \, \rho_T(\xi_T) \int d^3\vec{y} \, \rho_P(\vec{x} + \vec{y} + \vec{\xi}_t) t(e, \vec{y})[1 - C(\vec{y})] \tag{5.5}$$

Note that equation (5.5) reduces to equation (5.2) if there are no correlation effects ($C = 0$).

5.2.2. Abrasion theory.

From Bleszynski and Sander (1979), the cross section for abrading n-projectile nucleons is

$$\sigma_n = \binom{A_P}{n} \int d^2\vec{b} \left[1 - P(\vec{b})\right]^n P(\vec{b})^{A_F} \tag{5.6}$$

where $P(\vec{b})$ is the probability for not removing a nucleon in the collision and A_F, the residual fragment mass number, is

$$A_F = A_P - n \tag{5.7}$$

The total absorption cross section

$$\sigma_{\mathrm{abs}} = \int d^2\vec{b} \left[1 - P(\vec{b})^{A_P}\right] \tag{5.8}$$

348

is obtained by summing σ_n over all values of n. In the eikonal approximation, the absorption cross section is

$$\sigma_{\text{abs}} = \int d^2\vec{b} \left(1 - \exp\left\{-2 \, \text{Im} \, \left[\chi(\vec{b})\right]\right\}\right) \tag{5.9}$$

where $\text{Im} \, (\chi)$ denotes the imaginary part of χ. Comparing equation (5.8) and equation (5.9) gives

$$P(\vec{b}) = \exp\left\{\frac{-2 \, \text{Im} \, \left[\chi(\vec{b})\right]}{A_P}\right\} \tag{5.10}$$

Substituting for the eikonal phase function gives

$$P(\vec{b}) = \exp\left[-A_T\sigma(e)I(\vec{b})\right] \tag{5.11}$$

with

$$I(\vec{b}) = [2\pi B(e)]^{-3/2} \int dz \int d^3\vec{\xi}_T \, \rho_T(\vec{\xi}_T)$$

$$\times \int d^3\vec{y} \, \rho_P \left(\vec{b} + \vec{z} + \vec{y} + \vec{\xi}_T\right) \exp\left[\frac{-y^2}{2B(e)}\right]$$

$$\times [1 - C(\vec{y})] \tag{5.12}$$

Values for $\sigma(e)$ and $B(e)$, the nucleon-nucleon cross section and slope parameter, were taken from compilations.

The Glauber theory result (Bleszynski and Sander, 1979) is

$$P(\vec{b}) = 2\pi \int D_P(\vec{s}) \exp\left[-A_T\sigma_{NN} \, D_T(\vec{s} + \vec{b})\right] s \, ds \tag{5.13}$$

where $D(\vec{s})$, the single-particle densities summed along the beam direction, is given by

$$D(\vec{s}) = \int_{-\infty}^{\infty} \rho(\vec{s} + \vec{z}) \, dz \tag{5.14}$$

The main advantages of equation (5.11) over equation (5.13) are its improved convergence and the added symmetry feature that the projectile and target are treated on an equal basis.

Substituting equation (5.11) into equation (5.6) gives

$$\sigma_n = \binom{A_P}{n} \int d^2\vec{b} \left\{1 - \exp\left[-A_T\sigma(e)I(\vec{b})\right]\right\}^n$$

$$\times \exp\left[-A_T A_F\sigma(e)I(\vec{b})\right] \tag{5.15}$$

5.2.3. Collision parameters. From Bohr and Mottelson (1969), the correlation function in the Fermi gas model is

$$C(y) = \frac{1}{4} \frac{3j_1^2(k_F y)}{k_F y} \tag{5.16}$$

where $k_F = 1.36$ fm^{-1}. For analytic simplicity, equation (5.16) is replaced by a simple Gaussian function. Expanding equation (5.16) in a power series gives

$$\frac{3j_1^2(k_F y)}{k_F y} = 1 - \left(\frac{k_F^2 y^2}{10} \right) + 0(k_F^4 y^4) \tag{5.17}$$

For small values of $k_F y$, where correlations are most important in actual nuclei, we note that

$$\exp \left(\frac{-k_F^2 y^2}{10} \right) \approx 1 - \left(\frac{k_F^2 y^2}{10} \right) \tag{5.18}$$

Thus, for computations in this work, we use

$$C(y) \approx \frac{1}{4} \exp \left(\frac{-k_F^2 y^2}{10} \right) \tag{5.19}$$

Determinations of σ_n require the use of nuclear single-particle matter densities ρ for the nuclei in the collision. For the ^{20}Ne projectiles, matter densities are extracted from Woods-Saxon charge data of Knight et al. (1981) as described by Wilson and Townsend (1981). For the ^{96}Mo target, the matter density was found from the three-parameter Gaussian charge density data (De Jager, De Vries, and De Vries, 1974) by assuming that the charge density is given by the folded integral of a Gaussian proton charge density ρ_p, with the unknown nuclear matter density ρ_m according to

$$\rho_c(\vec{r}) = \int \rho_p(\vec{r}') \rho_m(\vec{r} + \vec{r}') \, d^3 r' \tag{5.20}$$

Inserting the Gaussian for $\rho_p(\vec{r})$, simplifying the ensuing expression (Wilson and Costner, 1975), and using a two-point Gauss-Hermite quadrature formula to evaluate the result yield

$$\rho_c(r) \approx \frac{1}{2} \left[\rho_m \left(r + \frac{r_p}{3^{1/2}} \right) + \rho_m \left(r - \frac{r_p}{3^{1/2}} \right) \right] \tag{5.21}$$

for determining the ^{96}Mo matter density. In equation (5.21), $r_p \approx 0.87$ fm is the proton rms radius (Borkowski et al., 1975). For the ^{12}C target, which has a harmonic well charge density, a matter density expression can be analytically extracted. Taking the Fourier transform of equation (5.20) and using the convolution theorem yield the following well-known result:

$$F_c(q) = F_p(q) F_m(q) \tag{5.22}$$

where for a Gaussian proton charge density,

$$F_p(q) = \exp\left(\frac{-q^2 r_p{}^2}{6}\right) \tag{5.23}$$

The ^{12}C harmonic well charge density (De Jager, De Vries, and De Vries, 1974)

$$\rho_c(r) = \rho_0 \left[1 + \alpha \left(\frac{r}{a}\right)^2\right] \exp\left(\frac{-r^2}{a^2}\right) \tag{5.24}$$

has a form factor (Townsend, 1982)

$$F_c(q) = \rho_0 \pi^{3/2} a^3 \left(1 + \frac{3\alpha}{2} - \frac{\alpha q^2 a^2}{4}\right) \exp\left(\frac{-q^2 a^2}{4}\right) \tag{5.25}$$

Values for parameters α and a are also given by De Jager, De Vries, and De Vries (1974).

Using equations (5.23) and (5.25) in equation (5.22) gives the matter density form factor $F_m(q)$. Taking the inverse transform of this $F_m(q)$ gives a ^{12}C harmonic well matter density

$$\rho_m(r) = \left(\frac{\rho_0 a^3}{8s^3}\right)\left(1 + \frac{3\alpha}{2} - \frac{3\alpha a^2}{8s^2} + \frac{\alpha a^2 r^2}{16s^4}\right) \exp\left(\frac{-r^2}{4s^2}\right) \tag{5.26}$$

with

$$s^2 = \frac{a^2}{4} - \frac{r_p{}^2}{6} \tag{5.27}$$

These density results are displayed in figures 5.1 through 5.3.

5.2.4. Results. Abrasion cross sections for ^{20}Ne-^{12}C collisions at 2.1 GeV/ nucleon obtained from equation (5.15) are given in table 5.1. Also listed are predicted cross sections when the Pauli correlation effects are ignored. From these results, the correlation effects have little or no effect on the abrasion cross sections for $n \leq 12$ because peripheral processes are the greatest contributors to

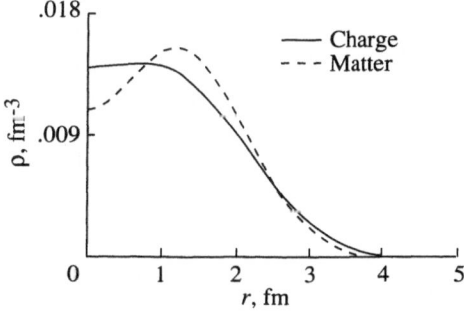

Figure 5.1. Harmonic well charge and matter density distributions for ^{12}C.

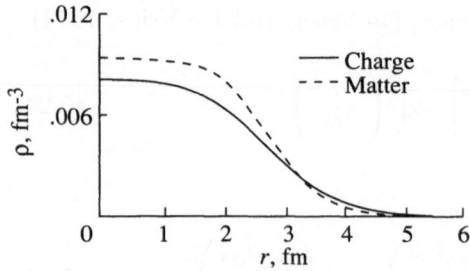

Figure 5.2. Woods-Saxon charge and matter density distribution for ^{20}Ne.

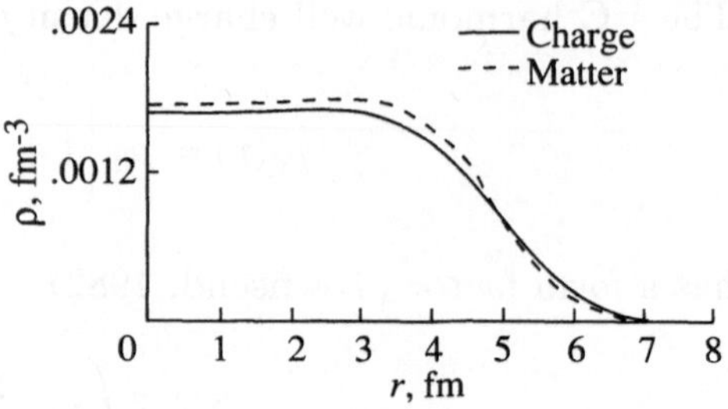

Figure 5.3. Three-parameter Gaussian charge and matter density distributions for ^{96}Mo.

Table 5.1. Optical Model Abrasion Cross Sections for ^{20}Ne-^{12}C Collisions at 2.1 GeV/Nucleon

	σ_n, mb	
n	Pauli correlations	No Pauli correlations
1	248	248
2	134	134
3	95	95
4	76	75
5	64	64
6	57	56
7	52	51
8	48	48
9	45	45
10	43	43
11	42	42
12	40	42
13	38	42
14	33	41
15	27	39
16	18	33
17	10	25
18	4	14
19	1	6
20	0.1	1

these abrasions. As n increases, greater overlap between the colliding nuclear volumes is required, and the importance of correlation effects increases. They are most important when there is complete overlap between the colliding volumes ($n = 20$). On the other hand, summing the abrasion cross sections to give a total absorption cross section demonstrates that correlation effects only reduce σ_{abs} by ≈ 6 percent (1076 mb versus 1144 mb) for this collision. These abrasion results are also displayed in figure 5.4 with the recent experimental results of Stevenson, Martinis, and Price (1981). Because the experimental results are given in relative probabilities (RP) rather than cross sections, theoretical relative probabilities were calculated from

$$\text{RP} = \frac{\sigma_n}{\sigma_{abs} - 0.5\sigma_1} \tag{5.28}$$

where the denominator correction $0.5\sigma_1$ accounts for the missing ^{19}Ne fragments that were discriminated out experimentally as discussed by Stevenson, Martinis, and Price (1981). Additionally, for $n = 1$, the relative probabilities were determined by setting the numerator in equation (5.28) equal to $0.5\sigma_1$ to again account for the missing ^{19}Ne fragments. Finally, the theoretical RP, which are discrete numbers, were folded with the finite detector resolution ($\sigma \approx$ 1.5 amu) to yield the displayed curves. As shown, the agreement between theory and experiment when correlation effects are included is excellent. The slight disagreement for small residual fragment masses may be caused by the approximations used in the correlation function.

To test the sensitivity of the abrasion results to the shape of the nuclear density distributions, relative probabilities were determined for the Ne + C collision by using a Woods-Saxon density for the neon projectile and two different distributions for the carbon target: a Woods-Saxon and a harmonic well. Correlation effects were not included. The superiority of the more exact harmonic well density is obvious in figure 5.5, where the theoretical predictions and the experimental results are presented.

Table 5.2 lists abrasion cross sections for ^{20}Ne-^{96}Mo collisions, at 2.1 GeV/ nucleon, obtained from equation (5.15). Pauli correlation effects are included in the results. The relative probabilities, obtained from equation (5.28), are plotted in figure 5.6 with the experimental results of Stevenson, Martinis, and Price (1981). The agreement between theory and experiment for this collision pair is good but not as good as was obtained in the Ne-C collision. The discrepancy may be caused by inaccuracies in the correlation function approximation and/or the ^{96}Mo matter density distribution approximation because the theoretical RP clearly overshoot the experimental values for small residual mass fragments ($A_F < 6$).

5.3. Simple Ablation Model

The quantum mechanical abrasion model using the optical model approximation was so successful that further development (Townsend et al., 1984) seems warranted. The obvious starting point is to use a simple compound nuclear evaporation decay model. Such calculations require specification of the initial compound nuclear state defined by the mass, charge, and excitation energy.

Table 5.2. Optical Model Abrasion Cross Sections
for ^{20}Ne-^{96}Mo Collisions at 2.1 GeV/Nucleon

n	σ_n, mb
1	380
2	207
3	148
4	119
5	101
6	90
7	82
8	77
9	73
10	71
11	70
12	71
13	72
14	75
15	81
16	90
17	105
18	135
19	201
20	294

5.3.1. Prefragment charge distributions. Since the abraded nucleons consist of protons and neutrons, which are not identical, a prescription for calculating the charge dispersions of the prefragments is needed to calculate final, isotope, and/or elemental production cross sections caused by the fragmentation process. Two such methods are used in the fragmentation theory described in this work. The method of Oliveira, Donangelo, and Rasmussen (1979) treats the neutron and proton distributions as completely uncorrelated. The cross section for forming a particular prefragment of mass A_j and charge Z_j is then given in terms of the

$$\sigma_{\text{abr}}(Z_j, A_j) = \frac{\binom{N}{n}\binom{Z}{z}}{\binom{A_P}{m}}\sigma_m \tag{5.29}$$

where z out of the original Z projectile nucleus protons is abraded along with n out of the original N projectile neutrons. Note that

$$A_P = N + Z \tag{5.30}$$

and

$$m = n + z \tag{5.31}$$

with

$$Z_j = Z - z \tag{5.32}$$

354

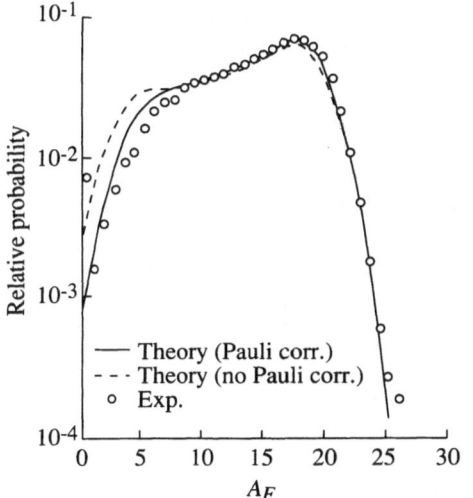

Figure 5.4. Ne-C abrasion results. Experimental results are from Stevenson, Martinis, and Price (1981).

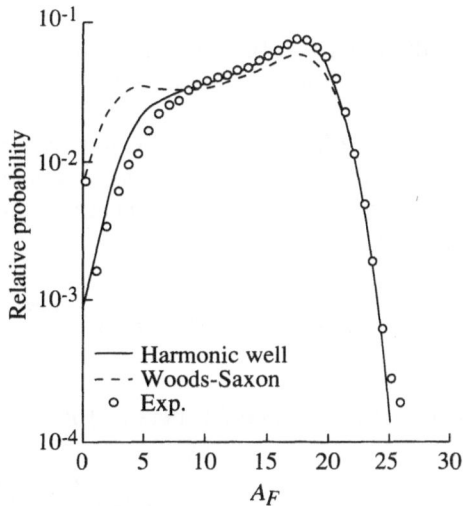

Figure 5.5. Theoretical Ne-C abrasion predictions. Experimental results are from Stevenson, Martinis, and Price (1981).

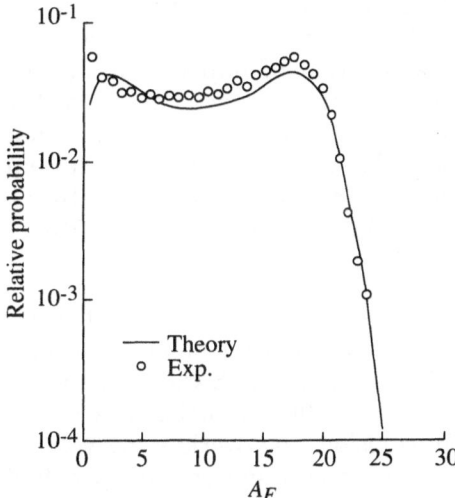

Figure 5.6. Ne-Mo abrasion results. Experimental results are from Stevenson, Martinis, and Price (1981).

and

$$A_j = A_P - m \tag{5.33}$$

This hypergeometric distribution is based on the assumption that there is no correlation at all between neutron and proton distributions. Therefore, unphysical results such as abrading all neutrons or protons from a nucleus while leaving the remaining fragment intact could occur.

As an alternative to the hypergeometric distribution, Morrissey et al. (1978) proposed a charge dispersion model based upon the zero-point vibrations of the giant dipole resonance of the projectile nucleus. In this model, equation (5.29) becomes

$$\sigma_{\text{abr}}(Z_j, A_j) = N_j(2\pi\alpha_Z{}^2)^{-1/2} \; \exp\left\{\frac{-[Z_j - A_j(Z/A_P)]^2}{2\alpha_Z{}^2}\right\} \sigma_m \tag{5.34}$$

where the variance (dispersion) is

$$\alpha_Z = 2.619 \left(\frac{u}{A_P}\right)^{1/2} \frac{Z}{A_P} \frac{dm}{db}(1+u)^{-3/4} \tag{5.35}$$

with

$$u = \frac{3J}{Q(A_P)^{1/3}} \tag{5.36}$$

In the droplet model of the nucleus, the coefficients J and Q have the nominal values of 25.76 and 11.9 MeV, respectively. The rate of change of the number of nucleons removed as a function of impact parameter (dm/db) is calculated numerically by using the geometric abrasion model of Bowman, Swiatecki, and Tsang (1973). The normalization factor N_J ensures that a given value of A_j, the discrete sum over all allowed values of Z_j, yields unity for the dispersion probabilities. This overall normalization is a new feature of this work and is not included in the original model of Morrissey et al. (1978).

5.3.2. Prefragment excitation energies.
The excitation energy of the projectile prefragment following abrasion of m nucleons is calculated from the clean-cut abrasion formalism of Bowman, Swiatecki, and Tsang (1973) and Gosset et al. (1977). For this model, the colliding nuclei are assumed to be uniform spheres of radii R_i $(i = P, T)$. In collision, the overlapping volumes shear off so that the resultant projectile prefragment is a sphere with a cylindrical hole gouged out of it. The excitation energy is then determined by calculating the difference in surface area between the misshapen sphere and a perfect sphere of equal volume. This excess surface area Δ is given by Gosset et al. (1977) as

$$\Delta = 4\pi R_P{}^2 \left[1 + P - (1 - F)^{2/3}\right] \tag{5.37}$$

where the expressions for P and F differ, depending upon the nature of the collision (peripheral versus central) and the relative sizes of the colliding nuclei.

356

For the case where $R_T > R_P$, we have

$$P = 0.125(\mu\nu)^{1/2} \left(\frac{1}{\mu} - 2\right) \left(\frac{1-\beta}{\nu}\right)^2$$
$$- 0.125 \left[0.5(\mu\nu)^{1/2} \left(\frac{1}{\mu} - 2\right) + 1\right] \left(\frac{1-\beta}{\nu}\right)^3 \qquad (5.38)$$

and

$$F = 0.75(1-\nu)^{1/2} \left(\frac{1-\beta}{\nu}\right)^2 - 0.125 \left[3(1-\nu)^{1/2} - 1\right] \left(\frac{1-\beta}{\nu}\right)^3 \qquad (5.39)$$

with

$$\nu = \frac{R_P}{R_P + R_T} \qquad (5.40)$$

$$\beta = \frac{b}{R_P + R_T} \qquad (5.41)$$

and

$$\mu = \frac{1}{\nu} - 1 = \frac{R_T}{R_P} \qquad (5.42)$$

Equations (5.38) and (5.39) are valid when the collision is peripheral (i.e., the two nuclear volumes do not completely overlap). In this case, the impact parameter b is restricted such that

$$R_T - R_P \le b \le R_T + R_P \qquad (5.43)$$

If the collision is central, then the projectile nucleus volume completely overlaps the target nucleus volume ($b < R_T - R_P$), and all the projectile nucleons are abraded. In this case, equations (5.38) and (5.39) are replaced by

$$P = -1 \qquad (5.44)$$

and

$$F = 1 \qquad (5.45)$$

and there is no ablation of the projectile because it was destroyed by the abrasion.

For the case where $R_P > R_T$ and the collision is peripheral, equations (5.38) and (5.39) become (Morrissey et al., 1978)

$$P = 0.125(\mu\nu)^{1/2} \left(\frac{1}{\mu} - 2\right) \left(\frac{1-\beta}{\nu}\right)^2 - 0.125 \left\{0.5 \left(\frac{\nu}{\mu}\right)^{1/2} \left(\frac{1}{\mu} - 2\right)\right.$$
$$\left. - \frac{[(1/\nu)(1-\mu^2)^{1/2} - 1][(2-\mu)\mu]^{1/2}}{\mu^3}\right\} \left(\frac{1-\beta}{\nu}\right)^3 \qquad (5.46)$$

and

$$F = 0.75(1 - \nu)^{1/2} \left(\frac{1 - \beta}{\nu}\right)^2 - 0.125\left\{\frac{3(1 - \nu)^{1/2}}{\mu}\right.$$

$$\left. - \frac{\left[1 - (1 - \mu^2)^{3/2}\right]\left[1 - (1 - \mu)^2\right]^{1/2}}{\mu^3}\right\}\left(\frac{1 - \beta}{\nu}\right)^3 \tag{5.47}$$

where the impact parameter is restricted such that

$$R_P - R_T \leq b \leq R_P + R_T \tag{5.48}$$

For a central collision $(b < R_P - R_T)$ with $R_P > R_T$, equations (5.46) and (5.47) become

$$P = \left[\frac{1}{\nu}(1 - \mu^2)^{1/2} - 1\right]\left[1 - \left(\frac{\beta}{\nu}\right)^2\right]^{1/2} \tag{5.49}$$

and

$$F = \left[1 - (1 - \mu^2)^{3/2}\right]\left[1 - \left(\frac{\beta}{\nu}\right)^2\right]^{1/2} \tag{5.50}$$

For the excess surface area obtained from equation (5.37), the excitation energy is given by

$$E_{\text{exc}} = \Delta E_s \tag{5.51}$$

where E_s, the nuclear surface energy coefficient (Bowman, Swiatecki, and Tsang, 1973; Gosset et al., 1977) obtained from the liquid drop model of the nucleus, is 0.95 MeV/fm^2.

5.3.3. Ablation factors (EVAP-4).

Depending upon the excitation energy, the excited prefragment may decay by emitting one or more nucleons (protons or neutrons), composites (deuterons, tritons, ^{3}He, or α-particles), or gamma rays. The probability α_{ij} for formation of a particular final fragment of type i as a result of the de-excitation of a prefragment of type j is obtained from the EVAP-4 computer code (Guthrie, 1970) by treating the prefragment as a compound nucleus with an excitation energy given by equation (5.51). The final fragmentation cross section for projectile of the type i isotope is then given by

$$\sigma_F(Z_i, A_i) = \sum_j \alpha_{ij}\, \sigma_{\text{abr}}(Z_j, A_j) \tag{5.52}$$

where $\sigma_{\text{abr}}(Z_j, A_j)$ is obtained from equation (5.29) or equation (5.34). The elemental production cross sections are obtained by summing over all isotope contributions as

$$\sigma_F(Z) = \sum_A \sigma_F(Z, A) \tag{5.53}$$

5.3.4. Fragmentation results.

As an illustrative application of the theory, element production cross sections for fragments of calcium $(Z = 20)$ and heavier

Table 5.3. Elemental Production Cross Sections for
Reaction ^{56}Fe + ^{12}C → Z + X

[Incident kinetic energy, 1.88 GeV/nucleon]

Element produced	Elemental production cross sections, mb		
	Hypergeometric	Giant dipole resonance	Westfall et al., 1979
Fe	161	209	
Mn	321	308	181 ± 27
Cr	156	142	124 ± 13
V	126	124	100 ± 11
Ti	90	88	87 ± 11
Sc	69	69	54 ± 9
Ca	77	78	78 ± 11

Table 5.4. Elemental Production Cross Sections for
Reaction ^{56}Fe + ^{108}Ag → Z + X

[Incident kinetic energy, 1.88 GeV/nucleon]

Element produced	Elemental production cross sections, mb		
	Hypergeometric	Giant dipole resonance	Westfall et al., 1979
Fe	296	262	
Mn	381	446	280 ± 23
Cr	226	230	218 ± 21
V	150	149	117 ± 15
Ti	126	128	124 ± 15
Sc	101	100	104 ± 13
Ca	102	112	118 ± 14

elements were calculated for ^{56}Fe projectiles at an incident kinetic energy of
1.88 GeV/nucleon and collided with stationary target nuclei of ^{12}C, ^{108}Ag, and
^{208}Pb. These reactions were chosen for analysis because of the availability of
experimental data for comparison purposes (Westfall et al., 1979) and because
relativistic ^{56}Fe nuclei are among the dominant high charge and energy (HZE)
particles of radiobiological significance for manned spaceflight.

Tables 5.3 through 5.5 display the elemental production cross sections obtained
for carbon, silver, and lead targets by using both the hypergeometric (eq. (5.29))
and giant dipole resonance (eq. (5.34)) dispersion expressions. Also displayed are
the experimental results of Westfall et al. (1979). Except for the cross section

for Mn production (carbon and silver targets) and V production (silver and lead targets), the agreement between theory and experiment is quite good. When compared with the predictions obtained with the hypergeometric distribution assumption of equation (5.29), the use of the giant dipole resonance expression for charge dispersion (eq. (5.34)) appears to yield slightly improved overall agreement between theory and experiment.

Figures 5.7 through 5.9 display the elemental production cross sections obtained from equation (5.34), for the giant dipole resonance dispersion (GDR) along with the experimental data from Westfall et al. (1979). Also displayed, for comparison, are the predictions from the semiempirical relations of Silberberg, Tsao, and Shapiro (1976). For the semiempirical relations, the unmodified predictions are displayed. Also displayed are the fragmentation cross sections obtained by renormalizing to ensure mass and charge conservation. Details of the renormalization can be found in Wilson et al. (1984). For the carbon target (Bevington, 1969), χ^2 for the giant dipole resonance predictions is 31.6, which is larger than the 19.4 obtained using the Silberberg-Tsao (ST) methods. For the GDR, most of the χ^2 comes from the Mn overestimate. If that point is excluded, χ^2 is reduced from 31.6 to 9.4. The comparative results for Ca, Sc, Ti, V, and Cr are in better agreement with the experiment for the silver target; the χ^2 for GDR is 57.3 (5.2 if the Mn datum is excluded), whereas the χ^2 for ST is 32.4 (9.5 if the Mn datum is excluded). For the lead target, the χ^2 for GDR is 4.4 (1.9 if the V datum is excluded), compared with the χ^2 for ST of 52.4 (2.3 if the Mn underestimate is excluded). In general, the overall agreement between theory and experiment for the abrasion-ablation model is satisfactory when considering its simple nature.

To illustrate further the results of the model, cross sections for the production of sulfur, phosphorous, silicon, and aluminum isotopes caused by the fragmentation of ^{40}Ar projectiles at 213 MeV/nucleon by carbon targets are shown in figure 5.10. These theoretical predictions were obtained with the hypergeometric distribution.

Also shown are the experimental data from Viyogi et al. (1979). In general, the agreement is surprisingly good, considering the simple nature of the calculations. Partial production cross sections for these same isotopes were also calculated with the GDR distribution. In general, those cross sections were less accurate when compared with the experiment than the ones obtained from the hypergeometric distribution. Typical results are shown as dashed lines in figure 5.10 for the sulfur and silicon isotopes.

In previous heavy ion transport work (Wilson, 1983; Wilson et al., 1984), the improved agreement between theory and experiment for Bragg (depth-dose) curves was obtained by using ST fragmentation parameters modified to scale by velocity (rather than total kinetic energy) and renormalized to conserve fragment charge and mass. As shown in figures 5.7 through 5.9, the modifications (labeled VR) do improve the ST predictions for the predominant, near-projectile mass fragments (in this case, Mn) but yield substantial overestimates for the fragmentation cross sections for the lighter mass fragments. Simple corrections to the ST parameters,

Table 5.5. Elemental Production Cross Sections for
Reaction ^{56}Fe + ^{208}Pb → $Z + X$

[Incident kinetic energy, 1.88 GeV/nucleon]

Element produced	Elemental production cross sections, mb		
	Hypergeometric	Giant dipole resonance	Westfall et al., 1979
Fe	345	302	
Mn	445	521	509 ± 40
Cr	267	268	242 ± 25
V	175	174	142 ± 20
Ti	152	151	148 ± 22
Sc	121	119	111 ± 17
Ca	116	129	144 ± 22

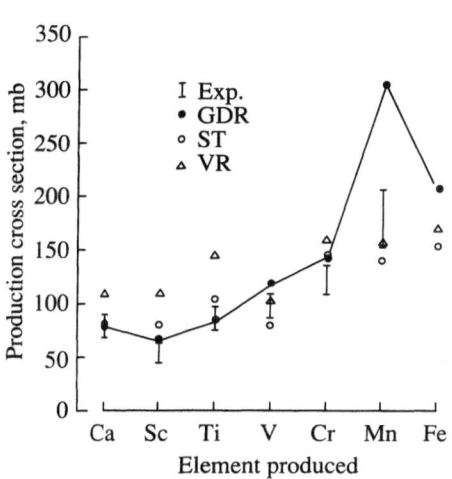

Figure 5.7. Elemental production cross sections for iron projectile nuclei fragmenting in carbon targets.

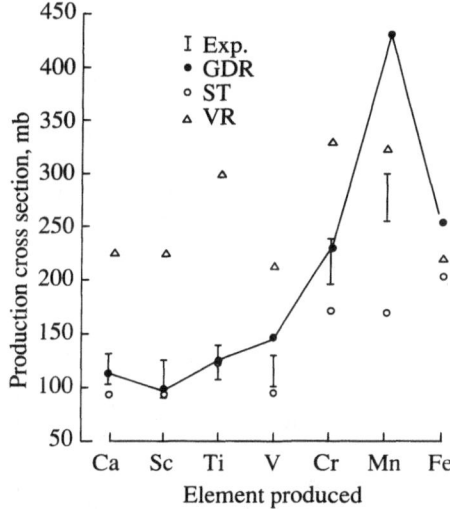

Figure 5.8. Elemental production cross sections for iron projectile nuclei fragmenting in silver targets.

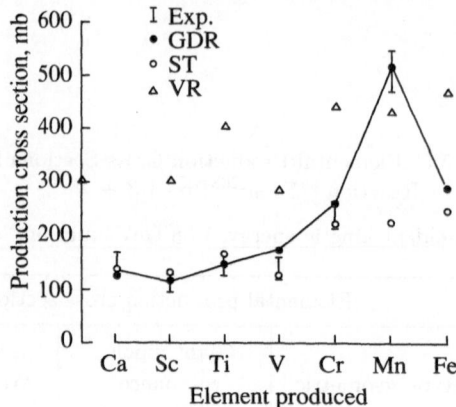

Figure 5.9. Elemental production cross sections for iron projectile nuclei fragmenting in lead targets.

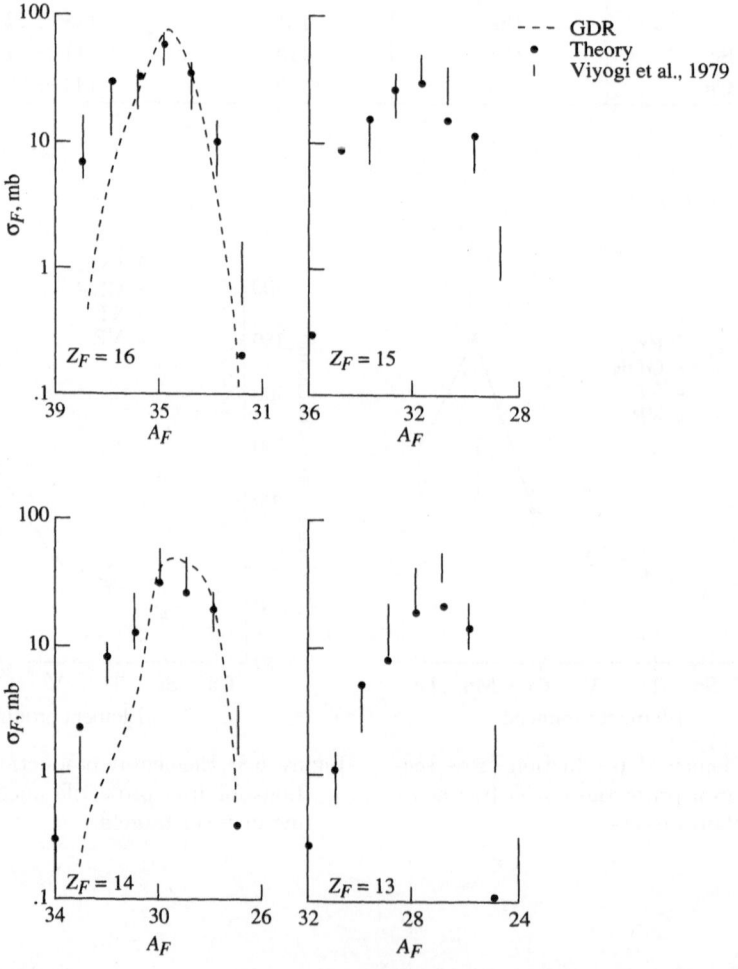

Figure 5.10. Production cross sections for isotopes of sulfur, phosphorous, silicon, and aluminum produced by the fragmentations of ^{40}Ar projectiles at 213 MeV/nucleon in carbon targets.

such as renormalization, are apparently adequate for gross total-dose comparisons (Wilson et al., 1984). However, only certain fragments may be biologically significant. Therefore, these corrections may be inadequate for the more pertinent shielding problems such as the accurate predictions of individual fragment species production. Clearly, the need remains for a comprehensive and accurate HZE particle fragmentation theory, of which the work just described is a beginning.

Improvements to this simple abrasion-ablation model should center on extending the GDR charge dispersion method to incorporate the actual quantum mechanical abrasion formalism rather than using the geometric model approximation of Bowman, Swiatecki, and Tsang (1973). Improved methods for estimating the prefragment excitation energy spectrum should also be developed (Khan, 1989). Finally, an alternative to the EVAP-4 ablation code should be tried, such as an intranuclear cascade code (Morrissey et al., 1979); or the development of other methods to describe the ablation step should be undertaken (Townsend et al., 1986a and 1986b; Cucinotta et al., 1987).

5.4. Abrasion Dynamics

One possible limitation of the abrasion-ablation model described in sections 5.2 and 5.3 is the use of the geometric model in estimating the prefragment excitation energies. We now look at an alternative method (Khan, 1989) of estimating the prefragment system parameters closely related to the work of Fricke (1985).

5.4.1. Method of calculation. The coupled-channel Schrödinger equation for composite particle scattering, which relates the entrance channel to all the excited states of the target and projectile, was derived by assuming large, incident projectile kinetic energies and closure of the accessible eigenstates (Wilson, 1975 and chapter 3). The equation is written as

$$\left(\nabla^2 + k^2\right)\psi_{n\mu}(\vec{x}) = 2m\, A_P A_T \left(A_P + A_T\right)^{-1} \sum_{n'\mu'} V_{n\mu,n'\mu'}(\vec{x})\psi_{n'\mu'}(\vec{x}) \qquad (5.54)$$

where the subscripts n and μ (with and without primes) label the projectile and target eigenstates; m is the nucleon mass; A_P and A_T are the mass numbers of the projectile and target; $\vec{k}$ is the incident projectile momentum relative to the center of mass; and $\vec{x}$ is the projectile position vector relative to the target. As for the nucleon-nucleon scattering t-matrix $t_{\alpha j}$ and the internal state vectors of the projectile $g_n^P(\vec{\xi}_P)$ and target $g_u^T(\vec{\xi}_T)$, the potential matrix can be expressed as

$$V_{n\mu,n'\mu'}(\vec{x}) = \left\langle g_n^P, g_\mu^T \middle| V_{\text{opt}}\left(\vec{\xi}_P, \vec{\xi}_T, \vec{x}\right) \middle| g_{n1}^P, g_{\mu'}^T \right\rangle \qquad (5.55)$$

where

$$V_{\text{opt}}\left(\vec{\xi}_P, \vec{\xi}_T, \vec{x}\right) = \sum_{\alpha j} t_{\alpha j} \qquad (5.56)$$

This same formalism can be used to investigate relativistic heavy ion collision momentum transfers. Within the context of eikonal scattering, the solution to the

Schrödinger equation

$$H \, \psi \left(\vec{x}, \vec{\xi}_P, \vec{\xi}_T \right) = E \, \psi \left(\vec{x}, \vec{\xi}_P, \vec{\xi}_T \right) \tag{5.57}$$

at high energies is

$$\psi \left(\vec{x}, \vec{\xi}_P, \vec{\xi}_T \right) = (2\pi)^{-3/2} \, \exp \left[-\frac{i}{v} \int_{-\infty}^{z} V_{\mathrm{opt}} \left(\vec{x}, \vec{\xi}_P, \vec{\xi}_T \right) dz' \right]$$
$$\times \, g_n^P(\vec{\xi}_P) \, g_\mu^T(\vec{\xi}_T) \, \exp(i\vec{k} \cdot \vec{x}) \tag{5.58}$$

where v is the velocity. The total momentum of the projectile is then given by the matrix element involving the sum of the projectile single-nucleon momentum operators as

$$\vec{P}_{\mathrm{tot}} = \left\langle \psi \left| -i \sum_{\alpha=1}^{A_P} \vec{\nabla}_{P,\alpha} \right| \psi \right\rangle \tag{5.59}$$

where the subscript P on the gradient operator denotes that the gradient is to be taken with regard to the projectile internal coordinates $\vec{\xi}_P$. Equation (5.59) actually denotes a momentum matrix $\vec{P}_{n\mu,n'\mu'}$ in analogy with equation (5.55). Therefore, substituting equation (5.58) into equation (5.59) yields

$$\vec{P}_{n\mu,n'\mu'} = \left\langle g_n^P(\vec{\xi}_P) \, g_\mu^T(\vec{\xi}_T) \left| \exp(-iS) \left(-i \sum_{\alpha=1}^{A_P} \vec{\nabla}_{P,\alpha} \right) \exp(iS) \right| g_{n'}^P(\vec{\xi}_P) \, g_{\mu'}^T(\vec{\xi}_T) \right\rangle \tag{5.60}$$

where

$$S = \frac{1}{v} \int_{-\infty}^{z} V_{\mathrm{opt}} \left(\vec{x}', \vec{\xi}_P, \vec{\xi}_T \right) dz' \tag{5.61}$$

With the chain rule for differentiation, equation (5.60) can be further expressed as

$$\vec{P}_{n\mu,n'\mu'} = \vec{P}_o + \left\langle g_n^P g_\mu^T \left| \left(-\sum_{\alpha}^{A_P} \vec{\nabla}_{P,\alpha} S \right) \right| g_{n'}^P g_{\mu'}^T \right\rangle \tag{5.62}$$

where the incident projectile momentum before the collision is

$$\vec{P}_o = \left\langle g_n^P g_\mu^T \left| \left(-i \sum_{\alpha=1}^{A_P} \vec{\nabla}_{P,\alpha} \right) \right| g_{n'}^P g_{\mu'}^T \right\rangle \tag{5.63}$$

The total momentum transfer to the projectile is then given by

$$\vec{Q}_{n\mu,n'\mu'} = \vec{P}_{n\mu,n'\mu'} - \vec{P}_o = \left\langle g_n^P g_\mu^T \left| \left(-\sum_{\alpha} \vec{\nabla}_{P,\alpha} S \right) \right| g_{n'}^P g_{\mu'}^T \right\rangle \tag{5.64}$$

For high-energy collisions, dominant scattering processes occur near the forward directions, because the momentum transferred is small when compared with the incident momentum of the projectile; hence, couplings between excited states are small and can be neglected (Wilson 1975). The total momentum transfer to the projectile is then approximated by

$$\vec{Q} = \vec{Q}_{00,00} = \left\langle g_o^P g_o^T \left| \left(-\sum_\alpha \vec{\nabla}_{P,\alpha} S \right) \right| g_o^P g_o^T \right\rangle \qquad (5.65)$$

In terms of projectile and target number densities and the constituent-averaged two-nucleon transition amplitude $\tilde{t}$, equation (5.65) becomes

$$\vec{Q}(\vec{b}) = -A_P A_T \int d^3\vec{\xi}_P \, \rho_P(\vec{\xi}_P) \int d^3\vec{\xi}_T \, \rho_T(\vec{\xi}_T)$$
$$\times \left[\vec{\nabla}_P \int_{-\infty}^{\infty} \tilde{t}(\vec{b} + \vec{z}' + \vec{\xi}_P - \vec{\xi}_T) \, \frac{dz'}{v} \right] \qquad (5.66)$$

where the integration limit in the longitudinal direction has been extended to infinity. The momentum transfer in equation (5.66) is therefore only a function of the impact parameter of the collision. The projectile and target number densities (ρ_P and ρ_T) are normalized to unity as

$$\int \rho(\vec{x}) \, d^3\vec{x} = 1 \qquad (5.67)$$

The constituent-averaged, two-nucleon transition amplitude is obtained from the impulsive, first-order t-matrix used in our previous studies (Wilson 1975; Wilson and Townsend, 1981; Townsend, 1981 and 1982; Townsend et al., 1986a and 1986b) of nucleus-nucleus collisions as

$$\tilde{t}(e, \vec{x}) = -\left(\frac{e}{m} \right)^{1/2} \sigma(e) \left[\alpha(e) + i \right] \left[2\pi B(e) \right]^{-3/2} \exp\left[\frac{-x^2}{2B(e)} \right] \qquad (5.68)$$

where e is the two-nucleon kinetic energy in its center-of-mass frame, $\sigma(e)$ is the nucleon-nucleon total cross section, $\alpha(e)$ is the ratio of the real-to-imaginary part of the forward-scattering amplitude, and $B(e)$ is the nucleon-nucleon slope parameter. Values for these parameters taken from various compilations are given in Wilson and Townsend (1981) and Townsend (1982).

The dynamic momentum transfer to the projectile, given by equation (5.66), results from interactions with the target. Note that it is a complex quantity that is consistent with the use of a complex optical potential (Rodberg and Thaler, 1967). The real part of the momentum transfer, which comes from the real part of the complex optical potential, is the contribution arising from elastic scattering. It is purely transverse. The imaginary component, which comes from the absorptive part of the complex optical potential, arises mainly from absorption and inelastic scattering processes. At high energies, the latter are mainly breakup (fragmentation) reactions because these account for over 95 percent of the total reaction cross section. Physically, this imaginary component represents attenuation of the incident wave front in analogy with the usual discussions for a complex index of refraction in an absorptive medium (Rodberg and Thaler,

1967). Concomitant with this attenuation of the incident wave by these absorptive processes, there is a loss of momentum from the wave front in the beam direction. This longitudinal momentum transfer (loss) is interpreted as arising from the imaginary component of $\vec{Q}$. From equation (5.66), the transverse component is

$$Q_\perp = -A_P A_T \int d^3\vec{\xi}_P \, \rho_P(\vec{\xi}_P) \int d^3\vec{\xi}_T \, \rho_T(\vec{\xi}_T)$$
$$\times \vec{\nabla}_P \int_{-\infty}^{\infty} \mathrm{Re} \left[\tilde{t}(\vec{b} + \vec{z}' + \vec{\xi}_P - \vec{\xi}_T) \right] \frac{dz'}{v} \qquad (5.69)$$

and the longitudinal component is

$$Q_\parallel = -A_P A_T \int d^3\vec{\xi}_P \, \rho_P(\vec{\xi}_P) \int d^3\vec{\xi}_T \, \rho_T(\vec{\xi}_T)$$
$$\times \left\{ \vec{\nabla}_P \int_{-\infty}^{\infty} \mathrm{Im} \left[\tilde{t}(\vec{b} + \vec{z}' + \vec{\xi}_P - \vec{\xi}_T) \right] \frac{dz'}{v} \right\} \qquad (5.70)$$

Calculated momentum transfers obtained with equations (5.69) and (5.70) are displayed in figure 5.11 for $^{16}\mathrm{O}$ at 2.1 GeV/nucleon colliding with a beryllium target. These calculations use the harmonic well nuclear densities from our previous work (Townsend, 1982 and 1983; Townsend et al., 1984). From the figure, two features are readily apparent. First, the longitudinal momentum transfer is larger than the transverse; this indicates the primarily absorptive nature of the nuclear collision at this energy. Second, the predicted momentum transfer decreases rapidly with increasing impact parameter. This decrease is discussed further in sections 5.4.3 and 5.4.4, but its occurrence is not surprising because the nuclear optical potential decreases rapidly with increasing separation of the colliding nuclei.

5.4.2. Results. The collisional momentum transfers computed with the model described in section 5.4.1 can be related to experimentally measured, heavy ion fragment momentum downshifts/widths through considerations of energy and momentum conservation. As has been formulated by Goldhaber (1974) and Wong (1981), a momentum transfer in any direction Q_j modifies the width h_j of the fragment momentum distribution in that direction by

$$(h_j')^2 = h_j^2 + \frac{F^2 \, Q_j^2}{A^2} \qquad (5.71)$$

and the mean by

$$\vec{P}_j' = \vec{P}_j + \frac{F}{A} \vec{Q}_j \qquad (5.72)$$

From equation (5.72), the longitudinal momentum downshift is given by

$$\Delta P_\parallel = P_\parallel' - P_\parallel = \frac{F}{A} Q_\parallel, \qquad (5.73)$$

366

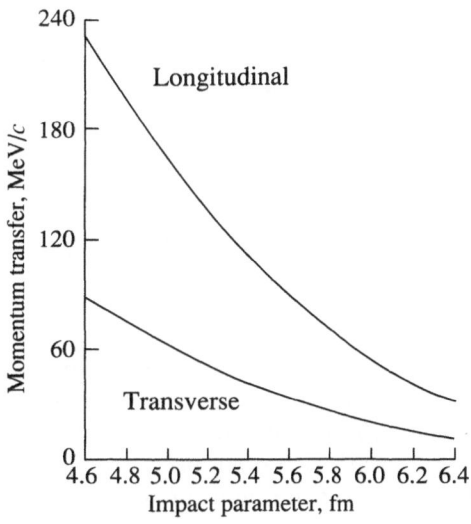

Figure 5.11. Momentum transfer to ^{16}O projectile as function of impact parameter for oxygen colliding with beryllium target at 2.1 GeV/nucleon.

where $Q_\parallel$ is the magnitude of the longitudinal momentum transfer (obtained from eq. (5.70)), F is the fragment mass number, and A is the initial mass number of the fragmenting nucleus. Recalling that $Q_\parallel$ is a function of impact parameter, an appropriate method for choosing the impact parameter for each fragmentation channel is necessary. Recently, a semiempirical abrasion-ablation fragmentation model (NUCFRAG) was proposed by Wilson, Townsend, and Badavi (1987b). Although it assumes simple uniform density distributions for the colliding ions and a zero-range (delta function) interaction, it does include frictional spectator interactions (FSI) and agrees with experimental cross-section data to the extent that they agree among themselves. Also, and most importantly for this work, it is easily modified to yield impact parameters for each fragmentation channel. Hence, the procedure for evaluation of equations (5.71) and (5.73) is to extract impact parameters from NUCFRAG for each nucleon removal corresponding exactly to $\Delta A = 1,\ 2,\ 3, \ldots$. These most probable impact parameters are then inserted in equations (5.69) and (5.70) to obtain the corresponding momentum transfers for use in evaluating equations (5.71) and (5.73) because NUCFRAG uses uniform densities; uniform densities are also used in evaluating equations (5.69) and (5.70). In addition, the zero-range interaction in NUCFRAG is simulated for numerical integration purposes in equations (5.69) and (5.70) through the use of a very narrow Gaussian form for the t-matrix given by equation (5.68). This narrow Gaussian is the same width for all collision pairs and therefore is not an arbitrarily adjusted parameter. We have checked the validity of using the "most probable" impact parameter in the calculations by actually computing the momentum transfers averaged over a range of impact parameters from NUCFRAG corresponding to $\Delta A - 0.5$ to $\Delta A + 0.5$. The differences between the estimates using averaged and most probable values are negligible (Khan, 1989).

Representative calculations for momentum downshifts as a function of fragment mass number are displayed in figure 5.12 for ^{16}O projectiles colliding with targets of Be, C, Al, Cu, Ag, and Pb at 2.1 GeV/nucleon. These momentum downshifts are target averaged by using simple arithmetic averaging. For comparison, the target-averaged experimental data of Greiner et al. (1975) are also displayed. For display and comparison purposes, the theory is also averaged over all isotopes contributing to each fragment mass number using

$$(\Delta P_{\parallel})_{\text{av}} = \frac{\sum_i \sigma_i \left(\Delta P_{\parallel}^i\right)}{\sum_i \sigma_i} \tag{5.74}$$

where σ_i is the experimental production cross section for the ith fragment isotope. Reasonable agreement is obtained for the heavier fragments when comparing the theoretical estimates to the experimental data. When considering the simplified form of the nuclear fragmentation model used in these calculations and the overall sensitivity of the calculated momentum transfer to the choice of impact parameter, the agreement is rather good. Improved agreement is expected if impact parameters from a fragmentation model using realistic nuclear densities and interactions were used. This is especially true for collisions involving lighter ions, such as carbon, oxygen, and beryllium, which are poorly represented by simple uniform nuclear distributions.

Figure 5.13 displays transverse momentum widths as a function of fragment mass number for ^{139}La fragmenting in carbon targets at 1.2 GeV/nucleon. The experimental data are taken from Brady et al. (1988). Again, impact parameters from NUCFRAG are used as inputs into the momentum transfer expressions (eqs. (5.69) and (5.70)). For consistency with the use of these impact parameters, a narrow Gaussian t-matrix and uniform nuclear densities were again used in the momentum transfer calculations. From figure 5.13, the agreement is much better than in figure 5.12 and probably reflects that a uniform nuclear density distribution is a more reasonable approximation for a heavy nucleus like lanthanum than for light nuclei such as oxygen.

5.4.3. Estimating collision impact parameters. So far in this work, we have used collision impact parameters as inputs into a momentum transfer computational model, which in turn, has yielded estimates of heavy ion fragment momentum downshifts/widths for comparison with experimental data. However, this procedure can be reversed and the model used to estimate collision impact parameters from measured momentum downshifts for relativistic collisions. Let F be the fragment mass number with measured longitudinal momentum downshift $\Delta P_{\parallel}$ produced in a relativistic collision between a projectile nucleus (mass number A) and some target. Then, from equation (5.73), the longitudinal momentum transfer to the projectile from the target is

$$Q_{\parallel} = \frac{A}{F} \Delta P_{\parallel} \tag{5.75}$$

The collision impact parameter can then be estimated from equation (5.70) by computing $Q_{\parallel}$ as a function of impact parameter (e.g., in fig. 5.11) and using $Q_{\parallel}$ from equation (5.75) as the entry. To illustrate, consider a collision involving

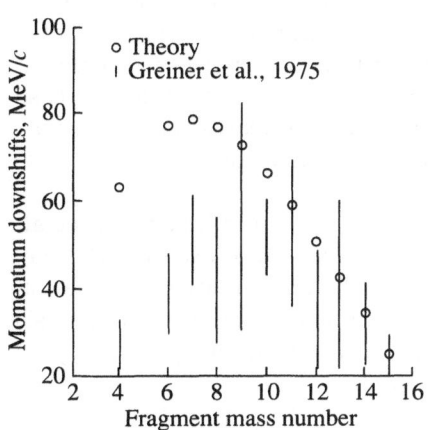

Figure 5.12. Target-averaged longitudinal momentum downshifts as function of projectile fragment mass number for ^{16}O colliding with Be, C, Al, Cu, Ag, and Pb targets at 2.1 GeV/nucleon.

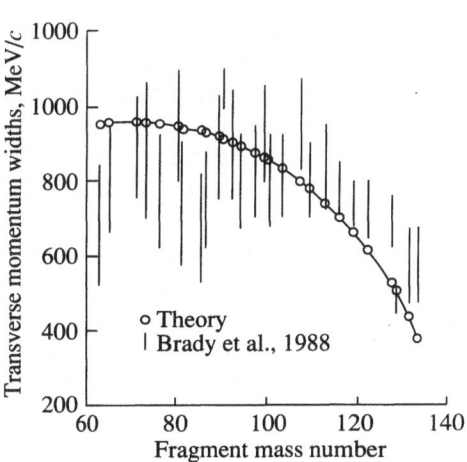

Figure 5.13. Transverse momentum widths as function of projectile fragment mass number for ^{139}La colliding with a carbon target at 1.2 GeV/nucleon.

oxygen colliding with a beryllium target at 2.1 GeV/nucleon. The calculated momentum transfer using realistic nuclear densities is displayed in figure 5.11. If the measured (hypothetical) momentum downshift for the ^{14}N fragment is 35 ± 7 MeV/c, then equation (5.75) yields a longitudinal momentum transfer of 40 ± 8 MeV/c. From figure 5.11, the corresponding range of impact parameters is 6.1–6.4 fm. A similar procedure incorporating measured momentum distribution widths and equations (5.71) and (5.69) or (5.70) could also be used to estimate collision impact parameters. These proposed methods for estimating collision impact parameters are similar in concept to the use of heavy fragment yields in the quantum molecular dynamics approach of Aichelin and collaborators (Aichelin et al., 1988).

5.4.4. Remarks.

Beginning with composite particle multiple-scattering theory, an optical model description of collision momentum transfer in relativistic heavy ion collisions was derived. General expressions for transverse and longitudinal momentum transfers, which use a finite-range, two-nucleon interaction and realistic nuclear densities, were presented. The theory was used as input into the Goldhaber (1974) formalism to estimate heavy ion fragment momentum downshifts for relativistic oxygen and transverse momentum widths for relativistic lanthanum projectiles. The main new feature of this work was the interpretation of the imaginary component of the momentum transfer as the longitudinal collision momentum transfer. Finally, the use of the model as a mechanism for estimating collision impact parameters was described.

The present theory is mainly applicable at intermediate or high energies because of the use of eikonal wave functions and the impulse approximation. At lower energies (below several hundred MeV/nucleon), the validity of straight-line trajectories and the assumption of a constant projectile velocity are questionable. Therefore, revisions to the model are necessary to compare theory with experiment at lower energies. In particular, deceleration corrections to the constant velocity assumption are being developed. For incident energies greater than 1 GeV/nucleon, first-order deceleration corrections are small (<1 percent). As the incident energy decreases, however, the first-order corrections increase significantly (over 50 percent) at 100 MeV/nucleon; this indicates that higher order terms must be included (Khan, 1989). Work on this is in progress.

In addition to the described work on abrasion-ablation models, Khan et al. (1988) have examined contributions of direct knockout and excitation decay contributions in ^{12}C fragmentation (Webb et al., 1987). The t-matrix formulation of Norbury, Townsend, and Deutchman (1985) has received additional analysis (Cucinotta et al., 1987) but requires more fundamental development.

5.5. Direct Reaction Processes

Reaction mechanisms discussed in sections 5.2 through 5.4 have dealt with the outcome for spectator constituents in the reaction. In this section we look at collision participants and the direct breakup and knockout of particles from the projectile or target nucleus including transitions to excited nuclear states. The calculations follow closely the work of Cucinotta (1988) and Cucinotta et al. (1988).

5.5.1. *Exclusive inelastic scattering.* The scattering amplitude matrix in the eikonal approximation is given by Wilson (1975) as

$$f(\vec{q}) = -\frac{ik}{2\pi} \int \exp\left(-i\vec{q} \cdot \vec{b}\right) \left\{\exp\left[i\chi(\vec{b})\right] - 1\right\} d^2b \qquad (5.76)$$

where the eikonal phase shift matrix is related to the coupled-channel optical potential

$$\chi(\vec{b}) = \frac{-i}{2k} \int_{-\infty}^{\infty} \mathbf{U}(\vec{b}, z)\, dz \qquad (5.77)$$

where $\mathbf{U}$ is the optical potential relating all the states of the interacting systems. The elastic scattering amplitude is found to be (Cucinotta et al., 1988)

$$f_{\text{elas}}(\vec{q}) = \frac{-ik}{2\pi} \int \exp(-i\vec{q} \cdot \vec{b})\, [\exp(i\chi_{\text{opt}}) \cos \Upsilon - 1]\, d^2b \qquad (5.78)$$

where the inelastic amplitudes are

$$f_{00,n\mu}(q) = \frac{k}{2\pi} \int \exp(-i\vec{q} \cdot \vec{b})\, \exp(i\chi_{\text{opt}}) \frac{\sin \Upsilon}{\Upsilon} \chi_{00,n\mu}\, d^2b \qquad (5.79)$$

where

$$\Upsilon^2 = \sum_{(n,\mu)\neq(0,0)} \chi_{00,n\mu} \chi_{n\mu,00} \qquad (5.80)$$

The usual coherent approximation and the DWBA (distorted-wave Born approximation) are found for $\Upsilon \to 0$ in equations (5.78) and (5.79), respectively.

The function Υ as given in equation (5.80) is directly related to the pair correlation function. This can be seen as follows (Cucinotta et al., 1988):

$$\Upsilon^2 = \left(\frac{\mu_r}{2k}\right)^2 \left(\frac{1}{2\pi}\right)^4 \int d^2q\, d^2q' \exp(-i\vec{q}\cdot\vec{b})\exp(-i\vec{q}'\cdot\vec{b})$$

$$\times \sum_{\mu \text{ or } n\neq 0}^{\infty} F_{0n}(\vec{q})F_{0n}(\vec{q}')\,F_{0\mu}(-\vec{q})\,F_{\mu 0}(-\vec{q}') \qquad (5.81)$$

$$\times\, t(\vec{q})\,t(\vec{q}')$$

From Kerman, McManus, and Thaler (1959), we have the following sum rule on the form factors

$$\sum_{n\neq 0}^{\infty} F_{0n}(\vec{q})\,F_{n0}(\vec{q}') = \frac{-1}{A}F_{00}(q)\,F_{00}(\vec{q}')$$

$$+ \frac{1}{A}F_{00}(\vec{q}+\vec{q}') + \left(1-\frac{1}{A}\right)C_{00}(\vec{q},\vec{q}') \qquad (5.82)$$

where A is the mass number of the nucleus in question, and $C_{00}(\vec{q},\vec{q}')$ is the Fourier transform of the pair correlation function. Analytic models for $C_{00}(\vec{q},\vec{q}')$ are under investigation. In section 5.5.3, we consider a numerical study of long-range correlations involving partial summation of the infinite sum that appears in equation (5.81) for ^{12}C.

Finally, the first- and second-order solutions to the eikonal coupled-channel scattering amplitudes were found by approximating the form of χ. We expect that higher order solutions, though more difficult, could be found by approximating the form of higher powers of χ.

5.5.2. Physical inputs. As a numerical study, we compare the first- and second-order eikonal coupled-channel solutions of p for ^{12}C and ^{4}He on ^{12}C scattering. The 2^+ at 4.65 MeV, 0^+ at 7.66 MeV, 3^- at 9.65 MeV, and 4^+ at 14.1 MeV excited states of ^{12}C are considered. An advantage of the bordered interaction matrix is that the eikonal phase matrix elements may be obtained through knowledge of form factors measured in electron scattering experiments, so that no excited-state wave function is needed as inputs. This would not be true for couplings between the off-diagonal elements. The charge form factors for the ground and first three excited states have been parameterized by Saudinos and Wilkin (1974) and Viollier (1975) in the form

$$F_{\text{charge}}(q) = Bq^m(1-Cq^2)\exp(-dq^2) \qquad (5.83)$$

where the parameters B, C, d, and m are listed in table 5.6. Table 5.6 also lists the form factor for excitation of the 4^+ state at 14.1 MeV of ^{12}C, which we have parameterized to the data of Nakada, Torizuka, and Horikawa (1971).

Table 5.6. Form Factors

(a) ^{12}C

E, MeV	J^P	m	B	C	d, fm^{-2}
0	0^+	0	1.0	0.296	0.7
4.43	2^+	2	0.24	0.13	0.57
7.65	0^+	2	0.167	0	0.99
9.67	3^-	3	0.134	0	0.77
14.1	4^+	4	0.00392	0	0.64

(b) ^{4}He

$$C_1 = 1.098$$
$$C_2 = 0.098$$
$$d_1 = 0.72$$
$$d_2 = 3.6$$

The matter form factors are obtained from the charge form factors in the following equation (Überall, 1971):

$$F_A(q) = \frac{F_{\text{charge}}(q)}{F_p(q)F_{cm}(q)} \tag{5.84}$$

where $F_p(q)$ is the proton charge form factor given by

$$F_p(q) = \exp\left(\frac{-r_p^2 q^2}{6}\right) \tag{5.85}$$

where $r_p = 0.87$ fm, and $F_{cm}(q)$ is a center-of-mass correction of the form

$$F_{cm}(q) = \exp\left(\frac{q^2 a_0^2}{4A}\right) \tag{5.86}$$

with

$$a_0^2 = \frac{\langle r^2 \rangle - r_p^2}{\frac{Z-2}{Z} + \frac{3(A-1)}{2A}} \tag{5.87}$$

where $\langle r^2 \rangle$ is the root-mean-square radius of the nucleus. For the ground state of ^{4}He, we use the parameterization of Auger, Gillespie, and Lombard (1976),

$$F_{4_{\text{He}}} = C_1 \exp(-d_1 q^2) - C_2 \exp(-d_2 q^2) \tag{5.88}$$

where C_1, C_2, d_1, and d_2 are listed in table 5.6.

The two-body amplitude is assumed to contain only a central piece of the usual form

$$t(\bar{q}) = -\sqrt{\frac{e}{m}}\sigma(e)[\alpha(e) + i] \exp\left(-\frac{1}{2}B(e)q^2\right) \tag{5.89}$$

Table 5.7. Two-Body Amplitude Parameters

[Isospin-averaged parameters of Ray (1979)]

T_{lab}, MeV	$B(e)$, fm^2	$\sigma(e)$, fm^2	$\alpha(e)$
340	0.62	3.03	0.28
800	0.20	4.3	−0.056
1000	0.21	4.3	−0.26

where the energy-dependent parameters $\sigma(e)$, $\alpha(e)$, and $B(e)$ are taken from Ray (1979) and given in table 5.7. The Gaussian forms for the form factors and two-body amplitude assumed in the calculations allow us to obtain analytic solutions for all eikonal phase matrix elements needed as inputs for our calculations.

5.5.3. Results and discussion. The first- and second-order scattering solutions and experimental data (Blanpied et al., 1981; Bertini et al., 1973; Chaumeaux et al., 1976) for elastic and inelastic scattering of p on ^{12}C at 800 MeV and 1000 MeV and for ^{4}He on ^{12}C at 340 MeV/nucleon are shown in figures 5.14 through 5.24. For p-^{12}C elastic scattering (figs. 5.14 and 5.18), the coherent approximation (dashed line, first-order) and bordered matrix (solid line, second-order) results are nearly the same in the region of the forward peak where single scattering dominates. This was implied by Wilson (1975) on theoretical arguments

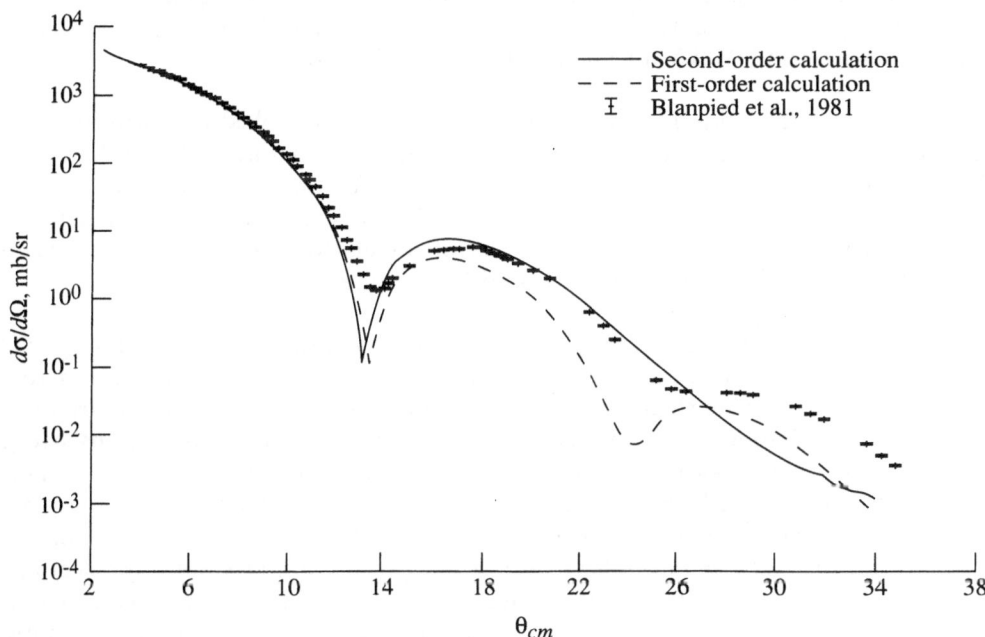

Figure 5.14. Theoretical and experimental elastic angular distributions for p-^{12}C scattering at 800 MeV.

repeated in chapter 3. We include coulomb effects only in an approximate way assuming a point coulomb interaction. A more exact treatment is needed to completely fill in the first minimum. (See, for example, Chaumeaux et al., 1976; Glauber and Matthiae, 1970.) Here, spin effects may also be important as noted by Saudinos and Wilkin (1974) and Ahmad (1975). The effect of coupling the elastic channel to low-lying excited states is seen in the second maximum (figs. 5.14 and 5.18) where the bordered matrix agrees well, whereas the coherent approximation underestimates the data both at 800 and 1000 MeV. The sensitivity to the number of channels included in the second-order calculations can be seen in figure 5.19, where the dashed line includes only the 2^+ state; the long-dash–short-dash line, the 2^+ and 0^+ states; and the solid line, the $2^+, 0^+, 3^-$, and 4^+ states. At larger angles, agreement with the data is poor. Here, the validity of the eikonal approximation is suspect, and the momentum transfers being probed are beyond the region where the phenomenological fits to the form factors and two-body amplitudes are made. For the second-order solutions, the effects of channel truncation, including the neglect of short-range correlations in the Υ-function, may be more important at larger angles.

Calculations of the excitation of the $2^+, 0^+$, and 3^- states in ^{12}C by 800 and 1040 MeV protons are shown in figures 5.15 through 5.17 and 5.20 through 5.22, respectively. The dashed line is the DWBA, and the solid line is the bordered matrix (second-order) solution. For all excited states, the DWBA and bordered matrix give similar results in the region of the first and second maxima. Although the bordered matrix contains all couplings to second order for the elastic channel, the cascades between excited states, which are neglected, should be considered a second-order effect for inelastic transitions. These cascades would be more important in the region of the second maximum. In the region of the third maximum, we do see better agreement for the bordered matrix solutions as compared with the DWBA for all transitions considered.

In figures 5.23 and 5.24 we show calculations for elastic scattering and excitation of the 0^+ state of ^{12}C for ^{4}He on ^{12}C collisions at 340 MeV/nucleon. The experimental results of Chaumeaux et al. (1976) do not report the forward peak with the data beginning at approximately 5°. No correlation effect is included for ^{4}He in the calculations. The importance of correlations is expected to increase for the lightest nuclei (Feshbach, Gal, and Hüfner, 1971).

In table 5.8, total cross sections are given for all channels considered for p-^{12}C scattering at 340, 800, and 1000 MeV. The total of the cross sections $\sigma(\text{tot})$ is calculated by the optical theorem, and the reaction cross section $\sigma(\text{re})$ is taken as the difference between the total and total elastic cross sections $\sigma(\text{el})$. The first- and second-order results are nearly the same for all channels. This agreement is expected because our angular distributions show almost complete agreement between the two solutions in the forward angles where most of the cross section occurs. In table 5.8, we also sum the excited-state cross sections $\sigma(2^+)$, $\sigma(0^+)$, $\sigma(3^-)$, $\sigma(4^+)$ for the bound-excited (BE) states calculated, $\sigma(\text{be})$. We note that $\sigma(\text{be})$ represents only a small fraction (<5 percent) of the total reaction cross section. This is an indication that the neglect of the bound-excited states in the abrasion model is a good approximation, although the importance of the giant dipole resonance state should be estimated.

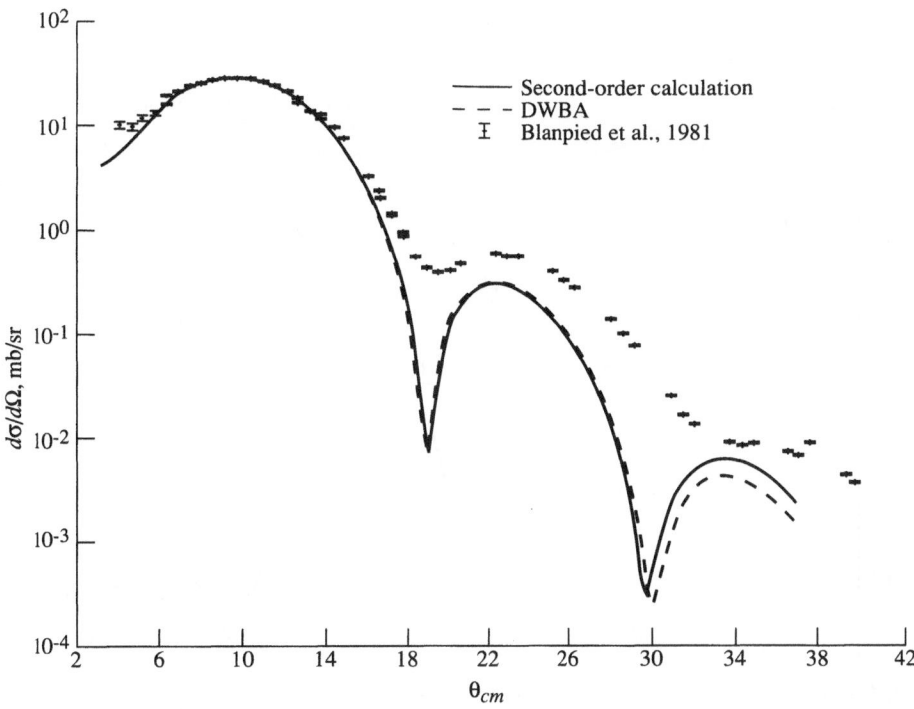

Figure 5.15. Theoretical and experimental inelastic angular distributions for excitation of 2^+ state in ^{12}C by 800-MeV protons.

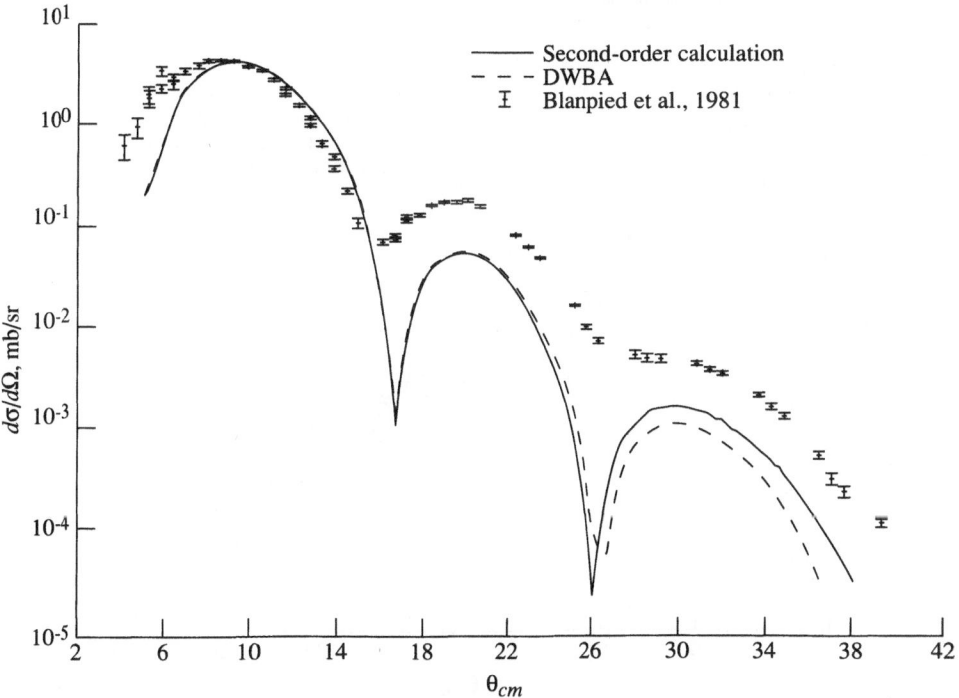

Figure 5.16. Theoretical and experimental inelastic angular distributions for excitation of 0^+ state in ^{12}C by 800-MeV protons.

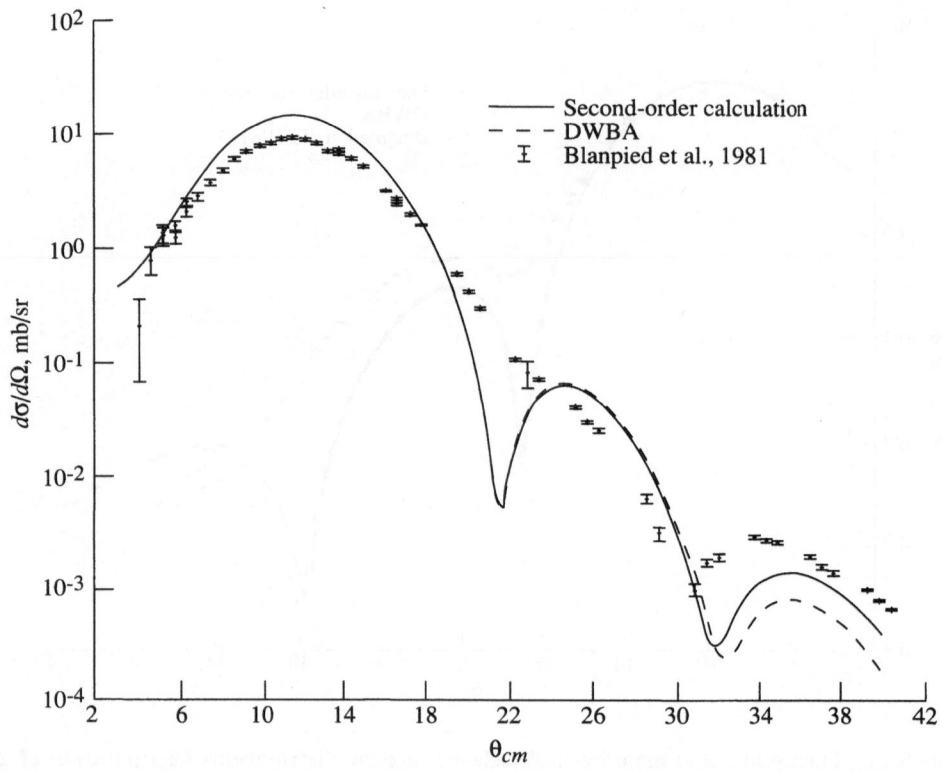

Figure 5.17. Theoretical and experimental inelastic angular distributions for excitation of 3^- state in ^{12}C by 800-MeV protons.

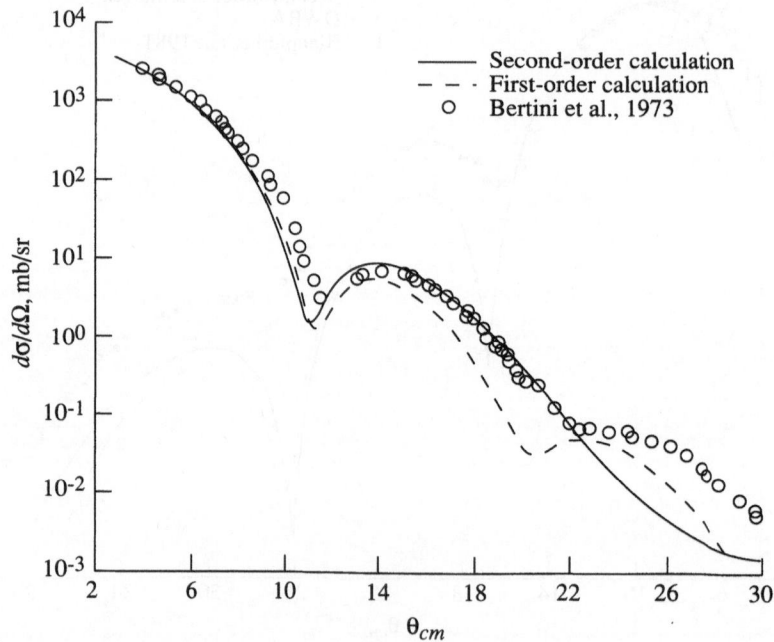

Figure 5.18. Theoretical and experimental elastic angular distributions for excitation for p-^{12}C scattering at 1040 MeV.

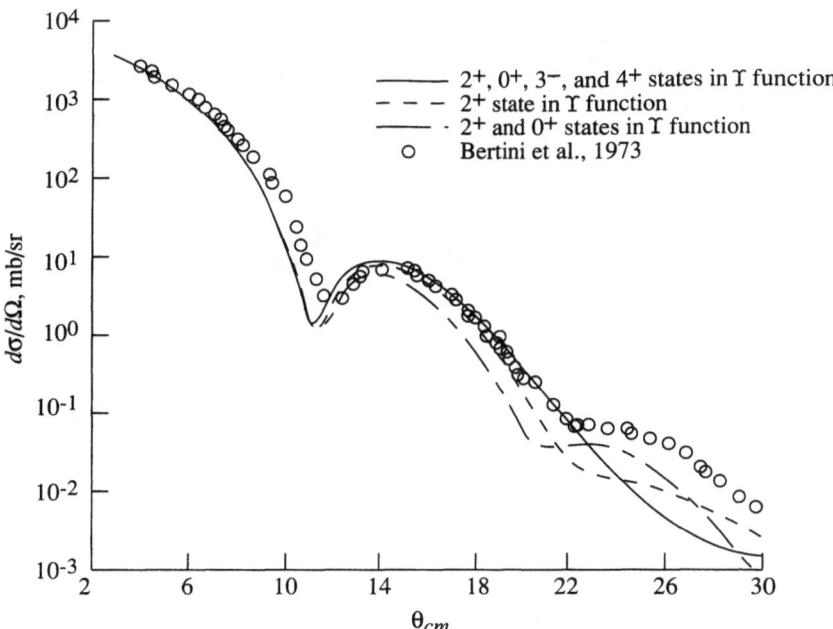

Figure 5.19. Effects of channel truncation in second-order calculations for p-^{12}C elastic scattering at 1040 MeV.

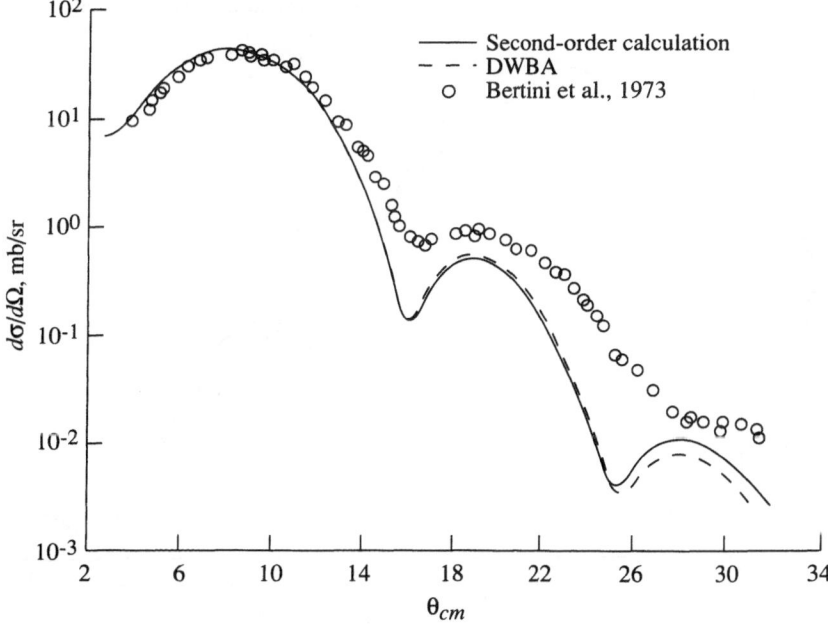

Figure 5.20. Theoretical and experimental inelastic angular distributions for excitation of 2^+ state in ^{12}C by 1040-MeV protons.

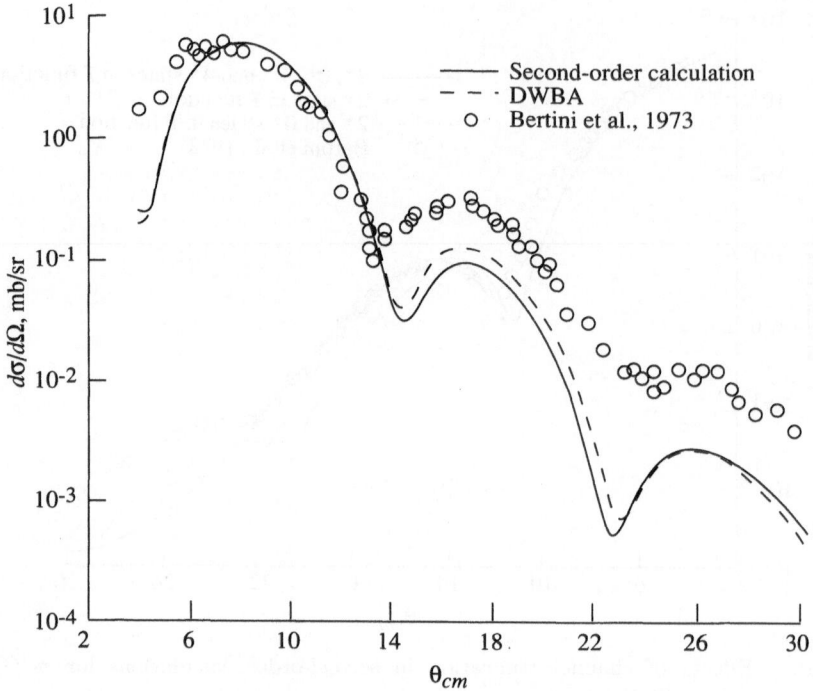

Figure 5.21. Theoretical and experimental inelastic angular distributions for excitation of 0^+ state in ^{12}C by 1040-MeV protons.

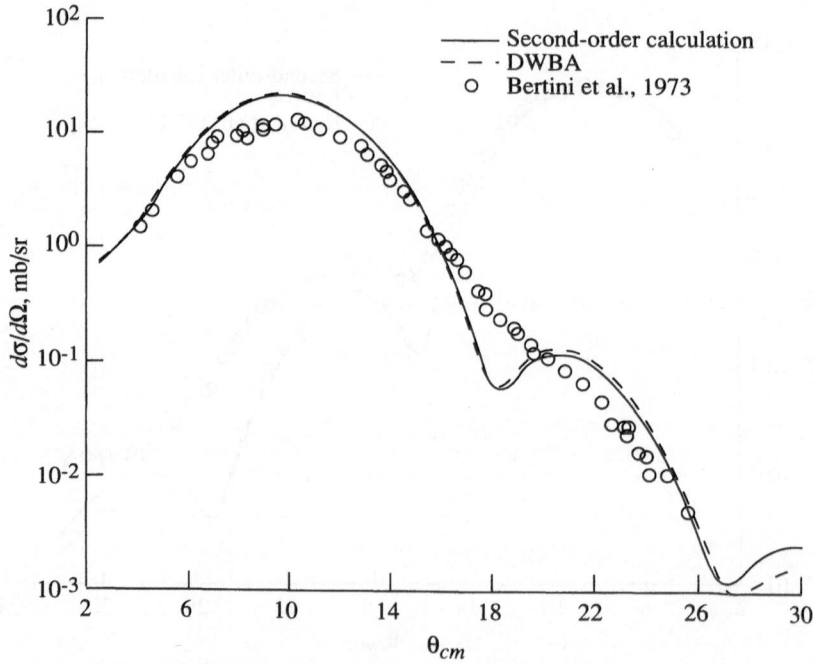

Figure 5.22. Theoretical and experimental inelastic angular distributions for excitation of 3^- state in ^{12}C by 1040-MeV protons.

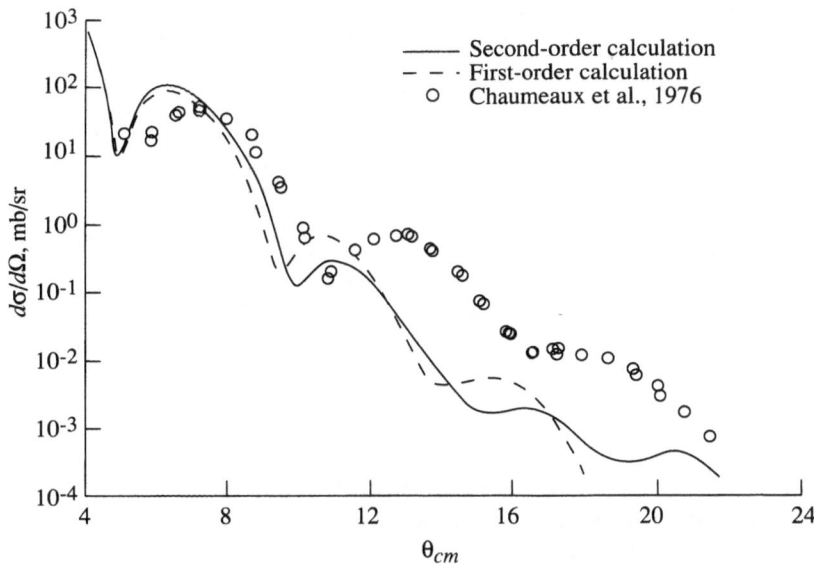

Figure 5.23. Theoretical and experimental elastic angular distributions for α-^{12}C scattering at 340 MeV.

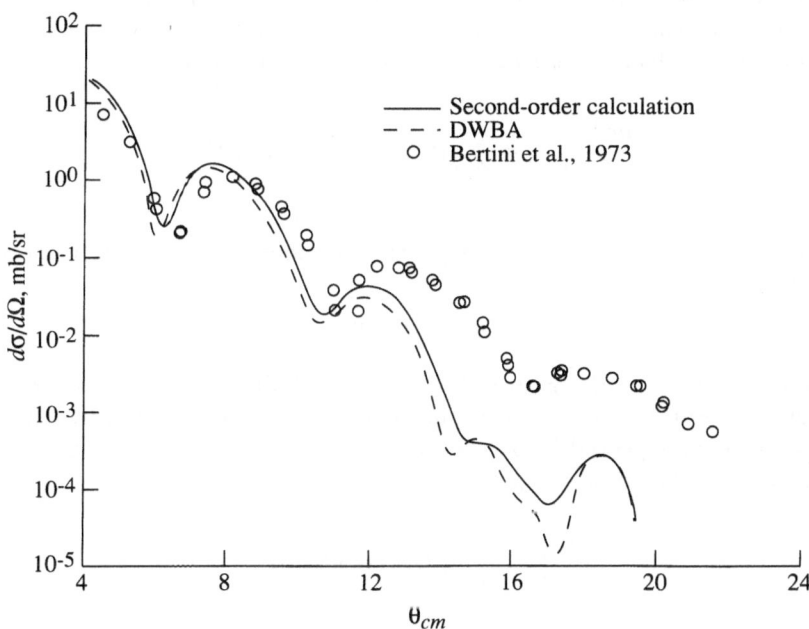

Figure 5.24. Theoretical and experimental inelastic angular distributions for excitation of 0^+ state in α-^{12}C scattering at 340 MeV/nucleon.

Table 5.8. Total Channel Cross Sections for p on ^{12}C

Cross section	$T_{\text{lab}} = 340$ MeV		$T_{\text{lab}} = 800$ MeV		$T_{\text{lab}} = 1000$ MeV	
	Coherent	Bordered	Coherent	Bordered	Coherent	Bordered
σel, mb	54.1	53.5	92.5	91.3	103.5	102.1
σ(be), mb	3.5	3.5	7.1	7.1	7.5	7.5
$\sigma(2^+)$, mb	2.2	2.2	4.1	4.1	4.3	4.3
$\sigma(0^+)$, mb	0.3	0.3	0.4	0.4	0.4	0.4
$\sigma(3^-)$, mb	1.0	1.0	2.5	2.5	2.6	2.6
$\sigma(4^+)$, mb	0.03	0.03	0.1	0.1	0.1	0.1
σ(re), mb	220.9	220.3	238.3	237.2	223.8	223.0
σ(tot), mb	275.0	273.8	330.8	328.5	327.3	325.1

5.5.4. Inclusive inelastic scattering. Cucinotta et al. (1990) consider nucleus-nucleus scattering at high energies for the case where an inclusive measurement of the projectile final state is made,

$$P + T \rightarrow P + X \qquad (5.90)$$

with P and T labeling the projectile and target, respectively, and X being some final state of the target that is not measured. In equation (5.90), the projectile scatters without fragmenting, and meson production is not considered. In the overall center-of-mass (CM) frame, with the projectile and target states denoted by $|m_P\rangle$ and $|\nu_T\rangle$, respectively, the angular distribution for equation (5.90) is found by summing the nuclear-scattering operator over all final states of the target,

$$\left. \frac{d\sigma^P}{d\Omega} \right)_{\text{tot}} = \sum_{\nu_T} |\langle \nu_T 0_P | \hat{f}(\vec{q}) | 0_T 0_P \rangle|^2 \qquad (5.91)$$

where $\hat{f}$ is the scattering operator and $\vec{q}$ is the momentum transfer to the projectile defined by

$$\vec{q} \equiv \vec{k} - \vec{k}_F \qquad (5.92)$$

In equation (5.92), $\vec{k}$ and $\vec{k}_F$ are the initial and final projectile wave vectors, respectively. In equation (5.91), the phase space is approximated by a two-body phase space that is expected to be accurate at high energies. Equation (5.91) can be separated into elastic and inelastic contributions given by

$$\left. \frac{d\sigma}{d\Omega} \right)_{\text{el}} = |\langle 0_P 0_T | \hat{f}(\vec{q}) | 0_P 0_T \rangle|^2 \qquad (5.93)$$

and

$$\left. \frac{d\sigma^P}{d\Omega} \right)_{\text{in}} = \sum_{\nu_T \neq 0} |\langle 0_P \nu_T | \hat{f}(\vec{q}) | 0_P 0_T \rangle|^2 \qquad (5.94)$$

The summation in equation (5.94) includes all excited states, bound and continuum, of the target. This infinite summation can be reduced to a single matrix element through the use of closure on the target states:

$$\sum_{\nu_T \neq 0} |\nu_T\rangle\langle\nu_T| = 1 - |0_T\rangle\langle 0_T| \tag{5.95}$$

Inserting equation (5.95) into equation (5.94), we find

$$\left.\frac{d\sigma^P}{d\Omega}\right)_{\text{in}} = \left.\frac{d\sigma^P}{d\Omega}\right)_{\text{tot}} - \left.\frac{d\sigma}{d\Omega}\right)_{\text{el}} \tag{5.96}$$

where

$$\left.\frac{d\sigma^P}{d\Omega}\right)_{\text{tot}} = \langle 0_T|\langle 0_P|\hat{f}(\vec{q})|0_P\rangle\langle 0_P|\hat{f}^+(\vec{q})|0_P\rangle|0_T\rangle \tag{5.97}$$

The great advantage of equation (5.96) over equation (5.94) is that only the ground-state wave function of the target is required.

A second reaction that we consider is complete inelastic scattering

$$P + T \rightarrow X + Y \tag{5.98}$$

where the projectile and target are both left in excited states. The angular distribution for equation (5.98) is given by

$$\left.\frac{d\sigma^{PT}}{d\Omega}\right)_{\text{in}} = \sum_{\nu_T \neq 0} \sum_{n_P \neq 0} |\langle\nu_T n_P|\hat{f}(\vec{q})|0_P 0_T\rangle|^2 \tag{5.99}$$

which is written, using closure on both the target and projectile states, as

$$\left.\frac{d\sigma^{PT}}{d\Omega}\right)_{\text{in}} = \left.\frac{d\sigma}{d\Omega}\right)_{\text{tot}} + \left.\frac{d\sigma}{d\Omega}\right)_{\text{el}} - \left.\frac{d\sigma^P}{d\Omega}\right)_{\text{tot}} - \left.\frac{d\sigma^T}{d\Omega}\right)_{\text{tot}} \tag{5.100}$$

where

$$\left.\frac{d\sigma}{d\Omega}\right)_{\text{tot}} = \left\langle 0_P 0_T\left|\left(|\hat{f}(\vec{q})|^2\right)\right|0_P 0_T\right\rangle \tag{5.101}$$

Equation (5.100) may be written as

$$\left.\frac{d\sigma^{PT}}{d\Omega}\right)_{\text{in}} = \left.\frac{d\sigma}{d\Omega}\right)_{\text{tot}} - \left.\frac{d\sigma}{d\Omega}\right)_{\text{el}} - \left.\frac{d\sigma^P}{d\Omega}\right)_{\text{in}} - \left.\frac{d\sigma^T}{d\Omega}\right)_{\text{in}} \tag{5.102}$$

The distributions given by equations (5.96) through (5.102) are evaluated when models for the nuclear-scattering operator and ground-state wave functions are introduced.

5.5.5. Correlations and inclusive scattering. The effects of short-range dynamical correlations and Pauli blocking in the nuclear wave function will be most pronounced in the inelastic distribution at small and medium momentum transfers. In order to include these effects in the inelastic scattering distributions, we consider the eikonal coupled-channel (ECC) model. Assuming correct or equivalent kinematics, the ECC can be considered the matrix representation of the Glauber amplitude. In the ECC, the matrix of scattering amplitudes for all possible projectile-target transitions is given by Cucinotta et al. (1989) as

$$f(\vec{q}) = \frac{ik}{2\pi} \int d^2b \, \exp(i\vec{q} \cdot \vec{b}) \left\{ \exp\left[i\chi(\vec{b})\right] - 1 \right\} \tag{5.103}$$

where bold-faced quantities represent matrices and the elements of χ are written as

$$\langle m_P \mu_T | \hat{\chi}(\vec{b}) | n_P \nu_T \rangle = \frac{1}{2\pi \, k_{NN}} \sum_{\alpha j} \int d^2q \, \exp(i\vec{q} \cdot \vec{b})$$

$$\times F^{(1)}_{m_p n_p}(-\vec{q}) \, G^{(1)}_{\mu_T \nu_T}(\vec{q}) \, f_{NN}(\vec{q}) \tag{5.104}$$

where F and G are projectile and target form factors. Assuming that the off-diagonal terms in χ are small compared with the diagonal terms (see chapter 3 and Wilson, 1975), we separate χ into diagonal (χ_D) and off-diagonal, χ_O terms as

$$\chi(\vec{b}) = \chi_D(\vec{b}) + \chi_O(\vec{b}) \tag{5.105}$$

We further assume that the nuclear density in the excited states is approximately the same as the ground state, such that the elements of the diagonal matrix χ_D are all taken as the elastic element,

$$\chi(\vec{b}) = \frac{A_P A_T}{(2\pi \, k_{NN})} \int d^2q \, F^{(1)}(-\vec{q}) \, G^{(1)}(\vec{q}) \, \exp(i\vec{q} \cdot \vec{b}) \tag{5.106}$$

To treat off-diagonal scattering, we expand f in powers of χ_O

$$f(\vec{q}) = \frac{-ik}{2\pi} \int d^2b \, \exp\left[i\vec{q} \cdot \vec{b} + i\chi_D(\vec{b})\right] \sum_{m=1} \frac{[i\chi_O(\vec{b})]^m}{m!} \tag{5.107}$$

The inclusive distribution for the projectile then follows as:

$$\left. \frac{d\sigma^P}{d\Omega} \right)_{in} = \left| \frac{ik}{2\pi} \right|^2 \int d^2b \, d^2b' \, \exp\left[i\vec{q} \cdot (\vec{b} - \vec{b}')\right] \, \exp\left\{ i \left[\chi(\vec{b}) - \chi^+(\vec{b}')\right] \right\}$$

$$\times \sum_{v_T \neq 0} \left[T_S(\vec{b}, \vec{b}') + T_D(\vec{b}, \vec{b}') + \ldots \right] \tag{5.108}$$

where the single inelastic scattering terms are

$$T_S(\vec{b}, \vec{b}') = \langle 0_P \, 0_T | \hat{\chi}(\vec{b}) | 0_P \nu_T \rangle \langle \nu_T 0_P | \hat{\chi}^+(\vec{b}) | 0_P 0_T \rangle \tag{5.109}$$

and the double inelastic scattering terms are

$$T_D(\vec{b}, \vec{b}') = -\frac{1}{4} \sum_{\mu_T \neq 0} \sum_{n_p = 0} \left\langle 0_P 0_T \middle| \hat{\chi}(\vec{b}) \middle| \mu_T n_p \right\rangle \left\langle \mu_T n_p \middle| \hat{\chi}(\vec{b}) \middle| 0_P \nu_T \right\rangle$$

$$\times \sum_{\mu_T' \neq 0} \sum_{n_p' = 0} \left\langle \nu_T 0_P \middle| \hat{\chi}^+(\vec{b}') \middle| \mu_T' n_p' \right\rangle \left\langle \mu_T' n_p' \middle| \hat{\chi}^+(\vec{b}') \middle| 0_P 0_T \right\rangle \quad (5.110)$$

Each term in the inelastic scattering expansion of equation (5.108) can be reduced through use of closure to terms involving matrix elements of one-, two-, ..., n-body operators over the ground state and thus includes the effects of two or more particle correlations. Details are given by Cucinotta et al. (1990).

5.5.6. Model calculations. We now consider the evaluation of the inelastic distributions in equation (5.106). Ignoring spin effects, we use an isospin-averaged, two-body amplitude given by

$$f_{\mathrm{NN}}(\vec{q}) = \frac{\sigma(\alpha + i)k_{\mathrm{NN}}}{4\,\pi} \exp\left(\frac{-Bq^2}{2}\right) \quad (5.111)$$

where the energy dependent parameters σ, B, and α are listed in table 5.9. For the projectile, we use a one-body form factor

$$F^{(1)}(\vec{q}) = \exp\left(\frac{-R_P^2 q^2}{4}\right) \quad (5.112)$$

where R_P is the matter radius of the projectile. For the target one-body form factor, we use the harmonic well form of Townsend and Wilson (1985)

$$G^{(1)}(\vec{q}) = (1 - C_T q^2) \exp\left(\frac{-R_T^2 q^2}{4}\right) \quad (5.113)$$

where R_T is the target matter radius and

$$C_T = \frac{\gamma_T R_T^2}{4(1 + \frac{3}{2}\gamma_T)} \quad (5.114)$$

where values of γ_T are given by Townsend and Wilson (1985)

Correlation effects are included in the two-particle density through the approximate form given by Moniz and Nixon (1971) as

$$\rho^{(2)}(\vec{x}, \vec{y}) = \rho^{(1)}(\vec{x})\rho^{(1)}(\vec{y}) \left\{ 1 - \exp\left[\frac{-(\vec{x} - \vec{y})^2}{2\ell_c^2}\right] \right\} \quad (5.115)$$

Table 5.9. Parameters For Nucleon-Nucleon Scattering Amplitude

Reaction	σ, fm^2	B, fm^2	α
α-α at 642 MeV/nucleon	3.93	0.13	-0.39
α-^{12}C at 3.64 GeV/nucleon	4.2	0.28	-0.43
p-^{16}O at 1 GeV/nucleon	4.3	0.26	-0.22
$\bar{\alpha}$-A_T at 1 GeV/nucleon	4.3	0.26	-0.23

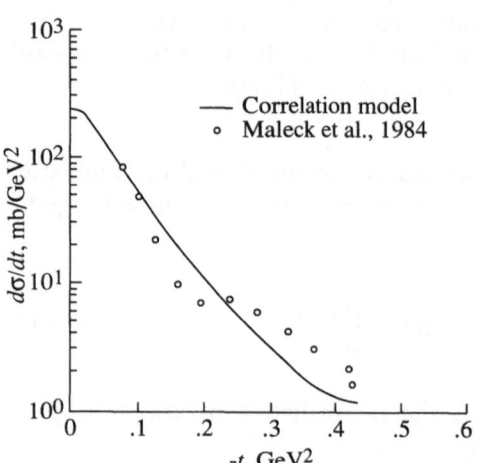

Figure 5.25. Inclusive α cross section on helium target.

Figure 5.26. Inclusive α cross section on carbon target.

where ℓ_c is an effective correlation length, $\ell_c = 0.7$ fm. For comparison with experimental results, the inclusive invariant distribution is written as

$$\left. \frac{d\sigma^P}{dt} \right)_{in} \approx \frac{\pi}{k^2} \left. \frac{d\sigma^P}{d\Omega} \right)_{in} \qquad (5.116)$$

with

$$t \approx -q^2 \qquad (5.117)$$

In figure 5.25, we show the correlation model and the experimental results of Maleck, Picozza, and Satta (1984) for α-α scattering at 642 MeV/nucleon. The correlation model produces good agreement over the region of momentum transfers studied.

Experimental results of Ableev et al. (1982) for total inclusive scattering of α-particles on ^{12}C at 3.64 GeV/nucleon are shown in figure 5.26. The solid line represents the sum of inelastic and elastic contributions. Agreement with the data is fair, whereas calculations underestimate the data at larger values of t. Correlation effects in elastic scattering have been shown to increase the cross section in this region by a substantial amount (Cucinotta et al., 1989; Cucinotta, 1988) such that a second-order elastic scattering model should lead to improved agreement. The dominance of elastic scattering at small values of t, as seen in figure 5.26, indicates that the model is sufficient when total scattering distributions are considered.

5.6. Coulomb Dissociation

The coulomb cross section for producing state X was given as a multipole series as

$$\sigma_{\text{em}}(X) = \sum_{\pi\ell} \int_{E_o(X)}^{\infty} \sigma_{\gamma}^{\pi\ell}(E, X) \, N^{\pi\ell}(E) \, dE \qquad (5.118)$$

where $\sigma_{\gamma}^{\pi\ell}(E, X)$ is the photonuclear cross section and $N^{\pi\ell}(E)$ is the virtual photon density produced by the passing ion. The virtual photon densities $N^{\pi\ell}(E)$ are known, and the corresponding photonuclear cross sections are the primary uncertainties.

5.6.1. Electric dipole transitions. The $E1$ virtual photon density experienced by a passing ion is (Norbury and Townsend, 1990a)

$$N^{E1}(E) = \frac{1}{E}\frac{2}{\pi}Z_t^2\alpha\frac{1}{\beta^2}\left\{ xK_0(x)\,K_1(x) - \frac{1}{2}x^2\beta^2\left[K_1^2(x) - K_0^2(x)\right] \right\} \qquad (5.119)$$

where E is the photon energy, Z_t the nuclear charge of the target, β is the velocity in units of c, α the fine structure constant, and $K_0(x)$ and $K_1(x)$ are the modified Bessel functions of the second kind.

The parameter x is given as

$$x = \frac{Eb_{\text{min}}}{\gamma\beta\hbar c} \qquad (5.120)$$

where γ is the usual relativistic factor and b_{min} is the minimum impact parameter taken as (Norbury et al., 1988)

$$b_{\text{min}} = R_{0.1}(A_T) + R_{0.1}(A_P) \qquad (5.121)$$

where the 10-percent charge density radius measured in fm is given by (Norbury et al., 1988)

$$R_{0.1}(A) = (1.18A^{1/3} + 0.75) \qquad (5.122)$$

The photonuclear cross sections are assumed to be of the form

$$\sigma_{\gamma}^{E1}(E, X) = g_X^{E1}(E)\sigma_{\text{abs}}^{E1} \qquad (5.123)$$

with $g_X^{E1}(E)$ assumed to be energy independent. The $E1$ absorption cross section is taken to be the giant electric dipole resonance (GDR), which is

$$\sigma_{\text{abs}}^{E1}(E) = \frac{\sigma_m}{1 + [(E^2 - E_{\text{GDR}}^2)/E^2\Gamma^2]} \qquad (5.124)$$

where E_{GDR} is the resonance energy, Γ is the resonance width, and σ_m is given by (Levinger, 1960)

$$\sigma_m = \frac{\sigma_{\mathrm{TRK}}}{\pi(\Gamma/2)} \tag{5.125}$$

with the Thomas-Reiche-Kuhn cross section in MeV-mb given by

$$\sigma_{\mathrm{TRK}} = \frac{60NZ}{A} \tag{5.126}$$

where N, Z, and A are neutron numbers, proton numbers, and nucleon numbers, respectively. The resonance energy is given by (Westfall et al., 1979)

$$E_{\mathrm{GDR}} = \hbar c \left[\frac{m^* R_o^2}{8J} \left(1 + u - \frac{1 + \epsilon + 3u}{1 + \epsilon + u} \epsilon \right) \right]^{-1/2} \tag{5.127}$$

with

$$u = \frac{3J}{Q' A^{1/3}} \tag{5.128}$$

and

$$R_o = r_o A^{1/3} \tag{5.129}$$

where $\epsilon = 0.0768, Q' = 17$ MeV, $J = 36.8$ MeV, $r_o = 1.18$ fm, and m^* is 0.7 times the nucleon rest mass. The resonance width in MeV is approximately

$$\Gamma = \left\{ \begin{array}{ll} 10 & (A < 50) \\ 4.5 & (A \geq 50) \end{array} \right\} \tag{5.130}$$

The branching ratios g_x satisfy

$$\sum_x g_x(E) = 1 \tag{5.131}$$

The proton branching ratio was found by Westfall et al. (1979) to be

$$g_p = \min \left[\frac{Z}{A}, \ 1.95 \exp(-0.075Z) \right] \tag{5.132}$$

If all other processes are assumed to emit neutrons, we may write

$$\sum_s \sigma_\gamma(E, sn) = \sigma_\gamma(E, n) + \sigma_\gamma(E, 2n) + \ldots$$

$$= (1 - g_p)\sigma_{\mathrm{abs}}^{E1}(E) \tag{5.133}$$

with the total photoneutron production cross section as

$$M(E) \sum_s \sigma_\gamma(E, sn) = \sum_s s\, \sigma_\gamma(E, sn)$$

$$= \sum_s s g_{sn} \sigma_{\mathrm{abs}}^{E1}(E) \tag{5.134}$$

and further assume that the processes emitting more than two neutrons are negligible yields

$$M(E)\,(1 - g_p) \approx g_n + 2g_{2n} \tag{5.135}$$

which yields

$$g_{2n} = [M(E) - 1](1 - g_p) \tag{5.136}$$

and

$$g_n = 1 - g_p - g_{2n} \tag{5.137}$$

The average multiplicity was shown to be

$$M(E) = 1 + (1 - f_d)\left[1 - \left(1 + \frac{E_{\text{sec}}}{\Theta}\right)\exp\left(-\frac{E_{\text{sec}}}{\Theta}\right)\right] \tag{5.138}$$

where

$$E_{\text{sec}} = E - E_o(\gamma, 2n) \tag{5.139}$$

and the nuclear temperature (Blatt and Weisskopf, 1952) is

$$\Theta = \left[\left(\frac{A}{8E^*}\right)^{1/2} - \frac{5}{4E^*}\right]^{-1} \tag{5.140}$$

where the excitation energy

$$E^* = E - E_o(\gamma, n) \tag{5.141}$$

See Norbury and Townsend (1990a) for details.

5.6.2. Results. Reasonable values of coulomb dissociation are obtained with experiments (Heckman and Lindstrom, 1976; Mercier et al., 1984) by neglecting the $2n$ channel (Norbury, Townsend, and Badavi, 1988) by setting $M(E) = 1$ as shown in tables 5.10 and 5.11. Further analysis is given by Norbury (1989). A fully parameterized computer program is available for generating one nucleon removal cross section (Norbury, Townsend, and Badavi, 1988). Preliminary work on evaluating the $2n$ photonuclear cross sections is hopeful but not complete (Norbury and Townsend, 1990a; Cucinotta, Norbury, and Townsend, 1988).

5.6.3. Electric quadrupole transitions. The $E2$ virtual photon density experienced by a passing ion is (Bertulani and Baur, 1988)

$$N^{E2}(E) = \frac{1}{E}\frac{2}{\pi}Z_T^2\,\alpha\frac{1}{\beta^4}\left\{2(1 - \beta^2)K_1^2(x) + x(2 - \beta^2)K_0(x)K_1(x)\right.$$

$$\left. - \frac{x^2}{2\beta^4}\left[K_1^2(x) - K_0^2(x)\right]\right\} \tag{5.142}$$

where E is the photon energy, Z_T is the nuclear charge of the target, β is the velocity of the projectile in units of c, α is the fine structure constant, and $K_0(x)$

Table 5.10. Calculated Total Electromagnetic Reaction Cross Sections
for ^{12}C And ^{16}O Incident Upon Various Targets

Projectile	Energy, GeV/nucleon	Target	Final state	σ_{EM}, mb (a)	σ_{EM}, mb (present work)
^{12}C	2.1	^{208}Pb	^{11}C + n	50 + 18	68
			^{11}B + p	50 ± 25	68
	1.05		^{11}C + n	38 ± 24	43
			^{11}B + p	50 ± 26	43
^{16}O	2.1		^{15}O + n	50 + 25	99
			^{15}N + p	97 ± 17	99
^{12}C	2.1	^{108}Ag	^{11}C + n	22 ± 12	26
			^{11}B + p	20 ± 12	26
	1.05		^{11}C + n	22 ± 12	17
			^{11}B + p	25 ± 20	17
^{16}O	2.1		^{15}O + n	26 ± 13	37
			^{15}N + p	29 ± 19	37
^{12}C	2.1	^{64}Cu	^{11}C + n	10 ± 6	11
			^{11}B + p	4 ± 8	11
	1.05		^{11}C + n	10 ± 7	7.4
			^{11}B + p	5 ± 8	7.4
^{16}O	2.1		^{15}O + n	10 ± 7	16
			^{15}N + p	14 ± 9	16
^{12}C	2.1	^{27}Al	^{11}C + n	0 ± 3	2.5
			^{11}B + p	0 ± 3	2.5
	1.05		^{11}C + n	1 ± 3	1.8
			^{11}B + p	1 ± 3	1.8
^{16}O	2.1		^{15}O + n	0 ± 3	3.6
			^{15}N + p	0 ± 0	3.6
^{12}C	2.1	^{12}C	^{11}C + n	0 ± 1	0.58
			^{11}B + n	0 ± 3	0.58
	1.05		^{11}C + n	0 ± 2	0.43
			^{11}B + p	0 ± 1	0.43
^{16}O	2.1		^{15}O + n	0 ± 2	0.83
			^{15}N + p	0 ± 3	0.83

[a]This column represents the measurements (isotope averaged) of Heckman and Lindstrom (1976). See Mercier et al. (1984).

Table 5.11. Calculated Total Electromagnetic Reaction Cross Sections for
Various Projectiles Incident Upon ^{197}Au

Projectile	Energy, GeV/nucleon	Final state	σ_{EM}, mb (a)	σ_{EM}, mb (present work)
^{12}C	2.1	^{196}Au + n	66 ± 20	39
^{20}Ne	2.1		136 ± 21	104
^{40}Ar	1.8		420 ± 120	299
^{56}Fe	1.7		680 ± 160	588

^{a} This column represents the data of Mercier et al. (1984).

Table 5.12. EM Cross Sections for Reaction ^{89}Y(projectile, X)^{88}Y

Projectile	$R_{0.1}(P)$ fm	Energy, GeV/nucleon	σ_{expt}, mb (a)	σ_{ww}, mb	σ_{E1}, mb	σ_{E2}, mb	$\sigma_{E1} + \sigma_{E2}$, mb
^{12}C	3.30	2.1	9 ± 12	12	12	1	13
^{20}Ne	4.00	2.1	43 ± 12	32	31	3	34
^{40}Ar	4.72	1.8	132 ± 17	90	88	9	97
^{56}Fe	5.24	1.7	217 ± 20	175	171	16	187

^{a} This column represents the data of Mercier et al. (1984).

and $K_1(x)$ are the modified Bessel functions of the second kind. The parameter x is given as

$$x = \frac{E b_{min}}{\gamma \beta \hbar c} \tag{5.143}$$

where γ is the usual Lorentz factor and b_{min} is the minimum impact parameter taken herein as the sum of the 10-percent radii of the target and projectile. The photonuclear cross section is assumed to be the isoscalar component of the electric giant quadrupole resonance (GQR) given by Norbury (1990b) as

$$\sigma_{E2}(E) = \frac{\sigma_{EWSR} E_{GQR}^2}{1 + (E^2 - E_{GQR}^2)^2/\Gamma^2 E^2} \tag{5.144}$$

where the energy-weighted sum rule (EWSR) cross section in μb/MeV is

$$\sigma_{EWSR} = f \frac{0.22 Z A^{2/3}}{\pi \Gamma/2} \tag{5.145}$$

The data of Mercier et al. (1986) were analyzed by Norbury (1990a) with results shown in table 5.12. In this table, the 10-percent charge radius used for ^{89}Y is 6.02 fm and the GQR parameters are $f = 0.55$, $\Gamma = 3.2$ MeV, and

$E_{\mathrm{GQR}} = 13.8\,\mathrm{MeV}$. The $^{89}\mathrm{Y}(\gamma, n)$ threshold is at 11.0 MeV. Calculations are made for Weizsäcker-Williams theory (σ_{WW}), and individual $E1$ and $E2$ multipole cross sections are calculated. The cross section $\sigma_{E1} + \sigma_{E2}$ is in reasonable agreement with experiment. All calculations use the minimum impact parameter given by $b_{\min} = R_{0.1}(A_P) + R_{0.1}(A_T)$.

5.6.4. *Nuclear versus coulomb contributions.*

An extensive comparison of the electromagnetic theory (Norbury, 1989, 1990a, 1990b, and 1990c) with experiment (Hill et al., 1988) showed that significant discrepancies still remained at higher energies. These have been attributed to an incorrect subtraction of the nuclear contribution to the total experimental cross section; thus an incorrect experimental electromagnetic cross section (Benesh, Cook, and Vary, 1989) resulted. Calculations were then performed (Benesh, Cook, and Vary, 1989; Norbury and Townsend, 1990b) of both the nuclear and electromagnetic cross sections and were compared with the total measured cross sections. Much more satisfactory agreement was then obtained. Thus, the electromagnetic and nuclear contributions to single-nucleon removal seem reasonably well understood; however, recent comparisons with exclusive data for nucleon removal from $^{28}\mathrm{Si}$ indicate that some work remains to be done in understanding how to calculate excitation energies relevant to single-nucleon removal at higher energies (14 GeV/nucleon).

Future work should be directed to understanding nuclear and coulomb contributions to the removal of a few nucleons.

5.7. Semiempirical Data Base

Even though the accuracy of the data for specific reactions improves, a reasonable means of representing data in computational procedures is still a challenge. We have avoided a point representation of the data since large multidimensional arrays will eventually rival computer storage. Mostly, we use empirical methods built on some theoretical base which describes approximately the systematic variation of reaction cross sections and add a few adjustable parameters or interpolation and extrapolation procedures.

5.7.1. *Nucleon nonelastic spectrum.*

The nonelastic differential cross sections (the inelastic process in which the nucleus is raised to an excited level is ignored) use the results of Bertini's MECC-7 (Anon., 1968) program. The nucleon multiplicities are given in tables 5.13 and 5.14. We have required the multiplicities to be monotonic in energy, and thus the values in parentheses, which were obtained by scaling from lower and higher energies, are correct values and are used in the calculations. The results below 400 MeV were taken from Alsmiller, Barish, and Leimdorfer (1968), and the results for carbon, calcium, bromine, cesium, and holmium above 400 MeV are obtained by interpolation. The nonelastic spectra are represented as

$$f(E, E') = \sum_{i=1}^{3} \frac{N_i}{\alpha_i} \frac{\exp(-E'/\alpha_i)}{1 - \exp(-E'/\alpha_i)}$$
$$+ \frac{N_Q}{E'} \left\{ 1 + \exp\left[-20(1 - E/E')\right] \right\} \tag{5.146}$$

The first term of the summation represents the evaporation peak so that N_1 (the number of evaporation nucleons) is taken from table 5.13 and the spectral parameter α_1 (in GeV) is taken from Ranft (1980)

$$\alpha_{1p} = \begin{cases} (0.019 + 0.0017E')(1 - 0.001A_T) & (E' < 5 \text{ GeV}) \\ 0.027(1 - 0.001A_T) & (E' \geq 5 \text{ GeV}) \end{cases} \quad (5.147)$$

$$\alpha_{1n} = \begin{cases} (0.017 + 0.0017E')(1 - 0.001A_T) & (E' < 5 \text{ GeV}) \\ 0.023(1 - 0.001A_T) & (E' \geq 5 \text{ GeV}) \end{cases} \quad (5.148)$$

The second term is taken from Ranft (1980) to represent the low-energy cascade particles as

$$n_{2p} = \begin{cases} 0.0035\sqrt{A_T} & (E' \leq 0.1 \text{ GeV}) \\ 0.007\sqrt{A_T}[0.5 + 1(1 + \log_{10} E')^2] & (0.1 < E' < 5 \text{ GeV}) \\ 0.0245\sqrt{A_T} & (E' \leq 5 \text{ GeV}) \end{cases} \quad (5.149)$$

$$n_{2n} = \begin{cases} 0.0042\sqrt{A_T} & (E' \leq 0.1 \text{ GeV}) \\ 0.007\sqrt{A_T}[0.6 + 1.3(1 + \log_{10} E')^2] & (0.1 < E' < 5 \text{ GeV}) \\ 0.032\sqrt{A_T} & (E' \leq 5 \text{ GeV}) \end{cases} \quad (5.150)$$

with the corresponding spectral parameters

$$\alpha_{2p} = \begin{cases} (0.11 + 0.01E')(1 - 0.001A_T) & (E' < 5 \text{ GeV}) \\ 0.16(1 - 0.001A_T) & (E' \geq 5 \text{ GeV}) \end{cases} \quad (5.151a)$$

$$\alpha_{1n} = \begin{cases} (0.1 + 0.01E')(1 - 0.001A_T) & (E' < 5 \text{ GeV}) \\ 0.15(1 - 0.001A_T) & (E' \geq 5 \text{ GeV}) \end{cases} \quad (5.151b)$$

The third term in the summation is the balance of cascade particles after the inclusion of the quasi-elastic contribution.

The quasi-elastic contribution is estimated by including the nuclear attenuation following the quasi-elastic event. The proton quasi-elastic cross section is

$$\left. \begin{array}{l} \sigma_{Q,pp} = Z_T\sigma_{pp} + (A_T - Z_T)\sigma_{np} \\ \sigma_{Q,pn} = (A_T - Z_T)\sigma_{np} \end{array} \right\} \quad (5.152)$$

and similarly for neutrons

$$\left. \begin{array}{l} \sigma_{Q,nn} = (A_T - Z_T)\sigma_{nn} + Z_T\sigma_{np} \\ \sigma_{Q,np} = Z_T\sigma_{np} \end{array} \right\} \quad (5.153)$$

Table 5.13. Number of Evaporation Nucleons Produced in Nuclear Collisions
[Values in parentheses are modified and used in the code]

	Number of nucleons produced at—					
	25 MeV	200 MeV	400 MeV	1000 MeV	2000 MeV	3000 MeV
$A_T = 12$:						
$p \to p$	0.51	0.54	0.50	0.72	0.75	0.84
$p \to n$	0.026	0.32	0.35	0.79	0.79	0.79
$n \to p$	0.052	0.30	0.35	0.73	0.73	0.80
$n \to n$	0.43	0.57	0.52	0.77 (0.71)	0.71 (0.71)	0.73
$A_T = 16$:						
$p \to p$	0.62	0.73	0.71	0.84	0.89	0.98 (0.93)
$p \to n$	0.87	0.36	0.441	0.11 (0.87)	0.93 (0.87)	0.82 (0.87)
$n \to p$	0.12	0.47	0.53	0.86	0.86	0.89
$n \to n$	0.55	0.60	0.59	0.79	0.79	0.81
$A_T = 27$:						
$p \to p$	0.54	0.99	1.03	1.36	1.49	1.86
$p \to n$	0.37	0.61	0.62	1.29	2.03 (1.92)	1.52 (1.92)
$n \to p$	0.14	0.78	0.82	1.29	1.60	1.74
$n \to n$	0.75	0.76	0.71	1.34	1.51	1.60
$A_T = 40$:						
$p \to p$	0.50	1.03	1.06	1.74	2.32	2.93
$p \to n$	0.53	1.12	1.24	2.63	3.36	3.64
$n \to p$	0.12	0.74	0.84	1.60	2.29	2.67
$n \to n$	0.89	1.39	1.44	2.76	3.25	3.54
$A_T = 65$:						
$p \to p$	0.18	0.75	0.91	2.11	3.15	4.00
$p \to n$	1.04	2.33	2.65	3.97	4.79	5.37
$n \to p$	0.03	0.49	0.66	1.90	2.98	3.61
$n \to n$	1.46	2.77	2.90	4.17	4.99	5.49
$A_T = 80$:						
$p \to p$	0.10	0.60	1.07	2.2	3.18	4.89
$p \to n$	1.29	2.20	3.18	3.72	5.07	6.77
$n \to p$	0.02	0.53	0.79	1.87	2.91	4.53
$n \to n$	1.58	3.19	3.43	4.07	5.35	6.91
$A_T = 100$:						
$p \to p$	0.03	0.46	1.28	2.96	4.56	5.78
$p \to n$	1.53	1.97	3.72	5.46	7.04	8.17
$n \to p$	0.004	0.59	0.96	2.71	4.27	5.44
$n \to n$	1.67	3.60	3.97	5.63	7.31	8.33
$A_T = 132$:						
$p \to p$	0.01	0.61	1.03	2.68	4.51	6.32
$p \to n$	1.91	4.11	5.25	8.76	11.34	12.31
$n \to p$	0.001	0.47	0.81	2.51	4.47	5.98
$n \to n$	1.96	4.73	5.59	8.93	10.6	12.42
$A_T = 164$:						
$p \to p$	0.003	0.42	0.76	2.38	4.68	6.86
$p \to n$	2.17	5.79	7.07	12.09	15.7	16.45
$n \to p$	0.003	0.28	0.58	2.30	4.68	6.52
$n \to n$	2.26	5.96	7.07	12.3	14.6	16.51
$A_T = 207$:						
$p \to p$	0.001	0.21	0.44	2.23	5.19	7.39
$p \to n$	2.29	7.22	9.24	15.3	17.81	20.6
$n \to p$	0.00	0.10	0.30	2.10	4.88	7.05
$n \to n$	2.29	7.38	9.53	15.6	18.2	20.6

Table 5.14. Number of Cascade Nucleons Produced in Nuclear Collisions

	Number of nucleons produced at—					
	25 MeV	200 MeV	400 MeV	1000 MeV	2000 MeV	3000 MeV
$A_T = 12$:						
$p \to p$	0.58	1.43	1.63	1.95	2.15	2.48
$p \to n$	0.41	0.86	0.93	1.42	1.66	2.08
$n \to p$	0.42	0.90	0.92	1.43	1.65	1.91
$n \to n$	0.56	1.42	1.69	1.95	2.27	2.57
$A_T = 16$:						
$p \to p$	0.56	1.41	1.72	2.05	2.39	2.60
$p \to n$	0.38	0.90	0.98	1.47	1.86	2.19
$n \to p$	0.38	0.91	0.96	1.49	1.85	2.01
$n \to n$	0.54	1.43	1.70	2.05	2.52	2.70
$A_T = 27$:						
$p \to p$	0.46	1.38	1.67	2.29	2.86	3.19
$p \to n$	0.34	0.97	1.16	1.86	2.54	3.25
$n \to p$	0.32	0.93	1.01	1.69	2.28	2.71
$n \to n$	0.49	1.48	1.81	2.42	3.22	3.71
$A_T = 40$:						
$p \to p$	0.40	1.33	1.69	2.32	3.01	3.53
$p \to n$	0.30	1.04	1.24	2.46	3.52	4.48
$n \to p$	0.28	0.89	1.08	1.79	2.51	3.06
$n \to n$	0.45	1.49	1.88	2.99	4.13	4.83
$A_T = 65$:						
$p \to p$	0.30	1.21	1.69	2.35	3.16	3.87
$p \to n$	0.28	1.09	1.46	3.06	4.49	5.72
$n \to p$	0.21	0.86	1.08	1.88	2.75	3.41
$n \to n$	0.40	1.53	2.00	3.55	5.03	5.95
$A_T = 80$:						
$p \to p$	0.27	1.18	1.57	2.32	3.18	3.95
$p \to n$	0.25	1.08	1.45	3.27	4.92	6.35
$n \to p$	0.19	0.81	1.04	1.86	2.78	3.54
$n \to n$	0.36	1.51	1.98	3.78	5.40	6.64
$A_T = 100$:						
$p \to p$	0.25	1.15	1.55	2.29	3.20	4.04
$p \to n$	0.22	1.06	1.52	3.47	5.35	6.98
$n \to p$	0.17	0.78	1.08	1.84	2.44	3.67
$n \to n$	0.31	1.47	2.03	3.96	5.76	7.33
$A_T = 132$:						
$p \to p$	0.20	1.00	1.46	2.21	3.17	3.87
$p \to n$	0.20	1.11	1.57	3.31	5.20	7.91
$n \to p$	0.13	0.70	1.00	1.79	2.69	3.52
$n \to n$	0.28	1.45	2.10	3.86	6.86	8.29
$A_T = 164$:						
$p \to p$	0.16	0.90	1.36	2.13	3.15	3.69
$p \to n$	0.18	1.11	1.60	3.16	5.06	8.86
$n \to p$	0.11	0.63	0.88	1.72	2.55	3.39
$n \to n$	0.26	1.42	2.11	3.56	7.94	9.25
$A_T = 208$:						
$p \to p$	0.14	0.82	1.27	2.05	7.74	3.51
$p \to n$	0.16	1.03	1.71	2.97	7.23	9.77
$n \to p$	0.09	0.58	0.87	1.67	2.41	3.24
$n \to n$	0.23	1.36	2.10	3.36	7.63	10.21

The corresponding multiplicities are taken as

$$N_{Q,jk} = \frac{\exp\left(-0.05\sqrt{A_T}\right)\sigma_{Q,jk}}{\sum_\ell \sigma_{Q,j\ell}} \qquad (5.154)$$

where the exponential factor accounts for the attenuation of the quasi-elastic particles before they escape the nucleus. The balance of the cascade particles is contained in N_3 as

$$N_3 = N_c - N_2 - N_Q \qquad (5.155)$$

with an assumed spectral coefficient given by

$$\alpha_3 = \frac{\alpha_2}{0.7} \qquad (5.156)$$

Results of the present formalism and the calculations of Bertini, Guthrie, and Culkowski (1972) are shown in figures 5.27 to 5.40. Some further improvements in this parameterization need to be made.

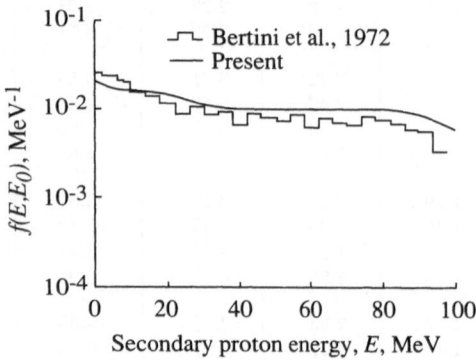

Figure 5.27. Nucleon cascade spectrum for protons produced by 100-MeV protons on oxygen.

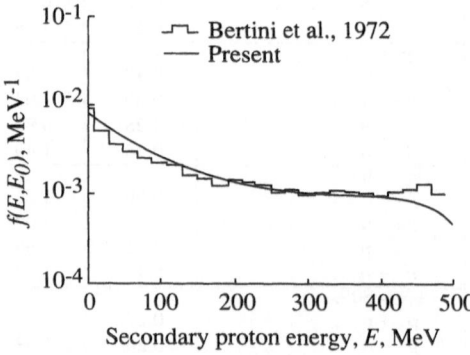

Figure 5.28. Nucleon cascade spectrum for protons produced by 500-MeV protons on oxygen.

5.7.2. Light-fragment spectrum.
The light-fragment yields per event are given in table 5.15 as obtained from Bertini's MECC-7 (Anon., 1968) calculations. These results are extrapolated and interpolated in energy and mass number. The corresponding mean energies are given in table 5.16. The mean energies are used in Ranft's formula for nucleons and are similarly used for the light ions.

5.7.3. Fragmentation cross sections.
The local distribution of ions and radicals produced in ionizing radiation events is known to be an indicator of biological response. Such distributions for high-energy nuclear radiation vastly altered by local nuclear-reaction events have been studied in nuclear emulsion (Van Allen, 1952; Lord, 1951) and are regular components in risk assessments in high-energy neutron and proton radiation fields (Alsmiller, Armstrong, and Coleman, 1970; Foelsche et al., 1974). Risk assessments have generally depended on the results of calculational models of these reactions because the detailed study of such reactions was largely inaccessible to experimental study until the advent of high-energy heavy ion beams.

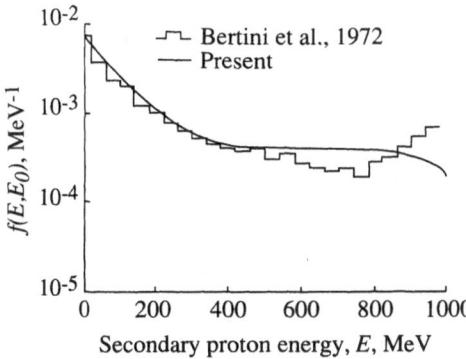

Figure 5.29. Nucleon cascade spectrum for protons produced by 1000-MeV protons on oxygen.

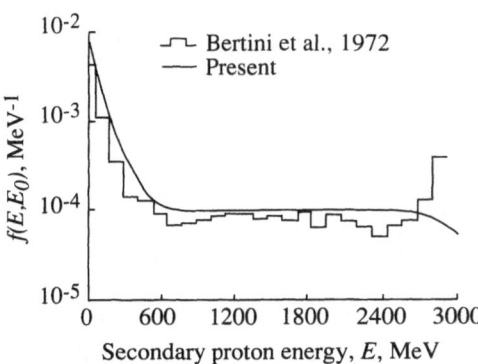

Figure 5.30. Nucleon cascade spectrum for protons produced by 3000-MeV protons on oxygen.

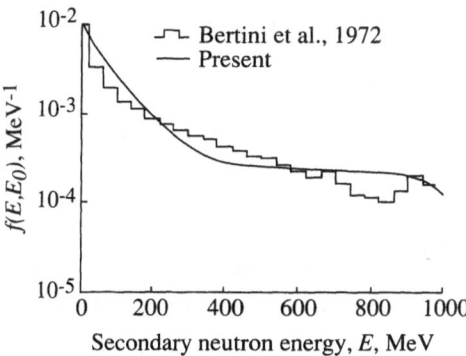

Figure 5.31. Nucleon cascade spectrum for neutrons produced by 1000-MeV protons on oxygen.

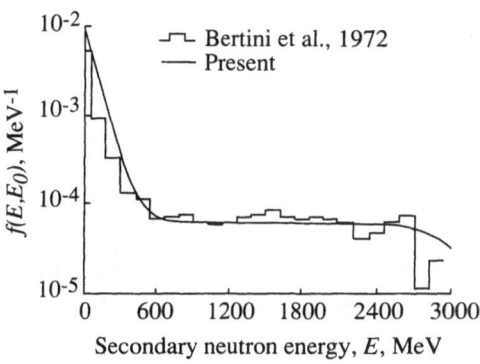

Figure 5.32. Nucleon cascade spectrum for neutrons produced by 3000-MeV protons on oxygen.

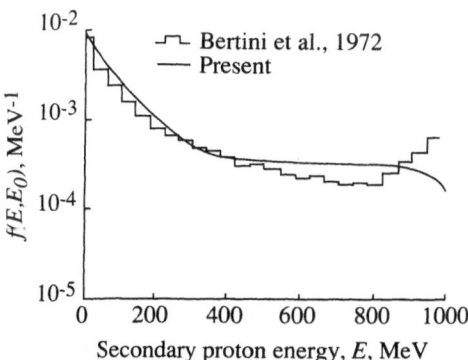

Figure 5.33. Nucleon cascade spectrum for protons produced by 1000-MeV protons on aluminum.

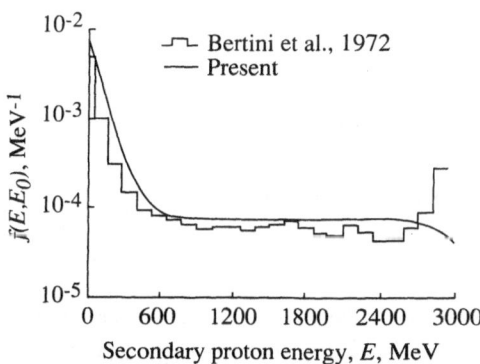

Figure 5.34. Nucleon cascade spectrum for protons produced by 3000-MeV protons on aluminum.

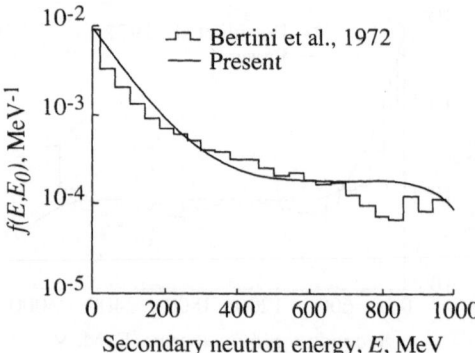

Figure 5.35. Nucleon cascade spectrum for neutrons produced by 1000-MeV protons on aluminum.

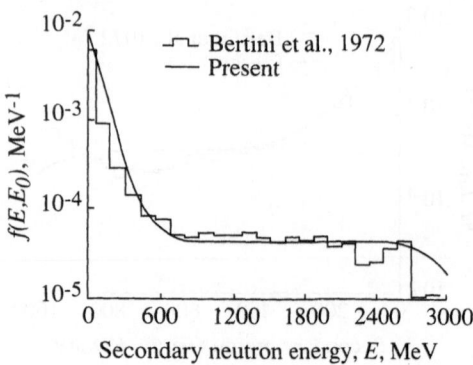

Figure 5.36. Nucleon cascade spectrum for neutrons produced by 3000-MeV protons on aluminum.

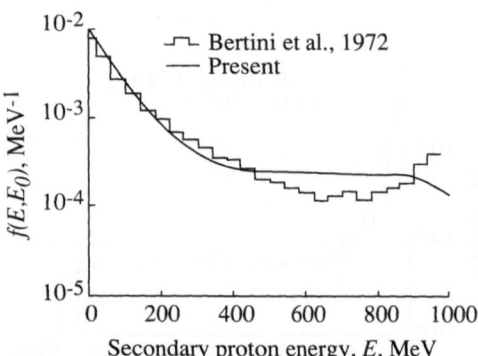

Figure 5.37. Nucleon cascade spectrum for protons produced by 1000-MeV protons on lead.

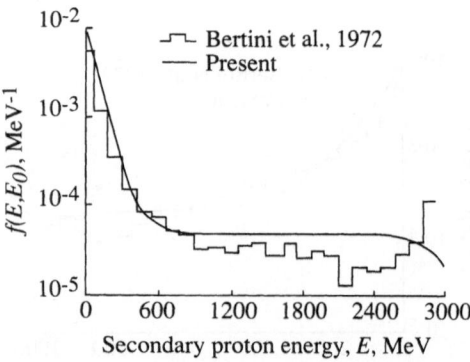

Figure 5.38. Nucleon cascade spectrum for protons produced by 3000-MeV protons on lead.

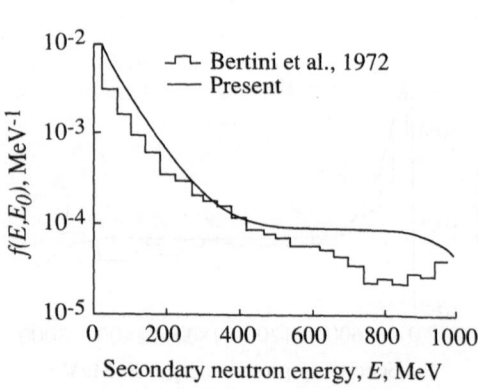

Figure 5.39. Nucleon cascade spectrum for neutrons produced by 1000-MeV protons on lead.

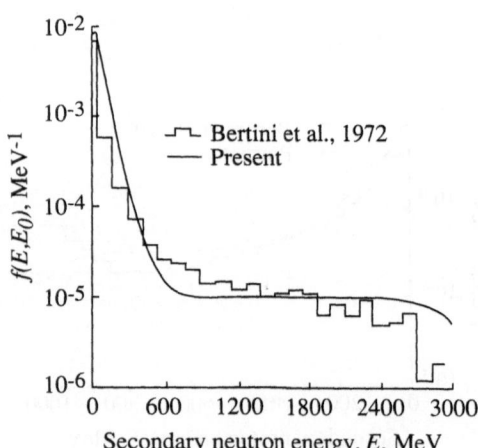

Figure 5.40. Nucleon cascade spectrum for neutrons produced by 3000-MeV protons on lead.

Table 5.15. Evaporated Ion Yields From Nucleon-Nucleus Collisions

[Values in parentheses are for proton reactions]

	Ion yields at—			
	500 MeV	1000 MeV	2000 MeV	3000 MeV
$A_T = 16$:				
d	0.111 (0.094)	0.199 (0.237)	0.257 (0.265)	0.304 (0.311)
t	0.022 (0.029)	0.024 (0.025)	0.033 (0.025)	0.029 (0.029)
he	0.018 (0.034)	0.035 (0.043)	0.037 (0.052)	0.037 (0.048)
α	0.664 (0.400)	0.720 (0.696)	0.666 (0.624)	0.640 (0.667)
$A_T = 27$:				
d	0.126 (0.130)	0.245 (0.269)	0.380 (0.396)	0.442 (0.433)
t	0.028 (0.023)	0.048 (0.052)	0.063 (0.065)	0.072 (0.069)
he	0.042 (0.035)	0.067 (0.074)	0.073 (0.091)	0.083 (0.092)
α	0.370 (0.400)	0.550 (0.566)	0.597 (0.582)	0.577 (0.577)
$A_T = 65$:				
d	0.150 (0.171)	0.379 (0.390)	0.748 (0.766)	0.935 (0.087)
t	0.031 (0.035)	0.075 (0.068)	0.145 (0.145)	0.177 (0.191)
he	0.013 (0.014)	0.039 (0.056)	0.112 (0.124)	0.166 (0.177)
α	0.124 (0.137)	0.231 (0.231)	0.373 (0.377)	0.431 (0.441)
$A_T = 100$:				
d	0.174 (0.183)	0.456 (0.475)	1.01 (1.02)	1.44 (1.48)
t	0.028 (0.029)	0.080 (0.081)	0.207 (0.192)	0.269 (0.273)
he	0.012 (0.017)	0.055 (0.060)	0.162 (0.185)	0.249 (0.262)
α	0.158 (0.156)	0.320 (0.339)	0.490 (0.467)	0.549 (0.540)
$A_T = 207$:				
d	0.131 (0.152)	0.536 (0.565)	1.51 (1.57)	2.54 (2.54)
t	0.038 (0.037)	0.152 (0.163)	0.415 (0.424)	0.641 (0.644)
he	0.001 (0.002)	0.017 (0.017)	0.112 (0.106)	0.211 (0.239)
α	0.053 (0.063)	0.195 (0.210)	0.527 (0.515)	0.751 (0.746)

The first detailed relativistic heavy ion beam experiments were performed by the Heckman group (Heckman, 1975; Greiner et al., 1975; Lindstrom et al., 1975) at the Lawrence Berkeley Laboratory (LBL), in which beams of carbon and oxygen were fragmented on a series of targets ranging from hydrogen to lead. The momentum distribution of the projectile fragments relative to the projectile rest frame was measured for all the isotopes produced. These results will be analyzed to ascertain relevant biological factors with their corresponding implications on radiation risk assessment in high-energy nucleonic radiation fields. An ion fragmentation model will be recommended for use in radiological protection and studies.

Individual nuclear constituents are ejected in the collision of high-energy neutrons and protons by direct collision (Serber, 1947). The remaining nuclear structure is left in an excited state that seeks an equilibrium minimum-energy configuration through particle emission (Rudstam, 1966). This is the basis of Rudstam's study of the systematics of spallation products produced in such collisions in which he assumes that the resultant isotopes are distributed in a bell-shaped distribution near the nuclear stability line. The total change in nuclear mass and the dependence on the incident projectile energy are treated empirically in Rudstam's formalism.

The fragment charge distribution for a given fragment mass A_F is given as

$$f(Z_F) = \exp\left(\rho A_F - r \left| Z_F - sA_F + vA_F^2 \right|\right) \tag{5.157}$$

Table 5.16. Mean Energies of Light Nuclear Fragments
Produced in Nucleon-Nucleus Collisions

[Values in parentheses are for proton reactions]

	Mean energies at—			
	500 MeV	1000 MeV	2000 MeV	3000 MeV
$A_T = 16$:				
n	5.55 (6.19)	7.91 (7.89)	9.55 (9.81)	11.1 (9.80)
p	6.10 (6.40)	8.33 (8.69)	9.71 (10.2)	10.3 (11.2)
d	8.53 (7.64)	12.2 (10.7)	14.9 (14.8)	16.3 (13.08)
t	6.40 (7.83)	10.6 (10.4)	12.5 (9.74)	13.7 (10.1)
he	12.1 (8.76)	11.8 (11.2)	11.1 (13.1)	12.9 (10.3)
α	9.36 (6.24)	12.6 (12.3)	13.1 (14.6)	13.6 (13.8)
$A_T = 27$:				
n	5.08 (5.09)	7.34 (7.48)	9.91 (10.5)	11.6 (12.0)
p	6.87 (6.90)	8.61 (8.92)	11.1 (11.9)	13.5 (13.7)
d	9.57 (9.42)	10.8 (11.2)	14.3 (14.8)	17.2 (17.4)
t	9.16 (9.54)	10.8 (11.1)	13.0 (13.9)	16.6 (13.7)
he	10.5 (10.8)	12.5 (12.8)	13.4 (14.1)	14.4 (14.5)
α	12.7 (13.4)	13.2 (13.6)	13.8 (13.8)	14.5 (14.6)
$A_T = 65$:				
n	4.24 (4.32)	5.67 (5.70)	7.92 (7.91)	9.67 (9.58)
p	8.25 (8.30)	9.66 (9.76)	12.1 (12.3)	14.4 (14.2)
d	9.88 (10.1)	13.5 (11.8)	13.8 (14.2)	15.6 (15.9)
t	10.0 (10.5)	11.7 (11.6)	13.7 (13.8)	15.1 (15.9)
he	14.6 (14.1)	16.4 (16.2)	17.5 (19.3)	19.5 (19.2)
α	12.7 (13.4)	13.2 (13.6)	13.8 (13.8)	14.5 (14.6)
$A_T = 100$:				
n	3.90 (3.90)	5.13 (5.16)	7.11 (7.04)	8.61 (8.74)
p	9.63 (9.62)	11.0 (11.0)	12.9 (13.2)	14.6 (14.7)
d	11.0 (11.1)	12.5 (12.6)	14.4 (15.0)	16.1 (16.0)
t	11.3 (11.7)	12.6 (13.0)	14.7 (14.3)	15.5 (16.5)
he	17.8 (18.7)	13.2 (13.6)	13.8 (13.8)	14.5 (14.6)
α	16.5 (16.5)	16.8 (16.9)	17.5 (17.5)	17.6 (17.6)
$A_T = 107$:				
n	3.28 (3.27)	4.37 (4.33)	5.83 (5.78)	6.90 (6.95)
p	12.5 (12.5)	12.2 (13.4)	14.9 (14.9)	16.2 (16.3)
d	13.2 (13.2)	14.4 (14.2)	16.0 (16.8)	17.4 (17.8)
t	13.6 (13.8)	5.0 (15.3)	16.6 (16.8)	17.4 (17.8)
he	24.1 (27.0)	26.2 (26.5)	28.0 (27.8)	29.1 (28.5)
α	25.3 (25.7)	26.0 (26.3)	26.4 (26.3)	25.9 (26.4)

where the coefficients show a slight energy and fragment-mass dependence as

$$r = 11.8 A_F^{-0.45} \tag{5.158}$$

$$s = 0.486 \tag{5.159}$$

$$v = 3.8 \times 10^{-4} \tag{5.160}$$

$$\rho = \begin{cases} 20E^{-0.77} & (E < 2100 \text{ MeV}) \\ 0.056 & (E \geq 2100 \text{ MeV}) \end{cases} \tag{5.161}$$

where E is the nucleon energy. The complete Rudstam cross section is given by

$$\sigma(A_F, Z_F) = \frac{F_1 F_2 \rho A_T^{-0.3} \; f(Z_F)}{D} \tag{5.162}$$

Table 5.17. Present Correction Factors
for Rudstam's Formula

ΔA	Correction factor for—	
	^{12}C	^{16}O
1	1.3	1.5
2	0.5	1.0
3	0.3	1.0
4	0.1	1.0
5	1.0	1.5
6	0.35	0.5
7		0.5
8		0.1
9		2.5
10		1.0

where

$$D = 1.79 \left[\exp(\rho A_T) \left(1 - \frac{0.3}{\rho A_T} \right) - \frac{0.3}{A_T} + \frac{0.3}{\rho A_T} \right] \tag{5.163}$$

$$F_1 = 5.18 \; \exp(-0.25 + 0.0074 A_T) \tag{5.164}$$

$$F_2 = \begin{cases} \exp(1.73 - 0.0071E) & (E < 240 \text{ MeV}) \\ 1 & (E \geq 240 \text{ MeV}) \end{cases} \tag{5.165}$$

We have applied a simple mass-dependent correction factor to Rudstam's formula as shown in table 5.17 and renormalized his cross sections to the total absorption cross section. Many corrective factors have been added to Rudstam's formalism by Silberberg, Tsao, and Shapiro (1976). Estimates have also been made by Guzik (1981) for some of the isotopes produced in connection with cosmic-ray propagation studies with some attempts at experimental verification (Guzik et al., 1985).

From a nuclear model point of view, isotope production at low energy results from the formation of a compound nuclear state that decays through particle emission. At higher energies, the direct ejection of particles from the nucleus becomes important, and intranuclear cascades represented as sequences of two-body scatterings within the nucleus with Pauli blocking are the usual means of evaluation. Subsequent to the cascade, the residual nucleus is assumed to be in thermal equilibrium and seeks to minimize its internal energy through particle emission.

The measurement of isotope production cross sections at proton accelerators does not allow the direct observation of the fragment products. Customary measurements used γ or β counting techniques to identify the isotopes produced. Stable and short-lived isotopes produced in the reactions were either not observed or their number was greatly distorted by loss through decay.

This is particularly true for light-mass targets such as those that are important to biological health considerations. Consequently, the fragmentation of carbon and oxygen nuclei by protons remained shrouded in experimental obscurity until the advent of heavy ion accelerators.

One of the earliest experiments performed at the Lawrence Berkeley Laboratory Bevatron, when the ions of carbon and oxygen could be accelerated to relativistic energies, used detectors able to measure the energy and charge of an ion beam in conjunction with a bending magnet for momentum analysis (Heckman, 1975). In this way, the density in phase space was measured for each isotope produced in collision with a fixed target.

The isobar cross section (σ_{LBL}) measured by Lindstrom et al. (1975) for 2.1 GeV/nucleon oxygen fragmentation on hydrogen targets is given in table 5.18 along with the results of the Bertini MECC-7 code (Anon., 1968), Rudstam (1966), and Silberberg, Tsao, and Shapiro (1976). Note that the Rudstam results contain the correction factors from table 5.17 and are renormalized as described previously.

The oxygen-fragmentation cross sections represented by three parametric forms are shown in figures 5.41 to 5.45 along with the Bertini results and various experiments. The baryon-15 isobaric cross sections in figure 5.41 show that experiments favor the curve of Silberberg, Tsao, and Shapiro, 1976. Although the Bertini model provides an overestimate, the other parametric curves provide improved estimates compared with the Bertini code. The baryon-14 isobaric experimental cross sections are in reasonable agreement with the three parametric curves as well as with the Bertini model as seen in figure 5.42. Again, the experiments show no clear advantage of one parametric curve over another for the baryon-13 cross section as seen in figure 5.43, although the Bertini results appear somewhat low. We show experimental results for baryon numbers between 9 and 14 of the LBL Group (Lindstrom et al., 1975) in table 5.18. Clearly, the equally good agreement for the Rudstam parameterization and the parameterization of Silberberg, Tsao, and Shapiro is obtained for baryon numbers 12, 11, and 10. The Bertini cross section is far too low to represent the cross section for baryon-11. The baryon-9 cross sections are shown in figure 5.44. (The results of Yiou are reported in Guzik, 1981.) The cross sections of Silberberg, Tsao, and Shapiro are favored. The baryon-7 cross sections are shown in figure 5.45. At energies below 300 MeV, the baryon-7 results of Silberberg, Tsao, and Shapiro are favored.

The measurements of Lindstrom et al. (1975) for relativistic carbon beams are shown along with the results from Rudstam and Silberberg, Tsao, and Shapiro in table 5.19 for two beam energies. The good agreement with the results of Silberberg, Tsao, and Shapiro is no surprise, because their parameterization was fit to these experimental data sets.

Table 5.18. Oxygen Fragmentation Cross Sections

	Fragmentation cross section, σ, mb, from—			
	Bertini	LBL		NRL
A_F	(a)	(b)	Present	(c)
16	7.0	0.02	8.7	
15	85.1	61.5	61.0	59.4
14	39.0	35.4	32.6	32.2
13	13.9	22.8	29.7	17.7
12	28.1	34.1	27.9	36.0
11	5.0	24.4	31.4	19.9
10	9.1	12.7	12.0	11.0
9	1.0	5.2	7.1	12.1
8	0.2	1.23	2.1	14.7
7	1.1	22.2	27.8	19.4
6	3.8	13.9	18.0	16.7
Total	193.3	235.5	258.3	239.1

[a] Anon., 1968.

[b] Heckman et al., 1975.

[c] Silberberg, Tsao, and Shapiro, 1976.

5.7.4. Heavy-fragment spectrum. Following the direct ejection of nucleons in nuclear collision, the nucleus is left in a highly excited state that decays through particle emission. From a sudden approximation point of view, as proposed by Serber (1947), the momentum distribution of the decay particles is governed by the Fermi distribution before collision. The collective momentum of decay products and nuclear fragments is thus derived on the basis of combinatorial rules on the random ways in which a given fragment mass can be formed from the nucleon distributions before collision. The formulation of Goldhaber (1974) is physically meaningful and simplistic. The momentum distribution is Gaussian in momentum space with a momentum width parameter given by

$$\sigma_p = \sigma_0 \left[\frac{A_F(A_T - A_F)}{(A_T - 1)} \right]^{1/2} \tag{5.166}$$

where σ_0 is the usual mean fermi momentum of the struck nucleus. However, the σ_0 of nuclear fragmentation is found to be about 25-percent smaller than that observed in electron scattering experiments (Greiner et al., 1975). The mean Fermi momentum is a slowly varying function of nuclear mass.

A slight modification of Goldhaber's results is found to adequately represent the experimental results of Greiner et al. (1975) given by

$$\sigma_p = 0.8b \left[\frac{4\delta_A}{20(A_T - 1)} \right]^{1/2} \tag{5.167}$$

Table 5.19. Carbon Fragmentation Cross Sections

(a) ^{12}C at 1000 MeV/nucleon

| A_F | Fragmentation cross section, σ, mb, from— | | |
	LBL (a)	Present	NRL (b)
12	0.1	6.7	0
11	55.3	63.2	69.0
10	22.7	28.0	22.0
9	5.8	10.0	15.2
8	1.4	4.8	26.0
7	18.9	21.7	20.7
6	12.4	14.7	16.9
Total	116.6	149.1	169.8

[a] Heckman et al., 1975.

[b] Silberberg, Tsao, and Shapiro, 1976.

(b) ^{12}C at 2000 MeV/nucleon

| A_F | Fragmentation cross section, σ, mb, from— | | |
	LBL (a)	Present	NRL (b)
12	0.09	6.2	0
11	57.0	60.4	58.5
10	22.7	27.8	20.5
9	6.20	10.4	14.2
8	1.6	5.2	24.1
7	20.49	24.4	19.9
6	14.8	17.2	16.7
Total	122.9	151.6	153.9

[a] Heckman et al., 1975.

[b] Silberberg, Tsao, and Shapiro, 1976.

where the parameters b and δ_A are given, respectively, by

$$b = \min\left(112A_T^{1/2}, 260\right) \tag{5.168}$$

and

$$\delta_A = \left\{ \begin{array}{ll} 0.45 & (A_T = A_F) \\ A_T - A_F & \text{(Otherwise)} \end{array} \right\} \tag{5.169}$$

A comparison of formulas (5.167) through (5.169) with experiments and the parameterization of Greiner et al. is given in table 5.20. Clearly, the present formulas are quite accurate.

The spectral distributions of the nuclear fragments in the rest frame of the struck nucleus before collision are given by

$$\frac{d\sigma_f}{dE} = \frac{\sigma_f}{(2\pi E_0^3)^{1/2}} E^{1/2} \exp\left(\frac{-E}{2E_0}\right) \tag{5.170}$$

where σ_f is the fragmentation cross section and the energy parameter is

$$E_0 = \frac{3\sigma_p^2}{2A_F} \tag{5.171}$$

The average energies $\bar{E}(= 3E_0)$ of various fragments obtained by equations (5.167) through (5.171) and the results of the Bertini model are presented in table 5.21. Generally, the average energies predicted by the Bertini model are reasonably accurate, although some specific isotopes differ by a factor of 2 or more.

5.7.5. Energy-transfer cross section. The energy-loss spectrum $\psi_j(\vec{x}, \vec{\Omega}, E)$ of the ion fragment j may be written as (Wilson, 1977)

$$\psi(\vec{x}, \vec{\Omega}, E) \approx A_j \zeta_j(\vec{x}) \int_E^{E_\gamma} \left(\frac{m}{2\pi\sigma_p^2}\right)^{3/2} \sqrt{2E'} \exp\left(\frac{-mE'}{\sigma_p^2}\right) dE' \tag{5.172}$$

where A_j is the fragment mass number, $\zeta_j(\vec{x})$ is the fragment source, and E_γ is related to the distance to the boundary along the direction $\vec{\Omega}$ as given elsewhere by Wilson (1977). For distances far from the boundary, one may take $E_\gamma = \infty$. The cumulative energy-loss spectrum far from the boundary ($E_\gamma = \infty$) is

$$D_j(\vec{x}, E) = 4\pi \int_E^{\infty} \psi_j(\vec{x}, \vec{\Omega}, E') \, dE' \tag{5.173}$$

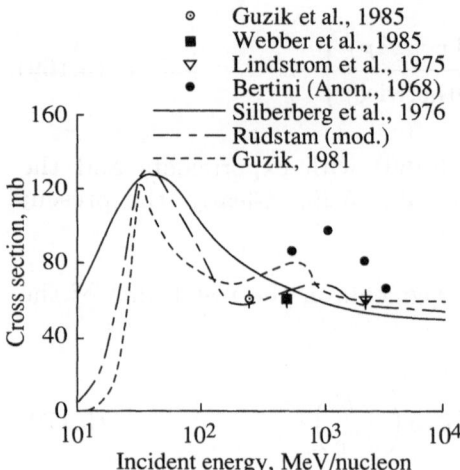

Figure 5.41. Oxygen fragmentation cross sections for baryon-15 isobars in proton collisions.

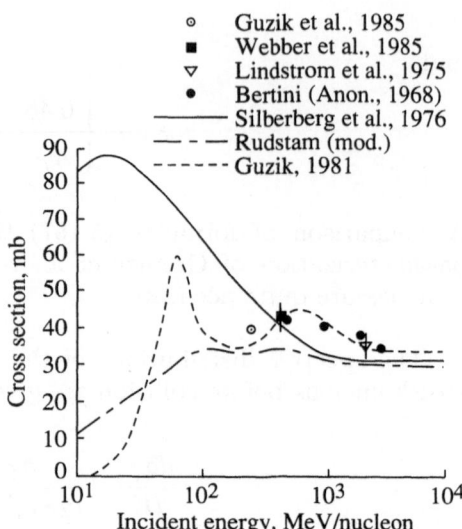

Figure 5.42. Oxygen fragmentation cross sections for baryon-14 isobars in proton collisions.

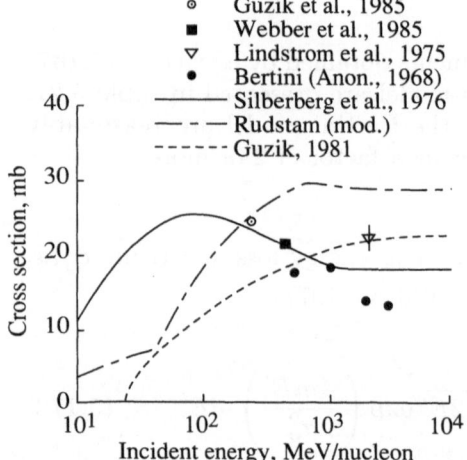

Figure 5.43. Oxygen fragmentation cross sections for baryon-13 isobars in proton collisions.

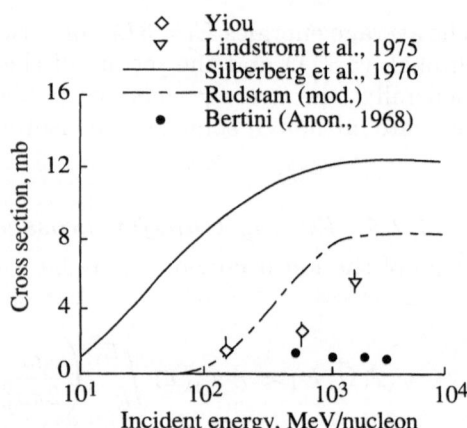

Figure 5.44. Oxygen fragmentation cross sections for baryon-9 isobars in proton collisions.

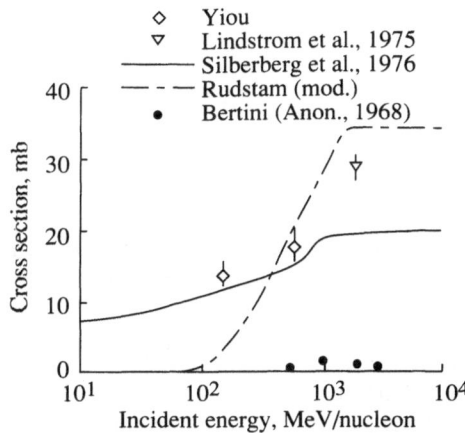

Figure 5.45. Oxygen fragmentation cross sections for baryon-7 isobars in proton collisions.

Table 5.20. σ_p for ^{16}O Fragments Produced by 2.1-GeV Protons

| Fragment | σ_p, MeV/c, from— | | |
	Experiments	Present work	Greiner et al., 1975
^{15}O	94 ± 3	80.0	83.8
^{14}O	99 ± 6	109.5	113.1
^{13}O	143 ± 14	129.2	133.5
^{16}N	54 ± 11	55.0	
^{15}N	95 ± 3	80.0	
^{14}N	112 ± 3	109.5	113.0
^{13}N	134 ± 2	129.2	133.5
^{12}N	153 ± 11	143.4	148.1
^{15}C	125 ± 19	80.0	82.8
^{14}C	125 ± 3	109.5	113.10
^{13}C	130 ± 3	129.2	133.5
^{12}C	120 ± 4	143.36	148.09
^{11}C	162 ± 5	153.45	158.5
^{10}C	190 ± 9	160.3	165.6
^{13}B	166 ± 10	129.2	133.5
^{12}B	163 ± 8	143.4	148.1
^{11}B	160 ± 2	153.5	158.5
^{10}B	175 ± 7	160.3	165.6
^{8}B	175 ± 22	165.5	171.0
^{11}Be	197 ± 20	153.5	158.5
^{10}Be	159 ± 6	160.0	165.0
^{9}Be	166 ± 7	164.24	169.66
^{7}Be	166 ± 2	164.24	169.66
^{9}Li	188 ± 15	164.24	169.66
^{8}Li	170 ± 13	165.4	171.0
^{7}Li	163 ± 4	164.24	169.66
^{6}Li	141 ± 7	160.0	169.66
^{6}He	167 ± 20	160.0	165.0

Table 5.21. Average Recoil Energy $\overline{E}$ of ^{16}O Fragments
Produced by 2.1-GeV Protons

Fragment	Average energy, $\overline{E}$, MeV, from—		
	Bertini (Anon., 1968)	Present results	Experiments
^{16}F	2.65	1.01	
^{15}F	4.19	.69	
^{16}O	1.05	1.012	1.01
^{15}O	.52	.69	.88
^{14}O	1.82	1.37	1.12
^{13}O	4.24	2.05	2.51
^{16}N	1.11	1.01	.30
^{15}N	.63	.69	.96
^{14}N	1.12	1.37	1.42
^{13}N	1.84	2.05	2.20
^{12}N	3.85	2.74	3.11
^{11}N	5.95	3.42	3.64
^{14}C	1.62	1.34	1.78
^{13}C	1.97	2.05	2.07
^{12}C	2.64	2.74	1.91
^{11}C	4.70	3.42	3.81
^{10}C	5.58	4.11	5.76
^{9}C	4.41	4.79	5.10
^{13}B	2.35	2.05	3.38
^{12}B	3.43	2.74	3.53
^{11}B	4.33	3.71	3.42
^{10}B	4.79	4.11	4.89
^{9}B	1.19	4.79	
^{10}Be	4.53	4.11	4.03
^{9}Be	8.76	4.79	4.89
^{10}Li	4.61	4.11	
^{9}Li	2.26	4.79	6.27
^{8}Li	4.41	5.48	5.76
^{7}Li	4.75	6.16	6.06
^{6}Li	5.76	6.85	5.29

Table 5.22. Fragment Energy-Transfer Cross Sections

| A_F | Energy-transfer cross sections, $\overline{E}\sigma$, MeV-mb, from— | | |
	Bertini	Greiner et al., 1975	Present
16	5.04	0.0006	0.26
15	60.6	56.9	56.4
14	48.8	48.3	62.9
13	37.6	48.3	62.9
12	85.8	68.2	55.8
11	37.9	99.1	117.9
10	52.8	62.0	58.6
9	6.5	25.7	35.1
8	2.5	7.1	12.1
7	6.11	121.7	152.4
6	31.4	73.4	95.1
Total	375.1	614.1	694.2

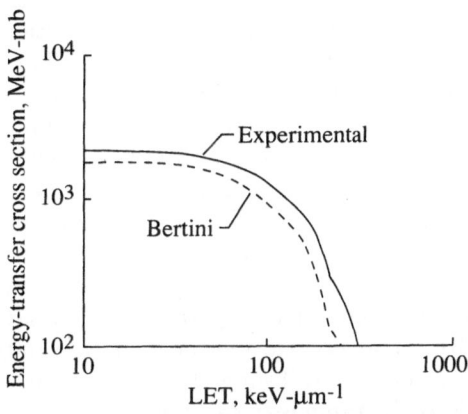

Figure 5.46. Linear energy-transfer (LET) cross section for fragmenting oxygen nucleus in water.

for which the distribution in linear energy transfer (LET) of energy deposit can be found. The total energy absorbed is given by

$$D(\vec{x}) = \sum_j D_j(\vec{x}, 0)$$

$$\approx \sum_j E_j \sigma_j \rho \phi \tag{5.174}$$

where E_j is the average energy of the fragment j, σ_j is the fragmentation cross section, ρ is the target density, and ϕ is the effective nucleon flux initiating the fragmentation events. The energy-transfer cross section of the various fragment components is $E_j \sigma_j$ and is shown in table 5.22 for the Rudstam parameterization (present results), Bertini data, and experiments of the Heckman group (Greiner et al., 1975) for comparison. Equations (5.172) through (5.174) also provide a basis for resolving the energy-transfer cross section into its various LET components. The LET components of equation (5.173) are shown in figure 5.46 for $\rho = \phi = 1$ for all contributions with a fragment charge greater than 1. The two curves shown in the figure are for the Bertini data and the experiments of the Heckman group. Results obtained with our modified Rudstam formalism and the parameterized momentum distributions are virtually indistinguishable from the curve based on the LBL experiments. The results shown in figure 5.46 clearly show that estimates of exposure from heavy ion recoil nuclei in tissue based on Bertini cross sections are generally low.

5.7.6. Heavy ion fragmentation model.

In the abrasion-ablation fragmentation model, the projectile nuclei, moving at relativistic speeds, collide with stationary target nuclei. In the abrasion step, those portions of the nuclear volumes that overlap are sheared away by the collision. The remaining projectile piece, called a prefragment or primary residue, continues its trajectory with essentially its precollision velocity. As a result of the dynamics of the abrasion process, the prefragment is highly excited and subsequently decays by the emission of gamma radiation and/or nuclear particles. This step is the ablation stage. The resultant isotope, sometimes referred to as a secondary product, is the nuclear fragment whose cross section is measured. The abrasion process can be analyzed with classical geometric arguments (Bowman, Swiatecki, and Tsang, 1973) or methods obtained from formal quantum scattering theory (Townsend et al., 1986a and 1986b). The ablation stage can be analyzed from geometric arguments (Bowman, Swiatecki, and Tsang, 1973) or more sophisticated methods based upon Monte Carlo or intranuclear cascade techniques (Gosset et al., 1977; Hüfner, Schäfer, and Schürmann, 1975; Morrissey et al., 1978; Guthrie, 1970). Predictions of fragmentation cross sections can also be made with the approximate semiempirical parameterization formulas of Silberberg, Tsao, and Shapiro (1976). The present data base uses the method of Wilson, Townsend, and Badavi (1987b).

The amount of nuclear material stripped away in the collision of two nuclei is taken as the volume of overlap region times an average attenuation factor. The relevant formula for the constituents in the overlap volume in the projectile is given by the following formula:

$$\Delta_{\text{abr}} = FA_P \left[1 - \frac{1}{2} \exp\left(\frac{-C_P}{\lambda}\right) - \frac{1}{2} \exp\left(\frac{-C_T}{\lambda}\right) \right] \tag{5.175}$$

where C_P and C_T are the maximum chord lengths of the intersecting surface in the projectile and the target, respectively, and the expressions for F differ depending on the nature of the collision (peripheral versus central) and the relative sizes of the colliding nuclei. The value for F is given by equations (5.38) through (5.50). The charge ratio of removed nuclear matter is assumed to be that of the parent nucleus.

The surface distortion excitation energy of the projectile prefragment following abrasion of m nucleons is calculated from the clean-cut abrasion formalism of Bowman, Swiatecki, and Tsang (1973). For this model, the colliding nuclei are assumed to be uniform spheres of radii R_i $(i = P, T)$. In the collision, the overlapping volumes shear off so that the resultant projectile prefragment is a sphere with a cylindrical hole gouged out of it. The excitation energy is then determined by calculating the difference in surface area between the misshapen sphere and a perfect sphere of equal volume. This excess surface area ΔS is given by Gosset et al. (1977) as

$$\Delta S = 4\pi R_p^2 \left[1 + P - (1 - F)^{2/3} \right] \tag{5.176}$$

where the expressions for P and F differ depending upon the nature of the collision (peripheral versus central) and the relative sizes of the colliding nuclei which were given in section 5.3.2. (See eqs. (5.38) through (5.50).)

The excitation energy associated with surface energy is well-known to be 0.95 MeV/fm^2 for near equilibrium nuclei so that

$$E'_s = 0.95\Delta S \tag{5.177}$$

for small surface distortions. When large numbers of nucleons are removed in the abrasion process, equation (5.177) is expected to be an underestimate to the actual excitation. We therefore introduce an excess excitation factor in terms of the number of abraded nucleons Δ_{abr} as

$$f = 1 + \frac{10\Delta_{\text{abr}}}{A_P} + \frac{25\Delta^2_{\text{abr}}}{A_P^2} \tag{5.178}$$

which approaches 1 when the impact parameter is large but increases the excess excitation when large portions of the nuclei are removed in the collisions and when grossly misshapened nuclei are formed. The total excitation energy is then

$$E_s = E'_s f \tag{5.179}$$

which reduces to equation (5.177) for small Δ_{abr}. We assume that all fragments with a mass of 5 are unbound, that 90 percent of the fragments with a mass of 8 are unbound, and that 50 percent of the fragments with a mass of 9 (^{9}B) are unbound.

A secondary contribution to the excitation energy is the transfer of kinetic energy of relative motion across the intersecting boundary of the two ions. The rate of energy loss of a nucleon when it passes through nuclear matter (Westfall et al., 1979) is taken at 13 MeV/fm, and the energy deposit is assumed to be symmetrically dispersed about the azimuth so that 6.5 MeV/nucleon-fm at the interface is the average rate of energy transfer into excitation energy. This energy is transferred in single particle collision processes, and on half of the events, the energy is transferred to excitation energy of the projectile and the remaining half of the events leaves the projectile excitation energy unchanged. The first estimate of this contribution is to use the length of the longest chord C_1 in the projectile surface interface. This chord length is the maximum distance traveled by any target constituent through the projectile interior. The number of other target constituents in the interface region may be found by estimating the maximum chord C_t transverse to the projectile velocity which spans the projectile surface interface. The total excitation energy from excess surface and spectator interaction is then

$$E'_x = 13C_1 + \frac{1}{3}\, 13C_1(C_t - 1.5) \tag{5.180}$$

where the second term only contributes if $C_t > 1.5$ fm. We further assume that the effective longitudinal chord length for these remaining nucleons is one third the maximum chord length.

The decay of highly excited nuclear states is dominated by heavy particle emission. In the present model, we assume that a nucleon is removed for every 10 MeV of excitation energy as

$$\Delta_{\mathrm{abl}} = \frac{(E_s + E_x)}{10 \text{ MeV}} \tag{5.181}$$

In accordance with the previously discussed directionality of the energy transfer, E_x is double valued as

$$E_x = \left\{ \begin{array}{ll} E'_x & \left(P_x = \frac{1}{2}\right) \\ 0 & \left(P_{\bar{x}} = \frac{1}{2}\right) \end{array} \right\} \tag{5.182}$$

where P_j is the corresponding probability of occurrence of each value in collisions.

The number of nucleons removed through the abrasion-ablation process is given as a function of impact parameter as

$$\Delta A = \Delta_{\mathrm{abr}}(b) + \Delta_{\mathrm{abl}}(b) \tag{5.183}$$

The values of ΔA for carbon projectiles on a copper target and for copper projectiles on a carbon target are shown in figure 5.47. In each case, the dashed curve corresponds to $E_x = 0$, whereas the solid curve corresponds to $E_x = E'_x$ as given by equation (5.180). A real collision would be given by a statistical distribution between the limits shown by these two curves. The average event

410

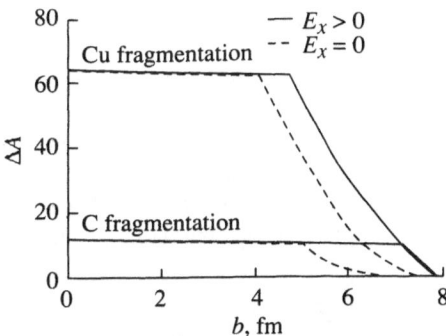

Figure 5.47. Nucleon removal number as function of impact parameter in carbon-copper collisions.

is calculated as if the two extremes occurred with equal probability, as noted in equation (5.182).

The nuclear fragmentation parameters herein are approximated according to the abrasion-ablation model of Bowman, Swiatecki, and Tsang (1973). The cross section for removal of ΔA nucleons is estimated as

$$\sigma(\Delta A) = \pi b_2^2 - \pi b_1^2 \qquad (5.184)$$

where b_2 is the impact parameter for which the volume of intersection of the projectile contains Δ_{abr} nucleons and the resulting excitation energies release an additional Δ_{abl} nucleons at the rate of 1 nucleon for every 10 MeV of excitation such that

$$\Delta_{\mathrm{abr}}(b_2) + \Delta_{\mathrm{abl}}(b_2) = \Delta A - \frac{1}{2} \qquad (5.185)$$

and similarly for b_1

$$\Delta_{\mathrm{abr}}(b_1) + \Delta_{\mathrm{abl}}(b_1) = \Delta A + \frac{1}{2} \qquad (5.186)$$

The charge distributions of the final projectile fragments are strongly affected by nuclear stability. We expect that the Rudstam (1966) charge distribution for a given $\sigma(\Delta A)$ to be reasonably correct as

$$\sigma(A_F, Z_F) = F_1 \exp\left[-R \left| Z_F - S A_F + T A_F^2 \right|^{3/2} \right] \sigma(\Delta A) \qquad (5.187)$$

where $R = 11.8/A_F^D$, $D = 0.45$, $S = 0.486$, and $T = 3.8 \times 10^{-4}$ according to Rudstam and F_1 is a normalizing factor such that

$$\sum_{Z_F} \sigma(A_F, Z_F) = \sigma(\Delta A) \qquad (5.188)$$

The Rudstam formula for $\sigma(\Delta A)$ was not used because the ΔA dependence is too simple and breaks down for heavy targets (Townsend et al., 1984; Townsend, Wilson, and Norbury, 1985).

The charge of the removed nucleons ΔZ is calculated according to charge conservation

$$Z_P = Z_F + \Delta Z \tag{5.189}$$

and is divided among the nucleons and α-particles according to the following rules. The abraded nucleons are those removed from that portion of projectile in the overlap region with the target. Therefore, the abraded nucleon charge is assumed to be proportional to the charge fraction of the projectile nucleus as

$$Z_{\text{abr}} = \frac{Z_P \Delta_{\text{abr}}}{A_P} \tag{5.190}$$

This, of course, ignores the charge separation caused by the giant dipole resonance model of Morrissey et al. (1978). The charge release in the ablation is then given as

$$Z_{\text{abl}} = \Delta Z - Z_{\text{abr}} \tag{5.191}$$

which simply conserves the remaining charge.

The α-particle is known to be unusually tightly bound in comparison with other arrangements of nucleons. Because of this unusually tight binding of the α-particle, the helium production is maximized in the ablation process

$$N_\alpha = \text{int}\left(\frac{Z_{\text{abl}}}{2}\right) \tag{5.192}$$

where $\text{int}(x)$ denotes the integer part of x. The number of protons produced is given by charge conservation as

$$N_P = \Delta Z - 2N_\alpha \tag{5.193}$$

Similarly, neutral conservation requires the number of neutrons produced to be

$$N_n = \Delta A - N_p - 4N_\alpha \tag{5.194}$$

The fragments with masses of 2 and 3 are ignored.

The calculation is performed for $\Delta A = 1$ to $\Delta A = A_P - 1$, for which the cross section associated with $\Delta A > A_P - 0.5$ is missed. These are, of course, the central collisions for which it is assumed that the projectile disintegrates into single nucleons if $R_P < R_T$ as

$$N_P = Z_P \tag{5.195}$$

$$N_n = A_P - Z_P \tag{5.196}$$

and is ignored otherwise. The energetic target fragments as well as the mesonic components are being ignored. The peripheral collisions with $\Delta A < 0.5$ are also missing. Most important in these near collisions will be the coulomb dissociation process studied by Norbury and Townsend (1986).

Only the nuclear radius for use in the model is yet undefined. The nuclear absorption cross sections are taken as energy independent and are approximated by Townsend and Wilson (1986) as

$$\sigma(A_1, A_2) = \pi r_0^2 \left(A_1^{1/3} + A_2^{1/3} - 0.2 - A_1^{-1} - A_2^{-1} \right)^2 \qquad (5.197)$$

where $r_0 = 1.26$ fm. Equation (5.197) is an accurate representation of the high-energy cross sections. The choice of nuclear radius as

$$R = 1.26 A^{1/3} \qquad (5.198)$$

is consistent with equation (5.197) when the peripheral collisions ($\Delta A < 0.5$) are taken into account. This completes the description of the basic fragmentation model in present use.

In the present evaluation, we look only to elemental fragmentation cross sections for which most of the experimental data have been obtained. This is also motivated by the crudeness of the present model which is not expected to be completely accurate. Even so, the quality of the experimental data base is uncertain with experiments of different groups differing by a factor of 2, in general, and differing even more for specific isotopes.

The first comparison is with the experiments of Heckman (1975) with ^{12}C ion beams at 1.05 GeV/nucleon on the series of targets extending from hydrogen to lead as shown in table 5.23. The present calculations are shown as values in parentheses. The calculated values for hydrogen targets are those of Rudstam. Note that all values are within 20 percent of the experiments with few exceptions (namely, fragments from hydrogen targets and the neutron removal cross section in copper and lead targets).

The charge removal cross sections for several projectiles on carbon targets are given in table 5.24. The agreement between the present model and the Lawrence Berkeley Laboratory groups (Heckman, 1975; Westfall et al., 1979) is quite good. Our results tend to be low compared with the experiments of Webber et al. (1983a and 1983b) and Guerreau et al. (1983). The model can be adjusted once experimental differences are resolved.

The elemental fragmentation cross section of iron projectiles on several targets is shown in table 5.25. Again, reasonable agreement is found generally with a few examples of relatively large errors. The bracketed quantities at the bottom of the table are the coulomb dissociation cross sections for forming manganese. These are to be added to the nuclear fragmentation cross sections for manganese in parentheses before comparing with experimental values.

Comparing the model cross sections with the experimental data set reveals that 92 percent of the calculated cross sections are within 50 percent of the measured values. If we reduce the error band to 30 percent, we will find 81 percent of the cross sections are in agreement to within this level. Among the least accurate are the iron on hydrogen target data which again is Rudstam's theory and the cross

Table 5.23. Fragmentation Cross Sections of Carbon Beams
at 1.05 GeV/nucleon in Various Targets

[Quantities in parentheses are present theory]

Fragment	Carbon cross section,[a] mb, in target of—				
	H (b)	Be	C	Cu	Pb
Li	23 ± 2 (34)	51 ± 2 (54)	52 ± 3 (61)	71 ± 5 (81)	103 ± 14 (113)
Be	17 ± 1 (22)	35 ± 1 (32)	35 ± 1 (33)	47 ± 2 (48)	71 ± 6 (63)
B	50 ± 4 (42)	81 ± 4 (86)	78 ± 3 (100)	119 ± 8 (138)	203 ± 32 (185)
C	28 ± 3 (10)	49 ± 3 (39)	50 ± 4 (44)	86 ± 8 (57)	139 ± 22 (79)

[a] Heckman, 1975.

[b] Values in parentheses in this column are those of modified Rudstam (1966).

Table 5.24. Charge Removal Cross Sections of Various
Projectiles on Carbon Targets

[Quantities in parentheses are present theory;
number in brackets is energy in GeV/nucleon]

ΔZ	Charge removal cross section, mb, of projectile of—					
	C [2.1] (a)	O [2.1] (a)	O [0.9] (b)	Ne [0.47] (b)	Ar [0.21] (c)	Fe [1.88] (d)
0	50 ± 4 (40)	45 ± 2 (45)	------	------ (40)	------ (132)	------ (64)
1	78 ± 3 (100)	105 ± 4 (101)	176 ± 5	129 ± 3 (90)	------ (151)	181 ± 27 (157)
2	35 ± 1 (33)	116 ± 6 (93)	164 ± 5	214 ± 3 (98)	154 ± 26 (85)	124 ± 13 (110)
3	52 ± 2 (61)	50 ± 2 (65)	55 ± 3	155 ± 3 (75)	122 ± 16 (72)	100 ± 11 (87)
4		36 ± 1 (24)	27 ± 2	140 ± 3 (65)	144 ± 19 (64)	87 ± 11 (76)
5		65 ± 3 (47)	------	74 ± 2 (54)	81 ± 15 (59)	54 ± 9 (62)
6				33 ± 1 (19)	112 ± 15 (51)	78 ± 11 (67)
7				------ (40)	90 ± 3 (50)	52 ± 7 (57)
8					92 ± 13 (44)	55 ± 9 (52)
9					65 ± 11 (42)	53 ± 7 (49)
10					83 ± 13 (37)	54 ± 10 (45)
11					------ (35)	59 ± 10 (42)
12						57 ± 10 (39)
13						83 ± 11 (36)
14						------ (35)

[a] Heckman, 1975.

[b] Webber et al., 1983a and 1983b.

[c] Guerreau et al., 1983.

[d] Westfall et al., 1979.

sections of Webber et al. Note that our model agrees with experiments to the extent that the experimentalists agree among themselves for the same projectile-target combinations. From this point of view, little progress can be made in

Table 5.25. Fragmentation Cross Section of Iron Projectiles at 1.88 GeV/nucleon in Various Targets

[Quantities in parentheses are values from present model; values in square brackets are coulomb dissociation cross sections for forming manganese]

Z_F	Cross section, mb, of iron projectiles in—									
	H (a)	Li	Be	C	S	Cu	Ag	Ta	Pb	U
13	25±10 (19)	50±5 (33)	50±7 (34)	83±11 (36)	78±18 (45)	179±27 (57)	112±19 (64)	81±14 (72)	191±34 (74)	307±79 (77)
14	31±9 (22)	54±5 (36)	75±8 (37)	57±10 (39)	106±14 (47)	72±11 (61)	158±20 (67)	115±20 (75)	119±22 (77)	169±28 (80)
15	22±10 (26)	57±6 (39)	57±8 (38)	59±10 (42)	50±8 (51)	88±15 (63)	64±13 (70)	133±20 (79)	78±16 (81)	176±34 (84)
16	37±24 (30)	56±6 (43)	63±8 (45)	54±10 (45)	74±12 (54)	56±11 (67)	96±13 (74)	109±17 (82)	116±19 (85)	116±22 (88)
17	36±18 (36)	38±4 (44)	54±7 (48)	53±7 (49)	66±14 (58)	86±13 (71)	79±14 (79)	101±18 (87)	90±19 (90)	133±22 (93)
18	31±9 (41)	55±6 (49)	54±7 (51)	55±9 (52)	74±13 (62)	95±15 (76)	84±14 (83)	100±18 (94)	73±15 (95)	113±19 (98)
19	36±9 (47)	56±5 (53)	65±7 (55)	52±7 (57)	55±21 (66)	88±14 (81)	79±11 (89)	111±20 (98)	90±19 (101)	105±15 (104)
20	47±11 (55)	64±6 (59)	68±7 (60)	78±11 (67)	97±14 (72)	98±14 (87)	118±14 (96)	107±17 (106)	144±22 (109)	143±19 (112)
21	62±11 (65)	67±6 (64)	77±8 (66)	54±9 (62)	91±13 (79)	100±15 (95)	104±13 (104)	129±18 (116)	111±17 (118)	153±21 (122)
22	22±13 (77)	75±6 (71)	83±9 (74)	87±11 (76)	64±10 (89)	101±14 (106)	124±16 (116)	152±19 (129)	148±22 (132)	95±16 (136)
23	60±11 (88)	88±7 (83)	88±9 (84)	100±11 (87)	86±12 (103)	121±15 (121)	117±15 (132)	150±19 (146)	142±20 (151)	181±27 (156)
24	80±13 (101)	98±7 (101)	111±9 (105)	124±13 (110)	128±16 (126)	149±16 (146)	218±21 (161)	206±22 (176)	242±25 (181)	208±22 (190)
25	127±24 (119)	141±18 (147)	156±21 (154)	181±27 (157)	250±22 (182)	219±20 (208)	280±23 (228)	457±34 (250)	509±40 (256)	646±43 (260)
	[0]	[1]	[1]	[3]	[15]	[42]	[97]	[211]	[258]	[316]

^{a} Values in parentheses in this column are those of Rudstam (1966).

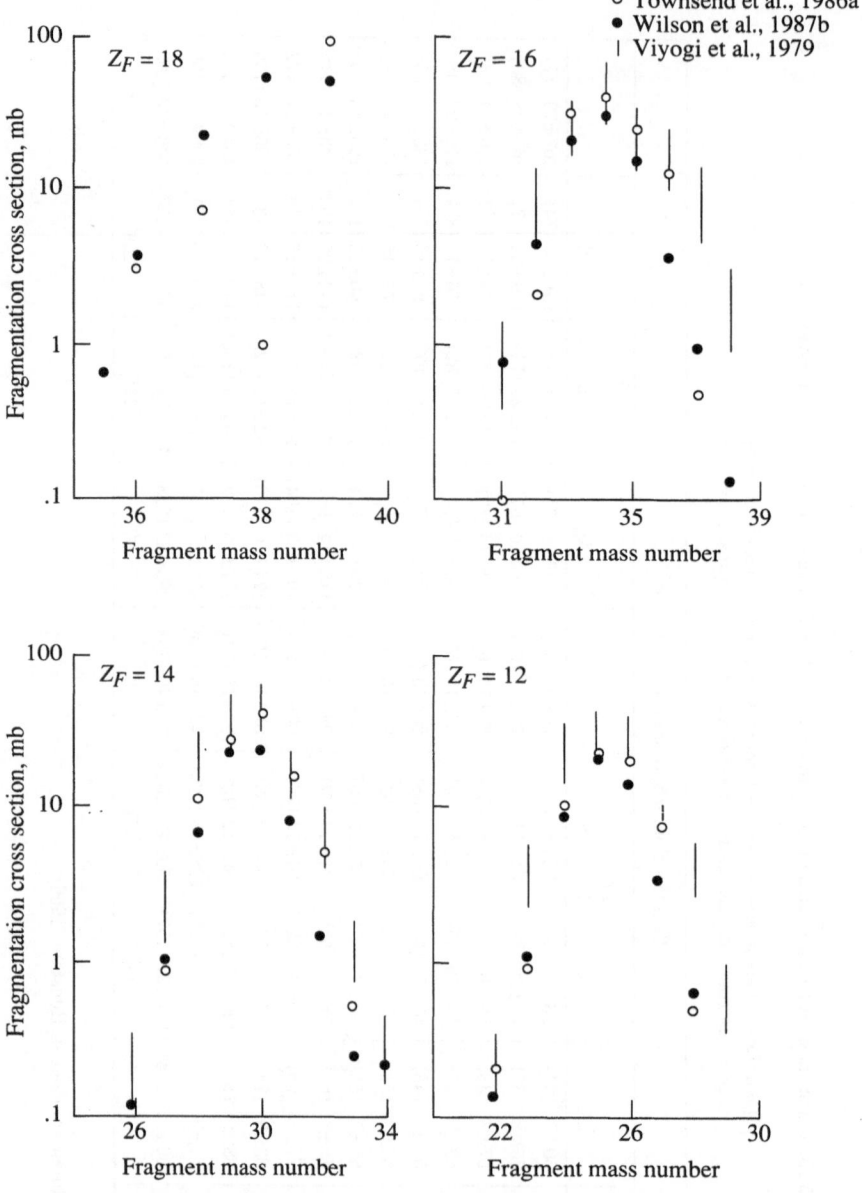

Figure 5.48. Representative argon-carbon fragmentation cross sections.

improving the model until the experimental situation is clarified. Clearly, the model of Silberberg, Tsao, and Shapiro (1976), which includes many corrections to Rudstam's formulas, is preferred for hydrogen targets.

The semiempirical model for argon fragmentation on carbon is shown with the quantum mechanical optical model calculation in figure 5.48. Also shown are experimental data of Viyogi et al. (1979). Reasonable agreements are seen between the two models except for neutron removal where there are no data yet to resolve the difference (Townsend and Wilson, 1989).

5.8. Summary

The current empirical data base represented in this chapter constitutes the nuclear data over which the current radiation transport codes are written. The adequacy of the data base depends on whether the important transport quantities are accurately represented. This issue is further addressed in chapters 8 and 9.

5.9. References

Anon., 1968: MECC-7 *Intranuclear Cascade Code, 500-*MeV *Protons on O-16.* I4C Analysis Codes (Programmed for H. W. Bertini). Available from Radiation Shielding Information Center, Oak Ridge National Lab.

Ableev, V. G.; Bodyagin, V. A.; Vorob'ev, G. G.; Zaporozhets, S. A.; Nomofilov, A. A.; Piskunov, N. M.; Sitnik, I. M.; Strokovskii, E. A.; Strunov, L. N.; Tarasóv, A. V.; Filipkovski, A.; Khristova, I. U.; and Sharov, V. I., 1982: Scattering of 17.9-GeV/c α Particles by C, Al, and Cu Nuclei. *Sov. J. Nuclear Phys.,* vol. 36, no. 5, pp. 698–703.

Ahmad, I., 1975: An Analysis of Some Proton-Nucleus Scattering Data at 1 GeV. *Nuclear Phys.,* vol. A247, no. 3, pp. 418–440.

Aichelin, J.; Peilert, G.; Bohnet, A.; Rosenhauer, A.; Stöcker, H.; and Greiner, W., 1988: Quantum Molecular Dynamics Approach to Heavy Ion Collisions: Description of the Model, Comparison With Fragmentation Data, and the Mechanism of Fragment Formation. *Phys. Review C,* vol. 37, no. 6, pp. 2451–2468.

Alsmiller, R. G., Jr.; Barish, J.; and Leimdorfer, M., 1968: Analytic Representation of Nonelastic Cross Sections and Particle-Emission Spectra From Nucleon-Nucleus Collisions in the Energy Range 25 to 400 MeV. *Protection Against Space Radiation,* NASA SP-169, pp. 495–515.

Alsmiller, R. G., Jr.; Armstrong, T. W.; and Coleman, W. A., 1970: The Absorbed Dose and Dose Equivalent From Neutrons in the Energy Range 60 to 3000 MeV and Protons in the Energy Range 400 to 3000 MeV. *Nuclear Sci. & Eng.,* vol. 42, no. 3, pp. 367–381.

Auger, J. P.; Gillespie, J.; and Lombard, R. J., 1976: Proton-^{4}He Elastic Scattering at Intermediate Energies. *Nuclear Phys.,* vol. A262, no. 3, pp. 372–388.

Badavi, Forooz F.; Townsend, Lawrence W.; Wilson, John W.; and Norbury, John W., 1987: An Algorithm for a Semiempirical Nuclear Fragmentation Model. *Comput. Phys. Commun.,* vol. 47, pp. 281–294.

Benesh, C. J.; Cook, B. C.; and Vary, J. P., 1989: Single Nucleon Removal in Relativistic Nuclear Collisions. *Phys. Review C,* vol. 40, third ser., no. 3, pp. 1198–1206.

Bertini, Hugo W.; Guthrie, Miriam P.; and Culkowski, Arline H., 1972: *Nonelastic Interactions of Nucleons and π-Mesons With Complex Nuclei at Energies Below 3* GeV. ORNL-TM-3148, U.S. Atomic Energy Commission.

Bertini, R.; Beurtey, R.; Brochard, F.; Bruge, G.; Catz, H.; Chaumeaux, A.; Durand, J. M.; Faivre, J. C.; Fontaine, J. M.; Garreta, D.; Gustafsson, C.; Hendrie, D.; Hibou, F.; Legrand, D.; Saudinos, J.; and Thiron, J., 1973: Angular Distribution of 1.04 GeV Protons Scattered by ^{12}C, ^{58}Ni, ^{208}Pb. *Phys. Lett.,* vol. 45B, no. 2, pp. 119–122.

Bertulani, Carlos A.; and Baur, Gerhard, 1988: Electromagnetic Processes in Relativistic Heavy Ion Collisions. *Phys. Rep.,* vol. 163, nos. 5 & 6, pp. 299–408.

Bevington, Philip R., 1969: *Data Reduction and Error Analysis for the Physical Sciences.* McGraw-Hill Book Co., Inc.

Blanpied, G. S.; Hoffmann, G. W.; Barlett, M. L.; McGill, J. A.; Greene, S. J.; Ray, L.; Van Dyck, O. B.; Amann, J.; and Thiessen, H. A., 1981: Large Angle Scattering of 0.8 GeV Protons From ^{12}C. *Phys. Review C,* vol. 23, third ser., no. 6, pp. 2599–2605.

Blatt, John M.; and Weisskopf, Victor F., 1952: *Theoretical Nuclear Physics.* John Wiley & Sons, Inc.

Bleszynski, M.; and Sander, C., 1979: Geometrical Aspects of High-Energy Peripheral Nucleus-Nucleus Collisions. *Nuclear Phys. A,* vol. 326, nos. 2–3, pp. 525–535.

Bohr, Aage; and Mottelson, Ben R., 1969: *Nuclear Structure. Volume I—Single-Particle Motion.* W. A. Benjamin, Inc.

Borkowski, F.; Simon, G. G.; Walther, V. H.; and Wendling, R. D., 1975: On the Determination of the Proton RMS-Radius From Electron Scattering Data. *Z. Phys. A,* vol. 275, no. 1, pp. 29–31.

Bowman, J. D.; Swiatecki, W. J.; and Tsang, C. F., 1973: *Abrasion and Ablation of Heavy Ions.* LBL-2908, Lawrence Berkeley Lab., Univ. of California.

Brady, F. P.; Christie, W. B.; Romero, J. L.; Tull, C. E.; McEachern, B.; Webb, M. L.; Young, J. C.; Crawford, H. J.; Greiner, D. E.; Lindstrom, P. J.; and Sann, H., 1988: Large p_T From the Fragmentation of 1.2-GeV/Nucleon ^{139}La Nuclei. *Phys. Review Lett.,* vol. 60, no. 17, pp. 1699–1702.

Chaumeaux, A.; Bruge, G.; Bauer, T.; Bertini, R.; Boudard, A.; Catz, H.; Couvert, P.; Duhm, H. H.; Fontaine, J. M.; Garreta, D.; Lugol, J. C.; Layly, V.; and Schaeffer, R., 1976: Scattering of 1.37 GeV α-Particles by ^{12}C. *Nuclear Phys.,* vol. A267, no. 3, pp. 413–424.

Cucinotta, Francis A.; Norbury, John W.; Khandelwal, Govind S.; and Townsend, Lawrence W., 1987: *Doubly Differential Cross Sections for Galactic Heavy-Ion Fragmentation.* NASA TP-2659.

Cucinotta, Francis A., 1988: Theory of Alpha-Nucleus Collisions at High Energies. Ph.D. Thesis, Old Dominion Univ.

Cucinotta, Francis A.; Khandelwal, Govind S.; Maung, Khin M.; Townsend, Lawrence W.; and Wilson, John W., 1988: *Eikonal Solutions to Optical Model Coupled-Channel Equations.* NASA TP-2830.

Cucinotta, Francis A.; Norbury, John W.; and Townsend, Lawrence W., 1988: *Multiple Nucleon Knockout by Coulomb Dissociation in Relativistic Heavy-Ion Collisions.* NASA TM-4070.

Cucinotta, F. A.; Khandelwal, G. S.; Townsend, L. W.; and Wilson, J. W., 1989: Correlations in α-α Scattering and Semi-Classical Optical Models. *Phys. Lett.,* vol. B223, no. 2, pp. 127–132.

Cucinotta, Francis A.; Townsend, Lawrence W.; Wilson, John W.; and Khandelwal, Govind S., 1990: *Inclusive Inelastic Scattering of Heavy Ions and Nuclear Correlations.* NASA TP-3026.

De Jager, C. W.; De Vries, H.; and De Vries, C., 1974: Nuclear Charge- and Magnetization-Density-Distribution Parameters From Elastic Electron Scattering. *At. Data & Nuclear Data Tables,* vol. 14, no. 5/6, pp. 479–508.

Feshbach, Herman; Gal, Avraham; and Hüfner, Jörg, 1971: On High-Energy Scattering by Nuclei—II. *Ann. Phys.,* vol. 66, no. 1, pp. 20–59.

Foelsche, Trutz; Mendell, Rosalind B.; Wilson, John W.; and Adams, Richard R., 1974: *Measured and Calculated Neutron Spectra and Dose Equivalent Rates at High Altitudes; Relevance to SST Operations and Space Research.* NASA TN D-7715.

Franco, Victor; and Varma, Girish K., 1978: Collisions Between Composite Particles at Medium and High Energies. *Phys. Review C,* vol. 18, no. 1, pp. 349–370.

Fricke, Scott Herbert, 1985: A Theoretical Study of Anomalous Projectile Fragments Produced in Relativistic Heavy Ion Interactions. Ph.D. Thesis, Univ. of Minnesota.

Glauber, R. J.; and Matthiae, G., 1970: High-Energy Scattering of Protons by Nuclei. *Nuclear Phys.,* vol. B21, no. 1, pp. 135–157.

Goldhaber, A. S., 1974: Statistical Models of Fragmentation Processes. *Phys. Lett.,* vol. 53B, no. 4, pp. 306–308.

Gosset, J.; Gutbrod, H. H.; Meyer, W. G.; Poskanzer, A. M.; Sandoval, A.; Stock, R.; and Westfall, G. D., 1977: Central Collisions of Relativistic Heavy Ions. *Phys. Review,* vol. 16, ser. C, no. 2, pp. 629–657.

Greiner, D. E.; Lindstrom, P. J.; Heckman, H. H.; Cork, Bruce; and Bieser, F. S., 1975: Momentum Distributions of Isotopes Produced by Fragmentation of Relativistic ^{12}C and ^{16}O Projectiles. *Phys. Review Lett.,* vol. 35, no. 3, pp. 152–155.

Guerreau, D.; Borrel, V.; Jacquet, D.; Galin, J.; Gatty, B.; and Tarrago, X., 1983: Isotopic Distributions of Projectile-Like Fragments in 44 MeV/u ^{40}Ar Induced Reactions. *Phys. Lett.,* vol. 131B, no. 4,5,6, pp. 293–296.

Guthrie, Miriam P., 1970: *EVAP-4: Another Modification of a Code To Calculate Particle Evaporation From Excited Compound Nuclei.* ORNL-TM-3119, U.S. Atomic Energy Commission.

Guzik, T. Gregory, 1981: The Low-Energy Galactic Cosmic Ray Carbon, Nitrogen, and Oxygen Isotopic Composition. *Astrophys. J.,* no. 2, pt. 1, pp. 695–710.

Guzik, T. G.; Wefel, J. P.; Crawford, H. J.; Greiner, D. E.; Lindstrom, P. J.; Schimmerling, W.; and Symons, T. J. M., 1985: Implications of New Measurements of ^{16}O $+ p \rightarrow$ ^{12,13}C, ^{14,15}N for the Abundances of C, N Isotopes at the Cosmic Ray Source. *19th International Cosmic Ray Conference,* OG Sessions, Volume 2, NASA CP-2376, pp. 80–83.

Heckman, H. H., 1975: *Heavy Ion Fragmentation Experiments at the Bevatron.* NASA CR-142589.

Heckman, Harry H.; and Lindstrom, Peter J., 1976: Coulomb Dissociation of Relativistic ^{12}C and ^{16}O Nuclei. *Phys. Review Lett.,* vol. 37, no. 1, pp. 56–59.

Hill, John C.; Wohn, F. K.; Winger, J. A.; and Smith, A. R., 1988: Electromagnetic Dissociation for High-Z Projectiles at Ultrarelativistic Energies. *Phys. Review Lett.,* vol. 60, no. 11, pp. 999–1001.

Hüfner, J.; Schäfer, K.; and Schürmann, B., 1975: Abrasion-Ablation in Reactions Between Relativistic Heavy Ions. *Phys. Review,* vol. 12, ser. C, no. 6, pp. 1888–1898.

Kerman, A. K.; McManus, H.; and Thaler, R. M., 1959: The Scattering of Fast Nucleons From Nuclei. *Ann. Phys. (N.Y.),* vol. 8, no. 4, pp. 551–635.

Khan, Ferdous; Khandelwal, Govind S.; Wilson, John W.; Townsend, Lawrence W.; and Norbury, John W., 1988: Excitation Decay Contribution of Projectile and Projectile Fragments to (^{12}C, ^{11}B+P) Cross Section at 2.1 A GeV With ^{12}C Targets. *Proceedings of the 8th High Energy Heavy Ion Study,* J. W. Harris and G. J. Wozniak, eds., LBL-24580, Lawrence Berkeley Lab., pp. 440–449.

Khan, Ferdous, 1989: Nuclear Fragmentation Energy and Momentum Transfer Distributions in Relativistic Heavy Ion Collisions. Ph.D. Thesis, Old Dominion Univ.

Knight, E. A.; Singhal, R. P.; Arthur, R. G.; and Macauley, M. W. S., 1981: Elastic Scattering of Electrons From ^{20,22}Ne. *J. Phys. G: Nuclear Phys.,* vol. 7, no. 8, pp. 1115–1121.

Levinger, Joseph S., 1960: *Nuclear Photo-Disintegration.* Oxford Univ. Press.

Lindstrom, P. J.; Greiner, D. E.; Heckman, H. H.; Cork, Bruce; and Bieser, F. S., 1975: *Isotope Production Cross Sections From the Fragmentation of ^{16}O and ^{12}C at Relativistic Energies.* LBL-3650 (NGR-05-003-513), Lawrence Berkeley Lab., Univ. of California.

Lord, J. J., 1951: The Altitude and Latitude Variation in the Rate of Occurrence of Nuclear Disintegrations Produced in the Stratosphere by Cosmic Rays. *Phys. Review,* vol. 81, second ser., no. 6, pp. 901–909.

Maleck, A.; Picozza, P.; and Satta, L., 1984: Inclusive Inelastic Cross Section for α-α Scattering at High Energies. *Phys. Lett.,* vol. 136B, no. 5/6, pp. 319–321.

Mercier, M. T.; Hill, J. C.; Wohn, F. K.; and Smith, A. R., 1984: Electromagnetic Dissociation of ^{197}Au by Relativistic Heavy Ions. *Phys. Review Lett.,* vol. 52, no. 11, pp. 898–901.

Mercier, M. T.; Hill, John C.; Wohn, F. K.; McCullough, C. M.; Nieland, M. E.; Winger, J. A.; Howard, C. B.; Renwick, S.; Matheis, D. K.; and Smith, A. R., 1986: Electromagnetic Dissociation of ^{59}Co, ^{89}Y, and ^{197}Au Targets by Relativistic Heavy Ions to $Z = 26$. *Phys. Review C,* vol. 33, third ser., no. 5, pp. 1655–1667.

Moniz, E. J.; and Nixon, G. D., 1971: High Energy Coherent Processes With Nuclear Targets. *Ann. Phys. (N.Y.),* vol. 67, no. 1, pp. 58–97.

Morrissey, D. J.; Marsh, W. R.; Otto, R. J.; Loveland, W.; and Seaborg, G. T., 1978: Target Residue Mass and Charge Distributions in Relativistic Heavy Ion Reactions. *Phys. Review,* vol. 18, ser. C, no. 3, pp. 1267–1274.

Morrissey, D. J.; Oliveira, L. F.; Rasmussen, J. O.; Seaborg, G. T.; Yariv, Y.; and Fraenkel, Z., 1979: Microscopic and Macroscopic Model Calculations of Relativistic Heavy-Ion Fragmentation Reactions. *Phys. Review Lett.*, vol. 43, no. 16, pp. 1139–1142.

Nakada, A.; Torizuka, Y.; and Horikawa, Y., 1971: Determination of the Deformation in ^{12}C From Electron Scattering. *Phys. Review Lett.*, vol. 27, no. 11, pp. 745–748.

Norbury, John W.; Townsend, Lawrence W.; and Deutchman, Philip A., 1985: *A T-Matrix Theory of Galactic Heavy-Ion Fragmentation.* NASA TP-2363.

Norbury, John W.; and Townsend, Lawrence W., 1986: *Electromagnetic Dissociation Effects in Galactic Heavy-Ion Fragmentation.* NASA TP-2527.

Norbury, John W.; Cucinotta, F. A.; Townsend, L. W.; and Badavi, F. F., 1988: Parameterized Cross Sections for Coulomb Dissociation in Heavy-Ion Collisions. *Nuclear Instrum. & Methods Phys. Res.*, vol. B31, no. 4, pp. 535–537.

Norbury, John W.; Townsend, Lawrence W.; and Badavi, Forooz F., 1988: *Computer Program for Parameterization of Nucleus-Nucleus Electromagnetic Dissociation Cross Sections.* NASA TM-4038.

Norbury, John W., 1989: Nucleon Emission Via Electromagnetic Excitation in Relativistic Nucleus-Nucleus Collisions: Reanalysis of the Weizsäcker-Williams Method. *Phys. Review C,* vol. 40, third ser., no. 6, pp. 2621–2628.

Norbury, John W., 1990a: Charge Dependence and Electric Quadrupole Effects on Single-Nucleon Removal in Relativistic and Intermediate Energy Nuclear Collisions. *Phys. Review C,* vol. 42, third ser., no. 5, pp. 2259–2262.

Norbury, John W., 1990b: Electric Quadrupole Excitations in the Interactions of ^{89}Y With Relativistic Nuclei. *Phys. Review C,* vol. 41, third ser., no. 1, pp. 372–373.

Norbury, John W., 1990c: Electric Quadrupole Excitations in Relativistic Nucleus-Nucleus Collisions. *Phys. Review C,* vol. 42, third ser., no. 2, pp. 711–715.

Norbury, John W.; and Townsend, Lawrence W., 1990a: *Calculation of Two-Neutron Multiplicity in Photonuclear Reactions.* NASA TP-2968.

Norbury, John W.; and Townsend, Lawrence W., 1990b: Single Nucleon Emission in Relativistic Nucleus-Nucleus Reactions. *Phys. Review C,* vol. 42, third ser., no. 4, pp. 1775–1777.

Oliveira, Luiz F.; Donangelo, Raul; and Rasmussen, John O., 1979: Abrasion-Ablation Calculations of Large Fragment Yields From Relativistic Heavy Ion Reactions. *Phys. Review C,* vol. 19, ser. C, no. 3, pp. 826–833.

Ranft, J., 1980: The FLUKA and KASPRO Hadronic Cascade Codes. *Computer Techniques in Radiation Transport and Dosimetry,* Walter R. Nelson and Theodore M. Jenkins, eds., Plenum Press, pp. 339–371.

Ray, L., 1979: Proton-Nucleus Total Cross Sections in the Intermediate Energy Range. *Phys. Review C,* vol. 20, third ser., no. 5, pp. 1857–1872.

Rodberg, Leonard S.; and Thaler, R. M., 1967: *Introduction to the Quantum Theory of Scattering.* Academic Press Inc.

Rudstam, G., 1966: Systematics of Spallation Yields. *Zeitschrift fur Naturforschung,* vol. 21a, no. 7, pp. 1027–1041.

Saudinos, Jean; and Wilkin, Colin, 1974: Proton-Nucleus Scattering at Medium Energies. *Annual Review of Nuclear Science,* Volume 24, Emilio Segrè, J. Robb Grover, and H. Pierre Noyes, eds., Annual Reviews, Inc., pp. 341–377.

Serber, R., 1947: Nuclear Reactions at High Energies. *Phys. Review,* vol. 72, no. 11, pp. 1114–1115.

Silberberg, R.; Tsao, C. H.; and Shapiro, M. M., 1976: Semiempirical Cross Sections, and Applications to Nuclear Interactions of Cosmic Rays. *Spallation Nuclear Reactions and Their Applications,* B. S. P. Shen and M. Merker, eds., D. Reidel Publ. Co., pp. 49–81.

Stevenson, J. D.; Martinis, J.; and Price, P. B., 1981: Measurement of the Summed Residual Projectile Mass in Relativistic Heavy-Ion Collisions. *Phys. Review Lett.,* vol. 47, no. 14, pp. 990–993.

Townsend, Lawrence W., 1981: *Optical-Model Abrasion Cross Sections for High-Energy Heavy Ions.* NASA TP-1893.

Townsend, Lawrence W., 1982: *Harmonic Well Matter Densities and Pauli Correlation Effects in Heavy-Ion Collisions.* NASA TP-2003.

Townsend, L. W., 1983: Abrasion Cross Sections for ^{20}Ne Projectiles at 2.1 GeV/Nucleon. *Canadian J. Phys.,* vol. 61, no. 1, pp. 93–98.

Townsend, Lawrence W.; Wilson, John W.; Norbury, John W.; and Bidasaria, Hari B., 1984: *An Abrasion-Ablation Model Description of Galactic Heavy-Ion Fragmentation.* NASA TP-2305.

Townsend, Lawrence W.; and Wilson, John W., 1985: *Tables of Nuclear Cross Sections for Galactic Cosmic Rays—Absorption Cross Sections.* NASA RP-1134.

Townsend, L. W.; Wilson, J. W.; and Norbury, J. W., 1985: A Simplified Optical Model Description of Heavy Ion Fragmentation. *Canadian J. Phys.,* vol. 63, no. 2, pp. 135–138.

Townsend, L. W.; and Wilson, J. W., 1986: Energy-Dependent Parameterization of Heavy-Ion Absorption Cross Sections. *Radiat. Res.,* vol. 106, pp. 283–287.

Townsend, L. W.; Wilson, J. W.; Cucinotta, F. A.; and Norbury, J. W., 1986a: Comparison of Abrasion Model Differences in Heavy Ion Fragmentation: Optical Versus Geometric Models. *Phys. Review,* vol. 34, ser. C, no. 4, pp. 1491–1494.

Townsend, Lawrence W.; Wilson, John W.; Cucinotta, Francis A.; and Norbury, John W., 1986b: *Optical Model Calculations of Heavy-Ion Target Fragmentation.* NASA TM-87692.

Townsend, L. W.; and Wilson, J. W., 1989: Nuclear Cross Sections for Estimating Secondary Radiations Produced in Spacecraft. *High-Energy Radiation Background in Space,* American Inst. of Physics, pp. 177–191.

Überall, Herbert, 1971: *Electron Scattering From Complex Nuclei*, Part A. Academic Press, Inc.

Van Allen, James A., 1952: The Nature and Intensity of the Cosmic Radiation. Chapter XIV of *Physics and Medicine of the Upper Atmosphere*, Clayton S. White and Otis O. Benson, Jr., eds., Univ. of New Mexico Press (Albuquerque), pp. 239–266.

Viollier, R. D., 1975: High-Energy Proton-Nucleus Scattering and Correlations. *Ann. Phys.*, vol. 93, no. 1–2, pp. 335–368.

Viyogi, Y. P.; Symons, T. J. M.; Doll, P.; Greiner, D. E.; Heckman, H. H.; Hendrie, D. L.; Lindstrom, P. J.; Mahoney, J.; Scott, D. K.; Van Bibber, K.; Westfall, G. D.; Wieman, H.; Crawford, H. J.; McParland, C.; and Gelbke, C. K., 1979: Fragmentations of ^{40}Ar at 213 MeV/Nucleon. *Phys. Review Lett.*, vol. 42, no. 1, pp. 33–36.

Webb, M. L.; Crawford, H. J.; Engelage, J.; Baumgartner, M. E.; Greiner, D. E.; Lindstrom, P. J.; Olson, D. L.; and Wada, R., 1987: Probing the Direct Step of Relativistic Heavy Ion Fragmentation: (^{12}C, ^{11}B + p) at 2.1 GeV/Nucleon With C and CH$_2$ Targets. *Phys. Review C*, vol. 36, third ser., no. 1, pp. 193–202.

Webber, W. R.; Brautigam, D. A.; Kish, J. C.; and Schrier, D., 1983a: Fragmentation of ∼ 500 MeV/nuc Ne and O Nuclei in CH$_2$, C and H Targets—Charge and Isotopic Cross Sections. *18th International Cosmic Ray Conference—Conference Papers*, OG Sessions, Vol. 2, Tata Inst. of Fundamental Research (Colaba, Bombay), pp. 202–205.

Webber, W. R.; Brautigam, D. A.; Kish, J. C.; and Schrier, D., 1983b: Fragmentation of 710 and 1050 MeV/nuc Fe Nuclei in CH$_2$ and C Targets—Isotopic Cross Sections for H Targets. *18th International Cosmic Ray Conference—Conference Papers*, OG Sessions, Vol. 2, Tata Inst. of Fundamental Research (Colaba, Bombay), pp. 198–201.

Westfall, G. D.; Wilson, Lance W.; Lindstrom, P. J.; Crawford, H. J.; Greiner, D. E.; and Heckman, H. H., 1979: Fragmentation of Relativistic ^{56}Fe. *Phys. Review*, vol. 19, ser. C, no. 4, pp. 1309–1323.

Wilson, J. W., 1974a: Multiple Scattering of Heavy Ions, Glauber Theory, and Optical Model. *Phys. Lett.*, vol. B52, no. 2, pp. 149–152.

Wilson, John W., 1974b: Proton-Deuteron Double Scattering. *Phys. Review C*, vol. 10, no. 1, pp. 369–376.

Wilson, John W., 1975: Composite Particle Reaction Theory. Ph.D. Diss., College of William and Mary in Virginia.

Wilson, John W.; and Costner, Christopher M., 1975: *Nucleon and Heavy-Ion Total and Absorption Cross Section for Selected Nuclei*. NASA TN D-8107.

Wilson, John W., 1977: *Analysis of the Theory of High-Energy Ion Transport*. NASA TN D-8381.

Wilson, J. W.; and Townsend, L. W., 1981: An Optical Model for Composite Nuclear Scattering. *Canadian J. Phys.*, vol. 59, no. 11, pp. 1569–1576.

Wilson, John W., 1983: *Heavy-Ion Transport in the Straight Ahead Approximation.* NASA TP-2178.

Wilson, John W.; Townsend, L. W.; Bidasaria, H. B.; Schimmerling, Walter; Wong, Mervyn; and Howard, Jerry, 1984: ^{20}Ne Depth-Dose Relations in Water. *Health Phys.,* vol. 46, no. 5, pp. 1101–1111.

Wilson, J. W.; Townsend, L. W.; and Badavi, F. F., 1987a: Galactic HZE Propagation Through the Earth's Atmosphere. *Radiat. Res.,* vol. 109, no. 2, pp. 173–183.

Wilson, John W.; Townsend, Lawrence W.; and Badavi, F. F., 1987b: A Semiempirical Nuclear Fragmentation Model. *Nuclear Instrum. & Methods Phys. Res.,* vol. B18, no. 3, pp. 225–231.

Wong, Cheuk-Yin, 1981: Peripheral and Central Collisions at Around 100 MeV per Nucleon. *5th High Energy Heavy Ion Study,* LBL-12652 (Contract W-7405-ENG-48), Lawrence Berkeley Lab., Univ. of California, pp. 61–86.

6. TRANSPORT THEORY

6.1. Introduction

The 1912 experiments of V. Hess to study the decline of terrestrial radiation in the atmosphere led to the discovery of cosmic rays (Hess and Eugster, 1949). The next two decades saw the study of the increase of cosmic rays with altitude and decreasing atmospheric shielding. Even after 30 years of study, a meaningful theory of the propagation of the nucleonic component came only after sufficient understanding of the nuclear force and nuclear theory. Thus, development began with the historic paper of Bethe, Korff, and Placzek (1940), which concerned atmospheric neutrons; these results, although incomplete, remain substantially correct today. The detection several years later of neutrons in coincidence with atmospheric air showers (Cocconi, Cocconi-Tongiorgi, and Greisen, 1948) and cloud chamber data with evaporation stars leads one to suspect that moderate energy neutrons are part of a normal air shower event as assumed by Bethe, Korff, and Placzek (1940). The subsequent work of Cocconi, Cocconi Tongiorgi, and Widgoff (1950) on atmospheric cascades begins to place the whole subject of air showers on firm ground. Yet to be added to the understanding of air showers is the discovery of the heavy ion component (Freier et al., 1948) and the related propagation equations.

Early works in setting up the galactic ion transport equations ignored energy loss by ionization of the medium. Peters (1958) used a one-dimensional equilibrium solution ignoring ionization energy loss and radioactive decay to show that the light ions of the galactic cosmic rays have their origin in the breakup of heavy particles in interstellar space. Davis (1960) showed that one-dimensional propagation is simplistic and that leakage at the galactic boundary must be included. Ginzburg and Syrovatskii (1964) argued that the leakage can be approximated as a superposition of nonequilibrium one-dimensional solutions.

In distinction to cosmic-ray studies that accentuated the nuclear reactions and ignored ionization energy loss, the early space shielding studies (mainly concerned with solar proton events and trapped radiation) ignored nuclear reaction effects and treated only the ionization energy loss (Shaefer, 1959; Foelsche, 1959; Dye and Noyes, 1960). Such studies were mainly limited by the available nuclear data. The hope for comprehensive nuclear data began with a study by M. L. Goldberger (1948) in which a two-dimensional, intranuclear cascade calculation by a young student named G. F. Chew was made with random number tables and a mechanical calculator. (G. F. Chew cast his vote for nuclear democracy at the first heavy ion conference at Lawrence Berkeley Laboratory 25 years later.) Detailed development of the intranuclear cascade method awaited the introduction of large-scale scientific computers (Metropolis et al., 1958) which, when developed, had a tremendous impact on the space-radiation program (Bertini, 1962; Alsmiller, 1967). A series of Monte Carlo and deterministic transport codes began to emerge using the new nuclear models (Alsmiller, 1967; Dye, 1962; Lambiotte, Wilson, and Filippas, 1971; Wilson and Lamkin, 1975). A relatively complete set of shielding codes was then available for determining shield requirements for protection from space protons. Now heavy ion transport required future development.

Heavy ion transport was important for understanding the origin of galactic cosmic rays as in the early works of Peters (1958), Davis (1960), and Ginzburg and Syrovatskii (1964). In these early works, the complications introduced by ionization energy loss were ignored. Even later papers would ignore or simplify the energy loss term. The "solution" to the steady-state equations is given as a Volterra equation by Gloeckler and Jokipii (1969) which is solved to first order in the fragmentation cross sections ignoring energy loss. They provide an approximation to the first-order solution with ionization energy loss included but are valid only at relativistic energies. Lezniak (1979) gives an overview to cosmic-ray propagation and derives a Volterra equation including the ionization energy loss, which he refers to as a solution "only in the iterative sense," and evaluates only the unperturbed term. No attempt is made to evaluate the first-order perturbation term or higher order terms. The main interest among cosmic-ray physicists has been in solution to, at most, first order in the fragmentation cross sections since path lengths in interstellar space are on the order of 3 to 4 g/cm^2. Clearly, higher order terms cannot be ignored in accelerator or space shielding transport problems (Wilson, 1977a and 1983; Wilson et al., 1984). Aside from this simplification, the cosmic-ray studies discussed previously have neglected the complicated three-dimensional nature of the fragmentation process.

Several approaches to the solution of high-energy, heavy ion propagation, including the ionization energy loss, have been developed over the last 20 years. All but one (Wilson, 1977a) have assumed the straight ahead approximation and velocity conserving fragmentation interactions. Only two (Wilson, 1977a; Wilson et al., 1984) have incorporated energy-dependent nuclear cross sections. The approach by Curtis, Doherty, and Wilkinson (1969) for a primary ion beam represented the first-generation secondary fragments as a quadrature over the collision density of the primary beam. Allkofer and Heinrich (1974) used an energy multigroup method in which an energy-independent fragment transport approximation was applied within each energy group after which the energy group boundaries were moved according to continuous slowing-down theory $(-dE/dx)$. Chatterjee, Tobias, and Lyman (1976) solved the energy-independent fragment transport equation with primary collision density as a source and neglected higher order fragmentation. The primary source term extended only to the primary ion range from the boundary. The energy-independent transport solution was modified to account for the finite range of the secondary fragment ions. Wilson (1977b) derived an expression for the ion transport problem to first order (first-collision term) and gave an analytic solution for the depth-dose relation. Wilson (1977a) further examined the more common approximations used in solving the heavy ion transport problem. The effects of conservation of velocity on fragmentation and the straight ahead approximation are found to be negligible for cosmic-ray applications. Solution methods for representing the energy-dependent nuclear cross sections are developed (Wilson, 1977a). Letaw, Tsao, and Silberberg (1983) approximate the energy-loss term and ion spectra by simple forms for which energy derivatives are evaluated explicitly (even if approximately). The resulting ordinary differential equations in position are solved analytically in a manner that is similar to the method of Allkofer and Heinrich (1974). This approximation assumes a separable solution and results in a decoupling of motion in space and a change in energy. In Letaw's formalism, the energy shift is replaced by an effective attenuation (separable) factor. Wilson (1983) adds the next higher order (second collision) term to his previous analytic expansion (Wilson, 1977b). This term was found to be very important in describing 670 MeV/nucleon ^{20}Ne beams. The three-term expansion of Wilson (1983) was modified to include the effects of energy variation of the nuclear cross sections (Wilson et al., 1984). The integral form of the transport equation was further used to derive a numerical marching procedure to solve the cosmic-ray transport problem (Wilson and Badavi, 1986). This method can easily include the energy-dependent nuclear cross sections within the numerical procedure. Comparison of the numerical procedure (Wilson and Badavi, 1986) with an analytical solution to a simplified problem (Wilson and Townsend, 1988) validates the numerical procedure to about 1-percent accuracy. Several solution techniques and analytic methods have been developed for testing future numerical solutions to the transport equations (Ganapol, Townsend, and Wilson, 1989; Ganapol et al., 1991). More recently, an analytic solution for the laboratory ion-beam transport problem has been derived assuming a straight ahead approximation, velocity conservation at the interaction site, and an energy-independent nuclear cross section (Wilson et al., 1989a and 1989b).

In the above overview of past developments, the applications were split into two separate categories: a single-ion species with a single energy at the boundary versus a broad host of elemental types with a broad, continuous energy spectrum. Techniques requiring a representation of the spectrum over an array of energy values require vast computer storage and computation speed for the laboratory beam problem to maintain sufficient energy resolution. On the other hand, analytic methods (Wilson, 1977a; Wilson and Badavi, 1986) are probably best applied in a marching procedure, which again has within it a similar energy resolution problem. This is a serious limitation, because we require a final code for cosmic-ray shielding that has been validated by laboratory experiments. In this chapter, we examine past developments and new methods in an attempt to overcome these difficulties. Our final objective, as always, is to develop a set of self-contained codes for use in an engineering design environment.

6.2. Transport Formalism

A massive ion, after entering a region filled with ordinary matter, interacts with orbital electrons, thus causing ionization and excitation of the medium. Because of the large mass difference between the ion and these orbital electrons, only a small amount of the ion energy can be transferred in a collision with a single electron. Because of the long range of the coulomb force and the large percentage of the material volume being occupied by electrons, the electron interactions can, to a good approximation, be treated as a continuous slowing-down process over any finite path length. Although the energy lost by an ion over some fixed path length fluctuates about a mean value, this fluctuation amounts to no more than a few percent (Janni, 1982a and 1982b; Schimmerling et al., 1986) and is of no importance in the study of space radiation (Alsmiller, Barish, and Scott, 1969). In the following paragraphs, continuous slowing-down theory will be assumed throughout, and the relevant quantity is the average energy loss per unit path length, denoted by $S_j(E)$, where E is the ion energy and j denotes the ion type.

The mean-free path for nuclear collisions is large (more than a centimeter); by comparison, the mean-free path for collision of the ion with electrons is small. Although collisions with electrons result only in a small transfer of energy compared with the total ion kinetic energy, the nuclear collision generally alters (loss of mass and charge) the ion and the struck nucleus, with many secondary particles being produced. The secondary particles produced as fragments of the primary heavy ion will have longer ranges and free paths causing much greater penetration. As the secondaries undergo additional nuclear reactions, more secondaries, which penetrate deeper into the material, are produced. This process produces the transition effect observed for cosmic rays. The purpose here is to develop the theoretical understanding of the transport of such radiations in extended materials.

The massive particle transport equations are derived by balancing the change in particle flux as it crosses a small volume of material with the gains and losses caused by nuclear collision (Wilson and Lamkin, 1975; see also chapters 1 and 2 for details). The resulting equations for a homogeneous material are given by

$$\left[\vec{\Omega} \cdot \nabla - \frac{1}{A_j} \frac{\partial}{\partial E} S_j(E) + \sigma_j(E) \right] \phi_j \left(\vec{x}, \vec{\Omega}, E \right)$$
$$= \sum_k \int dE' \, d\vec{\Omega}' \, \sigma_{jk} \left(E, E', \vec{\Omega}, \vec{\Omega}' \right) \phi_k \left(\vec{x}, \vec{\Omega}', E' \right) \tag{6.1}$$

where $\phi_j \left(\vec{x}, \vec{\Omega}, E \right)$ is the flux of ions of type j with atomic mass A_j at $\vec{x}$ with motion along $\vec{\Omega}$ and energy E in units of MeV/nucleon, $\sigma_j(E)$ is the corresponding macroscopic cross section, $S_j(E)$ is the linear energy transfer (LET), and $\sigma_{jk} \left(E, E', \vec{\Omega}, \vec{\Omega}' \right)$ is the production cross section for type j particles with energy E and direction $\vec{\Omega}$ by the collision of a type k particle of energy E' and direction $\vec{\Omega}'$. The term on the left side of equation (6.1) containing $S_j(E)$ is a result of the continuous slowing-down approximation, whereas the remaining terms of equation (6.1) are seen to be the usual Boltzmann terms. The solutions to equation (6.1) exist and are unique in any convex region for which the inbound flux of each particle type is specified everywhere on the bounding surface. If the boundary is given as the loci of the two-parameter vector function $\vec{\gamma}(s,t)$ for which a generic point on the boundary is given by $\vec{\Gamma}$, then the boundary condition is specified by requiring the solution of equation (6.1) to meet

$$\phi_j(\vec{\Gamma}, \vec{\Omega}, E) = \psi_j(\vec{\Gamma}, \vec{\Omega}, E) \tag{6.2}$$

for each value of $\vec{\Omega}$ such that

$$\vec{\Omega} \cdot \vec{n}(\vec{\Gamma}) < 0 \tag{6.3}$$

where $\vec{n}(\vec{\Gamma})$ is the outward-directed unit normal vector to the boundary surface at the point $\vec{\Gamma}$ and ψ_j is a specified boundary function.

The fragmentation of the projectile and target nuclei is represented by the quantities $\sigma_{jk}\left(E, E', \vec{\Omega}, \vec{\Omega}'\right)$, which are composed of three functions:

$$\sigma_{jk}\left(E, E', \vec{\Omega}, \vec{\Omega}'\right) = \sigma_k(E')\, \nu_{jk}(E')\, f_{jk}\left(E, E', \vec{\Omega}, \vec{\Omega}'\right) \tag{6.4}$$

where $\nu_{jk}(E')$ is the average number (which we loosely refer to as multiplicity) of type j particles being produced by a collision of a type k of energy E', and $f_{jk}\left(E, E', \vec{\Omega}, \vec{\Omega}'\right)$ is the probability density distribution for producing particles of type j of energy E into direction $\vec{\Omega}$ from the collision of a type k particle with energy E' moving in direction $\vec{\Omega}'$. For an unpolarized source of projectiles and unpolarized targets, the energy-angle distribution of reaction products is a function of the energies and cosine of the production angle relative to the incident projectile direction. The secondary multiplicities $\nu_{jk}(E')$ and secondary energy-angle distributions are the major unknowns in ion transport theory.

Information on the multiplicity $\nu_{jk}(E')$ was obtained in the past through experiments with galactic cosmic rays as an ion source, and the fragmentation of the ions on target nuclei was observed in nuclear emulsion (Cleghorn, Freier, and Waddington, 1968). Such data are mainly limited by not knowing the identity of the initial or secondary ions precisely and by relatively low counting rates of each ion type. The heavy ion acceleration by machine makes a reduction in the uncertainty possible because large count rates can be obtained with known ion types. In addition, the target nuclei in accelerator experiments can conveniently be other than nuclear emulsion, and accurate detector techniques with modern electronic processing are greatly improving the experimental data base. (See chapter 5.) In addition, the accelerator experiments are providing information on the spectral distribution $f_{jk}\left(E, E', \vec{\Omega}, \vec{\Omega}'\right)$ which has not been available before (Heckman et al., 1972).

The spectral distribution function is found to consist of two terms that describe the fragmentation of the projectile and the fragmentation of the struck nucleus as follows (Heckman, 1975; Raisbeck and Yiou, 1975):

$$\begin{aligned}
\sigma_{jk}\left(E, E', \vec{\Omega}, \vec{\Omega}'\right) = \sigma_k(E') \Big[& \nu_{jk}^P(E')\, f_{jk}^P\left(E, E', \vec{\Omega}, \vec{\Omega}'\right) \\
& + \nu_{jk}^T(E')\, f_{jk}^T\left(E, E', \vec{\Omega}, \vec{\Omega}'\right) \Big]
\end{aligned} \tag{6.5}$$

where ν_{jk}^P and f_{jk}^P depend only weakly on the target and ν_{jk}^T and f_{jk}^T depend only weakly on the projectile. Although the average secondary velocities associated with f^P are nearly equal to the projectile velocity, the average velocities associated with f^T are near zero. During experiments, Heckman (1975) observed that

$$f_{jk}^P\left(E, E', \vec{\Omega}, \vec{\Omega}'\right) \approx \left[\frac{m}{2\pi\left(\sigma_{jk}^P\right)^2}\right]^{3/2} \sqrt{2E} \exp\left[-\frac{(\vec{p}-\vec{p}')^2}{2\left(\sigma_{jk}^P\right)^2}\right]$$

$$\approx \left[\frac{m}{2\pi\left(\sigma_{jk}^P\right)^2}\right]^{3/2} \sqrt{2E} \exp\left[-\frac{\left(\vec{\Omega}\sqrt{2mE}-\vec{\Omega}'\sqrt{2mE'}\right)^2}{2\left(\sigma_{jk}^P\right)^2}\right] \quad (6.6)$$

where $\vec{p}$ and $\vec{p}'$ are the momenta per unit mass of j and k ions, respectively, and

$$f_{jk}^T\left(E, E', \vec{\Omega}, \vec{\Omega}'\right) \approx \left[\frac{m}{2\pi\left(\sigma_{jk}^T\right)^2}\right]^{3/2} \sqrt{2E} \exp\left[-\frac{\vec{p}^2}{2\left(\sigma_{jk}^T\right)^2}\right] \quad (6.7)$$

where σ_{jk}^P and σ_{jk}^T are related to the root-mean-square (rms) momentum spread of secondary products. These parameters depend only on the fragmenting nucleus. Feshbach and Huang (1973) suggested that the parameters σ_{jk}^P and σ_{jk}^T depend on the average square momentum of the nuclear fragments as allowed by Fermi motion. A precise formulation of these ideas in terms of a statistical model was obtained by Goldhaber (1974).

6.3. Approximation Procedures

6.3.1. Neglect of target fragmentation. Using equations (6.5), (6.6), and (6.7) in the evaluation of the source term $\zeta_j\left(\vec{x}, \vec{\Omega}, E\right)$ of equation (6.1) results in

$$\zeta_j\left(\vec{x}, \vec{\Omega}, E\right) = \sum_k \int dE'\, d\vec{\Omega}'\, \sigma_k(E')\, \phi_k\left(\vec{x}, \vec{\Omega}', \vec{E}'\right)$$

$$\times \left[\nu_{jk}^P(E')\, f_{jk}^P\left(E, E', \vec{\Omega}, \vec{\Omega}'\right) + \nu_{jk}^T\, f_{jk}^T\left(E, E', \vec{\Omega}, \vec{\Omega}'\right)\right]$$

$$\equiv \zeta_j^P\left(\vec{x}, \vec{\Omega}, E\right) + \zeta_j^T\left(\vec{x}, \vec{\Omega}, E\right) \quad (6.8)$$

where, as before, the superscripts P and T refer to fragmentation of the projectile and target, respectively. The target term is seen to be

$$\zeta_j^T\left(\vec{x}, \vec{\Omega}, E\right) = \sum_k \left[\frac{m}{2\pi\left(\sigma_{jk}^T\right)^2}\right]^{3/2} \sqrt{2E} \exp\left[\frac{-mE}{\left(\sigma_{jk}^T\right)^2}\right]$$

$$\times \int d\vec{\Omega}' \int_E^\infty dE'\, \nu_{jk}^T(E')\, \sigma_k(E')\phi_k\left(\vec{x}, \vec{\Omega}', E'\right) \quad (6.9)$$

which is negligibly small for

$$E \gg \frac{\left(\sigma_{jk}^T\right)^2}{m} \tag{6.10}$$

Thus, for calculating the flux at high energy,

$$\zeta_j\left(\vec{x}, \vec{\Omega}, E\right) \approx \zeta_j^P\left(\vec{x}, \vec{\Omega}, E\right) \tag{6.11}$$

6.3.2. Space radiations. A convenient property of space radiations is that they are nearly isotropic. This fact, coupled with the forward peaked spectral distribution, leads to substantial reductions in the source term as follows:

$$\zeta_j^P\left(\vec{x}, \vec{\Omega}, E\right) \approx \sum_k \int dE' \, d\vec{\Omega}' \sigma_k(E') \, \nu_{jk}^P(E') \left[\frac{m}{2\pi\left(\sigma_{jk}^P\right)^2}\right]^{3/2} \sqrt{2E'}$$

$$\times \exp\left[-\frac{\left(\vec{\Omega}\sqrt{2mE} - \vec{\Omega}'\sqrt{2mE'}\right)^2}{2\left(\sigma_{jk}^P\right)^2}\right] \phi_k\left(\vec{x}, \vec{\Omega}', E'\right) \tag{6.12}$$

Assuming that $\phi_k\left(\vec{x}, \vec{\Omega}', E'\right)$ is a slowly varying function of $\vec{\Omega}'$, one may seek an expansion about the sharply peaked maximum of the exponential function. Such an expansion is made by letting

$$\vec{\Omega}' = \vec{\Omega} + (\cos\theta - 1)\vec{\Omega} + \vec{e}_\phi \sin\theta \tag{6.13}$$

where

$$\cos\theta = \vec{\Omega} \cdot \vec{\Omega}' \tag{6.14}$$

and

$$\vec{e}_\phi = \frac{\vec{\Omega} \times \vec{\Omega}'}{\left|\vec{\Omega} \times \vec{\Omega}'\right|} \tag{6.15}$$

with which the flux may be expanded as

$$\phi_k\left(\vec{x}, \vec{\Omega}', E'\right) = \phi_k\left(\vec{x}, \vec{\Omega}, E'\right)$$
$$+ \left[\frac{\partial}{\partial\Omega} \phi_k\left(\vec{x}, \vec{\Omega}, E'\right)\right] \cdot \left[(\cos\theta - 1)\vec{\Omega} + \vec{e}_\phi \sin\theta\right] + \dots \quad (6.16)$$

Substituting equation (6.16) into equation (6.12) and simplifying results in

$$\zeta_j^P\left(\vec{x}, \vec{\Omega}, E\right) \approx \sum_k \int dE' \, \sigma_k(E') \nu_{jk}^P(E')$$

$$\times \left[\frac{m}{2\pi\left(\sigma_{jk}^P\right)^2}\right]^{3/2} \frac{\sqrt{2}}{\sqrt{E'}} \exp\left[-\frac{\left(\sqrt{2mE} - \sqrt{2mE'}\right)^2}{2\left(\sigma_{jk}^P\right)^2}\right]$$

$$\times \left\{\phi_k\left(\vec{x}, \vec{\Omega}, E'\right) - \left[\vec{\Omega} \cdot \frac{\partial}{\partial\vec{\Omega}} \phi_k\left(\vec{x}, \vec{\Omega}, E'\right)\right] \left[\frac{\left(\sigma_{jk}^P\right)^2}{2m\sqrt{EE'}}\right] + \dots\right\}$$
$$(6.17)$$

The leading term of equation (6.17) is clearly a good approximation to the source term whenever

$$\frac{2mE}{\left(\sigma_{jk}^P\right)^2} \gg \frac{\vec{\Omega} \cdot \frac{\partial}{\partial\vec{\Omega}} \phi_k\left(\vec{x}, \vec{\Omega}, E'\right)}{\phi_k\left(\vec{x}, \vec{\Omega}, E'\right)} \quad (6.18)$$

Note that the leading term is equivalent to assuming that secondary ions are produced only in the direction of motion of the primary ions. In the case of space radiations that are nearly isotropic, relation (6.18) is easily met, and neglect of higher order terms in equation (6.17) results in the usual straight ahead approximation. If the radiation is highly anisotropic, then relation (6.18) is not likely to apply. Validity of the straight ahead approximation was discovered empirically by Alsmiller et al. (1965) and Alsmiller, Irving, and Moran (1968) for the case of proton transport and is discussed further in the next chapter.

6.3.3. Velocity conserving interaction. Customarily, in cosmic ion transport studies (Curtis and Wilkinson, 1972), the fragment velocities are assumed to be equal to the fragmenting ion velocity before collision. Derived below is the order of approximation resulting from such an assumption. Assuming that the projectile energy E' is equal to the secondary energy plus a positive quantity ϵ,

$$E' = E + \epsilon \quad (6.19)$$

and that ϵ will contribute to equation (6.17) only over a small range above zero energy, substituting equation (6.19) into equation (6.17) and expanding the integrand results in

$$\zeta_j^P\left(\vec{x},\vec{\Omega},E\right) = \sum_k \sigma_k(E)\,\nu_{jk}^P(E)\left\{\phi_k\left(\vec{x},\vec{\Omega},E\right)\left[1 - \sqrt{\frac{\left(\sigma_{jk}^P\right)^2}{\pi m E}}\right]\right.$$

$$+ \left[E\frac{\partial}{\partial E}\,\phi_k\left(\vec{x},\vec{\Omega},E\right)\right]\sqrt{\frac{\left(\sigma_{jk}^P\right)^2}{\pi m E}}$$

$$\left.- \left[\vec{\Omega}\cdot\frac{\partial}{\partial\vec{\Omega}}\phi_k\left(\vec{x},\vec{\Omega},E\right)\right]\frac{\left(\sigma_{jk}^P\right)^2}{2mE} + \dots\right\} \tag{6.20}$$

Because $\sqrt{\frac{\left(\sigma_{jk}^P\right)^2}{\pi m E}} \ll 1$ at those energies at which most nuclear reactions occur, the assumption of velocity conservation is clearly inferior to a straight ahead approximation but may be adequate for space radiations where the variation of $\phi_k\left(\vec{x},\vec{\Omega},E\right)$ with energy is sufficiently smooth. That is,

$$E\frac{\partial}{\partial E}\,\phi_k\left(\vec{x},\vec{\Omega},E\right) \approx \phi_k\left(\vec{x},\vec{\Omega},E\right)$$

Although the validity of the velocity conserving approximation is usually accepted without question in transport applications, it is clearly an inferior approximation to the straight ahead approximation which is often held suspect.

6.3.4. Decoupling of target and projectile flux. Equation (6.1) with the use of equation (6.8) may be rewritten as

$$B_j\,\phi_j\left(\vec{x},\vec{\Omega},E\right) = \sum_k F_{jk}^T\,\phi_k\left(\vec{x},\vec{\Omega},E\right) + \sum_k F_{jk}^P\,\phi_k\left(\vec{x},\vec{\Omega},E\right) \tag{6.21}$$

where the differential operator is given by

$$B_j = \left[\vec{\Omega}\cdot\nabla - \frac{1}{A_j}\,\frac{\partial}{\partial E}\,S_j(E) + \sigma_j(E)\right] \tag{6.22}$$

and the integral operator $\left(F_{jk} = F_{jk}^T + F_{jk}^P\right)$ is given by

$$F_{jk}\,\phi_k\left(\vec{x},\vec{\Omega},E\right) = \int dE'\,d\vec{\Omega}'\,\upsilon_{jk}\left(E,E',\vec{\Omega},\vec{\Omega}'\right)\,\phi_k\left(\vec{x},\vec{\Omega}',E'\right) \tag{6.23}$$

Defining the flux as a sum of two terms

$$\phi_j\left(\vec{x},\vec{\Omega},E\right) = \phi_j^T\left(\vec{x},\vec{\Omega},E\right) + \phi_j^P\left(\vec{x},\vec{\Omega},E\right) \tag{6.24}$$

433

allows the following separation:

$$B_j \, \phi_j^P \left(\vec{x}, \vec{\Omega}, E\right) = \sum_k F_{jk}^P \, \phi_k^P \left(\vec{x}, \vec{\Omega}, E\right) + \sum_k F_{jk}^P \, \phi_k^T \left(\vec{x}, \vec{\Omega}, E\right) \qquad (6.25)$$

$$B_j \, \phi_j^T \left(\vec{x}, \vec{\Omega}, E\right) = \sum_k F_{jk}^T \, \phi_k^P \left(\vec{x}, \vec{\Omega}, E\right) + \sum_k F_{jk}^T \, \phi_k^T \left(\vec{x}, \vec{\Omega}, E\right) \qquad (6.26)$$

As noted in connection with equations (6.8) through (6.11), the source term on the right-hand side of equation (6.26) is small at high energies and one may assume that

$$\phi_j^T \left(\vec{x}, \vec{\Omega}, E\right) \approx 0 \qquad (6.27)$$

for $E \gg \left(\sigma_{jk}^T\right)^2 \big/ m$. As a result of equation (6.27) and the fact that the ion range is small compared with its mean-free path at low energy, one obtains

$$B_j \, \phi_j^P \left(\vec{x}, \vec{\Omega}, E\right) \approx \sum_k F_{jk}^P \, \phi_k^P \left(\vec{x}, \vec{\Omega}, E\right) \qquad (6.28)$$

$$B_j \, \phi_j^T \left(\vec{x}, \vec{\Omega}, E\right) \approx \sum_k F_{jk}^T \, \phi_k^P \left(\vec{x}, \vec{\Omega}, E\right) \qquad (6.29)$$

The advantage of this separation is that once equation (6.28) is solved by whatever means necessary, then equation (6.29) can be solved in closed form. The solution of equation (6.29) is accomplished by noting that the inwardly directed flux ϕ_j^T must vanish on the boundary, with the result that

$$\phi_j^T \left(\vec{x}, \vec{\Omega}, E\right) \approx \sum_k \int_E^{E_\gamma} dE' \, \frac{A_j \, P_j(E')}{P_j(E) S_j(E)} \int dE'' \, d\vec{\Omega}' \, \sigma_{jk}^T \left(E', E'', \vec{\Omega}, \vec{\Omega}'\right)$$
$$\times \, \phi_k^P \left\{\vec{x} + [R_j(E) - R_j(E')] \, \vec{\Omega}, \vec{\Omega}', E''\right\} \qquad (6.30)$$

where $E_\gamma = R_j^{-1}[d + R_j(E)]$, with d denoting the projected distance to the boundary.

Using equations (6.5) and (6.7) in equation (6.30) yields

$$\phi_j^T \left(\vec{x}, \vec{\Omega}, E\right) \approx \int_E^{E_\gamma} dE' \, \frac{A_j P_j(E')}{P_j(E) \, S_j(E)} \left[\frac{m}{2\pi \left(\sigma_{jk}^T\right)^2}\right]^{3/2} \sqrt{2E'} \exp\left[\frac{-mE'}{\left(\sigma_{jk}^T\right)^2}\right]$$
$$\times \, \zeta_j^T \left\{\vec{x} + [R_j(E) - R_j(E')] \, \vec{\Omega}\right\} \qquad (6.31)$$

where

$$\zeta_j^T \left(\vec{x}\right) = \sum_k \int dE' \, d\vec{\Omega}' \, \sigma_k(E') \, \nu_{jk}^T(E') \, \phi_k^P \left(\vec{x}, \vec{\Omega}', E'\right) \qquad (6.32)$$

and σ_{jk}^T has been assumed to be a slowly varying function of projectile type k and projectile energy E. If the range of secondary type j ions is small compared with their mean-free path lengths and the mean-free paths of the fragmenting parent ions ℓ_k, that is,

$$R_j \left[\frac{\left(\sigma_{jk}^T \right)^2}{m} \right] \ll \ell_k \tag{6.33}$$

then the integral of equation (6.31) may be simplified as

$$\phi_j^T \left(\vec{x}, \vec{\Omega}, E \right) \approx \frac{A_j}{S_j(E)} \zeta_j^T \left(\vec{x} \right) \int_E^{E_\gamma} \left[\frac{m}{2\pi \left(\sigma_{jk}^T \right)^2} \right]^{3/2} \sqrt{2E'} \exp \left[\frac{-mE'}{\left(\sigma_{jk}^T \right)^2} \right] dE' \tag{6.34}$$

which may be reduced into terms of known functions. Thus,

$$\phi_j^T \left(\vec{x}, \vec{\Omega}, E \right) \approx \frac{A_j}{S_j(E)} \zeta_j^T \left(\vec{x} \right) \frac{1}{2\pi \sqrt{\pi}} \left\{ \Gamma \left[\frac{3}{2}, \frac{mE}{\left(\sigma_{jk}^T \right)^2} \right] - \Gamma \left[\frac{3}{2}, \frac{mE_\gamma}{\left(\sigma_{jk}^T \right)^2} \right] \right\} \tag{6.35}$$

in terms of the incomplete gamma function. One can show that equation (6.35) is equivalent to

$$\phi_j^T \left(\vec{x}, \vec{\Omega}, E \right) \approx \frac{A_j}{S_j(E)} \zeta_j^T \left(\vec{x} \right) \frac{1}{2\pi} \left\{ \frac{1}{2} \operatorname{erfc} \left[\sqrt{\frac{mE}{\left(\sigma_{jk}^T \right)^2}} \right] - \frac{1}{2} \operatorname{erfc} \left[\sqrt{\frac{mE_\gamma}{\left(\sigma_{jk}^T \right)^2}} \right] \right.$$
$$\left. + \sqrt{\frac{mE}{\pi \left(\sigma_{jk}^T \right)^2}} \exp \left[\frac{-mE}{\left(\sigma_{jk}^T \right)^2} \right] - \sqrt{\frac{mE_\gamma}{\pi \left(\sigma_{jk}^T \right)^2}} \exp \left[\frac{-mE_\gamma}{\left(\sigma_{jk}^T \right)^2} \right] \right\} \tag{6.36}$$

At points sufficiently removed from the boundary such that

$$R_j^{-1}(d) \gg \frac{\left(\sigma_{jk}^T \right)^2}{m} \tag{6.37}$$

equation (6.36) may be reduced to

$$\phi_j^T \left(\vec{x}, \vec{\Omega}, E \right) \approx \frac{A_j}{S_j(E)} \zeta_j^T \left(\vec{x} \right) \frac{1}{2\pi} \left\{ \frac{1}{2} \operatorname{erfc} \left[\sqrt{\frac{mE}{\left(\sigma_{jk}^T \right)^2}} \right] + \sqrt{\frac{mE}{\pi \left(\sigma_{jk}^T \right)^2}} \exp \left[\frac{-mE}{\left(\sigma_{jk}^T \right)^2} \right] \right\} \tag{6.38}$$

The solution of equation (6.28) will now be further examined.

6.3.5. Back-substitution and perturbation theory. One approach to the solution of equation (6.28) results from the tendency of the multiple-charged ions

to be destroyed in nuclear reactions. Thus,

$$F_{jk}^P \equiv 0 \qquad\qquad (j \geq k) \qquad\qquad (6.39)$$

This means that there is a maximum j such that

$$B_J \, \phi_J^P \left(\vec{x}, \vec{\Omega}, E \right) = 0 \qquad\qquad (6.40)$$

where J is the largest j. Furthermore,

$$B_{J-1} \, \phi_{J-1}^P \left(\vec{x}, \vec{\Omega}, E \right) = F_{J-1,J}^P \, \phi_J^P \left(\vec{x}, \vec{\Omega}, E \right) \qquad\qquad (6.41)$$

and, in general,

$$B_{J-N} \, \phi_{J-N}^P \left(\vec{x}, \vec{\Omega}, E \right) = \sum_{k=1}^{N-1} F_{J-N,J-k}^P \, \phi_{J-k}^P \left(\vec{x}, \vec{\Omega}, E \right) \qquad\qquad (6.42)$$

for $N < J - 1$. Note that equations (6.41) and (6.42) constitute solvable problems. The singly charged ions satisfy

$$B_1 \phi_1^P \left(\vec{x}, \vec{\Omega}, E \right) = F_{1,1}^P \, \phi_1^P \left(\vec{x}, \vec{\Omega}, E \right) + \sum_{k=2}^{J} F_{1,k}^P \, \phi_k^P \left(\vec{x}, \vec{\Omega}, E \right) \qquad\qquad (6.43)$$

which, unlike equations (6.40) to (6.42), is an integral-differential equation that is difficult to solve directly. Equation (6.43) is solvable by perturbation theory, and the resultant series is known to converge rapidly for intermediate and low energies (Wilson and Lamkin, 1975; Wilson et al., 1989a and 1989b; Wilson and Townsend 1988). Note that equations (6.40) and (6.42) are also obtained from perturbation theory as applied to equation (6.28) at the outset. Thus, the perturbation series is expected to converge after the first J plus a few terms.

6.4. References

Allkofer, O. C.; and Heinrich, W., 1974: Attenuation of Cosmic Ray Heavy Nuclei Fluxes in the Upper Atmosphere by Fragmentation. *Nuclear Phys. B,* vol. B71, no. 3, pp. 429–438.

Alsmiller, R. G., Jr.; Irving, D. C.; Kinney, W. E.; and Moran, H. S., 1965: The Validity of the Straightahead Approximation in Space Vehicle Shielding Studies. *Second Symposium on Protection Against Radiations in Space,* Arthur Reetz, Jr., ed., NASA SP-71, pp. 177–181.

Alsmiller, R. G., 1967: High-Energy Nucleon Transport and Space Vehicle Shielding. *Nuclear Sci. & Eng.,* vol. 27, no. 2, pp. 158–189.

Alsmiller, R. G., Jr.; Irving, D. C.; and Moran, H. S., 1968: Validity of the Straightahead Approximation in Space-Vehicle Shielding Studies, Part II. *Nuclear Sci. & Eng.,* vol. 32, no. 1, pp. 56–61.

Alsmiller, R. G., Jr.; Barish, J.; and Scott, W. W., 1969: The Effects of Multiple Coulomb Scattering and Range Straggling in Shielding Against Solar-Flare Protons. *Nuclear Sci. & Eng.,* vol. 35, no. 3, pp. 405–406.

Bertini, H. W., 1962: Monte Carlo Calculations for Intranuclear Cascades. *Proceedings of the Symposium on the Protection Against Radiation Hazards in Space,* Book 2, TID-7652, United States Atomic Energy Commission, pp. 433–522.

Bethe, H. A.; Korff, S. A.; and Placzek, G., 1940.: On the Interpretation of Neutron Measurements in Cosmic Radiation. *Phys. Review,* vol. 57, second ser., no. 7, pp. 573–587.

Chatterjee, A.; Tobias, C. A.; and Lyman, J. T., 1976: Nuclear Fragmentation in Therapeutic and Diagnostic Studies With Heavy Ions. *Spallation Nuclear Reactions and Their Applications,* B. S. P. Shen and M. Merker, eds., D. Reidel Publ. Co., pp. 169–191.

Cleghorn, T. F.; Freier, P. S.; and Waddington, C. J., 1968: The Energy Dependence of the Fragmentation Parameters and Mean Free Paths of Cosmic-Ray Nuclei With $Z \geq 10$. *Canadian J. Phys.,* vol. 46, no. 10, pp. S572–S577.

Cocconi, G.; Cocconi-Tongiorgi, V.; and Greisen, K., 1948: Neutrons in the Penetrating Showers of the Cosmic Radiation. *Phys. Review,* vol. 74, second ser., no. 12, pp. 1867–1868.

Cocconi, G.; Cocconi Tongiorgi, V.; and Widgoff, M., 1950: Cascades of Nuclear Disintegrations Induced by the Cosmic Radiation. *Phys. Review,* vol. 79, second ser., no. 5, pp. 768–780.

Curtis, S. B.; Doherty, W. R.; and Wilkinson, M. C., 1969: *Study of Radiation Hazards to Man on Extended Near Earth Missions.* NASA CR-1469.

Curtis, S. B.; and Wilkinson, M. C., 1972: The Heavy Particle Hazard—What Physical Data Are Needed? *Proceedings of the National Symposium on Natural and Manmade Radiation in Space,* E. A. Warman, ed., NASA TM X-2440, pp. 1007–1015.

Davis, Leverett, Jr., 1960: On the Diffusion of Cosmic Rays in the Galaxy. *Proceedings of the Moscow Cosmic Ray Conference,* International Union of Pure and Applied Physics (Moscow), pp. 220–225.

Dye, D. L.; and Noyes, J. C., 1960: Biological Shielding for Radiation Belt Particles. *J. Astronaut. Sci.,* vol. VII, no. 3, pp. 64–70.

Dye, David L., 1962: Space Proton Doses at Points Within the Human Body. *Proceedings of the Symposium on the Protection Against Radiation Hazards in Space,* Book 2, TID-7652, United States Atomic Energy Commission, pp. 633–661.

Feshbach, H.; and Huang, K., 1973: Fragmentation of Relativistic Heavy Ions. *Phys. Lett.,* vol. 47B, no. 4, pp. 300–302.

Foelsche, T., 1959: Estimate of the Specific Ionization of Heavy Cosmic Ray Primaries in Tissue or Water. *J. Astronaut. Sci.,* vol. VI, no. 4, pp. 57–62.

Freier, Phyllis; Lofgren, E. J.; Ney, E. P.; and Oppenheimer, F., 1948: The Heavy Component of Primary Cosmic Rays. *Phys. Review,* vol. 74, second ser., no. 12, pp. 1818–1827.

Ganapol, Barry D.; Townsend, Lawrence W.; and Wilson, John W., 1989: *Benchmark Solutions for the Galactic Ion Transport Equations: Energy and Spatially Dependent Problems.* NASA TP-2878.

Ganapol, Barry D.; Townsend, Lawrence W.; Lamkin, Stanley L.; and Wilson, John W., 1991: *Benchmark Solutions for the Galactic Heavy-Ion Transport Equations With Energy and Spatial Coupling.* NASA TP-3112.

Ginzburg, V. L.; and Syrovatskii, S. I. (H. S. H. Massey, transl., and D. Ter Haar, ed.), 1964: *The Origin of Cosmic Rays.* Macmillan Co.

Gloeckler, G.; and Jokipii, J. R., 1969: Physical Basis of the Transport and Composition of Cosmic Rays in the Galaxy. *Phys. Review Lett.,* vol. 22, no. 26, pp. 1448–1453.

Goldberger, M. L., 1948: The Interaction of High Energy Neutrons and Heavy Nuclei. *Phys. Review,* vol. 74, no. 10, pp. 1269–1277.

Goldhaber, A. S., 1974: Statistical Models of Fragmentation Processes. *Phys. Lett.,* vol. 53B, no. 4, pp. 306–308.

Heckman, H. H.; Greiner, D. E.; Lindstrom, P. J.; and Bieser, F. S., 1972: Fragmentation of ^{14}N Nuclei at 29 GeV: Inclusive Isotope Spectra at $0°$. *Phys. Review Lett.,* vol. 28, no. 14, pp. 926–929.

Heckman, H. H., 1975: *Heavy Ion Fragmentation Experiments at the Bevatron.* NASA CR-142589.

Hess, Victor F.; and Eugster, Jakob, 1949: *Cosmic Radiation and Its Biological Effects,* Second ed., Revised and Augmented. Fordham Univ. Press.

Janni, Joseph F., 1982a: Proton Range-Energy Tables, 1 keV–10 GeV—Energy Loss, Range, Path Length, Time-of-Flight, Straggling, Multiple Scattering, and Nuclear Interaction Probability. Part 1. For 63 Compounds. *At. Data & Nuclear Data Tables,* vol. 27, nos. 2/3, pp. 147–339.

Janni, Joseph F., 1982b: Proton Range-Energy Tables, 1 keV–10 GeV—Energy Loss, Range, Path Length, Time-of-Flight, Straggling, Multiple Scattering, and Nuclear Interaction Probability. Part 2. For Elements $1 \leq Z \leq 92$. *At. Data & Nuclear Data Tables,* vol. 27, nos. 4/5, pp. 341–529.

Lambiotte, Jules J., Jr.; Wilson, John W.; and Filippas, Tassos A., 1971: Proper-3C: *A Nucleon-Pion Transport Code.* NASA TM X-2158.

Letaw, John; Tsao, C. H.; and Silberberg, R., 1983: Matrix Methods of Cosmic Ray Propagation. *Composition and Origin of Cosmic Rays,* Maurice M. Shapiro, ed., D. Reidel Publ. Co., pp. 337–342.

Lezniak, J. A., 1979: The Extension of the Concept of the Cosmic-Ray Path-Length Distribution to Nonrelativistic Energies. *Astrophys. & Space Sci.,* vol. 63, no. 2, pp. 279–293.

Metropolis, N.; Bivins, R.; Storm, M.; Turkevich, Anthony; Miller, J. M.; and Friedlander, G., 1958: Monte Carlo Calculations on Intranuclear Cascades. I. Low-Energy Studies. *Phys. Review,* vol. 110, second ser., no. 1, pp. 185–203.

Peters, B., 1958: The Nature of Primary Cosmic Radiation. *Progress in Cosmic Ray Physics,* J. G. Wilson, ed., Interscience Publ., Inc., pp. 191–242.

Raisbeck, G. M.; and Yiou, F., 1975: Production Cross Sections of Be Isotopes in C and O Targets Bombarded by 2.8 GeV α Particles: Implications for Factorization. *Phys. Review Lett.*, vol. 35, no. 3, pp. 155–159.

Schaefer, Hermann J., 1959: *Radiation and Man in Space.* Volume 1 of *Advances in Space Science,* Frederick I. Ordway III, ed., Academic Press, Inc., pp. 267–339.

Schimmerling, Walter; Rapkin, Marwin; Wong, Mervyn; and Howard, Jerry, 1986: The Propagation of Relativistic Heavy Ions in Multielement Beam Lines. *Med. Phys.*, vol. 13, no. 2, pp. 217–228.

Townsend, L. W.; and Wilson, J. W., 1988: Nuclear Cross Sections for Hadronic Transport. *Trans. American Nuclear Soc.*, vol. 56, pp. 277–279.

Wilson, John W.; and Lamkin, Stanley L., 1975: Perturbation Theory for Charged-Particle Transport in One Dimension. *Nuclear Sci. & Eng.*, vol. 57, no. 4, pp. 292–299.

Wilson, John W., 1977a: *Analysis of the Theory of High-Energy Ion Transport.* NASA TN D-8381.

Wilson, J. W., 1977b: Depth-Dose Relations for Heavy Ion Beams. *Virginia J. Sci.*, vol. 28, no. 3, pp. 136–138.

Wilson, John W., 1983: *Heavy Ion Transport in the Straight Ahead Approximation.* NASA TP-2178.

Wilson, John W.; Townsend, L. W.; Bidasaria, H. B.; Schimmerling, Walter; Wong, Mervyn; and Howard, Jerry, 1984: ^{20}Ne Depth-Dose Relations in Water. *Health Phys.*, vol. 46, no. 5, pp. 1101–1111.

Wilson, John W.; and Badavi, F. F., 1986: Methods of Galactic Heavy Ion Transport. *Radiat. Res.*, vol. 108, pp. 231–237.

Wilson, John W.; and Townsend, L. W., 1988: A Benchmark for Galactic Cosmic-Ray Transport Codes. *Radiat. Res.*, vol. 114, no. 2, pp. 201–206.

Wilson, John W.; Lamkin, Stanley L.; Farhat, Hamidullah; Ganapol, Barry D.; and Townsend, Lawrence W., 1989a: *A Hierarchy of Transport Approximations for High Energy Heavy (HZE) Ions.* NASA TM-4118.

Wilson, John W.; Townsend, Lawrence W.; Nealy, John E.; Chun, Sang Y.; Hong, B. S.; Buck, Warren W.; Lamkin, S. L.; Ganapol, Barry D.; Khan, Ferdous; and Cucinotta, Francis A., 1989b: BRYNTRN: *A Baryon Transport Model.* NASA TP-2887.

7. DOSE APPROXIMATION IN ARBITRARY CONVEX GEOMETRY

7.1. Introduction

A convenient property of energetic heavy-charged particles, when passing through matter, is that the primaries and their secondary particles remain relatively confined to the primary beam axis. As a consequence, the particle beam in matter is not strongly affected by near boundaries, and the problem of calculating dose in a complicated geometric object is greatly simplified (Wilson and Khandelwal, 1974; Schimmerling et al., 1986). Furthermore, the small beam width is a useful expansion parameter to develop a series that converges rapidly for most practical dose calculations. The final result relates dose at any point in an arbitrary convex region to an integral over solutions to the straight ahead approximation of the Boltzmann equation for normal incidence on a semi-infinite slab.

Energetic massive-charged particles constitute much of the radiation environment that man is subjected to in space. An attendant problem for radiation shield design is the evolution of the resultant dose in complex geometric objects, such as the human body. Although simple calculations neglecting nuclear reactions are often made (Cucinotta and Wilson, 1985; Santoro et al., 1986; Seltzer, 1980), several researchers have demonstrated that nuclear reaction effects are generally important (Shen, 1963; Alsmiller, 1967; Armstrong and Bishop, 1971). Incorporation of nuclear reaction effects requires solution of the Boltzmann transport equation subject to appropriate boundary conditions (Alsmiller, 1967).

The imposition of complicated boundary conditions often limits one's ability to obtain solutions to the transport problem even in the restricted sense of numerical solution. Even for such general techniques as the Monte Carlo method, the complexity introduced by complicated boundaries leads to prohibitively long calculations on present-day computers so that typical calculations approximate complicated objects such as the human body by slabs (Armstrong and Bishop, 1971), cylinders, or spheres (Santoro et al., 1986) of uniform soft tissue. In fact, the extensive calculations of dose in tissue slabs constitute most of our understanding of radiation dose in humans, although the relations between dose rates in a slab and dose rates in a particular body organ are at best poorly understood (Neufeld and Wright, 1972; Langley and Billings, 1972).

There is an approximate form of transport theory in which the transport equation is solvable for complex boundaries. The use of the "straight ahead" approximation reduces consideration to one-dimensional transport or a sort of ray tracing. The principal simplification of the straight ahead approximation is that lateral boundaries do not enter the solution along a given ray. The disadvantages of the method are that the effects of lateral dispersion on the buildup of secondaries are neglected and the errors of the approximation are generally not known (Alsmiller, 1967).

In this chapter, we derive an expansion for the solution to the transport equation in two dimensions subject to boundary conditions given for an arbitrary convex region. The expansion parameter is taken as the lateral dispersion, which is so small that the lowest order expansion term is the dominant contribution. When the expansion is applied to the straight ahead theory, only the first term is nonzero because lateral dispersion is zero in this case. The advantages of using the first term of the expansion of the transport solution instead of the straight ahead theory are that the effects of lateral dispersion on the buildup of secondaries are taken into account and the errors associated with the approximation are known. On the basis of general principles, the present expansion always provides an overestimate of dose (conservative estimate), and in almost all circumstances this overestimate is small. The extension to three dimensions is conceptually the same but with more complicated algebra.

7.2. High-Energy Transport

The fields of type j particles are represented by the functions $\phi_j(\vec{x}, E, \vec{\Omega})$, which denote the particle fluence of energy E that crosses a plane normal to the direction of motion $\vec{\Omega}$ per unit solid angle as seen at a position $\vec{x}$. The fields satisfy the steady linear Boltzmann transport equation

$$\left[\vec{\Omega} \cdot \vec{\nabla} + \sigma_j(E) - \frac{\partial}{\partial E} S_j(E)\right] \phi_j(\vec{x}, E, \vec{\Omega})$$
$$= \zeta_j(\vec{x}, E, \vec{\Omega}) + \sum_k \int d\Omega' \int dE' \sigma_{jk}(E, E', \vec{\Omega} \cdot \vec{\Omega}') \phi_k(\vec{x}, E', \vec{\Omega}') \quad (7.1)$$

where $S_j(E)$ is the stopping power (energy loss per unit path length), $\sigma_j(E)$ is the total macroscopic cross section, $\zeta_j(\vec{x}, E, \vec{\Omega})$ is the source of type j particles, and the cross section for type j secondary particles of energy E produced by collisions of type k particles of energy E' with the medium is represented by $\sigma_{jk}(E, E', \vec{\Omega} \cdot \vec{\Omega}')$. The Boltzmann equation admits solution in a closed region for which the inward-directed flux is specified on the boundary $\vec{\Gamma}$. The boundary condition is that

$$\phi_j(\vec{\Gamma}, E, \vec{\Omega}) = \psi_j(\vec{\Gamma}, E, \vec{\Omega}) \quad (7.2a)$$

for all $\vec{\Omega}$ such that $\vec{n} \cdot \vec{\Omega} < 0$, where $\vec{n}$ is the outward directed unit normal and $\psi_j(\vec{\Gamma}, E, \vec{\Omega})$ is a specified boundary function determined from exterior sources. Homogeneous boundary conditions can be used by including an equivalent boundary source on the right side of equation (7.1) given by

$$\xi_j(\vec{\Gamma}, E, \vec{\Omega}) = -\vec{n} \cdot \vec{\Omega} \psi_j(\vec{\Gamma}, E, \vec{\Omega}) \quad (7.2b)$$

A useful property of equation (7.1) is the linearity that allows superposition of solutions. Thus, the solution for a sum of sources is the sum of solutions for each source. As will be seen, there is a great advantage in considering the solution for an arbitrary source as a superposition of solutions for unidirectional monoenergetic point sources.

Presently, the energy absorbed per unit mass or dose is of interest. The dose at a point $\vec{x}$ is taken as

$$D(\vec{x}) = \frac{1}{d} \sum_j \int d\vec{\Omega}' \int dE' \; \phi_j(\vec{x}, E', \vec{\Omega}') \; S_j(E') \qquad (7.3)$$

where d is the mass density. Since the fields $\phi_j(\vec{x}, E', \vec{\Omega}')$ are linear functions of the source, we may rewrite equation (7.3) as

$$D(\vec{x}) = \sum_k \int_\Gamma dl \int d\vec{\Omega} \int dE \; R_k(\vec{x}, E, \vec{\Omega}, \vec{\Gamma}) \; \xi_k[\vec{\Gamma}(l), E, \vec{\Omega}] \qquad (7.4)$$

when all sources lie on the boundary $\vec{\Gamma}$ and dl is the incremental boundary surface. From the linearity of equation (7.1), the relation between the fields and the source is

$$\phi_j(\vec{x}, E', \vec{\Omega}') = \sum_k \int dE \int d\vec{\Omega} \int dl \; G_{jk}(\vec{x}, \vec{\Gamma}, E', E, \vec{\Omega}', \vec{\Omega}) \; \zeta_k(\vec{\Gamma}, E, \vec{\Omega}) \qquad (7.5)$$

where G_{jk} is Green's function. The dose response becomes

$$R_k(\vec{x}, E, \vec{\Omega}, \vec{\Gamma}) = \frac{1}{d} \int dE' \int d\vec{\Omega}' \sum_j G_{jk}(\vec{x}, \vec{\Gamma}, E', E, \vec{\Omega}', \vec{\Omega}) \; S_j(E') \qquad (7.6)$$

The function $R_k(\vec{x}, E, \vec{\Omega}, \vec{\Gamma})$ is the dose at a point $\vec{x}$ caused by a point source of type k particles of energy E located at the point $\vec{\Gamma}$ on the boundary and directed toward $\vec{\Omega}$. Note that R_k in equation (7.6) is a solution for homogeneous boundary conditions of equation (7.1) when taken with equation (7.4); R_k is a function of the geometric shape of the bounding surface.

We see that the boundaries enter the solution given by equation (7.4) in two essential ways:

1. The boundary enters the dose calculation by equation (7.4) as an integral over the equivalent boundary source.

2. The homogeneous boundary condition enters through the function R_k.

The properties of the function R_k are discussed in the next section, where it will be argued that although R_k depends on the boundaries, this dependence is weak and, therefore, negligible. Supportive evidence of this claim will be presented, and some numerical results will be discussed. Based on these properties, an approximate form of R_k will be derived that results in simplification of equation (7.4) by relating it to flux-to-dose-rate conversion factors for radiation incident normally on a slab.

7.3. Properties of the Dose Response Function

In this section we first discuss the dose response when all boundaries are far removed. A specific general form for the response is determined. The effects of near boundaries are then discussed, especially in the context of calculating $D(\vec{x})$ given by equation (7.4). The discussion is limited primarily to response to proton sources, but some aspects of high-energy neutrons will also be discussed. The methods are even more appropriate for ions heavier than protons.

Protons of energy less than a few hundred MeV interact with dense matter predominantly through energy loss in collisions with electrons in the surrounding material. The probability of nuclear reaction before stopping is rather small, and relatively few secondaries are produced when reactions do occur. The paths of the primary particles are confined to a small cylinder about the initial direction, with the deviations in the paths resulting predominantly from multiple coulomb scattering through very small angles.

At several hundred MeV, nuclear reactions are important, and most primary particles will suffer nuclear reaction before stopping. The most energetic secondaries are confined in a narrow cone about the initial direction (this cone narrows with increasing primary energy as shown by Shen (1965)) and are closely confined to the initial beam axis over at least the first mean-free path (≈ 1 m in water). Low-energy secondary-charged particles are stopped near their point of production, and only the low-energy neutrons are able to migrate far from the beam axis. At large distances from the beam, only a net outward flux of low-energy neutrons is observed; this flux decreases exponentially because of absorptive processes in the medium. These processes are shown schematically in figure 7.1.

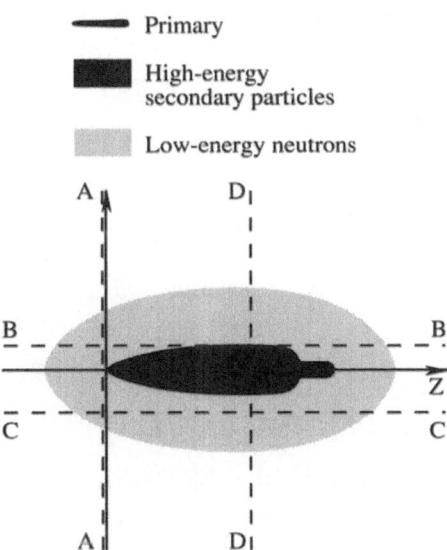

Figure 7.1. Schematic representation of the response to a point source of monoenergetic protons.

At very high energies, nuclear reactions tend to dominate, and the distinction between incident neutrons and protons all but disappears. A principal mode of transferring energy to the medium is through the multitude of secondary particles produced in nuclear reaction.

The dose response will now be parameterized in such a way to exhibit the above properties. We represent the dose response of a unidirectional point source of protons of energy E at the origin directed along the z-axis of an infinite medium as

$$R(z, E, r) = R_N(z, E) \, g[\delta(E, z), r] \qquad (7.7)$$

where r is the radial coordinate normal to the z-axis. This is shown schematically in figure 7.1. The parameter $\delta(E, z)$ is the root-mean-square value of the beam width. The amount of shading in the figure illustrates the magnitude of the dose. The functions on the right side of equation (7.7) are chosen so that $g(\delta, r)$ gives the lateral profile of the beam with

$$\int_{-\infty}^{\infty} r^{2n} \, g(\delta, r) \, dr = C_{2n} \delta^{2n} \qquad (7.8)$$

and $C_o = 1$ so that $R_N(z, E)$ is the fluence-to-dose conversion factor for a uniform source incident normal to the $z = 0$ plane. We now consider the changes in the dose response if a boundary is moved to some position near the beam.

First, consider placing a void to the left of the $z = 0$ plane, which is boundary A in figure 7.1. This means that radiation will no longer backscatter from the material to the left of boundary A and the doses with and without the boundary are related by

$$R(E, z, r) = R^{A}(E, z, r) + \epsilon^{A}(z, r) \qquad (7.9)$$

The fact that the dose with the boundary far removed is an overestimate of the dose with the boundary A is denoted by

$$\epsilon^{A}(0, r) \geq \epsilon^{A}(z, r) \geq 0 \qquad (7.10)$$

If we use the approximation of R^{A} as

$$R^{A}(E, z, r) \approx R(E, z, r) \qquad (7.11)$$

then the order of magnitude of the overestimate of dose is given by

$$
\begin{aligned}
\int_{-\infty}^{\infty} & [R(E, z, r) - R^{A}(E, z, r)] \, dr \\
&= [R_N(E, z) - R_N^{A}(E, z)] \\
&\leq [R_N(E, 0) - R_N^{A}(E, 0)] \\
&\approx 2 \times 10^{-3} \, R_N(E, 0) \qquad (7.12)
\end{aligned}
$$

as estimated (actually overestimated) from the numerical results of Irving et al. (1965) and Alsmiller et al. (1965). The placement of a void to the right of boundary D parallel to A located downstream from the source would be expected to produce similar effects.

A more difficult question concerns the effects of lateral boundaries, shown as C or B in figure 7.1. Clearly, as long as the lateral boundaries are removed from the heavily shaded areas, the effects are negligible. Thus, if the distance from the beam axis to the boundary is greater than the beam width δ, then the lateral boundaries will not greatly affect the interior solution.

Beam profiles for 400-MeV protons and neutrons have been calculated by Wright, Hamm, and Turner (1971) and are shown in figures 7.2 and 7.3; their results show that $\delta \leq 1$ cm in tissue. The purpose of the next section is to derive an approximate form for equation (7.4), which makes use of the beam width being small, that is applicable to the transport of protons, high-energy neutrons, and heavy ions.

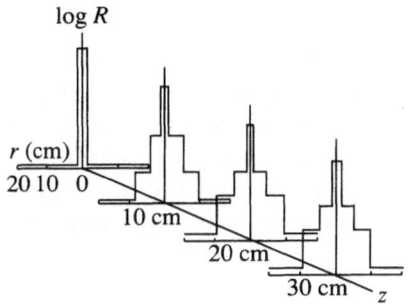

Figure 7.2. Beam profiles of a point source of 400-MeV protons calculated by Wright, Hamm, and Turner (1971).

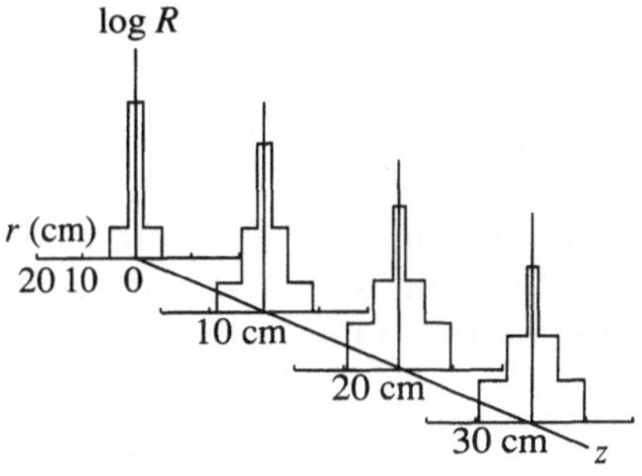

Figure 7.3. Beam profiles of a point source of 400-MeV neutrons calculated by Wright, Hamm, and Turner (1971).

7.4. Expansion for Two-Dimensional Transport

We consider the transport solution in a closed, convex, two-dimensional region. The boundary is defined by a vector function $\vec{\Gamma}(l)$, where l is the path length along the boundary curve. The unit tangent to the curve is

$$\vec{t}(l) = \frac{d\vec{\Gamma}(l)}{dl} \tag{7.13}$$

and points in the direction of increasing l. The outward-directed unit normal is

$$\vec{n}(l) = -\rho(l)\frac{d\vec{t}(l)}{dl} \tag{7.14}$$

where $\rho(l)$ is the radius of curvature. The dose Δ at an interior point $\vec{x}$ caused by a boundary source $\zeta(\vec{\Gamma}, E, \vec{\Omega})$ is given by

$$\Delta(\vec{x}, E, \vec{\Omega}) = \int_{\Gamma} R^{\Gamma}[E, z(\vec{\Gamma}, \vec{x}, \vec{\Omega}), r(\vec{\Gamma}, \vec{x}, \vec{\Omega})]\zeta(\vec{\Gamma}, E, \vec{\Omega})\, dl \tag{7.15}$$

where the arguments of R^{Γ} are

$$z(\vec{\Gamma}, \vec{x}, \vec{\Omega}) = \vec{\Omega} \cdot (\vec{\Gamma} - \vec{x}) \tag{7.16}$$

$$r(\vec{\Gamma}, \vec{x}, \vec{\Omega}) = \vec{\Omega}_{\perp} \cdot (\vec{\Gamma} - \vec{x}) \tag{7.17}$$

445

where $\vec{\Omega}_\perp$ is a unit vector perpendicular to $\vec{\Omega}$. We now seek means whereby the above integral can be approximated.

We recall that the effects of boundaries on $R^\Gamma(E, z, r)$ are expected to be small so that the replacement

$$R^\Gamma(E, z, r) \to R(E, z, r) \tag{7.18}$$

provides an overestimate of the dose and the error is negligible so that

$$\Delta(E, x, \Omega) \approx \int_\Gamma R[E, z(\vec{\Gamma}, \vec{x}, \vec{\Omega}), r(\vec{\Gamma}, \vec{x}, \vec{\Omega})]\zeta(\vec{\Gamma}, E, \vec{\Omega})\, dl \tag{7.19}$$

Note that equation (7.19) is a considerable simplification over equation (7.15). The replacement of R^Γ by R means that the transport equation needs to be solved only once to determine the function R, which is then applicable to all problems through equation (7.19). Now we will make use of the property that $R(E, z, r)$ has a maximum at $r = 0$ and drops precipitously away from the maximum. To do this, we first make a translation along the bounding curve to the point where

$$r(\vec{\Gamma}, \vec{x}, \vec{\Omega}) = \vec{\Omega}_\perp \cdot (\vec{x} - \vec{\Gamma}) = 0 \tag{7.20}$$

label the point $l_x(\vec{\Omega})$, and define a new variable

$$s = l - l_x(\vec{\Omega}) \tag{7.21}$$

and the vector function

$$\vec{\gamma}(s) = \vec{\Gamma}(l) - \vec{\Gamma}[l_x(\vec{\Omega})] \tag{7.22}$$

We can then rewrite

$$z(\vec{\Gamma}, \vec{x}, \vec{\Omega}) = z_x(\vec{\Omega}) - \vec{\Omega} \cdot \vec{\gamma}(s) \tag{7.23}$$
$$r(\vec{\Gamma}, \vec{x}, \vec{\Omega}) = -\vec{\Omega}_\perp \cdot \vec{\gamma}(s) \tag{7.24}$$

and the integral as

$$\Delta(\vec{x}, E, \vec{\Omega}) = \int_{-s_1(\vec{\Omega})}^{s_2(\vec{\Omega})} R[E, z_x(\vec{\Omega}) - \vec{\Omega} \cdot \vec{\gamma}(s), -\vec{\Omega}_\perp \cdot \vec{\gamma}(s)]$$
$$\times \zeta[\vec{\Gamma}_x + \vec{\gamma}(s), E, \vec{\Omega}]\, ds \tag{7.25}$$

where the limits of integration are given by solutions of

$$\vec{t}(-s_1) \cdot \vec{\Omega} = -1 \tag{7.26}$$
$$\vec{t}(s_2) \cdot \vec{\Omega} = 1 \tag{7.27}$$

where $\vec{t}(s)$ is the unit tangent at s and is directed toward increasing s. We can further simplify equation (7.25) by transforming the integral over path length s to an integral over the lateral dimension r.

Consider the expansion

$$r = -\vec{\Omega}_\perp \cdot \vec{t}(0)s + \frac{1}{2\rho_o}\vec{\Omega}_\perp \cdot \vec{n}(0)s^2 + \dots \tag{7.28}$$

which, when inverted, reads ($\vec{n}_o \cdot \vec{\Omega} = \vec{\Omega}_\perp \cdot \vec{t}_o$ and where subscript o denotes evaluation at $r = s = 0$):

$$s = \frac{-r}{\vec{n}_o \cdot \vec{\Omega}} + \frac{1}{2\rho_o}\frac{\vec{\Omega}\cdot\vec{t}_o}{\vec{n}_o \cdot \vec{\Omega}}\frac{r^2}{(\vec{n}_o \cdot \vec{\Omega})^2} + \dots \tag{7.29}$$

so that

$$z(\vec{\Gamma},\vec{x},\vec{\Omega}) = z_x(\vec{\Omega}) + \Big(\frac{\vec{\Omega}\cdot\vec{t}_o}{\vec{\Omega}\cdot\vec{n}_o}\Big)r - \frac{1}{2\rho_o}$$
$$\times \frac{(\vec{\Omega}\cdot\vec{t}_o)^2 - (\vec{\Omega}\cdot\vec{n}_o)^2}{(\vec{\Omega}\cdot\vec{n}_o)^4}r^2 + \dots \tag{7.30}$$

and

$$\vec{\gamma}(s) = -\vec{t}_o\Big(\frac{r}{\vec{n}_o \cdot \vec{\Omega}}\Big) + \frac{1}{2\rho_o}\Big(\vec{t}_o\frac{\vec{\Omega}\cdot\vec{t}_o}{\vec{\Omega}\cdot\vec{n}_o} - \vec{n}_o\Big)\Big(\frac{r}{\vec{n}_o \cdot \vec{\Omega}}\Big)^2 + \dots \tag{7.31}$$

With a change of variables,

$$\Delta(\vec{x},E,\vec{\Omega}) = \int_{-r_1(\vec{\Omega})}^{r_2(\vec{\Omega})} R\Big[E, z_x + \Big(\frac{\vec{\Omega}\cdot\vec{t}_o}{\vec{\Omega}\cdot\vec{n}_o}\Big)r - \frac{1}{2\rho_o}\frac{(\vec{\Omega}\cdot\vec{t}_o)^2 - (\vec{\Omega}\cdot\vec{n}_o)^2}{(\vec{\Omega}\cdot\vec{n}_o)^2}r^2, r\Big]$$
$$\times \phi\Big\{\vec{\Gamma}_x - \vec{t}_o\Big(\frac{r}{\vec{n}_o \cdot \vec{\Omega}}\Big) + \frac{1}{2\rho_o}\Big[\vec{t}_o\frac{\vec{\Omega}\cdot\vec{t}_o}{\vec{n}_o \cdot \vec{\Omega}} - \vec{n}_o\Big]\Big(\frac{r}{\vec{n}_o \cdot \vec{\Omega}}\Big)^2$$
$$+ \dots, \vec{\Omega}, E\Big\}\Big[1 - \frac{1}{\rho_o}\frac{\vec{\Omega}\cdot\vec{t}_o}{(\vec{\Omega}\cdot\vec{n}_o)^2}r + \dots\Big]\,dr \tag{7.32}$$

where $r_1(\vec{\Omega})$ and $r_2(\vec{\Omega})$ are the distances from $\vec{x}$ to the boundary along the direction perpendicular to $\vec{\Omega}$. Expanding the integrand of equation (7.32) and using equations (7.7) and (7.8) give

$$\Delta(x,E,\Omega) = \Big[\int_{-\infty}^{\infty} - \int_{r_2(\vec{\Omega})}^{\infty} + \int_{-r_1(\vec{\Omega})}^{-\infty}\Big] R_N[z_x(\vec{\Omega}),E]\phi(\vec{\Gamma}_x,\vec{\Omega},E)\,g(\delta,r)$$

$$\times \Big[1 + \frac{R_N'}{R_N}\Big(\frac{(\vec{\Omega}\cdot\vec{t}_o)}{(\vec{\Omega}\cdot\vec{n}_o)}\Big)r + \frac{\vec{t}_o\cdot\vec{\nabla}\phi}{\vec{n}_o \cdot \vec{\Omega}}r + \frac{1}{\rho_o}\frac{\vec{t}_o\cdot\vec{\Omega}}{(\vec{\Omega}\cdot\vec{n}_o)^2}r + O(r^2)\Big]\,dr$$

$$\approx R_N[z_x(\vec{\Omega}),E]\,\phi(\vec{\Gamma}_x,\vec{\Omega},E)[1 + O(\delta^2)]$$

$$- \Big(\int_{r_2}^{\infty} + \int_{-\infty}^{-r_1}\Big) R_N[z_x(\vec{\Omega}),E]\,\phi(\vec{\Gamma}_x,\vec{\Omega},E)\,g(\delta,r)\,dr$$

$$\approx R_N[z_x(\vec{\Omega}),E]\phi(\vec{\Gamma}_x,\vec{\Omega},E)\{1 + O(\delta^2) - \delta[g(\delta,r_1) + g(\delta,r_2)]\} \tag{7.33}$$

447

where the last term in equation (7.33) is inferior to terms $O(\delta^2)$, i.e.,

$$O(\delta^2) >> \delta g(\delta, r_i)$$

for $r_i >> \delta$. The solution for all sources of different E and $\vec{\Omega}$ is then found from equation (7.33) as

$$D(\vec{x}) = \int_0^\infty \int_\Omega R_N[z_x(\vec{\Omega}), E] \ \phi(\vec{\Gamma}_x, \vec{\Omega}, E)[1 + O(\delta^2)] \ d\vec{\Omega} \ dE \qquad (7.34)$$

which can be evaluated by using a simple approximate quadrature. The error terms in equation (7.34) contain the quantities

$$\frac{\delta}{\rho_o}, \ \frac{R_N' \delta}{R_N}, \ \frac{\vec{\nabla}\phi \cdot \vec{t}_o \delta}{\phi}, \ \frac{R_N'' \delta^2}{R_N}, \ \frac{\vec{\nabla}^2 \phi \delta^2}{\phi}$$

in the form of second-order products in δ.

Generally, δ is quite small. This is the basis for the straight ahead approximation; and in the limit as $\delta \rightarrow 0$, the first term in equation (7.34) is the sole surviving term when applied to the straight ahead theory. In the real situation, δ is nonzero and the first term in equation (7.34) will still give satisfactory results even when straight ahead theory is inadequate.

The greatest importance of equation (7.34) is that the fluence-to-dose conversion factors for normal incidence on a semi-infinite slab are all the information required to compute doses in any arbitrary convex object. These fluence-to-dose conversion factors are well-known for neutrons and protons of several energies for depths up to 30 cm in soft tissue. An interpolation to other energies and an extrapolation to large depths are given by Wilson and Khandelwal (1974 and 1976) and have been extended for tissue systems shielded by aluminum. (See chapter 8.)

Perhaps the main limitation in using equation (7.34) is the assumption that the lateral distance from the dose point to the boundary is large compared with the beam width; i.e., $r_i \leq \delta$ so that $\delta g(\delta, r_i) >> O(\delta^2)$. This could always be corrected by a knowledge of the lateral profile of the beam where the expression

$$D(\vec{x}) = \int_0^\infty \int_\Omega \int_{-r_1(\vec{\Omega})}^{r_2(\vec{\Omega})} R_N[z_x(\vec{\Omega}), E] \ \phi(\vec{\Gamma}_x, \vec{\Omega}, E) \ g(\delta, r) \ dr \ d\vec{\Omega} \ dE + O(\delta^2) \quad (7.35)$$

is used in place of equation (7.34). The main limitation is that $g(\delta, r)$ is not sufficiently well-known in the literature. There is some question as to the importance of such terms, as has been shown by Wilson and Khandelwal (1974) with a numerical example. Note also that the result of equation (7.34) is always an overestimate to equation (7.35) and thus is always conservative. The conservativeness of equation (7.34) follows because the integrand of equation (7.35) is positive definite. Finally, we note that equation (7.34) or (7.35) is the desired approximate form for equation (7.4). The accuracy is associated with the beam width δ, and errors are sufficiently small in many applications to the transport of protons, heavy ions, and high-energy neutrons. As a check on the formalism expressed by equation (7.34), we have calculated the fluence-to-dose conversion factors for isotropic protons on a 30-cm tissue slab and compared the results with "exact" Monte Carlo results. The comparison can be made from table 7.1. Equation (7.34) is seen to be accurate for dose points near the boundary ($x = 0$) as well as at great depths. The condition under which equation (7.35) provides a more accurate result is not yet clear. The calculation of the function $R_N(x, E)$ for protons and heavy ions and their validation is the main topic of the rest of this report.

Table 7.1. Comparison of Isotropic Incident Proton Conversion Factors[a]
for a 30-cm Tissue Slab

[Present approximations compared with "exact" Monte Carlo results]

x, cm	Exact	Approximate	Error, percent
E = 100 MeV			
1	2.78	2.65	−4.7
3	2.92	2.45	−16
5	2.15	1.92	11
7	0.56	.90	60
8	≈0.01	0	
E = 200 MeV			
1	1.66	1.70	2.4
5	1.72	1.62	−6.0
10	1.48	1.41	−4.6
15	1.14	1.16	1.7
20	.81	.79	−2.5
E = 300 MeV			
1	1.29	1.29	0
5	1.34	1.35	.8
10	1.28	1.26	−1.6
15	1.18	1.14	−3.4
20	1.03	1.01	−1.9
25	.89	.88	−1.1
E = 400 MeV			
1	1.16	1.17	0.9
5	1.20	1.27	5.9
10	1.14	1.22	7.0
15	1.17	1.13	−4.1
20	1.10	1.04	−5.5
25	1.01	.94	−6.9

[a]Units are chosen as 10^{-10} Gy/(proton cm^{-2}).

7.5. References

Alsmiller, R. G., Jr.; Irving, D. C.; Kinney, W. E.; and Moran, H. S., 1965: The Validity of the Straightahead Approximation in Space Vehicle Shielding Studies, *Second Symposium on Protection Against Radiations in Space*, Arthur Reetz, Jr., ed., NASA SP-71, pp. 177–181.

Alsmiller, R. G., Jr., 1967: High-Energy Nucleon Transport and Space Vehicle Shielding. *Nuclear Sci. & Eng.*, vol. 27, no. 2, pp. 158–189.

Armstrong, T. W.; and Bishop, B. L., 1971: Calculation of the Absorbed Dose and Dose Equivalent Induced by Medium-Energy Neutrons and Protons and Comparison With Experiment. *Radiat. Res.*, vol. 47, no. 3, pp. 581–588.

Cucinotta, Francis A.; and Wilson, John W., 1985: *Computer Subroutines For Estimation of Human Exposure to Radiation in Low Earth Orbit*. NASA TM-86324.

Irving, D. C.; Alsmiller, R. G., Jr.; Kinney, W. E.; and Moran, H. S., 1965: The Secondary-Particle Contribution to the Dose From Monoenergetic Proton Beams and the Validity of Current-to-Dose Conversion Factors. *Second Symposium on Protection Against Radiations in Space*, Arthur Reetz, Jr., ed., NASA SP-71, pp. 173–176.

Langley, R. W.; and Billings, M. P., 1972: A New Model For Estimating Space Proton Dose to Body Organs. *Nuclear Technol.*, vol. 15, no. 1, pp. 68–74.

Neufeld, Jacob; and Wright, Harvel, 1972: Radiation Levels and Fluence Conversion Factors. *Health Phys.*, vol. 23, no. 2, pp. 183–186.

Santoro, R. T.; Alsmiller, R. G., Jr.; Barnes, J. M.; and Corbin, J. M., 1986: *Shielding of Manned Space Stations Against Van Allen Belt Protons: A Preliminary Scoping Study*. ORNL/TM-10040 (Contract No. IDOD3070-0000), Oak Ridge National Lab.

Schimmerling, Walter; Rapkin Marwin; Wong, Mervyn; and Howard, Jerry, 1986: The Propagation of Relativistic Heavy Ions in Multielement Beam Lines. *Med. Phys.*, vol. 13, no. 2, pp. 217–228.

Seltzer, Stephen, 1980: SHIELDOSE: *A Computer Code for Space-Shielding Radiation Dose Calculations*. NBS Tech. Note 1116, U.S. Dep. of Commerce.

Shen, S. P., 1963: Nuclear Problems in Radiation Shielding in Space. *Astronaut. Acta,* vol. IX, Fasc. 4, pp. 211–274.

Shen, B. S. P., 1965: Some Experimental Data on the Nuclear Cascade in Thick Absorbers. *Second Symposium on Protection Against Radiations in Space*, Arthur Reetz, Jr., ed., NASA SP-71, pp. 357–362.

Wilson, John W.; and Khandelwal, G. S., 1974: Proton Dose Approximation in Arbitrary Convex Geometry. *Nuclear Technol.,* vol. 23, no. 3, pp. 298–305.

Wilson, John W.; and Khandelwal, Govind S., 1976: Proton-Tissue Dose Buildup Factors. *Health Phys.*, vol. 31, no. 2, pp. 115–118.

Wright, H. A.; Hamm, R. N.; and Turner, J. E., 1971: Effect of Lateral Scattering on Absorbed Dose From 400 MeV Neutrons and Protons. *International Congress on Protection Against Accelerator and Space Radiation*, J. Baarli and J. Dutrannois, eds., CERN 71-16, Volume 1, European Organization for Nuclear Research, pp. 207–219.

8. NUCLEON TRANSPORT METHODS

8.1. Introduction

Understanding the interaction of energetic charged particles with bulk matter is important for determining shield quality (Shen, 1963; Alsmiller, 1967), dosimeter design (Khandelwal and Wilson, 1973), and radiology (Armstrong, 1972; Armstrong and Bishop, 1971), as well as for astrophysics (Murzin and Sarycheva, 1970; Shen 1967), and solar system studies (Reedy and Arnold, 1972). Detailed studies of charged-particle transport with regeneration have been hampered by terms arising from energy loss caused by atomic ionization and excitation (Haffner, 1967). Generally, charged-particle transport calculations are made by using Monte Carlo methods (Alsmiller, 1967), where errors of the procedure are difficult to evaluate and a degree of insight is lost in the handling of large amounts of numerical data. Clearly, an approximation procedure that allows an error test that is both easily mechanized and related directly to the particle fields is desirable.

Analytical methods have been developed to represent solutions to the Boltzmann equation for charged-particle transport. These results are obtained by using an approximate Boltzmann equation for which analytical solution is possible. For example, approximations ignoring energy loss through ionization have been used in studying the high-energy nucleonic cascade development in the Earth's atmosphere (Cocconi, Cocconi Tongiorgi, and Widgoff, 1950). Energy loss caused by ionization is generally important, and only the high-energy secondaries are adequately treated. The so-called Passow approximations have found utility in recent studies (O'Brien, 1971). The restrictions on the Passow approximations are that all secondaries are produced in the forward direction (straight ahead approximation), energy loss caused by ionization is neglected, and the secondary-particle-production spectra must be proportional to E_s^α, where α is constant.

The solution of the Boltzmann equation including ionization energy loss has been inadequately treated. Alsmiller (1967) has given approximate expressions without estimates of their validity, but with a note of caution that the result is adequate only over very small distances compared with one mean-free path. Neither has a general treatment of the solution been given nor have questions of convergence been answered.

In the present chapter, we consider the solution of the Boltzmann equation for charged-particle transport and answer questions on the convergence of the solution technique (Wilson and Lamkin, 1975). For the present, we assume one-dimensional transport (which is a reasonable approximation for nucleon transport, see Alsmiller (1967) and Wright et al. (1969)) in which production of neutral secondaries is ignored. The equations are otherwise realistic and, as we will see, provide a good first approximation to proton transport. Herein we solve the charged-particle Boltzmann equation as coupled perturbation equations (in the context of slowing-down theory). The inverse of the Boltzmann operator is derived and the original Boltzmann equation is replaced by a set of quadratures. The implementation is discussed and convergence of the perturbation series is examined. The final results are compared with those obtained by Monte Carlo methods (Wright et al., 1969). Implications on the interaction of high-energy protons with tissue are then discussed.

8.2. Charged-Particle Transport

8.2.1. Energy-independent cross sections.
The Boltzmann equation for proton transport (neglecting coupling to other particle fields) in the straight ahead approximation is given as

$$\left(\frac{\partial}{\partial x} - \frac{\partial}{\partial E} S(E) + \sigma\right) \phi(x, E) = \int_E^\infty f(E, E') \phi(x, E') \, dE' \tag{8.1}$$

where $S(E)$ is the proton stopping power, σ is the macroscopic proton cross section that we take as energy independent, and $f(E, E')$ is the secondary-particle-production cross section. The production cross section satisfies

$$\int_0^{E'} f(E, E') \, dE = m\sigma \tag{8.2}$$

where m is the average number of protons produced per event.

The differential operator of equation (8.1) can be simplified by making a nonlinear transformation on E as

$$r = \int_0^E \frac{dE'}{S(E')} \tag{8.3}$$

which follows from the method of characteristics in partial differential equations theory (Wilson and Lamkin, 1975). By further mapping the particle field $\phi(x, E)$ and production cross section as

$$\psi(x, r) = S(E) \, \phi(x, E) \tag{8.4}$$

$$\bar{f}(r, r') = S(E) \, f(E, E') \tag{8.5}$$

we may rewrite equation (8.1) as

$$\left[\frac{\partial}{\partial x} - \frac{\partial}{\partial r} + \sigma\right] \psi(x, r) = \int_r^\infty \bar{f}(r, r') \, \psi(x, r') \, dr' \tag{8.6}$$

The characteristic coordinate along which the solution propagates is simply $x - r$, and the solution may be written as a line integral along this coordinate. The result may be written as

$$\psi(x, r) = \exp(-\sigma x) \, \psi(0, x + r)$$

$$+ \int_0^x dz \exp(-\sigma z) \int_{r+z}^\infty dr' \, \bar{f}(r + z, r') \, \psi(x - z, r') \tag{8.7}$$

where the boundary condition is

$$\psi(0, r) = S(E) \, \phi(0, E) \tag{8.8}$$

Solution to equation (8.7) may be written in terms of quadratures as follows

$$\psi_0(x, r) = \exp(-\sigma x)\, \psi(0, r + x) \tag{8.9}$$

$$\psi_n(x, r) = \int_0^x dz \exp(-\sigma z) \int_{r+z}^\infty dr'\, \bar{f}(r + z, r')\, \psi_{n-1}(x - z, r') \tag{8.10}$$

with

$$\psi(x, r) = \sum_{i=0}^\infty \psi_i(x, r) \tag{8.11}$$

which is recognized as the Neumann series.

The secondary source spectra were discussed in chapter 5 and are of the form

$$\bar{f}(r, r') \approx a \exp(-\alpha r) + c \exp[\gamma(r - r')] \tag{8.12}$$

where the coefficients α and γ are slowly varying functions of r'.

The first term corresponds to knockout nucleons and evaporation particles so that

$$\alpha r' \gg 1 \tag{8.13}$$

The second term corresponds to the quasi-elastic scattered primary such that

$$\gamma r' \ll 1 \tag{8.14}$$

Typically, $\alpha \approx 1 \to 10$ cm^2/g while $\gamma \approx 10^{-3} \to 10^{-2}$ cm^2/g. As a result of linearity of equation (8.7), we will solve for each term of equation (8.12) separately and evaluate the cross term last.

8.2.2. Discrete spectrum. The solution for a discrete spectrum at the boundary is developed as follows. The first term is given as

$$\psi_0(x, r) = \exp(-\sigma x)\, \delta(r_o - r - x) \tag{8.15}$$

where r_o corresponds to the energy at the boundary. The quasi-elastic transported particles may be solved directly (Wilson et al., 1988 and 1989) as

$$\psi_{c_1}(x, r) = x \exp(-\sigma x) c \exp[\gamma(r_o - r - x)] \tag{8.16}$$

with

$$\psi_{c_n}(x, r) = \frac{1}{n!} x^n \exp(-\sigma x) \frac{c^n}{(n-1)!} (r_o - r - x)^{n-1} \exp[-\gamma(r_o - r - x)] \tag{8.17}$$

It is easily shown that

$$\psi_c(x,r) = \sum_{n=1}^{\infty} \psi_{c_n}(x,r)$$

$$= \exp[-\sigma x - \gamma(r_o - r - x)]\sqrt{\frac{cx}{r_o - r - x}}$$

$$\times I_1\left[2\sqrt{cx(r_o - r - x)}\right] \tag{8.18}$$

where $I_1(Z)$ is a modified Bessel function of first order (Wilson et al., 1988). Equation (8.18) can also be derived by using Laplace transforms (Ganapol et al., 1991). The total flux is found for each term as

$$\Phi_{c_n}(x) = \int_0^{r_o-x} \psi_n(x,r)\,dr \tag{8.19}$$

so that

$$\Phi_0(x) = \exp(-\sigma x) \tag{8.20}$$

as expected, and

$$\Phi_{c_1}(x) = \frac{c}{\gamma}x\,\exp(-\sigma x)\{1 - \exp[-\gamma(r_o - x)]\} \tag{8.21}$$

$$\Phi_{c_2}(x) = \frac{1}{2}\,\frac{c^2}{\gamma^2}x^2\,\exp(-\sigma x)\left\{1 - [1 + \gamma(r_o - x)]\,\exp[-\gamma(r_o - x)]\right\} \tag{8.22}$$

At very high energies ($r_o \to \infty$), the total flux values become

$$\Phi_{c_n}(x) = \frac{1}{n!}\left(\frac{c}{\gamma}x\right)^n\,\exp(-\sigma x) \tag{8.23}$$

as expected because $c \approx m_Q\sigma\gamma$, where m_Q is the quasi-elastic multiplicity (Wilson et al., 1989).

The low-energy term of the secondary spectrum can likewise be treated. The first term is

$$\psi_{a_1}(x,r) = \frac{a}{\alpha}\,\exp(-\sigma x)\left\{\exp(-\alpha r) - \exp[-\alpha(r+x)]\right\}U(r_o - x - r)$$

$$= \frac{a}{\alpha}\,\exp(-\sigma x)\left[F_a(r+x) - F_a(r)\right] \tag{8.24}$$

where $F_a(r)$ is the cumulative spectrum. The total fluence contribution is

$$\Phi_{a_1}(x) = \frac{a}{\alpha^2}\,\exp(-\sigma x)\left\{1 + \exp(-\alpha r_o) - \exp(-\alpha x) - \exp[-\alpha(r_o - x)]\right\} \tag{8.25}$$

455

which shows initial linear growth for small x to some maximum value limited by the $\exp(-\sigma x)$ factor if αr_o is large or by $1 - \exp[-\alpha(r_o - x)]$ if αr_o is small compared with σr_o. If αr_o is large, the secondary particle fluence rises quickly to equilibrium with the primary beam for $x \gg \alpha^{-1}$ and slowly declines as the primary beam attenuates according to

$$\Phi_{a_1}(x) \approx \frac{a}{\alpha^2} \exp(-\sigma x) \tag{8.26}$$

The next secondary spectral term is

$$\psi_{a_2}(x, r) = \frac{a^2}{2\alpha^3} \exp(-\sigma x) \exp(-\alpha r)[1 + \exp(-2\alpha x) - 2\exp(-\alpha x)]$$
$$\times \left\{ \exp(-\alpha r) - \exp[-\alpha(r_o - x)] \right\} \tag{8.27}$$

The fluence grows for small x as

$$\psi_{a_2}(x, r) \approx \frac{3}{4} \frac{a^2}{\alpha} x^2 \exp(-\sigma x) \exp(-2\alpha r) \tag{8.28}$$

and approaches a value given by

$$\psi_{a_2}(x, r) \approx \frac{a^2}{2\alpha^3} \exp(-\sigma x) \exp(-2\alpha r) \tag{8.29}$$

where equilibrium with the primary beam is established. The total flux is

$$\Phi_{a_2}(x) = \frac{a^2}{4\alpha^4} \exp(-\sigma x)[1 + \exp(-2\alpha x) - 2\exp(-\alpha x)]\left\{ 1 + \exp[-2\alpha(r_o - x)] \right.$$
$$\left. - 2\exp[-\alpha(r_o - x)] \right\} \tag{8.30}$$

The equilibrium flux is

$$\Phi_{a_2}(x) \approx \frac{a^2}{4\alpha^4} \exp(-\sigma x) \tag{8.31}$$

and can be compared with the first perturbation term for which

$$\frac{\Phi_{a_2}(x)}{\Phi_{a_1}(x)} \leq 10^{-2} \tag{8.32}$$

for most materials. The series for the $\psi_a(x, r)$ converges very rapidly so that

$$\psi_a(x, r) \approx \psi_{a_1}(x, r) + \psi_{a_2}(x, r) \tag{8.33}$$

to better than 1 percent accuracy, and neglecting the second term gives results within 1 percent.

We now evaluate the cross terms of the series. The low-energy secondary particle flux produced by quasi-elastic secondaries is given by

$$\psi_{ac}(x,r) = \frac{a}{\alpha^2}\frac{c}{\gamma}\exp(-\sigma x)\exp(-\alpha r)\Big[(\alpha x - 1) + \exp(-\alpha x)\Big]$$
$$\times \Big\{1 - \exp[-\gamma(r_o - x - r)]\Big\} \qquad (8.34)$$

for which the total integrated flux is

$$\Phi_{ac}(x) = \frac{a}{\alpha^2}\frac{c}{\gamma\alpha(\alpha-\gamma)}\exp(-\sigma x)[(\alpha x - 1) + \exp(-\alpha x)]$$
$$\times \Big((\alpha-\gamma)\{1 - \exp[-\alpha(r_o - x)]\}$$
$$-\alpha\exp[-\gamma(r_o - x)]\{1 - \exp[-(\alpha-\gamma)](r_o - x)\}\Big) \qquad (8.35)$$

Because $\alpha \gg \gamma$, we find that for high-energy primaries (i.e., $r_o \to \infty$) the total flux rises quadratically from the boundary and rapidly approaches equilibrium with the quasi-elastic secondaries (eq. (8.21)) as seen in

$$\Phi_{ac}(x) \approx \frac{a}{\alpha^2}\frac{c}{\gamma}x\exp(-\sigma x) \qquad (8.36)$$

Similar results can be found for the higher order quasi-elastic secondary terms. The quasi-elastic scattering of low-energy secondaries is similarly derived as

$$\psi_{ca}(x,r) = \frac{a}{\alpha}\frac{c}{\alpha+\gamma}\exp(-\sigma x)\left\{\frac{1 - \exp[-(\alpha-\gamma)x]}{\alpha-\gamma} - \frac{1}{\gamma}[1 - \exp(-\gamma x)]\right\}$$
$$\times\Big\{\exp(-\alpha r) - \exp[-\alpha(r_o - x)]\exp[-\gamma(r_o - x - r)]\Big\} \qquad (8.37)$$

The total integrated flux is

$$\Phi_{ca}(x) = \frac{a}{\alpha}\frac{c}{\alpha+\gamma}\exp(-\sigma x)\left\{\frac{1 - \exp(-\alpha-\gamma)x}{\alpha-\gamma} - \frac{1}{\gamma}[1 - \exp(-\gamma x)]\right\}$$
$$\times\left(\frac{1}{\alpha}\{1 - \exp[-\alpha(r_o - x)]\}\right.$$
$$\left.-\frac{1}{\gamma}\{\exp[-\alpha(r_o - x)] - \exp[-(\alpha+\gamma)(r_o - x)]\}\right) \qquad (8.38)$$

For high energy and $x \gg \alpha^{-1}$,

$$\Phi_{ca}(x) \approx \frac{a}{\alpha^2}\frac{c}{\alpha^2}\exp(-\sigma x) \qquad (8.39)$$

It is clear that equilibrium is established with the primary beam caused by the short range of the first-generation, low-energy secondary particles (i.e., $\alpha \gg \gamma$). Note that $\Phi_{ca}(x)$ is inferior to $\Phi_{ac}(x)$ once equilibrium is established ($x \gg \alpha^{-1}$) and is therefore small as $x \to 0$, thus making the cross term of equation (8.36) the main contribution.

8.2.3. Continuous spectrum. We consider a continuous energy spectrum at the boundary given by an exponential so that

$$\psi_0(x, r) = \exp(-\sigma x) \exp[-\beta(r + x)] \tag{8.40}$$

where typical values of β are $\alpha \gg \beta \gg \gamma$. The next terms of the quasi-elastic series are given by

$$\psi_{c_n}(x, r) = \frac{1}{n!} x^n \exp(-\sigma x) \left(\frac{c}{\gamma + \beta}\right)^n \exp[-\beta(x + r)] \tag{8.41}$$

Clearly, the quasi-elastic series may be summed as

$$\sum_{n=0}^{\infty} \psi_{c_n}(x, r) = \exp[cx/(\gamma + \beta)] \exp(-\sigma x) \exp[-\beta(x + r)] \tag{8.42}$$

The total flux associated with each term is

$$\Phi_n(x) = \frac{1}{n!} \left(\frac{cx}{\gamma + \beta}\right)^n \frac{1}{\beta} \exp(-\sigma x - \beta x) \tag{8.43}$$

The total flux

$$\Phi_c(x) = \frac{1}{\beta} \exp[cx/(\beta + \gamma)] \exp(-\sigma x) \exp(-\beta x) \tag{8.44}$$

converges by virtue of $\beta \gg \gamma$.

The low-energy secondaries may likewise be evaluated to give

$$\psi_{a_1}(x, r) = \frac{a}{\alpha} \exp(-\sigma x) \exp[-\beta(x + r)] \left\{\exp(-\alpha r) - \exp[-\alpha(r + x)]\right\} \tag{8.45}$$

The total flux is given as

$$\Phi_{a_1}(x) = \frac{a}{\alpha} \frac{1}{\alpha + \beta} \exp(-\sigma x) \exp(-\beta x)[1 - \exp(-\alpha x)] \tag{8.46}$$

The flux grows linearly at small $x \ll \alpha^{-1}$ and approaches equilibrium with the primary beam for $x \gg \alpha^{-1}$. The equilibrium value is

$$\Phi_{a_1}(x) \approx \frac{a}{\alpha} \frac{1}{\alpha + \beta} \exp(-\sigma x) \exp(-\beta x) \tag{8.47}$$

The next term is given as

$$\psi_{a_2}(x, r) = \frac{a}{\alpha} \frac{a}{\alpha + \beta} \exp(-\sigma x) \exp\left[-\beta(x + r)\right]$$

$$\times \left\{ \frac{1}{2\alpha} \exp(-2\alpha r)\left[1 - \exp(-2\alpha x)\right] \right.$$

$$\left. - \frac{1}{\beta} \exp(-\alpha x - 2\alpha r)\left[1 - \exp(-\beta x)\right] \right\} \tag{8.48}$$

for which the corresponding integrated flux is

$$\Phi_{a_2}(x) = \frac{a}{\alpha} \frac{a}{\alpha + \beta} \frac{1}{2\alpha + \beta} \exp(-\sigma x) \exp(-\beta x)$$

$$\times \left\{ \frac{1}{2\alpha}[1 - \exp(-2\alpha x)] - \frac{1}{\beta}[1 - \exp(-\beta x)] \exp(-\alpha x) \right\} \tag{8.49}$$

The flux again approaches equilibrium for $x \gg \alpha^{-1}$ as

$$\Phi_{\alpha_2}(x) \approx \frac{1}{2} \frac{a^2}{\alpha^2} \frac{\beta}{\alpha + 2\beta} \frac{1}{\beta} \exp(-\sigma x) \exp(-\beta x) \tag{8.50}$$

As in the discrete spectrum case,

$$\frac{\Phi_{a_2}(x)}{\Phi_{a_1}(x)} < 0.01 \tag{8.51}$$

The cross term representing the low-energy secondaries produced by the quasi-elastic scatter particles is

$$\psi_{ac}(x, r) = \frac{a}{\beta} \frac{c}{\gamma + \beta} \exp(-\sigma x - \alpha r - \beta x)$$

$$\times \left\{ \frac{\alpha x - \beta x - 1 + \exp\left[-(\alpha - \beta)x\right]}{(\alpha - \beta)^2} \right.$$

$$\left. - \frac{\exp(-\beta r)\left[\alpha x - 1 + \exp(-\alpha x)\right]}{\alpha^2} \right\} \tag{8.52}$$

The corresponding total flux is

$$\Phi_{ac}(x) = \frac{a}{\alpha\beta} \frac{c}{\gamma + \beta} \exp(-\sigma x - \beta x)$$

$$\times \left\{ \frac{\alpha x - \beta x - 1 + \exp[-(\alpha - \beta)x]}{(\alpha - \beta)^2} - \frac{\alpha x - 1 + \exp(-\alpha x)}{\alpha(\alpha + \beta)} \right\} \tag{8.53}$$

The flux rises quadratically for small x to an equilibrium value for $x \gg \alpha^{-1}$ as

$$\Phi_{ac}(x) \approx \frac{a}{\alpha + \beta} \frac{c}{\gamma + \beta} \frac{x}{(\alpha - \beta)^2} \exp(-\sigma x - \beta x) \tag{8.54}$$

and remains in equilibrium with the $\psi_{c_1}(x, r)$ term throughout the rest of the median. The cross term representing the quasi-elastic scattering of the first-generation, low-energy flux is

$$\psi_{ca}(x, r) = \frac{a}{\alpha^2} \frac{c}{\gamma + \beta + \alpha} \exp(-\sigma x) \exp(-\beta x) \exp\left[-(\alpha + \beta)r\right]$$
$$\times \left[1 - (\alpha x - 1) \exp(-\alpha x)\right] \tag{8.55}$$

The corresponding integrated flux is

$$\Phi_{ca}(x) = \frac{a}{\alpha^2} \frac{c}{\gamma + \alpha + \beta} \frac{\exp(-\sigma x - \beta x)}{\alpha + \beta} \left[1 - (\alpha x - 1) \exp(-\alpha x)\right] \tag{8.56}$$

The flux rises quadratically at small x and approaches an equilibrium value for $x \gg \alpha^{-1}$ as

$$\Phi_{ca}(x) \approx \frac{a}{\alpha^2} \frac{c}{\gamma + \beta + \alpha} \frac{\exp(-\sigma x - \beta x)}{\alpha + \beta} \tag{8.57}$$

and remains in equilibrium with the surviving flux of primary particles. Clearly, the Φ_{ca} flux is inferior to the Φ_{ac} flux at all values of x.

These results are useful in developing an understanding of the role of various contributions in the particle-production terms of equations (8.10) or (8.6). They likewise serve as a means of developing numerical procedures and testing those procedures for error propagation. We will further discuss this role in a latter section. The fact that the lower energy solutions converge quickly and remain in equilibrium with the primary beam has played an important role in the study of target-induced reaction and nuclear recoil effects in biological materials (Wilson, Townsend, and Buck, 1986; Wilson, Townsend, and Khan, 1989; Wilson, Shinn, and Townsend, 1990; Shinn, Wilson, and Ngo, 1990) and electronic materials (Ngo et al., 1989; Wilson, Stith, and Stock, 1983,; Wilson, Walker, and Outlaw, 1984).

8.3. Buildup Factors

In passing through tissue, energetic protons interact mostly through ionization of atomic constituents by transferring small amounts of momentum to orbital electrons. Although nuclear reactions are far less numerous, their effects are magnified because of the large momentum transferred to the nuclear particles and the struck nucleus itself. Unlike the secondary electrons formed through atomic ionization by interaction with the primary protons, the radiations resulting from nuclear reactions are mostly heavily ionizing and generally have large biological effectiveness. Many of the secondary particles of nuclear reactions are sufficiently energetic to promote similar nuclear reactions and thus cause a buildup of secondary radiations. The description of such processes requires a solution of the transport equation. The approximate solutions for transporting protons in 30-cm-thick slabs of soft tissue for fixed incident energies have been found (Armstrong and Bishop, 1971; Wright et al., 1969; Alsmiller, Armstrong, and Coleman, 1970; Armstrong and Chandler, 1970; Snyder et al., 1969; Turner et al., 1964; Wright, Hamm, and Turner, 1971; Zerby and Kinney, 1965). The results of such calculations are dose conversion factors for relating the primary monoenergetic proton fluence to dose or dose equivalent as a function of position in a tissue slab.

Whenever the radiation is spatially uniform, the dose at any point x in a convex object may be calculated by (Wilson and Khandelwal, 1974)

$$D(\vec{x}) = \int_0^\infty \int_\Omega R_n[z_x(\vec{\Omega}), E] \, \phi(\vec{\Omega}, E) \, d\vec{\Omega} \, dE \tag{8.58}$$

where $R_n(z, E)$ is the dose at depth z for normal incident protons of energy E on a tissue slab; $\phi(\vec{\Omega}, E)$ is the differential proton fluence along direction $\vec{\Omega}$; and $z_x(\vec{\Omega})$ is the distance from the boundary along $\vec{\Omega}$ to the point $\vec{x}$. It has been shown that equation (8.58) always overestimates the dose but is an accurate estimate when the ratio of the proton beam divergence (caused by nuclear reaction) to the radius of curvature of the body is small. Equation (8.58) is a practical prescription for introducing nuclear reaction effects into calculations of dose in geometrically complex objects, such as a human body. The main requirement is that the dose conversion factors for a tissue slab be adequately known for a broad range of energies and depths. Such a description is obtained through the use of buildup factors.

8.3.1. Simplified theory of proton buildup factors. The Boltzmann equation for proton transport in straight ahead approximation is given as

$$\left[\frac{\partial}{\partial x} - \frac{\partial}{\partial E} S(E) + \sigma \right] \phi(x, E) = \int_E^\infty f(E, E') \, \phi(x, E') \, dE' \tag{8.59}$$

where $S(E)$ is the proton stopping power, σ is the macroscopic interaction cross section which we presently take as energy independent, and $f(E, E')$ is the production cross section for secondary protons. Using the definitions

$$r = \int_0^E \frac{dE'}{S(E')} \tag{8.60}$$

$$\psi(x, r) = S(E)\phi(x, E) \tag{8.61}$$

$$\bar{f}(r, r') = S(E)f(E, E') \tag{8.62}$$

allows equation (8.59) to be rewritten as

$$\psi(x, r) = \exp(-\sigma x) \, \psi(0, x + r)$$
$$+ \int_0^x dz \exp(-\sigma z) \int_{r+z}^\infty dr' \, \bar{f}(r + z, r') \, \psi(x - z, r') \tag{8.63}$$

where the boundary condition is

$$\psi(0, r) = S(E) \, \phi(0, E) \tag{8.64}$$

The secondary-particle-production cross section is normalized as

$$\int_0^{E'} f(E, E')\, dE = m\sigma \tag{8.65}$$

where m is the average number of protons produced per nuclear event. Although m and σ are, in reality, functions of E', our current interest is in monoenergetic boundary conditions as

$$\phi(0, E) = \delta(E - E_0) \tag{8.66}$$

and we take m and σ evaluated at the beam energy E_0. The corresponding boundary condition on ψ is

$$\psi(0, r) = \delta(r - r_o) \tag{8.67}$$

The high-energy-production cross section is an exponential function of $E' - E$ and is used to approximate equation (8.62) as

$$f(r, r') \approx b \exp\left[-\gamma(r' - r)\right] \tag{8.68}$$

The normalization in equation (8.65) requires

$$b = \frac{m\sigma\gamma}{1 - \exp(-\gamma r_o)} \tag{8.69}$$

and $\gamma \approx 0.01$ cm^2/g. Equation (8.63) may be solved by perturbation theory to obtain

$$\psi(x, r) = \sum_{i=0}^{\infty} \psi_i(x, r) \tag{8.70}$$

where

$$\psi_0(x, r) = \exp(-\sigma x)\, \delta(r + x - r_o) \tag{8.71}$$

$$\psi_1(x, r) = \exp(-\sigma x) \int_0^x \bar{f}(r + z, r_o - x + z)\,dz \tag{8.72}$$

$$\psi_{n+1}(x, r) = \int_0^x dz\, \exp(-\sigma z) \int_{r+z}^{\infty} dr'\, \bar{f}(r + z, r')\, \psi_n(x - z, r') \tag{8.73}$$

which may be reduced using equation (8.69). For example,

$$\psi_1(x, r) = x \exp(-\sigma x)\, \bar{f}(r, r_o - x) \tag{8.74}$$

$$\psi_2(x, r) = \frac{1}{2}x^2 \exp(-\sigma x)\, b(r_o - x - r)\, \bar{f}(r, r_o - x) \tag{8.75}$$

The successive contributions to dose may now be calculated

$$D_0(x) = \int_0^\infty \exp(-\sigma x)\,\delta(r + x - r_o)\,dE = S\big[\varepsilon(r_o - x)\big]\exp(-\sigma x) \qquad (8.76)$$

$$D_1(x) = \bar{S}_1\big[\varepsilon(r_o - x)\big] m\sigma x \exp(-\sigma x) \qquad (8.77)$$

$$D_2(x) = \bar{S}_2\big[\varepsilon(r_o - x)\big] m^2 \frac{\sigma^2 x^2}{2}\exp(-\sigma x) \qquad (8.78)$$

where $\bar{S}_1$ and $\bar{S}_2$ are spectral averages of stopping power in which $\bar{S}_1(\varepsilon) = O(\varepsilon)$ and $\bar{S}_2(\varepsilon) = O(\varepsilon^2)$ for small ε. The total dose is then

$$D(x) = S\big[\varepsilon(r_o - x)\big]\exp(-\sigma x) + \sum_{i=1}^\infty \frac{1}{i!}(m\sigma x)^i \exp(-\sigma x)\,\bar{S}_i\big[\varepsilon(r_o - x)\big] \qquad (8.79)$$

The dose buildup factor is then defined as

$$B(x, E_o) = 1 + \frac{\sum_{i=1}^\infty \frac{1}{i!}(m\sigma x)^i\,\bar{S}_i[\varepsilon(r_o - x)]}{S[\varepsilon(r_o - x)]} \qquad (8.80)$$

With the property that

$$\lim_{x \to r_o} B(x, E_o) = 1 \qquad (8.81)$$

which follows here from the neglect of the coupling between the proton and neutron fields.

Wilson and Khandelwal (1974) assumed that the buildup factor had the following form

$$B(x, E_0) = (A_1 + A_2 x + A_3 x^2)\exp(-A_4 x) \qquad (8.82)$$

where A_4 was chosen to satisfy equation (8.81). It is seen that the choice of A_4 is not governed by the nuclear cross section but rather by the result that $\bar{S}_i(\varepsilon) \approx O(\varepsilon^i)$ for small ε. The presence of neutron production in the medium modifies the conclusion that equation (8.81) is valid.

8.3.2. Buildup in an external shield. We now consider the problem of buildup in an external shield. We assume that an equivalent distance in the shield can be defined so that the stopping power in equivalent distance units of the shield and exposed media are equal. Consequently, the Boltzmann equations of the two media differ only in their nuclear cross sections as

$$\left[\frac{\partial}{\partial x} - \frac{\partial}{\partial E}S(E) + \sigma_s\right]\phi(x, E) = \int_E^\infty f_s(E, E')\,\phi(x, E')\,dE' \qquad (8.83)$$

$$\left[\frac{\partial}{\partial x} - \frac{\partial}{\partial E}S(E) + \sigma\right]\phi(x,E) = \int_E^\infty f(E,E')\,\phi(x,E')\,dE' \qquad (8.84)$$

For a monoenergetic beam on the boundary of the shield, the solution is given as

$$\psi_0(x,r) = \exp(-\sigma_s x)\,\delta(r_o - x - r) \qquad (8.85)$$

$$\psi_1(x,r) = x\exp(-\sigma_s x)\,\bar{f}_s(r,r_o - x) \qquad (8.86)$$

The particles appearing at the media interface provide the boundary condition of the exposed medium; thus,

$$\psi_0(0,r) = \exp(-\sigma_s t_s)\,\delta(r_o - t_s - r) + t_s\exp(-\sigma_s t_s)\,\bar{f}_s(r,r_o - t_s) + \ldots \quad (8.87)$$

where t_s is the shield thickness. To evaluate the proton field in the exposed media we may use equation (8.63) and the above boundary value in equation (8.87) to obtain

$$\begin{aligned}
\psi(x,r) = {} & \exp(-\sigma x)\exp(-\sigma_s t_s)\,\delta(r_o - t_s - x - r) \\
& + \exp(-\sigma x)t_s\exp(-\sigma_s t_s)\,\bar{f}_s(r + x, r_o - t_s) \\
& + \exp(-\sigma_s t_s)x\exp(-\sigma x)\,\bar{f}(r, r_o - t_s - x) + \ldots \qquad (8.88)
\end{aligned}$$

Using equation (8.68), we may rewrite equation (8.88) as

$$\begin{aligned}
\psi(x,r) = {} & \exp(-\sigma_s t_s - \sigma x)\,\delta(r_o - t_s - x - r) + \big[t_s\,\bar{f}_s(r,r_o - t_s - x) \\
& + x\,\bar{f}(r,r_o - t_s - x)\big]\exp(-\sigma_s t_s - \sigma x) + \ldots \qquad (8.89)
\end{aligned}$$

Similar to equation (8.79), we have

$$\begin{aligned}
D_s(t_s,x,E_o) = {} & \exp(-\sigma_s t_s - \sigma x)S\big[\varepsilon(r_o - t_s - x)\big] \\
& + \Big\{m_s\sigma_s t_s\bar{S}_{s1}\big[\varepsilon(r_o - t_s - x)\big] \\
& + m\sigma x\bar{S}_1\big[\varepsilon(r_o - t_s - x)\big]\Big\}\exp(-\sigma_s t_s - \sigma x) + \ldots \quad (8.90)
\end{aligned}$$

It is clear that equation (8.90) may be written as

$$\begin{aligned}
D_s(t_s,x,E_o) = {} & D(t_s + x, E_o) + \big[\exp(-\sigma_s t_s) - \exp(-\sigma t_s)\big]D(x,E_s) \\
& + t_s\exp(-\sigma x)\Big\{m_s\sigma_s\,\bar{S}_{s1}\big[\varepsilon(r_o - t_s - x)\big]\exp(-\sigma_s t_s) \\
& - m\sigma\,\bar{S}_1\big[\varepsilon(r_o - t_s - x)\big]\exp(-\sigma t_s)\Big\} \\
\approx {} & D(t_s + x, E_o) + t_s\Big\{(\sigma - \sigma_s)D(x,E_s) \\
& + \exp(-\sigma x)(m_s\sigma_s - m\sigma)S_1\big[\varepsilon(r_o - t_s - x)\big]\Big\} + \ldots \quad (8.91)
\end{aligned}$$

where $E_s = \varepsilon(r_o - t_s)$. One may define the shield buildup factor relative to the exposed media as

$$D_s(t_s, x, E_o) = \left\{ \frac{1 + t_s[(\sigma - \sigma_s)D(x, E_s) + \exp(-\sigma x)(m_s\sigma_s - m\sigma)\,\bar{S}_1]}{D(t_s + x, E_o)} + \cdots \right\}$$
$$\times\, D(t_s + x, E_o) \tag{8.92}$$

Clearly, the coefficient of t_s in equation (8.92) is reduced if $\sigma \approx \sigma_s$, and furthermore it is reduced to a small contribution as $m \to m_s$. This occurs because the spectral distribution functions have $\gamma \approx \gamma_s$ for all materials since γ is largely determined from the proton-proton scattering amplitude.

8.3.3. Tissue buildup factors. We now consider the implementation of the above ideas into a practical formulation of proton buildup factors in tissue. Available information on conversion factors is for discrete energies from 100 MeV to 1 TeV in rather broad energy steps for depths from 0 to 30 cm in semi-infinite slabs of tissue (Alsmiller, Armstrong, and Coleman, 1970; Armstrong and Chandler, 1970; Turner et al., 1964; Zerby and Kinney, 1965). The nuclear reaction data used for high-energy nucleons are usually based on Monte Carlo estimates (Bertini, 1963 and 1969; Bertini and Guthrie, 1971) with low-energy neutron reaction data taken from experimental observation. The quality factor as defined by ICRP 26 (Anon., 1977) is used for protons. The quality factor for heavier fragments and the recoiling nuclei is arbitrarily set to 20 which is considered conservative (Wilson, Shinn, and Townsend, 1990), but the average quality factor obtained by calculation is comparable with estimates obtained through observations made in nuclear emulsion (Schaefer and Sullivan, 1970).

To fully use equation (8.58), a parameterization of the conversion factor was introduced by Wilson and Khandelwal (1974) that allowed reliable interpolation and extrapolation from known values. A refinement and an extension of that work are now discussed.

The conversion factor $R_n(z, E)$ is composed of two terms representing the dose caused by the primary beam protons and the dose caused by secondary particles produced in nuclear reactions. Thus,

$$R_n(z, E) = R_p(z, E) + R_s(z, E) \tag{8.93}$$

where the conversion factor of the primary dose equivalent is

$$R_p(z, E) = \frac{P(E)\,Q_F[S(E_r)]\,S(E_r)}{P(E_r)} \tag{8.94}$$

with the reduced energy given by

$$E_r = \varepsilon[R(E) - z] \tag{8.95}$$

and with the usual quality factor Q_F defined as a function of linear energy transfer (LET), with LET denoted here by the symbol S, and total nuclear survival probability for a proton of energy E given by

$$P(E) = \exp\left[-\int_0^E \frac{\sigma(E')\,dE'}{S(E')}\right] \tag{8.96}$$

where macroscopic cross section $\sigma(E)$ for tissue as calculated by Bertini is given by Alsmiller et al. (1972). The $R(E)$ is the usual range-energy relation for protons in tissue, and $\varepsilon(x)$ is the inverse of $R(E)$. The proton total optical thickness given by

$$\tau(E) = \int_0^E \frac{\sigma(E')\,dE'}{S(E')} \tag{8.97}$$

is given in table 8.1 for purposes of numerical interpolation. In the case of conversion factors for absorbed dose, $R_p(z, E)$ is taken as

$$R_p(z, E) = \frac{P(E)\,S(E_r)}{P(E_r)} \tag{8.98}$$

The representation of the conversion factors is simplified by rewriting equation (8.93) as

$$R_n(z, E) = \left[1 + \frac{R_s(z, E)}{R_p(z, E)}\right] R_p(z, E) \equiv B(z, E)\,R_p(z, E) \tag{8.99}$$

where $B(z, E)$ is recognized as the dose buildup factor. The main advantage of introducing the buildup factor into equation (8.99) is that unlike $R_n(z, E)$, the buildup factor is a smoothly varying function of energy at all depths in the slab and can be approximated by the simple function

$$B(z, E) = (A_1 + A_2 z + A_3 z^2)\exp(-A_4 z) \tag{8.100}$$

where the parameters A_i are understood to be energy dependent. The parameters A_i are found by fitting equation (8.100) to the values of the buildup factors as estimated from the Monte Carlo calculations of proton conversion factors. The resulting coefficients are shown in table 8.2. The coefficients for 100-, 200-, and 300-MeV protons were obtained with the Monte Carlo data of Turner et al. (1964). The values at 400, 730, 1500, and 3000 MeV were obtained from the results of Alsmiller, Armstrong, and Coleman (1970). The 10-GeV entry was obtained from the calculations of Armstrong and Chandler (1970). Some values noted in table 8.2 were obtained by interpolating between data points or smoothly extrapolating to unit buildup factor at proton energies near the coulomb barrier for tissue nuclei ($\approx$12 MeV). The coefficients are found for all energies to 10 GeV by using second-order Lagrange interpolation between the values shown in table 8.2. The resulting buildup factors are shown in figures 8.1 and 8.2 in comparison with the Monte Carlo results, where the error bars were determined by drawing smooth limiting curves so as to bracket the Monte Carlo values and to follow the general functional dependence. The uncertainty limits should, therefore, be interpreted as approximately 2σ limits, rather than 1σ ranges generally used in expressing uncertainty limits.

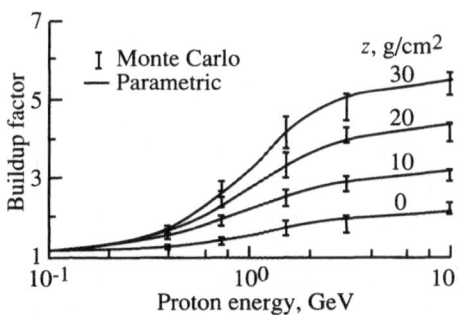

Figure 8.1. Dose buildup factor for several depths in tissue as function of incident proton energy.

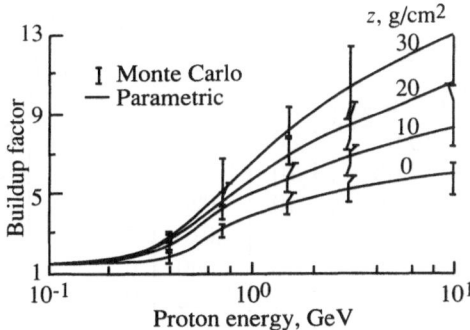

Figure 8.2. Dose equivalent buildup factor for several depths in tissue as function of incident proton energy.

Table 8.1. Total Tissue Optical Thickness for Protons

E, GeV	$\tau(E)$	E, GeV	$\tau(E)$
0	0	1.3	6.57
.01	.0033	1.5	8.03
.025	.0171	1.7	9.52
.05	.0510	2.0	11.76
.1	.135	2.2	13.27
.15	.239	2.4	14.78
.2	.362	2.6	16.29
.25	.501	2.8	17.79
.3	.655	3.0	19.29
.35	.822	4.0	26.62
.4	1.004	5.0	3.81
.5	1.429	6.0	40.84
.7	2.471	7.0	47.75
.9	3.743	8.5	57.91
1.1	5.143	10.0	67.85

In figure 8.3 the dose as a function of depth is shown in comparison to measurements of Baarli and Goebel at CERN (Switzerland) (Properties of High-Energy Beams From a 600-MeV Synchrocyclotron. Presented at XI International Congress of Radiology (Rome), Sept. 1965). Also shown are the Monte Carlo values interpolated between 400 and 730 MeV. The uncollided primary proton contribution is shown separately. The dose equivalent is likewise shown in figure 8.4. The extreme importance of secondary radiation is clearly shown.

Within the space program, one has shield material that is mostly aluminum. We are therefore interested in attenuating space radiation by the appropriate

Table 8.2. Buildup-Factor Parameters

E, GeV	Dose equivalent				Dose			
	A_1	A_2	A_3	A_4	A_1	A_2	A_3	A_4
[a]0.03	1.00	0	0	0	1.00	0	0	0
[a].06	1.20	0	0	.0300	1.07	.010	0	.010
.10	1.40	.020	0	.0300	1.10	.040	0	.026
[a].15	1.50	.070	0	.0385	1.12	.060	0	.031
.20	1.60	.090	0	.0400	1.15	.062	0	.032
.30	1.70	.110	0	.0330	1.20	.068	0	.026
.40	1.90	.130	0	.0228	1.24	.071	0	.0228
.73	3.40	.156	.00035	.0150	1.40	.090	.0001	.0150
[a]1.2	4.32	.167	.00145	.0130	1.67	.094	.0008	.0122

[a]Denotes interpolated values.

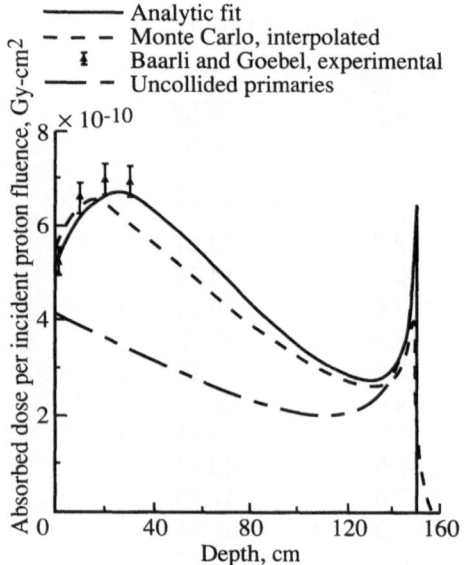

Figure 8.3. Proton depth-dose relation: analytic fit (nuclear effects), Monte Carlo, experiments, and primary protons for 592-MeV protons.

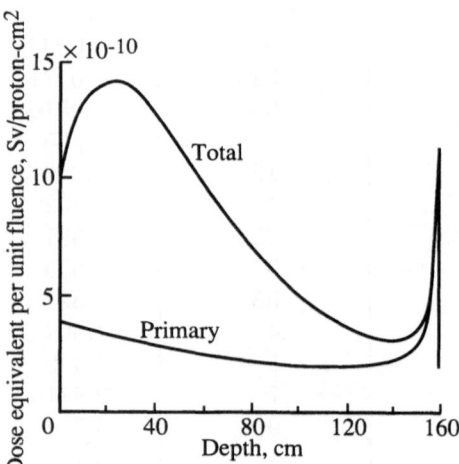

Figure 8.4. Proton depth-dose equivalent relation including nuclear effects for 600-MeV protons.

amount of aluminum before the radiation enters the astronaut's body. As a first step, we replace the appropriate aluminum thickness z_s (given in g/cm²) by a range of equivalent thickness of tissue $\hat{z}_s$ for 50-MeV protons as has been the custom in space-radiation protection as

$$\hat{z}_s = \frac{R_{\text{tiss}}(50)}{R_{\text{Al}}(50)} z_s \equiv \rho z_s \tag{8.101}$$

For definiteness we note that $\rho \approx 0.787$. The conversion factor of the primary dose equivalent is then

$$R_p(z + \hat{z}_s, E) = \exp\left[-\left(\rho^{-1}\sigma_{Al} - \sigma_{tiss}\right)\hat{z}_s\right]$$
$$\times \frac{P(E)\,Q_F[S(E_r)]\,S(E_r)}{P(E_r)} \tag{8.102}$$

where the reduced energy is

$$E_r = \varepsilon\left[R(E) - z - \hat{z}_s\right] \tag{8.103}$$

and the exponential factor corrects $P(E)$ by the appropriate aluminum-tissue-combined attenuation factor. The primary absorbed dose is identical in form to equation (8.102) except that $Q_F(S)$ is equal to unity. Note that σ_{Al} and σ_{tiss} are taken presently as the asymptotic macroscopic cross sections where energy dependence is negligible. The complete conversion factors are

$$R_n(z + \hat{z}_s, E) = R_p(z + \hat{z}_s, E) + R_s(z + \hat{z}_s, E) \tag{8.104}$$

where $R_s(z + \hat{z}_s, E)$ is the contribution including secondary particles. We rewrite equation (8.104) as

$$R_n(\hat{z}_s + z, E) = B_\Delta(\hat{z}_s, E)\,R_{tiss}(z + \hat{z}_s, E) \tag{8.105}$$

where $B_\Delta(\hat{z}_s, E)$ is an aluminum buildup factor relative to tissue which is unity for $\hat{z}_s = 0$ and $E \ll 100$ MeV. The aluminum factor has been found (units for E are GeV and for $\hat{z}_s$ are g/cm^2) to be reasonably approximated by

$$B_\Delta(\hat{z}_s, E) = 1 + \frac{0.02\hat{z}_s E}{(1 + E)}\exp(-0.022\hat{z}_s) \tag{8.106}$$

for the dose equivalent and by

$$B_\Delta(\hat{z}_s, E) = 1 + \frac{0.02\hat{z}_s E}{6(1 + E)}\exp(-0.01\hat{z}_s) \tag{8.107}$$

for the absorbed dose. Equation (8.56) is rewritten as

$$D(\vec{x}) = \int_0^\infty \int_\Omega R_n[\hat{z}_s(\vec{\Omega}), z(\vec{\Omega}), E]\,\phi(\vec{\Omega}, E)\,d\vec{\Omega}\,dE \tag{8.108}$$

where $\hat{z}_0(\vec{\Omega})$ is the aluminum thickness distribution (Atwell et al., 1989) about the dose point $\vec{x}$ and $z(\vec{\Omega})$ is the astronaut self-shielding distribution about the dose point (Billings and Yucker, 1973). This method has proven very useful in estimating space proton exposures.

8.4. Numerical Methods

In the rest of this chapter, we consider numerical methods for estimating solutions to the Boltzmann equation. We first consider the numerical solution of the charged-particle transport equation with energy-independent nuclear parameters as a test bed for numerical methods. We then propose a general method for fully coupled, neutron-proton transport in the straight ahead approximation using the nuclear data bases of chapters 4 and 5 which resulted in the code system (Wilson et al., 1989) known as the BRYNTRN (baryon transport code.) The numerical convergence and the comparison with Monte Carlo derived results are studied.

8.4.1. Energy-independent proton model.
The Boltzmann equation for proton transport in the straight ahead approximation is given as

$$\left[\frac{\partial}{\partial x} - \frac{\partial}{\partial E} S(E) + \sigma\right] \phi(x, E) = \int_E^\infty f(E, E') \, \phi(x, E') \, dE' \tag{8.109}$$

where $S(E)$ is the proton stopping power, σ is the macroscopic interaction cross section that we presently take as energy independent, and $f(E, E')$ is the production secondary-particle spectrum. Using the definitions

$$r = \int_0^E \frac{dE'}{S(E')} \tag{8.110}$$

$$\psi(x, r) = S(E) \, \phi(x, E) \tag{8.111}$$

and

$$\bar{f}(r, r') = S(E) \, f(E, E') \tag{8.112}$$

allows equation (8.109) to be written as

$$\left[\frac{\partial}{\partial x} - \frac{\partial}{\partial r} + \sigma\right] \psi(x, r) = \int_r^\infty \bar{f}(r, r') \, \psi(x, r') \, dr' \tag{8.113}$$

The advantage of equation (8.113) over equation (8.109) is that derivatives of $\phi(x, E)$ with respect to E display large variations at low energy and are difficult to approximate numerically, whereas $\psi(x, r)$ is well behaved at all values of r and approaches a constant at small values of r.

8.4.2. First-order explicit methods.
The boundary condition is specified at $x = 0$ and first-order explicit methods imply a forward difference formula in x to propagate the solution from the boundary. We assume an x-grid denoted by x_i separated by distance h and an r-grid denoted by r_j separated by distance Δ. A backward difference along the boundary yields

$$\frac{1}{h} \left(\psi_{i+1,j} - \psi_{i,j}\right) - \frac{1}{\Delta} \left(\psi_{i,j} - \psi_{i,j-1}\right) + \sigma\psi_{ij} = \zeta_{ij} \tag{8.114}$$

where $\psi_{i,j}$ is taken as zero for $j < 0$ corresponding to a negative residual range.

This explicit procedure yields

$$\psi_{i+1,j} = \left(1 + \frac{h}{\Delta} - \sigma h\right) \psi_{i,j} - \frac{h}{\Delta} \, \psi_{i,j-1} + h\zeta_{ij} \tag{8.115}$$

Clearly, a stable method must propagate the boundary as an energy shifted and attenuated beam of particles. Note that the lowest energy point may be solved as

$$\psi_{i+1,0} = \left(1 + \frac{h}{\Delta} - \sigma h\right)^{i+1} \psi_{0,0} \qquad (8.116)$$

when the secondary source terms are set to zero. A stable solution requires

$$1 > \sigma h > \frac{h}{\Delta} \qquad (8.117)$$

for which Δ must be chosen greater than the nuclear mean-free path. Such a requirement $(\Delta > \sigma^{-1})$ resulting in poor energy resolution will not allow an adequate representation of typical boundary conditions resulting in large numerical errors.

A second explicit method uses a forward difference along the boundary as

$$\frac{1}{h}\left(\psi_{i+1,j} - \psi_{i,j}\right) - \frac{1}{\Delta}\left(\psi_{i,j+1} - \psi_{ij}\right) + \sigma\psi_{ij} = \zeta_{ij} \qquad (8.118)$$

and is represented in a stepping procedure as

$$\psi_{i+1,j} = \left(1 - \frac{h}{\Delta} - h\sigma\right)\psi_{ij} + \frac{h}{\Delta}\psi_{i,j+1} + \zeta_{ij}h \qquad (8.119)$$

It is clear that for some values of J the values of $\psi_{0,j}$ are zero for $j > J$ and the boundary propagation of $\psi_{i,J}$ is

$$\psi_{i,J} = \left(1 - \frac{h}{\Delta} - h\sigma\right)^{i}\psi_{0,J} \qquad (8.120)$$

which converges if $\frac{h}{\Delta} + h\sigma < 1$. For numerical accuracy, $h \ll \Delta$ and $h \ll \sigma^{-1}$ are also required. The first condition is particularly hard to meet because Δ becomes rapidly small at low energies which makes the low-energy spectrum difficult to calculate without special procedures.

8.4.3. "Linearized" methods. A method was proposed and received considerable use in which the stopping-power term was "linearized" in such a way that analytical methods could be applied and numerical stability issues circumvented. This requires an assumed form for the flux as

$$\phi(x, E) \approx \frac{c}{E^\alpha} \qquad (8.121)$$

and stopping power as

$$S(E) = nk\left(\frac{k}{E}\right)^{\frac{1}{n}-1} \qquad (8.122)$$

The energy derivative in equation (8.109) is then approximated as

$$\frac{\partial}{\partial E}[S(E)\,\phi(x,E)] \approx -\frac{n\alpha - n + 1}{r(E)}\,\phi(x,E) \tag{8.123}$$

after which equation (8.109) is written as

$$\left[\frac{\partial}{\partial x} + \frac{n\alpha - n + 1}{r(E)} + \sigma\right]\phi(x,E) = \int_E^\infty f(E,E')\,\phi(x,E')\,dE' \tag{8.124}$$

which may be solved analytically because E enters only as a parameter. Note that the range-energy relations enter as an effective attenuation similar to the finite-difference approximation (eq. (8.120)). Although equation (8.124) contains no instability, there are large inherent errors as discussed elsewhere (Wilson and Badavi, 1986).

8.4.4. Unconditionally stable numerical methods. The differential operator of equation (8.113) may be inverted to yield

$$\psi(x,r) = \exp(-\sigma x)\,\psi(0, r + x) + \int_0^x dz\,\exp(-\sigma z) \int_{r+z}^\infty dr'\,\bar{f}(r+z, r')\,\psi(x-z, r') \tag{8.125}$$

where the boundary condition is

$$\psi(0, r) = S(E)\,\phi(0, E) \tag{8.126}$$

A numerical algorithm for equation (8.125) is found by noting that

$$\begin{aligned}
\psi(x + h, r) = {}& \exp(-\sigma h)\,\psi(x, r + h) \\
& + \int_0^h dz\,\exp(-\sigma z) \int_r^\infty dr'\,\bar{f}(r+z, r'+z) \\
& \times \psi(x + h - z, r' + z)
\end{aligned} \tag{8.127}$$

which can be simplified by using

$$\psi(x + h - z, r) \approx \exp\left[-\sigma(h - z)\right]\psi(x, r + h - z) + O(h) \tag{8.128}$$

which yields

$$\begin{aligned}
\psi(x + h, r) \approx {}& \exp(-\sigma h)\,\psi(x, r + h) \\
& + \exp(-\sigma h) \int_0^h dz \int_r^\infty dr'\,\bar{f}(r+z, r'+z)\,\psi(x, r' + h)
\end{aligned} \tag{8.129}$$

with the order of h^2, where h is the step size. Equation (8.129) is accurate for distances such that $\sigma h \ll 1$ and may be used to relate the spectrum at some point x to the spectrum at $x + h$. Therefore, one may begin at the boundary ($x = 0$) and propagate the solution to any arbitrary interior point using equation (8.129).

Several advantages are seen in the above method. First, the range-energy relations enter the solution exactly. Second, the method introduces no extraneous unstable roots that arose in earlier methods by not treating the range-energy relation accurately. The inherent stability will tend to dampen any errors committed at the boundary or generated in the interior. Truncation errors enter the solution of equation (8.129), and their generation and propagation will now be considered.

8.5. Error Analysis of Unconditionally Stable Methods

There are two immediate questions regarding the use of equation (8.129) in the solution of charged-particle transport: (1) What are the relative errors in numerical implementation, and (2) how do these errors propagate into the solution domain? Although these two questions cannot be dealt with entirely independently, we first consider relative errors and then study their propagation.

8.5.1. Local relative error. Numerical interpolation in BRYNTRN was motivated by the observation that the high-energy spectrum for most space radiation varies as $E^{-\alpha}$ as noted in equation (8.121). Similarly, for $r_i \leq r \leq r_{i+1}$ we used

$$\psi(x, r) \approx a_i (r_i/r)^{\alpha_i} \tag{8.130}$$

where

$$a_i = \psi(x, r_i) \tag{8.131}$$

$$\alpha_i = \frac{\ln\left[\frac{\psi(x, r_{i+1})}{\psi(x, r_i)}\right]}{\ln(r_i/r_{i+1})} \tag{8.132}$$

and we define $\Delta = r_2 - r_1$. We evaluate the relative error for equation (8.130) for typical space spectra.

The galactic cosmic-ray spectrum is given approximately as

$$\psi_{\text{GCR}}(r) \approx \frac{1}{1 + r^2} \tag{8.133}$$

for which (with $i = 1$)

$$a_1 = (1 + r_1^2)^{-1} \tag{8.134}$$

$$\alpha_1 = \frac{\ln\left\{1 + \left[2r_1\,\Delta + \Delta^2)/(1 + r_1^2\right]\right\}}{\ln\left[1 + (\Delta/r_1)\right]} \tag{8.135}$$

and we assume that $\Delta \ll 1$. We evaluate the spectrum at the midpoint $r_m = r_1 + \frac{1}{2}\Delta$ and compare. First, note that

$$\ln\left[\psi_{\text{GCR}}(r_m)\right] = -\ln\left(1 + r_1^2 + \Delta r_1 + \frac{1}{4}\Delta^2\right) \tag{8.136}$$

is the exact value compared with

$$\ln \left[a_1 \left(\frac{r_1}{r_m} \right)^{\alpha_1} \right] = -\ln \left(1 + r_1^2 \right) - \frac{\ln \left(1 + \frac{2r_1 \Delta + \Delta^2}{1 + r_1^2} \right) \ln \left(1 + \frac{1}{2} \frac{\Delta}{r_1} \right)}{\ln \left(1 + \frac{\Delta}{r_1} \right)} \tag{8.137}$$

The error ε is examined in the following three limits:

For $r_i \gg 1$,

$$\varepsilon \approx +\frac{1}{4} \frac{\Delta^2}{r_1^2} \tag{8.138}$$

for $\Delta \ll r_i \ll 1$,

$$\varepsilon \approx \frac{1}{4} \Delta^2 \tag{8.139}$$

and for $\Delta \approx r_1 \ll 1$,

$$\varepsilon \approx 0.17 r_1 \Delta \tag{8.140}$$

It is clear from these limits that accuracy is easy to maintain for large values of r_1, but errors are progressively greater for lower values of r_1. One obvious problem with equation (8.130) is its concave shape between grid values leading to discontinuous derivatives.

Many solar-flare particle events are exponential rigidity spectra as

$$\phi \approx \exp \left[\frac{-P(E)}{P_o} \right] \tag{8.141}$$

where the momentum is

$$P(E) = \sqrt{E(E + 2Mc^2)} \tag{8.142}$$

with Mc^2 denoting the rest energy. In analogy we consider a trial spectrum

$$\phi = \exp \left\{ \frac{-[r(r+1)]^{1/2}}{r_o} \right\} \tag{8.143}$$

and the interpolating function of equation (8.130).

Consequently,

$$a_1 = \exp \left\{ \frac{-[r(r+1)]^{1/2}}{r_o} \right\} \tag{8.144}$$

$$\alpha_1 = \frac{\sqrt{(r_1 + \Delta)(r_1 + 1 + \Delta)} - \sqrt{r_1(r_1 + 1)}}{r_o \ln[1 + (\Delta/r_1)]} \tag{8.145}$$

and assume that $\Delta \ll 1$. At the midpoint we have

$$\ln \phi = \frac{-\left[\left(r_1 + \frac{\Delta}{2}\right)\left(1 + r_1 + \frac{\Delta}{2}\right)\right]^{1/2}}{r_0} \tag{8.146}$$

which is the exact value to be compared with

$$\ln\left[a_1 \left(\frac{r}{r_m}\right)^{\alpha_1}\right] = \frac{-[(r_1 + \Delta)(1 + r_1 + \Delta)]^{1/2}}{r_0}$$

$$- \frac{\left[\sqrt{(r_1 + \Delta)(1 + r_1 + \Delta)} - \sqrt{r_1(r_1 + 1)}\right] \ln\left(1 + \frac{\Delta}{2r_1}\right)}{r_0 \ln\left(1 + \frac{\Delta}{r_1}\right)} \tag{8.147}$$

The error ε is examined in the limits as before:

For $r_1 \gg 1$,

$$\varepsilon \approx -\frac{3}{2} \frac{\Delta}{r_0} \tag{8.148}$$

and for $\Delta \leq r_1 \ll 1$,

$$\varepsilon \approx -\frac{3}{4} \frac{\Delta}{r_0 \sqrt{r_1}} \tag{8.149}$$

Clearly, one requirement for high accuracy is $\Delta \ll r_0$, but the error still increases as r_1 becomes small.

An alternate choice for an interpolating function is

$$\psi(r) = a_i \exp\left[-b_i(r - r_i)\right] \qquad (r_i \leq r \leq r_{i+1}) \tag{8.150}$$

This function has the qualitative feature of being convex as are most space spectra. As before,

$$a_i = \psi(r_i) \tag{8.151}$$

with b_i given as

$$b_i = \frac{-\ln\left[\psi(r_2)/\psi(r_1)\right]}{r_2 - r_1} \tag{8.152}$$

Analysis shows that the assumed GCR spectrum is always correct to $O(\Delta^2)$ which is taken to be small. For the rigidity spectrum, the error is

$$\varepsilon \approx \frac{\Delta}{r_0} \qquad (r_1 \gg 1 \gg \Delta) \tag{8.153}$$

$$\varepsilon \approx \frac{\sqrt{r_1}}{2r_o} \Delta \qquad (1 \gg r_1 \gg \Delta) \qquad (8.154)$$

$$\varepsilon = \frac{\Delta}{2\sqrt{r_1} r_o} \qquad (1 \gg r_1 \approx \Delta) \qquad (8.155)$$

The errors for the interpolating function (8.150) are on the same order as the earlier interpolating function (8.130), except that their coefficient is a factor of 2 to 3 smaller and the error is now an overestimate.

8.5.2. Error propagation. In consideration of how errors are propagated in the use of equation (8.129), the error is introduced locally by calculating $\psi(x, r+h)$ over the range (energy) grid over which it was defined as

$$\psi(x + h, r_i) = \exp(-\sigma h) \, \psi(x, r_i + h) \qquad (8.156)$$

We denote the truncation error introduced into equation (8.156) as

$$\psi(x, r_i + h) = \psi_{\text{int}}(x, r_i + h) + \varepsilon_i(h) \qquad (8.157)$$

After the kth step from the boundary, the numerical solution is

$$\psi(kh, r_i) = \exp(-\sigma h) \, \psi_{\text{int}}\left[(k-1)h, r_i + h\right] + \sum_{\ell=0}^{k-1} \exp\left[-\sigma(k-\ell)h\right] \varepsilon_\ell(h) \quad (8.158)$$

Suppose that $0 \le \varepsilon_\ell(h) \le \varepsilon(h)$ for all values of ℓ, then the propagated error is bound by

$$\varepsilon_{\text{prp}}(h) = \sum_{\ell=0}^{k-1} \exp\left[-\sigma(k-\ell)h\right] \varepsilon_\ell(h) \le \varepsilon(h) \sum_{\ell=0}^{k-1} \exp\left[-\sigma(k-\ell)h\right] \qquad (8.159)$$

We note that

$$\sum_{\ell=0}^{k-1} \exp(-\sigma kh) \exp(\sigma h\ell) \approx \frac{1}{h\sigma} \left[1 - \exp(-\sigma kh)\right] \qquad (8.160)$$

Clearly, the propagated error on the kth step is bound by

$$\varepsilon_{\text{prp}}(h) < \frac{\varepsilon(h)}{h\sigma} \left[1 - \exp(-\sigma kh)\right] \qquad (8.161)$$

where $\varepsilon(h)$ is the maximum error per step. The propagated error grows each step to a maximum value of $\varepsilon(h)/h\sigma$ and would require $\varepsilon(h)$ to be on the order of $O(h^2)$ for good convergence. The asymptotic bound for the propagated error is

$$\varepsilon_{\text{prp}} \le \varepsilon(h) \sum_{\ell=0}^{\infty} \exp(-\sigma\ell h) = \varepsilon(h) \frac{\exp(-\sigma h)}{1 - \exp(-\sigma h)} \qquad (8.162)$$

emphasizing again the need to control the error as $h\sigma \to 0$. It is clear that the higher order techniques are required to control error propagation as found in recent studies (Shinn et al., 1991).

8.5.3. Numerical procedure. We now consider numerical methods for the integral portion of equation (8.127). We will make use of the form of the interaction given by equation (8.12) for which an analytic solution has already been obtained. As shown in connection with equation (8.129), equation (8.127) may be rewritten as

$$\psi(x+h,r) = \exp(-\sigma h) \; \psi(x,r+h)$$

$$+ \exp(-\sigma h) \int_0^h dz \int_r^\infty dt \; a \exp\left[-\alpha(r+z)\right] \psi(x,t+h)$$

$$+ \exp(-\sigma h) \int_0^h dz \int_r^\infty dt \; c \exp\left[\gamma(r-t-Q)\right] \psi(x,t+h)$$

$$+ O(h^2) \tag{8.163}$$

where Q represents the average energy shift of the projectile producing the secondary particles across the interval x to $x+h$. In principle, $a, \alpha, c,$ and γ are dependent on projectile energy as well and would be evaluated using the same value of Q (Wilson et al., 1989). With the analytic forms in equation (8.163), we may perform the integrals as

$$\psi(x+h,r) = \exp(-\sigma h) \; \psi(x,r+h)$$

$$+ \exp(-\sigma h) \int_r^\infty dt \; \frac{a}{\alpha}\left[\exp(-\alpha r) - \exp\left[-\alpha(r+h)\right]\right]\psi(x,t+h)$$

$$+ \exp(-\sigma h) \int_r^\infty dt \; hc \exp\left[\gamma(r-t)\right] \psi(x,t+h) + O(h^2) \tag{8.164}$$

The integral terms of equation (8.164) can be written in terms of the cumulative secondary spectra denoted as

$$F_a(r,t) = \int_0^r a \, \exp(-\alpha z) \; dz \tag{8.165}$$

and

$$F_c(r,t) = \int_0^r c \exp[\gamma(z-t)] \, dz \tag{8.166}$$

In particular,

$$\frac{a}{\alpha}\left\{\exp(-\alpha r) - \exp\left[-\alpha(r+h)\right]\right\} = F_a(r+h,t+Q) - F_a(r,t+Q) \tag{8.167}$$

$$ch \exp[\gamma(r - t - Q)] = [F_c(r + h, t + Q) - F_c(r, t + Q)]$$
$$\times \left[1 - \frac{1}{2}\gamma(h - 2Q) + O(h^2)\right] \qquad (8.168)$$

which may now be substituted into equation (8.164) to obtain

$$\psi(x + h, r) = \exp(-\sigma h)\, \psi(x, r + h)$$
$$+ \exp(-\sigma h) \int_r^\infty dt\, [F_a(r + h, t + Q) - F_a(r, t + Q)]\, \psi(x, t + h)$$
$$+ \exp(-\sigma h) \int_r^\infty dt\, [F_c(r + h, t + Q) - F_c(r, t + Q)]\, \psi(x, t + h)$$
$$+ O(h - 2Q) + O(h^2) \qquad (8.169)$$

The second-order accuracy is maintained only if Q is chosen at the midpoint of the interval (i.e., $Q = \frac{1}{2}h$). Additional details of this analysis can be found in Wilson et al. (1989). The propagation equation is implemented as

$$\psi(x + h, r) = \exp(-\sigma h)\, \psi(x, r + h)$$
$$+ \int_r^\infty dt\, \bar{F}\left(h, r, t + \frac{h}{2}\right)\, \psi(x, t + h) \qquad (8.170)$$

where

$$\bar{F}(h, r, t) = \int_0^h \bar{f}(r + z, t)\, dz$$
$$\equiv F(r + h, t) - F(r, t) \qquad (8.171)$$

and is related to the cumulative energy spectrum by

$$F(r, t) = \int_0^{\varepsilon(r)} f(E, E')\, dE \qquad (8.172)$$

where $\varepsilon(r)$ is the energy associated with the residual range r and $E' = \varepsilon(t)$.

8.6. Coupled Baryon Transport Methods

The coupled baryon transport equations are of the form

$$\left[\frac{\partial}{\partial x} - \nu_j \frac{\partial}{\partial E} S(E) + \sigma_j(E)\right] \phi_j(x, E) = \sum_k \int_0^\infty f_{jk}(E, E')\, \phi_k(x, E')\, dE'$$
$$(8.173)$$

where ν_j is the range scaling parameter, $S(E)$ is the stopping power of the protons, $\sigma_j(E)$ is the total cross section, $\phi_j(x, E)$ is the differential flux spectrum of type j

baryons, and $f_{jk}(E, E')$ is a differential energy cross section for redistribution of particle type and energy. Utilizing the definitions

$$r = \int_0^E \frac{dE'}{S(E')} \tag{8.174}$$

$$\psi_j(x, r) = S(E)\,\phi_j(x, E) \tag{8.175}$$

and

$$\bar{f}_{jk}(r, r') = S(E)\,f_{jk}(E, E') \tag{8.176}$$

allows equation (8.173) to be written as

$$\left[\frac{\partial}{\partial x} - \nu_j \frac{\partial}{\partial r} + \sigma_j(r)\right] \psi_j(x, r) = \sum_k \int_r^\infty \bar{f}_{jk}(r, r')\,\psi_k(x, r')\,dr' \tag{8.177}$$

which may be rewritten as (Wilson and Lamkin, 1975; Wilson and Badavi, 1986)

$$\psi_j(x, r) = \exp\left[-\zeta_j(r, x)\right] \psi_j(0, r + \nu_j x)$$

$$+ \sum_k \int_0^x \int_r^\infty \exp\left[-\zeta_j(r, z)\right] \bar{f}_{jk}(r + \nu_j z, r')$$

$$\times \psi_k(x - z, r')\,dr'\,dz \tag{8.178}$$

where the exponential is the integrating factor with

$$\zeta_j(r, t) = \int_0^t \sigma_j(r + \nu_j t')\,dt' \tag{8.179}$$

If the interactions are such that

$$\bar{f}_{jk}(r, r') = \delta_{jk}\,g(r - r') \tag{8.180}$$

where g denotes the appropriate spectral function, then the solutions to equation (8.177) are of the form

$$\psi_j(x, r) = \chi(x, r + \nu_j x) \tag{8.181}$$

To demonstrate how remarkable equation (8.181) is, we note that if $\chi(x, r)$ is the solution to the neutron transport equation ($\nu_n = 0$), then $\chi(x, r + \nu_p x)$ is the solution to the proton transport problem independent of the functional form chosen for the stopping power.

Rather simple numerical procedures follow from equation (8.179). Noting that the first-order nature of equation (8.173) allows $\psi_j(x, r)$ to be taken as a boundary condition for propagation to larger values of x, one may approximate equation (8.179) as

$$\psi_j(x+h,r) = \exp\left[-\zeta_j(r,h)\right] \psi_j(x,r+\nu_j h)$$

$$+ \sum_k \int_0^h \int_r^\infty \exp\left[-\zeta_j(r,z)\right] \bar{f}_{jk}(r+\nu_j z, r')$$

$$\times \psi_k(x+h-z,r') \, dz \, dr' \tag{8.182}$$

which may be used to develop a numerical stepping procedure. Equation (8.182) has provided the basis for a number of new transport codes for baryons of mass number greater than or equal to 1. (See Wilson and Lamkin (1975); Wilson and Badavi (1986); Wilson et al. (1989); Shinn et al. (1990).) These codes are now being extended to couple with the meson fields and with the negative baryon number fields.

If h is sufficiently small such that

$$\sigma_j(r') \, h \ll 1 \tag{8.183}$$

then, according to perturbation theory (Wilson and Lamkin, 1975)

$$\psi_k(x+h-z,r') \approx \exp[-\zeta_k(r,h-z)]\psi_k[x,r'+\nu_k(h-z)] \tag{8.184}$$

which may be used to approximate the above integral of equation (8.182).

For many cases of practical interest (e.g., accelerator studies), monoenergetic particle beams are used, and separation of the singular terms from the solution becomes convenient. The initial beam of type J particles of energy E_0 (where $r_o = R(E_o)$) is taken as

$$\psi_j(0,r) = \delta_{jJ}\, \delta(r_0 - r) \tag{8.185}$$

and the solution is written by the replacement

$$\psi_j(x,r) \Rightarrow \psi_{j0}(x,r) + \psi_j(x,r) \tag{8.186}$$

The corresponding singular terms are

$$\psi_{k0}(x,r) = \exp[-\zeta_k(r,x)]\, \delta(r_0 - r - \nu_k x)\delta_{kj} \tag{8.187}$$

The regular terms of equation (8.182) for $k = p$ may be written as

$$\psi_p(x+h,r) = \exp\left[-\zeta_p(r,h)\right]\psi_p(x,r+h)$$

$$+ \int_0^h dz \exp[-\zeta_p(r,z)] \sum_j \int_{r+z}^\infty \bar{f}_{pj}(r+z,r')$$

$$\times [\psi_{j0}(x+h-z,r') + \psi_j(x+h-z,r')] \, dr' \tag{8.188}$$

and the regular terms for $k = n$ are

$$\psi_n(x+h,r) = \exp[-\sigma_n(r)h]\psi_n(x,r)$$
$$+ \int_0^h dz \exp[-\sigma_n(r)z] \sum_j \int_r^\infty \bar{f}_{nj}(r,r')$$
$$\times [\psi_{j0}(x+h-z,r') + \psi_j(x+h-z,r')]\,dr' \qquad (8.189)$$

The singular contribution under the integrals of equations (8.188) and (8.189) can be evaluated with equation (8.187), and the approximations in equations (8.183) and (8.184) can be applied to find

$$\psi_p(x+h,r) = \exp\left[-\sigma_p(r)h\right]\psi_p(x,r+h)$$
$$+ \exp\left\{-[\sigma_p(r)+\sigma_p(r'_o)]\frac{h}{2}\right\}\,\overline{F}_{pp}(h,r,r'_o)\,\delta_{pj}\,\exp\left[-\zeta_p(r'_o,x)\right]$$
$$+ \exp\left\{-[\sigma_p(r)+\sigma_n(r_o)]\frac{h}{2}\right\}\,\overline{F}_{pn}(h,r,r_o)\,\delta_{nj}\,\exp\left[-\sigma_n(r_o)x\right]$$
$$+ \int_r^\infty \exp\left\{-\left[\sigma_p(r)+\sigma_p\left(r'+\frac{h}{2}\right)\right]\frac{h}{2}\right\}$$
$$\times \overline{F}_{pp}\left(h,r,r'+\frac{h}{2}\right)\psi_p(x,r'+h)\,dr'$$
$$+ \int_r^\infty \exp\left\{-\left[\sigma_p(r)+\sigma_n\left(r'+\frac{h}{2}\right)\right]\frac{h}{2}\right\}$$
$$\times \overline{F}_{pn}\left(h,r,r'+\frac{h}{2}\right)\psi_n\left(x,r'+\frac{h}{2}\right)\,dr' \qquad (8.190)$$

and

$$\psi_n(x+h,r) = \exp\left[-\sigma_n(r)\,h\right]\psi_n(x,r)$$
$$+ h\,\bar{f}_{np}(r,r'_o)\,\exp\left\{-[\sigma_n(r)+\sigma_p(r'_o)]\frac{h}{2}\right\}\delta_{pj}\,\exp\left[-\zeta_p(r'_o,x)\right]$$
$$+ h\bar{f}_{nn}(r,r'_o)\,\exp\left\{-[\sigma_n(r)+\sigma_n(r_o)]\frac{h}{2}\right\}\delta_{nj}\,\exp\left[-\sigma_n(r_o)x\right]$$
$$+ h\int_r^\infty \exp\left\{-[\sigma_n(r)+\sigma_p(r')]\frac{h}{2}\right\}\bar{f}_{np}(r,r')\,\psi_p\left(x,r'+\frac{h}{2}\right)\,dr'$$
$$+ h\int_r^\infty \exp\left\{-[\sigma_n(r)+\sigma_n(r')]\frac{h}{2}\right\}\bar{f}_{nn}(r,r')\,\psi_n(x,r')\,dr'$$
$$\qquad (8.191)$$

where $r'_o = r_o - x - \frac{h}{2}$ and $\overline{F}$ is related to the cumulative spectrum F as given by

$$\overline{F}_{ij}(h,r,r') = \int_0^h \bar{f}_{ij}(r+z,r')\,dz \qquad (8.192)$$

with

$$F_{ij}(r, r') = \int_0^{\varepsilon(r)} f_{ij}(E, E') \, dE \qquad (8.193)$$

$\varepsilon(r)$ is the energy associated with residual range r, and $E' = \varepsilon(r')$. Equations (8.190) and (8.191) are evaluated by establishing an x-grid at which $\psi_j(x_m, r)$ is evaluated, where h is the distance between each successive evaluation. The integral over r' is accomplished by establishing an r-grid (and the corresponding E-grid) and using

$$\int_{r_n}^{\infty} g(r_n, r') \, \psi_j(x_m, r') \, dr' \approx \sum_{\ell=n}^{\infty} g_n(r_n, \bar{r}_\ell) \int_{r_\ell}^{r_{\ell+1}} \psi_j(x_m, r) \, dr' \qquad (8.194)$$

where $\bar{r}_\ell = (r_\ell + r_{\ell+1})/2$, and the series terminates at the highest value of ℓ in the r-grid. There is a spatially dependent discontinuity in the proton flux spectrum that requires right- and left-hand interpolation and integration. These discontinuities have been treated in the computational procedures.

8.7. Results and Discussion

Because the buildup factors are functions of both energy and thickness, the first step in verifying such a method is to compare the result at various fixed (discrete) energies. Comparisons were made (Wilson et al., 1989) between BRYNTRN and Monte Carlo results for monoenergetic protons at various energies and were in reasonable agreement considering the numerical difficulty involved in discrete energy calculations with BRYNTRN. For the buildup-factor method, comparisons made with Monte Carlo are shown in figures 8.5 and 8.6 and with experiment in figure 8.7.

The dose and dose equivalent calculated as functions of depth in tissue with and without aluminum shield are shown in figures 8.5 and 8.6 for normal incident protons at discrete energies of 400, 660, 730, 1500, and 3000 MeV. The limited Monte Carlo results with 0 g/cm^2 shielding are obtained from Alsmiller, Armstrong, and Coleman (1970). The calculated values with the buildup-factor method are seen to be in reasonable agreement despite the crudeness in the buildup parameters chosen. Although there are no Monte Carlo data available with shielding at discrete energy, the doses calculated with 30 g/cm^2 of aluminum shield by the buildup-factor method are also presented in the figures for qualitative comparison. In general, the dose is increased because of the presence of the shield (transition effect discussed in chapter 1). The increase in dose over those with no shielding results from neutrons produced in aluminum, especially in the first few centimeters of the tissue. For protons at the lowest energy (400 MeV), the Bragg peak appears at 55 cm in depth as the protons approach their limiting range as they travel through the shield and tissue.

Figure 8.7 shows a comparison of the experimental data of Baarli and Goebel at CERN, the buildup-factor method, and the interpolated Monte Carlo result (Alsmiller, Armstrong, and Coleman, 1970; Turner et al., 1964; Wright, Hamm,

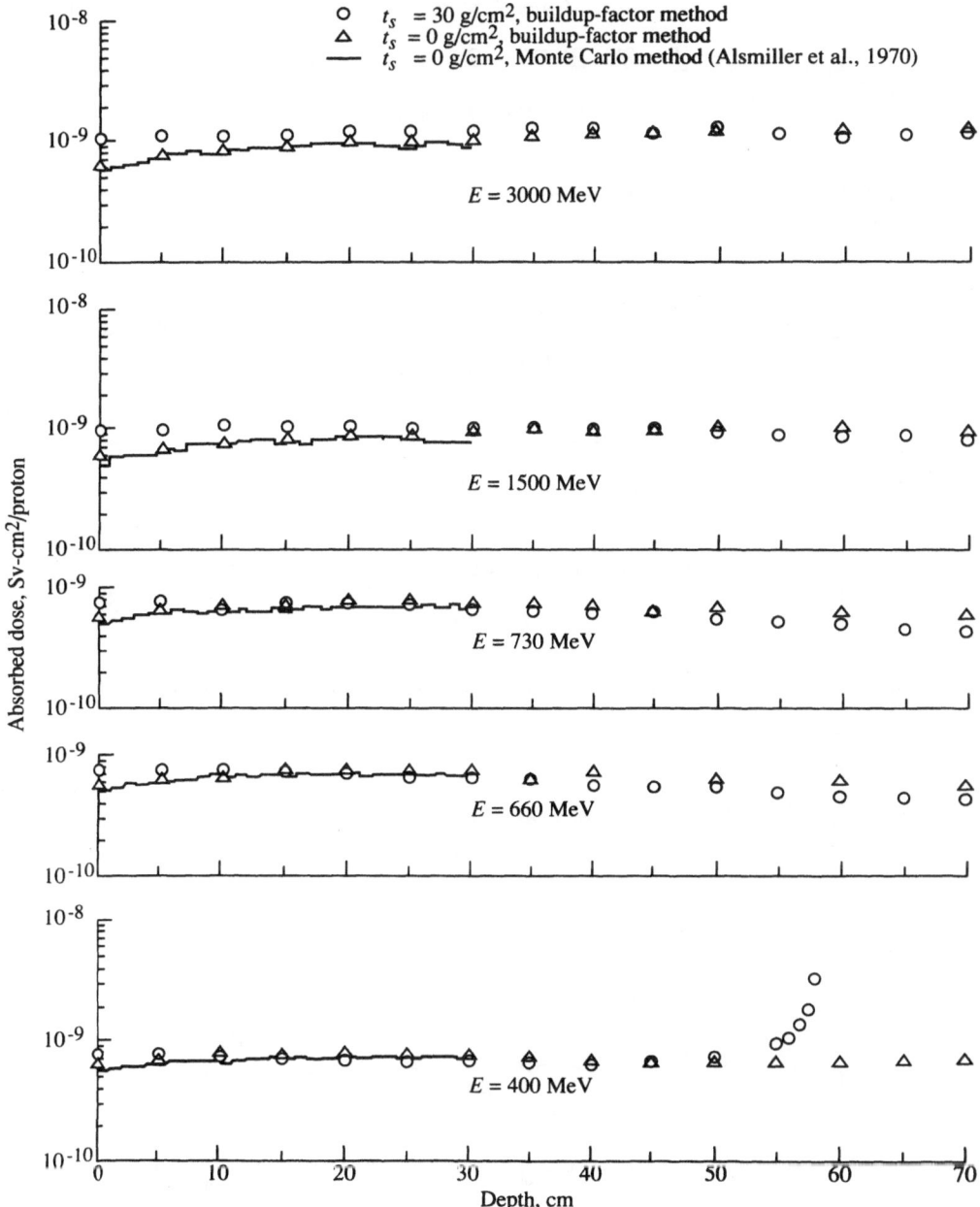

Figure 8.5. Dose in tissue (with or without aluminum shield) exposed to normal incident protons at various discrete energies.

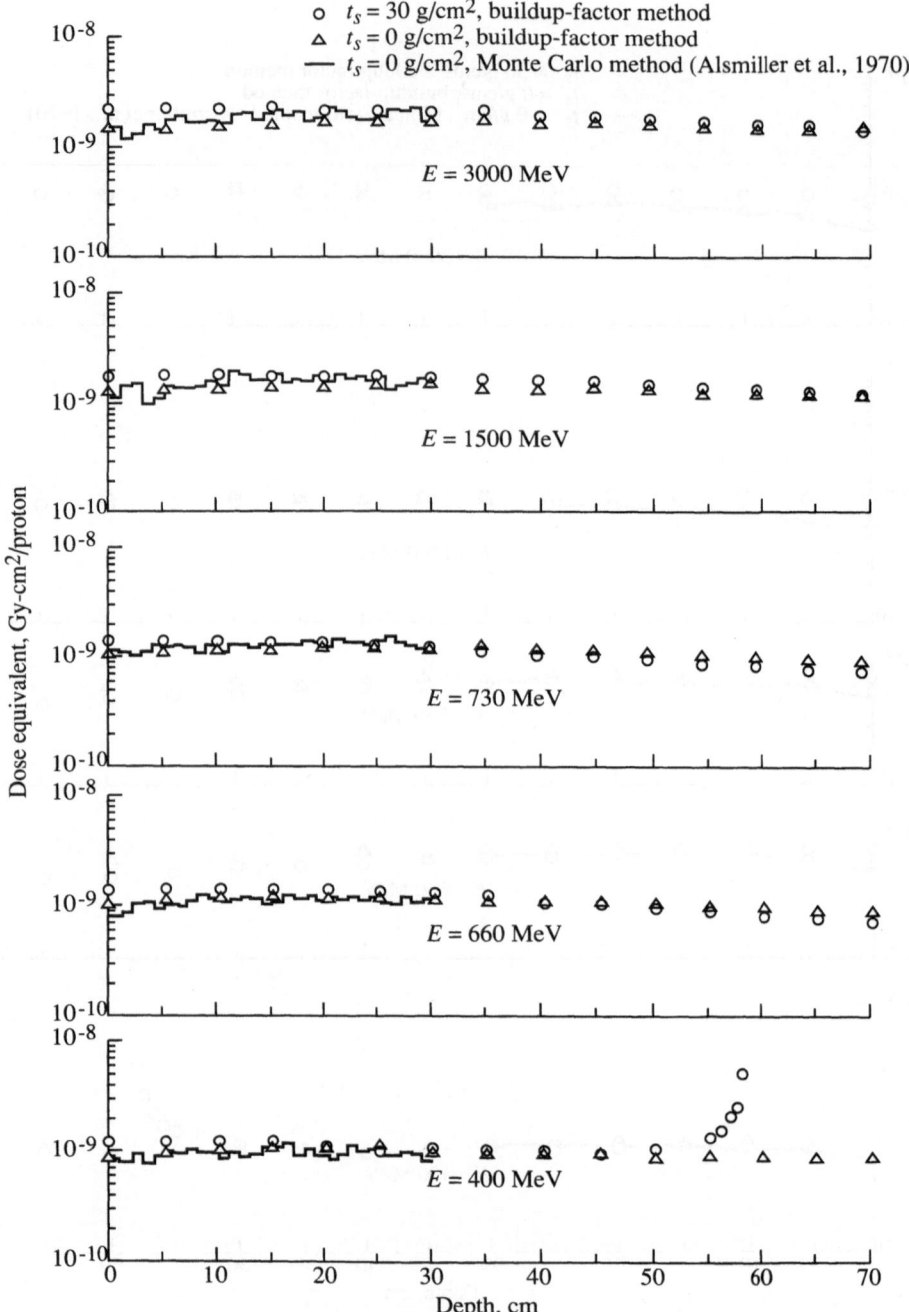

Figure 8.6. Dose equivalent in tissue (with or without aluminum shield) exposed to normal incident protons at various discrete energies.

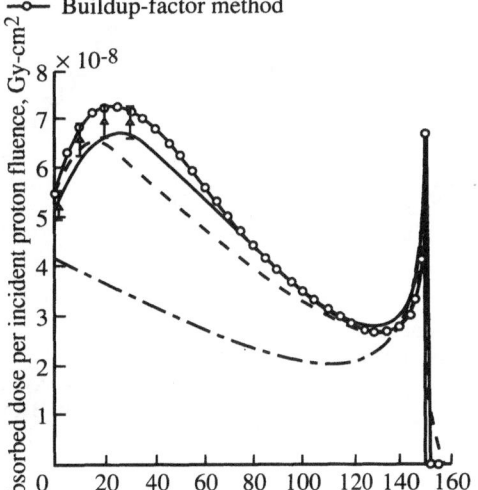

Figure 8.7. Comparison of predicted dose in tissue with experimental data for normal incident protons at 592 MeV.

and Turner, 1971) for the absorbed dose in tissue that is exposed to a proton beam of 592 MeV. Also shown are the earlier buildup-factor calculations (Wilson and Khandelwal, 1974) for uncollided primary and total absorbed doses. Observe that the dose from the buildup, which is the difference between the total and uncollided primary, is substantial. The usual Bragg peak is also obvious for both the analytical and experimental results. The buildup-factor calculations are approximately within the uncertainties of interpolated Monte Carlo values and are in reasonable agreement with the experimental data.

To verify both BRYNTRN and the buildup-factor method in case of a continuous energy spectrum of incident protons, dose calculations were made (figs. 8.8 and 8.9) for shielded tissue being exposed to a typical solar-flare spectrum for which Monte Carlo results were available (Scott and Alsmiller, 1967 and 1968). The flare spectrum taken from Scott and Alsmiller (1967) is of the Webber (1966) form and is exponential in rigidity with characteristic rigidity $P_o = 100$ MV and normalized to 10^9 protons/cm^2 with energy greater than 30 MeV. Only the portion of the spectrum between 50 and 400 MeV was considered for the Monte Carlo calculation (Scott and Alsmiller, 1967 and 1968). Nevertheless, for the current calculations, the high-energy cutoff at 400 MeV was ignored, and very small differences of a few percent were found (figs. 8.8(a) and 8.9(a)) because the spectrum contains very few highly penetrating energetic protons, which may become significant only at depths beyond current interest. The tissue had an aluminum shield thickness of 20 g/cm^2 (fig. 8.8) and an iron shield thickness of 20 g/cm^2 (fig. 8.9).

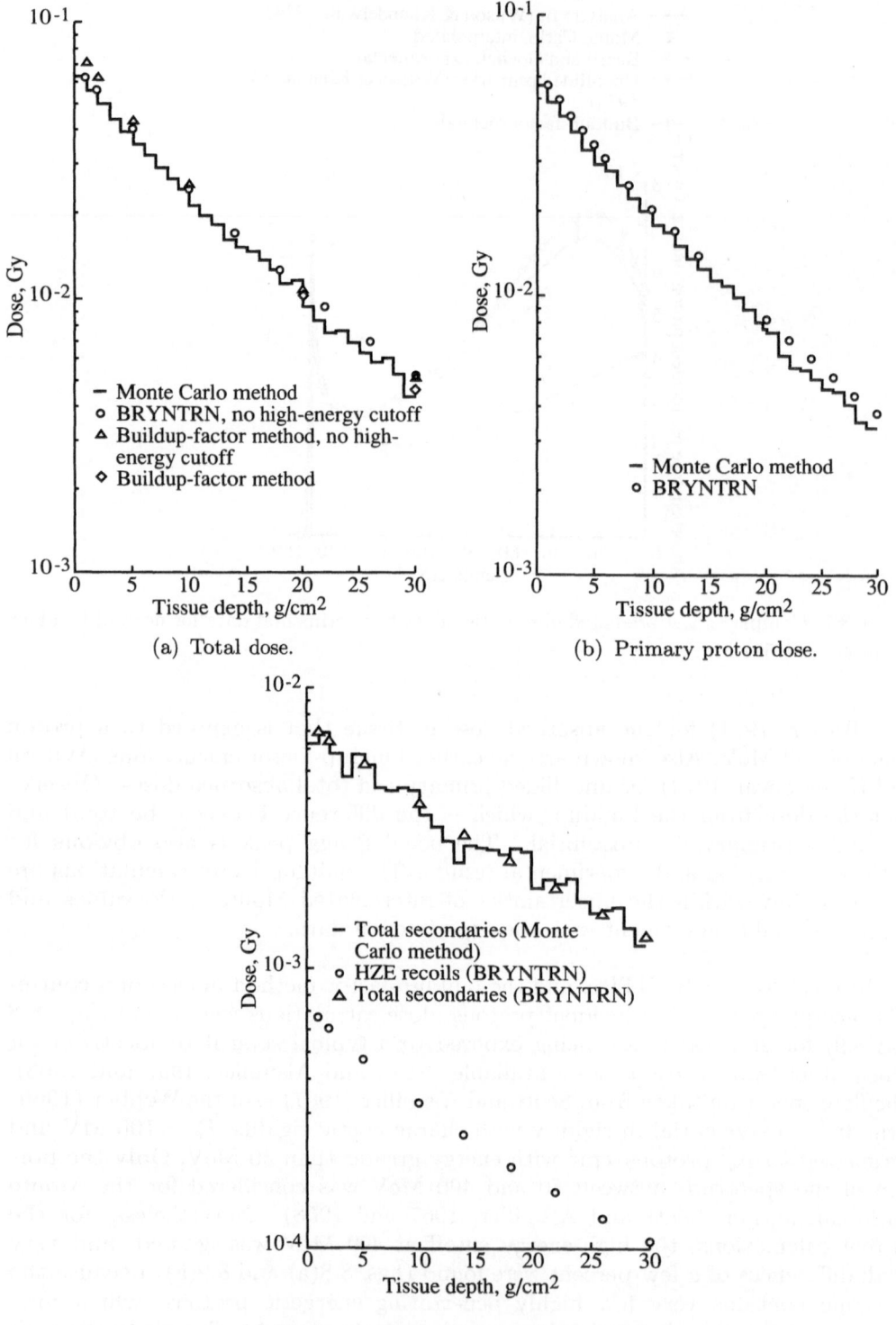

(a) Total dose.

(b) Primary proton dose.

(c) Total secondary and heavy ion recoil dose.

Figure 8.8. Various quantities of doses in tissue behind 20 g/cm² of aluminum shield to normal incidence of a solar-flare proton spectrum of Webber form with rigidity equal to 100 MV.

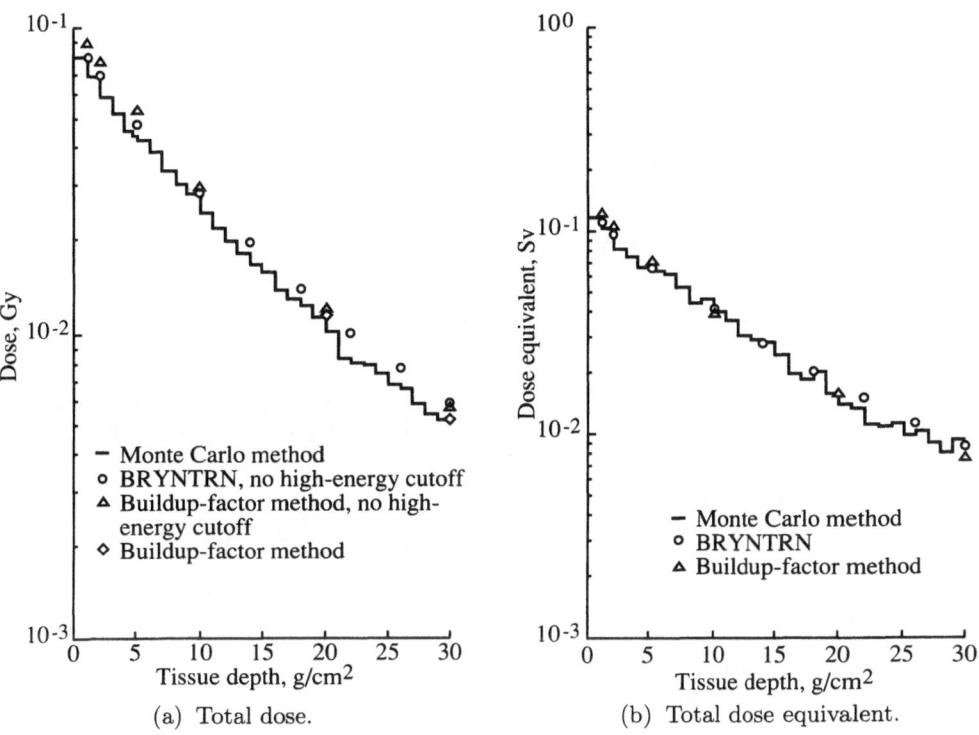

(a) Total dose.   (b) Total dose equivalent.

Figure 8.9. Various quantities of doses in tissues behind 20 g/cm^2 of iron shield to normal incidence of a solar-flare spectrum of Webber form with rigidity equal to 100 MV.

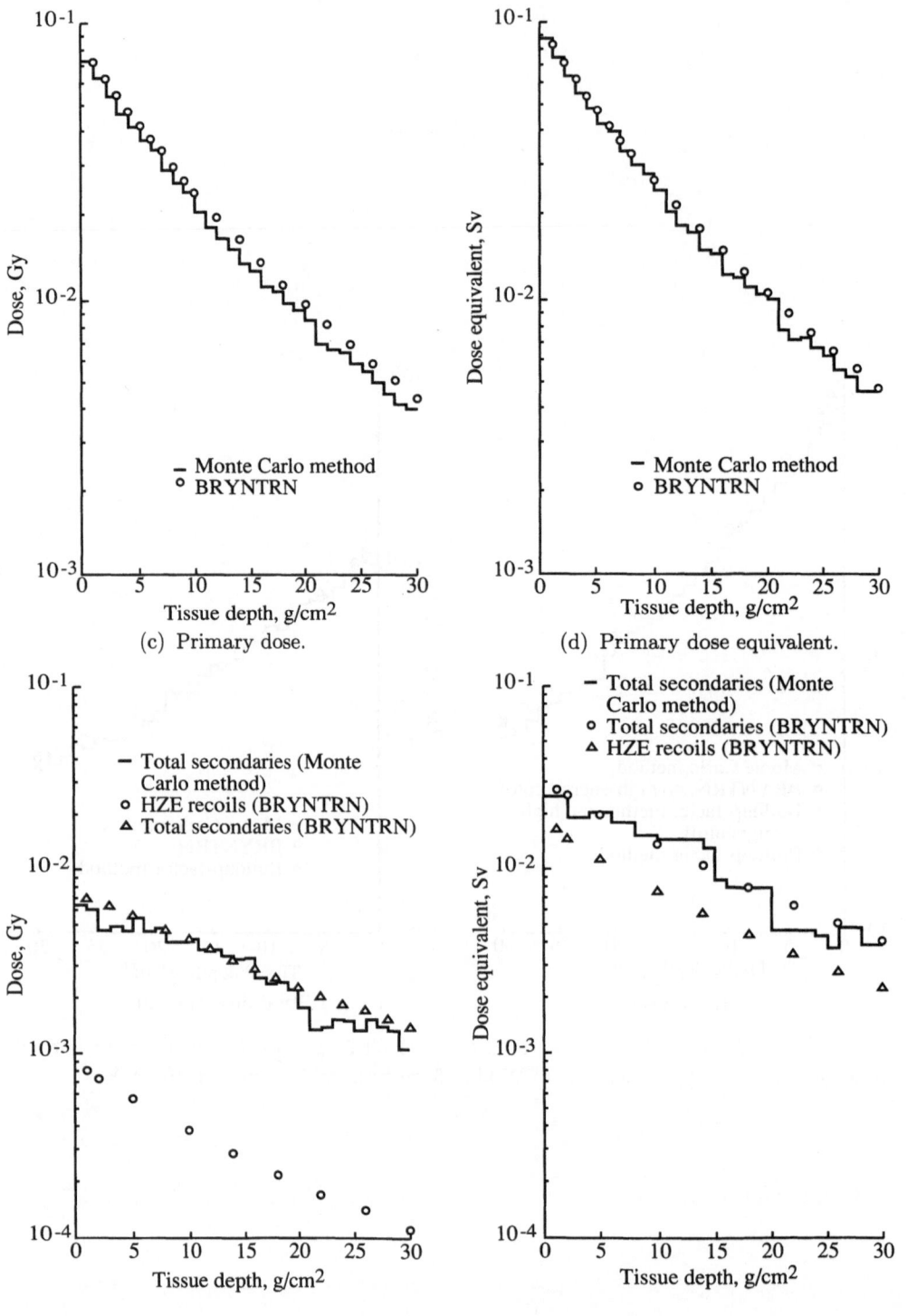

(c) Primary dose.

(d) Primary dose equivalent.

(e) Total secondary and heavy ion recoil doses.

(f) Total secondary and heavy ion recoil dose equivalents.

Figure 8.9. Concluded.

The buildup-factor results presented for iron, however, were obtained with the same values of buildup parameters chosen for aluminum. This is allowable because the nuclear reaction cross sections are roughly the same for both materials at the energies of interest (below 400 MeV). The total doses by the buildup-factor method are seen to be in good agreement (within 5–10 percent) with both Monte Carlo and BRYNTRN results. (See figs. 8.8(a) and 8.9(e)–(f).) (Note that the total secondary dose is the sum of the secondary proton and secondary neutron dose of Scott and Alsmiller (1967 and 1968).) The heavy ion recoil dose and dose equivalent of BRYNTRN show that the actual physical dose from heavy recoils may not be important, but their contribution to the dose equivalent can be significant because of the large quality factor.

Dose calculations are also made for a continuous spectrum that contains more high-energy protons. Because no Monte Carlo results are available in the high-energy range, the February 1956 solar-flare event is chosen for comparison between BRYNTRN and the buildup-factor method. The solar-flare spectrum given as the integral fluence form in protons/cm^2 is

$$\phi_P(> E) = 1.5 \times 10^9 \exp\left(-\frac{E - 10}{25}\right) + 3 \times 10^8 \exp\left(-\frac{E - 100}{320}\right) \qquad (8.195)$$

where E is the energy in MeV. The results shown in figure 8.10 are for the dose and dose equivalent in tissue with 0, 10, and 30 g/cm^2 of aluminum shielding. The agreement between these two deterministic methods is seen to be reasonably good. With the future improvement to include neutron coupling, the buildup-factor method, which is gaining in computational efficiency for flare-dose analysis, would probably be favored in parametric studies of spacecraft shield design.

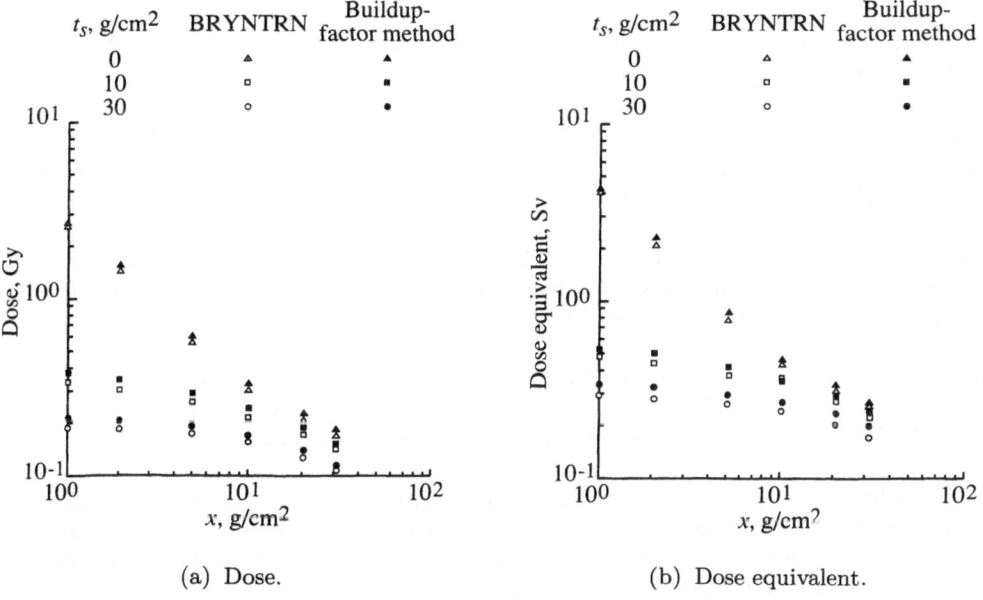

(a) Dose.   (b) Dose equivalent.

Figure 8.10. Dose in tissue behind various thicknesses of aluminum shield to normal incidence of the February 1956 solar-flare event.

8.8. Concluding Remarks

A comparison has been made of the calculated doses in tissue behind various thicknesses of shielding with exposure to various proton spectra for the buildup-factor method, BRYNTRN, and the Monte Carlo method. The results are found to be in reasonable agreement (within 5–10 percent), but with some overestimation by the buildup factors when the effect of neutron production in the shield is significant. Future improvement to include neutron coupling in the buildup-factor theory should alleviate this shortcoming. Impressive agreement for various components of doses is obtained between BRYNTRN and the Monte Carlo calculation. This is not surprising in the sense that both use some form of the Bertini cross-section data, but it does reflect on the contributions from corrections for the straight ahead approximation since the Monte Carlo calculation is a fully three-dimensional code. The straight ahead errors are, as expected, reasonably small. (See chapter 7.)

8.9. References

Alsmiller, R. G., Jr., 1967: High-Energy Nucleon Transport and Space Vehicle Shielding. *Nuclear Sci. & Eng.,* vol. 27, no. 2, pp. 158–189.

Alsmiller, R. G., Jr.; Armstrong, T. W.; and Coleman, W. A., 1970: The Absorbed Dose and Dose Equivalent From Neutrons in the Energy Range 60 to 3000 MeV and Protons in the Energy Range 400 to 3000 MeV. *Nuclear Sci. & Eng.,* vol. 42, no. 3, pp. 367–381.

Alsmiller, R. G., Jr.; Santoro, R. T.; Barish, J.; and Claiborne, H. C., 1972: *Shielding of Manned Space Vehicles Against Protons and Alpha Particles.* ORNL-RSIC-35, U.S. Atomic Energy Commission.

Anon., 1977: *Recommendations of the International Commission on Radiological Protection.* ICRP Publ. 26, Pergamon Press.

Armstrong, T. W.; and Chandler, K. C., 1970: *Calculation of the Absorbed Dose and Dose Equivalent From Neutrons and Protons in the Energy Range From 3.5 GeV to 1.0 TeV.* ORNL-TM-3758, U.S. Atomic Energy Commission.

Armstrong, T. W.; and Bishop, B. L., 1971: Calculation of the Absorbed Dose and Dose Equivalent Induced by Medium-Energy Neutrons and Protons and Comparison With Experiment. *Radiat. Res.,* vol. 47, no. 3, pp. 581–588.

Armstrong, T. W., 1972: *Radiation Transport Codes for Potential Applications Related to Radiobiology and Radiotherapy Using Protons, Neutrons, and Negatively Charged Pions.* ORNL-TM-3816 (NASA Order H-38280A), Oak Ridge National Lab.

Atwell, William; Beever, E. Ralph; Hardy, Alva C.; and Cash, Bernard L., 1989: A Radiation Shielding Model of the Space Shuttle for Space Radiation Dose Exposure Estimations. *Advances in Nuclear Engineering Computation and Radiation Shielding,* Volume 1, Michael L. Hall, ed., American Nuclear Soc., Inc., pp. 11:1–11:12.

Bertini, Hugo W., 1963: Low-Energy Intranuclear Cascade Calculation. *Phys. Review,* vol. 131, second ser., no. 4, pp. 1801–1821.

Bertini, Hugo W., 1969: Intranuclear-Cascade Calculation of the Secondary Nucleon Spectra From Nucleon-Nucleus Interactions in the Energy Range 340 to 2900 MeV and Comparisons With Experiment. *Phys. Review,* vol. 188, second ser., no. 4, pp. 1711–1730.

Bertini, Hugo W.; and Guthrie, Miriam P., 1971: News Item—Results From Medium-Energy Intranuclear-Cascade Calculation. *Nuclear Phys.,* vol. A169, no. 3, pp. 670–672.

Billings, M. P.; and Yucker, W. R., 1973: *The Computerized Anatomical Man (CAM) Model.* NASA CR-134043.

Cocconi, G.; Cocconi Tongiorgi, V.; and Widgoff, M., 1950: Cascades of Nuclear Disintegrations Induced by the Cosmic Radiation. *Phys. Review,* vol. 79, second ser., no. 5, pp. 768–780.

Ganapol, Barry D.; Townsend, Lawrence W.; Lamkin, Stanley L.; and Wilson, John W., 1991: *Benchmark Solutions for the Galactic Heavy-Ion Transport Equations With Energy and Spatial Coupling.* NASA TP-3112.

Haffner, James W., 1967: *Radiation and Shielding in Space.* Academic Press, Inc.

Khandelwal, G. S.; and Wilson, J. W., 1973: Proton Dosimeter Design for Distributed Body Organs. *Nuclear Technol.,* vol. 20, no. 1, pp. 64–67.

Murzin, V. S.; and Sarycheva, L. I., 1970: *Cosmic Rays and Their Interactions.* NASA TT F-594.

Ngo, Duc M.; Wilson, John W.; Buck, Warren W.; and Fogarty, Thomas N., 1989: *Nuclear-Fragmentation Studies for Microelectronic Applications.* NASA TM-4143.

O'Brien, K., 1971: Cosmic-Ray Propagation in the Atmosphere. *Nuovo Cimento,* vol. 3A, no. 3, pp. 521–547.

Reedy, R. C.; and Arnold, J. R., 1972: Interaction of Solar and Galactic Cosmic-Ray Particles With the Moon. *J. Geophys. Res.,* vol. 77, no. 4, pp. 537–555.

Schaefer, Hermann J.; and Sullivan, Jeremiah J., 1970: *Nuclear Emulsion Recordings of the Astronauts' Radiation Exposure on the First Lunar Landing Mission Apollo* XI. NASA CR-115804.

Scott, W. Wayne; and Alsmiller, R. G., Jr., 1967: *Comparisons of Results Obtained With Several Proton Penetration Codes.* ORNL-RSIC-17 (Contract No. W-7405-eng-26, NASA Order R-104(10)), Oak Ridge National Lab.

Scott, W. Wayne; and Alsmiller, R. G., Jr., 1968: *Comparisons of Results Obtained With Several Proton Penetration Codes—Part II.* ORNL-RSIC-22 (Contract No. W-7405-eng-26, NASA Order R-104(10)), Oak Ridge National Lab.

Shen, S. P., 1963: Nuclear Problems in Radiation Shielding in Space. *Astronaut. Acta,* vol. IX, Fasc. 4, pp. 211–274.

Shen, B. S. P., ed., 1967: *High-Energy Nuclear Reactions in Astrophysics—A Collection of Articles.* W. A. Benjamin, Inc.

Shinn, Judy L.; Wilson, John W.; Nealy, John E.; and Cucinotta, Francis A., 1990: *Comparison of Dose Estimates Using the Buildup-Factor Method and a Baryon Transport Code (BRYNTRN) With Monte Carlo Results.* NASA TP-3021.

Shinn, J. L.; Wilson, J. W.; and Ngo, D. M., 1990: Risk Assessment Methodologies for Target Fragments Produced in High-Energy Nucleon Reactions. *Health Phys.*, vol. 59, no. 1, pp. 141–143.

Shinn, Judy L.; Wilson, John W.; Weyland, Mark; and Cucinotta, Francis A., 1991: *Improvements in Computational Accuracy of* BRYNTRN *(A Baryon Transport Code)*. NASA TP-3093.

Snyder, W. S.; Wright, H. A.; Turner, J. E.; and Neufeld, Jacob, 1969: Calculations of Depth-Dose Curves for High-Energy Neutrons and Protons and Their Interpretation for Radiation Protection. *Nuclear Appl.*, vol. 6, no. 4, pp. 336–343.

Turner, J. E.; Zerby, C. D.; Woodyard, R. L.; Wright, H. A.; Kinney, W. E.; Snyder, W. S.; and Neufeld, J., 1964: Calculation of Radiation Dose From Protons to 400 MeV. *Health Phys.*, vol. 10, no. 11, pp. 783–808.

Webber, William R., 1966: *An Evaluation of Solar-Cosmic-Ray Events During Solar Minimum.* D2-84274-1, Boeing Co.

Wilson, John W.; and Khandelwal, G. S., 1974: Proton Dose Approximation in Arbitrary Convex Geometry. *Nuclear Technol.*, vol. 23, no. 3, pp. 298–305.

Wilson, John W.; and Lamkin, Stanley L., 1975: Perturbation Theory for Charged-Particle Transport in One Dimension. *Nuclear Sci. & Eng.*, vol. 57, no. 4, pp. 292–299.

Wilson, John W.; Stith, John J.; and Stock, L. V., 1983: *A Simple Model of Space Radiation Damage in* GaAs *Solar Cells.* NASA TP-2242.

Wilson, John W.; Walker, Gilbert H.; and Outlaw, R. A., 1984: Proton Damage in GaAs Solar Cells. *IEEE Trans. Electron Devices,* vol. ED-31, no. 4, pp. 421–422.

Wilson, John W.; and Badavi, F. F., 1986: Methods of Galactic Heavy Ion Transport. *Radiat. Res.*, vol. 108, pp. 231–237.

Wilson, John W.; Townsend, Lawrence W.; and Buck, Warren W., 1986: On the Biological Hazard of Galactic Antinuclei. *Health Phys.*, vol. 50, no. 5, pp. 666–667.

Wilson, John W.; Townsend, Lawrence W.; Ganapol, Barry D.; and Lamkin, Stanley L., 1988: Methods for High Energy Hadronic Beam Transport. *Trans American Nuclear Soc.*, vol. 56, pp. 271–272.

Wilson, John W.; Townsend, Lawrence W.; and Khan, Ferdous, 1989: Evaluation of Highly Ionizing Components in High-Energy Nucleon Radiation Fields. *Health Phys.*, vol. 57, no. 5, pp. 717–724.

Wilson, John W.; Townsend, Lawrence W.; Nealy, John E.; Chun, Sang Y.; Hong, B. S.; Buck, Warren W.; Lamkin, S. L.; Ganapol, Barry D.; Khan, Ferdous; and Cucinotta, Francis A., 1989: BRYNTRN: *A Baryon Transport Model.* NASA TP-2887.

Wilson, John W.; Shinn, Judy L.; and Townsend, Lawrence W., 1990: Nuclear Reaction Effects in Conventional Risk Assessment for Energetic Ion Exposure. *Health Phys.*, vol. 58, no. 6, pp. 749–752.

Wright, Harvel A.; Anderson, V. E.; Turner, J. E.; Neufeld, Jacob; and Snyder, W. S., 1969: Calculation of Radiation Dose Due to Protons and Neutrons With Energies From 0.4 to 2.4 GeV. *Health Phys.*, vol. 16, no. 1, pp. 13–31.

Wright, H. A.; Hamm, R. N.; and Turner, J. E., 1971: Effect of Lateral Scattering on Absorbed Dose From 400 MeV Neutrons and Protons. *International Congress on Protection Against Accelerator and Space Radiation,* J. Baarli and J. Dutrannois, eds., CERN 71-16, Volume 1, European Organization for Nuclear Research, pp. 207–219.

Zerby, C. D.; and Kinney, W. E., 1965: Calculated Tissue Current-to-Dose Conversion Factors For Nucleons Below 400 MeV. *Nuclear Instrum. & Methods,* vol. 36, no. 1, pp. 125–140.

9. HIGH CHARGE AND ENERGY (HZE) TRANSPORT

9.1. Introduction

Propagation of galactic ions through matter has been studied for the past 40 years as a means of determining the origin of these ions. Peters (1958) used the one-dimensional equilibrium solution ignoring ionization energy loss and radioactive decay to show that the light ions have their origin in the breakup of heavy particles. Davis (1960) showed that one-dimensional propagation is simplistic and that leakage at the galactic boundary must be taken into account. Ginzburg and Syrovatskii (1964) argued that the leakage can be approximated as a superposition of nonequilibrium one-dimensional solutions. The "solution" to the steady-state equations is given as a Volterra equation by Gloeckler and Jokipii (1969), which is solved to first order in the fragmentation cross sections by ignoring energy loss. They provide an approximation to the first-order solution with ionization energy loss included that is only valid at relativistic energies. Lezniak (1979) gives an overview of cosmic-ray propagation and derives a Volterra equation including the ionization energy loss which he refers to as a solution "only in the iterative sense" and evaluates only the unperturbed term. No attempt is made to evaluate either the first-order perturbation term or higher order terms. The main interest among cosmic-ray physicists has been in first-order solutions in the fragmentation cross sections, since path lengths in interstellar space are on the order of 3–4 g/cm^2. Clearly, higher order terms cannot be ignored in accelerator or space shielding transport problems (Wilson, 1977a, 1977b, and 1983; Wilson et al., 1984). Aside from this simplification, the cosmic-ray studies discussed above have neglected the complicated three-dimensional nature of the fragmentation process.

Several approaches to the solution of high-energy heavy ion propagation including the ionization energy loss have been developed (Wilson, 1977a, 1977b, and 1983; Wilson et al., 1984, 1989a, and 1987b; Wilson and Badavi, 1986; Wilson and Townsend, 1988; Curtis, Doherty, and Wilkinson, 1969; Allkofer and Heinrich, 1974; Chatterjee, Tobias, and Lyman, 1976; Letaw, Tsao, and Silberberg, 1983; Ganapol, Townsend, and Wilson, 1989; Townsend, Ganapol, and Wilson, 1989) over the last 20 years. All but one (Wilson, 1977a) have assumed the straight ahead approximation and velocity conserving fragmentation interactions. Only two (Wilson, 1977a; Wilson, et al., 1984) have incorporated energy-dependent nuclear cross sections. The approach by Curtis, Doherty, and Wilkinson (1969) for a primary ion beam represented the first-generation secondary fragments as a quadrature over the collision density of the primary beam. Allkofer and Heinrich (1974) used an energy multigroup method in which an energy-independent fragmentation transport approximation was applied within each energy group after which the energy group boundaries were moved according to continuous slowing down theory $(-dE/dx)$. Chatterjee, Tobias, and Lyman (1976) solved the energy-independent fragment transport equation with primary collision density as a source and neglected higher order fragmentation. The primary source term extended only to the primary ion range from the boundary. The energy-independent transport solution was modified to account for the finite range of the secondary fragment ions.

Wilson (1977b) derived an expression for the ion transport problem to first order (first collision term) and gave an analytic solution for the depth-dose relation. Wilson (1977a) examined the more common approximations used in solving the heavy ion transport problem. The effects of conservation of velocity on fragmentation and the straight ahead approximation are found to be negligible for cosmic-ray applications. Solution methods for representing the energy-dependent nuclear cross sections are developed (Wilson, 1977a). Letaw, Tsao, and Silberberg (1983) approximate the energy loss term and ion spectra by simple forms for which energy derivatives are evaluated explicitly (even if approximately). The resulting ordinary differential equations in position are solved analytically similar to the method of Allkofer and Heinrich (1974). This approximation results in a decoupling of motion in space and a change in energy. In Letaw's formalism, the energy shift is replaced by an effective attenuation factor. Wilson adds the next higher order (second collision) term (Wilson, 1983). This term was found to be very important in describing ^{20}Ne beams at 670 MeV/nucleon. The three-term expansion of (Wilson, 1983) was modified to include the effects of energy variation of the nuclear cross sections (Wilson et al., 1984). The integral form of the transport equation (Wilson 1977a) was further used to derive a numerical marching procedure to solve the cosmic-ray transport problem (Wilson and Badavi, 1986). This method can easily include the energy-dependent nuclear cross sections within the numerical procedure. Comparison of the numerical procedure (Wilson and Badavi, 1986) with an analytic solution to a simplified problem (Wilson and Townsend, 1988) validates the solution technique to about 1 percent accuracy. Several solution techniques and analytic methods have been developed for testing future numerical solutions to the transport equation (Ganapol, Townsend, and Wilson, 1989; Townsend, Ganapol, and Wilson, 1989). More recently, an analytic solution for the laboratory ion beam transport problem has been derived assuming a straight ahead approximation, velocity conservation at the interaction site, and energy-independent nuclear cross sections (Wilson et al., 1989a).

In the previous overview of past developments, the applications split into two separate categories according to a single ion species with a single energy at the boundary versus a broad host of elemental types with a broad, continuous energy spectrum. Techniques requiring a representation of the spectrum over an array of energy values require vast computer storage and computation speed for the laboratory beam problem to maintain sufficient energy resolution. On the other hand, analytic methods (Wilson, 1977a and 1977b; Wilson and Badavi, 1986) are probably best applied in a marching procedure (Wilson and Badavi, 1986), which again has within it a similar energy resolution problem. This is a serious limitation because we require a final High Charge and Energy (HZE) Code for cosmic-ray shielding that has been validated by laboratory experiments.

In this chapter, we begin with the most simplified assumptions for which the problem may be solved completely. Solutions to a more complete theory may then be compared with prior results as limiting cases. In this way, the more complete but approximate analysis will have some basis for evaluating the accuracy of the solution method. The lowest order approximation will be totally energy independent. The next most complicated solution to be considered herein will have energy-independent nuclear cross sections but will treat the energy-dependent atomic/molecular processes and the energy spread of the primary beam. A fully energy-dependent theory must await further development, although some terms are discussed.

9.2. Energy-Independent Flux

If the ion beam is of sufficiently high energy that the energy shift caused by atomic/molecular collisions brings none of the particles to rest in the region of interest, then

$$\left[\frac{\partial}{\partial x} + \sigma_j\right] \phi_j(x) = \sum_k m_{jk}\sigma_k \ \phi_k(x) \tag{9.1}$$

where $\phi_j(x)$ is the flux of type j ions, σ_j is the nuclear absorption cross section, and m_{jk} is the fragmentation parameter for producing type j ions from type k. The solution for a given incident ion type J is given in terms of a set of g-functions as follows:

$$g(j_1) = \exp(-\sigma_{j_1} x) \tag{9.2}$$

$$g(j_1, j_2, \ldots, j_n, j_{n+1}) = \frac{g(j_1, j_2, \ldots, j_{n-1}, j_n) - g(j_1, j_2, \ldots, j_{n-1}, j_{n+1})}{\sigma_{j_{n+1}} - \sigma_{j_n}} \tag{9.3}$$

for which the solution for the type j ion flux is written as

$$\phi_j^{(0)}(x) = \delta_{jJ} \ g(j) \tag{9.4}$$

$$\phi_j^{(1)}(x) = m_{jJ}\sigma_J \ g(j, J) = m_{jJ}\sigma_J \frac{\exp(-\sigma_j x) - \exp(-\sigma_J x)}{\sigma_J - \sigma_j} \tag{9.5}$$

$$\phi_j^{(2)}(x) = \sum_k m_{jk}\sigma_k \ m_{kJ}\sigma_J \ g(j, k, J) \tag{9.6}$$

$$\phi_j^{(3)}(x) = \sum_{k,l} m_{jk}\sigma_k \ m_{kl}\sigma_l \ m_{lJ}\sigma_J \ g(j, k, l, J) \tag{9.7}$$

with

$$\phi_j(x) = \sum_i \phi_j^{(i)}(x) \tag{9.8}$$

This solution is equivalent to that derived by Ganapol, Townsend, and Wilson (1989). We now consider some applications of this formalism. The cross-section data base is discussed by Townsend, Wilson, and Bidasaria (1983a and 1983b).

9.2.1. Neon beam transport. We first note that for ^{20}Ne incident on water, ^{19}Ne and ^{19}F have only one contributing term in equation (9.8). This is shown in figure 9.1. Also shown in figure 9.1 are the fluxes of various isotopes of secondary ion fragments. The effect of successive terms of equation (9.8) is shown in table 9.1 for the ^{15}O flux. It is clear from the table that the fourth and higher order collision terms are completely negligible and that third collision terms are a rather minor contribution. Hence, a three-term expansion as we have used in the past (Wilson, 1977a and 1983; Wilson et al., 1984) appears justified. The relative magnitude of the terms contributing to the ^{7}Li flux generated by the ^{20}Ne beam is presented in table 9.2. The fourth collision term is negligible at small penetration distances and small, but not negligible, at distances greater than 30 cm. The greater penetrating power of the lighter mass fragments is demonstrated in figure 9.2. Also note the difference in solution character caused by the importance of the higher order term.

9.2.2. Iron beam transport. We first note that for ^{56}Fe incident on water, ^{55}Fe and ^{55}Mn have only one contributing term in equation (9.8). The ^{54}Mn has two terms, and the slight difference in solution character can be seen in figure 9.3. Results for ^{52}V are also shown. The convergence rate of equation (9.8) is demonstrated in table 9.3. Again, the fourth collision term is negligible, whereas the three-term expansion we have used before seems quite accurate at these depths for these ions. Distinguished from prior results, the ^{16}O flux has significant contributions from higher order terms for depths beyond 20 cm, as seen in table 9.4. Clearly, a more complete theory with higher order terms is required than the one previously used for ion beams of particles heavier than ^{20}Ne. The different solution character of the lighter mass fragments is clearly demonstrated in figure 9.4.

9.3. Monoenergetic Ion Beams

When moving through extended matter, heavy ions lose energy through interaction with atomic electrons along their trajectories. On occasion, they interact violently with nuclei of the matter and produce ion fragments moving in the forward direction and low-energy fragments of the struck target nucleus. The transport equations for the short range target fragments can be solved in closed form in terms of collision density (Wilson, 1977a; Wilson et al., 1984). Hence, the projectile fragment transport is the interesting unsolved problem. In previous work, the projectile ion fragments were treated as if all went straightforward (Wilson, 1977b and 1983; Wilson et al., 1984 and 1989a; Wilson and Badavi, 1986; Wilson and Townsend, 1988; Curtis, Doherty, and Wilkinson, 1969; Allkofer and Heinrich, 1974; Chatterjee, Tobias, and Lyman, 1976; Letaw, Tsao, and Silberberg, 1983; Ganapol, Townsend, and Wilson, 1989; Townsend, Ganapol, and Wilson, 1989). We continue with this assumption herein, noting that an extension of the beam fragmentation model to three dimensions is being developed (Shavers, 1988; Schimmerling et al., 1986).

With the straight ahead approximation and the target secondary fragments neglected (Wilson, 1977a and 1977b; Wilson, 1983; Wilson et al., 1984), the transport equation may be written as

$$\left[\frac{\partial}{\partial x} - \frac{\partial}{\partial E} \widetilde{S}_j(E) + \sigma_j \right] \phi_j(x, E) - \sum_{k>j} m_{jk} \sigma_k \ \phi_k(x, E) \tag{9.9}$$

where $\phi_j(x, E)$ is the flux of ions of type j with atomic mass A_j at x moving along the x-axis at energy E in units of MeV/nucleon, σ_j is the corresponding macroscopic nuclear absorption cross section, $\widetilde{S}_j(E)$ is the change in E per unit

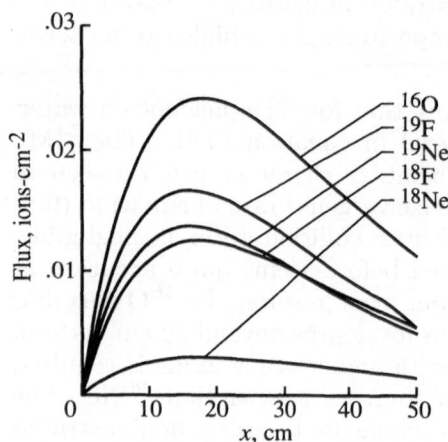

Figure 9.1. Ion fragment flux of various isotopes as function of depth in water for ^{20}Ne incident beam.

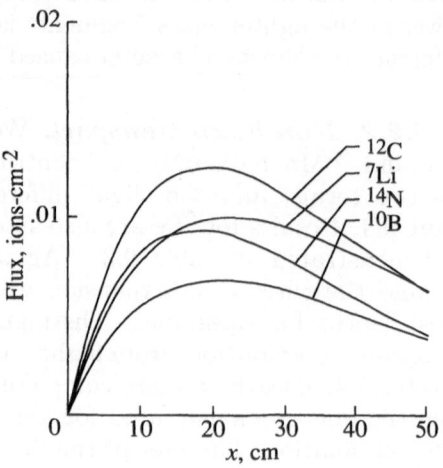

Figure 9.2. Flux of light ion fragments as function of depth in water for ^{20}Ne incident beam.

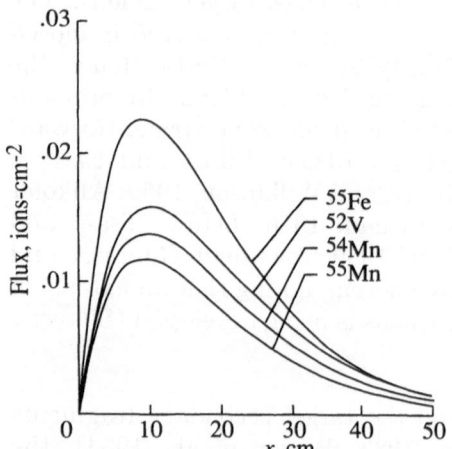

Figure 9.3. Ion fragment flux of various isotopes as function of depth in water for ^{56}Fe incident beam.

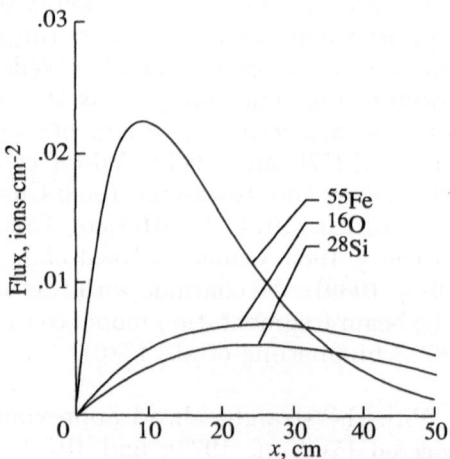

Figure 9.4. Flux of light ion fragments as function of depth in water for ^{56}Fe incident beam.

Table 9.1. Normalized Contributions to ^{15}O Flux From Successive
Collision Terms for ^{20}Ne Transport in Water

Fragment term	^{15}O flux at x of—				
	10 cm	20 cm	30 cm	40 cm	50 cm
$\phi^{(1)}$	1.00E0	1.00E0	1.00E0	1.00E0	1.00E0
$\phi^{(2)}$	1.01E−1	2.01E−1	3.02E−1	4.03E−1	5.04E−1
$\phi^{(3)}$	2.63E−3	1.05E−2	2.36E−2	4.18E−2	6.52E−2
$\phi^{(4)}$	3.31E−5	2.52E−4	8.58E−4	2.03E−3	3.95E−3

Table 9.2. Normalized Contributions to ^{7}Li Flux From Successive
Collision Terms for ^{20}Ne Transport in Water

Fragment term	^{7}Li flux at x of—				
	10 cm	20 cm	30 cm	40 cm	50 cm
$\phi^{(1)}$	1.00E0	1.00E0	1.00E0	1.00E0	1.00E0
$\phi^{(2)}$	1.62E−1	3.20E−1	4.72E−1	6.18E−1	7.58E−1
$\phi^{(3)}$	1.15E−2	4.53E−2	9.98E−2	1.73E−1	2.63E−1
$\phi^{(4)}$	4.02E−4	3.16E−3	1.04E−2	2.39E−2	4.53E−2

Table 9.3. Normalized Contributions to ^{52}V Flux From Successive
Collision Terms for ^{56}Fe Transport in Water

Fragment term	^{52}V flux at x of—				
	10 cm	20 cm	30 cm	40 cm	50 cm
$\phi^{(1)}$	1.00E0	1.00E0	1.00E0	1.00E0	1.00E0
$\phi^{(2)}$	7.91E−2	1.52E−1	2.37E−1	3.15E−1	3.94E−1
$\phi^{(3)}$	2.37E−3	9.48E−3	2.13E−2	3.79E−2	5.91E−2
$\phi^{(4)}$	2.24E−5	1.73E−4	5.93E−4	1.41E−3	2.75E−3

Table 9.4. Normalized Contributions to ^{16}O Flux From Successive
Collision Terms for ^{56}Fe Transport in Water

Fragment term	^{16}O flux at x of—				
	10 cm	20 cm	30 cm	40 cm	50 cm
$\phi^{(1)}$	1.00E0	1.00E0	1.00E0	1.00E0	1.00E0
$\phi^{(2)}$	5.87E−1	1.12E0	1.59E0	2.00E0	2.36E0
$\phi^{(3)}$	1.86E−1	7.08E−1	1.49E0	2.46E0	3.56E0
$\phi^{(4)}$	3.06E−2	2.63E−1	9.44E−1	2.33E0	4.72E0

distance, and m_{jk} is the fragmentation parameter for ion j produced in collision by ion k. The range of the ion is given as

$$R_j(E) = \int_0^E \frac{dE'}{\widetilde{S}_j(E')} \tag{9.10}$$

The solution to equation (9.9) is found subject to boundary specification at $x = 0$ and arbitrary E as

$$\phi_j(0, E) = F_j(E) \tag{9.11}$$

Usually, $F_j(E)$ is called the incident beam spectrum.

It follows from Bethe's theory that

$$\widetilde{S}_j(E) = \frac{A_p Z_j^2}{A_j Z_p^2} \, \widetilde{S}_p(E) \tag{9.12}$$

for which

$$\frac{Z_j^2}{A_j} \, R_j(E) = \frac{Z_p^2}{A_p} \, R_p(E) \tag{9.13}$$

The subscript p refers to proton. Equation (9.12) is quite accurate at high energy and only approximately true at low energy. At low energy, equation (9.12) is modified by electron capture by the ion which effectively reduces its charge, higher order Born corrections to Bethe's theory, and nuclear stopping at the lowest energies. Herein, the parameter ν_j is defined as

$$\nu_j = \frac{Z_j^2}{A_j} \tag{9.14}$$

so that

$$\nu_j \, R_j(E) = \nu_k \, R_k(E) \tag{9.15}$$

Equations (9.14) and (9.15) are used in the subsequent development, and the energy variation in ν_j is neglected. The limits of assumed constant ν_j hold only for $E > 10$ MeV/nucleon (Schimmerling et al., 1986). The inverse function of $R_j(E)$ is defined as

$$E = R_j^{-1} \left[R_j(E) \right] \tag{9.16}$$

and subsequently plays a fundamental role. For the purpose of solving equation (9.9), define the coordinate transformation (Wilson, 1977a and 1983),

$$\left. \begin{array}{l} \eta_j \equiv x - R_j(E) \\[2mm] \xi_j \equiv x + R_j(E) \end{array} \right\} \tag{9.17}$$

and new functions

$$\left.\begin{array}{l} \chi_j(\eta_j, \xi_j) \equiv \tilde{S}_j(E)\, \phi_j(x, E) \\[2mm] \overline{\chi}_k(\eta_j, \xi_j) \equiv \chi_k(\eta_k, \xi_k) \end{array}\right\} \tag{9.18}$$

where

$$\left.\begin{array}{l} \xi_j + \eta_j = \xi_k + \eta_k \\[2mm] \eta_j - \xi_j = \dfrac{\nu_k}{\nu_j}(\eta_k - \xi_k) \end{array}\right\} \tag{9.19}$$

for which equation (9.9) becomes

$$\left(2\frac{\partial}{\partial \eta_j} + \sigma_j\right)\chi_j(\eta_j, \xi_j) = \sum_k m_{jk}\sigma_k \frac{\nu_j}{\nu_k}\, \overline{\chi}_k(\eta_j, \xi_j) \tag{9.20}$$

where the σ_j is assumed to be energy independent. Solving equation (9.20) by using line integration with the integrating factor,

$$\mu_j(\eta_j, \xi_j) = \exp\left[\frac{1}{2}\sigma_j(\xi_j + \eta_j)\right] \tag{9.21}$$

results in

$$\chi_j(\eta_j, \xi_j) = \exp\left[-\frac{1}{2}\sigma_j(\xi_j + \eta_j)\right]\chi_j(-\xi_j, \xi_j)$$

$$+ \frac{1}{2}\int_{-\xi_j}^{\eta_j} \exp\left[\frac{1}{2}\sigma_j(\eta' - \eta_j)\right]\sum_k m_{jk}\sigma_k \frac{\nu_j}{\nu_k}\, \chi_k(\eta_k', \xi_k')\, d\eta' \tag{9.22}$$

where

$$\left.\begin{array}{l} \eta_k' = \dfrac{\nu_k + \nu_j}{2\nu_k}\eta' + \dfrac{\nu_k - \nu_j}{2\nu_k}\xi_j \\[3mm] \xi_k' = \dfrac{\nu_k - \nu_j}{2\nu_k}\eta' + \dfrac{\nu_k + \nu_j}{2\nu_k}\xi_j \end{array}\right\} \tag{9.23}$$

and the boundary condition (eq. (9.11)) is written as

$$\chi_j(-\xi_j, \xi_j) = \tilde{S}_j\left[R_j^{-1}(\xi_j)\right]\, F_j\left[R_j^{-1}(\xi_j)\right]$$

Consider a Neumann series for equation (9.22) for which the first term is

$$\chi_j^{(0)}(\eta_j, \xi_j) = \exp\left[-\frac{1}{2}\sigma_j(\eta_j + \xi_j)\right]\, \tilde{S}_j\left[R_j^{-1}(\xi_j)\right]\, F_j\left[R_j^{-1}(\xi_j)\right] \tag{9.24}$$

and the second term is

$$\chi_j^{(1)}(\eta_j, \xi_j) = \frac{1}{2}\int_{-\xi_j}^{\eta_j} \exp\left[\frac{1}{2}\sigma_j(\eta' - \eta_j)\right]\sum_k m_{jk}\sigma_k \frac{\nu_j}{\nu_k}\exp\left[-\frac{1}{2}\sigma_k(\eta_k' - \xi_k')\right]$$

$$\times \tilde{S}_k\left[R_k^{-1}(\xi_k')\right]\, F_k\left[R_k^{-1}(\xi_k')\right]\, d\eta' \tag{9.25}$$

501

An expression for $\chi_j^{(2)}(\eta_j, \xi_j)$ is derived once equation (9.25) is reduced, and higher order terms can be found by continued iteration of equation (9.22). These expressions (eqs. (9.24) and (9.25)) are now simplified for a monoenergetic beam of type M ions.

The boundary condition is now taken as

$$F_j(E) = \delta_{jM} \, \delta(E - E_o) \tag{9.26}$$

where δ_{jM} is the Kronecker delta, $\delta(\ )$ is the Dirac delta, and E_o is the incident beam energy. Thus,

$$\chi_j(-\xi_j, \xi_j) = \tilde{S}_j\left[R_j^{-1}(\xi_j)\right] \, \delta_{jM} \, \delta\left[R_j^{-1}(\xi_j) - E_o\right] = \delta_{jM} \, \delta[\xi_j - R_j(E_o)] \tag{9.27}$$

for which $\chi_j^{(0)}$ becomes

$$\chi_j^{(0)}(\eta_j, \xi_j) = \delta_{jM} \exp\left[-\frac{1}{2}\sigma_j(\eta_j + \xi_j)\right] \, \delta[\xi_j - R_j(E_o)] \tag{9.28}$$

and $\chi_j^{(1)}$ becomes

$$\chi_j^{(1)}(\eta_j, \xi_j) = \frac{1}{2} \int_{-\xi_j}^{\eta_j} m_{jM}\sigma_M \frac{\nu_j}{\nu_M} \exp\left[\frac{1}{2}\sigma_j(\eta' - \eta_j) - \frac{1}{2}\sigma_M(\eta'_M + \xi'_M)\right]$$

$$\times \, \delta[\xi'_M - R_M(E_o)] \, d\eta' \tag{9.29}$$

where ξ'_M is given by equations (9.23) for $k = M$. The contribution to the integral (eq. (9.29)) occurs at

$$\eta' = \frac{2\nu_M}{\nu_M - \nu_j} R_M(E_o) - \frac{\nu_M + \nu_j}{\nu_M - \nu_j}\xi_j \tag{9.30}$$

provided that η' lies on the interval $-\xi_j < \eta' < \eta_j$ so that

$$\chi_j^{(1)}(\eta_j, \xi_j) = \frac{m_{jM}\sigma_M\nu_j}{|\nu_M - \nu_j|} \exp\left[-\frac{1}{2}\sigma_M(\xi_j + \eta') - \frac{1}{2}\sigma_j(\eta_j - \eta')\right] \tag{9.31}$$

The simplified form in equation (9.31) may now be used to calculate the next iteration of equation (9.22):

$$\chi_j^{(2)}(\eta_j, \xi_j) = \frac{1}{2} \sum_k m_{jk}\sigma_k m_{kM}\sigma_M \frac{\nu_j}{|\nu_M - \nu_k|} \int_{-\xi_j}^{\eta_j} \exp\left[-\frac{1}{2}\sigma_M(\xi''_k + \tilde{\eta})\right.$$

$$\left. -\frac{1}{2}\sigma_k(\eta''_k - \tilde{\eta}) - \frac{1}{2}\sigma_j(\eta_j - \eta'')\right] d\eta'' \tag{9.32}$$

where

$$\left.\begin{array}{l} \eta_k'' = \dfrac{\nu_k + \nu_j}{2\nu_k}\eta'' + \dfrac{\nu_k - \nu_j}{2\nu_k}\xi_j \\[3mm] \xi_k'' = \dfrac{\nu_k - \nu_j}{2\nu_k}\eta'' + \dfrac{\nu_k + \nu_j}{2\nu_k}\xi_j \end{array}\right\} \tag{9.33}$$

and

$$\widetilde{\eta} = \frac{2\nu_M}{\nu_M - \nu_k}\, R_M(E_o) - \frac{\nu_M + \nu_k}{\nu_M - \nu_k}\xi_k'' \tag{9.34}$$

with the requirement that $-\xi_k'' < \widetilde{\eta} < \eta_k''$. The inverse of the transformation is now applied to obtain from equation (9.28)

$$\phi_j^{(0)}(x, E) = \frac{1}{\widetilde{S}_j(E)}\exp(-\sigma_j x)\,\delta_{jM}\,\delta\big[x + R_j(E) - R_M(E_o)\big] \tag{9.35}$$

and from equation (9.31)

$$\phi_j^{(1)}(x, E) = \frac{1}{\widetilde{S}_j(E)}m_{jM}\sigma_M\frac{\nu_j}{|\nu_M - \nu_j|}\exp\Big\{-\frac{1}{2}\sigma_j\big[x - R_j(E) - \eta'\big]$$

$$-\frac{1}{2}\sigma_M\big[x + R_j(E) + \eta'\big]\Big\} \tag{9.36}$$

so long as

$$\frac{\nu_M}{\nu_j}\big[R_M(E_o) - x\big] < R_j(E) < \frac{\nu_M}{\nu_j}R_M(E_o) - x \tag{9.37}$$

Otherwise, $\phi_j^{(1)}(x, E)$ is zero. After a complicated but straightforward manipulation, a similar result may be obtained from equation (9.32) for $\phi_j^{(2)}(x, E)$.

In reducing equation (9.32), it is useful to define

$$x_M = \frac{1}{2}\left(\xi_k'' + \widetilde{\eta}\right) \tag{9.38}$$

$$x_k = \frac{1}{2}\left(\eta_k'' - \widetilde{\eta}\right) \tag{9.39}$$

$$x_j = \frac{1}{2}\left(\eta_j - \eta''\right) \tag{9.40}$$

and make a change in variables as

$$\phi_j^{(2)}(x, E) = \frac{1}{\widetilde{S}_j(E)}\sum_k m_{jk}\sigma_k m_{kM}\sigma_M\frac{\nu_j}{|\nu_M - \nu_k|}$$

$$\times \int_{x_{jl}}^{x_{ju}} \exp(-\sigma_M x_M - \sigma_k x_k - \sigma_j x_j)\,dx_j \tag{9.41}$$

503

where the integral is understood to be nonzero only in the physically allowed regions as presently explained. One may easily demonstrate

$$x_M + x_k + x_j = x \tag{9.42}$$

and

$$\nu_M x_M + \nu_k x_k + \nu_j x_j = \nu_M \, R_M(E_o) - \nu_k \, R_k(E) \tag{9.43}$$

for which the parametric solution is given as

$$x_M = \frac{\nu_M \, R_M(E_o) - \nu_k[R_k(E) + x] + (\nu_k - \nu_j)x_j}{\nu_M - \nu_k} \tag{9.44}$$

$$x_k = \frac{\nu_M[R_M(E) + x] - \nu_M \, R_M(E_o) - (\nu_M - \nu_j)x_j}{\nu_M - \nu_k} \tag{9.45}$$

The requirement that x_M and x_k be bounded by the interval 0 to $x - x_j$ yields

$$\left\{ \begin{array}{c} 0 \\ \dfrac{\nu_k \, [R_k(E) + x] - \nu_M \, R_M(E_o)}{\nu_k - \nu_j} \end{array} \right\} \le x_j \le \left\{ \begin{array}{c} x \\ \dfrac{\nu_M \, [R_M(E) + x] - \nu_M \, R_M(E_o)}{\nu_M - \nu_j} \end{array} \right\} \tag{9.46}$$

as the appropriate limits for the integral in equation (9.41) when $\nu_M > \nu_k > \nu_j$. In the braces in equation (9.46), we always choose the most restrictive value for the limit. The requirement of equation (9.46) also implies the result that

$$R_M^{-1} \, [R_M(E_o) - x] \le E \le R_k^{-1} \left[\frac{\nu_M \, R_M(E_o) - \nu_j x}{\nu_k} \right] \tag{9.47}$$

as the range over which the result of equation (9.41) is not zero. If $\nu_k > \nu_M > \nu_j$, then

$$\left\{ \begin{array}{c} 0 \\ \dfrac{\nu_M \, [R_M(E) + x] - \nu_M \, R_M(E_o)}{\nu_M - \nu_j} \end{array} \right\} \le x_j \le \left\{ \begin{array}{c} x \\ \dfrac{\nu_k \, [R_k(E) + x] - \nu_M \, R_M(E_o)}{\nu_k - \nu_j} \end{array} \right\} \tag{9.48}$$

As a result of equation (9.48),

$$R_k^{-1} \left[\frac{\nu_M}{\nu_k} \, R_M(E_o) - x \right] \le E \le R_M^{-1} \left[R_M(E_o) - \frac{\nu_j}{\nu_M} x \right] \tag{9.49}$$

If $\nu_M > \nu_j > \nu_k$, it follows that

$$
0 \le x_j \le
\begin{Bmatrix}
x \\[6pt]
\dfrac{\nu_M \left[R_M(E) + x - R_M(E_o)\right]}{\nu_M - \nu_j} \\[10pt]
\dfrac{\nu_M R_M(E_o) - \nu_k\, R_k(E) - \nu_k x}{\nu_j - \nu_k}
\end{Bmatrix}
\tag{9.50}
$$

where the lesser of the three values in the braces is used as the upper limit of x_j for which the integral of equation (9.32) is not zero. As a result of equation (9.50),

$$
R_M^{-1}\left[R_M(E_o) - x\right] \le E \le R_k^{-1}\left[\frac{\nu_M}{\nu_k}\, R_M(E_o) - x\right]
\tag{9.51}
$$

The integral in equation (9.41) may now be evaluated as

$$
\phi_j^{(2)}(x, E) = \sum_k \frac{\sigma_{jk}\sigma_{kM}\nu_j}{\widetilde{S}_j(E)\,|\nu_M - \nu_k|\Delta_{jkM}} \left[\exp(-\sigma_M x_{Ml} - \sigma_k x_{kl} - \sigma_j x_{jl})\right.
$$

$$
\left. - \exp(-\sigma_M x_{Mu} - \sigma_k x_{ku} - \sigma_j x_{ju})\right]
\tag{9.52}
$$

where x_{Mu}, x_{ku}, x_{Ml}, and x_{kl} are the values of equations (9.44) and (9.45) evaluated at the corresponding upper and lower limits of x_j and

$$
\Delta_{jkM} = \sigma_j + \left[\frac{(\nu_k - \nu_j)}{(\nu_M - \nu_k)}\sigma_M - \frac{(\nu_M - \nu_j)}{(\nu_M - \nu_k)}\sigma_k\right]
\tag{9.53}
$$

Higher order terms are similarly derived.

The total integral flux associated with each term may be evaluated as

$$
\Phi_j^{(1)}(x) = \int_0^\infty \phi_j^{(1)}(x, E)\, dE
\tag{9.54}
$$

One may easily show that

$$
\int_0^\infty \phi_j^{(1)}(x, E)\, dE = \frac{\sigma_{jM}\left[\exp(-\sigma_j x) - \exp(-\sigma_M x)\right]}{\sigma_M - \sigma_j}
\tag{9.55}
$$

in agreement with equation (9.5). Furthermore,

$$
\int_0^\infty \phi_j^{(2)}(x, E)\, dE = \sum_k \frac{\sigma_{jk}\sigma_{kM}}{|\nu_M - \nu_k|\Delta_{jkM}} \left\{ \left(\frac{\nu_k - \nu_j}{\sigma_k} \quad \sigma_j\right)\right.
$$

$$
\times \left[\exp(-\sigma_j x) - \exp(-\sigma_k x)\right]
$$

$$
\left. - \left(\frac{\nu_M - \nu_j}{\sigma_M - \sigma_j}\right)\left[\exp(-\sigma_j x) - \exp(-\sigma_M x)\right] \right\}
\tag{9.56}
$$

which agrees with equation (9.6) as $\nu_k \rightarrow \nu_M$. This relation of equation (9.56) and equation (9.6) has been used previously (Wilson, 1983).

9.3.1. Total flux comparisons. The results of equations (9.36) and (9.52) are integrated numerically over their entire energy spectrum and given along with values from corresponding energy-independent solutions in table 9.5. The primary beam was taken as ^{20}Ne at 1380 MeV/nucleon. Clearly, the energy-dependent solutions appear quite accurate.

Table 9.5. Total Flux From Energy-Independent Solution and Numerically
Integrated Differential Spectrum

[Values in parentheses are from energy-independent solution]

Fragment	Term	Flux, cm^{-2}, at water depth x of—			
		5 cm		20 cm	
^{18}F	$\phi^{(1)}$	0.00727	(0.00717)	0.01148	(0.01140)
	$\phi^{(2)}$	0.00018	(0.00018)	0.00114	(0.00114)
^{17}O	$\phi^{(1)}$	0.00729	(0.00729)	0.01173	(0.01174)
	$\phi^{(2)}$	0.00017	(0.00017)	0.00112	(0.00112)
^{16}O	$\phi^{(1)}$	0.01350	(0.01349)	0.02193	(0.02202)
	$\phi^{(2)}$	0.00029	(0.00029)	0.00191	(0.00190)
^{15}N	$\phi^{(1)}$	0.00470	(0.00481)	0.00796	(0.00796)
	$\phi^{(2)}$	0.00032	(0.00033)	0.00220	(0.00220)
^{13}C	$\phi^{(1)}$	0.00511	(0.00521)	0.00894	(0.00887)
	$\phi^{(2)}$	0.00032	(0.00033)	0.00224	(0.00224)
^{12}C	$\phi^{(1)}$	0.00668	(0.00682)	0.01173	(0.01178)
	$\phi^{(2)}$	0.00056	(0.00056)	0.00398	(0.00398)
^{11}B	$\phi^{(1)}$	0.00417	(0.00417)	0.00735	(0.00732)
	$\phi^{(2)}$	0.00036	(0.00036)	0.00259	(0.00259)

9.3.2. Monoenergetic beam results. The fluorine spectral flux at various depths in a water column is shown in figure 9.5. The primary beam was ^{20}Ne ions at 600 MeV/nucleon corresponding to a range of 30 cm. There is a clear structure caused by the fluorine isotopes shown in the spectrum. The most energetic ions are ^{19}F. The ^{18}F and ^{17}F spectral components are clearly resolved. Only the ^{19}F is able to penetrate to the largest depth represented in figure 9.5 (35 cm). A similar, but more complicated, isotopic structure is seen in the oxygen spectra of figure 9.6. The greater number of oxygen isotopes contributing has a smoothing effect on the resultant spectrum. This effect is even more clearly seen in figure 9.7 for the nitrogen isotopes. Some of the smoothness results from the higher order term $\phi^{(2)}$ in the perturbation expansion. The boron flux of figure 9.8 shows very little isotopic structure. Qualitatively, similar results are obtained for an iron beam of the same range (30 cm) as shown in figures 9.9 to 9.13.

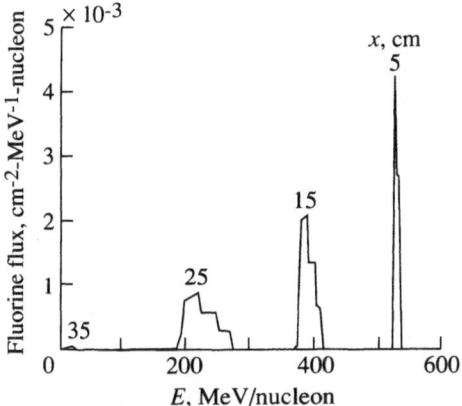

Figure 9.5. Fluorine flux spectrum
produced by ^{20}Ne beam at
600 MeV/nucleon in water column
at various depths.

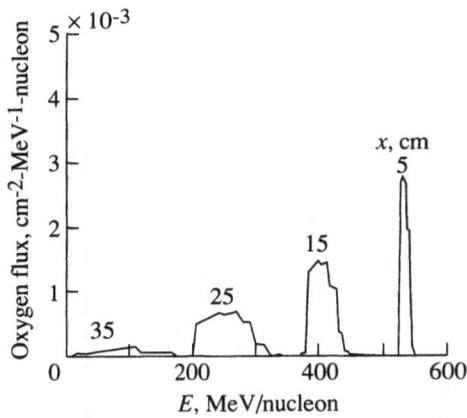

Figure 9.6. Oxygen flux spectrum
produced by ^{20}Ne beam at
600 MeV/nucleon in water column
at various depths.

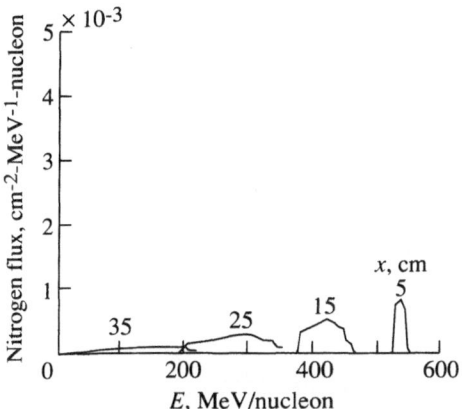

Figure 9.7. Nitrogen flux spectrum
produced by ^{20}Ne beam at
600 MeV/nucleon in water column
at various depths.

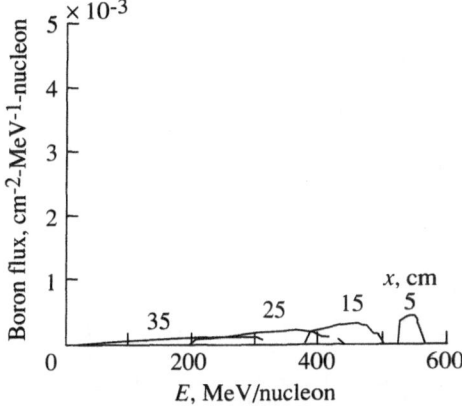

Figure 9.8. Boron flux spectrum
produced by ^{20}Ne beam at
600 MeV/nucleon in water column
at various depths.

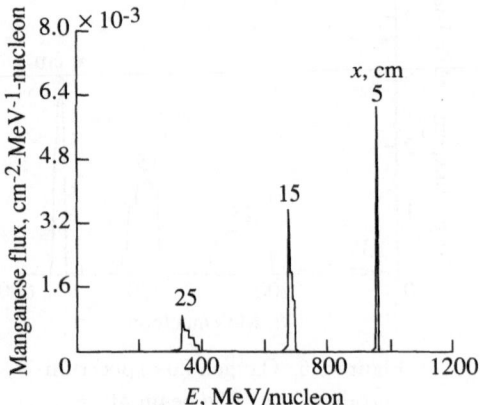

Figure 9.9. Manganese flux spectrum produced by ^{56}Fe beam at 1090 MeV/nucleon in water column at various depths.

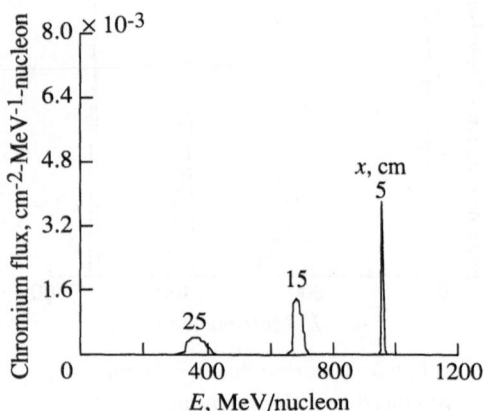

Figure 9.10. Chromium flux spectrum produced by ^{56}Fe beam at 1090 MeV/nucleon in water column at various depths.

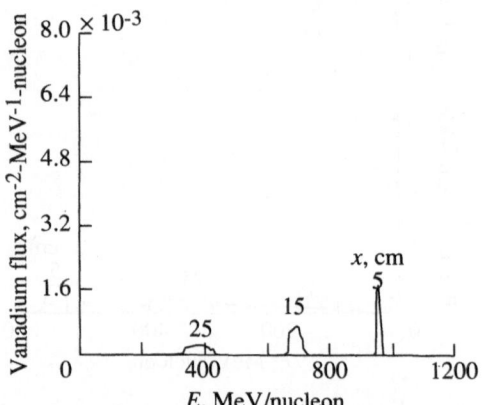

Figure 9.11. Vanadium flux spectrum produced by ^{56}Fe beam at 1090 MeV/nucleon in water column at various depths.

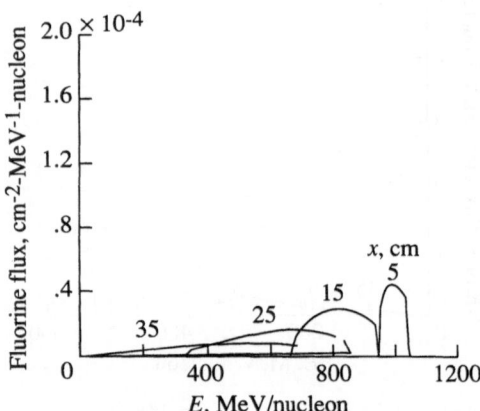

Figure 9.12. Fluorine flux spectrum produced by ^{56}Fe beam at 1090 MeV/nucleon in water column at various depths.

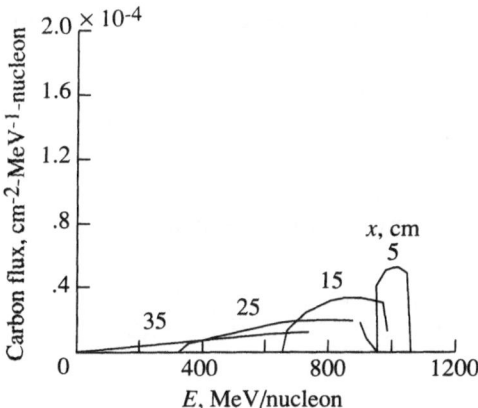

Figure 9.13. Carbon flux spectrum produced by ^{56}Fe beam at 1090 MeV/nucleon in water column at various depths.

9.4. Realistic Ion Beams

In section 9.3, we assumed that a monoenergetic beam was present at the boundary. We now take the incident ion beam flux to be

$$\phi_j(0, E) = \frac{1}{\sqrt{2\pi}\Delta} \exp\left[-\frac{(E - E_o)^2}{2\Delta^2}\right] \tag{9.57}$$

where E_o is the nominal beam energy and Δ is related to the half-width at half-maximum. The full solution is then found as a superposition of results from section 9.3. The uncollided flux is found to be

$$\phi_M^{(0)}(x, E) = \frac{\widetilde{S}_M(E_m)}{\widetilde{S}_M(E)} \exp(-\sigma_M x) \frac{1}{\sqrt{2\pi}\Delta} \exp\left[\frac{(E_o - E_m)^2}{2\Delta^2}\right] \tag{9.58}$$

where $R_M(E_m) = R_M(E) + x$. One similarly arrives at

$$\phi_j^{(1)}(x, E) = \frac{\sigma_{jM}\nu_j}{\widetilde{S}_j(E)\,|\nu_M - \nu_j|} \exp\left\{-\frac{1}{2}\sigma_j\left[x - R_j(E) - \eta_o'\right] - \frac{1}{2}\sigma_M\left[x + R_j(E) + \eta_o'\right]\right\}$$

$$\times \frac{1}{2}\left[\mathrm{erf}\left(\frac{E_u - E_o}{\sqrt{2}\Delta}\right) - \mathrm{erf}\left(\frac{E_l - E_o}{\sqrt{2}\Delta}\right)\right] \tag{9.59}$$

where

$$E_l = R_M^{-1}\left\{\frac{\nu_j}{\nu_M}\left[R_j(E) + x\right]\right\} \tag{9.60}$$

$$E_u = R_M^{-1}\left\{\frac{\nu_j}{\nu_M}R_j(E) + x\right\} \tag{9.61}$$

$$\eta_o' = \frac{2\nu_M}{\nu_M - \nu_j}R_M(E_o) - \frac{\nu_M + \nu_j}{\nu_M - \nu_j}\left[R_j(E) + x\right] \tag{9.62}$$

509

The second collision contribution to the ion energy spectrum is similar:

$$\phi_j^{(2)}(x, E) = \sum_k \frac{\sigma_{jk}\sigma_{kM}\nu_j}{\tilde{S}_j(E)\,|\nu_M - \nu_k|\,\Delta_{jkM}} \left[\exp(-\sigma_M x_{Ml} - \sigma_k x_{kl} - \sigma_j x_{jl})\right.$$

$$\left. - \exp(-\sigma_M x_{Mu} - \sigma_k x_{ku} - \sigma_j x_{ju})\right]$$

$$\times \frac{1}{2}\left[\operatorname{erf}\left(\frac{E_u - E_o}{\sqrt{2}\Delta}\right) - \operatorname{erf}\left(\frac{E_l - E_o}{\sqrt{2}\Delta}\right)\right] \tag{9.63}$$

where

$$E_l = \begin{cases} R_M^{-1}\left[\dfrac{\nu_k R_k(E) + \nu_j x}{\nu_M}\right] & (\nu_M > \nu_k > \nu_j) \\[2ex] R_M^{-1}\left[R_M(E) + \dfrac{\nu_j x}{\nu_M}\right] & (\nu_k > \nu_M > \nu_j) \\[2ex] R_M^{-1}\left[\dfrac{\nu_k(R_k(E) + x)}{\nu_M}\right] & (\nu_M > \nu_j > \nu_k) \end{cases} \tag{9.64}$$

$$E_u = \begin{cases} R_M^{-1}\left[R_M(E) + x\right] & (\nu_M > \nu_k > \nu_j) \\[2ex] R_M^{-1}\left[\dfrac{\nu_k(R_k(E) + x)}{\nu_M}\right] & (\nu_k > \nu_M > \nu_j) \\[2ex] R_M^{-1}\left[R_M(E) + x\right] & (\nu_M > \nu_j > \nu_k) \end{cases} \tag{9.65}$$

and x_M and x_k evaluated at the upper and lower limit values of x_j are obtained from equations (9.44) and (9.45).

The elemental flux spectra were recalculated for ^{20}Ne ions at 600 MeV/nucleon with a 0.2-percent energy spread assumed for the primary beam. The resulting fluorine flux is shown in figure 9.14. Although the spectral results are quite similar to the monoenergetic beam case, there is a considerable smoothing of the total spectrum. Similar results are obtained for the oxygen flux as well in figure 9.15. In distinction, the nitrogen and carbon spectra show only slight isotopic structure as seen in figures 9.16 and 9.17. Qualitatively similar results are obtained for the ^{56}Fe realistic beam as shown in figures 9.18 through 9.22.

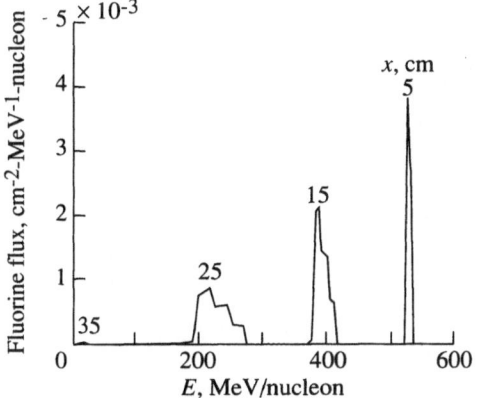

Figure 9.14. Fluorine flux spectrum produced by ^{20}Ne beam at 600 MeV/nucleon with 0.2-percent energy spread in water column at various depths.

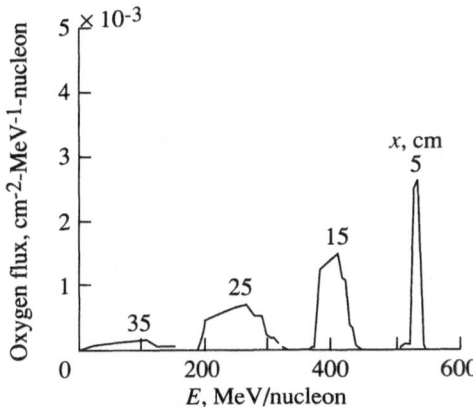

Figure 9.15. Oxygen flux spectrum produced by ^{20}Ne beam at 600 MeV/nucleon with 0.2-percent energy spread in water column at various depths.

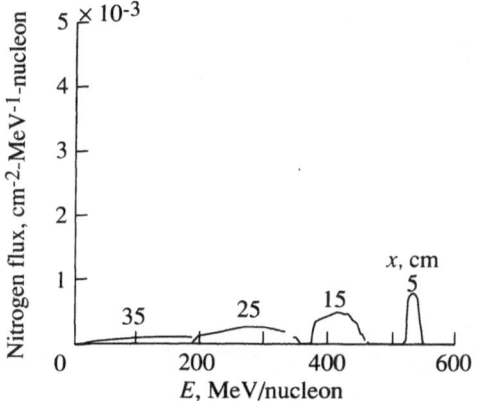

Figure 9.16. Nitrogen flux spectrum produced by ^{20}Ne beam at 600 MeV/nucleon with 0.2-percent energy spread in water column at various depths.

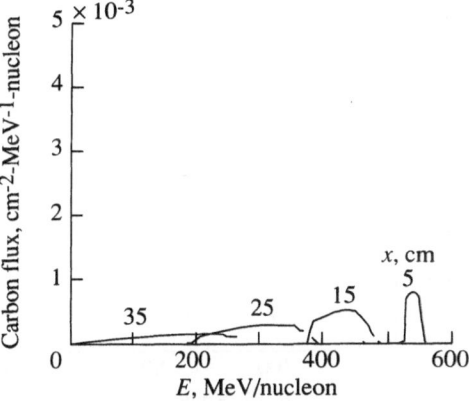

Figure 9.17. Carbon flux spectrum produced by ^{20}Ne beam at 600 MeV/nucleon with 0.2-percent energy spread in water column at various depths.

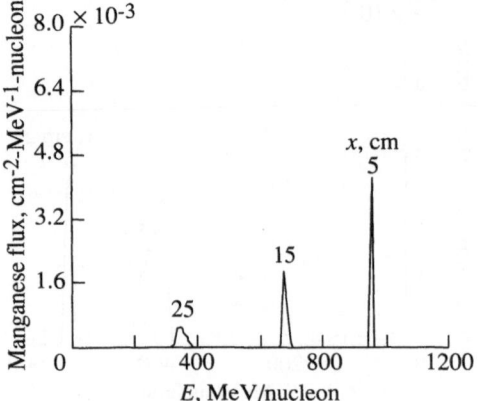

Figure 9.18. Manganese flux spectrum produced by ^{56}Fe beam at 1090 MeV/nucleon with 0.2-percent energy spread in water column at various depths.

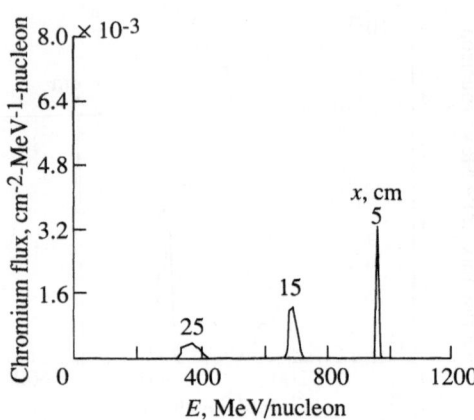

Figure 9.19. Chromium flux spectrum produced by ^{56}Fe beam at 1090 MeV/nucleon with 0.2-percent energy spread in water column at various depths.

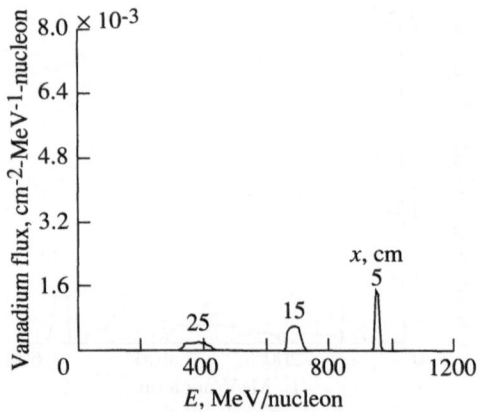

Figure 9.20. Vanadium flux spectrum produced by ^{56}Fe beam at 1090 MeV/nucleon with 0.2-percent energy spread in water column at various depths.

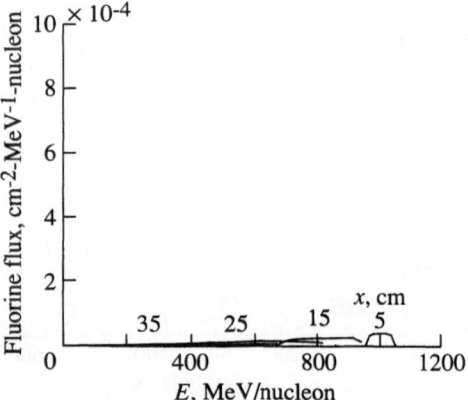

Figure 9.21. Fluorine flux spectrum produced by ^{56}Fe beam at 1090 MeV/nucleon with 0.2-percent energy spread in water column at various depths.

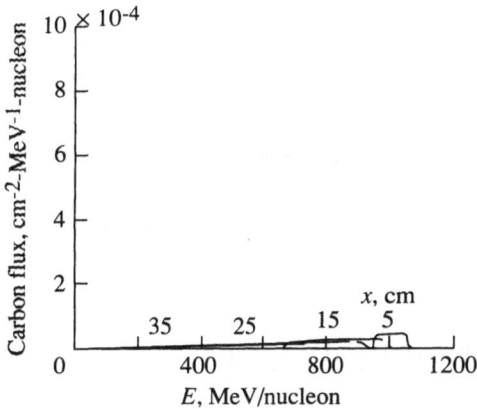

Figure 9.22. Carbon flux spectrum produced by ^{56}Fe beam at 1090 MeV/nucleon with 0.2-percent energy spread in water column at various depths.

9.5. Approximate Spectral Solutions

In sections 9.3 and 9.4, the spectral solutions of the secondary ion flux were derived to second-order collision terms. Such a three-term expansion is not always an adequate representation of the transport solution. In this section, we derive approximate expressions for the perturbation series. Clearly, the more accurate results would be used to the order to which they are known, and the higher order terms would be taken to the approximate expressions of this section.

9.5.1. Approximate monoenergetic beams. The uncollided beam solution is taken as

$$\phi_j^{(0)}(x, E) = \phi_j^{(0)}(x) \frac{1}{\widetilde{S}_j(E)} \delta[x + R_j(E) - R_j(E_o)] \qquad (9.66)$$

which is equal to the result in equation (9.35). The first-order collision term is approximated by noting that the energy dependence of the exponent of equation (9.36) is slowly varying in energy resulting in (Wilson, 1977a, 1977b, and 1983; Wilson et al., 1984)

$$\phi_j^{(1)}(x, E) \approx \frac{\phi_j^{(1)}(x)}{E_{ju} - E_{jl}} \qquad (9.67)$$

where

$$E_{jl} = R_j^{-1} \left\{ \frac{\nu_M}{\nu_j} [R_M(E_o) - x] \right\} \qquad (9.68)$$

$$E_{ju} = R_j^{-1} \left\{ \frac{\nu_M}{\nu_j} R_M(E_o) - x \right\} \qquad (9.69)$$

Similarly,

$$\phi_j^{(2)}(x, E) = \sum_k \frac{\sigma_{jk}\sigma_{kM} \ g(j, k, M)}{E_{ju} - E_{jl}} \qquad (9.70)$$

where

$$E_{jl} = \begin{cases} R_M^{-1}\left\{R_M(E_o) - x\right\} & (\nu_M > \nu_k > \nu_j) \\ R_k^{-1}\left\{\dfrac{\nu_M}{\nu_k}\,R_M(E_o) - x\right\} & (\nu_k > \nu_M > \nu_j) \\ R_M^{-1}\left\{R_M(E_o) - x\right\} & (\nu_M > \nu_j > \nu_k) \end{cases} \qquad (9.71)$$

and

$$E_{ju} = \begin{cases} R_k^{-1}\left\{\dfrac{\nu_M\,R_M(E_o) - \nu_j x}{\nu_k}\right\} & (\nu_M > \nu_k > \nu_j) \\ R_M^{-1}\left\{R_M(E_o) - \dfrac{\nu_j x}{\nu_M}\right\} & (\nu_k > \nu_M > \nu_j) \\ R_k^{-1}\left\{\dfrac{\nu_M}{\nu_k}\,R_M(E_o) - x\right\} & (\nu_M > \nu_j > \nu_k) \end{cases} \qquad (9.72)$$

Higher order terms $(n > 2)$ are taken as

$$\phi_j^n(x, E) \approx \frac{\phi_j^{(n)}(x)}{E_{ju} - E_{jl}} \qquad (9.73)$$

where E_{ju} and E_{jl} are given by equations (9.71) and (9.72). In all the expressions for $\phi_j^{(n)}$ given by equations (9.67), (9.70), and (9.73), the flux values are taken as zero unless

$$E_{jl} \leq E \leq E_{ju} \qquad (9.74)$$

The approximate monoenergetic beam solutions are given in figures 9.23 through 9.26 and should be compared with the solutions found in section 9.3. The ^{17}O flux at 20 cm of water is shown in figure 9.23 as contributed by the first collision term. The trapezoidal (solid) curve is the exact solution for the first collision term derived in section 9.3. The rectangular (dashed) curve is the approximate first collision term of equation (9.67). Terms for other fragment spectra are similar to those shown in figure 9.23. The solution for the second collision contribution to the ^{17}O flux at 20 cm of water is shown in figure 9.24. The nearly rectangular solution (dashed curve) is the approximation given by equation (9.70). A triangular spectral function of the same energy interval could yield improved results. The spectra of fragments which are much lighter than the primary beam are more accurately represented by the approximate solutions as seen in figures 9.25 and 9.26. This improvement results from the greater number of terms in the summation of equation (9.52). This leads us to believe that the higher order terms in the perturbation series can be adequately represented by the approximation in equation (9.73). This is especially true because higher order terms in many applications are only small corrections.

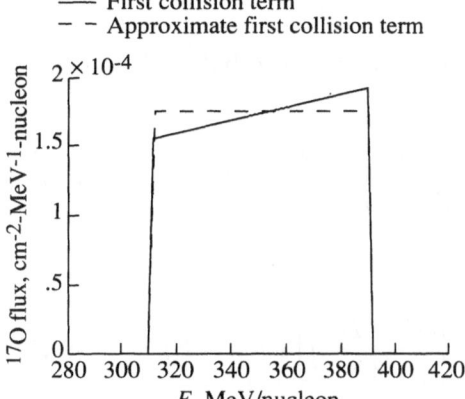

Figure 9.23. ^{17}O flux spectral term
according to first collision term
and approximate first collision term.

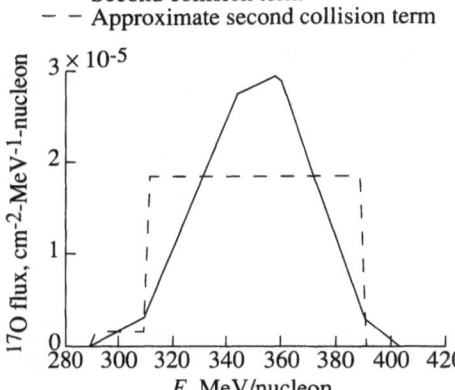

Figure 9.24. ^{17}O flux spectral term
according to second collision term
and approximate second collision term.

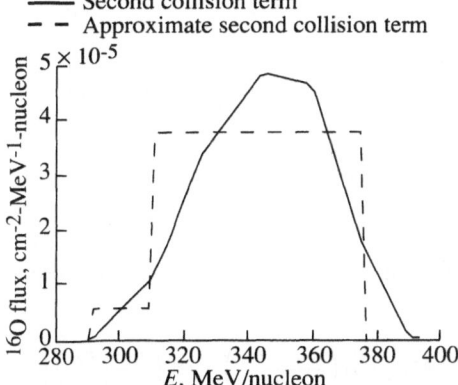

Figure 9.25. ^{16}O flux spectral term
according to second collision term
and approximate second collision term.

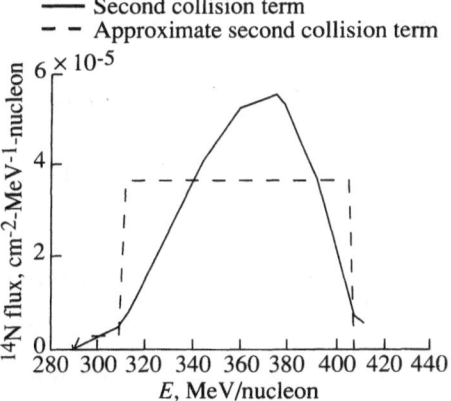

Figure 9.26. ^{14}N flux spectral term
according to second collision term
and approximate second collision term.

9.5.2. Approximate realistic beams. Approximate solutions for realistic ion beams may be found by using a superposition of the approximate monoenergetic beam solutions. The incident ion beam is taken as

$$\phi_j(0, E) = \frac{1}{\sqrt{2\pi}\Delta} \exp\left[-\frac{(E - E_o)^2}{2\Delta^2}\right] \delta_{jM} \qquad (9.75)$$

where E_o is the nominal beam energy and Δ is related to the half-width at half-maximum. The first term is then, as before,

$$\phi_M^{(0)}(x, E) = \frac{\widetilde{S}_M(E_m)}{\widetilde{S}_M(E)} \exp(-\sigma_M x)\frac{1}{\sqrt{2\pi}\Delta} \exp\left[-\frac{(E_o - E_m)^2}{2\Delta^2}\right] \qquad (9.76)$$

where $R_M(E_m) = R_M(E) + x$. One similarly arrives at

$$\phi_j^{(1)}(x, E) = \phi_j^{(1)}(x)\frac{1}{2}\left[\mathrm{erf}\left(\frac{E_u - E_o}{\sqrt{2}\Delta}\right) - \mathrm{erf}\left(\frac{E_l - E_o}{\sqrt{2}\Delta}\right)\right](E_{ju} - E_{jl})^{-1} \quad (9.77)$$

where

$$E_u = R_M^{-1}\left\{\frac{\nu_j}{\nu_M}R_j(E) + x\right\} \qquad (9.78)$$

$$E_l = R_M^{-1}\left\{\frac{\nu_j}{\nu_M}[R_j(E) + x]\right\} \qquad (9.79)$$

and E_{ju} and E_{jl} are given by equations (9.68) and (9.69). Additional computation yields

$$\phi_j^{(2)}(x, E) = \sum_k \sigma_{jk}\sigma_{kM}\, g(j, k, M)\frac{1}{2}\left[\mathrm{erf}\left(\frac{E_u - E_o}{\sqrt{2}\Delta}\right) - \mathrm{erf}\left(\frac{E_l - E_o)}{\sqrt{2}\Delta}\right)\right]$$

$$\times (E_{ju} - E_{jl})^{-1} \qquad (9.80)$$

where E_{ju} and E_{jl} are given by equations (9.71) and (9.72), and E_l and E_u are given in equations (9.64) and (9.65). The remaining higher order terms are taken as

$$\phi_j^{(n)}(x, E) = \sum_{j_1,\ldots,j_{n-1}} \sigma_{jj_{n-1}},\ldots,\sigma_{j1,M}\, g(j, j_{n-1},\ldots,j_1, M)\frac{1}{2}\left[\mathrm{erf}\left(\frac{E_u - E_o}{\sqrt{2}\Delta}\right) - \mathrm{erf}\left(\frac{E_l - E_o}{\sqrt{2}\Delta}\right)\right]$$

$$\times (E_{ju} - E_{jl})^{-1} \qquad (9.81)$$

where E_{jl} and E_{ju} are given by equations (9.68) and (9.69), and E_u and E_l are given by equations (9.78) and (9.79).

These approximate equations for realistic ion beams are given in figures 9.27 to 9.30 and should be compared with the more exact formulas given in section 9.4.

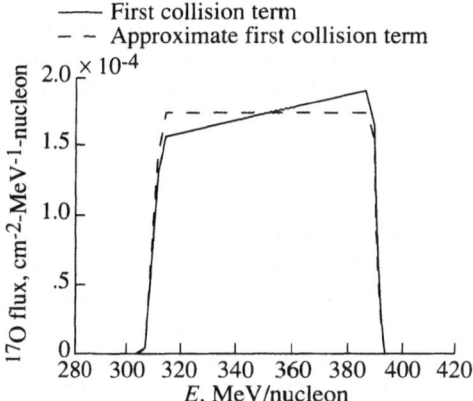

Figure 9.27. ^{17}O flux spectral term
for energy spread ^{20}Ne beam
according to first collision term
and approximate first collision term.

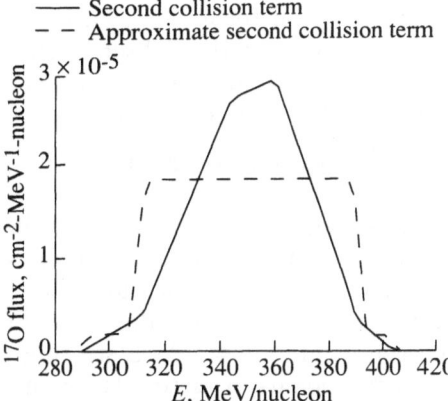

Figure 9.28. ^{17}O flux spectral term
for energy spread ^{20}Ne beam
according to second collision term
and approximate second collision term.

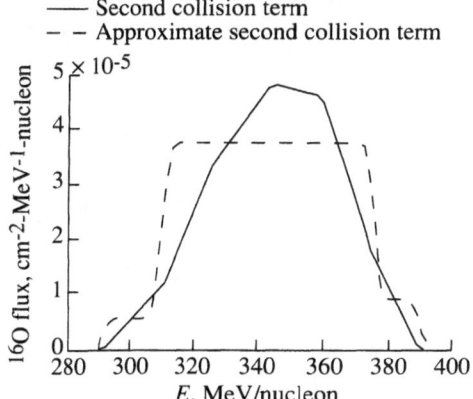

Figure 9.29. ^{16}O flux spectral term
for energy spread ^{20}Ne beam
according to second collision term
and approximate second collision term.

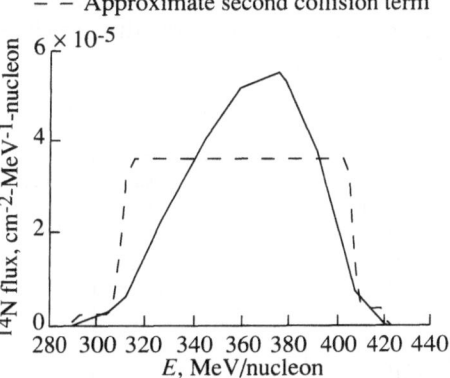

Figure 9.30. ^{14}N flux spectral term
for energy spread ^{20}Ne beam
according to second collision term
and approximate second collision term.

The primary ion beam is taken as ^{20}Ne at 600 MeV/nucleon with a 0.2-percent energy spread. The ^{17}O flux first collision term is shown in figure 9.27 for the two formalisms. The effect of the beam energy spread is seen as a rounding of the spectrum at the edges compared with the monoenergetic case in figure 9.23. The second collision term is shown in figure 9.28. The approximate second collision term improves for the lighter fragments as seen in figures 9.29 and 9.30. Higher order collision terms are expected to be more accurate because of the large number of combinations of contributing ion terms.

9.6. Recommended Methods

An energy-independent theory has been used to show that the perturbation expansion up to the double collision term is adequate for all fragments whose mass is near that of the projectile. This is why the three-term expansion was able to explain the Bragg curve data for ^{20}Ne beams in water with reasonable accuracy (Wilson et al., 1984). As a starting point for the calculation of the transition of heavy ion beams in materials, the use of the three-term expansion can be further corrected by use of the approximate higher order terms given in section 9.5. As an example of such a procedure, we give results for ^{20}Ne beams at 600 MeV/nucleon in water. The results are shown in figures 9.31 to 9.36 as successive partial sums of the perturbation series. The solid line is the first collision term. The dashed curve includes the double collision terms. The long-dash–short-dash curve includes the triple collision term and can hardly be distinguished from the long-dash–double-short-dash curve which includes the quadruple collision terms. The results for penetration to 20 cm of water are shown in figures 9.31 to 9.36. The monoenergetic beam results for ^{17}O, ^{16}O, and ^{12}C are given in figures 9.31 to 9.33, respectively. The double collision term is seen to be always an important contribution. The triple collision term shows some importance for ^{12}C, whereas higher order terms are negligible. Similar results are shown in figures 9.34 to 9.36 for an energy spread of 0.2 percent.

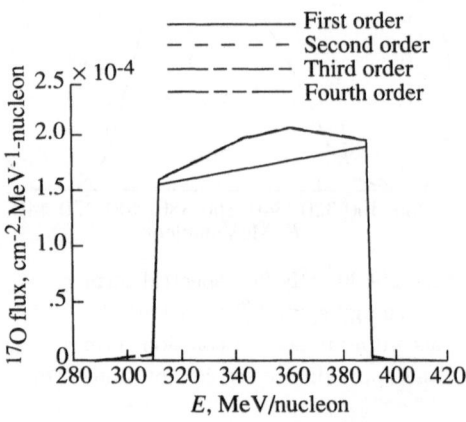

Figure 9.31. Sequence of approximations of ^{17}O flux spectrum after 20 cm of water for first-, second-, third-, and fourth-order theories.

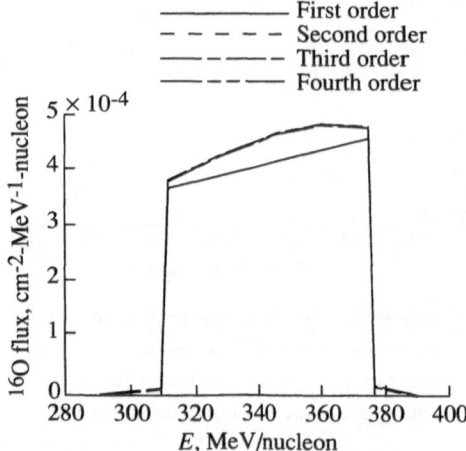

Figure 9.32. Sequence of approximations of ^{16}O flux spectrum after 20 cm of water for first-, second-, third-, and fourth-order theories.

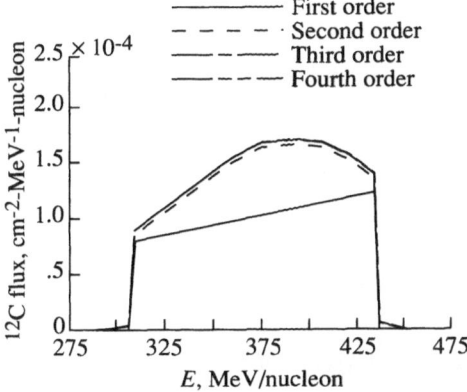

Figure 9.33. Sequence of approximations of ^{12}C flux spectrum after 20 cm of water for first-, second-, third-, and fourth-order theories.

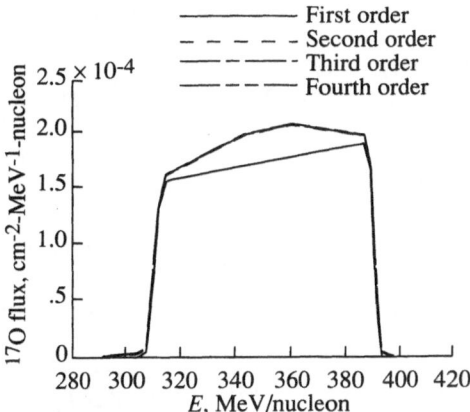

Figure 9.34. Sequence of approximations for energy spread solution of ^{17}O flux spectrum after 20 cm of water for first-, second-, third-, and fourth-order theories.

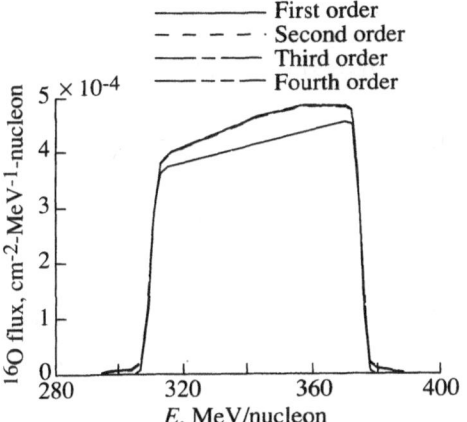

Figure 9.35. Sequence of approximations for energy spread solution of ^{16}O flux spectrum after 20 cm of water for first-, second-, third-, and fourth-order theories.

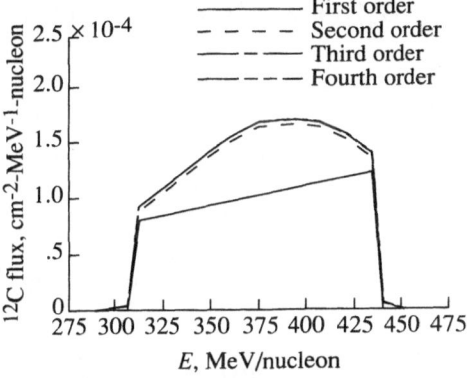

Figure 9.36. Sequence of approximations for energy spread solution of ^{12}C flux spectrum after 20 cm of water for first-, second-, third-, and fourth-order theories.

9.7. Impulse Response

One form of Green's function is the impulse response corresponding to a δ-like source term at the boundary. We therefore seek a solution of

$$\left[\frac{\partial}{\partial x} - \frac{\partial}{\partial E}\tilde{S}_j(E) + \sigma_j\right] G_{jM}(x, E; E') = \sum_k m_{jk}\sigma_k\, G_{kM}(x, E; E') \qquad (9.82)$$

where the boundary condition is

$$G_{jM}(0, E; E') = \delta_{jM}\delta(E - E') \qquad (9.83)$$

for which any arbitrary transport solution may be written as

$$\phi_j(x, E) = \sum_M \int_0^\infty G_{jM}(x, E; E')\, F_M(E')\, dE' \qquad (9.84)$$

where $F_M(E')$ is the flux at the boundary. The solution to equations (9.82) and (9.83) is straightforward, even if tedious (Wilson, 1977a, 1977b, and 1983; Wilson et al., 1989a) and is arrived at by using the method of characteristics (Wilson and Lamkin, 1975). The solution is expressed as a series as

$$G_{jM}(x, E; E') = \sum_i G_{jM}^{(i)}(x, E; E') \qquad (9.85)$$

where

$$G_{jM}^{(0)}(x, E; E') = \frac{1}{\tilde{S}_j(E)}\exp(-\sigma_j x)\delta_{jM}\delta\left[x + R_j(E) - R_M(E')\right] \qquad (9.86)$$

and

$$G_{jM}^{(1)}(x, E; E') = \frac{1}{\tilde{S}_j(E)}m_{jM}\,\sigma_M\frac{\nu_j}{|\nu_M - \nu_j|}\exp\left\{-\frac{1}{2}\,\sigma_j\left[x - R_j(E) - \eta'\right]\right.$$

$$\left. -\frac{1}{2}\sigma_M\left[x + R_j(E) + \eta'\right]\right\} \qquad (9.87)$$

so long as

$$\frac{\nu_M}{\nu_j}\left[R_M(E') - x\right] < R_j(E) < \frac{\nu_M}{\nu_j}R_M(E') - x \qquad (9.88)$$

where

$$\eta' = \frac{2\nu_M}{\nu_m - \nu_j}R_M(E') - \frac{\nu_M + \nu_j}{\nu_M - \nu_j}\left[R_j(E) + x\right] \qquad (9.89)$$

otherwise, $G_{jM}^{(1)}(x, E; E')$ is zero. After a complicated but straightforward manipulation, a similar result may be obtained for $G_{jM}^{(2)}(x, E; E')$ as

$$G_{jM}^{(2)}(x, E; E') = \sum_k \frac{\sigma_{jk}\sigma_{kM}\nu_j}{\widetilde{S}_j(E)|\nu_M - \nu_k|\Delta_{jkM}} \left[\exp\left(-\sigma_M x_{Ml} - \sigma_k x_{kl} - \sigma_j x_{jl}\right)\right.$$

$$\left. - \exp(-\sigma_M x_{Mu} - \sigma_k x_{ku} - \sigma_j x_{ju})\right] \tag{9.90}$$

where

$$\Delta_{jkM} = \sigma_j + \left(\frac{\nu_k - \nu_j}{\nu_M - \nu_j}\sigma_M - \frac{\nu_M - \nu_j}{\nu_M - \nu_k}\sigma_k\right) \tag{9.91}$$

and x_{Mu}, x_{ku}, x_{Ml}, and x_{kl} are values of x_M and x_k evaluated at the corresponding upper and lower limits of x_j and

$$x_M = \frac{\nu_M \; R_M(E') - \nu_k[R_k(E) + x] + (\nu_k - \nu_j)x_j}{\nu_M - \nu_k} \tag{9.92}$$

$$x_k = \frac{\nu_M[R_M(E) + x] - \nu_M \; R_M(E') - (\nu_M - \nu_k)x_j}{\nu_M - \nu_k} \tag{9.93}$$

The requirements that x_M and x_k be bounded by the interval 0 to $x - x_j$ yields

$$\left\{\begin{array}{c} 0 \\ \dfrac{\nu_k \, [R_k(E) + x] - \nu_M \; R_M(E')}{\nu_k - \nu_j} \end{array}\right\} \le x_j \le \left\{\begin{array}{c} x \\ \dfrac{\nu_M \, [R_M(E) + x] - \nu_M \; R_M(E')}{\nu_M - \nu_j} \end{array}\right\} \tag{9.94}$$

as the appropriate limiting values in equation (9.90) when $\nu_M > \nu_k > \nu_j$. In the braces, we always choose the most restrictive value for the limit. The requirement of equation (9.94) also implies the result that

$$R_M^{-1}[R_M(E') - x] \le E \le R_k^{-1}\left[\frac{\nu_M \; R_M(E') - \nu_j x}{\nu_k}\right] \tag{9.95}$$

as the range over which the result of equation (9.90) is not zero. In the event that $\nu_k > \nu_M > \nu_j$, then

$$\left\{\begin{array}{c} 0 \\ \dfrac{\nu_M \, [R_M(E) + x] - \nu_M \; R_M(E')}{\nu_M - \nu_j} \end{array}\right\} \le x_j \le \left\{\begin{array}{c} x \\ \dfrac{\nu_k \, [R_k(E) + x] - \nu_M \; R_M(E')}{\nu_k - \nu_j} \end{array}\right\} \tag{9.96}$$

As a result of equation (9.96),

$$R_k^{-1}\left[\frac{\nu_M}{\nu_k}R_M(E') - x\right] \le E \le R_M^{-1}\left[R_M(E') - \frac{\nu_j}{\nu_M}x\right] \tag{9.97}$$

In the event that $\nu_M > \nu_j > \nu_k$, it follows that

$$0 \le x_j \le \begin{Bmatrix} x \\ \dfrac{\nu_M \left[R_M(E) + x - R_M(E')\right]}{\nu_M - \nu_j} \\ \dfrac{\nu_M \, R_M(E') - \nu_k \, R_k(E) - \nu_k \, x}{\nu_j - \nu_k} \end{Bmatrix} \tag{9.98}$$

where the lesser of the three values in the braces is used as the upper limit of x_j for which $G^{(2)}$ of equation (9.90) is not zero. As a result of equation (9.98),

$$R_M^{-1}\left[R_M(E') - x\right] \le E \le R_k^{-1}\left[\frac{\nu_M}{\nu_k} R_M(E') - x\right] \tag{9.99}$$

Higher order terms are similarly derived. Approximate expressions have been obtained (Wilson et al., 1988) as

$$G_{jM}^{(n)}(x, E; E') = \sum_{k, j_1, \cdots, j_{n-1}} \sigma_{jk} \frac{\sigma_{kj_1} \cdots \sigma_{j_{n-2}M} \; g(j, k, j_1 \cdots, j_{n-2}, M)}{E_{uj} - E_{lj}} \tag{9.100}$$

where

$$E_{jl} = \begin{Bmatrix} R_M^{-1}\left\{R_M(E') - x\right\} & (\nu_M > \nu_k > \nu_j) \\ R_k^{-1}\left\{\frac{\nu_M}{\nu_k} R_M(E') - x\right\} & (\nu_k > \nu_M > \nu_j) \\ R_M^{-1}\left\{R_M(E') - x\right\} & (\nu_M > \nu_J > \nu_k) \end{Bmatrix} \tag{9.101}$$

and

$$E_{ju} = \begin{Bmatrix} R_k^{-1}\left\{\left[\nu_M \, R_M(E') - \nu_j \, x\right]\nu_k^{-1}\right\} & (\nu_M > \nu_k > \nu_j) \\ R_M^{-1}\left\{R_M(E') - \frac{\nu_j \, x}{\nu_M}\right\} & (\nu_k > \nu_M > \nu_j) \\ R_k^{-1}\left\{\frac{\nu_M}{\nu_k} \, R_M(E') - x\right\} & (\nu_M > \nu_j > \nu_k) \end{Bmatrix} \tag{9.102}$$

and the g functions of $n + 1$ arguments are defined as

$$g(j_1) = \exp(-\sigma_{j_1} \, x) \tag{9.103}$$

$$g(j_1, j_2, \cdots j_n, j_{n+1}) = \frac{g(j_1, j_2, \cdots, j_{n-1}, j_n) - g(j_1, j_2, \cdots, j_{n-1}, j_{n+1})}{\sigma_{j_{n+1}} - \sigma_{j_n}} \tag{9.104}$$

The expression for $G_{jM}^{(n)}$ given by equation (9.100) is taken as zero unless

$$E_{jl} \le E \le E_{ju} \tag{9.105}$$

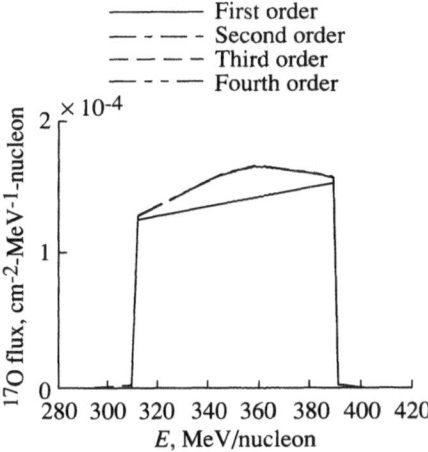

Figure 9.37. Sequence of approximations of ^{17}O flux spectrum after 20 cm of water for first-, second-, third-, and fourth-order (which cannot be distinguished) theories.

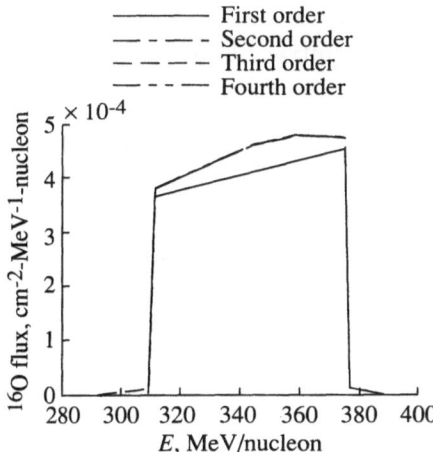

Figure 9.38. Sequence of approximations of ^{16}O flux spectrum after 20 cm of water for first-, second-, third-, and fourth-order (which cannot be distinguished) theories.

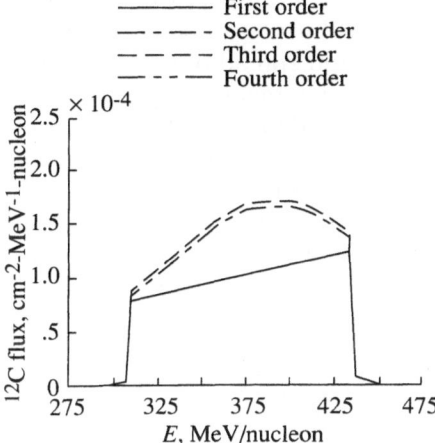

Figure 9.39. Sequence of approximations of ^{12}C flux spectrum after 20 cm of water for first-, second-, third-, and fourth-order (which cannot be distinguished) theories.

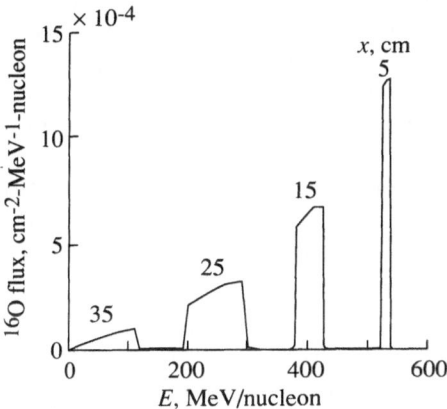

Figure 9.40. Green's function for ^{16}O flux spectrum response to 600-MeV/nucleon ^{20}Ne flux at boundary.

Portions of the Green's function are shown for incident ^{20}Ne beams at $E' = 600$ MeV/nucleon at $x = 20$ cm in figures 9.37 to 9.39. The contribution from $G^{(1)}$ is shown as the solid curve, $G^{(1)} + G^{(2)}$ is shown as the long-dash–short-dash curve, and $G^{(1)} + G^{(2)} + G^{(3)}$ is shown as the dashed curve. The long-dash–double-short-dash curve representing the inclusion of $G^{(4)}$ in the sum cannot be distinguished signifying convergence to a high degree of accuracy. A fuller presentation of the Green's function for ^{16}O fragments is given in figure 9.40, and a presentation of the Green's function for ^{12}C is given in figure 9.41. The results in figures 9.40 and 9.41

are found by summing the terms to $G^{(4)}$. From the present result, the solution for any arbitrary boundary condition may be found by using equation (9.84).

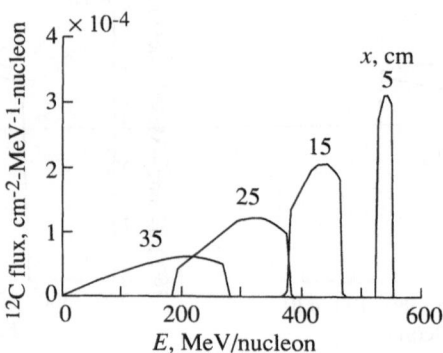

Figure 9.41. Green's function for ^{12}C flux spectrum response to a 600-MeV/nucleon ^{20}Ne flux at boundary.

Although the present formalism presents a closed-form solution for the more common form of the HZE propagation problem, many tasks remain before it is adequately solved. The inclusion of energy-dependent nuclear cross sections is known to be very important in obtaining accurate solutions to some problems (Townsend and Wilson, 1988a). Treating the momentum spread of the fragments is more complicated for the higher-order terms. The inclusion of the light fragment spectra is a difficult challenge (Wilson et al., 1989b). Finally, the three-dimensional aspects of the problem have only partially been treated (Shavers, 1988; Schimmerling et al., 1986). Even these shortcomings of the HZE propagation problem remain without the mention of uncertainties in nuclear cross sections (Townsend and Wilson, 1988a and 1988b) or atomic/molecular cross sections. Clearly much work remains.

9.8. Galactic Ion Transport

In the present section, we expand on the methods developed earlier for nucleon transport (Wilson and Lamkin, 1975) by combining analytic and numerical tools. The galactic cosmic-ray ion transport problem is transformed to an integral along the characteristic curve of that particular ion. As a result of the conservation of velocity in fragmentation, the perturbation series (Wilson and Lamkin, 1975) is replaced by a simple numerical procedure. The resulting method reduces the difficulty associated with the low-energy discretization and the restriction to a definite form for the stopping power. The resulting numerical computation is simple and nondemanding from computer requirements and yet gives superior results compared with other methods.

In the present work, we use the straight ahead approximation and neglect the target secondary fragments (Wilson, 1977a and 1983). The transport equation may be written as

$$\left[\frac{\partial}{\partial x} - \frac{\partial}{\partial E} \tilde{S}_j(E) + \sigma_j \right] \phi_j(x, E) = \sum_{k>j} m_{jk} \sigma_k \phi_k(x, E) \tag{9.106}$$

where $\phi_j(x, E)$ is the flux of ions of type j with atomic mass A_j at x moving along the x axis at energy E in units of MeV/nucleon; σ_j is the corresponding macroscopic nuclear absorption cross section; $\tilde{S}_j(E)$ is the change in E per unit distance; and m_{jk} is the multiplicity of ion j produced in collision by ion k. We recall the result of equation (9.22) as

$$\chi_j(\eta_j, \xi_j) = \exp\left[-\frac{1}{2}\sigma_j(\xi_j + \eta_j)\right] \chi_j(-\xi_j, \xi_j)$$

$$+ \frac{1}{2} \int_{-\xi_j}^{\eta_j} \exp\left[\frac{1}{2}\sigma_j(\eta' - \eta_j)\right] \sum_k m_{jk}\sigma_k \frac{\nu_j}{\nu_k} \chi_k(\eta_k', \xi_k') \, d\eta' \quad (9.107)$$

where

$$\eta_k' = \frac{\nu_k + \nu_j}{2\nu_k}\eta' + \frac{\nu_k - \nu_j}{2\nu_k}\xi_j$$

and

$$\xi_k' = \frac{\nu_k - \nu_j}{2\nu_k}\eta' + \frac{\nu_k + \nu_j}{2\nu_k}\xi_j$$

Defining

$$\psi_j(x, r_j) = \chi_j(\eta_j, \xi_j) \quad (9.108)$$

one may show

$$\psi_j(x, r_j) = \exp(-\sigma_j x) \, \psi_j(0, r_j + x)$$

$$+ \int_0^x dz \, \exp(-\sigma_j z) \sum_k m_{jk}\sigma_k \frac{\nu_j}{\nu_k} \, \psi_k\left(x - z, r_k + \frac{\nu_j}{\nu_k}z\right) \quad (9.109)$$

Furthermore, it is easy to show that

$$\psi_j(x + h, r_j) = \exp(-\sigma_j h) \, \psi_j(x, r_j + h)$$

$$+ \int_0^h dz \, \exp(-\sigma_j z) \sum_k m_{jk}\sigma_k \frac{\nu_j}{\nu_k} \, \psi_k\left(x + h - z, r_k + \frac{\nu_j}{\nu_k}z\right) \quad (9.110)$$

It is clear from equation (9.109) that

$$\psi_k(x + h - z, r_k) = \exp[-\sigma_k(h - z)] \, \psi_k(x, r_k + h) + O(h - z) \quad (9.111)$$

which upon substitution into equation (9.110) yields

$$\psi_j(x + h, r_j) = \exp(-\sigma_j h)\psi_j(x, r_j + h)$$

$$+ \int_0^h dz \, \exp(-\sigma_j z) \sum_k m_{jk}\sigma_k \frac{\nu_j}{\nu_k}$$

$$\times \exp[-\sigma_k(h - z)] \, \psi_k\left(x, r_k + \frac{\nu_j}{\nu_k}z + h - z\right) \quad (9.112)$$

which is correct to order h^2. This expression may be further approximated by

$$\psi_j(x+h, r_j) = \exp(-\sigma_j h) \; \psi_j(x, r_j + h)$$

$$+ \sum_k m_{jk} \sigma_k \frac{\nu_j}{\nu_k} \left[\frac{\exp(-\sigma_j h) - \exp(-\sigma_k h)}{\sigma_k - \sigma_j} \right] \psi_k \left(x, r_k + \frac{\nu_j}{\nu_k} h \right) \quad (9.113)$$

which is accurate to $O\left[(\nu_k - \nu_j)\, \nu_j^{-1} h \right]$. Equation (9.113) is the basis of the Galactic Cosmic-Ray (GCR) Transport Code, HZETRN (Wilson and Badavi, 1986; Wilson et al., 1988; Wilson, Townsend, and Badavi, 1987a).

There are several quantities of interest that are now given. The integral fluence is given as

$$\phi_j(x, >E) = \int_{R_j(E)}^{\infty} \psi_j(x, r) \; dr \quad (9.114)$$

The energy absorption per gram is

$$D_j(x, >E) = \int_E^{\infty} A_j \psi_j[x, R_j(E)] \; dE \quad (9.115)$$

with the dose equivalent given as

$$H_j(x, >E) = A_j \int_E^{\infty} Q_F \; \psi_j[x, R_j(E)] \; dE \quad (9.116)$$

These quantities are used in shield design studies for protection against galactic cosmic rays.

9.8.1. Galactic cosmic-ray propagation. The semiempirical fragmentation cross sections (Wilson, Townsend, and Badavi, 1987b) for iron nuclei colliding with atmospheric nuclei are presented along with values obtained by the original model of Bowman, Swiatecki, and Tsang (1973), the results of the parameterization by Silberberg, Tsao, and Letaw (1983), and the experiments (for carbon targets) of Westfall et al. (1979) in figure 9.42. Quite reasonable values are obtained for these elemental fragmentation cross sections although considerable uncertainty exists in the neutron removal cross section. The light fragment production cross sections for the three models are shown in table 9.6. Clearly, vast differences in estimates of light fragment production exist between the three models. Unfortunately, there are no experimental data to resolve these differences.

The galactic cosmic-ray propagation in the Earth's atmosphere was calculated according to the present propagation model and shown with the fluence measurements of Webber and Ormes (1967) in figure 9.43. The differential spectrum of each species was taken as $1/(1000 + E)^{2.5}$ and elemental distributions according to Adams, Silberberg, and Tsao (1981) and Silberberg, Tsao, and Letaw (1983). These data (table 1 of Tsao et al., 1983) were renormalized within the usual categories of L $(3 < Z < 5)$, M $(6 < Z < 9)$, LH $(10 < Z < 14)$, and H $(20 < Z < 28)$ at the top of the atmosphere with extrapolations of the data of Webber and Ormes

(1967). The renormalized incident flux was used to calculate the results shown in figure 9.43.

Table 9.6. Light Fragment Production From Iron Nuclei Cross Sections

	Iron nuclei cross sections, mb, from—		
Element	Bowman, Swiatecki, and Tsang, 1973	Silberberg, Tsao, and Shapiro, 1976	Present
Li	10.6	130	19.2
Be	40.3	107	6.9
B	46.0	81	20.2
C	20.8	72	22.0

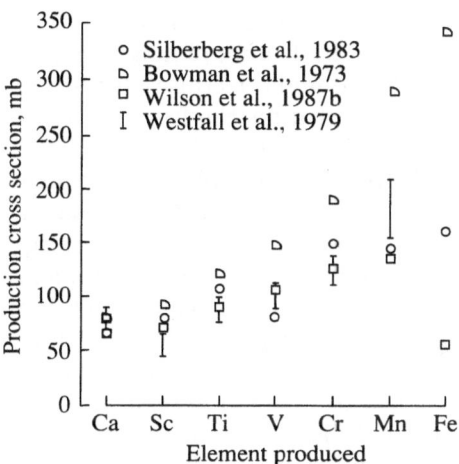

Figure 9.42. Fragment production cross sections of three models and experiments of Westfall et al. (1979).

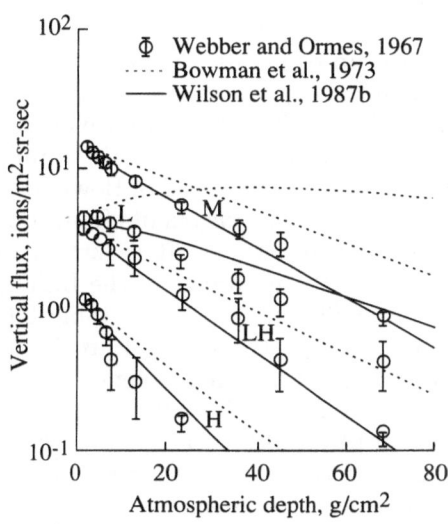

Figure 9.43. Cosmic-ray transition curves for ion energies above 360 MeV/nucleon at high latitudes for two fragmentation models and experiments of Webber and Ormes (1967). Input flux (Wilson et al., 1987b) was renormalized at top of atmosphere.

9.8.2. Discussion of results. It is clear that the original fragmentation model of Bowman, Swiatecki, and Tsang (1973) is oversimplified, a fact of which they have been fully aware. It appears that the present corrections to their model will allow "reasonable" results in the propagation of these broad categories. It is difficult to say as to what accuracy the calculations are performed because various compensating errors can be committed without changing the present result. This is especially important in view of the uncertainty in the neutron removal cross section as noted by comparing the three values of iron fragmentation cross sections in figure 9.42. Such an error in removal cross section could be hidden in the measurements of Webber and Ormes by either a change in total absorption or various charge changing processes, which cannot be observed in the data of Webber and Ormes (1967).

The linear energy-transfer (LET) spectra of normal incident galactic ions are derived from equation (9.114) with

$$\phi_j(x,>S) = \phi_j(x,>E_{m_j}) - \phi_j(x,>E_{S_j}) \tag{9.117}$$

where E_{m_j} corresponds to the energy when $S_j(E)$ is maximum and E_{S_j} is the energy when $S_j(E_{S_j}) = S$. The LET spectra are calculated for the present Langley Research Center nuclear model (NUCFRAG) and are shown in figure 9.44 for several atmospheric depths and geomagnetic cutoff of 1.48 GeV/c. The nearly discontinuous steps in the LET spectra are located at the highest LET for each ion type and are located according to the ion charge squared. There is an observable steepening of the LET spectra as a function of atmospheric depth associated with the breakup of highly ionizing heavy nuclei into smaller nuclear fragments. The flux with LET greater than 200 MeV-cm^2/g is reduced by 1 order of magnitude in penetrating to 60 g/cm^2 (approximately 63 000 ft), the highest cruise altitude of present day commercial aircraft. Similar results are shown for the fragmentation cross sections of the model of Bowman, Swiatecki, and Tsang (1973) in figure 9.45. It is clear in comparing results from the two models that considerable disagreement remains between them. It is further observed that considerable disagreement is expected in the midrange of the LET spectra for the cross sections of Silberberg, Tsao, and Shapiro (1976) in table 9.6. Clearly, a full solution to the problem of the LET spectra in the Earth's atmosphere must await the solution of the nuclear fragmentation problem. Meanwhile, the nuclear fragmentation model of the Langley Research Center presented in chapter 5 and used herein contains the principal physical mechanisms involved and is shown to give fair agreement with the data of Webber and Ormes (1967) as well as the fragmentation cross sections measured for iron beams on carbon targets by Westfall et al. (1979). For further comparison of this model with experiment, see Wilson, Townsend, and Badavi (1987a).

9.9. Analytic Benchmarks

In the present section, we address the question of GCR transport code validation. Ideally, validation should be accomplished with detailed transport data obtained from carefully planned and controlled experiments; unfortunately, there exists a paucity of such data. Although useful for comparison purposes, the atmospheric propagation measurements used previously (Wilson, Townsend, and Badavi, 1987a) are clearly not definitive because they consist of integral fluences of as many as 10 different nuclear species combined into a single datum. Although limited quantities of HZE dosimetry measurements from manned space missions (e.g., Skylab) are also available (Benton, Henke, and Peterson, 1977), numerous assumptions concerning the relationships between dosimeter locations and spacecraft shield thicknesses and geometry must be made to estimate astronaut doses using GCR codes. Because many of these assumptions may involve inherently large uncertainties (a factor of 2 or greater), it becomes difficult to attribute any differences to particular assumptions or approximations that may have been used in the analyses. Without definitive GCR transport measurements with which to compare code predictions, other methods of validation must be considered. As noted by Wilson (1983) and Wilson et al. (1989a), there are several different versions of HZE transport codes available. When used with the same input spectra, interaction parameters, and boundary conditions, all should yield comparable results. The history of transport code development, however, suggests otherwise. For this reason, a realistic, nontrivial exact analytic solution to the simplified Boltzmann equation used to describe HZE transport has been formulated as an absolute standard for code comparison purposes.

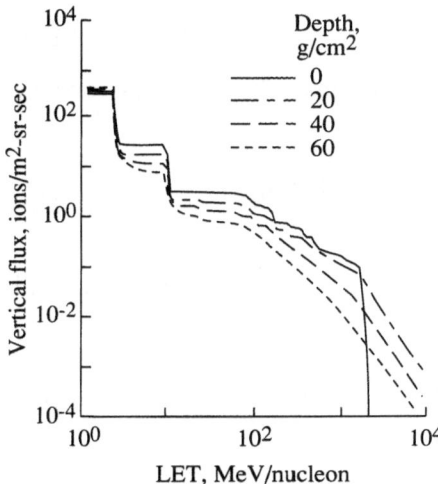

Figure 9.44. Cosmic-ray LET spectra at rigidity cutoff of 1.48 GV/c according to LaRC fragmentation model.

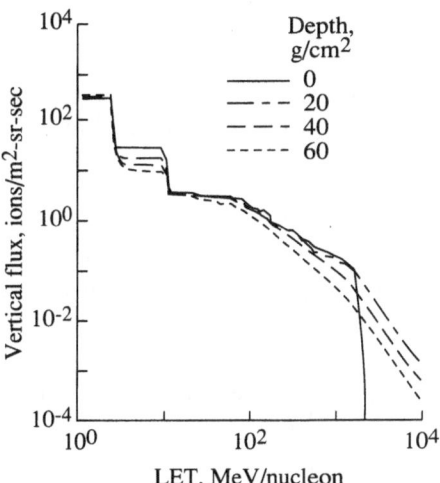

Figure 9.45. Cosmic-ray LET spectra at rigidity cutoff of 1.48 GV/c according to fragmentation model of Bowman, Swiatecki, and Tsang (1973).

For the benchmark problem, the incident spectrum is limited to a single ion type ($j = J$). Because the GCR spectrum for a typical ion is of the form

$$F(E) \sim E^{-\alpha} \tag{9.118}$$

where $\alpha \approx 2.5$, we choose the energy spectrum to be of similar functional form as

$$F_j(E) = \frac{\delta_{jJ}}{[R_J(E)]^2 \widetilde{S}_j(E)} \tag{9.119}$$

Defining the characteristic variables as

$$\eta_j = x - R_j(E) \tag{9.120}$$

and

$$\xi_j = x + R_j(E) \tag{9.121}$$

equation (9.106) can be solved by the method of characteristics (Wilson, 1977a; Wilson and Townsend, 1988) to give

$$\psi_J(x, E) = \frac{\exp(-\hat{\sigma}_J x)}{[\nu_J x + R_J(E)]^2} \tag{9.122}$$

where

$$\psi_J(x, E) \equiv \tilde{S}_j(E) \, \phi_J(x, E) \tag{9.123}$$

and

$$\hat{\sigma}_J = \sigma_J(1 - m_{JJ}) \tag{9.124}$$

This is the trivial solution for the incident beam species. For $j < J$ (secondary fragments), it can be shown that

$$\psi_j(x, E) = \sigma_J m_{jJ} \frac{\nu_j}{\nu_J} \, I_j(x, E) \, \exp\left[\frac{-(\hat{\sigma}_j \eta_j + \hat{\sigma}_J \xi_j)}{2}\right] \tag{9.125}$$

where in terms of the exponential integral function $E_2(x)$ (see Abramowitz and Stegun, 1964),

$$I_j(x, E) = \frac{\exp\left[-b(\nu_J + \nu_j)\xi_j/2\right]}{\nu_J - \nu_j} \left[\frac{E_2(b\nu_j \xi_j)}{\nu_j \xi_j} - \frac{E_2(b\nu_J \xi_J)}{\nu_J \xi_J}\right] \tag{9.126}$$

for $j = J - 1$ and

$$b = \frac{\hat{\sigma}_J - \hat{\sigma}_j}{\nu_J - \nu_j} \tag{9.127}$$

Clearly, equations (9.126) and (9.127) are true for all values of j if $m_{kj} = 0$ for all values of $j < J$ (i.e., if the secondary fragments do not further fragment).

9.9.1. Benchmark results.

The benchmark solution was calculated for an incident iron beam ($J = 26$) in an aluminum target, for which the input parameters are $\hat{\sigma}_{26} = 0.04568 \text{ cm}^2/\text{g}$, $\hat{\sigma}_{25} = 0.04260 \text{ cm}^2/\text{g}$, and $m_{25,26}\sigma_{26} = 0.00403 \text{ cm}^2/\text{g}$. Results of the GCR transport code simulation of this benchmark for the propagating incident iron beam and secondary manganese ($j = 25$) ions and the exact analytic predictions obtained from equations (9.122) and (9.125) are given in tables 9.7 and 9.8, respectively. It is clear from these tabulated results that the numerical solution methods developed previously (Wilson and Badavi, 1986; Wilson, Townsend, and Badavi, 1987a) are accurate in solving equation (9.106) for GCR transport to within about 1 percent. This indicates that any limitations to solving GCR transport problems accurately must focus upon the simplifying approximations used to obtain equation (9.106) as well as upon unresolved issues concerning the need to include the effects of multiple coulomb scattering, fragment momentum dispersion and, perhaps most importantly, the nature and quality of the input cross-section data bases. To illustrate this point, we are aware of only one heavy ion transport code (Wilson et al., 1984) which uses energy-dependent cross sections. Recent studies, however, suggest that fully energy-dependent cross sections may be important for some transport code applications (Townsend and Wilson, 1988a).

Table 9.7. Benchmark Numerical Simulation and Analytic Solution for Iron Ions as Function of Ion Depth[a] and Energy Into Aluminum Absorber

E, MeV/nucleon	$\psi_{Fe}(0, E)$ Numerical	$\psi_{Fe}(0, E)$ Analytic	$\psi_{Fe}(10, E)$ Numerical	$\psi_{Fe}(10, E)$ Analytic	$\psi_{Fe}(20, E)$ Numerical	$\psi_{Fe}(20, E)$ Analytic
0.0198	1.394E5	1.394E5	4.334E−5	4.382E−5	6.942E−6	7.044E−6
0.1147	1.692E4	1.692E4	4.334E−5	4.381E−5	6.942E−6	7.044E−6
1.090	9.217E2	9.217E2	4.333E−5	4.379E−5	6.942E−6	7.043E−6
10.07	1.062E1	1.062E1	4.321E−5	4.360E−5	6.932E−6	7.027E−6
100.1	9.310E−3	9.310E−3	3.699E−5	3.718E−5	6.400E−6	6.478E−6
1 059	5.089E−6	5.089E−6	2.014E−6	2.019E−6	8.741E−7	8.799E−7
10 490	2.970E−8	2.970E−8	1.833E−8	1.833E−8	1.132E−8	1.132E−8

[a]Depth is given in g/cm^2.

Table 9.8. Benchmark Numerical Simulation and Analytic Solution for Secondary Manganese Ions as Function of Ion Depth[a] and Energy Into Aluminum Absorber

E, MeV/nucleon	$\psi_{Mn}(10, E)$ Numerical	$\psi_{Mn}(10, E)$ Analytic	$\psi_{Mn}(20, E)$ Numerical	$\psi_{Mn}(20, E)$ Analytic
0.0198	1.772E−6	1.780E−6	5.704E−7	5.768E−7
0.1147	1.772E−6	1.780E−6	5.704E−7	5.768E−7
1.090	1.772E−6	1.779E−6	5.704E−7	5.767E−7
10.07	1.767E−6	1.771E−6	5.696E−7	5.753E−7
100.1	1.504E−6	1.503E−6	5.242E−7	5.291E−7
1 059	7.797E−8	7.806E−8	6.880E−8	6.918E−8
10 490	7.004E−10	7.004E−10	8.728E−10	8.728E−10

[a]Depth is given in g/cm^2.

9.9.2. Remarks. The need to develop suitable benchmarks for use in validating and comparing existing galactic cosmic-ray transport codes has been described and an exact nontrivial analytic benchmark solution presented. This benchmark solution was then used to establish computational accuracy for a previously published cosmic-ray transport code to within 1 percent. Finally, remaining unresolved issues in GCR transport were briefly described.

9.10. Methods for Energy-Dependent Cross Sections

The HZE transport methods presented herein are an extension of those presented elsewhere (Wilson, 1977a, 1977b, and 1983) and are based upon an analytical solution to the transport equation. In principle, these methods allow for the calculations of absorbed dose caused by fragments of any species in each interaction generation for any arbitrary sequence of absorber layers. The present theory makes several approximations. In particular, projectile fragmentation parameters are obtained from semiempirical formulas (Silberberg, Tsao, and Shapiro, 1976; Silberberg, Tsao, and Letaw, 1983). Target fragmentation is neglected, and the energy loss of charged particles is accounted for by using the continuous slowing down approximation. The present calculations are one-dimensional; however, an extension to three dimensions is planned. Concurrently, an extensive experimental program is in progress at the Lawrence Berkeley Laboratory, BEVALAC, to measure the radiation fields of relativistic nuclei (Schimmerling, Curtis, and Vosburgh, 1977; Schimmerling et al., 1986 and 1989).

In this section, we compare our results with a Bragg ionization curve, obtained at the Lawrence Berkeley Laboratory, BEVALAC, for a neon beam at 670 MeV/nucleon. This example is used to present basic transport data in water. It also illustrates the need for obtaining a more accurate description of basic fragmentation parameters. In addition to the interests of the manned space program, the results are of importance to the radiation safety of high altitude aircraft (Wilson, 1981) and for radiation therapy using heavy ion beams (Schimmerling et al., 1986).

9.10.1. Depth-dose relations.
The transport equation in the straight ahead approximation and neglecting target secondary fragments (Wilson, 1977a, 1977b, and 1983) may be written as

$$
\left[\frac{\partial}{\partial x} - \frac{\partial}{\partial E} \widetilde{S}_j(E) + \sigma_j(E) \right] \phi_j(x, E) = \sum_{k>j} m_{jk}(E)\, \sigma_k(E) \phi_k(x, E) \qquad (9.128)
$$

where $\phi_j(x, E)$ is the flux of ions of type j with atomic mass A_j at x with motion along the x axis and energy E in units of MeV/nucleon, $\sigma_j(E)$ is the corresponding macroscopic nuclear absorption cross section, $\widetilde{S}_j(E)$ is the change in E per unit distance, and $m_{jk}(E)$ is the fragmentation parameter of ion j produced in collision by ion k. The form of the operator on the left-hand side of equation (9.128) is derived in chapter 6 (Wilson and Lamkin, 1975). Note that two terms arise from the energy differential operator. The first term arises from the scattering of particles to lower energy when traversing a distance Δx. The second term arises from contraction of the energy interval caused by the nonlinear relation between space and energy. The solution to equation (9.128) is to be found subject to boundary specification at $x = 0$ and arbitrary E as

$$
\phi_j(0, E) = F_j(E) \qquad (9.129)
$$

and is usually called the incident beam spectrum.

The transport equation (9.128) is solved by the method of characteristics by using an iterative procedure (Wilson, 1983). The resultant series is used to evaluate the dose given by

$$D(x) = \sum_j \int_o^\infty dE \, S_j(E) \, \phi_j(x, E) \tag{9.130}$$

which is evaluated for a monoenergetic beam of energy E_o. The stopping power is $S_j = A_j \tilde{S}_j$. The solution of the homogeneous equation resulting from setting the right-hand side of equation (9.128) to zero yields

$$D^{(0)}(x) = \frac{S_J(E_x) \, P_J(E_o)}{P_J(E_x)} \tag{9.131}$$

where

$$E_x = R_J^{-1}[R_J(E_o) - x] \tag{9.132}$$

is the residual energy and the P_j factors account for nuclear attenuation of the primary beam with

$$P_j(E) = \exp\left[-\int_0^E \frac{\sigma_j(e) \, de}{\tilde{S}_j(e)}\right] \tag{9.133}$$

The first perturbation to the homogeneous solution yields an additional contribution

$$D^{(1)}(x) \approx \sum_j A_j \nu_j (E_{u_j} - E_{l_j}) \left[\frac{m_{jJ}(E_o) \, \sigma_J(E_o) P_j(E_o)}{P_j(E_{l_j})}\right.$$
$$\left. - \frac{m_{jJ}(E_{lj}) \, \sigma_J(E_{lj}) P_J(E_o)}{P_J(E_{l_j})}\right] \{[\sigma_j(E_o) - \sigma_j(E_o)]x\}^{-1} \tag{9.134}$$

where the energy spanned by these secondary ions is given by the "lower limit,"

$$E_{l_j} = R_j^{-1}\left\{\frac{\nu_J}{\nu_j}[R_J(E_o) - x]\right\} \tag{9.135}$$

and "upper limit,"

$$E_{u_j} = R_j^{-1}\left[\frac{\nu_J}{\nu_j} R_J(E_o) - x\right] \tag{9.136}$$

A second perturbation of the homogeneous solution yields an additional contribution to the dose given by

$$D^{(2)}(x) \approx \sum_{jk} m_{jk} \sigma_k m_{kJ} \sigma_J \frac{\Lambda_j \nu_j (E'_{u_j} - E'_{l_i})}{(\nu_j - \nu_k)(\sigma_J - \sigma_k)x}$$
$$\times \left[\frac{\exp(-\sigma_j x) - \exp(-\sigma_J x)}{\sigma_J - \sigma_j} - \frac{\exp(-\sigma_j x) - \exp(-\sigma_k x)}{\sigma_k - \sigma_j}\right] \tag{9.137}$$

where the energy range spanned by these tertiary ions is given by the upper limit,

$$E'_{u_j} = R_j^{-1} \left[\frac{\nu_J}{\nu_j} R_J(E_o) - x \right] \tag{9.138}$$

and the corresponding lower limit,

$$E'_{l_j} = R_j^{-1} \left\{ \frac{\nu_J}{\nu_j} [R_J(E_o) - x] \right\} \tag{9.139}$$

where m and σ of equation (9.137) are evaluated at E_o. The results of equations (9.138) and (9.139) are understood to be zero whenever the right-hand sides are negative. Equations (9.131) through (9.139) can be applied to various shield materials of uniform composition. Each specific application requires knowledge of the appropriate transport coefficients $S_j(E), \sigma_j$, and m_{jk}.

9.10.2. Nuclear absorption. The nuclear absorption cross section σ_k is calculated from a quantum mechanical model of the heavy ion reaction (Wilson, 1975; Wilson and Costner, 1975). Appropriate solutions of the coupled-channel equations for high-energy composite particle scattering are used to calculate the elastic scattered amplitude from which total and absorption cross sections are derived (Wilson and Townsend, 1981; Townsend, Wilson, and Bidasaria, 1983a and 1983b).

The proton cross sections are shown in figures 9.46 and 9.47 for the oxygen constituent of water. The cross sections of ^{12}C projectiles at 2.1 GeV/nucleon onto various targets are shown in figure 9.48 with corresponding results for ^{16}O projectiles in figure 9.49. Additional values for ^{12}C projectiles (Townsend, 1982) are shown in table 9.9 along with experimental values from Jaros et al. (1978), Heckman et al. (1978), Cheshire et al. (1974), and Skrzypczak (1980). It is clear from the present results (Townsend, Wilson, and Bidasaria, 1983a and 1983b) that these cross sections are quite superior to our previous calculations (Wilson and Costner, 1975) and are in good agreement with the limited experimental data. Most nuclear cross sections appear better than 5 percent accurate regarding absorption mean-free paths, except for carbon nuclei where differences may be as large as 10 percent. Although little data are yet available on neon beams, the single datum ($A_T = 20$) in table 9.9 lends confidence that the cross-section data set presented in table 9.10, for light and medium mass ion mean-free paths in water, is quite accurate. Clearly, a great deal more experimental data are required for a complete evaluation. For convenience, we define a fundamental parameter associated with the absorption of a given ion in coming to rest. The average extinction coefficient is defined as

$$O_j(E) = \frac{\int_0^E \sigma_j(e) \, de / \tilde{S}_j(e)}{R_j(E)} \tag{9.140}$$

so that the survival probability is

$$P_J(E) = \exp\left[-O_j(E) \, R_j(E)\right] \tag{9.141}$$

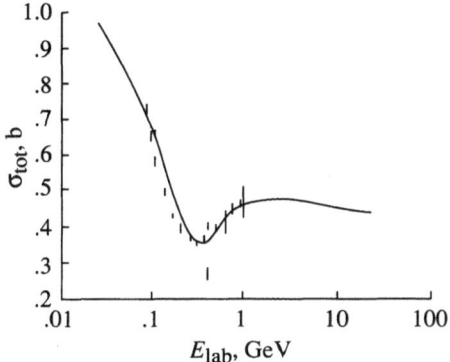

Figure 9.46. Total nuclear cross
section for nucleons on oxygen
targets.

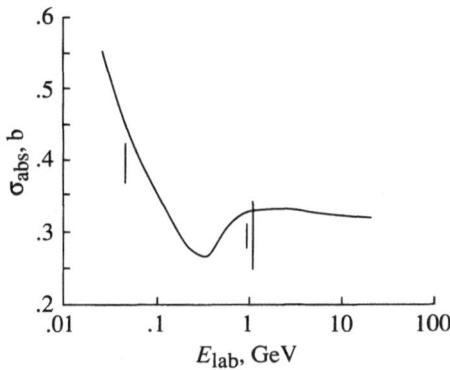

Figure 9.47. Absorption nuclear cross
sections for nucleons on oxygen
targets.

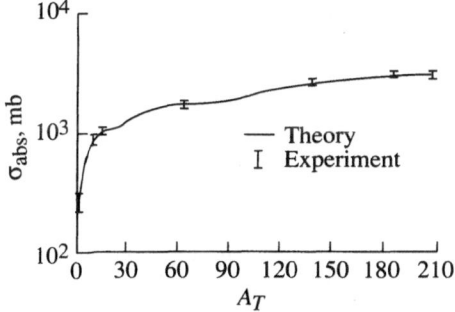

Figure 9.48. Absorption cross
sections for ^{12}C projectiles at
2.1 GeV/nucleons onto various
targets.

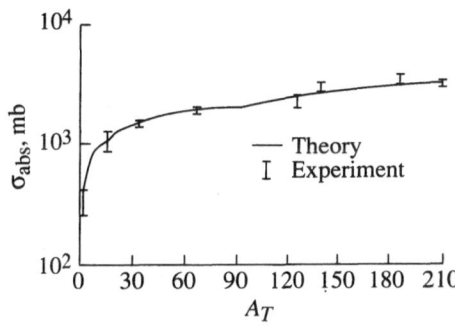

Figure 9.49. Absorption cross
sections for ^{16}O projectiles at
2.1 GeV/nucleons onto various
targets.

The average extinction coefficients for several projectiles in water are shown in table 9.11. Values at intermediate projectile mass numbers can be found by numerical interpolation.

Table 9.9. Absorption Cross Sections for ^{12}C Projectiles Colliding With
Various Target Nuclei

| A_T | σ_{abs}, mb | | σ_{exp}, mb |
	Townsend, 1982	Wilson and Badavi, 1986	
0.87 GeV/nucleon			
12	819	763	[a]939 ± 49
2.1 GeV/nucleon			
1	237	246	[a]269 ± 14 [b]258 ± 21
12	839	781	[a]888 ± 50 [b]826 ± 23
16	990	820	[b]1022 ± 25
64	1727	1656	[b]1730 ± 36
138	2519	2447	[c]2600 ± 100
184	2924		[c]3000 ± 100
208	3047	2969	[b]2960 ± 65
3.6 GeV/nucleon			
12	836	779	[d]780 ± 30
20	1059	902	[d]1040 ± 60
64	1723	1653	[d]1700 ± 90

[a]Jaros et al., 1978.
[b]Heckman et al., 1978.
[c]Cheshire et al., 1974.
[d]Skrzypczak, 1980.

Table 9.10. Macroscopic Nuclear Absorption Cross Sections in Water

E, MeV/nucleon	σ_{abs}, m^{-1}										
	^{1}H	^{4}He	^{7}Li	^{9}Be	^{12}C	^{16}O	^{20}Ne	^{27}Al	^{40}A	^{56}Fe	^{64}Cu
25	2.57	4.18	6.72	7.41	7.35	8.80	9.57	9.93	13.33	15.10	15.60
50	1.84	3.66	5.35	5.98	6.10	7.33	7.99	8.50	11.32	13.00	13.60
75	1.53	3.01	4.76	5.36	5.54	6.67	7.30	7.87	10.46	12.10	12.70
100	1.34	2.73	4.30	4.87	5.10	6.16	6.77	7.36	9.79	1.140	1.200
200	1.12	2.35	3.66	4.19	4.48	5.43	6.00	6.64	8.81	1.040	1.100
300	1.06	2.25	3.48	4.00	4.31	5.23	5.79	6.43	8.53	10.10	10.70
400	1.09	2.28	3.52	4.04	4.36	5.28	5.84	6.48	8.59	10.20	10.80
600	1.30	2.52	3.87	4.42	4.74	5.72	6.31	6.93	9.16	10.80	11.40
1 000	1.43	2.67	4.10	4.68	4.98	6.01	6.61	7.22	9.52	1.115	1.180
2 000	1.43	2.72	4.17	4.75	5.05	6.09	6.70	7.31	9.64	11.30	11.90
4 000	1.39	2.70	4.12	4.69	5.01	6.03	6.64	7.27	9.57	11.20	11.80
10 000	1.37	2.68	4.07	4.64	4.97	5.97	6.58	7.21	9.49	11.10	11.74

9.10.3. Nuclear fragmentation parameters. The basic fragmentation parameters for ions onto hydrogen targets are those of Silberberg, Tsao, and Shapiro (1976). These have been augmented by light fragment production cross sections of Bertini (Anon., 1968). The extension of the fragmentation on hydrogen targets to an arbitrary target nucleus is by a multiplicative scale factor (Silberberg, Tsao, and Shapiro, 1976) determined from the measured carbon fragmentation data at 1.05 GeV/nucleon (Lindstrom, et al., 1975) on various target nuclei. Silberberg, Tsao, and Shapiro suggested that the appropriate energy for evaluation of the hydrogen target fragmentation parameters is the total kinetic energy of the target ion as seen in the projectile rest frame. A second procedure was used herein in which the relative target velocity, rather than total kinetic energy, was assumed to be the appropriate parameter for evaluation of the hydrogen fragmentation cross sections (Schimmerling, Curtis, and Vosburgh, 1977), after which renormalization was used to ensure mass and charge conservation (velocity renormalized, VR). It is seen from figure 9.50 that reasonable estimates of the Bragg curve are obtained from these three terms provided by the present theory when the velocity-scaled renormalized parameters are used. Values for these fragmentation parameters (m_{ij} of equation (9.128)) are listed in table 9.12.

Contributions of the homogeneous term and the first and second perturbations with the velocity-scaled renormalized parameters are shown separately in figure 9.50 and show the rapid convergence of the series for the first 20 cm. The generally greater penetrability of successive generations of ion fragments indicates the need to consider the third perturbation term for depths beyond 30 cm.

Table 9.11. Average Extinction Coefficients for Ions in Water

E, MeV/nucleon	Extinction coefficient, m^{-1}										
	^{1}H	^{4}He	^{7}Li	^{9}Be	^{12}C	^{16}O	^{20}Ne	^{30}Si	^{40}Ar	^{50}V	^{60}Ni
25	2.57	4.18	6.72	7.41	7.34	8.79	9.56	10.42	13.30	14.55	15.35
50	2.24	4.00	6.11	6.77	6.79	8.15	8.87	9.79	12.46	13.67	14.52
75	1.94	3.62	5.55	6.19	6.28	7.54	8.22	9.19	11.65	12.82	13.69
100	1.73	3.31	5.13	5.75	5.89	7.08	7.74	8.73	11.04	12.19	13.05
200	1.35	2.72	4.25	4.81	5.04	6.09	6.70	7.72	9.71	10.81	11.67
300	1.21	2.50	3.89	4.44	4.70	5.69	6.28	7.31	9.17	10.24	11.09
400	1.16	2.41	3.74	4.27	4.56	5.52	6.10	7.13	8.93	10.00	10.85
600	1.18	2.40	3.72	4.25	4.55	5.51	6.08	7.11	8.90	9.98	10.83
1 000	1.28	2.52	3.88	4.43	4.74	5.73	6.32	7.34	9.17	10.26	11.12
2 000	1.38	2.64	4.06	4.63	4.93	5.95	6.55	7.57	9.46	10.55	11.42
4 000	1.40	2.68	4.11	4.69	4.99	6.02	6.63	7.66	9.55	10.65	11.51
10 000	1.38	2.69	4.10	4.67	4.99	6.00	6.61	7.65	9.53	10.62	11.48

Table 9.12. ^{20}Ne Fragmentation Parameters in Water for Velocity Scaling and Total Fragment Charge $\bar{Z}_F$

$\bar{Z}_F$	^{20}Ne fragmentation parameter at E, MeV/nucleon, of—				
	10	31.6	100	316	1000
0	1.313	0.968	0.962	1.289	2.160
1	3.141	2.176	1.975	2.377	3.181
2	0.971	0.713	0.587	0.525	0.652
3	0.035	0.023	0.031	0.046	0.082
4	0.042	0.026	0.037	0.050	0.079
5	0.003	0.013	0.035	0.061	0.074
6	0.005	0.026	0.080	0.121	0.111
7	0.020	0.048	0.090	0.106	0.083
8	0.193	0.225	0.289	0.267	0.183
9	0.303	0.268	0.217	0.195	0.158
10	0.008	0.158	0.122	0.077	0.060
11	0.043	0.009	0.002	0.000	0.000
$\bar{Z}_F$. . . .	10.4	10.2	10.2	10.2	10.2

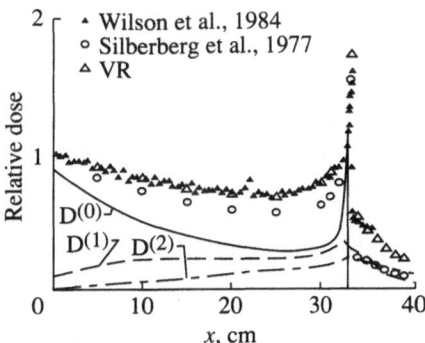

Figure 9.50. Energy deposition in water by ^{20}Ne ions at 670 MeV/nucleon.

It should be emphasized that the Bragg curve does not provide adequate information for radiation protection because of the probable large relative biological effectiveness (RBE) differences of the secondary components. The Bragg curve, however, is a readily accessible experimental quantity and any meaningful theory must reliably produce it. Since any variation from the Bragg curve may have important biological consequences caused by differences in RBE factors, a full evaluation requires more complete sets of experimental data. Clearly, evaluation of the next perturbation term is important, but an adequate knowledge of the specific fragmentation parameters is ultimately required to reduce the remaining error. This point is discussed further in section 9.11.

9.11. Laboratory Validation

Concurrent with the development of heavy ion transport codes and associated nuclear and atomic data bases are laboratory experiments with detailed measurements of ion type and energy within absorber layers of various materials. A detailed description of the experimental setup is given by Schimmerling et al. (1989) along with a description of experimental results for ^{20}Ne beams in water. The results were compared with a first collision transport code (HZESEC) developed by Curtis, Doherty, and Wilkinson (1969) and further improved by Schimmerling, Curtis, and Vosburgh (1977) by using the analysis code described by Shavers (1988) and Shavers et al. (1990). Generally, the transport calculations were found to be accurate to within 30 percent for depths less than $\approx$15 cm, and the effects of tertiary particles (not included in HZESEC) become significant at greater depths with differences up to a factor of 2. This is consistent with our earlier results (Wilson et al., 1984) in figure 9.50. The theory (HZESEC) based on the fragmentation cross sections of Silberberg, Tsao, and Shapiro generally underestimate the fragment fluence which is again consistent with our earlier conclusions (Wilson et al., 1984). A later version of this fragmentation shows a significant increase in the production of C, O, and F, whereas the production of N was reduced 7 percent.

The experimental methods and the corresponding elements of the theoretical model are depicted in figure 9.51. The experiment shown in the top view has the beam incident from the right passing through a multiwire proportional counter (B2WC2) before scattering on a lead foil which spreads the beam. The beam is bent with a bending magnet (B2M1) to align with the experiment axis where it first passes through a beam-monitoring scintillation counter (J). The beam focus is monitored by a second multiwire proportional counter (MWPC) before passing through the monitoring ion chamber (IC1). The variable water column is the transport media being studied. The Bragg curve behind the water column is determined by using the ion chamber (IC2) for various water thicknesses. The time of flight column (TOF1 and TOF2) is enclosed by a vacuum chamber to minimize scattering losses in the detector system. The T_1 detector begins the timing clock which is stopped by a signal from T_2. The differential energy detector (D_1–D_3) responses are proportional to the particle charge squared. A second time of flight measurement is made with T_2, T_3, and T_4 for the most energetic particles. The theoretical model is represented by two parts. The nuclear fragmentation is represented by a one-dimensional transport code as depicted in the lower part of figure 9.51. Because of the large dimension of the experimental arrangement (several meters), even a small scattering is magnified in the large distances involved. The angular scattering is represented as a multiple-scattering acceptance model defining a solid angle of acceptance for the various (geometry-defining) detectors of the apparatus.

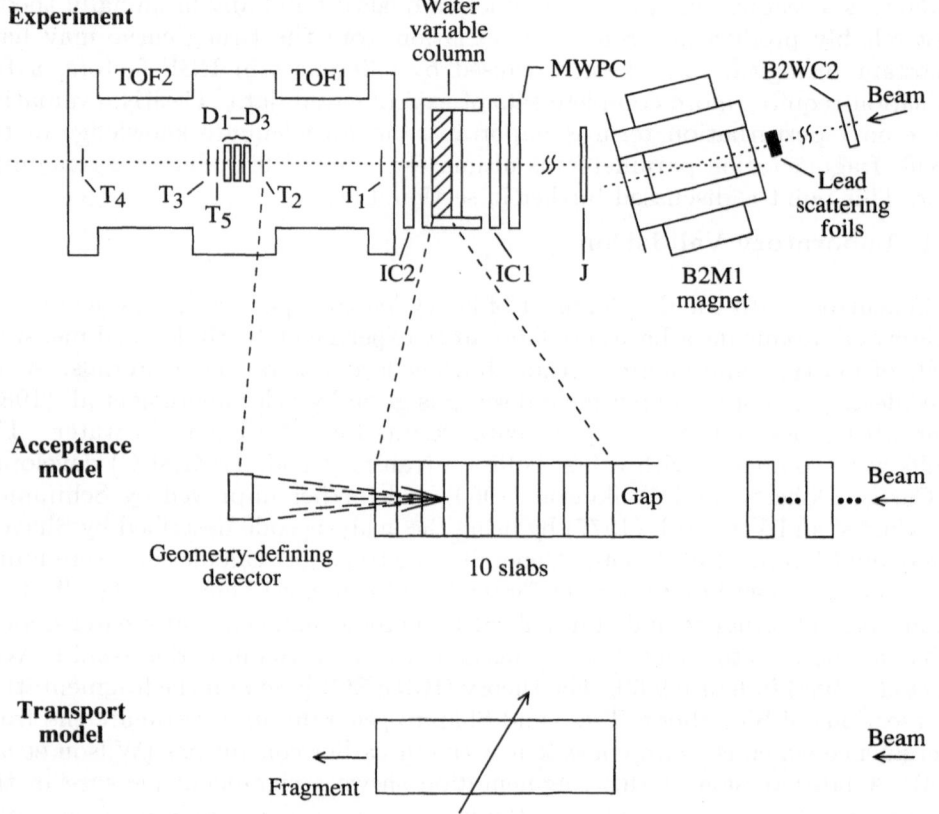

Figure 9.51. The fragmentation of ^{20}Ne at 670 MeV/nucleon as function of depth in water (from Shavers et al., 1990).

The acceptance functions for the experimental arrangement which accounts for the three-dimensional aspects of the transport (in part) were evaluated by the PROPAGATE code developed by Schimmerling et al. (1986) and is based on their fundamental studies in multiple scattering (Wong et al., 1990). The analysis of the experimental data (Schimmerling et al., 1989; Schimmerling, 1990) was accomplished with the code (Shavers, 1988; Shavers et al., 1990) and equations (9.58), (9.59), (9.63), and (9.81) for the transport of an incident primary beam with a narrow Gaussian energy spread. A post-1984 version of the fragmentation model of Silberberg, Tsao, and Shapiro was used as opposed to an older version used in HZESEC (circa 1973).

The measured neon fluence spectra are shown with the calculated values in figure 9.52. The theoretical result multiplied by the acceptance of the detector at each value of LET is plotted as a histogram. In the plateau region of the Bragg curve, the agreement is within 1 percent of the maximum value. The differences at high LET are caused by neon isotopes which were not evaluated. The nuclear mean free path for the experiment was 16.5 g/cm^2 compared with the LBLBEAM value of 16.0 g/cm^2. Figure 9.53 shows the first generation term for carbon fragments (dashed curve) and experimental data. Each isotope of carbon had applied its own acceptance. The fluence predictions are within 30 percent at water thicknesses up to 30 g/cm^2. For increasing water depths and decreasing fragment mass, the number of possible reaction channels and higher-order interactions become more likely. The tertiary fluence contribution (solid curve) is included in figure 9.53, with the acceptance factor of the first generation term for each isotope being assumed. The integral fluence is shown in figure 9.54. Also shown are the results of HZESEC used in a previous analysis of the data (Shavers et al., 1990). The circles denote the experimental data, the solid line is the result of HZESEC (Shavers et al., 1990), the short-dashed line is the first generation term of LBLBEAM, and the long-dashed line is LBLBEAM including tertiary particles.

The threshold effect of the detector can be seen in the plots for the Be and B data, where the measured number of Be fragments falls significantly below all predictions for absorber thicknesses less than 25 g/cm^2; B seems to deposit signals above threshold at thicknesses greater than approximately 15 g/cm^2.

The predictions of HZESEC are systematically lower than those of LBLBEAM, even for the lighter fragments, where differences in stopping-power calculations are not significant. One way in which such behavior could arise is if one of the models did not properly account for the lower mass isotopes. Such an effect may be a consequence of the fragmentation cross sections used by LBLBEAM, obtained from the latest semiempirical fits of Silberberg, Tsao, and Shapiro, which are larger than the older values used by HZESEC.

The difference between the first generation and the second generation predictions of LBLBEAM shows the expected effect of tertiary particles, which increases for lighter particles and thicker absorbers; however, the tertiary prediction of LBLBEAM is systematically greater than the data. One likely reason for this behavior must lie with the angular distribution of the fragments, which is not taken into account in the straight ahead approximation used by the present version of the transport code. Two effects contribute to the angular distribution of fragments. One effect is the compounding of the angular distributions of fragments emitted in nuclear interactions, which results in fewer fragments emitted into the narrow solid angle of the detector. The other effect is that the velocity

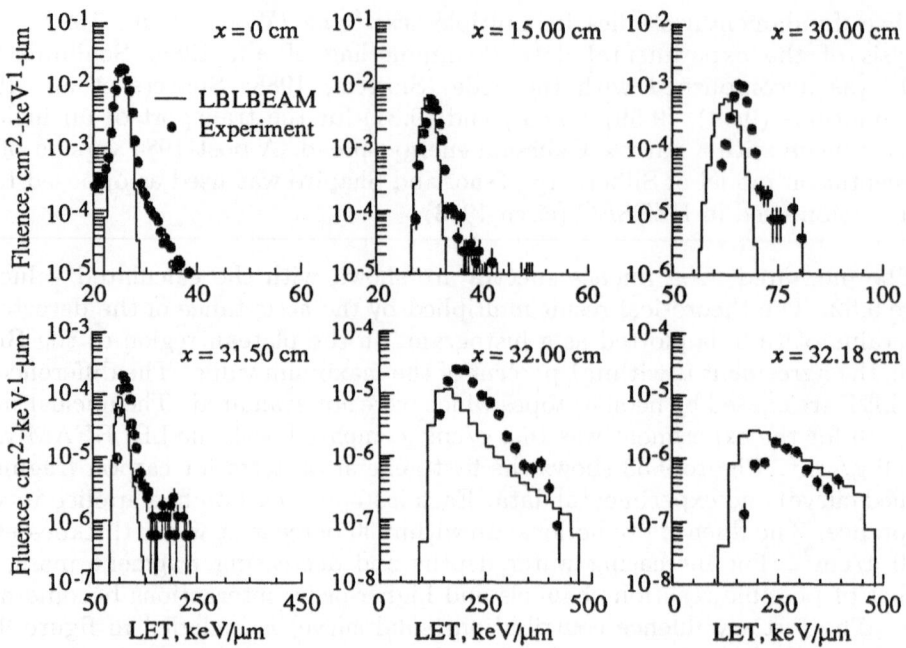

Figure 9.52. Differential fluence for neon nuclei incident on various thicknesses of water.

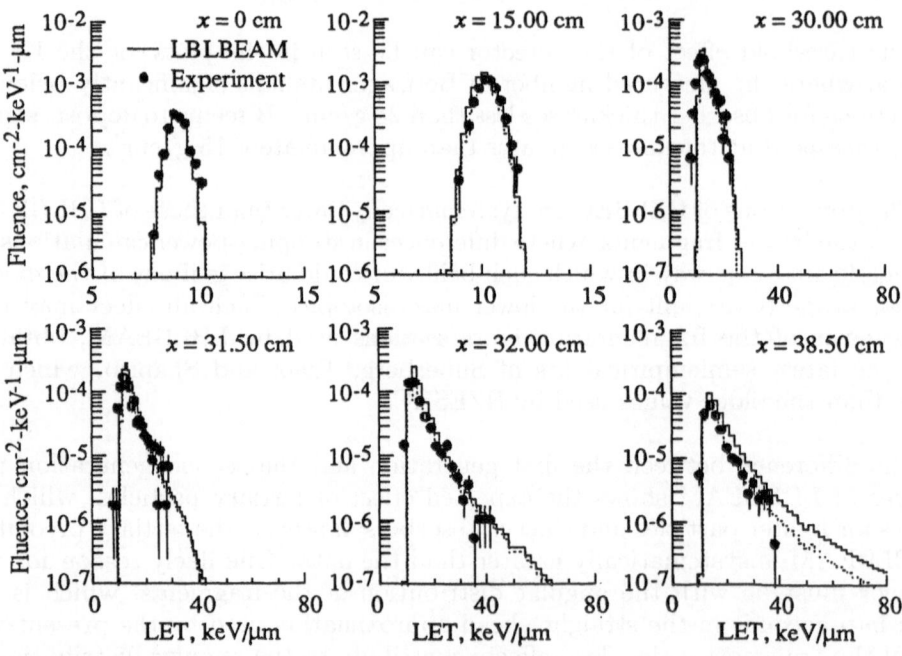

Figure 9.53. Differential fluence for carbon nuclei produced by neon nuclei incident on various thicknesses of water.

of tertiary particles emerging from the absorber is no longer uniquely tied to the position at which it is produced and, therefore, the scattering calculation used for the acceptance is based on an erroneous amount of scattering material.

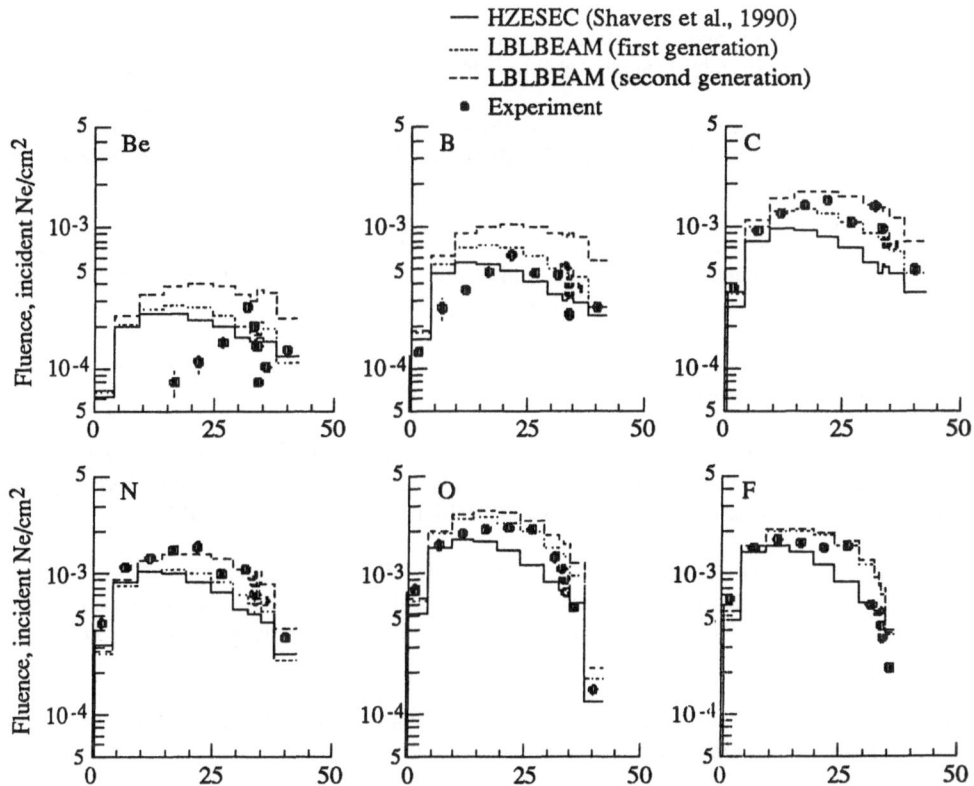

Figure 9.54. Integral fluence for all fragments produced by neon nuclei incident on various thicknesses of water as function of water column thickness.

It is not immediately apparent that the more recent values of the cross sections are more accurate than the older values. The first-generation predictions of LBLBEAM may also be too large, if the geometric effects discussed do not result in significant differences between predicted secondary and tertiary particles. If this is the case, then the predictions of LBLBEAM would be expected to be below the measured data by an amount equivalent to the added contribution of tertiary particles, which is approximately the difference seen by HZESEC.

9.12. References

Abramowitz, Milton; and Stegun, Irene A., eds., 1964: *Handbook of Mathematical Functions With Formulas, Graphs, and Mathematical Tables.* John Wiley & Sons, Inc. (Reprinted with corrections Dec. 1972.)

Adams, J. H., Jr.; Silberberg, R.; and Tsao, C. H., 1981: *Cosmic Ray Effects on Microelectronics, Part I—The Near-Earth Particle Environment.* NRL Memo. Rep. 4506-Pt. I, U.S. Navy. (Available from DTIC as AD A103 897.)

Allkofer, O. C.; and Heinrich, W., 1974: Attenuation of Cosmic Ray Heavy Nuclei Fluxes in the Upper Atmosphere by Fragmentation. *Nuclear Phys. B,* vol. B71, no. 3, pp. 429–438.

Anon., 1968: *MECC-7 Intranuclear Cascade Code, 500-MeV Protons on O-16.* I4C Analysis Codes (Programmed for H. W. Bertini). Available from Radiation Shielding Information Center, Oak Ridge National Lab.

Benton, E. V.; Henke, R. P.; and Peterson, D. D., 1977: Plastic Nuclear Track Detector Measurements of High-LET Particle Radiation on Apollo, Skylab, and ASTP Space Missions. *Nuclear Track Detect.,* vol. 1, no. 1, pp. 27–32.

Bowman, J. D.; Swiatecki, W. J.; and Tsang, C. F., 1973: *Abrasion and Ablation of Heavy Ions.* LBL-2908, Lawrence Berkeley Lab., Univ. of California.

Chatterjee, A.; Tobias, C. A.; and Lyman, J. T., 1976: Nuclear Fragmentation in Therapeutic and Diagnostic Studies With Heavy Ions. *Spallation Nuclear Reactions and Their Applications,* B. S. P. Shen and M. Merker, eds., D. Reidel Publ. Co., pp. 169–191.

Cheshire, D. L.; Huggett, R. W.; Johnson, D. P.; Jones, W. V.; Rountree, S. P.; Verma, S. D.; Schmidt, W. K. H.; Kurz, R. J.; Bowen, T.; and Krider, E. P., 1974: Fragmentation Cross Sections of 2.1-GeV/Nucleon ^{12}C and ^{16}O Ions. *Phys. Review,* vol. 10, ser. D, no. 1, pp. 25–31.

Curtis, S. B.; Doherty, W. R.; and Wilkinson, M. C., 1969: *Study of Radiation Hazards to Man on Extended Near Earth Missions.* NASA CR-1469.

Davis, Leverett, Jr., 1960: On the Diffusion of Cosmic Rays in the Galaxy. *Proceedings of the Moscow Cosmic Ray Conference,* International Union of Pure and Applied Physics (Moscow), pp. 220–225.

Ganapol, Barry D.; Townsend, Lawrence W.; and Wilson, John W., 1989: *Benchmark Solutions for the Galactic Ion Transport Equations: Energy and Spatially Dependent Problems.* NASA TP-2878.

Ginzburg, V. L.; and Syrovatskii, S. I. (H. S. H. Massey, transl., and D. Ter Haar, ed.), 1964: *The Origin of Cosmic Rays.* Macmillan Co.

Gloeckler, G.; and Jokipii, J. R., 1969: Physical Basis of the Transport and Composition of Cosmic Rays in the Galaxy. *Phys. Review Lett.,* vol. 22, no. 26, pp. 1448–1453.

Heckman, H. H.; Greiner, D. E.; Lindstrom, P. J.; and Shwe, H., 1978: Fragmentation of ^{4}He, ^{12}C, ^{14}N and ^{16}O Nuclei in Nuclear Emulsion at 2.1 GeV/Nucleon," *Phys. Review,* vol. 17, ser. C, no. 5, pp. 1735–1747.

Jaros, J.; Wagner, A.; Anderson, L.; Chamberlain, O.; Fuzesy, R. Z.; Gallup, J.; Gorn, W.; Schroeder, L.; Shannon, S.; Shapiro, G.; and Steiner, H., 1978: Nucleus-Nucleus Total Cross Sections for Light Nuclei at 1.55 and 2.89 GeV/c per Nucleon. *Phys. Review*, vol. 18, ser. C, no. 5, pp. 2273–2292.

Letaw, John; Tsao, C. H.; and Silberberg, R., 1983: Matrix Methods of Cosmic Ray Propagation. *Composition and Origin of Cosmic Rays*, Maurice M. Shapiro, ed., D. Reidel Publ. Co., pp. 337–342.

Lezniak, J. A., 1979: The Extension of the Concept of the Cosmic-Ray Path-Length Distribution to Nonrelativistic Energies. *Astrophys. & Space Sci.*, vol. 63, no. 2, pp. 279–293.

Lindstrom, P. J.; Greiner, D. E.; Heckman, H. H.; Cork, Bruce; and Bieser, F. S., 1975: *Isotope Production Cross Sections From the Fragmentation of ^{16}O and ^{12}C at Relativistic Energies.* LBL-3650 (NGR-05-003-513), Lawrence Berkeley Lab., Univ. of California.

Peters, B., 1958: The Nature of Primary Cosmic Radiation. *Progress in Cosmic Ray Physics*, J. G. Wilson, ed., Interscience Publ., Inc., pp. 191–242.

Schimmerling, Walter; Curtis, Stanley B.; and Vosburgh, Kirby G., 1977: Velocity Spectrometry of 3.5-GeV Nitrogen Ions. *Radiat. Res.*, vol. 72, no. 1, pp. 1–17.

Schimmerling, Walter; Rapkin, Marwin; Wong, Mervyn; and Howard, Jerry, 1986: The Propagation of Relativistic Heavy Ions in Multielement Beam Lines. *Med. Phys.*, vol. 13, no. 2, pp. 217–228.

Schimmerling, Walter; Miller, Jack; Wong, Mervyn; Rapkin, Marwin; Howard, Jerry; Spieler, Helmut G.; and Jarret, Blair V., 1989: The Fragmentation of 670A MeV Neon-20 as a Function of Depth in Water. *Radiat. Res.*, vol. 120, pp. 36–71.

Schimmerling, Walter, 1990: Ground-Based Measurements of Galactic Cosmic Ray Fragmentation in Shielding. Paper presented at the 28th Plenary Meeting of COSPAR (The Hague, The Netherlands).

Shavers, Mark R., 1988: Heavy-Ion Transport Code Calculations and Comparison With Experiment for a 670A MeV Accelerated Neon Beam Interacting With Water. M. E. Thesis, Univ. of Florida.

Shavers, Mark R.; Curtis, Stanley B.; Miller, Jack; and Schimmerling, Walter, 1990: The Fragmentation of 670A MeV Neon-20 as a Function of Depth in Water. II. One-Generation Transport Theory. *Radiat. Res.*, vol. 124, pp. 117–130.

Silberberg, R.; Tsao, C. H.; and Shapiro, M. M., 1976: Semiempirical Cross Sections, and Applications to Nuclear Interactions of Cosmic Rays. *Spallation Nuclear Reactions and Their Applications*, B. S. P. Shen and M. Merker, eds., D. Reidel Publ. Co., pp. 49–81.

Silberberg, R.; Tsao, C. H.; and Letaw, John R., 1983: Improvement of Calculations of Cross Sections and Cosmic-Ray Propagation. *Composition and Origin of Cosmic Rays*, Maurice M. Shapiro, ed., D. Reidel Publ. Co., pp. 321–336.

Skrzypczak, E., 1980: Cross-Sections for Inelastic ^{4}He and ^{12}C-Nucleus Collisions at 4.5 GeV/c/N Incident Momentum. *Proceedings of the International Conference on Nuclear Physics, Volume 1, Abstracts*, LBL-11118, (Contract No. W-7405-ENG-48), Lawrence Berkeley Lab., Univ. of California, p. 575.

Townsend, Lawrence W., 1982: *Harmonic Well Matter Densities and Pauli Correlation Effects in Heavy Ion Collisions.* NASA TP-2003.

Townsend, Lawrence W.; Wilson, John W.; and Bidasaria, Hari B.; 1983a: *Heavy-Ion Total and Absorption Cross Sections Above 25 MeV/Nucleon.* NASA TP-2138.

Townsend, Lawrence W.; Wilson, John W.; and Bidasaria, Hari B., 1983b: *Nucleon and Deuteron Scattering Cross Sections From 25 MeV/Nucleon to 22.5 GeV/Nucleon.* NASA TM-84636.

Townsend, L. W.; and Wilson, J. W., 1988a: An Evaluation of Energy-Independent Heavy Ion Transport Coefficient Approximations. *Health Phys.*, vol. 54, no. 4, pp. 409–412.

Townsend, L. W.; and Wilson, J. W., 1988b: Nuclear Cross Sections for Hadronic Transport. *Trans. American Nuclear Soc.*, vol. 56, pp. 277–279.

Townsend, Lawrence W.; Ganapol, Barry D.; and Wilson, John W., 1989: Benchmark Solutions for Heavy Ion Transport Code Validation. *Abstracts of Papers for the Thirty-Seventh Annual Meeting of the Radiation Research Society and Ninth Annual Meeting of the North American Hyperthermia Group,* Radiation Research Soc., p. 126.

Tsao, C. H.; Silberberg, R.; Adams, J. H., Jr.; and Letaw, J. R., 1983: Cosmic Ray Transport in the Atmosphere—Dose and LET-Distributions in Materials. *IEEE Trans. Nuclear Sci.*, vol. NS-30, pp. 4398–4404.

Webber, W. R.; and Ormes, J. F., 1967: Cerénkov-Scintillation Counter Measurements of Nuclei Heavier Than Helium in the Primary Cosmic Radiation. 1. Charge Composition and Energy Spectra Between 200 MeV/Nucleon and 5 beV/Nucleon. *J. Geophys. Res.—Space Phys.*, vol. 72, no. 23, pp. 5957–5976.

Westfall, G. D.; Wilson, Lance W.; Lindstrom, P. J.; Crawford, H. J.; Greiner, D. E.; and Heckman, H. H., 1979: Fragmentation of Relativistic ^{56}Fe. *Phys. Review*, vol. 19, ser. C, no. 4, pp. 1309–1323.

Wilson, John W., 1975: Composite Particle Reaction Theory. Ph.D. Diss., College of William and Mary in Virginia.

Wilson, John W.; and Costner, Christopher M., 1975: *Nucleon and Heavy-Ion Total and Absorption Cross Section for Selected Nuclei.* NASA TN D-8107.

Wilson, John W.; and Lamkin, Stanley L., 1975: Perturbation Theory for Charged-Particle Transport in One Dimension. *Nuclear Sci. & Eng.*, vol. 57, no. 4, pp. 292–299.

Wilson, John W., 1977a: *Analysis of the Theory of High-Energy Ion Transport.* NASA TN D-8381.

Wilson, J. W., 1977b: Depth-Dose Relations for Heavy Ion Beams. *Virginia J. Sci.*, vol. 28, no. 3, pp. 136–138.

Wilson, J. W., 1981: Solar Radiation Monitoring for High Altitude Aircraft. *Health Phys.*, vol. 41, no. 4, pp. 607–617.

Wilson, J. W.; and Townsend, L. W., 1981: An Optical Model for Composite Nuclear Scattering. *Canadian J. Phys.*, vol. 59, no. 11, pp. 1569–1576.

Wilson, John W., 1983: *Heavy Ion Transport in the Straight Ahead Approximation.* NASA TP-2178.

Wilson, John W.; Townsend, L. W.; Bidasaria, H. B.; Schimmerling, Walter; Wong, Mervyn; and Howard, Jerry, 1984: ^{20}Ne Depth-Dose Relations in Water. *Health Phys.*, vol. 46, no. 5, pp. 1101–1111.

Wilson, John W.; and Badavi, F. F., 1986: Methods of Galactic Heavy Ion Transport. *Radiat. Res.*, vol. 108, 1986, pp. 231–237.

Wilson, J. W.; Townsend, L. W.; and Badavi, F. F., 1987a: Galactic HZE Propagation Through the Earth's Atmosphere. *Radiat. Res.*, vol. 109, no. 2, pp. 173–183.

Wilson, John W.; Townsend, Lawrence W.; and Badavi, F. F., 1987b: A Semiempirical Nuclear Fragmentation Model. *Nuclear Instrum. & Methods Phys. Res.*, vol. B18, no. 3, pp. 225–231.

Wilson, John W.; and Townsend, L. W., 1988: A Benchmark for Galactic Cosmic-Ray Transport Codes. *Radiat. Res.,* vol. 114, no. 2, pp. 201–206.

Wilson, John W.; Townsend, Lawrence W.; Ganapol, Barry D.; and Lamkin, Stanley L., 1988: Methods for High-Energy Hadronic Beam Transport. *Trans. American Nuclear Soc.,* vol. 56, pp. 271–272.

Wilson, John W.; Lamkin, Stanley L.; Farhat, Hamidullah; Ganapol, Barry D.; and Townsend, Lawrence W., 1989a: *A Hierarchy of Transport Approximations for High Energy Heavy (HZE) Ions.* NASA TM-4118.

Wilson, John W.; Townsend, Lawrence W.; Nealy, John E.; Chun, Sang Y.; Hong, B. S.; Buck, Warren W.; Lamkin, S. L.; Ganapol, Barry D.; Khan, Ferdous; and Cucinotta, Francis A., 1989b: BRYNTRN: *A Baryon Transport Model.* NASA TP-2887.

Wong, Mervyn; Schimmerling, Walter; Phillips, Mark H.; Ludewigt, Bernhard A.; Landis, Donald A.; Walton, John T.; and Curtis, Stanley B., 1990: The Multiple Coulomb Scattering of Very Heavy Charged Particles. *Med. Phys.*, vol. 17, no. 2, pp. 163–171.

10. TARGET FRAGMENT AND NUCLEAR RECOIL TRANSPORT

10.1. Introduction

The interaction of energetic charged particles with matter encompasses a broad host of physical processes of both atomic and nuclear origins. For biological systems, the patterns of local-event distributions are of essential concern because biological response is dominated by physical/chemical changes in localized sensitive volumes (Rossi, 1959). In principle, one can specify the fields within the material from which specific-event rates at the sensitive sites can be evaluated. Lacking from such a description are spatially and temporally correlated events which may bear some importance in some sensitive systems. In practice, however, the field quantities are never specified completely, thus leading to questions about the adequacy of such a reduced expression.

In most field calculations stopping power is customarily used without regard to how the energy lost to the medium is actually mediated. Inherent in such a description is that the details concerning the track structure formed by the passage of the particles is unimportant to system response. A suggestion long ago by Schaefer (1952) and experimental evidence now testify to the importance of track structure to biological consequences (Katz et al., 1971; Watt, 1989). The field of microdosimetric techniques has developed in response to the need to quantify the track structure of field quantities (Rossi, 1959). The importance of track structure is dependent on site size (Chatterjee, Maccabee, and Tobias, 1973) and is most effective for sites that are less than 1 μm.

As an energetic charged particle passes through a region of material, it will suffer many atomic/molecular interactions to which only small amounts of energy are given to ionization/excitation at each interaction site. Secondary electrons and photons propagate the energy from the initial loss site and cause a broadening of the particle track (Katz et al., 1971; Chatterjee, Maccabee, and Tobias, 1973; Kellerer and Chmelevsky, 1975; Paretzke, 1988). In this way, the passing particle can affect a localized volume, even though the path is remote to the localized volume itself. Occasionally, the passing energetic particle undergoes a nuclear reaction in which a large amount of its kinetic energy is given to the nucleus of the struck atom. Often, several nuclear disintegration fragments (nuclear stars) are produced of sufficient energy to form well-defined tracks emanating from the interaction site. These fragments may also affect localized volumes remote to the initial trajectory of the particle.

In the past it was assumed that the radiobiological effectiveness (RBE) was related to the linear energy transfer (LET). Thus, risk assessment used LET-dependent quality factors (QF's) evaluated directly from field quantities. However, if biological response is governed by the energy deposit in small localized volumes, then biological risk is related to energy deposits within such volumes (Watt, 1989). The local energy deposited is related to lineal energy transfer, and RBE is not LET-dependent but rather depends on the lineal energy (energy absorbed by a small volume divided by the average chord of the volume). Indeed, recommended

QF's in the future (e.g., ICRU 40 (Anon., 1986)) may well be assumed to be a function of lineal energy and not LET. Thus, a current need is then to understand more fully the relationship between field quantities and microdosimetric quantities. If microdosimetry is the way of the future, such an understanding will be essential to the future assessment of radiation risk in such a complex radiation field as the space environment.

In the present work, we endeavor to evaluate the field quantities for the nuclear disintegration products formed in nuclear reactions in tissue. We consider explicitly the localized energy deposit associated with microdosimetric measurement. Effects of secondary electron fields are not considered herein but will be considered separately. The effect of secondary electrons was considered by Kellerer and Chmelevsky (1975), who showed that α-particles below 10 MeV and ^{16}O less than 12 MeV have negligible δ-ray contributions for site sizes 1 μm and larger. It is clear from their analysis that application to target fragmentation products in tissue limits the site size of the present work to the order of 1 μm or larger. We also address several related issues of target recoil and fragmentation effects on biological systems (Wilson, Townsend, and Badavi, 1987).

10.2. Target Fragment Transport

In chapter 6 we derived an expression for ion fluence in a region bounded by a surface $\vec{\Gamma}$ as (Wilson, 1977)

$$\phi_j(\vec{x}, \vec{\Omega}, E) = \frac{S_j(E_\gamma)\ P_j(E_\gamma)}{S_j(E)\ P_j(E)}\ \phi_j(\vec{\Gamma}_{\Omega,x}, \vec{\Omega}, E_\gamma)$$

$$+ \sum_k \int_E^{E\gamma} dE'\ \frac{A_j\ P_j(E')}{S_j(E)\ P_j(E)}\ \int_{E'}^{\infty} dE'' \int d\vec{\Omega}'\ r_{jk}(E', E'', \vec{\Omega}, \vec{\Omega}')$$

$$\times\ \phi_k\left\{\vec{x} + [R_j(E) - R_j(E')]\vec{\Omega}, \vec{\Omega}', E''\right\} \tag{10.1}$$

where $E_\gamma = R_j^{-1}[\rho - d + R_j(E)]$, $\rho = \vec{\Omega} \cdot \vec{x}$, and $d = \vec{\Omega} \cdot \vec{\Gamma}_{\Omega,x}$. The integral over E' is a summation over the collisional source distribution from the boundary $(E' = E_\gamma)$ to the point $\vec{x}$ $(E' = E)$. We approximate equation (10.1) in a perturbation series by taking

$$\phi_k(\vec{x}', \vec{\Omega}', E'') \approx \frac{S_k(E_\gamma'')\ P_k(E_\gamma'')}{S_k(E'')\ P_k(E'')}\ \phi_k(\vec{\Gamma}_{\Omega',x'}, \vec{\Omega}', E_\gamma'') \tag{10.2}$$

where

$$\vec{x}' = \vec{x} + [R_j(E) - R_j(E')]\vec{\Omega}$$

$$\rho' = \vec{\Omega}' \cdot \vec{x}'$$

$$d' = \vec{\Omega}' \cdot \vec{\Gamma}_{\Omega'x'}$$

$$E_\gamma'' = R_k^{-1}[\rho' - d' + R_k(E'')]$$

We specialize to a unidirectional monoenergetic beam at the boundary as

$$\phi_k(\vec{\Gamma}, \vec{\Omega}, E) = \delta_{sM}\delta(E - E_p)\delta(\vec{\Omega} - \vec{\Omega}_p) \tag{10.3}$$

for which equation (10.1) may be simplified to

$$\phi_j(\vec{x}, \vec{\Omega}, E) = \exp[-\sigma_j(\rho - d)]\frac{S_j(E\gamma)}{S_j(E)}\delta_{jM}\ \delta(E\gamma - E_p)\ \delta(\vec{\Omega} - \vec{\Omega}_p)$$

$$+ \int_E^{E_\gamma} dE'\ \frac{A_j}{S_j(E)}\ \exp\{-\sigma_j[R_j(E') - R_j(E)]\}$$

$$\times \exp[-\sigma_m(\rho' - d')]\sigma_{jM}(E', E_p'', \vec{\Omega}, \vec{\Omega}_p) \tag{10.4}$$

where $E_p'' = R_k^{-1}[R_k(E_p) - \rho' - d']$. If we restrict ourselves to a small volume of material $(\sigma_j(\rho - d) < 0.01)$, then

$$\phi_j(\vec{x}, \vec{\Omega}, E) = \frac{S_j(E_\gamma)}{S_j(E)}\delta_{jM}\ \delta(E_\gamma - E_p)\ \delta(\vec{\Omega} - \vec{\Omega}_p)$$

$$+ \int_E^{E_\gamma} dE'\ \frac{A_j}{S_j(E)}\ [\sigma_{jM}^P(E', E_p'', \vec{\Omega}, \vec{\Omega}_p) + \sigma_{jM}^T(E', E_p'', \vec{\Omega}, \vec{\Omega}_p)] \tag{10.5}$$

where the projectile and target fragment terms are shown separately. The projectile term has a contribution only at energies near E_p'' whereas the target term has a contribution only for $E << E_p''$. At high energy we have

$$E_\gamma \approx E + \frac{1}{A_j}S_j(E)\ (\rho - d) \tag{10.6}$$

in which case

$$\phi_j(\vec{x}, \vec{\Omega}, E) \approx \delta_{jM}\ \delta(E - E_p)\ \delta(\vec{\Omega} - \vec{\Omega}_p)$$

$$+ \sigma_{jM}^P(E, E_p, \vec{\Omega}, \vec{\Omega}_p)\ (\rho - d)$$

$$+ \int_E^{E_\gamma} dE'\ \frac{A_j}{S_j(E)}\ \sigma_{jM}^T(E', E_p, \vec{\Omega}, \vec{\Omega}_p) \tag{10.7}$$

The absorbed-energy density is then

$$D(\vec{x}) = \sum_j \int_0^\infty dE \int d\vec{\Omega}\ S_j(E)\ \phi_j(\vec{x}, \vec{\Omega}, E)$$

$$= S_M(E_p) + \sum_j\ S_j(E_p)\ m_{jM}^P\ \sigma_m(\rho - d)$$

$$+ \sum_j \int_0^\infty A_j\ dE \int_E^{E_\gamma}\ dE'\ \sigma_{jM}^T(E', E_p, \vec{\Omega}, \vec{\Omega}_p) \tag{10.8}$$

The first term clearly dominates the second. The fact that the third term is nonnegligible results from the large stopping power of the low-energy fragments represented by the third term compared with $S_M(E_p)$, which also results in all their energy being deposited locally. It is clear on this basis that equation (10.7) may be reduced to

$$\phi_j(\vec{x}, \vec{\Omega}, E) \approx \delta_{jM}\ \delta(E - E_p)\ \delta(\vec{\Omega} - \vec{\Omega}_p)$$

$$+ \int_E^{E_\gamma} dE' \frac{A_j}{S_j(E)}\ \sigma_{jM}^T(E', E_p, \vec{\Omega}, \vec{\Omega}_p) \qquad (10.9)$$

Accordingly, the high-energy beam exposure of a small object can be treated by evaluating the direct ionization of the primary particles and the transport of low-energy fragments produced uniformly throughout the volume. This is represented in equation (10.9). We now consider some applications of target fragment transport.

10.3. Neutron KERMA

An interesting quantity concerning nuclear recoil and fragments is the neutron KERMA. The KERMA is a physically meaningful quantity at low-neutron energies where the kinetic energies of the produced charged particles are sufficiently low that they come to rest near their production site. Thus, equilibrium is easily established and the dose and KERMA are almost numerically equal. The kinetic energies of protons produced in tissue by high-energy neutrons are deposited far from the production site, and equilibrium is difficult to establish for high-neutron energies so that KERMA is numerically greater than dose. The KERMA is shown for various tissue constituents in figures 10.1 to 10.4 with an explanatory key for these figures given in chart A. Also shown are results from others (MacFarlane et al., 1978; Dimbylow, 1982; Caswell and Coyne in Anon., 1977a; Herling et al., 1981; Wells, 1979; Brenner, 1983; Alsmiller and Barish, 1976). Clearly, the KERMA from our data base is reasonable and reflects on the veracity of our charged-particle data base, which is based on the same models.

Chart A. Key to figures 1–4

Symbol	Source
⊢—⊣	MacFarlane et al., 1978
□	Dimbylow, 1982
○	Caswell and Coyne in Anon., 1977a
▽	Herling et al., 1981
⋆	Wells, 1979
⊙	Brenner, 1983
△	Alsmiller and Barish, 1976
●	Present data

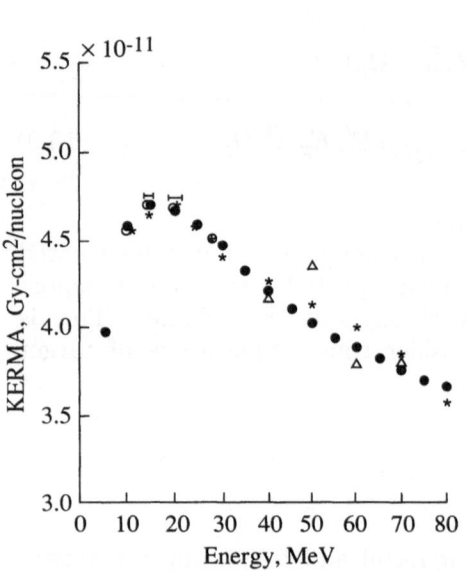

Figure 10.1. Neutron KERMA in hydrogen.
See key given in chart A.

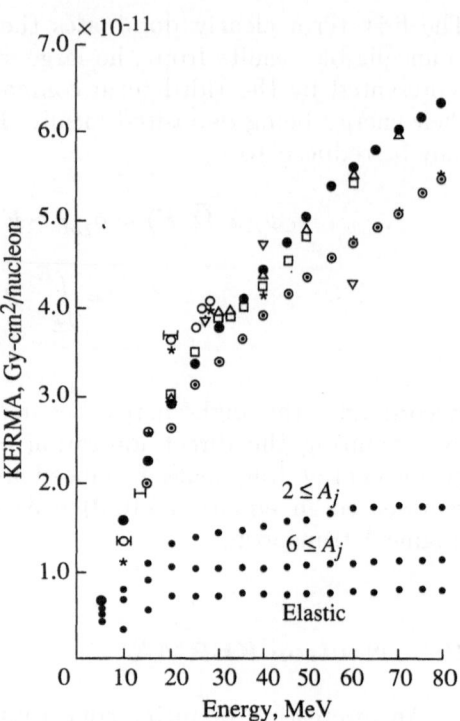

Figure 10.2. Neutron KERMA in
carbon. See key given in chart A.

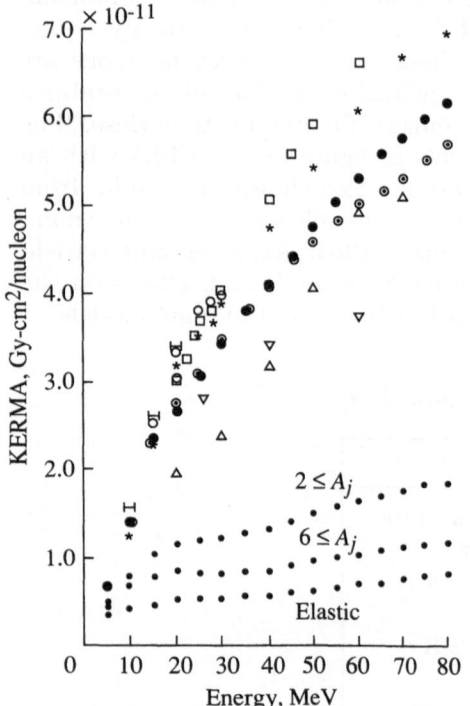

Figure 10.3. Neutron KERMA in nitrogen.
See key given in chart A.

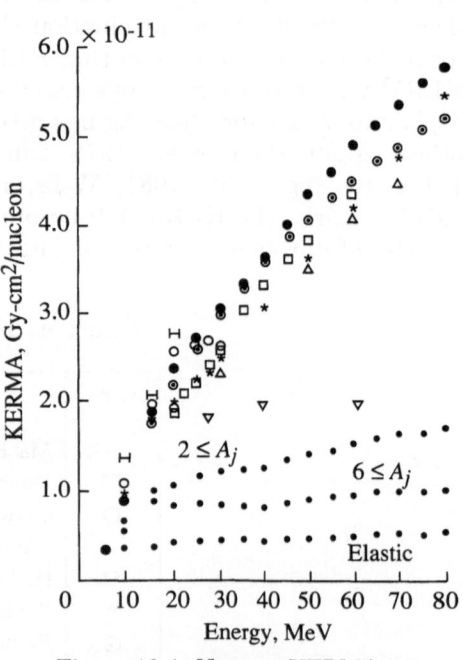

Figure 10.4. Neutron KERMA in
oxygen. See key given in chart A.

10.4. Effects in Conventional Risk Assessment

Biological risks are related to the local energy deposited by the passage of energetic ions. The ionization energy loss is on the order of $0.2Z^2$ keV/μ for a passing relativistic ion of charge Z. Some ions produce nuclear reactions in which 10 to 100 MeV are released (per event) locally as secondary nuclear fragments. The average energy transfer rate is $0.05A^{2/3}$ keV/μ, where A is the ion atomic weight. Because the quality factor of the fragments is usually taken as 20, the risk associated with direct atomic ionization is on the order of the risk associated with the nuclear events for incident low-charge ions ($Z < 5$), although the risk is dominated by the direct ionization for high charge ($Z \gg 5$). At a sufficiently low energy, the direct ionization always dominates the biological risk independent of the ion charge and mass. In this chapter, we quantify these various contributions to biological risk using quality factors presently in force (ICRP 26 (Anon., 1977b)) and evaluate the effects of newly proposed quality factors (ICRU 40 (Anon., (1986)).

10.4.1. Theoretical considerations. We consider a volume of tissue through which a monoenergetic ion fluence $\phi_z(E_p)$ of energy E_p has passed and then evaluate the energy absorbed by the media in the passage. There are several processes by which the ion gives up energy to the media: electronic excitation/ionization, nuclear coulomb elastic scattering, nuclear elastic scattering, and nuclear reaction. The electronic excitation/ionization is contained in the stopping power that is evaluated by methods discussed in relation to equation (10.8) (Wilson et al., 1984). The nuclear coulomb elastic scattering is highly peaked at low momentum transfer, and the energy transfers per event of a few hundred electron volts or less are typical (Wilson, Stith, and Stock, 1983). The nuclear elastic scattering energy transfer is on the order of 1 MeV or less and can be neglected in comparison with reactive processes. A model for proton-induced reactions in tissue constituents has been given elsewhere (Wilson, Townsend, and Khan, 1989) and will provide the basis for the present evaluation.

The secondary-particle radiation fields $\phi_j(E)$ are given as

$$\phi_j(E) = \frac{1}{S_j(E)} \int_E^\infty \zeta_j(E') \, dE' \tag{10.10}$$

where $S_j(E)$ is the stopping power and $\zeta_j(E')$ denotes the particle source energy distributions given as

$$\zeta_j(E') = \rho \sigma_j(E_p) \, f_j(E') \, \phi_z(E_p) \tag{10.11}$$

where ρ is the nuclear density, $\sigma_j(E_p)$ is the fragmentation cross section, and $f_j(E)$ is the fragment spectrum as discussed elsewhere (Wilson, Townsend, and Khan, 1989). The total absorbed energy is approximately

$$D_Z(E_p) = S_Z(E_p) \, \phi_Z(E_p) + \sum_j \int_0^\infty S_j(E) \, \phi_j(E) \, dE \tag{10.12}$$

Equation (10.12) may be written as

$$D_Z(E_p) = S_Z(E_p)\,\phi_Z(E_p) + \sum_j E_j\,\sigma_j(E_p)\,\rho\,\phi_Z(E_Z) \qquad (10.13)$$

where E_j is the average energy associated with each spectral distribution $f_j(E)$. Similarly, the dose equivalent is

$$H_Z(E_p) = Q_Z(E_p)\,S_Z(E_p)\,\phi_Z(E_p)$$

$$+ \sum_j \bar{Q}_{F_j} E_j\,\sigma_j(E_p)\,\rho\,\phi_Z(E_p) \qquad (10.14)$$

where $\bar{Q}_{F_j}$ is the spectral-averaged quality factor of the jth secondary particle (Shinn, Wilson, and Ngo, 1990). The sum over j will include the usual "evaporation" products, including the low-energy protons.

We now evaluate equation (10.13) for the conventional LET-dependent quality factor $Q(L)$ (ICRP 26 (Anon., 1977b)) and the lineal-energy-dependent (y) quality factor $Q(y)$ recently proposed (ICRU 40 (Anon., 1986)). To implement the $Q(y)$, we used appendix B of ICRU 40 (Anon., 1986), in which the lineal-energy distributions are assumed to be linearly dependent on y at a fixed LET. Some of the problems of this assumption have been discussed by Townsend, Wilson, and Cucinotta (1987), which we circumvent herein by assuming Q to be greater than or equal to 1. The spectral-averaged quality factors of the conventional method (ICRP 26 (Anon., 1977b)) and proposed method (ICRU 40 (Anon., 1986)) are shown in table 10.1 for the various isotopes produced in ^{16}O reactions. The proposed values are generally greater than the conventional values, except for the heavier fragments where the proposed values are substantially smaller. The conventional average quality factors show, generally, a weak isotope dependence, whereas the proposed average quality factors show a strong isotope dependence, with neutron-rich isotopes being the most biologically damaging.

10.4.2. Results and discussion. The dose equivalent per unit fluence of incident ions of charge Z and energy per nucleon E_p are shown in figures 10.5 and 10.6. Figure 10.5 is based on current quality factors (ICRP 26 (Anon., 1977b)), and figure 10.6 is based on newly proposed quality factors (ICRU 40 (Anon., 1986)). The proton-induced fragmentation cross sections are taken from chapters 4 and 5. The proton cross sections are velocity scaled according to the proposed factorization model of Lindstrom et al. (1975), as modified by Silberberg, Tsao, and Shapiro (1976). The limitations of this model, as discussed elsewhere (Wilson et al., 1984), do not concern us here because the ^{16}O and ^{12}C data were used in the original derivation by Lindstrom et al. (1975), retained in subsequent modifications of Silberberg, Tsao, and Shapiro (1976), and adequately represent the ^{16}O and ^{12}C data. The problems with this scaling model arise for nuclear fragmentation predictions far removed from projectile-target combinations used in fitting the model parameters. For example, there are no light fragment data for iron fragmentation for which the model could be fit (Wilson, Townsend, and Badavi, 1987); such experiments are currently in progress. The 50-percent increase

Table 10.1. Spectral-Averaged Quality Factor for Individual Isotopes
Produced in 1-GeV, Proton-Induced Reactions in ^{16}O

Z_j	A_j	E_j, MeV	$\bar{Q}_{Fj}$ ICRP 26 (a)	$\bar{Q}_{Fj}$ ICRU 40 (b)
1	1	8.69	2.73	3.71
1	2	10.70	4.09	5.87
1	3	10.40	5.20	7.72
2	3	11.20	12.38	19.52
2	4	12.30	13.90	21.88
3	6	6.85	19.25	23.19
3	7	6.16	19.50	22.38
4	9	4.79	19.77	15.14
4	10	4.11	19.73	14.95
5	9	4.79	19.81	11.47
5	10	4.11	19.78	11.54
5	11	3.71	19.75	11.88
5	12	2.74	19.70	12.58
6	11	3.71	19.79	9.96
6	12	2.74	19.75	10.71
6	13	2.05	19.68	11.96
6	14	1.34	19.56	14.11
6	15	0.69	19.17	18.30
7	13	2.05	19.74	10.53
7	14	1.37	19.64	12.61
7	15	0.69	19.35	16.73
8	15	0.69	19.45	15.50

[a]ICRP 26 (Anon., 1977b).
[b]ICRU 40 (Anon., 1986).

we proposed for the ^{20}Ne data is within the uncertainty generally regarded for the Silberberg, Tsao, and Letaw (1983) parameterization. (Also see Mathews (1983).)

The nuclear contribution to the dose equivalent increases rapidly at the lowest energies as new channel thresholds are passed with increasing energy opening new reaction mechanisms. The small variation seen between 20 MeV per nucleon and 300 MeV per nucleon is related to nuclear transparency (Townsend, Wilson, and Bidasaria, 1982). The new inelastic channels open above the pion production threshold cause a rapid rise in dose equivalent above 300 MeV per nucleon. The fractional contribution of nuclear reaction to the total dose equivalent is shown in table 10.2 for the two quality factors and ion types shown in figures 10.5 and 10.6 at 10 GeV per nucleon. The nuclear contribution to dose equivalent for ^{12}C and heavier ions is less than 5 percent. Nuclear contributions for lighter ions can be substantial and as high as 70 percent.

Table 10.2. Fractional Contribution of Nuclear Reactions to the Total Dose Equivalent
at 10 GeV/nucleon for the ICRP 26 and ICRU 40 Quality Factors

Report (a)	Projectile						
	^{1}H	^{4}He	^{3}Li	^{9}Be	^{12}C	^{28}Si	^{56}Fe
ICRP 26	0.59	0.43	0.27	0.17	4.2×10^{-2}	2.5×10^{-3}	3.8×10^{-4}
ICRU 40	.70	.51	.34	.21	4.7×10^{-2}	2.1×10^{-3}	3.5×10^{-4}

[a]ICRP 26 (Anon., 1977b); ICRU 40 (Anon., 1986).

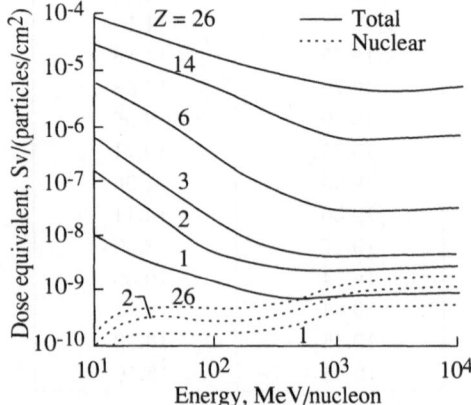

Figure 10.5. Dose equivalent of various
ion types including nuclear reactions
for the ICRP 26 quality factor
(Anon., 1977b).

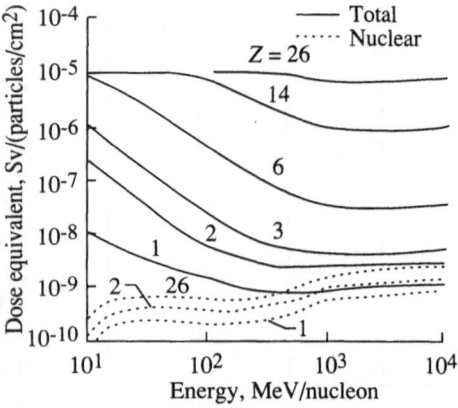

Figure 10.6. Dose equivalent of various
ion types including nuclear reactions
for the ICRU 40 quality factor
(Anon., 1986).

The average quality factors including nuclear reaction effects are shown in
figures 10.7 and 10.8. The nuclear effects are clearly seen as the rise in average
quality factor at high energies, especially for the light ions. The increase in average
quality factors for high-energy protons of ICRU 40 (Anon., 1986) compared with
the value obtained for ICRP 26 (Anon., 1977b) is only 25 percent. Consequently,
the earlier estimates of Alsmiller, Armstrong, and Coleman (1970), in which the
quality factor of 20 for all nuclear fragments of mass greater than 1 nucleon was
used, are expected to remain slightly conservative with respect to biological risk,
even if the quality factor for ICRU 40 (Anon., 1986) is enforced. One interesting
note with respect to figures 10.7 and 10.8 is that the average quality factor for
high-energy protons is on the order of that for high-energy carbon ions.

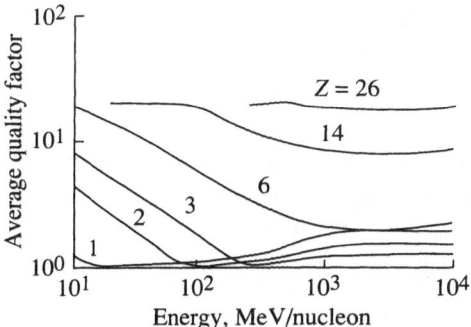

Figure 10.7. Average quality factor of various ion types including nuclear reactions for the ICRP 26 quality factor (Anon., 1977b).

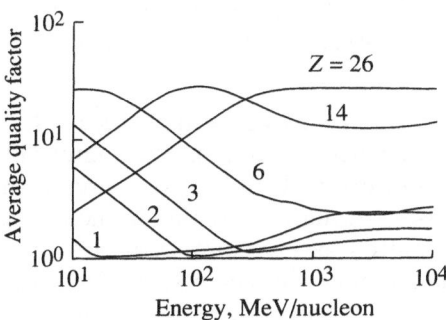

Figure 10.8. Average quality factor of various ion types including nuclear reactions for the ICRU 40 quality factor (Anon., 1986).

10.5. Bone-Tissue Interface Effects

The transport of high-energy nuclei through biological media is greatly affected by nuclear fragmentation events that produce both high- and low-energy sources of nuclear fragmentation product ions, depending on the mechanisms of projectile or target fragmentation, respectively. Fragmentation channels dominate the nuclear cross section at high energies causing broad distributions in ion mass and energy. For high-energy nucleons, the direct ionization is small and the high LET tissue fragments are expected to contribute significantly to their biological effectiveness. The low-energy target fragments with energies of a few MeV are more localized and simpler to treat because of the reduced importance of delta rays and energy-loss straggling (Kellerer and Chmelevsky, 1975). Although energy distributions for charged secondaries in tissue have been studied previously (Wilson, Townsend, and Khan, 1989), the importance of interface effects on the distribution of ions (Cucinotta, Hajnal, and Wilson, 1990) has not been considered in the past. The energy deposition in tissue from fragmentation products produced in bone may be of importance for determining carcinogenic risk because, for example, the epithelium cells near this interface within trabecular cavities are the relevant tissue for induction of bone tumors (Spiers, 1969; ICRP 11 (Anon., 1968b)).

Tissue sites within bone are randomly distributed throughout the calcified matrix, with typical marrow cavities (Beddoe, Darley, and Spiers, 1976) in trabecular bone having dimensions between 50 and 500 μm. Particle ranges for high Z recoils are much smaller than the cavity dimensions, with average energies of heavy target fragments being on the order of several MeV (Greiner et al., 1975). The interface geometry may have an important effect for the more penetrating proton recoils produced in the fragmentation event.

In this section we present solutions for fragment transport in the interface region within a plane geometry model for an incident beam of high-energy nucleons. The model presented is general and may be used to consider other interface effects, e.g., near the connective tissues located in bone. Fragment source spectra are modeled using the high-energy fragmentation model (Wilson, Townsend, and

Khan, 1989) and numerical results given for bidirectional transport. The effects of the interface geometry will be considered in future work.

10.5.1. Fragment transport.

We consider the transport of nuclear fragments with charge Z in the vicinity of an interface of two distinct regions denoted by subscripts 1 and 2. We assume that a high-energy flux density of nucleons ϕ passes through the region inducing nuclear reactions. We assume the region to be of limited extent compared with the range of the incident particle, such that the change in ϕ over the region of interest is negligible. The fragment sources ζ_{1j} and ζ_{2j} within the regions are defined by Wilson (1977) as

$$\zeta_{1j}(E) = \frac{d\sigma_{1j}}{dE}\rho_1\phi \tag{10.15}$$

$$\zeta_{2j}(E) = \frac{d\sigma_{2j}}{dE}\rho_2\phi \tag{10.16}$$

where the subscript j labels the nuclear fragments; $d\sigma_{1j}/dE$ and $d\sigma_{2j}/dE$ are the fragmentation energy distributions for the two regions; and ρ_1 and ρ_2 are the densities of the two media. Constituent densities for tissue and bone found by Santoro, Alsmiller, and Chandler (1974) are given in table 10.3. The source terms are used in the Boltzmann equation for fragment transport in the two media, which is solved using the method of characteristics with the appropriate boundary condition at the interface. The flux density in medium 1 along a ray passing through the interface is then found as

$$\phi_{1j}(x, E) = \frac{1}{S_{1j}(E)}\frac{S_{1j}(E_{1jb})}{S_{2j}(E_{1jb})}\int_{E_{1jb}}^{E_{2jab}}\zeta_{2j}(E')\,dE'$$
$$+ \frac{1}{S_{1j}(E)}\int_E^{E_{1jb}}\zeta_{1j}(E')\,dE' \tag{10.17}$$

where x is a point in region 1 through which the ray passes, b is the distance in medium 1 along the ray, a is the chord of region 2 along the ray, and S_{1j} and S_{2j} are the fragment stopping powers in the two regions. The energy limits in equation (10.17) are given by

$$E_{1jb} = R_{1j}^{-1}\left[R_{1j}(E) + b\right] \tag{10.18}$$

and

$$E_{2jab} = R_{2j}^{-1}\left[R_{2j}(E_{1jb}) + a\right] \tag{10.19}$$

where the functions $R_{1j}(E)$ and $R_{2j}(E)$ are the range relations for fragments j, and R_{1j}^{-1} and R_{2j}^{-1} are the energies associated with the range r in each region defined by Wilson (1977) as

$$E = R_{1j}^{-1}\left[R_{1j}(E)\right] \tag{10.20}$$

The dose at depth x is given by

$$D_{1j}(x) = \int_0^\infty S_{1j}(E) \; \phi_{1j}(x, E) \; dE \tag{10.21}$$

which is evaluated numerically using the modified stopping powers (Wilson et al., 1984) of the Ziegler (1980) handbook.

Table 10.3. Densities for Tissue and Bone

Element	ρ_1, tissue, cm^{-3} (a)	ρ_2, bone, cm^{-3} (b)
H	6.265×10^{22}	7.074×10^{22}
C	9.398×10^{21}	2.851×10^{22}
N	1.342×10^{21}	2.148×10^{21}
O	2.551×10^{22}	2.855×10^{22}
P		2.518×10^{21}
Ca		4.086×10^{21}

[a] ρ_1, tissue = 1.0 g/cm.
[b] ρ_2, bone = 1.85 g/cm.

The spectral distribution for the production of a target fragment j is parameterized as

$$\frac{d\sigma_{1j}}{dE} = \frac{\sigma_{1j}}{\left(2\pi E_{01j}^3\right)^{1/2}} \sqrt{E} \; \exp\left(\frac{-E}{2E_{01j}}\right) \tag{10.22}$$

where σ_{1j} is the total fragmentation cross section for the production of j, and the energy parameter E_{01j} is related to the average energy of the fragment $\bar{E}_{01j}$ by

$$E_{01j} = \frac{1}{3}\bar{E}_{01j} \tag{10.23}$$

The fragmentation cross sections used here are taken from the empirical parameterizations of Silberberg, Tsao, and Shapiro (1976) for $Z > 2$ ions and from the MECC-7 computer code for $Z = 1, 2$ ions in Anon. (1968a). The average energy of the fragmentation products is related to the momentum width measured in high-energy-ion fragmentation experiments (Greiner et al., 1975) that are fitted by an empirical model by Wilson, Townsend, and Khan (1989). These models for the fragmentation cross sections and average energies were applied to 1-GeV proton collisions on ^{12}C, ^{14}N, ^{16}O, ^{31}P, and ^{40}Ca, with results for ^{12}C, ^{14}N, and ^{40}Ca listed in table 10.4. Numerical values for ^{16}O were given in Wilson, Townsend, and Khan (1989).

We now pursue an analytic reduction of equation (10.21). Over a limited energy interval, the range of a fragment, which is fully ionized, can be approximated by a simple power relation

$$R_{1j}(E) = \alpha E^{p_j} \tag{10.24}$$

Values for the parameter p_j are determined by considering the energy dependence for the ion ranges over the region of interest, corresponding to the spectral distributions, and α is fixed by the range energy relations. The distinct composition of the two media results in a host of nuclear fragments (the recoil ions) with varied energy spectra at x in such a way that distinct values of p_j, denoted p_{1j} and p_{2j}, should be used. Some modifications were made to improve the agreement with the numerical results of equation (10.21), with the resulting values shown in table 10.5. Listed in parentheses in table 10.5 are parameters for hydrogen and helium ions used to calculate the tissue contribution for the absorbed dose in bone. With the approximation of equation (10.24), the integral in equation (10.21) may be reduced to

$$
\begin{aligned}
D_{1j}(x) = \phi\sigma_{1j}\rho\Bigg\{ & 3E_{01j} - E_{01j}\left[3 + \frac{\nu_b}{2E_{01j}}\right]\exp\left(\frac{-\nu_b}{2E_{01j}}\right) \\
& + \frac{2\sqrt{2}\,p_{1j}b\,E_{01j}}{R_{1j}(E_{01j})}\left[\sqrt{\pi}\,\mathrm{erfc}\left(\sqrt{\frac{\nu_b}{2E_{01j}}}\right) + \frac{1}{2}\Gamma\left(\frac{3}{2},\frac{\nu_b}{2E_{01j}}\right)\right]\Bigg\} \\
& + \phi\sigma_{2j}\rho\,\frac{R_{2j}(E_{02j})}{R_{1j}(E_{02j})}\Bigg\{ E_{02j}\left(3 + \frac{\nu_b}{2E_{02j}}\right)\exp\left(\frac{-\nu_b}{2E_{02j}}\right) \\
& - \frac{2\sqrt{2}\,p_{2j}b\,E_{02j}}{R_{1j}(E_{02j})}\left[\sqrt{\pi}\,\mathrm{erfc}\left(\sqrt{\frac{\nu_b}{2E_{02j}}}\right) + \frac{1}{2}\Gamma\left(3/2,\frac{\nu_b}{2E_{01j}}\right)\right] \\
& - E_{02j}\left(3 + \frac{\nu_{ab}}{2E_{02j}}\right)\exp\left(\frac{-\nu_{ab}}{2\,E_{02j}}\right) \\
& + \frac{2\sqrt{2}\,p_{2j}(b+a_1)\,E_{02j}}{R_{1j}(E_{02j})}\left[\sqrt{\pi}\,\mathrm{erfc}\left(\sqrt{\frac{\nu_{ab}}{2E_{02j}}}\right) + \frac{1}{2}\Gamma\left(\frac{3}{2},\frac{\nu_{ab}}{2E_{02j}}\right)\right]\Bigg\}
\end{aligned}
\tag{10.25}
$$

where $\nu_b = R_{1j}^{-1}(b)$ and $a_1 = R_{1j}\left[R_{2j}^{-1}(a)\right]$. The ambiguity in expressing equation (10.25) was resolved by requiring the first term to vanish linearly as $b \to 0$ and the second term to vanish linearly as $a_1 \to 0$. This analytical reduction of the dose contribution for charged secondaries is expected to be extremely useful when the angular dependence is considered, where a similar form is obtained. Results for bidirectional transport along a single ray are given in the next section.

10.5.2. Results and discussion. The fragment transport model is now applied for the case of bidirectional transport for 1-GeV incident protons at the bone-tissue interface. We consider the interface at $x = 0$ and assume that $a \gg x$ such that $E_{2jab} \to \infty$. The source term is then divided into forward and backward contributions with

$$
\zeta_{1j}^{+}(E) = \zeta_{1j}^{-}(E) = \frac{1}{2}\zeta_{1j}(E)
\tag{10.26}
$$

and similarly for ζ_{2j}^{+}. We assume a unit incident fluence of protons. The source term moving to the left in tissue from tissue fragments is assumed to contribute

Table 10.4. Fragmentation Cross Sections and Fragment Average
Energies for p-Nucleus Scattering at 1 GeV

Z_j	A_j	σ, mb	E_j, MeV
\multicolumn{4}{c}{$A_T = 12; Z_T = 6$}			
1	1	490.1	8.5
1	2	42.5	7.6
1	3	4.5	7.4
2	3	7.7	8.0
2	4	118.5	8.8
3	5	10.9	8.9
3	6	12.2	7.6
3	7	8.3	6.3
4	6	4.3	7.6
4	7	11.9	6.3
4	8	21.9	5.1
4	9	2.6	3.8
5	8	3.6	5.1
5	9	12.1	3.8
5	10	17.2	2.5
5	11	41.5	1.3
6	10	3.9	2.5
6	11	27.5	1.3
\multicolumn{4}{c}{$A_T = 14; Z_T = 7$}			
1	1	559.4	8.6
1	2	56.8	9.2
1	3	6.0	8.9
2	3	10.3	9.6
2	4	153.2	10.5
3	5	10.2	9.6
3	6	11.5	8.6
3	7	7.8	7.5
4	6	4.1	8.6
4	7	11.1	7.5
4	8	11.4	6.4
4	9	2.4	5.4
5	8	3.2	6.4
5	9	10.8	5.4
5	10	15.3	4.3
5	11	10.2	3.2
6	10	3.3	4.3
6	11	13.0	3.2
6	12	40.3	2.1
6	13	13.8	1.1
7	12	2.6	2.1
7	13	10.9	1.1

Table 10.4. Continued

Z_j	A_j	σ, mb	E_j, MeV
$A_T = 16; Z_T = 8$			
1	1	630.4	8.7
1	2	72.9	10.7
1	3	7.7	10.4
2	3	13.2	11.2
2	4	203.0	12.3
3	5	9.6	10.2
3	6	10.8	9.3
3	7	7.3	8.4
4	6	3.8	9.3
4	7	10.5	8.4
4	8	10.7	7.4
4	9	2.3	6.5
5	8	2.9	7.4
5	9	9.6	6.5
5	10	7.6	5.6
5	11	9.1	4.6
6	10	2.8	5.6
6	11	11.0	4.6
6	12	34.3	3.7
6	13	14.4	2.8
6	14	2.1	1.9
7	12	2.2	3.7
7	13	3.9	2.8
7	14	28.0	1.9
7	15	41.5	.9
8	14	3.5	1.9
8	15	28.5	.9
$A_T = 31; Z_T = 15$			
1	1	1150.6	9.0
1	2	140.9	11.3
1	3	26.9	11.2
2	3	36.1	13.2
2	4	265.5	10.1
3	5	5.7	12.1
3	6	6.9	11.6
3	7	5.2	11.1
4	6	2.1	11.6
4	7	6.2	11.1
4	8	7.0	10.7
4	9	1.7	10.2
5	8	1.1	10.7
5	9	3.9	10.2
5	10	3.5	9.7
5	11	4.8	9.3

Table 10.4. Continued

Z_j	A_j	σ, mb	E_j, MeV
$A_T = 40; Z_T = 20$			
1	1	1396.0	9.1
1	2	186.8	11.4
1	3	34.6	11.3
2	3	40.8	14.0
2	4	271.6	11.0
3	5	4.9	12.5
3	6	5.9	12.1
3	7	3.7	11.8
4	6	1.9	12.1
4	7	5.3	11.8
4	8	5.0	11.4
4	9	1.1	11.1
5	8	.7	11.4
5	9	2.3	11.1
5	10	2.2	10.7
5	11	1.7	10.4
6	10	.4	10.7
6	11	1.2	10.4
6	12	3.4	10.0
6	13	1.8	9.6
6	14	.9	9.3
7	12	.3	10.0
7	13	.6	9.6
7	14	2.8	9.3
7	15	2.6	8.9
7	16	.8	8.6
8	14	.5	9.3
8	15	1.6	8.9
8	16	4.9	8.6
8	17	2.9	8.2
8	18	1.5	7.9
9	17	2.0	8.2
9	18	4.0	7.9
9	19	4.2	7.5
9	20	1.4	7.1
10	19	2.0	7.5
10	20	6.9	7.1
10	21	5.1	6.8
10	22	2.9	6.4
11	21	2.6	6.8
11	22	5.8	6.4
11	23	7.2	6.1
11	24	2.6	5.7

Table 10.4. Concluded

Z_j	A_j	σ, mb	E_j, MeV
\multicolumn{4}{c}{$A_T = 40; Z_T = 20$}			
12	23	2.6	6.1
12	24	9.9	5.7
12	25	8.2	5.4
12	26	5.2	5.0
13	25	3.1	5.4
13	26	7.8	5.0
13	27	11.7	4.6
13	28	4.8	4.3
13	29	1.5	3.9
14	27	3.0	4.6
14	28	13.0	4.3
14	29	13.3	3.9
14	30	9.6	3.6
14	31	1.9	3.2
15	29	3.5	3.9
15	30	9.9	3.6
15	31	18.4	3.2
15	32	8.7	2.9
15	33	3.0	2.5
16	31	3.2	3.2
16	32	15.5	2.9
16	33	19.9	2.5
16	34	15.9	2.1
16	35	3.6	1.8
17	33	3.4	2.5
17	34	10.5	2.1
17	35	24.5	1.8
17	36	14.3	1.4
17	37	5.7	1.1
18	35	2.8	1.8
18	36	15.9	1.4
18	37	25.1	1.1
18	38	23.7	.7
19	37	18.0	1.1
19	38	21.3	.7
19	39	40.1	.4
20	39	18.9	.4

Table 10.5. Range-Energy Parameters

Z	p_1	p_2 (a)
1	0.10	0.1 (0.06)
2	.20	.2 (0.13)
3,4	0	.1
5,6	.15	0
>6	.25	0

[a]Values in parentheses denote absorbed dose in bone.

one-half the equilibrium value of the absorbed dose, which is given by

$$D_{1j}^{-}(x) \approx \bar{E}_{01j}\, \sigma_{1j}\, \rho_1 \qquad (10.27)$$

with a similar contribution for the absorbed dose in bone. Thus, each medium is assumed to be semi-infinite in extent.

The differential flux density at several depths in tissue is shown in figure 10.9 for low Z fragment components ($Z = 1, 2$) and in figure 10.10 for the high Z components ($Z > 2$), with the dashed lines showing the bone contribution to the flux, the dotted lines showing the tissue term, and the solid lines showing the total flux. From figure 10.9, the low Z fragments from bone entering tissue are seen to be appreciable for at least 1000 μm of tissue with a significant contribution from high-energy recoil and fragmentation particles ($E > 25$ MeV). In figure 10.10, the high Z fragments from bone are seen to contribute only over a small spatial region (<20 μm) but they dominate the low- and high-energy portions of the spectra nearer to the interface.

In figures 10.11 and 10.12, the absorbed doses are shown in the region of the interface for the low Z and high Z fragments, respectively. The dotted line denotes the tissue contribution; the dashed line, the bone contribution; the solid line, the total absorbed dose. Both figures 10.11 and 10.12 show contributions to the absorbed dose from fragmentation events in adjacent media near the interface. There is also a buildup in dose close to the interface in tissue that is not present in bone. For the low Z ions, this buildup can be attributed to the higher production multiplicities for light ions produced from potassium and calcium present in bone; and for the high Z ions, this buildup is caused by $Z > 8$ ions produced from fragmentation on phosphorous and calcium. In tissue the interface effect on the absorbed dose is appreciable for only a few tens of a micrometer for high Z ions, and it is substantial to about 1000 μm for low Z ions. Conversion of the depth-dose curves to distribution in linear energy transfer (LET) will give a better indication of the importance of the high Z ions near the interface.

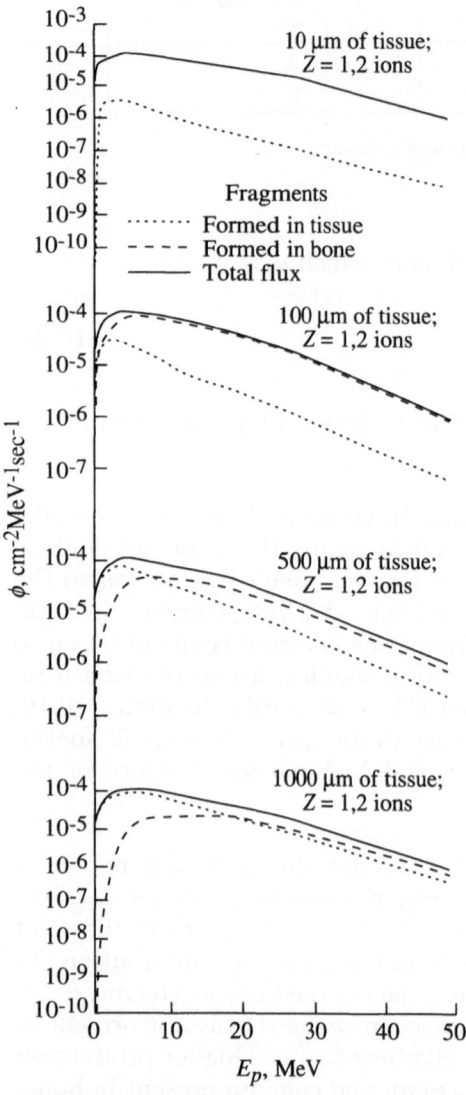

Figure 10.9. Differential flux for $Z = 1,2$ ion fragments at several depths in tissue.

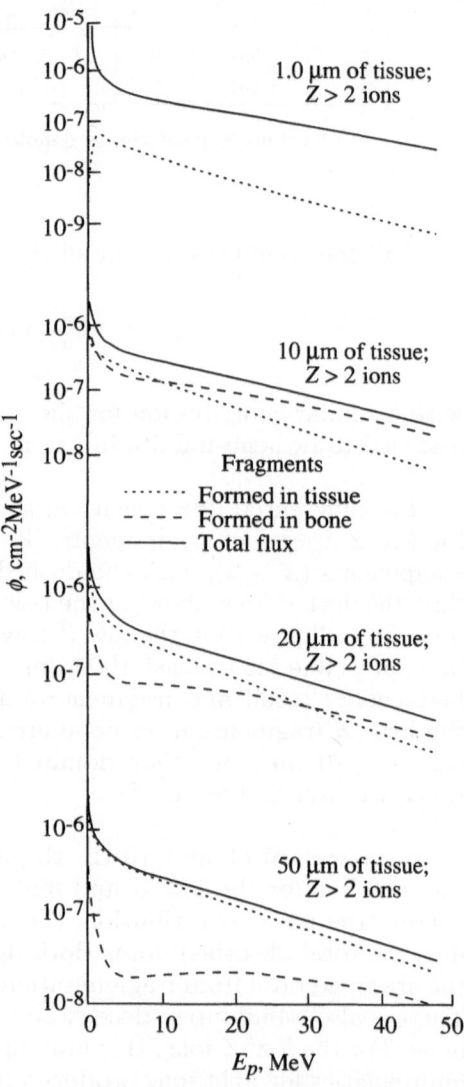

Figure 10.10. Differential flux for all fragments with $Z > 2$ at several depths in tissue.

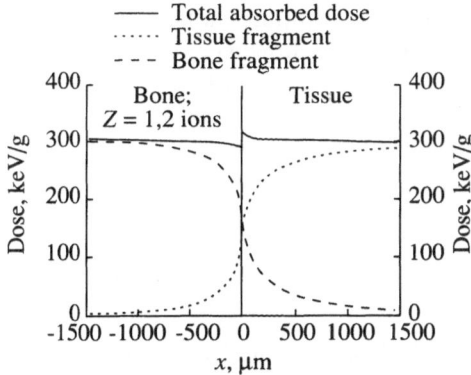

Figure 10.11. Absorbed depth dose in the region of the bone-tissue interface for $Z = 1,2$ fragments.

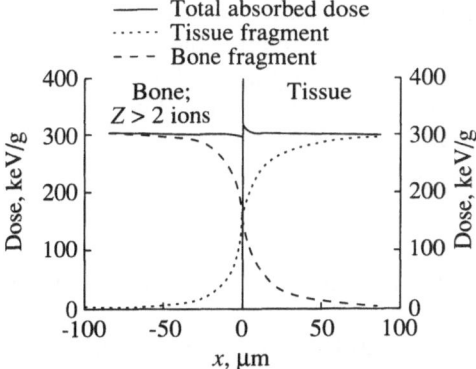

Figure 10.12. Absorbed depth dose in the region of the bone-tissue interface for all fragments with $Z > 2$.

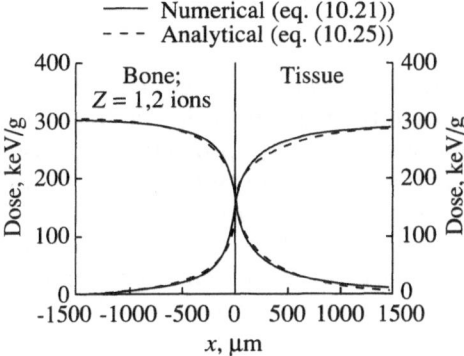

Figure 10.13. Comparison between numerical and analytical results for absorbed depth dose for $Z = 1,2$ fragments.

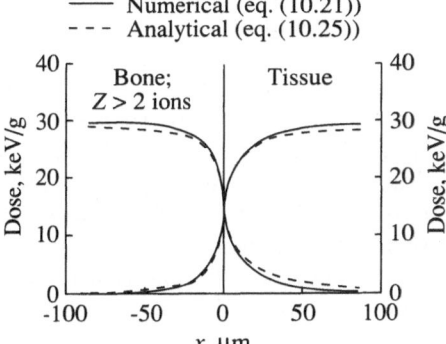

Figure 10.14. Comparison between numerical and analytical results for absorbed depth dose for $Z > 2$ fragments.

In figures 10.13 and 10.14, the analytic approximation of equation (10.25) to the absorbed dose (shown by dashed lines) is shown in comparison with the result obtained from numerical integration of equation (10.21) (shown by solid lines). The analytic results are seen to be quite accurate, converging to the equilibrium values slightly slower than the numerical result. The differences seen are within the accuracy of the fragmentation cross-section model used. The values for p_{1j} and p_{2j} listed in table 10.5 should show a slight dependence on the energy of the incident nucleon.

10.5.3. Conclusions. Solutions for the transport of fragmented target recoils produced by high-energy nucleons are found in the region of the interface between two distinct media. The differential flux and absorbed dose of the nuclear fragments are considered. A simple analytic formula for the absorbed dose is found in an approximate range-energy model. Interface effects were found to be important at the tissue-bone interface in a simple bidirectional transport model, showing a large enhancement in energy deposition from fragments formed in adjacent media. The bone-tissue interface geometry should not be important for high Z fragments based on the results of past work; however, this geometry may become important for the more penetrating low Z fragments and should be a topic of future study. The evaluation of LET distributions near this interface should also be considered to assess the biological effectiveness of the individual fragment contributions.

10.6. Effects on Harderian Tumorigenesis

A quantitative understanding of the contributions of nuclear fragments to biological injury is required for an unambiguous interpretation of the relationship between relative biological effectiveness (RBE) and energy deposition as measured by linear energy transfer (LET). In the study of biological response, concomitant with exposure by particles from projectile fragmentation, there are contributions to biological injury by low-energy, target-nuclear fragments produced within the biological medium.

The effects of target fragmentation on the dose delivered by charged particles along their range can be regarded as small relative to those of projectile fragmentation for heavy-ion beams because fragments of high-energy projectiles are also of high energy in the target rest frame and dissipate this energy over their entire range. This is not true for protons that have no "projectile" fragments below the π-production thresholds. Although these facts have been known for a long time, there has not been a quantitative evaluation of the target fragment contributions to the exposure of biological systems. This section presents such an evaluation.

Biological injury is related to the local energy deposited by the passage of energetic ions. As noted in section 10.4, the ionization energy loss is on the order of $0.2Z^2$ keV/μm for a traversing (i.e., not stopping) high-energy ion of charge Z. The average nuclear energy-transfer rate is $0.05A^{2/3}$ keV/μm, where A is the ion atomic mass (i.e., the number of nucleons). Because the RBE factor of the fragments may be 20 or more, the biological response associated with direct atomic ionization is on the order of that associated with the nuclear events for incident low-charge ions ($Z < 5$) while the response is dominated by direct ionization for particles of high charge ($Z \gg 5$). In the present section, we quantify these various contributions by using, as an illustrative example, the dependence of RBE on LET obtained by Fry et al. (1985) for the induction of tumors in the Harderian gland of mice. This method is similar to that used by Schimmerling et al. (1987) for spermatogonial cell survival.

10.6.1. Radiation response of Harderian gland. The dose response of Harderian tumorigenesis has been measured for ^{60}Co γ and several ions (Fry et al., 1985; Fry, 1986). The approximate response is given as

$$P = 2.5 + 50 \left[1 - \exp\left(\frac{-D}{D_o}\right) \right] \tag{10.28}$$

where the percentage of tumor prevalence P at 600 days is given in terms of a radiosensitivity parameter D_o and the dose D. Tumor prevalence was scored by Fry et al. (1985) at 600 days after exposure as opposed to incident rates scored on postmortem. The D_o of equation (10.28) gives the exponential slope of the nonincidence curve of the at-risk population. The spontaneous tumor rate is the leading coefficient representing 2.5 percent. Note that only one-half the animals appear susceptible to tumor induction, or else the radiation has inactivated pretumorous cells before expression (Fry, 1986). The value of D_o depends on the ion type through which the RBE is found as ratios to the reference radiation value of D_o. The radiation types, their LET's, RBE, and radiosensitivity D_o are given in table 10.6. The measurements of Fry et al. (1985) are all given on a common equivalent dose basis in figure 10.15. It can be seen in figure 10.15 that the dose response curve of equation (10.28) represents the data to within 25 percent. It is easily shown that a small increase in RBE for ^{20}Ne and a small decrease in RBE for ^{12}C would bring all the data points to within 18 percent of equation (10.28).

Table 10.6. Harderian Gland Radiosensitivity Parameters

Radiation type	LET, keV/μ	RBE	D_o, cGy
^{60}Co γ	0.2	1	769.5
^{4}He	18	5	153.9
^{12}C	80	11.2	68.7
^{20}Ne (distal)	190	28.5	27
^{40}Ar	6540	28.7	26.8

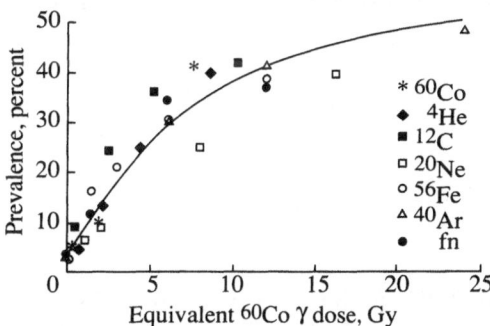

Figure 10.15. Dose response relationship for Harderian gland tumor from various radiations on an equivalent dose scale. The symbol fn refers to fission neutrons.

Equation (10.28) provides a basis for additivity. The probability of radiation-induced tumors within the population at risk is

$$P_r = 1 - \exp\left(\frac{-D}{D_o}\right) \tag{10.29}$$

for which the probability of being among the unaffected population is

$$Q = 1 - P_r = \exp\left(\frac{-D}{D_o}\right) \tag{10.30}$$

Consider two exposures with two different ion types. The probability of being unaffected by the first ion type is

$$Q_1 = \exp\left(\frac{-D_1}{D_{o1}}\right) \tag{10.31}$$

the probability that the unaffected population after exposure D_1 is unaffected by exposure D_2 is then

$$Q_2 = \exp\left(\frac{-D_2}{D_{o2}}\right) \tag{10.32}$$

so that the total unaffected population from the two exposures is

$$Q_{12} = Q_1 Q_2 = \exp\left[-\left(\frac{D_1}{D_{o1}} + \frac{D_2}{D_{o2}}\right)\right] \tag{10.33}$$

Therefore, the total radiation-induced prevalence is

$$P_{r_{12}} = 1 - Q_{12} = 1 - \exp\left[-\left(\frac{D_1}{D_{o1}} + \frac{D_2}{D_{o2}}\right)\right] = 1 - \exp\left(\frac{-M}{D_{o1}}\right) \tag{10.34}$$

where

$$M = D_1 + \frac{D_{o1}}{D_{o2}} D_2 \tag{10.35}$$

is the effective dose for the two types of radiation. The RBE with exposure 1 taken as reference radiation is

$$\text{RBE} = \frac{D_{o1}}{D_{o2}} \tag{10.36}$$

and the effective dose is

$$M = \sum_i \text{RBE}_i \, D_i \tag{10.37}$$

The values of D_{oi} and RBE_i found by Fry et al. (1985) are given in table 10.6 with estimates of corresponding LET values. Note that we have used the LET estimates of Fry et al. (1985) and Fry (1986) except for ^{4}He, where the LET of the most energetic ^{4}He ion at the exposure point is used.

For the fluence ϕ_i of monoenergetic particles of type i and energy E_p, the dose is approximated by $\phi_i(E_p) \, S_i(E_p)$, where $S_i(E_p)$ is the stopping power. For a distribution of particles with a spectrum of energies, the generalized effective dose M is given by

$$M = \sum_i \int \text{RBE}[L_i(E_p)] \, S_i(E_p) \, \phi_i(E_p) \, dE_p \tag{10.38}$$

after which the prevalence can be determined from equation (10.28) with $D_o = 770$ cGy for ^{60}Co γ.

There are two approaches to understanding the RBE relation to LET. The first starts with the observation that biological response is related to the absorbed energy (for example, as Fry et al. (1985) have found for the Harderian gland). As the density of ionization at the sites of energy deposit increases, the ionization becomes more effective in producing biological response, and RBE rises with increasing LET. At some value of LET, the energy deposition reaches a maximum efficiency after which RBE remains constant as shown in figure 10.16. (The dashed line indicates ionization density enhancement saturation (IDES).) Such a view is suggested by the ^{56}Fe and ^{40}Ar data of Fry et al. (1985) which show the same RBE, but this is not necessarily conclusive. The second view is that biological response is a function of particle fluence and that a cross section can be defined which relates biological response to fluence level. This second view can also be justified using the data of Fry et al. (1985), where the cross section depends on particle type. If biological response is fluence related, then RBE decreases with increasing LET. Experimentally, fluence-based limited response is observed only at high LET, which we refer to as the fluence-based limit (FBL). The fluence-based limit RBE is shown in figure 10.16 as the full curve above 190 keV/μm. Katz incorporates these two views into his track-structure model (Katz et al., 1971). For the present, we assume with Katz that low-LET radiation response is dose dependent (gamma-kill mode) and that the high-LET asymptotic RBE is proportional to the inverse LET (ion-kill mode).

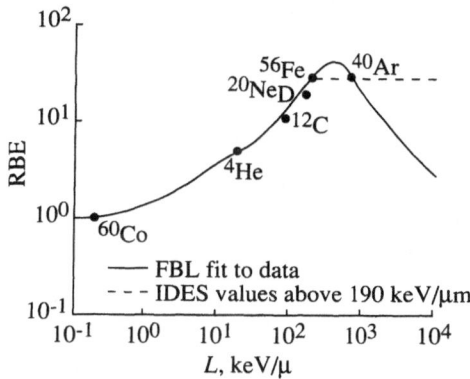

Figure 10.16. RBE measured by Fry et al. (1985) for various radiations.

The relation between RBE and LET for the various ions is shown in figure 10.16. The assumed functional form for the fluence-based limit (FBL) case is

$$\text{RBE} = 0.95 + \frac{a_1}{L} \left[1 + 2 \exp\left(\frac{-L}{14}\right) \right] \left[1 + \exp\left(-a_2 L^2 - a_3 L^3\right) \right] \qquad (10.39)$$

where

$$\left. \begin{array}{l} a_1 = 18\,720 \\ a_2 = 7.43 \times 10^{-6} \\ a_3 = 1.14 \times 10^{-8} \end{array} \right\} \qquad (10.40)$$

and L denotes LET in units of keV/μm. At high LET, the exponential terms become negligible and the response becomes fluence dependent above 500 keV/μm. For LET in the range above 190 keV/μm for the high-ionization density results in enhancement saturation (IDES), and the RBE remains constant as shown by the dashed line in figure 10.16.

10.6.2. Ion reactions in tissue. Again we consider a volume of tissue through which a monoenergetic ion fluence $\phi_Z(E_p)$ of energy E_p has passed and endeavor to evaluate the energy absorbed by the media in the passage. There are several processes by which the ions give up energy to the media: electronic excitation/ionization, nuclear coulomb scattering, nuclear elastic scattering, and nuclear reaction. The electronic excitation/ionization is contained in the stopping power, which is evaluated by the methods discussed elsewhere (Wilson et al., 1984). The nuclear coulomb elastic scattering is highly peaked at low momentum transfer, and energy transfers per event of a few hundred electron volts or less are typical. (See, for example, Wilson, Stith, and Stock (1983).) The nuclear elastic scattering energy transfer is on the order of 1 MeV or less and can be neglected in comparison with reactive processes in the energy range of interest here. A model for proton-induced reactions in tissue constituents, given elsewhere (Wilson, Townsend, and Khan, 1989), will be used here as well.

The secondary particle radiation field $\phi_j(E)$ is given by Wilson (1977) as

$$\phi_j(E) = \frac{1}{S_j(E)} \int_E^\infty \zeta_j(E') \, dE' \qquad (10.41)$$

where $S_j(E)$ is the stopping power and $\zeta_j(E')$ denotes the particle source energy distributions

$$\zeta_j(E) = \rho \sigma_j(E_p) \, f_j(E) \, \phi_Z(E_p) \qquad (10.42)$$

where ρ is the nuclear density, $\sigma_j(E_p)$ is the fragmentation cross section, and $f_j(E)$ is the fragment spectrum as discussed elsewhere (Wilson, Townsend, and Khan, (1989)). The total absorbed energy is approximately

$$D_Z(E_p) = S_Z(E_p) \, \phi_Z(E_p) + \sum_j \int_0^\infty S_j(E) \, \phi_j(E) \, dE \qquad (10.43)$$

Equation (10.43) may be written as

$$D_Z(E_p) = S_Z(E_p) \, \phi_Z(E_p) + \overline{E}_j \, \sigma_j(E_p) \, \rho \, \phi_p(E_p) \qquad (10.44)$$

where $\overline{E}_j$ is the average energy associated with each spectral distribution $f_j(E)$,

$$\overline{E}_j = \int_0^\infty E \, f_j(E) \, dE \qquad (10.45)$$

Similarly, the equivalent ^{60}Co γ dose is

$$M_Z(E_p) = \mathrm{RBE}_Z(E_p) \, S_Z(E_p) \, \phi_Z(E_p) + \sum_j \overline{\mathrm{RBE}}_{Fj} \overline{E}_j \, \sigma_j(E_p) \, \rho \, \phi_Z(E_p) \qquad (10.46)$$

where $\overline{\mathrm{RBE}}_{Fj}$ is the spectral average RBE factor of the jth-type secondary-fragment particle. The sum over j will include the usual "evaporation" products including the low-energy protons. We now evaluate equation (10.46) for the LET-dependent RBE factor found by Fry et al. (1985) and extrapolated by equation (10.39) for both the FBL model and the IDES model. The spectral-averaged RBE factors are shown in table 10.7 for the various isotopes produced in ^{16}O reactions.

Table 10.7 Average RBE Factors for Individual Isotopes Produced in 1-GeV Proton-Induced Reactions in ^{16}O

Z_j	A_j	E_j, MeV	$\overline{\mathrm{RBE}}_{Fj}$	
			IDES	FBL
1	1	8.69	3.67	3.67
1	2	10.70	4.83	4.83
1	3	10.40	5.73	5.73
2	3	11.20	14.09	14.37
2	4	12.30	16.39	16.79
3	6	6.85	27.45	32.89
3	7	6.16	28.22	34.60
4	9	4.79	29.22	37.82
4	10	4.11	29.12	37.45
5	9	4.79	29.32	32.54
5	10	4.11	29.25	32.06
5	11	3.71	29.18	31.88
5	12	2.74	29.03	32.07
6	11	3.71	29.26	27.52
6	12	2.74	29.18	28.32
6	13	2.05	29.00	29.69
6	14	1.34	28.67	31.70
6	15	.69	27.70	33.22
7	13	2.05	29.15	27.17
7	14	1.37	28.89	29.86
7	15	.69	28.16	33.01
8	15	.69	28.42	32.43

10.6.3. Results. The equivalent ^{60}Co γ dose per unit fluence of incident ions of charge Z and energy per nucleon E_p is shown in figures 10.17 and 10.18. Figure 10.17 is based on the RBE factor of the IDES model shown as the dashed line in figure 10.16. Figure 10.18 is based on the FBL model. The proton-induced fragmentation cross sections are taken from Wilson, Townsend, and Khan (1989). The proton cross sections are velocity scaled according to the proposed factorization model of Lindstrom et al. (1975) as modified by Silberberg, Tsao, and Shapiro (1976). The limitations of this model, as discussed elsewhere (Wilson, et al., 1984; Wilson, Townsend, and Badavi, 1987), do not concern us here because the ^{16}O and ^{12}C data, which were used in the original derivation by Lindstrom et al. (1975) and were retained in subsequent modifications of Silberberg, Tsao, and Shapiro (1976), adequately represent the scaling for ^{16}O and ^{12}C data. The problems with this scaling model arise for nuclear fragmentation predictions far removed from projectile-target combinations used in fitting the model parameters. For example, there are no light fragment data ($A_j < 28$) for iron fragmentation to which the model could be fit. Such experiments are currently in progress. The 50-percent increase that we proposed for the ^{20}Ne data (Wilson et al., 1984) is within the uncertainty generally regarded for the Silberberg, Tsao, and Shapiro (1976) parameterization (for example, Mathews, 1983; Silberberg, Tsao, and Letaw, 1983).

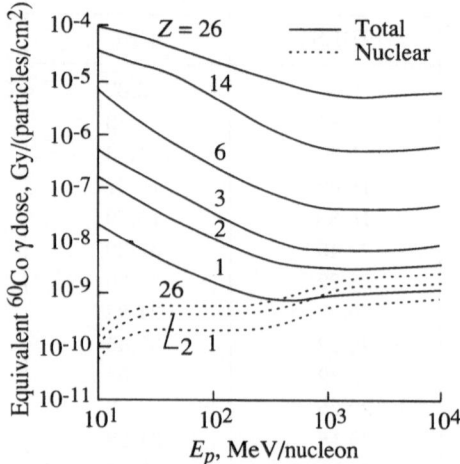

Figure 10.17. Equivalent ^{60}Co γ dose for various ion types including nuclear reaction effects with IDES.

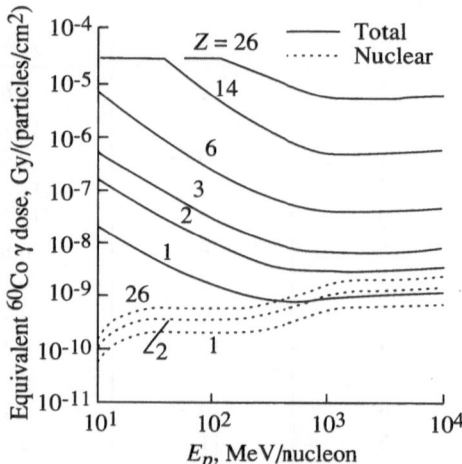

Figure 10.18. Equivalent ^{60}Co γ dose for various ion types including nuclear reaction effects for FBL.

The nuclear contribution to the equivalent ^{60}Co γ initially increases rapidly with increasing energy as new reaction-mechanism thresholds are passed. Only a small variation is seen between 30 MeV/nucleon and 300 MeV/nucleon and this is related to the nuclear transparency (Townsend, Wilson, and Bidasaria, 1982). New inelastic channels open above the pion production threshold and cause a rapid rise in the equivalent dose above 300 MeV/nucleon. Similar results for the fluence-based limit model are shown in figure 10.18. The fractional contribution of nuclear reaction effects for either RBE factor of figure 10.16 differs by a few percent, and representative values are shown in table 10.8 at 0.1, 1.0, 3.0, and 10.0 GeV/nucleon. The nuclear contribution to equivalent ^{60}Co γ doses for carbon and heavier ions is less than 5 percent. Nuclear contributions for lighter ions can be substantial and as high as 69 percent.

The average RBE factors including nuclear reaction effects are shown in figures 10.19 and 10.20. The nuclear effects are clearly seen as the rise in average RBE factor at high energies, especially for the light ions. The present results can now be used to better determine the relation of the ions between RBE and LET in future experiments.

Table 10.8. Fractional Contribution of Nuclear Reactions to the
Total Dose Equivalent at Typical Energies

E, GeV/ nucleon	^{1}H	^{4}He	^{3}Li	^{9}Be	^{12}C	^{28}Si	^{56}Fe
				Projectile ion			
0.1	0.12	0.03	0.01	4.90×10^{-3}	1.50×10^{-3}	8.75×10^{-5}	1.80×10^{-5}
1.0	.59	.35	.17	8.10×10^{-2}	2.50×10^{-2}	2.40×10^{-3}	2.70×10^{-4}
3.0	.67	.43	.22	.11	3.50×10^{-2}	3.50×10^{-3}	4.0×10^{-4}
10.0	.69	.44	.23	.11	3.60×10^{-2}	3.60×10^{-3}	4.0×10^{-4}

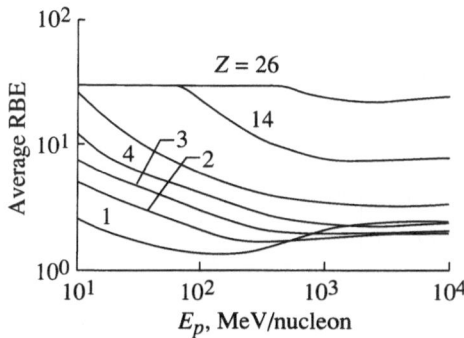

Figure 10.19. Average RBE of various ion types including contributions from nuclear reactions for the IDES model.

Figure 10.20. Average RBE of various ion types including contributions from nuclear reactions for the FBL model.

10.7. Effects on Cellular Track Models

The cellular track model of Katz has been described extensively (Katz et al., 1971; Katz, Sharma, and Homayoonfar, 1972; Katz, 1986). Here we outline its basic concepts and consider the extension to the mixed radiation field seen in space. The biological damage from passing ions is caused by delta-ray production. Cell damage is separated into a grain-count regime where damage occurs randomly along the ion path and the track-width regime, in which the damage is distributed like a "hairy rope." The response of the cells is described by four cellular parameters, two of which (m, the number of targets per cell, and D_o, the characteristic X-ray dose) are extracted from the response of the cellular system to X-ray or γ-ray irradiation. The other two (Σ_o, interpreted as the cross-sectional area of the cell nucleus within which the damage sites are located, and κ, a measure of size of the damage site) are found from survival measurements with track segment irradiations by energetic charged particles. The transition from the grain-count regime to the track-width regime takes place at $Z^{*2}/\kappa\beta^2$ on the order of 4, where Z^* is the effective charge and β, the velocity. The grain-count regime is at the lower values of $Z^{*2}/\kappa\beta^2$ and the track-width regime is at the higher values.

To accommodate for the capacity of cells to accumulate sublethal damage, two modes of inactivation are identified: ion kill (intratrack) and gamma kill (intertrack). For cells damaged by the passage of a single ion, the ion-kill mode occurs. The fraction of cells damaged in the ion-kill mode is taken as $P = \Sigma/\Sigma_o$, where Σ is the single-particle inactivation cross section and P is the probability of the damage in the ion-kill mode. Cells not damaged in the ion-kill mode can be sublethally damaged by the delta rays from the passing ion and then inactivated in the gamma-kill mode, by the cumulative addition of sublethal damage caused by delta rays from other passing ions. The surviving fraction of a cellular population N, whose response parameters are m, D_o, Σ_o, and κ, after irradiation by a fluence of particles F, is written as (Katz et al., 1971)

$$\frac{N}{N_o} = \pi_i \times \pi_\gamma \tag{10.47}$$

where the ion-kill survivability is

$$\pi_i = \exp\left(-\Sigma F\right) \tag{10.48}$$

and the gamma-kill survivability is

$$\pi_\gamma = 1 - \left[1 - \exp\left(\frac{-D_\gamma}{D_o}\right)\right]^m \tag{10.49}$$

The gamma-kill dose fraction is

$$D_\gamma = (1 - P)D \tag{10.50}$$

where D is the absorbed dose. The single-particle inactivation cross section is given by

$$\Sigma = \Sigma_o \left[1 - \exp\left(\frac{-Z^{*2}}{\kappa\beta^2}\right)\right]^m \tag{10.51}$$

where the effective charge number is

$$Z^* = Z \left[1 - \exp\left(\frac{-125\beta}{Z^{2/3}}\right)\right] \tag{10.52}$$

In the track-width regime where $P > 0.98$, we take $P = 1$.

For cell transformation, the fraction of transformed cells per surviving cell is

$$T = 1 - \frac{N'}{N'_o} \tag{10.53}$$

where N'/N'_o is the fraction of nontransformed cells, and a set of cellular response parameters for transformations m', D'_o, Σ'_o, and κ' is used. The RBE at a given survival level is given by

$$\text{RBE} = \frac{D_X}{D} \tag{10.54}$$

where

$$D_X = -D_o \ln \left[1 - \left(1 - \frac{N}{N_o} \right)^{1/m} \right]$$ (10.55)

is the X-ray dose at which this level is obtained. Equations (10.47) through (10.55) represent the cellular track model for monoenergetic particles. Mixed radiation fields have been considered previously in the Katz model. (See, for example, Katz, Sharma, and Homayoonfar, 1972.) Next, we consider placing the model in terms of the particle fields described previously.

10.7.1. Katz model and target fragments. The target fragmentation fields are found in closed form in terms of the collision density (Wilson, 1977) because these ions are of relatively low energy. Away from any interfaces, the target fields are in a local equilibrium and may be written as

$$\phi_\alpha(x, E_\alpha; E_j) = \frac{1}{S_\alpha(E_\alpha)} \int_{E_\alpha}^{\infty} \frac{d\sigma_{\alpha j}(E', E_j)}{dE'} \, \phi_j(x, E_j) \, dE'$$ (10.56)

where the subscript α denotes the target fragment type, $S_\alpha(E)$ denotes the stopping power, and E_α and E_j are in units of MeV.

The particle fields of the projectiles and target fragments determine the level and type of radiological damage at the endpoint of interest. The relationship between the fields and the cellular response is now considered within the Katz cellular track model.

The ion-kill term now contains a projectile term as well as a target fragment term as

$$(\Sigma F) = \Sigma_j(E_j) \, \phi_j(x, E_j) + \sum_\alpha \int_0^{\infty} dE_\alpha \, \phi_\alpha(x, E_\alpha; E_j) \, \Sigma_\alpha(E_\alpha)$$ (10.57)

while the corresponding gamma-kill dose becomes

$$D_\gamma = [1 - P_j(E_j)] \, S_j(E_j) \, \phi_j(x, E_j)$$
$$+ \sum_\alpha \int_0^{\infty} dE_\alpha [1 - P_\alpha(E_\alpha)] \, S_\alpha(E_\alpha) \, \phi_\alpha(x, E_\alpha; E_j)$$ (10.58)

Use of equations (10.56) and (10.57) allows one to define an effective cross section as

$$\Sigma_j^*(E_j) = \Sigma_j(E_j) + \sum_\alpha \int_0^{\infty} dE_\alpha \frac{\Sigma_\alpha(E_\alpha)}{S_\alpha(E_\alpha)} \int_{E_\alpha}^{\infty} dE' \, \frac{d\sigma_{\alpha j}(E', E_j)}{dE'}$$ (10.59)

The first term of equation (10.59) is caused by the direct ionization of the media by the passing ion of type j. The second term results from the target fragment produced in the media.

577

10.7.2. Results and discussion. Katz (Waligorski, Sinclair, and Katz, 1987) has obtained cellular parameters for survival and neoplastic transformations of C3H10T1/2 from the experiments of Yang et al. (1985) as given in table 10.9. We note the large uncertainties in the transformation data of Yang, which should lead to a similar uncertainty in the transformation parameters. Parameter sets were found from data for instantaneous and delayed plating of the cells after the irradiation. Here, only the delayed plating case is considered. General agreement with the measured RBE values (Waligorski, Sinclair, and Katz, 1987) was found using these parameter sets. The single-particle inactivation cross section neglecting target fragmentation of equation (10.59) is shown in figures 10.21 and 10.22 for cell death and transformation, respectively, as a function of the energy (given in MeV/nucleon) of the passing ion. The target fragmentation contribution (the second term in eq. (10.59)) for protons has been evaluated as shown in figures 10.23 and 10.24 and is shown as the dashed curve. The second term of equation (10.59) dominates over the proton direct ionization (dotted line) at high energy. For high-LET particles (low energy), the direct ionization dominates and target fragmentation effects become negligible. A simple scaling by $\sqrt{A_j}$ relates the proton target fragment term to ions of mass A_j. The resulting effective action cross sections for cell death and transformation are plotted in figures 10.25 and 10.26, respectively. We note that the low-energy ^{56}Fe component of the GCR spectra extends into the track width regime where $\Sigma > \Sigma_o$ and is not represented in the present calculation. The error introduced by the present calculation is small.

Table 10.9. Cellular Response Parameters for C3H10T1/2 Cells

Cell-damage type	m	D_o, cGy	Σ_o, cm^2	κ
Death	3	280	5.0×10^{-7}	750
Transformation	2	26 000	1.15×10^{-10}	750

10.8. Effects on Fluence-Related Risk Coefficients

The idea underlying the risk per unit fluence concept was introduced by Curtis, Dye, and Sheldon (1965). The fluence-related risk coefficient F_j was defined by S. B. Curtis[1] et al. as the probability of a given endpoint of interest (e.g., cancer) per unit fluence of type j particles passing through the organ. A first estimate of $F_j(L_j)$ can be found from the RBE values of Fry et al. (1985) and Fry (1986) as approximated by equation (10.39) using the aforementioned definition of Curtis et al.[1]

$$F_j(L_j) = \frac{\text{RBE}(L_j)\, L_j}{12.5 D_o} \tag{10.60}$$

---

[1] Curtis, S. B.; Townsend, L. W.; Wilson, J. W.; Powers-Risius, P.; Alpen, E. L.; and Fry, R. J. M.: Fluence-Related Risk Coefficients Using the Harderian Gland Data as an Example. Paper presented at the 28th Plenary Meeting of COSPAR (The Hague, The Netherlands), June 25–July 6, 1990.

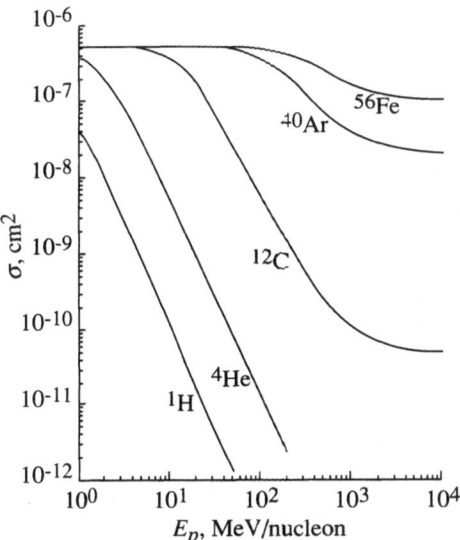

Figure 10.21. Cell-death cross sections for several ions in C3H10T1/2 cells according to the Katz model for direct ionization effects only.

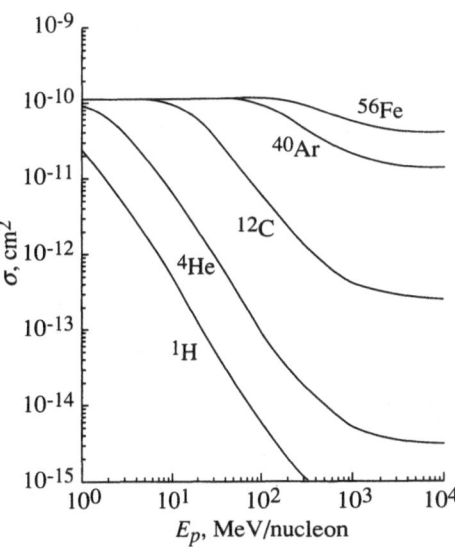

Figure 10.22. Cell-transformation cross sections for several ions in C3H10T1/2 cells according to the Katz model for direct ionization effects only.

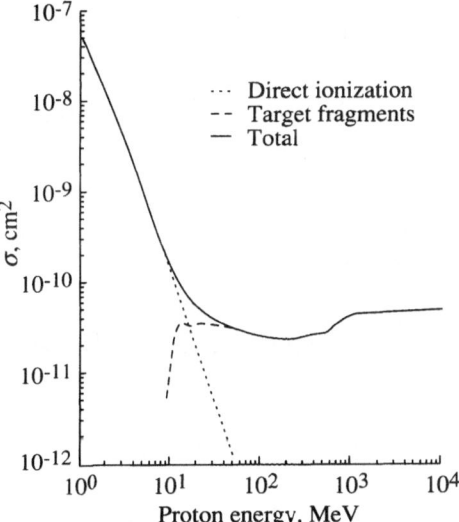

Figure 10.23. Cell-death cross sections including effects of nuclear reactions for protons in C3H10T1/2 cells according to the Katz model.

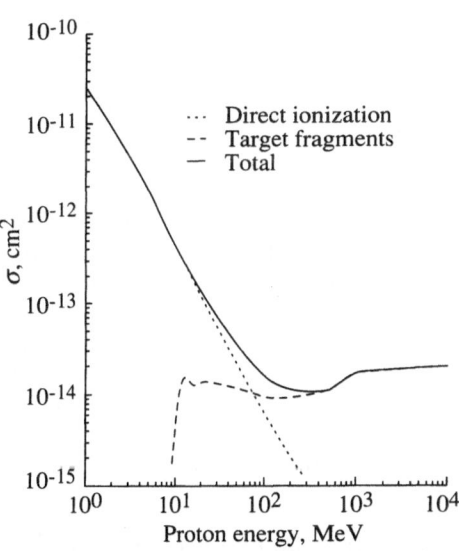

Figure 10.24. Cell-transformation cross sections including nuclear reaction effects for protons in C3H10T1/2 cells according to the Katz model.

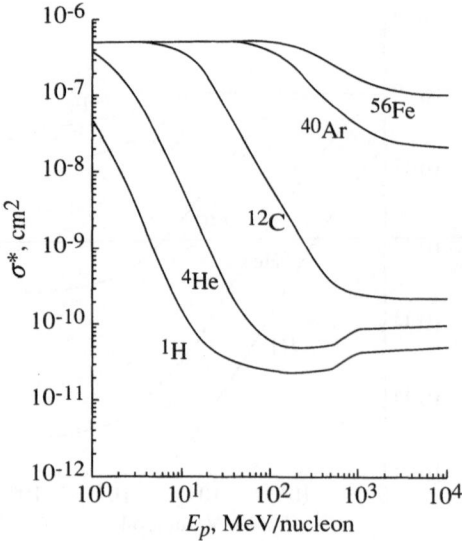

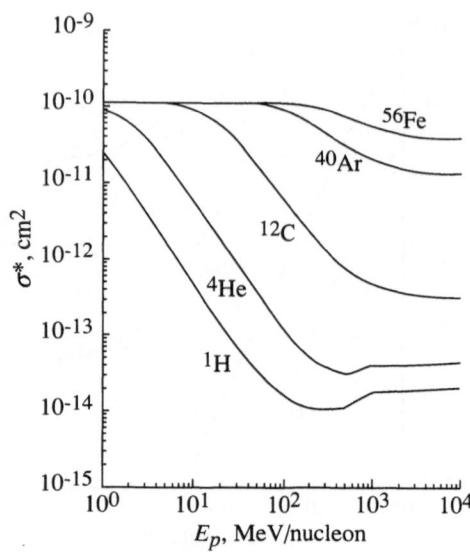

Figure 10.25. Effective cell-death cross sections including nuclear reaction effects for various ions in C3H10T1/2 cells.

Figure 10.26. Cell-transformation cross sections including nuclear reaction effects for various ions in C3H10T1/2 cells.

and represents the risk coefficient for direct ionization only because RBE was taken as unity for ^{60}Co γ rays. In addition to the ionization caused directly by primary and high-energy secondary nuclei from fragmentation of the primary ions, the nuclei constituting biological tissue (i.e., the "target" nuclei) will break up into lower energy and, in some cases, very highly ionizing target fragments. Target fragment fluences $\phi_j(E_j)$, produced by a passing energetic ion of energy E_j, are given by

$$\phi_j(E_j) = \frac{1}{L_j(E_j)} \int_{E_j}^{\infty} \sigma_j(E_i) \ f_j(E_j', E_i) \ \phi_i(E_i) \ dE_j' \qquad (10.61)$$

where $\sigma_j(E_i)$ is the macroscopic fragmentation cross section, $f_j(E_j', E_i)$ is the energy distribution of the jth fragment, and $\phi_i(E_i)$ is the fluence of passing ions of energy E_i. The total prevalence is given in terms of F_i as

$$P = F_i(L_i) \ \phi_i(E_i) + \sum_j \int_0^{\infty} F_j[(L_j(E_j)] \ \phi_j(E_j) \ dE_j \qquad (10.62)$$

It is useful to define an effective $F_i^*(L_i)$ using equation (10.62) as

$$F_i^*(L_i) = F_i(L_i) + \sum_j \int_0^{\infty} dE_j \frac{F_j[L_j(E_j)]}{L_j(E_j)} \int_{E_j}^{\infty} dE_j' \ \sigma_j(E_i) \ f_j(E_j', E_i) \qquad (10.63)$$

The effective F_i^* are shown in figure 10.27 as a function of particle energy for representative charge components with the target fragment contributions shown separately.

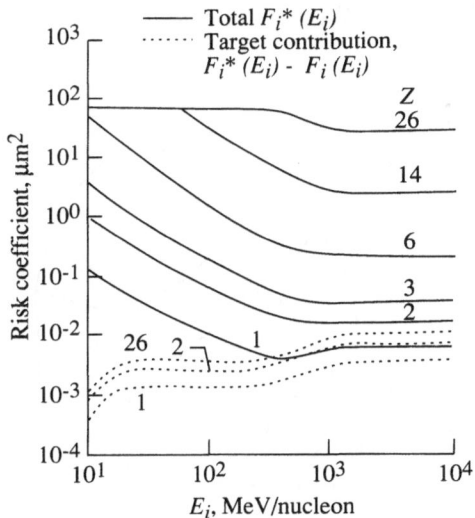

Figure 10.27. Fluence-related risk coefficients as a function of particle energy per nucleon for the total (ionization plus target fragmentation) contribution.

10.9. Effects on Hit-Size Effectiveness Spectrum

A somewhat different fluence-related concept has been introduced to overcome the problem of relating the absorbed dose to the dose equivalent (Bond et al., 1985). This method requires the measurement of the hit-size spectra within an appropriate sensitive volume size. The risk is estimated as a weighted average over the hit-size spectra. The measurement device is a gas proportional counter simulating a small tissue volume, and the weight function is found from radiobiological experimental data. Because we use continuous slowing down theory without regard to the transport of secondary electrons, we are limited to site sizes greater than 1 μm.

10.9.1. Microdosimetric quantities. Consider a convex region of tissue bounded by a surface Γ through which passes a flux of energetic nuclear particles displayed as dashed lines in figure 10.28. We currently examine the nuclear fragments they produce within the tissue volume denoted by starlike events in figure 10.28 and ignore the direct effects of the primary particles that we have treated elsewhere. The appropriate continuity equation is given as

$$\left[\vec{\Omega} \cdot \vec{\nabla} - \frac{\partial}{\partial E} S_Z(E) \right] \phi_Z \left(\vec{x}, \vec{\Omega}, E \right) = \zeta_Z(E) \tag{10.64}$$

where $S_Z(E)$ is the stopping power, $\phi_Z(\vec{x}, \vec{\Omega}, E)$ is the ion flux at $\vec{x}$ moving in direction $\vec{\Omega}$ with energy E (in MeV), and the volumetric ion source $\zeta_Z(E)$ is formed by the collision density of the passing energetic particles and is herein assumed to be uniform and isotropic. We assume the inward-directed flux of ions Z to be zero on the boundary. We note that the solution to equation (10.64) is related to the cumulative spectrum of ion fragments produced by the nuclear collision. This spectrum is readily related to the LET spectrum in tissue as has been derived elsewhere (Wilson et al., 1988). We now consider the more complicated problem of microdosimetry.

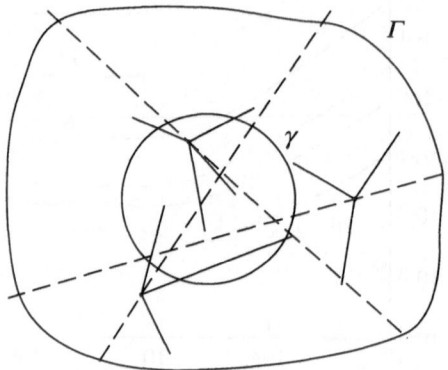

Figure 10.28. Nuclear-star events in a tissue region Γ in which a sensitive site is bound by γ.

The microdosimetric problem is to simply evaluate the energy fluctuations in a small specific region caused by the interaction of a tissue system with a radiation field. We will follow custom and assume the sensitive volume bound by a surface $\vec{\gamma}$ to be a small sphere of tissue of radius a embedded in a larger tissue matrix, as represented in figure 10.28. We represent the sensitive site boundary by the surface $\vec{\gamma}$ and consider the solution of equation (10.64) subject to the boundary condition

$$\phi_Z(\vec{\gamma}, \vec{\Omega}, E) = \frac{1}{S_Z(E)} \int_E^{E_b} \zeta_Z(E') \, dE' \qquad (10.65)$$

for values of $\vec{\Omega}$ such that $\vec{n} \cdot \vec{\Omega} < 0$, where $\vec{n}$ is the outward-directed normal to $\vec{\gamma}$. We assume that the boundary $\vec{\Gamma}$ is sufficiently removed that $E_b \to \infty$ for simplicity. Note that this is usually true because such limits are normally achieved over distances of a few hundred micrometers. For example, the average energy of the ^{7}Li fragments in tissue (Wilson, Townsend, and Khan, 1989) is 6.1 MeV. The field solution to equation (10.64) subject to the boundary condition in equation (10.65) is

$$\phi_Z(\vec{x}, \vec{\Omega}, E) = \frac{S_Z(E_a)}{S_Z(E)} \, \phi_Z(\vec{\gamma}, \vec{\Omega}, E_a) + \frac{1}{S_Z(E)} \int_E^{E_a} \zeta_Z(E') \, dE' \qquad (10.66)$$

which is the fragment field due to the surrounding tissue environment and ion sources within the sensitive volume itself.

Although the fields within the region $\vec{\gamma}$ may be evaluated directly by equations (10.65) and (10.66), it is not clear how this field quantity is related to the fluctuation of absorbed energy within the volume bound by $\vec{\gamma}$. Caswell (1966) solved the problem of absorbed-energy fluctuation for nuclear recoils from low-energy neutron elastic scattering assuming isotropic scattering. The ICRU 40 report (Anon., 1986) suggests a relation that assumes constant stopping powers across the region. We derive herein a general solution to the energy fluctuation problem by using equations (10.65) and (10.66) for a monoenergetic source for which any arbitrary spectral source may be found by superposition.

10.9.2. Response to an internal volumetric source. We first consider the term with internal sources. The particles produced within the sphere have a spectrum given by $\zeta_z(E)$, and as the particles leave the volume, the spectrum is modified to

$$\phi_Z(\vec{\gamma}, \vec{\Omega}, E) = \frac{1}{S_Z(E)} \int_E^{E_a} \zeta_Z(E') \, dE' \tag{10.67}$$

This spectral modification is related to the spectrum of energy absorbed (neglecting the diffusive role of the secondary electrons). We consider here a monoenergetic spectral source that can be modified for any arbitrary source spectrum by superposition. The source spectrum is taken as

$$\zeta_Z(E) = \frac{\rho\sigma}{4\pi} \phi \, \delta(E - E_p) \tag{10.68}$$

where σ is the nuclear production cross section, assumed herein to be energy independent, and ρ is the nuclear density. The solution is

$$\phi_Z(\vec{\gamma}, \vec{\Omega}, E) = \begin{cases} \frac{\rho\sigma\phi}{4\pi \, S_Z(E)} & (E \leq E_p \leq E_a) \\ 0 & (\text{Otherwise}) \end{cases} \tag{10.69}$$

where ϕ is the flux of energetic initiating particles and

$$E_a = R_Z^{-1} \left[R_Z(E) + 2a \, \cos \, \theta \right] \tag{10.70}$$

with θ being the colatitude at the local normal to $\vec{\gamma}$. The differential energy spectrum of particles leaving the surface $\vec{\gamma}$ is

$$\frac{df_{vi}}{dE} = 2\pi A \int_{\mu_o}^1 \phi_Z(\vec{\gamma}, \vec{\Omega}, E) \, \mu \, d\mu = \frac{A\rho\sigma\phi}{4 \, S_Z(E)} (1 - \mu_o^2) \tag{10.71}$$

where μ_o is the lower range of $\mu = \cos\theta$ subject to

$$\mu_o = \begin{cases} \frac{R_Z(E_p) - R_Z(E)}{2a} & (E < R_Z^{-1}([R_Z(E_p) - 2a])) \\ 1 & (\text{Otherwise}) \end{cases} \tag{10.72}$$

and A is the surface area of the sphere. The total number of fragments produced is

$$N_{\text{tot}} = 4\pi \int_0^\infty \int_{\vec{\gamma}} \zeta_Z(E) \, r^2 \, dr \, d\vec{\Omega} \, dE = \frac{4}{3}\pi a^3 \rho\sigma\phi \tag{10.73}$$

with which we may write for the normalized spectrum

$$\frac{dF_{vi}}{dE} = \frac{1}{N_{\text{tot}}} \frac{df_{vi}}{dE} = \frac{3}{4a \, S_Z(E)} \left\{ 1 - \left[\frac{R_Z(E_p) - R_Z(E)}{2a} \right]^2 \right\} \tag{10.74}$$

The spectrum extends over the range

$$R_Z^{-1} \left[R_Z(E_p) - 2a \right] \leq E \leq E_p \tag{10.75}$$

where the lower limit extends to zero if $R_Z(E_p)$ is less than $2a$ for which some particles will come to rest inside the volume. The spectral distributions of emitted ^{7}Li ions are indicated in figures 10.29 and 10.30 for several values of a between 1 and 20 μm.

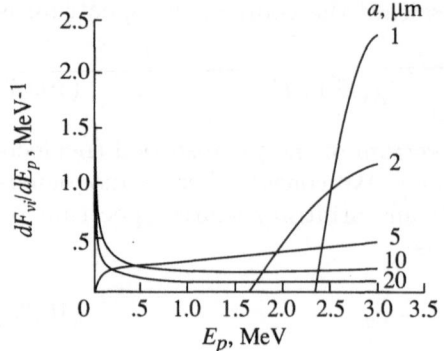

Figure 10.29. Exit spectrum of ^{7}Li due to internal 3-MeV sources in tissue spheres of radii between 1 and 20 μm.

Figure 10.30. Exit spectrum of ^{7}Li due to internal 6-MeV sources in tissue spheres of radii between 1 and 20 μm.

The spectrum of ^{7}Li ions emitted from the surface of the tissue spheres for a 3-MeV source energy is shown in figure 10.29. The results for the 1-μm sphere show minimal degradations of the source energy spectrum. The escaping energy spectrum for the larger spheres is beginning to approach the shape of the equilibrium spectrum of the ion flux spectrum of the surrounding tissue. Indeed, very little effect on the escaping spectrum is seen from the increased leakage in reducing the radius from 20 to 10 μm. The ^{7}Li range on the order of the sphere diameter and the spectrum is probably an equilibrium spectrum. Note that the spectrum in figure 10.29 is only from the escaping ions corresponding to the ion sources within 10 μm of the surface. Figure 10.30 shows similar results for 6-MeV ^{7}Li ions. Note that the 20-μm spectrum is still near equilibrium because the ion range (22 μm) is on the order of the sphere radius.

The integrated spectral function represents the fraction of escaping particles

$$\int_0^{E_p} \frac{dF_{vi}}{dE} dE = \frac{3}{4a} \int_{R_Z(E_p)-2a}^{R_Z(E_p)} \left\{ 1 - \left[\frac{R_Z(E_p) - R}{2a} \right]^2 \right\} dR$$

$$= \left\{ \begin{array}{ll} 1 & R_Z(E_p) > 2a \\ \frac{3}{4a} R_Z(E_p) \left[1 - \frac{R_Z(E_p)^2}{12a^2} \right] & (R_Z(E_p) \leq 2a) \end{array} \right\} \quad (10.76)$$

The energy loss spectrum is made up of two components: those generated with energy E_p that escape with energy E and those that do not escape. The energy losses ε and E are related as

$$\varepsilon = E_p - E \quad (10.77)$$

with the result that the distribution of ε and E is related as

$$\frac{dF_{va}}{d\varepsilon} = \frac{dF_{vi}}{dE} \tag{10.78}$$

so that

$$\frac{dF_{va}}{d\varepsilon} = \frac{3}{4a \; S_Z(E_p - \varepsilon)} \left\{ 1 - \left[\frac{R_Z(E_p) - R_Z(E_p - \varepsilon)}{2A} \right]^2 \right\} \tag{10.79}$$

thus giving the energy loss spectrum of escaping ions. The energy loss spectrum for ^{7}Li ions escaping the sphere is shown for three sphere radii in figures 10.31 to 10.33. A second contribution comes from ions that stop within the volume, giving the total energy loss spectrum as

$$\frac{dF_{ai}}{d\varepsilon} = \frac{dF_{va}}{d\varepsilon} + \left\{ 1 - \frac{3}{4a} R_Z(E_p) \left[1 - \frac{R_Z(E_p)^2}{12a^2} \right] \right\} \delta(\varepsilon - E_p) \tag{10.80}$$

Equation (10.80) is useful for calculating the absorption spectrum for an isolated sphere. The energy loss spectrum is related to the lineal energy spectrum by multiplying equation (10.80) by the average chord of the sensitive site. One may show with little effort that the terms in equation (10.80) are equivalent to Caswell's starter and insider results for recoil proton spectra, respectively. Indeed, Caswell's results are easily derived using equation (10.80). The energy loss spectrum of ^{7}Li ions produced externally to the surface $\vec{\gamma}$ is now considered.

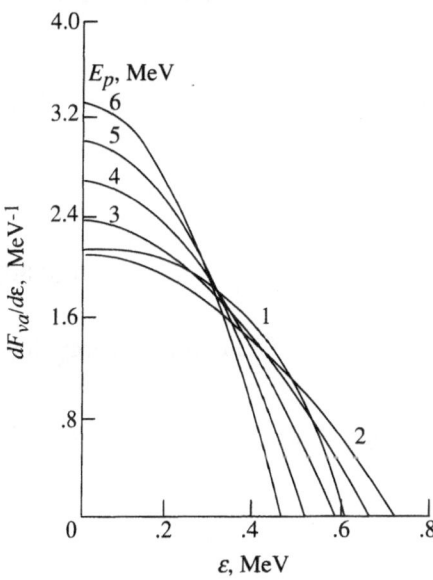

Figure 10.31. Energy loss spectrum of ^{7}Li ions in a 1-μm tissue sphere for internal source energies between 1 and 6 MeV.

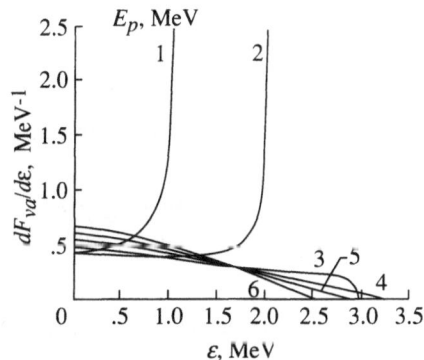

Figure 10.32. Energy loss spectrum of ^{7}Li ions in a 5-μm tissue sphere for internal source energies between 1 and 6 MeV.

10.9.3. Response to a surface source. To evaluate the surface terms, we first consider the response to a monoenergetic source term by taking the boundary function as

$$\phi_Z(\vec{\gamma}, \vec{\Omega}, E) = \frac{1}{2\pi} \delta(E - E') \tag{10.81}$$

for which the solution for surface sources is

$$\phi_Z(\vec{x}, \vec{\Omega}, E) = \frac{S_Z(E_a)}{2\pi \, S_Z(E)} \delta(E_a - E') \tag{10.82}$$

We evaluate equation (10.82) on the surface of the sphere (i.e., $\vec{x} = \vec{\gamma}$) and calculate the total spectrum of exiting particles as

$$\frac{df_s}{dE} = 2\pi A \int_0^1 \mu \, \phi_Z(\vec{\gamma}, \vec{\Omega}, E) \, d\mu$$

$$= \frac{A}{4a^2 \, S_Z(E)} \left\{ \begin{array}{ll} R_Z(E') - R_Z(E) & (R_Z^{-1}[R_Z(E') - 2a] \leq E \leq E') \\ 0 & \text{(Otherwise)} \end{array} \right\} \tag{10.83}$$

The total number of escaping ions is

$$N_{\text{tot}} = \int_0^\infty \frac{df_s}{dE} \, dE = \left\{ \begin{array}{ll} \frac{A}{2} & (R_Z(E') \geq 2a) \\ \frac{A}{2} \frac{[R_Z(E')]^2}{2a} & (R_Z(E') < 2a) \end{array} \right\} \tag{10.84}$$

We may now find the energy loss spectrum from this source. The total number of particles crossing into the volume is $A/2$ with individual energies of E'. For $R_Z(E') < 2a$, the number of stopping ions is

$$N_s = \frac{A}{2} \left\{ 1 - \left[\frac{R_Z(E')}{2a} \right]^2 \right\} \tag{10.85}$$

and the corresponding ion energy loss spectrum is

$$\frac{df_s}{d\varepsilon} = N_s \, \delta(E' - \varepsilon) \tag{10.86}$$

Note that there are no stopping ions if $R_Z(E') > 2a$. The total energy loss spectrum is

$$\frac{df_{sa}}{d\varepsilon} = \left. \frac{df_s}{dE} \right|_{E=E'-\varepsilon} + N_s \, \delta(E' - \varepsilon) \tag{10.87}$$

similar to our earlier result in equation (10.80).

10.9.4. Response to an external volumetric source. We now apply equation (10.87) to determine the response to external volumetric source terms present at the boundary of the sphere as given by equation (10.66). First, by using equations (10.83) and (10.87) we consider the energy loss spectrum for

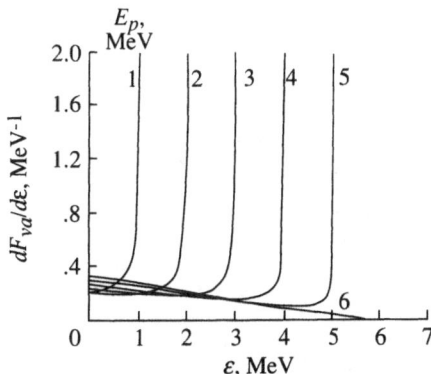

Figure 10.33. Energy loss spectrum of ^{7}Li ions in a 10-μm tissue sphere for internal source energies between 1 and 6 MeV.

the surface ion source term for ions that are able to escape the sphere. The energy loss spectrum is

$$\frac{df_{ve}}{d\varepsilon} = \frac{2\pi A}{4a^2} \int_\varepsilon^{E_p} \frac{R_Z(E') - R_Z(E' - \varepsilon)}{S_Z(E' - \varepsilon)} \frac{\rho\sigma\phi}{4\pi} \frac{}{S_Z(E')} dE'$$

$$= \frac{A}{2} \frac{\rho\sigma\phi}{4a^2} \int_{R_Z(\varepsilon)}^{R_Z(E_p)} \frac{R_Z(E') - R_Z(E' - \varepsilon)}{S_Z(E' - \varepsilon)} d\left[R_Z(E')\right] \qquad (10.88)$$

One can easily show that equation (10.88) produces Caswell's results for crossers produced by neutron collisions with hydrogen. The integral in equation (10.88) cannot be further reduced without assuming some functional form for $S_Z(E)$. Because the ions are of relatively low energy, we assume that $S_Z(E)$ is proportional to velocity as predicted by Fermi and Teller (1947) and Lindhard (1954). We find that

$$\frac{df_{ve}}{d\varepsilon} = \frac{A}{2} \frac{\rho\sigma\phi}{4a^2} \frac{R_Z(E_p - \varepsilon)}{S_Z(E_p - \varepsilon)} \{2 R_Z(E_p - \varepsilon) - [R_Z(E_p) - R_Z(\varepsilon)]\} \qquad (10.89)$$

The contribution to the energy loss spectrum from stopping ions is

$$\frac{df_{ve}}{d\varepsilon} = 2\pi \int_\varepsilon^{E_p} N_s(E') \, \delta(E' - \varepsilon) \frac{\rho\sigma\phi}{4\pi} \frac{}{S_Z(E')} dE'$$

$$= \frac{A}{4} \frac{\rho\sigma\phi}{S_Z(\varepsilon)} \left\{ 1 - \left[\frac{R_Z(\varepsilon)}{2a}\right]^2 \right\} \qquad (10.90)$$

for which the total energy loss spectrum from external volume sources is given by summing equations (10.89) and (10.90). It can be shown that equation (10.90) produces Caswell's results for the stoppers produced by low-energy neutron collisions with hydrogen.

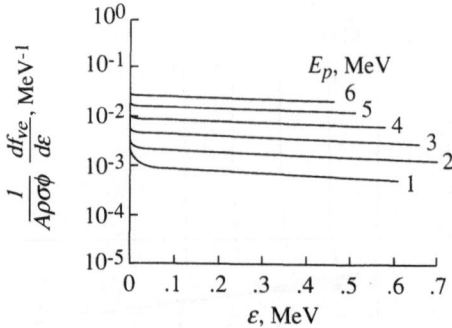

Figure 10.34. Energy loss spectrum of ^{7}Li ions in a 1-μm tissue sphere for external source energies between 1 and 6 MeV.

The energy loss spectrum in a 1-μm sphere caused by external volumetric sources in the tissue matrix is shown in figure 10.34. The maximum energy loss is indicative of the energy-dependent stopping power. Otherwise, the energy loss spectrum is nearly uniform in energy. A rather strong energy dependence is seen in the energy loss spectrum for the 5- and 10-μm spheres shown in figures 10.35 and 10.36, respectively. The low-energy spectrum is dominated by the low-energy particles entering the boundary $\vec{\gamma}$ and has insufficient energy to escape, as given by equation (10.90). The high-energy shoulder seen most clearly for the 2-MeV curve in figure 10.36 is caused by passing entirely through the region and escaping $\vec{\gamma}$ with a portion of the low-energy particles as given by equation (10.89).

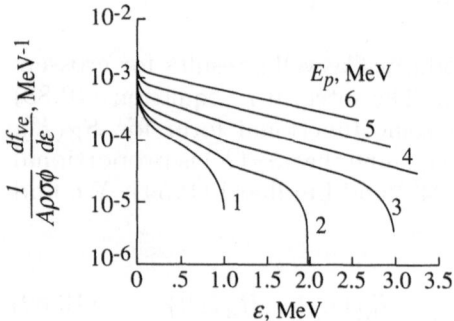

Figure 10.35. Energy loss spectrum of ^{7}Li ions in a 5-μm tissue sphere for external source energies between 1 and 6 MeV.

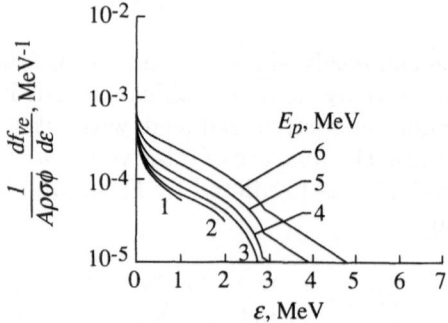

Figure 10.36. Energy loss spectrum of ^{7}Li ions in a 10-μm tissue sphere for external source energies between 1 and 6 MeV.

10.9.5. Remarks. A general method is found by which the energy absorption spectrum in a small spherical cavity is related to the source spectrum of the nuclear fragmentation event. The energy absorption spectrum may be found for arbitrary reaction source spectra. The current method can be used for evaluating the lineal energy distribution of the important 1-μm spherical dosimeter, which directly relates to the proposed quality factors (ICRU 40 (Anon., 1986)). The results of Caswell for the proton spectra from isotropic neutron scattering are easily derived from the present formalism. The current method can be easily applied to other geometries for any arbitrary source spectra. For example, the same methods are being used to analyze fragmentation events in thin surface barrier detectors.

The hit spectrum in electronic devices is related to the single-event upset problem. Application of these methods to electronic-related problems is given elsewhere (Ngo et al., 1989).

10.10. References

Alsmiller, R. G., Jr.; Armstrong, T. W.; and Coleman, W. A., 1970: *The Absorbed Dose and Dose Equivalent From Neutrons in the Energy Range 60 to 3000* MeV *and Protons in the Energy Range* 400 *to* 3000 MeV. ORNL TM-2924, U.S. Atomic Energy Commission.

Alsmiller, R. G., Jr.; and Barish, J., 1976: *Neutron Kerma Factors for* H, C, N, *and* O *and Tissue in the Energy Range 20 to 70* MeV. ORNL/TM-5702 (Contract No. W-7405-eng-26), Oak Ridge National Lab.

Anon., 1968a: *MECC-7 Intranuclear Cascade Code, 500-*MeV *Protons on* O-16. *I4C Analysis Codes* (Programmed for H. W. Bertini). Available from Radiation Shielding Information Center, Oak Ridge National Lab.

Anon., 1968b: *A Review of the Radiosensitivity of the Tissues in Bone.* ICRP Publ. 11, Pergamon Press.

Anon., 1977a: *Neutron Dosimetry for Biology and Medicine.* ICRU Rep. 26, International Commission on Radiation Units and Measurements.

Anon., 1977b: *Recommendations of the International Commission on Radiological Protection.* ICRP Publ. 26, Pergamon Press.

Anon., 1986: *The Quality Factor in Radiation Protection.* ICRU Rep. 40, International Commission on Radiation Units and Measurements.

Beddoe, A. H.; Darley, P. J.; and Spiers, F. W., 1976: Measurements of Trabecular Bone Structure in Man. *Phys. Med. & Biol.,* vol. 21, no. 4, pp. 589–607.

Bond, V. P.; Varma, M. N.; Sondhaus, C. A.; and Feinendegen, L. E., 1985: An Alternative to Absorbed Dose, Quality, and RBE at Low Exposures. *Radiat. Res.,* suppl. 8, vol. 104, pp. S-52–S-57.

Brenner, D. J., 1983: Neutron Kerma Values Above 15 MeV Calculated With a Nuclear Model Applicable to Light Nuclei. *Phys. Med. & Biol.,* vol. 29, no. 4, pp. 437–441.

Caswell, Randall S., 1966: Deposition of Energy by Neutrons in Spherical Cavities. *Radiat. Res.,* vol. 27, pp. 92–107.

Chatterjee, A.; Maccabee, H. D.; and Tobias, C. A., 1973: Radial Cutoff LET and Radical Cutoff Dose Calculations for Heavy Charged Particles in Water. *Radiat. Res.,* vol. 54, pp. 479–494.

Cucinotta, Francis A.; Hajnal, Ferenc; and Wilson, John W., 1990: Energy Deposition at the Bone-Tissue Interface From Nuclear Fragments Produced by High-Energy Nucleons. *Health Phys.*, vol. 59, no. 6, pp. 819–825.

Curtis, S. B.; Dye, D. L.; and Sheldon, W. R., 1965: Fractional Cell Lethality Approach to Space Radiation Hazards. *Second Symposium on Protection Against Radiations in Space*, Arthur Reetz, Jr., ed., NASA SP-71, pp. 219–223.

Dimbylow, P. J., 1982: Neutron Cross-Section and Kerma Value Calculations for C, N, O, Mg, Al, P, S, Ar and Ca From 20 to 50 MeV. *Phys. Med. & Biol.*, vol. 27, no. 8, pp. 989–1001.

Fermi, E.; and Teller, E., 1947: The Capture of Negative Mesotrons in Matter. *Phys. Review*, vol. 72, no. 5, pp. 399–408.

Fry, R. J. M.; Powers-Risius, P.; Alpen, E. L.; and Ainsworth, E. J., 1985: High-LET Radiation Carcinogenesis. *Radiat. Res.*, vol. 104, pp. S-188–S-195.

Fry, R. J. M., 1986: Radiation Effects in Space. *Adv. Space Res.*, vol. 6, no. 11, pp. 261–268.

Greiner, D. E.; Lindstrom, P. J.; Heckman, H. H.; Cork, Bruce; and Bieser, F. S., 1975: Momentum Distributions of Isotopes Produced by Fragmentation of Relativistic ^{12}C and ^{16}O Projectiles. *Phys. Review Lett.*, vol. 35, no. 3, pp. 152–155.

Herling, G. H.; Bassel, R. H.; Adams, J. H.; and Fraser, W. A., 1981: *Neutron-Induced Reactions in Tissue-Resident Elements.* NRL Rep. 8441, U.S. Navy. (Available from DTIC as AD A101 519.)

Katz, R.; Ackerson, B.; Homayoonfar, M.; and Sharma, S. C., 1971: Inactivation of Cells by Heavy Ion Bombardment. *Radiat. Res.*, vol. 47, pp. 402–425.

Katz, Robert; Sharma, S. C.; and Homayoonfar, M., 1972: The Structure of Particle Tracks. *Topics in Radiation Dosimetry, Supplement 1*, F. H. Attix, ed., Academic Press, Inc., pp. 317–383.

Katz, Robert, 1986: Biological Effects of Heavy Ions From the Standpoint of Target Theory. *Adv. Space Res.*, vol. 6, no. 11, pp. 191–198.

Kellerer, A. M.; and Chmelevsky, D., 1975: Criteria for the Applicability of LET. *Radiat. Res.*, vol. 63, pp. 226–234.

Lindhard, J., 1954: On the Properties of a Gas of Charged Particles. *Mat.-Fys. Medd.—K. Dan. Vidensk. Selsk.*, vol. 28, no. 8, pp. 1–58.

Lindstrom, P. J.; Greiner, D. E.; Heckman, H. H.; Cork, Bruce; and Bieser, F. S., 1975: *Isotope Production Cross Sections From the Fragmentation of ^{16}O and ^{12}C at Relativistic Energies.* LBL-3650 (NGR-05-003-513), Lawrence Berkeley Lab., Univ. of California.

MacFarlane, R. E.; Barrett, R. J.; Muir, D. W.; and Boicourt, R. M., 1978: *The NJOY Nuclear Data Processing System: User's Manual.* LA-7584-M (ENDF-272) (Contract W-7405-ENG. 34), Los Alamos Scientific Lab.

Mathews, G. J., 1983: Complete Fragment Yields From Spallation Reactions Via a Combined Time-of-Flight and $\Delta E - E$ Technique. *Composition and Origin of Cosmic Rays*, Maurice M. Shapiro, ed., D. Riedel Publ. Co., pp. 317–320.

Ngo, Duc M.; Wilson, John W.; Buck, Warren W.; and Fogarty, Thomas N., 1989: *Nuclear-Fragmentation Studies in Microelectronic Applications*. NASA TM-4143.

Paretzke, Herwig G., 1988: Problems in Theoretical Track Structure Research for Heavy Charged Particles. *Quantitative Mathematical Models in Radiation Biology*, Jürgen Kiefer, ed., Springer-Verlag, pp. 49–56.

Rossi, Harald H., 1959: Specification of Radiation Quality. *Radiat. Res.*, vol. 10, pp. 522–531.

Santoro, R. T.; Alsmiller, R. G., Jr.; and Chandler, K. C., 1974: *The Effects of Bone in the Use of Negatively Charged Pions in Cancer Radiotherapy*. ORNL-TM-4407 (Order NSF/RANN AG-399), Oak Ridge National Lab.

Schaefer, Hermann J., 1952: Exposure Hazards From Cosmic Radiation Beyond the Stratosphere and in Free Space. *J. Aviation Med.*, vol. 23, no. 4, pp. 334–344.

Schimmerling, Walter; Alpen, Edward L.; Powers-Risius, Patricia; Wong, Mervyn; DeGuzman, Randy J.; and Rapkin, Marwin, 1987: The Relative Biological Effectiveness of 670 MeV/A Neon as a Function of Depth in Water for a Tissue Model. *Radiat. Res.*, vol. 112, no. 3, pp. 436–448.

Shinn, J. L.; Wilson, J. W.; and Ngo, D. M., 1990: Risk Assessment Methodologies for Target Fragments Produced in High-Energy Nucleon Reactions. *Health Phys.*, vol. 59, no. 1, pp. 141–143.

Silberberg, R.; Tsao, C. H.; and Shapiro, M. M., 1976: Semiempirical Cross Sections, and Applications to Nuclear Interactions of Cosmic Rays. *Spallation Nuclear Reactions and Their Applications*, B. S. P. Shen and M. Merker, eds., D. Reidel Publ. Co., pp. 49–81.

Silberberg, R.; Tsao, C. H.; and Letaw, John R., 1983: Improvement of Calculations of Cross Sections and Cosmic-Ray Propagation. *Composition and Origin of Cosmic Rays*, Maurice M. Shapiro, ed., D. Reidel Publ. Co., pp. 321–336.

Spiers, F. W., 1969: Transition-Zone Dosimetry. *Radiation Dosimetry, Second ed.,* Volume III, Frank H. Attix and Eugene Tochilin, eds., Academic Press, pp. 809–867.

Townsend, L. W.; Wilson, J. W.; and Bidasaria, H. B., 1982: On the Geometric Nature of High-Energy Nucleus-Nucleus Reaction Cross Sections. *Canadian J. Phys.*, vol. 60, no. 10, pp. 1514–1518.

Townsend, Lawrence W.; Wilson, John; and Cucinotta, Francis A., 1987: A Simple Parameterization for Quality Factor as a Function of Linear Energy Transfer. *Health Phys.*, vol. 53, no. 5, pp. 531–532.

Waligorski, M. P. R.; Sinclair, G. L.; and Katz, R., 1987: Radiosensitivity Parameters for Neoplastic Transformations in C3H10T1/2 Cells. *Radiation Res.*, vol. 111, pp. 424–437.

Watt, D. E., 1989: On Absolute Biological Effectiveness and Unified Dosimetry. *J. Radiol. Prot.*, vol. 9, no. 1, pp. 33–49.

Wells, Alan H., 1979: A Consistent Set of KERMA Values for H, C, N, and O for Neutrons of Energies From 10 to 80 MeV. *Radiat. Res.*, vol. 80, pp. 1–9.

Wilson, John W., 1977: *Analysis of the Theory of High-Energy Ion Transport.* NASA TN D-8381.

Wilson, John W.; Stith, John J.; and Stock, L. V., 1983: *A Simple Model of Space Radiation Damage in GaAs Solar Cells.* NASA TP-2242.

Wilson, John W.; Townsend, L. W.; Bidasaria, H. B.; Schimmerling, Walter; Wong, Mervyn; and Howard, Jerry, 1984: ^{20}Ne Depth-Dose Relations in Water. *Health Phys.*, vol. 46, no. 5, pp. 1101–1111.

Wilson, J. W.; Townsend, L. W.; Badavi, F. F, 1987: Galactic HZE Propagation Through the Earth's Atmosphere. *Radiat. Res.*, vol. 109, no. 2, pp. 173–183.

Wilson, John W.; Chun, S. Y.; Buck, W. W.; and Townsend, L. W., 1988: High Energy Nucleon Data Bases. *Health Phys.*, vol. 55, no. 5, pp. 817–819.

Wilson, John W.; Townsend, Lawrence W.; Khan, Ferdous, 1989: Evaluation of Highly Ionizing Components in High-Energy Nucleon Radiation Fields. *Health Phys.*, vol. 57, no. 5, pp. 717–724.

Yang, Tracy Chui-Hsu; Craise, Laurie M.; Mei, Man-Tong; and Tobias, Cornelius A., 1985: Neoplastic Cell Transformation by Heavy Charged Particles. *Radiation Res.*, vol. 104, pp. S-177–S-187.

Ziegler, J. F., 1980: *Handbook of Stopping Cross-Sections for Energetic Ions in All Elements.* Volume 5 of *The Stopping and Ranges of Ions in Matter*, J. F. Ziegler, ed., Pergamon Press Inc.

11. SPACE-RADIATION EXPOSURE ISSUES

11.1. Introduction

Most of the work presented in previous chapters has been of a more fundamental nature rather than strictly space related exposure issues. That is, the main body of issues covered is as fundamental to radiobiology and radiation therapy as to space radiation. There are clear gaps in the presented methodologies and data bases. There is also a need to improve the atomic/molecular description in our work to include, first of all, improved stopping powers and to treat the fluctuation phenomena associated with slowing down as well as multiple scattering. The light fragment ($Z \leq 2$) production and transport need to treat the full energy spectra as well as the angular dependence. This is especially true not only for the laboratory related work but also for the angular scatter of neutrons in space computations. A heavy ion code for the laboratory analysis with energy-dependent nuclear cross sections has been developed (Wilson et al., 1984; Townsend and Wilson, 1988) but space-radiation codes have ignored the energy-dependent HZE cross sections. A laboratory code that includes light fragments in any realistic way does not exist. Greater knowledge of nuclear fragmentation processes and a corresponding transport theory are required.

In addition to these physical factors, there are unresolved biological issues. The differences in the radiosensitivity of various tissues within an individual as well as individual differences are generally assumed to result from repair mechanisms (Curtis, 1986; Fritz-Niggli, 1988). The work of Swenberg, Holwitt, and Speicher (1990) suggests these differences may result from the structural state of the DNA as well. Repair also affects the dose response for protracted exposure. Current radiation-protection guidelines use quality factors that are independent of dose rate (no time-modifying factors), which may be of unusual importance in the small dose rate exposures often experienced in space (NCRP 98, (Anon., 1989)). Clearly, well understood dose-rate-dependent models are needed (Scott and Ainsworth, 1980; Curtis, 1986; Anon., 1989). Furthermore, exposure received on a Lunar or Mars mission will involve heavy ion exposure for which many issues concerning appropriate relative biological effectiveness (RBE) factors (hence, quality factors) are yet unknown. The accumulated heavy ion exposure levels will be large and unprecedented in human experience. Although these issues may be studied in ground-based exposures with model biological systems, extrapolating to human exposure is difficult at best, and space stress factors such as microgravity are unknown possible modifying factors in radiobiological response.

In addition to radiobiological response issues is the need to evaluate dose nonuniformity caused by body self-shielding and dose gradients within the shielding structure. For example, tumor prevalence in the female breast is site specific even for relative uniform exposure (NCRP 85 (Anon., 1986a)). We are led to believe that the exposure of only sensitive sites may be effective in tumor formation. Conversely, exposure of insensitive sites is assumed to be noneffective, and nonuniformity of exposure is a critical issue. In this chapter, we make some preliminary assessments concerning these issues and examine a limited number of shielding strategies to mitigate these radiation effects.

11.2. Galactic Cosmic-Ray Exposure

The incident galactic cosmic-ray spectrum (Adams, 1987) for free space is propagated through the target material by using the accurate analytical/numerical solutions to the transport equation described in chapter 10. These solution methods have been verified (to within 2 percent accuracy) by comparison with exact, analytical benchmark solutions to the ion transport equation (Wilson and Townsend, 1988; Wilson et al., 1988).

These transport calculations include

1. Linear energy-transfer (LET) dependent quality factors from ICRP 26 (Anon., 1977)

2. Dose contributions from propagating neutrons, protons, α-particles, and heavy ions (high-energy, high-charge (HZE) particles)

3. Dose contributions that result from target nuclear fragments produced by all neutrons and primary protons and their secondaries

4. Dose contributions due to nuclear recoil in tissue

Major shortcomings of the calculations are as follows:

1. Except for tissue targets, mass number 2 and 3 fragment contributions are neglected

2. Target fragmentation contributions from HZE particles and their charged secondaries are neglected (although they are included for nucleons)

3. All secondary particles from HZE interactions are presently assumed to be produced with a velocity equal to that of the incident particle; for neutrons produced in HZE particle fragmentations, this is conservative

4. A quality factor of 20 is assigned to all multiple charged target fragments from the incident protons; to improve this approximation, one must calculate target fragment spectra correctly

5. Meson contributions to the propagating radiation fields are neglected

6. Nucleus-nucleus cross sections are not fully energy dependent (nucleon-nucleus cross sections are fully energy dependent)

For these shortcomings, items 3 and 4 are conservative; the remaining items, however, are not and alone probably result in a 15- to 30-percent underestimate of the exposure. As discussed elsewhere (Townsend, Wilson, and Nealy, 1989), the main sources of uncertainty are the input nuclear fragmentation model and the incident galactic cosmic-ray (GCR) spectrum. Taken together, they could easily impose a factor of 2 or more uncertainty in the exposure predictions.

11.2.1. Results. In the present results we use the ICRP 26 (Anon., 1977) quality factors, which are currently in force within the U.S. space program. Figure 11.1 displays dose equivalent (in units of sieverts per year) as a function of water shield thickness (in units of areal density, g/cm^2, or thickness, cm). Curves are displayed for solar minimum and solar maximum periods. The numerical values used in this figure are listed in table 11.1. Also listed in this table are values for the absorbed dose in centigrays per year as a function of water shield thickness. For all thicknesses considered, the dose and dose equivalent during solar maximum are less than half the dose equivalent during solar minimum, at least according to the current estimates derived from the CREME environmental model of Adams (1987). Figure 11.2 displays results for dose and dose equivalent behind an aluminum shield. Also shown are measurements with argon filled ion chambers at two shielded locations on the Prognoz satellite (Kovalev, Muratova, and Petrov, 1989). The results for the 1 g/cm^2 location are the most clear by experimental design. The mass distribution for the deeply shielded counter was poorly defined (Kovalev, Muratova, and Petrov, 1989); this uncertainty is denoted in the figure by parentheses around the data points. The solar maximum model predicted by CREME is clearly an underestimate. The solar minimum model appears in reasonable agreement with the Prognoz data. Therefore, we will restrict the present analysis to solar minimum periods, which are the most limiting for GCR exposures. This is

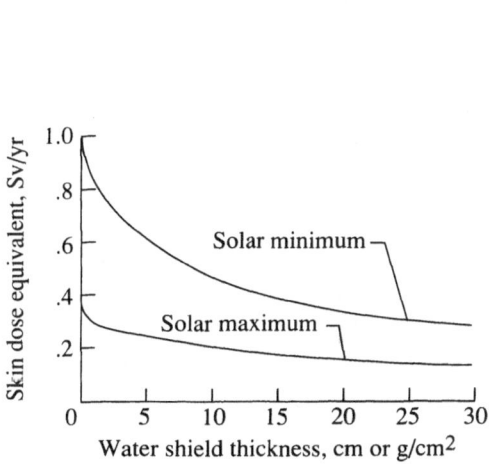

Figure 11.1. Dose equivalent resulting from galactic cosmic rays as function of water shield thickness.

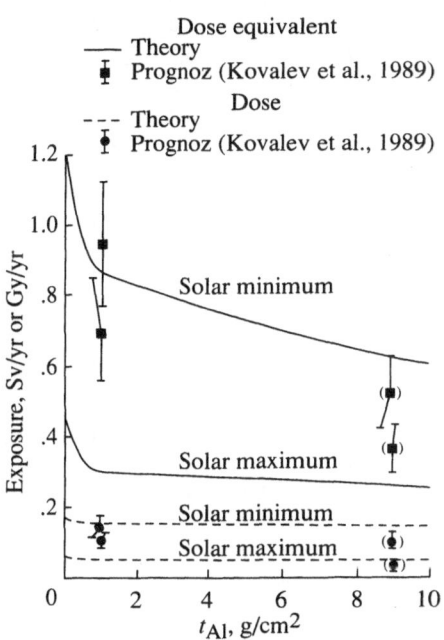

Figure 11.2. Deep space exposure behind aluminum shield. Parentheses denote depth of interior of Prognoz spacecraft.

Table 11.1. Galactic Cosmic-Ray Dose and Dose Equivalent in Tissue for Various
Water Shield Thicknesses

[All values are rounded to nearest 0.1]

Thickness, cm or g/cm^2	Solar maximum		Solar minimum	
	Dose, cGy/yr	Dose equivalent, cSv/yr	Dose, cGy/yr	Dose equivalent, cSv/yr
0	6.4	45.1	17.1	120.6
1	5.4	29.0	14.9	82.6
2	5.4	27.6	14.6	76.1
3	5.4	26.3	14.4	70.3
4	5.4	25.1	14.2	65.4
5	5.4	24.0	14.0	61.1
6	5.4	23.0	13.8	57.4
7	5.3	22.0	13.7	54.1
8	5.3	21.2	13.5	51.1
9	5.3	20.4	13.4	48.6
10	5.3	19.6	13.3	46.3
11	5.3	19.0	13.2	44.2
12	5.3	18.4	13.1	42.4
13	5.3	17.8	13.0	40.7
14	5.2	17.3	12.9	39.3
15	5.2	16.8	12.8	37.9
16	5.2	16.3	12.7	36.7
17	5.2	15.9	12.6	35.6
18	5.2	15.6	12.6	34.6
19	5.2	15.2	12.5	33.7
20	5.2	14.9	12.4	32.8
25	5.1	13.6	12.1	29.6
30	5.1	12.7	11.9	27.3
40	5.0	11.6	11.4	24.6
50	4.9	11.0	10.9	22.9

not meant to imply, however, that exposures during solar maximum periods are not important. On the contrary, the cumulative exposures that result from combined GCR and increased solar flare activity during solar maximum could potentially be significant. Analyses of these hazards are reported in chapter 12.

The actual compositions of the calculated radiation fields are displayed in tables 11.2 through 11.4; values for dose equivalent, dose, and particle flux are listed by particle type (neutrons, protons, etc.) and as a function of water thickness. The target fragment dose and dose equivalent contributions for incident protons and their secondaries, computed with BRYNTRN (Wilson et al., 1989), are displayed separately in these tables. The quality factor of target fragments $(A > 1)$ is assumed to be 20.

Estimates of the thicknesses of water shielding required to protect astronauts from GCR particles can be obtained from table 11.1 or figure 11.1. At present, there are no recommended exposure limits for deep space exploratory missions. Therefore, we will use the currently proposed annual limits for Space Station *Freedom* as guidelines as recommended by NCRP 98 (Anon., 1989). The annual limits are 3 Sv to the skin (0.01 cm depth), 2 Sv to the ocular lens (0.3 cm depth), and 0.5 Sv to the blood-forming organs (5 cm depth). Clearly, from table 11.1, none of these limits are exceeded during periods of

Table 11.2. Solar Minimum Galactic Cosmic-Ray Dose Equivalent in Tissue for Various Particle Types and Water Shield Thicknesses

[All values are rounded to nearest 0.1]

Thickness, cm or g/cm^2	Dose equivalent, cSv/yr, from—					
	Neutrons	Protons	Target fragments	α-particles	HZE	Total
0	0	9.7	0	7.0	102.5	119.2
1	0.3	6.6	5.9	3.4	66.4	82.6
2	0.6	7.0	5.9	3.2	59.3	76.0
3	0.9	7.4	5.8	3.1	53.1	70.3
4	1.2	7.7	5.8	3.0	47.7	65.4
5	1.4	8.0	5.8	2.9	43.0	61.1
6	1.7	8.2	5.8	2.8	38.9	57.4
8	2.1	8.6	5.7	2.6	32.1	51.1
10	2.6	9.0	5.6	2.4	26.7	46.3
15	3.5	9.6	5.4	2.0	17.4	37.9
20	4.3	10.0	5.2	1.6	11.7	32.8
25	4.9	10.2	5.0	1.3	8.0	29.4
30	5.4	10.4	4.8	1.0	5.2	26.8
40	6.2	10.4	4.4	0.7	2.8	24.5
50	6.6	10.2	4.0	0.5	1.4	22.7

Table 11.3 Solar Minimum Galactic Cosmic-Ray Dose in Tissue for Various Particle Types and Water Shield Thicknesses

[All values are rounded to nearest 0.1]

Thickness, cm or g/cm^2	Dose, cGy/yr, from—					
	Neutrons	Protons	Target fragments	α-particles	HZE	Total
0	0	6.2	0	3.0	7.8	17.0
1	0.1	6.0	0.3	2.7	5.8	14.9
2	0.1	6.4	0.3	2.6	5.3	14.7
3	0.2	6.6	0.3	2.5	4.9	14.5
4	0.2	6.8	0.3	2.4	4.5	14.2
5	0.3	7.1	0.3	2.3	4.1	14.1
6	0.4	7.2	0.3	2.2	3.8	13.9
8	0.5	7.5	0.3	2.0	3.2	13.5
10	0.5	7.8	0.3	1.9	2.8	13.3
15	0.7	8.3	0.3	1.6	1.9	12.8
20	0.9	8.6	0.3	1.3	1.4	12.5
25	1.1	8.7	0.3	1.1	1.0	12.2
30	1.2	8.8	0.2	0.9	0.8	11.9
40	1.3	8.8	0.2	0.6	0.4	11.3
50	1.4	8.6	0.2	0.4	0.2	10.8

Table 11.4. Solar Minimum Galactic Cosmic-Ray Flux for Various Particle Types and Water Shield Thicknesses

[All values are rounded to nearest 0.1]

Thickness, cm or g/cm^2	Flux, particles/cm^2/yr, from—			
	Neutrons	Protons	α-particles	HZE
0	0E7	1.3E8	1.2E7	1.4E6
1	0.4E7	1.3E8	1.2E7	1.3E6
2	0.8E7	1.3E8	1.2E7	1.2E6
3	1.2E7	1.3E8	1.1E7	1.2E6
4	1.6E7	1.4E8	1.1E7	1.1E6
5	1.9E7	1.4E8	1.0E7	1.1E6
6	2.2E7	1.4E8	1.0E7	1.0E6
8	2.9E7	1.4E8	0.9E7	0.9E6
10	3.5E7	1.4E8	0.9E7	0.8E6
15	4.7E7	1.4E8	0.7E7	0.6E6
20	5.8E7	1.4E8	0.6E7	0.5E6
25	6.7E7	1.4E8	0.5E7	0.4E6
30	7.4E7	1.4E8	0.4E7	0.3E6
40	8.4E7	1.3E8	0.3E7	0.2E6
50	9.0E7	1.3E8	0.2E7	0.1E6

solar maximum activity, as the unshielded (0 cm depth) dose equivalent is estimated to be less than 0.5 Sv. Similarly, during solar minimum periods, the estimated unshielded dose equivalent of 1.2 Sv does not exceed either exposure limits for the skin or the ocular lens. The dose equivalent at 5 cm depth, which yields an estimate of the exposure to the unshielded blood-forming organs (BFO), is 0.61 Sv, which exceeds the 0.5-Sv limit by 22 percent. To reduce this estimated exposure below 0.5 Sv requires about 3.5 g/cm^2 (3.5 cm) of water shielding in addition to the body self-shielding of 5 g/cm^2 (5 cm).

For comparison purposes, calculations of skin (0 cm depth) and BFO (5 cm depth) exposures behind various thicknesses of aluminum and liquid hydrogen shielding were made. The results are presented in tables 11.5 through 11.12. For aluminum, 6.5 g/cm^2 (2.4 cm) of shielding thickness is required to reduce the BFO dose equivalent below the annual limit. (See table 11.7.) For liquid hydrogen, 1 g/cm^2 (14 cm) of shielding is required. For relative comparison purposes, the BFO dose equivalent as a function of shield thickness (areal density) is plotted in figure 11.3 for these three materials. Clearly, shielding effectiveness per unit mass increases as the composition of the shield changes from heavier to lighter mass elements. For liquid hydrogen, an added advantage is the reduced neutron fluence that is caused by the absence of neutrons in the target composition and by the lack of target fragment contributions because of the elementary nature of hydrogen. From these results, for an allowed BFO exposure of 0.25 Sv/year, which corresponds to an uncertainty factor of 2 in a 0.5 Sv/year estimate, the mass ratios for the shielding are about 1:5:11 for LH$_2$:H$_2$O:Al. Obviously, for GCR shielding, the materials of choice are those composed of low atomic mass number constituents with significant hydrogen content.

Table 11.5. Solar Minimum Galactic Cosmic-Ray Dose Equivalent at 0 cm Deep in Tissue for Various Particle Types and Aluminum Shield Thicknesses

[All values are rounded to nearest 0.1]

Thickness,[a] g/cm^2	Dose equivalent, cSv/yr, from—					
	Neutrons	Protons	Target fragments	α-particles	HZE	Total dose equivalent
1	0.4	7.5	5.9	3.5	69.4	86.8
2	0.8	8.2	5.9	3.4	64.5	82.8
3	1.2	8.6	5.9	3.3	59.9	79.0
4	1.6	9.0	5.9	3.2	55.7	75.4
5	2.0	9.4	5.9	3.2	51.9	72.2
6	2.4	9.7	5.8	3.1	48.4	69.4
8	3.1	10.2	5.8	2.9	42.4	64.4
10	3.8	10.6	5.8	2.8	37.4	60.3
15	5.3	11.5	5.7	2.4	27.9	52.7
20	6.6	12.0	5.5	2.1	21.3	47.6
30	8.7	12.7	5.3	1.6	13.1	41.3

[a]1 g/cm^2 of aluminum is equivalent to 0.37 cm thickness.

Table 11.6. Solar Minimum Galactic Cosmic-Ray Dose at 0 cm Deep in Tissue
for Various Particle Types and Aluminum Shield Thicknesses

[All values are rounded to nearest 0.1]

Thickness,[a] g/cm^2	Dose, cGy/yr, from—					
	Neutrons	Protons	Target fragments	α-particles	HZE	Total dose
1	0.1	6.3	0.3	2.7	6.1	15.5
2	0.2	6.8	0.3	2.6	5.7	15.5
3	0.2	7.1	0.3	2.6	5.3	15.5
4	0.3	7.4	0.3	2.5	5.0	15.5
5	0.4	7.6	0.3	2.4	4.9	15.5
6	0.5	7.8	0.3	2.4	4.5	15.4
8	0.6	8.2	0.3	2.3	4.0	15.4
10	0.7	8.5	0.3	2.2	3.6	15.3
15	1.0	9.1	0.3	1.9	2.8	15.1
20	1.3	9.5	0.3	1.7	2.3	15.0
30	1.7	10.0	0.3	1.3	1.5	14.7

[a]1 g/cm^2 of aluminum is equivalent to 0.37 cm thickness.

Table 11.7. Solar Minimum Galactic Cosmic-Ray Dose Equivalent at 5 cm Deep in Tissue
for Varous Particle Types and Aluminum Shield Thicknesses

[All values are rounded to nearest 0.1]

Thickness,[a] g/cm^2	Dose equivalent, cSv/yr, from—					
	Neutrons	Protons	Target fragments	α-particles	HZE	Total dose equivalent
1	1.7	8.2	5.8	2.8	40.3	58.8
2	2.1	8.5	5.8	2.8	37.7	56.8
3	2.4	8.7	5.8	2.7	35.4	54.9
4	2.7	8.9	5.7	2.6	33.3	53.3
5	3.0	9.1	5.7	2.5	31.4	51.7
6	3.3	9.3	5.7	2.5	29.6	50.3
8	3.9	9.6	5.7	2.3	26.4	47.8
10	4.4	9.9	5.6	2.2	23.6	45.7
15	5.6	10.5	5.5	1.9	18.1	41.7
20	6.6	10.9	5.4	1.7	14.1	38.8
30	8.3	11.4	5.3	1.3	8.9	35.2
50	10.4	11.6	4.3	0.8	3.8	30.9

[a]1 g/cm^2 of aluminum is equivalent to 0.37 cm thickness.

Table 11.8. Solar Minimum Galactic Cosmic-Ray Dose at 5 cm Deep in Tissue for Various Particle Types and Aluminum Shield Thicknesses

[All values are rounded to nearest 0.1]

Thickness,[a] g/cm^2	Dose, cGy/yr, from—					
	Neutrons	Protons	Target fragments	α-particles	HZE	Total dose
1	0.4	7.2	0.3	2.2	3.8	14.0
2	0.4	7.4	0.3	2.2	3.7	14.0
3	0.5	7.6	0.3	2.1	3.5	13.9
4	0.6	7.7	0.3	2.1	3.3	13.9
5	0.6	7.8	0.3	2.0	3.1	13.9
6	0.7	8.0	0.3	2.0	3.0	13.9
8	0.8	8.2	0.3	1.9	2.7	13.8
10	0.9	8.4	0.3	1.8	2.5	13.8
15	1.3	8.7	0.3	1.6	2.0	13.8
20	1.3	9.0	0.3	1.4	1.6	13.6
30	1.7	9.3	0.3	1.1	1.1	13.3
50	2.1	9.3	0.2	0.7	0.5	12.7

[a]1 g/cm^2 of aluminum is equivalent to 0.37 cm thickness.

Table 11.9. Solar Minimum Galactic Cosmic-Ray Flux for Various Particle Types and Aluminum Shield Thicknesses

[All values are rounded to nearest 0.1]

Thickness,[a] g/cm^2	Flux, particles/cm^2/yr, from—			
	Neutrons	Protons	α-particles	HZE
0	0E7	1.3E8	1.2E7	1.4E6
1	0.6E7	1.3E8	1.2E7	1.3E6
2	1.2E7	1.3E8	1.2E7	1.3E6
3	1.8E7	1.3E8	1.2E7	1.2E6
4	2.4E7	1.4E8	1.1E7	1.2E6
5	2.9E7	1.4E8	1.1E7	1.1E6
6	3.4E7	1.4E8	1.0E7	1.1E6
8	4.4E7	1.4E8	1.0E7	1.0E6
10	5.4E7	1.4E8	1.0E7	1.0E6
15	7.6E7	1.4E8	0.9E7	0.8E6
20	9.5E7	1.4E8	0.8E7	0.7E6
30	12.5E7	1.4E8	0.6E7	0.5E6
50	16.3E7	1.4E8	0.4E7	0.3E6

[a]1 g/cm^2 of aluminum is equivalent to 0.37 cm thickness.

Table 11.10. Solar Minimum Galactic Cosmic-Ray Depth Dose Equivalent in Tissue
for Various Particle Types and Liquid Hydrogen Shield Thicknesses

[All values are rounded to nearest 0.1]

Thickness,[a] g/cm^2	Dose equivalent, cSv/yr, from—				Total dose equivalent
	Neutrons	Protons	α-particles	HZE	
Skin dose equivalent (0 cm depth)					
0	0	9.4	6.7	101.6	117.7
3	0.2	6.6	2.7	31.8	41.3
10	0.6	7.8	1.5	6.31	6.2
25	0.8	8.1	0.4	0.4	9.7
50	0.7	6.6	0.1	<0.1	7.4
75	0.6	4.8	<0.1	<0.1	5.4
100	0.4	3.3	<0.1	<0.1	3.8
BFO dose equivalent (5 cm depth)					
0	1.4	8.0	2.9	43.0	61.1
3	1.8	8.8	2.2	21.2	34.1
10	1.9	9.6	1.2	4.6	17.2
25	1.7	9.4	0.4	0.3	11.7
50	1.3	7.4	<0.1	<0.1	8.7
75	0.9	5.3	<0.1	<0.1	6.2
100	0.7	3.6	<0.1	<0.1	4.3

[a]1 g/cm^2 of LH_2 is equivalent to 14 cm thickness.

Table 11.11. Solar Minimum Galactic Cosmic-Ray Depth Dose in Tissue for Various
Particle Types and Liquid Hydrogen Shield Thicknesses

[All values are rounded to nearest 0.1]

Thickness,[a] g/cm^2	Dose, cGy/yr, from—				
	Neutrons	Protons	α-particles	HZE	Total dose
Skin dose equivalent (0 cm depth)					
0	0	6.2	3.0	7.8	17.0
3	0.1	6.4	2.2	3.2	11.9
10	0.1	7.5	1.2	0.9	9.7
25	0.2	7.7	0.4	0.1	8.4
50	0.2	6.2	<0.1	<0.1	6.5
75	0.1	4.5	<0.1	<0.1	4.7
100	0.1	3.1	<0.1	<0.1	3.2
BFO dose (5 cm depth)					
0	0.3	7.1	2.3	4.1	14.0
3	0.4	7.8	1.8	2.3	12.3
10	0.4	8.5	1.0	0.7	10.6
25	0.4	8.3	0.3	0.1	9.0
50	0.3	6.5	<0.1	<0.1	6.8
75	0.2	4.7	<0.1	<0.1	4.9
100	0.1	3.2	<0.1	<0.1	3.3

[a] 1 g/cm^2 of LH_2 is equivalent to 14 cm thickness.

Table 11.12. Solar Minimum Galactic Cosmic-Ray Flux for Various Particle Types and
Liquid Shield Thicknesses

[All values are rounded to nearest 0.1]

Thickness,[a] cm or g/cm^2	Flux, particles/cm^2/yr, from—			
	Neutrons	Protons	α-particles	HZE
0	0E6	1.3E8	1.2E7	1.39E6
3	2.8E6	1.4E8	9.9E6	9.49E5
10	6.8E6	1.4E8	5.8E6	4.24E5
25	9.6E6	1.3E8	1.8E6	8.10E4
50	8.9E6	1.0E8	0.3E6	0.50E4
75	7.0E6	0.7E8	0.4E5	0.30E3
100	5.3E6	0.5E8	0.6E4	0.20E2

[a] 1 g/cm^2 of LH_2 is equivalent to 14 cm thickness.

11.2.2. Discussion. Although the calculations are useful for estimating relative shield effectiveness to compare different materials, quantitatively the calculations should be considered preliminary estimates of actual shield mass requirements. Aside from the previously mentioned shortcomings related to neglecting meson production and target fragment contributions from interactions of HZE particles and the target medium, figure 11.3 shows that the dose equivalent is a slowly decreasing function of shield thickness. This is a result of secondary particle production processes whereby the heavier GCR nuclei are broken into nucleons and lighter nuclear fragments by nuclear and coulombic interactions with the shield material. This slow decrease in dose equivalent with increasing shield thickness means that relatively small uncertainties in predicted doses that arise from nuclear fragmentation model inaccuracies may yield large uncertainties in estimated shield thicknesses. A preliminary analysis of the nonlinear relationship between exposure uncertainty was presented by Townsend, Wilson, and Nealy (1989). The most startling finding was that a factor of 2 uncertainty in exposure amplified into an order of magnitude uncertainty in shield mass requirements. To further illustrate this, water shield mass increase (in percent) as a function of BFO exposure uncertainty (in percent) is listed in table 11.13. For the latter quantity, the calculated exposure is assumed to be smaller than the actual exposure by the percentage indicated; that is, the exposure is underestimated.

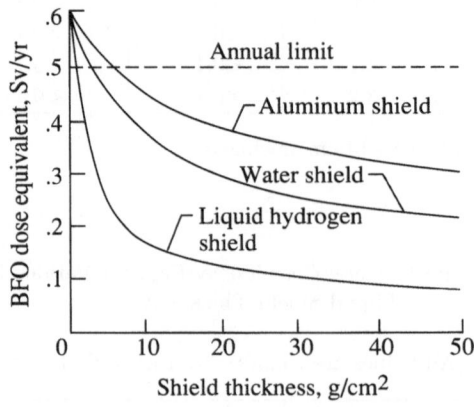

Figure 11.3. Dose equivalent in BFO as function of shield type and thickness.

Again we note that if the exposure is underestimated by a factor of 2 (the 50-percent entry), then the resultant shield mass must be increased by an order of magnitude (1000 percent). To account for the ≈15-percent uncertainty that results from the neglect of meson production and the incomplete treatment of target fragmentation, the shield mass must be doubled (increased by 100 percent). Similarly, possible inaccuracies in the input fragmentation cross sections could underestimate the exposures by as much as 20 to 30 percent (Townsend and Wilson, 1988) and result in potential shield mass increases by up to a factor of 4 (over a 400 percent increase). Clearly the *complete* development of an accurate and comprehensive transport code is needed, and uncertainties in the actual GCR environmental model and in the input nuclear fragmentation models need to be resolved through additional theoretical and experimental research as has been emphasized at Langley for many years. (See chapter 1.). Finally, we note that radiation exposure is cumulative and therefore requires consideration of contributions from all sources, including onboard nuclear power sources, solar particle events, and GCR's. Exposure to onboard sources reduces the allowed exposures from solar flares and cosmic rays and thereby increases the required shield thicknesses that are necessary to stay below the exposure limits.

Table 11.13. Increase in Water Shield Mass for Various Exposure Uncertainties

BFO exposure uncertainty,[a] percent	Increase in water shield mass, percent
10	43
15	100
20	129
30	414
40	614
50	1000

[a]Exposures assumed to be *underestimated* by the indicated percent.

11.3. GCR Component Breakdown

Although GCR's probably include every natural element, not all are important for space-radiation-protection purposes. For example, the elemental abundances for species heavier than iron (charge number, $Z > 26$) are typically 2 to 4 orders of magnitude smaller than iron (Adams, Silberberg, and Tsao, 1981), and therefore are of negligible importance in this regard. For elements lighter than iron, species with an odd charge number are significantly less abundant than their neighboring species with even charge numbers. This is readily seen in figure 11.4, which displays abundances (normalized to silicon = 100) for ions from helium ($Z = 2$) to iron. The data were taken from Simpson (1983). From this figure, the most abundant elements that are heavier than helium are carbon ($Z = 6$), oxygen ($Z = 8$), magnesium ($Z = 12$), silicon ($Z = 14$), and iron ($Z = 26$). Although neon ($Z = 10$) is about equal in abundance to iron, its much lower charge number suggests that its contribution to the total dose (nearly proportional to Z^2) should be much lower than that for iron. Therefore, we did not include neon in this analysis. Instead, we estimated component contributions to dose and dose equivalent for seven elemental GCR constituents: protons, helium, carbon, oxygen, magnesium, silicon, and iron.

The component analysis is performed by separately transporting each of the seven ion species (solar minimum abundances) through the aluminum shield. At each shield thickness (0, 2, and 10 g/cm^2), the dose and dose equivalent are computed for the incident ion species and all subsequent-generation collision products. The latter are categorized as HZE (all secondary ions having $Z > 2$), alpha particles ($Z = 2$), protons ($Z = 1$), and neutrons. The results are presented in tables 11.14 and 11.15. The entries labeled 0 g/cm^2 aluminum shield represent unshielded exposures and consist only of primary ion contributions. From these tables, we note that these seven ions constitute over 80 percent of the unshielded GCR dose and nearly 70 percent of the unshielded GCR dose equivalent. Although protons (hydrogen) account for nearly 90 percent of the incident flux, they only account for 36 percent of the unshielded total GCR dose and less than 10 percent of the total dose equivalent. Helium nuclei, which comprise nearly 10 percent of the incident flux, account for 18 percent of the total unshielded dose and only 6 percent of the unshielded total dose equivalent. The entire heavy ion component of the spectrum, which comprises about 2 percent of the incident GCR flux, accounts for nearly half of the unshielded dose (46 percent) and over 85 percent of the unshielded total dose equivalent. The largest single contribution to the heavy ion component is iron, which accounts for 9 percent of the total unshielded GCR dose and 26 percent of the total dose equivalent.

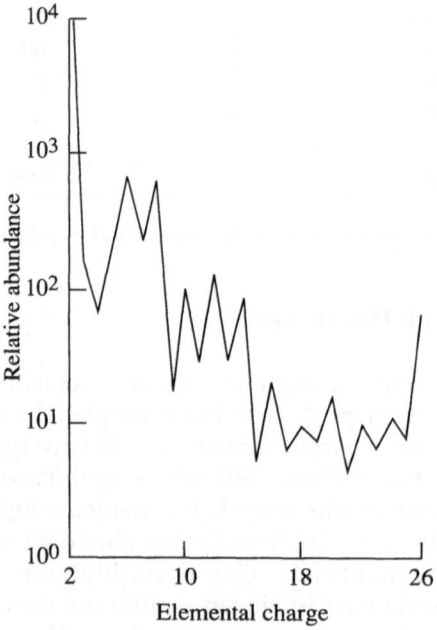

Figure 11.4. Elemental abundances.

Behind 2 g/cm^2 of aluminum shielding (a thin spacecraft), the main contributions to the dose equivalent (table 11.15) come from the incident ions. The total contribution from all secondaries is less than 2 percent of the primary contribution from incident protons and their secondaries. The HZE contribution to the secondary total dose equivalent (0.67 cSv) is nearly equal to the secondary neutron contribution (0.81 cSv). In these tables, no separate entry for secondary protons produced by primary protons is made because of the difficulty in extracting this information from the current version of the GCR transport code. Instead, the values listed for primary dose and dose equivalent represent the sum of primary and secondary proton contributions.

At 10 g/cm^2 aluminum shielding, the largest single contributor to GCR total dose and dose equivalent is hydrogen (protons) and its secondaries, which account for 58 percent of the dose and 24 percent of the dose equivalent. Surprisingly, the second largest contributor is iron and its secondaries, which accounts for 18 percent of the total GCR dose equivalent. Again from table 11.15, we note that the HZE

Table 11.14. Dose in Water by GCR Component for Several Aluminum Shield Thicknesses

Incident ion species	Primary ion dose, cGy	Dose from secondary particles, cGy				Ion total, cGy	Percent of total GCR
		HZE	α-particles	Protons	Neutrons		
Aluminum shield 0 g/cm^2 thick							
p	6.21				0	6.21	36
α	3.02			0	0	3.02	18
C	0.83	0	0	0	0	0.83	5
O	1.37	0	0	0	0	1.37	8
Mg	0.66	0	0	0	0	0.66	4
Si	0.69	0	0	0	0	0.69	4
Fe	1.56	0	0	0	0	1.56	9
Total	14.34	0	0	0	0	14.34	84
Aluminum shield 2 g/cm^2 thick							
p	6.70				0.12	6.82	45
α	2.62			0.03	0.03	2.68	18
C	0.69	<0.01	<0.01	<0.01	<0.01	0.69	5
O	1.13	0.01	<0.01	0.01	<0.01	1.15	8
Mg	0.53	0.01	<0.01	<0.01	<0.01	0.54	4
Si	0.54	0.01	<0.01	<0.01	<0.01	0.55	4
Fe	0.86	0.03	<0.01	<0.01	<0.01	0.89	6
Total	13.07	0.06	<0.01	<0.04	0.15	13.32	90
Aluminum shield 10 g/cm^2 thick							
p	8.17				0.55	8.72	58
α	2.13			0.21	0.13	2.47	16
C	0.48	0.01	<0.01	0.03	0.01	0.53	4
O	0.73	0.03	<0.01	0.03	0.01	0.80	5
Mg	0.31	0.02	<0.01	0.01	<0.01	0.34	2
Si	0.30	0.03	<0.01	0.01	<0.01	0.34	2
Fe	0.44	0.07	<0.01	0.01	0.01	0.53	4
Total	12.56	0.16	<0.01	0.30	0.71	13.73	91

Table 11.15. Dose Equivalent in Water by GCR Component for Several Aluminum Shield Thicknesses

Incident ion species	Primary Ion dose equivalent, cSv	Dose from secondary particles, cSv				Ion total, cSv	Percent of total GCR
		HZE	α-particles	Protons	Neutrons		
Aluminum shield 0 g/cm^2 thick							
p	9.73				0	9.73	8
α	6.96			0	0	6.96	6
C	4.94	0	0	0	0	4.94	4
O	11.11	0	0	0	0	11.11	9
Mg	8.01	0	0	0	0	8.01	7
Si	9.63	0	0	0	0	9.63	8
Fe	30.90	0	0	0	0	30.90	26
Total	81.28	0	0	0	0	81.28	68
Aluminum shield 2 g/cm^2 thick							
p	8.09				0.68	8.77	11
α	3.38			0.04	0.11	3.53	5
C	3.25	0.01	<0.01	0.01	0.01	3.28	4
O	7.85	0.05	<0.01	<0.01	<0.01	7.92	10
Mg	5.91	0.06	<0.01	<0.01	<0.01	5.97	8
Si	7.10	0.09	<0.01	<0.01	<0.01	7.19	9
Fe	17.04	0.46	<0.01	<0.01	<0.01	17.50	23
Total	52.62	0.67	<0.01	0.06	0.81	54.16	70
Aluminum shield 10 g/cm^2 thick							
p	10.15				2.99	13.14	24
α	2.75			0.30	0.54	3.59	7
C	2.12	0.04	<0.01	0.03	0.04	2.23	4
O	4.76	0.16	<0.01	0.04	0.05	5.01	9
Mg	3.23	0.18	<0.01	0.01	0.02	3.44	6
Si	3.71	0.24	<0.01	0.01	0.01	3.97	7
Fe	8.68	1.25	<0.01	0.02	0.02	9.97	18
Total	35.40	1.87	<0.01	0.41	3.67	41.35	75

contribution to the secondary total dose equivalent (1.87 cSv) is comparable with that which results from secondary neutrons (3.67 cSv).

Several other comments on the results displayed in tables 11.14 and 11.15 are appropriate. First, the secondary neutron and proton dose-equivalent contributions that arise from heavy ion fragmentations, although nearly equal, appear to be relatively small compared with the secondary HZE contribution. Second, the secondary alpha production from all sources appears negligible. Both of these findings may be in error because of the relatively simple treatment of light ion production used by the semiempirical nuclear fragmentation model (Wilson, Townsend, and Badavi, 1987) in the GCR transport code and the paucity of relevant experimental data needed to improve that treatment. Since the fragmentation model used does conserve charge and mass, the dose-equivalent contributions from these ions will probably not change enough to alter the major conclusions. For example, increasing secondary proton and alpha production from heavy ions by an order of magnitude would only increase the dose-equivalent contribution from secondary protons by 1.26 cSv and from alpha particles by less than 0.1 cSv. Finally, the results presented in the tables neglect target-fragment contributions. For heavy ions, the target-fragment contributions (Shinn, Wilson, and Nealy, 1990) to the total dose equivalent is small (less than 5 percent of the heavy ion total dose equivalent for ions heavier than carbon) and of no consequence for this study. For protons, the target-fragment contribution is about 5 cSv, which will increase the relative contribution from protons to the total dose and dose equivalent but will not alter the major conclusions of this chapter.

11.4. Quality Factors

The quality factor (QF) as defined in International Commission on Radiological Protection publication no. 26 (Anon., 1977) or proposed in the International Commission on Radiological Units and Measurements report no. 40 (Anon., 1986b) is not expected to be a valid method for assessing biological risk for deep space missions in which the HZE particles of the GCR are of major concern. No human data for cancer induction from HZE particles exist, and information on biological effectiveness is expected to be taken from experiments with animals and cultured cells (Sinclair, 1985). Experiments with cultured cells (Yang et al., 1985; Thacker, Stretch, and Stephens, 1979; Wulf et al., 1985) indicate that the relative biological effectiveness of HZE particles is dependent on particle type, energy, and level of fluence. Use of a single parameter, such as LET or lineal energy (Katz and Cucinotta, 1991), to determine radiation quality will therefore represent an extreme oversimplification for GCR risk assessment.

Katz has presented a theoretical model that predicts the correct RBE behavior as observed in recent experimental studies by using track segment irradiations with heavy ions on cultured mammalian cells (Katz et al., 1971; Katz, Sharma, and Homayoonfar, 1972; Waligorski, Sinclair, and Katz, 1987). Cells at risk in deep space will be subject to a complicated mixture of particles varying in composition with the amount and type of shielding surrounding them. The fluence levels in space are such that a single cell will probably be exposed to only a few ion encounters over an extended period. Katz has developed a model for the ion-kill mode of cell death or neoplastic transformation that corresponds to low-fluence exposures. The delta-ray (energetic electrons produced in ion collisions) radial dose distribution that surrounds the ion track is assumed to initiate the biological damage, and the cell response to the radiation field is parameterized by using target theory and results from gamma-ray and track segment irradiations. The level of damage for a mixed radiation field is determined by the cellular response parameters and the local flux of particles. The deterministic transport code for calculating the differential flux of ions behind natural and protective radiation shielding exposed to the GCR spectrum is used to calculate the biological damage to mammalian cell cultures expected for 1 year in deep space at solar minimum behind various depths of aluminum shielding by using the Katz cellular damage model. Cell death and neoplastic transformations for C3H10T1/2 cells (cultured mouse cells) are considered for typical levels of spacecraft shielding. The Katz parameters are given in table 11.16. The results of this study must be considered preliminary in that the transport code is in an early stage of development as discussed in section 11.2.

Table 11.16. Cellular Response Parameters for C3H10T1/2 Cells

Damage type	m	D_o, cGy	Σ_o, cm^2	κ
Killing	3	280	5.0E−7	750
Transformation	2	26 000	1.15E−10	750

11.4.1. Katz model. The surviving fraction of a cellular population N, whose response parameters are m, D_o, Σ_o, and κ, after irradiation by a fluence of particles F is written as

$$\frac{N}{N_o} = \pi_i \times \pi_\gamma \tag{11.1}$$

where the ion-kill survivability is

$$\pi_i = \exp\left(-\Sigma F\right) \tag{11.2}$$

and the gamma-kill survivability is

$$\pi_\gamma = 1 - \left[1 - \exp\left(\frac{-D_\gamma}{D_o}\right)\right]^m \tag{11.3}$$

The gamma-kill dose fraction is

$$D_\gamma = (1 - P)D \tag{11.4}$$

where D is the absorbed dose. The single-particle inactivation cross section is given by

$$\Sigma = \Sigma_o \left\{1 - \exp\left[\frac{-(Z^*)^2}{\kappa\beta^2}\right]\right\}^m \tag{11.5}$$

where the effective charge number is

$$Z^* = Z \left[1 - \exp \left(\frac{-125\beta}{Z^{2/3}} \right) \right] \tag{11.6}$$

In the track-width regime, where $P > 0.98$, we take $P = 1$.

For neoplastic cell transformation, the fraction of transformed cells per surviving cell is

$$T = 1 - \frac{N'}{N_o'} \tag{11.7}$$

where N'/N_o' is the fraction of nontransformed cells and a set of cellular response parameters for transformations m', D_o', Σ_o', and κ' is used. The RBE at a given survival level is given by

$$\text{RBE} = \frac{D_X}{D} \tag{11.8}$$

where

$$D_X = -D_o \ln \left[1 - \left(1 - \frac{N}{N_o} \right)^{1/m} \right] \tag{11.9}$$

is the X-ray dose at which this level is obtained. Equations (11.1) through (11.9) represent the cellular track model for monoenergetic particles. Mixed radiation fields have been considered previously in the Katz (1986) model.

The cellular track model was applied to predict the fraction of C3H10T1/2 cells killed or transformed for 1 year in deep space at solar minimum for typical spacecraft shielding. The GCR environment was taken from the Naval Research Laboratory code (Adams, Silberberg, and Tsao, 1981). Aluminum shielding was considered with a local region of tissue for the cell cultures. Tables 11.17 and 11.18 contain individual particle fluences and absorbed doses, respectively, for the protons, α-particles, $Z = 3$ to 9 ions, and $Z = 10$ to 28 ions as determined by the Langley GCR code. Results for the fraction of C3H10T1/2 cells killed and transformed for 1 year at solar minimum are listed in tables 11.19 and 11.20, respectively. The gamma-kill mode was of negligible importance in the calculations, which indicates that biological damage in deep space from GCR particles at the cellular level will indeed result from the action of single particles. The importance of the target terms in biological effects for low-LET protons and α-particles is apparent. The results also indicate that the HZE component of the GCR spectrum is most damaging for small shielding depths. At large depths, the HZE components undergo many fragmentations; this causes proton buildup with increasing shield depth. At large depths, the protons (and neutrons) dominate the biological effects. In comparing individual charge components, we see that the particles with high Z have a reduced effectiveness for the transformation endpoint.

Also listed in tables 11.19 and 11.20 are the RBE versus depth values for the two endpoints. Table 11.21 presents the current RBE values beside the average QF's taken from Townsend et al. (1990a) with the same transport code. That the RBE and QF are nearly equal at small depths is coincidental. We note that the QF is independent of the fluence level, which is not true for the Katz model.

Table 11.17. Flux Year at Solar Minimum Behind Aluminum Shielding

| x, g/cm^2 | Flux, particles/cm^2/yr, from— | | | |
	Protons	α-particles	Low Z (a)	High Z (b)
0	1.29E8	1.24E7	1.09E6	3.0E5
1	1.31E8	1.21E7	1.05E6	2.8E5
2	1.33E8	1.18E7	1.01E6	2.7E5
3	1.34E8	1.15E7	0.98E6	2.5E5
5	1.36E8	1.10E7	0.91E6	2.2E5
10	1.40E8	0.97E7	0.77E6	1.7E5
20	1.43E8	0.77E7	0.57E6	1.1E5

[a] Z = 3 to 9 ions.
[b] Z = 10 to 28 ions.

Table 11.18. Dose for Solar Minimum Behind Aluminum Shielding

| x, g/cm^2 | Dose, cGy/yr, from— | | | | |
	Protons	α-particles	Low Z (a)	High Z (b)	Total
0	6.2	3.0	2.8	5.0	17.1
1	6.3	2.7	2.5	3.6	15.1
2	6.8	2.6	2.4	3.3	15.1
3	7.1	2.6	2.3	3.1	15.0
5	7.6	2.4	2.1	2.7	14.8
10	8.5	2.1	1.7	2.0	14.3
20	9.5	1.7	1.1	1.1	13.4

[a] Z = 3 to 9 ions.
[b] Z = 10 to 28 ions.

Table 11.19. Fraction of C3H10T1/2 Cells Killed in Deep Space for 1 Year at
Solar Minimum Behind Aluminum Shielding

			Fraction of cells killed by—			
x, g/cm^2	Protons	α-particles	Low Z (a)	High Z (b)	Total	RBE
Including Target Fragments						
0	1.35E−2	0.46E−2	0.57E−2	2.08E−2	4.46E−2	7.1
1	0.76E−2	0.15E−2	0.43E−2	1.84E−2	3.18E−2	7.0
2	0.80E−2	0.14E−2	0.41E−2	1.69E−2	3.04E−2	6.9
3	0.83E−2	0.14E−2	0.38E−2	1.55E−2	2.90E−2	6.8
5	0.88E−2	0.14E−2	0.34E−2	1.32E−2	2.68E−2	6.7
10	0.95E−2	0.12E−2	0.25E−2	0.91E−2	2.22E−2	6.5
20	1.02E−2	0.09E−2	0.15E−2	0.49E−2	1.74E−2	6.2
Without target fragments						
0	0.84E−2	0.37E−2	0.55E−2	2.08E−2	3.79E−2	6.7
1	0.24E−2	0.06E−2	0.41E−2	1.83E−2	2.54E−2	6.5
2	0.28E−2	0.06E−2	0.39E−2	1.68E−2	2.41E−2	6.3
3	0.31E−2	0.06E−2	0.37E−2	1.55E−2	2.27E−2	6.2
5	0.35E−2	0.06E−2	0.33E−2	1.31E−2	2.04E−2	6.1
10	0.42E−2	0.05E−2	0.24E−2	0.91E−2	1.61E−2	5.7
20	0.49E−2	0.04E−2	0.14E−2	0.48E−2	1.15E−2	5.3

[a] Z = 3 to 9 ions.
[b] Z = 10 to 28 ions.

Table 11.20. Fraction of C3H10T1/2 Cells Transformed in Deep Space for 1 Year at
Solar Minimum Behind Aluminum Shielding

			Fraction of cells transformed by—			
x, g/cm^2	Protons	α-particles	Low Z (a)	High Z (b)	Total	RBE
Including target fragments						
0	5.2E−6	2.0E−6	3.1E−6	7.5E−6	1.78E−5	6.4
1	3.5E−6	1.0E−6	2.7E−6	6.7E−6	1.39E−5	6.4
2	3.7E−6	1.0E−6	2.6E−6	6.2E−6	1.35E−5	6.3
3	3.9E−6	0.9E−6	2.4E−6	5.7E−6	1.29E−5	6.3
5	4.2E−6	0.9E−6	2.2E−6	4.9E−6	1.22E−5	6.2
10	4.7E−6	0.8E−6	1.7E−6	3.5E−6	1.06E−5	6.0
20	5.2E−6	0.6E−6	1.1E−6	2.0E−6	.88E−5	5.7
Without target fragments						
0	3.2E−6	1.6E−6	3.1E−6	7.5E−6	1.53E−5	6.0
1	1.4E−6	0.6E−6	2.7E−6	6.7E−6	1.13E−5	5.8
2	1.6E−6	0.6E−6	2.5E−6	6.2E−6	1.09E−5	5.7
3	1.8E−6	0.6E−6	2.4E−6	5.7E−6	1.05E−5	5.6
5	2.1E−6	0.5E−6	2.1E−6	4.9E−6	0.97E−5	5.4
10	2.5E−6	0.5E−6	1.6E−6	3.5E−6	0.82E−5	5.2
20	3.0E−6	0.4E−6	1.0E−6	2.0E−6	0.64E−5	4.9

[a] Z = 3 to 9 ions.
[b] Z = 10 to 28 ions.

Table 11.21. Quality Factors and RBE for Cell Death and Transformation

[1 Year in deep space at solar minimum for aluminum shielding]

x, g/cm^2	Quality factor (a)	RBE (cell kill)	RBE (transformation)
0	7.1	7.1	6.4
1	5.6	7.0	6.4
2	5.3	6.9	6.3
3	5.1	6.8	6.3
5	4.7	6.7	6.2
10	3.9	6.5	6.0
20	3.2	6.2	5.7

[a]ICRP 26 (Anon., 1977).

The Katz model indicates a substantial increase in risk relative to the ICRP 26 (Anon., 1977) quality factors for greater amounts of shielding (Cucinotta et al., 1991). The RBE's show a simple scaling with exposure time for the GCR particles as shown in equations (11.8), (11.9), and (11.2) in which ion kill dominates. Here we find for

$$\frac{N}{N_o} \approx 1 \tag{11.10}$$

with

$$\Sigma F << 1 \tag{11.11}$$

that

$$RBE = \frac{D_o}{LET}\sigma^{1/m}F^{(-1+1/m)} \tag{11.12}$$

Then scaling the RBE as a function of duration in deep space to the 1-year value RBE$_1$, we find for a duration period of τ with $F = n\tau$ that

$$RBE(\tau) = (\frac{\tau}{\tau_1})^{(-1+1/m)}RBE_1 \tag{11.13}$$

such that a one-hit ($m = 1$) system RBE becomes fluence independent

$$RBE(\tau) = RBE_1 \tag{11.14}$$

for a two-hit system ($m = 2$)

$$RBE(\tau) = \frac{RBE_1}{\sqrt{\tau/\tau_1}} \tag{11.15}$$

and for a three-hit system ($m = 3$)

$$RBE(\tau) = \frac{RBE_1}{(\tau/\tau_1)^{2/3}} \tag{11.16}$$

Results of this scaling approximation agree with calculations that use equations (11.1) through (11.10), as shown in table 11.22, whereby values obtained with the approximations of equation (11.13) are shown in parentheses as scaled from the 1-year RBE values taken from table 11.21; results of calculations are shown without parentheses. The extremely large RBE values that would be obtained for small values of τ are caused by the choice of energetic photons as the reference radiation.

Table 11.22. RBE for Cell Death and Cell Transformations of C3H10T1/2 Cells for GCR Spectrum at Solar Minimum Behind Aluminum Shielding

[Values in parentheses scaled from 1 year value by using eq. (11.16)]

x, g/cm^2	1 month	1 year	2 years
0	33.2 (37.0)	7.1	4.8 (4.6)
1	33.2 (36.1)	7.0	4.7 (4.5)
3	32.4 (35.1)	6.8	4.5 (4.3)

x, g/cm^2	1 month	1 year	2 years
0	22.3 (22.2)	6.4	4.6 (4.5)
1	22.0 (22.2)	6.4	4.5 (4.5)
3	21.6 (21.8)	6.3	4.4 (4.4)

11.4.2. Remarks. A track structure model has been used with a deterministic GCR transport code to predict the fractions of cell death and neoplastic transformations for C3H10T1/2 cells in deep space behind typical spacecraft shielding. Results indicate that the level of damage from the GCR particles does not attenuate appreciably for large amounts of spacecraft shielding and that single particles acting in the ion-kill mode dominate the effects. The contribution from target fragments was important in assessing the biological effect of protons and alpha particles. The RBE values obtained in this fluence-dependent model were more severe than the ICRP 26 quality factors. A simple scaling law with the duration time in space accounted for the change in RBE with fluence for the uniform GCR background.

The resulting average RBE of our calculations for both cell killing and transformation are remarkably close, when we consider the large difference in radiosensitivity parameters for these endpoints and the huge difference in the fraction of affected cells. About 1000 times as many cells are killed as are transformed. Nevertheless, 90 percent of the cells survive the conditions calculated here, and of these about 1 or 2 in 100 000 are transformed. Yet this is not an insignificant fraction when we consider the number of cells per cubic centimeter in tissue and speculate about the number of cells transformed by radiation that might lead to cancer.

The cell population in tissue, about $10^9/cm^3$, suggests that after 1 year of exposure to GCR at solar minimum there would be about 10^4 transformed cells/cm^3 in tissue if in vitro and in vivo transformation parameters were equal. Additionally, we do not know the minimum number of transformed cells that can be injected into a mouse to induce a cancer. Clearly, priority must be assigned to the investigation of these questions. If one or two transformed cells per cm^3 were to lead to cancer, as in leukemia, we could not tolerate an exposure in which the transformation fraction exceeded 10^{-9}.

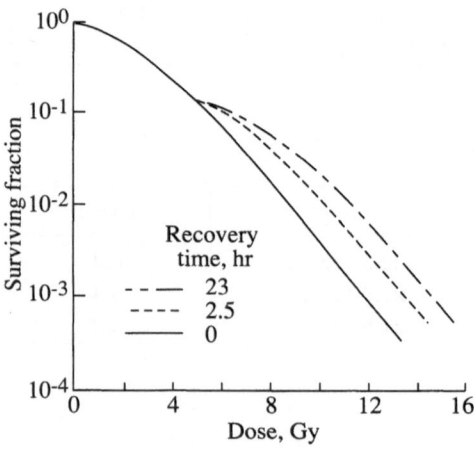

Figure 11.5. Fractional survival of cultured Chinese hamster cells for single exposure and exposure in two fractions (Elkind and Sutton, 1960).

11.5. Other Biological Effect Modifications

The rising RBE at low GCR dose results from the multitarget assumption in Katz theory that leads to the sigmoid behavior in the survival curve of low-LET radiation as opposed to the exponential relationship for high-LET radiation. The transition from sigmoid to exponential behavior is observed by Todd and Tobias (1974) to occur at 150 to 200 keV/μm for mammalian cells. Many also believe that the sigmoid behavior is related to repair mechanisms. This view is promoted by single-exposure and split-exposure experiments with a delay of 2.5 or 23 hours between fractions by using V79 hamster cells as shown in figure 11.5 (Elkind and Sutton, 1960). Repair is indicated by the sigmoid response of the second exposure after either the 2.5-hour repair period or especially the 23-hour repair period. Obviously, the RBE based on such a photon exposure protocol depends on the history of the radiation induced damage. Similar survival studies with confluent C3H10T1/2 mouse cells (G1) indicate no repair for this endpoint for high-LET radiations (Yang et al., 1986). As a result of operative repair mechanisms (sparing) for low-LET exposure (fig. 11.6) and the lack of repair for high-LET exposure (fig. 11.7), the corresponding RBE is dose rate dependent (Yang et al., 1986) as shown in figure 11.8. Also shown in figure 11.8 are the RBE values for neoplastic transformations. (Note, great liberty has been taken in connecting the data points). The increase in RBE at low dose rate is in part indicative of repair of the damage for low-LET radiation (fig. 11.9) but additional enhancement of high-LET exposure at low dose rate (presumably some misrepair mechanism) also contributes to cell transformations (fig. 11.10). If misrepair/repair plays a role, then this role should be observed in the delayed plating experiments of Yang et al. (1989) as shown in figure 11.11. Instead the delayed plating experiments show no transformation misrepair, but repair appears in cell survival data in distinction to the earlier low dose rate experiments with the same cell system.

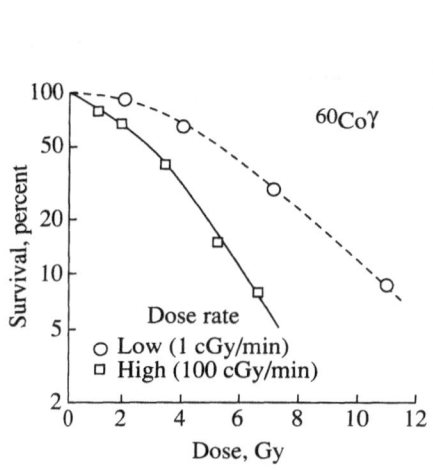

Figure 11.6. Survival fraction of confluent mouse cells at two dose rates displaying sparing at low dose rate (Yang, et al., 1986).

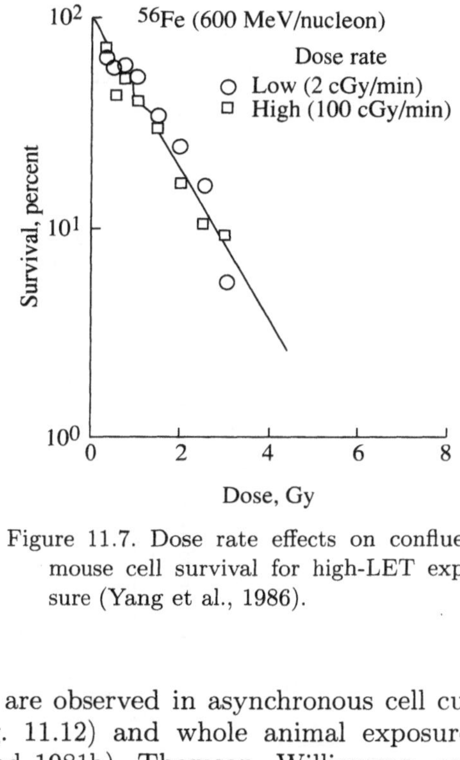

Figure 11.7. Dose rate effects on confluent mouse cell survival for high-LET exposure (Yang et al., 1986).

Similar dose rate enhancement effects are observed in asynchronous cell cultures by Hill et al. (1982 and 1985, fig. 11.12) and whole animal exposures as observed by Thomson et al. (1981a and 1981b), Thomson, Williamson, and Grahn (1983, 1985a, 1985b, and 1986), and Thomson and Grahn (1988 and 1989) (fig. 11.13). These effects are considered the result of cell cycle phenomena (Rossi and Kellerer, 1986; Brenner and Hall, 1990). The basic model assumes that some phases of the cell cycle are more affected by radiation exposure. This is clearly seen in the cell synchronous experiments which are shown in figure 11.14 (Sinclair, 1968). The model of dose rate enhancement assumes only one cell phase is effective in injury of only that fraction in the sensitive phase. At a later time, a different

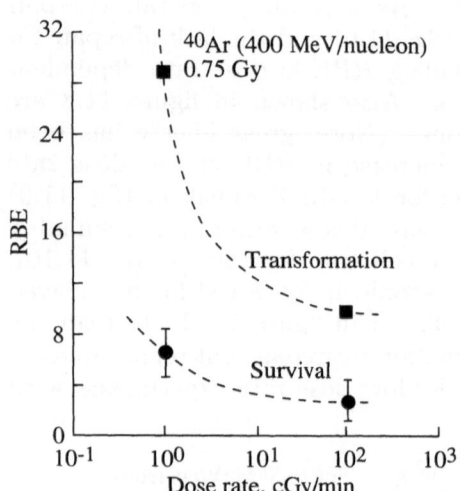

Figure 11.8. RBE as function of dose rate. The curves are to guide eye.

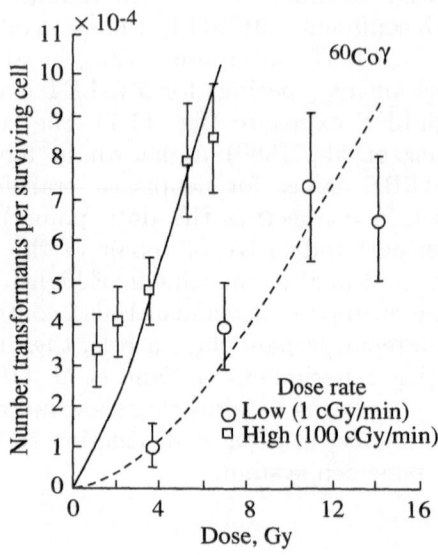

Figure 11.9. Repair processes for confluent mouse cell cultures exposed to γ-rays at low LET.

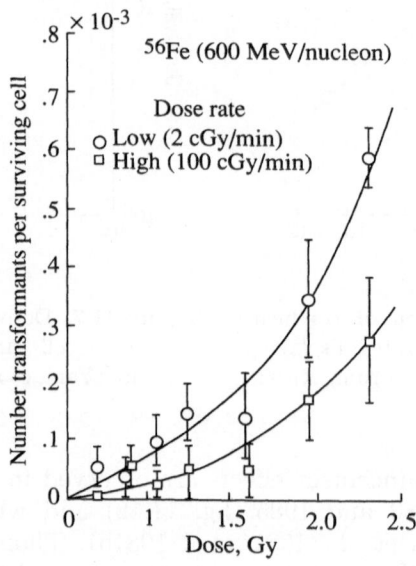

Figure 11.10. Cell transformation rate enhanced at low dose rate for high LET exposure with possible misrepair mechanism indicated.

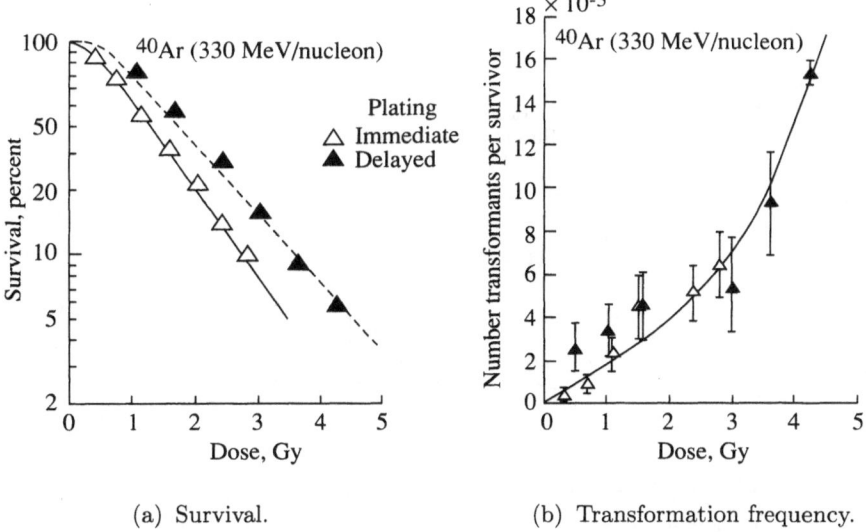

(a) Survival.

(b) Transformation frequency.

Figure 11.11. C3H10T1/2 cells irradiated by 330 MeV/u argon ions (Yang et al., 1986).

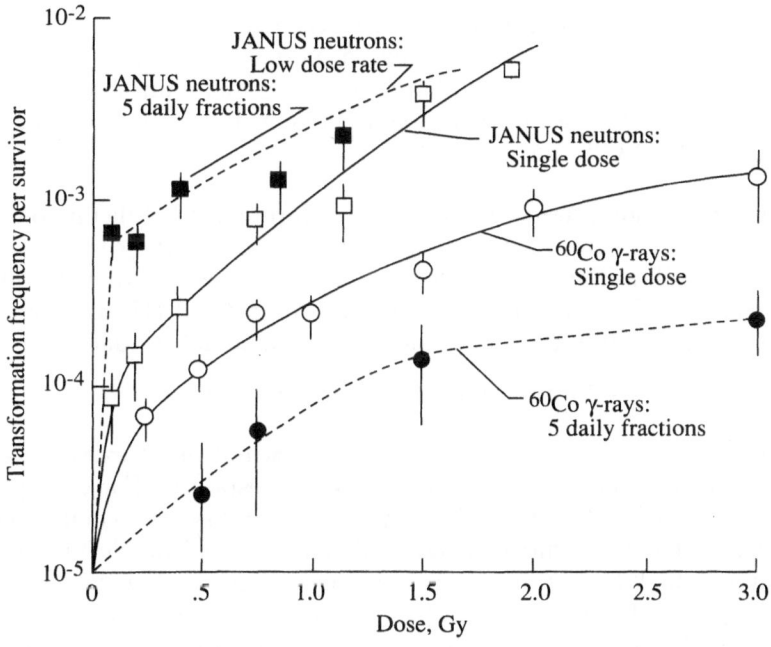

Figure 11.12. Transformation frequencies in C3H10T1/2 cells (Hill et al., 1982 and 1985).

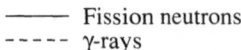

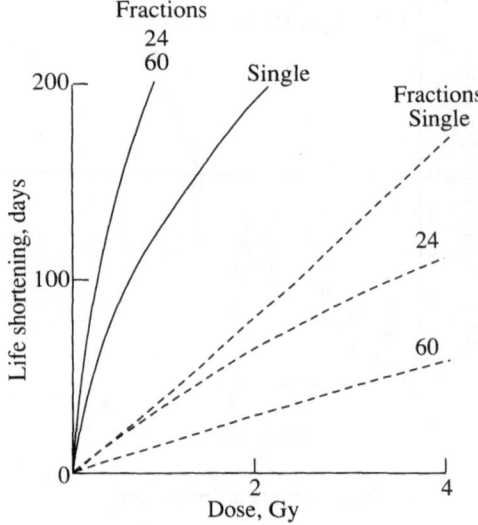

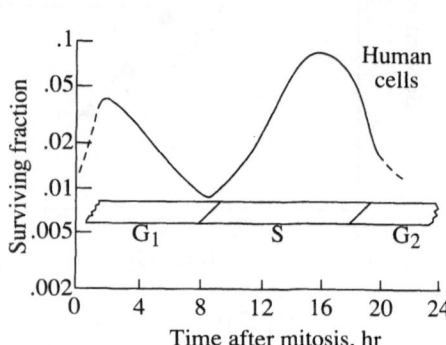

Figure 11.13. Life shortening in mice after single, 24, or 60 fractions of neutrons and after single, 24, or 60 fractions of γ-rays. Curves are fit to data of Thomson et al.

Figure 11.14. Survival of synchronized human cells exposed to 500 rads of 200 kV X-rays.

fraction of cells is in the appropriate phase thus providing two exposed groups of cells and an apparent enhancement. Such a model was exploited in the work of Brenner and Hall (1990). This explanation fails to explain the enhancement effects observed by Yang et al. (1986) in cell transformation in stationary phase (G1) confluent C3H10T1/2.

Clearly the risk to long-term GCR/SCR exposure will be difficult to evaluate because of the low dose rate, fractionated components, and the complex mixture of low- and high-LET radiations in space. Operative repair and cycle enhanced effects will require at least an intimate knowledge of the LET distributions at affected tissues or possibly more comprehensive track structure data.

11.6. Nuclear Models, Materials, and LET Spectra

As is clear from the previous discussion, the distribution of exposure components over LET is a primary indicator of biological response. For example, low-LET components allow certain biological repairs at low dose rates and a low RBE value, and high-LET components can show enhanced biological effects at low dose rates and generally high RBE values. There is clear evidence that the relative contributions to exposure from various LET components can be altered through the choice of shield material. The transmitted LET spectrum for an aluminum shield is shown in figure 11.15 with the transmitted LET spectrum for a liquid hydrogen shield in figure 11.16. Although a rather large shift in LET can be accomplished by choice of shield composition, an exact evaluation must await improved nuclear fragmentation cross sections since uncertainty in cross sections cause LET shifts of the same order of magnitude. These shifts can be seen by comparing LET spectra in the Earth's atmosphere by using the Bowman, Swiatecki, and Tsang (1973) fragmentation model shown in figure 11.17 with the LaRC fragmentation model (see chapter 5) shown in figure 11.18.

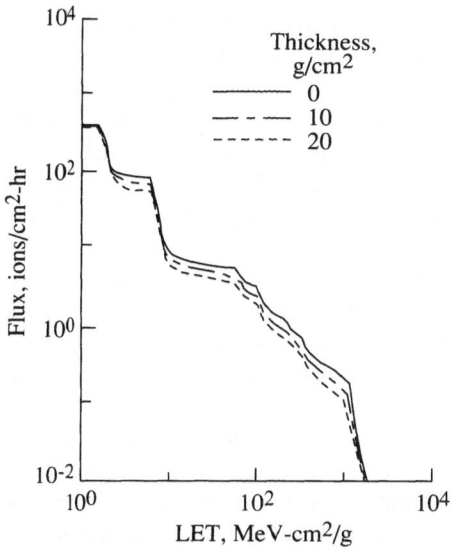

Figure 11.15. GCR integral LET spectra in aluminum for 30° orbit at altitude of 400 km.

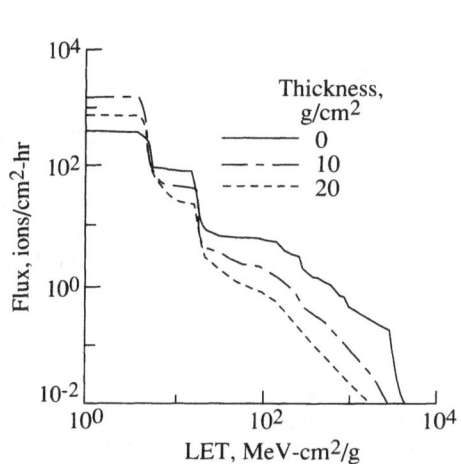

Figure 11.16. GCR integral LET spectra in hydrogen for 30° orbit at altitude of 400 km.

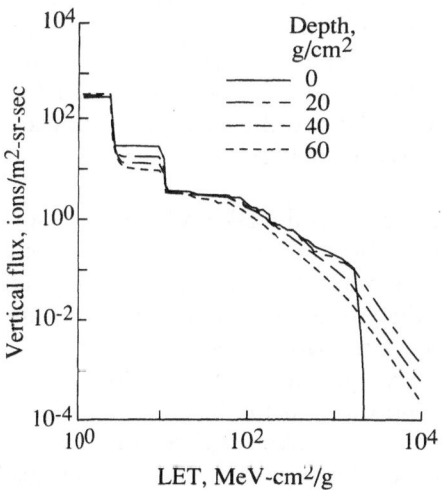

Figure 11.17. GCR integral LET spectra in Earth's atmosphere for fragmentation parameters of Bowman, Swiatecki, and Tsang, 1973.

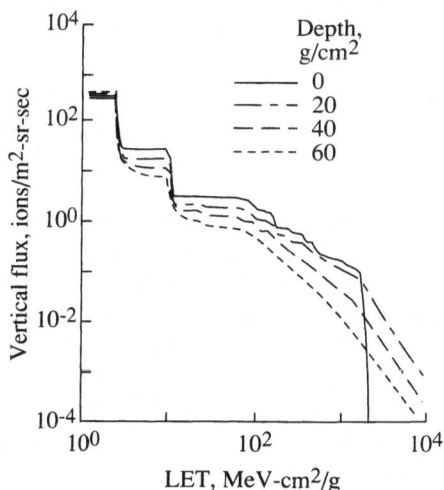

Figure 11.18. GCR integral LET spectra in Earth's atmosphere for fragmentation parameters of LaRC.

11.7. Human Geometry Factors

The significance of improving the accuracy (Shinn, Wilson, and Nealy, 1990) for predicting the dose and dose equivalent that astronauts will incur during future NASA missions has been demonstrated in several studies (Khandelwal and Wilson, 1974; Wilson and Denn, 1976; Townsend, Wilson, and Nealy, 1988; Nealy, Wilson, and Townsend, 1989). For example, Townsend, Wilson, and Nealy (1988) indicate that an increase of 20 percent in predicted BFO dose equivalent due to GCR's equates to a tripling of the required shield mass from 5 g/cm^2 to 16 g/cm^2 of water to meet the recommended annual BFO limit of 0.5 Sv. Large uncertainties are associated with the current dose estimate analysis, and every possible effort is needed to improve the accuracy to accomplish these missions in the most economical way without compromising the well-being of the astronauts.

One of the customary estimation practices that has been considered to be fairly reliable in the past is the use of the equivalent sphere model to obtain dose or dose equivalent to BFO. Langley and Billings (1972b) examined the feasibility of using a set of dosimetry spheres to monitor real-time organ doses received by astronauts under various space-radiation and vehicle conditions. They made comparisons between the doses calculated for the spheres and the detailed body geometry under a range of solar proton energy spectrum characteristics and also under various vehicle radiation-shielding thicknesses. The spectra were characterized by an assumed form described by Webber (1963). The optimal radii were determined for those spheres with the corresponding correlation constants that best represented the averaged organ doses under those assumed conditions. Although a moderate error of 18 percent for the correlation was found, one might question whether the accuracy will hold under less idealized particle spectral conditions.

11.7.1. Equivalent sphere model. This section reexamines the accuracy of the equivalent sphere model in approximating the BFO doses for more realistic conditions. The calculation made in a separate study (Simonsen et al., 1990) for the radiation transport through the atmosphere of Mars for the three largest solar flares observed in the last half century is extended here to include detailed BFO geometry. Comparisons are made for the dose equivalent to the various distributed BFO with the reported values based on the equivalent sphere model.

The Langley Research Center nucleon transport code BRYNTRN (Wilson et al., 1989) was used by Simonsen et al. (1990) to obtain dose and dose equivalent on the surface of Mars caused by large solar flares. The transport code was based on the straight ahead approximation, which reduces consideration to one-dimensional transport; the merits of this approximation have been discussed elsewhere (Alsmiller et al., 1965; Alsmiller, Irving, and Moran, 1968). An asymptotic expansion for the solution to the transport equation in two dimensions, subject to boundary conditions given for an arbitrary convex region, was derived by Wilson and Khandelwal (1974). The first term of the expansion was an accurate approximation of the dose and for the case of an isotropic proton fluence spectrum given by

$$D(\vec{x}) = 4\pi \int_0^\infty \phi(E) \int_0^\infty R(t, E) f_x(t) \; dt \; dE \qquad (11.17)$$

with

$$\int_0^\infty f_x(t) \; dt = 1 \tag{11.18}$$

where $R(t, E)$ is the fluence-to-dose conversion factor at the depth t for normal incidence protons on a slab and $f_x(t)$ is the areal density distribution function for the point $\vec{x}$. The quantity $f_x(t) \; dt$ is the fraction of the solid angle for which the distance to the surface from the point $\vec{x}$ lies between t and $t + dt$.

To simplify the computational task (that is, without making any change to the BRYNTRN code), equation (11.17) is rewritten as

$$D(\vec{x}) = 4\pi \int_0^\infty f_x(t) D_x(t) \; dt \tag{11.19}$$

with

$$D_x(t) = \int_0^\infty \phi(E) R(t, E) \; dE \tag{11.20}$$

where $D_x(t)$ is the dose or dose equivalent at depth t for normal incidence protons on a slab of tissue. With the areal density distribution function for BFO's given by the detailed geometry work described by Langley and Billings (1972a) and Billings and Yucker (1973), equation (11.19) can be calculated.

11.7.2. Results. The three solar flare spectra used for this study are those of February 1956, November 1960, and August 1972; however, Langley and Billings used a Webber (1963) form of integral spectra given by the inverse exponential of proton magnetic rigidity with a range of rigidity parameter P_o from 50 to 200 MV. Figure 11.19 shows these three flare spectra and the best fit to the earlier two events with the Webber form. The actual spectra, especially the high-energy range of the February 1956 event (Foelsche et al., 1974) are different from the analytical form of Webber.

The average dose equivalents at the surface of Mars caused by these three solar flare events are shown in figure 11.20 as a function of slab (water) thickness for the low density Mars atmosphere model (16 g/cm^2 CO$_2$ vertically) used in Simonsen et al. (1990). These average values of dose equivalent are obtained by summing the directional (anisotropic) dose equivalent over the solid angle and are used as $D_x(t)$ in this section. The calculated results from equation (11.19) are presented in table 11.23 for the five distributed compartments of the blood-forming organ. Also shown for comparison are the average BFO and 5-cm (water) depth dose equivalents.

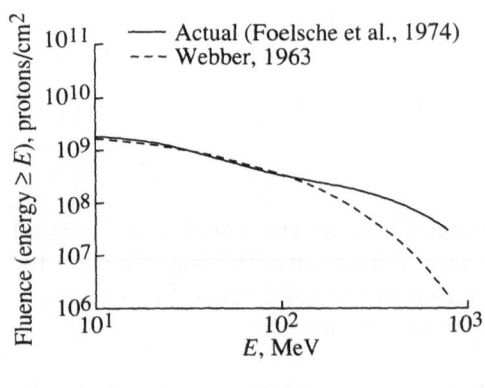

(a) February 1956 event.

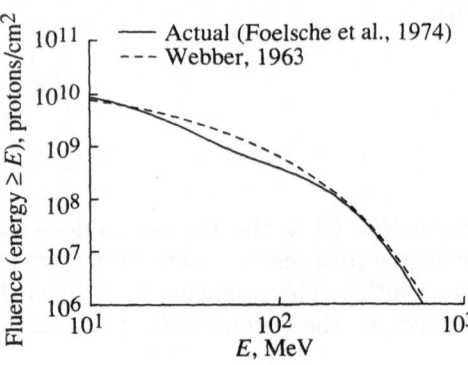

(b) November 1960 event.

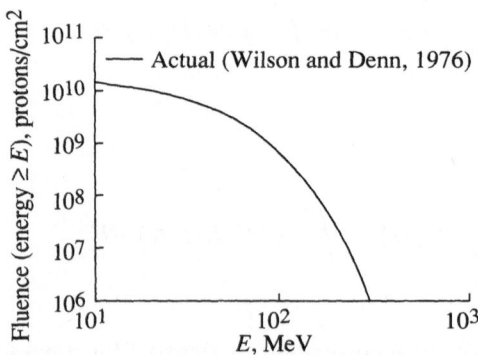

(c) August 1972 event. Webber spectrum fits actual event spectrum.

Figure 11.19. Fluence spectra for three major solar particle events.

It is customary (Space Science Board, 1970; Beck, Stokes, and Lushbaugh, 1972) to represent the average BFO exposure (dose or dose equivalent) with the 5-cm sphere based on the recommendation of the Space Science Board (1970). Conversely, the average BFO dose was found to be about half the 5-cm sphere dose in several analytical findings, such as the one from Langley and Billings (1972b). For the August 1972 event, the average BFO value for the detailed geometry (table 11.23) is fairly close (within 10 percent) to half of the value for a 5-cm sphere. However, the differences are larger for the other two flares, with 30 and 41 percent for November 1960 and February 1956 spectra, respectively. This wide discrepancy among these three events probably occurs because the two earlier flares contain more penetrating high-energy protons (fig. 11.19) and the actual spectra do not conform to the simple analytical form that Langley and Billings (1972b) used. Also, the 5-cm sphere dose is conservative for these three events.

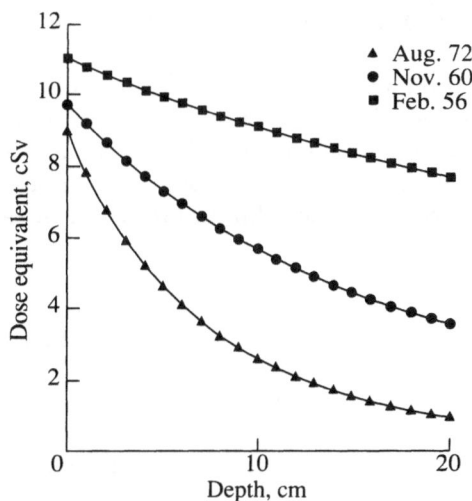

Figure 11.20. Dose equivalent at Mars surface as function of slab (water) thickness for low-density Mars atmosphere model.

Table 11.23. BFO Dose Equivalent on Mars Surface-Low Density Atmosphere Model

Solar flare event	Dose equivalents, cSv, for—						
	Arms	Legs	High trunk	Low trunk	Skull	Average BFO value	5-cm sphere
Feb. 1956	8.74	8.60	8.32	7.98	8.91	8.45	9.94
Nov. 1960	5.66	5.34	4.95	4.32	5.75	5.21	7.31
Aug. 1972	3.20	2.73	2.42	1.76	3.09	2.56	4.61

The equivalent sphere model of Billings and Langley is not accurate enough for precise, quantitative estimates of body doses and vehicle-shielding requirements in connection with future NASA mission studies. Furthermore, the 5-cm sphere dose recommended by the Space Science Board is always an overestimate and could lead to serious shielding penalties. This statement is based on the comparison made with the detailed body geometry calculation for BFO with actual solar flare spectra. The 5-cm equivalent sphere model of the BFO is shown to break down for more realistic spectra than the simple mathematical forms used in previous studies. Future works that involve actual exposure estimates or shield mass requirements should be extended to include all body geometries, including other critical organs, such as eyes, skin, and active BFO (Shinn, Wilson, and Nealy, 1990).

11.8. August 1972 Solar Particle Event Risk Assessment

At present no radiation exposure limits for astronauts on interplanetary missions have been established. However, it has been suggested that the dose equivalent limits recently recommended by the National Council on Radiation Protection and Measurements in their report 98 (Anon., 1989) be used as guidelines for planning purposes. These limits are listed in table 11.24.

Table 11.24. Ionizing Radiation Exposure Limits

[From NCRP 98 (Anon., 1989)]

Exposure interval	Dose equivalent, Sv, for—		
	Skin	Ocular lens	BFO
30 Days	1.5	1	0.25
Annual	3	2	0.50
Career	6	4	[a]1–4

[a]Varies with gender and age at initial exposure.

To assess the risk to astronauts on a mission outside the Earth's magnetosphere from the August 1972 solar particle event (SPE), the cumulative doses and dose equivalents as a function of time during the event were computed for the skin, ocular lens, and bone marrow. This event was chosen for analysis because it is one of the largest known SPE's and is the standard to which other events are compared. The integral proton fluence spectra as a function of energy and time are displayed in figure 11.21. These data, obtained from Interplanetary Monitoring Platform (IMP) satellite measurements during the SPE, were taken from figure 6 of Wilson and Denn (1976).

For purposes of analysis, two shielding thicknesses are considered: (1) a "nominal" spacecraft thickness of 2 g/cm^2 of aluminum and (2) a storm shelter shielded by 20 g/cm^2 of aluminum. The resultant skin doses and dose equivalents as a function of time are displayed in figure 11.22. The cumulative doses/dose equivalents for all organs are listed in table 11.25. For the nominal spacecraft, all limits in table 11.24, including the career limit, are exceeded by a substantial amount. For a total dose of about 11 Sv (1100 rem), skin erythema, and epilation (hair loss) are probable (Wilson and Denn, 1976). These are acute responses to the high exposure. If a storm shelter is provided, the skin dose equivalents are well below the recommended 30-day limit of 1.5 Sv and should pose no hazard to the crew.

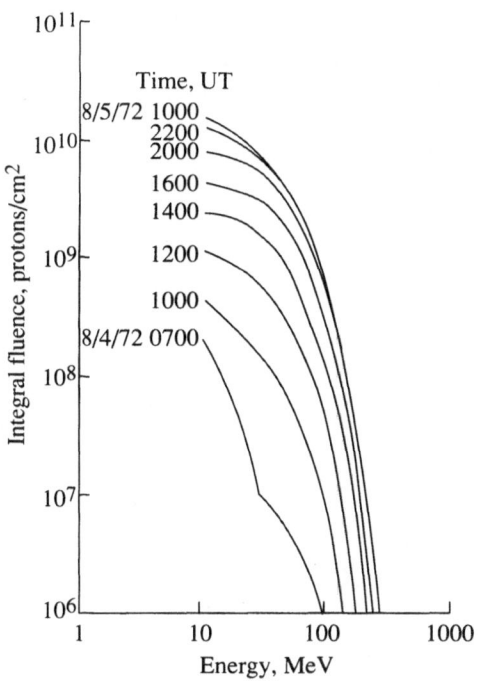

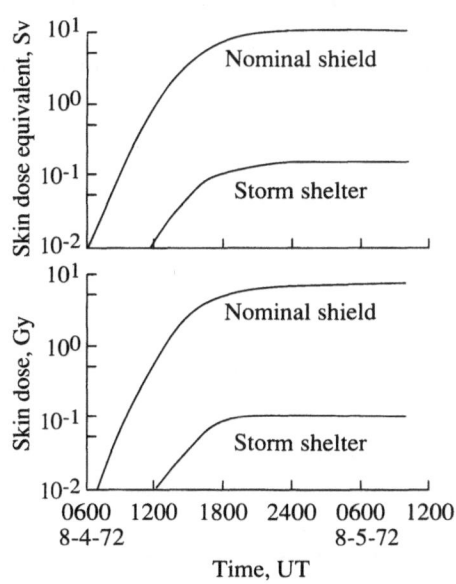

Figure 11.21. August 1972 solar particle event fluence as function of energy and time.

Figure 11.22. Skin dose and dose equivalent as function of time for August 1972 solar particle event. Results are displayed for nominal (2 g/cm² aluminum) and storm shelter (20 g/cm² aluminum) shielding configurations.

Table 11.25. Cumulative Doses and Dose Equivalents
for the August 1972 Solar Particle Event

Al shield thickness, g/cm²	Dose, Gy	Dose equivalent, Sv
Skin		
2	7.61	11.30
20	0.12	0.18
Ocular lens		
2	6.35	9.09
20	0.12	0.17
BFO		
2	0.92	1.24
20	0.04	0.07

Figure 11.23 displays the doses and dose equivalents to the ocular lens as a function of time. Again, for the nominal spacecraft, all limits in table 11.24 are significantly exceeded. Responses to an estimated dose equivalent to the eye in excess of 9 Sv include early erythema to the lid skin and an increased probability of cataract formation (Townsend et al., 1990a; Wilson and Denn, 1976). For the storm shelter configuration, the eye dose equivalent is well below the 30-day limit of 1.0 Sv.

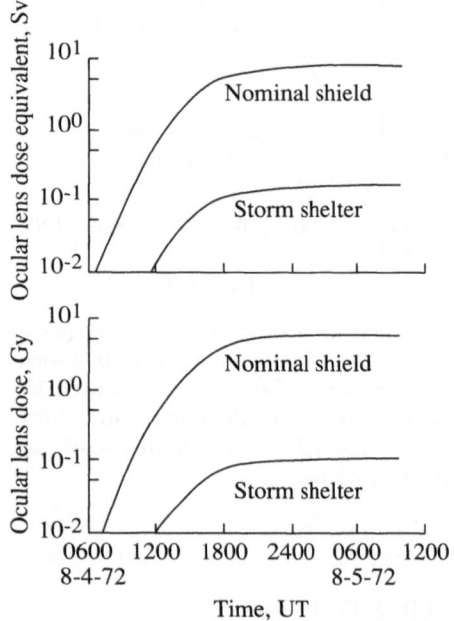

Figure 11.23. Ocular lens dose and dose equivalent as function of time for August 1972 solar particle event. Results are displayed for nominal (2 g/cm^2 aluminum) and storm shelter (20 g/cm^2 aluminum) shielding configurations.

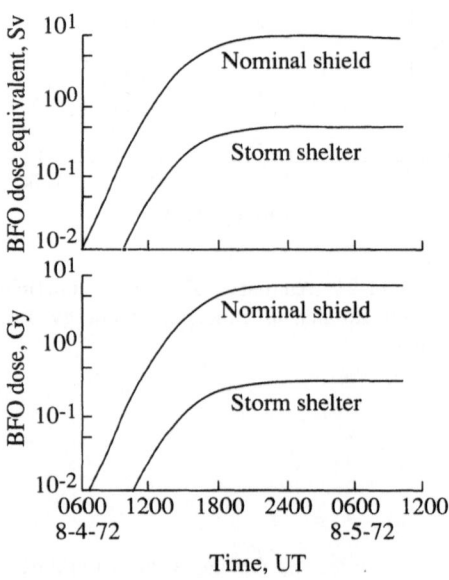

Figure 11.24. Bone marrow dose and dose equivalent as function of time for August 1972 solar particle event. Results are displayed for nominal (2 g/cm^2 aluminum) and storm shelter (20 g/cm^2 aluminum) shielding configurations.

Next, the doses and dose equivalents to the bone marrow, as a function of time, are displayed in figure 11.24. For the nominal spacecraft, the estimated dose equivalent of nearly 1.3 Sv is clinically significant. Blood-count changes will be detectable. Nausea and vomiting would be possible from damage to the intestinal lining (Wilson and Denn, 1976). For the storm shelter, the estimated dose equivalents are small (<70 mSv) and appear to pose no threat to the immediate health of the crew since they are well within the recommended 30-day limit.

Finally, mission directors need to recognize the critical nature of an SPE emergency and implement timely, appropriate, protective measures to ensure crew safety and health. To illustrate the rapidity with which such decisions may be required, table 11.26 presents an approximate time sequence of events beginning with the optical flare observation at 0621 UT on August 4, 1972. Radiation exposure references are to the nominal (2 g/cm^2 Al) configuration. Astronauts on extravehicular activity (EVA), who are essentially unshielded, would have considerably less time to seek shelter. During the peak intensity period (1400 UT to 1600 UT), the average dose equivalent rates were ≈ 1.5 Sv/hr (skin), ≈ 1.25 Sv/hr (eye), and ≈ 170 mSv/hr (bone marrow) behind nominal spacecraft shielding.

These dose equivalent rates are considerably lower than the usual rates used in radiotherapy of around 1 Gy/min. Repair processes in cells and tissues are known to have characteristic repair times of 30 to 120 minutes. Because the dose from this event will be received over a time period of about 12 to 18 hours, some damage that occurs early in the period will be repaired during the exposure period itself. Such a situation is known to decrease the resulting biological effect. Various theoretical approaches (Kellerer and Rossi, 1972; Thames, 1985; Curtis, 1986) have been developed to handle such a situation. A more extensive analysis, however, is beyond the scope of this section. For present purposes, and to be on the conservative side, we assume that the dose is acute (i.e., received fast enough so that repair during the exposure can be neglected).

Table 11.26. August 4, 1972, Event Sequence

Approximate time, UT	Event (Al shielding 2 g/cm^2 thick assumed)
0621	Optical flare observed
1300	30-day limit exceeded for skin and ocular lens
1400	30-day limit exceeded for BFO; annual limit exceeded for ocular lens
1500	Annual limit exceeded for skin
1600	Annual limit exceeded for BFO; career limit exceeded for ocular lens
1700	Career limit exceeded for skin

11.9. Hypothetical, Worst-Case SPE Scenario Results

Although the August 1972 SPE was the largest in terms of particle fluence, the overall spectrum was fairly soft (the fluence decreased rapidly with increasing particle energy) and therefore easily shielded. For more energetic events (with harder spectra), greater thicknesses of shielding are needed to achieve a significant percentage reduction in dose/dose equivalent. The SPE with the hardest spectrum to date was probably the February 1956 event. Fortunately, it had only one tenth the fluence of the August 1972 event and therefore probably would not have been mission or life threatening. Because the August 1972 SPE actually involved a series of large events over a period of several days (Wilson and Denn, 1976), it is interesting to hypothesize a possible combined August 1972 and February 1956 SPE scenario—with one of the events in the 1972 sequence having the hardness and fluence of the 1956 event (Townsend et al., 1990b). This combined spectrum is displayed in figure 11.25. The resultant doses and dose equivalents are listed in table 11.27. Acute effects expected with only nominal shielding include skin erythema, epilation, increased cataract formation in the lens, blood-count changes, nausea, and perhaps vomiting (NCRP 98 (Anon., 1989)). Again, quickly taking refuge in a storm shelter should provide adequate protection, although the 30-day limit for BFO exposure is exceeded.

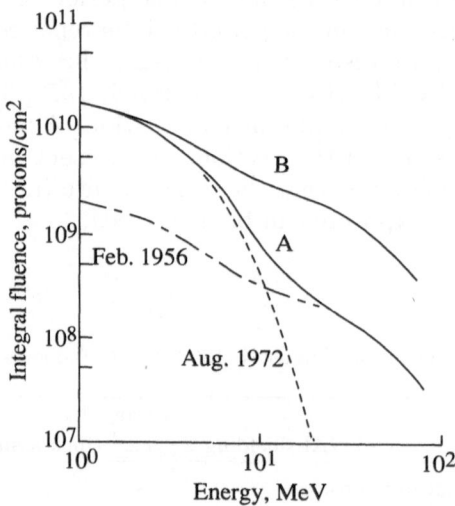

Figure 11.25. Fluence spectra for hypothetical worst-case solar particle events compared with February 1956 and August 1972 cases. Case A is combined August 1972 and February 1956 events. Case B is August 1972 event with February 1956 energy spectrum.

Another possible worst-case scenario would involve an event similar to the August 1972 event with the spectral hardness of the February 1956 event. The event fluence spectrum for this scenario is also depicted in figure 11.25. The resultant doses/dose equivalents for the nominal and storm shelter shielding chosen for this study are listed in table 11.28. For a nominal spacecraft, the effective whole body dose equivalent could be lethal. Even within a storm shelter, acute effects, such as skin erythema, vomiting, blood changes, and possibly even death, could occur since the bone marrow would receive over 2.6 Sv. However, such an event, or anything closely resembling it, has never been observed.

Table 11.27. Doses and Dose Equivalents for Combined
August 1972 and February 1956 Solar Particle Event

Al shield thickness, g/cm^2	Dose, Gy	Dose equivalent, Sv
Skin		
2	9.47	14.2
20	0.28	0.43
Ocular lens		
2	7.83	11.3
20	0.28	0.43
BFO		
2	1.18	1.64
20	0.20	0.31

Table 11.28. Doses and Dose Equivalents for 1972 Solar Particle
Event Fluence with a 1956 Solar Particle Event Spectral Hardness

Al shield thickness, g/cm^2	Dose, Gy	Dose equivalent, Sv
Skin		
2	0.3	15.5
20	1.99	3.02
Ocular lens		
2	8.95	13.0
20	2.00	3.04
BFO		
2	3.04	4.40
20	1.71	2.62

Using the coupled neutron-proton deterministic transport code, BRYNTRN, and the CAM model for the human geometry (Billings and Yucker, 1973), we have computed detailed exposure estimates for the bone marrow, ocular lens, and skin of astronauts on manned missions beyond the Earth's magnetosphere. Calculations were performed for crews protected by nominal (2 g/cm^2 aluminum) and heavily shielded (20 g/cm^2 aluminum) thicknesses of shielding for the August 1972 solar particle event, the largest ever recorded. We found that all current exposure limits, including career limits, would be exceeded for the ocular lens and skin if only nominal shielding is provided. For the bone marrow, 30-day and annual limits would also be exceeded. If the crew quickly sought refuge in a heavily shielded storm shelter, potential exposures could be maintained within current guidelines. Decisions to seek shelter must be made quickly, however, because limits are initially exceeded within 6 to 7 hours after the optical flare is observed on the solar disk. All exposure limits, except career BFO limits, would be exceeded within about 10 hours. Health effects from such exposures include skin erythema, epilation, blood changes, and cataract formation in the lens.

Two hypothetical worst-case SPE scenarios were also analyzed. The first scenario assumed a February 1956 event as a part of the sequence of flares that comprised the August 1972 event. For nominal spacecraft shielding, all exposure limits, including career BFO limits, would be exceeded. Acute health effects are identical to those previously mentioned for the August 1972 event. If a flare shelter is quickly used by the crew, only the 30-day limit for BFO would be exceeded. The time required to reach the shelter for such a high-energy event could be as short as 20 minutes after the optical flare is observed (Foelsche et al., 1974). The second scenario assumed the highly unlikely prospect that an SPE possessing the August 1972 fluence with the February 1956 spectral hardness could occur. For this hypothetical event, which has never been even approximately observed, the estimated crew exposures would be severe and possibly life threatening, even within a 20 g/cm^2 aluminum storm shelter.

11.10. Exposure of Female Breast

No regulatory dose limits are specifically assigned for the radiation exposure of female breasts during manned space flight. However, the relatively high radiosensitivity of the glandular tissue of the breasts and its potential exposure to solar-flare protons on short- and long-term missions mandate a priori estimation of the associated risks. In this section, a model for estimation of dose equivalent within the breast is developed to assess important exposure factors for future NASA missions.

The female breast and torso geometry is represented by a simple interim model. The proton dose-buildup factor procedure discussed in a previous chapter is used to estimate doses. A computer code has been developed that considers geomagnetic shielding, magnetic-storm conditions, spacecraft shielding, and body self-shielding. Inputs to the code include proton energy spectra, spacecraft orbital parameters, STS orbiter-shielding distribution at a given position, and a single parameter that allows for variation in breast size.

Virtually all breast cancers arise from the 15 to 20 glandular tissue lobes that exist within the connective tissue stroma. The stroma lies beneath a thin outer layer of skin and a subdermal layer of adipose tissue that is several millimeters to about 1 cm thick. Most breast cancers occur centrally and laterally in proportion to the amounts of glandular tissue in these volumes (NCRP 85 (Anon., 1986a)). The masses of the various types of tissue in the breast vary widely between individuals and with age.

11.10.1. Simplified breast geometry. We take as an interim geometry, a tissue-equivalent truncated sphere placed on a finite-tissue-equivalent slab (Shavers et al., 1991). The slab dimensions are taken as the mean dimensions of the trunk, and the dimensions for the truncated spheres are to be in accordance with individual geometry. Herein we consider small, medium, and large breast sizes by taking the sphere radius a to be taken as a parameter 7, 10, and 13 cm, respectively, with a base after truncation of 13.4, 19.6, and 25 cm, respectively; the $0g$ height (a–d) is then 5, 8, and 11 cm. The sensitive sites are assumed distributed and average exposure calculated along the axis of symmetry as a mean dose at depths of 0.5, 2, 4, and 7 cm. For convenience, we establish a spherical coordinate system centered at each dose point (t_o) and zenith outward along the axis of symmetry. The general geometry in any meridonal plane is shown in figure 11.26. The length l is a function of the azimuthal angle, ϕ. We neglect the shielding provided by the second breast. The chord length in the breast tissue is

$$\Delta_1(\theta) = \left[(a - t_o)^2 \cos^2 \theta + t_o(2\,a - t_o)\right]^{\frac{1}{2}} - (a - t_o) \cos \theta \qquad (11.21)$$

and the total chord length is

$$t(\theta) = \Delta_1(\theta) \qquad\qquad \left(0 \leq \theta \leq \frac{\pi}{2} + \varepsilon_1\right) \qquad (11.22)$$

where

$$\tan \varepsilon_1 = \frac{a - t_o - d}{l}$$

and θ is the polar angle. With the definition $\varepsilon = \theta - \frac{\pi}{2}$, the trunk chord length may be written in terms of the following function:

$$z = l \tan \varepsilon - (a - d - t_o) \qquad (11.23)$$

as

$$\Delta_2(\theta) = \left\{ \begin{array}{ll} 0 & (z \leq 0) \\ \frac{z}{\sin \varepsilon} & (0 \leq z \leq t) \\ \frac{t}{\sin \varepsilon} & (\text{Otherwise}) \end{array} \right\} \qquad (11.24)$$

The total thickness function is then

$$t(\theta) = \Delta_1(\theta) + \Delta_2(\theta) \qquad\qquad \left(0 \leq \theta \leq \frac{\pi}{2} + \varepsilon_2\right) \qquad (11.25)$$

where

$$\tan \varepsilon_2 = \frac{a - t_o - d}{\sqrt{a^2 - d^2}}$$

For larger values of θ,

$$t(\theta) = \left\{ \begin{array}{ll} \frac{z + a - d - t_o}{\sin \varepsilon} & (0 \leq z \leq t) \\ \frac{t + a - d - t_o}{\sin \varepsilon} & (t \leq z) \end{array} \right\} \qquad (11.26)$$

The azimuthal values of l will have local minima at $\phi = 0, \pi, \frac{\pi}{2}$, and $\frac{3}{2}\pi$ corresponding to l_1, l_2, l_3, and l_4. There are four boundaries in ϕ separating the local minima and related through

$$\frac{l_1}{\cos \phi_1} = \frac{l_2}{\cos \left(\frac{\pi}{2} - \phi_1\right)} \qquad (11.27)$$

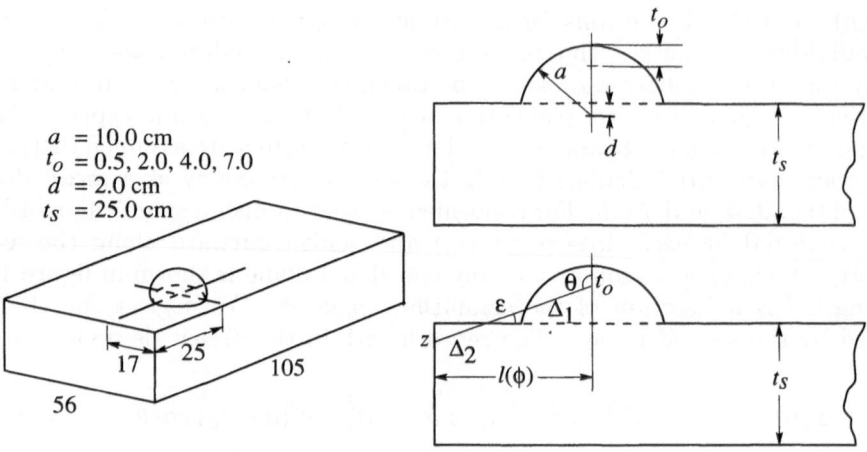

$a = 10.0$ cm
$t_o = 0.5, 2.0, 4.0, 7.0$
$d = 2.0$ cm
$t_S = 25.0$ cm

Figure 11.26. Breast geometry in meridional plane showing basic variables.

$$\frac{l_2}{\cos\left(\phi_2 - \frac{\pi}{2}\right)} = \frac{l_3}{\cos\left(\pi - \phi_2\right)} \tag{11.28}$$

$$\frac{l_3}{\cos\left(\phi_3 - \pi\right)} = \frac{l_4}{\cos\left(\frac{3}{2}\pi - \phi_3\right)} \tag{11.29}$$

$$\frac{l_4}{\cos\left(\phi_4 - \frac{3}{2}\pi\right)} = \frac{l_1}{\cos(2\pi - \phi_4)} \tag{11.30}$$

Then l as a function of azimuth is

$$l(\phi) = l_i \Big/ \cos\left[\phi - (i-1)\frac{\pi}{2}\right] \quad (\phi_{i-1} \le \phi \le \phi_1) \tag{11.31}$$

where $\phi_o \equiv \phi_4 - 2\pi$. The total chord length is then given by $t(\theta, \phi)$ or $t(\Omega)$, where numerical values are given by equations (11.21) through (11.31). Approximate values of l_i are 17, 80, 39, and 25 cm for which

$$\phi_1 \to \phi_4 = \{78°, 116°, 212.7°, 304.2°\}$$

The thickness t_s is taken as 25 cm. The dose response is given as usual by

$$R_B(E) = \int d\Omega \; R[E, t(\Omega)] \tag{11.32}$$

The present method has been incorporated into a computer code written for the Shuttle geometry (Wilson et al., 1990) for use in future mission analyses.

11.10.2. Results. The solar event of February 1956 was a large, high-energy event in which energetic particles up to several GeV were observed. As a relativistic particle event, the ground-level neutron monitor onset started about 20 minutes after the optical flare and peaked 20 minutes after onset as shown in figure 11.27. The intensities decayed in 2 to 3 hours after the event. In contrast, the event of August 1972 was a relatively soft spectrum but of high intensity. Onset was 4 hours with peak intensities reached a few hours after onset followed by slow decay over the next dozen hours (table 11.26). In terms of high-energy intensity and total proton fluence, these two events bracket most other large events.

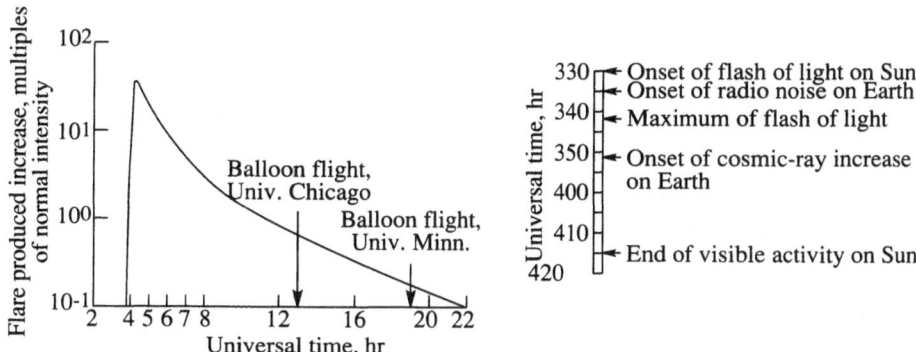

Figure 11.27. Cosmic-ray neutron surge at sea level during large solar event of Feb. 23, 1956.

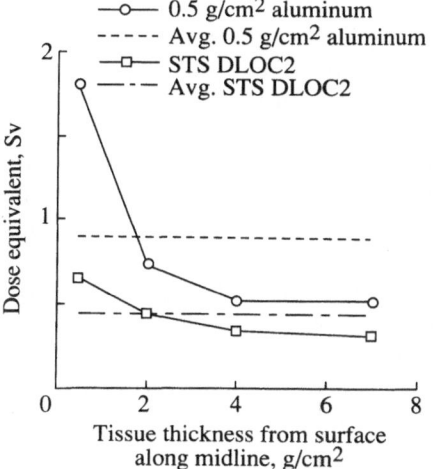

Figure 11.28. Dose equivalent exposure along breast centerline for Feb. 23, 1956.

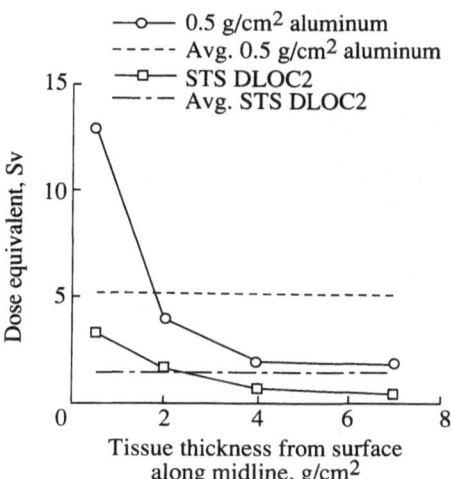

Figure 11.29. Dose equivalent exposure along breast centerline for Aug. 4, 1972.

A dose for a medium-size breast was calculated for these two events with two standard shield configurations. The first shield was an aluminum shield 0.5 g/cm^2 thick, which is representative of a hardened spacesuit, and the second was the least shielded region in the STS, which is representative of typical spacecraft shielding without the use of a storm shelter. The concern here is not so much the overall exposure, which varies greatly from event to event, but rather the dose distribution which may be important in assessing the exposure. Results are shown in figures 11.28 and 11.29. These results clearly show that the large variations in exposure occur over a large volume of breast tissue for either event for the aluminum shield 0.5 g/cm^2 thick. Even for a typical spacecraft configuration, large dose gradients exist within the breast tissues for the softer solar flare spectra.

The exposure is expected to be fairly uniform within a spacecraft for a high-energy spectrum that resembles the spectrum of the February 1956 event.

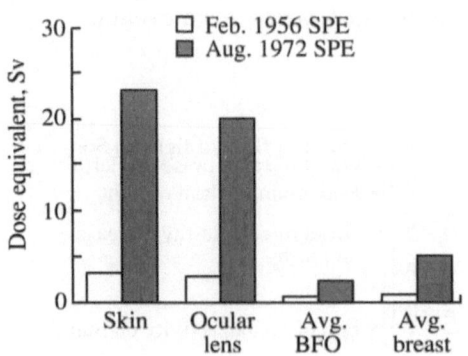

Figure 11.30. Organ dose equivalent for hardened space suit for two solar events.

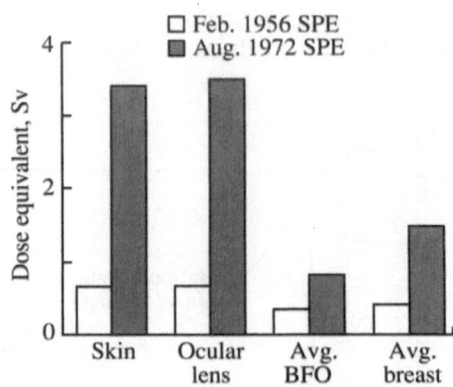

Figure 11.31. Organ dose equivalent for typical space vehicle for two solar events.

The average breast exposure and exposure values for other critical organs are presented in figures 11.30 and 11.31. A comparison of figures 11.30 and 11.31 clearly shows that the average breast exposure may be twice the exposure of the blood-forming organs, especially for low-energy, solar particle event spectra. This is a potentially important factor in the overall exposure budget.

Qualitatively, no great differences were observed in the dose gradient along the axis of symmetry for the three breast sizes; therefore average doses for the three breast sizes will be within 10 percent of the average among the sizes. This result occurs because dose distribution depends on the radius of curvature (Wilson and Khandelwal, 1974), which varies slowly with breast size. This result greatly simplifies the monitoring of individual exposure because the 10-cm radius model should provide adequate values for all. This is especially true if the astronauts are located inside the vehicle where breast-size effects are entirely negligible. However, even for extravehicular activity in heavy space suits (0.5 g/cm^2), this is a reasonably accurate approximation.

Exposure estimates for the female breast in future space missions can be made on the basis of a fixed, typical breast size ($a = 10$ cm). One should remember that dose variations within the sensitive volume can be large (a factor of 2 to 3), although breast size does not appear to be a sensitive factor. Further work in assessing the importance of this large dose variation should be made.

11.11. References

Adams, J. H., Jr.; Silberberg, R.; and Tsao, C. H., 1981: *Cosmic Ray Effects on Microelectronics. Part I—The Near-Earth Particle Environment.* NRL Memo. Rep. 4506-Pt. I, U.S. Navy. (Available from DTIC as AD A103 897.)

Adams, James H., Jr., 1987: *Cosmic Ray Effects on Microelectronics,* Part IV. NRL Memo. Rep. 5901 (Revised), Naval Research Lab.

Alsmiller, R. G., Jr.; Irving, D. C.; Kinney, W. E.; and Moran, H. S., 1965: The Validity of the Straightahead Approximation in Space Vehicle Shielding Studies. *Second Symposium on Protection Against Radiations in Space,* Arthur Reetz, Jr., ed., NASA SP-71, pp. 177–181.

Alsmiller, R. G., Jr.; Irving, D. C.; and Moran, H. S., 1968: Validity of the Straightahead Approximation in Space-Vehicle Shielding Studies, Part II. *Nuclear Sci. & Eng.,* vol. 32, no. 1, pp. 56–61.

Anon., 1977: *Recommendations of the International Commission on Radiological Protection.* ICRP Publ. 26, Pergamon Press.

Anon., 1986a: *Mammography—A User's Guide, Recommendations of the National Council on Radiation Protection and Measurements.* NCRP Rep. No. 85.

Anon., 1986b: *The Quality Factor in Radiation Protection.* ICRU Rep. 40, International Commission on Radiation Units and Measurements.

Anon., 1989: *Guidance on Radiation Received in Space Activities.* NCRP Rep. No. 98, National Council on Radiation Protection and Measurements.

Beck, W. L.; Stokes, T. R.; and Lushbaugh, C. C., 1972: Dosimetry for Radiobiological Studies of the Human Hematopoietic System. *Proceedings of the National Symposium on Natural and Manmade Radiation in Space,* E. A. Warman, ed., NASA TM X-2440, pp. 974–981.

Billings, M. P.; and Yucker, W. R., 1973: *The Computerized Anatomical Man (CAM) Model.* NASA CR-134043.

Bowman, J. D.; Swiatecki, W. J.; and Tsang, C. F., 1973: *Abrasion and Ablation of Heavy Ions.* LBL-2908, Lawrence Berkeley Lab., Univ. of California.

Brenner, D. J.; and Hall, E. J., 1990: The Inverse Dose-Rate Effect for Oncogenic Transformation by Neutrons and Charged Particles: A Plausible Interpretation Consistent With Published Data. *Int. J. Radiat. & Biol.* vol. 58, no. 5, pp. 745–758.

Cucinotta, Francis A.; Katz, Robert; Wilson, John W.; Townsend, Lawrence W.; Nealy, John E.; Shinn, Judy L., 1991: *Cellular Track Model of Biological Damage to Mammalian Cell Cultures From Galactic Cosmic Rays.* NASA TP-3055.

Curtis, Stanley B., 1986: Lethal and Potentially Lethal Lesions Induced by Radiation—A Unified Repair Model. *Radiat. Res.,* vol. 106, pp. 252–270.

Elkind, M. M.; and Sutton, Harriet, 1960: Radiation Response of Mammalian Cells Grown in Culture. I. Repair of X-Ray Damage in Surviving Chinese Hamster Cells. *Radiat. Res.,* vol. 13, nos. 1–6, pp. 556–593.

Foelsche, Trutz; Mendell, Rosalind, B.; Wilson, John W.; and Adams, Richard R., 1974: *Measured and Calculated Neutron Spectra and Dose Equivalent Rates at High Altitudes; Relevance to SST Operations and Space Research.* NASA TN D-7715.

Fritz-Niggli, Hedi, 1988: The Role of Repair Processes in Cellular and Genetical Response to Radiation. *Terrestrial Space Radiation and Its Biological Effects,* Percival D. McCormack, Charles E. Swenberg, and Horst Bücker, eds., Plenum Press, pp. 213–235.

Hill, C. K.; Buonaguro, F. M.; Myers, C. P.; Han, A.; and Elkind, M. M., 1982: Fission-Spectrum Neutrons at Reduced Dose Rates Enhance Neoplastic Transformation. *Nature,* vol. 298, pp. 67–69.

Hill, C. K.; Carnes, B. A.; Han, A.; and Elkind, M. M., 1985: Neoplastic Transformation Is Enhanced by Multiple Low Doses of Fission-Spectrum Neutrons. *Radiat. Res.,* vol. 102, pp. 404–410.

Katz, R.; Ackerson, R.; Homayoonfar, M.; and Sharma, S. C., 1971: Inactivation of Cells by Heavy Ion Bombardment. *Radiat. Res.,* vol. 47, pp. 402–425.

Katz, Robert; Sharma, S. C.; and Homayoonfar, M., 1972: The Structure of Particle Tracks. *Topics in Radiation Dosimetry,* Supplement 1, F. H Attix, ed., Academic Press, Inc., pp. 317–383.

Katz, Robert, 1986: Biological Effects of Heavy Ions From the Standpoint of Target Theory. *Adv. Space Res.,* vol. 6, no. 11, pp. 191–198.

Katz, Robert; and Cucinotta, F. A., 1991: RBE vs. Dose for Low Doses of High-LET Radiations. *Health Phys.,* vol. 60, no. 5, pp. 717–718.

Kellerer, Albrecht M.; and Rossi, Harald H., 1972: The Theory of Dual Radiation Action. *Current Top. Radiat. Res. Q.,* vol. 8, no. 2, pp. 85–158.

Kovalev, E. E.; Muratova, I. A.; and Petrov, V. M., 1989: Studies of the Radiation Environment Aboard Prognoz Satellites. *Nuclear Tracks & Radiat. Meas.,* vol. 16, no. 1, pp. 45–48.

Khandelwal, G. S.; and Wilson, John W., 1974: *Proton Tissue Dose for the Blood Forming Organ in Human Geometry: Isotropic Radiation.* NASA TM X-3089.

Langley, R. W.; and Billings, M. P., 1972a: Methods of Space Radiation Dose Analysis With Applications to Manned Space Systems. *Proceedings of the National Symposium on Natural and Manmade Radiation in Space,* E. A. Warman, ed., NASA TM X-2440, pp. 108–116.

Langley, R. W.; and Billings, M. P., 1972b: A New Model for Estimating Space Proton Dose to Body Organs. *Nuclear Technol.,* vol. 15, no. 1, pp. 68–74.

Nealy, John E.; Wilson, John W.; and Townsend, Lawrence W., 1989: Preliminary Analyses of Space Radiation Protection for Lunar Base Surface Systems. SAE Tech. Paper Ser. 891487.

Rossi, H. H.; and Kellerer, A. M., 1986: The Dose Rate Dependence of Oncogenic Transformation by Neutrons May Be Due to Variation of Response During the Cell Cycle. *Int. J. Radiat. & Biol.,* vol. 50, no. 2, pp. 353–361.

Scott, B. R.; and Ainsworth, E. J., 1980: State-Vector Model for Life Shortening in Mice After Brief Exposures to Low Doses of Ionizing Radiation. *Math. Biosci.,* vol. 49, pp. 185–205.

Shavers, Mark; Poston, John W.; Atwell, William; Hardy, Alva C.; and Wilson, John W., 1991: *Preliminary Calculation of Solar Cosmic Ray Dose to the Female Breast in Space Missions.* NASA TM-4235.

Shinn, Judy L.; Wilson, John W.; and Nealy, John E., 1990: *Reliability of Equivalent Sphere Model in Blood-Forming Organ Dose Estimation.* NASA TM-4178.

Simonsen, Lisa C.; Nealy, John E.; Townsend, Lawrence W.; and Wilson, John, W., 1990: *Radiation Exposure for Manned Mars Surface Missions.* NASA TP-2979.

Simpson, J. A., 1983: Elemental and Isotopic Composition of the Galactic Cosmic Rays. *Annual Review of Nuclear and Particle Science*, Volume 33, J. D. Jackson, Harry E. Gove, and Roy F. Schwitters, eds., Annual Reviews Inc., pp. 323–381.

Sinclair, Warren K., 1968: Cyclic X-Ray Responses in Mammalian Cells In Vitro. *Radiat. Res.*, vol. 33, pp. 620–643.

Sinclair, W. K., 1985: Experimental RBE Values of High LET Radiations at Low Doses and the Implications for Quality Factor Assignment. *Radiat. Prot. Dosim.*, vol. 13, no. 1–4, pp. 319–326.

Space Science Board, 1970: *Radiation Protection Guides and Constraints for Space-Mission and Vehicle-Design Studies Involving Nuclear Systems.* National Academy of Sciences, National Research Council.

Swenberg, Charles E.; Holwitt, Eric A.; and Speicher, James M., 1990: Superhelicity and DNA Radiation Sensitivity. SAE Tech. Paper Ser. 901349.

Thacker, John; Stretch, Albert; and Stephens, Miriam A., 1979: Mutation and Inactivation of Cultured Mammalian Cells Exposed to Beams of Accelerated Heavy Ions. II. Chinese Hamster V79 Cells. *Int. J. Biol.*, vol. 36, no. 2, pp. 137–148.

Thames, Howard D., 1985: 'An Incomplete-Repair' Model for Survival After Fractionated and Continuous Irradiations, *Int. J. Radiat. & Biol.*, vol. 47, pp. 317–339.

Thomson, John F.; Williamson, Frank S.; Grahn, Douglas; and Ainsworth, E. John, 1981a: Life Shortening in Mice Exposed to Fission Neutrons and γ Rays. I. Single and Short-Term Fractionated Exposures. *Radiat. Res.*, vol. 86, pp. 559–572.

Thomson, John F.; Williamson, Frank S.; Grahn, Douglas; and Ainsworth, E. John, 1981b: Life Shortening in Mice Exposed to Fission Neutrons and γ Rays. II. Duration-of-Life and Long-Term Fractionated Exposures. *Radiat. Res.*, vol. 86, pp. 573–579.

Thomson, John F.; Williamson, Frank S.; and Grahn, D., 1983: Life Shortening in Mice Exposed to Fission Neutrons and γ Rays. III. Neutron Exposures of 5 and 10 Rad. *Radiat. Res.*, vol. 93, pp. 205–209.

Thomson, John F.; Williamson, Frank S.; and Grahn, Douglas, 1985a: Life Shortening in Mice Exposed to Fission Neutrons and γ Rays. IV. Further Studies With Fractionated Neutron Exposures. *Radiat. Res.*, vol. 103, pp. 77–88.

Thomson, John F.; Williamson, Frank S.; and Grahn, Douglas, 1985b: Life Shortening in Mice Exposed to Fission Neutrons and γ Rays. V. Further Studies With Single Low Doses. *Radiat. Res.*, vol. 104, pp. 420–428.

Thomson, John F.; Williamson, Frank S.; and Grahn, Douglas, 1986: Life Shortening in Mice Exposed to Fission Neutrons and γ Rays. VI. Studies With the White-Footed Mouse, Peromyscus Leucopus. *Radiat. Res.*, vol. 108, pp. 176–188.

Thomson, John F.; and Grahn, Douglas, 1988: Life Shortening in Mice Exposed to Fission Neutrons and γ Rays. VII. Effects of 60 Once-Weekly Exposures. *Radiat. Res.*, vol. 115, pp. 347–360.

Thomson, John F.; and Grahn, Douglas, 1989: Life Shortening in Mice Exposed to Fission Neutrons and γ Rays. VIII. Exposures to Continuous γ Radiation. *Radiat. Res.*, vol. 118, pp. 151–160.

Todd, Paul; and Tobias, Cornelius A., 1974: Cellular Radiation Biology. *Space Radiation Biology and Related Topics*, Cornelius A. Tobias and Paul Todd, eds., Academic Press, Inc., pp. 141–196.

Townsend, L. W.; and Wilson, J. W., 1988: An Evaluation of Energy-Independent Heavy Ion Transport Coefficient Approximations. *Health Phys.*, vol. 54, no. 4, pp. 409–412.

Townsend, Lawrence W.; Wilson, John W.; and Nealy, John E., 1988: *Preliminary Estimates of Galactic Ray Shielding Requirements for Manned Interplanetary Missions*. NASA TM-101516.

Townsend, Lawrence W.; Wilson, John W.; and Nealy, John E., 1989: Space Radiation Shielding Strategies and Requirements for Deep Space Missions. SAE Tech. Paper Ser. 891433.

Townsend, Lawrence W.; Nealy, John E.; Wilson, John W.; and Simonsen, Lisa C., 1990a: *Estimates of Galactic Cosmic Ray Shielding Requirements During Solar Minimum*. NASA TM-4167.

Townsend, L. W.; Wilson, J. W.; Shinn, J. L.; and Curtis, S. B., 1990b: Human Exposure to Large Solar Particle Events in Space. Paper presented at the 28th Plenary Meeting of COSPAR (The Hague, The Netherlands).

Waligorski, M. P. R.; Sinclair, G. L.; and Katz, R., 1987: Radiosensitivity Parameters for Neoplastic Transformations in C3HT10T1/2 Cells. *Radiation Res.*, vol. 111, pp. 424–437.

Webber, W. R., 1963: *An Evaluation of the Radiation Hazard Due to Solar-Particle Events*. D2-90469, Aero-Space Div., Boeing Co.

Wilson, John W.; and Khandelwal, G. S., 1974: Proton Dose Approximation in Arbitrary Convex Geometry. *Nuclear Technol.*, vol. 23, no. 3, pp. 298–305.

Wilson, John W; and Denn, Fred M., 1976: *Preliminary Analysis of the Implications of Natural Radiations on Geostationary Operations*. NASA TN D-8290.

Wilson, John W.; Townsend, L. W.; Bidasaria, H. B.; Schimmerling, Walter; Wong, Mervyn; and Howard, Jerry, 1984: ^{20}Ne Depth-Dose Relations in Water. *Health Phys.*, vol. 46, no. 5, pp. 1101–1111.

Wilson, John W.; Townsend, Lawrence W.; and Badavi, F. F., 1987: A Semiempirical Nuclear Fragmentation Model. *Nuclear Instrum. & Methods Phys. Res.*, vol. B18, no. 3, pp. 225–231.

Wilson, John W.; and Townsend, L. W., 1988: A Benchmark for Galactic Cosmic-Ray Transport Codes. *Radiat. Res.*, vol. 114, no. 2, pp. 201–206.

Wilson, John W.; and Townsend, Lawrence W.; Ganapol, Barry; Chun, Sang Y.; and Buck, Warren W., 1988: Charged-Particle Transport in One Dimension. *Nuclear Sci. & Eng.,* vol. 99, no. 3, pp. 285–287.

Wilson, John W.; Townsend, Lawrence W.; Nealy, John E.; Chun, Sang Y.; Hong, B. S.; Buck, Warren W.; Lamkin, S. L.; Ganapol, Barry D.; Khan, Ferdous; and Cucinotta, Francis A., 1989: BRYNTRN: *A Baryon Transport Model.* NASA TP-2887.

Wilson, John, W.; Khandelwal, Govind S.; Shinn, Judy L.; Nealy, John E.; Townsend, Lawrence W.; and Cucinotta, Francis A., 1990: *Simplified Model for Solar Cosmic Ray Exposure in Manned Earth Orbital Flights.* NASA TM-4182.

Wulf, H.; Kraft-Weyrather, W.; Miltenburger, H. G.; Blakely, E. A.; Tobias, C. A.; and Kraft, G., 1985: Heavy-Ion Effects on Mammalian Cells: Inactivation Measurements With Different Cell Lines. *Radiat. Res.,* vol. 104, suppl. 8, pp. S-122–S-134.

Yang, Tracy Chui-Hsu; Craise, Laurie M.; Mei, Man-Tong; and Tobias, Cornelius A., 1985: Neoplastic Cell Transformation by Heavy Charged Particles. *Radiat. Res.,* vol. 104, pp. S-177–S-187.

Yang, Tracy Chui-Hsu; Craise, Laurie M.; Mei, Man-Tong; and Tobias, Cornelius A., 1986: Dose Protection Studies With Low- and High-LET Radiations on Neoplastic Cell Transformation In Vitro. *Adv. Space Res.,* vol. 6, no. 11, pp. 137–147.

Yang, Tracy Chui-Hsu; Craise, Laurie M.; Mei, Man-Tong; and Tobias, Cornelius A., 1989: Neoplastic Cell Transformation by High-LET Radiation: Molecular Mechanisms. *Adv. Space Res.,* vol. 9, no. 10, pp. (10)131–(10)140.

12. APPLICATION TO SPACE EXPLORATION

12.1. Introduction

The next major space endeavor after Space Station *Freedom* will be the human exploration of the Moon and Mars (The 90-Day Study (Anon., 1989b)). A critical aspect of these missions is the safety and health of the crew. One of the major health concerns is the damaging effects of ionizing space radiation (Parker and West, 1973). Once the crew leaves the Earth's protective environment, they will be bombarded by radiations of varying energies and ranges of intensity. The most harmful components of these radiations are trapped electrons and protons in the Van Allen belts, solar flare protons, and galactic cosmic rays. Adequate shielding will be required to protect the crew from this environment.

Astronaut doses incurred from the Van Allen belts are highly dependent on the time spent in the high flux regions of the belt and the state of fields at the time of exposure (Burrell, Wright, and Watts, 1968; Wilson and Cucinotta, 1984; Cucinotta and Wilson, 1985). Large temporal variations are observed in the outer zone for which dose incurred over a short time period may increase by an order of magnitude and more (Wilson and Denn, 1976 and 1977; Wilson, 1978). The nature of the energy spectrum is such that crew members in a typical shielded spacecraft can incur very large doses. However, moderate shielding (approximately 2 g/cm^2) and a single pass through the belts usually result in relatively small delivered doses (<1 cSv) under normal field conditions. These doses are of most concern for low Earth to geostationary orbit operations (Wilson and Denn, 1977; Wilson, 1978) and for spiraling trajectories through the belts. In either case, the large scale fluctuations are of great importance in determining shield requirements. Although galactic cosmic rays (GCR) are ever present, the low-energy GCR are deflected by the geomagnetic field (Wilson, 1978).

Outside the influence of the Earth's magnetic field, the astronauts will be constantly bombarded by galactic cosmic rays. The constant bombardment of these particles delivers a steady although low-level dose rate. The intensity of the GCR flux varies over the approximately 11-year solar cycle due to the interplanetary plasma resulting from the expanding solar corona. The maximum dose received occurs at solar minimum due to the lower solar plasma output. For the long-duration missions, this dose can become career limiting. Thus, the amount of shielding required to protect the astronauts will depend on the time within the solar cycle and duration of the mission.

Anomalously large solar proton events are relatively rare with one or two events per solar cycle. The largest flares observed in cycles 18 and 19 are the November 1948, February 1956, the May to July 1959 events series, and the November 1960 series. It was generally believed that the unusually large solar turbulence experienced in cycles 18 and 19 resulted in the largest events to be observed. However, the rather uneventful cycle 20 at the close of its activity produced an event on August 4, 1972, completely out of proportion to all expectations (Wilson and Denn, 1976; Wilson, 1978). Solar cycle 21 (1975–1986) proved relatively quiet with no unusually large events. However, with the onset of cycle 22, old concerns are confirmed with several large events occurring in the latter half of 1989. A solar flare event can be very dangerous if a spacecraft is inadequately shielded because flares can deliver a very high dose in a short period of time as was first made clear by the August 1972 event (Wilson and Denn, 1976). For relatively short-duration missions (2–3 months), the most important radiation hazard is the possibility of an unusually large solar proton event. The amount of shielding required for protection will depend on the nature of the energy spectrum and intensity of the flare. The means of setting shield requirements for such events are uncertain because there is no way yet of predicting either event size or spectrum.

Shielding must be provided to maintain crew doses to an acceptable level. Currently there are no limits established for exploratory class missions; however, it is recommended by NCRP 98 (Anon., 1989a, p. 163) that limits established for operations in low Earth orbit be used as *guidelines* for mission studies. The Space Station *Freedom* (SSF) limits are established (table 12.1) by the National Council on Radiation Protection and Measurement (NCRP 98 (Anon., 1989a)) and include dose equivalent limits for the blood-forming organ (BFO), ocular lens, and skin. For high-energy radiation from galactic cosmic rays and solar flare protons, the dose delivered to the BFO is the most important because of latent carcinogenic effects. Although other organs of the individual are at risk to cancer, only the blood-forming organ (BFO), ocular lens, and skin have been specifically limited (NCRP 98 (Anon., 1989a)). It is generally regarded that the BFO dose is a good indicator of whole-body exposure. Such notions are founded on ground level experience and need not apply to space radiation where large dose gradients are known to exist and have important consequences on risk assessment (Shinn, Wilson, and Nealy, 1990; Shavers et al., 1991). When detailed body geometry is not considered, the BFO dose is usually computed as the dose incurred at a depth of 5 cm in tissue as recommended by the Space Science Board (1970). Dose equivalent rate limits are established for short-term exposures (30 days), annual exposures, and total career exposure. These values are given in table 12.1. Note that dose equivalent is used for all limits although the quality factor mainly applies to carcinogenesis and mutagenesis (NCRP 98 (Anon., 1989a)). Short-term exposures are important when considering solar flare events because of their high dose rate (Committee on the Biological Effects of Ionizing Radiations, 1990 (BEIR V)). Doses received from GCR on long-duration missions are especially

Table 12.1. Dose Equivalent Limits Recommended for
U.S. Astronauts in Low Earth Orbit

Exposure time	Dose equivalent recommended limit, Sv, for—		
	BFO	Ocular lens	Skin
Career	[a]1–4	4	6
Annual	0.5	2	3
30 days	0.25	1	1.5

[a]Varies with age and gender.

important to total career limits, which are determined by the age and gender of the individual. For instance, career limits for typical male and female astronauts who are 30 years old at the time of their first exposure are 2 Sv and 1.4 Sv, respectively. The limit on whole-body exposure for a 3-year Mars mission would be 1.5 Sv using the BFO limit for SSF compared with the limit used in the Soviet Union space program of 4 Sv (Yablontsev, 1990). The appropriateness of the use of quality factors for GCR exposures is unknown (NCRP 98 (Anon., 1989a)).

Current mission scenarios for the Nation's Space Exploration Initiative are described in The 90-Day Study (Anon., 1989b). The final goal of the Initiative is to establish two permanent operational outposts on both the Moon and Mars. After a 3-day trip from Earth to the Moon, crew rotation times on the surface are described as starting with a 30-day stay, to a 6-month stay, to a 12-month stay, and finally growing to a 600-day stay. The flight time to Mars is estimated to take from 7 months to over a year each way. Crew rotations on the martian surface are described as starting with a 30-day stay, to a 90-day stay, up to a 600-day stay. Thus, an entire Mars mission is estimated to take anywhere from 500 to 1000 days round trip. Different shielding strategies will exist for each phase of each lunar and martian mission. Deep space shielding requirements for lunar transfer vehicles will differ greatly from those selected for the Mars vehicles because of the large differences in travel time. Likewise, planetary habitation shielding strategies utilizing local resources will differ greatly from the transfer vehicles. Habitation shielding on the lunar surface versus that on the martian surface will also differ greatly because of the differences in the environment and the protection provided by the martian atmosphere.

12.2. Space-Radiation Environment

The types of particle radiations that occur in space are summarized in figure 12.1. There are both temporal as well as spatial variations. For example, trapped particles exist only in the geomagnetic field where mirror points lie well above the atmosphere, the solar wind can only be seen outside the Earth's magnetosphere, the auroral electrons are trapped particles with mirror points in the atmosphere and are seen only in polar regions during geomagnetic disturbances, solar cosmic rays are rare transient events associated with solar flares, and so on. (See Wilson (1978) for more details.) The radiations with energies below 100 keV and the protons below 10 MeV are mainly important only from a material point of view—for example, thermal control coatings—and are considered biologically unimportant. The radiations of immediate importance for biological consideration are the trapped protons in the inner zone, the trapped electrons in both the inner and the outer zones, and solar flare protons. Galactic cosmic rays are also biologically important. They are of low intensity but many questions surround them because of their particular composition, and their biological action is potentially hazardous and not well understood experimentally. Data used in constructing figure 12.1 are taken from Noll and McElroy (Anon., 1975a), Foelsche (1963), McDonald (1963), Divine (Anon., 1975b), and Johnson (1965).

The impact of radiation on Earth orbital operations is shown in table 12.2 (Wilson, 1978). We see that imposed limits are very restrictive in some regions of space. Within the inner zone below 400 n.mi. are mostly protons and electrons.

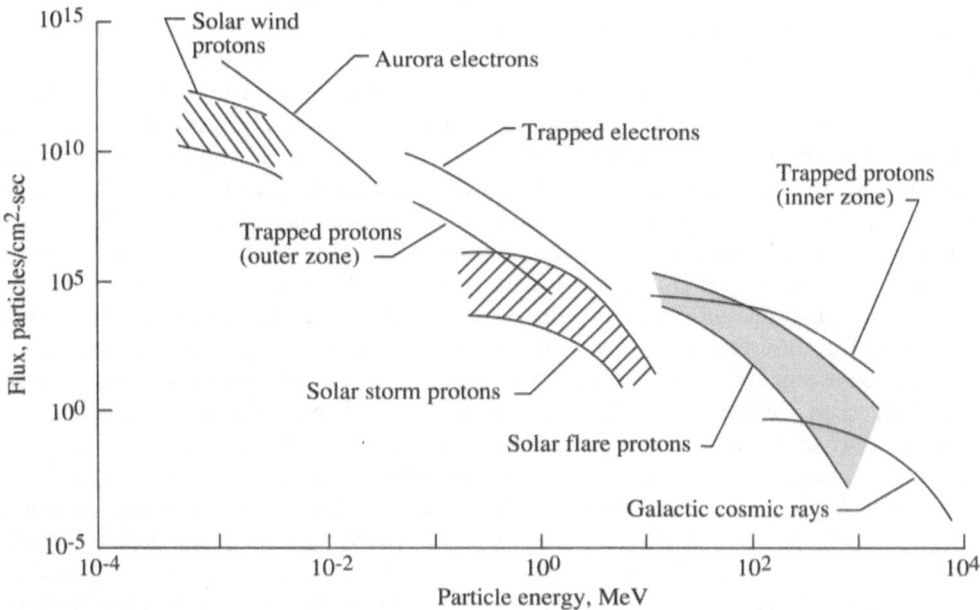

Figure 12.1. Space-radiation environment (Wilson, 1978).

Table 12.2. Impact of Radiation on Operations in Earth Orbit

Source	Particles	Aluminum shielding, g/cm^2	Number of days[a]	Limiting factor
Inner zone ($h < 400$ n.mi)	Protons Electrons	2	22	Testes
Outer zone ($h > 19\,000$ n.mi)	Electrons	6.7 1.4	90 90	Testes [b]Skin, lens, BFO
Solar cosmic rays	Proton Alpha	5 10	0.25 0.5	Lens, testes, skin BFO, lens, testes
Galactic cosmic rays	Proton Alpha Carbon ⋮ Iron	?	?	Nonregenerative tissue with unique function?

[a]Number of days to reach quarterly exposure limit.
[b]Personal shielding is assumed for testes.

Behind a shield 2 g/cm^2 thick, only about 22 days are required to reach the quarterly exposure limit at this altitude. The limiting biological factor is the testes which could be protected by personal shielding. As for the outer zone, which is important to the space solar power satellites and interplanetary transportation, the radiation is primarily electrons, and a shield constructed from aluminum on the order of 6.7 g/cm^2 is required to reach the quarterly exposure limits in 90 days. The limiting biological factor is again the testes. If the bremsstrahlung is eliminated by putting a high Z material on the innerside of the wall, the shield could be reduced substantially. Then a shield thickness of only 1.4 g/cm^2 is required to meet the quarterly exposure limit in 90 days and the limiting factors are skin, ocular lens, and the blood-forming organs. This shield 1.4 g/cm thick is, of course, an absolute minimum shield because there is no personal shielding that is practical for the organs involved. Interplanetary travel exposure is reduced by a rapid transit through this region but large short-term temporal variations need to be addressed (Wilson and Denn, 1977; Wilson, 1978). The solar cosmic rays consist mostly of protons and alphas with fewer other particles. Behind 5 g/cm^2, only about 6 hours are required to reach exposure limits, and the limiting factors are listed in table 12.2. For a shield of 10 g/cm^2, it takes about 0.5 day to reach exposure limits for the ocular lens and testes, which can be protected by using personal shielding. The galactic cosmic rays contain a little bit of everything, and the type of shielding required and the number of days to reach exposure limits are presently in question. Most probably the hazard will be associated with nonregenerative tissues which also have a unique function, carcinogenesis, and mutagenesis. Galactic heavy ions will probably be the ultimate limiting factor in space operations, but all these points are still open for debate since definitive biological data are still lacking (NCRP 98 (Anon., 1989a)). Conclusions are drawn from data taken from Burrell, Wright, and Watts (1968), Wilson and Denn (1976 and 1977), and Grahn (1973).

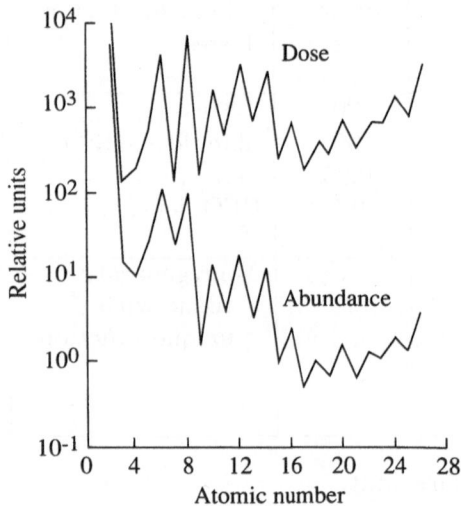

Figure 12.2. Galactic heavy ion intensities.

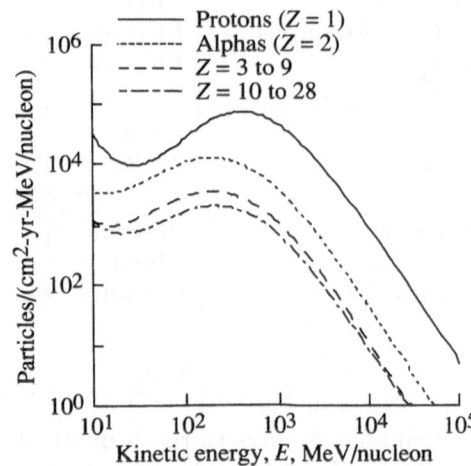

Figure 12.3. Galactic cosmic-ray ion spectra for solar minimum conditions (Adams, Silberberg, and Tsao, 1981).

12.2.1. Galactic cosmic rays. Figure 12.2 shows how the abundance of galactic cosmic rays falls off as the higher atomic numbers are reached (J. A. Simpson and M. Garcia-Munoz, University of Chicago).[1] The dose is proportional to the charge squared. The relative dose contribution is more nearly the same for different particle types; it doesn't follow that the less abundant types are necessarily negligible. The deep space differential spectra are shown in figure 12.3 for solar minimum conditions (Adams, Silberberg, and Tsao, 1981).

The galactic cosmic rays are affected by interaction with the Earth's magnetic field (fig. 12.4). Mainly the low rigidity particles are excluded from equatorial regions at low altitudes, whereas near the poles the particles may come in freely at all altitudes. Although the particles with low rigidity are seen at low altitude mainly near the polar region, the heavy ions are by far the most rigid particles (mass to charge ratio) of the galactic beam. Consequently, mostly protons are lost in the equatorial region having a greater proportion of heavy ions.

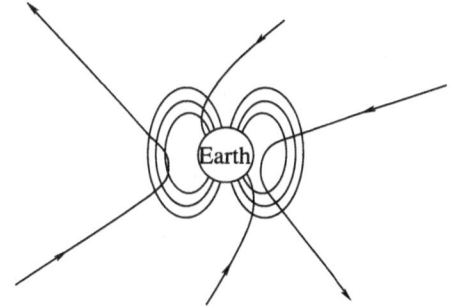

Figure 12.4. Galactic cosmic-ray interaction with geomagnetic field.

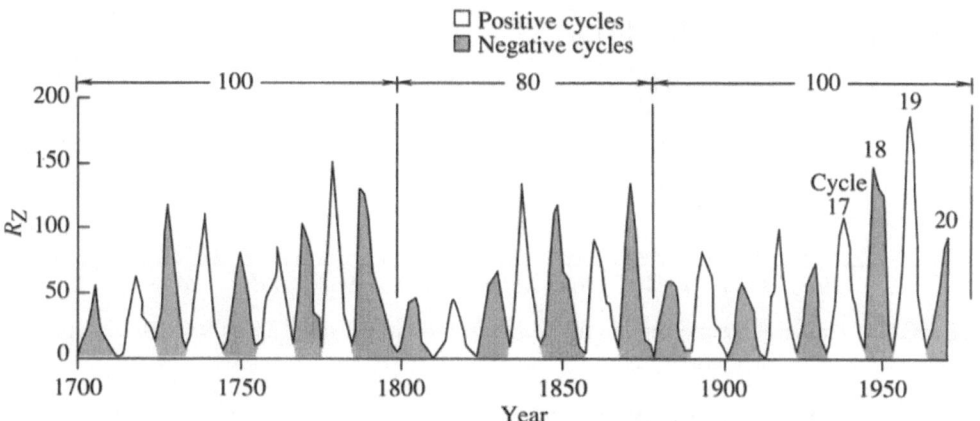

Figure 12.5. Annual smoothed sunspot numbers.

12.2.2. Solar activity. Most space radiations are affected by solar activity in one way or the other, either as their source or in some secondary effect. Figure 12.5 shows the annual smoothed sunspot numbers for the past few hundred years. The main feature is that the concept of a "typical" solar cycle is uncertain.

---

[1] Unpublished data measured aboard IMP-4 in 1970.

It is clearly illustrated that cycle 19 is one of the most extreme cycles in terms of sunspot numbers that we have ever seen (last full cycle at right). Cycle 20, shown in part just to the right of cycle 19, was pretty close to an average cycle and we should keep that in mind when we discuss solar cosmic rays later. These data were taken from Sleeper (1972).

One effect observed during solar activity was the fluctuation of the expanding solar corona. The galactic cosmic rays coming in from galactic space interact with this plasma and slow down. In figure 12.6 we show the amount of energy that the particles lose coming in from galactic space to Earth orbit represented as a potential function. It correlates reasonably well with sunspot number which is related to solar activity. These data were taken from O'Brien (1972). The effects of increased solar modulation during solar maximum are demonstrated in figure 12.7.

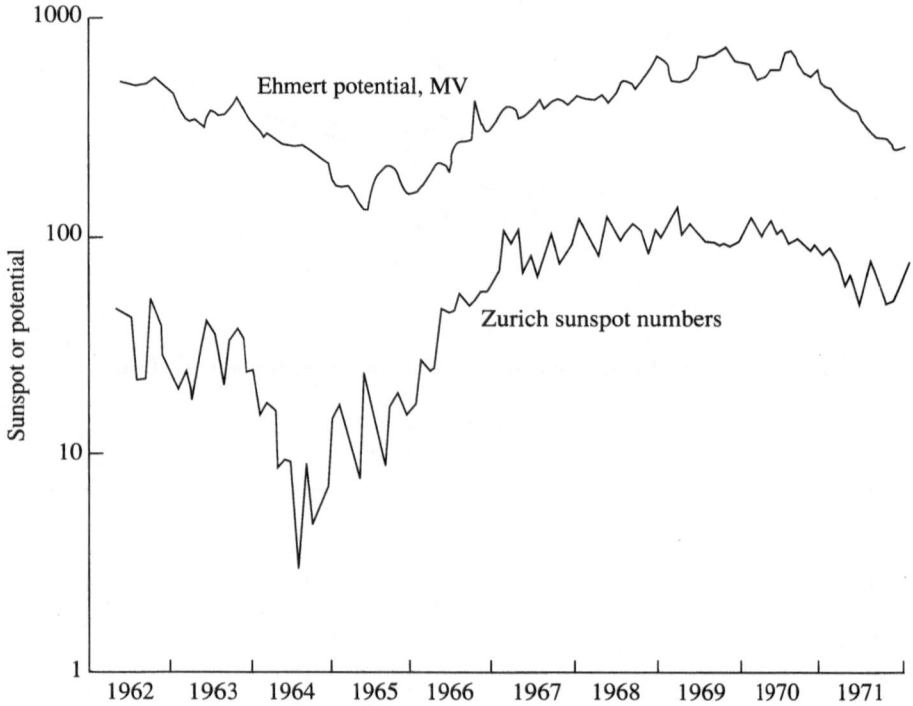

Figure 12.6. Galactic cosmic-ray modulation parameter and solar activity (O'Brien, 1972).

12.2.3. Solar cosmic rays. Occasional solar flares are associated with the Sun and solar activity. Plasma is ejected from a large number of optical flares, and this causes type IV radio bursts. During some of these flares (actually very few), there are particles that are ejected at high energy into interplanetary space. These high-energy particles are able to escape the solar magnetic fields only if the lines are open to the interplanetary region. The data shown in figure 12.8 were taken from Slutz et al. (1971), King (1974), and Blizard (1969). This figure shows the sunspot numbers during cycles 19 and 20 and plots of the proton fluence greater than 30 MeV. This is the total fluence of each individual particle event as

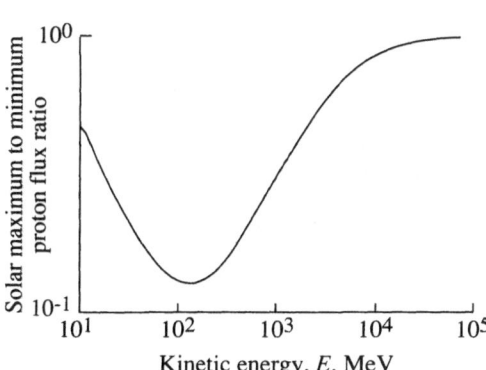

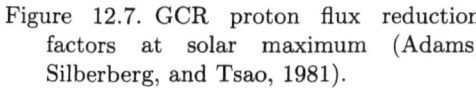

Figure 12.7. GCR proton flux reduction factors at solar maximum (Adams, Silberberg, and Tsao, 1981).

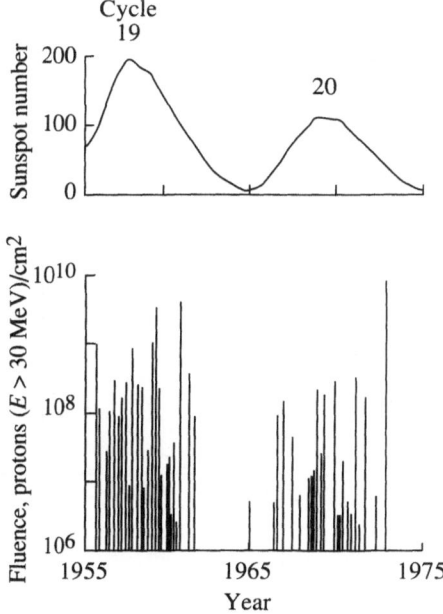

Figure 12.8. Solar activity and flare proton fluence.

a function of time of occurrence. There is a rough correlation between the number of particles and the degree of solar activity. Generally there are anywhere from one to perhaps five solar events which might be called major events during any particular cycle.

Some details of what happened during cycles 18, 19, and 20 can be found in figure 12.9. Here we show just the major events that occurred during these particular cycles; notice that, in general, the largest events happened during the ascending or descending phase of the solar cycle. Major events are usually absent during solar maximum and, of course, also during solar minimum. The events denoted by dashes are of little significance to manned space flight. Data were taken from Blizard (1969) and King (1974).

There is a rough correlation between the solar activity and the particle fluences that are observed in any given year. Plotted in figure 12.10 are proton yearly fluences as a function of the average yearly sunspot number during cycle 19 for protons of energy greater than 1 MeV (upper curve), greater than 10 MeV (middle curve), and greater than 40 MeV (lower curve). There is some general dependence of the fluence of particles associated with sunspot number, although there are significant deviations. These correlations are made for predictive purposes. If the sunspot numbers in the next cycle can be predicted, a correlation between sunspot number and particle fluence can be made. Then it is possible to make an estimate of what sort of exposure might be expected in the coming solar cycle. Data were taken from Webber (1966) and Curtis, Doherty, and Wilkinson (1969).

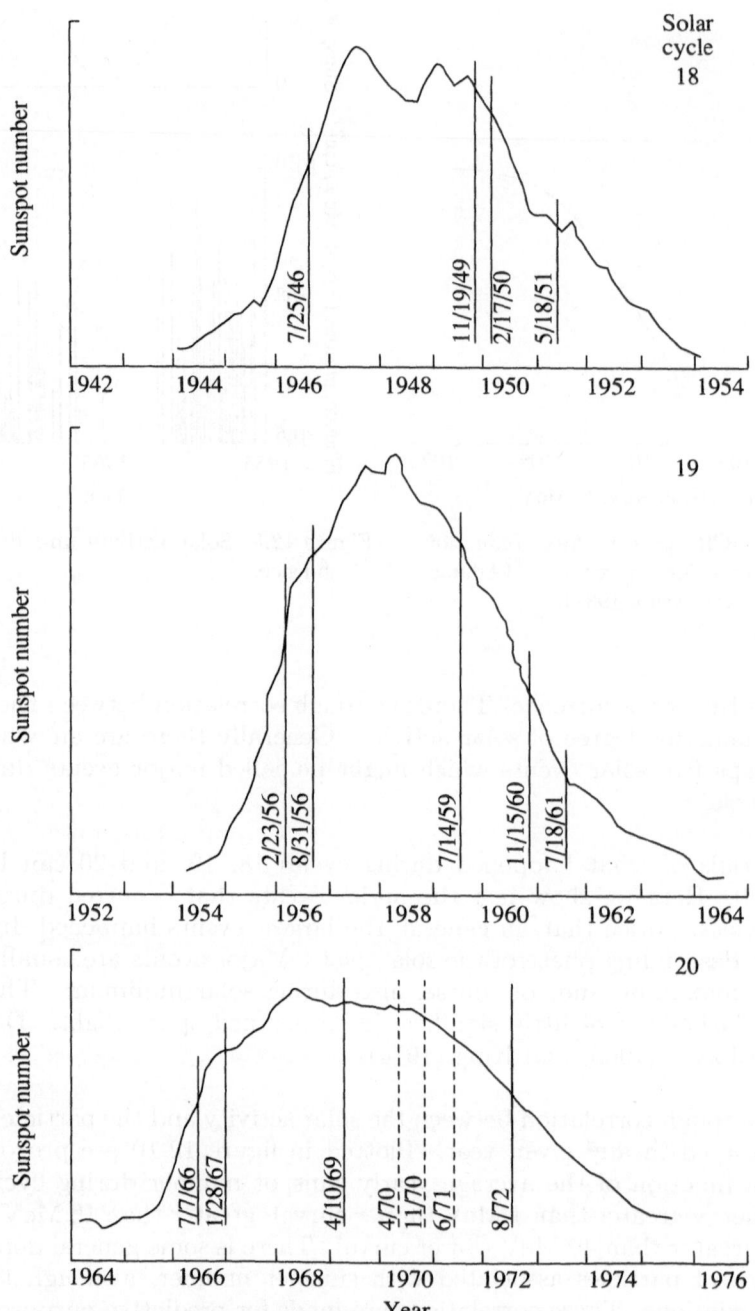

Figure 12.9. Major solar particle events of the last three solar cycles.

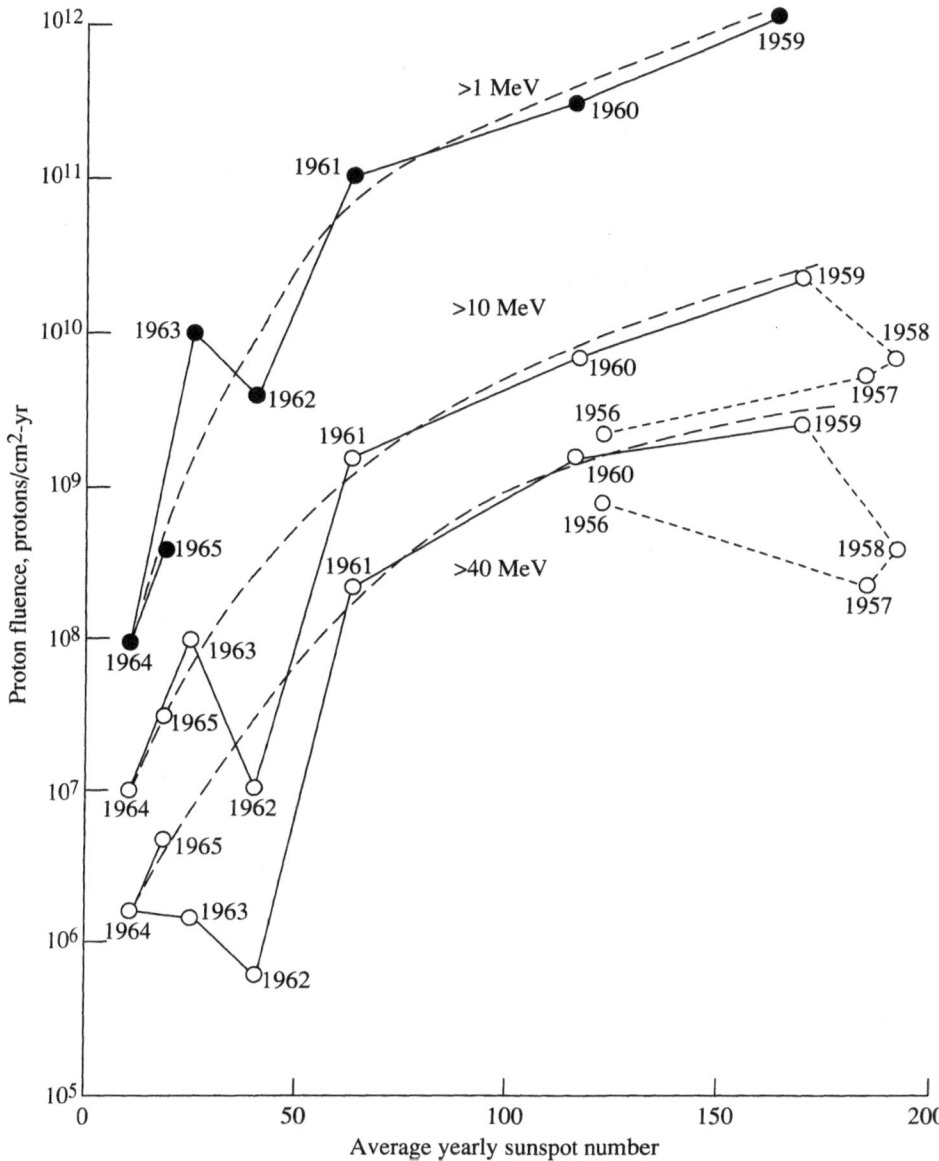

Figure 12.10. Solar cosmic-ray yearly fluence during cycle 19.

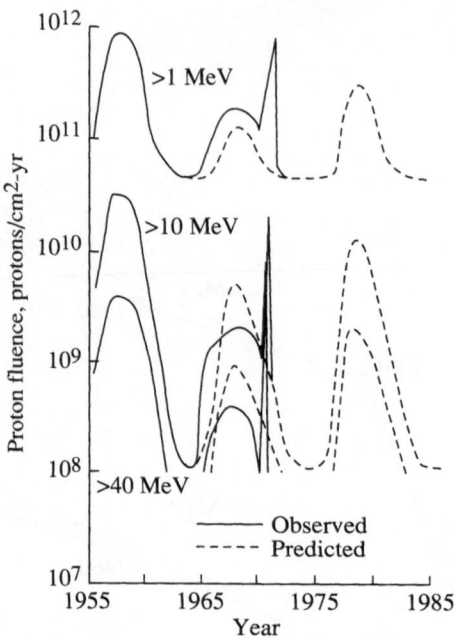

Figure 12.11. Yearly proton fluence at geostationary orbit.

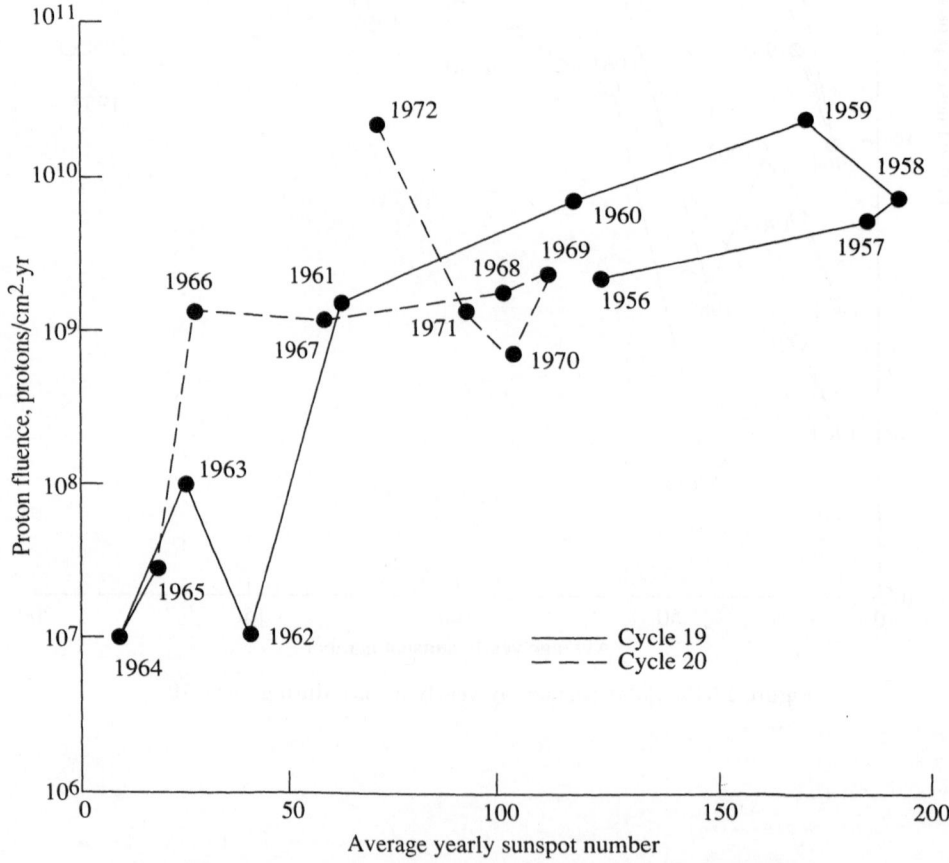

Figure 12.12. Solar cosmic-ray yearly fluence above 10 MeV.

A set of such predictions based on the correlations shown in figure 12.10 is shown in figure 12.11 for some predictive models on solar cycles. We plotted the calendar year as a function of the proton fluence and we show cycle 19 on which the correlations were based. The solid curves are the observed values going into cycle 20, and the dashed curves are the predicted values. The predictions were fairly accurate up to August 1972; after that there are rather large deviations from the predictive curve. In fact, while we thought that cycle 20 was going to be a rather mild cycle it turned out that the largest event, as far as space exposure is concerned, occurred during this rather ordinary cycle—consequently changing our thinking about the importance of solar flares (Wilson and Denn, 1976). We always thought solar flares were serious but we did not realize just how hazardous (potentially lethal) they were until August 1972. Data were taken from Curtis and Wilkinson (1971) and King (1974).

Figure 12.12 shows the data for energies above 10 MeV as a function of sunspot number from cycle 19 (figure 12.11) with the observation made during cycle 20 added to it. We see that the August 1972 event gives most of the contribution during that particular year. Keeping the correlation curve of figure 12.10 in mind and comparing the location of the 1962 event and 1972 event (or the 1962 year and the 1972 year), these correlations are accurate to within about a factor of 10 on the basis of the data we now have.

In particular, the event of November 1960 and the May–July event series of 1959 were previously thought to be the most serious events we had to design for in space operations. Now we find that the exposure from the August 1972 event is about three to four times greater than the two earlier events. Whether we will have a future event that will exceed the dose of the August 1972 event is an open question. Obviously, one would expect that an even larger event could occur with some smaller probability. How to assign the chance of occurrence is questionable.

The solar cosmic rays produced on the Sun must still travel to Earth. The transit time between the Sun and the Earth is typically 20 minutes for relativistic particles, but sometimes it takes up to a few hours depending on the spectrum and the interplanetary magnetic field configuration (fig. 12.13). The spectral distribution at Earth changes as a function of time because high-energy particles tend to arrive before lower energy particles. The angular distribution of the particles varies greatly from event to event. During some of the high-energy events, the particles tend to be directional early in the event and approach isotropy later as the lower energy particles arrive. Similar to galactic cosmic rays, the solar protons tend to be eliminated from equatorial regions of the Earth's magnetosphere. However, nearly all particles incident in polar regions are transmitted to low altitudes (fig. 12.13).

The integral fluence spectra of three major proton events observed during cycles 19 and 20 are shown in figure 12.14. It was previously thought that the November 1960 event was the most hazardous for space operations, and we were basing our designs on this limit. As the figure shows, the August 1972 event dominates at energies below 100 MeV, and it has changed our thinking about the limits for the most hazardous case. We originally considered the largest event observed in the cycle of greatest activity to be the worst-case event, i.e., November

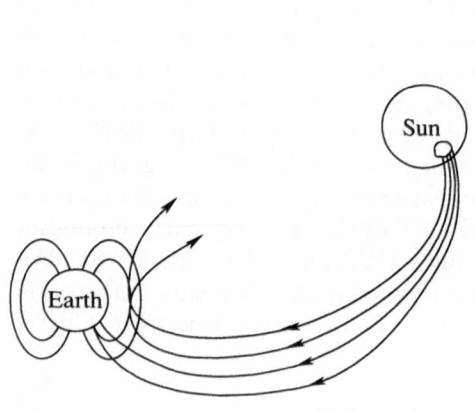

Figure 12.13. Solar cosmic-ray Earth interaction.

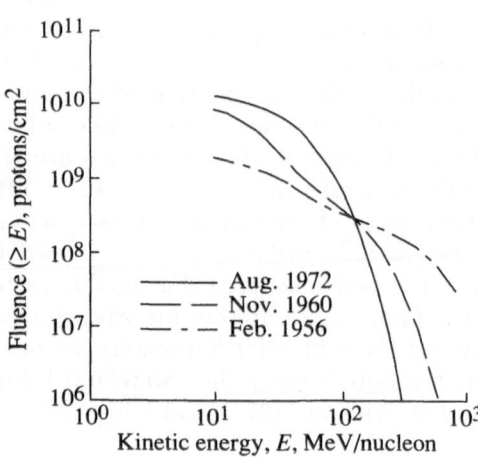

Figure 12.14. Proton fluence of three major solar events.

1960. Now a much larger event has occurred in a rather inactive cycle which destroys our logic. Someday the August 1972 event may well be overshadowed by some future event. These data were taken from Foelsche (1963) and King (1974).

The dose equivalent in the center of the sphere of radius r is shown in figure 12.15. Compare the August 1972 depth-dose relation to that of the February 1956 event. Clearly in the region about 1 to 20 g/cm^2, which is the important region for spacecraft shielding, the August 1972 event is (in places) an order of magnitude more serious than the February 1956 event. The November 1960 event lies about halfway between these two curves.

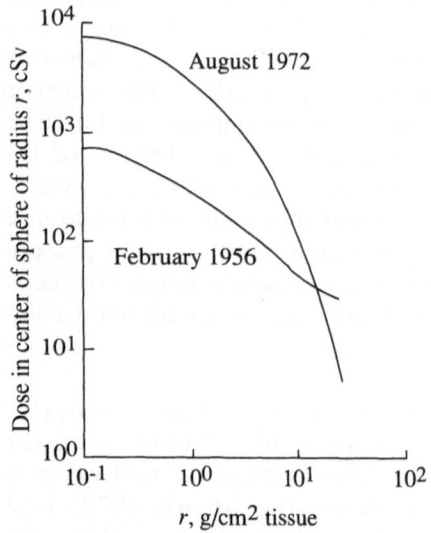

Figure 12.15. Dose equivalent from two major solar events (Wilson and Denn, 1976).

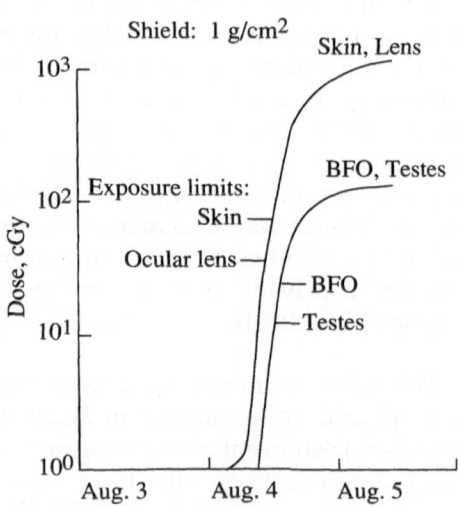

Figure 12.16. Dose from a major solar flare (Wilson and Denn, 1976).

654

Figure 12.16 presents the exposure limits and the dose behind a shield of 1 g/cm^2 during the August 1972 event. The dose limits were determined by calculating the effective average quality factor for the August 4, 1972 event. This average quality factor is about 1.3, and it is the value used in deriving this figure. Of course the average quality factor is spectrum dependent; 1.3 cannot be used for all events because it depends very much on the energy content of the event. For this particular quality factor the exposure limits are reached for the lens of the eye first and the skin later. Note that the dose greatly exceeds the allowed limits behind the shielding 1 g/cm^2 thick, which in the past has been a typical thickness for spacecraft shielding. These curves also take into account the body geometry (Khandelwal and Wilson, 1974; Billings and Yucker, 1973). Data were taken from Wilson and Denn (1976).

The 30-day exposure limits and also the time required to reach these exposure limits during the August 1972 event are shown in table 12.3. This is the time after the onset of particle emission—not the time after the optical flare is observed but rather the time after the particles are first seen arriving at Earth's orbit. Generally, if a person is very lightly shielded he still has about 2 to 4 hours to seek shelter. This is adequate time to move to a more protective region. At 10 g/cm^2 of tissue equivalent material, the dose limits to the BFO and skin are never reached. The limiting factors are the lens of the eye and the testes, and these can be taken care of by using personal shielding. Therefore, a shelter of about 10 g/cm^2 of a material like polyethylene (plus personal shielding) would be adequate protection from the August 1972 event. Data were taken from Wilson and Denn (1976).

Table 12.3. Time Required to Reach Exposure Limits for
August 4, 1972 Event

Shield thickness, g/cm^2	BFO, hr	Skin, hr	Ocular lens, hr	Testes,[a] hr
0.2	6.0	3.0	1.9	4.4
0.4	6.1	3.5	2.4	4.9
1	6.3	4.7	3.6	5.2
5	8.9	8.0	6.5	7.3
10	∞	∞	11.7	12.7

[a]Values are overestimated since the testes dose is taken to be the same as the BFO dose.

Solar cycle 21 (1975–1986) was relatively quiet with no flare events of these magnitudes recorded. The flares of cycle 21 may constitute the typical proton fluence within a solar cycle due to the more normally occurring smaller and medium size events. The proton fluxes due to flare events were measured by particle monitors onboard the Interplanetary Monitoring Platform satellites, IMP-7 and IMP-8. Fifty-five flares within solar cycle 21 had integral fluences greater that 10^7 protons/cm^2 for energies greater than 10 MeV. The other flares of lower fluence and energy would contribute negligibly to dose calculations.

Figure 12.17(a) shows the integral fluences of the 55 flares as they are distributed in time throughout the cycle, and figure 12.17(b) shows the fluence spectra for each of these flares (Goswami et al., 1988).

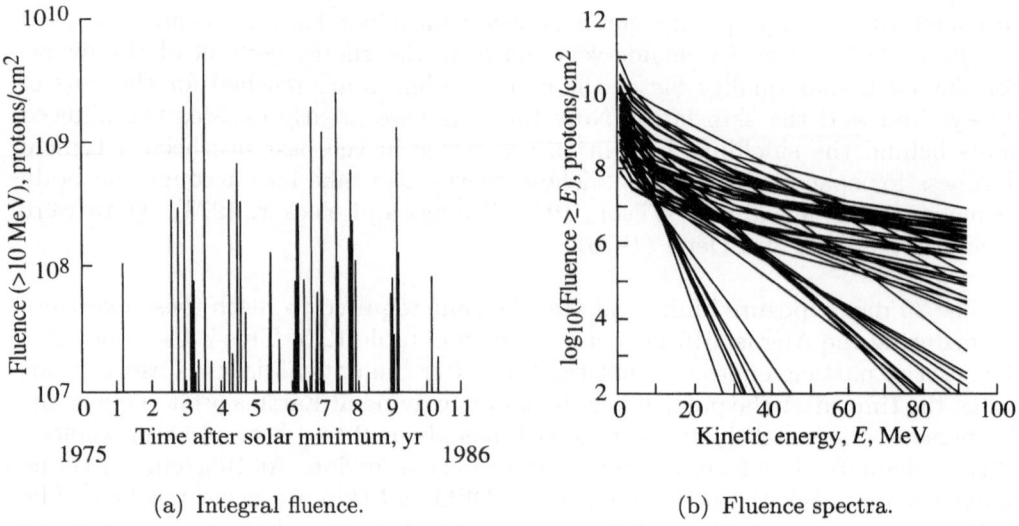

(a) Integral fluence.

(b) Fluence spectra.

Figure 12.17. Solar proton flares during solar cycle 21 (1975–1986) (Goswami et al., 1988).

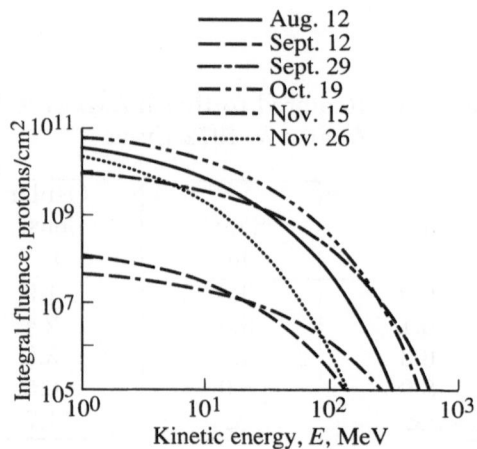

Figure 12.18. Six large solar flare integral fluences based on 1989 GOES-7 data (Sauer, Zwickl, and Ness, 1990).

With the onset of solar cycle 22 (1986–1997), several flares larger than any recorded in cycle 21 have already occurred in the months of August through December 1989. Six flares occurring in this time frame have been recorded by the GOES-7 satellite. Figure 12.18 shows the proton fluence energy spectra based on rigidity functions reported by Sauer, Zwickl, and Ness (1990). The magnitude of the October 1989 event is on the order of that of the August 1972 event and has heightened concern over flare shielding strategies. The addition of these six flares can provide a fairly realistic estimate of a flare environment that may be encountered during missions taking place in the 5 or 6 years of active Sun conditions.

12.2.4. Geomagnetic effects on orbital environment. Charged particles arriving at some location within the geomagnetosphere are deflected by the Lorentz force $e\vec{v} \times \vec{B}$ which prevents penetration for some directions of incidence and some energies. Such phenomena were extensively studied by Störmer (1930) for a dipole magnetic field which provides the basis for classifying the orbital trajectories of charged particles arriving at some location within the field. As a part of Störmer's theory, allowed trajectories with no connection to asymptotic trajectories exist; these are now recognized as trapping regions associated with Van Allen radiation. Numerical solutions to the charged particle equations of motion in a more realistic geomagnetic field model were introduced by McCracken (1962a, 1962b, and 1962c) and further advanced by Smart and Shea (1983), Shea and Smart (1983), and Shea, Smart, and Gentile (1983). Our purpose here is not to supplant the vastly detailed numerical work but to seek a simple analytic form to reasonably approximate the more general numerical solutions. The numerical work of Smart and Shea is indispensable.

The geomagnetic field can be reasonably approximated by a tilted dipole with moment $M = r_e^3 31\,500$ nT displaced from the Earth's center by 430 km or $0.068 r_e$, where $r_e = 6378$ km. The tilt angle is 11.7° at 69° W longitude. The magnetic quadrupole contributions are then about 10 percent at the surface and decrease to 5 percent at $2 r_e$. Higher order moments are even smaller. The motion of charged particles in the geomagnetic field was studied extensively by Störmer. We outline his methods herein. In spherical coordinates, Störmer showed that the azimuth angle ϕ is an ignorable coordinate possessing an integral for the particles trajectories such that

$$\cos \omega = \frac{\gamma}{mvr \sin \theta} - \left(\frac{ZeM}{mvc} \right) \frac{\sin \theta}{r^2} \tag{12.1}$$

where m is the mass of the particle, Ze is the charge, v is the velocity, c is the velocity of light, r is radial distance from the center of the field, θ is colatitude, γ is an integration constant, and ω is the angle between the velocity vector and the azimuthal direction. The allowed Störmer regions consist of the space for which

$$|\cos \omega| \equiv \left| \frac{\gamma}{mvr \sin \theta} - \left(\frac{ZeM}{mvc} \right) \frac{\sin \theta}{r^2} \right| \leq 1 \tag{12.2}$$

Further analysis of the condition in equation (12.2) shows stable trapping regions as well as the Störmer main cone of transmission given for $\gamma = 2mv(ZeM/mvc)^{1/2}$. The Störmer main cone is given (Kuhn, Schwamb, and Payne, 1965) by the solid angle element

$$\Omega = 2\pi(1 + \cos \omega) \tag{12.3}$$

657

which contains the allowed directions of arrival for particles of rigidity R (momentum per unit charge) given by

$$R = \frac{M}{c} \frac{\sin^4 \theta}{r^2 \left[1 + (1 - \sin^3 \theta \cos \omega)^{1/2}\right]^2} \tag{12.4}$$

Henceforth we replace the colatitude θ by the magnetic latitude λ_m and note that Ω varies from 0 to 4π reaching its half-value at $\omega = \pi/2$ including angles up to the vertical direction. The vertical cutoff model is expressed as

$$\Omega \approx 4\pi \, U \left[R - R_C \left(\lambda_m\right)\right] \tag{12.5}$$

where the vertical cutoff rigidity from equation (12.4) is

$$R_C \left(\lambda_m\right) = \frac{M}{4cr^2} \cos^4 \lambda_m \tag{12.6}$$

and $U(x)$ is the unit step function.

Not included in the above formalism are those trajectories which are cut off by the shadow cast by the solid Earth. The fraction of the solid angle covered by the shadow of the Earth is estimated with the assumption that the curvature of the local trajectories is large compared with the radius of the Earth (Kuhn, Schwamb, and Payne, 1965). Then the solid angle fraction is

$$\frac{\Omega_{\mathrm{sh}}}{4\pi} = \frac{1}{2} \left[1 + \cos\left(\sin^{-1}\frac{1}{r}\right)\right] \tag{12.7}$$

The corrected solid angle for the vertical cutoff model is then

$$\Omega = \Omega_{\mathrm{sh}} U \left[R - R_C \left(\lambda_m\right)\right] \tag{12.8}$$

which leaves the local solid angle open to transmission of charged particles of rigidity R at altitude r and geomagnetic latitude λ_m.

Spacecraft in low Earth orbit (LEO) are typically in circular orbits; this simplifies the analysis. The orbit plane is inclined with respect to the equatorial plane. Since the angular momentum (spin) of the Earth and the orbital angular momentum of the spacecraft are conserved, the angle between them is fixed and equal to the inclination angle i. The magnetic axis rotates with the Earth and therefore precesses about the rotational axis within a 24-hour period. The geographic location of the ascending node likewise moves around the geographic equator every 24 hours. The inclination of the orbit plane relative to the magnetic axis i_m likewise is periodic. If η is the geographic coordinate of the ascending node line, then

$$\cos i_m = \cos i \cos \theta_m + \sin i \sin \theta_m \cos \left(\eta - \phi_m - 90°\right) \tag{12.9}$$

where θ_m and ϕ_m are the magnetic north pole colatitude and longitude. The average transmission factor around this orbit $\overline{F}$ is then

$$\overline{F}(R, i, \eta) = \frac{\Omega_{\text{sh}}}{i_m} \int_0^{i_m} U\left[R - R_C(\lambda)\right] d\lambda = \frac{i_m - \lambda_m}{i_m} \Omega_{\text{sh}} \qquad (12.10)$$

where λ_m is the magnetic latitude with cutoff at R as given by

$$R_C(\lambda_m) = R \qquad (12.11)$$

We note that i_m goes through a maximum and minimum orbit corresponding to $\eta_{\max} = \phi_m - 90°$ and $\eta_{\min} = \phi_m + 90°$ for which $i_m = i + \theta_m$ and $i_m = |i - \theta_m|$, respectively, as we have shown elsewhere (Wilson et al., 1990).

We may also calculate the long-term average over many days of orbits by averaging equation (12.10) over the node angle η as

$$\overline{F}(R, i) = \frac{1}{\pi} \int_0^\pi \Omega_{\text{sh}} \frac{i_m - \lambda_m}{i_m} \, d\eta = \frac{1}{\pi} \int_{\phi_\lambda}^\pi \Omega_{\text{sh}} \frac{i_m - \lambda_m}{i_m} \, d\phi \qquad (12.12)$$

where

$$\cos i_m = \cos i \cos \theta_m + \sin i \sin \theta_m \cos \phi \qquad (12.13)$$

and

$$\phi_\lambda = \left\{ \begin{array}{ll} 0 & (\lambda_m \leq |i - \theta_m|) \\[2mm] \cos^{-1}\left[\frac{\cos \lambda_m - \cos i \cos \theta_m}{\sin i \sin \theta_m}\right] & (|i - \theta_m| \leq \lambda_m \leq i + \theta_m) \\[2mm] \pi & (\lambda_m > i + \theta_m) \end{array} \right\} \qquad (12.14)$$

Equation (12.12) may be rewritten as

$$\overline{F}(R, i) = \Omega_{\text{sh}} \left[\left(1 - \frac{\phi_\lambda}{\pi} \right) - \frac{1}{\pi} \int_{\phi_\lambda}^\pi \frac{\lambda_m}{i_m} \, d\phi \right] \qquad (12.15)$$

where the last integral is approximated by a numerical quadrature. The results of equation (12.15) are compared with the numerical calculations of Smart and Shea (1983) for 400-km (216-n.mi.) orbits at several inclinations in figure 12.19 for this centered dipole field model with $\theta_m = 11.7°$ (tilt angle) and longitude $\phi_m = -69°$ (69° W).

An important correction to the centered dipole field is the displacement of the geomagnetic dipole 430 km (232 n.mi.) from the Earth's center. Unfortunately, the formalism is very complicated, since the distance r from the dipole center is no longer constant even for a circular orbit. The offset dipole decreases the cutoffs in the Atlantic hemisphere defined by the meridional plane normal to the tilt direction and increases the cutoffs in the remaining hemisphere over the Pacific. We define two cutoff functions for centered dipole fields as

$$R_j(\lambda_m) = \frac{14.9}{\left(r + \delta_j\right)^2} \cos^4 \lambda_m \qquad (12.16)$$

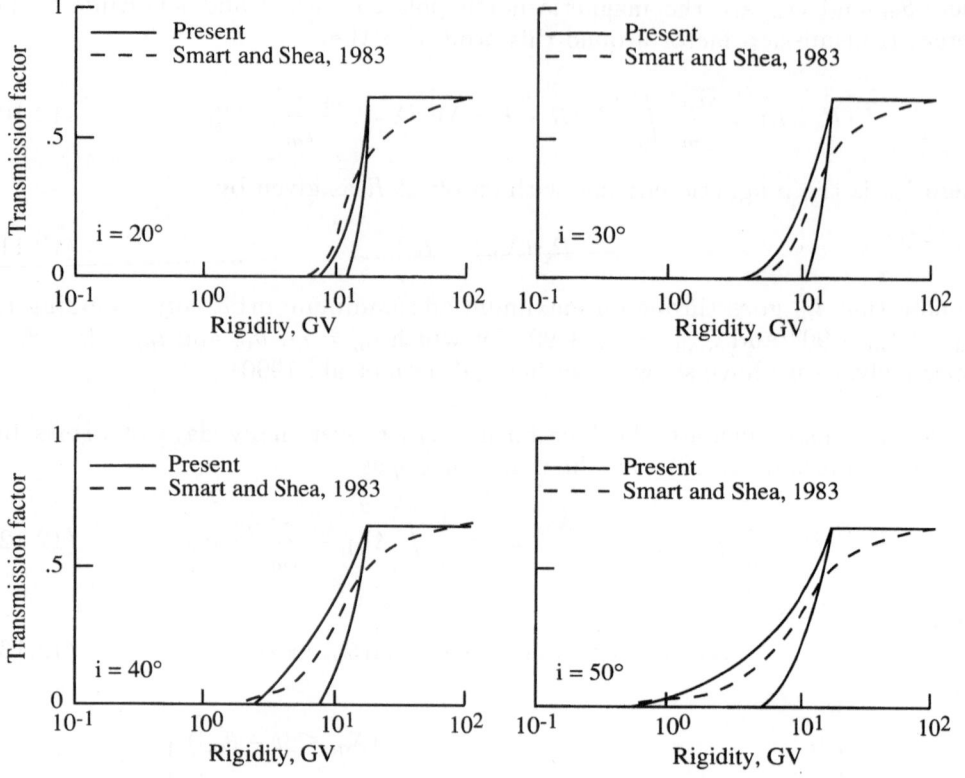

Figure 12.19. Dipole maximum and minimum cutoff model and numerical simulation of exact geomagnetic field model.

where $j = A$, $\delta_A = 593$ km (320 n.mi.) for the Atlantic hemisphere; $j = P$, $\delta_P = -504$ km (-272 n.mi.) for the Pacific; λ_m is the usual magnetic latitude which depends on the hemisphere; and r is the geocentric radius of the orbit. The value 14.9 GV is found from the value of the dipole moment of $r_e^3 31\,500$ nT, and values of δ_j were chosen to match the minimum equatorial cutoff in the Atlantic region of Shea and Smart (1983) and the maximum cutoff in the Pacific. The calculation of the orbit average transmission factor is as before except that the two hemispheres are considered separately as

$$\overline{F}(R, i, \eta) = \frac{1}{2}\Omega_{\text{sh}}\left[\frac{i_m - \lambda_A}{i_m} + \frac{i_m - \lambda_P}{i_m}\right] \tag{12.17}$$

where

$$R_A(\lambda_A) = R \tag{12.18}$$

and

$$R_P(\lambda_P) = R \tag{12.19}$$

Similarly, the long-term average of equation (12.12) is extended to each hemisphere as

$$\overline{F}(R, i) = \frac{\Omega_{\text{sh}}}{2\pi}\left[\int_{\phi_A}^{\pi}\left(\frac{i_m - \lambda_A}{i_m}\right)\,d\phi + \int_{\phi_P}^{\pi}\left(\frac{i_m - \lambda_P}{i_m}\right)\,d\phi\right] \tag{12.20}$$

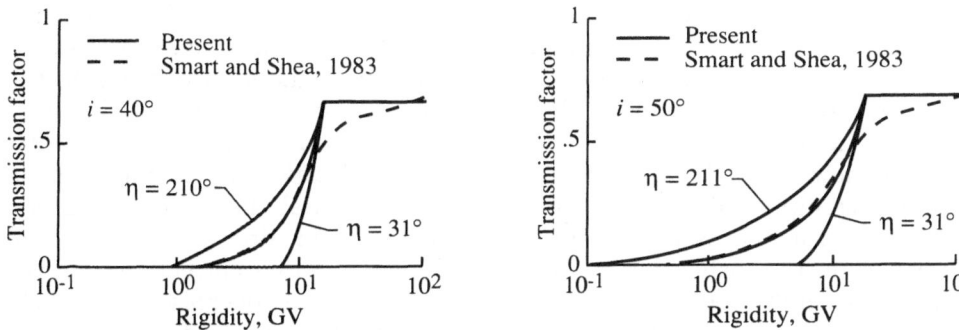

Figure 12.20. Offset dipole model average transmission factors with detailed calculations of Smart and Shea (1983) and maximum and minimum transmission factors.

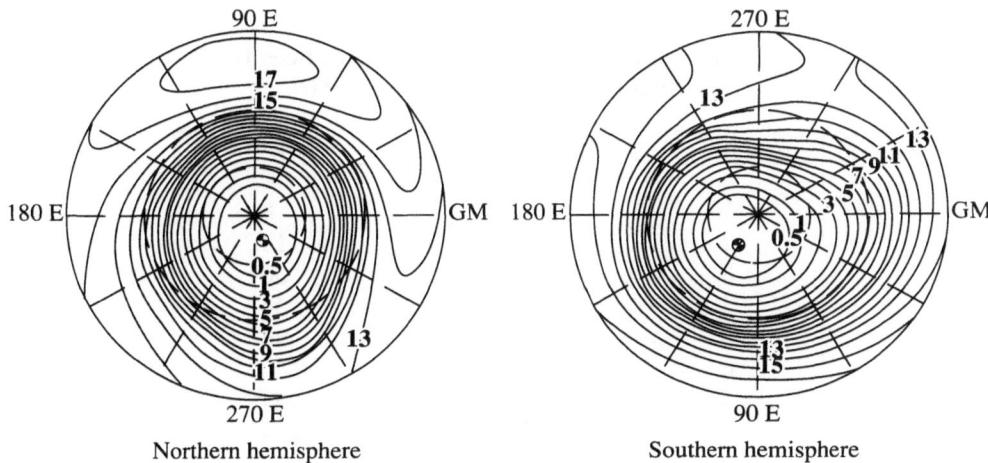

Figure 12.21. Vertical cutoff contours showing location of corresponding magnetic poles.

Table 12.4. Geographic Locations of Offset
Poles in Present Calculations

Magnetic pole	Longitude, ϕ_m, deg	Tilt, θ_m, deg
North	−69	16
South	121	22

where the integrals are evaluated as described for equation (12.15). The average transmission factors of equation (12.20) are compared with the numerical calculations of Smart and Shea (1983) in figure 12.20. The tilt angles of the poles (Johnson, 1965) are given in table 12.4 along with suitably chosen longitudes and are shown in relation to the vertical cutoff rigidities of Shea and Smart in figure 12.21.

During times of intense solar activity, the solar plasma emitted in solar flares and subflares advances outward and arrives at 1 AU from the Sun. If the Earth is locally present, the plasma interacts with the geomagnetic field in which the plasma pressure performs work on the local geomagnetic field. The initial impact produces hydromagnetic waves causing a general increase in geomagnetic intensity. As plasma flow is established, it generates large electric ring currents and a corresponding impressed magnetic storm field. In the initial phase (hydromagnetic wave), the storm field is parallel to the equatorial field after which the storm field reverses in the main phase of the storm caused by ring currents within the magnetopause and opposes the quiet field, causing a net decrease of the field strength. The main phase is followed by slow recovery to the quiet field conditions (Johnson, 1965).

Table 12.5. Relation of Magnetic Indices to
Magnetic Storm Field Strength

K_p	a_p	$|H_{st}|$, nT
0	0	0
1	4	8
2	7	14
3	15	30
4	27	54
5	48	96
6	80	160
7	132	264
8	207	414
9	400	800

The magnetic storm model used here assumes a uniform magnetic field impressed on the normal quiet field (Kuhn, Schwamb, and Payne, 1965). The storm field strength can be found from the change in the horizontal field component around the geomagnetic equator. We represent this field by H_{st}. Typical values of H_{st} in the main phase range from substorm values -10 nT to severe storms with -500 nT. On rare occasions, for very intense storms, the storm field exceeds -1000 nT.

Magnetic disturbances have been observed for many years, and various classification schemes for such disturbances have been proposed. The planetary magnetic index K_p is based on magnetometer measurements of 12 stations worldwide. The K_p index is related to a derived planetary index a_p and storm field strength by Bartels (Johnson, 1965) given in table 12.5.

The vertical cutoff rigidity as given by equation (12.16) is further modified to approximate the effects of geomagnetic disturbances. It was shown by Kuhn, Schwamb, and Payne (1965) that the appropriate equation is

$$R_C(\lambda_m) = \frac{14.9}{r^2} \cos^4 \lambda_m \left[1 + \frac{H_{st} r^3}{M} \left(\frac{4}{\cos^6 \lambda_m} - 1 \right) \right] \qquad (12.21)$$

for the centered dipole field. In the context of our approximation of the offset-tilted dipole field, we get

$$R_j(\lambda_m) = \frac{14.9}{(r + \delta_j)^2} \cos^4 \lambda_m \left[1 + \frac{H_{st} (r + \delta_j)^3}{M} \left(\frac{4}{\cos^6 \lambda_m} - 1 \right) \right] \qquad (12.22)$$

This vertical cutoff replaces equation (12.16) and applies to storm conditions. Note that the cutoff is zero whenever the result of equation (12.22) is negative. The corresponding transmission factor on the worst-case orbit ($\eta \approx 211°$) is shown in relation to the quiet field average transmission factors of Smart and Shea (1983) in figure 12.22.

12.2.5. Dose estimation.

In passing through tissue, energetic protons interact mostly through ionization of atomic constituents by the transfer of small amounts of momentum to orbital electrons. Although the nuclear reactions are far less numerous, their effects are magnified because of the large momentum transferred to the nuclear particles and the struck nucleus itself. Unlike the secondary electrons formed through atomic ionization by interaction with the primary protons, the radiations resulting from nuclear reactions are mostly heavy ionizing and generally have large biological effectiveness. Many of the secondary particles of nuclear reactions are sufficiently energetic to promote similar nuclear reactions and thus cause a buildup of secondary radiations. The description of such processes requires solution of the transport equation. The approximate solutions for the transition of protons in 30-cm-thick slabs of soft tissue for fixed incident energies have been made (Wilson and Khandelwal, 1976). The results of such calculations are dose conversion factors for relating the primary monoenergetic proton fluence to dose or dose equivalent as a function of position in a tissue slab.

Whenever the radiation is spatially uniform, the dose at any point x in a convex object may be calculated (Wilson and Khandelwal, 1974) by

$$D(\vec{x}) = \int_0^\infty \int_\Omega R_n \left[z_x(\vec{\Omega}), E \right] \phi(\vec{\Omega}, E) \, d\vec{\Omega} \, dE \qquad (12.23)$$

where $R_n(z, E)$ is the dose at depth z for normal incident protons of energy E on a tissue slab, $\phi(\vec{\Omega}, E)$ is the local differential proton fluence along direction $\vec{\Omega}$, and $z_x(\vec{\Omega})$ is the distance from the boundary along $\vec{\Omega}$ to point $\vec{x}$. It has been shown that equation (12.23) always overestimates the dose but is an accurate estimate when the ratio of the proton beam divergence due to nuclear reaction to the radius of curvature of the body is small (Wilson and Khandelwal, 1974). Equation (12.23) is a practical prescription for introducing nuclear reaction effects into calculations

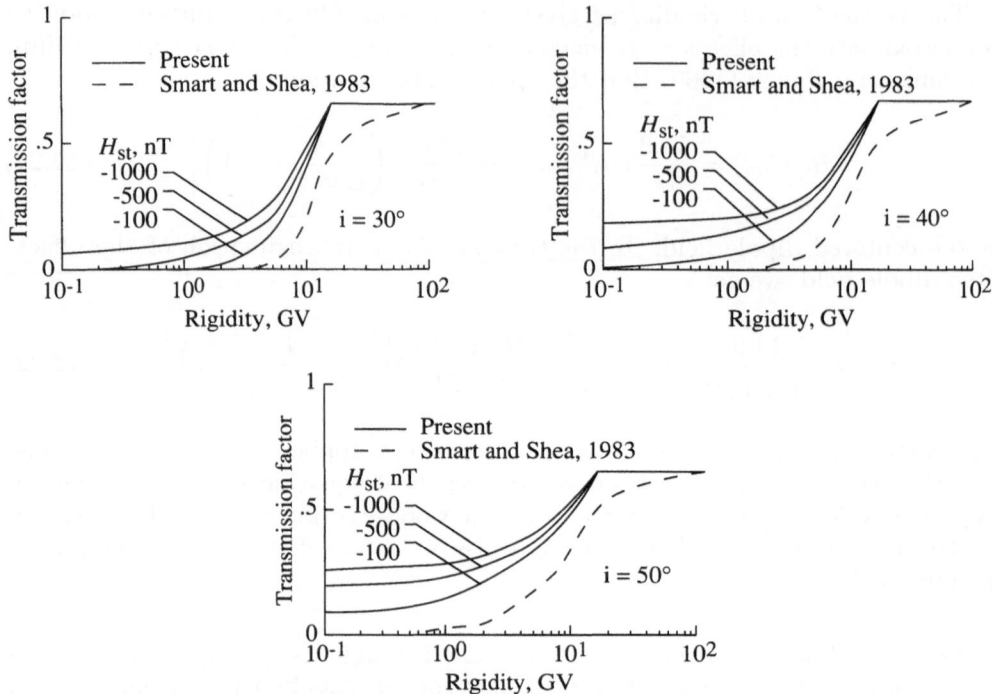

Figure 12.22. Maximum transmission factor for various storm fields and quiet time average transmission of Smart and Shea (1983). $\eta = 211°$.

of dose in geometrically complex objects such as the human body. The main requirement is that the dose conversion factors for a tissue slab be adequately known for a broad range of energies and depths. The dose conversion factors for tissue were derived by Wilson and Khandelwal (1976), and a correction for an aluminum shield is found in chapter 8. The spacecraft geometry is taken as an aluminum sphere of large radius.

12.2.6. Method for Shuttle geometry. In section 12.2.5, the calculation of astronaut (convex object) exposure in the center of a large aluminum sphere of arbitrary thickness was derived for a specific orbit with either the quiet geomagnetic field or with a geomagnetic disturbance. We denote that result by $D_{\text{sph}}(t_s)$, and it has a different value for each critical organ for which exposure within the aluminum sphere of thickness t_s is evaluated. Within the context of assumed isotropic radiation, the exposure at some location within the Shuttle (taken as a typical spacecraft geometry) is

$$D = \int_0^\infty D_{\text{sph}}(t_s) \, f(t_s) \, dt_s \qquad (12.24)$$

where $f(t_s)$ describes the mass distribution of the Shuttle structure assumed to be aluminum about that particular location. Physically, $f(t_s) \, dt_s$ is the solid angle fraction for which the areal density to the Shuttle surface lies between t_s and $t_s + dt_s$. The cumulative distribution of areal density is given by

$$F_C(t_s) = \int_0^{t_s} f(t_s)\, dt_s \qquad (12.25)$$

and is shown for two locations in the Shuttle (Atwell et al., 1989) in figure 12.23. Also shown in figure 12.23 are the following approximate functions:

$$f_1(t_s) = \begin{cases} \dfrac{0.176}{t_s} & (1 \le t_s \le 2) \\[2mm] \dfrac{0.113}{t_s} & (2 \le t_s \le 20) \\[2mm] \dfrac{0.353}{t_s} & (20 \le t_s \le 120) \end{cases} \qquad (12.26)$$

and

$$f_2(t_s) = \begin{cases} \dfrac{0.303}{t_s} & (1 \le t_s \le 6) \\[2mm] \dfrac{0.147}{t_s} & (6 \le t_s \le 132) \end{cases} \qquad (12.27)$$

where the functions are understood to be zero outside specified ranges. Formulas (12.24), (12.26), and (12.27) are used in conjunction with the methods described in sections 12.2.4 and 12.2.5 to estimate Shuttle exposure in the two locations, referred to by Atwell et al. (1989) as dosimeter locations 1 and 2, which are the most and the least shielded locations in the Shuttle crew compartment, respectively. The method can be easily expanded to include more astronaut organs and other Shuttle locations.

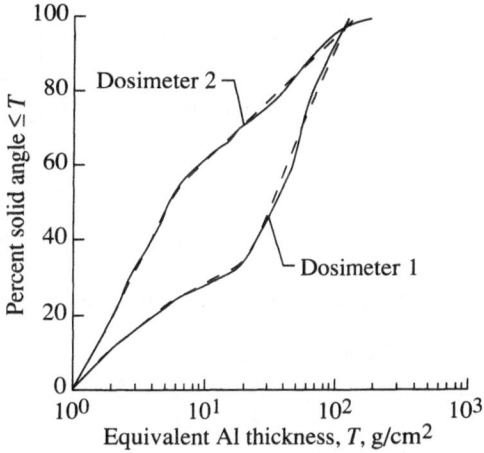

Figure 12.23. Mass distribution of two locations on Shuttle flight deck (Atwell et al., 1989).

12.2.7. Results. The maximum exposure limits in force for the Space Station *Freedom* are shown in table 12.1. The dose and dose equivalent to critical body organs for an aluminum shield 1 g/cm² thick are shown in tables 12.6 through 12.9 for various storm conditions (H_{st}). The exposures are shown for a worst exposed orbit ($\eta = 211°$) and the average over all η. The average is shown, since average transmission factors are calculated by several groups, and one may be tempted to use the transmission factor appropriate for the galactic cosmic-ray background (Curtis, Doherty, and Wilkinson, 1969; Adams, 1987). It is clear from the results in tables 12.6 through 12.9 that such use of average cutoffs provides exposure estimates which could be too small by a factor of 2 to 10. Such an underestimate is clearly unacceptable. Furthermore, if the current dose estimates are compared with values for transmission factors derived for a tilted concentric dipole field (Wilson et al., 1990), the dose values of the current field model are a factor of 3 to 5 higher. The eccentric field has two effects which lower the cutoffs. The offset displaces the South magnetic pole to lower latitudes and lowers the geomagnetic cutoff values over the Atlantic. The methods derived herein allow evaluating exposures as a function of the location of the line of nodes and should provide acceptable estimates of exposure.

Table 12.6. Skin Dose Behind Aluminum Shield 1 g/cm^2 Thick During February 25, 1956, November 12–13, 1960, and August 4, 1972, Events With Various Storm Fields

| Orbit inclination, deg | Skin dose, cGy, during— | | | | | | | | |
| | Feb. 1956 for H_{st}, nT, of— | | | Nov. 1960 for H_{st}, nT, of— | | | Aug. 1972 for H_{st}, nT, of— | | |
	-100	-500	-900	-100	-500	-900	-100	-500	-900
30, max	< 0.1	2.2	12.0	0	4.9	34.0	0	15.0	100.0
30, avg	< 0.1	0.3	2.6	0	0.5	7.1	0	1.6	22.0
40, max	2.9	19.0	28.0	6.3	53.0	79.0	19.0	170.0	240.0
40, avg	0.4	5.4	9.8	0.8	15.0	28.0	2.3	47.0	86.0
50, max	17.0	31.0	39.0	47.0	89.0	111.0	140.0	280.0	340.0
50, avg	4.8	13.0	18.0	13.0	36.0	51.0	39.0	110.0	160.0

Table 12.7. Skin Dose Equivalent Behind Aluminum Shield 1 g/cm^2 Thick During February 25, 1956, November 12–13, 1960, and August 4, 1972, Events With Various Storm Fields

| Orbit inclination, deg | Skin dose equivalent, cSv, during— | | | | | | | | |
| | Feb. 1956 for H_{st}, nT, of— | | | Nov. 1960 for H_{st}, nT, of— | | | Aug. 1972 for H_{st}, nT, of— | | |
	-100	-500	-900	-100	-500	-900	-100	-500	-900
30, max	< 0.1	3.3	17.0	0	7.0	49.0	0	20.0	140.0
30, avg	< 0.1	0.4	3.7	0	0.7	10.0	0	2.1	30.0
40, max	4.3	27.0	39.0	8.9	78.0	110.0	25.0	230.0	333.0
40, avg	0.7	7.7	14.0	1.1	22.0	40.0	3.1	63.0	120.0
50, max	24.0	44.0	54.0	68.0	130.0	160.0	200.0	380.0	470.0
50, avg	6.8	18.0	25.0	18.0	52.0	74.0	53.0	150.0	220.0

Table 12.8. BFO Dose Behind Aluminum Shield 1 g/cm^2 Thick During
February 25, 1956, November 12–13, 1960, and August 4, 1972,
Events With Various Storm Fields

Orbit inclination, deg	BFO dose, cGy, during—								
	Feb. 1956 for H_{st}, nT, of—			Nov. 1960 for H_{st}, nT, of—			Aug. 1972 for H_{st}, nT, of—		
	-100	-500	-900	-100	-500	-900	-100	-500	-900
30, max	< 0.1	1.0	3.3	0	0.9	4.4	0	1.8	10.0
30, avg	< 0.1	0.2	0.8	0	0.1	1.0	0	0.2	2.3
40, max	1.4	4.9	6.8	1.3	6.7	9.5	2.5	16.0	24.0
40, avg	0.3	1.5	2.5	0.2	2.0	3.4	0.4	4.7	8.8
50, max	4.6	7.6	9.3	6.2	11.0	13.0	15.0	27.0	33.0
50, avg	1.5	3.2	4.4	1.8	4.4	6.2	4.2	11.0	16.0

Table 12.9. BFO Dose Equivalent Behind Aluminum Shield 1 g/cm^2 Thick
During February 25, 1956, November 12–13, 1960, and August 4, 1972,
Events With Various Storm Fields

Orbit inclination, deg	BFO dose equivalent, cSv, during—								
	Feb. 1956 for H_{st}, nT, of—			Nov. 1960 for H_{st}, nT, of—			Aug. 1972 for H_{st}, nT, of—		
	-100	-500	-900	-100	-500	-900	-100	-500	-900
30, max	< 0.1	1.7	5.1	0	1.3	6.1	0	2.6	15.0
30, avg	< 0.1	0.3	1.2	0	0.2	1.3	0	0.3	3.3
40, max	2.3	7.6	10.0	1.8	9.2	13.0	3.6	23.0	34.0
40, avg	0.5	2.3	3.9	0.3	2.7	4.7	0.5	6.7	12.0
50, max	7.3	12.0	14.0	8.5	15.0	18.0	21.0	38.0	48.0
50, avg	2.3	5.0	6.9	2.5	6.1	8.6	6.0	15.0	22.0

From observing the levels of exposure in low inclination orbits ($i \approx 30°$), a significant exposure could clearly occur if particle arrivals coincided with a large magnitude ($K_p \approx 9$) magnetic disturbance. On the basis of the present analysis, a more in-depth study of potential solar flare exposure of the Space Station *Freedom* seems warranted. Such a study should include a review of the history of major geomagnetic disturbances in proximity to solar particle events, a review of alternate geomagnetic storm models, and a review of the specific Space Station *Freedom* shield geometry.

The exposure for Shuttle flight in a 400-km (216-n.mi.) orbit with a 50° inclination is shown in table 12.10 for the February 23, 1956, solar event spectrum as compiled by Foelsche et al. (1974). A magnetic storm was assumed to be in progress with an impressed field of -100 nT. The results shown in table 12.10 are for the long-term, average geomagnetic cutoffs, since these are directly comparable with the work of other geomagnetic models (Curtis, Doherty, and Wilkinson, 1969; Adams, 1987). We note, however, that actual exposure could be greatly different depending on the location of the line of nodes at the time of arrival of the high-energy flare particles.

Table 12.10. Human Exposure at Two Locations in Shuttle Crew Compartment for February 23, 1956, Event With $H_{st} = -100$ nT at 400 km and Orbit Inclination of $50°$

	Exposure in—					
	BFO		Skin		Lens	
Location	cGy	cSv	cGy	cSv	cGy	cSv
1	2.6	4.9	4.0	6.8	4.1	7.6
2	3.4	5.9	6.0	9.4	6.0	10.0

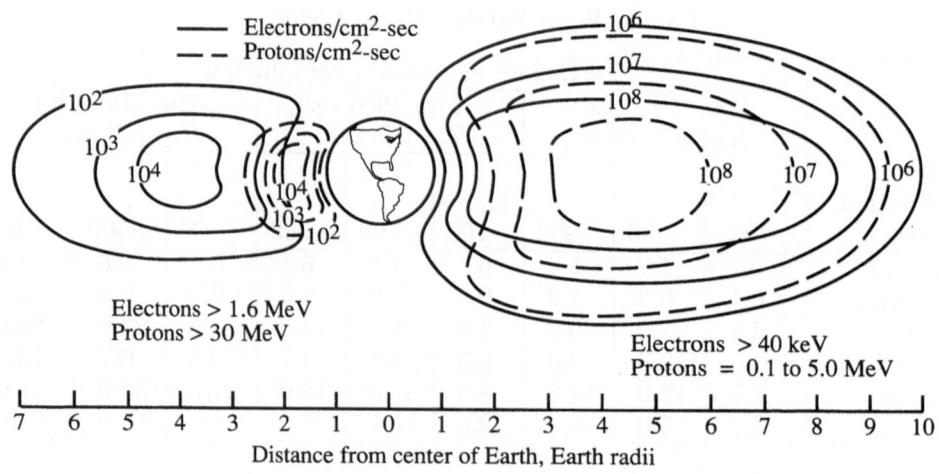

Figure 12.24. Near-Earth trapped radiation and solar proton environment (Parker and West, 1973).

12.2.8. Geomagnetically trapped radiations.

The trapped radiation, illustrated in figure 12.24, follows a helical path along the magnetic field lines between the mirror points. The location of the mirror points along the field line depends on the pitch angle at the magnetic equator and the energy of the particle. The greater the energy or the higher the pitch angle, the deeper the mirror points lie in the magnetic field. If the particle energy and pitch angle are sufficiently large, the mirror point is so deep that the particle interacts with the atmosphere and is lost from the particle population. For the inner zone it appears at least for the protons that the particle source is primarily neutrons which are produced in atmosphere by solar and galactic cosmic rays. The outer zones appear to be something like a pipe line with strong sources and strong sinks. The particles flow rapidly through these regions, and on the average they maintain a fairly high population density although the residence time is short. (See Singley and Vette, 1972.)

The Earth's magnetic field is not centered at the Earth's geographic center. Also, the main dipole moment, along the principal axis of the magnetic field, is tilted with respect to the Earth's rotational axis so that the geomagnetic field is not symmetrical with respect to geographic coordinates. The exposure as a function of altitude is expressed as maximum stay time before reaching the exposure limit (Wilson, 1978) in figure 12.25.

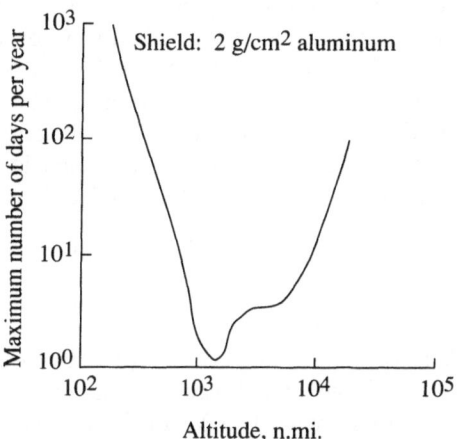

Figure 12.25. Limits imposed by trapped radiations on space operations within a 2 g/cm^2 aluminum shield.

12.3. Analysis for Deep Space Missions

The analyses presented here focus on the shielding requirements for GCR and different flare scenarios (Simonsen and Nealy, 1991). Shielding thicknesses selected for these missions should also reduce doses incurred from the Van Allen belts to a negligible amount providing long times are not spent in the belts.

12.3.1. Transport codes. The NASA Langley Research Center nucleon and heavy ion transport computer code HZETRN (Wilson et al., 1991) is used to predict the propagation and interactions of the deep space nucleons and heavy ions through various media. For large solar flare radiation, the baryon transport code BRYNTRN (Wilson et al., 1989) is used. For the galactic cosmic rays, an existing heavy ion transport code is integrated with the BRYNTRN code to include the transport of high-energy heavy ions up to atomic number 28 (Wilson and Badavi, 1986; Wilson, Townsend, and Badavi, 1987; Wilson et al., 1991). Both codes solve the fundamental Boltzmann transport equation in the one-dimensional, or straight ahead, approximation form:

$$\left[\frac{\partial}{\partial x} - \frac{\partial}{\partial E} S_j(E) + \sigma_j(E)\right] \phi_j(x, E) = \sum_{k>j} \int_E^\infty \sigma_{jk}(E, E') \phi_k(x, E') \, dE' \quad (12.28)$$

where the quantity to be evaluated, $\phi_j(x, E)$, is the flux of particles of type j having energy E at spatial location x. The solution methodology of this integrod-ifferential equation may be described as a combined analytical-numerical technique (Wilson, 1977). The accuracy of this numerical method has been determined to be within approximately 1 percent of exact benchmark solutions (Wilson and Townsend, 1988). The data required for solution consist of the stopping power S_j in various media, the macroscopic total nuclear cross sections σ_j, and the differential nuclear interaction cross sections σ_{jk}. The differential cross sections describe the production of type j particles with energy E by type k particles of energies $E' > E$. Detailed information on these data base compilations is described in chapters 4 and 5.

In addition to benchmark solution checks on the numerical precision of the code, comparisons with standard Monte Carlo type calculations have been made (Shinn et al., 1990). Samples of BRYNTRN results and results from the statistical Nucleon Transport Code (NTC) (Scott and Alsmiller, 1968) are shown in figure 12.26, where the dose values are given for a 30-cm tissue layer behind an aluminum shield of 20 g/cm^2. The input spectrum used is expressed analytically with the integral fluence F as a function of proton rigidity R:

$$F(>E) = C_o \exp\left[\frac{-R(E)}{R_o}\right] \tag{12.29}$$

with R_o equal to 100 MV and C_o chosen so that $F(30 \text{ MV})$ equals 10^9 protons/cm^2. Such a function is representative of a large proton event, and the BRYNTRN results show excellent agreement with the Monte Carlo calculations.

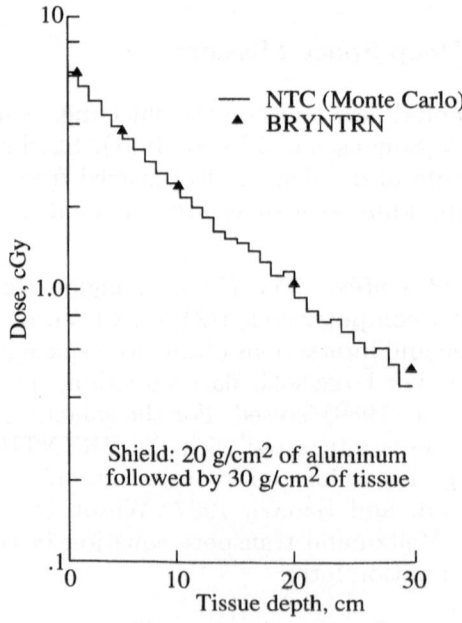

Figure 12.26. Comparison of results from BRYNTRN with equivalent Monte Carlo calculations (Shinn et al., 1990).

670

The present GCR code formulation is considered to be an interim version, since some features of the transport interaction phenomena have yet to be incorporated. These include improvements and additions to the existing nucleus-nucleus cross sections and their energy dependence and provisions for pion and muon contributions. Further improvements in target fragmentation treatment and computational efficiency are to be incorporated (Shinn and Wilson, 1991) even though computational execution times are already faster than counterpart statistical (Monte Carlo) calculations. These improvements should not greatly alter the current results, and the present interim version of the GCR code should provide a reasonable description of cosmic-ray particle fluxes and the corresponding dose predictions. The results included herein are preliminary and should be considered as current state-of-the-art "best estimates."

The absorbed dose D due to energy deposition at a given location x by all particles is calculated according to

$$D(x) = \sum_j \int_0^\infty S_j(E)\, \Phi_j(x, E)\, dE \tag{12.30}$$

The degree to which biological systems undergo damage by ionizing radiation is not simply proportional to this absorbed dose for all particle types. For human exposure, the dose equivalent is defined by introducing the quality factor Q which relates the biological damage incurred because of any ionizing radiation to the damage produced by soft X-rays. (See limitations on Q discussed in chapter 11.) In general, Q is a function of linear energy transfer, which in turn is a function of both particle type and energy. For the present calculations, the quality factors used are those defined by the International Commission on Radiological Protection (ICRP 26 (Anon., 1977)). The values of dose equivalent H are computed as

$$H(x) = \sum_j \int_o^\infty Q_j(E)\, S_j(E)\, \Phi_j(x, E)\, dE \tag{12.31}$$

These values are used to specify radiation exposure limits. (See table 12.1.) Strictly speaking, the limits in table 12.1 apply to low Earth orbit operations but are used as guidelines in the current analysis as suggested by NCRP 98 (Anon., 1989a).

12.3.2. Propagation data. The BRYNTRN code and the combined nucleon/heavy ion transport code are easily applied to various media. The GCR and solar flare energy distributions (figs. 12.3, 12.14, 12.17(b), and 12.18) are input for the code as the initial particle fluxes at the media boundaries. Results include slab calculations of the particle-flux energy distributions at various absorber amounts from which slab dose estimates as a function of absorber amount are determined. The slab calculations correspond to a monodirectional beam of particles normally incident on a planar layer of shield material. The dose at a specific slab-shield depth with normal incident radiation is equivalent to the dose in the center of a spherical shield of the same thickness in a field of isotropic radiation. (See chapter 7 for limitations.) This is depicted in figure 12.27. This relation was shown formally by Wilson and Khandelwal (1974).

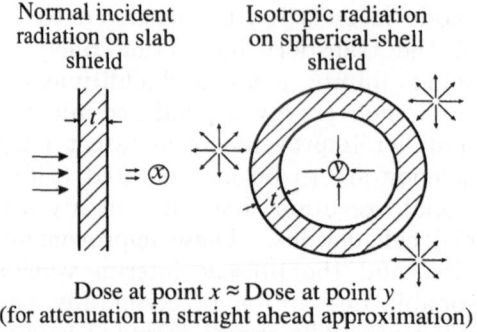

Normal incident radiation on slab shield

Isotropic radiation on spherical-shell shield

Dose at point x ≈ Dose at point y
(for attenuation in straight ahead approximation)

Figure 12.27. Calculation equivalence of slab shield and spherical-shell shield.

Basic propagation data have been generated for a variety of materials for both the GCR spectrum and different flare spectra. The propagation results are displayed as dose versus absorber amounts (g/cm^2) which can be converted to a linear thickness by dividing by density. Displaying results in this manner is helpful in comparing the shield effectiveness of various materials because equal absorber amounts for a given shielded volume will yield equal shield masses even though their linear thicknesses may differ.

For incident solar flare protons, the variation of dose with shield amount is sensitive to the energy characteristics (differential flux spectra). Figure 12.28 illustrates the BFO dose as a function of depth in aluminum followed by a 5-cm tissue layer for the three flares whose spectra are shown in figure 12.14. For these flares, the proton fluences have an approximate coincidence close to 100 MeV. Consequently, this behavior is reflected in a corresponding crossover of the dose-depth curves of figure 12.28, where the coincidence occurs at approximately 15 g/cm^2 of aluminum.

The combined fluences of the solar proton events occurring in the latter part of 1989 (fig. 12.18) have spectral characteristics similar to the August 1972 event. The BFO dose as a function of depth for several shield materials is shown in figure 12.29 for this flare scenario. On a per-unit-mass basis, water and lithium hydride have almost identical shield effectiveness properties for all shield thicknesses. Such similarities apply as well to media of low atomic weight and high hydrogen content (e.g., hydrocarbon polymers) which may be used as bulk shields. The curves for aluminum and lead are indicative of the decreasing relative effectiveness of higher atomic weight media. This effect can be attributed to the differences in proton stopping powers of the materials and to the greater numbers of secondary nucleons generated in the heavier materials. This effect is further exemplified by the results shown in figure 12.30, which shows the BFO dose-depth functions for the GCR spectra at solar minimum conditions. In addition to water and aluminum, results for liquid hydrogen (which may be used in an application to propellant tank structures) show the dramatic superiority of this material as a shield. This is largely caused by the greatly reduced generation of reaction products (nucleons and fragments) created by the GCR heavy ions traversing the hydrogen medium. For the very energetic GCR spectrum, most of the reduction in dose for all the materials shown occurs in the first 20 to 30 g/cm^2, with the magnitude of the dose gradient decreasing at larger thicknesses.

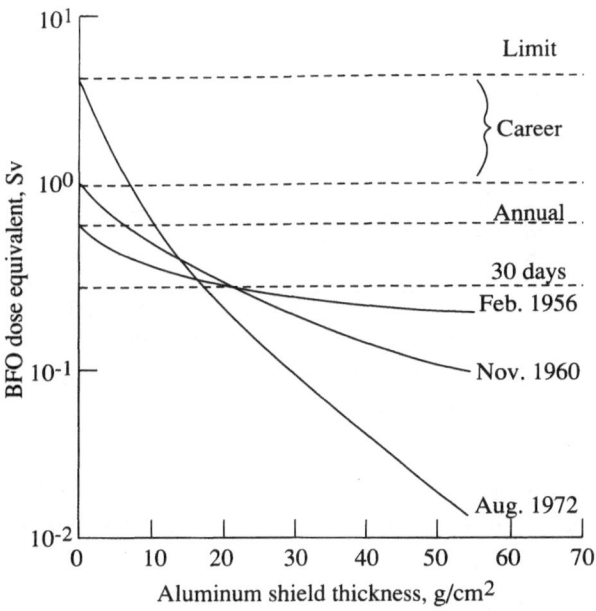

Figure 12.28. BFO dose equivalent as function of aluminum shield thickness for three large solar flare events (Townsend et al., 1989).

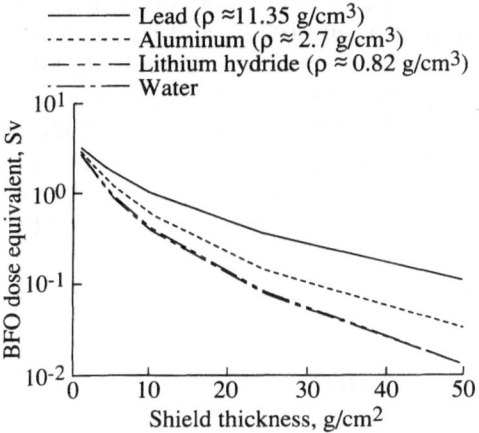

Figure 12.29. BFO dose equivalent versus depth functions for sum of 1989 flare fluences for four materials.

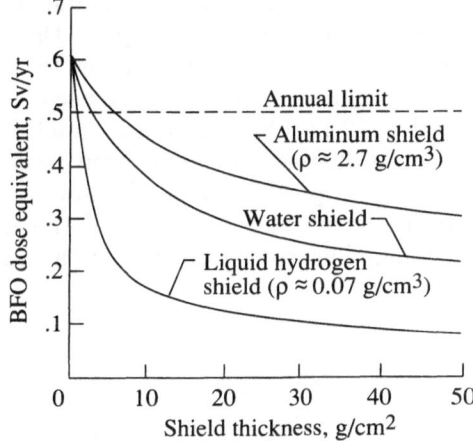

Figure 12.30. BFO dose equivalent as function of shield type and thickness resulting from galactic cosmic rays at solar minimum (Townsend et al., 1990a).

The differences between the GCR at solar minimum and maximum with respect to water shield thicknesses are shown in figure 12.31 (Townsend ct al., 1990a). The incurred dose equivalents between these two extremes are seen to differ by about a factor of 2 for shield amounts up to 30 g/cm^2. These results were computed for the GCR spectra at solar minimum and maximum as specified by the NRL CREME model. However, recent measurements (Kovalev, Muratova, and

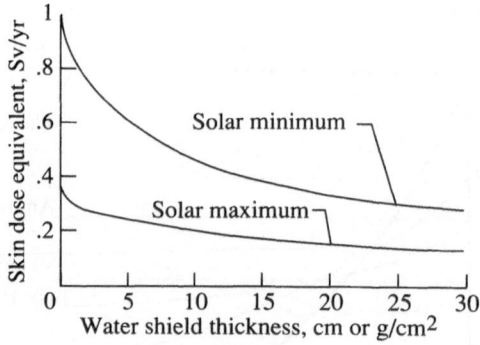

Figure 12.31. Dose equivalent as function of water shield thickness resulting from galactic cosmic
rays at solar minimum (Townsend et al., 1990a).

Petrov, 1989) made during the last solar cycle imply that the GCR intensity during
solar maximum may actually be greater than that prescribed in the NRL model.
(See chapter 11.)

We have dealt with the transport results for some of the more common
materials which may be fabricated and/or supplied as shield media. For habitats
on the Moon and Mars surfaces, the regolith (or soil) of a particular locale is
a convenient candidate for bulk shielding. In the analyses presented herein, the
regolith composition is modeled with the mass normalized concentrations of the
five most abundant elements found in the soil. The lunar model composition is
based on Apollo return samples (Dalton and Hohmann, 1972), and the martian
model composition is based on Viking lander data (Smith and West, 1983).
The normalized compositions used in the regolith shielding studies are given
in table 12.11 (Nealy, Wilson, and Townsend, 1988; Simonsen et al., 1990b).
Moderate changes in composition are found to have negligible effects on the overall
shielding properties (Nealy, Wilson, and Townsend, 1988 and 1989). As might be
expected from the similarity of the Mars and Moon constituents, the regolith
shielding characteristics are comparable.

Table 12.11. Composition of Lunar and Martian Regoliths

Regolith	Composition, percent of normalized mass	Density, g/cm^3
Lunar	52.6 SiO_2 19.8 FeO 17.5 Al_2O_3 10.0 MgO	0.8–2.15
Martian	58.2 SiO_2 23.7 Fe_2O_3 10.8 MgO 7.3 CaO	1.0–1.8

The results of BFO dose versus depth in lunar regolith are given for the three large flares of solar cycles 19 and 20 in figure 12.32. The regolith results are very similar to those for aluminum (fig. 12.28); this is not surprising, since the mean molecular weight of the lunar regolith is comparable with the atomic weight of aluminum. Figure 12.33 shows the calculated propagation data for the GCR at solar minimum conditions, with the contributions to the dose by neutrons, protons, α-particles, and two groups of heavier ions shown individually. For very thin layers, the heaviest ion group ($10 \leq Z \leq 28$) contributes over half the dose. For increasing thicknesses, the heavier ions fragment and react with target nuclei to produce particles of lower mass (ultimately nucleons) which then deliver the greater percentage of the dose. For the lunar soil, approximately 90 percent of the dose is estimated to result from nucleons (mostly secondaries) for shield layers greater than approximately 20 g/cm^2.

The exposures on Mars differ considerably from those on the moon because of the carbon dioxide atmosphere on Mars. Consequently, dose-depth functions are generated in carbon dioxide for the flare spectra of figure 12.14. These results are shown in figure 12.34. The shielding effectiveness per unit mass of carbon dioxide is greater than the effectiveness of either aluminum or regolith results as shown previously (figs. 12.28 and 12.32). This is particularly the case for shield amounts exceeding 25 to 30 g/cm^2 of material. A similar observation may be made for the GCR results for carbon dioxide (fig. 12.35) compared with the corresponding calculations for aluminum and lunar regolith (figs. 12.30 and 12.33). The basic carbon dioxide propagation data may be applied to the martian atmosphere when gas density as a function of altitude is specified.

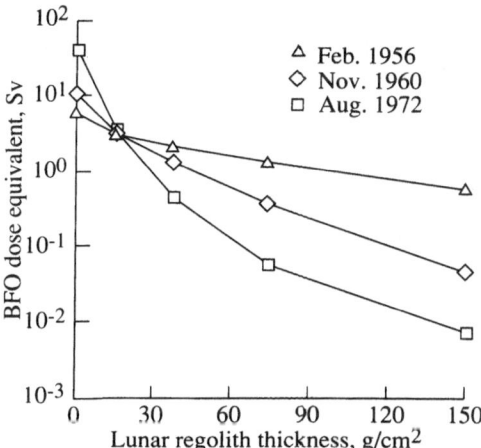

Figure 12.32. Predicted BFO dose equivalent for slab thickness between 0 and 150 g/cm^2 in simulated lunar regolith for three solar flare events (Nealy, Wilson, and Townsend, 1988).

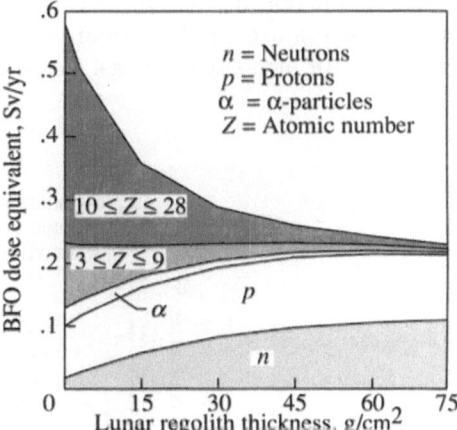

Figure 12.33. Annual BFO dose equivalent contribution from specified particle constituents as function of lunar regolith thickness for GCR at solar minimum (Nealy, Wilson, and Townsend, 1989).

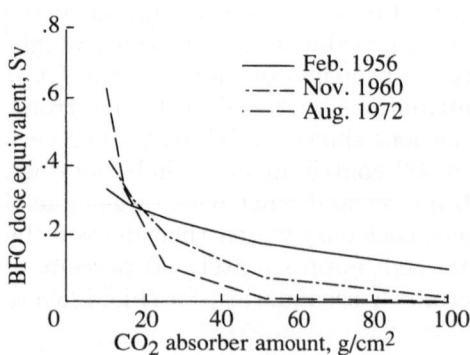

Figure 12.34. BFO dose equivalent as function of carbon dioxide absorber amounts for three solar flare events (Simonsen et al., 1990a).

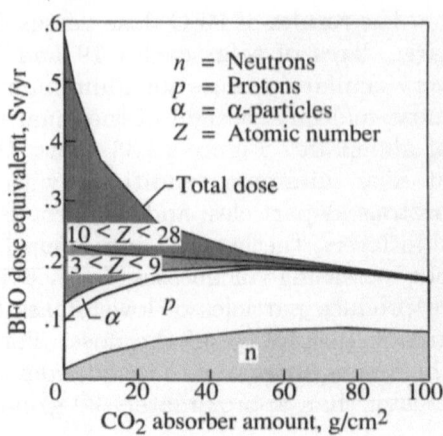

Figure 12.35. Annual BFO dose equivalent contributions from specified particle constituents as function of carbon dioxide absorber amount for GCR at solar minimum (Simonsen et al., 1990a).

When martian regolith is considered as a protective shield medium, the transport calculations must be made for the atmosphere-regolith thicknesses combined. In this case, the detailed flux-energy spectra emergent from a specified carbon dioxide amount must be used as input for the subsequent regolith calculation. Sample BFO dose results for such a procedure are given in figure 12.36, where fixed carbon dioxide amounts are used in conjunction with regolith layers. Two GCR cases and the energetic 1956 solar flare are included in the analysis. For moderate carbon dioxide absorber amounts, the dose reductions from additional regolith layers are small compared with the dose reduction occurring in the first few g/cm^2 of atmosphere (figs. 12.34 and 12.35).

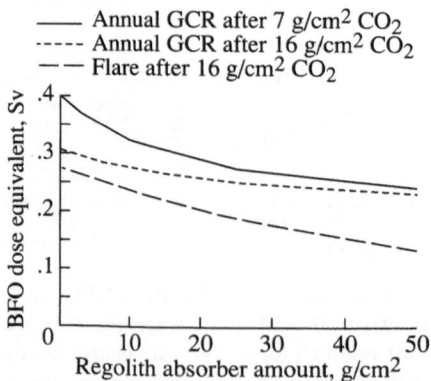

Figure 12.36. BFO dose equivalent as function of regolith absorber amount after transport through martian atmosphere in vertical direction (Simonsen et al., 1990b).

12.4. Description of Shield Assessment Results

When the computed propagation data for the GCR and solar flare protons in different materials are applied to specific shield geometries, the dose at a specific target point can be evaluated. To evaluate the dose at a particular point, the radiation from all directions must be determined. In deep space, radiation will surround the crew from the full 4π sr. However, on a planetary surface, only a solid angle of 2π is considered because the mass of the planet protects the crew from half the deep space radiation. The dose contribution attributed to particles arriving from a given direction is determined by the shield thickness encountered along its straight-line path to a specified target point. For shield assessments in these analyses, the absorber amounts and the corresponding dosimetric quantities are evaluated for zenith angles between $0°$ and $90°$ in $5°$ increments, and for azimuth angles of $0°$ to $360°$ also in $5°$ increments. The directional dose is then numerically integrated over the solid angle (4π for deep space, 2π for planetary applications) about a target point to determine the total dose at that point. For deep space calculations when a spherical shielded volume is considered, the slab dose calculations can be used directly. (See fig. 12.27). The dose estimates presented here conservatively estimate the skin dose as the dose at 0 cm depth and the BFO dose as the dose at 5 cm depth. When detailed body geometry is considered, for example, incorporating the thickness distributions of the Computerized Anatomical Man Model (Billings and Yucker, 1973), the estimated doses will generally be lower, with the amount of reduction being dependent on the energetic particle environment spectrum (Shinn, Wilson, and Nealy, 1990). Dose estimates using the propagation data for various materials are determined for the following shielded volumes: (1) interplanetary transportation vehicles, (2) lunar habitats, and (3) martian habitats.

12.4.1. Transportation vehicles. Unshielded BFO dose equivalents in deep space are substantial and could be lethal if an unusually large flare occurred. From galactic cosmic radiation at solar minimum, the unshielded astronaut would receive approximately 0.6 Sv/yr. The three large flare events of August 1972, November 1960, and February 1956 would have delivered unshielded doses of approximately 4.11, 1.10, and 0.62 Sv, respectively. The GCR dose is over the annual limit of 0.5 Sv/yr and the flare doses are significantly greater than the 30-day limit of 0.25 Sv. Clearly, both lunar and martian transportation vehicles must offer adequate protection. The protection for the short lunar travel time will most likely emphasize flare protection, whereas the protection required for the longer travel time required for Mars must consider both the GCR and the flares combined. The following analyses consider radiation protection for transportation vehicles required for various flare scenarios and for galactic cosmic radiation.

The normal-incidence slab calculations, presented in section 12.3, can be used to estimate the doses inside nearly spherical structures in an assumed isotropic radiation field. Results of such an application are presented in table 12.12 for the three large solar flare events (Townsend et al., 1989). The aluminum wall thicknesses required to reduce the incurred dose from large flares to the astronaut 30-day limits for ocular lens, skin, and blood-forming organs are estimated. Even though the individual flare spectra exhibit marked differences (fig. 12.14), the

Table 12.12. Aluminum Shield Thickness Required for Solar Flare
Protection to Remain Below the 30-Day Limit

[Data from Townsend et al., 1989]

Organ	Aluminum shield thickness for solar flare event in—					
	February 1956		November 1960		August 1972	
	g/cm^2	cm	g/cm^2	cm	g/cm^2	cm
Skin	1.3	0.5	2.5	1.0	7.5	2.8
Ocular lens	1.5	0.6	3.5	1.3	9.5	3.5
BFO	24.0	8.9	22.0	8.1	18.0	6.7

required shielding thicknesses range from approximately 18 to 24 g/cm^2 (7 to 9 cm) of aluminum. The shielding mass required can be reduced by approximately 15 to 30 percent using water as shielding with thicknesses of only 15 to 20 g/cm^2 required (Townsend et al., 1989). These shielding estimates include only a flare contribution and represent a minimum acceptable wall thickness. Rather than shielding an entire spacecraft with these wall thicknesses, the crew can be provided with a heavily shielded "shelter" for protection during a large flare event. More recent solar flare analyses have been done by Nealy et al. (1990a), Simonsen et al. (1991), and Townsend, Shinn, and Wilson (1991).

For long-duration missions, contributions from the GCR and the more numerous smaller flares should be considered. Dose evaluations throughout a complete solar cycle are made with the flare data (fig. 12.17) measured during solar cycle 21 between 1975 and 1986 (Nealy et al., 1990b). The GCR contribution is assumed to vary sinusoidally from peak values at solar minimum to the smallest dose rate at solar maximum. Normal-incidence slab calculations for the dose evaluations are made with effective water shield thicknesses. Water, both potable and waste, may be a likely shield material for long-duration missions since it will probably be available in large quantities. Water calculations can be used to simulate results for other media with low atomic weight and high hydrogen content. Consequently, reasonable shield mass requirements may be estimated on the basis of water transport results.

Figures 12.37 and 12.38 show sample BFO dose estimates from this study as a function of time within the solar cycle. In figure 12.37, the dose equivalent incurred for an effective water shield of 5 g/cm^2 is given for mission durations of 3, 12, and 36 months. The figure shows the dose integrated over mission duration time, with the flare contribution (according to solar cycle 21 distribution) appearing as deviations above the smooth sinusoidal curve, which would be seen for the GCR contribution alone. The results indicate that the flare contribution is not conspicuous in comparison with the more regularly varying GCR component. Note, however, that there were no unusually large events in cycle 21 as has been observed in cycles 18, 19, 20, and 22. For missions of duration longer than 1 year in cycle 21, the dose contributions due to the normally occurring solar flares may not be significant in comparison with the GCR (for shield amounts greater than

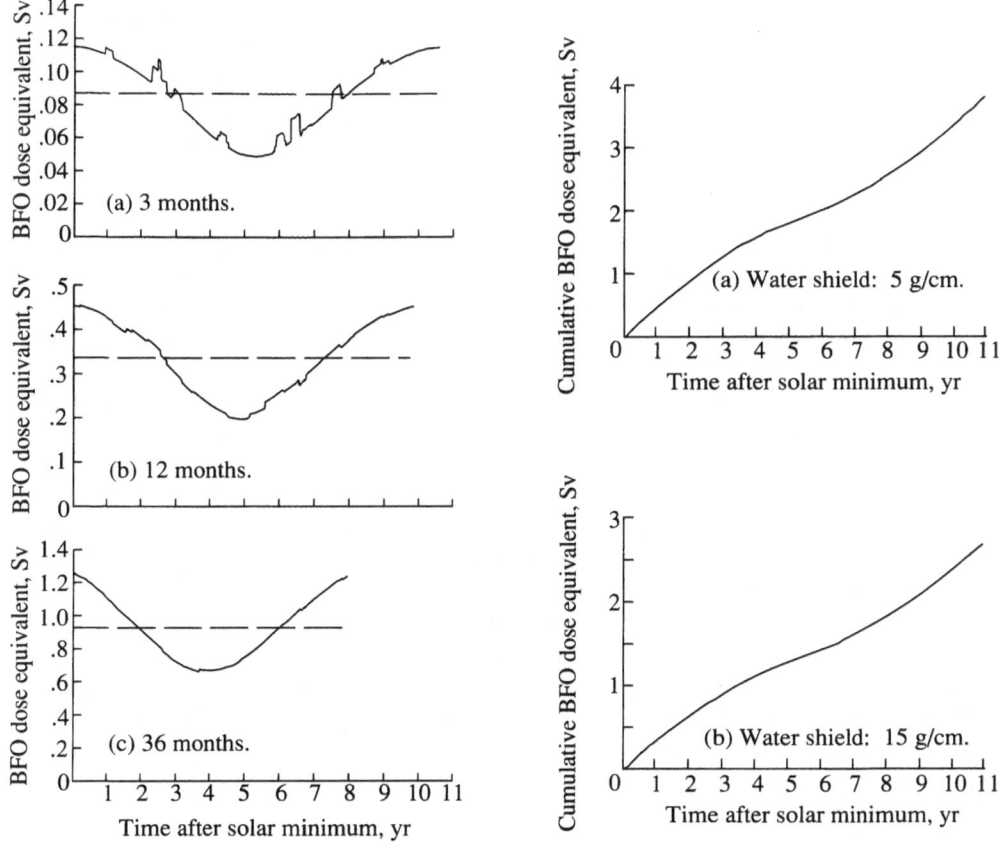

Figure 12.37. Free-space BFO dose equivalent incurred for water-slab shields 5 g/cm^2 thick for three mission lengths as function of time in cycle after solar minimum. Dashed lines indicate cycle average values (Nealy et al., 1990b).

Figure 12.38. Cumulative total BFO dose equivalent incurred throughout the 11-year solar cycle for water slab shields (Nealy et al., 1990b).

5 g/cm^2). If this is true, the cumulative dose is approximately proportional to the mission duration time.

The BFO dose received by crew members on a 3-month, 12-month, or 36-month mission starting in any portion of solar cycle 21 may be predicted from figure 12.37. For example, the final dose value on figure 12.37(c) of about 1.25 Sv represents the dose incurred for a mission beginning 8 years after solar minimum and lasting over the next 3 years. This plot (fig. 12.37(c)) also indicates that a 36-month mission beginning 4 years after solar minimum would result in a total incurred dose approximately 45 percent lower than would be received on a mission beginning at solar minimum.

Figure 12.38 illustrates the variation of the cumulative incurred dose equivalent throughout the entire solar cycle 21 for water shields 5 and 15 g/cm^2 thick. This type of representation is useful in estimating incurred dose for long-duration missions (2 years or more) which begin and end at arbitrary times within solar cycle 21. For example, from figure 12.38(a), the total BFO dose for a 5-year mission beginning at solar minimum is predicted to be approximately 1.80 Sv for shielding 5 g/cm^2 thick. However, if the 5-year mission begins 3 years after solar minimum, the total incurred dose is estimated to be approximately 1.35 Sv (2.6 Sv at year 8 minus 1.25 Sv at year 3).

The preceding results from the solar cycle 21 analysis do not include contributions from a rarely occurring giant solar proton event (e.g., the events of 1956, 1960, 1972, 1989), and such an event must be accounted for separately as circumstances warrant. For example, for a 1- or 2-year mission spanning the solar minimum, a large proton event would be highly unlikely, whereas during active Sun conditions, a larger (but still relatively small) probability exists that incurred doses would be considerably increased because of large flare episodes.

The results of the solar cycle 21 study indicate that a reasonably conservative radiation environment for exposure analysis may be derived from the solar minimum GCR flux with the inclusion of one large proton event. The BFO dose depth variation for such an environment consisting of the fluence of the 1972 large proton event in combination with the annual GCR contribution is given in figure 12.39. The BFO dose equivalent of 0.5 Sv is exceeded for water shield thicknesses less than about 18 g/cm^2. For shields thicker than 25 or 30 g/cm^2, the flare dose is insignificant. These propagation data can be used to estimate shield masses of various manned habitation modules.

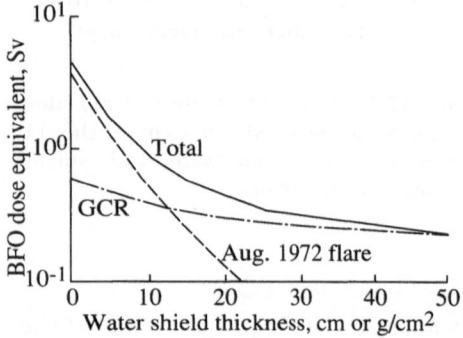

Figure 12.39. BFO dose equivalent as a function of water shield thickness for August 1972 flare and GCR at solar minimum (Simonsen and Nealy, 1991).

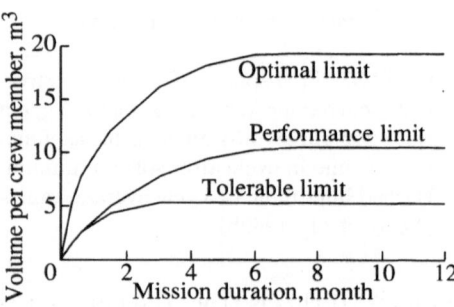

Figure 12.40. Guidelines for the determination of total habitable volume required per person in space module (NASA-STD-3000 (Anon., 1987)).

Guidelines for manned module volume requirements are graphically depicted in figure 12.40 (Anon., 1987). According to these guidelines, long-duration missions would require at least 10 m^3 per crew member as a performance limit and approximately 19 m^3 as an optimal limit. (Here, the tolerance limit volume is not considered to be applicable for normal operations on extended missions.) A four person crew is recommended for a manned Mars mission (The 90-Day Study (Anon., 1989b)) which implies a minimum habitable volume of approximately 42 m^3. If a cylindrical module is assumed, with diameter equal to length, the shield mass of the configuration may then be found as a function of dose delivered near the center of the module. Figure 12.41 shows the annual delivered dose due to GCR and the August 1972 flare as a function of cylinder wall mass. Again, equivalent water shield thicknesses are used in these estimates (fig. 12.39). If one considers an acceptable design criteria to be 50 to 70 percent of the maximum allowable dose, then shield masses on the order of 20 to 30 metric tons are required for the volume of 42 m^3. Estimates of shield mass will be greater if aluminum is assumed to be the shielding material because of the poorer shielding characteristics of aluminum. In some cases, the shield mass can be a significant fraction of the total mass of the candidate concepts of the Mars transportation vehicle (The 90-Day Study (Anon., 1989b)). However, the bulk shielding mass is not necessarily extra mass that must be provided but the total shielding required which can include the pressure vessel walls, water tanks, fuel tanks, and other components of the spacecraft.

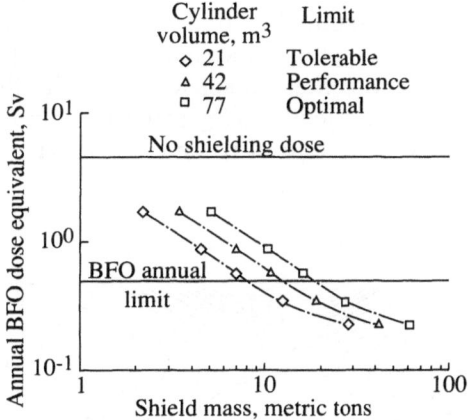

Figure 12.41. BFO dose equivalent incurred from the August 1972 flare and GCR (fig. 12.39) versus shield mass for cylindrical modules (Length/Diameter = 1.0) of various volumes based on the requirements of figure 12.40 for a four man crew (Simonsen and Nealy, 1991).

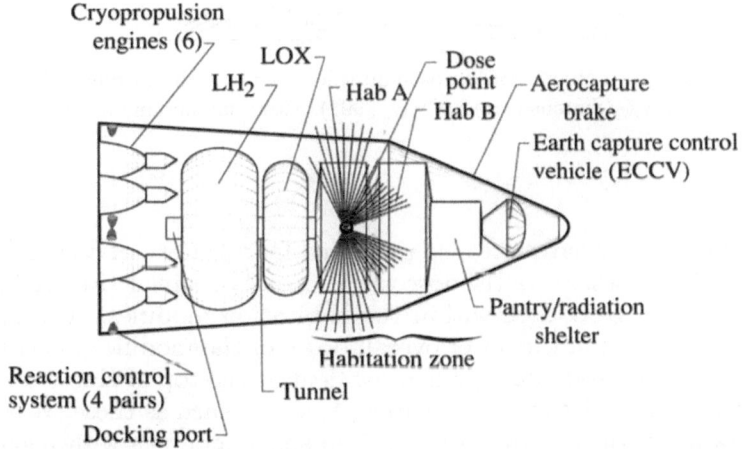

Figure 12.42. Configuration of martian piloted vehicle with sample directional dose patterns for a point inside of Hab A module (Simonsen and Nealy, 1991).

The basic propagation data generated in the form of slab dose estimates can also be used for more detailed dose analyses of specific shielded configurations. One such configuration, depicted in figure 12.42, is a concept of a manned Mars transfer vehicle developed by the Martin Marietta Corporation. This concept contains two cylindrical habitation modules (diameter of 7.6 m, length of 2.7 m). For the present calculation, the combination of components and bulk shielding for each habitat module is assumed to be equivalent to an effective water shield thickness of 5 g/cm^2. Also contributing to the shielding are the ECCV, pantry, and fuel tanks.

The directional dose due to GCR (at solar minimum) was calculated for an interior point in the center habitat modules. Figure 12.42 shows the axisymmetric directional dose pattern superimposed on the vehicle configuration outline. This pattern consists of vectors emanating from a target point with their lengths proportional to the annual GCR dose per unit solid angle. Although the radiation field outside the spacecraft is assumed to be isotropic, geometry effects cause the internal field to be highly anisotropic. In particular, very little radiation penetrates from solid angles subtending the fuel tanks, which in the illustrative calculation are assumed to be full. By numerically integrating the directional dose, the BFO dose in the center of Hab A is estimated to be 0.29 Sv/yr.

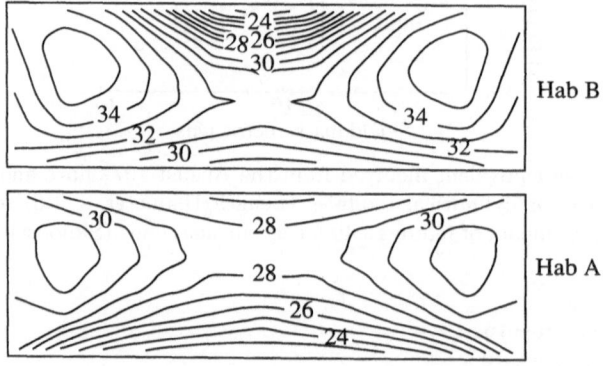

Figure 12.43. Variation in annual BFO dose equivalent for cylindrical habitation modules for galactic cosmic rays. (Simonsen and Nealy, 1991). Contour increments are 1 cSv/yr.

Total BFO dose estimates are also predicted for a variety of points within each module from which contours of the dose variation are obtained. Figure 12.43 shows the variation of the BFO dose within the habitation modules. The influence of the fuel tank is evident in the lower overall doses of the module closest to the fuel supply (Hab A). The large dose gradient evident at the top of Hab B is because of the thick walls of the adjacent flare shelter. Analyses such as these are expected to be of importance in the design stages of deep space modules with regard to such things as crew-quarters layout, placement of equipment, storage of consumables, and waste.

12.4.2. Lunar surface habitation. Once on the lunar surface, the radiation hazards of deep space will be less severe. Unshielded BFO dose estimates for the flare events of August 1972, November 1960, and February 1956 are approximately half those of deep space: 2.05 Sv, 0.55 Sv, and 0.31 Sv, respectively. These dose estimates are significantly higher than the 30-day limit of 0.25 Sv. The BFO dose incurred from the GCR at solar minimum is estimated to be approximately 0.3 Sv/yr, which is below the annual limit of 0.5 Sv/yr. However, the GCR dose in conjunction with medium to large flare event doses may reach the annual limit and become career limiting for long-duration missions. These values clearly show the need for radiation protection while on the lunar surface. Local resources, such as lunar regolith, will be available for use as protective shielding to cover habitats. In this section, several habitat configurations are considered with different regolith shielding thicknesses for protection.

Dose calculations inside candidate habitats are predicted with the computed propagation data for solar flares and the GCR shown in figures 12.32 and 12.33. A conservative estimate of the deep space environment is to assume the combination of GCR at solar minimum and one large proton event. From figures 12.32 and 12.33, the regolith slab dose estimates imply that a thickness of 50 cm (75 g/cm^2 with regolith density assumed to be 1.5 g/cm^3) will reduce the BFO dose equivalent to approximately 0.4 Sv for the sum of the GCR and one large flare (February 1956). With the 2π-sr shielding on the lunar surface, it is expected that, with a regolith layer of 50 cm, the annual dose for this environment is reduced to approximately 0.2 Sv. Thus, a shield thickness of 50 cm is selected for analysis to reduce dose levels to slightly less than one half the annual limit (or a design safety factor of approximately 2). Shield thicknesses of 75 cm and 100 cm are also selected for analysis to determine the extent to which additional shielding will further reduce annual doses.

Early lunar habitats are described as a Space Station *Freedom* derived module and an inflatable/constructible sphere (Alred et al., 1988). The Space Station *Freedom* module is assumed to be 4.6 m in diameter and 12.2 m in length as shown in figure 12.44(a). The module is assumed to be lengthwise on the lunar surface covered with either 50 cm (or 75 g/cm^2 with regolith density assumed to be 1.5 g/cm^3) or 100 cm of lunar regolith overhead. Along the sides, the regolith material is filled in around the cylinder to form a vertical wall up to the central horizontal plane. For the 50-cm layer, the shield thickness will vary from 230 cm to 50 cm from ground level up to this plane. The spherical habitat is 15.2 m in diameter and is modeled as a half-buried sphere with the portion above ground level shielded with a regolith layer of either 50 cm, 75 cm, or 100 cm (fig. 12.44(b)).

The integrated BFO dose estimates which would have been incurred from the three flare events with shield thicknesses of 75 g/cm^2 and 150 g/cm^2 are shown in table 12.13 (Nealy, Wilson, and Townsend, 1988). The values in the table represent the dose in the center of the habitat for each flare event. The dose distribution was also calculated throughout each habitat. For the cylindrical module, the general dose levels show little change for heights above and below the center plane. The radiation field maxima occur at about two thirds the distance between the center and end wall. For the spherical habitat the field maximum occurs above the center point at positions closer to the top, whereas doses in

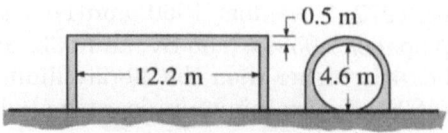

(a) Cylindrical module (side and end views).

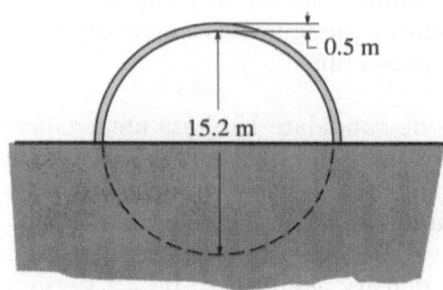

(b) Spherical module.

Figure 12.44. Modeled shielded configurations of candidate lunar habitation modules (Nealy, Wilson, and Townsend, 1989).

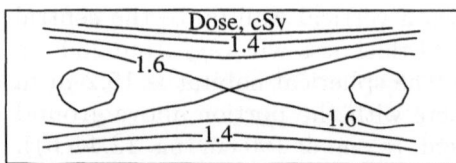

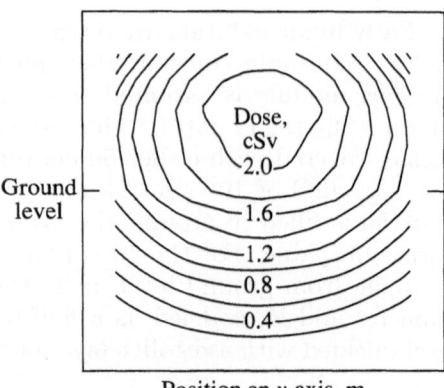

Position on y-axis, m

Figure 12.45. Variation in BFO dose equivalent from November 1960 solar flare within cylinder shielded overhead with regolith 75 g/cm^2 thick for central horizontal plane (Nealy, Wilson, and Townsend, 1988).

Figure 12.46. Variation in BFO dose equivalent from November 1960 flare event within a half-buried sphere shielded overhead with regolith 75 g/cm^2 thick (Nealy, Wilson, and Townsend, 1988).

Table 12.13. BFO Dose Comparisons for Three Large Solar Flares
for Lunar Habitats

[Data from Nealy, Wilson, and Townsend, 1988]

Flare data	Regolith thickness, cm	Predicted dose, cSv	
		Cylinder (center)	Sphere (center)
Feb. 1956	50	7.48	7.04
	100	2.70	2.94
Nov. 1960	50	1.60	1.90
	100	0.16	0.23
Aug. 1972	50	0.25	0.30
	100	0.03	0.04

the buried half are significantly reduced. The BFO dose variations within these habitats for the November 1960 flare event are shown in figures 12.45 and 12.46.

Dose predictions are also included for the GCR at solar minimum conditions. The maximum integrated BFO doses estimated in each habitat for various shield thicknesses are shown in table 12.14 (Nealy, Wilson, and Townsend, 1989). For the cylindrical habitat configuration, the dose variation throughout the configuration is relatively small (fig. 12.47). For the portion of the spherical habitat above ground level, the dose variation is also relatively small with a broad maximum dose rate observed directly above the center point of the sphere (approximately 0.11 to 0.12 Sv/yr). Below ground level, a large gradient in dose rate is shown in the downward direction, with values in the lower section decreasing to less than 5 cSv/yr (fig. 12.48). With overhead shielding of 112.5 g/cm^2, the dose rate maximum is reduced to 0.08 to 0.1 Sv/yr throughout the upper half of the sphere. This increased shielding is of even less significance in the regions below the ground where predicted doses approach the same low values as seen in the calculation for 75 g/cm^2. Relatively little reduction in dose (less than 20 percent) occurs for a 50-percent increase in layer thickness; this indicates that further substantial dose reductions would require very thick layers of material.

To make a conservative yearly estimate of dose, the crew is assumed to receive the dose delivered from the GCR and from one large flare (the February 1956 flare since it delivers the largest dose in the shielded module). If 75 g/cm^2 of regolith is selected for coverage, such a BFO dose in the cylindrical habitat is approximately 0.195 Sv/yr. Estimating the dose estimate in the spherical habitat is more complicated because of the large variation in dose throughout the habitat; however, the maximum dose estimated is approximately 0.19 Sv/yr. These dose estimates are well below the established guidelines for U.S. astronauts of 0.5 Sv/yr. The 30-day limits, with regard to the flares, remain below the 0.25 Sv limit. The skin doses, not presented in this analysis, are also well below the established 30-day and annual limits. These estimates have not taken into account the added shielding provided by the pressure vessel wall, supporting structures, or the placement of equipment in and around the module.

Table 12.14. GCR Integrated BFO Results for Lunar Habitats

[Data from Nealy, Wilson, and Townsend, 1989]

Habitat geometry	Regolith thickness		BFO dose equivalent Sv/yr
	cm	g/cm^2 (a)	
Cylindrical	50	75	0.12
Spherical	50	75	0.12
	75	112.5	0.10

[a]Regolith density of 1.5 g/cm^3 is assumed.

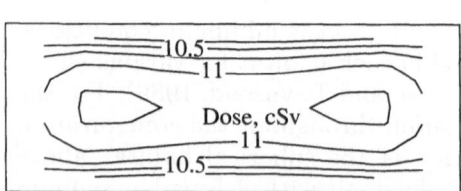

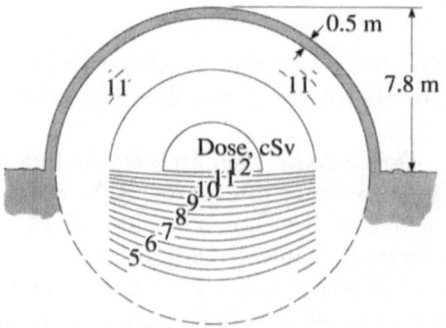

Figure 12.47. Variation in annual BFO dose equivalent from GCR variation within a cylinder shielded overhead with regolith 75 g/cm^2 thick for the central horizontal plane (Nealy, Wilson, and Townsend, 1989).

Figure 12.48. Variation in annual BFO dose equivalent from GCR within a half-buried sphere shielded overhead with regolith 75 g/cm^2 thick (Nealy, Wilson, and Townsend, 1989).

Shielding from solar flare events is essential on the lunar surface whether in the form of heavily shielded areas (i.e., flare shelters) or overall habitat protection for any mission duration. For longer stay times on the surface, the shielding from GCR becomes necessary to reduce the overall career exposure of the crew. A regolith shield thickness of 50 cm is estimated to provide adequate flare and GCR protection. However, before an optimum thickness and shielding strategy is selected, the complete mission scenario (including the lunar transport vehicle) needs to be studied in detail.

12.4.3. Martian surface habitation. The radiation environment on the martian surface is less severe than that found on the lunar surface. Although Mars is devoid of an intrinsic magnetic field strong enough to deflect charged particles, it does have a carbon dioxide atmosphere which will help protect surface crews from deep space radiative fluxes. Estimating the unshielded doses anticipated for crew members on the surface of Mars is more difficult than estimates made for the Moon in which deep space estimates are simply divided in half. On Mars, the protection provided by the atmosphere must be considered.

The amount of protection provided by the Mars atmosphere depends on the composition and structure of the atmosphere and altitude of the crew. In this analysis, the composition of the atmosphere is assumed to be 100 percent carbon dioxide. The Committee on Space Research has developed warm high- and cool low-density models of the atmospheric structure (Smith and West, 1983). The low-density model and the high-density model use surface pressures of 5.9 mbars and 7.8 mbars, respectively. The amount of protection provided by the atmosphere, in the vertical direction, at various altitudes is shown in table 12.15 (Simonsen et al., 1990a). In these calculations, a spherically concentric atmosphere is assumed such that the amount of protection provided increases with increasing zenith angle. Dose predictions at altitudes up to 12 km are included in the analysis because of the great deal of topographical relief present on the Mars surface.

Table 12.15. Martian CO_2 Atmospheric Protection in Vertical Direction

[Data from Simonsen et al., 1990a]

Altitude, km	Low-density model, g/cm^2	High-density model, g/cm^2
0	16	22
4	11	16
8	7	11
12	5	8

Dose estimates are predicted for the galactic cosmic radiation for the minimum of the solar activity cycle (fig. 12.3). The fluence spectra at 1 AU are used for the three large flares of August 1972, November 1960, and February 1956 (fig. 12.14). In the vicinity of Mars (approximately 1.5 AU), the fluence from these flares is expected to be less; however, there is still much discussion on the dependence of the radial dispersion of the flare with distance. Therefore, for the flare calculations in this analysis, the deep space fluence energy spectra at 1 AU have been conservatively applied to Mars. The surface doses at various altitudes in the atmosphere are determined from the computed propagation data for the GCR and the solar flare protons through carbon dioxide as shown in figures 12.34 and 12.35.

Integrated dose equivalent calculations were made for both the high density and the low-density atmosphere models at altitudes of 0, 4, 8, and 12 km. The corresponding skin and BFO dose estimates are shown in tables 12.16 and 12.17 (Simonsen et al., 1990a). A total yearly skin and BFO dose may be conservatively estimated as the sum of the annual GCR dose and the dose due to one large flare. At the surface, such an estimated skin dose equivalent is 0.21 to 0.24 Sv/yr and an estimated BFO dose equivalent is 0.19 to 0.22 Sv/yr (GCR plus Feb. 1956 event).

Table 12.16. Integrated Skin Dose Equivalents for Martian Atmospheric Models

[Data from Simonsen et al., 1990a]

Condition		Integrated skin dose equivalent, Sv, at altitude of—			
		0 km	4 km	8 km	12 km
Galactic cosmic	High-density	0.113	0.134	0.158	0.186
ray (annual)	Low-density	0.132	0.159	0.189	0.224
Aug. 1972	High-density	0.039	0.095	0.211	0.428
solar flare event	Low-density	0.09	0.219	0.462	0.826
Nov. 1960	High-density	0.064	0.10	0.148	0.211
solar flare event	Low-density	0.097	0.151	0.219	0.296
Feb. 1956	High-density	0.092	0.111	0.133	0.159
solar flare event	Low-density	0.11	0.134	0.162	0.191

Table 12.17. Integrated BFO Dose Equivalents for Martian Atmospheric Models

[Data from Simonsen et al., 1990a]

Condition		Integrated BFO dose equivalent, Sv, at altitude of—			
		0 km	4 km	8 km	12 km
Galactic cosmic	High-density	0.105	0.12	0.137	0.156
ray (annual)	Low-density	0.119	0.138	0.158	0.18
Aug. 1972	High-density	0.022	0.048	0.095	0.174
solar flare event	Low-density	0.046	0.099	0.185	0.303
Nov. 1960	High-density	0.050	0.075	0.106	0.144
solar flare event	Low-density	0.073	0.108	0.148	0.191
Feb. 1956	High-density	0.085	0.10	0.117	0.134
solar flare event	Low-density	0.099	0.118	0.136	0.153

At an altitude of 12 km, an estimated skin dose equivalent is 0.61 to 1.05 Sv/yr and an estimated BFO dose equivalent is 0.33 to 0.48 Sv/yr (GCR plus Aug. 1972 event). These dose predictions imply that the atmosphere of Mars may provide shielding sufficient to maintain the annual skin and BFO dose levels below the current U.S. astronaut limits of 3 and 0.5 Sv/yr, respectively.

The 30-day limits are important when considering the doses incurred from a solar flare event. The only 30-day limit exceeded is the BFO limit of 0.25 Sv for the August 1972 event at an altitude of 12 km. However, as seen in figure 12.34, the August 1972 flare is rapidly attenuated by matter, and a few g/cm^2 of additional shielding should reduce the anticipated dose below this limit. These dose predictions imply that the atmosphere of Mars may also provide sufficient shielding to maintain 30-day dose levels for the skin and BFO below the current U.S. astronaut limits of 1.5 and 0.25 Sv, respectively.

Mars exploration crews are likely to incur a substantial dose while in transit to Mars and perhaps from other radiation sources (e.g., nuclear reactors) which will reduce the allowable dose that can be received while on the surface. Therefore, additional shielding may be necessary to maintain short-term dose levels below limits or to help maintain career dose levels as low as possible. By utilizing local resources, such as martian regolith, shielding materials can be provided without excessive launch weight requirements from Earth.

The GCR particle flux and solar flare particle flux spectra obtained during the atmosphere calculations at altitudes of 0 and 8 km are now used as input conditions for regolith shield calculations. For a representative large solar flare contribution, the very penetrating spectrum of the February 1956 event is selected for further analysis. This event has the greatest flux of high-energy particles which results in the highest dose at the martian surface. The subsequently calculated particle flux versus energy distributions in the regolith can then be used to determine the dose at specified locations in the shield media. The dose contribution attributed to particles arriving from a given direction is now determined by the amount of CO_2 traversed and then the shield thickness encountered along its straight-line path to a specified target point. An example of some of the basic propagation data required is shown in figure 12.36.

One early martian habitat is described as a Space Station *Freedom* derived module, which is 8.2 m in length and 4.45 m in diameter (The 90-Day Study (Anon., 1989b)). The cylindrical module is assumed to be lengthwise on the martian surface with various thicknesses of martian regolith surrounding it. Another configuration assumes that the module is situated 2 m from a cliff 10-m high. (See fig. 12.49.)

A series of calculations are performed for various regolith thicknesses covering the module. Again, no consideration is given to the added shielding provided by the pressure vessel and internal equipment. The largest integrated dose equivalent in a vertical plane through the center of the cylinder is plotted versus an effective regolith thickness in figure 12.50 (Simonsen et al., 1990b). As shown in the figure, the regolith does not provide much additional protection from the GCR or the flare event than that already provided by the carbon dioxide atmosphere. The slope of

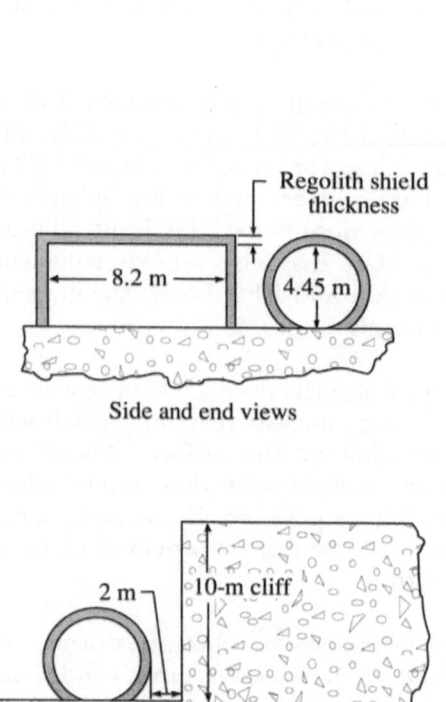

Regolith shield thickness

8.2 m

4.45 m

Side and end views

2 m

10-m cliff

End view near cliff

Figure 12.49. Cylindrical habitat module with regolith shielding for Mars (Simonsen et al., 1990b).

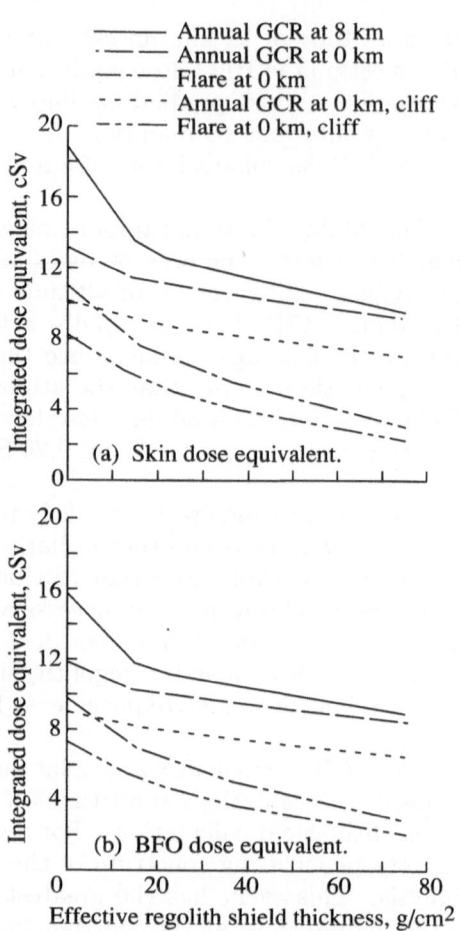

Annual GCR at 8 km
Annual GCR at 0 km
Flare at 0 km
Annual GCR at 0 km, cliff
Flare at 0 km, cliff

(a) Skin dose equivalent.

(b) BFO dose equivalent.

Effective regolith shield thickness, g/cm^2

Figure 12.50. Maximum dose equivalent in central cross-sectional plane of module as function of effective regolith shield thickness (Simonsen et al., 1990b).

each curve is relatively flat after 20 g/cm^2, with most of the dose reductions for the skin and BFO occurring in the first 20 g/cm^2. For 20 g/cm^2 of regolith protection, the annual BFO dose equivalent due to GCR is reduced from 0.119 to 0.1 Sv/yr at 0 km, and from 0.158 to 0.112 Sv/yr at 8 km. The annual skin dose equivalent is reduced from 0.132 to 0.11 Sv/yr at 0 km and from 0.189 to 0.126 Sv/yr at 8 km. For 20 g/cm^2 of regolith protection, the BFO dose equivalent due to the solar flare is reduced from 0.099 to 0.063 Sv/event at 0 km. The skin dose equivalent is reduced from 0.11 to 0.069 Sv/event.

For the GCR, the dose variation within the module in the radial direction is not large, approximately 5 to 20 percent for 15 to 50 cm of shielding, respectively. For the February 1956 solar flare event, the variation in dose equivalent is approximately 25 to 40 percent for 15 to 50 cm of shielding, respectively. In the axial direction, the dose estimates for both the GCR and the flare showed a variation of less than 1 percent; this suggests that the doses incurred in cylindrical habitats of other lengths would be comparable in magnitude.

A possible way to further reduce the dose equivalent received on the martian surface would be to locate the habitat next to a cliff. As shown in figure 12.50(b), the cliff further reduces the BFO dose equivalent by approximately 0.02 to 0.03 Sv/yr for the GCR at 0 km and by approximately 0.01 to 0.015 Sv/event for the February 1956 flare at 0 km. Similar decreases are also obtained for the skin dose (fig. 12.50(a)). The shielding provided by the cliff and atmosphere alone results in a BFO dose equivalent of 0.091 Sv/yr for the GCR and 0.074 Sv/event for the February 1956 event.

From this analysis, it is concluded that moderate thicknesses of martian regolith do not provide substantial additional protection to that already provided by the carbon dioxide atmosphere. If regolith is used as shielding material, the largest reduction in dose equivalent occurs in the first 20 g/cm^2 (or approximately 15 cm if a regolith density of 1.5 g/cm^3 is assumed). Thus, if additional protection using martian regolith is desired, a shield thickness on the order of 15 to 20 cm is recommended. If additional protection using 15 cm of martian regolith is provided at an altitude of 0 km, the annual skin and BFO dose equivalent will be reduced from 0.24 to 0.18 Sv/yr and from 0.22 to 0.16 Sv/yr, respectively (Simonsen et al., 1990b).

For radiation protection provided by regolith on the surface of Mars, mission planners and medical personnel must decide if the radiation doses anticipated warrant the added equipment and time required for crew members to "bury" themselves. For the shorter stay times of 30 to 90 days, the additional requirements placed on a Mars mission to cover a module may be unnecessary, especially if a flare shelter is provided. A logical alternative to massive shielding efforts is to take advantage of local terrain features found on the surface of Mars. Regolith shielding may become more attractive for the longer stay times of 600 days or for futuristic permanent habitation.

12.5. Issues and Concerns

Estimates and predictions of radiation exposure and incurred doses for space exploration missions usually require complex analysis techniques and involve uncertainties that are presently difficult to quantify. Some issues and concerns regarding radiation exposure estimates and shielding requirements are discussed in the following subsections.

12.5.1. Environment. Confidence in the estimates of incurred dose for lunar and martian missions is directly related to the accuracy and development of the current space-radiation environmental models. With regard to the charged-particle environmental models, only in some cases do enough data exist for estimates of uncertainties and natural variabilities. At the present time, no particular flare model has been established as a practical standard. However, a likely future candidate is the statistical model developed at the Jet Propulsion Laboratory (Feynman and Gabriel, 1990). The continued development, endorsement, and implementation of standard environmental models are important aspects of mission scenario analyses and shield design studies.

12.5.2. Transport codes. The accuracy of transport codes used to describe the propagation of particles through matter is another concern. Monte Carlo techniques are generally regarded as most faithfully representing the details of the complex processes involving high-energy radiation transport. In many cases, simpler and faster codes, which are far less costly and time consuming to implement, may be used to adequately describe the transport. The precision of such codes may be evaluated by comparisons with equivalent Monte Carlo calculations, or with exact benchmark solutions (when they can be found). Once the mathematical precision of a particular code is established, the ultimate accuracy of its prediction will depend on the interaction cross-section data base used as input for calculations. Presently, nucleon (neutrons and protons) interaction cross sections are relatively well-known for wide ranges of energy and target materials. However, data are very limited for interaction cross sections for the 20 to 25 heavy ion nuclei of importance for GCR exposure. Inevitably, data extrapolations and extensions by complex theoretical techniques are implemented in order to provide a comprehensive cross-section data base (Norbury and Townsend, 1986; Townsend and Wilson, 1985; Wilson and Badavi, 1986). This creates uncertainties in the transport calculations which are very difficult to quantify. (See chapter 11 for further discussion.)

12.5.3. Radiobiology. Standard dosimetric techniques used to evaluate health risks due to radiation exposures are presently being challenged, particularly with regard to latent effects due to the high-energy, low dose-rate exposure from the GCR heavy ions. Current methods for evaluating dose equivalents resulting from heavy ion exposure utilize biological effectiveness quality factors (Q) which are specified as functions of linear energy transfer (LET) of the projectile particles to the biological system being traversed (ICRP 26 (Anon., 1977)). Predictions of dose equivalent incurred in deep space from the GCR with the standard methods indicate that substantial shielding (20–50 g/cm^2) is required to reduce dose levels to an annual dose of 0.25 to 0.3 Sv (Townsend et al., 1990a). Such shield amounts are very massive when large habitation modules are involved. Thus, efforts are in progress toward better definition of risk assessment for GCR exposures. Newly proposed quality factors have been based on recent biological effects data (ICRU 40 (Anon., 1986)). Preliminary calculations with the latest Q values indicate that previous evaluations may have been somewhat, but not dramatically, conservative (Wilson, Shinn, and Townsend, 1990). Other recent studies have suggested abandoning the Q value/LET system (Katz, 1986) and formulating more detailed models of cell destruction and transformation using radiosensitivity parameters derived from biological experiments (Cucinotta et al., 1991). The current limitation of such models is the lack of methods to extrapolate from cell damage to expression at the organistic level, lack of comprehensive cell repair model, and the physiological factors. It is expected that such direct biophysical models would be a distinct improvement. However, evolution of such models is directly coupled to the available data bases for radiobiological effects, which are very limited for GCR-type radiations. Clearly, the relationship between heavy ion exposure and health risk is in need of better definition. For further discussion, see chapter 11.

12.5.4. Dosimetric measurement.

The preceding discussion naturally leads to additional questions concerning measurement and monitoring of incurred radiation doses. Present space flight dosimetry instrumentation includes dosimeters of both thermo-luminescent and proportional counter types, and they have been shown to be reliable and accurate for the Space Transportation System (STS) mission (Atwell, 1990). In general, STS dose rates are fairly low. For the 28.5° inclination orbits at altitudes between 250 and 350 km, the average dose rate is observed to be approximately 0.01 cGy/day (or 0.036 Gy/yr). Steady dose rates in deep space, even with thick shields, are expected to be substantially higher (factors of 5 to 10), with intermittent (solar flare) dose rates higher still. Further advancement in dosimetric instrumentation and techniques will be required to monitor the astronaut deep space exposures, with emphasis on active, as opposed to passive, dosimeters. In particular, since the GCR interactions with thick shields may produce a high yield of neutrons and precision in neutron dosimetry is currently considered to be rather poor (Paić, 1988), improvements are currently needed in this case.

12.5.5. Flare prediction.

The forecasting of large solar proton events is of vital importance for missions of long duration. Practically continuous monitoring of various aspects of solar activity (i.e., X-rays, radio emissions, sunspot number) during solar cycle 21 (1975–1986) and up to the present time has provided a valuable data base for flare-forecasting statistics. The approach to flare forecasting used at the NOAA Space Environment Laboratory during recent years is to examine the intensities of X-rays and radio emissions as they relate to the release of energetic particles from the solar surface. Estimates of the peak proton flux may also be made from these observations. For flare predictions during solar cycle 21, the number of events which occurred without prediction was about 10 percent of the total. This resulted primarily because the initial X-ray and radio bursts were not on the visible portion of the Sun (Heckman et al., 1984). The false-alarm rate was approximately 50 percent; therefore, further work in this area is needed. Other techniques combine high-resolution observations of sunspot group patterns and magnetic field configurations in conjunction with H_α-line emission. The prediction of occurrence with the use of these techniques is claimed to be up to several days in advance (Zirin and Liggett, 1987). This method appears to show promise, but more observations are required to demonstrate the practicality of its implementation on a routine basis. For missions of long duration, additional onboard instruments for active proton detection should also be available to indicate when the use of a well-shielded storm shelter is warranted.

12.5.6. Alternate shielding concepts.

Other topics of concern in the area of space-radiation shielding include the effectiveness of material types (or combination of material types) and alternate approaches to bulk shielding (e.g., magnetic and electromagnetic field deflection methods). As previously discussed in section 2.3.2, recent results indicate that hydrogenous materials of low atomic weight are substantially superior to heavy metals for energetic ion shielding. However, little has been done in the study of the behavior of combinations, for example, alternating layers of light and heavy materials. Further studies should also address structural details of shields: in particular, corrugatedlike panels and/or shadow shielding techniques may offer advantages over simple wall structures. One recent study has indicated that magnetic shielding is of little use for protection from GCR (Townsend et al., 1990b). However, the Townsend study also showed that for representative large proton flares, great reductions in exposure can be achieved; thus, the potential use of such a technique for flare protection may still be viable.

12.6. Concluding Remarks

Before astronaut dose estimates and subsequent shielding requirements can be determined for advanced missions to the Moon and Mars, many details of the missions must be specified. For instance, many items must be defined in order to determine specific shielding requirements: the transfer vehicle configuration, the habitat configuration, the length of time required to shield habitats with regolith, the career limits of the crew, the year of the mission (solar minimum or maximum conditions), the duration of the mission. Mars mission planning includes particular concerns such as whether any nuclear powered propulsion is envisioned, the location of the habitat on the martian surface, whether the crew will be spiraled through the Van Allen belts. Estimates must also be made as to where the Mars crew will spend their time en route to Mars; that is, how much of their time is anticipated to be spent in the more heavily shielded areas of the spacecraft as opposed to the less heavily shield areas. Even with the specific details of the mission defined, the final shield design must consider the many uncertainties associated with current state-of-the-art transport analysis.

Steps toward quantifying some of the issues involved with radiation protection for advanced manned missions to the Moon and Mars have been presented in this chapter. After the definition of the galactic cosmic-ray environment and the selection of various flare environment scenarios, deterministic transport codes were used to determine the transport and attenuation of the deep space radiative fluxes through different media. From these basic propagation data, conservative dose estimates and shielding requirements are determined for simple-geometry transfer vehicles and for possible lunar and martian habitat configurations. The results that have been presented are just part of the information required to determine radiation-protection requirements for each phase of a complete mission scenario. However, all this must await an improved understanding of biological response to heavy ion exposure and identification of important biological consequences which must be mitigated.

12.7. References

Adams, J. H., Jr.; Silberberg, R.; and Tsao, C. H., 1981: *Cosmic Ray Effects on Microelectronics. Part I—The Near-Earth Particle Environment.* NRL Memo. Rep. 4506-Pt. I, U.S. Navy. (Available from DTIC as AD A103 897.)

Adams, James H., Jr., 1987: *Cosmic Ray Effects on Microelectronics*, Part IV. NRL Memo. Rep. 5901 (Revised), Naval Research Lab.

Alred, J.; Bufkin, A.; Graf, J.; Kennedy, K.; Patterson, J.; Petro, A.; Roberts, M.; Stecklein, J.; and Sturm, J., 1988: Development of a Lunar Outpost: Year 2000–2005. *Lunar Bases & Space Activities in the 21st Century*, NASA, AIAA, Lunar & Planetary Inst., American Geophysical Union, American Nuclear Soc., American Soc. of Civil Engineers, Space Studies Inst., and National Space Soc., Paper No. LBS-88-240.

Anon., 1975a: *The Earth's Trapped Radiation Belts.* NASA Space Vehicle Design Criteria (Environment). NASA SP-8116.

Anon., 1975b: *Interplanetary Charged Particles.* NASA Space Vehicle Design Criteria (Environment). NASA SP-8118.

Anon., 1977: *Recommendations of the International Commission on Radiological Protection.* ICRP Publ. 26, Pergamon Press.

Anon., 1986: *The Quality Factor in Radiation Protection.* ICRU Rep. 40, International Commission on Radiation Units and Measurements.

Anon., 1987: *Man-Systems Integration Standards*—Volume 1. NASA CR-185625-VOL. 1.

Anon., 1989a: *Guidance on Radiation Received in Space Activities.* NCRP Rep. No. 98, National Council on Radiation Protection and Measurements.

Anon., 1989b: *Report of the 90-Day Study on Human Exploration of the Moon and Mars.* NASA Johnson Space Center.

Atwell, William; Beever, E. Ralph; Hardy, Alva C.; and Cash, Bernard L., 1989: A Radiation Shielding Model of the Space Shuttle for Space Radiation Dose Exposure Estimations. *Advances in Nuclear Engineering Computation and Radiation Shielding*, Volume 1, Michael L. Hall, ed., American Nuclear Soc., Inc., pp. 11:1–11:12.

Atwell, William, 1990: Astronaut Exposure to Space Radiation: Space Shuttle Experience. SAE Tech. Paper Ser. 901342.

Blizard, J. B., 1969: *Long Range Solar Flare Prediction.* NASA CR-61316.

Billings, M. P.; and Yucker, W. R., 1973: *The Computerized Anatomical Man (CAM) Model.* NASA CR-134043.

Burrell, M. O.; Wright, J. J.; and Watts, J. W., 1968: *An Analysis of Energetic Space Radiation and Dose Rates.* NASA TN D-4404.

Committee on the Biological Effects of Ionizing Radiations, 1990: *Health Effects of Exposure to Low Levels of Ionizing Radiation.* BEIR V, National Academy Press.

Cucinotta, Francis A.; and Wilson, John W., 1985: *Computer Subroutines for Estimation of Human Exposure to Radiation in Low Earth Orbit.* NASA TM-86324.

Cucinotta, Francis A.; Katz, Robert; Wilson, John W.; Townsend, Lawrence W.; Nealy, John E.; and Shinn, Judy L., 1991: *Cellular Track Model of Biological Damage to Mammalian Cell Cultures From Galactic Cosmic Rays.* NASA TP-3055.

Curtis, S. B.; Doherty, W. R.; and Wilkinson, M. C., 1969: *Study of Radiation Hazards to Man on Extended Near Earth Missions*. NASA CR-1469.

Curtis, S. B.; and Wilkinson, M. C., 1971: *Radiation Hazards to Man*. NASA CR-125592.

Dalton, Charles; and Hohmann, Edward, eds., 1972: *Conceptual Design of a Lunar Colony*. NASA CR-129164.

Feynman, Joan; and Gabriel, Stephen B., 1990: A New Model for Calculation and Prediction of Solar Proton Fluences. AIAA-90-0292.

Foelsche, T., 1963: Specific Solar Flare Events and Associated Radiaton Doses. *Space Radiation Effects*, ASTM Special Tech. Publ. No. 363, American Soc. for Testing & Materials, pp. 1–13.

Foelsche, Trutz; Mendell, Rosalind B.; Wilson, John W.; and Adams, Richard R., 1974: *Measured and Calculated Neutron Spectra and Dose Equivalent Rates at High Altitudes; Relevance to SST Operations and Space Research*. NASA TN D-7715.

Grahn, Douglas, ed., 1973: *HZE-Particle Effects in Manned Spaceflight*. Space Science Board, National Research Council, National Academy of Sciences.

Goswami, J. N.; McGuire, R. E.; Reedy, R. C.; Lal, D.; and Jha, R., 1988: Solar Flare Protons and Alpha Particles During the Last Three Solar Cycles. *J. Geophys. Res.*, vol. 93, no. A7, pp. 7195–7205.

Heckman, G.; Hirman, J.; Kunches, J.; and Balch, C., 1984: The Monitoring and Prediction of Solar Particle Events—An Experience Report. *Adv. Space Res.*, vol. 4, no. 10, pp. 165–172.

Johnson, Francis S., ed., 1965: *Satellite Environment Handbook*, Second ed. Stanford Univ. Press.

Khandelwal, G. S.; and Wilson, John W., 1974: *Proton Tissue Dose for the Blood Forming Organ in Human Geometry: Isotropic Radiation*. NASA TM X-3089.

Katz, Robert, 1986: Biological Effects of Heavy Ions From the Standpoint of Target Theory. *Adv. Space Res.*, vol. 6, no. 11, pp. 191–198.

King, Joseph H., 1974: Solar Proton Fluences for 1977–1983 Space Missions. *J. Spacecr. & Rockets*, vol. 11, no. 6, pp. 401–408.

Kovalev, E. E.; Muratova, I. A.; and Petrov, V. M., 1989: Studies of the Radiation Environment Aboard Prognoz Satellites. *Nuclear Tracks & Radiat. Meas.*, vol. 16, no. 1, pp. 45–48.

Kuhn, E.; Schwamb, F. E.; and Payne, W. T., 1965: Solar Flare Hazard to Earth-Orbiting Vehicles. *Second Symposium on Protection Against Radiations in Space*, Arthur Reetz, Jr., ed., NASA SP-71, pp. 429–434.

McCracken, K. G., 1962a: The Cosmic-Ray Flare Effect. 1. Some New Methods of Analysis. *J. Geophys. Res.*, vol. 67, no. 2, pp. 423–434.

McCracken, K. G., 1962b: The Cosmic-Ray Flare Effect. 2. The Flare Effects of May 4, November 12, and November 15, 1960. *J. Geophys. Res.*, vol. 67, no. 2, pp. 435–446.

McCracken, K. G., 1962c: The Cosmic-Ray Flare Effect. 3. Deductions Regarding the Interplanetary Magnetic Field. *J. Geophys. Res.*, vol. 67, no. 2, pp. 447–458.

McDonald, Frank B., ed., 1963: *Solar Proton Manual*. NASA TR R-169.

Nealy, John E.; Wilson, John W.; and Townsend, Lawrence W., 1988: *Solar-Flare Shielding With Regolith at a Lunar-Base Site*. NASA TP-2869.

Nealy, John E.; Wilson, John W.; and Townsend, Lawrence W., 1989: Preliminary Analyses of Space Radiation Protection for Lunar Base Surface Systems. SAE Tech. Paper Ser. 891487.

Nealy, John E.; Simonsen, Lisa C.; Sauer, Herbert H.; Wilson, John W.; and Townsend, Lawrence W., 1990a: *Space Radiation Dose Analysis for Solar Flare of August 1989*. NASA TM-4229.

Nealy, John E.; Simonsen, Lisa C.; Townsend, Lawrence W.; and Wilson, John W., 1990b: Deep Space Radiation Exposure Analysis for Solar Cycle XXI (1975–1986). SAE Tech. Paper Ser. 901347.

Norbury, John W.; and Townsend, Lawrence W., 1986: *Electromagnetic Dissociation Effects in Galactic Heavy-Ion Fragmentation*. NASA TP-2527.

O'Brien, Keran, 1972: The Cosmic Ray Field at Ground Level. Paper presented at the Second International Symposium on the Natural Radiation Environment (Houston, Texas).

Paić, Guy, ed., 1988: *Ionizing Radiation: Protection and Dosimetry*. CRC Press, Inc.

Parker, James F., Jr.; and West, Vita R., eds., 1973: *Bioastronautics Data Book*, Second ed. NASA SP-3006.

Sauer, Herbert H.; Zwickl, Ronald D.; and Ness, Martha J., 1990: *Summary Data for the Solar Energetic Particle Events of August Through December 1989*. Space Environment Lab., National Oceanic and Atmospheric Adm.

Scott, W. Wayne; and Alsmiller, R. G., Jr., 1968: *Comparisons of Results Obtained With Several Proton Penetration Codes—Part II*. ORNL-RSIC-22 (Contract No. W-7405-eng-26, NASA Order R-104(10)), Oak Ridge National Lab.

Shavers, Mark; Poston, John W.; Atwell, William; Hardy, Alva C.; and Wilson, John W., 1991: *Preliminary Calculation of Solar Cosmic Ray Dose to the Female Breast in Space Missions*. NASA TM-4235.

Shea, M. A.; and Smart, D. F., 1983: A World Grid of Calculated Cosmic Ray Vertical Cutoff Rigidities for 1980.0. *18th International Cosmic Ray Conference—Conference Papers*, MG Sessions, Volume 3, Tata Inst. of Fundamental Research (Colaba, Bombay), pp. 415–418.

Shea, M. A.; Smart, D. F.; and Gentile, L. C., 1983: The Cosmic Ray Equator Determined Using the International Geomagnetic Reference Field for 1980.0. *18th International Cosmic Ray Conference—Conference Papers*, MG Sessions, Volume 3, Tata Inst. of Fundamental Research (Colaba, Bombay), pp. 423–426.

Shinn, Judy L.; Wilson, John W.; and Nealy, John E., 1990: *Reliability of Equivalent Sphere Model in Blood-Forming Organ Dose Estimation*. NASA TM-4178.

Shinn, Judy L.; Wilson, John W.; Nealy, John E.; and Cucinotta, Francis A., 1990: *Comparison of Dose Estimates Using the Buildup-Factor Method and a Baryon Transport Code* (BRYNTRN) *With Monte Carlo Results*. NASA TP-3021.

Shinn, Judy L.; and Wilson, John W., 1992: *An Efficient Heavy Ion Transport Code:* HZETRN. NASA TP-3147.

Simonsen, Lisa C.; Nealy, John E.; Townsend, Lawrence W.; and Wilson, John W., 1990a: *Radiation Exposure for Manned Mars Surface Missions*. NASA TP-2979.

Simonsen, Lisa C.; Nealy, John E.; Townsend, Lawrence W.; and Wilson, John W., 1990b: Space Radiation Shielding for a Martian Habitat. SAE Tech. Paper Ser. 901346.

Simonsen, Lisa C.; and Nealy, John E., 1991: *Radiation Protection for Human Missions to the Moon and Mars*. NASA TP-3079.

Simonsen, Lisa C.; Nealy, John E.; Sauer, Herbert H.; and Townsend, Lawrence W., 1991: Solar Flare Protection for Manned Lunar Missions: Analysis of the October 1989 Proton Flare Event. SAE Tech. Paper Ser. 911351.

Singley, G. W.; and Vette, J. I., 1972: *A Model Environment for Outer Zone Electrons*. NASA TM X-69989.

Sleeper, H. P., Jr., 1972: *Planetary Resonances, Bi-Stable Oscillation Modes, and Solar Activity Cycles*. NASA CR-2035.

Slutz, Ralph J.; Gray, Thomas B.; West, Marie L.; Stewart, Frank G.; and Leftin, Margo, 1971: *Solar Activity Prediction*. NASA CR-1939.

Smart, D. F.; and Shea, M. A., 1983: Geomagnetic Transmission Functions for a 400 km Altitude Satellite. *18th International Cosmic Ray Conference—Conference Papers*, MG Sessions, Volume 3, Tata Inst. of Fundamental Research (Colaba, Bombay), pp. 419–422.

Smith, Robert E.; and West, George S., compilers, 1983: *Space and Planetary Environment Criteria Guidelines for Use in Space Vehicle Development*, 1982 Revision (Volume 1). NASA TM-82478.

Space Science Board, 1970: *Radiation Protection Guides and Constraints for Space-Mission and Vehicle-Design Studies Involving Nuclear Systems*. National Academy of Sciences, National Research Council.

Störmer, Carl, 1930: Periodische Elektronenbahnen im Felde eines Elementarmagneten und ihre Anwendung auf Brüches Modellversuche und auf Eschenhagens Elementarwellen des Erdmagnetismus. *Z. Astrophys.*, Bd. 1, pp. 237–274.

Townsend, Lawrence W.; and Wilson, John W., 1985: *Tables of Nuclear Cross Sections for Galactic Cosmic Rays—Absorption Cross Sections*. NASA RP-1134.

Townsend, Lawrence W.; Nealy, John E.; Wilson, John W.; and Atwell, William, 1989: Large Solar Flare Radiation Shielding Requirements for Manned Interplanetary Mission. *J. Spacecr. & Rockets*, vol. 26, no. 2, pp. 126–128.

Townsend, Lawrence W.: Nealy, John E.; Wilson, John W.; and Simonsen, Lisa C., 1990a: *Estimates of Galactic Cosmic Ray Shielding Requirements During Solar Minimum.* NASA TM-4167.

Townsend, L. W.; Wilson, J. W.; Shinn, J. L.; Nealy, J. E.; and Simonsen, L. C., 1990b: Radiation Protection Effectiveness of a Proposed Magnetic Shielding Concept for Manned Mars Missions. SAE Tech. Paper Ser. 901343.

Townsend, Lawrence W.; Shinn, Judy L.; and Wilson, John W., 1991: Interplanetary Crew Exposure Estimates for the August 1972 and October 1989 Solar Particle Events. *Radiat. Res.*, vol. 126, pp. 108–110.

Webber, William R., 1966: *An Evaluation of Solar-Cosmic-Ray Events During Solar Minimum.* D2-84274-1, Boeing Co.

Wilson, John W.; and Khandelwal, G. S., 1974: Proton Dose Approximation in Arbitrary Convex Geometry. *Nuclear Technol.*, vol. 23, no. 3, pp. 298–305.

Wilson, John W.; and Denn, Fred M., 1976: *Preliminary Analysis of the Implications of Natural Radiations on Geostationary Operations.* NASA TN D-8290.

Wilson, John W.; and Khandelwal, Govind S., 1976: Proton-Tissue Dose Buildup Factors. *Health Phys.*, vol. 31, no. 2, 1976, pp. 115–118.

Wilson, John W., 1977: *Analysis of the Theory of High-Energy Ion Transport.* NASA TN D-8381.

Wilson, John W.; and Denn, Fred M., 1977: *Implications of Outer-Zone Radiations on Operations in the Geostationary Region Utilizing the AE4 Environmental Model.* NASA TN D-8416.

Wilson, John W., 1978: Environmental Geophysics and SPS Shielding. *Workshop on the Radiation Environment of the Satellite Power System*, Walter Schimmerling and Stanley B. Curtis, eds., LBL-8581, UC-41 (Contract W-7405-ENG-48), Univ. of California, pp. 33–116.

Wilson, John W.; and Cucinotta, Frank, 1984: *Human Exposure in Low Earth Orbit.* NASA TP-2344.

Wilson, John W.; and Badavi, F. F., 1986: Methods of Galactic Heavy Ion Transport. *Radiat. Res.*, vol. 108, pp. 231–237.

Wilson, J. W.; Townsend, L. W.; and Badavi, F. F., 1987: Galactic HZE Propagation Through the Earth's Atmosphere. *Radiat. Res.*, vol. 109, no. 2, pp. 173–183.

Wilson, John W.; and Townsend, L. W., 1988: A Benchmark for Galactic Cosmic-Ray Transport Codes. *Radiat. Res.*, vol. 114, no. 2, pp. 201–206.

Wilson, John W.; Townsend, Lawrence W.; Nealy, John E.; Chun, Sang Y.; Hong, B. S.; Buck, Warren W.; Lamkin, S. L.; Ganapol, Barry D.; Khan, Ferdous; and Cucinotta Francis A., 1989: BRYNTRN: *A Baryon Transport Model.* NASA TP-2887.

Wilson, John W.; Shinn, Judy L.; and Townsend, Lawrence W., 1990: Nuclear Reaction Effects in Conventional Risk Assessment for Energetic Ion Exposure. *Health Phys.*, vol. 58, no. 6, pp. 749–752.

Wilson, John W.; Khandelwal, Govind S.; Shinn, Judy L.; Nealy, John E.; Townsend, Lawrence W.; and Cucinotta, Francis A., 1990: *Simplified Model for Solar Cosmic Ray Exposure in Manned Earth Orbital Flights.* NASA TM-4182.

Wilson, John W.; Chun, Sang Y.; Badavi, Forooz F.; Townsend, Lawrence W.; and Lamkin, Stanley L., 1991: HZETRN: *A Heavy Ion/Nucleon Transport Code for Space Radiations.* NASA TP-3146.

Yablontsev, N. N., 1990: Documents Setting Standards for Radiation Safety for Space Flight. *Kosm. Biol. Aviakosmicheskaya Meditsina*, vol. 24, no. 1, p. 55.

Zirin, Harold; and Liggett, Margaret A., 1987: Delta Spots and Great Flares. *Sol. Phys.*, vol. 113, nos. 1 & 2, pp. 267–283.

13. RADIATION SAFETY IN THE EARTH'S ATMOSPHERE

13.1. Introduction

When the possibility of high-altitude supersonic commercial aviation was first seriously proposed, Foelsche (1962) brought to light a number of concerns with respect to atmospheric radiation. Subsequently, a detailed study of the atmospheric radiation components at high altitudes was conducted from 1965 to 1971 at the Langley Research Center (LaRC) by Foelsche et al. (1974). In that study the major role of atmospheric neutrons in radiation exposure was uncovered. These studies utilized an instrument package consisting of tissue equivalent ion chambers, organic scintillator neutron spectrometers, and nuclear emulsion. A theoretical program to predict atmospheric radiation levels and to specifically extend the neutron spectrum into the range outside that measured by the scintillation spectrometer was also developed (Wilson et al., 1970). It was found that the neutron spectrum due to galactic cosmic rays was nearly independent of solar modulation. However, the neutron spectrum produced by solar cosmic rays was found to vary from event to event. An overview of that program is given by Foelsche (1977). The conclusion of this previous work was that high-altitude commercial aviation required special considerations for radiation protection (Wilson, 1981), whereas the worst-case flights for pre-1980 subsonic airlines were well within the exposure limits of the general population (Foelsche et al., 1974; Friedberg and Neas (in Anon., 1980)).

Three factors have changed since those studies: (1) increases in quality factors seem imminent (ICRU 40 (Anon., 1986); Sinclair[1]); (2) reduced exposure limits are being proposed; and (3) flight crews are logging greatly increased hours at altitude (Bramlitt, 1985; Wilson and Townsend, 1988 and 1989; Friedberg et al., 1989; Busick, 1989; Barish, 1990). At present, the Langley data base on biologically important components appears to be the most complete and comprehensive available, but updating with new quality factors and providing for easy use by the health physics community is required. Furthermore, concerns on atmospheric radiation exposure remain for such NASA-related aircraft as the National Aero-Space Plane, the Advanced Supersonic Transport, and the Hypersonic Transport.

A computer program called GREP (Galactic Radiation Exposure Program) was written by S. B. Curtis of the Boeing Company over 20 years ago. The dose was estimated using the ion chamber data of Neher (1961) and Neher and Anderson (1962) and the neutron data of Hess, Canfield, and Lingenfelter (1961). The code was modified for various aircraft trajectories between city pairs by Wallace and Sondhaus (1978). The inadequacy of the data base used was reviewed by Foelsche (1962) and Friedberg and Neas (in Anon., 1980). A similar code was written later using the theoretical model of O'Brien and McLaughlin (1972) for atmospheric radiation. In this later code (Friedberg et al., 1989) the O'Brien-McLaughlin theoretical model was used with neutron, proton, and pion dose equivalents increased by a factor of 2 in accordance with the recommendations

---

[1] Sinclair, W. K.: Recent Developments in Estimates of Cancer Risk From Ionizing Radiation. Paper presented at the 20th SAE International Conference on Environmental Systems (Williamsburg, VA), July 9–12, 1990.

of NCRP 91 (Anon., 1987). On the basis of analysis of the high-latitude data of Foelsche et al. (1974) using the newly proposed quality factors of ICRU 40 (Anon., 1986), the increase of quality factor by a factor of 2 does not seem fully justified (Wilson and Townsend, 1988; Shinn, Wilson, and Ngo, 1990) and should be substantially conservative in exposure estimates. In the present chapter we undertake a reanalysis of the quality factor increase using measured field quantities. Furthermore, the use of measured data within the Earth's atmosphere is not governed by the uncertainty associated with the galactic cosmic ray (GCR) spectrum and charge distribution (Wilson, Shinn, and Townsend, 1990) and the uncertainty in their propagation in the atmosphere (Wilson, Townsend, and Badavi, 1987). Indeed, the measured data are the primary sources of information on the atmospheric exposure levels and their meaning in terms of biological risk.

In the present chapter we use currently known data to generate radiation field values within the Earth's atmosphere as a function of time. These are compared where possible with values obtained by other methods. The data base will ultimately provide a test bed for the transport codes discussed in previous chapters. In addition, the newly developed codes will provide information on high linear energy-transfer (LET) components in the uppermost atmosphere of importance to future NASA missions such as the National Aero-Space Plane (NASP).

13.2. Solar Effects on Atmospheric Radiations

Very little ionizing radiation would be found in the Earth's atmosphere were it not for the presence of extraterrestrial energetic particulate radiation. These extraterrestrial particles have two primary sources. The first discovered source is the diffuse component originating from remote regions of our galaxy (galactic cosmic rays or GCR), and the second source is a more local and more directed source from our local star, Sol (solar cosmic rays or SCR). Both sources contain particles consisting mainly of protons and smaller amounts of heavier particles as elements stripped of their electrons. Both show time variations correlated with the natural solar cyclic processes most easily observed as magnetic disturbances in the solar surface called sunspots. The area of the solar disk covered by sunspots varies in cycles lasting from 10.5 to 13 years. Detailed records exist on solar observations for 21 complete cycles, and we are currently near the maximum of cycle 22. A more fragmentary record exists for several hundred years. The solar cycle affects the extraterrestrial particulate environment in two ways.

The solar plasma output (solar wind) increases during years of high solar activity associated with the expanding solar corona. The expanding plasma field entraps the local solar magnetic field at the time of ejection and transports it outward into the solar system. The GCR are denied access to the solar system according to the status of the interplanetary plasma, and they are modulated by the solar cycle. The modulation must wait for the plasma to fill a certain region of space (out to several astronomical units (AU)). The time delay depends on the solar wind velocity, and the time to reach equilibrium with the sunspot number depends on the rate of rise in solar activity. Because of these delays, the GCR intensity in relation to sunspot number shows a typical hysteresis effect. Such a curve using neutron monitor data (Freier and Waddington, 1965) is shown in figure 13.1

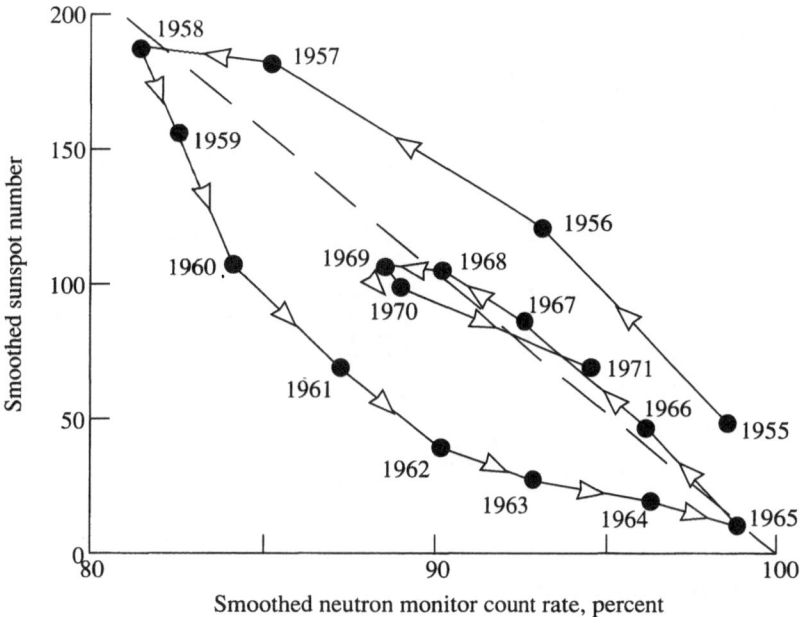

Figure 13.1. Regression plot between smoothed sunspot numbers and smoothed neutron monitor count rates of Mount Washington in New Hampshire and the Deep River neutron monitor (DRNM) in Canada.

for cycle 19 for 1955 to 1965. (A neutron monitor is a ground level device sensitive to neutrons produced in a lead converter by nuclear particles and has been in use for 50 years to observe cosmic-ray intensities.) If we imagine that the time delays are removed, we conjecture that all the data would lie on the correlation curve given as the dashed line. Note that the more slowly rising cycle 20 shown as 1965 to 1971 appears nearly in equilibrium with the modulation effects and closely approximates the correlation curve.

The magnetic irregularities observed as sunspots are responsible for the acceleration of the plasma of the solar surface. This occurs especially when magnetic regions coalesce into plages and their individual magnetic fields annihilate, thus generating large electric fields that accelerate the local plasma (solar flare) to very high energies (sometimes more than 15 GeV). Such particles escape the solar surface and propagate outward along the sectored magnetic field lines into interplanetary space. Such particles are sometimes seen arriving at the Earth. The events are most likely to cause local cosmic ray increases on the Earth if the solar flare occurred on the western limb. SCR arriving at Earth vary by orders of magnitude in intensity and spectral content and very few events are of importance to present-day commercial aircraft operation (operating below 50 000 ft).

13.3. Background Radiation Data Base

13.3.1. Radiation levels at high latitude. The low-level background radiation in the Earth's atmosphere is generated by the impact of the GCR on the top of the atmospheric layer. The incident GCR intensity varies in time

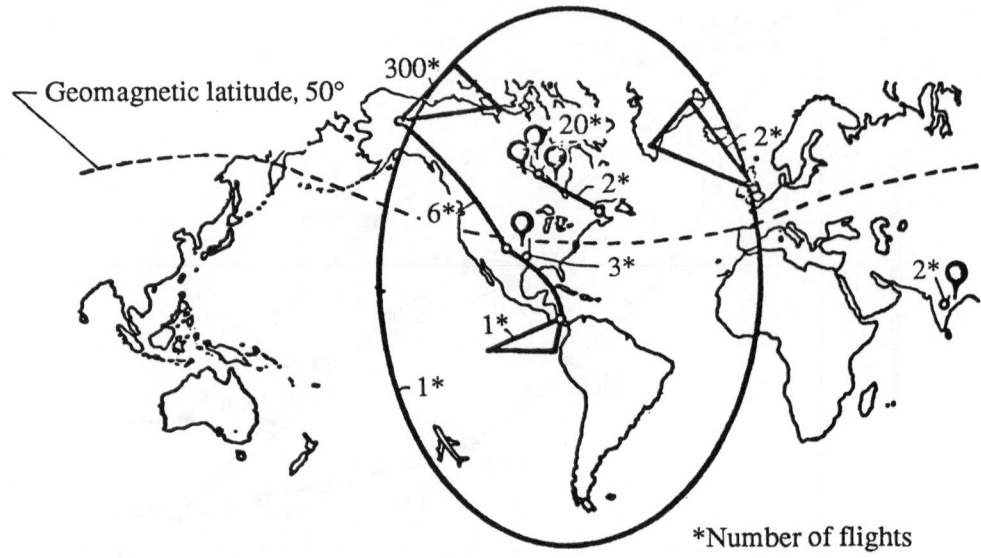

Figure 13.2. High-altitude radiation measurements made between 1965 and 1971.

because of solar modulation and over the surface because of exclusion of the lower energy particles as one approaches the equator by a deflection in the Earth's magnetic field. The representation of the time variations will be accomplished through keying the background to the high-latitude neutron monitor count rates (percent of maximum), and the geomagnetic effects are handled by expressing the environment in terms of the local geomagnetic vertical cutoff rigidity. (Rigidity is related to the radius of curvature of a charged particle in moving through a magnetic field.)

The NASA study is indicated in figure 13.2. The experimental package consisted of encapsulated NE-222 liquid scintillators for neutron spectral measurements throughout the fast neutron region and above, tissue equivalent ion chambers, and nuclear emulsion. The tissue equivalent ion chamber measures the overall radiation dose, and the neutron spectrometer and nuclear emulsion allow an assessment of the high LET components. There were 25 high-altitude balloon flights at various times in cycle 20 and at different latitudes. The balloon flights provide the best altitude survey of environmental quantities. There were over 300 airplane flights using General Dynamics/Martin RB-57F, Lockheed U-2, and Boeing 707 airplanes. The airplanes provided detailed latitude surveys, balloon calibration rendezvous flights, and flights during solar flare events. Because of limited funds, it is unfortunate that not all the data could be finally reduced and that we could not continue the flight program through August 1972 when an extremely large solar event occurred. A more detailed description of the experiment is given by Foelsche et al. (1974) and Korff et al. (1979).

Figure 13.3 shows the measurements made on a high-altitude balloon flight during galactic cosmic ray maximum at 69° N geomagnetic latitude. The instruments for this flight were only lightly shielded (less than 1 g/cm^2 of fiber glass and foam for thermal insulation). The features to be noted in figure 13.3 are the

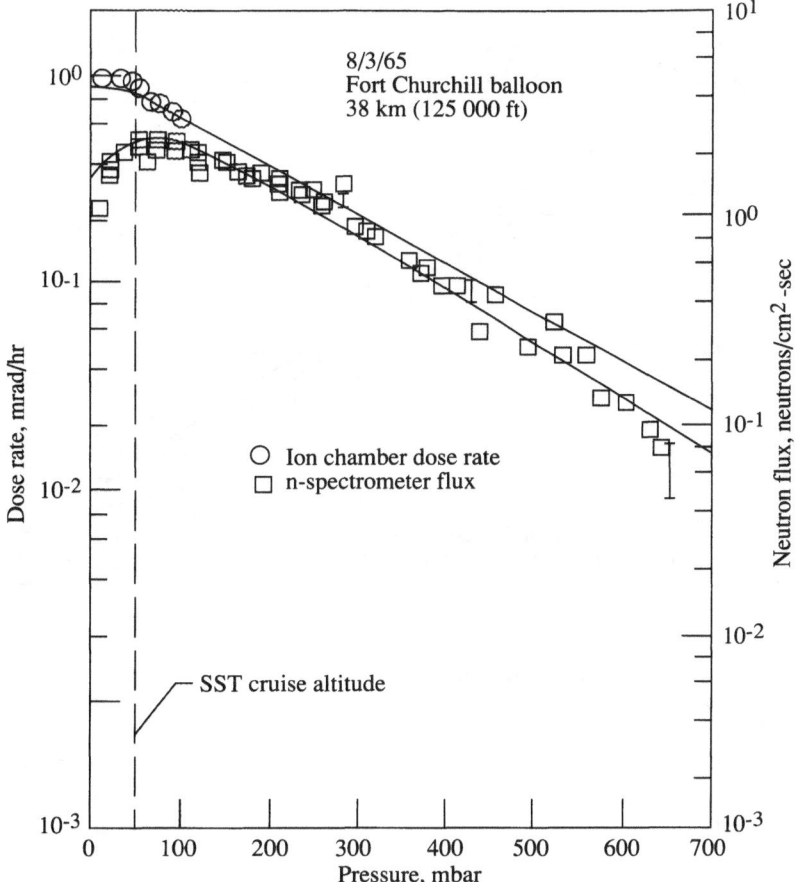

Figure 13.3. Galactic cosmic-ray maximum (August 3, 1965; 1 year after sunspot minimum; Fort Churchill, Canada; geomagnetic latitude ≈ 69°). Neutron flux from 1 to 10 MeV (right scale), and ion chamber dose rate (left scale) is a function of altitude.

broad maximum in the neutron flux, with peak at 60 to 70 millibars (mbar) and the leveling off of the ion chamber dose rate above 50 mbar (1 mbar ≈ 1 g/cm^2). Also shown is the present neutron model environment to be discussed and the model dose rate in tissue to be compared with ion chamber data.

Figure 13.4 shows data from a low-altitude balloon flight during galactic cosmic-ray maximum (1 month after the flight shown in figure 13.3) at 55° N geomagnetic latitude. Note that although the ionization dose rate is considerably reduced, the neutron flux has changed very little. These reductions are due to the increase in geomagnetic cutoff energies when going to lower latitudes. The proton cutoff at Fort Churchill, Canada, caused by solar modulation may have been on the order of a few hundred MeV because of the residual atmosphere (geomagnetic cutoff of 2 MeV), and the geomagnetic proton cutoff at St. Paul, Minnesota, during the magnetically quiet period at the beginning of September 1965 was approximately 800 MeV. The present neutron model environment and tissue dose rate agree well with the data.

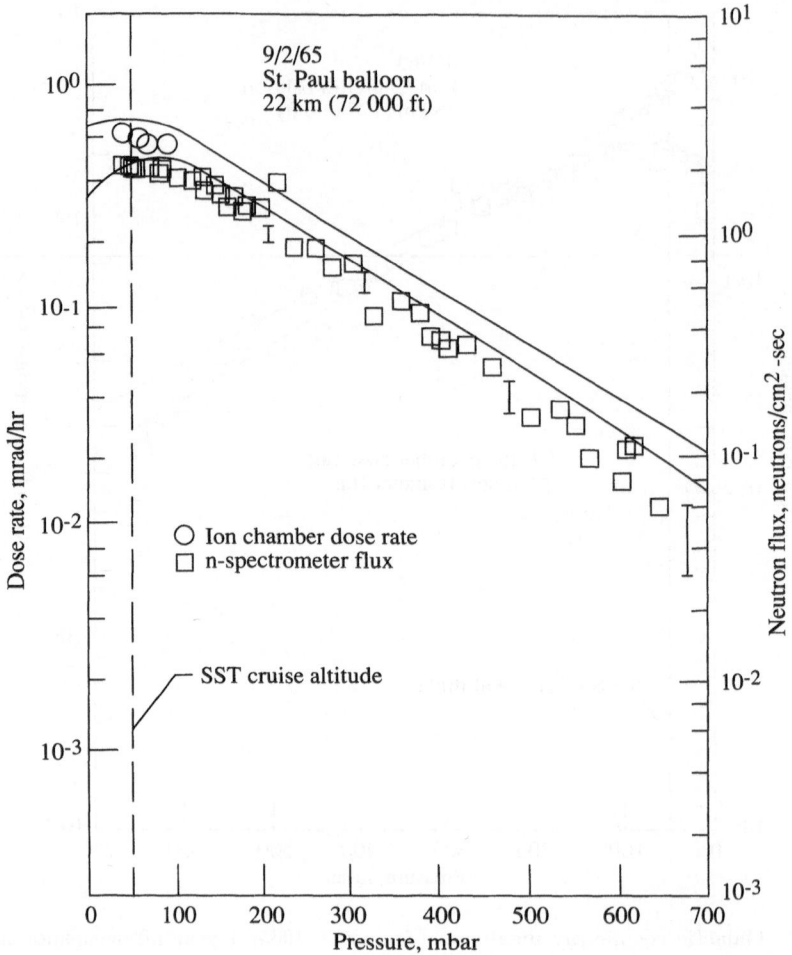

Figure 13.4. Galactic cosmic-ray maximum (September 2, 1965; St. Paul, Minnesota; geomagnetic latitude $\approx 55°$). Neutron flux from 1 to 10 MeV (right scale), and ion chamber dose rate (left scale) is a function of altitude. Compare with data in figure 13.3 at higher latitude.

Shown in figure 13.5 are data from a second flight above Minnesota. This flight differs from the one shown in figure 13.4 in that the ion chamber and neutron spectrometer were placed in a spherical shell of tissue equivalent material (phantom) 15 cm or 15 g/cm^2 thick. The ion chamber dose rate is not appreciably changed from the earlier flight. (See fig. 13.4.) The neutron flux has decreased significantly, and the neutron energy spectrum was found to be flatter in the range from 1 to 10 MeV. This reduction is due to the relatively large moderation of neutrons of energies below about 10 MeV by the hydrogen in the phantom, which outweighs the production of new neutrons by the calcium, carbon, and nitrogen in the phantom.

In figure 13.6 data are plotted for a flight from Fort Churchill, Canada, in a period of increased solar activity, which is typical for about 2 years after galactic

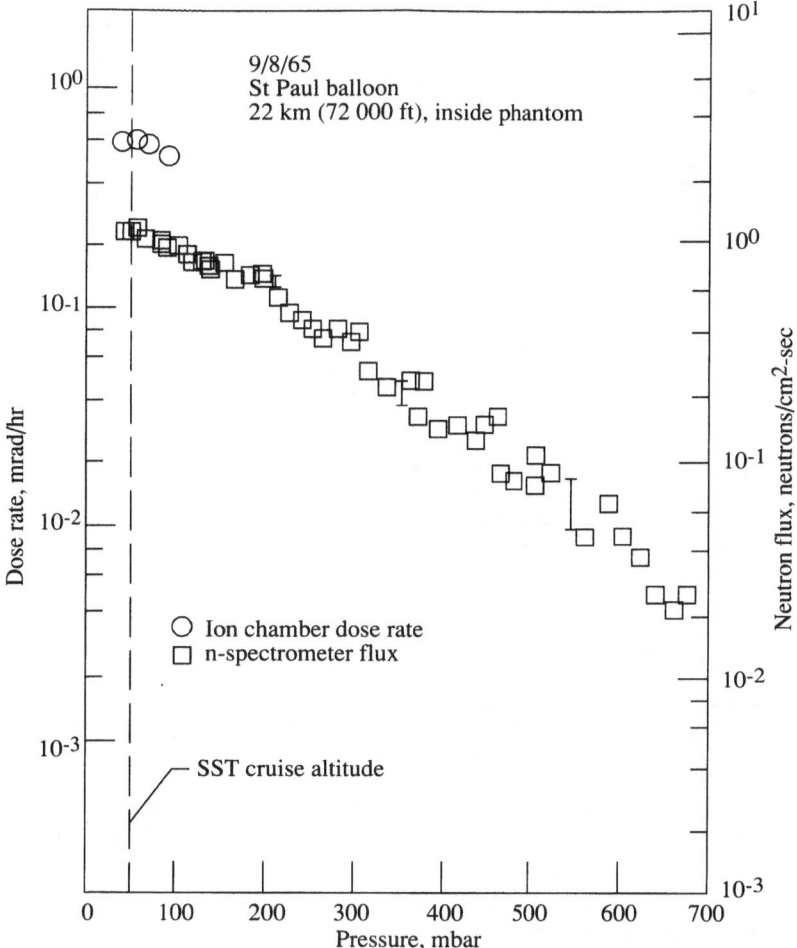

Figure 13.5. Galactic cosmic-ray maximum (September 8, 1965; St. Paul, Minnesota; geomagnetic latitude $\approx 55°$). In flights of figures 13.3 and 13.4, the sensors were lightly shielded (less than 1 g/cm^2 of fiber glass and foam). In this flight, the sensors were surrounded by tissue equivalent material, including calcium, of about 15 g/cm^2 thickness to obtain an approximate measurement of the neutron fluxes and ion chamber dose rates in the center of the human body.

cosmic-ray maximum. The ion chamber dose rate and the neutron flux decreased by about the same percentage during the 2 years. These decreases are due to a corresponding increase in the scattering power of the interplanetary magnetic fields. The solid line between 0 and 300 g/cm^2 in figure 13.6 is the altitude profile of neutron intensities 1 to 10 MeV as obtained from the theoretical nucleon cascade calculations described by Wilson et al. (1970). The neutron flux and tissue dose rate of the present model environment are also shown.

Table 13.1 contains neutron fluxes and spectral indices in the range from 1 to 10 MeV for flights during 1965 to 1968 at high latitudes at an atmospheric depth

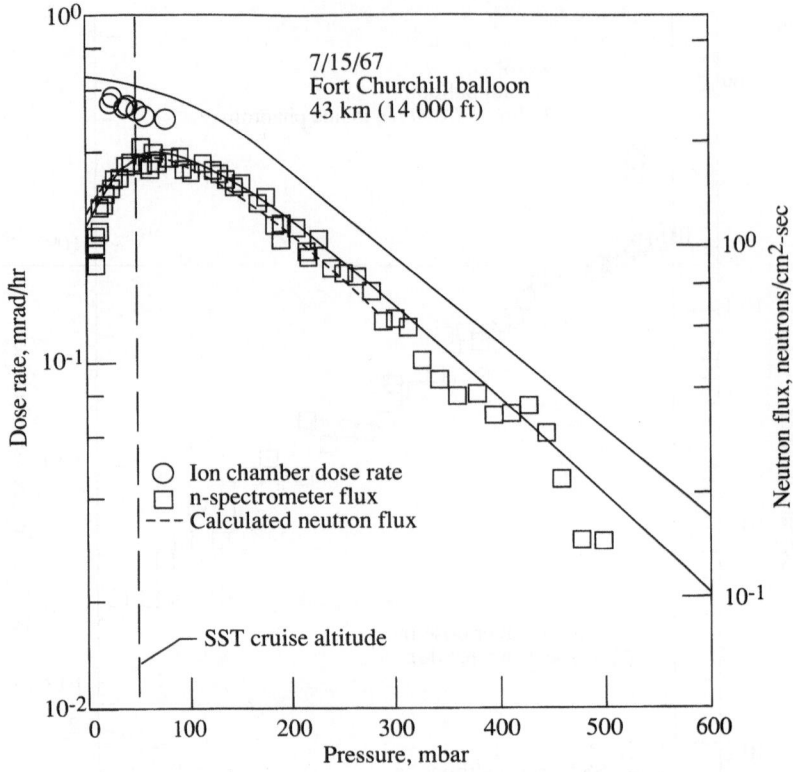

Figure 13.6. Galactic cosmic rays 2 years after galactic cosmic-ray maximum (July 15, 1967; Fort Churchill, Canada; geomagnetic latitude ≈ 69°). Compare with figure 13.3 for a flight at galactic cosmic-ray maximum. The neutron flux and ion chamber dose rate have both decreased about 25 to 30 percent at SST altitudes (solar modulation). The solid line is the altitude dependence obtained by theory.

Table 13.1. Neutron Flux (Integral Flux in Range From 1 to 10 MeV)
and Spectral Index (Differential Energy Spectrum $\approx AE^{-x}$) at SST Altitude

Date	Location[a]	Shielding[b]	Flux, neutrons/cm²-sec	Spectral index, x	DRNM
8/3/65[c]	Ft. Churchill	Air	2.46	1.26	7038
9/2/65	St. Paul	Air	2.16	1.29	6999
7/15/67[c]	Ft. Churchill	Air	1.81	1.23	6644
7/18/68[c]	Ft. Churchill	Air	1.52	1.16	6389
8/9/65[c]	Ft. Churchill	Air + phantom	1.35	.96	7004
9/8/65	St. Paul	Air + phantom	1.08	.86	7018

[a]Locations: Ft. Churchill, Canada, and St. Paul, Minnesota.
[b]Phantom denotes tissue equivalent material.
[c]Used in dose calculations.

of approximately 60 g/cm^2 or an altitude of 20 km (65 000 ft). These data were supplemented by theoretical extrapolations of the neutron spectra (see Wilson, Lambiotte, and Foelsche, 1969; Wilson et al., 1970; Wilson, 1972) in the range from 0.1 to 1 MeV according to the spectral shape calculated by Newkirk (1963) and in the range from 10 to 500 MeV according to the flat spectral slope ($E^{-1.2}$), first found by J. W. Wilson using Monte Carlo nucleon transport calculations on the basis of neutron production cross sections for incident protons up to 2 GeV energy of Bertini (1967) and semiempirical extrapolation to 10 GeV.

In figure 13.7, as an example for the neutron dose determinations, the neutron spectrum from galactic cosmic rays measured from 1 to 10 MeV at supersonic transport (SST) altitude on August 3, 1965, above Fort Churchill is extrapolated by the preceding method to lower energies (0.01 MeV) and higher energies (500 MeV); the results of the Monte Carlo calculations are shown by the horizontal dashes representing the neutron fluxes compiled in the corresponding energy bins. From this spectrum the dose and dose equivalent rates for hands and feet due to neutrons are obtained by summing the dose rates resulting from multiplying the flux in each energy interval by the corresponding flux-to-dose rate conversion factor for the extremities. The resulting dose rate is 1.23 μGy/hr (0.123 mrad/hr), and the corresponding dose equivalent rate is 7.72 μSv/hr (0.772 mrem/hr). In addition to the spectrum, the separate contributions in the different energy ranges to the dose and dose equivalent rates are indicated in figure 13.7 (linear scale). The neutrons of energies greater than 10 MeV are found to contribute (through recoil protons and stars) 35 percent to the total dose equivalent rate of neutrons. The neutrons of energy from 0.1 to 1 MeV, assumed to have an energy spectrum similar to that given by Newkirk, contribute about 27 percent. The unmeasured part of the spectrum thus contributes about 70 percent to the neutron dose equivalent rate in extremities.

The main results of these measurements on galactic cosmic rays are the determination of both the absolute values of the energetic secondary neutron fluxes (1 to 10 MeV) and the dose as measured with tissue equivalent ion chambers. The neutron spectrum, which was in doubt before the present experiments, especially for high latitudes and altitudes, was found to be of a flatter spectrum. The tissue equivalent ion chamber yielded the contributions of neutrons (via recoil protons) to the absorbed dose in tissue, which is not obtained in conventional metal-walled ion chambers. In addition, the actual measurements of neutron spectra and tissue dose rates inside a spherical body phantom experimentally relate the dose equivalents in thin tissue equivalent samples (corresponding to the extremities) to the depth dose equivalents in the human body, and thus confirm theoretical calculations (Foelsche et al., 1974).

The theoretical spectra have as yet to be normalized by adjusting the absolute intensities to the measured neutron spectra in the range from 1 to 10 MeV. The theoretical spectra are based on calculations for proton primaries and do not accurately take into account the α-particles and heavier nuclei that are present in galactic and solar cosmic rays because the secondary production cross sections in reactions with air have not been satisfactorily determined either theoretically or

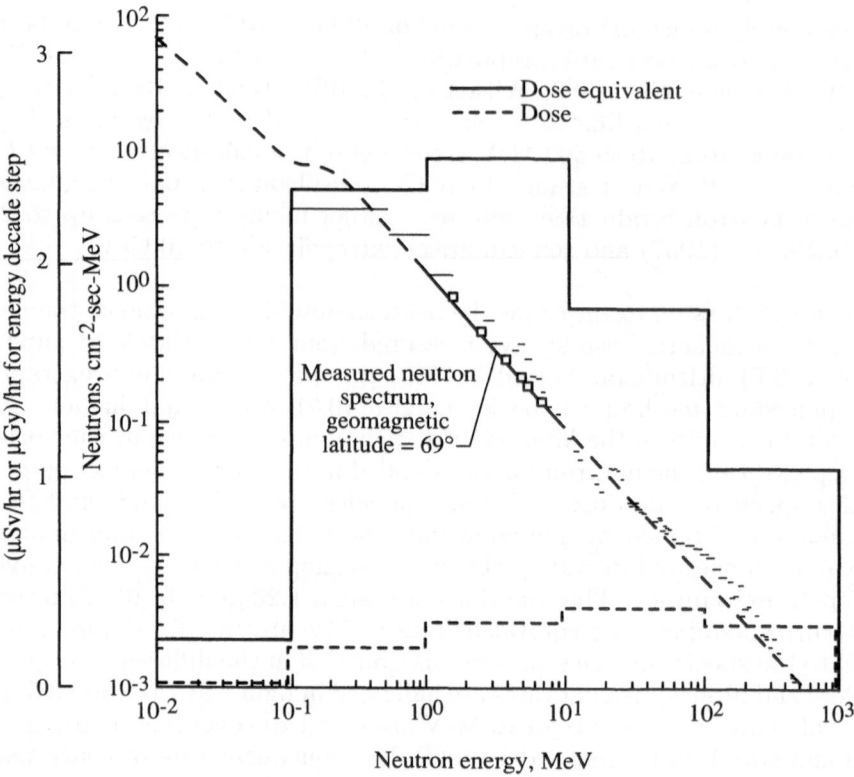

Figure 13.7. High-latitude (geomagnetic latitude $\approx 69°$) neutron spectrum measured at SST altitudes (≈ 50 g/cm^2) on August 3, 1965 (heavy solid line between 1 and 10 MeV), by Mendell in Korff et al. (1979) with its extension to lower and higher energies (heavy dashed curve) compared with the shape of the Monte Carlo spectrum (histogram, horizontal dashes). The linear scale is for the dose rates in extremities calculated from the NASA spectrum (Sv—heavy step curve; Gy—dashed step curve).

experimentally. For the present purposes, it is considered satisfactory to assume that only the intensity, and not the shape, of the neutron spectra at subsonic and supersonic jet altitudes is substantially changed by the heavier primaries.

The various contributions to the total dose equivalent rate as measured in high latitudes for the initial phases of the present solar cycle (1965 to 1968) are shown in figure 13.8. The different contributions in the figure correspond to the types of instrumentation with which the measurements were made. The tissue equivalent ion chamber measures only the energy deposited in a thin tissue sample (that is, the absorbed dose rate) by all radiation components. This measurement, however, does not provide the biologically equivalent dose rate since much of the dose equivalent rate is due to components with a quality factor Q_F greater than unity, such as proton recoils and heavily ionizing star prongs in tissue from neutron and charged-particle reactions with tissue nuclei. These contributions to the excess of the total dose equivalent over the corresponding absorbed dose given by the tissue equivalent ion chamber (Sv minus Gy) are referred to as the $Q_F - 1$ increments. They are derived by an analysis of the neutron spectrometer

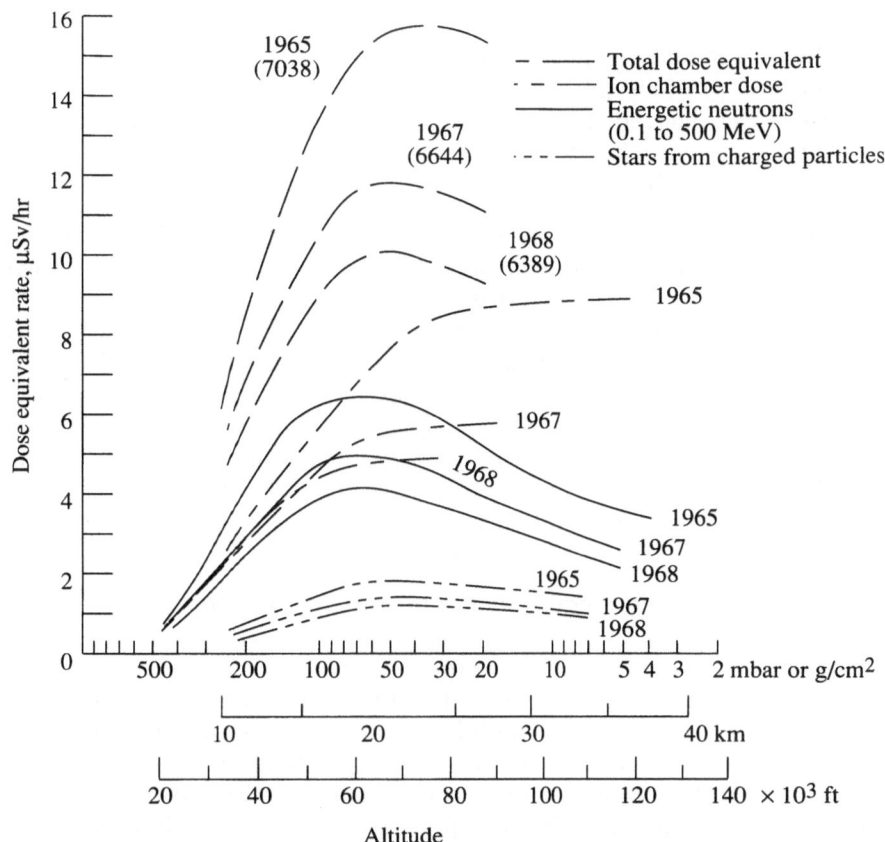

Figure 13.8. Galactic cosmic-ray dose equivalent rates for extremities (hands and feet) and approximately for eyes as a function of altitude at different phases of the solar cycle for high latitudes.

and tissue equivalent nuclear emulsion data. (The nuclear emulsion data from the British Royal Aerospace Establishment (R.A.E.) were used, as explained subsequently.) The components of the total dose equivalent rate, derived from measurements and shown in figure 13.8, are as follows:

(1) The tissue absorbed dose rate from all radiation components, that is, from charged primaries and secondaries, including mesons, gamma rays, neutrons (via recoil protons, heavy recoils, and neutron-produced stars), and stars produced by energetic charged particles, all of which are measured in the tissue equivalent ion chamber. As explained before, some of these components have, because of their large linear energy transfer, a quality factor greater than unity. This excess constitutes parts (2) and (3).

(2) The $Q_F - 1$ increment rate produced in tissue by energetic neutrons (0.1 to 500 MeV) via recoils and stars.

(3) The $Q_F - 1$ increment rate caused by stars in tissue produced by primary and secondary charged particles.

Component (2), the neutron $Q_F - 1$ increment rate, is found from the measured neutron fluxes in the range from 1 to 10 MeV extended to lower and higher energies as explained in a previous paragraph (see fig. 13.7) by subtracting the inferred dose rate from the inferred neutron dose equivalent rate. The neutron $Q_F - 1$ increment rate on August 3, 1965, at SST altitude is thus found, for example, for the extremities as 6.5 μSv/hr (i.e., 7.72 − 1.23, see fig. 13.7).

Component (3), the dose equivalent rate due to stars produced by charged particles, was derived from measurements at different altitudes in tissue equivalent emulsions by P. J. N. Davison of the British R.A.E., where the increment is referred to as "star damage energy." The total star $Q_F - 1$ increment derived by Davison included the contribution of neutron-produced stars already taken into account in (2); the star contribution from charged particles is approximately one-half of the total star $Q_F - 1$ increment at high altitudes (20 km (65 000 ft) to 34 km (100 000 ft)) and one-third of the total star $Q_F - 1$ increment at subsonic altitudes (11 km (37 000 ft)), as theoretical calculations indicate. This part is plotted in figure 13.8. The total dose equivalent rate is obtained by summing parts (1), (2), and (3).

The total extremity dose equivalent rate in figure 13.8 in high latitudes as a function of altitude exhibits a maximum at about 35 g/cm^2 (22 km or 75 000 ft) during galactic cosmic-ray maximum (approximately 1 to 2 years after sunspot minimum). The maximum decreases in magnitude and appears to move deeper into the atmosphere as the galactic cosmic-ray minimum is approached. This peak is mainly due to the broad maximum in the neutron fluxes at these altitudes. (See the neutron data in figs. 13.3 and 13.6 and the neutron contribution in fig. 13.8.) The absorbed energy measured in the ion chamber does not exhibit this peak. It may furthermore be noted that the neutron dose equivalent rate contributes about 50 percent to the total dose equivalent rate at these altitudes.

13.3.2. Radiation levels within the geomagnetic field. The latitude surveys were made mostly by airplanes so that the relation between airplane count rates and balloon count rates needs to be established (the effect of neutron production and moderation by the airplane structure). The airplane count rates are found to be 10 percent higher in balloon rendezvous flights with identical instruments as well as within the solar cycle for the same instrument (Korff et al., 1979) as shown in figure 13.9. The latitude surveys by balloons and airplanes are shown in figure 13.10 at the transition maximum and at 250 g/cm^2 for different phases of solar cycle 20. The curves in the figure are our approximation to the data given by

$$\phi_{1-10}(x_m, R, C) = 0.23$$
$$+ [1.1 + 0.167(C - 100)] \exp\left(\frac{-R^2}{81}\right) + \left[0.991 + 0.0501(C - 100) \right.$$
$$\left. + 0.4 \exp\left(\frac{(C - 100)}{3.73}\right) \right] \exp\left(\frac{-R^2}{12.96}\right) \tag{13.1}$$

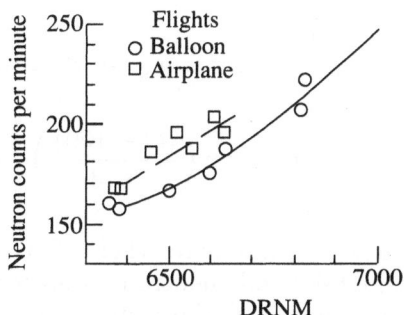

Figure 13.9. Count rates of neutron channels 2–7 in balloon flights and airplane flights. Effect of airplane material on neutron count rates is illustrated.

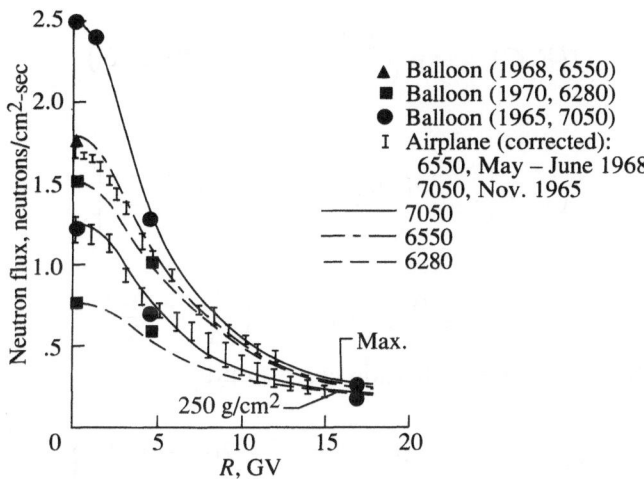

Figure 13.10. Fast neutron flux (in range from 1 to 10 MeV) at the transition maximum and at 250-g/cm^2 depth as a function of vertical cutoff rigidity R for various times in the solar cycle and DRNM count rates.

at the transition maximum and

$$\phi_{1-10}(250, R, C) = 0.17$$
$$+ [0.787 + 0.035(C - 100)]\exp\left(\frac{-R^2}{25}\right)$$
$$+ \left[-0.107 - 0.0265(C - 100) \right.$$
$$\left. + 0.612 \exp\left(\frac{(C - 100)}{3.73}\right) \right] \exp\left(\frac{-R^2}{139.2}\right) \qquad (13.2)$$

at depths of 250 g/cm^2 in the atmosphere where R is local cutoff rigidity (in units of GV) and C is the high-latitude neutron monitor count rate in percent of maximum. At depths below 250 g/cm^2, the neutrons attenuate with attenuation length (g/cm^2) given by

$$\lambda = 165 + 2R \qquad (13.3)$$

The flux at all altitudes is approximated as

$$\phi_{1-10}(x, R, C) = f(R, C)\exp\left(\frac{-x}{\lambda}\right) - F(R, C) \exp\left(\frac{-x}{\Lambda}\right) \qquad (13.4)$$

where

$$f(R, C) = \exp\left(\frac{250}{\lambda}\right) \phi_{1-10}(250, R, C) \qquad (13.5)$$

$$\Lambda = \lambda \left[1 - \frac{\phi_{1-10}(x_m, R, C) \exp\left(\frac{x_m}{\lambda}\right)}{f(R, C)} \right] \qquad (13.6)$$

and

$$F(R, C) = \frac{\Lambda}{\lambda} f(R, c) \exp\left(\frac{x_m}{\Lambda} - \frac{x_m}{\lambda}\right) \qquad (13.7)$$

where the transition maximum altitude corresponds to

$$x_m = 50 + \ln\left\{ 2000 + \exp\left[-2(C - 100) \right] \right\} \qquad (13.8)$$

The neutron environment model given by equations (13.1) to (13.8) is shown in figures 13.3, 13.4, and 13.6 in comparison with the experimental data. The flux from 1 to 10 MeV is converted to dose equivalent and dose using 3.14 μSv-cm^2-sec/hr (0.314 mrem-cm^2-sec/hr) and 0.5 μGy-cm^2-sec/hr (0.05 mrad-cm^2-sec/hr), respectively. The accepted quality factor within the U.S. is the one in ICRP 26 (Anon., 1977) and is used in the current estimates.

Unfortunately, not all ion chamber data or all nuclear emulsion data were reduced. For our present purpose we will use the argon-filled ion chamber data to represent the altitude, latitude, and solar cycle dependence and use the available tissue equivalent ion chamber data as a guide. The ion chamber data of GREP

Table 13.2. Ionization Rates in Air Measured by Argon-Filled Chambers[a] at Solar Minimum ($C = 98.3$ in 1965)

R, GV	Ion pairs, cm^{-3}-sec^{-1}, for air depths, g/cm^2, of—													
	30	40	50	60	70	80	90	100	120	140	200	245	300	1034
0	445.0	430.0	414.0	399.0	383.0	366.0	349.0	332.0	298.0	266.0	181.0	136.0	95.0	11.4
.01	445.0	430.0	414.0	399.0	383.0	366.0	349.0	332.0	298.0	266.0	181.0	136.0	95.0	11.4
.16	444.0	430.0	414.0	399.0	383.0	366.0	349.0	332.0	298.0	266.0	181.0	136.0	95.0	11.4
.49	411.8	404.3	394.4	382.0	369.0	354.8	339.4	325.0	292.3	264.5	181.0	136.0	95.0	11.4
1.97	325.0	333.0	340.0	335.0	330.0	312.5	308.0	300.0	285.0	264.0	181.0	134.0	95.0	11.4
2.56	300.0	305.0	310.0	305.0	300.0	290.0	285.0	280.0	255.0	230.0	173.0	126.0	95.0	11.4
5.17	185.0	195.0	208.0	208.0	208.0	208.0	208.0	208.0	195.0	185.0	135.0	103.0	75.0	10.6
8.44	127.6	137.0	145.0	150.2	153.8	155.8	156.0	154.6	149.7	142.2	111.3	87.0	66.6	10.4
11.70	85.0	92.0	98.0	100.0	102.0	105.0	107.0	110.0	108.0	105.0	80.0	77.0	60.0	10.0
14.11	70.0	75.0	82.0	85.0	89.0	93.6	95.0	100.0	98.0	95.0	78.0	68.8	50.0	10.0
17.00	66.3	73.8	80.0	84.8	88.5	91.1	92.6	93.5	93.4	90.5	75.0	62.3	48.0	10.0

[a]Experimental data extrapolated to provide estimates of ionization rates over a wide range of altitudes and geomagnetic cutoffs.

Table 13.3. Ionization Rates in Air Measured by Argon-Filled Chambers[a] at Solar Maximum ($C = 80$ in 1958)

R, GV	Ion pairs, cm^{-3}-sec^{-1}, for air depths, g/cm^2, of—													
	30	40	50	60	70	80	90	100	120	140	200	245	300	1034
0	264.6	267.5	267.0	265.0	258.0	252.0	243.0	235.0	216.3	197.0	145.0	109.2	78.8	11.4
.01	264.6	267.8	267.0	265.0	258.0	251.0	243.0	235.0	216.3	197.0	145.0	109.2	78.8	11.4
.16	264.0	264.9	265.0	264.0	257.0	250.0	243.0	233.0	215.0	197.0	145.0	109.2	78.8	11.4
.49	264.0	264.9	265.0	262.0	256.0	249.0	242.0	231.0	213.2	197.0	145.0	109.2	78.8	11.4
1.97	264.0	265.0	265.0	262.0	252.0	245.0	241.0	231.0	212.5	197.0	145.0	107.8	78.8	11.4
2.56	235.0	237.5	240.0	240.0	239.0	238.0	237.0	230.0	209.0	197.0	145.0	101.6	78.8	11.4
5.17	162.5	168.0	179.0	182.0	178.0	175.2	174.0	173.8	170.0	160.0	159.0	88.3	65.0	10.6
8.44	95.0	103.5	112.0	118.0	118.0	119.0	120.0	122.0	118.0	117.0	100.6	78.7	60.2	10.4
11.70	78.2	85.0	90.7	92.7	94.8	98.0	100.0	103.1	101.2	98.4	75.0	72.2	56.2	10.0
14.11	65.7	70.7	77.5	80.5	84.3	89.0	90.5	95.3	93.5	90.9	74.0	65.9	47.9	10.0
17.0	63.0	70.3	76.4	81.1	84.8	87.5	89.1	90.2	90.1	87.4	72.6	60.3	46.5	10.0

[a]Experimental data extrapolated to provide estimates of ionization rates over a wide range of altitudes and geomagnetic cutoffs.

is shown in table 13.2 for solar minimum ($C = 98.3$ in 1965) and in table 13.3 for solar maximum ($C = 80$ in 1958) as obtained for cycle 19. We note that the low-energy GCR had not fully recovered in the summer of 1965 with the result that the high-latitude ionization at high altitude is about 10 percent lower than that in 1954. Furthermore, the 1958 measurements near solar maximum covered only mid to high latitudes, and the low-latitude data in table 13.3 are likely to be about 10 percent too high at high altitudes. The ionization rates in tables 13.2 and 13.3 are the rates in air per atmosphere of pressure. They are directly converted to exposure units and absorbed dose in tissue. The comparison with the tissue equivalent ion chamber requires the addition of the neutron absorbed dose rates as shown in figures 13.3, 13.4, and 13.6 where good consistency between the two methods is demonstrated. Dose equivalent estimates require an estimate of the high LET components associated with charged particles and are found from the measurements in nuclear emulsion as shown in figure 13.8. The corresponding average quality factor for proton-produced stars is found from Davison's emulsion data as

$$Q_F = 1 + 0.35 \exp\left(\frac{-x}{416}\right) - 0.194 \exp\left(\frac{-x}{65}\right) \tag{13.9}$$

This average quality factor will be applied to ion chamber dose rate data.

13.3.3. *Comparison with other methods.*

This first comparison is with the dose equivalent rate meter of the Brookhaven National Laboratory (BNL). The BNL instrument is a tissue equivalent spherical proportional counter of 22-cm radius filled at 10 torr with tissue equivalent gas (corresponds to a 3-μm tissue site). The LET spectra are derived using the triangular relation to lineal energy which assumes that the chamber size is small compared with the range of the particles being detected. The ICRP 26 quality factor (Anon., 1977) is used to calculate dose equivalent rates. Because of the large chamber size at the chosen pressures, the high LET events are greatly distorted by the triangular assumption. For example, a 100-keV/μm proton moving along the diameter would register only a 75-keV pulse instead of the 300-keV pulse expected using the triangular distribution. The instrument-assigned quality factor would be 10 instead of the correct value of 20. Such distortions are worse for multiple charged ions. Therefore, the BNL instrument could significantly underestimate the average quality factor. With this limitation in mind, in table 13.4 we give the BNL measurements made in 1971 and 1972 along with results of the present model. Approximate neutron monitor count rates used in evaluating the present model are given for the Deep River neutron monitor (DRNM) in Canada. The dose rates are in very good agreement, but the quality factors of the present model are substantially larger than the values measured by the BNL instrument. The present results are shown in table 13.5 with the results of the British Aerospace/Aerospatiale Concorde dosimeter and the HARIS instrument. The Concorde dosimeter consists of three shielded Geiger counters and a BF_3 proportional counter for neutron detection. The HARIS instrument consists of a gas-filled tissue equivalent proportional counter and tissue equivalent ion chamber. It has some of the same limitations as the BNL instrument.

Geomagnetic latitude, °N	Date	Altitude, km	Areal density, g/cm^2	Absorbed dose rate, μGy/hr	Dose equivalent rate, μSv/hr
36.7	Aug. 29, 1972:	3.0	694	0.18 (0.18)	0.25 (0.30)
	DRNM ≈6950	6.1	460	.58 (0.57)	1.00 (1.12)
		9.1	303	1.38 (1.21)	2.50 (2.55)
		12.2	188	2.80 (2.32)	4.75 (4.82)
41.7	Aug. 30, 1972:	3.0	694	0.20 (0.19)	0.25 (0.36)
	DRNM ≈6950	6.1	460	.63 (0.59)	0.88 (1.29)
		9.1	303	1.50 (1.28)	2.45 (2.99)
		12.2	188	3.10 (2.57)	5.25 (5.76)
50.0	June 17, 1972:	3.0	694	0.22 (0.20)	0.35 (0.43)
	DRNM ≈6950	6.1	460	.61 (0.69)	1.05 (1.63)
		9.1	303	1.68 (1.59)	2.75 (3.95)
		12.2	188	3.45 (3.14)	5.80 (7.61)
58.0	July 18, 1972:	3.0	694	0.19 (0.20)	0.42 (0.45)
	DRNM ≈6950	6.1	460	.63 (0.69)	1.21 (1.72)
		9.1	303	1.70 (1.59)	2.90 (4.25)
		12.2	188	3.90 (3.35)	7.00 (8.52)
69.4	June 29, 1971:	9.1	303	1.59 (1.57)	2.34 (4.09)
69.6	DRNM ≈6850	11.6	204	2.88 (2.97)	4.89 (7.53)
68.4		15.2	116	4.89 (4.97)	7.93 (12.09)
67.3		18.3	71	6.03 (6.14)	10.39 (14.24)

Table 13.5. Present Results and Radiation Measurements Made With the Concorde
Instrument and the HARIS at 18.3 km and High Geomagnetic Latitude

Geomagnetic latitude, °N	Date (1969)	DRNM	Dose equivalent rates, μSv/hr, for—		
			Concorde	HARIS	Present
65–70	Nov. 3	6369	9	10.7	11.7–11.5
67	Nov. 17	6429	7–9	7.8	11.9
70	Nov. 19	6442	5–7	6.3	12.0
70	Nov. 21	6510	7–9	6.9	12.4
70	Nov. 21	6537	7–9	6.5	12.5

The present model is compared with several sources in table 13.6. Of particular note is the result of Schaefer's measurements with nuclear emulsion (Friedberg and Neas in Anon., 1980). In particular, the ionization tracks of all particles other than nuclear stars lead to 5.8 μGy/hr and 10.5 μSv/hr, which is in good agreement with the values from the BNL instrument of 6.0 μGy/hr and 10.4 μSv/hr as given in tables 13.4 and 13.6. The nuclear-star contributions from Schaefer's nuclear emulsion data give final values of 6.5 μGy/hr and 15.0 μSv/hr in excellent agreement with those of the present model of 6.1 μGy/hr and 14.2 μSv/hr, respectively. The importance of the nuclear-star contribution has also been noted by Friedberg and Neas in Anon. (1980). The present model appears to be in excellent agreement with measurements, which include the high LET components associated with nuclear-star contributions.

Table 13.6. Present Results Compared With Other Estimates of Galactic Radiation

Source of data	Dose equivalent rate, μSv/hr, for altitude (areal density) of—	
	11.0 km (255 g/cm^2)	18.3 km (71 g/cm^2)
At or near 55°N, solar average (DRNM = 6660)		
Present	5.9	12.1
O'Brien and McLaughlin (1972)	4.1	13.2
LaRC:		
Extremities	6.2	12.5
42°N, 43°N, at or near solar maximum (DRNM = 6950)		
Present	4.8	8.6
O'Brien and McLaughlin (1972)	3.3	9.8
BNL	3.9	
67°N–70°N, at or near solar maximum (DRNM = 6850)		
Present	6.7	14.2
BNL	4.1	10.4
Concorde		8.0
HARIS		9.0
Schaefer in Anon. (1980)		15.0

13.4. Global Dose Rate Estimates

The dose rate evaluation methods described herein are directly applicable to predictions, or estimates, of global dose rate patterns. The requirements for such an analysis include the geographic distribution of magnetic cutoff values and the global pressure fields at the altitudes of interest. The vertical magnetic cutoff values are taken here as the 1980.0 epoch data of Smart and Shea (1983) and are shown in figure 13.11. The atmospheric climatological data have been extracted from the Langley Research Center General Circulation Model (LaRC-GCM) of the Earth's atmosphere (Grose et al., 1987; Blackshear, Grose, and Turner, 1987), with enhanced resolution for vertical structure as given by R. S. Eckman et al. in presently unpublished work. For purposes of computations presented herein, the original latitude-longitude grid has been reduced to 10° increments by appropriate interpolation of the LaRC-GCM data; i.e., the latitude grid points are −90° S to +90° N and the longitudes progress from −180° W to +180° E in 10°

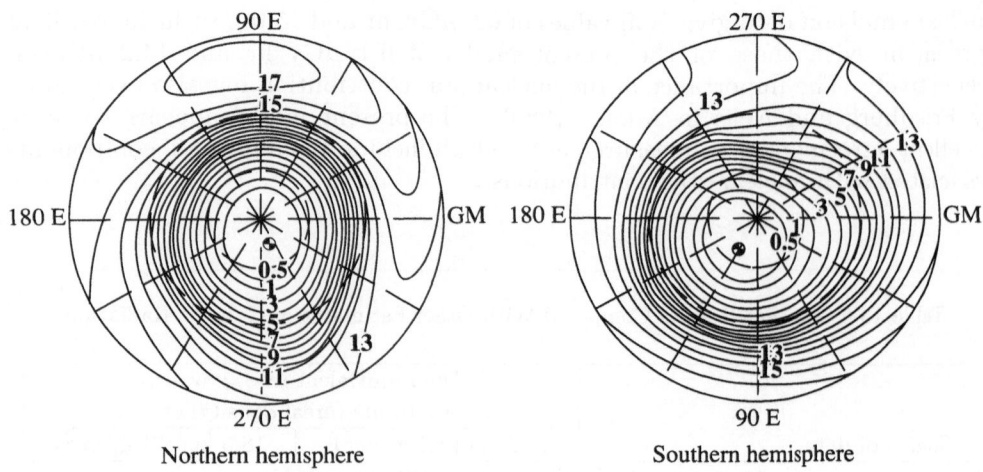

Figure 13.11. Contour of vertical geomagnetic cutoff values from data of Smart and Shea (1983). Contour increments are 1 GV except for the lowest (0.5 GV) contour. Magnetic pole locations are indicated.

steps. Of course, any desired pressure altitude distribution may be substituted; for example, the widely used NASA/Marshall Space Flight Center Global Reference Atmospheric Model (Justus et al., 1980) or other real-time meteorological fields can be used. The principal reasons for utilization of the LaRC-GCM data in the present calculations relate to the fact that both seasonal and north-south hemisphere asymmetries appear to be well represented in the upper troposphere and stratosphere (Grose et al., 1989), with the influence of surface topography taken into account.

The global dose rate analysis has been performed for three altitudes: 10, 14, and 18 km (approximately 33 000, 46 000, and 60 000 ft, respectively). Pressure distributions at these altitudes are shown in figures 13.12, 13.13, and 13.14, respectively. In these figures, the pressure distributions are shown in polar stereographic projection for both hemispheres and for both solstice conditions. The northern hemisphere winter solstice (the southern hemisphere summer) has been modeled as an average of pressure for the first 10 days in January as given by the LaRC-GCM. The corresponding opposite solstice conditions are represented by an average of the data for the first 10 days in July. Pressure contours are given in millibars (mbar). (Note that for the Earth's atmosphere, pressure in millibars is numerically within 2 percent of the overhead absorber amount in units of g/cm^2.) Some general salient features of the pressure distributions may be noted; for example, the equator-to-pole latitudinal gradient is much larger for each hemisphere in winter than in summer. As a consequence, a substantial seasonal variation in dose rates may result, particularly for high-latitude flights in the northern hemisphere.

13.4.1. ICRP 26 quality factors. Contour maps of the dose rates at solar minimum corresponding to the three altitude levels are presented in figures 13.15, 13.16, and 13.17. Dose rate values for the contours are expressed in units of cSv/1000 hr (equivalent to mrem/hr). For altitudes of 10 and 14 km, dose rates increase from low latitudes by approximately a factor of 3 as polar regions are approached. At the 18-km altitude ($\approx$60 000 ft), the equator-to-pole increase is on the order of a factor of 5, largely because of the increasing influence of the geomagnetic cutoffs relative to the atmospheric attenuation. Note that even for the 14-km altitude, dose rates at high latitudes in the polar winter are greater than 1 cSv/1000 hr. Such regions of the globe in the northern hemisphere encompass several common international flight paths. Two such routes (N.Y.–London and N.Y.–Tokyo), which are indicated in figure 13.18, have been overlaid on satellite-view projections of the globe showing contour maps of the northern hemisphere winter dose rate at the 18-km level. The orientations of the projections are such that the pertinent great circle routes appear as straight lines on the figures.

Figure 13.19 shows the winter dose rate contours at the 10-km altitude for the contiguous United States. (Note the change in scale from previous figures; dose rates in fig. 13.19 are given in mSv/1000 hr.) The U.S. dose rate differs by approximately a factor of 2 within the latitude belt of the nation. Dose rates for many of the northern states exceed 0.6 cSv/1000 hr (0.6 mrem/hr).

During the summer months at solar minimum, the dose rates for the northern-most flights in the contiguous U.S. drop to about 0.47 cSv/1000 hr. Crew members operating 1000 hr or more on northern U.S. routes at 10 km or more obviously exceed the allowable exposure for the general population. The dose rates at solar minimum at 14 km are greater by nearly a factor of 2 and represent a substantial fraction of the dose allowed a radiation worker (50 mSv/year) for a 1000 hr/year exposure. A more detailed analysis of commercial airplane operations is clearly needed. This is especially true if newly proposed increases in quality factors are enforced (ICRU 40 (Anon., 1986)).

13.4.2. Revised quality factors. All previous results were based on the quality factors of ICRP 26 (Anon., 1977). Average neutron quality factors have been derived using the proposed ICRU 40 quality factors in Anon. (1986) and will be used here to estimate expected future upward revisions. The contributions to neutron dose equivalent rates in neutron energy subintervals as presented in figure 13.7 are shown in the first three columns of table 13.7. The corresponding previous average quality factors for each subinterval are shown in column four. The newly proposed quality factors (ICRU 40 (Anon., 1986)) are averaged over each subinterval according to the neutron spectrum produced by GCR and then are applied to obtain new estimates of the neutron dose equivalent rates, as shown in columns five and six. The results change the dose equivalent by 60 percent. We do not increase the charged-particle star contribution since the assumed quality factor of 20 is a generous overestimate of nuclear-star contributions to dose equivalent (Wilson, Shinn, and Townsend, 1990).

The revised dose equivalent rates at solar minimum for 10-, 14-, and 18-km altitudes are shown for winter solstice and summer solstice in figures 13.20, 13.21, and 13.22, respectively. During solar minimum years, flight crews who spend over 1000 hr at altitudes in excess of 10 km (33 000 ft) in the northern polar region can receive up to 1 cSv per year in excess of that allowed for the general population.

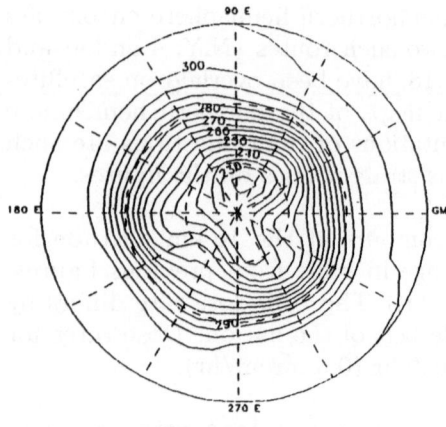

(a) Northern hemisphere winter
(January data).

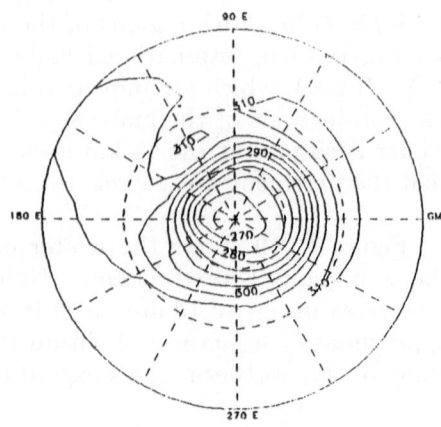

(b) Northern hemisphere summer
(July data).

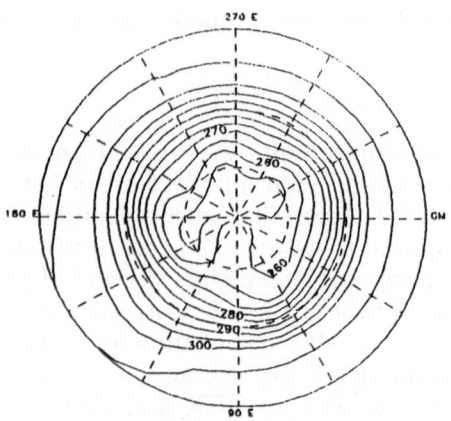

(c) Southern hemisphere winter
(July data).

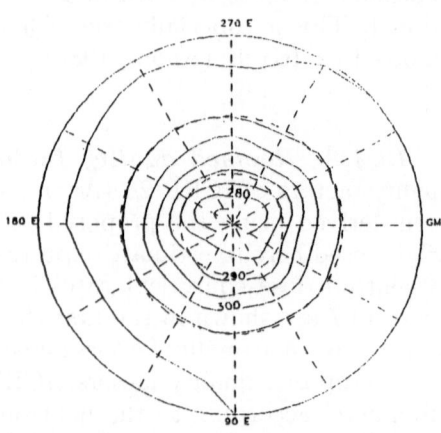

(d) Southern hemisphere summer
(January data).

Figure 13.12. Global pressure distribution for solstice conditions at 10-km altitude (approximately 33 000 ft). Contour increment is 5 mbar.

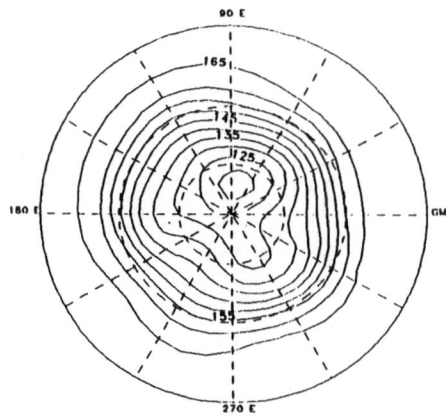

(a) Northern hemisphere winter
(January data).

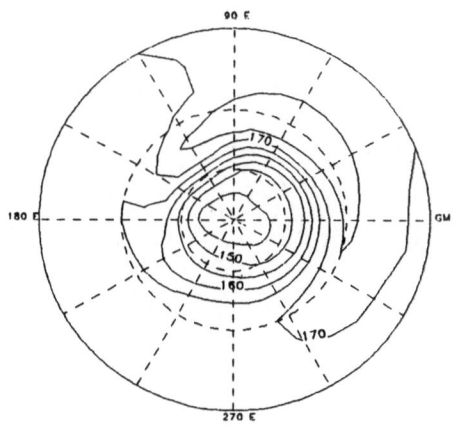

(b) Northern hemisphere summer
(July data).

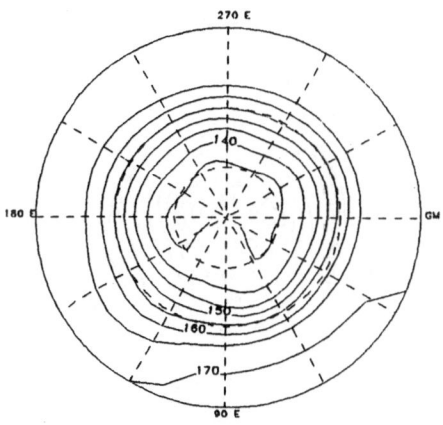

(c) Southern hemisphere winter
(July data).

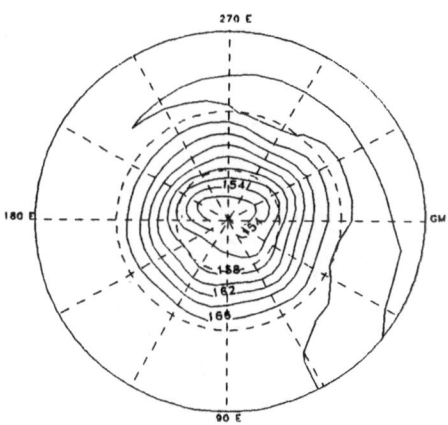

(d) Southern hemisphere summer
(January data).

Figure 13.13. Global pressure distribution for solstice conditions at 14-km altitude (approximately 46 000 ft). Contour increment is 5 mbar.

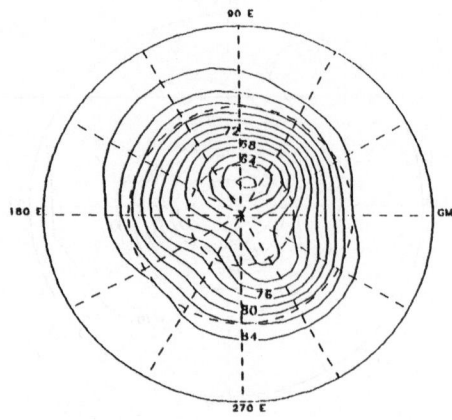

(a) Northern hemisphere winter
(January data).

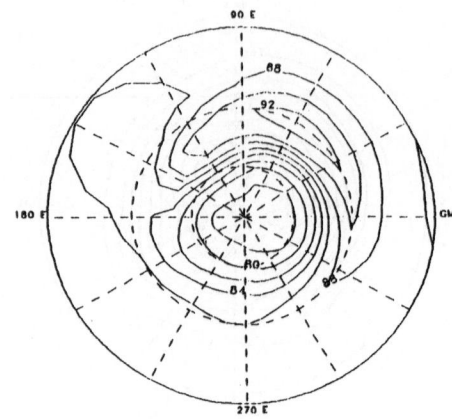

(b) Northern hemisphere summer
(July data).

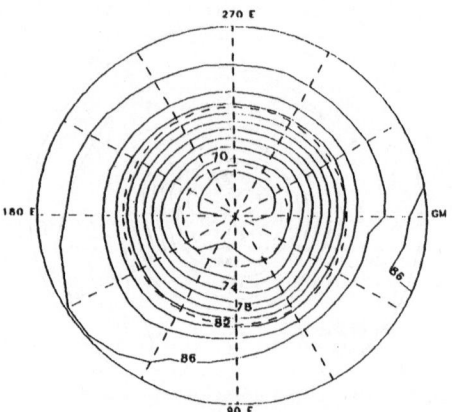

(c) Southern hemisphere winter
(July data).

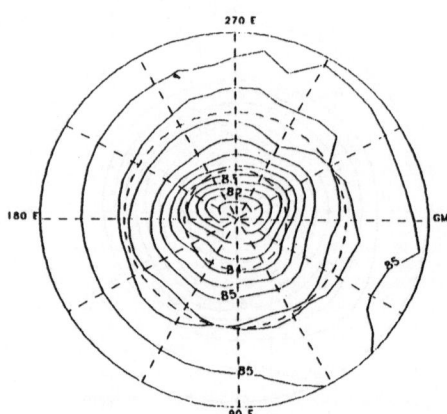

(d) Southern hemisphere summer
(January data).

Figure 13.14. Global pressure distribution for solstice conditions at 18-km altitude (approximately 60 000 ft). Contour increment is 2 mbar.

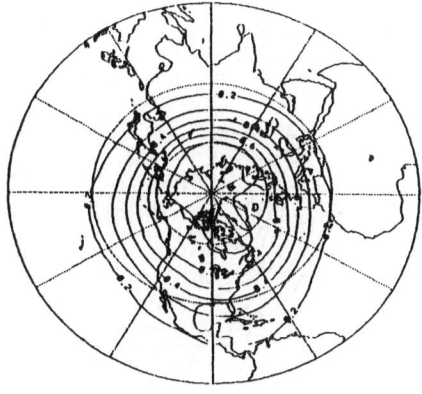

(a) Northern hemisphere winter
    (January data).

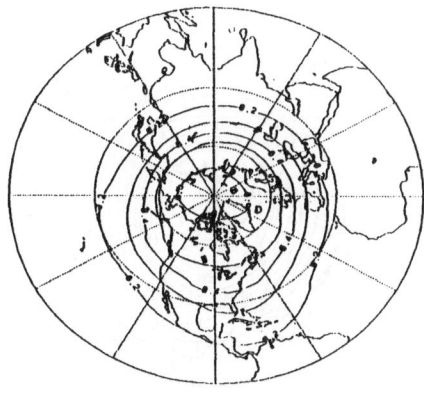

(b) Northern hemisphere summer
    (July data).

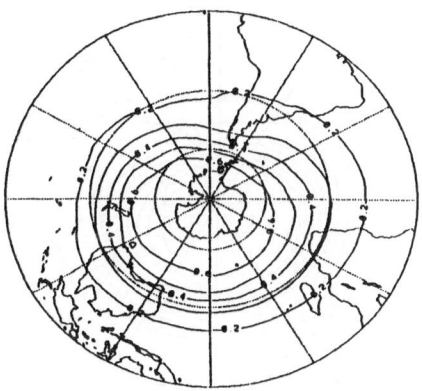

(c) Southern hemisphere winter
    (July data).

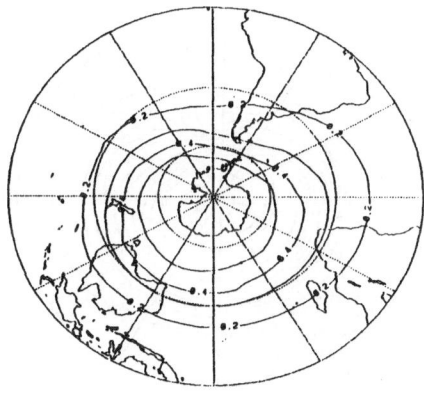

(d) Southern hemisphere summer
    (January data).

Figure 13.15. Contours of dose rate during solstice conditions at 10-km altitude (approximately 33 000 ft). Contour increment is 0.1 cSv/1000 hr.

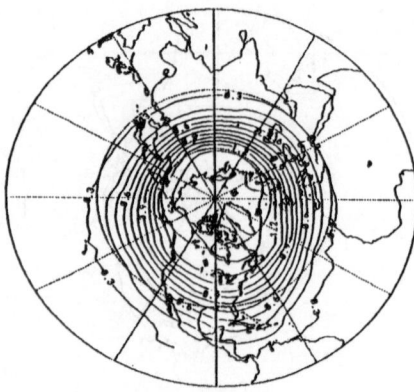

(a) Northern hemisphere winter
(January data).

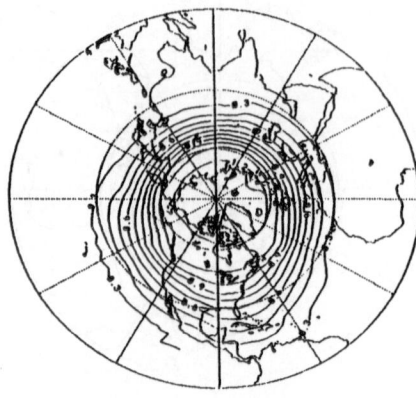

(b) Northern hemisphere summer
(July data).

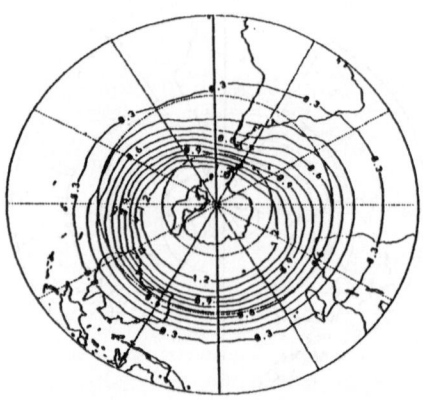

(c) Southern hemisphere winter
(July data).

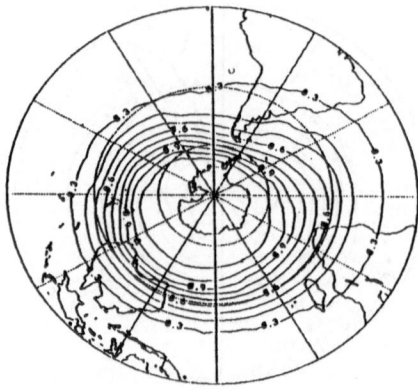

(d) Southern hemisphere summer
(January data).

Figure 13.16. Contours of dose rate during solstice conditions at 14-km altitude (approximately 46 000 ft). Contour increment is 0.1 cSv/1000 hr.

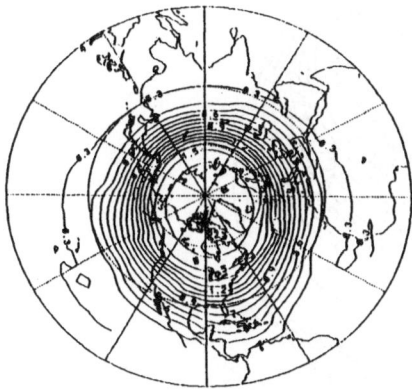

(a) Northern hemisphere winter
(January data).

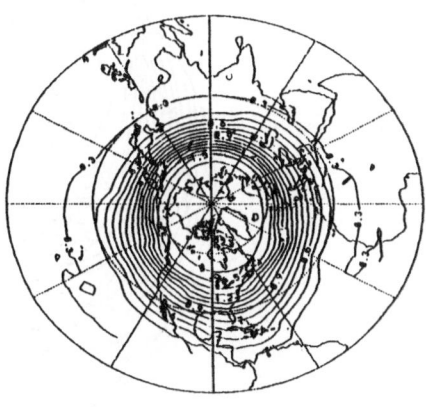

(b) Northern hemisphere summer
(July data).

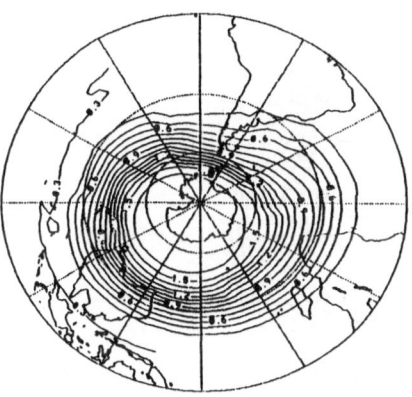

(c) Southern hemisphere winter
(July data).

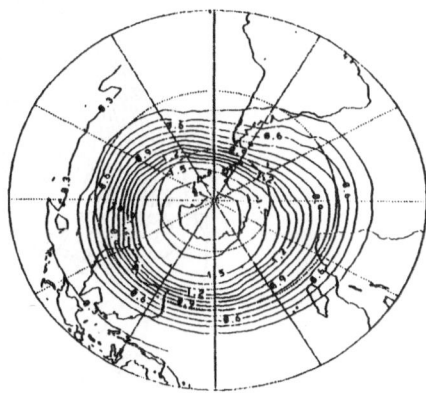

(d) Southern hemisphere summer
(January data).

Figure 13.17. Contours of dose rate during solstice conditions at 18-km altitude (approximately 60 000 ft). Contour increment is 0.1 cSv/1000 hr.

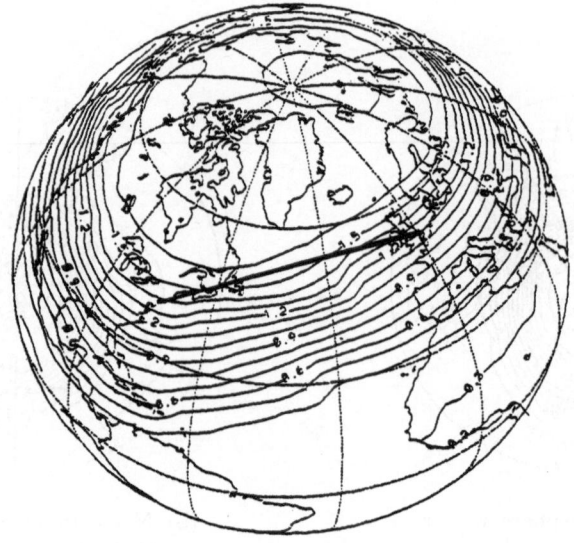

(a) New York–London.

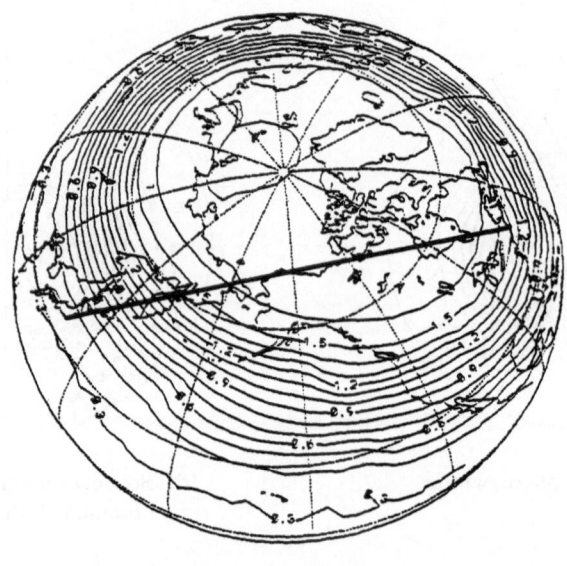

(b) New York–Tokyo.

Figure 13.18. International great circle routes for northern hemisphere winter at 18-km altitude. Dose rate contour increment is 0.1 cSv/1000 hr.

The exposure nearly doubles if the cruise altitudes are near 14 km (46 000 ft), and this increase represents a substantial fraction of the exposure allowed a radiation worker.

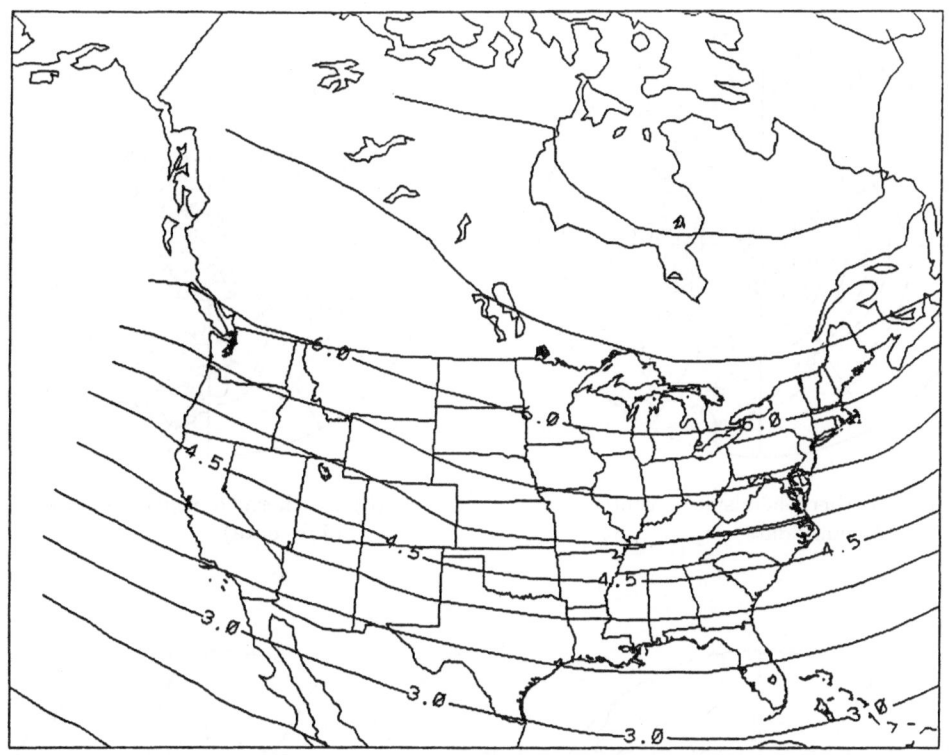

Figure 13.19. Winter solstice dose rate contours for continental United States at 10-km altitude (approximately 33 000 ft). Contour increment is 0.5 mSv/1000 hr.

Table 13.7. Dose and Dose Equivalent Rates in Neutron Energy Intervals
With the ICRP 26[a] and Proposed ICRU 40[b] Quality Factors

$$\left[\begin{array}{l} \Delta E:\ \text{neutron energy interval, MeV};\ \Delta D:\ \text{dose, } \mu\text{Gy/hr}; \\ \Delta H:\ \text{dose equivalent, } \mu\text{Sv/hr};\ Q:\ \text{quality factor} \end{array} \right]$$

ΔE	ΔD	ΔH_{ICRP}	Q_{ICRP}	Q_{ICRU}	ΔH_{ICRU}
0.1–1	0.20	2.34	11.7	19.4	3.88
1–10	.32	2.52	7.9	17.6	5.63
10–100	.39	1.81	7.0	7.0	1.81
100–1000	.31	1.04	3.4	3.4	1.04
0.1–1000	1.22	7.71	6.3	10.1	12.36

[a]ICRP 26 (Anon., 1977).
[b]ICRU 40 (Anon., 1986).

729

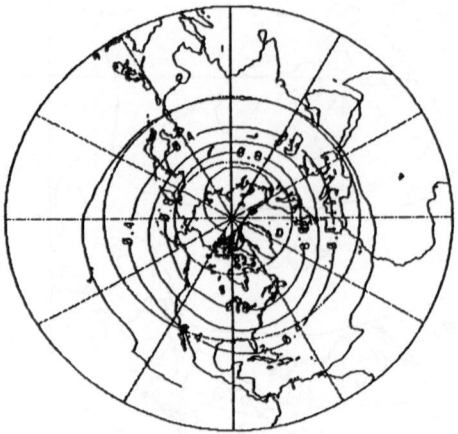

(a) Northern hemisphere winter
(January data).

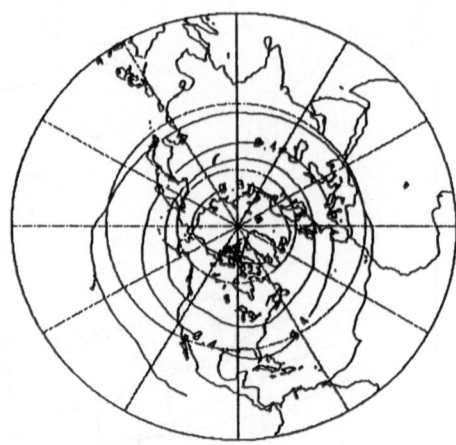

(b) Northern hemisphere summer
(July data).

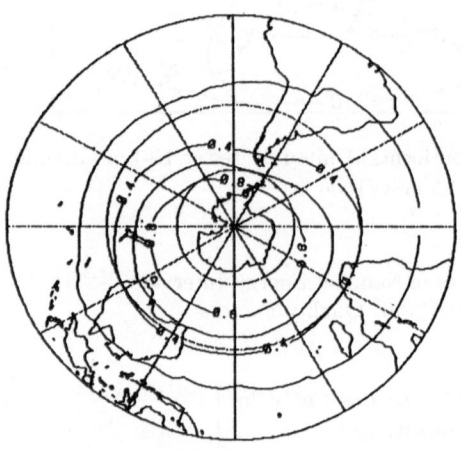

(c) Southern hemisphere winter
(July data).

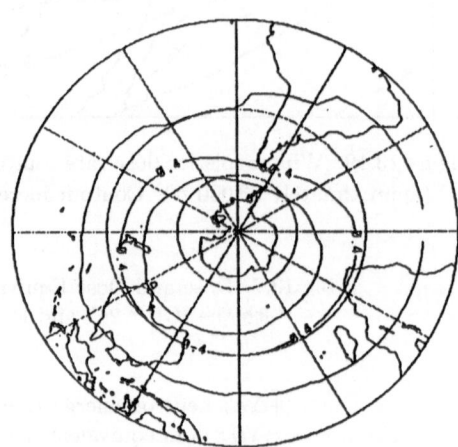

(d) Southern hemisphere summer
(January data).

Figure 13.20. Contours of dose rate during solstice conditions at 10-km altitude (approximately 33 000 ft) computed with ICRU 40 quality factors (Anon., 1986). Contour increment is 0.2 cSv/1000 hr.

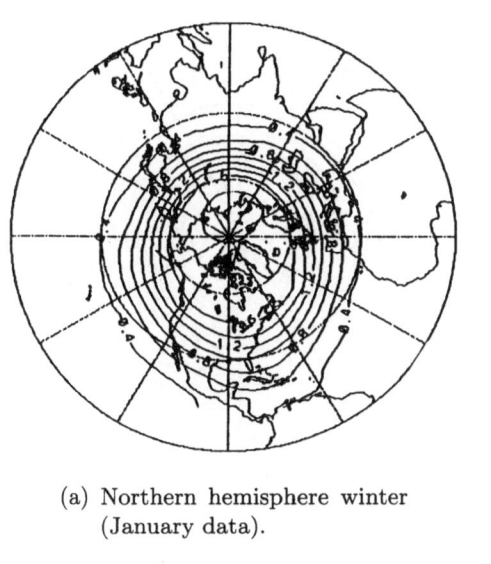

(a) Northern hemisphere winter
    (January data).

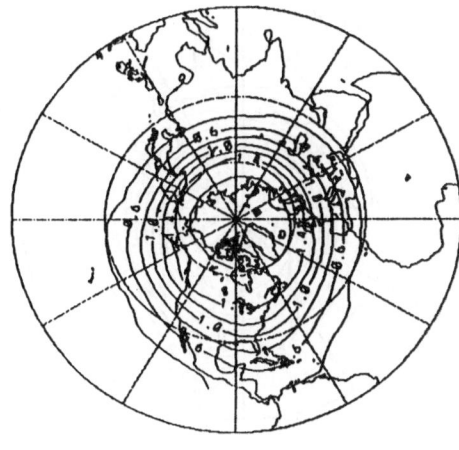

(b) Northern hemisphere summer
    (July data).

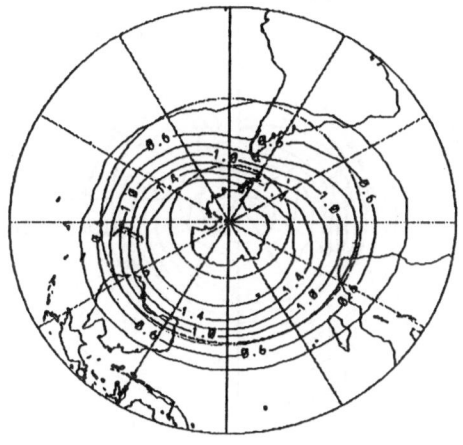

(c) Southern hemisphere winter
    (July data).

(d) Southern hemisphere summer
    (January data).

Figure 13.21. Contours of dose rate during solstice conditions at 14-km altitude (approximately 46 000 ft) computed with ICRU 40 quality factors (Anon., 1986). Contour increment is 0.2 cSv/1000 hr.

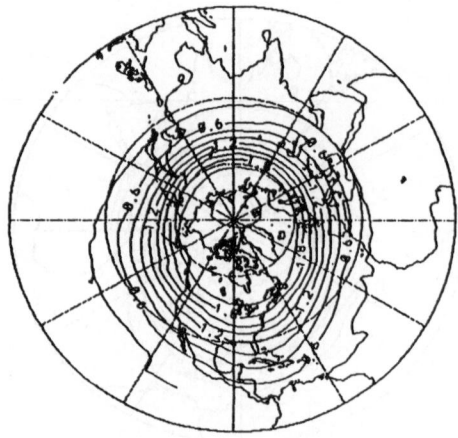

(a) Northern hemisphere winter
(January data).

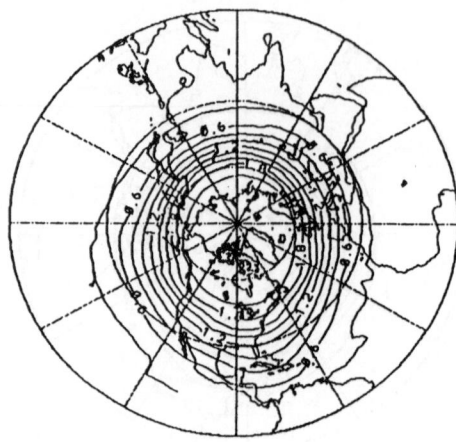

(b) Northern hemisphere summer
(July data).

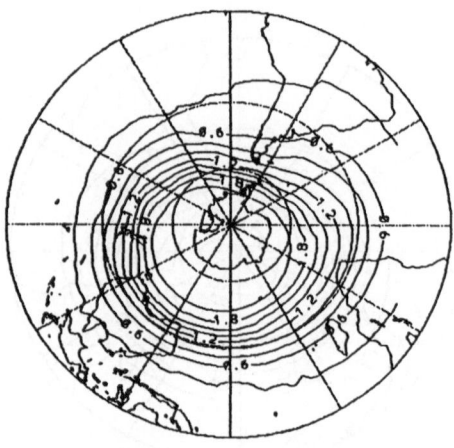

(c) Southern hemisphere winter
(July data).

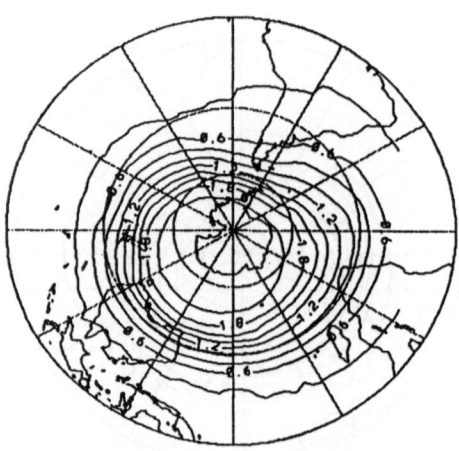

(d) Southern hemisphere summer
(January data).

Figure 13.22. Contours of dose rate during solstice conditions at 18-km altitude (approximately 60 000 ft) computed with ICRU 40 quality factors (Anon., 1986). Contour increment is 0.2 cSv/1000 hr.

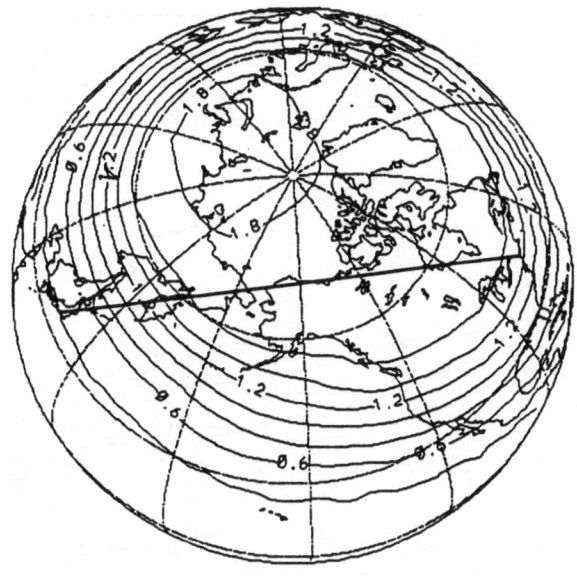

(a) New York–London.

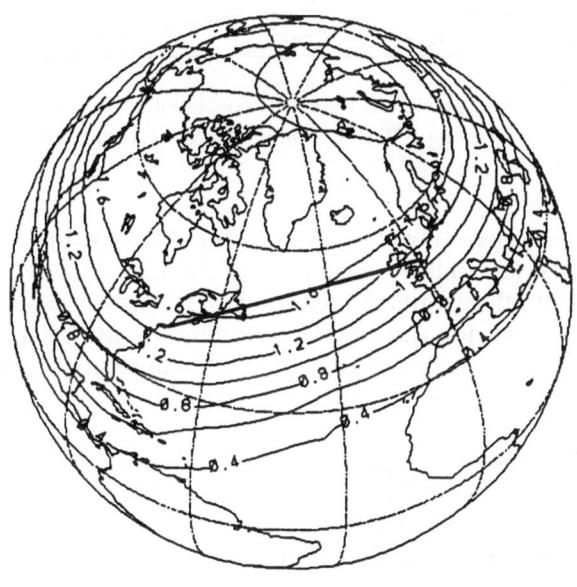

(b) New York–Tokyo.

Figure 13.23. International great circle routes for northern hemisphere at 14-km altitude. Dose rate contours are computed with ICRU 40 quality factors (Anon., 1986) and are in increments of 0.2 cSv/1000 hr.

International flights for New York–London and New York–Tokyo are shown in figure 13.23 at solar minimum during winter solstice for 14-km altitudes. Such

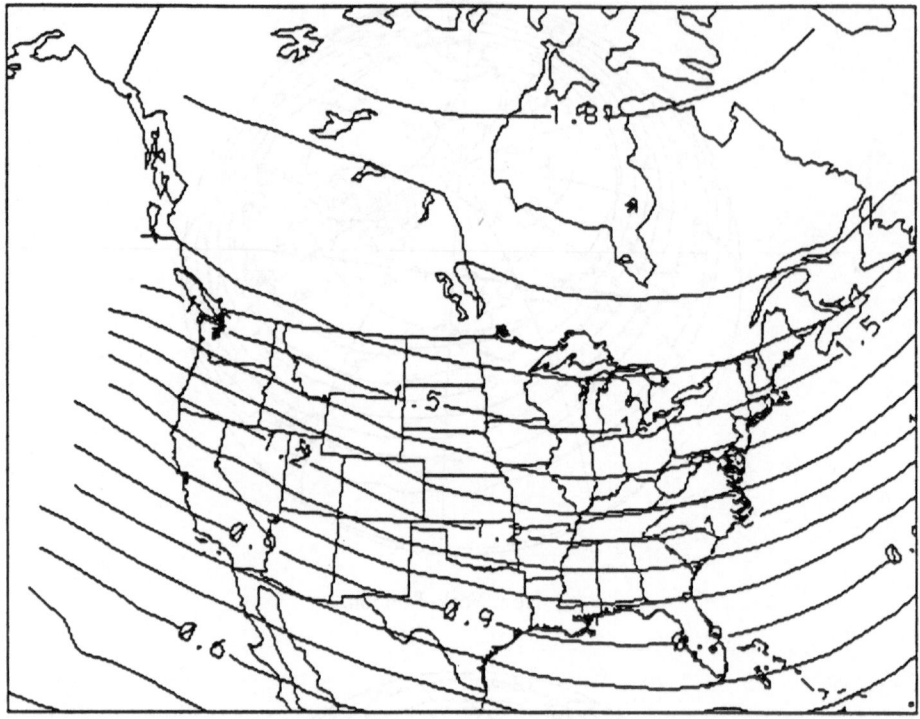

Figure 13.24. Winter solstice dose rate contours for contiguous United States at 14-km altitude (approximately 46 000 ft) using ICRU 40 quality factors (Anon., 1986). Contour increment is 0.1 cSV/1000 hr.

flights will accumulate more than 1.5 cSv/1000 hr. Clearly, high exposures relative to the general public are indicated and some thought to possible crew rotation (especially for potentially pregnant crew members) seems advisable. The winter dose rates at 14-km altitude during solar minimum can be quite high over the contiguous U.S. as shown in figure 13.24. Some counseling will be in order if the newly proposed quality factors are, in fact, adopted.

The revised Q_F dose equivalent rates are shown in figure 13.25 for solar maximum (in 1969) at the 14-km altitude. The dose rates decreased by 20 to 40 percent in reaching solar maximum in cycle 20. The decrease for cycle 19 (in 1958) would have been substantially greater.

13.5. Analysis for Selected Flight Paths

The environmental model described previously is used in conjunction with the background radiation data base to estimate incurred dose equivalent rates for several intercontinental and domestic routes at three altitudes (12, 14, and 17 km, corresponding approximately to 40 000, 46 000, and 56 000 ft, respectively). Several tables of results have been prepared (tables 13.8–13.12) in which the predicted average dose rates are presented for the selected flight paths as various input parameters are altered. The routes are specified as minimum distance (great circle) routes at constant altitude. Tables 13.8 and 13.9 give results computed for solar-minimum, northern hemisphere winter conditions, and they illustrate the effect of the proposed ICRU 40 quality factors (Anon., 1986) quality factors (table 13.8) as opposed to the generally lower ICRP 26 quantities (Anon., 1977) (table 13.9). The ICRU 40 quality factors result in dose equivalent values that are higher by 25 to 33 percent, with largest increases for the higher latitude routes. In order to examine the effect of active Sun conditions, tables 13.10 and 13.11 have been generated for solar maxima of cycles 19 and 20, respectively, in which the relative solar activity is shown to have an inverse relationship to the Deep River neutron monitor count in figure 13.1. It is seen that the decrease in dose rate during solar maximum may be quite large, especially at high latitudes. For example, the average rates at all altitudes considered for the New York–Tokyo route are reduced to almost half the predicted quiet Sun values (table 13.10 compared with table 13.8). A comparison of tables 13.10 and 13.11 for different solar maxima indicates the effects of variabilities between different cycles.

Finally, table 13.12 presents corresponding results for northern hemisphere summer solstice conditions, for which the pressure at a given altitude is usually

Table 13.8. Dose Equivalent Rate in January at Solar Cycle Minimum (DRNM = 7157) for ICRU 40 Quality Factors (Anon., 1986)

City pairs	Dose equivalent rate, cSv/1000 hr, at altitude of—		
	12 km (39 370 ft)	14 km (45 932 ft)	17 km (55 774 ft)
N.Y.–Tokyo	1.09	1.37	1.62
N.Y.–London	1.22	1.55	1.82
N.Y.–Seattle	1.20	1.56	1.86
Paris–Rio	.39	.49	.57
Paris–D.C.	1.17	1.49	1.74
Atlanta–L.A.	.79	1.02	1.22
Atlanta–N.Y.	1.00	1.29	1.56
Atlanta–S.F.	.85	1.10	1.31

Table 13.9. Dose Equivalent Rate in January at Solar Cycle Minimum (DRNM = 7157) for ICRP 26 Quality Factors (Anon., 1977)

City pairs	Dose equivalent rate, cSv/1000 hr, at altitude of—		
	12 km (39 370 ft)	14 km (45 932 ft)	17 km (55 774 ft)
N.Y.–Tokyo	0.82	1.04	1.24
N.Y.–London	.92	1.18	1.39
N.Y.–Seattle	.90	1.18	1.42
Paris–Rio	.31	.39	.46
Paris–D.C.	.88	1.13	1.33
Atlanta–L.A.	.60	.79	.94
Atlanta–N.Y.	.75	.98	1.19
Atlanta–S.F.	.64	.84	1.01

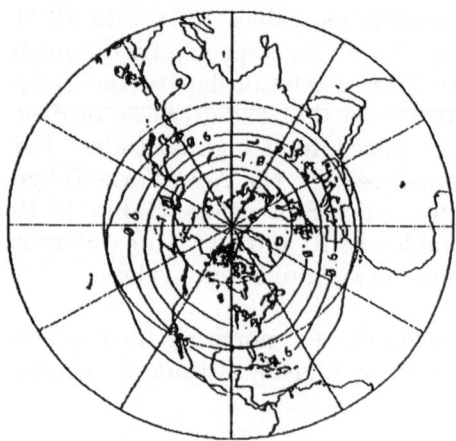
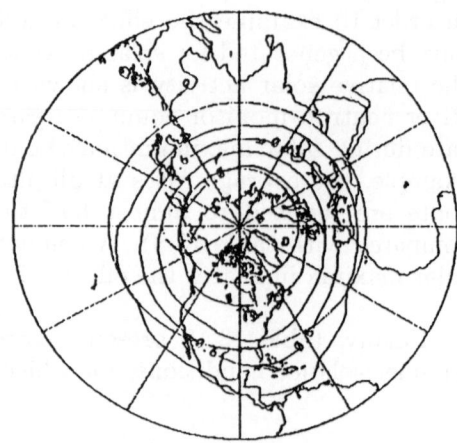

(a) Northern hemisphere winter (January data).

(b) Northern hemisphere summer (July data).

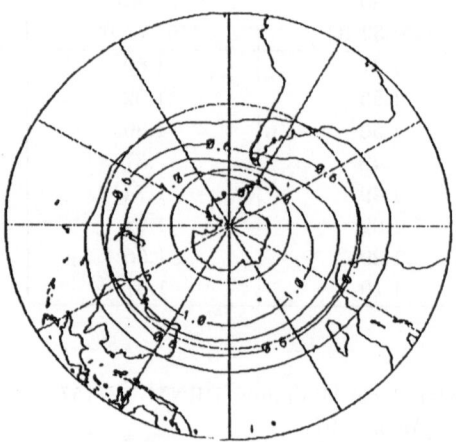
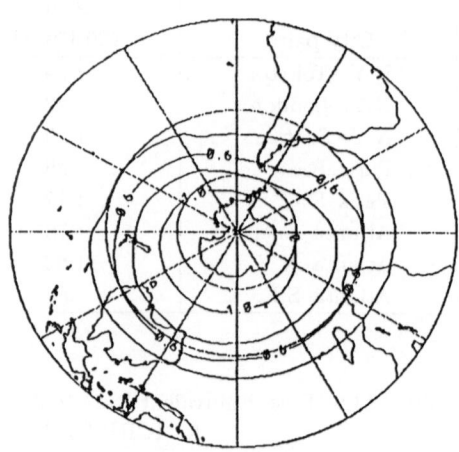

(c) Southern hemisphere winter (July data).

(d) Southern hemisphere summer (January data).

Figure 13.25. Contours of dose rate during solstice conditions at solar maximum activity for 14-km altitude computed with ICRU 40 quality factors (Anon., 1986). Contour increment is 0.2 cSV/1000 hr.

Table 13.10. Dose Equivalent Rate in January at Solar Cycle Maximum
(DRNM = 5700; cycle 19) for ICRU 40 Quality Factors (Anon., 1986)

City pairs	Dose equivalent rate, cSv/1000 hr, at altitude of—		
	12 km (39 370 ft)	14 km (45 932 ft)	17 km (55 774 ft)
N.Y.–Tokyo	0.63	0.77	0.86
N.Y.–London	.69	.85	.97
N.Y.–Seattle	.67	.84	.97
Paris–Rio	.35	.41	.46
Paris–D.C.	.68	.84	.95
Atlanta–L.A.	.59	.73	.82
Atlanta–N.Y.	.63	.80	.92
Atlanta–S.F.	.61	.76	.85

Table 13.11. Dose Equivalent Rate in January at Solar Cycle Maximum
(DRNM = 6280; cycle 20) for ICRU 40 Quality Factors (Anon., 1986)

City pairs	Dose equivalent rate, cSv/1000 hr, at altitude of—		
	12 km (39 370 ft)	14 km (45 932 ft)	17 km (55 774 ft)
N.Y.–Tokyo	0.75	0.96	1.12
N.Y.–London	.84	1.07	1.26
N.Y.–Seattle	.81	1.06	1.27
Paris–Rio	.35	.43	.50
Paris–D.C.	.81	1.04	1.22
Atlanta–L.A.	.62	.80	.96
Atlanta–N.Y.	.72	.94	1.14
Atlanta–S.F.	.66	.85	1.01

Table 13.12. Dose Equivalent Rate in July at Solar Cycle Minimum (DRNM = 7157)
for ICRU 40 Quality Factors (Anon., 1986)

City pairs	Dose equivalent rate, cSv/1000 hr, at altitude of—		
	12 km (39 370 ft)	14 km (45 932 ft)	17 km (55 774 ft)
N.Y.–Tokyo	0.92	1.24	1.57
N.Y.–London	1.04	1.41	1.77
N.Y.–Seattle	1.02	1.41	1.82
Paris–Rio	.37	.47	.56
Paris–D.C.	.99	1.35	1.70
Atlanta–L.A.	.70	.94	1.20
Atlanta–N.Y.	.86	1.18	1.52
Atlanta–S.F.	.75	1.02	1.29

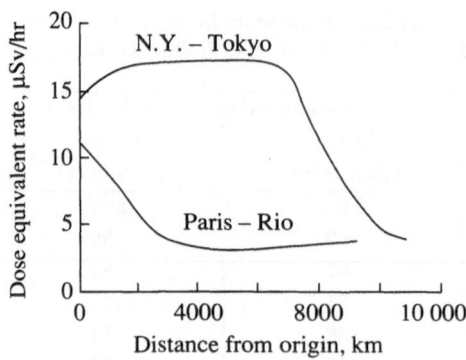

Figure 13.26. Dose equivalent rate profile along aircraft intercontinental flight trajectories at 17 km (56 000 ft).

higher than in winter. Consequently, dose equivalent rates are somewhat lower than those for the corresponding wintertime values, as may be seen by comparing tables 13.12 and 13.8. In addition, the seasonal variation is less for routes at low latitudes since annual pressure changes are not as pronounced.

As a final illustration of the range of dose rate variation along particular flight paths, the calculated dose equivalent rates at 14-km altitude for New York–Tokyo and for Paris–Rio de Janeiro are shown as a function of distance traveled in figure 13.26. The dose rate for New York–Tokyo is in excess of 1.5 cSv/1000 hr since much of the route is at high latitudes (>50° N). In contrast, the Paris–Rio calculation, for which most time is spent at tropical and subtropical latitudes, shows much lower dose rates. The average dose rate for the more northern route exceeds that of the low-latitude route by more than 2-1/2 times.

In the preceding discussion, dose equivalent rates have been expressed in cSv/1000 hr, or mrem/hr. If one assumes that crew members spend 1000 hr per year at altitude, the tabular results convert directly to annual incurred dose. When such a conversion is made, it is noteworthy that only the flight path at equatorial latitudes (Paris–Rio) is determined to be within the 0.5-cSv annual limit for the general populace. The more northern routes, especially at high altitudes, are often in excess of this limit by factors of 2 and sometimes 3.

13.6. References

Anon., 1977: *Recommendations of the International Commission on Radiological Protection.* ICRP Publ. 26, Pergamon Press.

Anon., 1980: *Cosmic Radiation Exposure During Air Travel.* FAA-AM-80-2, FAA Advisory Committee on Radiation Biology Aspects of the Supersonic Transport, U.S. Dep. of Transportation.

Anon., 1986: *The Quality Factor in Radiation Protection.* ICRU Rep. 40, International Commission on Radiation Units and Measurements.

Anon., 1987: *Recommendations on Limits for Exposure to Ionizing Radiation.* NCRP Rep. No. 91, National Council on Radiation Protection and Measurements.

Barish, R. J., 1990: Health Physics Concerns in Commercial Aviation. *Health Phys.,* vol. 59, no. 2, pp. 199–204.

Bertini, Hugo W., 1967: *Preliminary Data From Intranuclear-Cascade Calculations of 0.75-, 1-, and 2-GeV Protons on Oxygen, Aluminum, and Lead, and 1-rm GeV Neutrons on the Same Elements.* ORNL-TM-1996, U.S. Atomic Energy Commission.

Blackshear, W. T.; Grose, W. L.; and Turner, R. E., 1987: Simulated Sudden Stratospheric Warming; Synoptic Evolution. *Q. J. Royal Meteorol. Soc.,* vol. 113, no. 477, pp. 815–846.

Bramlitt, Edward T., 1985: Commercial Aviation Crewmember Radiation Doses. *Health Phys.,* vol. 49, no. 5, pp. 945–948.

Busick, D. D., 1989: Should Airline Crews Be Monitored for Radiation Exposure? Paper presented at Health Physics Society 34th Annual Meeting (Albuquerque, New Mexico).

Foelsche, Trutz, 1962: *Radiation Exposure in Supersonic Transports.* NASA TN D-1383.

Foelsche, Trutz; Mendell, Rosalind B.; Wilson, John W.; and Adams, Richard R., 1974: *Measured and Calculated Neutron Spectra and Dose Equivalent Rates at High Altitudes; Relevance to SST Operations and Space Research.* NASA TN D-7715.

Foelsche, Trutz, 1977: Radiation Safety in High-Altitude Air Traffic. *J. Aircr.,* vol. 14, no. 12, pp. 1226–1233.

Freier, P. S.; and Waddington, C. J., 1965: The Helium Nuclei of the Primary Cosmic Radiation as Studied Over a Solar Cycle of Activity, Interpreted in Terms of Electric Field Modulation. *Space Sci. Reviews,* vol. IV, no. 3, pp. 313–372.

Friedberg, W.; Faulkner, D. N.; Snyder, L.; Darden, E. B., Jr.; and O'Brien, K., 1989: Galactic Cosmic Radiation Exposure and Associated Health Risks for Air Carrier Crewmembers. *Aviation, Space, & Environ. Med.,* vol. 60, no. 11, pp. 1104–1108.

Grose, W. L.; Nealy, J. E.; Turner, R. E.; and Blackshear, W. T., 1987: Modeling the Transport of Chemically Active Constituents in the Stratosphere. *Transport Processes in the Middle Atmosphere,* G. Visconti and R. Garcia, eds., D. Reidel Publ. Co., pp. 229–250.

Grose, W. L.; Eckman, R. S.; Turner, R. E.; and Blackshear, W. T., 1989: Global Modeling of Ozone and Trace Gases. *Atmospheric Ozone Research and Its Policy Implications,* T. Schneider et al., eds., Elsevier, pp. 1021–1035.

Hess, W. N.; Canfield, E. H.; and Lingenfelter, R. E., 1961: Cosmic-Ray Neutron Demography. *J. Geophys. Res.*, vol. 66, no. 3, pp. 665–677.

Justus, C. G.; Fletcher, G. R.; Gramling, F. E.; and Pace, W. B., 1980: *The NASA/MSFC Global Reference Atmospheric Model—MOD 3 (With Spherical Harmonic Wind Model).* NASA CR-3256.

Korff, Serge A.; Mendell, Rosalind B.; Merker, Milton; Light, Edward S.; Verschell, Howard J.; and Sandie, William S., 1979: *Atmospheric Neutrons.* NASA CR-3126.

Neher, H. V., 1961: Cosmic-Ray Knee in 1958. *J. Geophys. Res.*, vol. 66, no. 12, pp. 4007–4012.

Neher, H. V.; and Anderson, Hugh R., 1962: Cosmic Rays at Balloon Altitudes and the Solar Cycle. *J. Geophys. Res.*, vol. 67, no. 4, pp. 1309–1315.

Newkirk, L. L., 1963: Calculation of Low-Energy Neutron Flux in the Atmosphere by the S_n Method. *J. Geophys. Res.*, vol. 68, no. 7, pp. 1825–1833.

O'Brien, Keran; and McLaughlin, James E., 1972: The Radiation Dose to Man From Galactic Cosmic Rays. *Health Phys.*, vol. 22, no. 3, pp. 225–232.

Shinn, J. L.; Wilson, J. W.; and Ngo, D. M., 1990: Risk Assessment Methodologies for Target Fragments Produced in High-Energy Nucleon Reactions. *Health Phys.*, vol. 59, no. 1, pp. 141–143.

Smart, D. F.; and Shea, M. A., 1983: Geomagnetic Transmission Functions for a 400 km Altitude Satellite. *18th International Cosmic Ray Conference—Conference Papers,* MG Sessions, Volume 3, Tata Inst. of Fundamental Research (Colaba, Bombay), pp. 419–422.

Wallace, Roger W.; and Sondhaus, C. A., 1978: Cosmic Radiation Exposure in Subsonic Air Transport. *Aviation, Space, & Environ. Med.*, vol. 49, no. 4, pp. 610–623.

Wilson, John W.; Lambiotte, Jules J.; and Foelsche, T., 1969: Structure in the Fast Spectra of Atmospheric Neutrons. *J. Geophys. Res.*, vol. 74, no. 26, pp. 6494–6496.

Wilson, John W.; Lambiotte, Jules J., Jr.; Foelsche, Trutz; and Filippas, Tassos A., 1970: *Dose Response Functions in the Atmosphere Due to Incident High-Energy Protons With Application to Solar Proton Events.* NASA TN D-6010.

Wilson, J. W., 1972: Production and Propagation of Atmospheric Neutrons. *Trans. American Nuclear Soc.*, vol. 15, no. 2, pp. 969–970.

Wilson, J. W., 1981: Solar Radiation Monitoring for High Altitude Aircraft. *Health Phys.*, vol. 41, no. 4, pp. 607–617.

Wilson, J. W.; Townsend, L. W.; and Badavi, F. F., 1987: Galactic HZE Propagation Through the Earth's Atmosphere. *Radiat. Res.*, vol. 109, no. 2, pp. 173–183.

Wilson, John W.; and Townsend, Lawrence W., 1988: Radiation Safety in Commercial Air Traffic: A Need for Further Study. *Health Phys.*, vol. 55, no. 6, pp. 1001–1003.

Wilson, J. W.; and Townsend, L. W., 1989: Errata—Radiation Safety in Commercial Air Traffic: A Need for Further Study. *Health Phys.*, vol. 56, no. 6, pp. 973–974.

Wilson, John W.; Shinn, Judy L.; and Townsend, Lawrence W., 1990: Nuclear Reaction Effects in Conventional Risk Assessment for Energetic Ion Exposure. *Health Phys.*, vol. 58, no. 6, pp. 749–752.

14. RADIATION EFFECTS IN ELECTRONIC MATERIALS

14.1. Introduction

The physical processes by which particles interact with electronic materials are the same as for any other material, namely, interaction with orbital electrons, elastic scattering with atomic nuclei of the material, and nonelastic reactions. It is the specific properties of these materials and the interaction in circuits which make the study of electronic materials of special interest. The interaction with orbital electrons raises the conductivity of the media until the charge released is collected or recombines with hole states or traps in the media. Such processes depend on external connections to the media. Elastic scattering with media nuclei depends on the binding potential of the surrounding medium, and a dislocation or series of dislocations can occur if the energy transfer is above the binding threshold. Such dislocations provide traps for the conduction electrons and holes. The nonelastic processes provide a release of kinetic energy as nuclear fragments produced by the reaction. The kinetic energy is given over to orbital electrons and elastic scattering with nuclei in the medium. The energy released in these nuclear reactions is given to orbital electrons and causes a temporary large increase in the conductivity. In the present chapter we will quantify these aspects of the interactions and treat two diverse applications of interest to the space program.

14.2. Gallium Arsenide Solar Cells

Gallium arsenide (GaAs) solar cells have received considerable attention because of their potential usefulness in high-power space-energy systems as well as special space-probe applications where high operating temperature is a limiting factor for silicon solar cells (Anon., 1977). However, space-radiation damage to the GaAs cells may be a limiting factor in Earth orbit above 2000 km and on interplanetary missions unless sufficient shielding is provided to keep damage levels within acceptable limits (Wilson, Stith, and Stock, 1983). Consequently, radiation damage studies have been made (Walker and Conway, 1978a and 1978b; Heinbockel, Conway, and Walker, 1980; Conway, Walker, and Heinbockel, 1981; Li et al., 1981; Loo, Knechtli, and Kamath, 1981; Kamath, 1981; Wilson et al., 1982; and Wilson, Stith, and Walker, 1982) on the effects of proton and electron irradiation, including defect characterization and annealing. Since damage effects are not generally additive, the combined effects of electron and proton exposure, as well as angular and spectral factors, are not known from the available experimental data base (Walker and Conway, 1978b; Loo, Knechtli, and Kamath, 1981; Kamath, 1981). To determine design parameters for a specific space environment, extensive laboratory testing or a model of the effects of the specific radiation components on the cell performance is required. Within the context of a detailed model, the question of additivity of specific radiation components can be adequately understood, and the cell performance can be evaluated under appropriate space environment conditions.

Earlier models for electron radiation damage assumed the defects to be produced uniformly throughout the cell volume and modeled the cell performance in terms of cell-averaged diffusion lengths of the minority carriers (Walker and

Conway, 1978a and 1978b; Heinbockel, Conway, and Walker, 1980). However, for low-energy protons, defects are not produced uniformly throughout the cell volume. Thus, there is a specific dependence of cell efficiency on proton energy. Consequently, the present report treats the geometric distribution of the displacement damage in detail, and cell performance is evaluated in terms of the cell-averaged minority-carrier recombination probability in diffusion to the cell junction. The average of the minority recombination probability over the cell active region weighted according to the solar-averaged photoabsorption rate is used to estimate the decrement in the short-circuit current.

14.2.1. Proton defect formation.

Atomic displacements caused by proton impact with atomic nuclei result in crystal defects as illustrated in figure 14.1. The formation rate of these defects is related to Rutherford's cross section (Dienes and Vineyard, 1957):

$$\sigma_D(E) = \frac{4\pi \, a_o^2 E_R^2 Z_2^2}{M_2 E} \left(\frac{1}{T_D} - \frac{1}{T_m} \right) \tag{14.1}$$

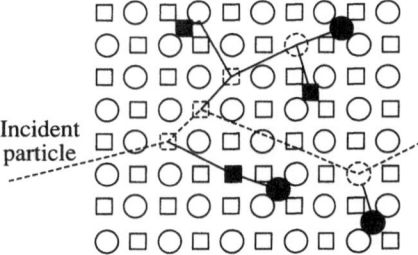

Figure 14.1. Defect formation by particulate radiation in a binary crystal. Defects shown are replacements, vacancies, and interstitials.

where a_o is Bohr's radius, E_R is Rydberg's constant, Z_2 is the atomic number of the struck nucleus, M_2 is the corresponding nuclear mass number, E is the proton kinetic energy, T_D is the energy required to displace the nucleus from its lattice site, and T_m is the maximum energy transfer in the collision; T_m is given by

$$T_m = \frac{4M_2}{(1 + M_2)^2} E \tag{14.2}$$

The displacement cross section and average energy transfer for protons in GaAs with $Z_2 \approx 32$ and $M_2 \approx 72.5$ are shown in figures 14.2 and 14.3. The threshold for displacement requires that $T_m > T_D$. The fact that $T_D \approx 9.5$ eV (Bauerlein, 1963) ensures that only close collisions result in displacement, so that screening corrections to the Rutherford formula are unimportant. If the atomic recoil energy is sufficiently large $(T \gg T_D)$, additional displacements can be produced by the recoiling nucleus before it comes to rest at an interstitial site. The average number of recoil displacements produced by one initiating proton collision event is given as a function of the maximum energy transfer by

$$\bar{\nu}_D(E) = \begin{cases} 1 + \dfrac{T_m}{2(T_m - T_D)} \log\left(\dfrac{T_m}{T_D}\right) & (T_m > 2T_D) \\[2ex] 1.0 & (2T_D > T_m > T_D) \end{cases} \tag{14.3}$$

with the assumption that half the recoil energy produces further displacements and the other half is dissipated in other processes. These quantities allow the calculation of the number of displacements produced per unit distance traveled by a proton of fixed energy.

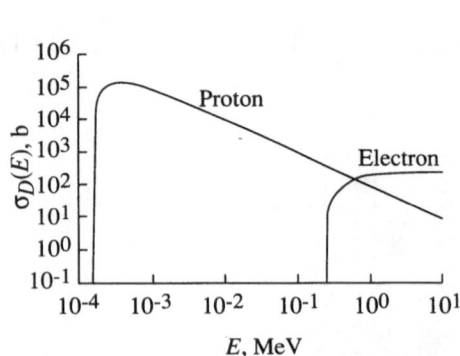

Figure 14.2. Displacement cross section for energetic protons and electrons. 1 barn = 1×10^{-28} m^2.

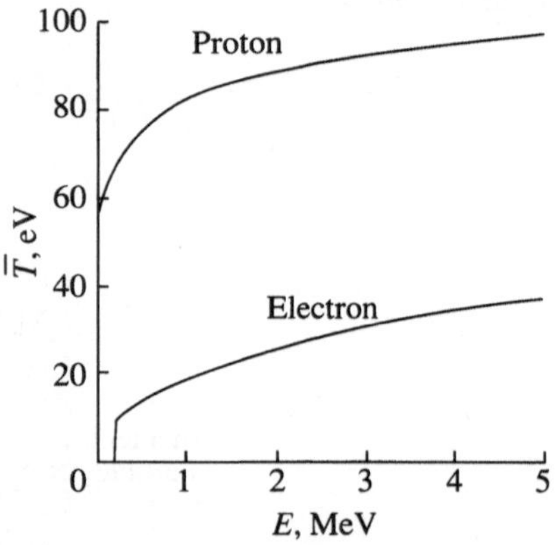

Figure 14.3. Average energy transferred to recoiling nucleus.

In passing through a crystal, most of the energy of a proton is transferred to orbital electrons (Andersen and Ziegler, 1977). The path length traveled in coming to rest is found by fitting the data of Andersen and Ziegler (1977) as

$$P(E) = 0.077E^{0.5} + 1.125 \times 10^{-4}E^{1.64} \tag{14.4}$$

where E is in keV and $P(E)$ is in μm. As derived from the slowing-down theory, a unique value of kinetic energy can be associated with each position along the trajectory of a proton. The proton energy as a function of the distance p yet to be traveled before coming to rest is given by

$$E = \frac{209.6p^{2.08}}{1 + 1.055p^{1.43}} \tag{14.5}$$

as determined by integrating the stopping-power data of Andersen and Ziegler (1977). In the process of coming to rest, the proton undergoes multiple scatterings from atomic nuclei, of which a few result in displacements. This process alters ever so slightly the direction of motion of the proton. The depth of penetration

$R(E)$ and path length $P(E)$ are approximately related by

$$P(E)^{-1}R(E) = \left[1 - \exp\left(-0.084E^{0.55}\right)\right] \tag{14.6}$$

This ratio is related to the average deviation in the direction of motion and is most important at low energies. The average depth of penetration and initial energy as related through equations (14.4) and (14.6) can be approximated by

$$R(E) = 0.0062E + 2.92 \times 10^{-5}E^{1.77} \tag{14.7}$$

There is no unique energy associated with a given depth of penetration because of multiple scattering. However, the average energy of protons which penetrate and stop at a depth x is

$$E = \frac{593x^{1.5}}{x + 3.71x^{0.5}} \tag{14.8}$$

The preceding quantities were used to determine the displacement density within a GaAs crystal.

A proton of energy E_o incident on the face of the crystal travels a distance

$$P_o = P(E_o) \tag{14.9}$$

before coming to rest. After traveling a distance p the energy will be reduced to

$$E = \frac{209.6\,(P_o - p)^{2.08}}{1 + 1.055\,(P_o - p)^{1.43}} \tag{14.10}$$

At this position p, the displacement mean-free path is

$$\ell_D(E) = \frac{1}{n\,\sigma_D(E)} \tag{14.11}$$

where n is the density of scattering centers in the crystal $\left(4.42 \times 10^{10}/\mu m^3\right)$, and $\sigma_D(E)$ is the displacement cross section averaged for GaAs ($M_2 = 72.5$ and $Z_2 = 32$). The average number of displacements per unit path length is

$$\xi_D(E) = \frac{\bar{\nu}_D(E)}{\ell_D(E)} \tag{14.12}$$

The use of equations (14.10) and (14.12) allows appropriate partitioning of the proton energy into electronic excitation and displacements everywhere along its path.

The number of displacements along the proton path is related to the displacement damage in the crystal. For a normally incident proton of energy E_o on the face of a crystal, the number of displacements along its path is given by equations (14.10) and (14.12). However, by the time its energy is reduced to E, it has penetrated to an average depth x given by

$$x = R(E_o) - R(E) \tag{14.13}$$

The path length and penetration depth are related to the average direction cosine (Janni, 1966) and are approximated here by solving the equation $\bar{\mu}(E) = dP(E)/dR(E)$ using equations (14.4) and (14.6). In terms of $\bar{\mu}(E)$, the average number of displacements per unit depth is

$$\frac{dD(E)}{dx} = \bar{\mu}(E)\,\xi_D(E) \tag{14.14}$$

where x is found from equation (14.13). The effects of multiple scattering are demonstrated in figure 14.4. The results of equation (14.14) for the average proton path due to multiple scattering (solid line) are compared with calculations neglecting multiple scattering (dashed line) according to equation (14.12). The difference between the two curves is a measure of the fluctuations caused by multiple scattering.

The total number of displacements formed along the path of a proton with initial energy E_o is

$$D(E_o) = \int_0^{E_o} \xi_D(E)\frac{dP(E)}{dE}\,dE \tag{14.15}$$

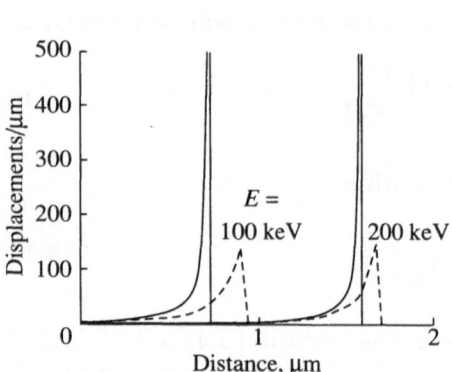

Figure 14.4. Displacement density for a single proton path.

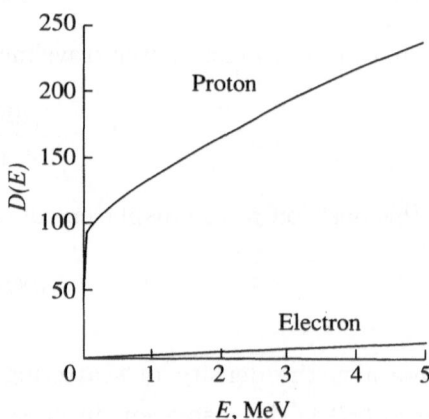

Figure 14.5. Total number of displacements formed in bringing a particle to rest in GaAs crystal.

The numerical evaluation of equation (14.15) as shown in figure 14.5 is approximated by (E_o in units of keV)

$$D(E_o) = \begin{cases} 0 & (E_o < 0.64) \\[2mm] 12.4 + 350.4\left(1 - 0.8236\,E_o^{0.016}\right)\log_{10}(E_o) & (0.64 < E_o < 20) \\[2mm] 47.83 + 20.48\left(1 + 3.246 \times 10^{-3}\,E_o^{0.721}\right)\log_{10}(E_o) & (20 < E_o) \end{cases} \tag{14.16}$$

where the effective threshold displacement energy for the proton is 0.64 keV. Equation (14.13) was also evaluated using the displacement theory of Lindhard, Scharff, and Schiott (LSS) (1963) as discussed by Peterson and Harkness (1976). The numerical values of the LSS theory and the present results are in good agreement. (See Wilson et al., 1982.)

14.2.2. Electron defect formation. One of the significant differences between proton and electron interaction is that relativistic effects must be included in the electron interaction. The Mott-McKinley-Feshbach (McKinley and Feshbach, 1948; see also Vook, 1968) relativistic electron scattering cross section leads to the expression

$$d\sigma = \pi r_c^2 Z_2^2 \left(\frac{1 - \beta^2}{\beta^4} \right) \left(\frac{T_m}{T^2} - \frac{\beta^2}{T} + \frac{\pi\alpha\beta}{T_m} \left(\frac{T_m}{T} \right)^{\frac{3}{2}} - \frac{\pi\alpha\beta}{T} \right) dT \qquad (14.17)$$

where r_c is the classical electron radius, Z_2 is the atomic number of the target atom, β is the ratio of the electron velocity to the speed of light, T is the energy transferred in the collision, and α equals $Z_2/137$. Integration yields the displacement cross section shown in figure 14.2 and given by

$$\sigma_D(E) = \pi Z_2^2 r_c^2 \left(\frac{1 - \beta^2}{\beta^4} \right)$$

$$\times \left[\frac{T_m}{T_D} - 1 - \beta^2 \, \log \left(\frac{T_m}{T_D} \right) + 2\pi\alpha\beta \left(\sqrt{\frac{T_m}{T_D}} - 1 \right) - \pi\alpha\beta \, \log \left(\frac{T_m}{T_D} \right) \right] \quad (14.18)$$

where T_D is the displacement threshold, the maximum energy transfer T_m is

$$T_m = \frac{2E}{M_2 c^2} \left(E + 2mc^2 \right) \qquad (14.19)$$

M_2 is the mass of the atom, m is the mass of the electron, and c is the velocity of light. In a collision between an electron of energy E and an atom, the atom acquires an energy in excess of T_D for cross section $\sigma_D(E)$.

A requirement for displacement of a nucleus is that $T > T_D$. The value of T_D used in deriving this model is 9.5 eV. The average energy transfer during a collision is

$$\overline{T}(E) = \frac{1}{\sigma_D(E)} \int_{T_D}^{T_m} T \frac{d\sigma_D(E)}{dT} dT$$

$$= \frac{T_m \, \log \left(\frac{T_m}{T_D} \right) - \beta^2 \left(T_m - T_D \right) + 2\pi\alpha\beta \left(T_m - \sqrt{T_m T_D} \right) - \pi\alpha\beta (T_m - T_D)}{\frac{T_m}{T_D} - 1 - \beta^2 \, \log \left(\frac{T_m}{T_D} \right) + 2\pi\alpha\beta \left(\sqrt{\frac{T_m}{T_D}} - 1 \right) - \pi\alpha\beta \, \log \left(\frac{T_m}{T_D} \right)} \quad (14.20)$$

Figure 14.3 illustrates the dependence on electron energy of the average energy transfer between an electron of initial energy E and a gallium or arsenic atom. If the energy transfer $T(E) \gg T_D$, additional atomic displacements can be

747

produced by the initial recoiling nucleus before it comes to rest at an interstitial or replacement site. The average number of recoils caused by one electron colliding with an atom is given as a function of the average energy transfer by

$$
\bar{\nu}_D(E) = \begin{cases} 1 & (T_D < \overline{T}(E) < 2T_D) \\ 1 + \frac{\overline{T}(E)}{2T_D} & (\overline{T}(E) > 2T_D) \end{cases} \tag{14.21}
$$

assuming half the recoil energy produces further displacements and assuming the other half is dissipated in other processes.

The displacement mean-free path is

$$
\ell_D(E) = \frac{1}{n\,\sigma_D(E)} \tag{14.22}
$$

where n is the density of scattering centers in the crystal $\left(4.42 \times 10^{22}/\text{cm}^3\right)$ and $\sigma_D(E)$ is the displacement cross section average for GaAs ($M_2 = 72.5$ and $Z_2 = 32$). The average number of displacements per unit path length produced by an electron of initial energy E is

$$
\xi_D(E) = \frac{\bar{\nu}_D(E)}{\ell_D(E)} = n\,\bar{\nu}_D(E)\,\sigma_D(E)
$$

$$
= \begin{cases} n\,\sigma_D(E) & (T_D < \overline{T}(E) < 2T_D) \\ n\,\sigma_D(E) + \frac{n}{2T_D}\,\sigma_D(E)\,\overline{T}(E) & (\overline{T}(E) > 2T_D) \end{cases} \tag{14.23}
$$

The total number of displacements produced along the path of an electron of initial energy E_o is

$$
D(E_o) = \int_0^{E_o} \frac{\xi_D(E)}{S(E)}\,dE = n\int_0^{E_o} \frac{\bar{\nu}(E)\,T_D(E)}{S(E)}\,dE \tag{14.24}
$$

where

$$
S(E) = \begin{cases} 0.381E^{0.084} & (260 < E < 1000 \text{ keV}) \\ 0.623 + 4.25 \times 10^{-5}E & (1000 < E < 10\,000 \text{ keV}) \end{cases} \tag{14.25}
$$

is the stopping-power formula (keV/μm) determined from data of Pages et al. (1972). Numerical evaluation of the displacement integral can be approximated by

$$
D(E) = \begin{cases} 0 & (E < 260 \text{ keV}) \\ -3.6 + 3.32 \times 10^{-3}E + 3.58\,\exp(-1.094 \times 10^{-3}E) & (260 < E < 10\,000 \text{ keV}) \end{cases} \tag{14.26}
$$

Figure 14.5 illustrates the dependence of atomic displacements in GaAs on initial electron energy as found from evaluation of equation (14.24).

In passing through the crystal, the electron is slowed down as it interacts with orbital electrons and atomic nuclei. Using data from Pages et al. (1972), the range of the electron in GaAs as a function of initial electron energy E is given by

$$R(E) = 0.4027 E^{1.16} - 5.95 \times 10^{-5} E^2 \qquad (14.27)$$

where $R(E)$ is in μm and E is in keV. The effects of multiple scattering are neglected in this formula for $R(E)$ because multiple scattering of electrons is relatively unimportant in the thin GaAs cells treated herein.

From the same data used in determining the range formula, a formula for the average energy of an electron that penetrates to a depth R and stops is

$$E = 2.217 R^{0.86} + \left(2.25 \times 10^{-5}\right) R^2 \qquad (14.28)$$

After penetrating to a depth of x within the crystal, the electron energy is given by

$$E_o(x) = 2.217 \left(R_o - x\right)^{0.86} + 2.25 \times 10^{-5} (R_o - x)^2 \qquad (14.29)$$

The effect of these radiation-induced defects on cell performance is discussed in the section 14.2.3.

14.2.3. Minority-carrier recombination. It is assumed that these radiation-induced displacements within the crystal form recombination centers for the minority carriers of the electron-hole pairs produced by photon absorption. A minority carrier, once formed, undergoes thermal diffusion (Hovel, 1975) until it is trapped and recombines or is separated at the junction. The root-mean-square distance traveled in moving to a position a distance L away from the source point is (Liverhant, 1960)

$$\bar{r} = \sqrt{6}\, L \qquad (14.30)$$

If σ_r is the recombination cross section and L is the distance along an arbitrary straight line path to the junction, the fractional loss of pairs due to recombination in reaching the junction along a fixed direction is

$$f(\mu) = \left\{ \begin{array}{ll} \left\{1 - \exp\left[-\int_{x_j}^{x} \sigma_r D_v(x) \sqrt{6}\frac{dx}{\mu}\right]\right\} & (x > x_j) \\[2ex] \left\{1 - \exp\left[-\int_{x}^{x_j} \sigma_r D_v(x) \sqrt{6}\frac{dx}{\mu}\right]\right\} & (x < x_j) \end{array} \right\} \qquad (14.31)$$

where μ is the cosine of the direction to the junction, and $D_v(x)$ is the displacement density. Averaging the fractional loss over all directions toward the junction

$$F(x) = \int_0^1 f(\mu)\, d\mu \qquad (14.32)$$

results in

$$F(x) = 1 - E_2\left(\sqrt{6}\ \sigma_r\left|\int_x^{x_j} D_v(x)\ dx\right|\right) \tag{14.33}$$

where summations over all spectral and angular components are implied. Note that $E_2(z)$ is the exponential integral of order 2.

The photoabsorption rate density at a depth x within the cell for the solar spectrum is

$$\rho(x) = K\gamma \exp(-\gamma x) \tag{14.34}$$

where K is the integrated flux in the absorption band and γ is the photoabsorption coefficient averaged over the solar spectrum $\left(\gamma \approx 1.4\ \mu\text{m}^{-1}\right)$. The rate at which the photocurrent is collected under short-circuit conditions is

$$I_{sc,0} = \int_0^t \eta_c(x)\,\rho(x)\,dx \tag{14.35}$$

where $\eta_c(x)$ is the normal or preirradiated collection efficiency and t is the depth of the active region. The normal collection efficiency is known in terms of diffusion lengths, lifetimes, and surface recombination rates of the minority carriers; electric fields; and cell dimensions (Loo et al., 1978).

To derive a simple expression for the short-circuit current in an irradiated cell, the following simplifying assumptions are made. First, it is assumed that the radiation-induced defects do not greatly alter the internal-cell electric fields. It is further assumed that the radiation defects mainly alter the cell operation through change in the minority-carrier lifetime in the bulk. Surface recombination plays only a secondary role for heteroface cells. (See Walker and Conway, 1978a and 1978b.) Viewing $\eta_c(x)$ as a probability of current collection of an electron-hole pair produced at x, it is further assumed that the normal collection efficiency and the recombination probability with radiation defects are statistically independent. This independence, which allows the postirradiation short-circuit current to be written as

$$I_{sc} = \int_0^t \eta_c(x)\,[1 - F(x)]\,\rho(x)\,dx \tag{14.36}$$

for which the fractional remaining current is

$$I_{sc}/I_{sc,0} = 1 - \left[\frac{\int_0^t \eta_c(x)\,F(x)\,\rho(x)\,dx}{\int_0^t \eta_c(x)\,\rho(x)\,dx}\right] \tag{14.37}$$

For a well-designed high-collection-efficiency solar cell, $\eta_c(x)$ is nearly spatially independent over the cell active volume, so that further simplification results in

$$I_{sc}/I_{sc,0} \approx 1 - \left[\frac{\int_0^t F(x)\,\rho(x)\,dx}{\int_0^t \rho(x)\,dx}\right] \tag{14.38}$$

which is used throughout the remainder of the present work.

14.2.4. Evaluation of defect spatial distribution. Central to the calculation of radiation effects as outlined in the preceding section is evaluation of the integral of the defect volume density. This integral is related to a cumulative defect function by

$$D_c(x) = \int_0^x D_v(x')\,dx' \tag{14.39}$$

This quantity may be evaluated for a fluence $\phi(E_o)$ of normally incident particles of energy E_o. This is accomplished by simply calculating the particle residual energy $E_o(x)$ after the particle penetrates to a depth x and noting that

$$D_c(x) = \{D(E_o) - D[E_o(x)]\}\,\phi(E_o) \tag{14.40}$$

Since $E_o(x)$ is the residual-energy function for normal incidence, the corresponding result for oblique incidence is

$$D_c(x) = \left\{ D(E_o) - D\left[E_o\left(\frac{x}{\cos\theta}\right)\right] \right\}\,\phi(E_o) \tag{14.41}$$

where θ is the angle of incidence to the normal of the surface. Generalizing for a spectrum of particles and isotropic incidence,

$$D_c(x) = 2\pi \int_0^\infty dE_o \int_0^1 d(\cos\theta) \left\{ D(E_o) - D\left[E_o\left(\frac{x}{\cos\theta}\right)\right] \right\}\,\phi(E_o) \tag{14.42}$$

where $4\pi\,\phi(E_o)$ is the omnidirectional differential fluence spectrum.

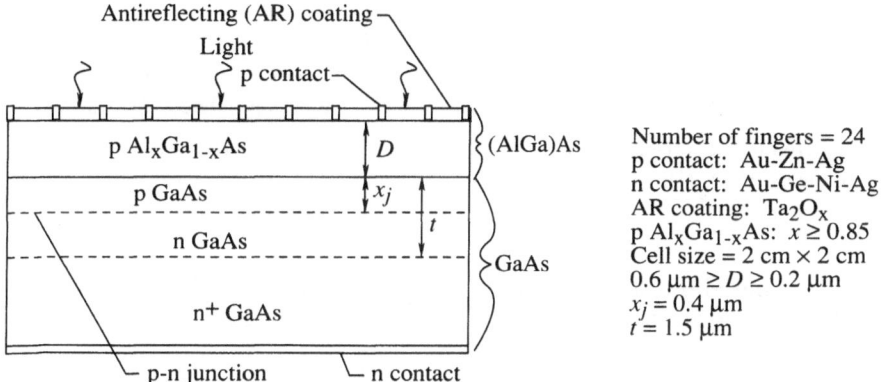

Figure 14.6. GaAs solar cell structure used in present model.

14.2.5. Comparison with experiment. The geometry of the solar cells used in experimental tests (Loo, Knechtli, and Kamath, 1981; Kamath, 1981) is shown in figure 14.6. The changes in the cell current collection efficiency as given by equation (14.38) were evaluated numerically and are shown in figure 14.7 for the solar cell parameters shown in figure 14.6. Since the protons follow neither the trajectory of the average proton nor the trajectory in which multiple-scattering is neglected (fig. 14.4), improvements were made by including the effects of multiple scattering. These effects were estimated by averaging with equal weight the cell damage for the two functions shown in figure 14.4, in which some effects of deviations about the average trajectory are included. It is clear that an understanding of the low-energy experimental data requires detailed modeling of multiple-scattering effects. The window thickness parameters which varied from cell to cell in experimental tests (Li et al., 1981; Kamath, 1981) were assumed to be governed by a uniform distribution in the present calculations. The model results averaged over the window thickness are compared with short-circuit current measurements (Loo, Knechtli, and Kamath, 1981; Kamath, 1981) of irradiated cells shown in figure 14.7. The best value of recombination cross section, in cm^2, for proton induced defects is

$$\sigma_r \approx 6 \times 10^{-14} \tag{14.43}$$

which is in fair agreement with the estimated average cross section ($\sigma_r \approx 1.06 \times 10^{-13}$ cm^2) determined from deep-level transient spectroscopy (Li et al., 1981).

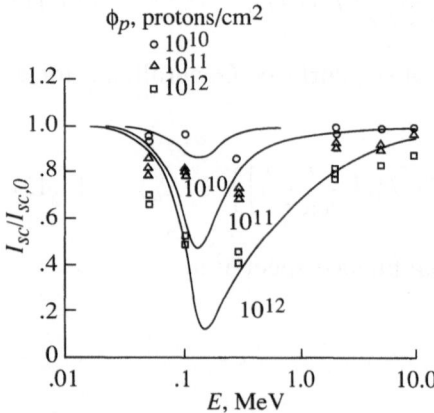

Figure 14.7. Reduced short-circuit efficiency for monoenergetic proton exposure at three fluence levels.

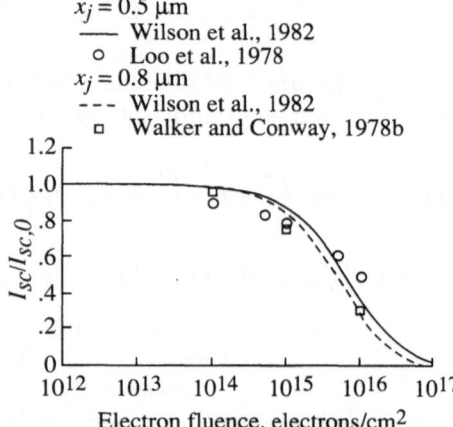

Figure 14.8. Reduced short-circuit current after 1 MeV of electron exposure for two junction depths.

The fractional short-circuit current remaining after 1 MeV of electron irradiation is shown as a function of electron fluence in figure 14.8. The recombination cross section, in cm^2, is

$$\sigma_r \approx 4 \times 10^{-14} \tag{14.44}$$

and calculations were made for two junction depths, namely 0.5 μm and 0.8 μm. Also shown in figure 14.8 are corresponding experimental data of Walker and Conway (1978b) and Loo et al. (1978). The reasonable consistency of the theory for vastly different particle types is gratifying.

14.2.6. Equivalent electron-fluence concept. It is customary in protection from mixed-radiation environments to develop concepts under which effects of radiations of different quality may be combined to ascertain the total effect on device performance. From an electronic device standpoint, the equivalent electron fluence is usually employed as the combinational rule. The equivalent electron fluence is defined as that fluence of electrons of fixed energy (usually 1 MeV) which produces the same effect on the device performance as a particle fluence of a particular type, energy, and fluence level. The fluence of electrons ϕ_e equivalent to a fluence of protons $\phi_p(E_p)$ of energy E_p is given by

$$R_p\left[\phi_p(E_p)\right] = R_e(\phi_e) \tag{14.45}$$

where R_p and R_e are the device response functions for proton and electron damage (Tada and Carter, 1977). If equation (14.45) is satisfied, the equivalent-fluence ratio may be defined as

$$r_f(E_p) = \frac{\phi_e}{\phi_p(E_p)} \tag{14.46}$$

and the main usefulness of the concept requires that $r_f(E_p)$ not depend on the magnitude of $\phi_p(E_p)$. The equivalence for solar cells is usually related through the minority-carrier diffusion length for which the equivalent-fluence ratio is expressed as the ratio of the damage coefficients (Wilson, Stith, and Walker, 1982; Tada and Carter, 1977). The combined effects of electron and proton exposure are then

$$R_{\text{tot}}\left[\phi_p(E_p), \phi_e\right] = R_e\left[\phi_e + r_f(E_p)\,\phi_p(E_p)\right] \tag{14.47}$$

where ϕ_e and $\phi_p(E_p)$ are the mixed environmental components. The strong energy dependence of the response to protons arising from spatial nonuniformity in cell damage brings into question the usefulness of the concept of equivalent electron fluence (Wilson, Stith, and Walker, 1982; Tada and Carter, 1977).

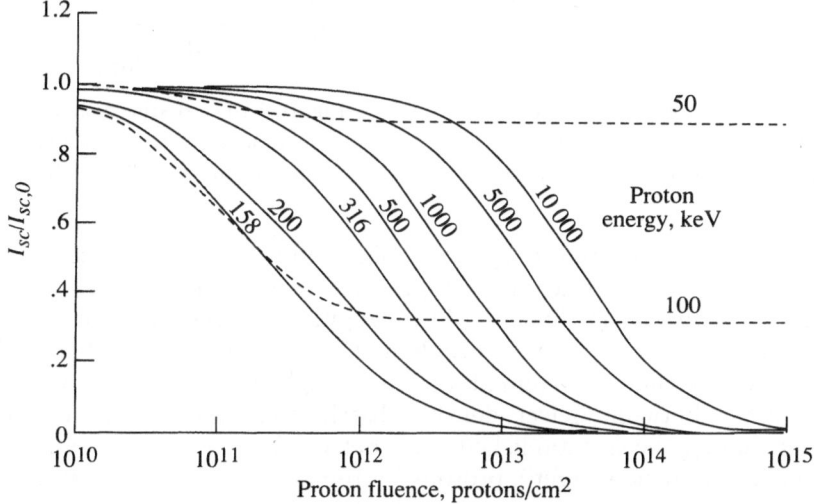

Figure 14.9. Reduced short-circuit current for monoenergetic proton exposure for 0.5-μm junction cells.

The remaining short-circuit current for 0.4-μm window cells and 0.5-μm junction cells as a function of proton energy and fluence is shown in figure 14.9. The equivalent-fluence ratio was calculated using equations (14.45) and (14.46) for 1-MeV electron-fluence levels $\phi_e = 1.7 \times 10^{15}$ electrons/cm^2, 6.8×10^{15} electrons/cm^2, and 2.3×10^{16} electrons/cm^2 at $I_{sc}/I_{sc,0} = 0.8, 0.5,$ and 0.2 (as shown in fig. 14.8 for the 0.5-μm junction cell). The resulting values of $r_f(E_p)$ are shown in figure 14.10 for each of the three fluence levels. For the equivalent-fluence concept to be useful, the three curves must coincide at all proton energies as they do above 500 keV. However, in the proton energy range 50 to 500 keV, where the cell is extremely sensitive, the usefulness of equivalent electron fluence is generally limited by the strong dependence of the equivalent-fluence ratio on the damage level. This has important consequences in terms of radiation testing, since the mixed environment generally must be simulated to ensure a valid test unless the bivariate equivalent-fluence ratio is adequately known. On the other hand, for a given (fixed) environment, test procedures could be established through the use of the present model, for a given cell type. Thus, an "equivalent" electron fluence could be established in the restricted sense of fixed environmental components.

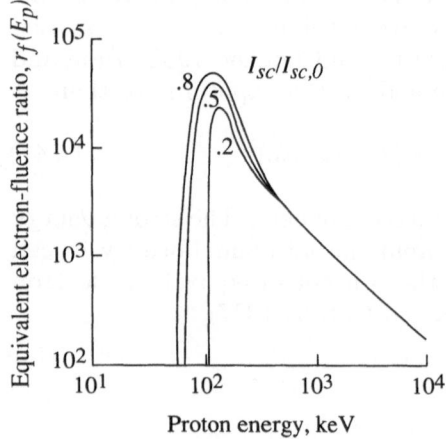

Figure 14.10. Equivalent electron-fluence ratio for a cell with 0.5-μm junction depth and a 0.4-μm $Al_xGa_{1-x}As$ window.

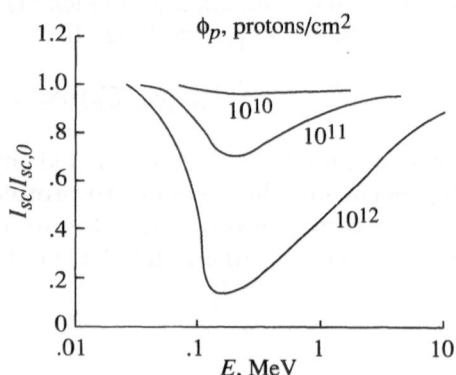

Figure 14.11. Reduced short-circuit current for isotropic incident protons at three fluence levels.

14.2.7. Angular isotropy effects. The radiation in space can, for most practical purposes, be considered isotropic, and most radiation models present data as the omnidirectional fluence. Such angular factors generally have great importance in radiation protection problems (Wilson, Stith, and Walker, 1982), and such effects within the context of this simple model are evaluated here. The relationship between defect density distribution within the cell and cell performance having been established, the defect density is now evaluated for isotropic-incidence monoenergetic protons by replacing $\phi(E_o)$ in equation (14.42) with a δ-function. Results are shown in figure 14.11. Clearly, angular isotropy effects show no major differences in cell sensitivity at all energies and fluence levels, although a general increase in radiation resistance at the lowest fluence levels is apparent. However, at the high fluence levels, the sensitivity is increased in the 200-keV to 1-MeV region. At higher energies ($E \gg 1$ MeV), angular factors are relatively less important because of the high penetrating power of the protons.

In general, the angular factors are helpful if fluence levels are sufficiently low that the reduced penetration of low-energy protons at oblique angles of incidence serves to provide the cell with added protection. At high fluence levels and fixed energy, the minority-carrier recombination rates near the end of the proton trajectories tend to saturate for normal incidence, whereas isotropic incidence tends to distribute these defects more uniformly over the cell. This uniform distribution increases their effectiveness for cell damage, which in turn accounts for the increased cell sensitivity for $E > 200$ keV (as shown in fig. 14.11 for $\phi_p = 10^{12}$ protons/cm^2). The spectral characteristics for performance evaluation in space applications must still be considered in the protection against space radiation.

Solar cell performance is likewise evaluated for isotropically incident 1-MeV electrons. The results are shown in figure 14.12 as a function of omnidirectional fluence level (0.5-μm junction depth and infinite backing is assumed). Comparison of figure 14.12 with figure 14.8 for normal incidence shows that an isotropically incident electron is equivalent to four normally incident electrons. Clearly, isotropic incidence is a most important factor for space-radiation testing.

14.2.8. Effects of space-radiation environment. Space missions to the fringes of the geomagnetic field and interplanetary missions experience the yearly solar-particle fluence during highly solar-active years (Foelsche, 1963) on the order of

$$\phi_p(E_p) \approx \frac{5 \times 10^{14}}{E_p} \tag{14.48}$$

where E_p is in keV and ϕ_p is in protons/cm^2. The remaining short-circuit current calculated from equations (14.38) and (14.48) as a function of cover glass thickness is shown in figure 14.13. It is clear that an unshielded cell would not survive a major solar event and requires a cover glass of about 25 μm to ensure performance levels to within 90 percent of their initial value.

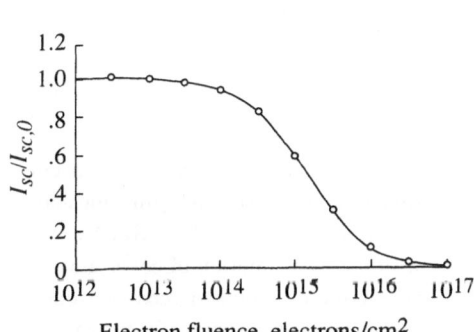

Figure 14.12. Reduced short-circuit current for isotropically incident 1-MeV electrons as a function of omnidirectional fluence.

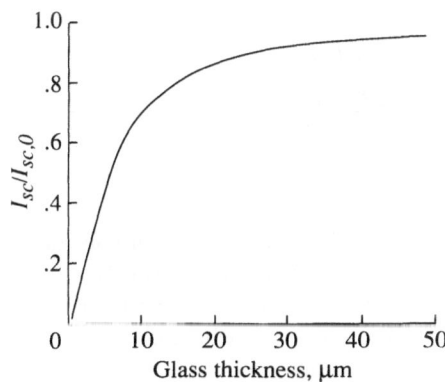

Figure 14.13. Reduced short-circuit current due to a large solar event of a cell with a 0.5-μm junction depth and a 0.5-μm Al$_{1-x}$Ga$_x$As window as a function of cover glass thickness.

The protons trapped at geosynchronous altitude ($L = 6.6$ Earth radii) are well approximated (Sawyer and Vette, 1976) by

$$\phi_p(E_p) = 2.5 \times 10^{14} \ \exp\left[-1.27 - 0.0072E_p + \frac{230}{E_p}\right] \qquad (14.49)$$

where ϕ_p is in protons/cm^2-yr. The corresponding yearly electron fluence (Singley and Vette, 1972) is

$$\phi_e(E_e) = 4.5 \times 10^{14} \ \exp\left(-2.832 \times 10^{-3} E_e\right) \qquad (14.50)$$

in electrons/cm^2-yr. The short-circuit current ratio is calculated for equivalent 1-, 5-, and 10-yr missions in the trapped environment with results shown in figure 14.14 as a function of cover glass thickness. Equations (14.48) to (14.50) are integrated flux and must be differentiated for use in equation (14.42). A 15-μm glass cover is required to stop the geosynchronous trapped protons. Cover glass thickness beyond 15 μm is ineffective for protection against the electron environment. The effects of the geosynchronous trapped environment are combined with a single large solar event in figure 14.15 for 1-, 5-, and 10-yr missions. Little improvement in cell protection is obtained by having a cover glass thickness in excess of about 30 μm. For a complete evaluation of solar cell performance, one needs to consider the production of color centers in the cover glass and their effect on cell performance.

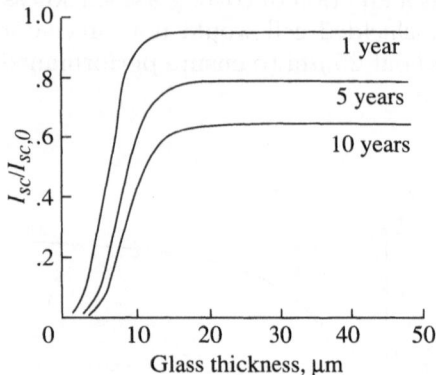

Figure 14.14. Reduced short-circuit current of a cell with 0.5-μm junction depth and a 0.5-μm Al$_{1-x}$Ga$_x$As window as a function of cover glass thickness in geosynchronous environment.

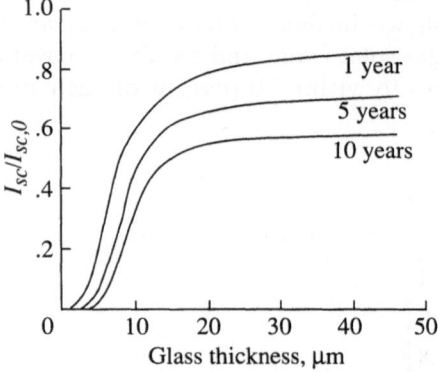

Figure 14.15. Reduced short-circuit current of a cell with a 0.5-μm junction depth and a 0.5-μm Al$_{1-x}$Ga$_x$As window as a function of cover glass thickness in combined geosynchronous and solar cosmic-ray environment.

14.3. Microscopic Defect Structures and Equivalent Electron-Fluence Concepts

The problem of additivity of exposure due to protons and electrons rests on the concept that proton damage and electron damage are in some sense equivalent.

When equivalence is valid then for any proton fluence causing damage in a cell, there is an equivalent electron fluence (usually 1 MeV) which causes the same damage level. In this way electron and proton damage can be added for total damage effect. In section 14.2.6 the issue of equivalent electron fluence was examined from the point of view of the macroscopic spatial density of defects produced in a GaAs shallow junction solar cell (Wilson and Stock, 1984). The calculations utilized a simple model of short-circuit current for the cell by Wilson, Walker, and Outlaw (1984) which has achieved considerable success in predicting experimental results (Wilson et al., 1982; Wilson, Stith, and Walker, 1982). More recently, further experiments have fully justified damage level for protons in the energy range below about 0.5 MeV (Anspaugh and Downing, 1984). Although it is not clear from these published experimental results as to what "high" and "low" damage levels were actually used in the experiment and that they measured the power decrement rather than the short-circuit current decrement, the comparison between theory and experiment shown in figure 14.16 is quite encouraging.

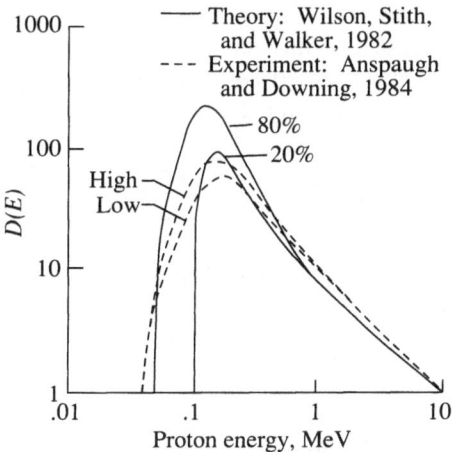

Figure 14.16. Proton damage coefficient.

In addition to the equivalent fluence for test exposure, one must take account of exposure time and temperature since a degree of self-healing of the cell is normally present. In this respect one may call to mind the experience with GaAs cells on NTS-2 for which annealing in flight is suspected (Walker, Statler, and Lambert, 1978). When such factors are fully considered, a reexamination of electron equivalency must be made since evidence exists which indicates that defect structures produced by proton exposure do not readily anneal (Anspaugh and Downing, 1981). Further study of space-radiation damage in which the chemistry of specific defects are included is clearly needed.

In the present section we will examine the question of equivalence in terms of the microscopic defect structures. The equivalent electron fluence concept is then said to hold only if the macroscopic and microscopic defect densities are reasonably represented under simplified test conditions. Implications as to minimum test requirements will be discussed.

14.3.1. Theory. Atomic displacements caused by particle impact with atomic nuclei result in crystal defects. The formation of the defects is related to the energy-transfer cross section, which is obtained from Rutherford's cross section for protons and Mott's cross section for electrons. The maximum energy transfer for protons is $T_m = 4M\,E_p/(1+M)^2$ where M is the atomic weight of the struck nucleus and E_p is the proton energy. The maximum energy transfer for an electron of energy E_e is similarly $T_m = 2E_e\left(E_e + 2m_ec^2\right)/M_c^2$ where m_ec^2 is the electron rest energy.

The minimum energy transfer to produce one displacement is $T_D \approx 9.5$ eV. At least one displacement is produced whenever $T \geq T_D$. When the recoil energy exceeds T_D, the nucleus is proficient in producing further recoils with half its energy and the remaining half is lost in electronic excitation. Hence, the total number of recoils is

$$\nu \approx 1 + \frac{T}{2T_D} \qquad (T \geq 2T_D) \qquad (14.51)$$

One may be tempted to take the integer part of ν, but this would be incorrect since ν as written is to be interpreted as a mean for many such events. We now introduce the cross section for producing ν or more defects as

$$\sigma_\nu(E) = \int_{T_\nu}^{T_m} \frac{d\sigma}{dT} dT \qquad (14.52)$$

where $T_\nu = 2(\nu - 1)T_D$. The probability that more than ν defects are produced in a given collision is then

$$P_\nu = \frac{\sigma_\nu(E)}{\sigma_D(E)} \qquad (14.53)$$

where $\sigma_D(E)$ is the total displacement cross section. Values for P_ν are shown in figure 14.17 as a function of proton energy for values of ν from 2 to 10. It is seen that an asymptotic value for P_ν is reached rather quickly ($E_p \approx 23$ keV).

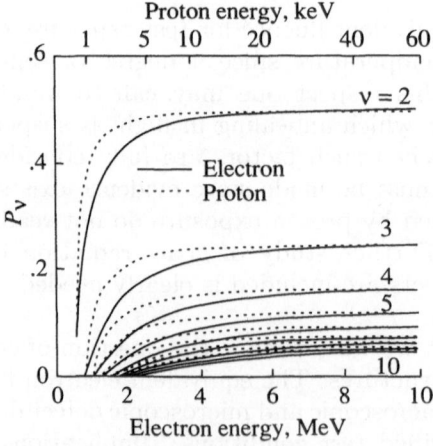

Figure 14.17 Defect number probabilities.

Values of P_ν are also shown in figure 14.17 as a function of electron energy and values of ν from 2 to 10. Note we have the relation between maximum energy transfer of the proton and electron (i.e., T_m (proton) $= T_m$ (electron)) so that

$$E_p = \frac{(1 + M)^2 \, E_e \left(E_e + 2m_e c^2 \right)}{2M^2 c^2} \tag{14.54}$$

Clearly the defect structures produced by protons and electrons are quite different for electron energies below 10 MeV. This is shown more clearly in table 14.1 where values of P_ν for protons and electrons of various energies are compared. The number of defects at a recoil site is nearly independent of proton energy, as shown in table 14.1. In distinction, the 1-MeV electron produced defects are vastly different. The number of defects at the recoil site of electrons approaches the proton values only as the electron energy exceeds 10 MeV.

Table 14.1. Defect Probabilities for Protons and Electrons

	Probability of forming ν defects—						
	For protons with energies E_p, MeV, of—				For electrons with energies E_e, MeV, of—		
ν	0.02	0.05	0.1	1	1	5	10
2	0.496	0.498	0.499	0.500	0.314	0.489	0.501
3	.243	.247	.249	.250	.046	.232	.249
4	.159	.164	.165	.167	.001	.145	.164
5	.117	.122	.123	.125		.103	.122
7	.075	.080	.082	.083		.060	.079
10	.047	.052	.054	.055		.033	.050

It is clear from the data presented that 1-MeV electron-induced defects appear as isolated events of two or three displacements. In distinction, proton-induced defects show a broad range of defect structures with appreciable numbers having more than five displacement sequences. It is believed that this is the main source of difference in annealing properties between proton and 1-MeV electron irradiation damage.

In order to provide a better understanding of the process of defect formation and kinetics, a binary-collision simulation code MARLOWE is employed (Robinson and Torrens, 1974). The GaAs unit cell is shown in figure 14.18. The primary recoil atom energy was taken as 20 eV corresponding to the average energy transferred by a 1-MeV electron and 90 eV representing the average energy transferred by the collision of a proton of a few MeV (see figure 14.3). This work is continuing at Virginia State University (John J. Stith).

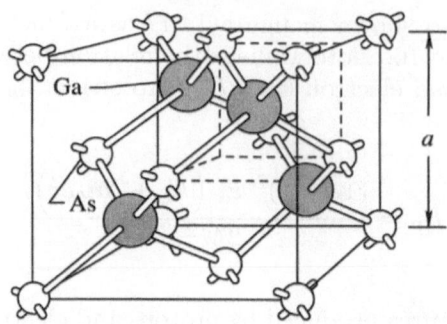

Figure 14.18. A unit cell of the GaAs crystal showing the corresponding lattice parameters.

The results from the computer simulations yielded information on the spatial distribution of the defect pairs (close pairs, near pairs, and distant pairs), as well as details on possible clusters of defects, such as multiple vacancies. Close pairs are vacancy-interstitial pairs that are separated by distances that are less than nearest neighbors separation; near pairs are vacancy-interstitial pairs that are separated by distances greater than nearest neighbors separation but less than the distance between second-nearest neighbors; and distant pairs are interstitial-vacancy pairs that are separated by distances greater than the distance between second-nearest neighbors. Information is also generated on improper replacements which are produced in irradiated binary crystals, such as gallium arsenide. These are a form of stable defect produced in the damaged crystal.

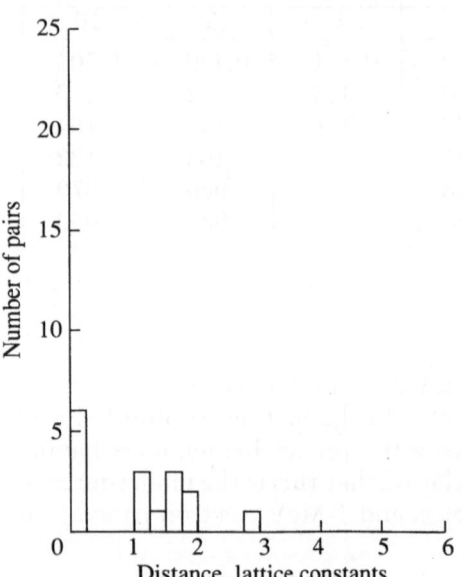

Figure 14.19. Distribution of interstitial-vacancy pairs for a 20 eV primary recoil atom.

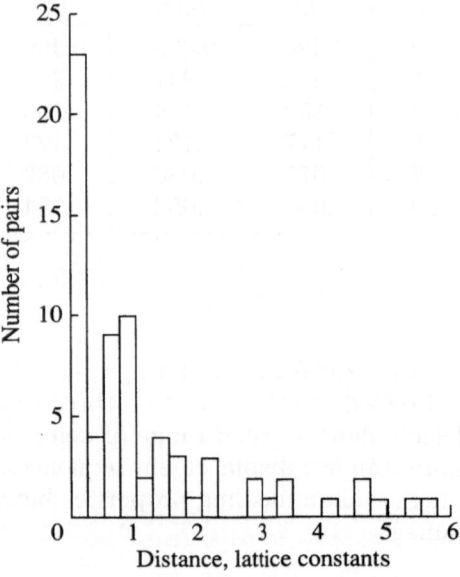

Figure 14.20. Distribution of interstitial-vacancy pairs for a 90 eV primary recoil atom.

The graphs in figures 14.19 and 14.20 show the distribution of the separations of the pairs for the 20- and 90-eV recoil energies, respectively. These distributions include close, near, and distant pairs. When the distributions on the two graphs are compared, it is clear that there are more interstitial-vacancy pairs for the higher-energy cascades than for the lower-energy cascades. This is to be expected. The results from two different cascades are displayed in figures 14.21 through 14.26. Figures 14.21 and 14.22 show the relative positions of the interstitials and the vacancies produced in the crystal. Close, near, and distant pairs are displayed. The circle represents a vacancy and the square represents an interstitial. In figures 14.23 and 14.24 the close pairs are not included since it is a good probability that the vacancies and interstitials that form the close pairs will combine (self-anneal). The triangle is used to represent defects caused by improper replacements that occur in the crystal. Figures 14.25 and 14.26 display improper replacements and the distant pairs. The distant pairs may still exist after annealing of the crystal and would represent, along with the improper replacements, stable defects within the crystal. When a proper pair combines, two defects are eliminated; but when an improper pair combines, two defects are reduced to one defect which is different from either of the two original defects. There is a sizable difference between the number of distant pairs for the high-energy primary recoil atoms over the number of distant pairs for the low-energy primary recoil atoms. It should also be noted that the high-energy primary recoil atom generates several subcascades, giving rise to more extensive damage structures within the crystal. This is qualitatively similar to the experimental results that demonstrated a high degree of difficulty in annealing proton radiation damage as compared with annealing electron radiation damage in gallium arsenide. Future work will concentrate on developing a defect kinetic model and correlation with deep level electron spectroscopic analysis of radiation produced defects.

14.3.2. Conclusions. The macroscopic defect density variation effects on equivalent electron fluence being well-established, the effect of microscopic defect structures reveals an additional requirement on the equivalent electron-fluence concept. It has been shown the 1-MeV electrons can never reproduce proton irradiation damage on the microscopic scale. This is the probable difference in annealing between electron and proton damaged cells observed by Anspaugh and Downing (1981). A full explanation must await further study on the chemical kinetics of defect structures. In any case, a minimum requirement will be the use of 10-MeV electrons to assure equivalence on the microscopic level of defect formation.

14.4. GaAs Model Refinements

The original model for GaAs solar cells was admittedly simplified (Wilson, Walker, and Outlaw, 1984) but still explained the main features of the proton-induced radiation damage response curves and the annealing characteristics of the cells (Stith and Wilson, 1985). Several modifications of the basic concepts were accomplished by various researchers.

The photon absorption coefficient exhibits strong wavelength dependence and has been used as a probe for study of the internal workings of GaAs photovoltaic systems. J. Y. Yaung (1984) incorporated photoabsorption properties into the short-circuit model to provide the spectral dependence of radiation damage. The resulting spectral response is shown in figure 14.27. When the spectral attenuation coefficient properly accounts for the depth dependence of the minority carrier

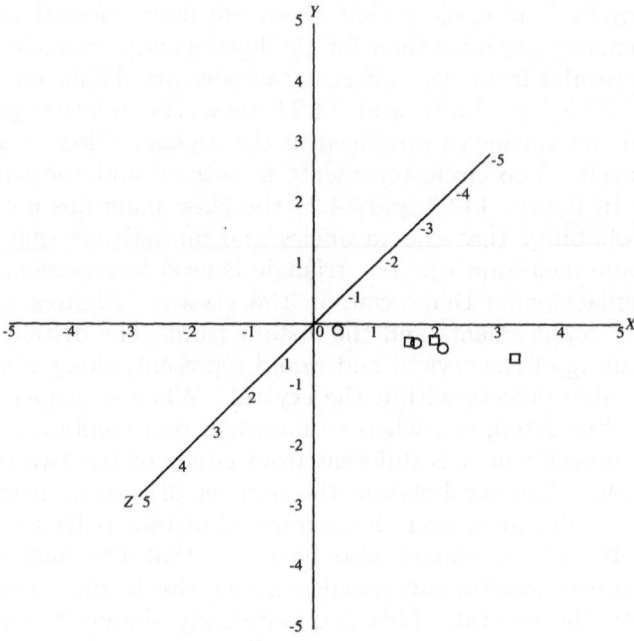

Figure 14.21. Vacancy and interstitial sites for a typical 20 eV primary atom recoil event. Coordinates in lattice constants. All pairs.

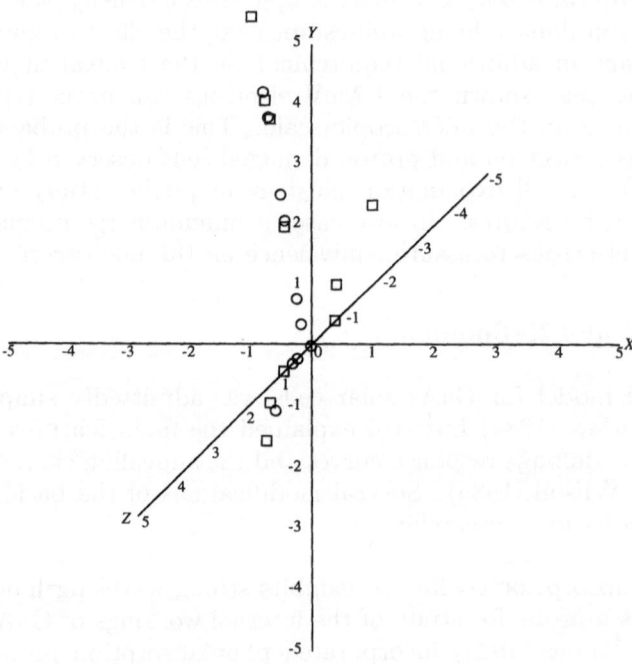

Figure 14.22. Vacancy and interstitial sites for a typical 90 eV primary atom recoil event. Coordinates in lattice constants. All pairs.

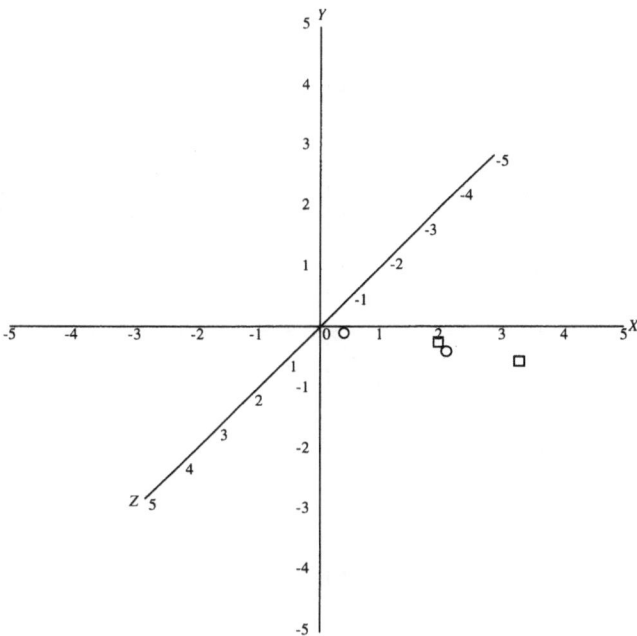

Figure 14.23. Vacancy and interstitial sites remaining after annealing of close pairs caused by 20 eV primary recoil atom. Coordinates in lattice constants. No close pairs.

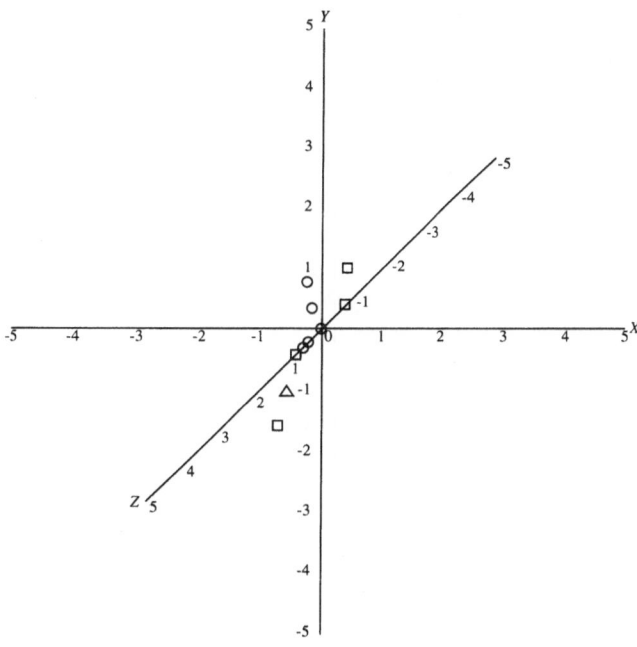

Figure 14.24. Vacancy and interstitial sites remaining after annealing of close pairs caused by 90 eV primary recoil atom. Coordinates in lattice constants. No close pairs.

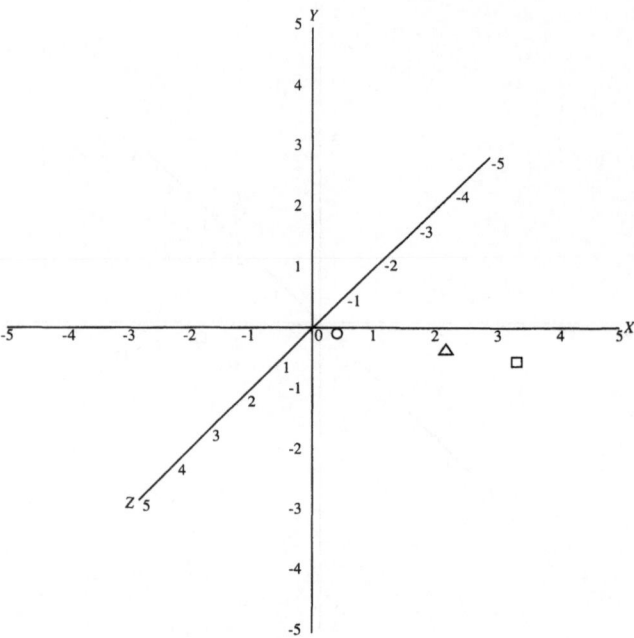

Figure 14.25. Improper replacements and distant pairs from 20 eV primary recoil event. Coordinates in lattice constants. No close pairs or near pairs.

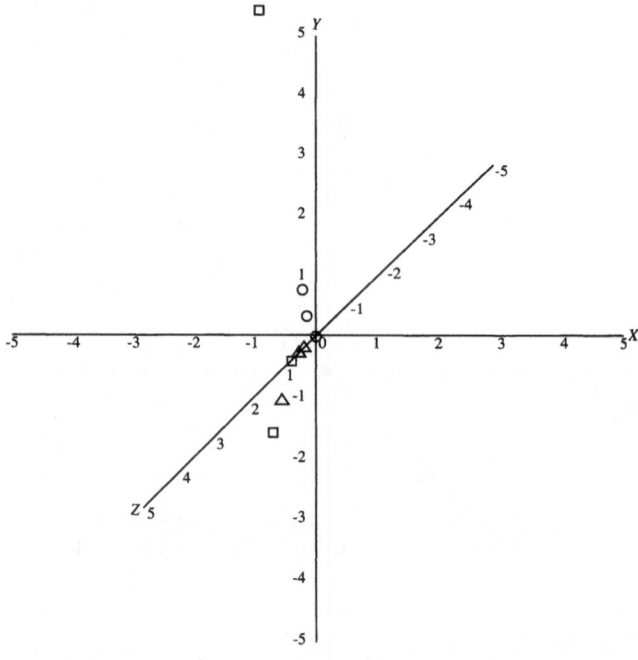

Figure 14.26. Improper replacements and distant pairs from 90 eV primary recoil event. Coordinates in lattice constants. No close pairs or near pairs.

source, the average damage response to low-energy protons is somewhat improved, as shown in figure 14.28. The spectral averaged model shows similar success in application to silicon solar cells, as seen in figures 14.29 and 14.30. In addition to the spectral dependence, Yeh, Li, and Loo (1985) added the recombination differences of the p- and n-material and find excellent agreement with their experimental data. (See fig. 14.31.)

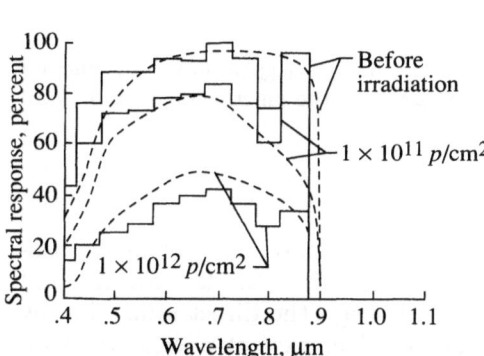

Figure 14.27. Experimental and simulated spectral responses for (AlGa)As-GaAs solar cell before and after proton irradiation (Proton energy = 290 keV).

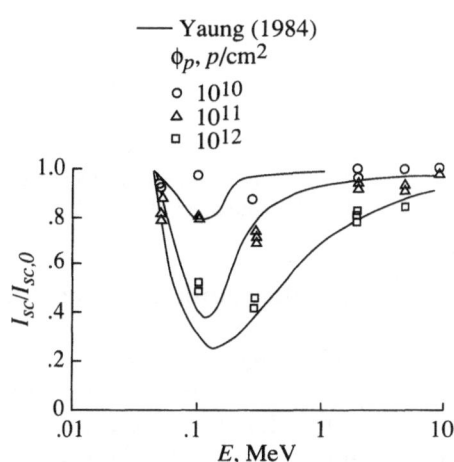

Figure 14.28. Final results of predicting I_{sc} damage on GaAs solar cell.

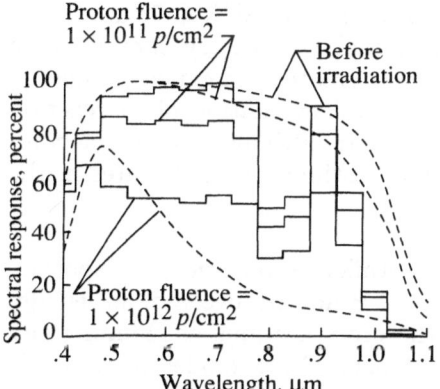

Figure 14.29. Experimental and Yaung's simulated spectral responses for silicon solar cell before and after proton irradiation (Proton energy = 290 keV).

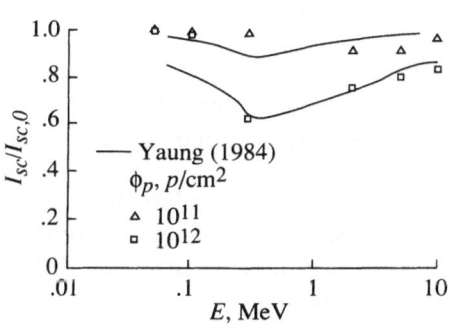

Figure 14.30. Final results of predicting I_{sc} damage on silicon solar cell.

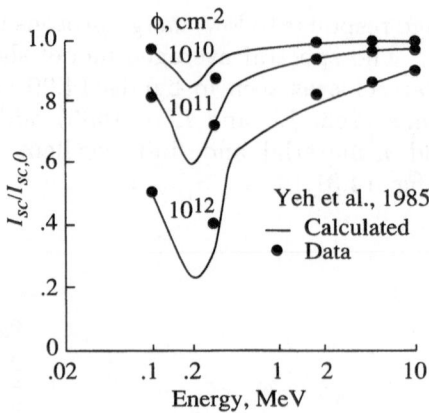

Figure 14.31. The calculated I_{sc} degradation ratio in the $Al_{0.85}Ga_{0.15}As$-GaAs p-n junction solar cell. The thickness of the window layer is 0.34 μm, and the junction depth is 0.5 μm.

14.5. Microelectronic Applications

The early suggestion that some spacecraft anomalies may result from the passage of the galactic ions through microelectronic circuits (Binder, Smith, and Holman, 1975) has now been well-established. Although the direct ionization by protons appears as an unlikely candidate, the recoil energy of nuclear-reaction products is suspected as a source of single-event upset (SEU) phenomena (Wyatt et al., 1979; Guenzer et al., 1980; Petersen, 1980). As a result, a number of fundamental experimental and theoretical studies were undertaken to better understand the phenomena. McNulty and coworkers examined the energy deposition of proton reaction products in Si by using surface-barrier detectors of various thicknesses for 2.5 to 200 μm (McNulty et al., 1980). They also developed a Monte Carlo code for theoretical evaluation of energy deposition for such products. (See McNulty et al., 1980; McNulty, Farrell, and Tucker, 1981.) A comparison of McNulty's work with the well-established medium energy cascade code (MECC-7) developed by Bertini and coworkers at the Oak Ridge National Laboratory showed some differences in predicted reaction products and even greater differences in energy spectral contribution. (See Hamm et al., 1981.) An evaluation of Si reaction products was likewise made by Petersen (1980), and, although no direct comparison was made with McNulty's experiments, an estimate of SEU rates in the trapped-proton environment was made.

Following these fundamental studies, more-detailed applications to specific-device geometries and parameters were made. Bradford evolved an energy deposition formalism (Bradford, 1982) using the cross sections of Hamm et al. (1981). McNulty et al. (1980) applied their Monte Carlo model to dynamic random access memory (DRAM) devices with reasonable success and discussed the implications of heavy ion SEU phenomena on proton-induced SEU events through secondary reaction processes (Bisgrove et al., 1986). The fundamental consideration is the evaluation of the energy deposited within the sensitive volume (depletion region) of the device in question as the result of a passing proton. The ionization due to the proton itself makes only a small contribution to the critical charge. Nuclear-reaction events usually produce several reaction products (a heavy fragment and several lighter particles, although a few heavy fragments may be produced simultaneously on some occasions), and all the resultant products can make important contributions to the deposited energy. Such nuclear-event products are, of course, correlated in both time and space.

There are three distinct approaches to a fundamental description of the energy deposition events. McNulty and coworkers developed a Monte Carlo code in which multiparticle events are calculated explicitly, including spatial and specific-event (temporal) correlation effects. Although this is the most straightforward way of treating the full detail, it is a complex computational task. A second class of methods begins with the volumetric source of collision events and calculates the SEU probability by using the chord-length distribution. (See Bradford, 1982; Fernald and Kerns, 1988.) Although in principle the correlation effects could be so incorporated, they appear to be ignored in both the cited references. A third approach in which linear energy-transfer (LET) distributions and chord-length distributions are used seems most appropriate for external sources. (See Petersen et al., 1982; Tsao et al., 1983.) This last approach applies if the LET distribution from external sources is constant over the sensitive volume, but its applicability to volumetric sources is questionable. At the very least, this approach ignores correlation effects.

Nuclear data bases for biological systems were examined by Wilson et al. (1988). The MECC-7 results underestimated by nearly a factor of 2 the energy-transfer cross section for multiple-charged ion products. In an analysis with greater detail, (Wilson et al., 1989), the Silberberg-Tsao (Tsao et al., 1983) fragmentation parameters were found to be superior to the MECC-7 results. The primary differences appear for the lighter of the multiple-charged fragments. Further comparison with experiments on Al targets shows both Monte Carlo nuclear models (McNulty's code OMNI as well as MECC-7) to underestimate production cross sections for products lighter than fluorine in proton-induced reactions. Although these intranuclear-cascade models are capable of representing multiparticle correlation, the inherent inaccuracies in predicting cross sections is a serious limitation.

In the present section, the effects of nuclear recoil on electronic devices are examined and the development of a formalism for application to specific-device parameters is begun. As a test of our methods as they develop, the results are compared with the experimental measurements of McNulty et al. (1980).

14.5.1. Microelectronic upsets. An electronic device is sensitive to the sudden introduction of charge into the active elements of its circuits. The amount of such charge that is sufficient to change the state of a logic circuit is called the critical charge. As shown in figure 14.32, there is a rough relationship between the critical charge Q_c and the device feature size L (Petersen et al., 1982). This relationship is as follows:

$$Q_c \approx 0.0156L^2 \tag{14.55}$$

where Q_c is measured in pC and L is measured in μm. Upsets in a device are then dependent on the charge produced in comparison to the critical charge.

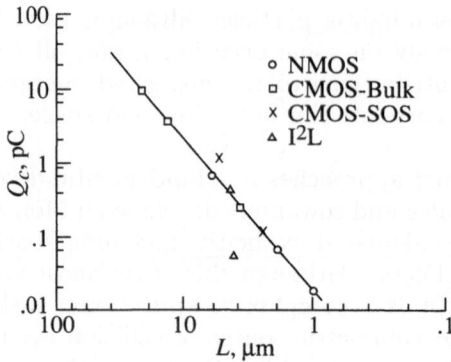

Figure 14.32. Critical charge as a function of feature size in several device types.

The charge released ΔQ in a material because of the passage of an energetic ion is related to the kinetic energy lost ΔE during the passage and is given by

$$\Delta Q = \frac{\Delta E}{22.5} \qquad (14.56)$$

where ΔQ has units pC and ΔE has units MeV. The energy lost by an ion in passing through a region is related to its stopping power $\left(-\frac{dE}{dx} = S_z(E')\right)$ in the medium. The distance traveled before coming to rest is

$$R_Z(E) = \int_0^E \frac{dE'}{S_z(E')} \qquad (14.57)$$

If an ion is known to come to rest in distance x, then its energy is found through the inverse of relation (14.57) as

$$E = R_Z^{-1}(x) \qquad (14.58)$$

Equation (14.58) is used to calculate energy loss. The energy loss by an ion of charge Z and energy E in passing through the active region of a device with collection length L_c is given by

$$\Delta E = E - R_Z^{-1}\left[R_Z(E) - L_c\right] \qquad (14.59)$$

where

$$L_c = W_{\text{epi}} + W_n \qquad (14.60)$$

In equation (14.60), W_{epi} is the epitaxial layer thickness and W_n is the width of the depletion region (Chern, Seitchik, and Yang, 1986). The energy loss depends on the particle isotope (i.e., ion mass) and angle of incidence. The range-energy relations described by Wilson et al. (1989) are utilized. As a practical matter to reduce numerical error inherent to numerical interpolation,

$$\Delta E = R_Z^{-1}\left[R_Z(E)\right] - R_Z^{-1}\left[R_Z(E) - L_c\right] \qquad (14.61)$$

is used in place of equation (14.59). The result of equation (14.59) depends on the global error (fixed at 1 percent) in the computer code, while equation (14.61) depends only on the local relative error (quite small). The charge introduced into the feature is given by equations (14.56) and (14.61). An example for a particular collection length of 2 μm is shown in figure 14.33 for each ion type. A simplified geometry is assumed in which the channel length and width and the collection length (fig. 14.34) are taken as equal to the feature size. The E, Z plane can be divided into regions for which

$$\Delta Q(E) \geq Q_c \tag{14.62}$$

The value of $\Delta Q(E)$ depends on the feature size L. (See eq. (14.55).) The ion mass for each value of Z was taken as the natural mass in arriving at the contour of constant ΔQ shown in figure 14.35. The average recoil energies from the fragmentation of ^{16}O and ^{28}Si produced by collision with high-energy protons (Wilson et al., 1989) are also shown in figure 14.35. The importance of a given fragment type for a given feature size for the device may be judged from the average recoil energies from the fragmentation of ^{16}O and ^{28}Si.

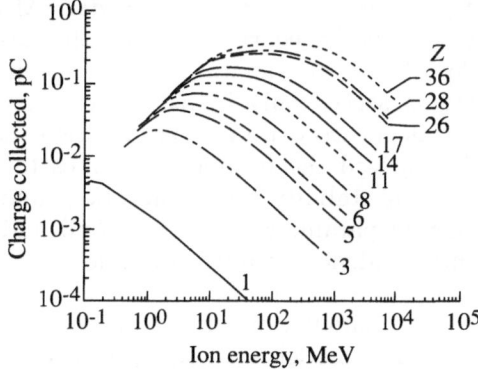

Figure 14.33. Charge collected as a function of ion energy with a collection length of 2 μm.

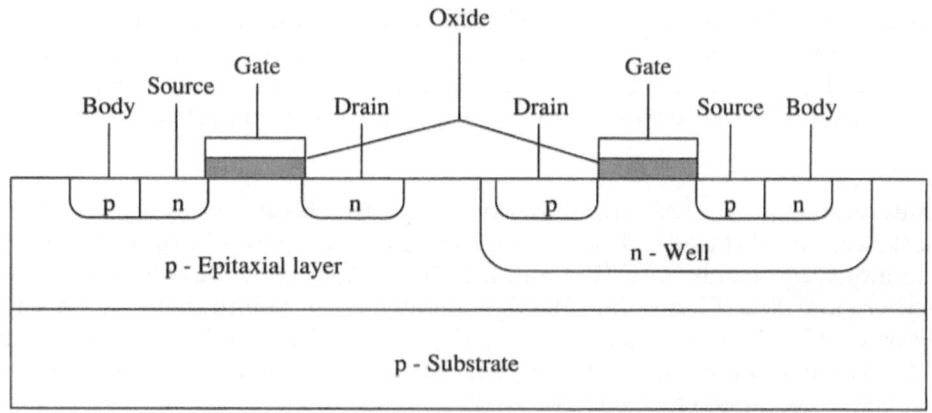

Figure 14.34. Cross section of bulk CMOS technology.

It is doubtful that any of the fragments produce upsets in the 4-μm and larger devices (note that simplified geometries have been used). Also, the lighter fragments of Li, He, and H are not suspected for SEU's in this simple geometry and figure 14.35 is applicable to incident cosmic-ray ions.

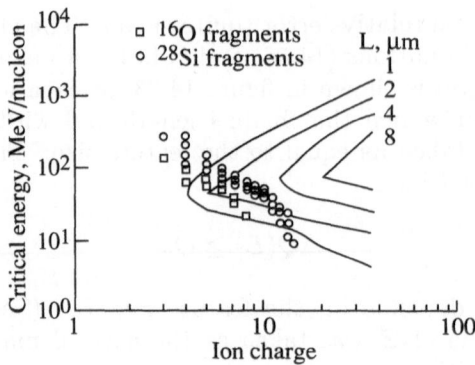

Figure 14.35. Critical energy as a function of ion charge for several feature sizes. Average recoil energies of fragments of silicon and oxygen are superimposed.

14.5.2 Nuclear-fragmentation cross sections.

Although nuclear fragmentation has been under study for nearly 50 years, the absolute cross sections still stir some controversy. The experimental problem is that the main-reaction products could be directly observed only in recent years and even now only in rather sophisticated experiments. Rudstam (1966) studied the systematics of nuclear fragmentation and supposed the fragment isotopes to be in a bell-shaped distribution about the nuclear stability line. Silberberg and Tsao (Tsao et al., 1983) continued the Rudstam parametric approach and added many correction factors as new experimental evidence became available.

Concurrently, Monte Carlo simulation of the Serber model (1947) and final decay through compound nuclear models showed some success (Hamm et al., 1981; Bertini, 1969). Even so, Monte Carlo simulation shows little success in predicting fragments whose mass is small compared with the original target nuclear mass (Wilson et al., 1988; Kwiatkowski et al., 1983). Of the various models for nucleon-induced fragmentation in ^{28}Si, the model of Silberberg and Tsao is probably the most reliable. The main limitation of their model is that only inclusive cross sections are predicted; particle correlations could prove important in predicting SEU.

Measurements of ^{27}Al fragmentation in proton beams have been made by Kwiatkowski et al. (1983). These experiments are compared in figure 14.36 with the Monte Carlo results of OMNI and MECC-7. Also shown are the results from Silberberg and Tsao (Tsao et al., 1983); generally, these results appear to be within a factor of 2 of the experiment. The model of Silberberg and Tsao (Tsao et al., 1983) is the only model which predicts significant contributions in the important range below the mass of carbon $A_F = 12$.

The spectrum of average recoil energy is calculated using the formalism of Wilson et al. (1989) and the Silberberg-Tsao cross sections and is shown for comparison with the spectrum according to the Bertini cross sections in figure 14.37. The Bertini cross sections are greatly underestimated above 24 MeV and greatly overestimated below 9 MeV. The Bertini results are typical for currently available intranuclear-cascade models. Experimental evidence indicates that even the Silberberg-Tsao values are too small above 6 MeV (Kwiatkowski et al., 1983).

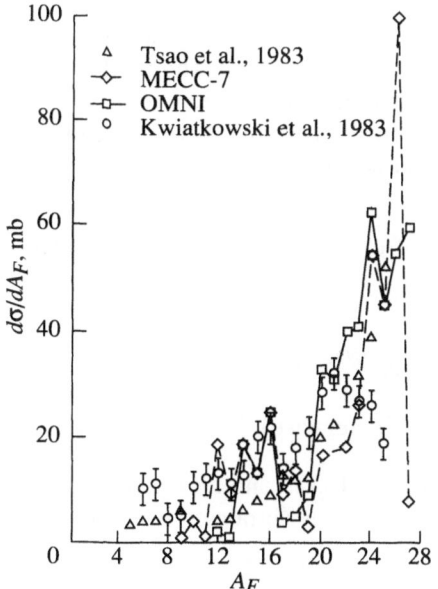

Figure 14.36. Fragmentation cross section for 180-MeV protons on Al targets calculated by various models compared with experimental measurements.

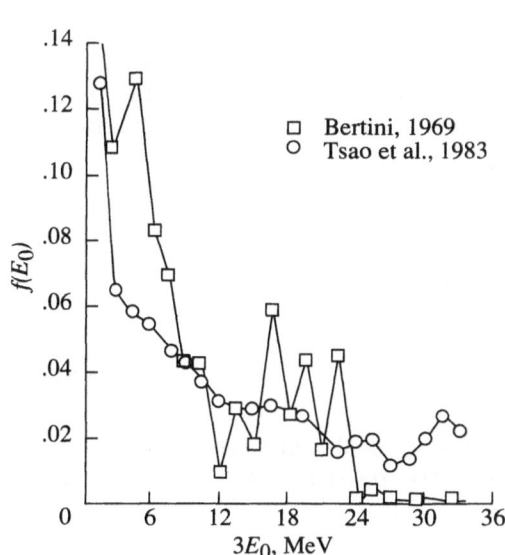

Figure 14.37. Spectrum of average energy predicted by Silberberg and Tsao cross sections compared with Bertini cross sections.

14.5.3. Nuclear recoil transport. The transport of the recoil fragments is described as follows:

$$\left[\vec{\Omega} \cdot \nabla - \frac{\partial}{\partial E} S_z(E)\right] \phi_z(\vec{x}, \vec{\Omega}, E) = \zeta_z(E) \tag{14.63}$$

where $\phi_z(\vec{x}, \vec{\Omega}, E)$ is the ion flux at $\vec{x}$ moving in direction $\vec{\Omega}$ with energy E and where $\zeta_z(E)$ is the ion-source density assumed to be isotropic and uniformly distributed through the media. The solution to equation (14.63) is in a closed region bounded by the surface $\vec{\Gamma}$ subject to the boundary condition

$$\phi_z(\vec{\Gamma}, \vec{\Omega}, E) = \psi_z(\vec{\Omega}, E) \qquad (\vec{n} \cdot \vec{\Omega} < 0) \tag{14.64}$$

where $\vec{n}$ is the outward-directed normal of the surface $\vec{\Gamma}$. The solution is found by using the method of characteristics (Wilson and Lamkin, 1975; Wilson, 1977) as

$$\phi_z(\vec{x}, \vec{\Omega}, E) = \frac{S_z(E_b)}{S_z(E)} \phi_z(\vec{\Gamma}, \vec{\Omega}, E_b) + \frac{1}{S_z(E)} \int_E^{E_b} \zeta_z(E') \, dE' \tag{14.65}$$

where $\vec{\Gamma}$ is the point on the boundary determined by projecting $\vec{x}$ along the direction $\vec{\Omega}$ and

$$E_b = R_Z^{-1} \left[R_Z(E) + b \right] \tag{14.66}$$

where

$$b = \vec{\Omega} \cdot (\vec{x} - \vec{\Gamma})$$ (14.67)

Equation (14.65) may be used to evaluate the spectrum of particles leaving the region that can be related to the spectrum of energy deposited in the media. An isolated sheet of silicon of thickness a, which is obviously similar to the McNulty surface-barrier detectors, is considered. The inward-directed flux at the boundary is then zero. We first consider a monoenergetic ion source

$$\zeta_z(E) = \frac{\sigma_z \phi}{4\pi} \delta(E - E')$$ (14.68)

for which

$$\phi_z(\vec{x}, \vec{\Omega}, E) = \frac{\sigma_z \phi}{4\pi S_z(E)} \left\{ \begin{array}{ll} 1 & (E \leq E' \leq E_b) \\ 0 & \text{(Otherwise)} \end{array} \right\}$$ (14.69)

where σ_z is the silicon-fragmentation cross section and ϕ is the flux of initiating energetic particles. The spectrum of ions leaving the sheet (ignoring edge effects) is

$$\frac{df_z}{dE} = 4\pi A \int_0^1 \mu \, \phi_z(\vec{\Gamma}, \vec{\Omega}, E) d\mu$$

$$= \frac{A\sigma_z \phi}{2 S_z(E)} \left\{ \begin{array}{ll} \frac{a^2}{[R_Z(E') - R_Z(E)]^2} & \left(0 \leq E \leq R_Z^{-1}\left[R_Z(E' - a)\right]\right) \\ 1 & \left(R_Z^{-1}\left[R_Z(E') - a\right] \leq E \leq E'\right) \\ 0 & (E' < E) \end{array} \right\}$$ (14.70)

where A is the area of the sheet and μ is the cosine of the colatitude with respect to the local surface normal. The total number of escaping particles is found by integrating the spectrum given by equation (14.70) and is

$$N_e = Aa\sigma_z \phi \left\{ \begin{array}{ll} \left[1 - \frac{a}{2 R_Z(E')}\right] & (a \leq R_Z(E')) \\ \frac{R_Z(E')}{2a} & (a > R_Z(E'')) \end{array} \right\}$$ (14.71)

From equation (14.71), the total number of ions which stop in the sheet is

$$N_S = Aa\sigma_z \phi \left\{ \begin{array}{ll} \left[\frac{a}{2 R_Z(E')}\right] & (a \leq R_Z(E')) \\ \left[1 - \frac{R_Z(E')}{2a}\right] & (a > R_Z(E')) \end{array} \right\}$$ (14.72)

Obviously, an ion produced with energy E' which leaves the sheet with energy E suffered an energy loss ε to the sheet given by

$$\varepsilon = E' - E \tag{14.73}$$

which we use to find the energy-loss spectrum as

$$\frac{df_{z\delta}}{d\varepsilon} = \left.\frac{df_z}{dE}\right|_{E=E'-\varepsilon} + N_S\,\delta(E' - \varepsilon) \tag{14.74}$$

Considering that equation (14.74) is the energy deposition in a sheet of area A and thickness a as the result of a monoenergetic volumetric source, the response to any arbitrary spectral source can be found by superposition.

14.5.4. Fragmentation energy-loss spectra. The fragmentation-source energy distribution (normalized to unity) is given as

$$\rho(E') = \sqrt{\frac{E'}{2\pi E_0^3}}\,\exp\left(\frac{-E'}{2E_0}\right) \tag{14.75}$$

where $3E_0$ is the mean-fragment energy and is given by Wilson et al. (1989) based on previous work by Goldhaber (1974)

The energy-loss spectrum is found by using equations (14.74) and (14.75) as

$$\frac{dF}{d\varepsilon} = \int_\varepsilon^\infty \left.\frac{df_{z\delta}}{d\varepsilon}\right|_{E=E'-\varepsilon} \rho(E')\,dE'$$

$$= \int_0^\infty \left.\frac{df_{z\delta}}{d\varepsilon}\right|_{E=E'+\varepsilon} \rho(E+\varepsilon)\,dE \tag{14.76}$$

The contribution from stopping ions is readily evaluated to give

$$\frac{dF}{d\varepsilon} = N_S(\varepsilon)\,\rho(\varepsilon) + \int_0^\infty \left.\frac{df_{z\delta}}{d\varepsilon}\right|_{E'=E+\varepsilon} \rho(E+\varepsilon)\,dE \tag{14.77}$$

where the second term of equation (14.77) requires more attention.

The energy-degradation function in the integral of equation (14.77) is given by equation (14.74). It is not clear how the integral in equation (14.77) is to be evaluated. As an approximate evaluation, the energy-degradation function is approximated by two or three line segments as given in equations (14.78) and (14.81).

If $R_Z(\varepsilon) > 2a$, then

$$\frac{df_z}{dE}\bigg|_{E'=E+\varepsilon} = \frac{A\sigma_z\phi}{2S_z(E)} \begin{cases} \frac{a^2}{R_Z^2(\varepsilon)} + \left[\frac{1}{4} - \frac{a^2}{R_Z^2(\varepsilon)}\right] \frac{R_Z(E)}{R_Z(E_2)} & (0 \leq E \leq E_2) \\[3mm] \frac{1}{4} + \frac{3[R_Z(E)-R_Z(E_2)]}{4[R_Z(E_1)-R_Z(E_2)]} & (E_2 \leq E \leq E_1) \\[3mm] 1 & (E_1 \leq E \leq \infty) \end{cases}$$

$$(14.78)$$

where E_2 is the solution of

$$R_Z(E_2) = R_Z(E_2 + \varepsilon) - 2a \qquad (14.79)$$

and E_1 is the solution of

$$R_Z(E_1) = R_Z(E_1 + \varepsilon) - a \qquad (14.80)$$

In the event that $R_Z(\varepsilon) < 2a$, then

$$\frac{df_z}{dE}\bigg|_{E'=E+\varepsilon} = \frac{A\sigma_z\phi}{2S_z(E)} \begin{cases} \frac{a^2}{R_Z^2(\varepsilon)} + \left[1 - \frac{a^2}{R_Z^2(\varepsilon)}\right] \frac{R_Z(E)}{R_Z(E_1)} & (0 \leq E \leq E_1) \\[3mm] 1 & (E_1 + E \leq \infty) \end{cases} \qquad (14.81)$$

with the understanding that E_1 is zero if $R_Z(\varepsilon) < a$. The second term of equation (14.77) is divided into three subintervals as follows:

$$I_1(\varepsilon) = \int_0^{E_2} \frac{df_z}{d\varepsilon}\bigg|_{E'=E+\varepsilon} \rho(E + \varepsilon)\, dE \qquad (14.82)$$

$$I_2(\varepsilon) = \int_{E_2}^{E_1} \frac{df_z}{d\varepsilon}\bigg|_{E'=E+\varepsilon} \rho(E + \varepsilon)\, dE \qquad (14.83)$$

$$I_3(\varepsilon) = \int_{E_1}^{\infty} \frac{df_z}{d\varepsilon}\bigg|_{E'=E+\varepsilon} \rho(E + \varepsilon)\, dE \qquad (14.84)$$

First, $I_1(\varepsilon)$ is zero unless $R_Z(\varepsilon) > 2a$, for which

$$I_1(\varepsilon) = \frac{A\sigma_z\phi}{2} \left\{ \frac{a^2}{R_Z^2(\varepsilon)} P(E_2, \varepsilon) + \left[\frac{1}{4} - \frac{a^2}{R_Z^2(\varepsilon)}\right] \frac{Q(E_2, \varepsilon)}{R_Z(E_2)} \right\} \qquad (14.85)$$

$$I_2(\varepsilon) = \frac{A\sigma_z\phi}{2}\left\{\frac{1}{4} - \frac{3}{4}\frac{R_Z(E_2)}{[R_Z(E_1) - R_Z(E_2)]}\right\}[P(E_1,\varepsilon) - P(E_2,\varepsilon)]$$

$$+ \frac{A\sigma_z\phi}{2}\frac{3}{4}\frac{Q(E_1,\varepsilon) - Q(E_2,\varepsilon)}{[R_Z(E_1) - R_Z(E_2)]} \tag{14.86}$$

$$I_3(\varepsilon) = \frac{A\sigma_z\phi}{2}\int_{E_1}^{\infty}\frac{\rho(E+\varepsilon)}{S_z(E)}\,dE \tag{14.87}$$

If $a \le R_Z(\varepsilon) \le 2a$, then E_2 and $I_1(\varepsilon)$ are zero and

$$I_2(\varepsilon) = \frac{A\sigma_z\phi}{2}\left\{\frac{a^2}{R_Z^2(\varepsilon)}P(E_1,\varepsilon) + \left[1 - \frac{a^2}{R_Z^2(\varepsilon)}\right]\frac{Q(E_1,\varepsilon)}{R_Z(E_1)}\right\} \tag{14.88}$$

When $R_Z(\varepsilon) \le a$, then $E_1 = E_2 = 0$, so that $I_1(\varepsilon)$ and $I_2(E)$ both vanish and

$$I_3(\varepsilon) = \frac{A\sigma_z\phi}{2}\int_{0}^{\infty}\frac{\rho(E+\varepsilon)}{S_z(E)}\,dE \tag{14.89}$$

In equations (14.85), (14.86), and (14.88), P and Q are given by

$$P(E_i,\varepsilon) = \int_{0}^{E_i}\frac{\rho(E+\varepsilon)}{S_z(E)}\,dE \tag{14.90}$$

$$Q(E_i,\varepsilon) = \int_{0}^{E_i}\frac{R_Z(E)\,\rho(E+\varepsilon)}{S_z(E)}\,dE \tag{14.91}$$

The integral of equation (14.90) may be approximated for values of $E_i \le \frac{1}{4}\varepsilon$ by

$$P(E_i,\varepsilon) \approx \frac{R_Z(E_0)}{\sqrt{2}}\rho(\varepsilon)\,\gamma\left(\frac{1}{2},\frac{E_i}{2E_0}\right) \tag{14.92}$$

where γ is an incomplete gamma function. For larger values of E_i $\left(\frac{1}{4}\varepsilon \le E_i \le 4\varepsilon\right)$, the integral may be taken as

$$P(E_i,\varepsilon) \approx \frac{R_Z(E_0)}{\sqrt{2}}\rho(\varepsilon)\left[\frac{1}{2}\gamma\left(\frac{1}{2},\frac{\varepsilon}{8E_0}\right) + \frac{1}{2}\gamma\left(\frac{1}{2},\frac{E_i}{2E_0}\right)\right.$$

$$\left. + \sqrt{\frac{2E_0}{\varepsilon}}\gamma\left(1,\frac{E_i}{2E_0}\right) - \sqrt{\frac{2E_0}{\varepsilon}}\gamma\left(1,\frac{\varepsilon}{8E_0}\right)\right] \tag{14.93}$$

Whenever $E_i > 4\varepsilon$, the integral is approximately

$$P(E_i, \varepsilon) \approx \frac{R_Z(E_0)}{\sqrt{2}} \rho(\varepsilon) \left[\frac{1}{2}\gamma\left(\frac{1}{2}, \frac{\varepsilon}{8E_0}\right) + \frac{1}{2}\gamma\left(\frac{1}{2}, \frac{4\varepsilon}{2E_0}\right) \right.$$

$$\left. + \sqrt{\frac{2E_0}{\varepsilon}}\gamma\left(1, \frac{E_i}{2E_0}\right) - \sqrt{\frac{2E_0}{\varepsilon}}\gamma\left(1, \frac{\varepsilon}{8E_0}\right) \right] \tag{14.94}$$

The integral in equation (14.91) may be approximated by

$$Q(E_i, \varepsilon) \approx \frac{R_Z^2(E_0)}{2E_0} \left[C(E_i + \varepsilon) - C(\varepsilon) \right] \tag{14.95}$$

where $C(\varepsilon)$ is the integral spectrum as follows:

$$C(E) = \int_0^E \rho(E')\, dE' \tag{14.96}$$

A useful check on the approximations involved is the strict requirement

$$I_1(\varepsilon) + I_2(\varepsilon) + I_3(\varepsilon) \leq \frac{A\sigma_z\phi}{2} \int_0^\infty \frac{\rho(E + \varepsilon)}{S_z(E)}\, dE \tag{14.97}$$

The total absorption spectrum is then

$$\frac{dF}{d\varepsilon} = N_S(\varepsilon)\,\rho(\varepsilon) + I_1(\varepsilon) + I_2(\varepsilon) + I_3(\varepsilon) \tag{14.98}$$

and is shown in figure 14.38 for detector thicknesses of 1 to 5 μm with $E_0 = 3.5$ MeV. Similar results are shown in figure 14.39 for detector thicknesses of 50 to 200 μm. In comparing figures 14.39 and 14.40, it is shown that the energy-loss spectrum is approaching the fragment-production spectrum as a becomes larger. The normalization is always

$$\int_0^\infty \frac{dF}{d\varepsilon} = 1 \tag{14.99}$$

which is satisfied by numerical evaluation to within 2 percent.

14.5.5. Results. Typical fragmentation cross sections calculated using the Silberberg-Tsao model are shown in table 14.2 for 125-MeV protons. The values of E_0 are taken from Wilson et al. (1989). The calculated response of the 2.5-μm detector is shown in figure 14.41; these values should be compared with the experiments of McNulty, Farrell, and Tucker (1981) and the values according to the Monte Carlo code of the McNulty group, which are also shown in figure 14.41. The peak value at zero energy is fixed by the total reaction cross section and total proton flux. It appears that the total reaction cross section of the McNulty code is too small. Otherwise, the present theory and the Monte Carlo code show nearly equivalent agreement with the experiments. Similar comments apply to the 4.2-μm detector response (fig. 14.42) with one exception. The energetic events above 20 MeV observed in experiments are well represented by the present theory

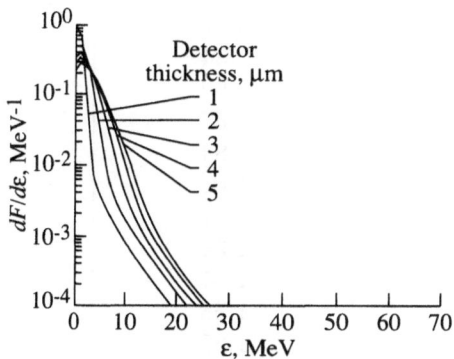

Figure 14.38. Total absorption spectrum for surface-barrier detector of 1 to 5 μm and $E_0 = 3.5$ MeV.

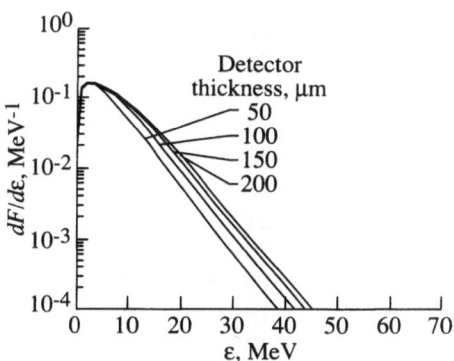

Figure 14.39. Total absorption spectrum for surface-barrier detector of 50 to 200 μm and $E_0 = 3.5$ MeV.

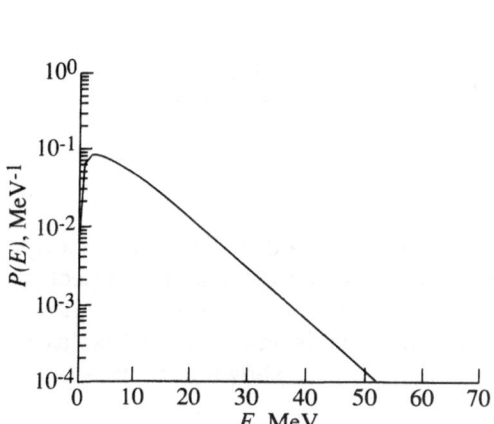

Figure 14.40. Energy-loss spectrum for $E_0 = 3.5$ MeV.

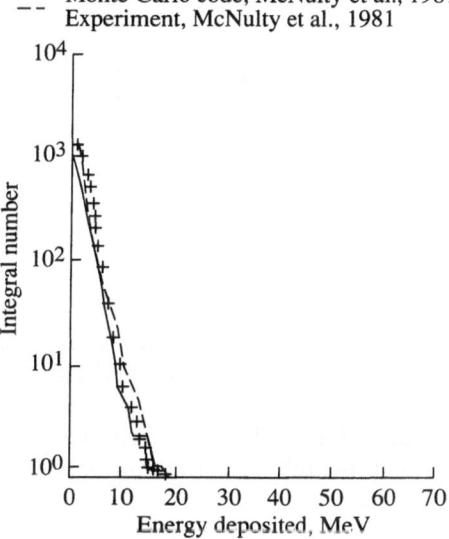

Figure 14.41. Response of a 2.5-μm surface-barrier detector to 125-MeV protons (2.14×10^8 protons).

Table 14.2. Cross-Section Parameters for Fragmentation of
^{28}Si by 125-MeV Protons

A_F	σ_F, mb	E_0, MeV
27	67.7	0.17
26	50.8	.34
25	44.5	.50
24	37.7	.67
23	24.7	.84
22	24.5	1.01
21	14.7	1.17
20	15.3	1.34
19	8.1	1.51
18	7.3	1.68
17	6.4	1.85
16	6.1	2.01
15	4.2	2.18
14	2.9	2.35
13	1.9	2.52
12	2.3	2.68
11	1.5	2.85
10	1.0	3.02
9	1.1	3.19
8	1.7	3.35
7	1.9	3.52
6	1.5	3.69
5	1.2	3.86
4	145.9	2.08
3	29.1	2.92
2	70.7	2.06
1	710.5	2.06

but, as expected, not by the Monte Carlo code (see fig. 14.37). This high-energy agreement between theory and experiment is observed for the 24.1-μm detector, but the Monte Carlo code again fails to predict the high-energy events, as shown in figure 14.43. The improved model of the present work is again clearly displayed for the 158-MeV experiments of McNulty et al. (1981), as shown in figures 14.44 and 14.45.

The inability of the Monte Carlo code to predict the most energetic fragments could be a serious limitation in predicting SEU in some devices. Although the Silberberg-Tsao cross sections for proton-induced reactions are not in complete agreement with some recent cross-section measurements, they still provide improved ability over Monte Carlo models. The methods of analysis used herein will be applied to specific-device geometries in the near future.

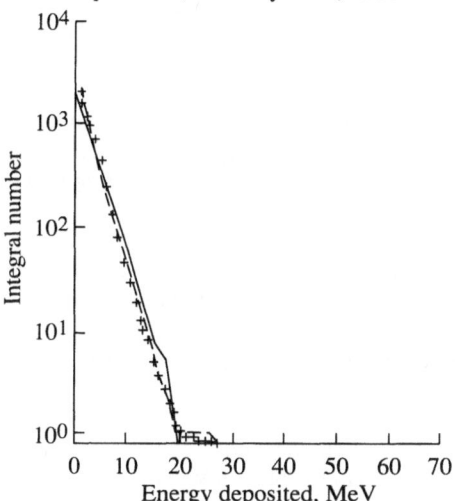

+ Theory: Ngo et al., 1989
— Monte Carlo code, McNulty et al., 1981
-- Experiment, McNulty et al., 1981

Figure 14.42. Response of a 4.2-μm surface-barrier detector to 125-MeV protons (2.14×10^8 protons).

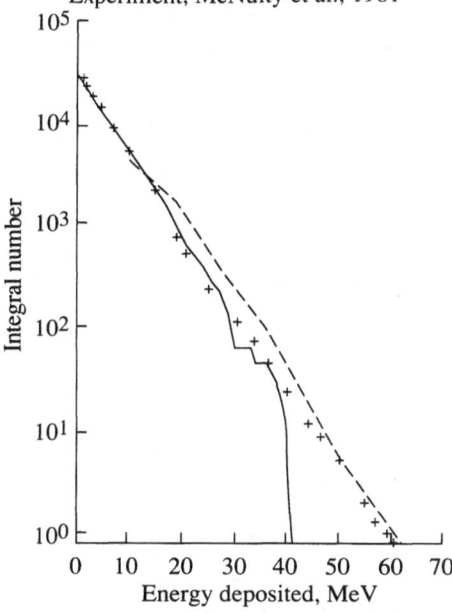

+ Theory: Ngo et al., 1989
— Monte Carlo code, McNulty et al., 1981
-- Experiment, McNulty et al., 1981

Figure 14.43. Response of a 24.1-μm surface-barrier detector to 125-MeV protons (6.42×10^8 protons).

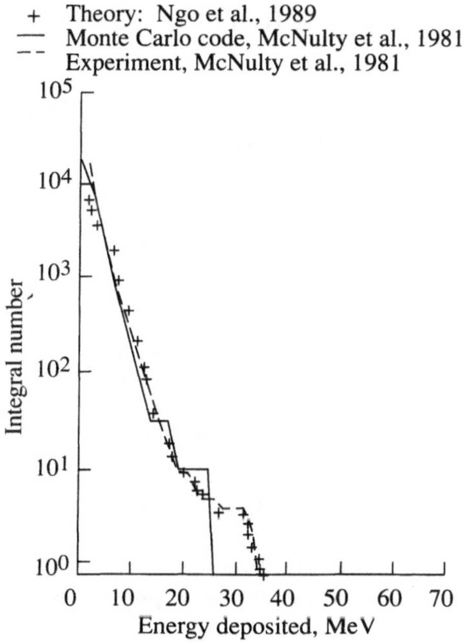

+ Theory: Ngo et al., 1989
— Monte Carlo code, McNulty et al., 1981
-- Experiment, McNulty et al., 1981

Figure 14.44. Response of a 2.5-μm surface-barrier detector to 158-MeV protons (3.9×10^9 protons).

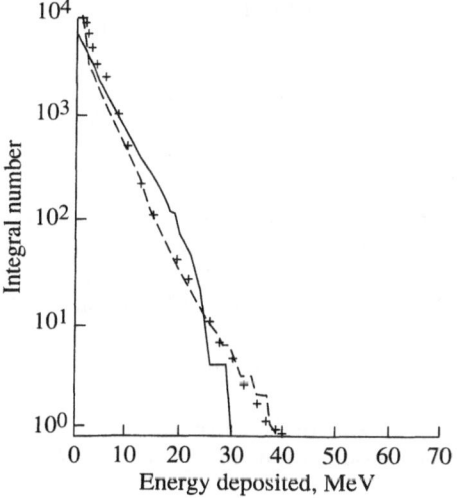

+ Theory: Ngo et al., 1989
— Monte Carlo code, McNulty et al., 1981
-- Experiment, McNulty et al., 1981

Figure 14.45. Response of a 8.7-μm surface-barrier detector to 158-MeV protons (3.9×10^8 protons).

779

14.6. References

Andersen, H. H.; and Ziegler, J. F., 1977: *Hydrogen—Stopping Powers and Ranges in All Elements. Volume 3 of The Stopping and Ranges of Ions in Matter*, J. F. Ziegler, organizer, Pergamon Press, Inc.

Anon., 1977: *Solar Power Satellite Concept Evaluation, Activities Report—July 1976 to June 1977. Volume I—Summary.* NASA TM-74820.

Anspaugh, B. E.; and Downing, R. G., 1981: Damage Coefficients and Thermal Annealing of Irradiated Silicon and GaAs Solar Cells. *The Conference Record of the Fifteenth IEEE Photovoltaic Specialists Conference—1981*, Inst. of Electrical and Electronics Engineers, Inc., pp. 499–505.

Anspaugh, B. E.; and Downing, R. G., 1984: *Radiation Effects in Silicon and Gallium Arsenide Solar Cells Using Isotropic and Normally Incident Radiation.* NASA CR-174007.

Bauerlein, R., 1963: Messung der Energie zur Verlagerung eines Gitteratoms durch Elektronen-stoss in $A^{III} B^V$-Verbindugen. *Z. Phys.*, vol. 176, no. 4, pp. 498–509.

Bertini, Hugo W., 1969: Intranuclear-Cascade Calculation of the Secondary Nucleon Spectra From Nucleon-Nucleus Interactions in the Energy Range 340 to 2900 MeV and Comparisons With Experiment. *Phys. Review*, vol. 188, second ser., no. 4, pp. 1711–1730.

Binder, D.; Smith, E. C.; and Holman, A. B., 1975: Satellite Anomalies From Galactic Cosmic Rays. *IEEE Trans. Nuclear Sci.*, vol. NS-22, no. 6, pp. 2675–2680.

Bisgrove, J. M.; Lynch, J. E.; McNulty, P. J.; Abdel-Kader, W. G.; Kletnieks, V.; and Kolasinski, W. A., 1986: Comparison of Soft Errors Induced by Heavy Ions and Protons. *IEEE Trans. Nuclear Sci.*, vol. NS-33, no. 6, pp. 1571–1576.

Bradford, John N., 1982: Microvolume Energy Deposition From High Energy Proton-Silicon Reactions. *IEEE Trans. Nuclear Sci.*, vol. NS-29, no. 6, pp. 2085–2089.

Chern, J. H.; Seitchik, J. A.; and Yang, P., 1986: Single Event Charge Collection Modeling in CMOS Multi-junctions Structure. *1986 International Electron Devices Meeting—Technical Digest*, IEEE Catalog No. 86CH2381-2, Inst. of Electrical and Electronics Engineers, Inc., pp. 538–541.

Conway, E. J.; Walker, G. H.; and Heinbockel, J. H., 1981: A Thermochemical Model of Radiation Damage and Annealing Applied to GaAs Solar Cells. *The Conference Record of the Fifteenth IEEE Photovoltaic Specialists Conference—1981*, IEEE Catalog No. 81CH1644-4, Inst. of Electrical and Electronics Engineers, Inc., pp. 38–44.

Dienes, G. J.; and Vineyard, G. H., 1957: *Radiation Effects in Solids.* Interscience Publ., Inc.

Fernald, Kenneth W.; and Kerns, Sherra E., 1988: Simulation of Proton-Induced Energy Deposition in Integrated Circuits. *IEEE Trans. Nuclear Sci.*, vol. 35, no. 1, pp. 981–986.

Foelsche, T., 1963: Specific Solar Flare Events and Associated Radiation Doses. *Space Radiation Effects*, ASTM Special Tech. Publ. No. 363, American Soc. for Testing & Materials, pp. 1–13.

Goldhaber, A. S., 1974: Statistical Models of Fragmentation Processes. *Phys. Lett.*, vol. 53B, no. 4. pp. 306–308.

Guenzer, C. S.; Allen, R. G.; Campbell, A. B.; Kidd, J. M.; Petersen, E. L.; Seeman, N.; and Wolicki, E. A., 1980: Single Event Upsets in RAMs Induced by Protons at 4.2 GeV and Protons and Neutrons Below 100 MeV. *IEEE Trans. Nuclear Sci.*, vol. NS-27, no. 6 pp. 1485–1489.

Hamm, R. N.; Rustgi, M. L.; Wright, H. A.; and Turner, J. E., 1981: Energy Spectra of Heavy Fragments From the Interaction of Protons With Communications Materials. *IEEE Trans. Nuclear Sci.*, vol. NS-28, no. 6, pp. 4004–4006.

Heinbockel, J. H.; Conway, E. J.; and Walker, G. H., 1980: Simultaneous Radiation Damage and Annealing of GaAs Solar Cells. *Fourteenth IEEE Photovoltaic Specialists Conference— 1980*, Inst. of Electrical and Electronics Engineers, Inc., pp. 1085–1089.

Hovel, Harold J., 1975: *Semiconductors and Semimetals. Volume 11—Solar Cells.* Academic Press. Inc.

Janni, Joseph, F., 1966: *Calculations of Energy Loss, Range, Pathlength, Straggling, Multiple Scattering, and the Probability of Inelastic Nuclear Collisions for 0.1- to 1000-MeV Protons.* AFWL-TR-65-150, U.S. Air Force. (Available from DTIC as AD 643 837.)

Kamath, G. S., 1981: GaAs Solar Cells for Space Application. *Proceedings of the 16th Intersociety Energy Conversion Engineering Conference. Volume I—Technologies for the Transition*, American Soc. of Mechanical Engineers, pp. 416–421.

Kwiatkowski, K.; Zhou, S. H.; Ward, T. E.; Viola, V. E., Jr.; Brever, H.; Mathews, G. J.; Gökmen, A.; and Mignerey, A. C., 1983: Energy Deposition in Intermediate-Energy Nucleon-Nucleus Collisions. *Phys. Review Lett.*, vol. 50, no. 21, pp. 1648–1651.

Li, S. S.; Chiu, T. T.; Schoenfeld, D. W.; and Loo, R. Y., 1981: Defect Characterization and Thermal Annealing Study of 200 KeV Proton Irradiated n-GaAs LPE Layers. *Defects and Radiation Effects in Semiconductors, 1980*, R. R. Hasiguti, ed., Inst. of Physics, pp. 335–340.

Lindhard, J.; Scharff, M.; and Schiott, H. E., 1963: Range Concepts and Heavy Ion Ranges (Notes on Atomic Collisions, II). *Mat.-Fys. Medd.—K. Dan. Vidensk. Selsk.*, vol. 33, no. 14, pp. 1–42.

Liverhant. S. E., 1960: *Nuclear Reactor Physics.* John Wiley & Sons, Inc.

Loo, R.; Goldhammer, L.; Anspaugh, B.; Knechtli, R. C.; and Kamath, G. S., 1978: Electron and Proton Degradation in (AlGa)As-GaAs Solar Cells. *Thirteenth IEEE Photovoltaic Specialists Conference—1978*, Inst. of Electrical and Electronics Engineers, Inc., pp. 562–570.

Loo, R.; Knechtli, R. C.; and Kamath, G. S., 1981: Enhanced Annealing of GaAs Solar Cell Radiation Damage. *The Conference Record of the Fifteenth IEEE Photovoltaic Specialists Conference—1981*, IEEE Catalog No. 81-CH1644-04, Inst. of Electrical and Electronics Engineers, Inc., pp. 33–37.

McKinley, William A., Jr.; and Feshbach, Herman, 1948: The Coulomb Scattering of Relativistic Electrons by Nuclei. *Phys. Review*, vol. 74, second ser., no. 12, pp. 1759–1763.

McNulty, Peter J.; Farrell, Gary E.; and Tucker, William P., 1981: Proton-Induced Nuclear Reactions in Silicon. *IEEE Trans. Nuclear Sci.*, vol. NS-28, no. 6, pp. 4007–4012.

McNulty, P. J.; Farrell, G. E.; Wyatt, R. C.; Rothwell, P. L.; Filz, R. C.; and Bradford, J. N., 1980: Upset Phenomena Induced by Energetic Protons and Electrons. *IEEE Trans. Nuclear Sci.*, vol. NS-27, no. 6, pp. 1516–1522.

Ngo, Duc M.; Wilson, John W.; Buck, Warren W.; and Fogarty, Thomas N., 1989: *Nuclear-Fragmentation Studies for Microelectronic Applications*. NASA TM-4143.

Pages, Lucien; Bertel, Evelyne; Joffre, Henri; and Sklavenitis, Laodamas, 1972: Energy Loss, Range, and Bremsstrahlung Yield for 10-keV to 100-MeV Electrons in Various Elements and Chemical Compounds. *At. Data*, vol. 4, no. 1, pp. 1–127.

Petersen, E. L., 1980: Nuclear Reactions in Semiconductors. *IEEE Trans. Nuclear Sci.*, vol. NS-27, no. 6, pp. 1494–1499.

Petersen, E. L.; Shapiro, P.; Adams, J. H., Jr.; and Burke, E. A., 1982: Calculation of Cosmic-Ray Induced Soft Upsets and Scaling of VLSI Devices. *IEEE Trans. Nuclear Sci.*, vol. NS-29, no. 6, pp. 2055–2063.

Peterson, N. L.; and Harkness, S. D., eds., 1976: *Radiation Damage in Metals*. American Soc. for Metals.

Robinson, Mark T.; and Torrens, Ian M., 1974: Computer Simulation of Atomic-Displacement Cascades in Solids in the Binary-Collision Approximation. *Phys. Review B*, vol. 9, no. 12, pp. 5008–5024.

Rudstam, G., 1966: Systematics of Spallation Yields. *Zeitschrift fur Naturforschung*, vol. 21a, no. 7, pp. 1027–1041.

Sawyer, Donald M.; and Vette, James I., 1976: *AP-8 Trapped Proton Environment for Solar Maximum and Solar Minimum*. NASA TM X-72605.

Serber, R.; 1947: Nuclear Reactions at High Energies. *Phys. Review*, vol. 72, no. 11, pp. 1114–1115.

Singley, G. Wayne; and Vette, James I., 1972: *The AE-4 Model of the Outer Radiation Zone Electron Environment*. NSSDC 72-06, NASA Goddard Space Flight Center.

Stith, John J.; and Wilson, John W., 1985: Microscopic Defect Structures and Equivalent Electron Fluence Concepts. *Eighteenth IEEE Photovoltaic Specialists Conference—1985*, IEEE Catalog No. 85CH2208-7, Inst. of Electrical and Electronics Engineers, Inc., pp. 1716–1717.

Tada, H. Y.; and Carter, J. R., Jr., 1977: *Solar Cell Radiation Handbook*. JPL Publ. 77-56 (Contract NAS7-100), California Inst. of Technology. (Available as NASA CR-155554.)

Tsao, C. H.; Silberberg, R.; Adams, J. H., Jr.; and Letaw, J. R., 1983: Cosmic Ray Transport in the Atmosphere—Dose and LET-Distributions in Materials. *IEEE Trans. Nuclear Sci.*, vol. NS-30, pp. 4398–4404.

Vook, Frederick L., ed., 1968: *Radiation Effects in Semiconductors*. Plenum Press.

Walker, D. H.; Statler, R. L.; and Lambert, R. J., 1978: Solar Cell Experiments on the NTS-2 Satellite. *The Conference Record of the Thirteenth IEEE Photovoltaic Specialists Conference—1978*, IEEE Catalog No. 78CH1319-3, Inst. of Electrical and Electronics Engineers, Inc., pp. 100–106.

Walker, Gilbert H.; and Conway, Edmund J., 1978a: Recovery of Shallow Junction GaAs Solar Cells Damaged by Electron Irradiation. *J. Electrochem. Soc.*, vol. 125, no. 10, pp. 1726–1727.

Walker, Gilbert H.; and Conway, Edmund J., 1978b: Short Circuit Current Changes in Electron Irradiated GaAℓAs/GaAs Solar Cells. *The Conference Record of the Thirteenth IEEE Photovoltaic Specialists Conference—1978*, Inst. of Electrical and Electronics Engineers, Inc., pp. 575–579.

Wilson, John W., 1977: *Analysis of the Theory of High-Energy Ion Transport.* NASA TN D-8381.

Wilson, John W.; and Lamkin, Stanley L., 1975: Perturbation Theory for Charged-Particle Transport in One Dimension. *Nuclear Sci. & Eng.*, vol. 57, no. 4, pp. 292–299.

Wilson, John W.; Stith, John J.; and Stock, L. V., 1983: *A Simple Model of Space Radiation Damage in GaAs Solar Cells.* NASA TP-2242.

Wilson, John W.; Stith, John J.; and Walker, Gilbert H., 1982: On the Validity of Equivalent Electron Fluence for GaAs Solar Cells. *Sixteenth IEEE Photovoltaic Specialists Conference—1982*, Inst. of Electrical and Electronics Engineers, Inc., pp. 1439–1440.

Wilson, John W.; and Stock, L. V., 1984: Equivalent Electron Fluence for Space Qualification of Shallow Junction Heteroface GaAs Solar Cells. *IEEE Trans. Electron Devices*, vol. ED-31, no. 5, pp. 622–625.

Wilson, John W.; Townsend, L. W.; Chun, S. Y.; and Buck, W. W., 1988: High Energy Nucleon Data Bases. *Health Phys.*, vol. 55, no. 5, pp. 817–819.

Wilson, John W.; Townsend, Lawrence W.; Nealy, John E.; Chun, Sang Y.; Hong, B. S.; Buck, Warren W.; Lamkin, S. L.; Ganapol, Barry D.; Khan, Ferdous; and Cucinotta, Francis A., 1989: BRYNTRN: *A Baryon Transport Model.* NASA TP-2887.

Wilson, John W.; Walker, Gilbert H.; and Outlaw, R. A., 1984: Proton Damage in GaAs Solar Cells. *IEEE Trans. Electron Devices*, vol. ED-31, no. 4, pp. 421-422.

Wilson, John W.; and Walker, Gilbert H.; Outlaw, R. A.; and Stock, L. V., 1982: A Model for Proton-Irradiated GaAs Solar Cells. *Sixteenth IEEE Photovoltaic Specialists Conference—1982*, Inst. of Electrical and Electronics Engineers, Inc., pp. 1441–1442.

Wyatt, R. C.; McNulty, P. J.; Toumbas, P.; Rothwell, P. L.; and Filz, R. C., 1979: Soft Errors Induced by Energetic Protons. *IEEE Trans. Nuclear Sci.*, vol. NS-26, no. 26, pp. 4905–4910.

Yaung, J. Y., 1984: Model of Solar Cell Proton Damage. *Space Photovoltaic Research and Technology—1983, High Efficiency, Radiation Damage, and Blanket Technology*, NASA CP-2314, pp. 56–62.

Yeh, C. S.; Li, S. S.; and Loo, R. Y., 1985: A Simple Model for Calculating the Displacement Damage in Electron and Proton Irradiated AlGaAs/GaAs/InGaAs (or Ge) Multijunction Solar Cells. *Eighteenth IEEE Photovoltaic Specialists Conference—1985*, IEEE Catalog No. 85CH2208-7, Inst. of Electrical and Electronics Engineers, Inc., pp. 657–662.

15. CONCLUDING REMARKS

15.1. Current Status

The final goal of the present research program is to provide the design engineer with analysis tools to adequately design future NASA space structures and to assure that acceptable risks are not exceeded. To validate code accuracy, we require these analysis tools to be compared with well-controlled experiments. Although we have made great progress toward this goal, still many difficult tasks remain before this goal is achieved.

The first step in progressing toward this goal is the development of transport codes and data bases for HZE and nucleonic components in the straight ahead approximation. An even more restrictive assumption than the straight ahead approximation is applied to the HZE fragments by assuming that the velocity is conserved in the interaction. The current code versions either apply in space (HZETRN, BRYNTRN) or in the laboratory (LABTRN, LBLTRN) exclusively. Hence, a code for space which can be validated in laboratory experiments is beyond our present capability. Even then the HZE cross sections are assumed to be energy independent for the space code HZETRN and the most general laboratory code LBLTRN. Although the laboratory code LABTRN does treat energy-dependent cross sections, it only allows evaluation of the absorbed dose within an absorber. Three generalizations of these codes are required: (1) the straight ahead approximation should be replaced with a two-stream approximation, (2) the full energy dependence of the nuclear cross sections should be added, and (3) the spectral components of the HZE fragmentation should be introduced. Even these additions to the current codes will not provide a complete description of the transport process. Such a complete description requires the introduction of mesons, antibaryons, and their decay, and reaction products, especially the electromagnetic cascades. The incompleteness of the present codes results in part from the data bases utilized. The generation of such a data base is in progress and our next immediate goal is to have a complete set of one-dimensional codes.

15.2. Future Goals

The first goal beyond the present work is to develop a complete set of one-dimensional codes that are able to evaluate biological response in an arbitrary shield geometry for engineering applications. After this initial goal has been accomplished, we will move onward toward fully three-dimensional codes by first generating a new data base for atomic/molecular collisions as well as a more complete nuclear data base. Such three-dimensional codes are particularly important for validation in laboratory experiments. These validated three-dimensional codes will provide the codes for future space engineering design.

ACKNOWLEDGMENTS

First I want to thank my wife, Delores Gelino Wilson, and son, John-Paul Wilson, for their patience during those periods when much of this work was being done at home and during the writing of the report. Also, I want to pay tribute to my mentor, teacher, and friend, Dr. Trutz Foelsche (deceased). The Langley Management deserves credit for encouraging this work both "in season" and "out of season." Special credit goes to Frank Hohl (deceased) and George P. Wood who rescued it; to Edmund J. Conway who encouraged it, even out of season; and to William M. Piland who helped restart it. Their foresight has at long last been vindicated. Thanks to the many friends and students who contributed to the work over these many years and are too numerous to mention, but whose names appear on many references cited herein, especially S. L. Lamkin, F. F. Badavi, C. M. Costner, D. M. Ngo, L. V. Stock, and S. Y. Chun. A special thank you to Professor Carl Carlson at the College of William and Mary in Virginia, Williamsburg, Virginia, for encouragement during a very difficult period of this research. Many thanks to the staff of the Langley Technical Editing Branch and the Langley Publications Section for their heroic efforts to complete this document for its intended use at the NATO Advanced Study Institute held in Amerção de Pera, Portugal. We acknowledge Dr. Frank Sultzman of the Life Sciences Division, NASA Office of Space Science and Applications, for his constant support of this effort. This work has been supported by the Life Sciences Division, NASA Office of Space Science and Applications; NASA Office of Aeronautics and Space Technology; Air Surgeon Generals Office, Federal Aviation Administration; the National Cancer Institute; and the Physics Department, Old Dominion University, Norfolk, Virginia.

HZE REACTIONS AND DATA-BASE DEVELOPMENT

Lawrence W. Townsend, Francis A. Cucinotta, and John W. Wilson

NASA Langley Research Center
Hampton, VA, U.S.A.

ABSTRACT

The primary cosmic rays are dispersed over a large range of linear energy transfer (LET) values and their distribution over LET is a determinant of biological response. This LET distribution is modified by radiation shielding thickness and shield material composition. The current uncertainties in nuclear cross sections will not allow the composition of the shield material to be distinguished in order to minimize biological risk. An overview of the development of quantum mechanical models of heavy ion reactions will be given and computational results compared with experiments. A second approach is the development of phenomenological models from semi-classical considerations. These models provide the current data base in high charge and energy (HZE) shielding studies. They will be compared with available experimental data. The background material for this lecture will be available as a review document of over 30 years of research at Langley but will include new results obtained over the last year.

INTRODUCTION

Whenever galactic cosmic rays (GCR) penetrate a spacecraft, their radiation fields change composition because of interactions with structural materials in their paths. These altered radiation fields, which depend upon target material geometry, thickness, and composition, are described by transport models which relate the transmitted fluxes to the incident fields. The main interaction processes involved in the transport of these radiation fields are ionization energy losses through collisions with atomic electrons, nuclear elastic and inelastic collisions, and nuclear fragmentation (breakup) and electromagnetic dissociation reactions. These latter processes are important because fragmentations result in the production of reaction products which alter the elemental and isotopic compositions of the transported radiation fields.

Propagation of these radiation fields is described by a Boltzmann equation derivable from mass and energy conservation. Its solutions yield particle fluxes and energies everywhere within and exiting the boundaries of the target material. For the HZE (high energy heavy ion) component of the galactic cosmic ray spectrum, the typically large ion kinetic energies allow changes in particle direction to be neglected and result in what is called the "straightahead approximation." The resulting one-dimensional Boltzmann equation is written

$$\left[\frac{\partial}{\partial x} + \sigma_i(E) - \frac{\partial}{\partial E}S_i(E)\right]\phi_i(x, E) = \sum_j \sigma_{ij}(E)\phi_j(x, E) \tag{1}$$

Biological Effects and Physics of Solar and Galactic Cosmic Radiation,
Part B, Edited by C.E. Swenberg *et al.*, Plenum Press, New York, 1993

In eq. (1), ϕ_i is the flux of type i ions at position x with motion along the x axis and energy E in units of A MeV, σ_i is the corresponding macroscopic nuclear absorption cross section in units of cm^{-1}, S_i is the change in E per unit distance (e.g., the stopping power per unit projectile mass), and σ_{ij} is the cross section, in units of cm^{-1}, for producing ion i from a collision by ion j. Accurate solutions to eq. (1) require accurate values for the nuclear absorption (reaction) and nuclear fragmentation (σ_{ij}) cross sections. Unfortunately, the experimental data bases for these cross sections are very sparse. Because of the large number of projectile ion-target ion combinations, the complexity of their reaction products, and the wide range of energies involved, it seems unlikely that the required cross section data bases will ever be obtained from experiments alone. Therefore, theoretical and semiempirical methods, validated and guided by experimental data, must be developed to provide the needed cross sections.

To illustrate the importance of these nuclear reaction processes, estimates of dose equivalent contributions to bone marrow from projectile ion fragments (secondaries and subsequent-generation fragmentation products) and from target nuclear fragments and recoils have been separately computed for several thicknesses of spacecraft aluminum shielding (Townsend, Cucinotta, and Wilson, 1991). These estimates were obtained using the Langley Research Center GCR transport computer code HZETRN (Wilson, Townsend, and Badavi, 1987a; Townsend, et al., 1991) and the computerized anatomical man (CAM) model of the human geometry (Billings and Yucker, 1973; Atwell, 1990). The input GCR spectra are the solar minimum spectra from the Naval Research Laboratory environmental model (Adams, 1986). The results, given in terms of percent contribution to the bone marrow annual total dose equivalent, are displayed in fig. 1. Note that projectile and target fragments (combined) contribute over half of the total dose equivalent behind a thick shield (30 g/cm^2 aluminum). Note also, for all shield thicknesses considered, that the projectile fragment contributions are a significantly greater percentage of the total than are the target fragment contributions.

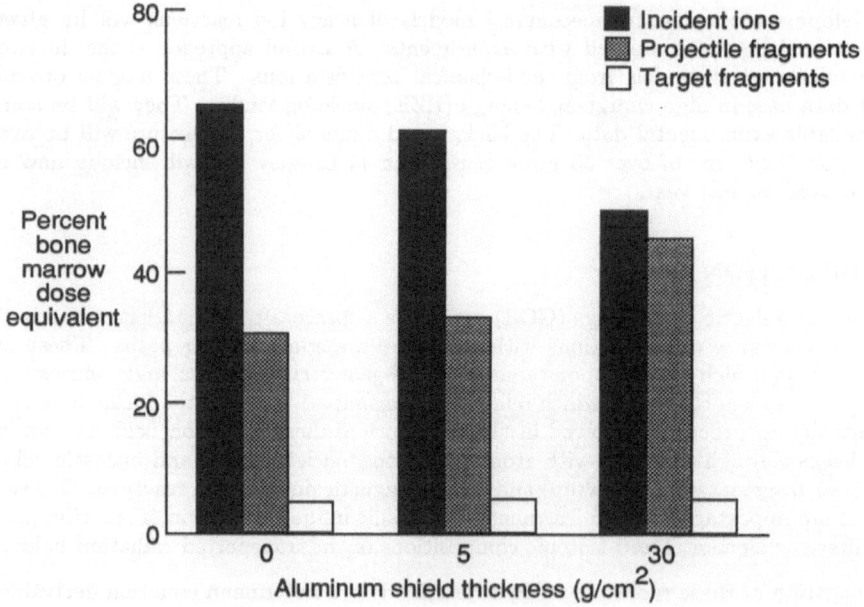

Fig. 1. Percentages of bone marrow dose equivalent resulting from incident galactic cosmic rays, projectile fragments and target fragments for several aluminum shield thickness.

Because of the paucity of experimental measurements of fragmentation cross sections for HZE particles colliding with spacecraft materials (e.g. aluminum), dose and dose equivalent uncertainties resulting from the use of particular nuclear fragmentation models cannot be established at this time. It is possible, however, to determine upper and lower bounds on absorbed dose and dose equivalent by investigating the physically limiting cases of the fragmentation events themselves. The minimum fragmentation event occurs whenever the collision removes only a single nucleon (neutron or proton) from the fragmenting nucleus. This we denote as a "high-LET" case because each fragmentation produces a heavy fragment which is nearly the same mass as the original nucleus. The maximum fragmentation event occurs whenever the collision causes the fragmenting nucleus to completely disintegrate into its constituent nucleons. This we denote as a "low-LET" case because there are no heavy fragments resulting from the breakup of the original nucleus. From a physical perspective, every other possible fragmentation event must be between these two extreme cases. Restricting every fragmentation event in the transport code to a high-LET one yields an upper bound on the estimated dose and dose equivalent. Conversely, limiting every fragmentation event to the low-LET case yields a lower bound. Figure 2 displays these bounds for the 0-cm depth dose equivalent in water, as a function of aluminum shield thickness. For comparison purposes, nominal values of dose equivalent using the NUCFRAG semiempirical fragmentation model (Wilson, Townsend, and Badavi, 1987b) are also displayed. For aluminum shield thickness between 20 g/cm^2 and 50 g/cm^2, differences between the upper and lower bounds exceed 50 percent. Bounds on the LET distributions are shown in fig. 3 where differences exceeding a factor of two are found for LET values greater than 10 keV/micron. Clearly, accurate laboratory-validated nuclear fragmentation models are necessary for precise evaluations of HZE particle shielding and dosimetry requirements.

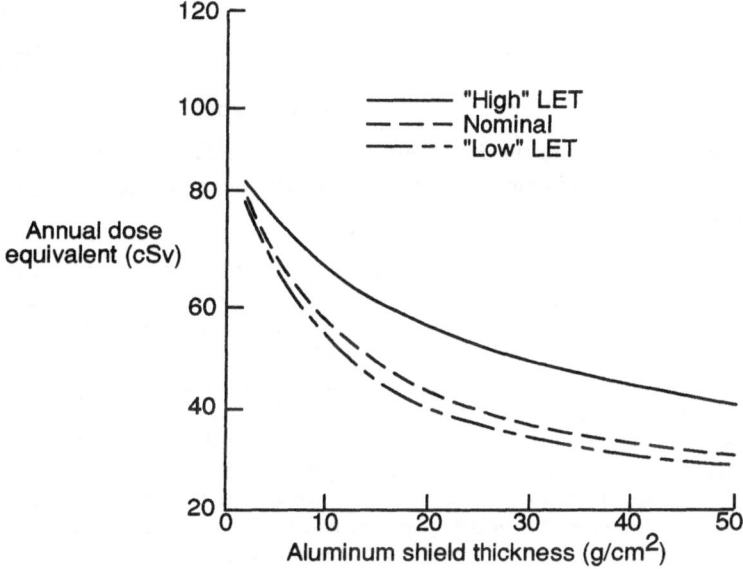

Fig. 2. Maximum variations in solar minimum galactic cosmic ray dose equivalent resulting from nuclear fragmentation uncertainties.

In the following sections, an overview of the Langley Research Center programs encompassing HZE reaction theory and data base development will be presented. Quantum mechanical optical model methods for estimating HZE reaction (absorption) cross sections will be described and representative results compared with experimental data and compared with energy-dependent cross section parameterizations developed for use directly in the transport codes. The main advantages of quantum mechanical models are their accuracy and generality. Excellent agreement between theory and experiment is obtained for any combination of projectile and target nuclei over all energies of interest for HZE particle shielding. The principal disadvantages are the complexity of the calculations and the extensive computer storage requirements if cross sections are stored in tabular form. As an alternative to these detailed quantum mechanical methods, energy-dependent parameterizations of nucleon-nucleus (Wilson et al., 1989) and nucleus-nucleus (Townsend and Wilson, 1986) absorption cross sections have been developed. These parameterizations typically agree to within approximately 10 percent of the detailed quantum mechanical calculations.

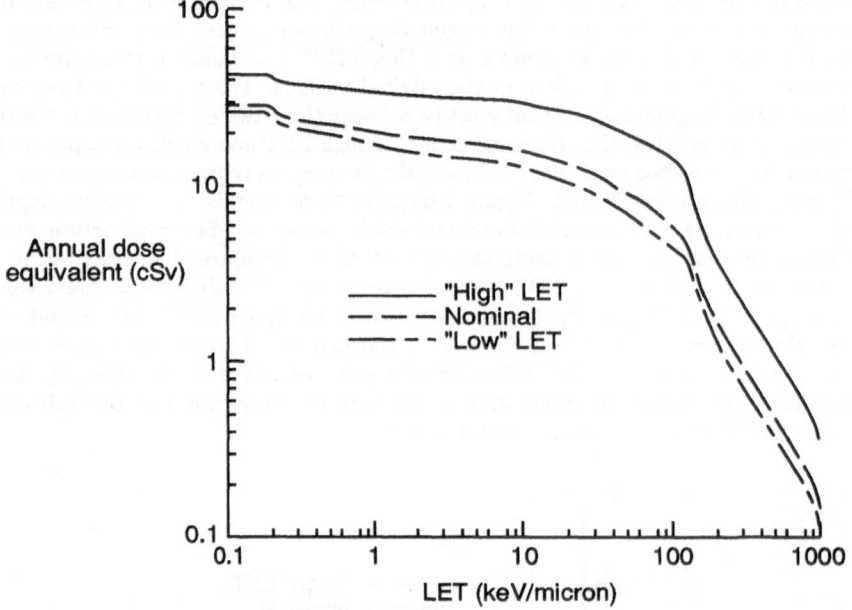

Fig. 3. Maximum variations in the linear energy transfer (LET) distribution due to nuclear fragmentation for solar minimum galactic cosmic rays after passage through 30g/cm^2 of aluminum shielding.

Unlike absorptive cross sections, the physics used to describe nuclear fragmentation is not as well understood, and the theoretical cross section estimates are not as accurate when compared with experimental measurements. Much of the sparse experimental data base is oriented toward studies of basic nuclear reaction mechanisms and not suitable for determining fragment production cross sections. Competing processes, such as fission and electromagnetic dissociation, further complicate the situation by contributing significantly to breakup cross section estimates for heavy nuclei and/or for high energies. Although fundamental theories based upon quantum mechanics are under development (Hufner, Schafer, and Schurmann, 1975; Townsend, 1983; Townsend, et al., 1986; Townsend, Norbury, and Khan, 1991), the calculations are too complex to repeatedly perform within a transport code. Details of these models and their shortcomings will be reviewed in subsequent sections of this work.

Recently, Langley has developed a semiempirical abrasion-ablation fragmentation model for use with the transport codes. In this work a description of the current version of the model will be presented. Comparisons of the theoretical fragmentation cross sections (optical and semiempirical models) with experimental data will also be presented. We will then close with a brief discussion of current research needs and what is being done to meet them.

NUCLEAR ABSORPTION CROSS SECTIONS

In terms of the microscopic nuclear absorption cross section $\sigma_{abs}(E, i, j)$, the macroscopic nuclear absorption cross section is given by

$$\sigma_i(E) = \sum_j \rho_j \sigma_{abs}(E, i, j) \tag{2}$$

where the ρ_j are the elemental constituent-number densities for the target, and (i, j) refers to the ith projectile nucleus —jth target nucleus collision pair.

Quantum Mechanical Model

During the decade of the 1970s and into the early 1980s, Langley developed a quantum mechanical optical model formalism (Wilson and Costner, 1975; Townsend and Wilson, 1985) based upon a microscopic optical potential approximation to the exact nucleus-nucleus multiple scattering series (Wilson, 1974; Wilson and Townsend, 1981). The methods use no arbitrarily-adjusted parameters and are applicable to any projectile nucleus colliding with any target nucleus at energies greater than 25 A MeV. Typically, when compared with experimental cross section measurements, the theoretical predictions are accurate to within 3 percent for energies greater than 80 A MeV and within 10 percent for energies as low as 30 A MeV. From eikonal scattering theory, the collision absorption cross sections are given by

$$\sigma_{abs} = 2\pi \int_0^\infty \left\{ 1 - \exp[-2\mathrm{Im}\chi(\vec{b})] \right\} b \, db \tag{3}$$

where the complex phase function, in terms of the reduced potential U, is

$$\chi(\vec{b}) = -\frac{1}{2k} \int_{-\infty}^\infty U(\vec{b}, z) dz \tag{4}$$

For composite particle scattering, the reduced potential is written as

$$U(\vec{x}) = 2mA_P A_T (A_P + A_T)^{-1} W(\vec{x}) \tag{5}$$

where m is the nucleon mass, A_P is the nuclear mass number of the projectile, and A_T is the nuclear mass number of the target. The nucleus-nucleus potential including Pauli correlation effects is (Townsend, 1982)

$$W(\vec{x}) = A_P A_T \int d^3\xi_T \, \rho_T(\vec{\xi}_T) \int d^3 y \rho_P(\vec{x} + \vec{y} + \vec{\xi}_T)$$
$$\times \, \tilde{t}(e, \vec{y})[1 - \tilde{C}(\vec{y})] \tag{6}$$

This potential was derived from an optical model potential approximation to the exact composite-particle multiple-scattering series and does not inherently depend on the eikonal approximation, although we are using it in that context. Other symbols in eqs. (3)–(6) are the impact parameter b, the projectile momentum k, and the nuclear densities of the colliding nuclei $\rho_i (i = P, T)$.

In equation (6), $\tilde{t}$ is the constituent-averaged energy-dependent two-body transition amplitude

$$\tilde{t}(e, \vec{y}) = \left(\frac{e}{m}\right)^{1/2} \sigma(e)[\alpha(e) + i][2\pi B(e)]^{-3/2} \exp\left[\frac{-y^2}{2B(e)}\right] \tag{7}$$

and the correlation function is taken to be

$$\tilde{C}(\vec{y}) = 0.25 \exp\left(\frac{-k_F^2 y^2}{10}\right) \tag{8}$$

For the analyses of this work, the Fermi momentum is assumed to be that of infinite matter, $k_F = 1.36 \text{ fm}^{-1}$.

The correct nuclear density distributions $\rho_j (j = P, T)$ to use in equation (6) are the nuclear ground state, single-particle number densities for the collision pair. Since these are not experimentally known, the number densities are obtained from their experimental charge density distributions by assuming that

$$\rho_c(\vec{r}) = \int \rho_p(\vec{r}')\rho_A(\vec{r} + \vec{r}')d^3r' \tag{9}$$

791

where ρ_c is the nuclear charge distribution, ρ_p is the proton charge distribution, and ρ_A is the desired nuclear single-particle density. All density distributions in equation (9) are normalized to unity. The proton charge distribution is taken to be the usual Gaussian form

$$\rho_p(\vec{r}) = \left(\frac{3}{2\pi r_p^2}\right)^{3/2} \exp\left(\frac{-3r^2}{2r_p^2}\right) \tag{10}$$

where $r_p = 0.87$ fm is the proton root-mean-square charge radius (Borkowski, et al., 1975).

When the projectile is a nucleon, equation (9) yields a delta function for ρ_A:

$$\rho_A(\vec{r} + \vec{r}') = \delta(\vec{r} + \vec{r}') \tag{11}$$

since ρ_c and ρ_p are identical.

For nuclei-lighter than neon ($A < 20$), the nuclear charge distribution is the harmonic well (HW) form given as (DeJager, DeVries, DeVries, 1974)

$$\rho_c(\mathbf{r}) = \rho_0 \left[1 + \gamma \left(\frac{r}{a}\right)^2\right] \exp\left(\frac{-r^2}{a^2}\right) \tag{12}$$

where ρ_0 is the normalization constant, r is the radial coordinate, and a and γ are charge parameters. Values for a and γ used herein are listed in table 1 of Townsend and Wilson (1985). Substituting equations (10) and (12) into equation (9) yields

$$\rho_A(\vec{r}) = \frac{\rho_0 a^3}{8s^3} \left(1 + \frac{3\gamma}{2} - \frac{3\gamma^2}{8s^2} + \frac{\gamma a^2 r^2}{16s^4}\right) \exp\left(\frac{-r^2}{4s^2}\right) \tag{13}$$

where

$$s^2 = \frac{a^2}{4} - \frac{r_p^2}{6} \tag{14}$$

For neon and heavier nuclei ($A > 20$), the nuclear charge distribution is taken to be the Woods-Saxon (WS) form given by

$$\rho_c(\vec{r}) = \frac{\rho_0}{1 + \exp[(r - R)/c]} \tag{15}$$

where R is the radius at half-density, and the surface diffuseness c is related to the nuclear skin thickness t_c through

$$c = \frac{t_c}{4.4} \tag{16}$$

Values for R and t_c used herein are listed in table 1 of Townsend and Wilson (1985). Inserting equations (10) and (15) into equation (9) yields, after some simplification, a number density ρ_A that is of the WS form (see eq. (15)) with the same R, but different overall normalization factor ρ_0 and surface thickness. The latter is given by

$$t_A = \frac{8.8 r_p}{3^{1/2}} \left[\ln\left(\frac{3\beta - 1}{3 - \beta}\right)\right]^{-1} \tag{17}$$

where

$$\beta = \exp\left(\frac{4.4 r_p}{t_c 3^{1/2}}\right) \tag{18}$$

with t_c denoting the charge skin thickness obtained by using equation (16) and the charge distribution surface diffuseness values listed in DeJager, DeVries and DeVries (1974).

The nucleon-nucleon cross sections, $\sigma(e)$, are obtained by performing a spline interpolation of values taken from various compilations (Benary, Price, and Alexander, 1970; Schopper, 1973, 1980).

Since scattering at these energies is mainly diffractive, the nucleon-nucleon slope parameters, $B(e)$, are those appropriate to purely diffractive scattering. They are given by (Ringia, et al., 1972)

$$B(e) = 10 + 0.5 \ln\left(\frac{s'}{s_0}\right) \tag{19}$$

where s' is the square of the nucleon-nucleon center-of-mass energy and $s_0 = 1\ (\text{GeV c}^{-1})^{-2}$. Values for the parameter $\alpha(e)$ are not required since only the imaginary part of $\tilde{t}$ is used in equation (3).

Using this formalism, absorption cross sections for nucleons, deuterons, and various heavy ions colliding with selected target nuclei have been computed and published in tabular form (Townsend and Wilson, 1985). Detailed comparisons with experimental data have also been made. Sample results are displayed in figures 4 through 6. Note the excellent agreement between theory and the experimental measurements.

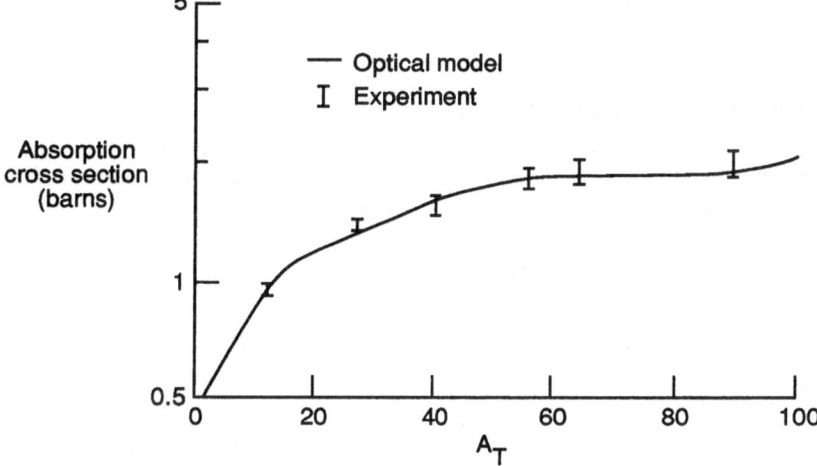

Fig. 4. Absorption cross sections, as a function of target mass number, for carbon projectiles at 83A MeV.

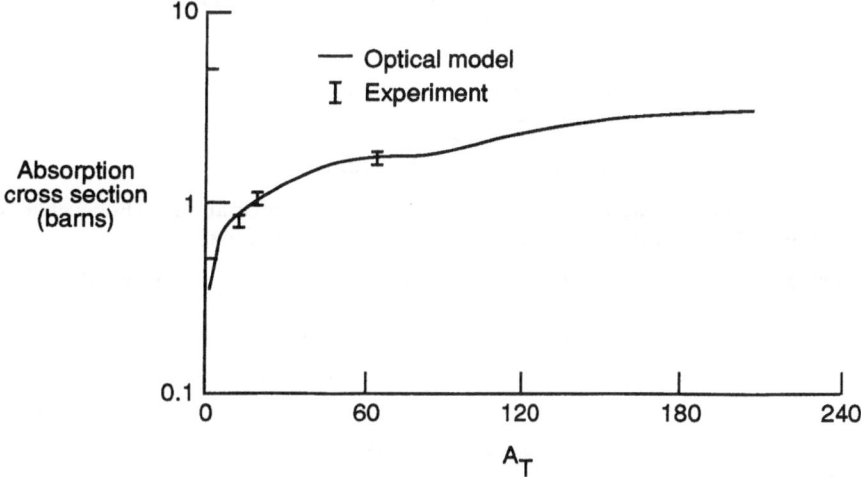

Fig. 5. Absorption cross sections, as a function of target mass number, for carbon projectiles at 3.6A GeV.

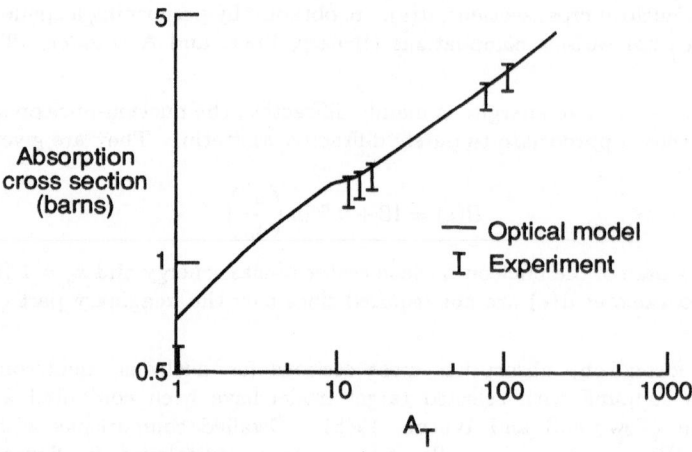

Fig. 6. Absorption cross sections, as a function of target mass number, for iron projectiles at 1.88A GeV. The experimental data were obtained with an emulsion target.

Cross Section Parameterizations

As an alternative to these complicated absorption cross section calculations, simple energy-dependent parameterizations have been developed. For nucleon-nucleus collisions, the absorption cross section is parameterized by (Wilson, et al., 1989).

$$\sigma_{abs} = f(E)\sigma_A(A_T) \tag{20}$$

where

$$\sigma_A(A_T) = 45A_T^{0.7}(1 - 0.018 \sin\theta_A), \tag{21}$$

and the angle θ_A is

$$\theta_A = 2.94 \ln(A_T) + 0.63 \sin[3.92 \ln(A_T) - 2.329] - 0.176. \tag{22}$$

The energy-dependent amplitude is

$$f(E) = 1 - (0.3E^{-0.22} + 0.76 \ e^{-E/135})(0.4 + 0.9 \ e^{-A_T/30}) \sin\theta_E, \tag{23}$$

where

$$\theta_E = 1.44 \quad (E < 25 \text{ MeV}), \tag{24}$$

and

$$\theta_E = 1.33 \ln(E) - 2.84 \quad \text{(otherwise)} \tag{25}$$

Figures 7 and 8 display sample results compared with experimental data (Bobchenko, et al., 1979) and the parameterization of Letaw, Silberberg and Tsao (1983).

For nucleus-nucleus collisions, the energy-dependent parameterization is (Townsend and Wilson, 1986)

$$\sigma_{abs}(A_P, A_T, E) = \pi r_o^2 \beta(E)[A_P^{1/3} + A_T^{1/3} - \delta(A_P, A_T, E)]^2 \tag{26}$$

where

$$\beta(E) = 1 + 5 \ E^{-1} \tag{27}$$

and

$$\delta(A_P, A_T, E) = 0.2 + A_P^{-1} + A_T^{-1} - 0.292 \ e^{-E/792} \cos(0.229E^{0.453}) \tag{28}$$

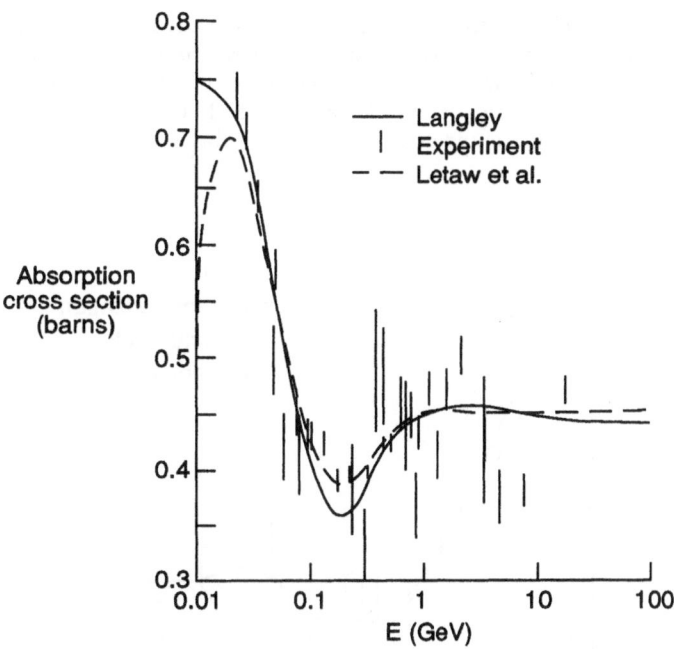

Fig. 7. The neutron-aluminum absorption cross section according to the Langley parameterization compared with Letaw et al. (1983) and various experiments.

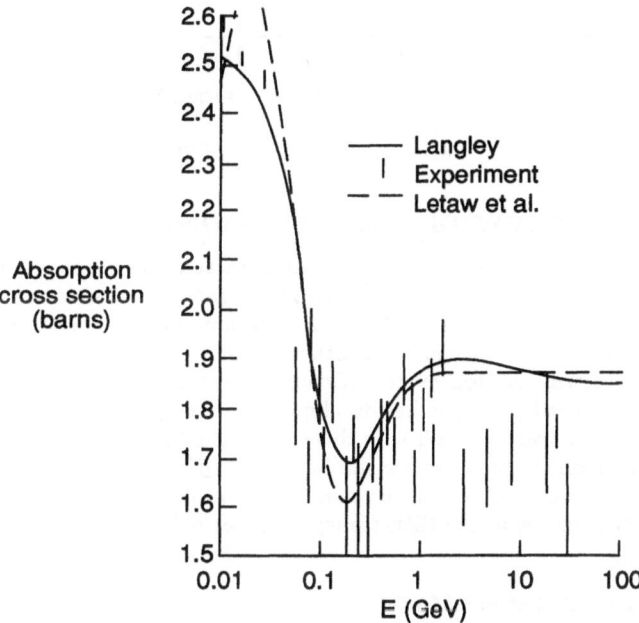

Fig. 8. The neutron-lead absorption cross section according to the Langley parameterization compared with Letaw et al. (1983) and various experiments.

with $r_0 = 1.26$ fm and E in units of MeV/nucleon (A MeV). Note that equation (28) reduces to the usual asymptotic form at high energies ($E \to \infty$)

$$\sigma_{abs} = \pi r_o^2 \left[A_P^{1/3} + A_T^{1/3} - \delta \right]^2 \qquad (29)$$

where δ is a constant independent of energy. Representative results for ^{12}C - ^{12}C collisions are displayed in figure 9. Also displayed are predictions from the optical model [Eq. (3)] and various experimental measurements (Kox, et al., 1987; Jaros, et al., 1978; Aksinenko, et al., 1980; Heckman, et al., 1978). Note that both calculations reliably reproduce the experimentally measured cross sections including the minimum near 300 A MeV.

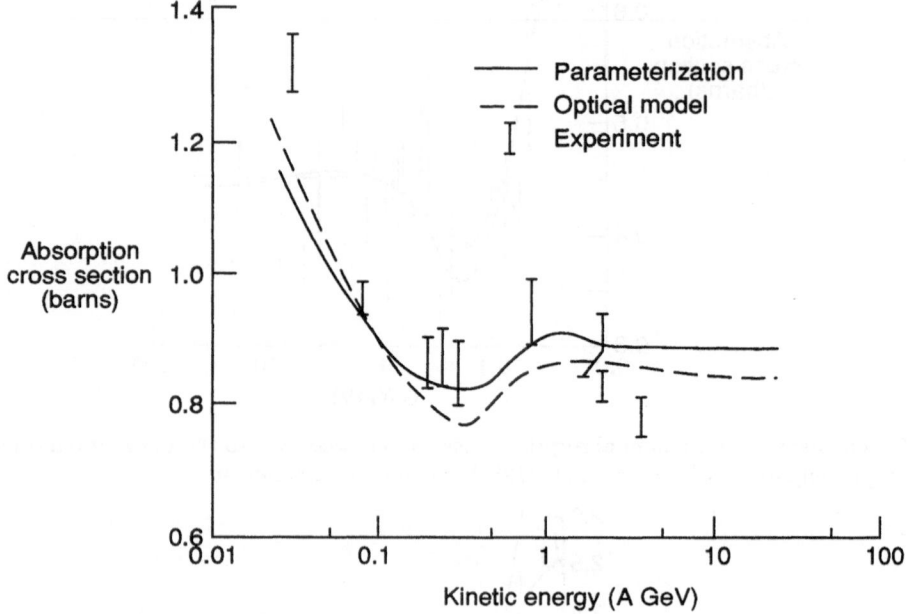

Fig. 9. Absorption cross sections for carbon-carbon collisions.

NUCLEAR FRAGMENTATION CROSS SECTIONS

Methods of calculating nuclear fragmentation cross sections for secondary particle production in HZE transport generally fall into one of two main categories: semiempirical parameterizations (Silberberg, Tsao, and Shapiro, 1976) or abrasion-ablation formulations based upon quantum-mechanical (Hufner, Schafer, and Schurmann, 1975; Townsend, 1983; Townsend, et al., 1986) or geometric collision models (Bowman, Swiatecki, and Tsang, 1973). As an alternative to these methods, Langley has developed a semiempirical abrasion-ablation model (Wilson, Townsend, and Badavi, 1987b). In this section the Langley semiempirical and quantum mechanical abrasion-ablation models will be described. In addition, the electromagnetic dissociation methods developed by Norbury and collaborators (Norbury and Townsend, 1986; Norbury, et al., 1988) for use in the HZE transport codes will be briefly described.

Quantum Mechanical Optical Model

In previous work (Townsend, 1983; Townsend, Wilson, and Norbury, 1985; Townsend, et al., 1986) the quantum mechanical optical model reaction theory was extended to investigate nuclear fragmentation within an abrasion-ablation formalism which includes contributions from frictional-spectator-interactions. The main goal of this effort is an attempt to develop a fundamental fragmentation theory which builds upon the successes of the absorption cross section theory, in particular its fundamental nature, broad generality, and excellent accuracy when compared with experimental data.

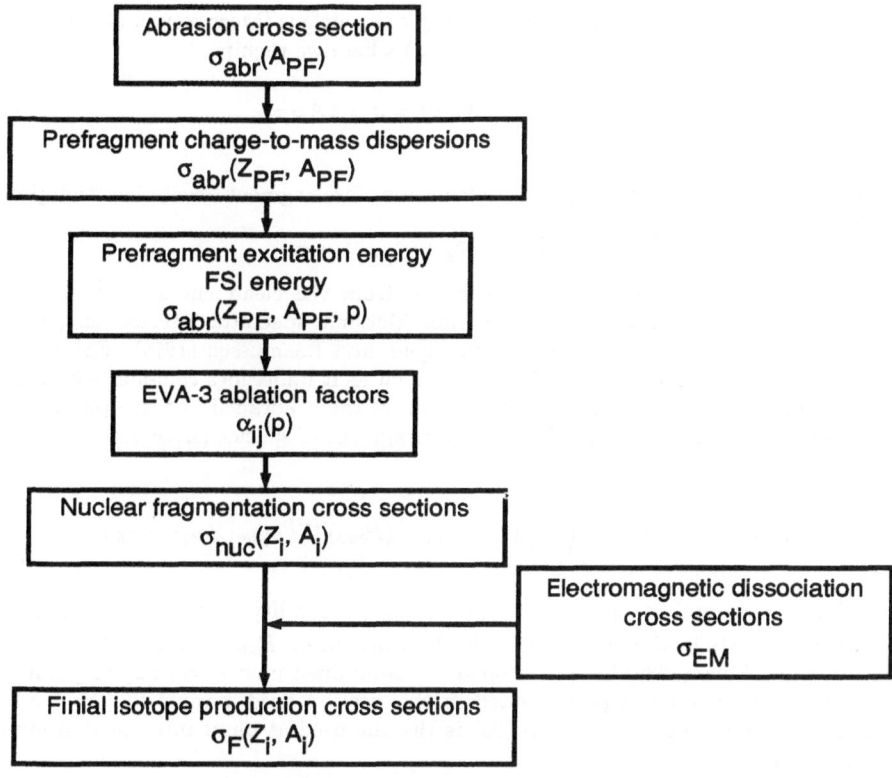

Fig. 10. Fragmentation calculations flow diagram.

As currently formulated, the computational steps in the Langley optical model fragmentation formalism are depicted in figure 10. For the initial (abrasion) stage of the fragmentation event, the cross section for abrading n out of A_P projectile nucleons to form a prefragment with mass number $A_{PF}(= A_P - n)$ is given by

$$\sigma_{abr}(A_{PF}) = \binom{A_P}{n} \int d^2\mathbf{b}[1 - P(\vec{\mathbf{b}})]^n P(\vec{\mathbf{b}})^{A_{PF}}, \tag{30}$$

where the probability, as a function of impact parameter, for not removing a projectile nucleon is

$$P(\vec{\mathbf{b}}) = \exp[-A_T\sigma(e)I(\vec{\mathbf{b}})], \tag{31}$$

with

$$I(\vec{\mathbf{b}}) = [2\pi B(e)]^{-3/2} \int dz \int d^3\xi_T \rho_T(\vec{\xi}_T)$$
$$\times \int d^3\mathbf{y}\rho_P(\vec{\mathbf{b}} + \vec{z} + \vec{y} + \vec{\xi}_T) \exp[-y^2/2B(e)]\, [1 - C(\vec{y})] \tag{32}$$

The symbols are identical to those used previously in equations (3)–(8). The nuclear density distributions are also described previously. Because the model is often used to describe few nucleon removal cross sections for peripheral collisions, the Pauli correlation correction is often neglected (Townsend, Norbury, Khan, 1991).

Since the abraded nucleons consist of protons and neutrons, a prescription for calculating the prefragment charge dispersion is needed. The most commonly used are the hypergeometric distribution (Oliveira, Donangelo, and Rasmussen, 1979) and one based upon the zero-point vibrations of the giant dipole resonance of the projectile nucleus (Morrissey, et al., 1978). The resultant abrasion cross section is $\sigma_{abr}(Z_{PF}, A_{PF})$ where we require

$$\sum_{Z_{PF}} \sigma_{abr}(Z_{PF}, A_{PF}) = \sigma_{abr}(A_{PF}) \tag{33}$$

To complete the abrasion stage of the calculation, prefragment excitation energies are estimated from

$$E_{exc} = E_s + E_{\text{FSI}} \tag{34}$$

where the surface energy term (E_s) is computed from the clean cut abrasion formalism (Gosset, et al., 1977) and the contribution from frictional-spectator-interactions (E_{FSI}) is estimated using the methods of Oliveira, Donangelo, and Rasmussen (1979). FSI results in the deposition of additional energy in the prefragment as it undergoes collisions with some of the abraded nucleons. If we assume that $p(\leq n)$ out of the n abraded nucleons participate in FSI, then the final abrasion cross section for a prefragment of species (Z_{PF}, A_{PF}) which has undergone p FSI's is given by

$$\sigma_{abr}(Z_{PF}, A_{PF}, p) = \binom{n}{p} (1 - P_{esc})^p (P_{esc})^{n-p} \sigma_{abr}(Z_{PF}, A_{PF}) \tag{35}$$

where $\binom{n}{p}$ is a binomial coefficient and P_{esc} is the probability that the abraded nucleon escapes without participating in an FSI. Early work used $P_{esc} = 0.5$. More recently, Benesh, Cook and Vary (1989) have developed a more refined model based upon geometrical considerations. The results to be presented in this section used a value $P_{esc} = 0.5$. The abrasion cross section given by equation (35) is the one used as input into the ablation part of the calculation.

Depending upon the excitation energy, the excited prefragment may decay by emitting one or more nucleons, composites, or gamma rays. The probability, $\alpha_{ij}(p)$, for formation of a particular final fragment of type i as a result of the deexcitation of a prefragment of type j which has undergone p frictional-spectator interactions, is obtained using the EVA-3 computer code (Morrissey, et al., 1979). The final nuclear fragmentation cross section for production of the type i isotope is given by

$$\sigma_{nuc}(Z_i, A_i) = \sum_j \sum_{p=0}^{m} \alpha_{ij}(p) \sigma_{abr}(Z_j, A_j, p), \tag{36}$$

where the summation over j accounts for the contributions to i from different prefragment species j, and the summation over p accounts for the effects of the different excitation energies resulting from FSI's. In order to compare these predictions with experiment, contributions resulting from electromagnetic dissociation must be included. At present only single nucleon cross section contributions to σ_{EM} are incorporated into the Langley model. Details of the methods for estimating σ_{EM} will be described in a subsequent section of this work.

Although completely general in applicability, and reasonably accurate for predicting heavy-fragment production cross sections, these optical model methods are of questionable accuracy for predicting light-ion production cross sections because of the restrictive nature of the peripheral collision assumption. In addition these calculational methods are far too complex to use directly in a transport code and appear to be better suited for generating data tables for parameterization or for other access as needed.

Semiempirical Model

In an attempt to maintain the physics inherent in the optical model while concomitantly reducing the complexity of the fragmentation calculations, a semiempirical nuclear fragmentation model was developed (Wilson, Townsend, and Badavi, 1987b). In this formalism, the geometric abrasion-ablation model of Bowman, Swiatecki, and Tsang (1973) was modified to incorporate FSI contributions through the use of a semiempirical correction to the abraded prefragment excitation energies.

In this method, the nuclear fragmentation cross sections for a final fragment with charge Z_f and mass A_f are obtained using

$$\sigma_{nuc}(Z_f, A_f) = F_1 \exp[-R|Z_f - SA_f + TA_f^2|^{3/2}]\sigma(\Delta A) \tag{37}$$

where (Rudstam, 1966) $R = 11.8A_F^{-0.45}$, $S = 0.486$, $T = 0.00038$, and F_1 is a normalizing factor such that

$$\sum_{Z_f} \sigma_{nuc}(Z_f, A_f) = \sigma(\Delta A) \tag{38}$$

which ensures charge and mass conservation. The Rudstam formula for $\sigma(\Delta A)$ is not used because its ΔA dependence is too simple and breaks down for heavy targets. Instead, the cross section for removal of ΔA nucleons is estimated using

$$\sigma(\Delta A) = \pi b_2^2 - \pi b_1^2 \tag{39}$$

where b_2 is the impact parameter at which Δ_{abr} nucleons are abraded by the collision and Δ_{abl} nucleons are ablated, in the subsequent prefragment deexcitation, such that

$$\Delta_{abr}(b_2) + \Delta_{abl}(b_2) = \Delta A - 1/2 \tag{40}$$

and similarly for b_1

$$\Delta_{abr}(b_1) + \Delta_{abl}(b_1) = \Delta A + 1/2 \tag{41}$$

The number of abraded nucleons is estimated from the geometric overlap volume and the mean free path in nuclear matter, λ, as

$$\Delta_{abr} = FA_p[1 - 0.5\exp(-C_p/\lambda) - 0.5\exp(-C_t/\lambda)] \tag{42}$$

where F is the fraction of the volume in the geometric overlap between projectile and target nuclei, and C_p and C_t are the maximum chord lengths of the intersecting surface in the projectile and target, respectively. Expressions for F, which differ depending on the relative sizes of the colliding nuclei and on the nature of the collision (peripheral versus central), are given elsewhere (Wilson, Townsend, and Badavi, 1987b; Gosset, et al., 1977). The value chosen for the mean free path in nuclear matter is $\lambda = 1.5$ fm. The number of ablated nucleons is computed from

$$\Delta_{abl} = (E_s + E_{FSI})/10 \text{ MeV} \tag{43}$$

which assumes that a nucleon is ablated for every 10 MeV of excitation energy. Varying the denominator in equation (43) from 8 to 12 MeV has little effect on the predicted cross sections. In equation (43) the surface energy contribution (E_s) is computed from the clean cut abrasion formalism (Gosset, et al., 1977). The frictional-spectator-interaction contributions (E_{FSI}) are estimated using methods detailed in Wilson, Townsend, and Badavi (1987b) and Badavi, et al. (1987).

The change of the nucleons removed by abrasion and ablation ΔZ is calculated according to charge conservation

$$Z_P = Z_F + \Delta Z \tag{44}$$

and is divided among the nucleons and alpha particles according to the following rules. The abraded nucleons are those removed from the portion of projectile in the overlap region with the target. Therefore, the abraded nucleon charge is assumed to be proportional to the charge fraction of the projectile nucleus as

$$Z_{abr} = Z_P \Delta_{abr}/A_P \tag{45}$$

The charge release in ablation is then given as

$$Z_{abl} = \Delta Z - Z_{abr} \tag{46}$$

which conserves the remaining charge.

Due to unusually tight binding of alpha particles in comparison to other nucleon arrangements, the helium production is maximized in the ablation process.

$$N_\alpha = int(Z_{abl}/2), \tag{47}$$

where $int(X)$ denotes the integer part of X. The number of protons produced is given by charge conservation as

$$N_P = \Delta Z - 2N_\alpha. \tag{48}$$

Similarly nucleon conservation requires the number of neutrons produced to be

$$N_N = \Delta A - N_P - 4N_\alpha. \tag{49}$$

The nuclear absorption cross section are taken as energy independent and approximated by

$$\sigma(A_1, A_2) = \pi r_o^2 (A_1^{1/3} + A_2^{1/3} - 0.2 - A_1^{-1} - A_2^{-1})^2, \tag{50}$$

where $r_o = 1.26$ fm. The choice of nuclear radius as

$$R(A) = 1.26 A^{1/3} \tag{51}$$

is consistent with equation (51) when peripheral collisions ($\Delta A < 0.5$) are taken into account. This is the version of the fragmentation model currently in use in the GCR transport code HZETRN. A separate computer code, NUCFRAG, incorporating this semiempirical fragmentation model and the electromagnetic dissociation model discussed in the next section has been developed (Badavi, et al., 1987).

Electromagnetic Dissociation

Because projectile and target nuclei may also breakup through their interacting electromagnetic fields, this contribution must be added to the nuclear fragmentation cross section to give the total fragmentation cross section as

$$\sigma_{Frag} = \sigma_{nuc} + \sigma_{EM} \tag{52}$$

An early study (Cucinotta, Norbury, Townsend, 1988) suggested that cross sections for multiple-nucleon removal by Coulomb dissociation were at least one order of magnitude smaller than single-nucleon removal cross sections. Therefore, the current version of NUCFRAG (Badavi, et al., 1987) includes only single nucleon removal cross sections for σ_{EM}. Recently, it has been noted that two-neutron removal cross section contributions to σ_{EM} may also be significant (Norbury and Townsend, 1990) and should be incorporated in future work.

Currently, methods for estimating electromagnetic dissociation cross sections use the Weiszacher-Williams method of virtual quanta (Westfall, et al., 1979; Norbury and Townsend, 1986; Norbury, et al., 1988). In this theory, the cross section is computed using

$$\sigma_{EM} = \int \sigma(E)N(E)dE \tag{53}$$

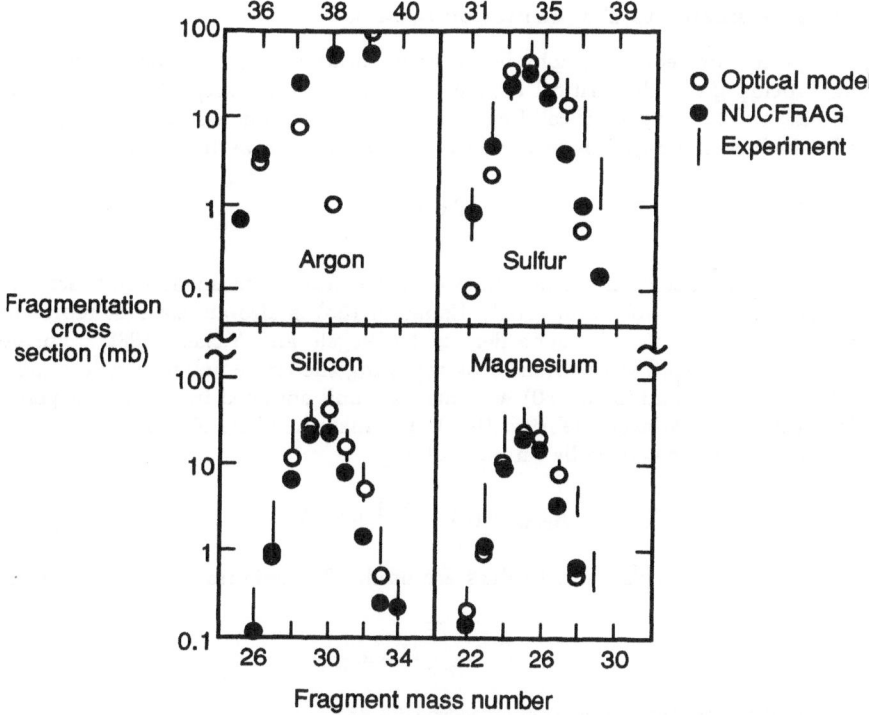

Fig. 11. Representative argon-carbon fragmentation cross sections at 213 A MeV.

where $N(E)$ is the virtual photon number spectrum, $\sigma(E)$ is the photonuclear cross section for neutron or proton knockout, and E is the virtual photon energy. The virtual photon number spectrum is given by (Westfall, et al., 1979)

$$N(E) = \frac{1}{E}\frac{2}{\pi}Z_t^2\alpha\frac{1}{\beta^2}\left\{xK_0(x)K_1(x) - \frac{1}{2}x^2\beta^2[K_1^2(x) - K_0^2(x)]\right\} \qquad (54)$$

where $N(E)$ is the number of virtual photons per unit energy E, Z_t is the number of protons in the target nucleus, β is the velocity of the target in units of c, and α is the electromagnetic fine structure constant. The $K_0(x)$ and $K_1(x)$ are modified Bessel functions of the second kind. The parameter x in Eq. (54) is defined as

$$x = \frac{Eb_{\min}}{\gamma\beta(\hbar c)} \qquad (55)$$

where γ is the usual relativistic factor and $b_{\min}$ is the minimum impact parameter. the dependence of $N(E)$ on the $b_{\min}$ in Eq. (54) comes from the fact that there is a finite maximum momentum transfer in the collision process. The photonuclear cross sections $\sigma(E)$ are usually obtained from experiment or a parameterization (Westfall, et al, 1979).

Electromagnetic dissociation contributions are significant for heavy nuclei and/or at high energies (Townsend and Wilson, 1989). Figure 11 displays sample predictions using the quantum-mechanical abrasion-ablation-FSI model and the semiempirical model NUCFRAG compared with the experimental data of Viyogi, et. al, (1979) for argon fragmenting on a carbon target. These cross section predictions are in reasonable agreement with the experimental measurements, although significant uncertainties in the data are evident from the error bars.

Recent Improvements to Fragmentation Models

After a period of reduced emphasis on fragmentation cross section studies, renewed attention is being given this critical area by the Langley Research Center team. Recent improvements to the semiempirical fragmentation model and associated computer code NUCFRAG include the use of an energy-dependent nucleon mean free path

$$\lambda = 16.6 \ E^{-0.26} \tag{56}$$

parameterized from phenomenological values obtained from experimental cross section measurements (Dymarz and Kohmura, 1982). Equation (56) is also an excellent representation of the non-local optical model results derived by Negele and Yazaki (1981). Further improvements to NUCFRAG were obtained by replacing the crude uniform nuclear radius approximation [eq. (51)] used in eq. (50) with nuclear radii obtained from recent experimental tabulations (DeVries, DeJager, DeVries, 1987). For mass numbers $A > 26$ the experimental rms radii are well-parameterized by (in units of fm)

$$R_{\rm rms} = 0.84 \ A^{1/3} + 0.55 \tag{57}$$

For $A \leq 26$ the actual experimental values are used. The uniform nuclear radii, computed from the relation

$$R = \sqrt{\frac{5}{3}} R_{\rm rms}, \tag{58}$$

are used to find the revised absorption cross section as

$$\sigma(A_1, A_2) = \pi(R_1 + R_2 - 0.504)^2 \tag{59}$$

which replaces eq. (50). Finally, a modification to the second-order correction to the surface excitation energy was incorporated. Typical results for elemental production cross sections are displayed in figs. 12 and 13. Also shown in fig. 12 are experimental measurements from Westfall, et. al, (1979) for incident iron nuclei at 1.88A GeV and from Cummings, et. al, (1990) for incident iron nuclei at 1.55A GeV colliding with a copper target. Note that the NUCFRAG predictions agree reasonably well with the older Westfall measurements, but typically tend to overestimate the recent measurements of Cummings and collaborators by approximately 25 to 50 percent.

Figure 13 displays elemental production cross sections for 1.65A GeV argon colliding with a KCl target. The experimental data are taken from Tull (1990). The differences between calculation and experiment for some elements are nearly a factor of two. Note also that there is significant scatter in the data which complicates efforts to improve the modeling. Planned improvements to NUCFRAG include a more realistic treatment of light ion production including an extension to incorporate deuteron, triton, and helion (^{3}He) production into the ablation stage. The single nucleon removal cross sections will be improved by incorporating the recent parameterization of Benesh, Cook, and Vary (1989).

LIGHT ION FRAGMENTATION

Within the current-generation of Langley Research Center space radiation transport codes, only nucleons and alpha particles are transported as light ions ($Z \leq 2$). There is no separate transport of deuterons, tritons and helions. In addition, the database of breakup cross sections for light ion production from HZE particles is somewhat crude. Because of the nature of the peripheral assumption in the quantum-mechanical abrasion-ablation formalism, it is not clear that such a formalism will ever produce accurate light-ion production cross sections. Finally, such a formalism may not even be applicable to the breakup of light ions such as alpha particles.

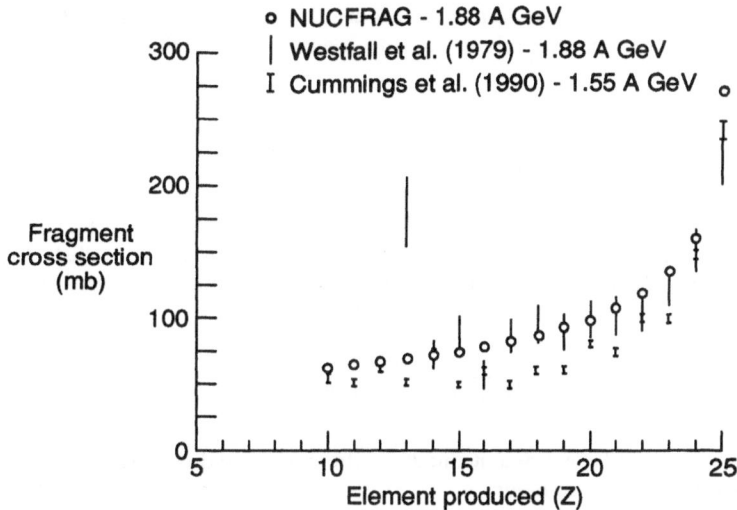

Fig. 12. Elemental production cross sections for iron-copper collisions.

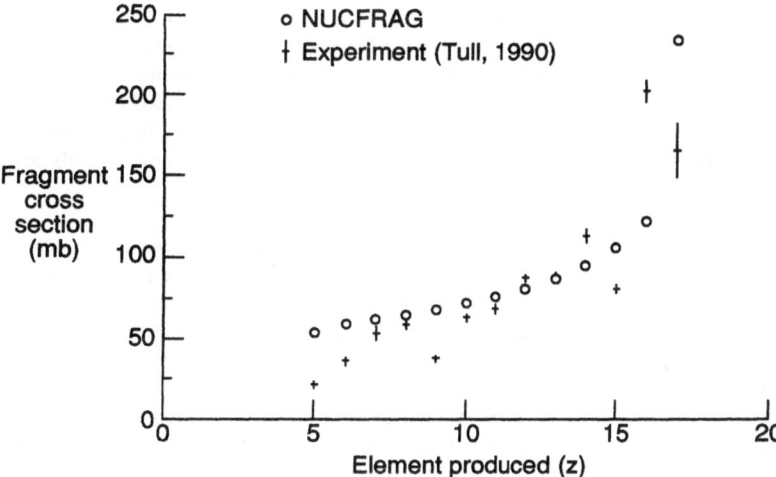

Fig. 13. Elemental production cross sections for Ar-KCl collisions at 1.65 A GeV.

To rectify this state of affairs, a comprehensive research program to develop a theory of light-ion breakup has been initiated (Cucinotta, 1988; Cucinotta, et. al, 1989; Cucinotta, et al., 1990; Cucinotta, et al., 1991). The initial emphasis is centered upon formulating a complete theory of alpha particle fragmentation in collisions with atomic nuclei. The theory is based upon an abrasion-ablation model where the abrasion step is described in terms of dispersion pole diagrams obtained from direct reaction theory (Shapiro, 1968). Corrections for final-state-interactions between the projectile fragments and for scattering by a second projectile fragment on the target (double-scattering) are also included (Cucinotta, et. al, 1991).

We start by considering the two-body dissociation of a projectile P fragmenting on a target T as

$$P + T \rightarrow a + b + X \tag{60}$$

where X is the final target state, and a and b are the projectile fragments. The transition matrix T_{fi} for this reaction is related to the momentum distribution for producing the fragment a by

$$\frac{d\sigma}{d\vec{p}_a} = \frac{V^3}{(2\pi)^5 \beta_{PT}} \int d\Omega_b K |T_{fi}|^2 \tag{61}$$

where β is the relative velocity in the initial state, K is the phase space factor, and V is the normalization volume; a summation over all final target states X is implied. In the overall center-of-mass frame (CM), assuming azimuthal symmetry about the beam direction, we have

$$K = \frac{p_b^2 E_b E_X}{p_b(E_b + E_X) + p_a E_b \, \cos(\theta_a + \theta_b)} \tag{62}$$

where p_i is the i-th particle's momentum; E_i is its energy; and θ_i is its emission angle. The transition matrix is written as a sum

$$T = \tilde{T}_s + \tilde{T}_p + T_D^1 + T_D^2 \tag{63}$$

where $\tilde{T}_s$ and $\tilde{T}_p$ represent distorted transition matrices for the spectator(s) and participant (p) contributions which factorize into a wavefunction for relative motion between the fragments a and b, and the interaction T-matrix for the fragment and target. These include final-state-interactions. By participants we mean the abraded nucleons; spectators refers to the prefragments left after the participants have been abraded. Either can be observed. The double scattering contributions (T_D^1 and T_D^2) presently neglect final-state-interactions. Specific expressions for these various transition matrices and additional details are available in Cucinotta, et. al, (1991).

As an example of this method, fig. 14 displays the predicted laboratory system angular distributions for ^{3}He production in the collision of 1A GeV alpha particles with protons. The experimental data are from Bizard, et. al, (1977). The importance of the participant and double-scattering terms is seen at all angles. We expect that the small differences between our calculations and the data at the smallest angles between 1.5° and 4.5° could be reduced if the phases between the various terms were treated correctly. We underestimate the data at the largest angles, the data at the smallest angles between 1.5° and 4.5° could be reduced if the phases between the various terms were treated correctly. We underestimate the data at the largest angles, which may be because of contributions not treated here, such as intermediate-state deuteron production, charge exchange, and pion production. Results for the total production cross section, obtained by integrating overall angles, are also in excellent agreement. Theory predicts 22.7 mb compared with the measured value of 24.1 ± 1.9 mb.

Planned improvements to this work include extending the methods to incorporate additional contributions from pion production, intermediate state deuteron production, and charge exchange. Modifications for application to heavier nuclear systems will be made by using the high-energy optical model (Wilson, 1975; Thies, 1979) to properly account for distortion and cascade effects and by treating the ablation step according to the methods of Norbury, Townsend, and Deutchman (1985).

SUMMARY AND FUTURE PLANS

The significance of high-energy heavy ion interactions to galactic cosmic ray dose estimates was illustrated by noting that nearly half of the bone marrow dose equivalent, behind typical thicknesses of spacecraft aluminum shielding, results from projectile and target fragments.

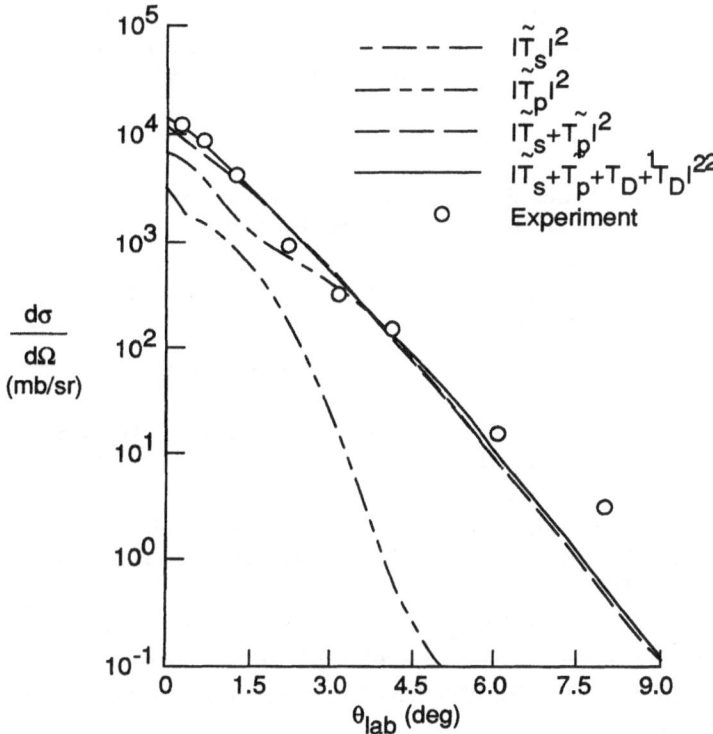

Fig. 14. Angular distribution for $\alpha + {}^1H \rightarrow {}^3He$ at 1.02 A GeV. Experimental data are from Bizard, et al. (1977).

Projectile fragments were also found to give a significantly larger dose equivalent contribution to the bone marrow than the contribution from target fragments. The potential effects of fragmentation cross section uncertainties on dose estimates were briefly addressed. Dose equivalent variations of over 50 percent are possible. Variations in LET distributions may even be larger and could exceed a factor of two for LET > 10 keV/micron.

An overview of the current quantum mechanical models of heavy ion reactions was presented and comparisons made with recent experimental measurements of heavy ion absorption and fragmentation cross sections. The excellent agreement between theory and experiment for total absorption cross sections was clearly evident. Comparisons between theoretical and experimental fragmentation cross sections indicated that much work remains to be done in developing a suitably accurate fragmentation theory from fundamental considerations. Improvements in the quality and quantity of experimental data are also needed in order to accomplish this goal. Future plans are to complete the light-ion breakup (alpha fragmentation) formalism and then attempt to extend the methods to heavier ions. Eventually, both total and differential fragmentation cross sections for any galactic cosmic ray ion of interest must be reliably predicted by the theory. After validation by comparison with high-quality experimental data, the theory will be used to generate a data base for parameterization or for use directly in the radiation transport codes.

At present, the total fragmentation cross sections used in the space radiation transport codes are obtained from phenomenological models derived from semi-classical considerations. The Langley semiempirical model, NUCFRAG, was described and future plans for its improvement discussed. Included, in these plans are a more realistic treatment of light ion production (^2H, ^3H, and ^3He) and the development of methods to estimate electromagnetic dissociation contributions to two-nucleon removal cross sections.

The current fragmentation models (quantum mechanical and semiempirical) both yield reasonable agreement with the existing, sparse, experimental data base. Significant improvements in the predictive accuracy of these models will require additional, high quality data for nuclei in the cosmic ray spectrum interacting with other target nuclei over a range of energies relevant to the GCR protection problem (100A MeV to 10A GeV).

REFERENCES

Adams, J. H., Jr., 1986: Cosmic Ray Effects on Microelectronics, Part IV. NRL Memo. Rep. 5901 (Revised), Naval Research Lab., Dec. 31, 1986.

Aksinenko, V. D., et. al, 1980: Streamer Chamber Study of the Cross Sections and Multiplicities in Nucleus-Nucleus Interactions at the Incident Momentum of 4.5 GeV/c per Nucleon. *Nucl. Phys.*, vol. A348, pp. 518–534.

Atwell, W., 1990: Astronaut Exposure to Space Radiation: Space Shuttle Experience. SAE Tech. Paper Ser. 901342, July 1990.

Badavi, F. F.; Townsend, L. W.; Wilson, J. W.; and Norbury, J. W., 1987: An Algorithm for a Semiempirical Nuclear Fragmentation Model. *Computer Phys. Comm.*, vol. 47, pp. 281–294.

Benary, O.; Price, L. R.; and Alexander, G., 1970: NN and ND Interactions (Above 0.5 GeV/c)—A Compilation. UCRL-20000 NN, Lawrence Radiation Lab., Univ. of California, Aug. 1970.

Benesh, C. J.; Cooke, B. C.; and Vary, J. P., 1989: Single Nucleon Removal in Relativistic Nuclear Collisions. *Phys. Review C*, vol. 40, no. 3, pp 1198-1206.

Billings, M. P.; and Yucker, W. R., 1973: The Computerized Anatomical Man (CAM) Model. NASA CR-13043.

Bizard, G.; Le Brun, C.; Berger, J.; Duflo, J.; Goldzahl, L.; Plouin, F.; Oostens, J.; Van Den Bossche, M.; Vu Hai, L.; Fabbri, F. L.; Picozza, P.; and Satta, L., 1977: ^{3}He Production in ^{4}He Fragmentation on Protons on 6.85 GeV/c. *Nucl. Phys.*, vol. A285, no. 3, pp. 461–468.

Bobchenko, B. M.; Buklei, A. E.; Viasov, A. V.; Vorob'ev, I. I.; Vorob'ev, L. S.; Goryainov, N. A.; Grishuk, Yu. G.; Gushchin, O. B.; Druzhinin, B. L.; Zhurkin, V. V.; Zavrazhnov, G. N.; Kosov, M. V.; Leksin, G. A.; Stolin, V. L.; Surin, V. P.; Fedorov, V. B.; Fominykh, B. A.; Shvartsman, B. B.; Shevchenko, S. V.; and Shuvalov, S. M., 1979: Measurement of Total Inelastic Cross Sections for Interaction of Protons With Nuclei in the Momentum Range from 5 to 9 GeV/c and for interaction of π^- Mesons with Nuclei in the Momentum Range From 1.75 to 6.5 GeV/c. *Soviet J. Nucl. Phys.*, vol. 30, no. 6, pp. 805–813.

Borkowski, F.; Simon., G. G.; Walther, V. H.; and Wendling, R. D., 1975: On the Determination of the Proton RMS-Radius From Electron Scattering Data. *Z. Phys. A*, vol. 275, no. 1, pp. 29–31.

Bowman, J. D.; Swiatecki, W. J.; and Tsang, C. F., 1973: Abrasion and Ablation of Heavy Ions. LBL-2908, Lawrence Buckley Lab., Univ. of California.

Cucinotta, F. A., 1988: Theory of Alpha-Nucleus Collisions at High Energies. Ph.D. Thesis, Old Dominion University.

Cucinotta, F. A.; Khandelwal, G. S.; Townsend, L. W.; and Wilson, J. W., 1989: Correlations in $\alpha - \alpha$ Scattering and Semi-Classical Optical Models. *Phys. Lett. B*, vol. 223, no. 2, pp. 127–132.

Cucinotta, F. A.; Norbury, J. W.; and Townsend, L. W., 1988: Multiple Knockout By Coulomb Dissociation in Relativistic Heavy Ion Collisions. NASA TM-4070.

Cucinotta, F. A.; Townsend, L. W.; Wilson, J. W.; and Khandelwal, G. S., 1990: Inclusive Inelastic Scattering of Heavy Ions and Nuclear Correlations. NASA TP-3026.

Cucinotta, F. A.; Townsend, L. W.; Wilson, J. W.; and Norbury, J. W.; 1991: Corrections to the Participant—Spectator Model of High-Energy Alpha-Particle Fragmentation. NASA TM-4262.

Cummings, J. R.; Binns, W. R.; Garrard, T. L.; Israel, M. H.; Klarmann, J.; Stone, E. C.; and Waddington, C. J., 1990: *Phys. Review C*, vol. 42, no. 6, pp. 2508–2529.

De Jager, C. W.; De Vries, H.; and De Vries, C., 1974: Nuclear Charge- and Magnetization-Density-Distribution Parameters From Elastic Electron Scattering. *At. Data & Nucl. Data Tables*, vol. 14, no. 5/6, pp. 479–508.

De Vries, H.; De Jager, C. W.; and De Vries, C., 1987: Nuclear Charge-Density Distribution Parameters from Elastic Electron Scattering. *At. Data & Nucl. Data Tables*, vol. 36, no. 3, pp. 495–536.

Dymarz, R.; and Kohmura, T., 1982: The Mean Free Path of Protons in Nuclei and the Nuclear Radius. Oxford Univ. Report 58/82.

Gosset, J.; Gutbrod, H. H.; Meyer, W. G.; Poskanzer, A. M.; Sandoval, A.; Stock, R.; and Westfall, G. D., 1977: Central Collisions of Relativistic Heavy Ions. *Phys. Review C*, vol. 16, no. 2, pp. 629–657.

Heckman, H. H.; Greiner, D. E.; Lindstrom, P. J.; and Shwe, H., 1978: Fragmentation of ^{4}He, ^{12}C, ^{14}N, and ^{16}O Nuclei in Nuclear Emulsion at 2.1 GeV/Nucleon. *Phys. Review C*, vol. 17, no. 5, pp. 1735–1747.

Hufner, J.; Schafer, K.; and Schurmann, B., 1975; Abrasion-Ablation in Reactions Between Relativistic Heavy Ions. *Phys. Review C*, vol. 12, no. 6, pp. 1888–1898.

Jaros, J.; Wagner, A.; Anderson, L.; Chamberlain, O.; Fuzesy, R. Z.; Gallup, J.; Gorn, W.; Schroder, L.; Shannon, S.; Shapiro, G.; and Steiner, H.; 1978: Nucleus-Nucleus Total Cross Sections for Light Nuclei at 1.55 and 2.89 GeV/c per Nucleon. *Phys. Review C*, vol. 18, no. 5, pp. 2273–2292.

Kox, S.; Gamp, A.; Perrin, C.; Arvieux, J.; Bertholet, R.; Bruandet, J. F.; Buenerd, M.; Cherkaoui, R.; Cole, A. J.; El-Masri, Y.; Longequeue, N.; Menet, J.; Merchez, F.; and Viano, J. B., 1987: Trends of Total Reaction Cross Sections for Heavy Ion Collisions in the Intermediate Energy Range. *Phys. Review C*, vol. 35, no. 5, pp. 1678–1691.

Letaw, J. R.; Silberberg, R.; and Tsao, C. H., 1983: Proton-Nucleus Total Inelastic Cross Sections: An Empirical Formula for E > 10 MeV. *Astrophys. J.*, Suppl ser., vol. 51, no. 3, pp. 271–276.

Morrissey, D. J.; Marsh, W. R.; Otto, R. J.; Loveland, W.; and Seaborg, G. T., 1978: Target Residue Mass and Charge Distributions in Relativistic Heavy Ion Reactions. *Phys. Review C*, vol. 18, no. 3, pp. 1267–1274.

Morrissey, D. J.; Oliveira, L. F.; Rasmussen, J. O.; Seaborg, G. T.; Yariv, Y.; and Fraenkel, Z., 1979: Microscopic and Macroscopic Model Calculations of Relativistic Heavy-Ion Fragmentation Reactions. *Phys. Review Lett.*, vol. 43, no. 16, pp. 1139–1142.

Negele, J. W.; and Yazaki, K., 1981: Mean Free Path in a Nucleus. *Phys. Review Lett.*, vol. 47, pp. 71–74.

Norbury, J. W.; Cucinotta, F. A.; Townsend, L. W.; and Badavi, F. F., 1988: Parameterized Cross Sections for Coulomb-Dissociation in Heavy Ion Collisions. *Nucl. Instrum. & Methods B*, vol. 31, no. 4, pp. 535–537.

Norbury, J. W.; and Townsend, L. W.; 1986: Electromagnetic Dissociation Effects in Galactic Heavy Ion Fragmentation, NASA TP-2527.

Norbury, J. W.; and Townsend, L. W.; 1990: Calculation of Two-Neutron Multiplicity in Photonuclear Reactions. NASA TP-2968.

Norbury, J. W.; and Townsend, L. W.; and Deutchman, P. A., 1985: A T-Matrix Theory of Galactic Heavy Ion Fragmentation, NASA TP-2363.

Oliveira, L. F.; Donangelo, R.; and Rasmussen, J. O., 1979: Abrasion-Ablation Calculations of Large Fragment Yields from Relativistic Heavy Ion Reactions. *Phys. Review C*, vol. 19, no. 3, pp. 826–833.

Ringia, F. E.; Dobrowolski, T.; Gustafson, H. R.; Jones, L. W.; Longo, M. J.; Parker, E. F.; and Cork, Bruce, 1972: Differential Cross Sections for Small-Angle Neutron-Proton and Neutron-Nucleus Elastic Scattering at 4.8 GeV/c. *Phys. Review Lett.*, vol. 28, no. 3, pp. 185–188.

Rudstam, G., 1966: Systematics of Spallation Yields. *Zeitshrift fur Naturforschung*, vol. 21a, no. 7, pp. 1027–1041.

Schopper, H., ed., 1973: Elastic and Charge Exchange Scattering of Elementary Particles. *Landolt-Börnstein Numerical Data and Functional Relationships in Science and Technology*, Group I, Volume 7, Springer-Verlag.

Schopper, H., ed., 1980: Elastic and Charge Exchange Scattering of Elementary Particles. *Landolt-Börstein Numerical Data and Functional Relationships in Science and Technology*. Group I, Volume 9, Springer-Verlag.

Shapiro, I. S., 1968: Some Problems in the Theory of Nuclear Reactions at a High Energies. *Sov. Phys. Uspekhi*, vol. 10, no. 4, pp. 515–535.

Silberberg, R.; Tsao, C. H.; and Shapiro, M. M., 1976: Semiempirical Cross Sections, and Applications to Nuclear Interactions of Cosmic Rays. *Spallation Nuclear Reactions and Their Applications*, B. S. P. Shen and M. Merker, eds., D. Reidel Publ. Co., c.1976, pp. 49–81.

Thies, M., 1979: Quantum Theory of Inelastic Multiple Scattering in Phase Space. *Ann. Phys. (NY)*, vol. 123, no. 2, pp. 411–441.

Townsend, L. W., 1982: Harmonic Well Matter Densities and Pauli Correction Effects in Heavy-Ion Collisions. NASA TM-2003.

Townsend, L. W., 1983: Abrasion Cross Sections for ^{20}Ne Projectiles at 2.1 GeV/Nucleon. *Canadian J. Phys.*, vol. 61, no. 1, pp. 93–98.

Townsend, L. W.; Cucinotta, F. A.; and Wilson, J. W., 1991: Interplanetary Crew Exposure Estimates for Galactic Cosmic Rays. *Radiat. Res.*, vol. 129, no. 1, pp. 48–52.

Townsend, L. W.; Norbury, J . W.; and Khan, F., 1991: Calculations of Hadronic Dissociation of ^{28}Si Projectiles at 14.6A GeV by Nucleon Emission. *Phys. Review C*, vol. 43, no. 5, pp. R2045–R2048.

Townsend, L. W.; and Wilson, J. W., 1985: *Tables of Nuclear Cross Sections for Galactic Cosmic Rays—Absorption Cross Sections.* NASA RP-1134.

Townsend, L. W.; and Wilson, J. W., 1986: Energy-Dependent Parameterization of Heavy-Ion Absorption Cross Sections. *Radiat. Res.*, vol. 106, pp. 283–287.

Townsend, L. W.; and Wilson, J. W., 1989: Nuclear Cross Sections for Estimating Secondary Radiations Produced in Spacecraft. Conference on the High Energy Radiation Background in Space (CHERBS '87), Sanibel Island, FL, November 3–5, 1987. AIP Conference Proceedings No. 186. Editors: A. C. Rester, Jr and J. I. Trombka (American Institute of Physics, New York, 1989), pp. 177–191.

Townsend, L. W.; and Wilson, J. W.; Cucinotta, F. A.; and Norbury, J. W., 1986: Comparison of Abrasion Model Differences in Heavy Ion Fragmentation: Optical Versus Geometric Models. *Phys. Review C*, vol. 34, no. 4, pp. 1491–1494.

Townsend, L. W.; and Wilson, J. W.; Cucinotta, F. A.; and Shinn, J. L., 1991: Galactic Cosmic Ray Transport Methods and Radiation Quality Issues. *Nucl. Tracks Radiat. Meas.*, 1992 (in press).

Townsend, L. W.; Wilson, J. W.; and Norbury, J. W., 1985: A Simplified Optical Model Description of Heavy Ion Fragmentation. *Canadian J. Phys.*, vol. 63, no. 2, pp. 135–138.

Tull, C. E., 1990: Relativistic Heavy Ion Fragmentation at HISS (Ph.D. Thesis). LBL-29718, Lawrence Berkeley Lab., Univ. of California.

Viyogi, Y. P.; Symons, T. J. M.; Doll, P.; Greiner, D. E.; Heckman, H. H.; Hendrie, D. L.; Lindstrom, P. J.; Mahoney, J.; Scott, D. K.; Van Bibber, K.; Westfall, G. D.; Wieman, H.; Crawford, H. J.; McParland, C.; and Gelbke, C. K., 1979: Fragmentations of ^{40}Ar at 213 MeV/Nucleon. *Phys. Review Lett.*, vol. 42, no. 1, pp. 33–36.

Westfall, G. D.; Wilson, Lance, W.; Lindstrom, P. J.; Crawford, H. J.; Greiner, D. E.; and Heckman, H. H., 1979: Fragmentation of Relativistic ^{56}Fe. *Phys. Review*, ser. C, vol. 19, no. 4, pp. 1309–1323.

Wilson, J. W., 1974: Multiple Scattering of Heavy Ions, Glauber Theory, and Optical Model. *Phys. Lett.*, vol. B52, no. 2, pp. 149–152.

Wilson, J. W.; 1975: *Composite Particle Reaction Theory*, Ph.D. Diss., College of William and Mary in Virginia. June 1975.

Wilson, J. W.; and Costner, C. M., 1975: *Nucleon and Heavy-Ion Total and Absorption Cross Section for Selected Nuclei.* NASA TN D-8107.

Wilson, J. W.; and Townsend, L. W., 1981: An Optical Model for Composite Nuclear Scattering. *Canadian J. Phys.*, vol. 59, no. 11, pp. 1569–1576.

Wilson, J. W.; Townsend, L. W.; and Badavi, F. F., 1987a: Galactic HZE Propagation Through the Earth's Atmosphere. *Radiat. Res.*, vol. 109, no. 2, pp. 173–183.

Wilson, J. W.; Townsend, L. W.; and Badavi, F. F., 1987b: A Semiempirical Nuclear Fragmentation Model. *Nucl. Instrum. & Methods B*, vol. 18, no. 3, pp. 225–231.

Wilson, J. W.; Townsend, L. W.; Nealy, J. E.; Chun, Sang, Y.; Hong, B. S.; Buck, Warren, W.; Lankin, S. L.; Ganapol, Barry D.; Kahn, Ferdous; and Cucinotta, F. A., 1989: BRYNTRN: *A Baryon Transport Model.* NASA TP-2887.

PLANS FOR A NEW GROUND BASED SPACE RADIATION RESEARCH FACILITY IN THE USA *

Peter Thieberger

Department of Physics
Brookhaven National Laboratory
Upton, New York 11973 U.S.A.

INTRODUCTION

The availability of energetic heavy ion beams at particle accelerator laboratories is essential to simulate components of the space radiation environment. Such ion beams are needed for radiobiological research, to assess radiation shielding requirements for space missions, to calibrate spacecraft radiation detectors, and to study space radiation effects on electronic devices and systems. Such ground-based studies are complementary to obviously more realistic, but also much more difficult and more expensive space-based research. The purity of the accelerator beam and their monoenergetic and unidirectional nature are indispensible to gain a detailed understanding of the effects caused by the complex radiation fields encountered in space.

The existing heavy ion accelerator facilities have all been built to perform research in nuclear and particle physics. Several of them have also been used for some of the applications which are of interest here, and others will be used for such applications in the future. Figure 1 shows a summary of maximum energies as functions of atomic mass number for all the heavy ion accelerators capable of energies higher than 100 MeV/amu. There are many lower energy heavy ion accelerators, some of which are also being used for space-related applications, but they will not be included in the present discussion.

Space-related tests and experiments performed at nuclear physics heavy ion accelerators are sometimes performed at beam lines not specifically designed for the application at hand. The requirements for most of these applications are, however, very different from those normally encountered in nuclear physics research. For example, while well focussed and intense beams are the norm for experiments in nuclear physics, large area, uniform, and relatively weak beams are usually required for space-related work. In addition, for all radiobiological research, special facilities and environmental control are required in relation to work with cells, tissues, and animals. To make optimum use of these accelerators for space-related research, it is indispensible to count with equipment and facilities specially designed for the specific applications.

The radiobiological facility at the Bevalac accelerator located at the Lawrence Berkeley Laboratory is the prime example of such a high energy heavy ion facility, designed for non-nuclear physics research (Alonso, 1989). The Bevalac was also the first, and for many years

* Supported by the U.S. Department of Energy under Contract #DE-ACO2-76CHOOO16.

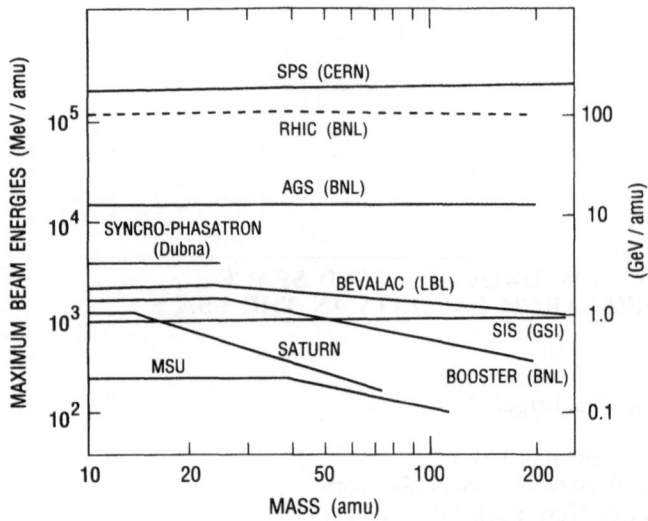

Figure 1. Maximum Beam Energies as Functions of Atomic Mass for All the Heavy Ion Accelerators with Energies Higher than 100 MeV/amu. Starting from the Top, SPS (CERN) Stands for Super Proton Synchrotron at Centre European de Recherches Nucleaires, Geneva, Switzerland. RHIC (BNL) and AGS (BNL) are Respectively the Relativistic Heavy Ion Collider Now under Construction at Brookhaven National Laboratory and the Alternating Gradient Synchrotron. The Syncro-Phasatron is a Weak Focussing Synchrotron Located at the Joint Institute for Nuclear Research, Dubna, Russia. The BEVALAC is a Linac Injected weak focussing synchrotron located at the Lawrence Berkeley Laboratory, and SIS is the Schwerionen Synchrotron Located at the Gesellschaft fur Schwerionenforschung, Darmstadt, Germany. The Booster (BNL) is a Newly Completed Synchrotron Which Will Inject the AGS and Will also Provide Beams for the New Space Research Facility. SATURN is the Synchrotron Located at the Laboratoire National Saturn in Saclay, France and MSU stands for the Michigan State University National Superconducting Cyclotron Laboratory.

the only, accelerator to reach heavy ion energies in excess of 1 GeV/amu. The only other accelerator facility in the USA where heavy ion energies of this magnitude and higher are now becoming available is at Brookhaven National Laboratory. Here we will briefly review the capabilities of the Brookhaven heavy ion accelerators and the plans for the construction of a new facility for space-related research.

THE BROOKHAVEN HEAVY ION ACCELERATORS

Research with heavy ions at Brookhaven, initiated about 20 years ago at the Tandem Facility (Thieberger, 1984), is now performed at progressively higher and higher energies achieved by the coupling of several accelerators. The four accelerators which will all eventually become part of the chain of heavy ion accelerators at Brookhaven (Thieberger, et al., 1988) are the Tandem Van de Graaff, the Booster Synchrotron, (Weng, 1990), the Alternating Gradient Synchrotron (AGS) and the Relativistic Heavy Ion Collider (RHIC) (Hahn, 1990). The maximum energies for typical beams from these accelerators are listed in Table 1.

The Tandem facility, where presently ongoing space-related research (Thieberger, et al., 1991) will continue, is also an injector for the other accelerators. Direct injection of Tandem beams into the AGS has provided 14.5 GeV/amu oxygen and silicon beams for several years. The recently completed Booster synchrotron will in the near future receive heavy ions from the Tandem for preacceleration before injection into the AGS. Thus heavier ions up to the mass of gold will become available at the AGS energies (see Table 1). Finally,

Table 1. Maximum Energies in MeV/amu of Typical Beams at the Brookhaven Heavy Ion Accelerators

	Protons	Oxygen	Iron	Gold
Tandem	30	7.5	2.5	1.0
Booster	1,500	1,500	1,250	350
AGS	29,000	14,500	13,500	12,500
RHIC	250,000*	125,000*	115,000*	100,000*

* These energies are for each of the two counterrotating beams. The center-of-mass energies in beam-beam collisions are twice these values for beams of equal mass.

the relativistic heavy ion collider will be charged with heavy ions from the AGS for further acceleration and storage to perform high energy heavy ion collision experiments.

The two Brookhaven high energy heavy ion accelerators which are of interest from the point of view of space-related research are the Booster synchrotron and the AGS. The relativistic heavy ion collider is designed for beam-beam collision experiments and is therefore not suitable for fixed target experiments. The AGS maximum energies exceed by far the energies corresponding to most of the cosmic ray flux, even though low intensity components of the cosmic ray spectrum reach even higher energies. The AGS can, and probably will, be used for some special space-related experiments or tests. The Brookhaven heavy ion accelerator of greatest interest for the simulation of interplanetary heavy ions is the Booster synchrotron because of the good overlap with energies of the main part of the cosmic ray flux.

PLANS FOR A NEW HEAVY ION SPACE RADIATION EFFECTS FACILITY

A new experimental facility will be constructed which will receive heavy ion beams from the Brookhaven Booster accelerator while this accelerator is simultaneously utilized for the nuclear physics program. Maximum energies, expected intensities, ranges, and surface LETs for the Booster synchrotron are listed in Table 2.

This accelerator was initially intended to only serve as an injector for the AGS and therefore no experimental facilities for the use of the Booster beams were constructed. Plans for such a facility (Thieberger, et al., 1991) have now been finalized and the construction of the new facility will probably start in 1993 and may be completed in 1996 or 1997. It will consist of three large irradiation rooms and an associated building with preparation rooms, laboratories, offices, and a control room.

The project includes the necessary accelerator modifications to extract the beam from the Booster at the appropriate point and to achieve slow extraction so as to deliver beam with an acceptable duty factor. The project also includes the upgrading of a second Tandem Van de Graaff injector and the construction of short beam transport to connect this injector to the injection line. This second injector is required to achieve simultaneous operation of the nuclear physics program and of the new space radiation research facility. A beam pulse from one of the heavy ion injectors will be accelerated in the Booster and then transferred to the AGS for further acceleration. Before the next beam pulse is required for the AGS, there is time for the Booster to accelerate ions from the other Tandem injector for delivery to the new facility. The ion species for the two applications will in general be different, and this is why two independent injectors are required. The shared use of the accelerators will increase beam

Table 2. Examples of Maximum Energies, Estimated Intensities, Ranges and Dose Rates for Heavy Ions from the BNL Booster

Ion	Energy (MeV/amu)	Average intensity (Ions/sec x 10^8)	Range in water (cm)	Surface dose-rate in water for a 10 cm^2 beam(Gy/min x 10^2)
$_6C^{12}$	1500	33	185	2.4
$_8O^{16}$	1500	24	139	3.1
$_9F^{19}$	1500	23	131	3.8
$_{14}Si^{28}$	1500	15	79	5.9
$_{15}P^{31}$	1500	10	77	4.5
$_{16}S^{32}$	1500	10	69	5.2
$_{17}Cl^{35}$	1500	9	67	5.2
$_{26}Fe^{56}$	1248	0.5	36	0.7
$_{29}Cu^{63}$	1127	6	29	10.5
$_{32}Ge^{70}$	1097	5	27	10.8
$_{35}Br^{79}$	917	6	19	16.0
$_{53}I^{127}$	620	4	7.4	27.4
$_{79}Au^{197}$	350	4	2.2	77.5

availability and reduce operating costs, thus benefiting all users. Once RHIC is operating in 1997, the shared use of the Booster will continue since collider injection will only be required once or twice a day. Each time this occurs, other activities will be interrupted for about half an hour.

Available beam spot areas will range from about 1cm x 1cm to about 50cm x 50cm, or even 1m x 1m, if necessary. The required beam uniformities over these areas will be achieved through a combination of beam focussing and beam sweeping techniques. Full beam characterization and dosimetry will be provided for users. There also are plans to develop unidimensional microbeams (e.g., 25 microns x 2 cm) taking advantage of the excellent beam quality (low emittance) of the Booster beams. Such microbeams are of interest for special and important experiments in radiobiology and for the detailed study of the response of microelectronic devices.

Beam energies will be continually variable from the maximum value listed in Table 2 down to a few MeV/amu. The energy selection will be achieved by adjusting the acceleration cycle and without the use of beam degraders, thus maximum beam purity will be maintained. In the support building adjacent to the irradiation rooms, temporary animal holding and preparation areas will be provided, as well as laboratory facilities for work with cell cultures and tissues. Other extensive laboratory facilities, fully certified animal holding facilities, and other services exist at BNL and can be made available to experimenters.

CONCLUSION

The planned heavy ion space radiation research facility will be designed taking into account the experiences at other facilities, and will also incorporate new ideas and advanced capabilities. Through the use of a modern accelerator and by sharing this use with the nuclear physics program, a facility will be developed which is expected to be reliable, cost effective, and user-friendly.

REFERENCES

Alonso, J.R., 1989, Intermediate Energy Heavy Ions: An Emerging Multidisciplinary Research Tool, Nucl. Instr. and Meth. B40/41: 921.

Hahn, H., 1990, The Relativistic Heavy Ion Collider at Brookhaven, Part. Accel. 32: 75.

Thieberger, P., 1984, The Brookhaven double MP facility: recent developments and plans for the future, Nucl. Instr. and Meth. 220: 45.

Thieberger, P., Barton, D.S., Benjamin, J., Chasman, C., Foelsche, H., and Wegner, H.E., 1988, Tandem

Injected Relativistic Heavy Ion Facility at Brookhaven, Present and Future, Nucl. Instr. and Meth. A268: 513.

Thieberger, P., Chanana, A.D., Foelsche, H.W., Lee, Y.Y., and Snead, C.L., 1991, Booster Applications Facility Report (phase II), BNL Report #52291.

Thieberger, P., Stassinopoulos, E.G., Van Gunten, O., and Zajic, V., 1991, Heavy ion beams for single-event research at Brookhaven - present and future, Nucl. Instr. and Meth. B56/57: 1251.

Weng, W.T., 1990, Progress and Status of the AGS Booster Project, Part. Accel. 27: 13.

RELATING SPACE RADIATION ENVIRONMENTS TO RISK ESTIMATES

Stanley B. Curtis

Lawrence Berkeley Laboratory
University of California
Berkeley, CA 94720 U.S.A.

INTRODUCTION

A number of considerations must go into the process of determining the risk of deleterious effects of space radiation to travelers on long missions. Among them are (1) determination of the components of the radiation environment (particle species, fluxes and energy spectra) which the travelers will encounter, (2) determination of the effects of shielding provided by the spacecraft and the bodies of the travelers which modify the incident particle spectra and mix of particles, and (3) determination of relevant biological effects of the radiation in the organs of interest. The latter can then lead to an estimation of risk from a given space scenario. Clearly, the process spans many scientific disciplines from solar and cosmic ray physics to radiation transport theory to the multistage problem of the induction by radiation of initial lesions in living material and their evolution via physical, chemical, and biological processes at the molecular, cellular, and tissue levels to produce the end point of importance.

This lecture will provide a bridge from the physical energy or LET spectra as might be calculated in an organ to the risk of carcinogenesis, a particular concern for extended missions to the moon or beyond to Mars. Topics covered will include (1) LET spectra expected from galactic cosmic rays, (2) probabilities that individual cell nuclei in the body will be hit by heavy galactic cosmic ray particles, (3) the conventional methods of calculating risks from a mixed environment of high and low LET radiation, (4) an alternate method which provides certain advantages using fluence-related risk coefficients (risk cross sections), and (5) directions for future research and development of these ideas.

RISKS OF RADIATION IN SPACE

We first consider the established and potential radiation risks to be expected on extended travel in space. It is accepted that, given adequate shielding against giant solar particle events, the most important radiation effects outside the earth's magnetic field are late effects: cancer, cataracts, and genetic effects caused by radiation produced both from solar particle events and from the galactic cosmic rays. The National Council on Radiation Protection and Measurements (NCRP) has published a report discussing various aspects of the space radiation risk assessment problem (Report #98, NCRP, 1989). It was concluded that the most important late effect was cancer. Recently, new assessments of cancer risk have been made from the Atomic Bomb survivor data base (Shimizu et al, 1990), and the BEIR V Committee (1990). The International Commission on Radiological Protection (ICRP) has

Biological Effects and Physics of Solar and Galactic Cosmic Radiation,
Part B, Edited by C.E. Swenberg *et al.*, Plenum Press, New York, 1993

817

recently published revised estimates of the risk of cancer mortality per unit of exposure. The value of 4% per Sv was given for the excess risk of cancer mortality for radiation received at low dose rate by adult members of the population (ICRP, 1991). This new value is roughly a factor of two higher than previously thought to apply. Based on these new estimates, the NCRP career Exposure Limits for near-earth orbit, suggested in Report #98, are presently under review.

Other potential late effects include cataractogenesis and possibly irreparable damage to the central nervous system accumulating from the low level of the penetrating highly ionizing component of the galactic cosmic rays (sometimes called HZE particles (Grahn, 1973)). It is not known whether the doses will be below the threshold dose for cataractogenesis and there is only speculation at present on the effects to the human central nervous system. Genetic effects to subsequent generations may become important if and when space travelers begin procreation after having been on an extended mission.

IMPORTANCE OF SPECIFIC ORGAN RISK ASSESSMENT

In the radiation exposures from large solar particle events that will sometimes occur in space flight, each organ of the body will receive a different amount of radiation due to the different amounts of shielding provided by the rest of the body. In addition, it has been determined that every organ or tissue has a different radiosensitivity for tumor induction. This is shown in Table 1 taken from ICRP Report #60 (1991). We note that some organs are over an order of magnitude more sensitive to cancer induction than others. Thus, in determining the overall cancer risk to the body from solar particle events, it is important to determine the exposure to each organ separately, estimate the average dose equivalent to the organ and add the contributions of risk to each organ to obtain the total risk of cancer induction in the body. In order for this to be a meaningful calculation, the external shielding (i.e., spacecraft, lunar habitat, etc.) must be very well defined, and the extent of modulation of the radiation field must be well understood.

Table 1. Probability of Fatal Cancer by Organ.

Tissue or organ	Probability of fatal cancer for working-age population (10^{-2} Sv^{-1})
Bladder	0.24
Bone marrow	0.40
Bone surface	0.04
Breast	0.16
Colon	0.68
Liver	0.12
Lung	0.68
Oesophagus	0.24
Ovary	0.08
Skin	0.02
Stomach	0.88
Thyroid	0.06
Remainder	0.40
Total	4.00

LET SPECTRA FROM GALACTIC COSMIC RAYS

Another important concern is the risk from the galactic cosmic radiation. A good review of our knowledge in cosmic ray abundances, their energy spectra, and their modulation through the eleven and twenty-two year solar cycles has been presented in other papers in this course. The question arises: what is the relative "importance" of the different species in the cosmic rays in determining the carcinogenic risk? A feeling for this can be obtained by examining the LET spectrum of the radiation in question. The fluence-LET spectrum is the distribution of LET's of the particles found in the radiation environment. Mathematically, the fluence-LET spectrum, f(L), is defined such that the total fluence (number of particles per unit area), F, can be written as the integral over all LET:

$$F = \int f(L) \; dL \tag{1}$$

where L is the LET_∞ or dE/dx (i.e., the stopping power) of the particles and f(L) is the fluence of particles having LET between L and L + dL.

The absorbed dose, D, can be written in terms of the fluence-LET spectrum as follows:

$$D = k \int f(L) \; L \; dL \tag{2}$$

where k is a constant linking fluence and dose units. We note that the integrand can also be defined as the distribution in LET of absorbed dose, D(L), the dose-LET spectrum, so that

$$D(L) = k \; f(L) \; L. \tag{3}$$

A representation of the dose-LET spectrum for the galactic cosmic rays at solar minimum is shown in Figure 1. The distribution D(L) has been multiplied by L so that equal areas under the curve have equal weights since the distribution is plotted on a logarithmic scale, i.e., linearly in the log of the LET. Several of the peaks in this distribution have been identified on the curve to show the relative importance of protons, helium, silicon, and iron ions in producing absorbed dose. The peak of low energy iron is a reflection of the fact that in free space the low energy iron ions having quite large LET values are still present. It should also be noted that the actual heights of the peaks as shown on the graph are of no significance since the curve should in theory go to infinity where the LET vs. energy relationship for each particle type reaches a maximum or minimum.

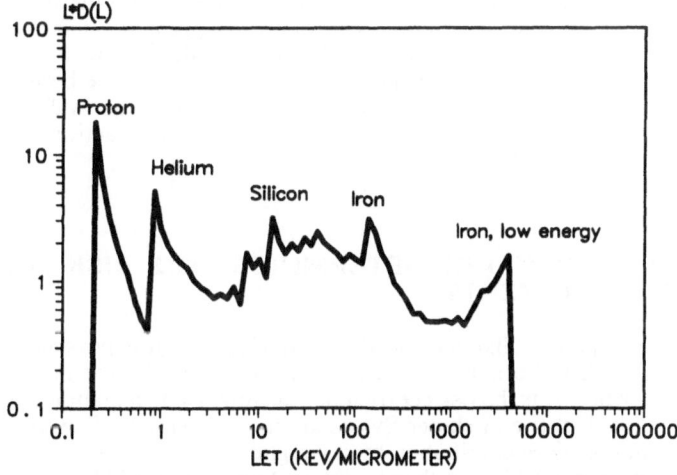

Figure 1. The differential dose-LET spectrum for galactic cosmic rays in free space at solar minimum. Several of the peaks are identified as being caused by specific components of the radiation. The distribution has been multiplied by L, the LET, in order to give equal weights to equal areas under the curve, since the LET scale is linear in the logarithm of the LET.

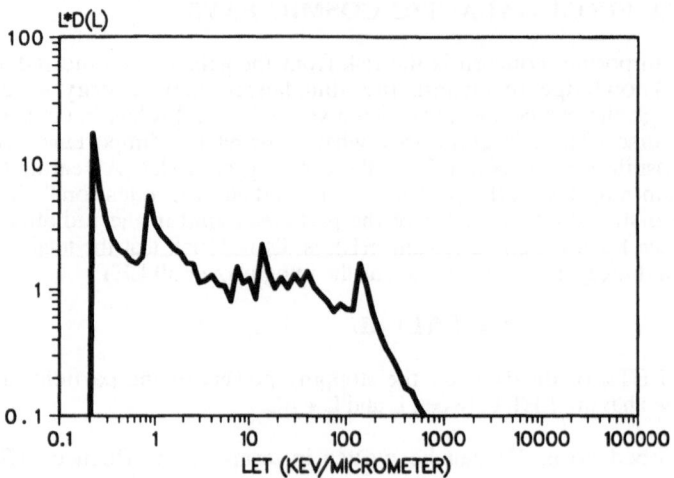

Figure 2. The differential dose-LET spectrum for galactic cosmic rays behind 10 g/cm² aluminum shielding at solar minimum. The "low-energy" iron peak at high LET is absent since that component is absorbed in the shielding. The distribution has been multiplied by the LET as Fig. 1.

There is a modification of this spectrum as the particles traverse matter. Coulomb effects (i.e., ionization of the atoms through which charged particles pass) will slow the particles down, thus changing their LET's and the absorbed dose they produce. Nuclear interactions will occur causing secondary particles which, in turn, will slow down and cause other nuclear interactions. The nuclear processes are very complex for all the species and energies involved in the galactic cosmic ray spectra. Computer codes have been developed to calculate the transport of the galactic spectra through matter, and a description of several of them has been presented at this course (see papers by J. W. Wilson). One output of such codes is the integral dose-LET distribution, i.e., the dose produced at the depth in question by particles above a given LET plotted as a function of LET. This distribution can be differentiated to give the differential dose-LET distribution at the depth of interest and plotted in the same manner as in Figure 1. Such a plot for 10 g/cm² of aluminum shielding is shown in Figure 2. The same peaks for the different ions are still seen even at this depth and the proton peak has remained constant, while the higher LET peaks are somewhat less important and the low energy, very high LET iron peak has totally disappeared. It should be noted that multiple production of nuclear secondaries has been accounted for in the calculation, but multiple coulomb and nuclear scattering and as well as straggling have been neglected, i.e., the "straight-ahead approximation" has been assumed. Note that the "valleys" at low LET have been filled up to some extent by the secondary nuclear interactions of the higher Z particles.

CONVENTIONAL METHOD OF DETERMINING THE RISK FROM MIXED RADIATION ENVIRONMENTS

Given an LET spectrum like the one shown in Figure 2, it is possible to calculate the dose equivalent, H. Then the risk of excess cancer mortality by the radiation is just the product of R and H, where R is the risk coefficient for low LET radiation expressed in risk per unit of dose equivalent (i.e., risk per Sv) as already discussed. The dose equivalent is found by multiplying the dose distribution by a weighting factor (called the Quality Factor, Q(L)), which is a function of LET, and then by integrating over all LET.

Mathematically, we have:

$$H = \int Q(L) \ D(L) \ dL \tag{4}$$

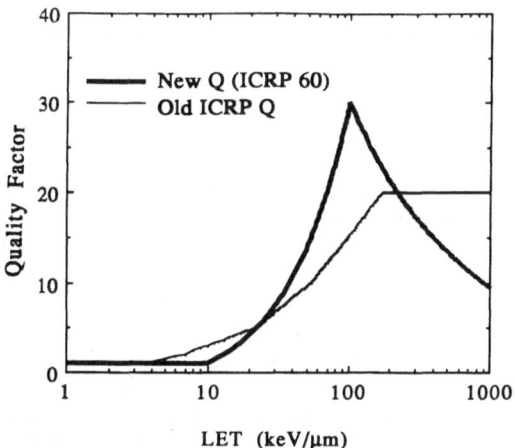

Figure 3. The new (ICRP, 1991) and old (ICRP, 1977) Quality Factors as determined by the International Commission on Radiological Protection are shown as a function of LET.

The Quality Factor is intended to weight the higher LET components of the radiation, in line with the experimentally found observation that, per unit of absorbed dose, high LET radiation is more effective in causing biological damage (thus presumably in causing human cancer) than low LET radiation. The functional dependence decided upon comes from examination at various LET's of experimental data on dose response curves of various biological systems deemed relevant to carcinogenesis and extrapolated to low dose and low dose rates or protracted exposures.

The ICRP has recently published a new dependence of Quality Factor on LET (ICRP, 1991). The new and old (ICRP, 1977) functions are plotted in Figure 3. The three main differences are that the new function (1) remains at unity until the LET reaches 10 keV/μm, (2) rises to a value of 30 at 100 keV/μm, and (3) decreases as the LET rises above 100 keV/μm.

Multiplication of the incident galactic cosmic ray dose distribution (under no shielding,

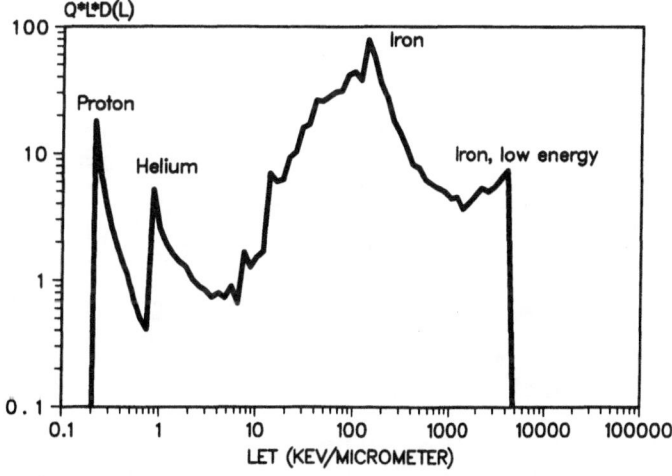

Figure 4. The differential dose-equivalent distribution for galactic cosmic rays in free space at solar minimum. This curve gives an idea of the relative importance of the various components of the radiation in causing risk. The distribution has been multiplied by the LET as in Fig. 1.

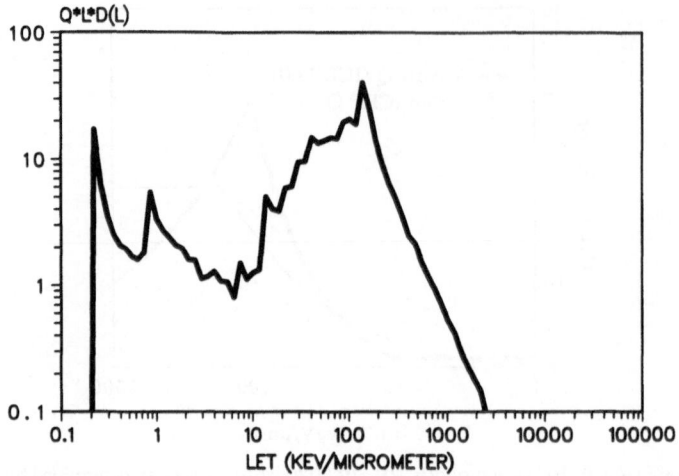

Figure 5. The differential dose-equivalent distribution for galactic cosmic rays behind 10 g/cm² aluminum shielding. The distribution has been multiplied by the LET as in Fig. 1.

cf. Figure 1) by the new Quality Factor yields the curve shown in Figure 4. Because a minimum ionizing iron ion has an LET of about 145 keV/μm, near the maximum of the Quality Factor vs. LET curve, we see a very prominant peak in the distribution around this value. The curve as calculated behind 10 g/cm² of aluminum shielding is given in Figure 5. Here again the contribution around 100 keV/μm from the iron and other high Z components is quite large. This indicates that high LET radiation will play a considerable role in determining the ultimate biological response even under fairly heavy shielding.

HIT FREQUENCIES OF CELL NUCLEI BY GALACTIC COSMIC RADIATION

It is of some interest to estimate the frequency with which the nuclei of typical cells within the bodies of space travelers might be hit by tracks of the galactic cosmic radiation.Calculations have been made for two specific simple shielding configurations at solar minimum conditions (Curtis and Letaw, 1989). Case a is a point at the center of a

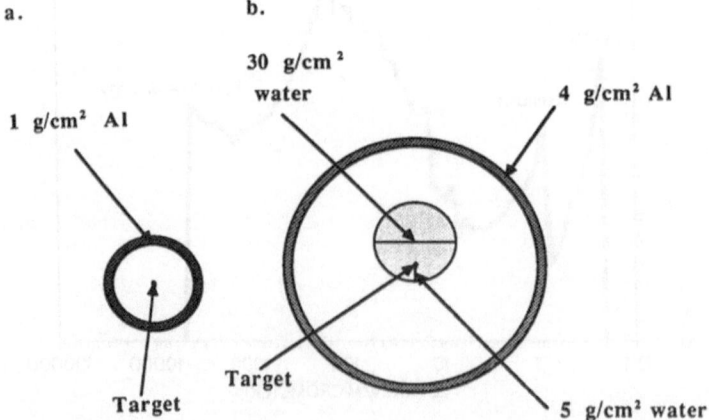

Figure 6. Description of the shielding configurations for the hit frequency calculations. Case a: the point of interest is the center of an aluminum shell 1 g/cm² thick; case b: the point of interest is the center of an aluminum shell 4 g/cm² thick and 5 g/cm² beneath the surface of a sphere of water 30 g/cm² in diameter.

spherical aluminum shell 1 g/cm² thick, and case b is a point at the center of a spherical aluminum shell 4 g/cm² thick and 5 cm inside the surface of a sphere of water 30 cm in diameter to approximate in a rough way the human body. The two shielding configurations are shown in Figure 6. An area of 100 μm² was chosen for the size of the "target" cell nucleus. This is a conservatively large estimation for the cross sectional area of many cell nuclei in the human body.

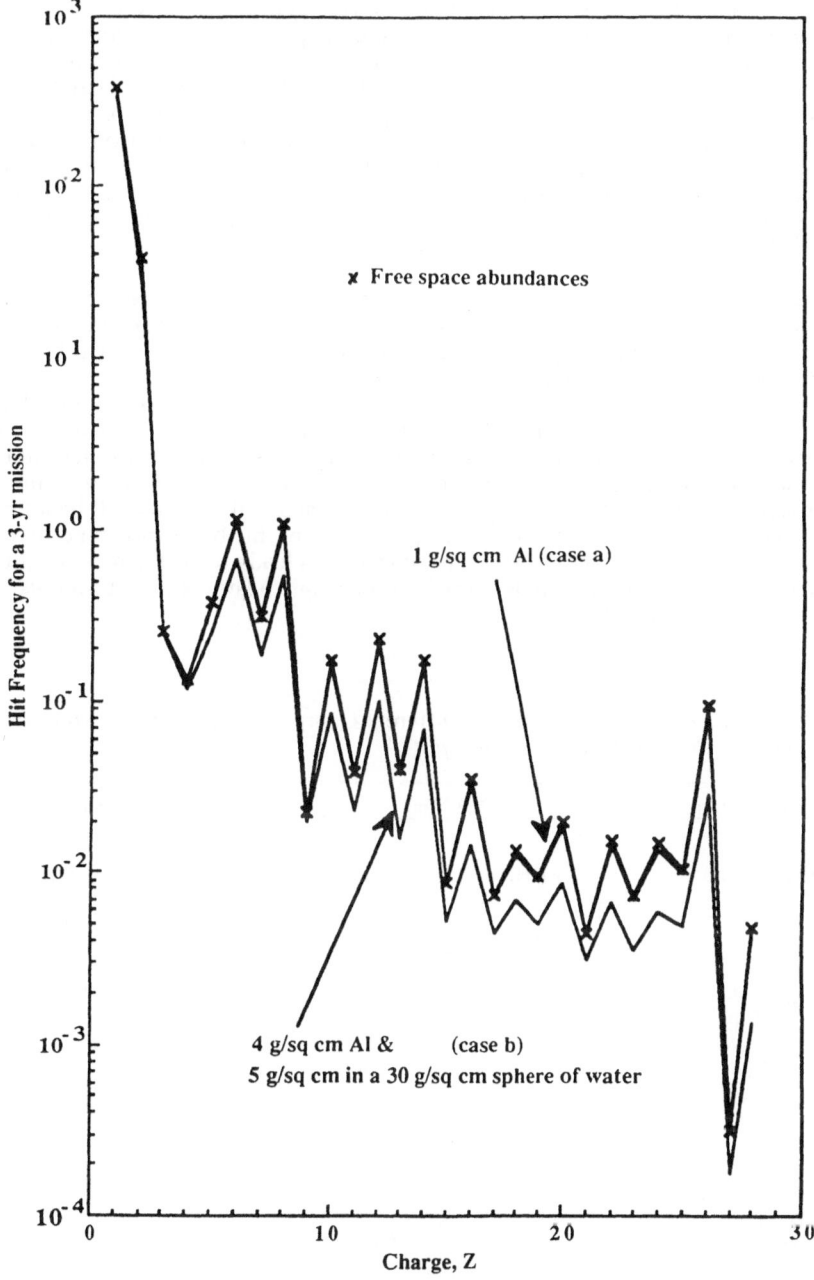

Figure 7. Frequencies of charged-particle hits caused by galactic cosmic rays in a 100-μm² area at the point of interest within the two shielding configurations (cases a and b defined in Fig. 6) plotted as a function of the charge of the particles for a 3-year mission outside the geomagnetosphere at solar minimum. The free-space galactic cosmic ray abundances converted to the same units are shown as x's.

The calculation uses the "straight-ahead" approximation and includes nuclear fragmentation. Hit frequencies for the two configurations are shown as a function of the charge of the particle in Figure 7 for a three-year mission outside the geomagnetosphere at solar minimum. The free-space abundances are plotted as x's in the figure. It is interesting to note that the peaks and valleys reflecting the relative free space abundances are almost indistinguishable from the more lightly shielded configuration (case a) and persist even for the more heavily shielded configuration (case b). We see, for instance, that each cell nucleus in the case b configuration would receive over a three year mission, on the average, roughly 400 proton hits, 40 helium ion hits, 0.7 carbon ion hit, 0.5 oxygen ion hit, and so on. From these data and assuming Poisson statistics (i.e., a random process) for particle arrivals, the percentages of cell nuclei hit at least once or at least twice behind the two shielding configurations can be calculated. These are shown for the more heavily shielded configuration in Table 2 for various charge groups. We note that 33% of cell nuclei will receive one or more hits during a three-year mission at solar minimum from particles with charge between 10 and 28. Some 80% of these will be single traversals only, that is, no other traversals of particles within that charge group. However, the probability is 0.86 that these cell nuclei will receive at least one hit from a particle in the charge 3-9 group and all will receive many hits from helium ions and protons.

Another way to present the results is in terms of the mean time between hits of ions with the same charge, i.e., the reciprocal of the mean frequencies. The frequencies can then be presented in terms of one hit per mean time between hits. This is shown for protons, helium ions, oxygen ions and iron ions in the second column of Table 3. In the third column are shown for comparison hit frequencies simply scaled to the human body by multiplying by the ratio of the two presumably relevant areas (0.3 m^2 for the body/100 μm^2 for the cell nucleus). Because of the assumptions in this calculation, the latter are very rough values. We conclude that although the bodies of the space travelers will be hit by many galactic cosmic ray particles during a mission lasting a year or longer, each cell nucleus will be hit by very few heavy ions with high charge, and the majority will be hit by at most one very heavy charged particle (with charge above 10). This conclusion leads us to the realization that for travel on extended deep space missions, the effects of single heavy particle traversals of cells may play very a important role.

Table 2. Percentage of Cell Nuclei Hit for Different Charge Groups at Solar Min. Assuming Poisson Statistics. (Cell Nuclear Area = 100 μm^2)[1]

Mission Duration	One Year		Three Years	
Charge Group	≥1 Hit	≥2 Hits	≥1 Hit	≥2Hits
3-9	49	14	86	59
10-16	10	0.51	27	4
17-25	1.6	0.001	4.8	0.12
26-28	0.99	0.005	2.9	0.04
10-28	12	0.8	33	6

[1] Behind 4g/cm^2 Al and 5 cm beneath surface of a 30 cm diameter water sphere(From Curtis and Letaw, 1989).

Table 3. Frequency of Hits (events) for Different Charge Species.

Particle Species (and charge)	Hits/cell nucleus [1] (Area ~100 μm²)	Hits/human body [2] (Area ~ 0.3 m²)
Protons (Z=1)	~ 1 per 3 days	~ 10^4 per second
He ions (Z=2)	~ 1 per month	~ 10^3 per second
Oxygen ions (Z=8)	~ 1 per 6 years	~ 20 per second
Iron ions (Z=26)	~ 1 per 100 years	~ 1 per second

[1] Behind 4 g/cm² Al and 5 cm beneath surface of a 30-cm diameter water sphere, (from Curtis and Letaw, 1989).

[2] Obtained simply by scaling to the larger area.

From the above discussion, we realize that it might be advantageous to split the dose equivalent, H, as calculated from Eq. (4), into its component parts, keeping the contributions from the different ion species separate. Thus, Eq. (4) can be rewritten:

$$H = \sum_{i=1}^{n} \int Q(L_i) \, D(L_i) \, dL_i \qquad (5)$$

where the summation is over the n different ion species in the cosmic ray spectrum. Remembering the relationship between the dose- and the fluence-LET spectra (Eq. 3), the risk then becomes:

$$R_c = R \, H = R \sum_{i=1}^{n} \int kQ(L_i) \, L_i \, f_i(L_i) \, dL_i \qquad (6)$$

INTRODUCTION OF AN ALTERNATIVE TO THE CONVENTIONAL APPROACH OF CALCULATING DOSE EQUIVALENT

It was pointed out some time ago that there is a way of defining risk without having first to estimate the dose equivalent. In 1966 it was suggested that a fractional cell lethality (FCL) be defined and calculated for astronauts caught in large solar particle events (Curtis et al, 1966). In this approach, fluence-LET spectra of protons, helium ions and heavier particles as found at depth within the body of a seated astronaut were multiplied by cell inactivation cross sections as a function of LET as measured with heavy ions on human kidney cells at the Berkeley Hilac (Todd, 1965), and then integrated over LET to obtain the number of inactivation hits. This yielded the survival, S, of cells at that depth, and the FCL was just 1 - S. The FCL was considered to be a measure of risk, since it was a direct measure of cells killed. Other similar ideas relating risk directly to a microdosimetric spectrum of energy deposited locally, (i.e., a y-spectrum) and defining a "hit-size effectiveness function" (HSEF) have been discussed in the literature (Bond et al, 1985, Sondhaus et al, 1990).

We now formally introduce the concept of a fluence-based risk coefficient (Curtis et al, 1991). It is defined as the risk per unit fluence so that the product of it and the fluence of a particular particle species yields the risk, for instance, the excess relative risk of cancer mortality, caused by that fluence. Since it has the units of risk/ (particle/unit area), this gives it the units of area and it is called a *risk cross section*. Under the assumptions that only single particle traversals are important (see above) and that the risk probability is small compared to unity for each particle type, the risks from all particle species are additive. If we denote the risk cross section for the ith particle species by σ_i (L_i), we can write for the risk:

$$R_c = \sum_{i=1}^{n} \int \sigma(L_i) \ f_i(L_i) \ dL_i \tag{7}$$

RELATIONSHIP BETWEEN RISK CROSS SECTION AND QUALITY FACTOR

Comparing Eqs. (6) and (7), we note that the risk cross section can be written in conventional terms as follows:

$$\sigma_i(L_i) = k \ R \ Q(L_i) \ L_i \tag{8}$$

As seen in Fig. 3, the Quality Factor is defined as a single-valued function of the LET. This means that any particle of a given LET will have the same value of the Quality Factor. It is well known that because of the different track structure (spatial distribution of energy loss around the track's trajectory) resulting from particles with different charge, biological effects of different particles with the same LET can be quite different. The definition of the risk cross section allows for this possibility in its formulation; this is not the case in the conventional treatment.

AN EXAMPLE OF USING THE CONCEPT

The prevalence of radiation-induced tumors in the Harderian gland of the mouse has been used as an example of how the concept of risk cross section can be used to calculate expected tumor prevalence during a space flight at solar minimum outside the geomagnetic field. From fluence-response curves obtained at the Berkeley BEVALAC for several beams of charged particles with well-defined LET's, the initial slope (i.e., the slope of the curve of prevalence as a function of fluence at very low fluence) was determined. This slope is the risk cross section for Harderian tumor prevalence. The slopes in units of μm^2 are plotted in Figure 8 (Alpen et al, submitted). The solid line is an analytical expression developed simply to possess characteristics thought to pertain (Curtis et al, 1992): an initial linear increase with LET (corresponding to the region where the Relative Biological Effectiveness, RBE, is expected to be 1), a supralinear region (corresponding to the region where the RBE is increasing to values considerably greater than 1), and a region of constant cross section or plateau at high LET (corresponding to the region of "saturation" or "overkill" where the RBE is decreasing). We note that there is a fairly good fit to the data, but there is no evidence whatsoever that the plateau implied by the analytical expression has been reached in the experimental data.

An additional factor has been included in the final calculations. It is well known that at high energy, nuclear interactions of protons and helium ions with target molecules in tissue can contribute local high LET events that rival or even surpass the contribution of direct ionization losses from these particles. This phenomenon has been studied (Shinn et al, 1989), and we have included target fragmentation in the calculation when the galactic cosmic ray spectra were integrated. The total risk cross section becomes:

$$\sigma_i^*(L_i) = \sigma_i(L_i) + \sigma_{targ \ frag} \tag{9}$$

The results when using galactic cosmic ray spectra at solar minimum calculated behind 1 g/cm^2 of aluminum shielding are given in Table 4. The results of tumor prevalence per year both including and excluding target fragmentation are shown. The contributions from protons

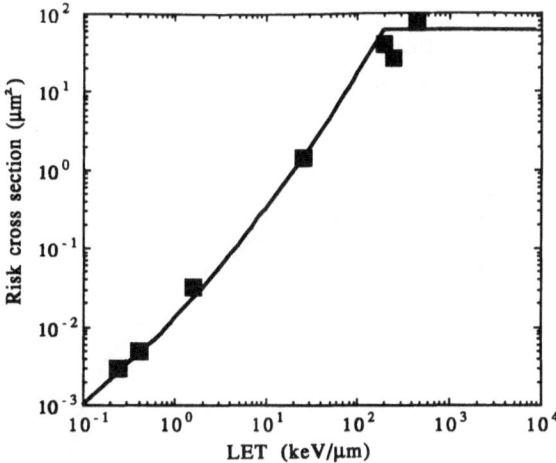

Figure 8. The risk cross sections for Harderian gland tumor prevalence as measured with monoenergetic beams from the Berkeley BEVALAC (Alpen et al, submitted) are shown as a function of LET. The curve is from an analytical expression as explained in the text (Curtis et al, 1992).

(Z=1), helium ions (Z=2), and two higher Z subgroups (Z=3-9 and 10-28) are presented separately. The prediction is that a 6% prevalence of Harderian tumors is expected in a space mission of one year outside the geomagnetosphere from the galactic cosmic radiation behind 1 g/cm² of aluminum shielding. Some 60% of the total comes from the Z = 10-28 charge group. The integral prevalance plotted as a function of LET (both for the total and for the contribution not including target fragmentation) is shown in Figure 9. The prevalence due to particles with LET greater than a given LET is plotted against the LET. The conclusion is

Table 4. Total Risk Cross Section at Solar Minimum Shielded by 1 g/cm² Aluminum.

| Charge Group | Prevalence per year | |
Z	Direct ionization only	With target fragmentation
1	0.0052	0.0092
2	0.0029	0.0039
3-9	0.0089	0.0101
10-28	0.0362	0.0367
TOTAL	0.0532	0.0599

From Curtis et al, 1992

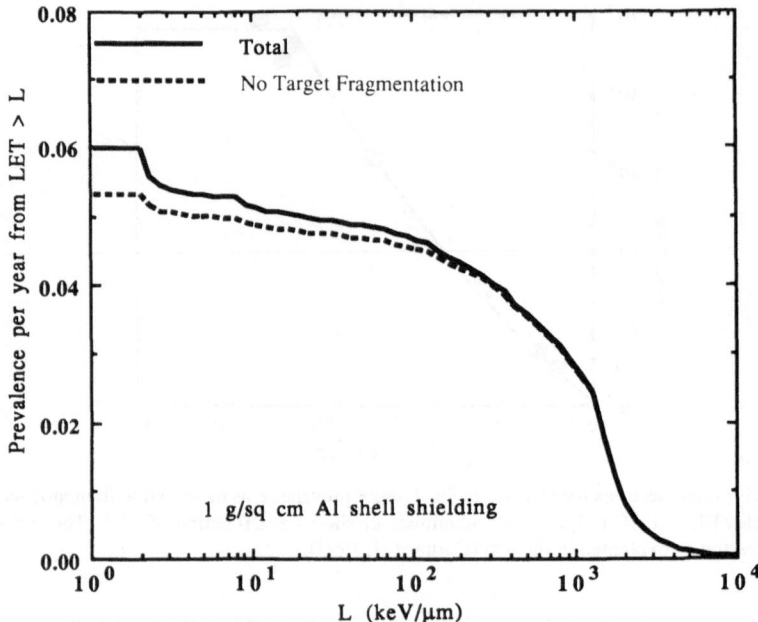

Figure 9. The integral probability distribution of Harderian gland tumor prevalence calculated for a year's mission at solar minimum behind 1 g/cm^2 aluminum shielding is plotted as a function of LET. It is the probability of tumor induction from LET's greater than a given LET plotted against the LET. Curves with and without inclusion of target fragmentation are shown for comparison (From Curtis et al, 1992).

clear that most (~ 80%) of the effect is caused by radiation with LET above 10 keV/μm for this thickness of shielding. We emphasize that this percentage will vary with shielding thickness as well as time through the solar cycle. This mode of evaluation thus results in the same conclusion as from the more traditional one: a large percentage of the biological effect is due to particles at high LET's.

CONCLUSIONS AND FUTURE DIRECTIONS

This discussion of radiation risks in space flight has emphasized the problem of determining the risk of cancer induction from the radiation environment that travelers will find outside the shielding confines of our earth's magnetic field, on a return excursion to the Moon, for example, or an exploratory trip to Mars. The conventional calculation of LET spectra produced by the galactic cosmic radiation behind a typical shielding thickness has been described. From such calculations, in which a weighting factor such as the Quality Factor weights the differential dose distribution in LET, it is possible to learn the relative importance of the different LET components constituting the radiation. We have seen that even under 10 g/cm^2 aluminum shielding, a considerable portion of the estimated risk (as suggested by the calculated distribution in LET of the dose equivalent) arises from components of LET greater than 10 keV/μm. In the future, it will be important to validate and improve the codes by which these transport calculations are made.

We have next approached the problem from a slightly different point of view by calculating the hit frequencies of cell nuclei at a point inside a simulated human body (a 30-cm diameter sphere of water) from the various charged components of the galactic cosmic radiation under well-defined shielding configurations. We have noted that most cell nuclei, if hit at all, will be hit by only one heavy highly charged particle during a long term mission of one to three year's duration. This emphasizes the need to study at ground level the biological effects which are considered relevant to the carcinogenic process of single traversals of cells by a high energy heavy particle. Such cells will also be hit by larger numbers of particles with lower charge over very long periods of time. Interactive effects of such hits should be studied to determine if they will provide anything more than a small second order modulation.

Finally, we have shown how a new concept related to particle fluence, the risk cross section, can be used to estimate risk and have pointed out several advantages for using such a concept for evaluating risk from the galactic cosmic radiation. The development of this idea paves the way toward a mechanistic understanding of radiation-induced carcinogenesis from charged particle radiation in terms of particle traversals of the cells at risk. Clearly, considerably more ground-based research is necessary on identifying the important changes inside a cell nucleus at the molecular level caused by a traversal of a high energy heavy particle track. Only in this way will we ultimately be able to better estimate the risk of these very low fluence, very high energy particles to the health of space travelers on long missions.

ACKNOWLEDGMENTS

The author acknowledges the calculations of integral LET spectra both in free space and behind aluminum shielding provided by Drs. L. W. Townsend and J. W. Wilson of the NASA Langley Research Center.

This work has been supported in part by the National Aeronautics and Space Administration, Order No. T-9310R and by the Office of Health and Environmental Research of the U. S. Department of Energy under Contract No. DE-AC03-76SF00098.

REFERENCES

BEIR, 1990. Health Effects of Exposure to Low Levels of Ionizing Radiation, Report of the BEIR V Committee, National Research Council, National Academy Press, Washington D.C..

Bond, V. P., Varma, M. N., Sondhaus, C. A., and Feinendegen, 1985. L. E., An alternative to absorbed dose, quality, and RBE at low exposures, Radiat. Res. 104: S-52 - S-57.

Curtis, S. B., Dye, D. L., and Sheldon, W. R., 1966. Hazard from highly ionizing radiation in space. Health Phys. 12: 1069-1075.

Curtis, S. B., and Letaw, J. R., 1989. Galactic cosmic rays and cell-hit frequencies outside the magnetosphere. Adv. Space Res. 9 (No. 10): 293-298.

Curtis, S. B., Townsend, L. W., Wilson, J. W., Powers-Risius, P., Alpen, E. L., and Fry, R. J. M., 1992. Fluence-related risk coefficients using the Harderian gland data as an example. Adv. Space Res., 12(2): 407-416.

Grahn, D. (ed.), 1973. HZE-Particle Effects in Manned Spaceflight, National Academy of Sciences, Washington, D.C..

ICRP, 1977. Recommendations of the International Commission on Radiological Protection, ICRP Publ. #26, Ann. of the ICRP, 1(3), Pergamon Press, Oxford.

ICRP, 1990 Recommendations of the International Commission on Radiaological Protection, ICRP Publ. #60, Ann. of the ICRP, 21, No.1-3, Pergamon Press, Oxford, 1991.

NCRP, 1989. Guidance on Radiation Received in Space Activities, NCRP Report No. 98, National Council on Radiation Protection and Measurements, Bethesda, MD.

Shimizu, Y., Kato, H., and Shull, W. J., 1990. Studies of the mortality of A-bomb survivors. 9. Mortality, 1950-1985: Part 2. Cancer mortality based on the recently revised doses (DS86). Radiat. Res. 121: 120-141.

Shinn, J. L., Wilson, J. W., and Ngo, D. M., Risk assessment methodologies for target fragments produced in high-energy nucleon reactions, Health Phys., 59: 141-143 (1990).

Sondhaus, C. A., Bond, V. P., and Feinendegen, L. E., 1990. Cell-oriented alternatives to dose, quality factor, and dose equivalent for low-level radiation. Health Phys., 59(1): 35-48.

Todd, P. , 1967. Heavy-ion irradiation of cultured human cells. Radiat. Res Suppl. 7: 196-207.

SOLAR PARTICLE DOSE RATE BUILDUP AND DISTRIBUTION IN CRITICAL BODY ORGANS

William Atwell and Mark D. Weyland

Rockwell International
Space System Division
Houston, Texas 77058 USA

Lisa C. Simonsen

NASA Langley Research Center
Hampton, Virginia 23665 USA

ABSTRACT

Human body organs have varying degrees of radiosensitivity as evidenced by radioepidemiologic tables. The major critical organs for both the male and female that have been identified include the lung, thyroid, stomach, and breast (female). Using computerized anatomical models of the 50th percentile United States Air Force male and female, we present the self-shielding effects of these various body organs and how the shielding effects change as the location (dose point) in the body varies.

Several major solar proton events from previous solar cycles and several events from the current 22nd solar cycle have been analyzed. The solar particle event rise time, peak intensity, and decay time vary considerably from event to event. Absorbed dose and dose equivalent rate calculations and organ risk assessment data are presented for each critical body organ. These data are compared with the current NASA astronaut dose limits as recommended by the National Council on Radiation Protection and Measurements.

INTRODUCTION

The purpose of this paper is to present the results of a preliminary investigation of the relative radiation risks to specific body organs, both male and female, from acute space radiation exposures due to major solar particle events. Extremely detailed computerized anatomical human models have been developed to aid in estimating these exposures. At the suggestion of the National Council on Radiation Protection and Measurements (NCRP) (Report No. 98, 1989), the National Aeronautics and Space Administration (NASA) has established radiation dose limits as part of the astronaut health care and protection plan. These dose limits have been adopted as part of the Space Shuttle program flight rules. Table 1 shows the current astronaut dose limits for the skin, eye, and a 5-cm depth. The depth dose is commonly referred to as the BFO (Blood-Forming Organ) exposure. These limits are gender-specific, although not organ-specific, and are based on age at first exposure. The recommended career limits are based on a three percent lifetime excess risk of fatal cancer from induced tumors.

Biological Effects and Physics of Solar and Galactic Cosmic Radiation,
Part B, Edited by C.E. Swenberg *et al.,* Plenum Press, New York, 1993

Table 1. NASA Astronaut Radiation Exposure Limits (cSv)

CONSTRAINT	DEPTH	EYE	SKIN
30-DAY	25	100	150
ANNUAL	50	200	300
CAREER	100 - 400[*]	400	600

[*]MALE: 200 + 7.5(AGE - 30) cSv, up to 400 cSv maximum

FEMALE: 200 + 7.5(AGE - 38) cSv, up to 400 cSv maximum

In addition, the National Institutes of Health (NIH) has published cancer morbidity and mortality data tables for specific body organs (Rall et al, 1985), and these data are incorporated in calculating radiation risk assessments for radiosensitive body organs. Major solar particle events (SPE's) that occur cannot be predicted with any significant confidence with regard to event magnitude and duration, and these represent a serious threat to space crews in high inclination, earth-orbiting missions or on flights beyond the geomagnetosphere. Two major solar particle events that occurred in August 1972, and more recently, October 1989 are used as the source terms in the analyses of radiation exposure and risk assessment in this paper.

We discuss the computerized anatomical male and female models, the shielding

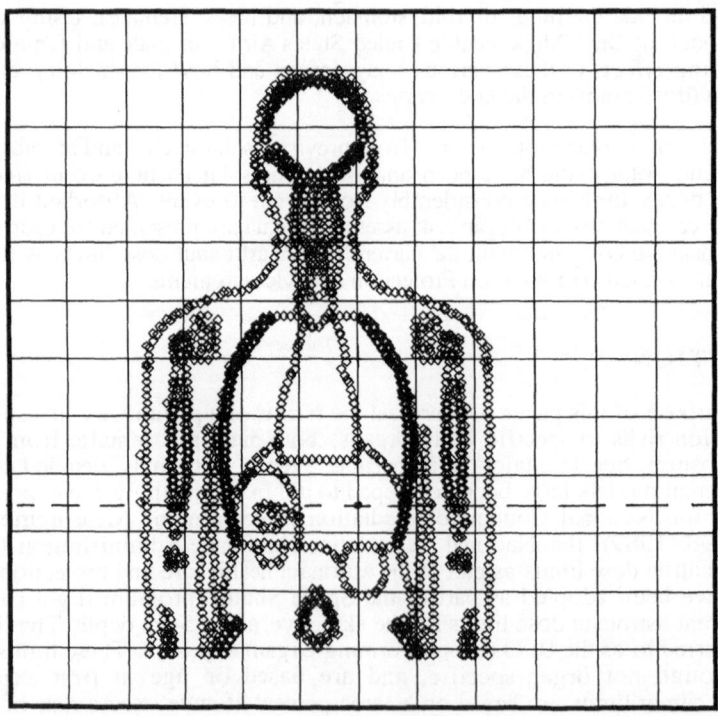

Figure 1. Cross-sectional Plot Through the Computerized Anatomical Man (CAM) - Stomach.

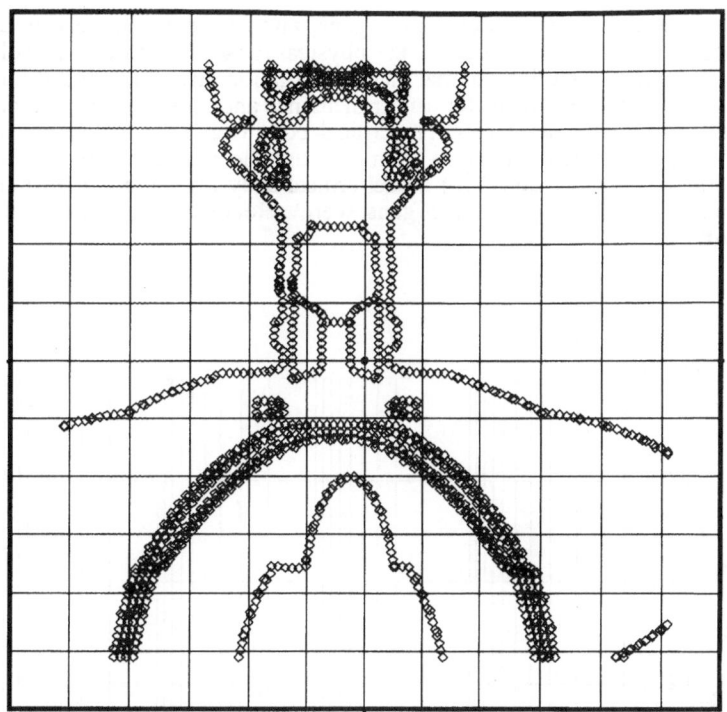

Figure 2. Cross-sectional Plot Through the Computerized Anatomical Man (CAM) - Thyroid.

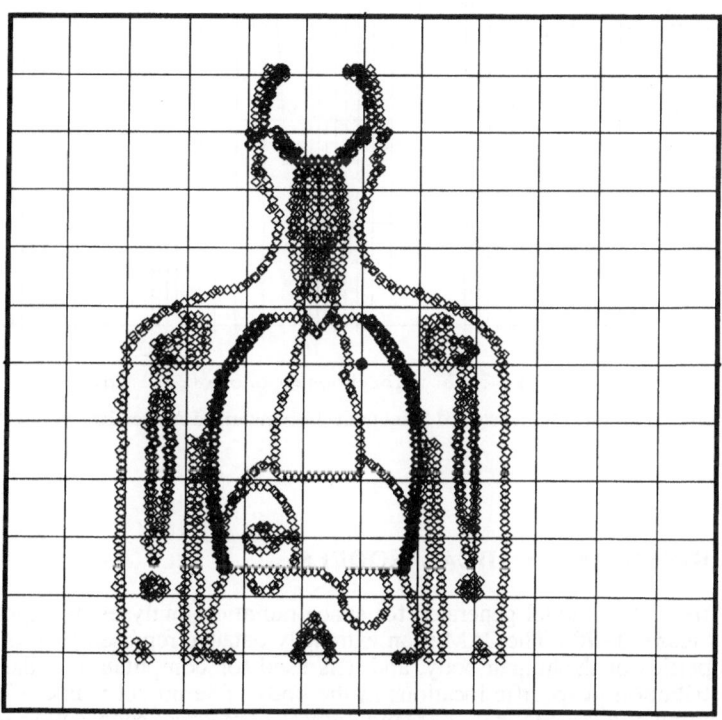

Figute 3. Cross-sectional Plot Through Computerized Anatomical Man (CAM) - Lung.

distributions of the individual critical body organs, and how these models are used with the radiation transport programs to calculate the physical (absorbed dose) and biological (dose equivalent) exposures given the external incident radiation environment and bulk shielding. The specific organ risk assessments for cancer incidence and mortality are then made based on NIH radioepidemiological data tables. The critical body organs considered in this paper for the computerized anatomical male (CAM) model are the stomach, thyroid, and lung; the computerized anatomical female (CAF) model critical body organs considered are the breast, thyroid, stomach, and lung. These organs were selected since they have the highest radiosensitivity.

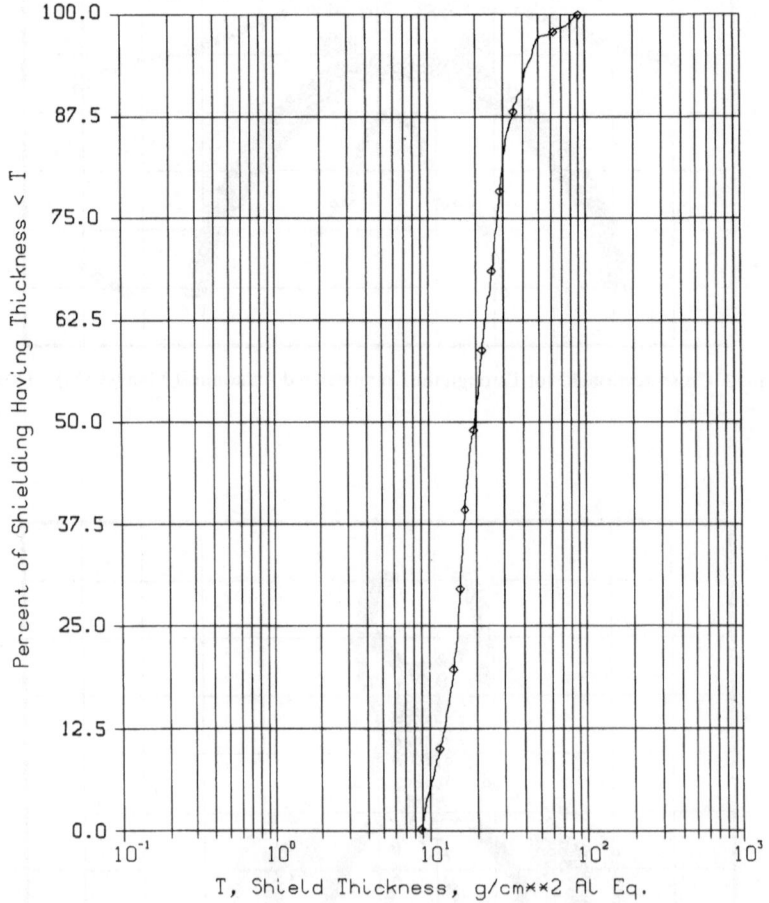

Figure 4.   Shield Distribution for a Selected Point in the Computerized Anatomical Man (CAM) - Stomach

COMPUTERIZED ANATOMICAL NODELS

The first CAM model generated for space radiation analyses and applications was developed by Kase (1970). The CAM is an extremely detailed representation of the radiation transport properties of the human body, and it is used for computation of the areal density shielding distribution at specific locations in the body. The model was configured in two versions: standing and seated. Over 2200 individual geometrical shapes were used to depict the external conformation, the skeleton, and the principal organs. The exterior dimensions are

834

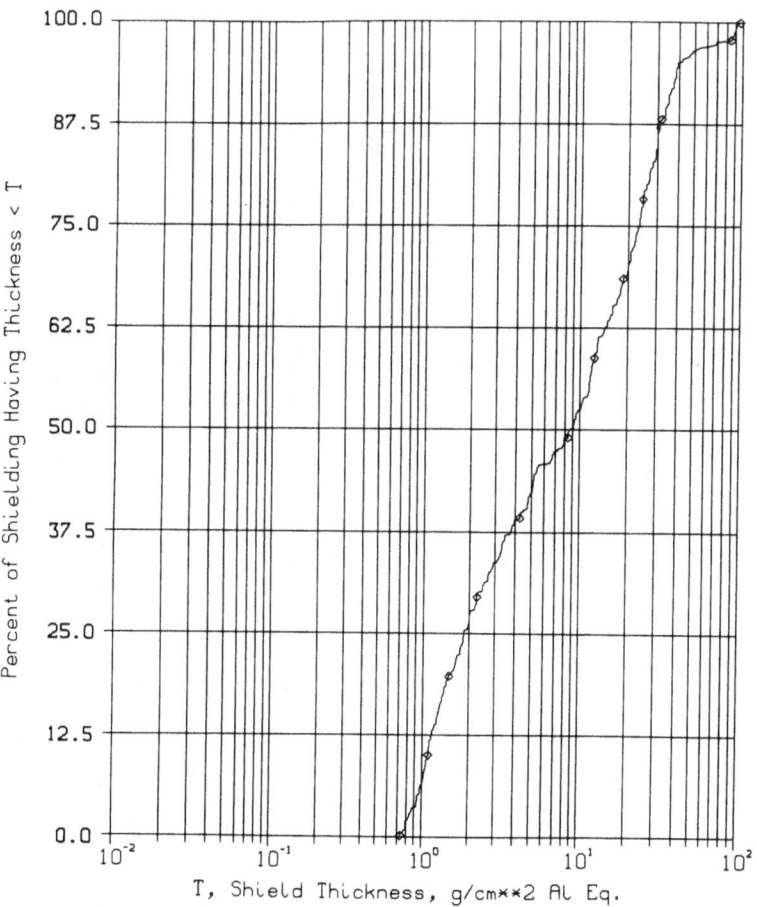

Figure 5. Shield Distribution for a selected Point in the Computerized Anatomical Man (CAM) - Thyroid.

those of the 50th percentile US Air Force man: weight equal to 72.855 kg (161.9 lb); height equal to 175.514 cm (69.1 in.). The skeleton and critical body organs were scaled from life-size models to conform to the exterior. The model includes variations of material density and fractional composition by weight due to the principal chemical elements contained in muscle, bone, bone marrow, and organ tissue.

Since a time constraint existed with task completion, the model was not fully verified. It was later discovered that numerous errors existed to the point that the model needed a complete "overhauling." Billings and Yucker (1973) were given the task of completely correcting and verifying the model. Over the past several years, it has been used very satisfactorily by the NASA Johnson Space Center, Space Radiation Analysis Group (SRAG) with only minor updates required. The computer program, CAMERA (see Billing and Yucker, 1973) is utilized to read in the CAM model to produce a number of user options. The SRAG primarily uses the CAMERA program and CAM model to produce a 512-ray, equal solid angle shielding distribution for any body location of interest. In addition, another attractive feature of the CAMERA program capability is the ability to produce cross-sectional computer plots in the x-, y-, and z-plane with the origin situated at the dose point. Figures 1-3 are cross-sectional plots through the CAM model stomach, thyroid, and lung, respectively; figures 4-6 are the corresponding organ shielding distributions converted to a standard material, aluminum. A number of dose points were taken for each organ to obtain some representative average distribution. Due to the lack of "body self-shielding," it is seen that the thyroid is a fairly "exposed" organ and thus susceptible to receiving larger radiation doses than the stomach or lung for a given external radiation source.

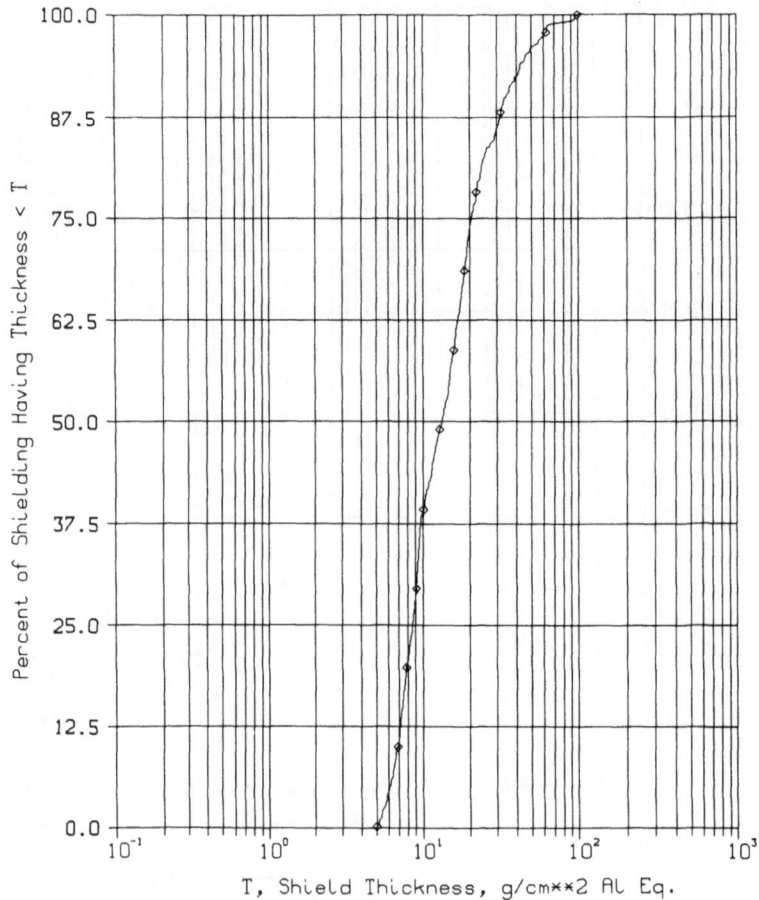

Figure 6.   Shield Distribution for a Selected Point in the Computerized Anatomical Man (CAM) - Lung.

The computerized anatomical female (CAF) model was developed by Yucker and Huston (1990) by modifying the CAM model and adding the female organs: breasts, uterus, and ovaries. By applying a scale factor of 0.92 to the CAM model geometry, a close approximation to the physical dimensions of the 50th percentile USAF female is obtained. The adipose tissue layer has also been included in the modelling of the female breasts. The thickness of this layer is taken as 0.5 cm of fat with a density of 0.918 g/cm^3 (seeNCRP, Report No. 85, 1986). The female organ specifications (size, location chemical composition, etc) were taken from reference ICPR, Report No. 23 (1973).

Figure 7 shows a cross-sectional plot through the breast. Figure 8 represents the shielding distribution for a location within the breast. The location selected is in the glandular tissue region which is the most vulnerable part of the breast to radiation risk (see NCRP,Report No. 85, 1986). Organ risk assessment will be discussed in a later section.

SOLAR PROTON EVENT EXPOSURES

Two of the largest solar proton events on record occurred in August 1972 and October 1989. We have used the free-space proton spectra from these two events to compute the CAM and CAF body organ radiation exposures using the shielding distributions discussed earlier. The integral proton spectra for these events are shown in figure 9.

Table 2.   Absorbed Doses and Dose Equivalents for the August 1972 Solar Event.

CAM	Absorbed Dose (cGy)				Dose Equivalent (cSv)			
	0.5	2.0 cm of H_2O	5.0	10.0	0.5	2.0 cm of H_2O	5.0	10.0
Stomach	26	19	10	4	36	25	14	6
Rt. Lung	72	48	23	8	97	64	31	11
Thyroid	726	342	107	26	1024	471	144	36
CAF	Absorbed Dose (cGy)				Dose Equivalent (cSv)			
	0.5	2.0 cm of H_2O	5.0	10.0	0.5	2.0 cm of H_2O	5.0	10.0
Stomach	33	23	12	4	44	31	16	6
Rt. Lung	84	55	26	9	112	74	35	12
Thyroid	739	349	110	27	1042	480	148	37
Breast	534	273	97	26	1054	371	130	35

Table 3.   Absorbed Doses and Dose Equivalents for the October 1989 Solar Event.

CAM	Absorbed Dose (cGy)				Dose Equivalent (cSv)			
	0.5	2.0 cm of H_2O	5.0	10.0	0.5	2.0 cm of H_2O	5.0	10.0
Stomach	27	21	14	8	36	28	19	11
Rt. Lung	53	40	25	12	70	53	33	17
Thyroid	361	160	66	26	538	220	87	34
CAF	Absorbed Dose (cGy)				Dose Equivalent (cSv)			
	0.5	2.0 cm of H_2O	5.0	10.0	0.5	2.0 cm of H_2O	5.0	10.0
Stomach	31	24	16	9	41	32	21	12
Rt. Lung	59	44	27	13	78	59	36	18
Thyroid	367	163	67	26	547	224	89	35
Breast	242	133	64	26	337	178	84	34

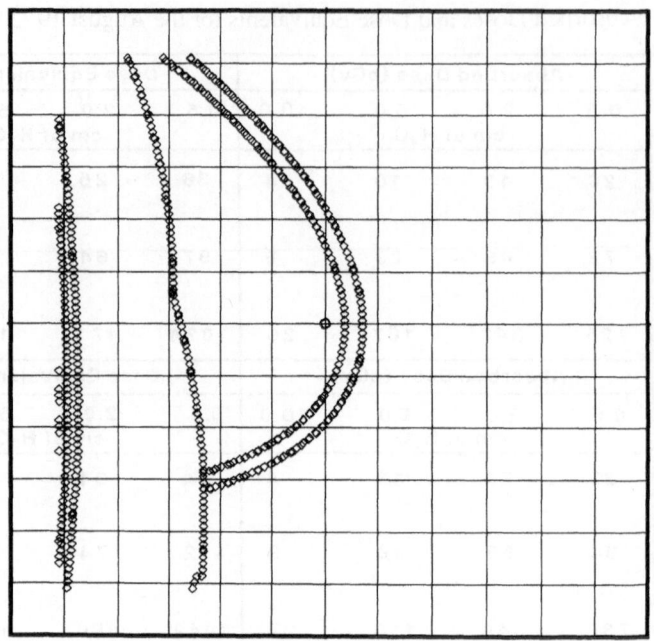

Figure 7. Cross-sectional Plot Through Computerized Anatomical Female (CAF)) - Breast.

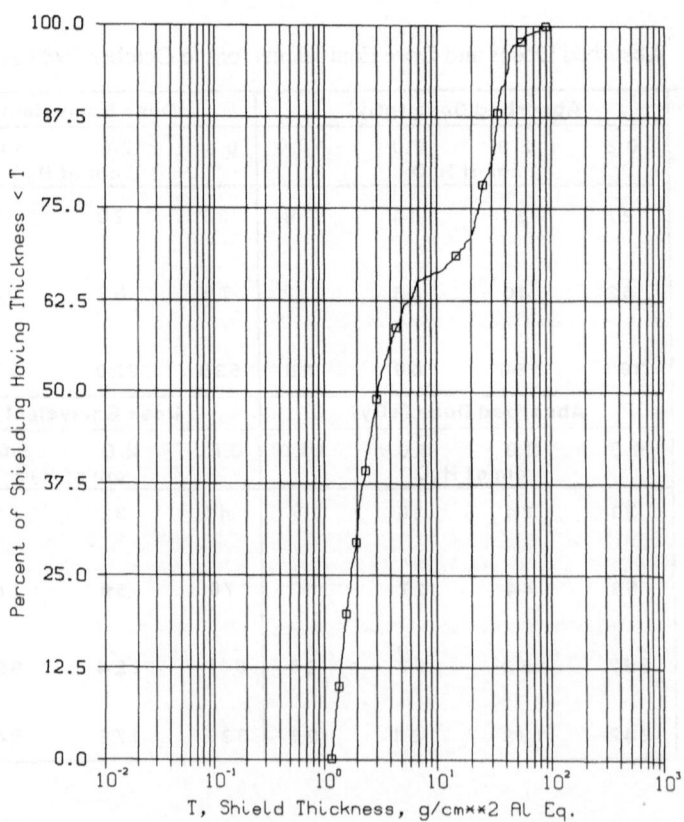

Figure 8. Shield Distribution for a Selected Point in the Computerized Anatomical Female (CAF) - Breast.

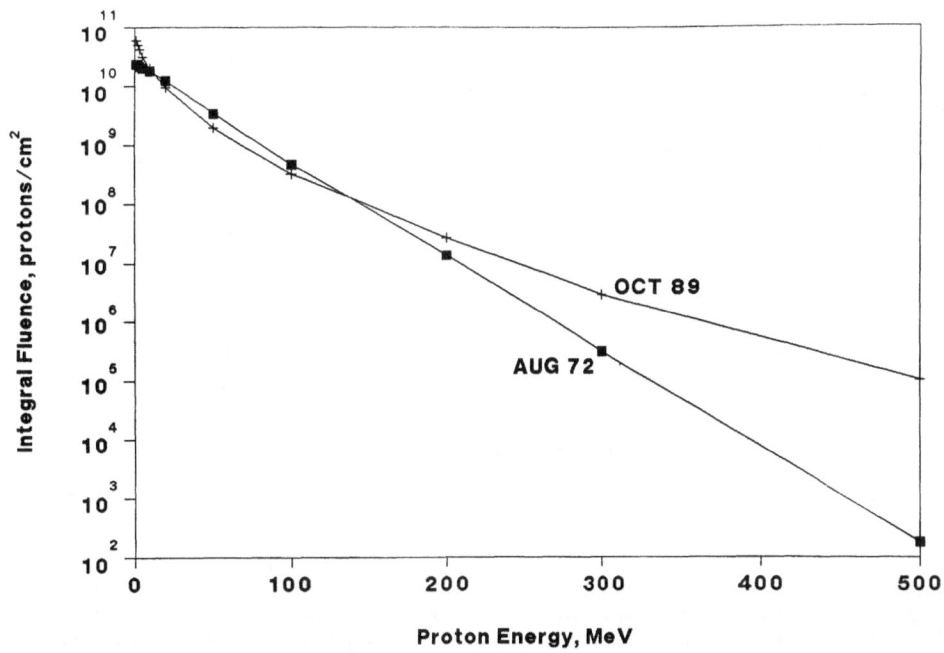

Figure 9.   Integral Fluence as a Function of Proton Energy for the Solar Proton Events of August 1972 and October 1989.

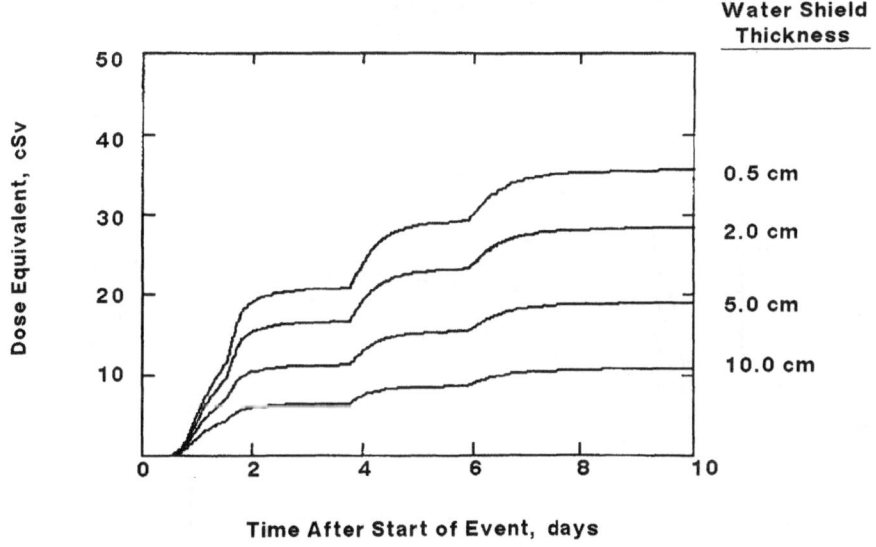

Figure 10.   Dose Rate Buildup in the CAM Stomach (October 1989 Solar Particle Event).

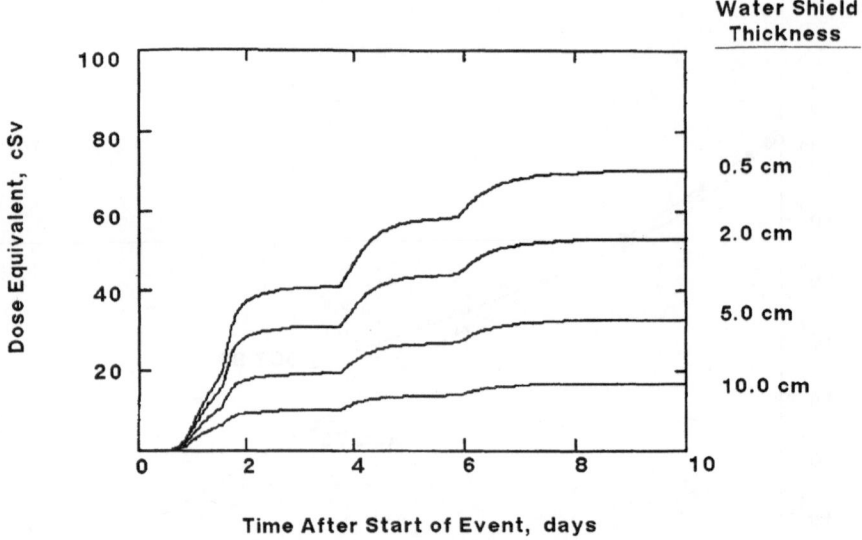

Figure 11.   Dose Rate Buildup in the CAM Right Lung (October 1989 Solar Particle Event).

Tables 2 and 3 show the computed exposures for the CAM and CAF body organs as a function of water shielding thickness. It is readily apparent that the thinly-shielded CAM thyroid and CAF thyroid and breast can receive very large exposures from these events. The CAM and CAF stomach and right lung experience comparable exposures with the CAF organs being slightly larger due to less self-shielding. The 30-day depth dose limit of 25 cSv (Table 1) is exceeded for less than 2 cm of water shielding. In fact, the annual limit of 50 cSv is exceeded for the right lung.

Cumulative dose rate buildup data for the October 1989 solar particle event are shown in figures 10-12 for the CAM stomach, right lung, and thyroid, respectively. Figures 13-16 show similar data for the CAF and include the breast. It is important to note that a significant fraction of the total event exposure is received within the first two days after particle onset. The obvious implication is the need for a prediction and warning system that minimizes a "false alarm" probability and provides for adequate time for the crew members to seek the

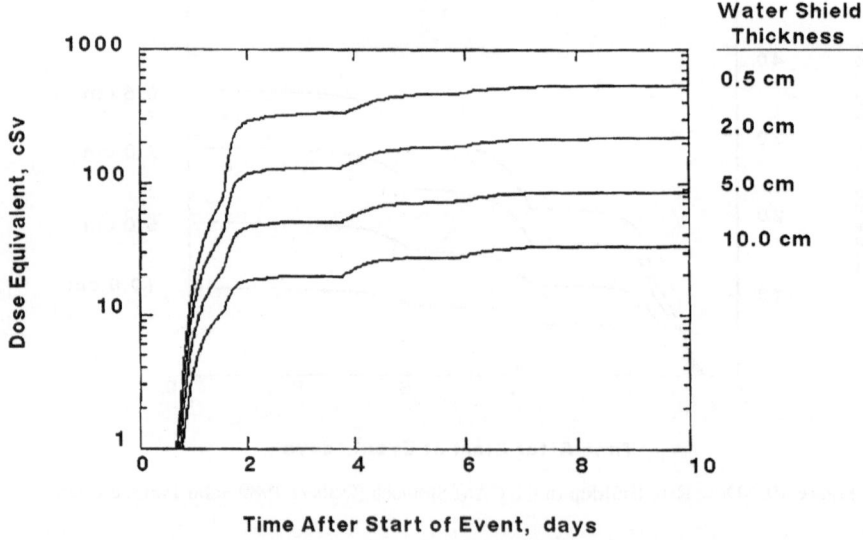

Figure 12.   Dose Rate Buildup in the CAM Thyroid (October 1989 Solar Particle Event).

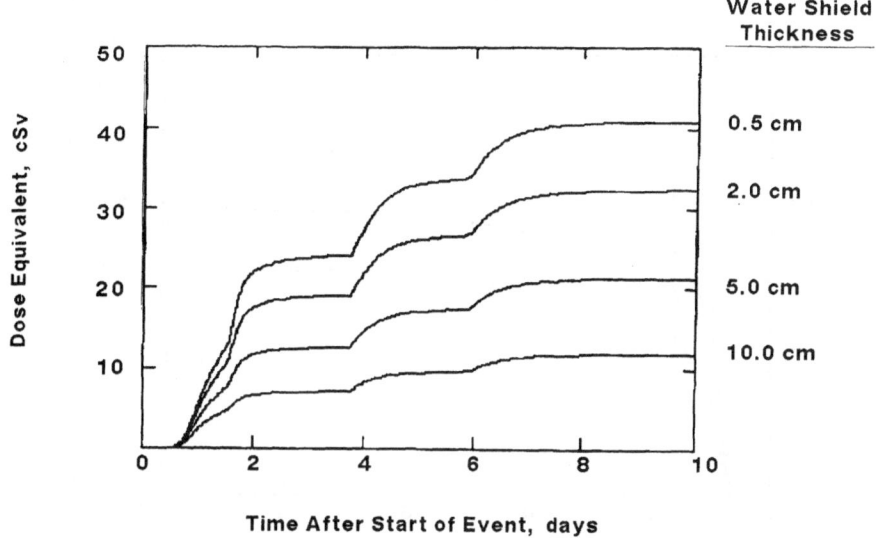

Figure 13. Dose Rate Buildup in the CAF Stomach (October 1989 Solar Particle Event).

protection of additional shielding or shelter to minimize their exposure. We are currently reviewing and refining the particle data from the August 1972 event and plan to perform a similar dose rate buildup analysis.

ORGAN RISK ASSESSMENT

In an earlier paper (Atwell et al, 1991) body organ risk assessments for the CAF were discussed. We have utilized the cancer morbidity and mortality data published by the NCRP (Report No. 85, 1989) and the NIH (Rall et al, 1985) to calculate radiation risks to specific male and female critical body organs for the total acute exposure from the August 1972 and October 1989 solar events. These excess risks are based on gender, age at first exposure, and

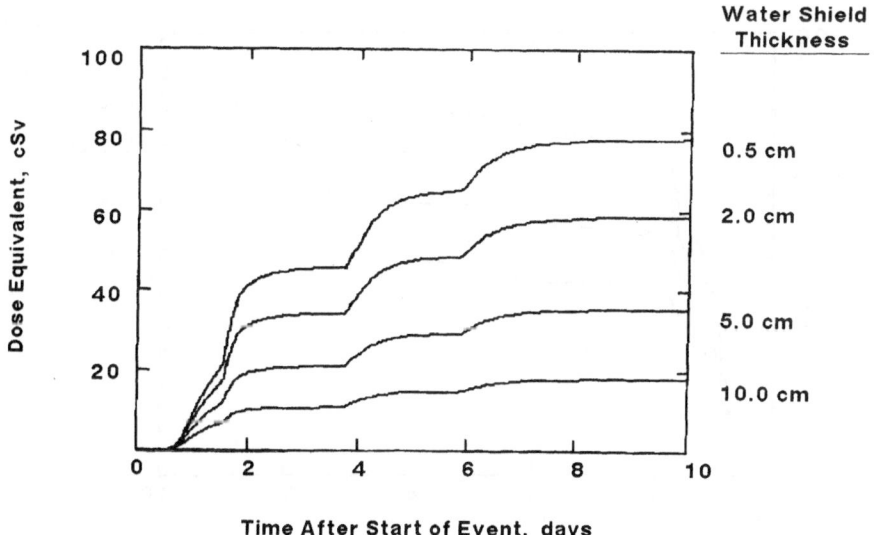

Figure 14. Dose Rate Buildup in the CAF Right Lung (October 1989 Solar Particle Event).

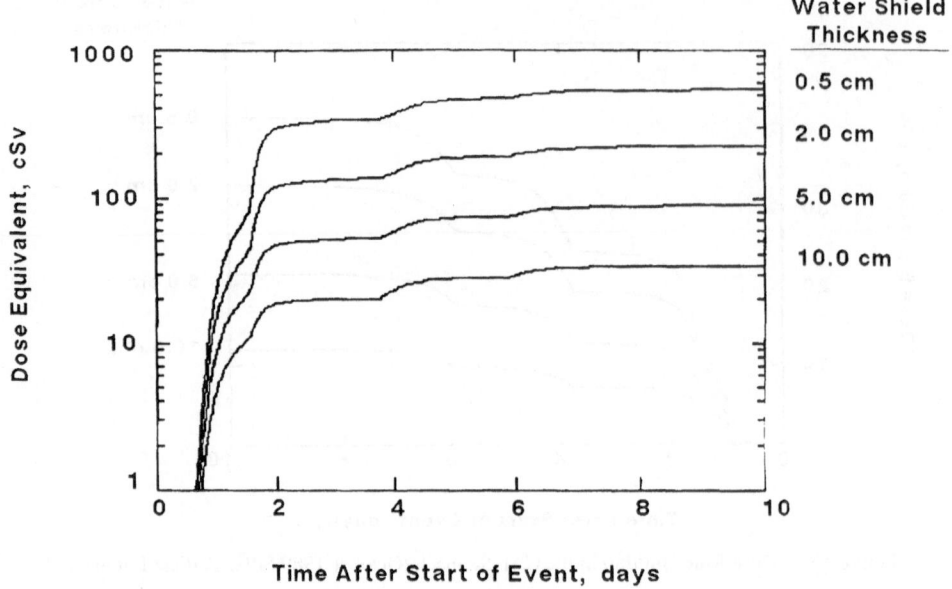

Figure 15.   Dose Rate Buildup in the CAF Thyroid (October 1989 Solar Particle Event).

body organ. As discussed in the NCRP, report No. 98 (1989), a linear-quadratic relationship exists between exposure and risk for all organs (sites) except for the thyroid and breast where a linear relationship is used. Table 4 shows several examples of the increased radiation risk of excess cancer incidence (morbidity) and cancer mortality per 1000 persons for these two SPE's for the CAM stomach and the CAF breast.

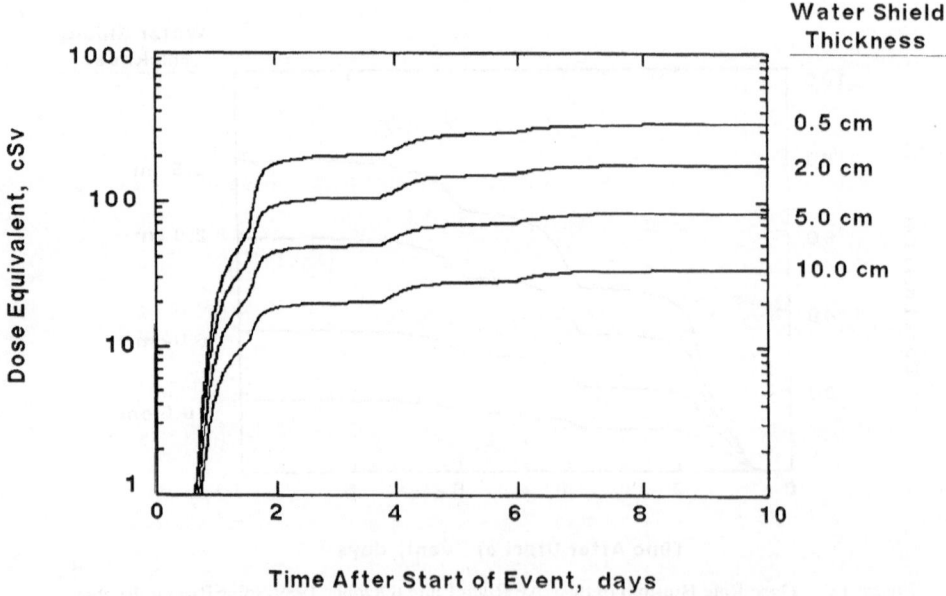

Figure 16.   Dose Rate Buildup in the CAF Breast (October 1989 Solar Particle Event).

Table 4.   Body Organ Risk Estimates From the August 1972 and October 1989 Solar Events.

CAM Stomach

	August 1972					October 1989			
Age	0.5	2.0	5.0	10.0	Age	0.5	2.0	5.0	10.0
	Water Shield (cm)					Water Shield (cm)			
Excess cancers/1000 people					Excess cancers/1000 people				
25	2.3	1.5	0.7	0.3	25	2.3	1.7	1.1	0.6
35	1.2	0.8	0.4	0.1	35	1.2	0.9	0.6	0.3
45	0.7	0.5	0.2	0.1	45	0.7	0.6	0.3	0.2
55	0.7	0.5	0.2	0.1	55	0.7	0.5	0.3	0.2
Excess cancer mortality/1000 people					Excess cancer mortality/1000 people				
25	1.6	1.1	0.5	0.2	25	1.6	1.2	0.8	0.4
35	0.8	0.5	0.3	0.1	35	0.8	0.6	0.4	0.2
45	0.5	0.3	0.2	0.1	45	0.5	0.4	0.2	0.1
55	0.5	0.3	0.2	0.1	55	0.5	0.4	0.2	0.1

CAF Breast

	August 1972					October 1989			
Age	0.5	2.0	5.0	10.0	Age	0.5	2.0	5.0	10.0
	Water Shield (cm)					Water Shield (cm)			
Excess cancers/1000 people					Excess cancers/1000 people				
25	329	116	41	11	25	1.5	56	26	11
35	203	71	25	7	35	65	34	16	6.6
45	36	13	5	1.2	45	12	6	3	1.2
55	18	6	2	0.6	55	6	3	1.4	0.6
Excess cancer mortality/1000 people					Excess cancer mortality/1000 people				
25	105	37	13	3.4	25	34	18	8.4	3.4
35	64	23	8	2.1	35	21	11	5.1	2.1
45	12	4	1.4	0.4	45	3.7	2	0.9	0.4
55	6	2	0.7	0.2	55	1.8	1	0.5	0.2

SUMMARY

Detailed male and female computerized anatomical models have been developed for use in space radiation analyses.

Capability exists for calculating shielding distributions at specific sites in specific critical body organs.

Absorbed dose and dose equivalent data have been presented for major solar proton events as a function of body organ and shielding material thickness.

Dose rate buildup data were presented and discussed for major solar particle events.

Excess cancer incidence and excess cancer mortality risk data have been presented and discussed as a function of gender, age, body organ, shield material thickness, and solar event.

REFERENCES

Atwell, William, Mark D. Weyland, and A.C. Hardy, "The Computerized Anatomical Female Model: Body Organ Risk Assessments," Second Annual Investigators Meeting on Radiation Research, Houston, TX, April 22-23, 1991.

Billings, M.P. and W.R. Yucker, "The Computerized Anatomical Man (CAM) Model," NASA CR-134043, 1973.

Kase, Paul G. , "Computerized Anatomical Model Man," Air Force Weapons Laboratory, Kirtland Air Force Base, NM, Tech. Report No. AFWL-TR-69-161, January 1970.

International Commission on Radiological Protection, "Report of the Task Group on Reference Man," Report No. 23, Pergamon Press, New York, NY, 1973.

National Council on Radiation Protection and Measurements, "Mammography - A User's Guide," NCRP Report No. 85, Bethesda, MD, March 1986.

National Council on Radiation Protection and Measurements, "Guidance on Radiation Received in Space Activities," NCRP Report No. 98, Bethesda, MD, July 31, 1989.

Rall, J.F. , G.W. Beebe, D.G. Hoel, S. Jablon, C.E. Land, O.F. Nygaard, A.C. Upton, and R.S. Yallow, "Report of the National Institutes of Health Ad Hoc Working Group to Develop Radioepidemiological Tables," DHHS Publication No. (NIH) 85-2748, U.S. Government Printing Office, Washington, D.C., 1985.

Yucker W.R. and S.L. Huston, "Computerized Anatomical Female. Final Report," McDonnell Douglas Corporation Report MDC H-6107, September 1990.

NUMERICAL SIMULATION OF 'DMSP' DOSIMETER RESPONSE

Thomas M. Jordan

Experimental and Mathematical Physics Consultants
Post Office Box 3191
Gaithersburg, Maryland 20885 USA

ABSTRACT

Four DMSP dosimeters were modeled for numerical simulation of radiation response. The modeling included the hemispherical aluminum dome, the solid state detector, and the tungsten base plate. Orbits were generated for 840 km and 98 degrees inclination and used with 1965 and 1985 magnetic field models and the AP8 and AE8 data sets to obtain solar minimum and solar maximum integral fluences for protons and electrons. Adjoint Monte Carlo methods were then used to simulate the transport of these environments in the geometric models of the dosimeters. Volume average dose calculations were used to calculate the response of the LOLET (less than 1 MeV deposited per particle) channels to electrons and secondary bremsstrahlung. Monte Carlo methods were used, in conjunction with a pulse height analysis, to obtain the proton response of the LOLET and HILET (1 to 10 MeV deposited per particle) channels. The HILET and LOLET responses obtained from these calculations are in good agreement with DMSP measurements for 1984-85.

INTRODUCTION

The Defense Meteorological Satellite Program (DMSP) satellite F7 carries a dosimeter package provided by the Air Force Geophysics Laboratory. The dosimeters provide accurate time-resolved dose measurements for an 840 km and 98 degree inclination orbit. These measurements provide a comparative basis for determining the accuracy of current trapped radiation models. The comparison with measurements uses Monte Carlo simulation of proton and electron transport in a three dimensional model of the dosimeters.

Dosimeter Characteristics

The four DMSP dosimeters have similar geometries consisting of a hemispherical aluminum dome covering a solid state detector which lies on a SiO_2 substrate. Both the dome and the substrate lie over a cylindrical tungsten plate which reduces the radiation levels from the backward hemisphere. The basic geometry characteristics (Hanser, 1991) of the dosimeters are listed in Table I. A cutout view of the geometry is shown in Figure 1.

The dosimeters are instrumented to obtain count rates in several broad channels. The LOLET channel measures particles that deposit, individually, less than 1 MeV. The HILET channel measures particles that drop between 1 and 10 MeV in the solid state detector. The VHLET channel measures particles that deposit at least 40 MeV. This channel applies

Biological Effects and Physics of Solar and Galactic Cosmic Radiation,
Part B, Edited by C.E. Swenberg *et al.*, Plenum Press, New York, 1993

845

Table 1. DMSP Dosimeters

```
          Characteristics of the DMSP Dosimeters

Dosimeter Number:  (1)     (2)     (3)     (4)
Dome (g/sqcm)      0.55    1.55    3.05    5.91
Inner Radius(mm)   10.2    11.5    11.4    11.6
Solid State Detector
   Area (sqcm)     .051    1.00    1.00    1.00
   Thick (microns) 398     403     390     384
   LOLET Channel     0.05 to 1.00 MeV Deposited
   HILET Channel     1.00 to 10.0 MeV Deposited
```

primarily to cosmic rays and proton nuclear events and is not discussed further here. Additional details of the dosimeter system and on orbit dose rates are given by Mullen et al., 1988.

Trapped Radiation Environment

The dosimeters fly in a 840 kilometer orbit with a 98.7 degree inclination. The trapped electron and proton environments for this orbit were obtained for two different years, 1965 and 1985. The trapped radiation environments were calculated using the SOFIP code and the AP8 (Sawyer and Vette, 1976) and AE8 (Space Science Data) environment models. For both years, solar minimum and solar maximum environments were obtained. The integral average daily fluxes for these conditions are given in Tables 2 (protons) and 3 (electrons).

The 1965 data are known to represent consistency between the measurements that constitute the AP8 and AE8 models and the magnetic fields that existed when the measurements were made. The 1985 data use the same AP8 and AE8 models but weaker magnetic fields when extrapolated to 1985. This combination does not properly account for absorption processes at lower altitudes, e.g., shuttle missions, but hopefully is reasonably accurate at the 840 km altitude.

Dose Attenuation Data

These environments were first used to generate one dimensional attenuation data for transport through simple aluminum shields. The transport of the protons, electrons, and

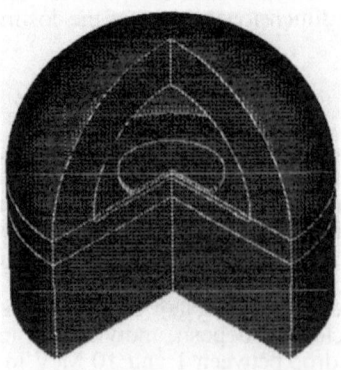

Figure 1. Dosimeter #2 Geometry.

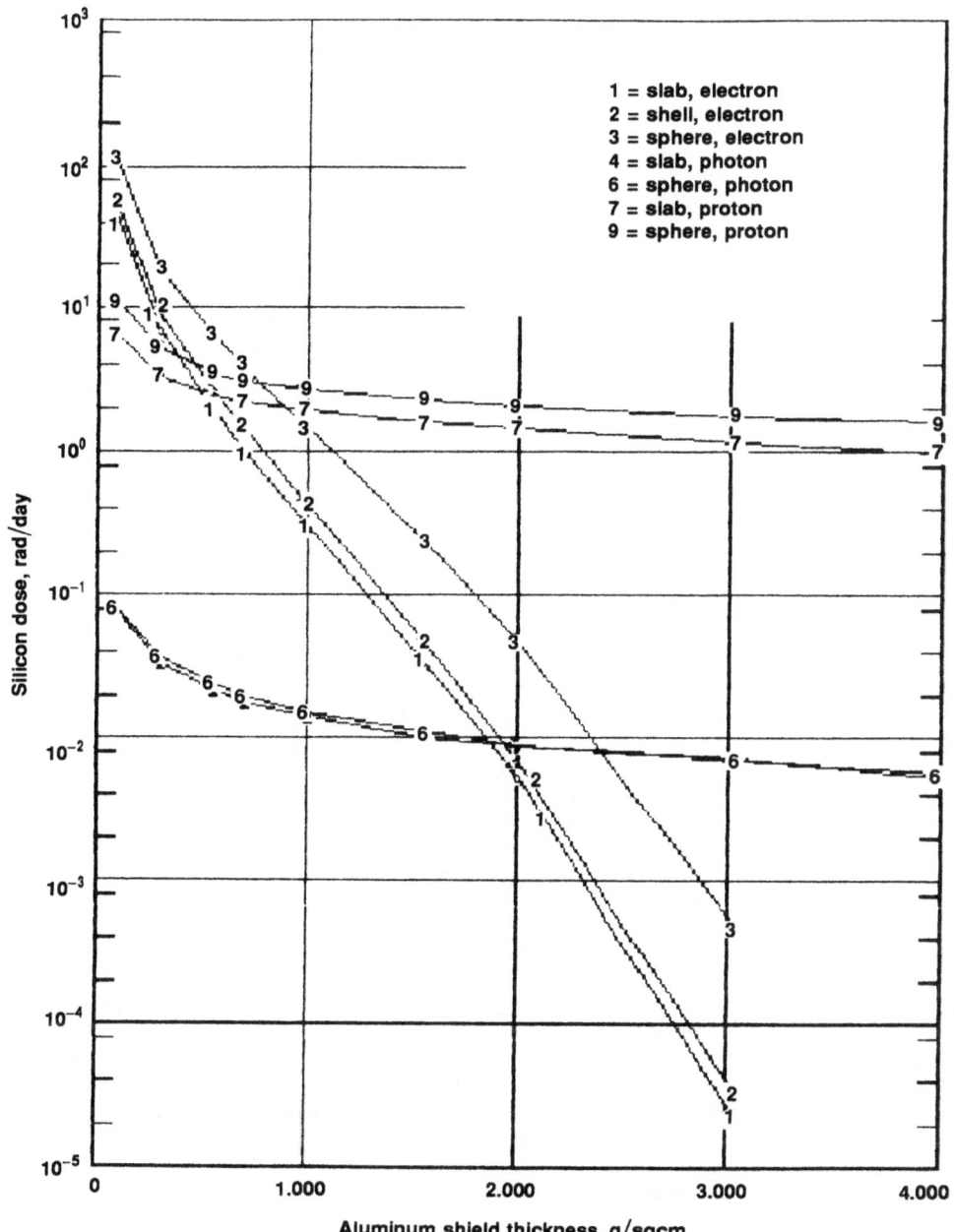

Figure 2. 1985 Solar Minimum Attenuation.

secondary bremsstrahlung was calculated using the NOVICE code(Jordan, 1991). The attenuation data is used for rough estimates of the detector response. These data can also be used in approximate ray-trace/sectoring calculations of detector responses as indicated in the appendix. The attenuation data for the 1985 solar minimum environment are shown in Figure 2.

There are several notable features of the one dimensional attenuation data. First, the primary electron dose dominates for shield thicknesses of less than 0.5 g/sqcm. Thereafter, the proton environment dominates. Secondary bremsstrahlung dose is at least two orders of

Table 2. Proton Spectra.

Integral Proton Spectra (#/sqcm/sec >E)			
E(MeV)	85Min	85Max	65Min	65Max
2	696	459	374	188
3	548	371	299	162
4	457	317	253	145
5	399	281	224	133
6	359	256	203	125
8	316	226	182	115
10	286	204	167	107
15	250	179	149	97.0
20	223	159	136	88.3
25	209	149	128	83.4
30	197	139	122	78.8
35	187	132	116	75.2
40	178	126	110	71.7
45	170	120	105	68.4
50	162	114	101	65.3
55	154	108	95.6	62.3
60	146	103	90.9	59.4
70	132	93.7	82.4	54.1
80	120	85.1	74.7	49.3
90	109	77.4	67.8	45.0
100	98.8	70.4	61.6	41.2
125	76.4	54.9	47.6	32.2
150	59.2	42.9	36.9	25.2
175	46.1	33.8	28.8	19.9
200	36.0	26.6	22.5	15.8
250	22.2	16.4	13.8	9.71
300	13.7	10.2	8.57	6.01
350	8.52	6.39	5.33	3.75
400	5.31	4.01	3.32	2.34
500	2.08	1.59	1.31	0.93

magnitude below the proton dose. Of particular interest are the two curves for electron dose in spherical geometry (curves 2 and 3). The higher curve gives the dose at the center of a solid sphere while the lower curve is for the dose at the center of a spherical shell of the same thickness. The difference in dose shown by these two curves indicates the magnitude of the errors that can be introduced by using only simple attenuation models for the prediction of electron dose in more complicated geometries.

Monte Carlo Calculations

The three dimensional geometry of the four dosimeters was used in a series of adjoint Monte Carlo calculations. In the adjoint method, particles are tracked backward from the detector (the solid state detector in these problems) through the geometry to the external environment. In the backtracking process, the particles speed up. The calculations for the low LET (linear energy transfer) particles (electrons and secondary bremsstrahlung) were performed as dose calculations averaged over the volumes of the solid state detectors. Separate runs were made for each detector and for the primary electrons and the secondary bremsstrahlung photons. The results of these calculations are tabulated in Table 4. The measured total LOLET response of the dosimeters is indicated at the bottom of this table for comparison.

It can be seen that the total electron response (electrons + secondary bremsstrahlung)of the dosimeters is appreciable only for the first dosimeter. For dosimeter 2, the total electron response is about a factor of eight less than the measured LOLET response,

Table 3. Electron Spectra.

Integral Electron Spectra (#/sqcm/sec >E)			
E(MeV)	85Min	85Max	65Min	65Max
0.1	266000	611000	151000	352000
0.2	101000	266000	599000	153000
0.3	48200	118000	30500	71000
0.4	27700	55200	18400	36500
0.5	16500	28700	11600	20800
0.6	11900	19900	8560	14900
0.7	8710	14200	6420	11000
0.8	6710	10800	5030	8510
0.9	5400	8680	4090	6910
1	4350	6990	3330	5620
1.25	2720	4519	2090	3640
1.5	1710	2940	1320	2380
1.75	1130	1920	872	1560
2	747	1260	578	1020
2.25	501	843	388	684
2.5	337	565	261	459
2.75	202	336	162	279
3	126	205	104	172
3.25	78.1	123	66.4	105
3.5	50.6	75.1	43.6	64.7
3.75	31.3	43.2	27.1	37.3
4	19.8	25.1	17.1	21.6
4.25	11.4	13.6	9.84	11.7
4.5	6.63	7.43	5.73	6.41
4.75	3.59	3.89	3.09	3.33
5	2.00	2.07	1.70	1.75
5.5	0.46	0.46	0.38	0.38
6	.070	.070	.054	.054
6.5	.0017	.0017	0	0
7	0	0	0	0

and for dosimeters 3 and 4, the measured response is at least a factor of 20 higher than the total electron dose. However, this disagreement is caused by response of the LOLET channel to protons rather than errors in the electron environment.

To demonstrate the effect of protons on the LOLET channel, Monte Carlo calculations were performed for the protons using options that provide pulse height information. Processing of the pulse height distribution then gives the dose rate from protons in both the HILET and LOLET channels of the dosimeters. These data are shown in Table 5 for both channels, and the measured data for the channels are also indicated.

The HILET channel agrees best with the Monte Carlo calculations using the 1985 solar maximum environment. However, 1985 is supposed to be at the end of a solar minimum heading for solar maximum around 1991. The agreement with the 1985 solar minimum calculation is not as good, the calculation is about 30 to 40 percent higher than the measurements. The calculations using the 1965 solar minimum protons agree a little better but are approximately 25 percent lower than the measured data.

Returning to the LOLET channel, Table 6 sums the contributions from protons, electrons, and bremsstrahlung in the LOLET channel and compares these data with measured data. First, it is seen that the calculations now compare very favorably with the measurements. The response in the first dosimeter is predominately electrons. The response

Table 4.   Electron and Bremsstrahlung Dose Rates.

```
Calculated LOLET Channel Electron Dose (Rad/Day)
  Dosimeter Number:   (1)      (2)      (3)      (4)
     1965min          1.21     .024    .0001    0.00
     1965max          2.15     .036    .0002    0.00
     1985min          1.56     .029    .0001    0.00
     1985max          2.65     .042    .0002    0.00

Calculated LOLET Bremsstrahlung Dose (Rad/Day)
  Dosimeter Number:   (1)      (2)      (3)      (4)
     1965min          .010     .0049   .0033    .0024
     1965max          .022     .010    .0065    .0057
     1985min          .016     .0073   .0048    .0034
     1985max          .033     .015    .0095    .0068

Calculated LOLET Electron+Bremsstrahlung(Rad/Day)
  Dosimeter Number:   (1)      (2)      (3)      (4)
     1965min          1.22     .029    .0034    .0024
     1965max          2.17     .046    .0067    .0047
     1985min          1.58     .036    .0050    .0034
     1985max          2.69     .057    .011     .0068

Measured LOLET          2.51    .34      .26      .27
```

of the other three dosimeters is predominately protons. For dosimeter 1, the measured data is right between the 1985 solar minimum and solar maximum predictions (and agrees best with 1965 solar maximum prediction). For dosimeters 2 through 4, the best agreement is with 1985 solar minimum calculation.

Table 5.   Proton Dose Rate in Hilet and Lolet Channels.

```
Calculated HILET Channel Proton Dose (Rad/Day)
  Dosimeter Number:   (1)      (2)      (3)      (4)
     1965min          0.99     0.72     0.57     0.39
     1965max          0.64     0.46     0.36     0.25
     1985min          1.68     1.17     0.93     0.63
     1985max          1.21     0.83     0.64     0.44

Measured HILET:       1.21     0.83     0.63     0.49

Calculated LOLET Channel Proton Dose (Rad/Day)
  Dosimeter Number:   (1)      (2)      (3)      (4)
     1965min          0.28     0.20     0.19     0.15
     1965max          0.19     0.13     0.12     0.10
     1985min          0.46     0.33     0.29     0.25
     1985max          0.33     0.23     0.21     0.18

Measured LOLET:       2.51     0.34     0.26     0.27
```

Table 6.   Total Dose Rate, for the Lolet Channel.

```
Calculated Total LOLET Dose (Proton+Elec) (Rad/Day)

Dosimeter Number:  (1)    (2)    (3)    (4)
      1965min      1.50   0.23   0.19   0.15
      1965max      2.36   0.18   0.13   0.10
      1985min      2.04   0.37   0.29   0.25
      1985max      3.02   0.29   0.22   0.19

Measured LOLET:    2.51   0.34   0.26   0.27
```

CONCLUSION

Standard trapped radiation models, AP8 and AE8, were used to generate proton and electron environments for the DMSP dosimeters. These environments were then used in a Monte Carlo simulation of dosimeter response. Good agreement was obtained between calculated and measured responses for both the LOLET and HILET channels. Agreement was much better than the factor of 2 to 5 that is often given for the uncertainty of the trapped particle environment.

REFERENCES

Hanser, F. Personal Communication, Aug 1991.

Jordan, T.M., 1991, "NOVICE: A Radiation Transport/Shielding Code," E.M.P. Consultants, Gaithersburg MD, .

Mullen, E.G., M. S. Gussenhoven, and D. A. Hardy, 1988, "The Space Radiation Environment at 840 Km", in "Terrestrial Space Radiation and its Biological Effects" edited by P. D. McCormack, C. E. Swenberg, and Horst Bucker, NATO, Series A: Life Sciences Vol. 154, Plenum Press,.

Sawyer, D.G. and J. I. Vette, 1976, "AP-8 Trapped Proton Environment for Solar Maximum and Solar Minimum," National Space Science Data Center, NASA Goddard Space Flight Center, Greenbelt MD, WDC-A-R&S 76-06.

"AE-8," undocumented data set distributed by the National Space Science Data Center, NASA Goddard Space Flight Center.

Stassinopoulos, E.G., J. J. Hebert, B. L. Butler, and J. L. Barth, 1979,"SOFIP: A Short Orbital Flux Integration Program," NASA/WDC-A-R&S 79-01, National Space Science Data Center.

Appendix A: Comparison of Sectoring and Monte Carlo Calculations

The attenuation data obtained for simple one dimensional geometry shields can be used for approximate three dimensional calculations of radiation levels. This appendix describes the approximation and gives results of an application to the DMSP dosimeter system. The NOVICE code, SIGMA processor, was used for the sectoring calculations.

The sectoring approximation starts with a three dimensional model of a space system, here a DMSP dosimeter. A detector point is defined within the system, here the center of the solid state detector. Discrete solid angle intervals are defined around this point typically using many intervals in azimuth and polar angle cosine. For each solid angle interval, a direction is defined and a ray trace is performed from the detector point along that direction to the outside of the system. The mass thickness along the ray is then computed and used in an interpolation of a tabulated 'dose versus thickness' data set. The average of the doses received in all the solid angle intervals gives the final estimate of the dose at the detector point.

Sectoring can include a variety of scaling procedures to improve accuracy. Scaling included in the calculations here are: different attenuation properties of materials for protons, electrons, and bremsstrahlung, and different bremsstrahlung production properties of the materials. These scaling procedures make sectoring fairly accurate for protons and secondary bremsstrahlung.

Most sectoring calculations interpolate solid sphere attenuation data. However, the solid sphere data can over estimate radiation levels for electrons. Therefore, a second approximation is also used for electrons where the straight ahead approximation does not apply. In this approximation, the minimum path along the ray is estimated by the product of the slant path and the cosine of the angle between the ray and the surface normal at material interfaces. This minimum path is then used to interpolate a combination of the slab and spherical shell attenuation data (which are quite similar for electrons). Typical dose results obtained by these sectoring approximations are tabulated in Table A-I. Comparisions with Monte Carlo calculations are quite good except for electrons, where the two sectoring models bracket the more accurate Monte Carlo calculations.

Table A-1.   Sectoring Calculations, DMSP Dosimeters, 1985 Solar Minimum.

Dosimeter Number:	(1)	(2)	(3)	(4)
Protons:				
Sectoring	2.18	1.62	1.31	0.98
Monte Carlo	2.14	1.50	1.22	0.88
Electrons:				
Sectoring				
Slant Path	2.36	.083		
Minimum Path	1.08	.019		
Monte Carlo	1.56	.029		
Bremsstrahlung:				
Sectoring	.012	.0066	.0046	.0033
Monte Carlo	.016	.0073	.0048	.0034

THE RELATIVE BIOLOGICAL EFFECTIVENESS OF ATTENUATED PROTONS

James B. Robertson[1], William C. Glisson[1], John O. Archambeau[2],
George B. Coutrakan[2], Daniel W. Miller[2], Michael F. Moyers[2],
Jeffrey F. Siebers[2], James M. Slater[2], and John F. Dicello[3]

[1]Department of Environmental Health, East Carolina University, Greenville,
NC 27858 (U.S.A.)
[2]Department of Radiation Medicine, Loma Linda University Medical Center,
Loma Linda, CA 92354 (U.S.A.)
[3]Department of Physics, Clarkson University, Potsdam, NY 13676 (U.S.A.)

ABSTRACT

The 250 MeV Synchrotron at Loma Linda University Medical Center has been in operation since October 1990. In this paper we will report the data collected for survival of V79 cells suspended in gelatin exposed to 100 MeV protons that are attenuated with varying thicknesses of polycarbonate plastic. The measured RBE for single-cell killing ranges from 1.05 to 1.20 as the protons are reduced in energy from 64 MeV. Survival profiles show that this RBE effect will result in a 2.5 fold difference in cell survival between high and low energy protons at a dose of 1000 cGy. Correspondingly, the microdosimetric spectra measured in the same beam offer a quantitative explanation for the observed variation in RBE. Both the biological and the microdosimetric results suggest that the number of intermediate lineal energy events increase as the protons are attenuated. These results suggest that while shielding will reduce the total absorbed dose due to a solar proton event, it might not be equally effective in reducing the radiobiological effect of the exposure.

INTRODUCTION

If we are to explore space beyond the protection of the earth's magnetic fields, the individuals involved will be exposed to solar and galactic cosmic radiation. The radiation exposure consists of the relatively constant galactic cosmic radiation spectrum which is about 85 percent protons and the transient radiation from solar particle events, where 90 percent or more of the particles are protons (Stassinopoulos, 1988; NCRP Report No. 98). Clearly, protons are the most frequently encountered particles in the free space radiation environment beyond the earth's radiation belts. While heavy ions are known to be effective biologically, and their contribution to chronic effects may be significant, the level of damage is still uncertain (Dicello, 1992). Protons from solar particle events pose the only natural acute radiation hazard that will be encountered during interplanetary space flight or during normal planetary and lunar base operations. This will also be the case for individuals living on a

Biological Effects and Physics of Solar and Galactic Cosmic Radiation,
Part B, Edited by C.E. Swenberg *et al.*, Plenum Press, New York, 1993

853

space station if its orbit included regions outside the area protected by the earth's magnetosphere. All designs for human habitation must include some form of radiation protection from solar flare protons. Since these protons are of mixed energies, it is important to study and understand the radiation biology of mixed energy fields of protons. As we have stated earlier, underestimation of the risk could lead to serious harm of the exposed individuals while overestimation of the risk will cause unacceptable inefficiency, and neither of these alternatives is tolerable (Robertson, et al., 1988).

Mixed energy proton fields with energies up to 250 MeV are available for such studies at the Loma Linda University Proton Treatment Facility. In this paper we report on the results obtained when cells suspended in gelatin are exposed to a mixed field of protons with energies ranging from 64 MeV to stopping protons.

MATERIALS AND METHODS

Cell Culture

Chinese hamster lung cells, V79, were maintained as exponentially growing monolayers at 37°C in a humidified 5 percent CO_2, 95 percent air atmosphere. The nutrient medium used was Eagles' minimal essential medium, MEM, (Gibco F-15) supplemented with 10 percent foetal calf serum (Hyclone Lot #1111946), 50 U/ml potassium penicillin and 50 μg/ml of streptomycin sulfate (Sigma #P7539). Cells to be irradiated with protons or x-rays were harvested from stock cultures and suspended in a gelatin solution that is a gel at 4°C and a liquid at 37°C. The use of gelatin solutions was first suggested by Palcic and Skarsgard (Skarsgard, 1974) and refined by Raju et al., 1978b. Briefly, a sterile gelatin solution was prepared and dialyzed for four days to remove low molecular weight toxins. A single-cell suspension of V79 cells was prepared by harvesting the exponentially growing cells with a 0.5 percent trypsin solution (Sigma Cat. #T 9395). The harvested cells are then counted and the appropriate number of cells are added to the gelatin. This gelatin cell suspension is then mixed and poured into polycarbonate tubes which are then sealed. The sealed tubes are chilled to 4°C and once set, they are packed in ice. The tubes are kept in an ice water bath prior to, during, and after irradiation. After irradiation the gels are extruded from the tubes and each cylinder of gel is cut into slices. The slices are then placed into 10 cm petri dishes (Corning #25020-100) containing 10 ml of nutrient medium, where the gel melts. These dishes are then incubated for seven days, stained with crystal violet, and scored. All colonies having more than 50 cells were counted as survivors.

Irradiation

Proton irradiations were performed on the eye beam line of the Loma Linda University Proton Treatment Facility. Protons are first accelerated in a synchrotron with a 2.2 second repetition rate. After a 300 ms extraction time, they are transported to one of several beam lines in various rooms. Protons for the horizontal eye beam line nozzle are typically accelerated to 100 MeV but their energy is reduced before exiting the nozzle by a polycarbonate rangeshifter to give the desired depth of penetration. For these experiments, the energy was reduced to 64 MeV giving a maximum range in water of 3.5 cm. The beam range was additionally modulated in depth by a rotating stepped polycarbonate wheel, thereby giving a constant value of dose (±2 percent) from the surface at a depth of 3.3 cm. Protons are scattered laterally by a lead foil giving a dose uniformly across the field of ±3 percent. The gel tube containing the cells was irradiated end on by the protons. The field was defined by a circular brass aperture 8.1 cm upstream of the tube yielding a projected size at the upstream end of the tube of 2.58 cm. Alignment of the cell tube was facilitated by a beamline light source as well as overhead and beamline lasers. Measurement of the dose in the beam was accomplished using a 0.05 cm³ tissue-equivalent thimble ionization chamber with a ⁶⁰Co exposure calibration factor traceable to the NIST. The dose to tissue was calculated from the ionization measurements using an N_{gas} type formalism similar to the AAPM Task Group 21 protocol (Nath et al., 1987). This method gave identical results to calculations based on the AAPM Heavy Charged Particle protocol (Lyman et al., 1986). The dose rate delivered to the cells varied from 7 to 12 Gy/min. on the various days of the experiment.

Cobalt irradiations were performed using an AECL Eldorado Model G cobalt irradiator. For these irradiations, the beam was incident upon the sides of the cell containing tubes. The centers of the tubes were placed at a depth of 8 cm in a water phantom with the distance from the source to the surface of the water being 41.8 cm. The field size at the tubes was 9.5 cm by 9.5 cm. Measurement of the dose in the beam was accomplished using a 0.5 cm^3 air-equivalent thimble ionization chamber with a ^{60}Co exposure calibration factor traceable to the NIST. The dose to tissue was calculated from the ionization measurements using the AAPM Task Group 21 protocol (Nath et al., 1987). The dose rate delivered to the cells averaged 0.966 Gy/min. The corresponding microdosemetric spectra have been measured as a function of depth in this beam in accordance with the AAPM protocol (Dicello et al., 1991).

RESULTS AND DISCUSSION

The observed surviving fractions are plotted as a function of dose and position within the gel in Figure 1. For the ^{60}Co exposures there is no qualitative difference in the radiation. There was, however, a quantitative variation of ±2 percent in the ^{60}Co field. This lack of qualitative differences in the radiation delivered to each slice manifests itself by having a line with a slope equal to zero fitting the data at each of the different dose levels. When one examines the proton data, one observes an entirely different picture. The mixed proton beam enters the gel at slice one and penetrates to slice eight. The number of stopping protons per slice increases as one goes from slice one to slice eight, which has the greatest number of stopping protons in its volume. This change in radiation quality is easily observed as a change in surviving fraction as a function of depth in the gel.

Survival curves can be calculated for each position in the mixed proton field and compared to survival curves for the cells exposed to the ^{60}Co gamma irradiation and RBEs determined. At a survival value of 0.01 the RBE ranges from 1.05 in the first slice up to 1.20 in slice number eight. This variation in response to the same dose of protons can be explained if one looks at how the energy is deposited throughout the gel. In slice one, about 90 percent of the dose is deposited by events where the lineal energy is less than 10 KeV per event. At the end of the gel (slice 8) only about 60 percent of the dose is deposited by events where the lineal energy per event is less than 10 KeV per event. Conversely, this means that at this point in the field (slice 8), about 40 percent of the dose is deposited in events where the lineal energy is more than 10 KeV per event. The increased RBE at the entrance of the gel is easily explained when the protons are compared to ^{60}Co where 99.9 percent of the dose is delivered with a lineal energy less than 10 KeV per event, and the RBE would be expected to increase as the percentage of events with a lineal energy greater than 10 MeV increases.

Numerous studies have demonstrated that stopping protons have elevated RBEs in cultured cells. Robertson et al. (1975) reported a RBE of 1.4 for stopping protons when the survival of H4 rat hepatoma cells was examined. Several other investigators have reported increased RBEs for low energy protons. Bird et al. (1980) reported RBEs of 1.4 and 1.8 for the survival of V79 cells in Late S and G1/S respectively for 10 KeV/μm (track average LET) protons. Perris reported an RBE of 3.1 for 3.8 MeV protons, when comparing the alpha/alpha x ratio of V79 cell survival curves. Belli et al. (1989) reported RBEs ranging from 3.0 to 7.3 in protons from energy 3.4 to 1.2 MeV. At lower proton energies the RBE - actually decreased to 6.3 and 4.2 for protons of 0.84 and 0.73 MeV. Again, these investigators used the alpha/alpha x ratio of V79 cell survival curves as the RBE. This increase in RBE for low energy protons has also been reported for cellular events that can lead to cell death. Matsubara et al. reported an increase in chromosome aberration frequency reaching a high of 2.0 at the end of a 70 MeV spread out Bragg peak.

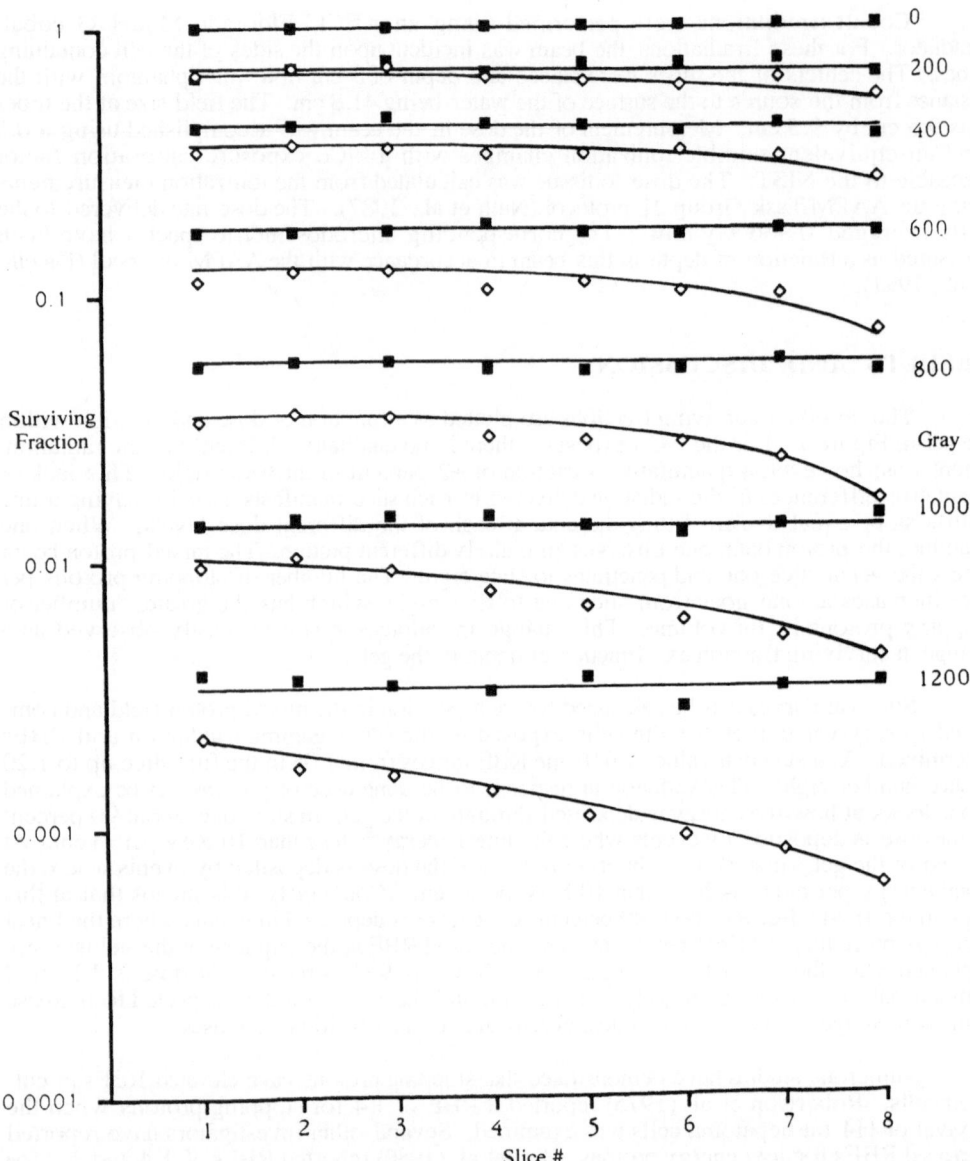

Figure 1. Cell Survival as a Function of Position. Solid Squares, Survival of V79 Cells Exposed to ^{60}Co Gamma Rays as a Function of Position in the Gel at the Dose Indicated. Open Diamonds, Survival of V79 Cells Exposed to Protons in the Gel as a Function of Position at the Doses Indicated. ^{60}Co Irradiated Cells Experience a Radiation Field of Uniform Quality While the Proton Irradiated Cells are Exposed to Radiation of Varying Quality as Described in the Text.

It is important to point out that many investigators have examined the relationship between RBE and high LET charged particles (see Raju et al., 1978a, 1978b). Belli et al. (1989) and Folkard et al. (1989) have pointed out that protons seem to have a higher RBE for a given LET than other particles. They, as well as Prise et al. (1990); Dicello et al. (1990), and Zaider et al. (1989) recognize that track average LET is not as important as how the energy is deposited in each energy loss event. We are continuing to collect and analyze data for mixed energy proton fields. We are using this data and the model of Zaider and Dicello (1978) to model radiobiological responses for mixed proton fields with greater accuracy than current methods, thereby allowing a better estimation of the risk of radiation exposure in space.

REFERENCES

Belli, M., Cherubini, R., Finotto, S., Moschini, G., Sapora, O., Simone, G., and Tabocchini, M.A. 1989. RBE-LET relationship for the survival of V79 cells irradiated with low energy protons. International Journal of Radiation Biology 55(1):93-104.

Bird, R.P., Rohrig, N., Colvett, R.D., Geard, C.R., and Marino, S.A. 1980. Inactivation of synchronized Chinese hamster V79 cells with charged-particle track segments. Radiation Research 82:277-289.

Dicello, J.F. 1992. HZE cosmic rays in space. Is it possible that they are not the major radiation hazard? Submitted for publication in Radiat. Prot. Dosimetry.

Dicello, J.F., Wasoiolek, M., and Zaider, M. 1991. Measured microdosimetric spectra of energetic ion beams of Fe, Ar, Ne, and C: Limitations of LET distributions and quality factor in space research and radiation effects. IEEE Trans. Nucl. Science NS-38, December (in press).

Dicello, J.F., Lyman, J.T., McDonald, J.C., and Verhey, L.J. 1990a. A portable system for microsimetric intercomparisons by task group #20 of the American Association of Physicists in Medicine (AAPM). Nuclear Inst. and Methods 45:724-729.

Dicello, J.F., Schillaci, M., and Lui, L. 1990b. Cross sections for pion, protein, and heavy-ion production from 800 MeV protons incident upon aluminum and silicon. Nucl Instr. Meth. B45:135-138.

Folkard, M., Prise, K.M., Vojnovic, B., Davies, S., Roper, M., and Michael, B.D. 1989. The irradiation of V79 mammalian cells by protons with energies below 2 MeV. Part I. Experimental arrangement and measurements of cell survival. International Journal of Radiation Biology 56:221-237.

Lyman, J.T., Awchalom, M., Berardo, P., Bichsel, H., Chen, G.T.Y., Dicello, J., Fessenden, P., Goitein, M., Lam, G., McDonald, J.C., Smith, A.R., Ten Haken, R., Verhey, L., and Zink, S. 1986. Protocol for heavy charged-particle therapy beam dosimetry. AAPM Report No. 16, American Institute of Physics. New York.

Matsubara, S., Ohara, H., Hiraoka, T., Koike, S., Ando, K., Yamaguchi, H., Kuwabara, Y., Hoshina, M., and Suzuki, S. 1990. Chromosome aberration frequencies produced by a 70-MeV proton beam. Radiation Research 123:182-191.

Nath, R., Anderson, L., Jones, D., Ling, C., Loevinger, R., Williamson, J., and Hanson, W. 1987. Specification of Brachytheraphy source strength. AAPM Report No. 21 American Institute of Physics. New York.

NCRP Report No. 98. 1989. Guidance on radiation received in space activities: National Council on Radiation Protection and Measurements, Bethesda, Maryland.

Perris, A., Pialoglou, P., Katsanos, A.A., and Sideris, E.G. 1986. Biological effectiveness of low energy protons. I. Survival of Chinese hamster cells. International Journal of Radiation Biology 50:1093-1101.

Prise, K.M., Folkdard, M., Davies, S., and Michael, B.D. 1990. The irradiation of V79 mammalian cells by protons with energies below 2 MeV. Part II. Measurement of oxygen enhancement ratios and DNA damage. International Journal of Radiation Biology 58:261-277.

Raju, M.R., Amols, H.I., Dicello, J.F., Howard, J., Lyman, J.T., Koehler, A.M., Graves, R., and Smathers, J.B. 1978a. A heavy particle comparative study. British Journal of Radiology 51:699-703.

Raju, M.R., Bain, E., Carpenter, S.G., Cox, R.A., and Robertson, J.B. 1978b. A heavy particle comparative study Part II: Cell survival versus depth. British Journal of Radiology 51:704-711.

Robertson, J.B., Koehler, A.M., Weideman, P.A., and McNulty, P.J. 1988. High energy proton induced mutations in cultured Chinese hamster cells. In Terrestrial Space Radiation and its Biological Effects (P.D. McCormack, C.E. Swenberg, and H. Bücker, eds), Plenum Press: New York.

Robertson, J.B., Williams, J.R., Schmidt, R.A., Little, J.B., Flynn, D.F., and Suit, H.D. 1975. Radiobiological studies of a high-energy modulated proton beam utilizing cultured mammalian cells. Cancer 35:1664-1677.

Skarsgard, L.D. 1974. Pretherapeutic research programmes at Pi-meson facilities. In Proceedings of the XIII International Congress of Radiology. International Congress Series No. 339. Radiology 2:447-454. Excerpta Medica, Amsterdam, the Netherlands.

Stassinopoulos, E.G. 1988. Earth's trapped and transient space radiation environment. In Terrestrial Space Radiation and its Biological Effects (P.D. McCormack, C.E. Swenberg, and H. Bücker, eds), Plenum Press: New York.

Zaider, M., Dicello, J.F., and Coyne, J.J. 1989. The effects of geometrical factors on microdosimetric probability distributions of energy deposition. Nucl. Instr. Meth. B40/41:1261-1265.

Zaider, M. and Dicello, J.F. 1978. RBEOER: A Fortran program for the computation of RBEs, OERs, survival ratios, and the effects of friclionation using the theory of dual resolution action. Alamo National Laboratory Report, No. LA-7196-MS, Los Alamos, NM.

RADIATION: WHAT DETERMINES THE RISK?

R.E.J. Mitchel and A. Trivedi

Health Sciences and Services Division
AECL Research
Chalk River, Ontario K0J 1J0 Canada

ABSTRACT

Radiation, like other DNA damaging agents, can initiate a series of cellular events responsible for cancer development. However, in any individual the risk of cancer arising from a carcinogen exposure is variable, and is not a fixed value dependent only on the dose of carcinogen. This variability in overall risk arises from variability in the probabilities of the intermediate steps of the multistep processes of carcinogenesis. Using cellular and animal model systems, we have shown that deliberate manipulation of these biological processes is possible, and that the risk of cancer from a fixed exposure to a carcinogen can be made to increase or decrease. We have also shown that such changes in risk can result from intervention at times long before or after that carcinogen exposure. These results indicate that the principles of radiation protection can be expanded. We suggest that in addition to offering protection against exposure, radiation protection can include the development of strategies for protection against the ultimate biological consequences of an exposure. Improved understanding of the biology of radiation responses may lead to techniques for deliberate intervention that could be particularly useful in long duration manned space flight.

INTRODUCTION

Historically, for radiation protection purposes, the risk of cancer arising from an exposure to ionizing radiation has been related only to the received dose. These risk estimates, and the exposure standards developed from them, are based primarily on epidemiological studies of excess cancer incidence in large groups of people such as the A-bomb survivors or those accidentally, medically or occupationally exposed to ionizing radiation. Many years after exposure, some of those exposed developed cancer potentially attributable to that exposure. Typically, the groups studied received high doses at high dose-rates, and of particular importance are the survivors of the atomic bombs exploded in Japan during World War II. Radiation risk estimates have recently undergone an upward revision, in part because of a recalculation of the radiation doses believed to have been absorbed by that group. Estimates derived from such high dose, high dose-rate exposures are also used to estimate carcinogenic risks associated with low doses and low dose-rates for both populations and individuals, although a dose and dose-rate reduction factor of two or more is usually used for protracted doses.

This approach provides ranges of values for cancer risk in large populations living under some set of circumstances which are undefined either at the time of a single or fractionated exposure or afterwards during the development of the cancer. It may not,

Biological Effects and Physics of Solar and Galactic Cosmic Radiation,
Part B, Edited by C.E. Swenberg *et al.*, Plenum Press, New York, 1993

however, provide adequate information on the risk for any particular individual living in a particular unique set of circumstances, nor does it necessarily produce valid estimates for other groups of people living under other environmental or exposure conditions. Current procedures for better estimating the risk of radiation induced cancer have placed more emphasis on increasingly accurate estimates of the dose received, while ignoring the potentially greater uncertainties resulting from individual variations in biological processes, working histories or personal habits; at best such factors have been averaged over large populations. The present approach tends to blur the importance of these biological factors and impairs the development of strategies which might modify them and reduce overall risk. Adequate assessment of risk at the individual level requires strategies that depend on knowledge of factors unique to the individual situation: the biologically effective dose after DNA repair has occurred, alterations in gene expression and certain aspects of cellular metabolism can all influence an individual's ultimate risk.

These considerations are particularly important in space flight where radiation protection procedures should consider each astronaut as a unique individual. It is conceivable that with a small defined group in a defined environment, it may be possible to devise procedures that manipulate some biological processes and so reduce the risk associated with their radiation exposure. We describe below some experiments showing that cancer risk can vary in a manner independent of the carcinogen exposure. These experiments show that the concept of biological manipulation of risk is feasible and that in the future it may be possible to devise strategies for deliberate intervention.

RISK OF CARCINOGENESIS

Current evidence indicates that cancer is a multistep process. While the initial DNA damage caused by ionizing radiation or other carcinogens is a necessary event, the extent of that initial damage is biologically remote from and only probabilistically related to the risk of cancer formation. All the stages in carcinogenesis depend on biological processes but cancer is only one of the possible consequences. Other biological responses to DNA damage are more likely. It is therefore the probabilities of the individual biological steps which occur subsequent to the initial DNA damage (during the "latent period") that determine the actual probability that a cancer will eventually result. The steps in this process, as they can be currently distinguished, are outlined in Figure 1. Since the overall risk of converting a normal cell into a cancer depends on the probability of each individual step, then if the probability of

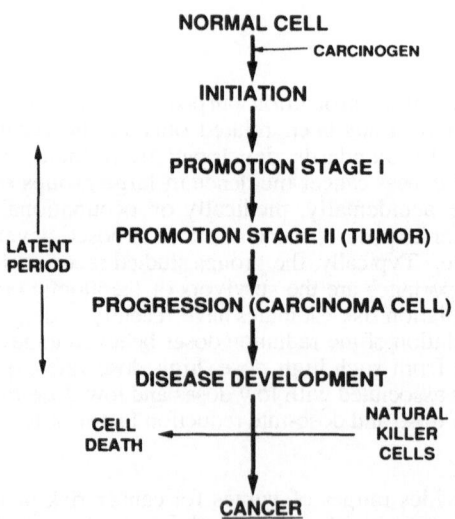

Figure 1. Identifiable Steps in the Multistep Process of Carcinogenesis.

any step increases, the overall risk will increase. However, if the probability of any step can be reduced the overall risk must also be reduced. If the probability of any step becomes zero, then overall risk must also be zero.

The events depicted in Figure 1 represent biological or biochemical processes. Some of these events are better understood than others but all can be demonstrated at either the cellular or whole animal level. It can be seen therefore, that if some or all of the biological processes can be manipulated or controlled, then the ultimate biological risk of a carcinogen exposure will not be a fixed value. We describe below the results of experiments, *in vitro* and *in vivo*, that substantiate this concept and point to potential new approaches to the interpretation and analysis of risk, and to the practice of radiation protection.

INITIATION

Of the steps shown in Figure 1, the best understood is the first, called "initiation" in mammalian carcinogenesis. It is believed to be equivalent to mutation. A mutation results not from the initial DNA damage *per se* but rather from the action or inaction of DNA repair systems on that damage. After the DNA has been damaged, by radiation for example, there are three possible general outcomes of repair: the cell dies, mutates or returns to normal (Fig. 2).

All DNA repair systems have an inherent, low probability of error and mutations result from these errors. Neither a return to normal nor cell death contribute to subsequent excess risk of carcinogenesis. The probability of any particular DNA damage resulting in any one of these outcomes depends upon the relative activities of the various DNA repair systems which recognize a particular type of DNA damage. If only one repair system acts on a particular type of damage, the observed mutation yield will vary with the absolute activity of that repair system. If, however, as is normally the case, more than one DNA repair system can act on a particular type of DNA damage and these repair systems differ in their error frequency ("error prone" vs. "error free"), then the mutation frequency (and hence its contribution to the overall risk) will vary with changes in their relative activity. This concept is diagrammed in Figure 3.

These ideas of variable risk after exposure to a fixed dose of carcinogen have been demonstrated by manipulating DNA repair processes and measuring changes in mutation frequency in a lower eukaryote, the yeast *Saccharomyces cerevisiae*. We have shown that a sublethal ionizing radiation exposure induced error-free recombinational repair capacity and that the extent of induction was proportional to dose and DNA damage (Mitchel and Morrison, 1984). Low LET radiation was more effective than high LET radiation at inducing this repair process (Boreham and Mitchel, 1991). This inducible process appears to respond to events which alter DNA topology or the enzymes (topoisomerases) that control that topology (Boreham et al., 1990). Besides recognizing and repairing DNA lesions generated

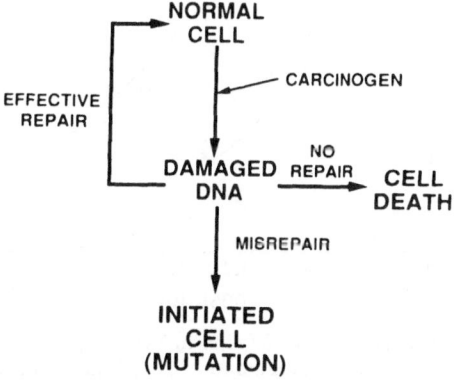

Figure 2. Initial Consequences of DNA Damage from a Carcinogen Exposure.

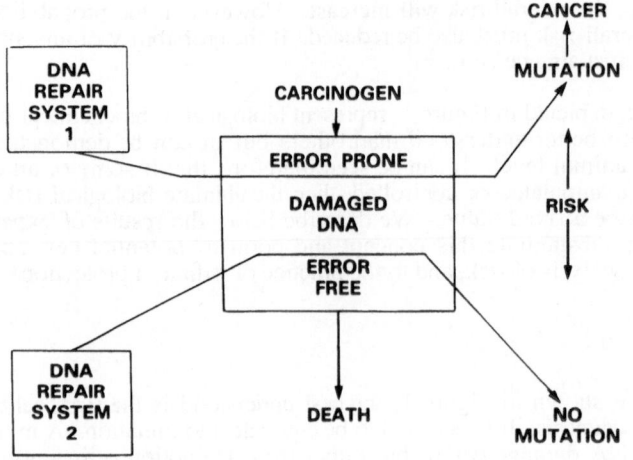

Figure 3. Variable Risks of Cancer Initiation Dependent Upon Relative DNA Repair Activities.

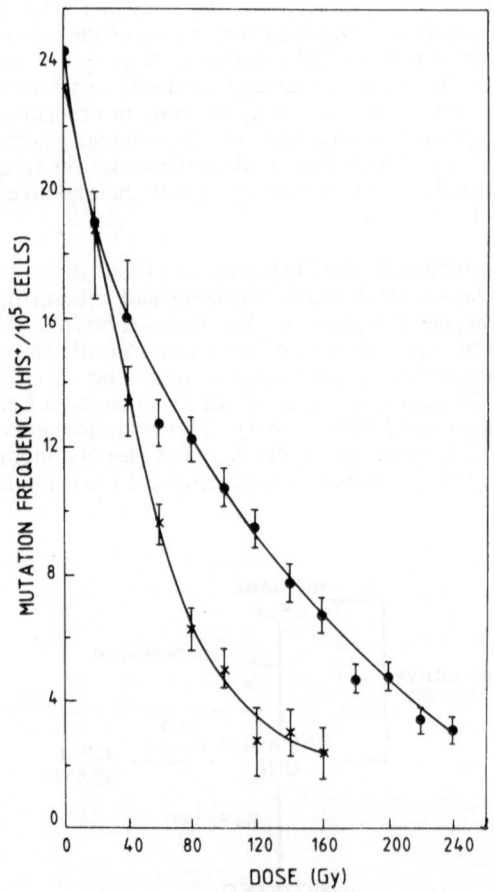

Figure 4. Gamma Radiation Protection Against Mutation in Yeast Exposed to the Chemical Mutagen N-methyl-N'-nitro-N-nitrosoguanidine (MNNG); Exposure to ^{60}Co-γ Radiation under Oxic (X) or Anoxic ($\bullet$) Conditions. Cell Survival was 100% at all Doses. (From Mitchel and Morrison, 1987).

by ionizing radiation, this repair system could also recognize and repair chemically produced DNA lesions that were normally repaired by error-prone (mutagenic) pathways. Induction of error-free recombinational repair allowed that repair system to compete with the error-prone repair process for those chemically generated DNA lesions. The result, shown in Figure 4, was that sublethal doses of radiation protected, in a dose dependent manner, against mutation by a fixed dose of a highly mutagenic chemical. In other words, this mutagenic chemical was made virtually non-mutagenic by increasing the probability that the pre-mutagenic DNA lesions would be repaired by an error-free process (Mitchel and Morrison, 1987).

The results quoted above demonstrate that the risk of mutation, from a fixed dose of carcinogen, can be made to decrease by an exposure to another agent. The other agent need not be a DNA damaging agent or a carcinogen. We have also shown, in yeast, that a heat stress prevented mutation by certain mutagens (Fig. 5). In this case, the evidence suggested that the heat-shock inhibited normal protein synthesis and shifted the cell instead to the heat-shock type of protein synthesis (Mitchel and Morrison, 1986). This had the effect of preventing induction of an error-prone type of DNA repair induced by the chemical and necessary for mutation, and instead forcing the cell to use other existing error-free DNA repair processes. The net effect of a heat-shock, however, was similar to that obtained with a radiation exposure. The risk of mutation was decreased by other cellular events occurring before or after exposure to the mutagen.

The above experiments both showed that the mutational risk from exposure of cells to a fixed dose of mutagen could be decreased by external stresses which allow or force the cell to process DNA damage in an error-free manner. Since however, the risk of mutation depends upon the relative probabilities of error-free versus error-prone repair of the existing DNA lesions, any event which shifts the balance toward error-prone repair would obviously have the opposite effect and increase mutation risk. Such a result was also demonstrated in our yeast experiments. Figure 6 shows the effect of pre-exposure of yeast to a small dose of

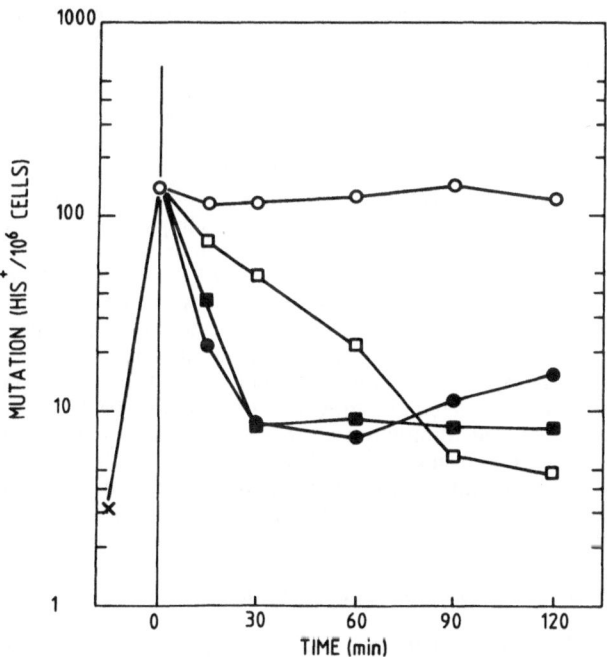

Figure 5. Heat- Shock Protection against Mutation in Yeast Exposed to the Chemical Mutagen N-methyl-N'-nitro-N-nitrosoguanidine (MNNG); Control, No Heat Exposure (o); Heat Shock at 38°C (●); Protein Synthesis Inhibition with Cycloheximide (◻); Heat-shock plus Cycloheximide (◼); Spontaneous Frequency (X). Cell Survival was 100% under all Conditions. (From Mitchel and Morrison, 1986).

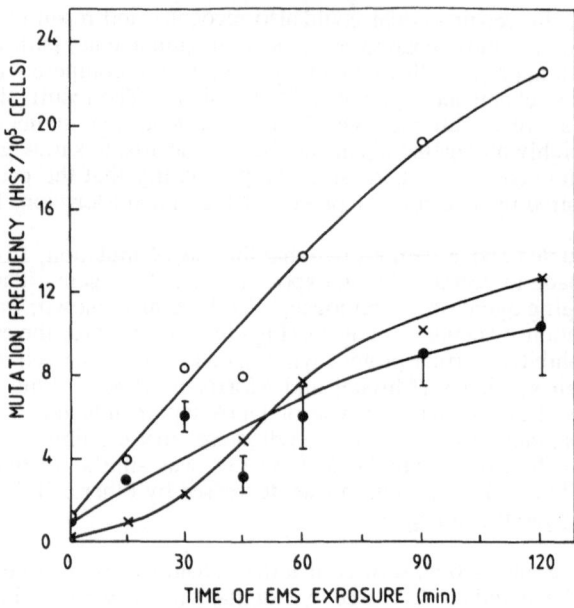

Figure 6. Amplification of Mutation Frequency in Yeast Exposed to Ethylmethanesulfonate (EMS). Error Prone Repair was stimulated by pre-exposure to a low dose of MNNG. Exposure to EMS alone (X); Exposure to MNNG followed by EMS (o); Excess EMS Mutation resulting from Error prone repair Stimulation (●). Cell Survival was 100% under all Conditions. (From Mitchel and Morrison, 1987).

a mutagen which induced the error-prone DNA process. Subsequent exposure to a second mutagen resulted in a frequency of mutation much greater than would normally be produced, because the DNA damage was preferentially processed via this enhanced error-prone process (Mitchel and Morrison, 1987).

The experiments described above clearly indicate that the risk of mutation in yeast can be altered by events which alter the balance of the various DNA repair processes. If the first step in the formation of cancer (which we call initiation, Fig. 1) is actually a mutation, and if DNA repair systems operate in a similar way in higher eukaryotes, then the same principles should apply in animal experiments. We have tested this hypothesis and shown that a heat or radiation stress, applied at or near the time of initiation of mice with a chemical mutagen, altered the risk of initiation and hence the probability of tumor formation. To demonstrate this, we used the two-stage skin tumor formation system in SENCAR mice (Slaga, 1984). This system produces skin tumors in mice where the dorsal skins have received a single application of a chemical tumor initiator, followed by multiple application of a chemical tumor promoter. This regimen accomplishes initiation and stages I and II of promotion (Fig. 1), producing visible skin tumors. We have shown (Mitchel et al., 1988) that a single heat stress applied near the time of chemical initiation increased the risk of tumor formation (Table 1). On the other hand, exposure to a single dose of β-radiation (Fig. 7), reduced the risk of tumor initiation by a chemical alkylating agent (Mitchel et al., 1990). This reduced risk was evident even if the radiation was given up to 6 days before or after exposure to the chemical initiator (Fig. 8). The reduced risk was likely due to radiation induction of a DNA repair system (alkyltransferase) capable of removing the DNA lesions, known to be induced in rodents exposed to ionizing radiation (Margison et al., 1985).

These results demonstrate that the activity of DNA repair systems, including those that recognize and repair radiation damage, can be temporarily altered in either direction (either inhibited or stimulated) by stresses like heat, radiation or exposure to DNA damaging chemicals. Changes in the relative activities of competing DNA repair systems (Fig. 3) can either increase or decrease the risk of tumor initiation. The changes in risk that we have

Table 1. Hyperthermia and the Risk of Tumor Initiation in Mouse Skin

INITIATOR	TREATMENT	RELATIVE TUMOR FREQUENCY
MNNG	I → P	1.0
	Δ I → P	1.4
DMBA	I → P	1.0
	I Δ → P	1.4

Δ = 44⁰ C, 30 min;   I = Initiation;   P = Promotion

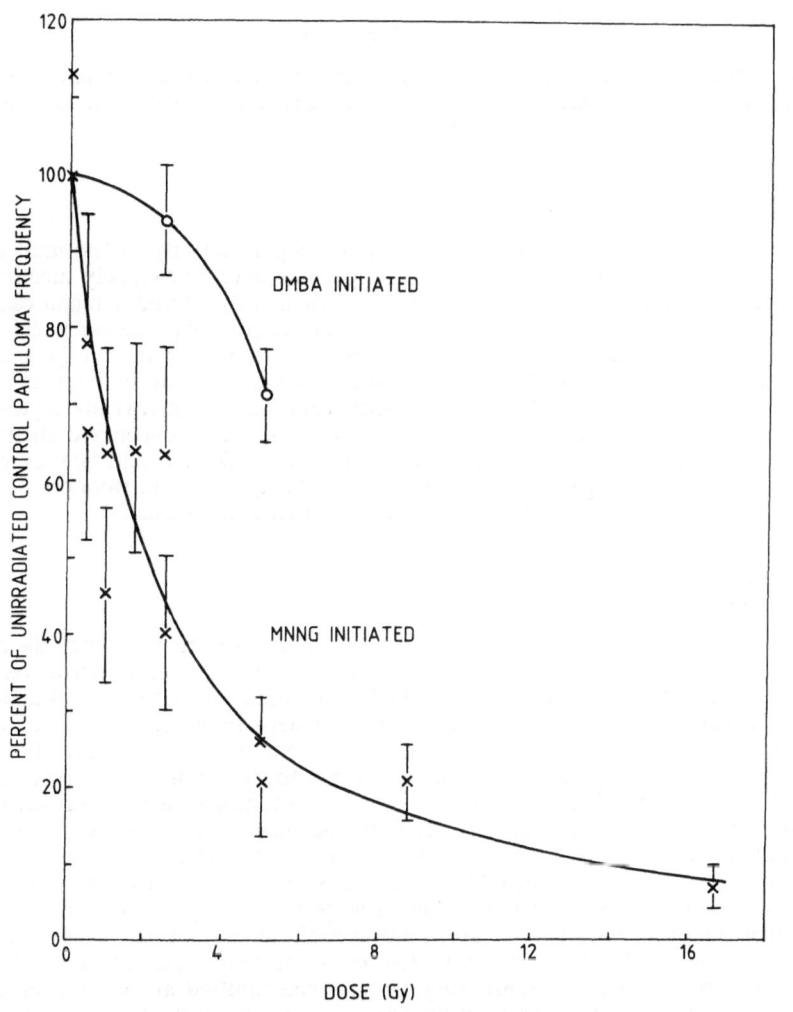

Figure 7. ß-Radiation-Induced Protection against Chemically-Initiated Skin Tumorigenesis in Mice. γ-radiation was delivered at the Time of Initiation and Mice were Subsequently Promoted with the Phorbol Ester TPA. (From Mitchel et al., 1990).

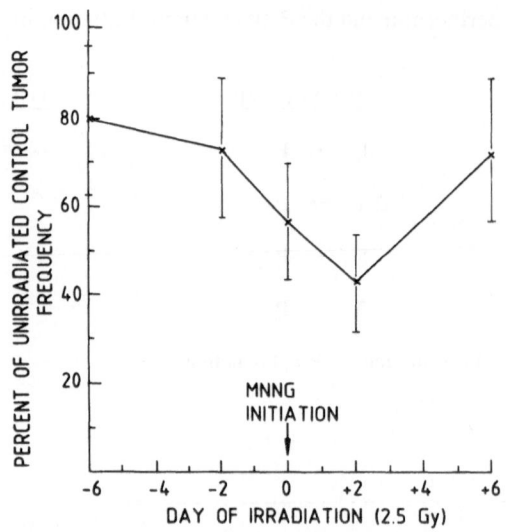

Figure 8. The Effect of the Time of Irradiation on the Radiation-Induced Protection against MNNG-Initiated Skin Tumorigenesis in Mice. β– radiation (2.5 Gy) was Delivered to the Skin at the Indicated Time before or after Initiation with MNNG. (From Mitchel et al. 1990).

observed resulted from changes in the level of DNA repair activity of "normal" cells and animals. It should be noted, however, that not all persons may have equally functional DNA repair processes to begin with. Such individuals, permanently altered in repair capacity for genetic reasons, are permanently at altered risk. Inducible repair processes which alter cell sensitivity to chromosomal aberrations resulting from radiation or chemical exposure are also known to exist in human cells (Wolff et al., 1988) but again not all individuals are equal. Recently, Bosi and Olivieri (1989) have shown that while most human lymphocytes exposed to low doses of radiation become less susceptible to induction of chromatid aberrations by subsequent high doses, lymphocytes from some individuals do not, and may even become more sensitive. Thus, changes in human cell repair capacity, resulting from prior exposures to genotoxins, may alter the actual level of risk for a particular individual.

PROMOTION

Once initiated, cells have the potential for further movement through the multistep process of carcinogenesis, but, as far as can be determined, can never return to a non-initiated (i.e., normal) state. They therefore remain available, if stimulated at some point in the future, to progress further down the chain to cancer. The next steps in the process, stage I and stage II of promotion (Fig. 1), are less well understood. Nonetheless, we have been able to demonstrate that the same agents (heat, radiation), when applied at the time of promotion, also alter the risk of tumor formation. Using the mouse skin tumor system, a heat stress applied to the skin near the time of each application of the chemical tumor promoter, or up to a day prior to the chemical, markedly reduced the risk of tumor formation (Mitchel et al., 1986). A similar result (Fig. 9) was obtained when the heat stress was applied only a very limited number of times at stage I of promotion and not at stage II, or vice versa (Mitchel et al., 1987) indicating that both processes can be affected by certain manipulations, thereby altering the risk of tumorigenesis. Note that the protective effect of heat stress applied at the time of promotion was opposite to its sensitizing effect when applied at the time of initiation, described above. Radiation exposure, at the time of tumor promotion, again had a different effect than heat stress on tumor risk. Chronic exposure of the initiated skin to ß-radiation did not produce tumors, indicating that ionizing radiation is not a complete tumor promoter. However, chronic ß-radiation exposure did act as a weak stage I promoter and increased tumor frequency (Table 2), perhaps by stimulating a naturally occurring process

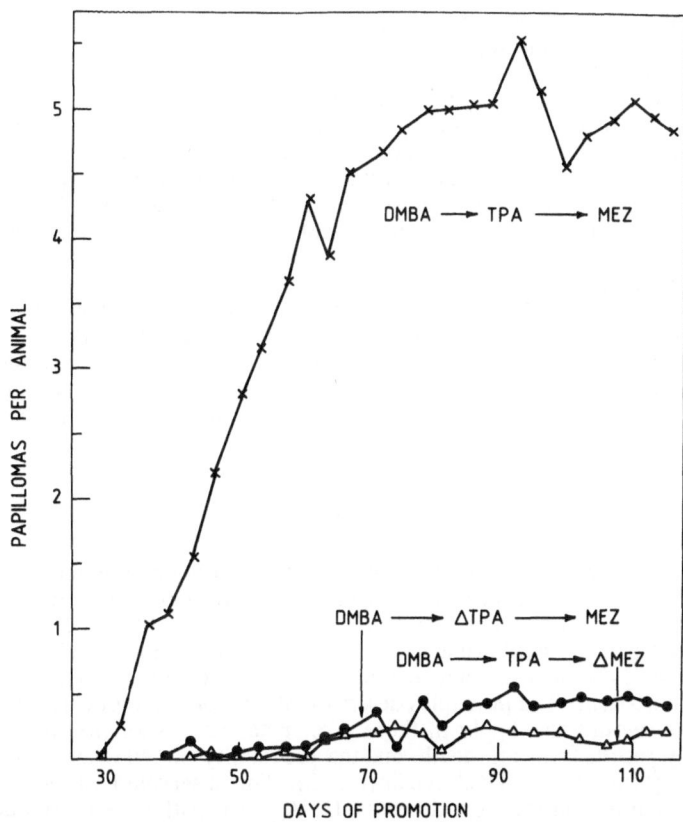

Figure 9. Inhibition of Stage I and Stage II Promotion in Mouse Skin by Hyperthermia (44°C, 30 min). Mouse Skin was Initiated with the Chemical Carcinogen DMBA; Stage I Promoted with the Phorbol Ester TPA and Stage II Promoted with Mezerein. Heat (Δ) was Applied only at Stage I or Stage II of Promotion. (From Mitchel et al. 1987).

(Mitchel and Trivedi, 1992). Chronic exposure to radiation, long after the tumor initiation event, therefore increases the risk that those initiated cells will form tumors.

IMMUNE SURVEILLANCE

The appearance of a clinically observable cancer, resulting from the progress of a cell

Table 2. Chronic Radiation As A Stage I Tumor Promotor

	TREATMENT		TUMORS/ANIMAL
Initiation	Stage I Promotion	Stage II Promotion	
DMBA	None	Mezerein for 13 weeks	0.85 ± 0.31
DMBA	No treatment for 13 weeks	Mezerein for 13 weeks	3.46 ± 0.16
DMBA	0.5 Gy β 2x/week for 13 weeks	Mezerein for 13 weeks	3.99 ± 0.23 P < 0.01

through the various steps of the multistep process of carcinogenesis, ultimately depends on the probability that the immune surveillance system will not detect and kill the changed cells. The importance of the natural killer cell component of this system was demonstrated by Warner and Dennert (1982) who produced a high frequency of leukemia in mice by exposure to whole body irradiation. The radiation exposure also depressed the immune surveillance system in these mice. However, when another group of mice received the same whole body dose, but also received transplanted natural killer cells, the leukemia frequency was reduced 10-fold. In our laboratory, the importance of the immune surveillance system was demonstrated at stage I of promotion, an early event after initiation (Fig. 1). Using the mouse skin tumor system, we have shown that immune system inhibition with cyclosporin at stage I of promotion doubled the tumor frequency when a stage II promoter was subsequently applied (Mitchel, unpublished). These results show that an impairment of the immune surveillance system can also alter the risk of this cancer, and that this risk could be altered at times soon after exposure to the initiating carcinogen. We have recently shown that heat stress induced changes in human natural killer cell activity varies between individuals, again indicating the importance of correctly assessing risk by treating each person as a unique individual (Yang and Mitchel, 1991).

RISK OF DEATH

The arguments presented above have been directed at demonstrating that the cancer risk arising from an exposure to a carcinogen like radiation can be altered by deliberate interventions into the intermediate steps of carcinogenesis. On earth, cancer probably constitutes the major hazard attributable to a radiation exposure. Such might not be the case in space, where periods of intense solar activity may produce radiation fields which threaten survival in the short-term. The principles described above for carcinogens, especially at early biological events, would also apply in those circumstances. Radiation-induced death of a person ultimately results from cell death, and the probability of cell death is a function of its maximum capacity for effective DNA repair (Fig. 2). The observation of increased sensitivity to killing by radiation resulting from defective DNA repair pathways is well documented in cells as diverse as those of bacteria and humans. Similarly, DNA repair processes induced by a primary exposure to DNA damaging agents have been shown to increase the resistance of cells to killing by ionizing radiation. DNA repair processes inducible by radiation exposure are well known in lower eukaryotes (e.g. Mitchel and Morrison, 1984) and are now being demonstrated in mammalian cells (Margison et al., 1985; Olivieri et al., 1984; Wolff et al., 1988). Such stress induced changes in radioresistance have also been extended to the whole animal level using stresses other than radiation. For example, recent experiments have shown that heavy metal stress substantially reduces the risk of death in mice subsequently exposed to normally lethal doses of low LET radiation (Matsubara et al., 1987). It was suggested that this increase in radioresistance was attributable to the metal-induced synthesis of metallothionein, a sulfydryl-rich protein believed to be an effective free radical scavenger. It should also be noted, however, that heavy metals have been shown to induce the cellular heat-shock response and the synthesis of heat-shock proteins in a manner analogous to a heat stress. We have described above how a heat-shock alters risk in both yeast and mice. In fact, an even more dramatic decrease in the risk of radiation induced death in mice was produced by a mild heat-shock given 24 h prior to a lethal dose of radiation, where all of the heat-shocked mice survived the exposure, while none of the control mice survived (Shen et al., 1991). In this case the authors attributed the increased radioresistance to heat-shock induced production of the cytokine IL-1. Indeed IL-1 has been shown to radioprotect cells in tissue culture. This result opens the possibility that cell-cell communication and control molecules like cytokines and growth factors could be used to temporarily alter the radiosensitivity of whole organisms. It is clear from our experiments, however, that such regulation can increase or decrease the risk, and that the same stress applied at different times can have different and opposite effects. Practical use of these techniques will require a sound understanding of all the parameters.

CONCLUSION

It is apparent that the cancer risk from exposure to a carcinogen, like radiation, should not be considered in isolation from other stresses and exposures. The risk of formation of a cancer resulting from a single or multiple carcinogen exposure is related only indirectly to the dose of the carcinogen, and can be highly variable depending to a large extent on events that may occur either before or after the exposure. The above results also demonstrate, however, that it may be biologically possible to establish intervention techniques to reduce the risk of at least some types of cancer in exposed individuals. We have shown that intervention is possible at many steps in the process of carcinogenesis.

In view of the results, the ideas of radiation protection can be considered in a new light. Heretofore, this concept has meant only protection against radiation exposure, but it is now possible to consider strategies for protection against the real hazard, that is the biological consequence of a radiation exposure, and even possibly to reduce those consequences by a selective intervention before or after an exposure has been received. The development of protocols that can be practically applied to reduce human risk certainly requires further research, but the general approaches discussed above can provide a basis for those investigations. Individuals living in a spacecraft, with only limited abilities to reduce their radiation exposure, might substantially benefit from such strategies of biological manipulation and could provide the impetus for such research. This approach has the additional distinct advantage of imposing virtually no engineering or hardware requirements and penalties.

ACKNOWLEDGEMENTS

This work was supported by AECL Research and CANDU Owner Group (COG).

REFERENCES

Boreham, D.R. and Mitchel, R.E.J., 1991, DNA lesions that signal the induction of radioresistance and DNA repair in yeast, Radiat. Res., 128: 19-28.

Boreham, D.R., Trivedi, A., Weinberger, P. and Mitchel, R.E.J., 1990, The involvement of topoisomerases and DNA polymerase I in the mechanism of induced thermal and radiation resistance in yeast. Radiat. Res., 123:203-212.

Bosi, A. and Olivieri, G., 1989, Variability of the adaptive response to ionizing radiations in humans, Mutation Res., 211:13-17.

Margison, G.P., Butler, J. and Hoey, B., 1985, O^6-Methylguanine methyltransferase activity is increased in rat tissues by ionizing radiation, Carcinogenesis, 6:1699-1702.

Matsuhara, J., Tajima, Y. and Karasawa, M., 1987, Metallothionein induction as a potent means of radiation protection in mice, Radiat. Res., 111:267-275.

Mitchel, R.E.J., Gragtmans, N.J. and Morrison, D.P., 1990, Beta-radiation-induced resistance to MNNG initiation of papilloma but not carcinoma formation in mouse skin, Radiat. Res., 121:180-186.

Mitchel, R.E.J. and Morrison, D.P., 1984, An oxygen effect for gamma-radiation induction of radiation resistance in yeast, Radiat. Res., 100:205-210.

Mitchel, R.E.J. and Morrison, D.P., 1986, Inducible error-prone repair in yeast. Suppression by heat shock, Mutation Res., 159:31-39.

Mitchel, R.E.J. and Morrison, D.P., 1987, Inducible DNA-repair systems in yeast: Competition for lesions. Mutation Res., 183:149-159.

Mitchel, R.E.J., Morrison, D.P. and Gragtmans, N.J., 1987, Tumorigenesis and carcinogenesis in mouse skin treated with hyperthermia during stage I or stage II of tumor promotion, Carcinogenesis, 8:1875-1879.

Mitchel, R.E.J., Morrison, D.P. and Gragtmans, N.J., 1988, The influence of a hyperthermia treatment on chemically induced tumor initiation and progression in mouse skin, Carcinogenesis, 9:379-385.

Mitchel, R.E.J., Morrison, D.P., Gragtmans, N.J. and Jevcak, J.J., 1986, Hyperthermia and phorbol ester tumor promotion in mouse skin, Carcinogenesis, 7:1505-1510.

Mitchel, R.E.J. and Trivedi, A., 1992, Chronic ionizing radiation exposure as a tumor promoter in mouse skin, Radiat. Res., 129:192-201.

Olivieri, G., Bodycote, J. and Wolff, S., 1984, Adaptive response of human lymphocytes to low concentrations of radioactive thymidine, Science, 223:594-597.

Slaga, T.J., 1984, Mechanisms involved in two-stage carcinogenesis in mouse skin, in: "Mechanisms of Tumor Promotion," Vol II, T.J. Slaga, ed., CRC Press, Boca Raton, Florida.

Shen, R.-N., Hornback, N.B., Shidnia, H., Wu, B., Lu, L. and Broxmyer, H.E., 1991, Whole body hyperthermia: a potent radioprotector in vivo, Int. J. Radiation Oncology Biol. Phys., 20:525-529.

Warner, J.F. and Dennert, G., 1982, Effects of a cloned cell line with NK activity on bone marrow transplants, tumor development and metastasis in vivo, Nature, 300:31-34.

Wolff, S., Afzal, V., Wiencke, J.K., Olivieri, G. and Michael, A., 1988, Human lymphocytes exposed to low doses of ionizing radiations become refractory to high doses of radiation as well as to chemical mutagens that induce double-strand breaks, Int. J. Radiat. Biol., 53:39-55.

Yang, H. and Mitchel, R.E.J., 1991, Hyperthermic inactivation, recovery and induced thermotolerance of human natural killer cell lytic function, Int. J. Hyperthermia, 7:35-49.

RADIATION PROTECTION FOR HUMAN INTERPLANETARY SPACEFLIGHT AND PLANETARY SURFACE OPERATIONS

Benton C. Clark

Planetary Sciences Laboratory (B0560)
Martin Marietta
Denver, CO 80201

ABSTRACT

Radiation protection issues are reviewed for five categories of radiation exposure during human missions to the moon and Mars: trapped radiation belts, galactic cosmic rays, solar flare particle events, planetary surface emissions, and on-board radiation sources. Relative hazards are dependent upon spacecraft and vehicle configurations, flight trajectories, human susceptibility, shielding effectiveness, monitoring and warning systems, and other factors. Crew cabins, interplanetary mission modules, surface habitats, planetary rovers, and extravehicular mobility units (spacesuits) provide various degrees of protection. Countermeasures that may be taken are reviewed relative to added complexity and risks that they could entail, with suggestions for future research and analysis.

INTRODUCTION

Radiation risk has been recognized almost since the beginning of the space era, with the discovery of the Earth's geomagnetically trapped radiation, the "Van Allen Belts", following immediately on the heels of Sputnik. Indeed, early rocket science emphasized radiation measuring instrumentation for studying cosmic rays.

Manned space flight has been restricted to altitudes below the trapped zones, although the Apollo missions required a sensibly quick traversal to prevent excessive exposure. The galactic cosmic rays (GCR) themselves are currently recognized as a serious potential threat to long-duration Mars missions because of the high biological damage, per unit of energy deposition, of their components and secondary products. Solar flare outbursts are well known to be capable of ejecting energetic particle streams into interplanetary space, causing solar particle events (SPE) that can in some cases produce lethal doses to unprotected astronauts. Two additional radiation environments are the emissions from planetary surfaces and from on-board radiation sources, but both of these can be shown to be negligible or controllable. Progressing from highest confidence to lowest, the following are all somewhat uncertain: the radiation fluxes, their interactions with spacecraft shields, and the biological effects they can produce. Although there is only now beginning to be a resurgence in the study of these problems, some progress has been made in the intervening two decades since Apollo, e.g., vigorous effort at calculational modeling of space radiation doses by the group at NASA's Langley Research Center (see Wilson et al., 1991; Simonsen and Nealy, 1991).

The following is an attempt to summarize the state of analysis as seen by the mission designer for human exploration of those celestial bodies whose surfaces are accessible, the moon and Mars. Much additional research is required, but some conclusions can be drawn at this time and concepts established for radiation protection approaches during human exploration beyond Earth.

Biological Effects and Physics of Solar and Galactic Cosmic Radiation,
Part B, Edited by C.E. Swenberg *et al.,* Plenum Press, New York, 1993

EFFECTIVE INTRINSIC SHIELDING (EIS)

Assessment of exposure risk must be made in the context of actual space mission conditions. For example, even a spacesuit will stop the low energy ultraviolet photons and solar wind ions. However, spacecraft crew modules are almost never intrinsically well shielded from the standpoint of most other ionizing radiations. Even if a nuclear fission reactor is utilized on the spacecraft for a nuclear propulsion or power system, at most a slab of special shielding material would be provided at the reactor, with the crew module in the "shadow" of the shield. No special shielding would be provided around the pressurized habitat itself. Furthermore, module hulls are typically quite thin. Even very large modules such as for Space Station Freedom (SSF) or the SEI Habitat design example shown in Figure 1 require only a 3 mm aluminum (Al) shell for pressure containment and structural integrity. The shielding thus provided is only 8 kg/m^2 (0.8 g/cm^2), about twice that of a spacesuit (wall thicknesses of the Apollo Lunar Excursion Module were extraordinarily thin, less than 5 kg/m^2 in some locations, to reduce mass of the spacecraft). Some additional shielding is available from material located outside the pressurized shells to provide thermal control and micrometeoroid protection, but this improves the total shielding by only a minor amount. Even the special orbital debris shields used by the space station add little effective radiation shielding, another 2-4 kg/m^2, but in any event are not required for spaceships operating far from the contaminated regions of low Earth orbit (LEO). Equipment and supplies will provide additional benefit, but not as much as sometimes assumed. For example, if the entire habitat mass is in the range of 10 to 20 t (t = tonne = 1000 kg = 2204.6 lbm) and if all internal structure and equipment could be spread uniformly along the walls, the additional shielding would be only 45 to 90 kg/m^2 for the original U.S. SSF module size (now being reduced by 40% in length). Against the space radiation challenges, this is "Shielding Lite." Furthermore, the mounting of equipment for functional purposes results in a highly heterogeneous intrinsic shielding environment, with mass thickness ranging from the minimum of 8 up to at least 800 kg/m^2. Equipment mass is not, however, totally negligible even though non-uniform. An "effective shielding" for a specified radiation threat is here defined as that single thickness which results in the same calculated dose as the more complex, realistic mass thickness distribution.

A second factor is that sensitive organs experience a variable intrinsic shielding just from the variation in path-lengths through the human body itself. Tissue fat, mineralized bone, and muscle provide relatively inert (biologically speaking) internal shields. For example, shielding of the abdomen varies from a low of less than 5 kg/m^2 for radiation particles normally incident at the navel up to 300 kg/m^2 when on a trajectory entering the body through the head or a foot. Even the so-called "skin dose" in space radiation is conventionally the abstraction of an infinitesimally small tissue-equivalent test volume with no body shielding. In actual fact, all skin except perhaps protruding ear lobes and extended digits are intrinsically shielded in the back-direction by 20 to 500 kg/m^2 of flesh and bone. Thus, in the human body each susceptible tissue or organ will actually be subjected to radiation through an intrinsic shielding path length distribution, determined for sensitive cells by the specific size and physique of the individual.

A third consideration, the proximity effect, is also a factor. Astronaut location affects the path length distribution through the intrinsic shielding. In omnidirectional fields, as are the GCR and predominate SPE fluxes, the closer the shielding to the body, the more effective it can be. Off-normal paths provide high mass thicknesses through moderate shields. As seen in Figure 2, the path lengths just inside a thick spherical shell are considerably greater than for a target located at the exact center of the sphere (note: the reader should be aware that in some conventional terminology, a slab shield is a parallel plate for which it is assumed that the radiation is normally incident and hence passes through a single path and a spherical shell is assumed to have the target at the center, which also produces only a single path length, even if the flux is anisotropic, e.g., Wilson et al., 1991, Figure 12.27. In real spacecraft geometries, however, neither case obtains and an actual path-length distribution, as shown for point (b) of Fig. 2, must be analyzed in detail). The result is that a crewmember could achieve greater efficiency in shielding against exposure to space environmental radiation by staying close to an equipment wall than by locating at the center of the module, further from the radiation-shield impingement interface.

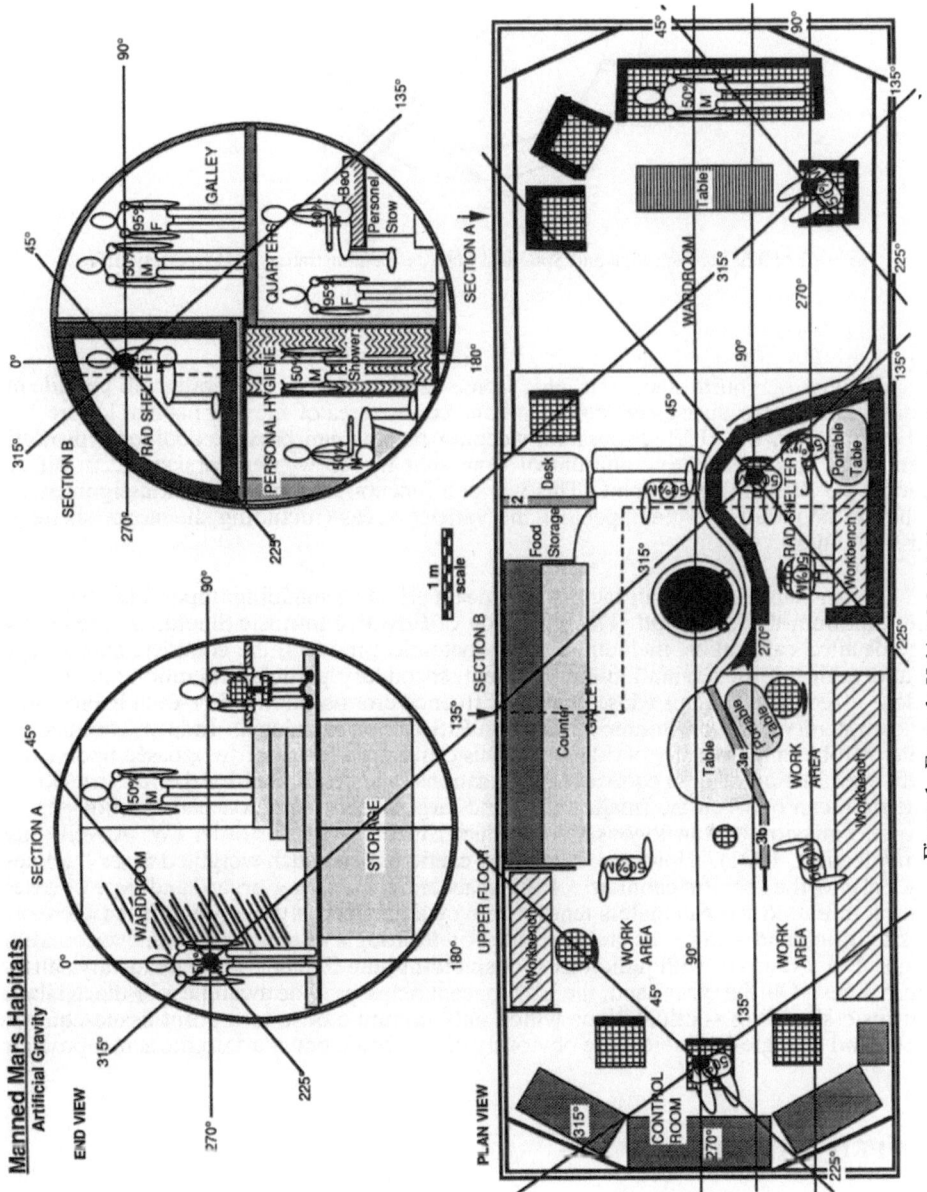

Figure 1. Example Habitat Module for Space Exploration

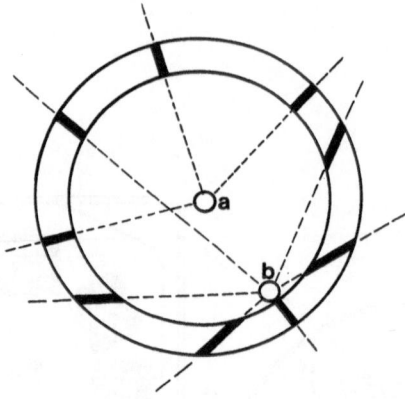

Figure 2. PathsThrough Slab and Spherical Shields, Demonstrating the Proximity Effect.

Finally, a fourth factor is that some areas of the crew habitat will provide more shielding than others, such as the command and control area of the SEI habitat, Figure 1. As will be shown in the GCR discussion, sleeping shields can provide additional protection. Depending upon the relative amount of time that the crew members spend in different locations, the EIS will be different. This will be a function not only of work assignments, but also habits acquired and the appeal of the various areas (including the storm shelter) for leisure activities.

When summed, the four factors provide an effective shielding importantly beyond that of the minimum-thickness wall. Determination of Effective Intrinsic Shielding (EIS) requires a very detailed calculation, including human anatomical models, the complete habitat layout, time allocation profiles, radiation environment, secondary particle generation and transport, and dose-effect relationships (dose/quality or fluence/cross-section). For each habitat design, crew complement, and environment model an EIS can be calculated. Indeed, for most cases the EIS will be in the vicinity of 50 kg/m², plus or minus a factor of two, based upon a quasi-objective evaluation of dose equivalent attenuations, and reinforced by the fact that 50 is the geometric mean between the smallest (10) and largest (250) common path thicknesses. For the Apollo mission, the analogous value deduced was also of this order (W. Atwell, private communication, 1991). However, it must be cautioned that such weighted mass thicknesses apply only to the specific geometry of the spacecraft, the target organ, and type of external radiation threat. Such parameters tend to convey a dangerously over-simplified view of the true situation, and cannot be a substitute for thorough re-analysis whenever conditions change. For example, with judicious spot shielding, the EIS can be dramatically shifted to higher values. On the other hand, the EIS concept helps provide awareness of the availability of intrinsic shielding -- calculations which only compute dose to a point tissue-equivalent mass behind a single-path shield are obviously only a crude boundary to the actual problem.

TRAPPED RADIATION BELTS

Other than the Earth, no other terrestrial planetary body has radiation belts. The hazard is thus peculiar to Earth operations, and includes both direct passage into deep space or orbital operations above the lower edge of the inner belt, about 700 kilometers above the surface. Even elliptical orbits with repeated passages through the belt can be acceptable, however, if perigee is below the inner edge and if apogee is high, preferably above 100,000 km. In this case, the integrated exposures are relatively low because of the phenomenon that most time in a highly eccentric orbit is spent near the apoapsis, outside the trapping zones. Belt passage for a 4-day orbit with apogee of 200,000 km takes only a few hours and the astronauts can take refuge in the solar flare radiation shelter during that time (Clark, 1990).

GALACTIC COSMIC RAYS (GCR)

The protons, alpha particles, and stripped heavy ions that impinge on our solar system comprise a low, but continuous flux of extremely penetrating ionizing radiation. The surface of the Earth is largely spared from this source because of the deflecting effect of Earth's unusually large intrinsic magnetic field and the huge 10,000 kg/m² shield provided by our atmosphere.

GCR During Interplanetary Mars Flight. Detailed calculations of rates for dose and dose-equivalent have been reported for the highest GCR fluxes (Wilson et al., 1991; Simonsen and Nealy, 1991; Letaw et al., 1987). Low-Z shields are shown to be more effective, per unit of mass thickness, than shields using elements of higher atomic number. For example, 100 kg/m² of path through pure hydrogen would reduce the tissue dose equivalent from the maximum unshielded value of 1.2 Sv/yr to less than 0.1 Sv/yr. In its liquid form, this would require 3.5 meters of the fluid, maintained at its cryogenic storage temperature of 20 K. Even though it is the ultimate shield, the use of liquid hydrogen becomes unrealistic from the standpoint of the enormous engineering difficulties that would arise from surrounding a room-temperature habitat with hard-cryogen tankage. Furthermore, the additional hazards associated with such a design could easily outweigh the gains in radiation risk reduction. For example, sudden loss of thermal insulating value could produce very low temperatures and/or thermally shock the habitat pressure vessel so badly that a rupture occurs. Hydrogen is also an explosion hazard, of course, if it mixes with the cabin atmosphere. Water is often suggested as a good low-Z shield for space radiation. A 250 kg/m² shield of H_2O will be three times less effective than hydrogen, producing a calculated dose equivalent rate of 0.3 Sv/yr. Using the same mass thickness of aluminum results in a 25% higher dose rate.

A pressure vessel wall thickness of 8 kg/m² Al would allow up to 0.9 Sv/yr skin dose rate. With an EIS of 50 kg/m² or more, provided by equipment, habitat, and self-shielding within the human body, the dose-rate is likely to be about 0.6 Sv/yr, i.e., twice that behind the thick low-Z shield.

When GCR particle fluxes are at their lowest, i.e., when solar activity and the repulsive effect of the associated interplanetary magnetic field is at its highest (Solar Max), the computed annual dose equivalents are only 40 to 45% of the above values. It is perhaps surprising that as a countermeasure to the space radiation hazard, the calculations indicate that simply flying Mars missions near or at Solar Max will reduce the GCR hazard with no additional shielding mass to a level as low as that behind a very massive shield of 250 kg/m² (2.5 tonnes for every 10 square meters of habitat wall area) when flying at Solar Min. Exactly how much mass penalty such shielding invokes depends, of course, upon the magnitude of the volume to be shielded. For a module as large as the original Space Station Freedom habitat module, a cylinder 4.5 m in diameter and 13.4 m long, the increased mass is 55 tonnes, compared to nominal starting mass of 10 to 20 t. The amount of mass that would have to be launched into low Earth orbit (LEO) would be considerably more because of the additional propellant needed for the trans-Mars injection propulsion. One measure of the economics of Mars missions is the gross mass of the spaceship when beginning the Earth departure rocket burn, i.e., analogous to gross lift-off weight for a rocket on Earth. The sensitivity of this IMLEO (initial mass in LEO) to habitat mass for missions to Mars has been studied for two example missions and plotted in Fig 3. Because the expense of launching mass into LEO can be anywhere from 1,000 to 10,000 $/kg, this would cause a significant addition to the total mission costs. These large amounts of "dead weight" shielding for interplanetary missions would almost certainly be considered to have an unjustifiable impact on the engineering of the spaceship and affect the general perception of the feasibility of Mars travel.

One popular suggestion for minimizing exposure to GCR has been to abandon the minimum energy Hohmann transfer trajectories, the so-called Conjunction Class missions which necessitate roundtrips approaching 1000 days, in favor of shorter trips utilizing Opposition Class trajectories, Fig. 4. The latter fly at least one of the transit legs toward the sun to speed rendezvous with Earth. In many cases a swingby of Venus to obtain a gravity assist can lower the energetic requirements on the propulsion system while still achieving

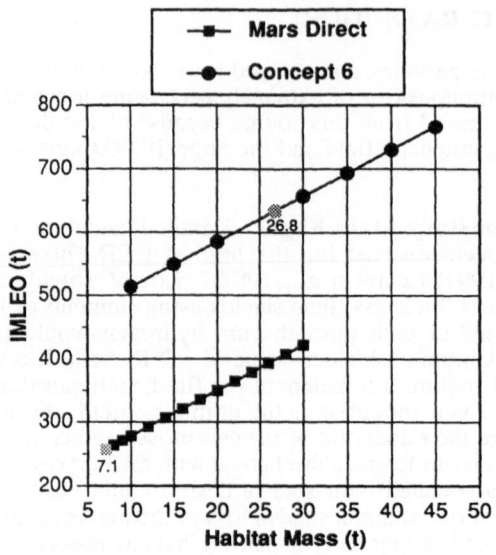

Figure 3. Mass in Orbit (IMLEO) for Mars Missions as Function of Habitat Mass

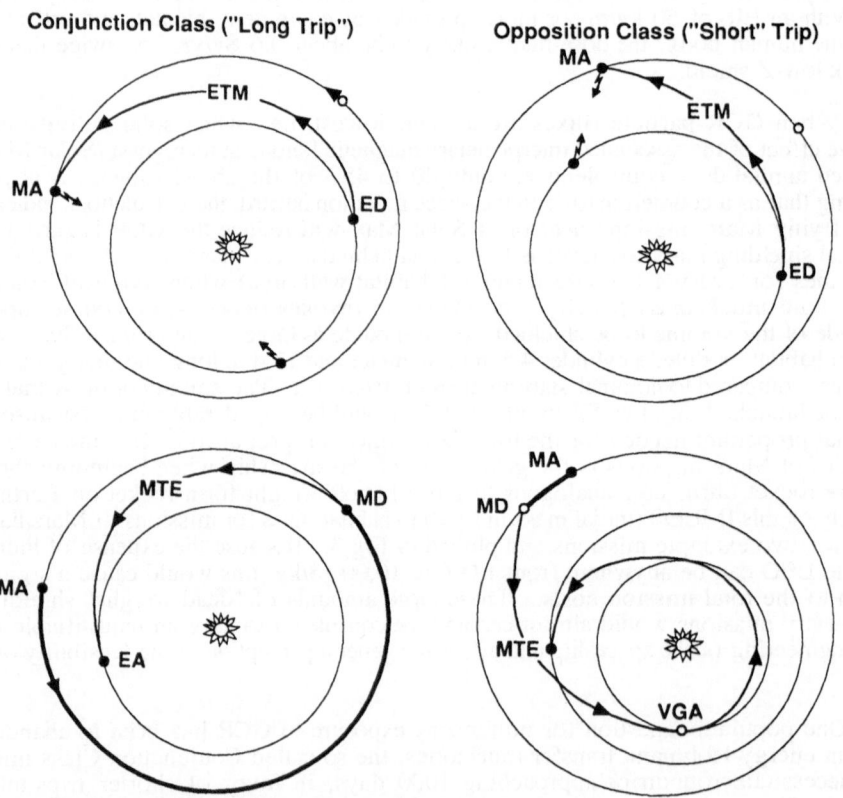

Figure 4. Typical Conjunction and Opposition Class Trajectories to Mars. (M = Mars, E = Earth, V = Venus, D = Departure, A = arrival, T = to, GA = Gravity assist)

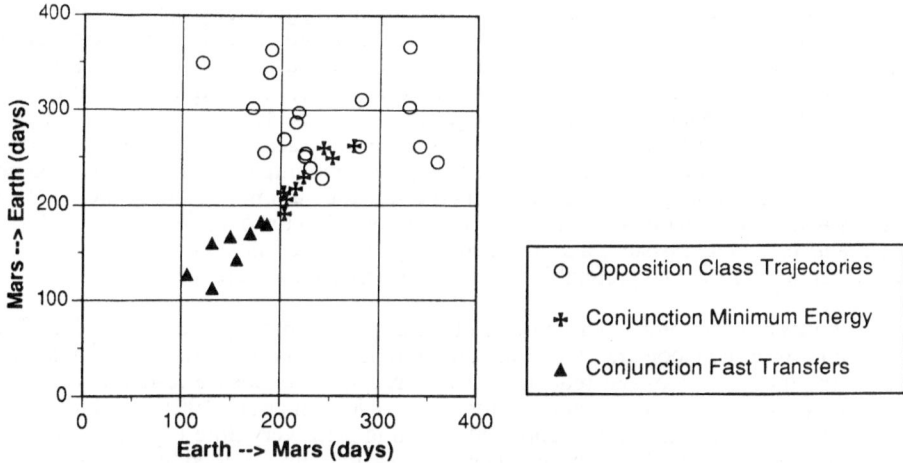

Figure 5. Transfer Times to Mars for Various Flight Trajectories and Energetics

round trips of 1.1 to 1.5 years. Nonetheless, the amounts of propellant needed for Conjunction trajectories are always much less than for Opposition. It has recently been shown (Clark et al., 1990) that for a modest penalty in added propellant to achieve higher departure velocities, much shorter transfer times can be obtained for the conjunction class than opposition missions, as amply demonstrated in Fig. 5. Note that for these "fast transfer" conjunctions, transit flight times can be less than six months in nearly all cases and even less than four months in certain specific cases. In addition to spending more time at Mars, the GCR exposure is minimized, because time spent on the surface will reduce exposures considerably. An optimum strategy would be to pre-place a Mars orbiting station, which could not only perform the functions of communications, observatory, consumables cache, and SolWatch, but also could be a large habitat which is heavily-shielded for GCR protection. Astronauts could then rendezvous with the orbiting station and spend their orbital stay at Mars in a more robustly shielded environment.

GCR Hazard on the Martian Surface. The martian atmosphere is composed mostly of CO_2 and although thin by terrestrial standards nonetheless provides significant shielding of the surface (Simonsen et al., 1990; Clark and Mason, 1990). At the reference level, the pressure is 6.1 millibars, which for Mars is an areal mass thickness of 160 kg/m2. Although

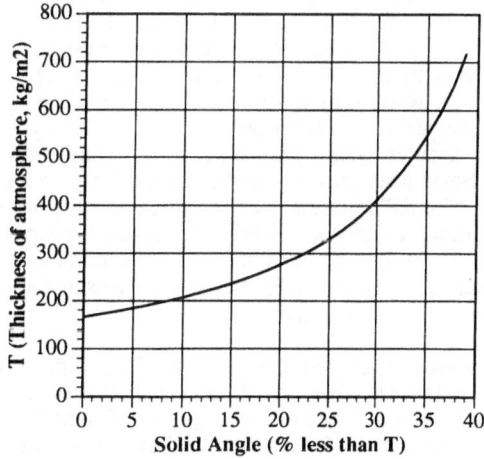

Figure 6. Integral Path Thickness Distributions Through the Martian Atmosphere

not massive, it is particularly effective because the shielding geometry approaches that of a semi-infinite slab. Very little solid angle is found in the zenith direction and most particles and their secondaries must reach their target through a much larger slant-path. Figure 6 is a solid-angle weighted path distribution through the atmosphere. Consequently, it may not be necessary to shield habitats on Mars with regolith as is expected for the moon. Indeed, Simonsen et al. (1990) calculate a worst-case GCR dose equivalent of 0.13 Sv per year on the surface of Mars. It should be acknowledged that the topography on Mars is exaggerated compared to the scale height of the atmosphere, and the summit of the highest volcanic mountain is at an atmospheric depth of only about 30 kg/m^2. However, landings on Mars are aerobrake and parachute assisted, and it is extremely likely that most candidate landing sites will be chosen to be near, and preferably below, the reference pressure level.

GCR on Lunar Missions. The doses from galactic cosmic rays during the 2 to 5 day journey to and from the moon are trivial, amounting to less than one millisievert. Once on the surface of the moon, a two-fold reduction in the free space exposure levels of 0.2 to 0.5 Sv/year can be further reduced another factor of two (Wilson et al., 1991) by shielding the habitats with 0.75 m, of lunar soil. Additional shielding helps very little (op. cit.), so that the nominal long-term exposure for crew members inside a regolith-shielded lunar base habitat is expected to be 0.05 to 0.1 Sv per year (Nealy et al., 1989). For 180-day tours of duty (TOD), this will be less than the new recommended limit for 5-year exposure of 0.1 Sv for Occupational Workers (ICRP-60, 1991). However, such limits cannot apply to Mars missions and are well below the currently accepted special limits for U.S. astronauts (0.5 Sv/yr and 1 to 4 Sv career total, NCRP-98, 1991). For shorter TOD's of 90 d, and especially during Solar Max, regolith shielding on the moon may be unnecessary for protecting against GCR.

SOLAR PARTICLE EVENTS (SPE)

Sudden brightenings on the solar disk, known as solar flares, can lead to the ejection of energetic plasmas, producing magnetic storms and particulate radiation in interplanetary space. These SPE's release high fluxes of protons and much lesser quantities of ions and electrons, which become experienced as mostly an omnidirectional flux at the observer.

SPE During Interplanetary Mars Flight. The solar flare hazard has long been recognized for Mars. Even in recent times, it has been suggested that Mars missions be conducted at times of Solar Minimum. From a project standpoint, however, this would be a difficult constraint. Mars missions encompass roundtrip travel times of 1.2 to 2.5 years, and can be only initiated at about 2.2 year intervals. In the worst case, planning would have to span nearly a five year period just to assure a primary and one fall-back launch opportunity. However, major SPE's do not occur only at the peak of Solar Max and the historical record suggests that assuming a five-year period at Solar Min with no hazardous flares would be risky. Another problem is that it is not possible to predict the solar cycle reliably beyond a decade or so because of its varying periodicity; indeed, the latest target Mars mission date of 2019 could be either Solar Min or Solar Max (Heckman, 1992). Furthermore, mission project office will set Design Requirements which will provide a worst-case design SPE, likely to be a single model regardless of position within the solar cycle. For this reason and the fact that the GCR is a greater hazard than a properly counter-measured SPE, we have previously suggested that Solar Max may be the best time to fly to Mars (Clark and Mason, 1990; Clark et al., 1990). The SPE hazard is different depending upon the trajectory taken. Although the Opposition Class provides nearly two-fold less total trip time, and hence a lower exposure probability, it must fly one or both legs to solar distances as close as 0.7 A.U. According to Shea and Smart (1992), the SPE flux will be less at the orbit of Mars than near the orbit of Venus by factors of 4 to 6 or more.

It has also long been recognized that a solar flare radiation shelter would be a prerequisite for Mars missions. Of what could such a shelter be constructed? A common suggestion has been to utilize the large quantity of H_2O which will be required for missions to Mars. Water, being low-Z, makes an efficient shield, being fluid, and could fill in gaps and voids. There are several problems, however, with this suggestion. First, it could not be the water that is planned for consumption -- the astronauts could not be allowed to drink away their shield. Second, Mars missions will not, in fact, need to take a large reservoir of liquid water. Rather, humans in a closed environment are a net producer of water from the metabolic conversion of food and breathing oxygen to CO_2 and H_2O. In addition, frozen and thermostabilized foods contain large amounts of water. Even dried food contains water of hydration. Thus, as the mission proceeds, water will be produced and recycled by the humidity control system and the hygienic extraction of water from urine. Possibly, a surplus of water will build up, but only a small supply would be available initially. In addition, in conditions of weightlessness liquids in unfilled containers tend to form large, random, and unpredictable voids which would allow radiation shielding holes unless actively controlled.

Since it is obviously highly undesirable to require inert shielding mass, some other source of material needs to be found. Various limitations, such as center of mass, functionality, packaging efficiency, human factors, and other constraints prohibit imposition of arbitrary requirements for placing all spacecraft equipment around the radiation shelter. Instead, we have suggested that consumables (e.g., food, medicines, household cleaning items, and maintenance supplies) be located in storage compartments situated in thick but otherwise hollow walls around the shelter (Clark et al., 1990; Clark and Mason, 1990). The quantities of food and miscellaneous supplies for supporting a Mars mission are typically quite large, about 10 tonnes for a three-year mission of five crew persons to the red planet. Obviously, as food is consumed, the resulting trash and human waste products must be packaged, compacted and re-stored in the walls to maintain the shielding mass. Normally on such a mission these waste materials would be dumped overboard into interplanetary space or left in Mars orbit. In addition to consumables, certain passive items such as spare parts, backup equipment, and tools which can be stored anywhere can be strategically placed around the shelter area.

SPE Hazard on the Martian Surface. As discussed above, the martian atmosphere provides excellent shielding against solar flare particles. According to Simonsen et al. (1990), doses at the reference surface level do not exceed 0.11 Sv for the largest known SPEs.

SPE Hazard during flight to the Moon. Although the transit time is just a few days, the danger of being caught in a severe SPE while inside a very lightweight transfer vehicle is of increasing concern. Either the entire cabin will have to be shielded, or a storm shelter of some sort provided. In this case, there will be no appreciable complement of consumables, spare parts, or other passive equipment which could provide the shielding. The shielding material will have to be dead weight unless supplies being transported to a base can be packaged in useful configurations or spacecraft equipment can be strategically placed.

SPE Hazard at the Lunar Base. Assuming that a regolith shield is provided over the Base habitat to protect against GCR radiation, such shielding also will protect against solar flare particle events. When the Base is initially under development, however, there may be a period where visits are sufficiently infrequent or short in duration that a GCR shield is not yet required. To handle these situations, we have developed a rapidly implementable shelter utilizing a drop-down compartment and lunar regolith. The compartment, inflatable or rigid, is first extended to contact the surface, Fig. 7. Astronauts then load pre-molded stackable forms (approximately the size of waste-paper baskets) with soil and array them around the compartment, much as a brick mason would but using contact fasteners, such as Velcro. Additional soil is bagged and taken inside the habitat to fill a sliding cover to complete the SPE storm shelter. It is estimated that this shelter could be completed in a single astronaut EVA work shift.

SPE Hazard to Long-Range Lunar Exploration. At the lunar surface, as compared with the martian surface, there is neither the shielding effect of an atmosphere nor the increased distance from the sun. The solar flare hazard is thus much greater on the moon than on Mars. Furthermore, until a planetary scale SolWatch program can be set up to monitor flare activity beyond the western limb of the sun, only the hemisphere visible from Earth can be directly viewed. In a pressurized lunar rover, with windows and wall perhaps only 10-30 kg/m^2 thick (1-3 g/cm^2), a large flare SPE could lead to ≥ 3 Sv, resulting in eventual radiation sickness, incapacitation, or even death. Several such events have occurred in solar cycles in the past four decades that they could be so monitored (Shea and Smart, 1992). Many remedies have been suggested. Foremost has been forecasting to avert EVA and rover operations during dangerous flare activity. Unfortunately, the ability to predict days or weeks in advance remains totally unreliable for this purpose (G. Heckman and D. Smart, private communications this conference). Even advance warning on time scales of hours or minutes suffer from many false positives and, worse, occasionally fail to predict an SPE before fluxes arrive at Earth (Heckman, 1992). Whatever the countermeasure, it must be a rapid solution, taking one-hour or less. A return to Base is essentially out of the question as a practical solution because the vehicle could be days away.

Figure 7 Lunar Habitat with Rapid-Construction SPE Storm Shelter

One suggestion has been for the astronauts to rapidly suit-up, abandon their rover, and use the lunar regolith for shelter. They could, for example, construct a trench, bridge it with a preconstructed roof, and load with soil. They could seek refuge under the rover storage tanks (assuming they are not near depletion!) and pile regolith around the sides. It has even been suggested they use a shaped charge to explode a cylindrical hole in the regolith. All of these concepts would require a special spacesuit with sustainability at least three times the current 6 to 8 hour limit. If the rover were pressurized to the one bar that is operational for Shuttle, planned for Space Station Freedom, and cited for a Lunar Base, the crew could suffer the bends when conducting a quick EVA without prebreathing pure oxygen for several hours. Finally, the success of such a strategy depends strongly upon the physical accessibility of loose soil where the rover is currently located.

Of course, the rover could have built-in shielding, taking a major mass penalty. Or, the crew could have a very small internal shelter, utilizing consumables and wastes, as with the Mars interplanetary spaceship, but also supplemented by pre-bagged or briquetted lunar soil. Batteries and equipment boxes could also be placed around such an interior shelter. Obviously, a lunar rover will pay a mass penalty compared to its martian counterpart because of this additional shielding mass which must be transported. But offsetting this will be the more than two-fold lower gravity and the more trafficable surface of the moon to lessen propulsion requirements.

PLANETARY SURFACE EMISSIONS

Prior to spacecraft missions to the planets, there were no direct measurements of surface compositions, particularly in terms of the naturally occurring radioactive elements -- potassium, uranium, thorium (K, U, Th) -- and their daughter products. On Earth, these elements are all enriched in the crust compared to deeper regions because of their geochemical lithophilic tendency, partitioning into uprising siliceous material during differentiation of the planet and formation of the mantle and core. Mars and the moon have much less differentiated cores. On Mars, the K content is very low in the soil, less than 0.25% by weight (Clark et al., 1982) and less than 0.5% in all but a few lunar soils (Adler, 1986). This is to be compared to 1-2% for most soils on Earth (Mason, 1966). Uranium plus thorium on Earth typically totals 15 ppm, but ranging from 3 to 50 ppm depending upon the sedimentary and igneous rock formations at a given site. These elements total less than 5 ppm at most sites on the moon, reaching 15 ppm only at the Apollo 14 site. Although there are as yet no conclusive measurements for U+Th on Mars, the available data from orbital remote sensing and relevant meteorites predict a sum of less than 0.5 ppm (Banin et al., 1992).

In addition to radiation doses due to natural soil emissions, there will be nuclear reaction products from the interaction of galactic cosmic rays with the soil, and in the case of Mars, also with the atmosphere. These product emissions include neutrons, gamma rays (from neutron inelastic and capture reactions, as well as decay of activated radionuclides), and protons and heavier ions. The actual levels for gamma rays and neutrons will be measured directly by the GRS instrument on the Mars Observer spacecraft in polar martian orbit. Gamma ray levels have already been measured directly at the moon (for a review, see Adler, 1986). Natural background dose rates at the surface of the Earth are typically 0.001 Sv/yr, with direct radioactive emissions accounting for about 20% of the total (Mayneord and Hill, 1969).

As a result of these considerations, the dose equivalent on Mars and the moon due to the planetary surface material is to be expected to be completely negligible, being less than 1% of the hazard posed by the incident GCR. As mentioned elsewhere, the planetary surfaces always are beneficial because of their 2-pi shielding and the availability of local soil to cover habitats or shelters.

ON-BOARD ARTIFICIAL RADIATION SOURCES

Man-made radiation sources will also be part of human missions to the planets. Radioactive isotopes have been built into many space instruments. Apollo used radioisotopic thermoelectric generators (RTG), whose energy source is plutonium-emitted alpha particles, to power deployed instruments on the moon. A nuclear thermal rocket is a prime candidate for the propulsion system for Mars exploration. The SP-100 thermoelectric nuclear reactor power system may be needed for supporting a lunar base during the lunar night, and with suitable modifications could be used on the martian surface. Since astronauts will be exposed to natural radiation, should these artificial radiation sources be allowed to provide equivalent doses? If there were no natural background, the ALARA principle (as low as reasonably achievable) would be invoked. It is suggested that this remain the case because the actual hazard from space radiation is even more difficult to determine because of the latitude in quality factors and the risk aversion procedures will tend to err on the conservative side (i.e., highest plausible quality factors). For controlling exposures due to economic activities, the U.S. OSHA regulation 29 1910.96 currently limits exposures to 0.03 Sv in any given calendar quarter, with 0.0125 Sv/qtr for the yearly average, and total cumulative exposure of 0.1 to 0.15 Sv, for 38 to 48 year old personnel, respectively. These standards could be placed upon all astronaut activities independent of the space background radiation, which could have its own limits. If so, then designs employing these artificial sources must be such that these regulations are satisfied. The design engineer must use distance and shielding to achieve satisfactory results, and also demonstrate that ALARA is being followed. This approach constrains nuclear reactor locations and shielding. As another example, the use of RTGs as a power source for manned rovers becomes considerably complicated. As seen in Fig. 8, the dose equivalent delivered in a 14 day sortie mission by a rover can exceed 0.03 Sv, the quarterly limit, even if a one kWe RTG cluster is located more than 4 meters from the

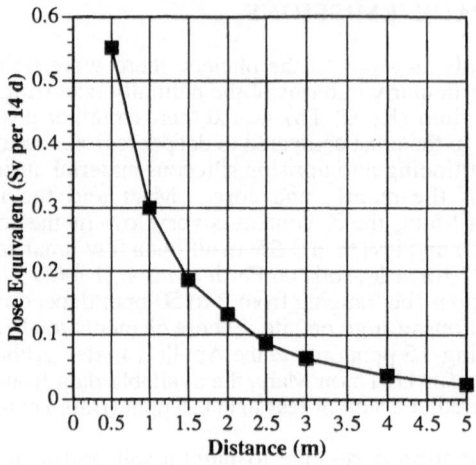

Figure 8. Dose From RTG Power Source

crew cabin. Shielding could be added, but at considerable mass penalty because the emissions are primarily neutrons and energetic gamma rays.

COUNTERMEASURES

A large number of issues arise when considering how to negate or minimize the effects of radiation using practical design options. Shielding is the first choice of countermeasure (CM), but shielding the entire habitat to 240 kg/m² for interplanetary flight to Mars, as suggested by Letaw et al. (1987), would result in very high mass penalties to the mission -- from 50% to 100% increases for typical mission trajectories and staging profiles (Clark and Mason, 1990).

Weightlessness might aggravate sensitivity to some radiation effects, but the data which indicate this possibility are presently only minimal. Providing artificial gravity for the astronauts would presumably eliminate such a concern.

Chemical prophylaxis by administration of certain drugs to the astronauts would probably never be tolerated, unless perhaps for a life-threatening emergency. With adequate shielding against SPE radiation, the possibility of acute radiation effects is eliminated and the chief hazards are carcinogenesis, genetic damage, and cataractogenesis. Of these, the danger of inducing carcinoma or other cancers is currently considered the most serious. Even if mutation of reproductive cells is found to be unacceptably likely, astronauts could have gametes stored on Earth until return or could elect not to reproduce further. Pregnancy on space missions is obviously to be forbidden, not only because of radiation, but the general physiological stress and hazardous nature of the mission itself. Cataracts are of very low probability during the mission, can be treated by several methods, and will eventually afflict virtually everyone fortunate enough to become aged. Much remains uncertain as to the effects of pharmacological agents upon physiological systems adapted to weightlessness, especially when subjected to chronic psychological and physical stress. Vitamins and other mild agents may be acceptable, but the efficacy of anti-oxidants in the human system as a CM to radiation carcinogenesis is essentially beyond validation.

Magnetic Shielding. On many occasions over the past three decades, it has been proposed that if a strong magnetic field were created around the crew cabin, the impinging radiation particles, being charged, would be deflected sufficiently that the dose could be substantially reduced or eliminated. Although correct in principle, this CM concept has always suffered from several drawbacks: (1) the magnetic moment required is enormous and would cause large inductive currents whenever it changed, a condition that could be especially catastrophic if a field collapse occurred; (2) superconductors would be needed to avoid totally impractical regimes for energy maintenance of the field, but such conductors need cryogenic temperatures (recent advances in high-temperature superconductors has been one of the main motivations for a resurgence of this concept); (3) all equipment operating within the protected zone would have to be tested and qualified for the field effects; and (4) the health and safety issues of allowing astronauts to be exposed to large electromagnetic fields is reason alone that such an approach would probably not be seriously considered by project management.

Tailored Shielding. It is interesting and potentially useful to recognize that of the 25 or more tissues that are susceptible to radiation-induced carcinogenesis, all except four -- the brain and the distributed tissues: blood, lymph, and skin -- belong exclusively to the trunk of the body. Partial body shielding could therefore be considered in an effort to reduce mass and the inconvenience of a garment shield. We have therefore considered as a shield the thick, elongated vest shown in Fig. 9, which for a mass of 150 to 300 kg, depending upon body size, could provide 200 kg/m² protection to the critical body organs and much of the distributed tissue. Goggles or a helmet could be added to provide protection to the eyes or head, for additional mass penalties of 5 to 50 kg. If the habitat is in a force field, whether due to planetary gravity or centrifugal acceleration (artificial gravity), such personal body shielding cannot be used because of its oppressive weight. Even for application in weightlessness, certain problems would arise in restriction of motion, and the material of the garment may have to be so special that it is useless for any other purpose.

As a supplementary approach, astronauts could receive special protection when in bed from a sliding "cocoon" shield which envelopes head-to-hips, as shown in Fig. 10. Astronauts could spend as much as one-third of their time in such cocoon shields just sleeping or resting in their crew quarters. In the fetal position, with legs retracted, the entire body is shielded except for a small fraction of solid angle, where the flux must first traverse the legs. With legs extended, the torso is still protected by the legs and shield. As with the solar flare shelter, these cocoon shields could have walls which incorporate consumables and inactive equipment, as well as the personal effects of the astronauts themselves. Water, aluminum,

Figure 9. Garment Shield Distributed Tissues.

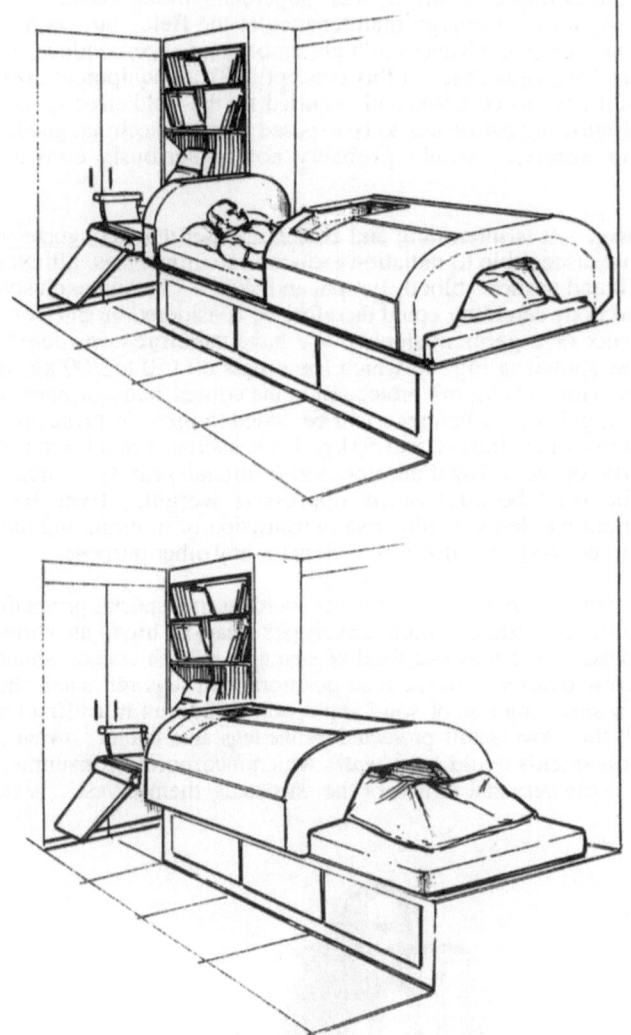

Figure 10. Sleeping "Cocoon" Shield.

hydrogen, or supplies could make up the shield. It should be pointed out that the total mass of a shield is directly proportional to the mass thickness per unit area only under the condition that the thickness of the shield is small compared to the scale of the volume being protected. As is more or less obvious by inspection in Figure 11, the large thickness of a low density material such as hydrogen results in an inordinately high volume simply because of the fact that volume is cubic with the radius from the center of the shielded volume whereas thickness is linear with radius. This actually results in much higher mass for the very low density LH2 even though the areal thickness is less. For example, a water shield could be provided for about one metric tonne or an Al shield for about 20% less, but a liquid hydrogen shield if it had the same 200 kg/m^2 minimum thickness would be several tonnes because of the density-volume effect. Due to the proximity effect in omnidirectional flux fields, the low density shielding is more effective and the areal mass thickness, hence total mass, could be somewhat reduced.

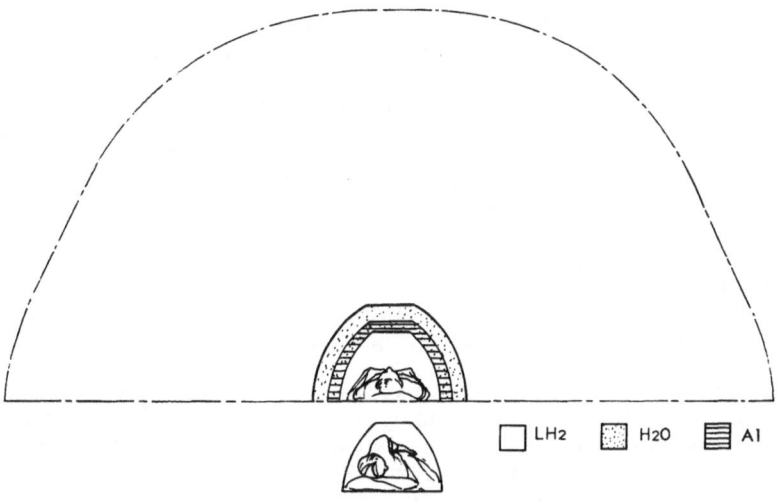

Figure 11. Cocoon Shields of Various Materials for Same Mass Thickness

Solar Monitoring. It is clear that both for space flights and for operations on planetary surfaces, it would be prudent to avoid EVA or extended rover sorties during times at which the risk of SPE is high. Monitoring solar dynamic activity and flare development will be a high priority. One dilemma is that flares located just over the west limb of the sun are still connected to the region of Earth by solar magnetic field lines and thus can produce SPEs without being observed. Ideally, a number of satellites should be equally spaced in heliocentric circular orbit so as to allow observation of both hemispheres of the sun at all times. This goal probably cannot be justified from a cost-benefit standpoint because of the unreliability of the predictions, but there are other strategies that can be used. For example, a SolWatch satellite placed in high Earth orbit would continually monitor the sun (emplacement of such a station on the lunar surface is both more costly and also not effective unless placed at several widely-spaced locations, because of lunar night). A similar SolWatch monitoring capability must also be provided on Mars spaceships, allowing observation of different portions of the sun. If SolWatch instruments could be included on satellites orbiting Mars and Venus, the coverage could be greatly expanded, although there might remain a small amount of time when unfavorable interplanetary geometries would not provide total coverage of the sun.

Individual Astronaut Susceptibility to Radiation Risk. Calculation of astronaut risk assumes uniformity in extrapolation from exposures of the general population and laboratory research on animals, mammalian tissue cultures, and even bacteria and other microorganisms. Actual data on human radiation response rests upon the recent re-analysis of Hiroshima and Nagasaki atomic bomb survivors and to a much lesser degree upon radiation treatments of two types of illness (ICRP-60, 1991), and nuclear accidents. There is no likely new data source that can become available, and it would be unthinkable to purposely expose astronauts to space radiation other than when required to execute deep space missions. In particular, the concept of exposing astronauts in orbit around the moon as part of a Mars mission precursor is ethically untenable because of exposure to damage without compensatory reward.

Likewise, it could be deemed unethical to allow individual astronauts who are more susceptible than normal to radiation damage to become exposed. At the very least, for Mars missions or long-term lunar assignments, persons who have family histories with high incidences of cancer should be considered for exclusion, radiation risk or not. Advanced tissue testing techniques may permit screening of individuals for either genotypic or phenotypic pre-disposition to oncogene activation, cell transformation, or other cancer etiological factor. In principle, it might be possible to identify individuals with greater than average resistance to radiation damage, just as astronauts and cosmonauts have always been selected on criteria far from the norm for a number of other attributes.

From the standpoint of lifetime risk, older persons have a lesser increase in probability of development of fatal radiation-induced cancer because their inherent risk is already elevated simply because of the age effect. Females have been previously identified as having greater risk to space radiation, primarily due to higher susceptibility of the reproductive organs. Exposure limits for age and gender thereby have spanned a range of four-to-one (NCRP-98, 1989, Table 1.1). One might deduce that to reduce radiation risk, only obese, elderly males should be allowed to travel to Mars, a conclusion that is obviously unacceptable (save, perhaps, to some fat old men) on many different grounds. This reductio ad absurdum is a graphic example of what often happens less obviously when disciplines are allowed to dictate their own requirements independent of other, higher-level project or systems needs.

Medical Treatment. The risk of fatal cancer can be mitigated to an important degree by diligent monitoring for first occurrence and by skilled, advanced intervention at the earliest detection. Exposed astronauts will certainly be automatically provided this enhanced care. For this reason alone, the risk factors may be somewhat exaggerated since they are derived for persons with average health care and generally no reasons for special diligence. Furthermore, the median latency period for most cancers is 15 to 25 years, with a minimum of 5 to 10 years. The major exception is leukemia, which can occur as soon as 2 years, but more typically 8 years, after exposure (ICRP-60, 1991). With human missions to Mars not likely to commence until the year 2009, and perhaps as late as 2019, the first exposures will not begin for another 17 to 27 years from the present. Combined with the latency periods, the cancers in question would not result for about 20 to 50 years from now (leukemia possibly earlier). What can be accomplished in the next two to five decades in medical progress in the detection and treatment of cancers is utterly beyond prediction. It is certainly to be hoped that outright cures become available in many cases and that prognoses for successful recovery become much improved. It is ironic that because radiation-induced cancers are indistinguishable from "spontaneous" ones, if an astronaut succumbs to fatal cancer (as one-third are predicted by the actuarial tables) there will be no possibility of determining whether the space flight was the actual cause or whether it is a "natural" illness. Mars astronauts will be a hopelessly small sample size to be applicable to the statistical tests normally used to detect radiation carcinogenesis.

CONCLUSIONS, OBSERVATIONS, AND RECOMMENDATIONS

(1) Intrinsic shielding exists for all space radiation exposures because of shielding by the equipment, structure of the vehicle, and the astronaut's body. The weighted intrinsic shielding is of the order of 50 kg/m^2, but must be calculated for each space mission's environment and spacecraft design.

(2) Intrinsic shielding is more effective if astronauts keep proximate to a thick equipment wall than locate in the center of a habitat.

(3) Lowest-Z material has the lowest mass thickness but not necessarily the lowest total mass for a given shielding effectiveness. Liquid hydrogen shields are theoretically best, but impractical and entail other hazards.

(4) The greatest radiation hazard for missions to Mars is the galactic cosmic radiation (GCR). The greatest radiation threat for missions to the moon is the solar particle event (SPE).

(5) It is better to fly to Mars during maximum solar activity (Solar Max) because GCR fluxes are reduced during this time. A lightly shielded habitat during Solar Max may be as safe as a heavily-shielded one flown during Solar Min.

(6) SPEs are shielded very effectively by a radiation storm shelter in the interplanetary space vehicle and by the 160 kg/m^2 thick martian atmosphere when on the surface.

(7) Radiation storm shelters for Mars missions should have compartmented walls, i.e., be designed to store consumables, waste products, tools, spare parts and equipment, thereby causing very little weight penalty to the mission.

(8) Sleeping cocoons and/or personnel radiation vests can help reduce the hazard of GCR. The sleep-shields could incorporate consumables; the vests probably could not.

(9) Transfers to and from Mars can be faster using Conjunction class trajectories rather than Opposition Class. At Mars, refuge can be found on the martian surface or in a well-shielded orbital station. Thus, radiation exposures can actually be less for the longer roundtrip trajectories than the short roundtrips (Opposition's).

(10) Short roundtrip trajectories, in flying closer to the sun, must expect higher fluxes and fluences from any SPEs which occur.

(12) For missions to the moon where personnel tours of duty are kept short (less than 90 days), shielding by piling lunar soil on the habitats to protect against GCR may not be necessary, but a SPE shelter would be needed.

(13) SPEs have many complex features. They produce mostly omnidirectional fluxes, requiring 4-pi shielding. They are difficult to predict solely from remote observations of the sun, and false negatives are possible which means that dosimetric monitoring and emergency procedures for rapid access to shelter are necessary.

(14) Solar monitoring should not be performed from lunar transfer vehicles or the surface of the moon, but rather from high Earth orbit, from a Mars orbiting platform, from all enroute Mars spaceships, and perhaps also from a Venus orbiter.

(15) Providing an efficient storm shelter in the small and lightweight lunar transfer vehicle is a major challenge.

(16) It is better operate on the moon during Solar Min, when SPEs are less frequent. The end of this decade would be an excellent time to begin a lunar base.

(17) Rovers away from a lunar base will be very vulnerable to SPEs. Lunar rovers may therefore be much more massive than martian rovers.

(18) The hazard of GCR must be ultimately evaluated from systematic laboratory research, because (a) it is unethical to purposely expose lunar astronauts to GCR to seek effects, and (b) even if one or more Mars astronauts ultimately have cancer, it will be statistically impossible to assign blame to the flight itself.

(19) Very high altitude, long-term balloon flights in Antarctica could provide a research platform for biological exposures to GCR at a fraction of the cost of recoverable satellites.

(20) Pharmacological agents as radiation countermeasures are almost certainly unacceptable to the astronauts and project medical management. Magnetic field protection is also probably untenable for these and other reasons.

(21) Individual astronaut sensitivity to radiation may have to be taken into consideration, including age, gender, and genetic make-up. Risk models favor elderly males with no familial background of cancer disease. However, other factors will mitigate these selection criteria.

(22) Human variability and the eventual medical improvements in reducing cancer fatality could permit higher dose exposures to become acceptable for exceptional situations like early Mars exploration.

(23) Human exposure in deep space remains a major problem; equipment hardening for deep space radiation is not new and has been solved many times for unmanned spacecraft missions to the planets and comet Halley.

ACKNOWLEDGEMENTS

Sensitivity calculations of habitat mass vs IMLEO were provided by S. A. Geels. The lunar quick-shelter concept was originated by P. S. Thompson and the author. L. W. Mason calculated the martian atmospheric solid angle weighted path distribution and has collaborated on a number of aspects of the radiation hazard to humans in space.

REFERENCES

Adler, I. 1986. The Analysis of Extraterrestrial Materials. Chemical Analysis series, Vol. 81. Wiley-Interscience.

Banin, A., Clark, B. C., and Waenke, H. 1992. Surface Chemistry and Mineralogy. Chap. 4.4 in The Mars Book, Ed. by H. Kieffer, B. Jakosky. Univ. of Arizona Press.

Clark, B. C. 1990. Benefits of HEPO as staging locations for human interplanetary exploration. IAF-90-339. 41st Congress of the Int'l Astron. Federation, Dresden.

Clark, B. C. and Mason, L. W. 1990. The radiation show-stopper to Mars missions: A solution. Presented at the Case for Mars IV, Boulder, CO. To be published by AAS. Proceedings in preparation.

Clark, B. C., Baird, A. K., Weldon, R. J., Tsusaki, D. M., Schnabel, L., and Candelaria, M. P. 1982. Chemical Composition of Martian Fines. J. Geophys. Res., 87, 10059-10067.

Clark, B. C., Baker, D. A., Geels, S. A., Redd, L., Thompson, P. S., Willcockson, W. H., and Zubrin, R. M. 1990. Manned Mars System Study, Final Report. Martin Marietta, MCR-90-537 for NAS8-37126.

Heckman, G. 1992. Prediction of solar particle events for exploration class missions. This conference.

ICRP-60. 1991. Radiation Protection; 1990 Recommendations of the ICRP. Pergamon Press. ISBM 0-08-04144-4.

Letaw, J. R., Silberberg, R., and Tsao, C. H. 1987. Radiation hazards on space missions. Nature, 330: 709-12.

Mason, B. 1966. Principles of Geochemistry. Wiley and Sons, NY.

Mayneord, W. V. and Hill, C. R. 1969. Natural and man-made background radiation. In Radiation Dosimetry, Vol. III. Ed by F. H. Attix and E. Tochilin, Academic Press, NY.

NCRP-98. 1989. Guidance on Radiation Received in Space Activities. Rept. No. 98, National Council on Radiation Protection and Measurements. R. J. M. Frye, Chair. ISBN 0-929600-04-5.

Nealy, J. E., Wilson, J. W., and Townsend, L. W. 1989. Preliminary analyses of space radiation protection for lunar base surface systems. SAE 891487.

Simonsen, L. C., Nealy, J. E., Townsend, L. W. and Wilson, J. W. 1990. Radiation exposure for manned Mars surface missions. NASA TP-2979.

Simonsen, L. C. and Nealy, J. E. 1991. Radiation protection for human missions to the moon and Mars. NASA TP-3079.

Shea, M. A., and Smart, D. A. 1992. History of energetic solar particles for the past three solar cycles; this conference.

Wilson, J. W., Townsend, L. W., Schimmerling, W., Khadelwal, G. S., Khan, F., Nealy, J. E., Cucinotta, F. A., Simonsen, L. C., Shinn, J. L., and Norbury, J. W. 1991. Transport methods and interactions for space radiations. NASA Ref. Publ. 1257, NASA/Langley Research Center.

EXPERIENCE, LESSONS LEARNED AND METHODOLOGY IN THE DESIGN OF SPACE SYSTEMS TO ACCOMMODATE TOTAL DOSE AND SEU EFFECTS

James H. Trainor

NASA, Goddard Space Flight Center
Greenbelt, MD, 20771, U.S.A.

ABSTRACT

We now have considerable experience with successfully designing science systems to function properly, even during and after exposure to ionizing radiation approaching one Megarad, and during large solar particle events with substantial high Z fluxes such as the events that happened in August, 1972 and in the fall of 1989. Nevertheless, with changing device dimensions and properties, newer technologies and new applications, new problems arise. Some recent basic research has shed light on these problems and pointed the way to better diagnostic tests. In the case of single event upsets, clearly one cannot always accurately predict upset probabilities simply by knowing the quiescent LET threshold and the asymptotic cross-section of the device.

A summary will be given as to how designers at the Goddard Space Flight Center are proceeding with the new information, with more sophisticated tests and with new devices. Examples will be given from the successful designs for instruments to measure the cosmic radiation on the Pioneer 10/11 and Voyager 1/2 spacecraft, as well as the extension of these instruments into the designs now being implemented for the 1990's. Additionally, since the potential designs for human shelters within possible space stations or planetary spacecraft have been a topic of interest at this conference, a brief discussion will be given of consequences of the materials selection for such shelters. Given the discussions that occurred in earlier papers at this conference, some early and straightforward measurements could supply data that would be much more useful to the radiobiologists than the primary cosmic ray spectra would be. Final remarks will concern how the presently known information, together with results from the flight and ground portions of the CRRES mission can lead to less risk and lower costs for future space systems.

INTRODUCTION

While extensive work has been done in measurements and analysis of nuclear radiation effects on materials, electronics devices, sensors and various life forms, in practice, the approach to the design of systems for space has often been a simplified approach using approximate sensitivities and active areas, a somewhat conservative model of the radiation to be encountered, and a safety factor of some sort. Not all managers, designers or flight projects approach these considerations in the same way. There is considerable variation in the end result. It is my observation that most experienced designers and managers are usually too conservative concerning the ionizing dose imparted to electronic parts. While this eliminates these problems in flight, this approach leads to other project risks, usually ones of schedule

Biological Effects and Physics of Solar and Galactic Cosmic Radiation,
Part B, Edited by C.E. Swenberg *et al.*, Plenum Press, New York, 1993

889

delay and increased costs. A discussion will be given of more realistic and successful approaches to these problems, together with recent examples of real problems in flight that resulted from inadequate designs. The inadequate designs can result from not knowing the performance of the newer electronic part or sensor in the "infrequent" radiation environment you may find the system exposed to once or twice per decade, for instance. The large solar nuclear particle events in March, 1991 followed by more than 18 years the previous, large event in August of 1972. I can only surmise that the damage that occurred to the power conversion circuitry on several commercial satellites in geosynchronous orbit as a result of the 1991 events [Allen, 1991] was due to lack of proper parts testing or failure to account for the possibility of such large events with high ionizing doses and large fluences of high Z nuclei.

SENSORS

Experienced investigators of phenomena in space have usually spent large efforts understanding both how their sensors change after exposure to integrated dose of ionizing radiation, as well as the sensor reaction to a single, highly ionizing event. These sensors often have relatively large active dimensions compared with transistors within electronic parts, and the effects on the sensors can be quite different. Most such investigators have flown a succession of evolving instruments, which have come to be very competent and tolerant systems. Usually they have had to do laboratory tests with the nuclear species of concern and with a representative spread of energies. Tests performed to some dose level with gamma rays from some nuclear radiation source is often inadequate and may be misleading.

Shielding often helps and sometimes does not help, especially for very high energy phenomena. Shielding for energetic electrons can be difficult when one has an aperture, because the electrons scatter so efficiently off the aperture material and off whatever is inside the aperture.

The selection of one particular sensor over another competing type is often dictated by the relative response of the sensor to the background nuclear radiation in space. The phenomena involved range from darkening of the particular glass in the window of the photomultiplier being used, for instance, to degradation in the sensitivity of the sensors because of damage resulting from high sensor currents (due to exposure to high nuclear particle fluxes while the sensor was biased), or to direct interference with the measurement. For certain solid state detectors when low energy nucleons are involved, it can be important as to the face of the sensor that is exposed (Coleman et al, 1968a and 1968b) or if certain crystalline lattice directions are exposed to the nucleons entering through the aperture (Dearnaley, 1964). Charge coupled devices and arrays (CCD's) are currently devices of great interest as sensors in the visible, infrared, ultraviolet and x-ray portions of the electromagnetic spectrum, but detailed testing is required to evaluate the variety of devices now available. A good example of such work is the paper in this NATO proceedings (Schott, 1992).

Measurements of the Radiation Environment

Several science teams from a variety of countries and agencies have measured and characterized the quiescent radiation in our solar system, as well as the transient events, both over the 11-year solar cycle. Excellent spectra are obtained element by element, and even by isotope (Pissarenko, 1992, for instance), although one still has the problem of even crudely predicting the occurrence of the large solar events. A particular instrumental example that is certainly adequate for measurements supportive of the radiobiology discussed at this institute is the instrument known as the Cosmic Radiation Subsystem (CRS) (Stilwell et al, 1979) for the NASA Voyager missions to the planets and the outer solar system, hopefully to penetrate the boundary of our solar system at perhaps ≈ 100 Astronomical Units from the Sun. As Voyager investigators, we used seven nuclear particle telescopes to make detailed measurements of the primary galactic and solar cosmic ray spectra, as well as measurements of the trapped radiation spectra and fluxes at the several planetary encounters. It would be similarly good at measuring the actual spectra inside the Space Station Freedom or in the radiation shelter on a Mars mission, for instance. Inside space vehicles generally or behind shielding of a few grams per cm^2 or more, the spectra are dominated by spallation secondaries

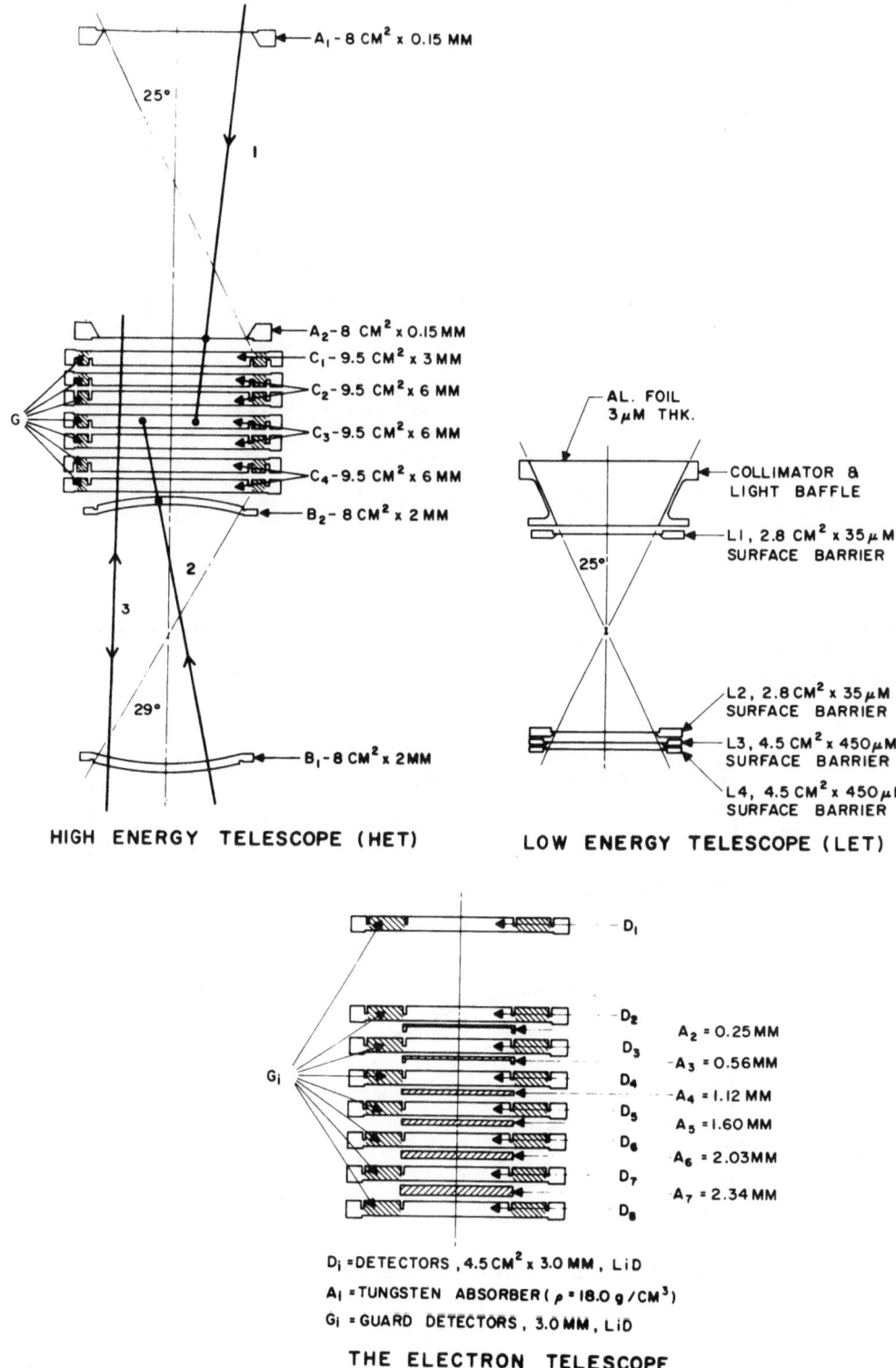

Figure 1. Cross-sectional Schematic of the HET, LET and TET Detector Systems of the Voyager Cosmic Ray Instrument (Stilwell, 1979).

Figure 2.   Photograph of the Flight Hardware of the Voyager Cosmic Ray Instrument.

from nuclear interactions between the incident nuclear particles and the local material. If one has a sufficiently detailed model of the surrounding material of the vehicle or the shield, accurate calculations can be made of the resultant spectra (see Wilson et al, 1992). It would

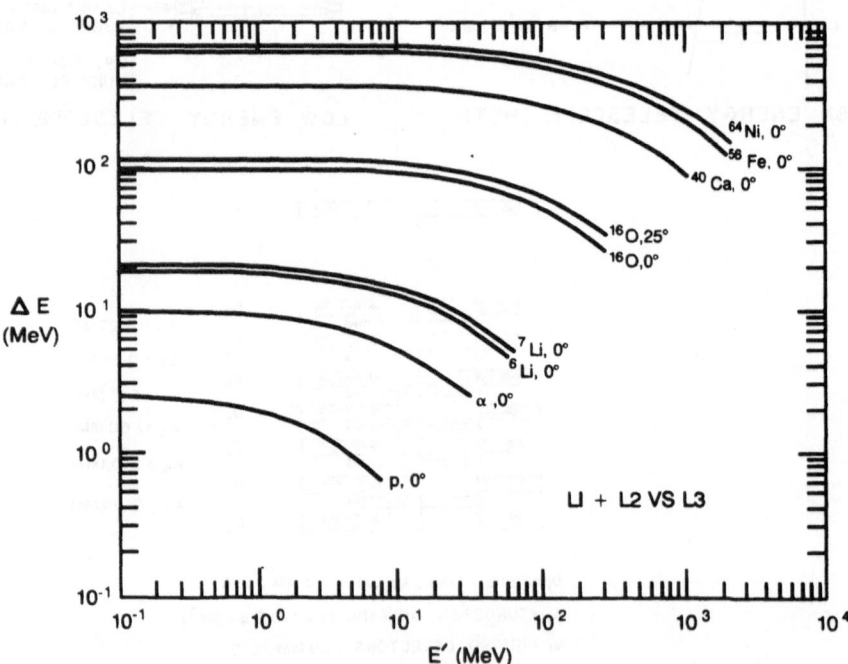

Figure 3.   A Response Curve for LET Telescope. Calibrations at Nuclear Particle Accelerators Result in Such Families of Curves. The Sum of the Energy Losses in L1 and L2 are Plotted against the Energy Deposited in L3. These Curves were Generated under the Constraint of No Event in L4. Such Telescopes have a Built-in System Calibration, since the End Point of Each Curve can be Readily Tracked during Operation in Space.

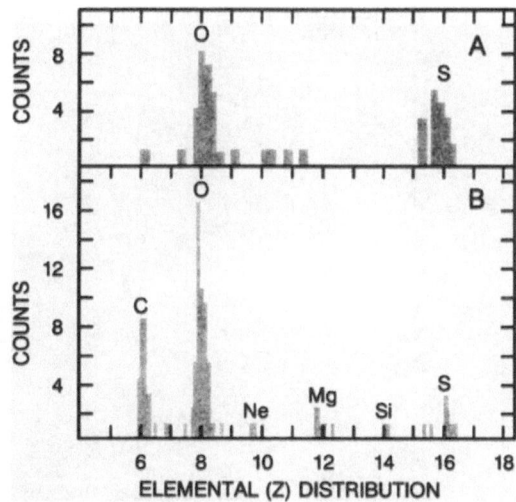

Figure 4.   These Plots Show the Raw Data of Number of Counts in Each Mass Bin Arrived at by Summing Data Points over the Appropriate Energy Range from a Data Plot Similar to : Figure 3. The Lower B Plot is for Data Taken between 10.6 and 25 Radii at Jupiter, while the Upper A Plot is for Data Taken from 10.1 to 10.6 Radii. The Growth of the Energetic Oxygen and Sulfur as the Planet is Neared is Apparent. These Nuclei Represent the Energization of the Atoms within the Sulfur Dioxide Propellant from the Volcanoes on the Moon Io near 5 Radii in the Jovian System (7.8 to 10.5 MeV/Nucleon).

also be straightforward and even necessary to have an instrument of the CRS type in the inhabited vehicle to simply measure the spectra directly and routinely, as well as to provide essential information for the crew decisions in the event of a large event.

The CRS instrument (Stilwell et al, 1979) as flown consisted of the three nuclear particle telescopes shown in cross-section in Fig. 1. The individual elements are totally depleted solid state detectors of both the silicon surface barrier and lithium-drifted silicon types. The set of seven telescopes consisted of two of the HET shown, four of the LET in a vierbeiner arrangement and one TET unit, all as shown in the photograph of the flight instrument in Fig. 2. This instrument measures the energy spectra and elemental composition from hydrogen through iron over the energy range from ≈ 1 to 500 MeV per nucleon, and resolves isotopes for the elements from hydrogen through sulfur over the energy range from ≈ 2 to 75 MeV per nucleon. The TET measures the energy spectra of electrons from ≈ 3 to 110 MeV. Fig. 3 shows the response of a LET telescope to energetic nucleons, while Fig. 4 shows the elemental separation obtained under difficult circumstances deep within the radiation belts at Jupiter. The sophisticated electronics system was designed to survive more than 500 Kilorad, contains several hundred thousand transistors, and has system integral linearity of less than 0.2% per channel in a 12-bit analog-to-digital system. Systems of better resolution and larger size are possible today, but they are more complex, carry more risk and one doesn't need them for the application discussed here.

Electronics Parts: Ionizing Dose

Spaceflight projects should have a realistic prediction of the dose to be expected for their mission. Models exist for the Earth's radiation belts that are adequate in the main radiation belts or the strong trapping regions. Below the main radiation belts in the orbits flown by the Space Shuttle and planned for Space Station Freedom, the trapped particle fluxes are small and quite variable. Fig. 5 shows a cross-sectional cut through the Earth's magnetosphere along a noon-midnight meridian (Heikkala, 1974). Geosynchronous satellites on the sun side of the Earth are outside the trapped radiation belts but fully exposed to the large solar events summarized in Fig. 6 (McGuire, 1986). While on the night side or magnetotail side of the magnetosphere, the satellite is in the outer radiation belts.

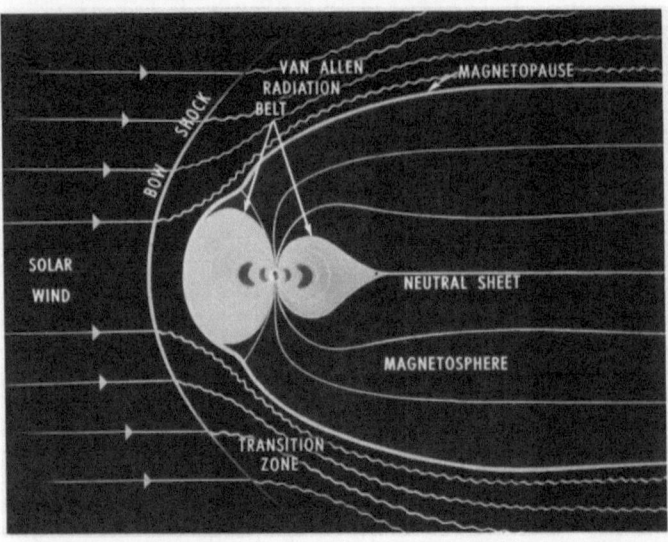

Figure 5 A Drawing due to Prof. Heikkala Showing a Noon-midnight Meridional Cut across a Model of the Earth's Magnetosphere in the Solar Wind.

If the orbit or trajectory allows the effluence of a large solar particle event to directly reach the spacecraft, then one should pragmatically allow for one or more of the largest events known during an average mission lifetime (5 to 10 years), rather than use some average fluence per year over a "typical" solar cycle as some models do. The consequences are to increase the expected mission dose, perhaps require some shielding and most likely to restrict the usage of some part types. These are usually straightforward tasks for an experienced team. Experienced engineers and physicists exist in many countries and agencies. I can only recommend that investigators search out these people and follow their recommendations, since the effects can not only be disastrous, but they can also be subtle and insidious.

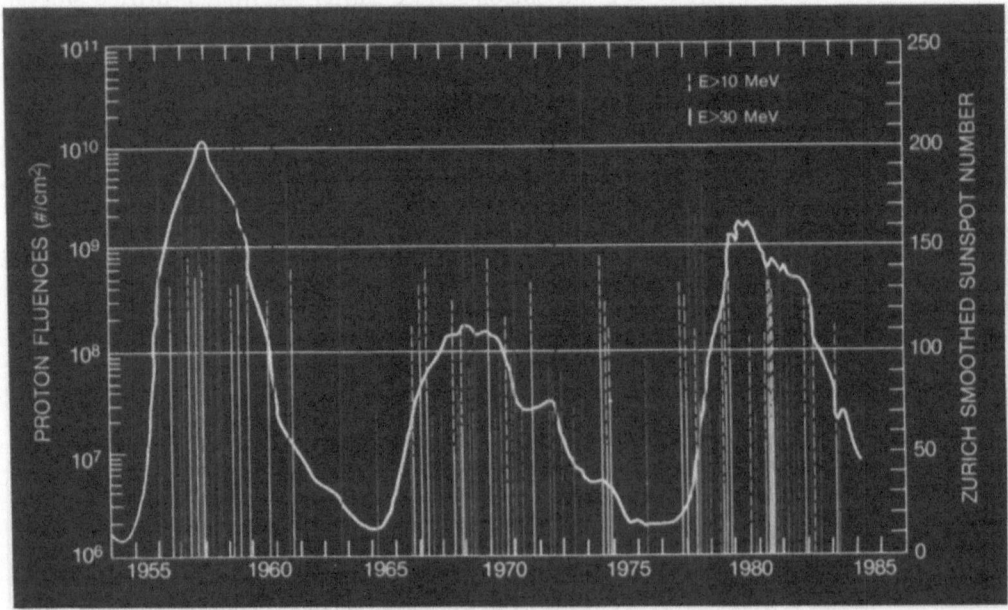

Figure 6. Summary Data for Three Solar Cycles Showing the Proton Fluences Associated with Large Solar Events on the Date of Occurrence as Well as the Smoothed Sunspot Numbers (McGuire, 1986).

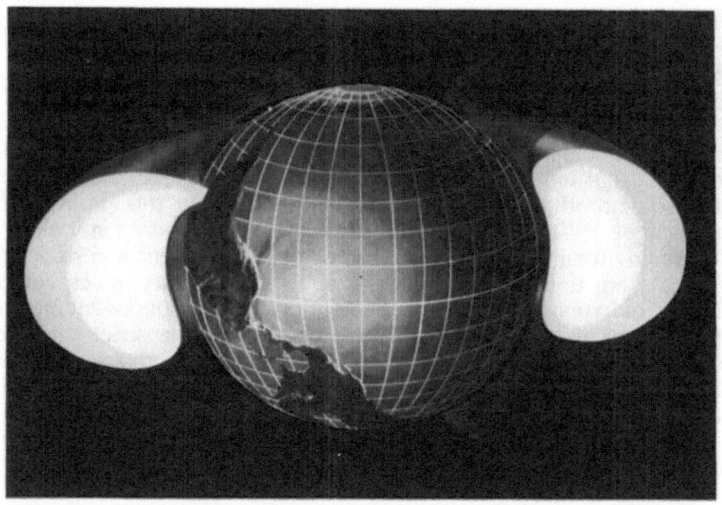

Figure 7. A Drawing due to E. G. Stassinopoulos which Shows the inner Van Allen Radiation Belts and Demonstrates how They Dip down to the Top of the Atmosphere due to the Anomaly in the Earth's Magnetic Field Located in the South Atlantic Region.

Most Earth-orbiting, low altitude satellites have orbital inclinations greater than 10 degrees, and so have to contend with the trapped radiation that the satellites will intercept in a broad region over the South Atlantic known as the South Atlantic Anomaly (SSA). This is shown schematically in Fig. 7. An anomaly in the Earth's magnetic field brings magnetic field lines carrying the trapped energetic nuclear particles right down to the top of the atmosphere. Many NASA missions are in orbits with inclinations near 28 degrees and accumulate the majority of their radiation dose in the SSA. At orbital altitudes of 900 Km. and above, the average dose is fairly predictable. At orbital altitudes of 350 Km. the average dose is much lower, but it is also highly variable from day to day and is especially dependent upon geomagnetic activity. At low altitudes the problems are more one of interference with sensitive astronomical detectors, for instance, rather than a problem with dosage. At low altitudes in a highly polar orbit, one has to additionally consider the passage through the ends of the outer radiation belts at geomagnetic latitudes from 40 to 60 degrees approximately, as

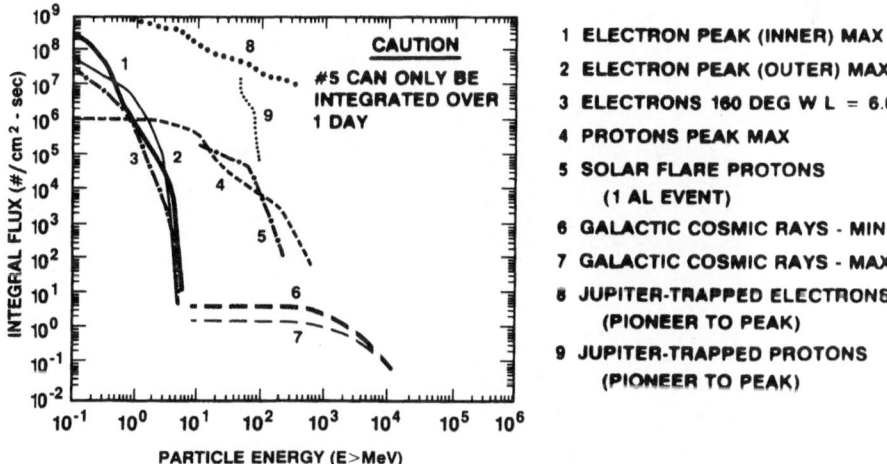

Figure 8. A compendium by E. G. Stassinopoulos of published Data for Cosmic Radiation and for TrappedElectrons and Protons at the Earth and Jupiter.

well as the solar particle events and cosmic radiation that can come directly into the atmosphere over the polar regions. In low altitude, moderate inclination orbits, the Earth's magnetic field is an effective shield for all but the most energetic solar particles and for many of the galactic cosmic rays.

Fig. 8 shows a compendium of published data by E. G.Stassinopoulos for electrons and protons in the Earth's and Jupiter's radiation belts, as well as for galactic cosmic radiation and an energetic solar proton event. Note that the ordinate for this plot of integral spectra covers eleven decades. While the scope of this presentation does not allow a thorough discussion of these environments as concerns radiobiological effects, it is sufficient to say that plausible solutions seem to exist for near-Earth and interplanetary travel, but that extended time in the Earth's radiation belts or in an energetic solar particle event requires a heavily shielded shelter for the crew and other biological specimens. The penetrating electron fluxes at Jupiter are so high, and contain electrons of such high energy that plausible solutions do not exist today for a fly-by trajectory or an orbit at 10 Jupiter radii from the radiobiological point of view, for example. Both because of the high energies of many of the Jupiter electrons, as well as the manner in which electrons interact with matter, the required shields are massive,and spacecraft weight becomes prohibitive.

The picture shown in Fig. 9 shows the two sides of a digital electronic "card" and the overall packaging of the digital assembly for the Pioneer 10/11 Cosmic Radiation Telescope built at the Goddard Space Flight Center. This circuitry is special for its low magnetic signature (interconnect is via a stitch-weld technique using alloy-180 ribbon wire) and for its first use of true LSI circuitry in space in 1972. Furthermore, it is of interest here because our engineers and physicists were able to radiation-harden these custom devices to about 600 Krads. This means that about one-half of these devices failed when irradiated to 600 Krads in our Cobalt-60 irradiation facility. The observed effects in these PMOS devices are dose and dose rate dependent, including the annealing effects observed. We determined by test that irradiating at a rate of 100 Krad overnight or, equivalently at ≈6 Krad per hour, followed by measurements some hours later, led to minimization both of dose rate effects and annealing effects that would occur after our measurements. On the Pioneer 10 flight to and by Jupiter,

Figure 9.   Photograph of Electronic Circuit Cards and the Packaging for the Custom PMOS LSI Circuitry Developed for the Data System for the NASA Goddard Space Flight Center Cosmic Radiation Experiment on Pioneer 10 and 11.

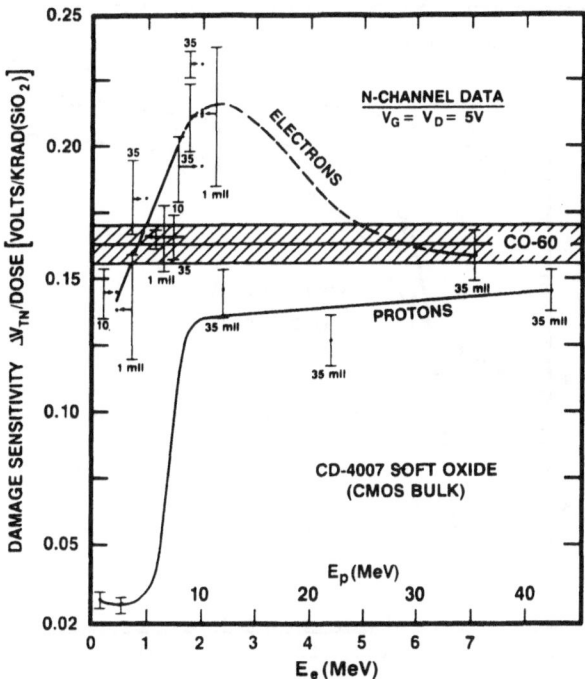

Figure 10. Published Figure [Brucker et al, 1982] Showing the Relative Damage Sensitivities of a CMOS CD-4007 Integrated Circuit to Electrons, Protons and Gamma Rays as a Function of Energy. (Dependence of Damage Sensitivity on Radiation Type and Particle Energy).

this subsystem actually experienced an exposure to nearly 600 Krads. Of the 32 such integrated circuits in the package, 3 had failed as the spacecraft exited the intense radiation belts at Jupiter. One device recovered (annealing) within a few hours, and another recovered within a few days. The third device never recovered. The point to be made is that radiation testing has to be done in a rational manner. You must test a reasonable number of integrated circuits made from wafers across your production run. You must irradiate them either at the rate expected in the flight exposure (for high dose missions) or at rates that ensure that secondary and annealing effects do not dominate the result.

A final topic needs to be addressed in concluding the discussion of ionizing dose effects. This concerns the equivalency of testing with Cobalt-60 gamma-rays vs. testing with electrons or protons. Fig. 10 shows interesting data on the relative-damage effects on a CMOS integrated circuit as a function of energy of the irradiating particle [Brucker et al, 1982]. For most orbits and trajectories one is subject to such a wide distribution in energy of electrons and protons that the equivalence of an ionizing dose is probably a good approximation independent of the cause of the dose. If one's orbit or application is such that electron fluences at energies of a few MeV are emphasized, at the least, one needs to be careful and conservative. Ongoing analysis [Danchenko, 1992] of CMOS circuits exposed on the CRRES satellite seem to show that the equivalency of ionizing dose due to charged particles in space and due to gamma rays in ground test is not so good as thought for these parts.

Single Event Upsets (SEU's)

As device dimensions for the active elements of integrated circuits have approached ≈1 micron and smaller, the inherent capacitance within the circuitry has grown smaller, and the required charge delivered to a junction to cause a change of state of the digital circuit is also correspondingly smaller. When the change of state is caused by charge resulting from the passage or stoppage of an energetic nucleon in the active region, it is conventionally called a

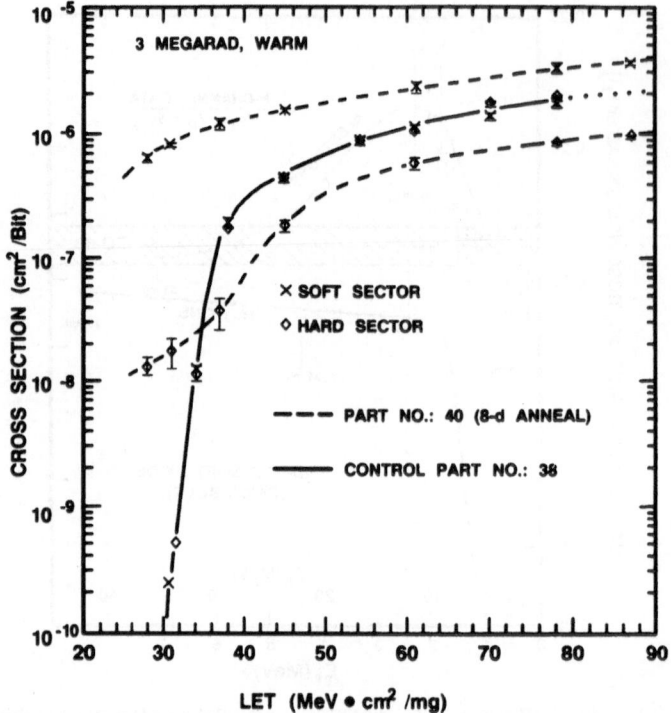

Figure 11. SEU Cross section vs. LET of the Incident Particle for a Radiation-hardened Harris 6504 static RAM. Data is Shown for the Sensitivity of Parts after Irradiation by Gamma Rays to 3 Megarad for Both Hard Sectors and Soft Sectors. The Control Plot is for the Same Parts before Being Irradiated to 3 Megarads.

single event upset (SEU). It is apparent from this simplified discussion that one sufficiently ionizing particle could cause the upset, just as well as several lesser ionizing particles in good time coincidence into the same sensitive region. It is usual in the characterization of such parts to demonstrate the sensitivity by a plot of the device effective cross section for the SEU as a function of rate of energy loss by the incident nucleon. Conventionally this rate of energy loss is labeled LET (linear energy transfer) in units of MeV per mg. per cm2. Nuclear physicists recognize this LET as what they also call stopping power or dE/dx.

Such a plot of cross section vs. LET is shown in Fig. 11 (Stassinopoulos et al, 1989) for a relatively hard CMOS part, the Harris 6504 RH static RAM. It also demonstrates that there can be substantial effects on the SEU sensitivity due to a prior or ongoing ionizing dose of radiation. The control part was not irradiated to 3 Mrad prior to the SEU measurements. It shows a typical response curve with a well defined threshold LET and an asymptotic cross section at high LET. As received it is an excellent part. With a LET threshold above 30 MeV per mg per cm2, the physics requires a nuclear ion heavier than the iron group to cause an upset with this device. Even then, the ion would have to be in a very narrow energy range incident on the chip to deposit the required charge for the upset. For practical purposes, this device is immune to upsets by the cosmic radiation. After irradiation to 3 Mrad, however, the LET threshold shifts to much lower values of LET with large cross sections. Those devices are readily upset by nuclei in the iron group and below. One would expect occasional upsets in such a chip due to galactic cosmic rays, and large numbers of upsets in a spacecraft memory during a large solar cosmic ray event where large numbers of high Z nuclei are present. The hard sector and soft sector refer to the fact that these RAM chips were irradiated with fixed patterns stored in the sectors of the memory. SEU sensitivity was then measured with the same pattern stored (hard sector) or with the complement stored (soft sector). The large differences are apparent. One's test results may well be dependent upon whether the test was done on a static device or on a device being dynamically used while being irradiated. The authors point out that a system in geosynchronous orbit with minimum shielding may experience an ionizing dose of 1 to 2 Mrad in a 5- to 10- year period, so that even for such a good part, the problem is real and the possible consequences can be serious.

With Fig. 11 in mind, it is instructive to point out that for Earth satellites in low altitude orbits with low to moderate inclinations, the Earth's magnetic field is an effective shield for high LET nuclear particles. At the equator, the physics requires a proton of kinetic energy of $\approx$10 GeV coming from the optimum direction to penetrate to the top of the atmosphere. Such a particle has a LET less than 1 Mev per mg per cm^2 and cannot cause an upset directly in any device yet tested. An SEU could result from the interaction of any energetic nuclei with atoms of high Z material near the chip, such that the spallation products of such an interaction would impact the active elements of the chip. While the efficiency for such an occurrence is not large, this effect may well be the dominant effect for circuitry in low to moderate altitude orbits in the inner proton belts.

The same physics applies to a radiation shelter within the Space Station Freedom, a Mars exploration spacecraft or a lunar surface shelter. The shelter can be very effective in reducing the total dose from a large solar particle event, but if it is not thick compared with the range of the nucleons, then some of the nucleons will interact will the shielding material near the sheltered volume, resulting in fluxes of high Z, high LET spallation products into the sheltered volume. The mass distribution of these secondaries will largely depend upon the composition of the shelter material, and so one would expect that an inner shield material of a low Z material such as a carbon composite to be preferable to aluminum. Such high LET nuclear particles are of obvious radiobiological concern.

As a practical matter, the measured LET thresholds for most digital integrated circuits fall in the range of 3 to 10 MeV per mg per cm^2, with some as low as $\approx$ 1 MeV per mg per cm^2 and some higher than 30 Mev per mg per cm^2. For most orbits, designs using parts with LET in the range 1 to 5 MeV per mg per cm^2 will result in substantial numbers of SEU's that the designers will have to cope with. Given the appropriate LET curves and an orbital description, many groups around the world can calculate the expected response. Recent results from the CRRES satellite (Ray et al, 1991) show that simple calculations using a "step function" approach with a simple, sharp LET threshold and a fixed cross section above the threshold can lead to errors in predicted response of up to two decades. A calculation which properly integrates the expected energy spectra over the appropriate LET response curve can give very good results. An excellent example concerns the measured response of a large RAM memory within the Total Ozone Measurement System (TOMS) developed by a group at NASA's Goddard Space Flight Center (Way, 1992). This TOMS instrument was launched in 1991 on a Russian Meteor-3 satellite in a high inclination orbit at $\approx$1200 Km altitude. The RAM memory is made up of Hitachi 256K devices totaling 57 megabits, and it functions effectively as a lightweight, small recorder to store data for downloading at both NASA and Russian ground stations. These RAM circuits have a very low LET threshold, but tests have shown [Koga et al, 1988] that the resultant upsets are all upsets in a single bit of a word. The TOMS instrument incorporates an error detection and correction circuit, such that all upsets are noted and corrected. After several months it is apparent that about 350 upsets per day occur in this memory, largely in the South Atlantic Anomaly, and that all have been corrected. This upset rate had been predicted with better than 10% accuracy by Stassinopoulos (1990). This is an important fact, because it means that under the right circumstances, one can successfully use a high density, relatively inexpensive, modern part with a low LET threshold.

The same Hitachi 62256 RAM (32K x 8 bits = 256K bits) is being used as the building element of a 25.2 Megabyte memory for data storage within a microprocessor data system on the SAMPEX Small Explorer (SMEX) mission to be launched in the summer of 1992 by NASA's Goddard Space Flight Center (LaBel, 1992). The design of the 80386-based data system uses hardened, expensive RAM for the smaller program memory, but uses the less expensive but more sensitive, high density RAM for the data storage. Again, EDAC circuitry is expected to correct any SEU's or to flag any uncorrected word. There is a large cost savings for the system, since the hardened chips cost about $400 each and have a long lead time, while the selected chip is readily available at about $25 each. This

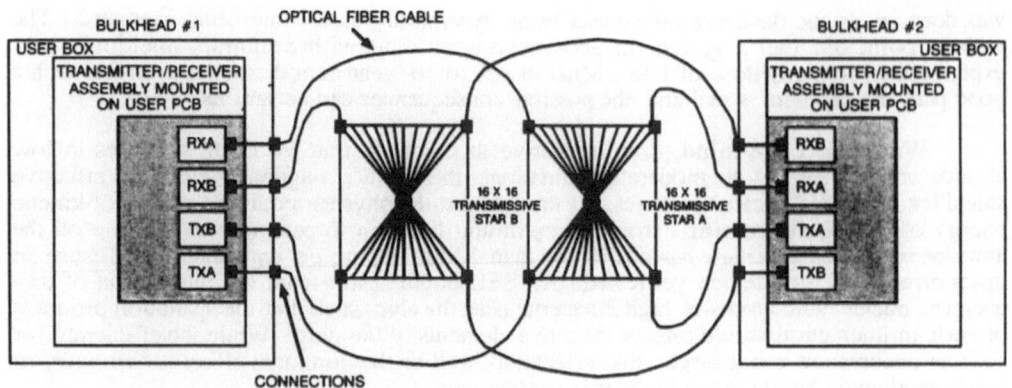

Figure 12. Functional Drawing Showing the Components and Interconnection of the Fiber Optic Data Bus for the Data System for the SMEX Satellite Missions.

results in a cost savings of nearly one million dollars for parts for this one subsystem on just the first satellite in the SMEX series.

Fiber Optic Data Bus Implementation

I have chosen the SMEX satellite's onboard, fiber optic, digital data bus (LaBel, 1992) as an example of the design and testing required for parts and components for which no flight history exists. The data bus uses the specifications and protocol of MIL-STD-1773. It allows the interconnection of the spacecraft sensors, control systems and the central processor, as well as allowing control of and access to the processors and large storage memory by the scientific instruments. It considerably reduces the complexity of the wiring harness, and obviates the need for much laborious hand work in building up such a custom harness. The largest advantage and the original reason for investigating such a design was the substantial shortening of the integration and test processes in the buildup and integration of the various subsystems and instruments on the spacecraft, as well as the potential weight savings. Much of this integration time has traditionally been used to discover and correct

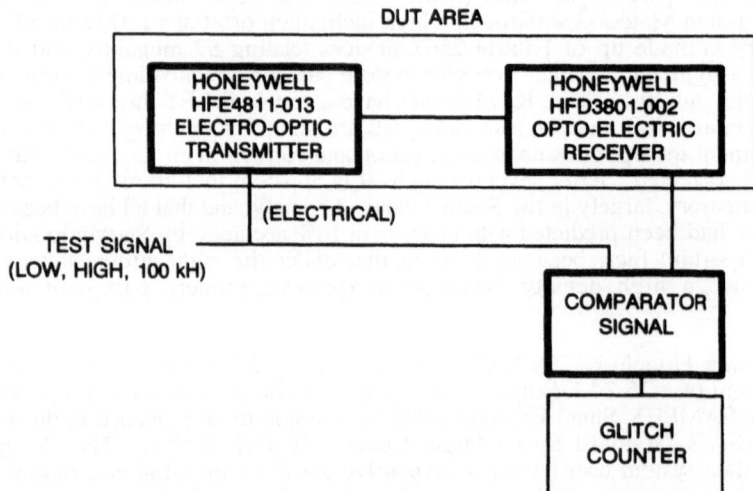

Figure 13. Fiber Optic System Test Setup (1773 RAD Effect). Glitches as Short as 13 Nanoseconds can be Measured.

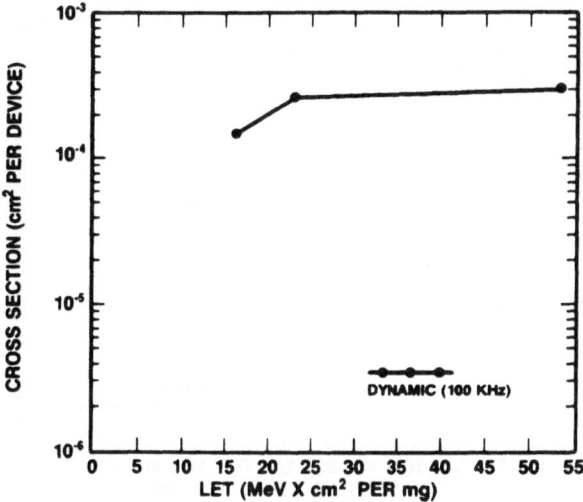

Figure 14.   Glitch Cross Sections for the Transmitter Circuits due to SEU's as a Function of the LET of the Incident High Z Nucleons (1773 Rad Effect).

components electrical interferences, cross talking, etc., between the various subsystems and boxes. Most of these effects can be traced to coupling due to currents flowing in the wiring harnesses. If one eliminates the wiring harnesses, except for the shielded leads bringing dc power to each subsystem, most of the electrical interference problems should not occur. Indeed, at this time the promise has been already realized at a large cost savings to this and subsequent projects. Such an approach will probably be the basic design for most new scientific and operational Earth satellites in the future.

Despite the advantages just outlined, there are serious concerns to be addressed prior to commitment to such an approach. The principal lack of detailed information was with nuclear radiation effects, especially any SEU effects. There were many new parts,

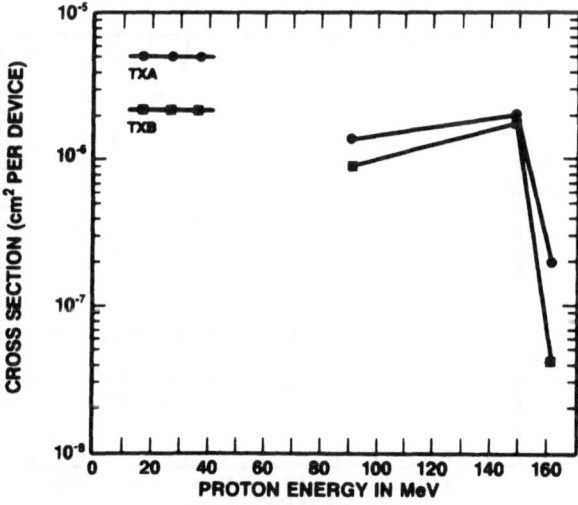

Figure 15.   Glitch Cross Sections for the Transmitters in One Package as a Function of Incident Proton Energy (1773 Rad Effect).

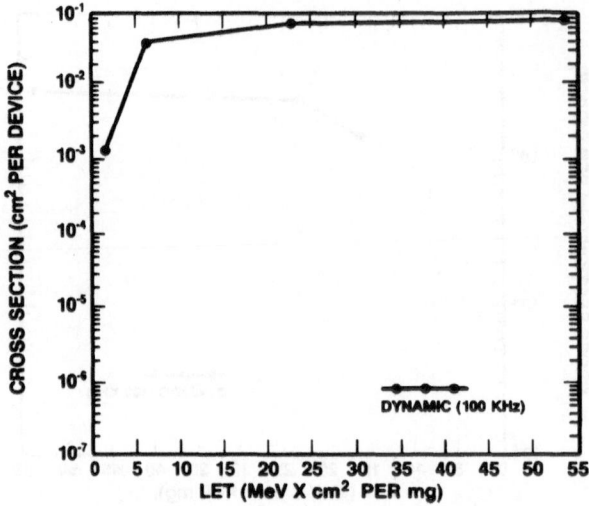

Figure 16. Glitch Cross Sections for the Worst Case Receiver as a Function of the LET of the Incident High Z Nucleons.

component and materials to be individually tested, eventually leading to a full systems test whereby the sensitivity of the system to upsets in individual parts was tested, followed by the performance of a tests where the system performance was carefully measured while all chips of the system were irradiated with high LET particles at the same time. These tests of this system for the SAMPEX mission was excellent, showing that the system is virtually immune to upset.

Most of the Figures and results which follow are the work of K. A. LaBel (1992) as given in his paper at the RADECS 91 Conference. Fig. 12 outlines the essentials of the fiber optic data bus itself. It consists of the transmitters, receivers, optical fibers and optical star couplers. Note that the transmitter and receiver functions are fully redundant. The testing of the finally selected elements are discussed in the paper, but the real difficulties were with the transmitters and receivers. While these devices are in wide use on the ground, in the oceans

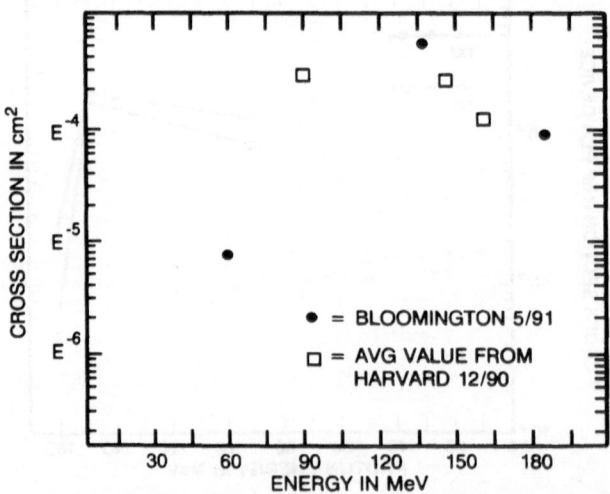

Figure 17. Glitch Cross Sections for Both Receivers in One Package as a Function of Incident Proton Energy. Data is Shown for Tests which were Repeated at Two Different Facilities (1773 Rad Effect).

and on complex vehicles and aircraft, we know of no previous use in space. In the radiation testing, most devices failed in one or more ways, but the devices identified in Fig. 13 meet the requirements of many of our missions. After one Megarad they still are fully functional with all important parameters within specification. Fig. 13 schematically shows the test that was run in a vacuum chamber while high LET particles were incident on the devices. Any difference between the transmitted and received data words is noted and counted for glitches as small as 13 nanoseconds.

Fig. 14 glitch cross section for a transmitter as a function of LET for heavy ions. Fig. 15 shows the glitch cross section as a function of proton energy for the two redundant transmitters within a circuit package. Fig. 16 shows the same kind of data for the worst case receiver. It is obviously a very sensitive part with a low LET threshold and a very large cross section. The passage of high LET particles during the missions will lead to many glitches in the received data words due to this part. Fig. 17 shows the response of receivers to high energy protons in tests a two facilities. While the response is moderate, the persistent falloff at higher energies was unexpected and is unexplained. The effect is real, since the test has been run carefully multiple times at both facilities.

What has not been demonstrated in these test shown here is the total performance of this master/slave type bus under MIL-STD-1773. The protocol includes encoding, error detection, retransmission, etc. In practice, there were no errors noted in system performance while individual circuits were irradiated with high LET nucleons and while all circuits were simultaneously irradiated with high LET nucleons. The individual circuits had small to large response to the nucleons individually, however, the total system response was unaffected by those SEU's. This partially demonstrates the degree to which one has to seriously know the parts one plans to use in their designs. This fiber optic bus system is part of the first SMEX satellite planned for launch in June, 1992 as part of the SAMPEX mission.

REFERENCES

Allen, J. H., "Technological Effects at Earth and in Space of Solar Activity in March, 1991", AGU 1991 Fall Meeting, Abstract SH11A-8, Dec. 9, 1991.

Brucker, G. J., E. G. Stassinopoulos, O. Van Gunten, L. S. August, and T. M. Jordan, "The Damage Equivalence of Electrons, Protons, and Gamma Rays in MOS Devices", IEEE Trans. Nucl. Sci. NS-29, No. 6, Dec. 1982.

Coleman, J., D. Love, J. Trainor and D. Williams, "Low Energy Proton Damage Effects in Silicon Surface Barrier Detectors", IEEE Trans. Nucl. Sci. NS-15, No. 1, 1968a.

Coleman, J., D. Love, J. Trainor and D. Williams, "Effects of Damage by 0.8 MeV to 5.0 MeV Protons in Silicon Surface Barrier Detectors", IEEE Trans. Nucl. Sci. NS-15, No. 3, 1968b.

Danchenko, V., private communication, 1991.

Dearnaley, G., "The Channelling of Ions Through Silicon Detectors", IEEE Trans. Nucl. Sci. NS-11, No. 3, 1964.

Heikkala, Prof., A sketch of a model presented to E. G. Stassinopoulos in the 1970's.

Koga, R., W. A. Kolasinski, J. V. Osborn, J. H. Elder and R. Chitty, "SEU Test Techniques for 256K Static RAMs and Comparisons of Upsets by Heavy Ions and Protons", IEEE Trans. Nucl. Sci. NS-35, No. 6, 1988.

LaBel, K. A., "A Spacecraft Fiber Optic Data System — Radiation Effects", Proceedings of the RADECS 91 Conference, in publication, 1992.

McGuire, R. E., National Space Science Data Center, 1986.

Pissarenko, N. F., Radiation Environments During Long Space Missions (Mars) due to Galactic Cosmic Rays", Proceedings of this NATO Advanced Study Institute, 1993.

Ray, K. P., M. S. Gussenhoven, E. G. Mullen, "Microelectronic Performance in Space: New Measurements from the CRRES Satellite", AGU Spring 1991 Meeting, Abstract SM32A-10, May 29, 1991.

Schott, J. U., "Time Resolving Detector Systems for Radiobiological Investigations on Effects of Single Heavy Ions", Proceedings of this NATO Advanced Study Institute, 1993.

Stassinopoulos, E. G. and T. M. Barth, "Space Radiation Analysis for SMEX", NASA GSFC Document X-600-89-1, Jan. 1989a.

Stassinopoulos, E. G., G. J. Brucker, O. Van Gunten and H. S. Kim, "Variation in SEU Sensitivity of Dose-Imprinted CMOS SRAMS", IEEE Trans. Nucl. Sci. NS-36, No. 6, 1989b.

Stilwell, D. E., W. D. Davis, R. M. Joyce, F. B. McDonald, J. H. Trainor, W. E. Althouse, A. C. Cummings, T. L. Garrard, E. C. Stone and R. E. Vogt, "The Voyager Cosmic Ray Instrument", IEEE Trans. Nucl. Sci. NS-26, No. 1, 1979.

Way, S. H., NASA GSFC memorandum to J. H. Trainor, Feb. 10, 1992.

Wilson, J. W., L. W. Townsend, W. Schimmerling, J. E. Nealy, "Space Radiation Protection", Proceedings of this NATO Advanced Study Institute, 1993.

RADIOLOGICAL OPERATIONAL SCENARIO FOR A PERMANENT LUNAR BASE

Percival D. McCormack

National Naval Medical Center
Bethesda, MD, U.S.A.

ABSTRACT

An operational scenario for a lunar base is postulated based on 30 lunar base personnel and 2 year tours of duty plus stipulated numbers of EVA's and sorties in the lunar rover vehicles. It is also postulated that the main shielding material for the lunar base units (habitats, laboratories, etc.) will be lunar regolith.

Using the solar minimum period as the basis, total accumulated dose equivalents for the galactic cosmic radiation over the two year period are computed at various shielding depths. Depths of regolith of over 20 g/cm^2 are sufficient to reduce the total dose equivalents to well under the present limits. It is also postulated that there is a significant possibility during a two year tour that a lunar base personnel would be exposed to a 1972 type large solar particle event.

Based on such a scenario, a total regolith depth of at least two meters would be optimal. However, this has implications with respect to the structural strength required of the lunar base units. In the absence of the development of more effective composite shielding, some increased risk of carcinogenesis, mutagenesis and/or cataractogenesis may have to be accepted.

The second arm of the radiological health strategy - continuous and all-encompassing radiation dosimetry - is also discussed in some detail. It is also emphasized that monitoring of the base personnel for genetic mutations and chromosomal aberrations must be part of the radiological health program in the lunar base.

The objective of minimizing the risk to the lunar base personnel must be accompanied by the objective of providing an estimation of the risk to each person prior to a two year tour and the measured risk after completion of the two year tour.

INTRODUCTION

The next major step in the United States manned space program will be a return to the moon and the construction of a permanent human base there. This will occur in the not too distant future-- probably before the end of the first decade of the next century. A major concern for the lunar base personnel is the radiation environment. The radiological health strategy is based on two major countermeasures, (1) risk calculation (any given scenario is dependent on knowledge of the radiation environment and it's biological effects) and (2)

Biological Effects and Physics of Solar and Galactic Cosmic Radiation,
Part B, Edited by C.E. Swenberg *et al.,* Plenum Press, New York, 1993
905

comprehensive dosimetry. These are the main drivers for the present manned space radiation research program.

The moon is the logical stepping stone for the permanent expansion of mankind into space and will be an ideal testing ground for closed biospheres and support systems for manned space missions.

The design of a lunar base will be influenced by many factors, but two factors of special importance are: (1) the cosmic radiation (solar and galactic) and (2) the lunar surface material or regolith (for mining of minerals and helium 3; for radiation shielding - exclusively the aspect dealt with in this paper; and for base structural purposes).

The lunar base habitat structures will be covered by a layer of regolith. Activities outside the habitats will involve various kinds of rover vehicles and personnel at work in space suits. Activities of personnel in donned space suits and outside habitats and vehicles will be termed EVA (extra-vehicular activities) in this work. Activities of personnel in powered vehicles will be deemed as involving LRVs (lunar rover vehicles) with their own internal environment and inside which the lunar base personnel - LBP - can work in "shirt sleeves." The lunar compatible EVA suit will be made of hard material (probably metal), will have a normal atmosphere inside and will be radiation resistant. This is in effect the AX-5 suit developed at the NASA Ames Research Center. With respect to the radiation risk to LBP, the solar and galactic exposures (doses) will be considered separately.

Solar particle radiation arriving at the lunar surface is comprised predominantly of energetic protons (with energies up to about 300 MeV). The major radiation fluences will be random in occurrence and are called anomalously large (AL) events. These events last for about 48 hours and occur approximately once in an 11 year cycle (solar maximum). LBP during their two year tours on the moon could expect to experience one such event. All solar particle event (SPE's) are potentially lethal to fully exposed humans.

Galactic cosmic radiation (GCR) is relatively steady in intensity, varying between Solar Min. and Solar Max. The flux is composed predominantly of very energetic (>1000 MeV) heavy ions (HZE particles), the most common of which is iron. These particles are much more carcinogenic than protons (quality factors in the ratio of 1.2:20). They form a much greater shielding problem due to the production of heavily ionizing knock-on protons and secondary neutrons.

The Operational Phase of a lunar base will be considered here, involving some 30 base personnel (LBP) on individual tours of two years duration. Personnel activities will include: 1).Living and working at the main base (habitats,recreational areas,clinic/hospital,laboratories and observatories, workshops). This is classed as Intravehicular Activities (IVA). 2). Up to 300 kilometer two day sorties in Lunar Rover Vehicles (LRV's). Also classed as IVA but an 8 hour EVA will be included. 3). EVA's from the main lunar base units; limited to distances such that return to the base and LRV, doffing suits and indressing to storm shelters can be achieved in one hour at most. This latter time is taken as advance warning time for SPE build-up.

GUIDELINES AND ASSUMPTIONS

1. Habitats and laboratories will be covered with a layer of regolith as a shield against ionizing cosmic radiation and micrometeorites. Regolith consists of the oxides SiO_2, MgO, Al_2O_3 and FeO, with an overall density of 1.5 g/cm³. This value for density is based on analyses of Apollo lunar return samples taken from different landing sites.

2. The cosmic ray environment used for calculations will be for solar minimum conditions when particle fluxes are highest.

3. Solar Particle Event will be based on the August 1972 Anomalously Large Event characteristics. A 1956 event actually had a "harder" particle spectra, but the fluences were lower.

Table 1. Ionization Radiation Exposure Limits

Exposure Interval	Dose Equivalent (Sv)		
	Skin	Eye	Blood-Forming Organs
30 Days	1.5	1	0.25
Annual	3	2	0.50
Career	6	4	1 - 4 *

*Dependent upon gender and age at initial exposure.

4. It will be postulated that the LRV's, lunar base units and EVA suits have basic shielding of 1.5 g/cm² aluminum equivalent. The primary shield of the base habitats, laboratories and workshops will be a layer of regolith at a depth to be discussed later in this paper.

5. Storm shelters for anomalously large SPE's will be provided in the habitats, labs, workshops, and in the LRVs. Metallic shielding will be used for the shelters (solar protons are attenuated quicker and produce less biologically damaging secondary fluences than HZE particles).

6. Radiation exposure limits (in Sieverts), as laid down by the NCRP for target organ doses and the low earth orbit environment, will be used (Table 1). The one year exposure limits will be multiplied by a factor of 2, for the two year tours considered in the chosen scenario. This will maintain the 3 x 10⁻² excess risk for cancer induction used as a basis for the dose limits. It is realized that due to the greater carcinogenic potential of the galactic heavy ions, which form the major environmental radiation source on the lunar surface, these limits may well be revised downwards in the future.

7. All LBP will carry personal, passive (x2) and active (x1) dosimeters, with alarms and meters. These devices will supply the data required to track the exposure status of all personnel. Readings will be taken at regular intervals (once per month) and stored in the central data base. After each EVA, LRV sortie and solar eruptions, readings will be taken and the exposure status of each LBP updated.

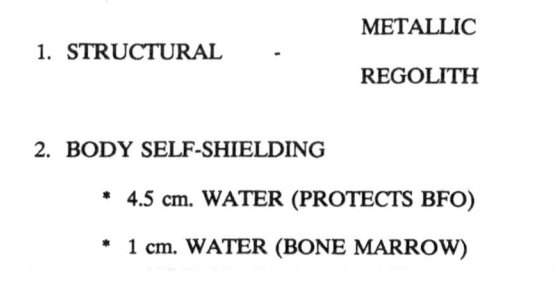

Figure 1. Shielding Configuration.

Table 2.   Cumulative GCR Dose Equivalents Over Two Year Period on Lunar Surface

	Dose Equivalents (Sieverts)	
Operation	Skin	BFO
EVA	.08	.05
LRV	.31	.17
Lunar Base	.58	.46
Totals	.97	.68

8.   Environmental dosimeters, active and passive, with visual and audible alarm systems, will be placed at designated locations inside and outside the main base habitats, labs, workshops, and LRV's. Dosimeters will also be located in the interior of each storm shelter and at predetermined depths through the regolith shield. Each EVA suit will have externally mounted dosimeters (one active and one passive).

Predicted Doses - GCR

The shielding configuration in the lunar base will be that described in Figure 1. Shielding in the LRV and EVA suits will not have the regolith component.

To arrive at estimates of predicted equivalent dose, an operational scenario is required. The following scenario will be chosen as representative:

1. LRV sorties: One two-day sortie per week and a maximum of 80 in two years. At a mean vehicle speed of 20 kilometers per hour, this would allow ranges of up to about 300 kilometers from the lunar base and would include an eight hour EVA sortie.Total hours of exposure to GCR would amount to almost 4,000 hours in the two year tour.

2. EVA's: One 8-hour EVA per week, for a maximum of 90 in two years, leads to a total EVA exposure time of 720 hours.

Table 3.   Quality Factors Used

Component	QF
Primary Heavy Ion	$1 \rightarrow 20$
Proton	1.2
α^{++}	3.0
Nuclear Recoil	20
Neutron	10
Gamma	1

Table 4. Tissue Weighting Factors

Tissue	F
Gonads	.25
Breast	.15
Red Bone Marrow	.12
Lung	.12
Thyroid	.03
Bones	.03
Remainder	.03

3. Lunar Habitat/Labs/Workshops: The residual exposure time in the lunar base amounts to 12,800 hours during a two year tour. It will be postulated that the primary shielding will be as follows: A. LRV - 1.5 g/cm^2 aluminum equivalent. B. Lunar Base - 1.5 g/cm^2 aluminum equivalent plus a minimum of 20 g/cm^2 regolith. C. EVA - 1.5 g/cm^2 aluminum equivalent over all areas of extremities, torso, and head.

Based on the data of Silberberg et al. (1984) and Wilson and Townsend et al. (1989) and with the above operational scenario and primary shielding, the expected dose equivalent in Sieverts (1 Sievert = 100 rem), during Solar Min conditions, are shown in Table 2 for two target organs (skin, and blood forming organs or BFO dose).

The total for the Lunar Base includes a contribution of about .07 Sievert due to secondary neutron production in the regolith shield. With a total risk for cancer of 2 x 10^{-2}/Sievert; a dose equivalent of .68 Sv leads to a risk of 1.36 x 10^{-2} - still well within the stipulated lifetime excess risk of cancer mortality of 3 x 10^{-2}.

Table 5. Quality Factors vs LET_∞

LET ∞ in Water (KEV/μm)	Q
$\leq$ 3.5	1
7	2
23	5
53	10
$\geq$ 175	20

Table 6. Annual Dose Equivalent of Neutron Dose in Lunar Regolith

Depth (G/cm^2)	Annual Dose (Sieverts)
0	.015
10	.03
20	.05
100	.13
200	.12
300	.08
400	.05
500	.02

Tables 3, 4, and 5 provide information on the bases of these calculations in terms of Quality Factors, Tissue Weighting Factors, and the variation Quality Factor with LET, respectively. Table 6 illustrates the annual neutron dose equivalent at various depths of regolith (Lingenfelter et al, 1972). It is seen that the neutron contribution to the dose will peak in the region of 200 g/cm2 (133 cm) and then decrease; becoming insignificant at depths of over three meters. Table 7 shows how the total dose equivalent (over the two year period) would decrease with regolith depth. At 500 gm/cm2 (333 cms) the skin dose would be as low as 0.1 Sievert and the BFO dose would be 0.07 Sievert. This, however, has a significant impact on the load on the lunar base structure. The force calculations (in the lunar 1/6 g environment) assume semi-cylindrical structures of 100 feet in length and 10 feet in radius. At 500 g/cm2 regolith, the force amounts to well over half a ton per square foot.

Predicted Doses - SPE's

As stated earlier, the chosen scenario will involve one such event in any two year period, and the August 1972 event will be taken as a basis for estimation.. There will be no EVA exposure of significance during a SPE as alarm systems will allow LBP on EVA sorties to return within one hour of the solar event precursor, to the storm shelters. The LRV storm shelters will have shielding of 20 g/cm2 aluminum equivalent. In the main base this will be in addition to the 1.5 g/cm2 Al equivalent in the covering structure, plus at least 20 g/cm2 of regolith.

Table 7. Variation of 2-Year Total Dose with Regolith Depth

Regolith Depth (cm)	Dose Equivalent (Sv)		Pressure on Habitat Structure (lbs./sq. ft. at $^1/_6$ g)
	Skin	BFO	
13.3 (20 g/cm^2)	.97	.68	42
50 (75 g/cm^2)	.57	.45	160
333 (500 g/cm^2)	.10	.07	1050

Table 8. August 4, 1972 Event Sequence

Approximate Time (UT)	Event (2 g/cm^2 Al Shielding assumed)
0621	Optical flare observed
1300	30-day limits exceeded for skin and eye
1400	30-day limit for BFO exceeded; annual eye limit exceeded
1500	Annual limit for skin exceeded
1600	Annual limit for BFO exceeded; career limit for eye exceeded
1700	Career limit for skin exceeded

Table 9. Cumulative Doses and Dose Equivalents for the August 1972 Solar Particle Event.

Shielding	Skin		Ocular Lens		Bone Marrow	
	Dose (Gy)	Dose Eq.(Sv)	Dose (Gy)	Dose Eq.(Sv)	Dose (Gy)	Dose Eq.(Sv)
2 g/cm^2 aluminum	7.61	11.30	6.35	9.09	0.92	1.24
20 g/cm^2	0.12	0.18	0.12	0.17	0.04	0.07

Fluence, dose data and shielding predictions have been given by Letaw and Clearwater (1986). Solar flare protons fall in the same energy range as trapped protons (inner zone), ranging from about 50 MeV to 400 MeV. Table 8 shows the event sequence for the 1972 event. Table 9 shows the cumulative dose equivalents for the '72 event at 2 g/cm^2 and

Table 10. Total Dose Equivalent Rate vs Shielding Depth- '72 Type SPE.

Aluminum Depth (G/cm^2)	Total Dose Equivalent Rate (Sieverts/Hour)
0	43.5
10	10.2
20	2.6
40	.39
70	.15

Table 11. Early Effects of Acute Radiation Exposure (Dose in rem in ≤ 1 Day)

Effect	ED_{10}	ED_{50}	ED_{90}
Anorexia	40	100	240
Nausea	50	170	320
Vomiting	60	215	380
Diarrhea	90	240	390
Death (in 20 - 60 days)	220	285	350

20 g/cm^2 Al shielding. Only BFO doses will be considered here and the calculations include 5 g/cm^2 body self-shielding. Table 10 shows the total dose equivalent rate in Sieverts per hour for shielding up to 70 g/cm^2 Al equivalent.

At the main base and within the storm shelters the total dose equivalents received is estimated to be less than 0.05 Sieverts (5 rem). In the LRV storm shelters, with 20 g/cm^2 Al equivalent, the total dose equivalent expected from a '72 like SPE, would be approximately 0.4 Sieverts (40 rem). The effective period of exposure at the rates given in Table 10 would be 15.5 hours. This correspondes to the worst case and is expected to occur if the SPE lasts the full two-days of a LRV sortie.

With a possible acute dose of .4 Sv in 48 hours, anorexia could be expected in one or more of the crew - see Table 11.

Means to treat acute radiation exposure should be available in the LRV storm shelters (anti-nausea medications; anti-radiation medication, etc.).Following such an acute exposure, evacuation of those LBP affected should be transported back to earth as soon as possible.

Dosimetry

The second vital component of the radiological health system in a lunar base is the dosimetry system. There are two main objectives of this system: (a) radiation protection and accumulated dose control for LBP and (b) the accumulation of dosimetric data: to predict future doses, to validate shielding and dose calculations, and to better understand the lunar radiation environment. The complete lunar dosimetry system must be able to record the temporal pattern of absorbed dose rate and dose equivalent rate, identify particle type and especially HZE particles, record the integrated personal dose equivalents, display personal dose equivalent rates, and have audible/visible alarms when preset dose equivalents and rates are exceeded. The system must include warnings of increased solar activity, based on: 1. satellite observations,   2. ground-based astronomical observations on the earth and moon, 3. lunar-based XR-image observations of the sun. Predictive computer algorithms based on these observations and models of solar physics will be developed to give maximum early warning of SPE's at the lunar surface.

Guiding principles for the dosimetry system are:

a. LBP will wear dosimeters at all times - three per person: one on the head and two on the torso. One of these will be an active, integrating solid-state device, coupled to visible and audible alarms.

b. dosimetry measurements in the LRV's, in the EVA suits and at the lunar main base, will be both exterior and interior to allow accurate checking of shielding calculations and effectiveness. Low LET (mainly protons) and high LET (mainly HZE particles, knock-on

Particles	Energy Range (MeV)	Fluence Rate (cm^{-2} Day^{-1})
Protons	10 → 500	→ 10^5 (transients to 10^8)
Neutrons	→ 100	→ 4 X 10^5
Electrons	→ 6	~ 10^5
α - Particles	→ 10^5	→ 10^5 (ALE)
HZE Particles (Z > 2)	10^3 → 10^6	→ 10^5
Photons	→ 6	→ 4 X 10^5

Figure 2.   Radiation Field.

protons, neutrons and alpha-particles) radiations will be identified and fluences measured. Slow and fast neutron fluences will be distinguished and measured also.

c.   all active personal dosimeters will be telemetrically linked to the lunar base main computer.

Maximum considerations must be given to the robustness, radiation hardening, power consumption, volume and mass of the dosimeter units, as well as to their dosimetric properties.Figure 2 summarizes the radiation field as it is known on the lunar surface. Figure 3 is a schematic outlining the main measurement components of the dosimetry system.

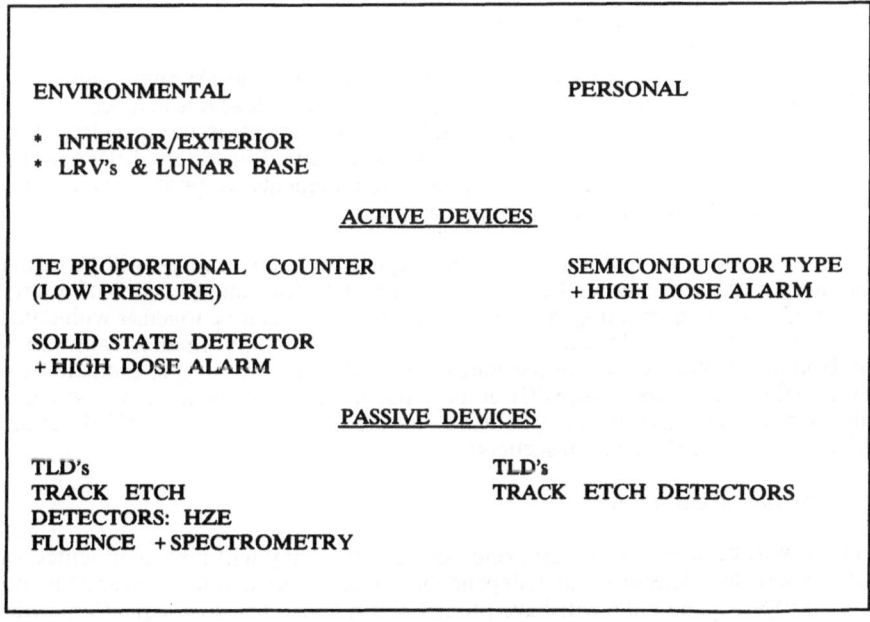

Figure 3.   Dosimetry

Dosimeter Types and Technology

Field dosimetry supplemented with computer calculations will determine the biologically equivalent doses. Both the energy deposited and the LET distribution of this energy must be determined. Tissue doses must be measured at 0.07 mm, 3 mm and 10 mm depth to enable the estimation of skin dose, eye lens dose and effective dose equivalents. Dose quantities to be measured in 1) personal monitoring are: - individual dose equivalent superficial, H_s; and - individual dose equivalent penetrating, H_p; 2) environmental monitoring are: - directional dose equivalent, H^1; and - ambient dose equivalent, H_t.

Dosimetry units must be correlated with the expected radiation fluxes. To satisfy these requirements, different types of active and passive dosimeters will be necessary for environmental and personal monitoring. The main characteristics are: - suitable dynamic range and - approximate dose equivalent proportionality of integrative response for the range of particle types and fluences.

Passive Dosimeters (LRV's, Lunar Base and Personal)

The presently preferred passive dosimeter to record absorbed dose of ionizing radiation (protons, electrons, alpha-particles and HZE particles, and for the photon and neutron fields) is a TLD (thermo-luminescent device). This gives a response proportional to the tissue absorbed dose at a defined depth for all ionizing radiations present on or below the lunar surface.

The TLD's used to measure skin and depth absorbed dose will be 'packaged' with a layered plastic polymer dosimeter in order to record the fluence and type of HZE particle. Track counting in each layer will determine the flux of particles above some known value of LET. One of the laboratories in the lunar base will be devoted exclusively to the preparation, calibration, and reading of each of these devices on a periodic basis.

Each LBP will wear two passive dosimeters: one on the torso and one on the head. Other passive dosimeters will be mounted at strategic locations in order to measure the fluences inside the LRV's and lunar base units.

Active Environmental Dosimeter

Two types of devices appear suitable:

1. Tissue Equivalent Proportional Counters with Electronic Pulse Height Analyzers. This device will allow for measurement of tissue absorbed dose and dose rate at a depth of 10mm. Both the absorbed dose and dose rate versus LET and the ambient tissue equivalent dose and dose rate will be recorded. These instruments will be mounted inside the LRV's and in the lunar base units and at selected exterior locations. Measurements are processed electronically for transmission by telemetry to the base main computer.

2. A set of solid state detectors (MOS) for high dose rate warnings, will be utilized to detect the rapid increase in photon fluence rate which provides an indication of increasing solar activity. Computer monitoring of the response of these detectors, together with data from lunar base observations of solar activity and X-ray output, plus information from the SESC system in Boulder, Colorado, will be integrated to provide early warning of anomalously high solar activity (flare) and impending SPE's at the lunar surface. This will allow ample time for countermeasures to be taken by the LBP (discontinue EVA; ingress to LRV or lunar base units; retirement of all LBP to storm shelters).

Active Personal Dosimeters

These will be worn by all LBP, one per person. They will also be mounted on the EVA suits. Solid state detectors with independent power supply and low power consumption must be developed. The units will have pre-set alarm levels for dose equivalent and dose

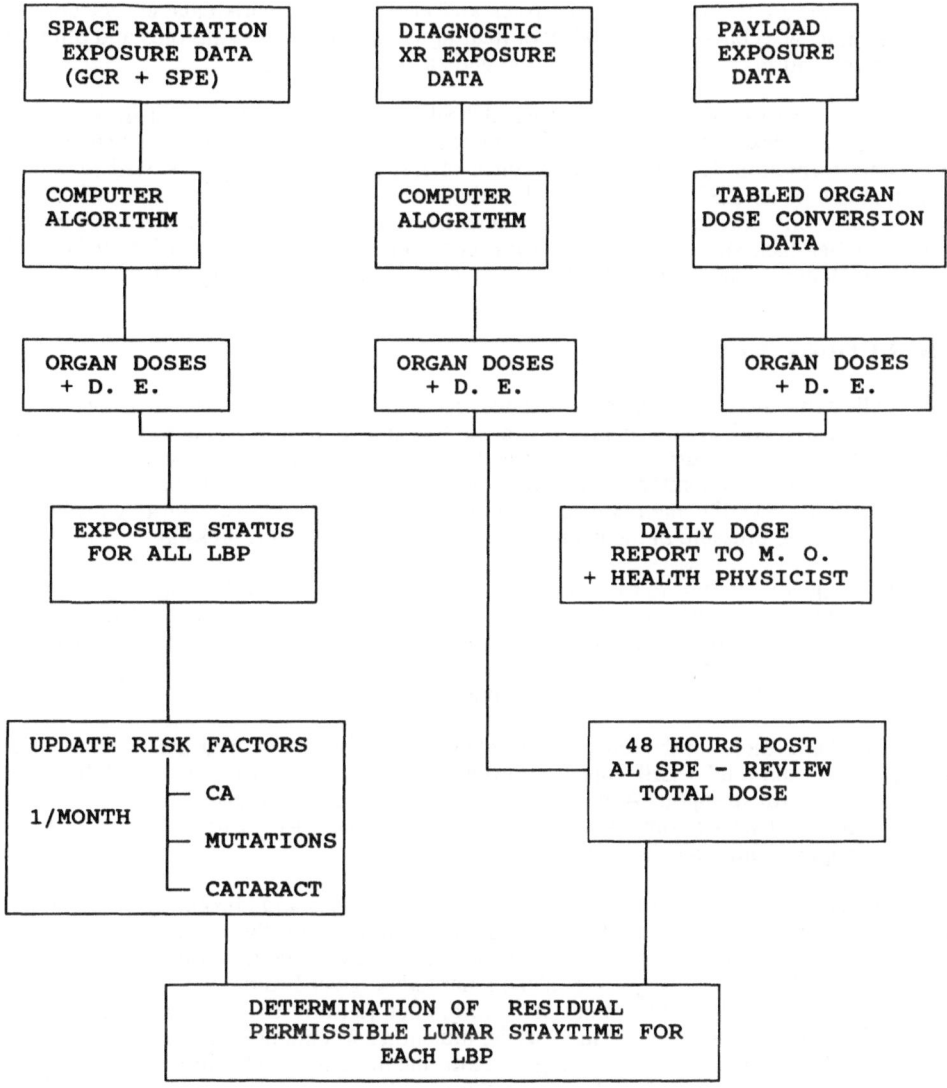

Figure 4.   Lunar Radiological Health Monitoring Flow Diagram.

equivalent rate with audio/visible alarms. The outputs will be telemetered to the lunar base main computer.

Neutron Fluence Measurement

Due to the build-up of secondary neutrons in the lunar surface layer or regolith, it will be important to measure low energy and high energy neutron fluences. The fission-foil type passive dosimeters are satisfactory for low energy neutrons however active devices are still necessary. A high energy neutron fluence dosimetry currently is being developed.

Radiological Health Strategy

It will be mandatory to have a radiation health physicist as a permanent member of the LBP. Also the medical officer present will have a subspecialty in environmental medicine, including radiological health.The main based medical facility should have the capability of hematologic analyses, bone marrow biopsies and analysis, and early cataract detection. In addition it must also have the capability of treating acute radiation symptoms and allow for the IV administration of radioprotectant drugs and parenteral feeding.

The central feature of the radiological health system will be the real-time radiation exposure history; risk factor calculation for cancer induction; measurement of genetic changes and cataract formation detection. The medical officer and health physicist will review the radiation status of all LBP once per month and after every SPE. Such reviews will also be performed after any accidental exposure to nuclear devices used as power sources for medical studies . After each such review, decisions on the member return to earth will be made. Figure 4 illustrates a schematic of the lunar radiological health monitoring system.

CONCLUDING REMARKS

Following the setting up of a manned international microgravity laboratory in near earth orbit at the end of this century, the next phase of manned space operations will undoubtedly involve a return to the moon and the establishment of a permanent lunar base consisting of habitats, laboratories (including observatories) and workshops. EVA will not be an option both from the main base and from LRV's. Critical decisions must be made on: 1. the atmospheric pressure in the habitats, LRV's and 2. EVA suit pressure.

The objective will be to avoid any risk of 'bends' and this could involve working at an R value of 1.4 or less in order to eliminate venous bubble formation. The 'hard suit' required to achieve this also is beneficial for radiation protection. Overall radiation shielding in the suit of at least 1.5 g/cm^2 is required to ensure that large skin doses (from high energy electrons, for example) will not be received by LBP during EVA sorties.

A regolith shield for the lunar base of at least 75 g/cm^2 is required and a regolith depth of 400 g/cm^2 would render LBP safe from even the most intense SPE. As noted this has an impact on structural strength. A decision early on in the design phase is essential. In the absence of more advanced composite shielding, one may have to opt for a lighter shield and therefore a higher radiation risk for the LBP.

Biological monitoring of the crew for genetic mutations and chromosomal aberrations must be developed and automated for routine use. In view of the build-up of secondary neutrons in the regolith, further development of active and passive neutron dosimeters are of higher priority. The overall objective of the lunar radiological health system will be to: 1). Present to each lunar base personnel prior to his/her journey to the moon,predicted risk estimates for radiation induced carcinogenesis, mutation and cataractogenesis and 2).Present to each lunar base personnel on returning to earth, the measured risk estimates for these end points. Shielding is a good countermeasure but it alone cannot eliminate risk. Efficient dosimetry will allow for the accurate estimation of risk. It is in effect a secondary countermeasure, but equal importance.

REFERENCES

Letaw, J. H. and Clearwater, S. H. 1986, Radiation Shielding Requirements on Long Duration Missions. Severn Communications Report 86-2.

Lingenfelter, R. E., Canfield, E. H. and Hampel, V. E., 1972, The Lunar Neutron Flux Persisted. Earth Planetary Sci Lett 16: 355-369.

Silberberg, R., Tsao, C. H., Adams, J. H. and Letaw, J. R., 1984, Radiation Doses and LET Distributions of Cosmic Rays, Radiation Research, 98: 209-226.

Townsend, L. W., Wilson, J. W. and Nealy, J. E., 1989,. Space Radiation Shielding Strategies and Requirements for Deep Space Missions;Proceedings; 19th Intersociety Conference on Environmental System(SAE Tech Paper, No. 891433), San Diego, CA.

SPACE RADIATION ISSUES WITHIN THE SPACE EXPLORATION INITIATIVE (SEI)

Thomas E. Ward

Brookhaven National Laboratory
Upton, L.I., New York 11973
and
The Synthesis Group
Arlington, Va., 22202
and
Office of Space
Department of Energy
Washington, D.C., 20585

INTRODUCTION

One of the more important considerations of manned space-flight, outside the Earths magnetoshere with exploration and habitation of the lunar and Martian surfaces, is the radiation hazard. Specifically, the risk of high levels of radiation, due to Galactic Cosmic-Rays (GCR) and Solar Particle Events (SPE) during long duration manned missions of 2-3 years length, must be quantitaively assessed. Current limits of space radiation to astronauts (NCRP, 1989) could easily be exceeded on long missions, if shielding requirements are not met for spacecraft and habitats. Life-threatening solar flares occurring during a solar maximum are readily detected but not reliably predicted. Quantitative shielding estimates (Simonsen and Nealy, 1991a) for spacecraft indicate that 10-20 metric tons (mt) of water equivalent are required for shielding from a large solar flare event. It is therefore necessary and prudent to define the passive shielding requirements quantitatively with minimum uncertainties in order to reduce the overall vehicle mass. Additionally, the engineering design that utilizes materials, fuels, and cargo for supplemental shielding to a storm shelter will further reduce the weight penalty for radiation protection shielding.

The worst case assumptions that incorporate present uncertainties as limits in a calculation grossly overestimate the shieling requirements by factors of two or more and lead to unacceptable shielding mass requirements for SEI (Ward, 1991). A quantitative estimate with an overall combined unceratainty of 15-25% requires reducing the present large uncertainties in the specific areas of environmental source terms, physics database, transport, modeling and radiobiological response. This goal is an enormous challenge to the technical and scientific community, but can be accomplished with a well correlated ground- and space-based research program.

RADIATION LIMITS

The recommendations of the National Council on Radiation Protection and Measurement (NCRP, 1989) are presently used as NASA guidelines for the radiation exposure limits for astronauts in low Earth orbit only. Limits or levels for SEI have not yet been determined. NASA

Biological Effects and Physics of Solar and Galactic Cosmic Radiation,
Part B, Edited by C.E. Swenberg *et al.,* Plenum Press, New York, 1993

917

has developed a research program to investigate the SEI radiation limits. The radiation research program involves planning to ensure As Low As Reasonably Achievable (ALARA) radiation exposure with the caveat that the radiation risks be kept in proper perspective with other risks of space-flight. Space radiation exposure to crewmembers can be reduced with shielding, but not entirely eliminated. The occupational hazard and career limits are based on a 3% lifetime risk to lethal cancer and are give in Table 1. It should be noted that approximately 20% of all deaths in the general population are due to cancer.

Table 1. Career Whole Body Dose Equivalent Limits Based on a Lifetime Excess Risk of Cancer Mortality of 3%.

Age(years)	Female(rem)*	Male(rem)*
25	100	150
35	175	250
45	250	320
55	300	400

(*) 100 rem=1Sv

The shorter term limits are given in Table 2 where the life threatening limits are set by the dose to the Blooding Forming Organs (BFO), the BFO being the self-shielding dose for 5 cm tissue depth. All shielding estimates will be made in this paper for the protection of BFO.

Table 2. Short-term Dose Eqivalent Limits and Career Limits.

Time Period	BFO(rem)*	Lens of Eye(rem)*	Skin(rem)*
30 day	25	100	150
annual	50	200	300
career	see Table 1	400	600

The NCRP is presently re-examining the database for establishing GCR limits based on particle fluences and a re-evaluation of the radiation basis obtained from BEIR V (1989).The present guidelines may well change after this re-examinination.

RADIATION ENVIRONMENT

The GCR is the major long-term space radiation component to be considered outside the magnetosphere. It has a relatively constant solar modulated intensity and a well characterized energy and particle type spectrum. Recently, a workshop entitled"Cosmic Radiation: Constraints on Space Exploration" was held at the 22nd International Cosmic Ray Conference (16 August, 1991). The results of the workshop and subsequent efforts by the cosmic ray community has been able to reduce the GCR flux uncertainty from 40-50% down to about 10%. However, the largest uncertainty level in dose response is due to the high Z-element (HZE) radiobiological response. Only 10% of the required radiobiological database has been produced in the 20 years since the National Academy of Sciences recommended a substantial effort in this research area (NAS/NRC 1973). The GCR components are comprised of protons (87%), helium (12%) and heavy-ions (1%) (Simpson, 1983: Menwalt, 1990). However, the relative ionizing power, which is proportional to Z-squared, together with the particle intensity and biological quality factor (Q) makes HZE Fe damage double that of protons and helium particles combined (Schimmerling, 1990a). The all pervasive GCR radiation burden represents a long-term excess cancer mortality risk unlike that of large and intense SPE which are catastrophic life-threatening events.

The solar flares or SPE are lower in energy on the average than the GCR, easier to shield against, variable in energy and intensity from event to event and highly unpredictable as to

Table 3. Large Solar Particle Events (SPE).

Date	Total BFO Dose(rem)*
23 Feb. 1956	60
9 Feb. 1958	30
10 May 1959	48
10,14 July 1959	114
12,14 Nov. 1960	102
4, 7 Aug. 1972	498
Sept.-Oct. 1989	200

(*) 3 gm/sq.cm aluminum shielding, 100 rem= 1 Sv

their occurrence. The frequency of SPE follows the well known 11 year solar cycle. During the 3-4 year solar minimum few if none large SPE are expected based on the past 3 solar cycles (Jordan and Stassinopoulos, 1989). Large events (>few billion particles/ sq.cm) that are life threatening occur only near solar maximum within a 5-7 year span. Table 3 list some of the more hazardous large events (Korvaleev, 1989), where the 1989 flares are included for comparison. These large SPE would all be life threatening to astronauts caught during unprotected EVA in space or on the lunar surface with no shielding.

Ordinary events (<few hundred million particles/sq. cm) can occur throughout the solar cycle, but are not life-threatening. The largest single integral intensity SPE was the August, 1972 flare, whereas the highest energy/intensity, and therefore the most penetrating SPE, was the February, 1956 event. There is no model that can predict the occurrence, energy, or relative intensity of SPE. They are readily detected, increasing in intensity over several hours, and therefore can be monitored to provide an early warning to the astronauts prior to a large event. A worst case scenario, a combination of the 1956 high energy event and the 1972 high intensity event, can be used to assess the radiation health risk and shielding requirements for large flares. The radiobiological Q values for protons of the SPE spectral range is much better known than that of HZE GCR. In addition, an extensive database on large animals for extrapolation to humans has been aquired since the late 1960's. Furthermore, the physics database, transport, and modeling of intermediate energy protons are well characterized. Total uncertainties of about 50% for SPE allow for quantitative passive shielding calculations that better quantify the risk and requirements for acute life-threatening events than for long-term GCR cancer risks.

PHYSICS DATABASE, TRANSPORT AND MODELING

Three groups have developed integrated analytical models for use in predicting radiation effects due to GCR and SPE. These are the NASA Langley group (Simonsen and Nealy, 1990a; Townsend, et.al.,1989; and Wilson, et.al., 1989), the NRL group (Adams, et.al., 1986; Letaw, Silberberg and Tsao, 1989 and 1990) and the NASA Goddard effort (Jordan and Stassinopoulos, 1989). A comparison of these efforts on a slab geometry configuration can be shown to be in overall agreement to within about 20%. The proton database is much better developed than that of helium or HZE particle interactions. However, high energy range thick transport data above 800 MeV proton energy is lacking. In general, uncertainties due to projectile or target fragmentation processes, secondary or tertiary interactions and a lack of well correlated data on complex thick targets restricts our present ability to reduce uncertainties to an acceptable level of 15-25%. Validation of the integrated database, transport and modeling codes requires a comprehensive ground based research program. An estimate of the ground based accelerator beam time required for the effort is given in Table 4 along with the radiobiology beam time requirements.

RADIOBIOLOGICAL RESPONSE

The largest uncertainties with regard to a quantitative risk assessment of radiation hazards lies with the radiobiological database. The experiments are often complex and difficult. The

Table 4. Accelerator Beam Time Requirements for the SEI Space Radiation Effects Program (First 10 years).

Research Area	Total Hours
Physics Database	
Central Collisions	3,000
Peripheral Collisions	3,000
Materials and Electronics	2,000
Neutron Yields	10,000
Radiobiology	
Cells	5,000
Tissue	5,000
Samll Animals	6,000
Large Animals	4,000
Total	38,000 hrs

radiation risks incurred in space environments can only be validated in well correlated ground and space-based experiments (Schimmerling and Sulzman, 1990b). The fact that the GCR is comprised of all particles at all energies and incident at all angles simultaneously precludes exact ground-based simulations. Other factors that must be considered for study are dose-rate effects since SPE can have dose-rates that affect thresholds and repair; and synergisms between microgravity, thresholds and repair. Considerable effort is required to correlate the endpoint matrices for cellular, tissue, organs, small animal and large animal extrapolations to human beings. The total accelerator beam time required is given above in Table 4.

RADIATION SHIELDING REQUIREMENTS

Radiation sources to be encountered on a long duration spaceflight are listed in Table 5.

Table 5. Space Radiation Sources and Levels Expected During Long Duration Missions with Minimum Shielding.

Source	BFO Dose*
GCR	
Solar Minimum	64 rem/yr
Solar Maximum	24 rem/yr
SPE	
Ordinary Events	14 rem/yr
Large Events	498 rem/yr

(*) assumes 3 gm/sq.cm aluminum shielding, 100 rem= 1 Sv;
(Letaw, 1989), (Simonsen and Nealy, 1990a) and (Simonsen, et. al., 1990b)

The large SPE of August, 1972 would produce lethality in 50% of the subjects in 60 days. Clearly a storm shelter capable of reducing these radiation effects is required. Using the analytical models referenced above (Simonsen and Nealy, 1990a; Simonsen, et. al., 1990b; and Letaw, et. al., 1989 and 1990) with shield effectiveness ratios of Al(1.0), CH_2(1.4), H_2O(1.6), Fe(0.68) and Pb(0.44) one obtains an optimum shield thickness of 16 gm/ sq.cm water equivalent in order to limit the total burden to below the limits set in Table 2 by the NCRP. A weight savings of 10 mt translates into a Mars mission cost savings of about $1Billion. It is therefore imperative that a well correlated, low uncertainty and high level of confidence data-base be established in order to assure proper radiation protection safety and reduce overall mission costs.

RECOMMENDATIONS

The research program for the radiation health protection is necessarily complex due to the multidisciplinary nature of the effort which includes integrating solar physics, nuclear physics, theory (modeling), radiobiology, and probability risk assessment. Impacts due to radiation issuses on spacecraft design, habitats and mission planning must be assessed with a high level of confidence to meet SEI milestones of our return to the moon in 2003-2005 and prior to the 2014-2018 Mars mission. Specifically, the recommendations are as follows:

1. The need for quantitative 25% uncertainties of potential hazards.

2. A need for additional experimental data on biological response, light and heavy-ion(HZE) physics, interactions on complex range thick targets, and the development of predictive transport and radiation risk assessment models.

3. A need for development of predictive models of the temporal solar cycle modulated GCR and SPE sources.

4. The need for a ground based space radiation effects facility at an accelerator capable of simulating SPE (protons and helium) and GCR (primarily HZE) particles with appropriate energies.

5. The need for space based science integration facilities for radiobiological research that can provide validation and verification of predictive models of space radiation risks.

In 1971 NASA requested an overall assessment of the agency radiobiological research program from which the 1973 report "HZE Particle Effects in Manned Spaceflight" (Grahn, 1973) was issued. Specifically, the National Research Council Radiobiological Advisory Panel recommended the first four points listed above. Those recommendations are still applicable today since less than 10% of the required radiobiological database has been aquired since 1973! SEI milestones require that a lunar shielding database be established with some confidence by the year 2000. In order to meet the milestone, a program capable of producing 50-60% of the database over the next 7-8 years must be implemented. The Mars mission requires a 80-90% database by the year 2005 to enable full validation of the predictive models with low uncertainties (15-25%) and high (90%) confidence level. The initiative will need substantial funding increase over the present budgets based on past program funding levels and research returns noted above. The implementation of the program requires a multiagency effort between NASA, DOE, DOD and others in order to utilize existing facilities and scientific manpower in a timely and efficient manner. The new National Exploration Program Office, to be headed by NASA but including DOE and DOD, will be instituted by Presidential Directive in 1992 and will coordinate the multiagency effort.

REFERENCES

Adams, J.H., 1986. Cosmic Ray Effects on Microelectronics, Part IV NRL Report 5901, Wash.,D.C.

Adams, J.H., Editor,1991. Galactic Cosmic Radiation: Constraints on Space Exploration, NRL Report 209-4154, Wash., D.C.

Grahn, D., Editor,1973. HZE-Particle Effects in Manned Spaceflight, NRC/NAS.

Jordan, T. and Stassinopoulos,1989, Effective Radiation Reduction in Space Station and Missions Beyond The Magnetiosphere, Adv. Space Res. 9: 261.

Korvaleev, E.E., Muratova, I.A. and Petrov, V.M., 1989, Nucl. Tracks Radiat.Meas. 16: 45.

Letaw, J.R., Silberberg, R. and Tsao, C.H., 1989, Radiation Hazards on Space Missions outside the Magnetosphere, Adv. Space Res. 9: 286.

Letaw, J.R., Silberberg, R., and Tsao, C.H.,1990. Natural Radiation on Manned Mars Mission, preprint.

Menwalt, R.A., 1988, in Interplanetary Part. Environmet, JPL Publ.No. 88-28, pg. 121.

Menwalt, R.A., 1990. Radiation Hazards to Man in Space, SRL Technical Report 90: 3.

NCRP Report No. 98, 1989. Guidance on Radiation Recieved in Space Activities, published bty the National Council on Radiation Protection and Measurements, Bethesda, Md..

Schimmerling, W., 1990a. The HZE Radiation Problem, preprint.

Schimmerling, W. and Sulzman, F., 1990b. The Radiation Health Program and SEI, preprint.

Simonsen, L. and Nealy, J., 1990a. Radiation Protection for Human Missions to the Moon and Mars, preprint.

Simonsen, L., Nealy, J., Townsend, L., and Wilson, J., 1990b. Radiation Exposure for Manned Mars Surface Missions, NASA Technical Report No.2979.

Simpson, J.A.,1983. Ann.Rev.Part. Sci. 33: 323.

Townsend, L., Wilson, J. and Nealy, J.,1989. Space Radiation Shielding Strategies and Requirements for Deep Space Missions, SAE paper No. 891433.

Ward, T.E.,1991. Space Radiation Hazards and the Space Exploration Initiative, NASA Workshop on Ionizing Radiation Environment, Models and Methods. Marshall Space Flight Center Presentation.

Wilson, J.,et. al., 1989. BRYNTRN: A Baryon Transport Model, NASA Technical Paper No. 2887.

PARTICIPANTS

Dr. Shamim Ahmad
Dept. od Life Sciences
Nottingham Polytechnic
Clifton Lane
Nottingham, NG11 8NS England

Mr. William Atwell
Rockwell International Space System
        Division
555 Gemini
Houston, Texas 77058, USA

Dr. B. Baican
Institut für Kernphysik
Universität Frankfurt
August Euler Str. 6
6000 Frankfurt/Main 90 Germany

Ms. Kira Bacal
Baylor College of Medicine
7575 Cambridge
2403
Houston, Texas 77054, USA

Dr. Rudolf Beaujean
Institut fur Reine und Angewandte
Kernphysik, Universität Keil
Olshausensstr. 40/60
23 Kiel 1, Germany

Mr. Dave Beckett
The University of Alabama
135 Angela Drive
#1308
Madison, Alabama 35758, USA

Dr. Breckow
TUV Reinland e. V
Fachbereich Kerntechnic, Am Grauen Stein
P. O. Box 101750
5000 Köln 91, Germany

Prof. Dr. J. J. Broerse
Institute of Applied Radiobiology &
        Immunology
Lange Kleiweg 151
NL-2288 HV Rijswijk, The Netherlands

Prof. A. H. Bücker
DLR Institut für Fluggmediczin
Abt. Biophysik
Linder Höhe
5000 Köln 90, Germany

Dr. A. Cajigas
Dept. of Pathology
Allbert Einstein College of Medicine,F-523N
1300 Morris Park Avenue
Bronx, NY 10561, USA

Prof. K. Carr
School of Biomedical Science
Queens University of Belfast
Belfast, Northern Ireland

Dr. Benton Clark
Martin Marietta (B0560)
P. O. Box 179
Denver, Colorado 80201, USA

Dr. Ann B. Cox
USAF Armstrong Laboratory
AL/OEDR
Brooks AFB
Texas 78235-5301, USA

Prof. J. Crepeau
University of Utah
Dept. Mechanical Engineering
Salt Lake City, Utah 84112, USA

Dr. Stan B. Curtis
Lawrence Berkeley Laboratory
University of California
Berkeley, California 94707, USA

Dr. Tsvetan P. Dachev
Solar Terrestrial Influences Laboratory
Bulgarian Academy of Sciences
Acad. G. Bonnchev Str. Block 3
Sofia 1113, Bulgaria

Dr. Dietrich Doll
Institut für Strahlenbiologie
UKE Hamburg 20; Martinistr. 52
2000 Hamburg 70, Germany

Dr. Mildred Donlon
AFRRI/DNA
Bethesda, Maryland,200889-5145
USA

Dr. Jürgen Fellinger
Techn. Universität Dresden
Institut fur Strahlenschutzphysik
Mommsenstr. 13
8027 Dresden, Germany

Dr. M. Frankenberger-Schwager
G S F Frankfurt
Institut für Biophysikalische
Strahlenforschung
Paul-Ehrlich-Str. 20,
6000Frankfurt/Main 70   Germany

Prof. Nicholas Geacintov
Dept. of Chemistry
New York University
New York, N. Y. 10003, USA

Dr. Henry Gerstenberg
AFPRI/ DNA
Bethesda, Maryland
20889-5145, USA

Mr. Frederick Hanser
Head, Research Branch
Radiation Physics Dept.
Pharmaceutics Inc.
221 Crescent St.,Waltham, MA 02154
USA

Dr. Thomas Heck
Universität des Saarlandes Fachrichtung 3.6
Theor. Medizin und Biophysik
(Bau 76)
6650 Homburg/Saar, Germany

Mr. Gary Heckman
National Oceanic & Atmospheric
        Administration, R/E/SE$_2$
325 Broadway
Boulder, Colorado 80303, USA

Dr. C. Heilmann
Centres de Recherches Nucleaires
23 Rue du Loess
67200 Strassbourg
France

Dr. J. Heilman
Abt. Biophysik, GSI
Planckstr. 12
Postfach 11 05 52
6100 Darmstadt 11, Germany

Dr. Ludwig Hieber
Institut für Strahlenbiologie
G S F
Ingolstaedter Landstrasse 1
8042 Neuherberg, Germany

Dr. Gerda Horneck
DLR Institut für Flugmedizin
Abt. Biophysik
Linder Höhe
5000 Köln 90, Germany

Mr. Stephen Hosselet
New York University
A.J. Lanza Labs.
Long Meadow Road
Tuxedo, New York 10987, USA

Mr. Thomas Jordan
Experimental & Mathematical Physics
Consultants
P.O. Box 3191
Gaithersburg, Maryland 20885, USA

Prof. Robert Katz
University of Nebraska
Physics Dept.
Lincoln, Nebraska 68588-0111, USA

Prof. J. Kiefer
Strahlenzentrum
Universität Giessen
Leihgesterner Weg 217
6300 Giessen, Germany

Dr. Michael Kost
Strahlenzentrum
Universität Giessen
Leihgesterner Weg 217
6300 Giessen, Germany

Prof. E. E. Kovalev
Research Center of Spacecraft
Radiation Safety, Apt 149
Leningradsky Prospeckt 69
Moscow 125057, USSR

Dr. S. Kozubek
DLR Institut für Flugmedizin
Abt. Biophysik
Linder Höhe
5000 Köln90, Germany

Prof. John Lett
Radiobiological Health Science Dept.
Colorado State University
Fort Collins, Colorado
80523, USA

Dr. Ch. Lücke-Huhle
Kernforschungszentrum Karlsruhe-Genetik
  und Toxikologie
P.O. Box 3640
7500 Karlsruhe 1, Germany

Dr. Percival McCormack
US Navy, National Naval Medical Center
Branch Clinic Washington Navy Yard
Washington, DC 20065, USA

Prof. P. McNulty
Dept. of Physics and Astronomy
Clemson University
Clemson, S. C.
29634, USA

Dr. J. H. Miller
Pacific Northwest Laboratory
P8-47
Richland, WA
99352, USA

Ms. Beatriz Momingues
Rua Elias Garcia #372-D
20 Esquerdo
27000 Amadora
Portugal

Dr. Lawrence Myers
AFRRI
Bethesda, MD
20889-5145, USA

Dr. Nazmi T. Okumusoglu
Physics Department
19 Mayis University
Samsun, Turkey

Prof. Y. Pak
GATA (Military Medical School)
Radiation Oncology
Ankara
Turkey

Dr. D. Philpott
NASA Ames Research Center
MS 239-14
Moffett Field, California
94035, USA

Prof. N. F. Pissarenko
Space Research Institute
IKIAN USSR
Profsojusznaja 84/32
Moscow 11178810 , USSR

Prof. Martin Pope
New York University
Dept. of Chemistry
24 Waverly Place
New York, NY 10003, USA

Prof. D. Pross
Strahlenzentrum
Universität Gießen
Leihgesterner Weg 217
6300 Gießen , Germany

Dr. B. Rabin
AFRRI
Behavioral Science Dept.
Bethesda, Maryland 20889-5145
USA

Dr. B. Reiss
St. George Krankenhaus
Strahlentherapie
Lohmühlen Str. 5
2000 Hamburg 1 , Germany

Dr. G. Reitz
DLR Institut für Flugmedizin
Abt. Biophysik
Linder Höhe
5000 Köln 90, Germany

Prof. J. B. Robertson
Dept. Environmental Health
East Carolina University
Greenville, North Carolina, USA

Dr. W. Rüther
Sonderforschungsbereich
Environtologie und Raumfahrtmedizin
Pilgrimstein 2
3550 Marburg/Lahn , Germany

Dr. D. M. Rust
John Hopkins University
Applied Physics Laboratory
Laurel, Maryland
20723-6099, USA

Prof. Leo Salter
Chemistry Department
University of Natal
King George V Avenue
Durban 4001, South Africa

Dr. E. Schopper
Institut für Kernphysik
 Universität Frankfurt
August Euler Str. 6
6000 Frankfurt/Main 90, Germany

Dr. J. U. Scott
DLR Institut für Flugmedizin
Abt. Biophysik
Linder Höhe
5000 Köln 90, Germany

Prof. D. Schulte-Frohlinde
MPI für Strahlenchemie
Stiftsstraße 34-36
4330 Mülheim/Ruhr, Germany

Mr. Mark Shavers
Texas A & M University
College Station, Texas
 77843-3133, USA

Dr. M. A. Shea
Space Physics Division (PHG)
Geophysics Directorate (OL-AA/PL)
Hanscom AFB
Bedford, Massachusetts 07131, USA

Dr. D. F. Smart
Geophysics Directorate PHG (OL-AA/PL)
Hanscom AFB
Bedford, Massachusetts 07131

Dr. E. G. Stassinopoulos
NASA Goddard Space Flight Center
Code 900, Earth Sciences Directorate
Greenbelt, Maryland 20771, USA

Ms. Lisa Steimel
DLR Institut für Flugmedizin
Abt. Biophysik
Linder Höhe
5000 Köln 90, Germany

Dr. J. J. Steinberg
Pathology/Radiation Oncology & Biology
Albert Einstein College of Medicine
1300 Morris Park Ave.
Bronx, New York 10461, USA

Mr. Michael Story
M. D. Anderson Cancer Center
University of Texas
1515 Holcombe Blvd.
Houston, Texas 77031, USA

Mr. Johannes Swenberg
California Institute of Technology
MS 128-95
Pasadena, California 91125, USA

Dr. Charles E. Swenberg
AFRRI/RBD
Bethesda, Maryland 20889-5145, USA

Dr. P. Thieberger
Physics Dept.
Bldg. 901A
Brookhaven National Laboratory
Upton, New York 11973, USA

Dr. J. Trainor
NASA Goddard Space Flight Center
Associate Director
Greenbelt, Maryland 20771, USA

Dr. A. Trivedi
AECI Research, Chalk River Laboratories
Chalk River, Ontario
Canada K05 130

Prof. Yurdanur Tulunay
The Middle East Technical University
ODTU, Dept of Aeronautical Engineering
06531 Ankara, Turkey

Dr. Alexander Vasilenko
 Institute of Botany
Ukrainian Academy of Sciences
2, Repina Str.
252601 KIEV-GSP-1, USSR

Dr. T. E. Ward
U. S. Dept of Energy
1000 Independent Ave. S. W.
Forrestal Bldg, Rm EA 155,S-1
Washington, D.C. 20585, USA

Mr. Jörg Wehner
DLR Institut für Flugmedizin
Abt. Biophysik
Linder Höhe
5000 Köln 90, Germany

Mr. G. Williams
Radiological Health Science Dept.
Colorado State University
Fort Collins, Colorado 80521, USA

Dr. J. W. Wilson
493
NASA Langley Research Center
Research Laboratory
Hampton, VA 23665, USA

Prof. B. Worgul
Eye Radiation and Environmental
         Research Laboratory
Columbia University
630 West 168th St.
New York, NY 10032, USA

GLOSSARY

amplification: The production of many DNA copies from one region of DNA.

Angstrom (Å): A distance used to specifiy the wavelength of X-ray emission. One Angstrom is 10^{-10} meters.

anisotropy: The ratio of the maximum to the average particle flux distribution as a function of angle.

anneal: A verb denoting the return towards the original condition or to a more homogeneous condition.

Archimedean spiral: A mathematical curve resulting from a linear angular rotation with increasing distance.

Astronomical Unit (AU): An astronomical unit is the average distance from the sun to the earth, 149.6 million km.

atomic charge: The number of protons contained in an atomic nucleus. Usually designated by the symbol Z. Hydrogen is $Z = 1$, Iron has an atomic charge of $Z=26$.

Background nuclear radiation: For low altitude spacecraft orbits, it would include low leakage fluxes from the Earth's radiation belts as well as albedo from nuclear reactions in the atmosphere in addition to galactic cosmic radiation.

bremsstrahlung: The electromagnetic energy (X-rays and gamma rays) emitted by a charged particle which is being accelerated or decelerated. Only electron bremsstrahlung is generally important.

carcinogen: A substance that causes cancer.

Carcinogenesis: The overall multistep process by which a normal cell becomes a cancer cell.

CCD: Charge Coupled Device. A semiconductor sensor which is composed of a planar array of sensitive volumes which are sensitive to photons of various wavelengths or are sensitive to the passage of charged particles. Readout of the contents is via charge- coupling technique on the semiconductor chip.

cell clone or cell colony: A group of genetically identical cells derived by division from a common ancestor.

chromosome: A linear arrangement of genes and the other DNA with associated protein.

chromosome rearrangement: A chromosome aberration involving new juxtapositions of chromosome parts.

chromosphere: The portion of solar atmosphere between the photosphere and the corona; one-thousand-kilometer wide outer layer of the solar atmosphere where temperature rises rapidly from 5000°C to 20000°C. Characterized by bright pink emission from atomic hydrogen.

chromospheric brightening: A literal description of the "solar flare" which is optically visible as a brightening of a small area of the solar chromosphere.

cis control: The ability of a gene to affect genes next to it on the same chromosome.

CMOS: Complementary metal-oxide semiconductor. The abbreviation CMOS usually denotes a form of complementary electronic digital logic that uses both N-channel and P-channel transitors for fast, low power circuitry.

Cobalt-60: A radioactive isotope of Cobalt that is commonly used as a source of gamma rays. It has a half life of 5.26 years and produces gamma rays at energies of 1.173 and 1.332 MeV.

conjunction class trajectory: Earth to Mars minimum energy interplanetary transfer trajectory, typically having the characteristic that at the time of arrival, Mars is near conjunction (behind the sun).

corona: The portion of the solar atmosphere above the chromosphere; X-ray emitting outermost layer. Temperature is 10^6 ° C.

coronal mass ejection: Transient disturbance of the solar corona in which a massive coronal streamer or loop leaves the sun.

coronal propagation: A concept developed to explain the apparent transport of solar particles around the sun.

cosmic ray modulation: The variation of the observed cosmic ray intensity as a function of the solar cycle. The cosmic ray intensity is observed to vary approximately inversely with the solar activity cycle.

CRRES: An abbreviation for the Chemical Release and Radiation Effects Satellite.

CRS: An abbreviation for the Cosmic Ray Subsystem, the cosmic radiation and magnetospheric radiation belt experiments on the Voyager 1 and 2 spacecraft to the outer planets and our outer solar system.

cross section: In the case of electronic parts or sensors, this refers to the active or susceptible cross-sectional area of the device exposed to the nuclear radiation.

differential flux: The particle flux in a specific energy interval. The differential directional flux is usually specified in units of $(cm^2 \, s \, ster \, MeV)^{-1}$. The differential omnidirectional flux is usually specified in units of $(cm^2 \, s \, MeV)^{-1}$.

diploid: A cell having two sets of chromosomes.

DNA : (deoxyribonucleic acid) A double chain of linked nucleotides; the fundamental substance of which genes are composed.

dosimetry: The measurement of the energy deposited in matter by radiation.

effective intrinsic shielding (EIS): The thickness of a single shield path length which results in the same calculated dose as the more complex, realistic mass thickness distribution that is found in spacecraft and the human body. The EIS is different for each type of radiation environment, spaceship design, and body organ.

electron volt(eV): The amount of energy an electron gains through a potential drop of one volt. $1 \, eV = 1.602 \times 10^{-19}$ joules

endotoxin: A toxic substance formed by bacteria.

erg: A cgs unit of energy. 10^7 ergs = one joule.

Error-Free repair: An enzymatic DNA repair system that is relatively accurate and unlikely to create coding errors during the process of repairing DNA damage.

Error-Prone repair: An enzymatic DNA repair system that is relatively prone to creating coding errors in DNA during the process of repairing DNA damage.

Extra-vehicular activity: Human activity occurring outside man-made or natural enclosures.

favorable propagation path: A concept suggesting that the Archimedean spiral path from the earth to the sun would connect to a specific solar longitude. It is based on the concept that charged particles travel along the interplanetary magnetic field which is transported out from the sun. For an idealized constant speed solar wind flow, if the interplanetary magnetic field is "frozen" in the plasma, then the result would form an Archimedean spiral.

fibril: A dark long (~ 10000 kilometers), thin (700 - 1000 kilometers) streak in the solar chromosphere. Usually indicates the direction of the magnetic field.

filament: A dense cloud of chromospheric material suspended in the solar corona.

first ionization potential: The energy required to remove the first electron from an electrically neutral atom. (The ionization potential is usually given in electron volts.)

flare: A transient brightening in the solar atmosphere. It usually occurs near sunspots.

fluence: The time integrated flux usually expressed in units of particles or photons per cm^2.

flux: The number of particles or photons passing through a unit area in unit time. This is usually expressed as the number per cm^2 per second. It may be used for directional or omnidirectional fluxes, usually specified in units of $(cm^2 \text{ s ster})^{-1}$ or $(cm^2 \text{ s})^{-1}$.

gamma ray: Short wavelength energetic photons.

gel electrophoresis: A technique for separating the components of a mixture of molecules in an electric field within a gel.

gene: The unit of inheritance; a portion of DNA molecule that codes for a specific protein.

glitch: A transient or permanent change of state. It can apply to both analog and digital circuitry and optoelectronic devices.

Golgi saccules: Saccules formed within the Golgi apparatus making up the flattened saccules of the Golgi apparatus.

gyration radius: The radius of a particle making a curved path in a magnetic field gyrating about a line of force.

heliocentric: A measurement system with its origin at the center of the sun.

heliocentric angular distance: A measurement of angular distance from the center of the sun.

heliographic: A measurement pertaining to the sun.

heliographic coordinate system: A coordinate system for measurement on the sun.

helilongitude: A measurement of longitude on the sun.

heliosphere: The portion of space where the plasma outflow from the sun dominates the characteristic of space.

heliospheric boundary: The distance (> 100 AU) where the solar wind flow is hypothesized to change from super-sonic flow to sub-sonic flow.

heterokaryon: A culture of cells composed of two different nuclear types in a common cytoplasm.

hydrogen-alpha line: An absorption line in the photospheric emission spectrum. This "dark" line is well suited for observation of brightening in the solar chromosphere, i.e. , a solar flare.

hybridize: To anneal nucleic acid strands from different sources.

hypokinesia: Reduced activity.

ionization: The production of electron and cation pairs in matter from energy deposited by radiation.

interplanetary magnetic field: The magnetic field in interplanetary space. The interplanetary magnetic field is transported out from the sun via the solar wind.

interplanetary shocks: An abrupt change in velocity or density that is moving faster than the wave propagation speed in interpanetary space.

interplanetary turbulence spectrum: The spectrum of change in the variation in the interplanetary magnetic field.

interstitium: A small interval, space, or gap in a tissue or structure.

Krad: One thousand rads. The rad is the unit of absorbed dose defined as 100 ergs per gram of absorber. Note : 100 rads equal one gray (Gy), the S. I. unit equals to 1 joule per kg.

LET: Abbreviation for linear energy transfer. Also seen as L, rather than LET. LET refers to local rate of energy deposition along the particle track often expressed as MeV cm^2/g. Physicists also refer to it as dE/dx and stopping power. LET is a nonlinear function of particle energy with the highest values of LET occurring as the particle slows to stopping.

L shell: The equatorial radius, in earth radii, of a dipole magnetic field line for trapped radiation.

LSI: This abbreviation refers to large scale integrated semiconductor circuits containing 1000 or more equivalent gate circuits.

magnetic storm: The disturbance of the earth's magnetic field, caused by solar activity, which also affects some components of the trapped radiation.

magnetogram: Map of the magnetic fields in the solar photosphere or chromosphere.

magnetosphere: The primary dipole magnetic field region around a planet.

Mars orbiting service station (MOSS): A facility placed into orbit around Mars to provide various functions to aid human exploration, including communications, observation, consumable stores, and well shielded, large volume habitation.

mass-energy equivalence: The energy realized by conversion of mass into energy. The mass-energy equivalence of a proton is 938.323 MeV.

mean free path: The distance between scattering events in interplanetary particle propagation.

megarad: One million rads. Also written as Mrad. See Krad.

metachromatic: Staining differently with the same dye, different elements of a tissue take on different colors when a certain dye is applied.

MeV: One million (10^6) electron volts. A measure of energy.

microburst: A transient flare of radio, EUV or X-ray emission lasting less than 10 seconds.

microwave frequencies: Radio wave frequencies in the range of about 100 MHZ to 10000 MHZ.

mucopolysaccharide (MPS): A group of polysaccharides which contains hexosamine, which may be combined with protein and which, when dispersed in water, form many of the mucins.

mutagenesis: The process by which cellular DNA is damaged, and that damage is misrepaired by a DNA repair system resulting in a mutation (a heritable change in the genetic code).

nucleotide: The basic building block of nucleic acid, composed of a nitrogen base, a sugar and a phosphate group.

oncogene : A gene that causes cancer.

opposition class trajectory: Fast round trip trajectories to Mars, typically with the initial transfer arriving at Mars while the two planets are in opposition (near each other, on the same side of the sun).

origin of replication: The point of specific sequence at which DNA replication is initiated.

perfusate: The liquid collected after being passed through the blood vessels.

periodic acid Schiff (PAS): A strong oxidizing agent used on tissue sections to liberate aldehydes from polysaccharides. The sites of liberation can be stained by the Schiff reaction.

photodiode: A type of silicon solid state detector which is used to detect light. It can also be used to detect charged particles.

photosphere: The visible portion of sun in white light, yellow layer of the solar atmosphere. The temperature is 6000° C. Also the limit of seeing down through the solar atmosphere in white light.

Pioneer 10/11: Two small , spinning NASA spacecrafts launched in March, 1972 and April, 1973 respectively, on trajectories to Jupiter, Saturn and the outer reaches of our solar system. Both spacecrafts are powered by ≈100 watt nuclear thermoelectric power supplies and are in relatively good condition more than 20 years after the launch of Pioneer 10.

plasma: An ionized gas.

PMOS: A type of metal oxide semiconductor transistor which uses a channel doped with p-type material.

polytenization: An endomitotic process by which giant chromosomes are produced (no cell division occurs and the multiple DNA sets remain bound within the initial chromosome).

quality factor (Q): Thr relative biological effectiveness of different radiation types. Electrons, X-rays and gamma rays have Q = 1, while other radiation types can have significantly higher values.

radian: A unit of angular measure. There are 2π radians in a circle. A radian is approximately 57.29578 degrees.

radiation dose: The energy deposited in matter by radiation. The true physical dose is the Gray (Gy = 1 j/kg), while the effective biological dose is the Sievert (Sv = Q x Gy).

radioisotope thermoelectric generator (RTG): A power source based upon conversion of the heat generated by decay of a radioisotope (typically, plutonium alpha source) to electrical energy. Synonym: Nuclear Battery.

RAM: The abbreviation for random access memory, a necessary component of modern microprocessors and computers. This semiconductor digital storage device is composed of a mass of individual storage devices organized as a large number of words of fixed bit length. As an example a RAM device might be composed of 32K words of 8 bits each, totaling 256K bits. Its name is derived from the fact that one can select any of 32K words in this example and read out its bits, as contrasted with the earlier implementation of memories as large shift registers which had to be read out serially.

recombination: Any process in a cell that generates a new gene or chromosomal combination.

reconnection: Hypothetical process by which magnetic fields in plasma weaken with attendant conversion of energy into heat, radiation and plasma motion.

regolith: The loose, unconsolidated layer of rock and fine material (soil) distributed on planetary surfaces and overlying the bedrock.

replication: DNA synthesis.

restriction enzyme: An endonuclease that will recognize specific target nucleotide sequences in DNA and breaks the DNA chain at those sites.

retrovirus: An RNA virus that replicates by first being converted into double-stranded DNA.

rigidity: The rigidity of a charged particle is its momentum per unit charge. Rigidity is usually specified in units of MV (10^6 volts) or GV (10^9 volts).

rigidity-energy conversion: Mathematical formulas for converting rigidity into kinetic energy.

Ruthenium red: An electron opaque dye that combines with mucopolysaccharides making them visible in the electron microscope.

SAMPEX: A mission composed of nuclear particle and plasma investigations on the first SMEX satellite. SAMPEX was launched successfully on a Scout rocket in June, 1992.

SEU: Abbreviation for single event upset. It refers to the change of state of a digital circuit caused by charge resulting from the passage or stoppage of an energetic nucleon in a c t i v e region(s) of the device.

shocks: see interplanetary shocks.

SMEX: Abbriviation for Small Explorer, the program name for modern series of small and less expensive NASA spacecraft with launched weights of 300 to 450 pounds (135 to 205 Kg.). They are to be launched on Scout or Pegasus launch vehicles.

solar atmosphere: The gaseous envelope of the sun above the solar photosphere.

solar chromosphere: see chromosphere.

solar cycle: The solar cycle refers to the solar activity cycle behavior, usually represented by the number of sunspots visible on the solar photosphere. The average solar cycle length is 11.4 years.

solar flare: The English term "solar flare" is the name given to the sudden energy release of about 10^{23} ergs of energy in a relatively small volume of solar atmosphere.

solar photosphere: see photosphere.

solar max.: Maximum activity phase of the sun as evidenced by relative sunspot number, accompanied by enhanced solar magnetic fields and higher probability of solar flare eruptions and SPEs. It usually lasts about 6-8 years, depending on actual length of cycle.

solar min.: Minimum solar activity phase. It usually lasts about 4-5 years.

solar wind: Outflow of coronal plasma into interplanetary space. The wind speed is 200-900 kilometers per second.

solar X-rays: X-ray emission from the sun.

solid state detector: A device, usually a silicon or germanium crystal, which can produce charge pulses from radiation energy deposits. The responses are linear, with the charge pulse amplitude being directly proportional to the energy deposited in sensitive volume of detector.

solWatch: A group of scientific instruments chosen for monitoring solar activity to predict and detect the occurrence of flares and SPEs.

somatic: All cells pertaining to the body, except the germ cells.

South Atlantic Anomaly (SAA): A depression in the Earth's magnetic field which results in magnetic field lines carrying the nuclear particles of the Van Allen radiation belts to the top of the atmosphere. This region is generally centered over the South Atlantic ocean, adjacent to the coast of South America.

spallation secondaries: Spallation is a term from nuclear physics which refers to the ejection from a nucleus of protons, neutrons, often in the form of light nuclei, as a result of collision of a target nucleus by an energetic nucleon. The ejecta are the spallation secondaries.

SPE: Solar particle events. Solar flares result from the ejection of energetic charged particles into interplanetary space.

steradian: A unit of spherical angular measures. There are 4π steradian in a sphere. Abbreviation: ster.

stopping power: The energy loss rate over distance of a charged particle in matter represented as dE/dx. Often also called LET (see LET).

sub-sonic flow: In this context, plasma flowing slower than the local speed of sound in the medium.

sunspot cycle: Periodic variation in intensity or number of various manifestations of solar activity, e.g., sunspots. The period is approximately eleven years.

supercoil: A closed double-standed DNA molecule that is twisted on itself.

super-sonic flow: In this context, plasma flowing faster than the local speed of sound in the medium.

TOMS: Abbreviation for Total Ozone Mapping Spectrometer, a space instrument designed, managed and operated by Goddard Space Flight Center. It measures the column density of ozone in the Earth's atmosphere. Presently two units are in operation in space, one on the old Nimbus 7 spacecraft and a new unit launched on August 15, 1991 on the Russian Meteor-3 spacecraft.

trapped radiation: Charged particles, usually protons and electrons, which are confined to a volume by closed magnetic field lines of a solar or galactic body.

Tumorigenesis: That part of multistep process of carcinogenesis by which a normal cell becomes a tumor cell with uncontrolled cell division, but may or may not yet have acquired the invasive characteristics of a cancer cell.

unequal crossover: The exchange of chromosome parts between homologs that are not perfectly aligned.

Van Allen radiation belts: The radiation belts surrounding the Earth. They are composed largely of energetic protons and electrons and some heavier nuclei spiraling around the Earth's magnetic field lines. The radiation belts are named after Professor James A. Van Allen of University of Iowa, their discoverer.

vector magnetograph: A photoelectric instrument for mapping all three components of the magnetic fields in the solar atmosphere.

Voyager 1/2: Two large, 3-axis stabilized spacecraft launched by NASA in 1977 on a mission to the outer planets and the outer solar system. These spacecraft have nuclear thermoelectric power supplies and probably have a reasonable chance of reaching the boundary of our solar system with interstellar space.

well connected: A concept suggesting that the Archimedean spiral path from the specified position in space would connect to a specific solar longitude. It is based on the concept that charged particles travel along the interplanetary magnetic field which is transported out from the sun along these spiral field lines.